MERRIST WOOD COLLEGE

D0719677

12 FEB 2002

RESOURCES CENTRE

library@merristwood.ac.uk

Merrist Wood College
LIBRARY
Worplesdon, Guildford, Surrey. GU3 3PE

This book is due for return on or before the last date shown below.
You may renew by phone on 01483 884022

TEMPORARY REFERENCE
FROM: 12/2/02
TO: 14/4/02

1 DAY
LOAN

TEMPORARY REFERENCE
FROM:
TO:

15 MAR 2016

TEMPORARY REFERENCE
FROM: 14.05.03
TO: 28.05.03

18 JAN 2007

19/2/07

5/3/07

7. OCT 08

30 MAY 2012

Environmental Science

Working with the Earth

EIGHTH EDITION

G. TYLER MILLER, JR.

President, Earth Education and Research
Adjunct Professor of Human Ecology
St. Andrews Presbyterian College

12 FEB 2002
363.7

Brooks/Cole
Thomson Learning™

Australia • Canada • Mexico • Singapore • Spain • United Kingdom • United States

Publisher: *Jack Carey*
Project Development Editor: *Kristin Milotich*
Editorial Assistant: *Dan Lombardino*
Production Editor: *Tessa McGlasson Avila*
Production Service: *Electronic Publishing Services Inc., NYC*
Manuscript Editor: *Electronic Publishing Services Inc., NYC*
Permissions Editor: *Mary Kay Hancharick*
Interior Illustration: *Electronic Publishing Services Inc., NYC*
Cover Design: *Vernon T. Boes*
Cover Photo: *George Grall/N.G.S. Image Collection*
Photo Researcher: *Linda L. Rill*
Print Buyer: *Vena Dyer*

Typesetting: *Electronic Publishing Services Inc., NYC*
Printing and Binding: *Transcontinental Printing, Inc.*
Title Page Photos: *Tiger © Gunter Zeisler/Bruce Coleman Ltd.; Coral reef © Karl and Jill Wallin/FPG International*
Part Opening Photos: Part I: *Composite satellite view of Earth. © Tom Van Sant/The Geosphere Project.* Part II: *Endangered green sea turtle. © David Fleetham/Tom Stack & Associates.* Part III: *Forest in Czechoslovakia killed by acid deposition and other air pollutants. Silvestris Fotoservice/NHPA.* Part IV: *Highly endangered Florida panther. © John Cancalosi/Peter Arnold.* Part V: *Wind farm in California. Courtesy, U.S. Windpower.*

COPYRIGHT © 2001 by Brooks/Cole
A division of Thomson Learning
The Thomson Learning logo is a trademark used herein under license.

For more information, contact:
BROOKS/COLE
511 Forest Lodge Road
Pacific Grove, CA 93950 USA
www.brookscole.com

All rights reserved. No part of this work may be reproduced, transcribed or used in any form or by any means—graphic, electronic, or mechanical, including photocopying, recording, taping, Web distribution, or information storage and/or retrieval systems—without the prior written permission of the publisher.

For permission to use material from this work, contact us by
Web: www.thomsonrights.com
fax: 1-800-730-2215
phone: 1-800-730-2214

Printed in Canada

10 9 8 7 6 5 4 3 2 1

Library of Congress Cataloging-in-Publication Data

Miller, G. Tyler (George Tyler), [date]
Environmental science: working with the Earth/G. Tyler Miller, Jr.—8th ed.
p. cm.
Includes bibliographical references and index.
ISBN 0-534-37613-4 (text: acid-free paper)—ISBN 0-534-37614-2 (international: acid-free paper)
1. Environmental sciences. 2. Human ecology 3. Environmental protection I Title.

GE105.M.544 2001
333.'2—dc21 99-042986

All photos for the Guest Essays are the courtesy of the individuals featured.

THIS BOOK IS PRINTED ON ACID-FREE RECYCLED PAPER

PREFACE

For Instructors and Students

How Did I Become Involved with Environmental Problems? In 1966 I heard a scientist give a lecture on the problems of overpopulation and environmental abuse. Afterward I went to him and said, "If even a fraction of what you have said is true, I will feel ethically obligated to give up my research on the corrosion of metals and devote the rest of my life to research and education on environmental problems and solutions. Frankly, I don't want to believe a word you have said, and I'm going into the literature to try to prove that your statements are either untrue or grossly distorted."

After 6 months of study I was convinced of the seriousness of these problems. Since then, I have been studying, teaching, and writing about them. This book summarizes what I have learned in over three decades of trying to understand environmental principles, problems, connections, and solutions.

What Is My Philosophy of Education? In our lifelong pursuit of knowledge, I believe we should do three things. The first is to question everything and everybody, as any good scientist does.

Second, each of us should develop a list of principles, concepts, and rules to serve as guidelines in making decisions, and continually evaluate and modify this list on the basis of experience. This is based on my belief that the key goal of education is to learn how to sift through mountains of facts and ideas to find the few that are most useful and worth knowing. We need to be *wisdom seekers*, not information vessels. This requires a firm commitment to learning how to think logically and critically. This book is full of facts and numbers, but they are useful only to the extent that they lead to an understanding of key ideas, scientific laws, concepts, principles, and connections.

Third, I believe in interacting with what I read as a way to sharpen my critical thinking skills. I do this by marking key sentences and paragraphs with a highlighter or pen. I put an asterisk in the margin next to something I think is important and double asterisks next to something that I think is especially important. I write comments in the margins, such as *Beautiful*, *Confusing*, *Misleading*, or *Wrong*. I fold down the top corner of pages with highlighted passages and the top and bottom corners of especially important pages. This way, I can flip through a book and quickly review the key passages. I urge you to interact in such ways with this book.

Why Is There a Need for More Scientific Content? This book is designed for introductory courses on environmental science. It treats environmental science as an *interdisciplinary* study, combining ideas and information from *natural sciences* (such as biology, chemistry, and geology) and *social sciences* (such as economics, politics, and ethics) to present a general idea of how nature works and how things are interconnected. It is a study of *connections in nature.*

Since its first edition in 1975 this book has led the way in using scientific laws, principles, models, and concepts to help us **(1)** understand environmental and resource problems and possible solutions, and **(2)** how these concepts, problems, and solutions are connected. I have introduced only the concepts and principles necessary for understanding the material in the book, and I have tried to present them simply but accurately.

A number of people have criticized environmental science teachers and textbooks for **(1)** not having enough basic science content, **(2)** treating key environmental issues (such as global warming, ozone depletion, and biodiversity loss) superficially, **(3)** giving an unbalanced, biased view of environmental issues, and **(4)** focusing mostly on bad environmental news without giving the good environmental news.

I generally agree with these criticisms and believe that *there is a need to expand the fundamental scientific content of environmental science courses*. In each new edition of this book I have done this. This latest edition further expands the treatment of the scientific content by adding an additional chapter on basic ecology (see changes in this edition listed on p. v).

Since the first edition of this book I have also *emphasized in-depth treatment of key environmental issues.* Many instructors have written me thanking me for giving students an in-depth, balanced, up-to-date, and easy-to-understand analysis of major environmental problems such as global warming and ozone depletion (Chapter 11) and loss of biodiversity (a major integrating theme of this book).

To help ensure that the material is accurate and up to date, I have consulted more than 10,000 research sources in the professional literature and about the same number of Internet sites. I have also benefited from the more than 200 experts and teachers (see list on pp. vii–viii) who have provided detailed reviews of this and my other three books in this field.

How Have I Attempted to Eliminate Bias? There are at least two sides to all controversial issues. The challenge for an author is to give a fair and balanced view of opposing viewpoints without injecting personal bias. This allows students to make up their own minds about important issues. Studying a subject as important as environmental science and ending up with no conclusions, opinions, and beliefs means that both the teacher and student have failed. But such conclusions should be based on using critical thinking to evaluate opposing ideas.

In this edition I continue to scour every sentence for possible bias. A few examples of my efforts to give a balanced presentation of opposing viewpoints are **(1)** the pros and cons of reducing birth rates (pp. 213–15), **(2)** the Pro/Con boxes on genetic engineering (pp. 196–97), and Genetically Modified Food, (pp. 396-97), **(3)** Section 11-3 on global warming, **(4)** Section 11-6 on ozone depletion, **(5)** discussion of acid deposition (pp. 242–46), **(6)** the asbestos problem (pp. 248-49), **(7)** problems with the Superfund law (pp. 379), and **(8)** pros and cons of pesticides (Sections 16-2 and 16-3).

Bias is subtle, however, and there is still room for improvement. I invite instructors and students to write me and point out any remaining bias.

How Do I Deal with the Bad News/Good News Dilemma? Critics have charged environmentalists with wallowing in doom and gloom and not reporting or rejoicing in the many improvements in environmental quality that have taken place over the past several decades. I generally agree with this criticism and in this book I emphasize the good environmental news along with the bad environmental news in Chapter 1 (Table 1-1, pp. 24-5) and throughout the book.

Rosy optimism and gloom-and-doom pessimism are traps that usually lead to denial, indifference, and inaction. I have tried to avoid these two extremes and give a realistic yet hopeful view of the future. This book is filled with technological and scientific advances that have led to environmental improvements (for example, see the Individuals Matter boxes on pp. 281, 311, 351, 532, and 541) and the openings to Chapters 15 and 20. It also has stories of people who have acted to help sustain the earth's life-support systems for us and for all life (for example, see the Individuals Matter boxes on pp. 43, 177, 306, 446, 447, and 452, the openings to Chapters 3 and 17, and the Guest Essay on p. 406). It's an exciting and challenging time to be alive as we struggle to enter into a new, more cooperative relationship with the planet that is our only home.

What Are Some Key Features of This Book? This book is *concept oriented* and *solutions oriented.* The book is divided into five major parts (see Brief Contents, p. x). After Chapter 1 and the scientific principles and concepts in Part II have been covered, the rest of the book can be used in almost any order. In addition, most chapters and many sections within these chapters can be moved around or omitted to accommodate courses with different lengths and emphases. I also present a range of solutions to environmental problems proposed by a variety of analysts.

Since its first edition this book has consistently used simple systems analysis to categorize proposed solutions to environmental problems as either **(1)** *input* (prevention) solutions such as pollution prevention and waste reduction or **(2)** *output* (cleanup) solutions such as pollution control and waste management. Both approaches are needed, but so far most emphasis has been on output or management solutions. There is a growing awareness of the need to put more emphasis on input or prevention approaches—another major theme of this book.

Each chapter begins with a brief *Earth Story,* a case study designed to capture interest and set the stage for the material that follows. In addition to these 20 case studies, 53 other case studies are found throughout the book (some in special boxes and others within the text); they provide a more in-depth look at specific environmental problems and their possible solutions. Thirteen *Guest Essays* present an individual researcher's or activist's point of view, which is then evaluated through Critical Thinking questions.

Other special boxes found in the text include **(1)** *Pro/Con boxes* that present both sides of controversial environmental issues, **(2)** *Connections boxes* that show connections in nature and among environmental concepts, problems, and solutions, **(3)** *Solutions boxes* that summarize a variety of solutions (some of them controversial) to environmental problems proposed by various analysts, **(4)** *Spotlight boxes* that highlight and give insights into key environmental problems and concepts, and **(5)** *Individuals Matter boxes* that describe what people have done to help solve environmental problems. To encourage critical thinking and integrate it throughout the book, all boxes (except Individuals Matter) end with Critical Thinking questions.

This book is an integrated study of environmental problems, connections, and solutions. The eight integrating themes in this book are **(1)** *biodiversity and natural capital,* **(2)** *sustainability,* **(3)** *connections in nature,* **(4)** *pollution prevention and waste reduction,* **(5)** *population and exponential growth,* **(6)** *energy and energy efficiency,* **(7)** *solutions to environmental problems,* and **(8)** *the importance of individual action.*

I hope you will start by looking at the brief table of contents (p. x) to get an overview of this book. Then I suggest that you look at the Concepts and Connections diagram inside the back cover, which shows the major components and relationships found in environmental science. In effect, it is a map of the book.

The book's 422 illustrations are designed to present complex ideas in understandable ways and to relate learning to the real world. They include 322 full-color diagrams (48 of them new to this edition, and 40 of them maps) and 100 carefully selected color photographs.

I have not cited specific sources of information. This is rarely done for an introductory-level text in any field and it would interrupt the flow of the material. Instead, the readings listed at the end of the book for each chapter provide backup for almost all the information in this book and serve as springboards to further information and ideas. This edition also has a greatly expanded and improved interactive World Wide Web site **(www.brookscole.com/biology)** that can be used as a source of further information and ideas.

Students and teachers also have access to InfoTrac, a fully searchable online database with access to complete environmental articles from over 700 periodicals. I have put two practice exercises at the bottom of two pages in each chapter to help users learn how to navigate this valuable source of information.

Instructors wanting a longer or a shorter book covering this material with a different emphasis and organization can use one of my two other books written for various types of environmental science courses: *Living in the Environment*, 11th edition (815 pages, Brooks/Cole, 2000) or *Sustaining the Earth: An Integrated Approach*, 4th edition (356 pages, Brooks/Cole, 2000, a shorter book with a different integrated approach).

What Are the Major Changes in the Eighth Edition? Major changes include the following:

■ Updated and revised material throughout the book.

■ Increased coverage of basic ecological concepts. The number of chapters on ecological principles has been increased from three to four and the order of topics in some of these chapters has been changed to improve scientific content and flow.

■ Greatly increased *scientific content*, with the addition or expansion of material on principles of ecological sustainability, overloading of the nitrogen cycle, island biogeography, population ecology, ecological restoration, infectious disease, survival ability of cockroaches, and conservation biology.

■ Addition or expansion of many topics, including the information revolution, green taxes, failure of Biosphere 2, food supply and demand, sustainable fishery management, possible signs of global warming, organic farming and global warming, solar village in Columbia (Gaviotas), the spread of genetic resistance of bacteria to antibiotics, controversy over reducing air pollution emissions of ultrafine particles, using honeybees to detect toxic pollutants, our ecological footprints, the termite from hell, genetically modified food, and solutions to the excessive snow goose population.

■ Consolidation of environmental economics, politics, and ethics into one chapter (Chapter 2) instead of two chapters.

■ Addition of 40 new figures and 8 new photos.

■ Addition of comprehensive Review Questions at the end of each chapter.

■ Revision and updating of *EnviroLink*, an instructor presentation tool.

■ *WebTutor in Web CT,* an online study tool.

■ *Online Regional Articles* (www.upcloser.com).

■ *Thomson Learning Course software.*

■ *CNN Videos for Environmental Science.*

■ Booklet on *Essential Study Skills for Science Students.*

■ Revision and updating of the *Instructor's Manual and Test Items* booklet.

Welcome to Controversy and Challenge Despite considerable research, we still know little about how nature works at a time when we are altering nature at an accelerating pace. This uncertainty, as well as the complexity and importance of these issues to current and future generations of humans and other species, makes many of these issues highly controversial.

Another source of disagreement is that science advances through controversy and careful scrutiny about its results until there is a general consensus about their validity. What is important is not what the experts disagree on (the frontiers of scientific knowledge that are still being developed, tested, and argued about), but what they generally agree on—the *scientific consensus*—about concepts, problems, and possible solutions.

Controversy also arises because environmental science is an interdisciplinary blend of natural and social sciences that sometimes questions the ways we view and act in the world around us.

Study Aids Each chapter begins with a few general questions to reveal how it is organized and what students will be learning. When a new term is introduced and defined, it is printed in boldface type. A glossary of all key terms is located at the end of the book.

Each chapter and box (except Individuals Matter boxes) ends with **(1)** a set of Review Questions covering all of the material in the chapter as a study guide for students and **(2)** a set of questions to encourage students to think critically and apply what they have learned to their lives. The Critical Thinking questions are followed by several projects that individuals or groups can carry out. Many additional projects are given in the *Instructor's Manual* available with this book.

Readers who become especially interested in a particular topic can consult the further readings for each chapter, given in the back of the book. Appendix 1 contains a list of important publications and some key environmental organizations and government and international agencies.

Students can also access World Wide Web for material in the book marked with the icon . Access to the Brooks/Cole Biology Resource Center is at www. brookscole.com/biology. Users can click on a chapter in the Hypercontents listed and find resources that couldn't be listed in the book. These resources are updated constantly. In the Web site for this text, you will also find Cool Events, Critical Thinking Questions, Tips on Surfing, Interactive Quizzes for each chapter, and much more. Happy surfing.

Help Me Improve This Book Let me know how you think this book can be improved; if you find any errors, bias, or confusing explanations please send them to Jack Carey, Biology Publisher, Brooks/Cole Publishing Company, 10 Davis Drive, Belmont, CA 94002. He will forward them to me. Most errors can be corrected in subsequent printings of this edition, rather than waiting for a new edition.

Annenberg/CPB Television Course This textbook is being offered as part of the Annenberg/CPB Project television series *Race to Save the Planet*, a 10-part public broadcasting series and a college-level telecourse examining the major environmental questions facing the world today. The series takes into account the wide spectrum of opinion about what constitutes an environmental problem and discusses the controversies about appropriate remedial measures. It analyzes problems and emphasizes the successful search for solutions. The course develops a number of key themes that cut across a broad range of environmental issues, including sustainability, the interconnections of the economy and the ecosystem, short-term versus long-term gains, and the trade-offs involved in balancing problems and solutions. A study guide and a faculty guide, both available from Brooks/Cole Publishing Company, integrate the telecourse and this text.

For further information about available television course licenses and duplication licenses, contact PBS Adult Learning Service, 1320 Braddock Place, Alexandria, VA 22314-1698 (1-800-ALS-AL5-8).

For information about purchasing videocassettes and print material, contact the Annenberg/CPB Collection, P.O. Box 2284, South Burlington, VT 05407-2284 (1-800-LEARNER).

Acknowledgments I wish to thank the many students and teachers who responded so favorably to the 7 previous editions of *Environmental Science*, the 11 editions of *Living in the Environment*, and the 4 editions of *Sustaining the Earth* and who corrected errors and offered many helpful suggestions for improvement. I am also deeply indebted to the reviewers, who pointed out errors and suggested many important improvements in this book. Any errors and deficiencies left are mine.

The members of the talented production team, listed on the copyright page, have made vital contributions as well. My thanks also go to **(1)** copyeditor Carol Anne Peschke, **(2)** production editors Brooks Ellis and Tessa Avila, **(3)** Brooks/Cole's hard-working sales staff, and **(4)** Kristin Milotich and Daniel Lombardino for their help and efficiency.

I also thank **(1)** C. Lee Rockett and Kenneth J. Van Dellen for developing the *Laboratory Manual* to accompany this book and **(2)** the people who have translated this book into five different languages for use throughout much of the world.

My special thanks go Jane Heinze-Fry for her outstanding work on concept mapping, *Critical Thinking and the Environment: A Beginner's Guide*, and the *Internet Booklet*. My thanks also go to Richard K. Clements for his excellent work on the *Instructor's Manual*.

My deepest thanks go to Jack Carey, Biology Publisher at Brooks/Cole, for his encouragement, help, 34 years of friendship, and superb reviewing system. It helps immensely to work with the best and most experienced editor in college textbook publishing.

G. Tyler Miller, Jr.

Guest Essayists and Reviewers

Guest Essayists The following are authors of Guest Essays: **(1) Lester R. Brown**, president, Worldwatch Institute; **(2) Alberto Ruz Buenfil**, environmental activist, writer, and performer; **(3) Robert D. Bullard**, professor of sociology and director of the Environmental Justice Resource Center at Clark Atlanta University; **(4) Lois Marie Gibbs**, director, Center for Health, Environment, and Justice; **(5) Garrett Hardin**, professor emeritus of human ecology, University of California, Santa Barbara; **(6) Paul Hawken**, environmental author and business leader; **(7) Jane Heinze-Fry**, author, teacher, and consultant in environmental education; **(8) Amory B. Lovins**, energy policy consultant and director of research, Rocky Mountain Institute; **(9) Lester W. Milbrath**, director of the research program in environment and society, State University of New York, Buffalo; **(10) Peter Montague**, director, Environmental Research Foundation; **(11) Norman Myers**, consultant in environment and development; **(12) David W. Orr**, professor of environmental studies, Oberlin College; and **(13) David Pimentel**, professor of entomology, Cornell University.

Cumulative Reviewers Barbara J. Abraham, Hampton College; Donald D. Adams, State University of New York at Plattsburgh; Larry G. Allen, California State University, Northridge; Susan Allen-Gil, Ithaca College; James R. Anderson, U.S. Geological Survey; Mark W. Anderson, University of Maine; Kenneth B. Armitage, University of Kansas; Gary J. Atchison, Iowa State University; Marvin W. Baker, Jr., University of Oklahoma; Virgil R. Baker, Arizona State University; Ian G. Barbour, Carleton College; Albert J. Beck, California State University, Chico; W. Behan, Northern Arizona University; Keith L. Bildstein, Winthrop College; Jeff Bland, University of Puget Sound; Roger G. Bland, Central Michigan University; Grady Blount II, Texas A&M University, Corpus Christi; Georg Borgstrom, Michigan State University; Arthur C. Borror, University of New Hampshire; John H. Bounds, Sam Houston State University; Leon F. Bouvier, Population Reference Bureau; Daniel J. Bovin, Universitè Laval; Michael F. Brewer, Resources for the Future, Inc.; Mark M. Brinson, East Carolina University; Dale Brown, University of Hartford; Patrick E. Brunelle, Contra Costa College; Terrence J. Burgess, Saddleback College North; David Byman, Pennsylvania State University, Worthington Scranton; Lynton K. Caldwell, Indiana University; Faith Thompson Campbell, Natural Resources Defense Council, Inc.; Ray Canterbery, Florida State University; Ted J. Case, University of San Diego; Ann Causey, Auburn University; Richard A. Cellarius, Evergreen State University; William U. Chandler, Worldwatch Institute; F. Christman, University of North Carolina, Chapel Hill; Preston Cloud, University of California, Santa Barbara; Bernard C. Cohen, University of Pittsburgh; Richard A. Cooley, University of California, Santa Cruz; Dennis J. Corrigan; George Cox, San Diego State University; John D. Cunningham, Keene State College; Herman E. Daly, The World Bank; Raymond F. Dasmann, University of California, Santa Cruz; Kingsley Davis, Hoover Institution; Edward E. DeMartini, University of California, Santa Barbara; Charles E. DePoe, Northeast Louisiana University; Thomas R. Detwyler, University of Wisconsin; Peter H. Diage, University of California, Riverside; Lon D. Drake, University of Iowa; David DuBose, Shasta College; T. Edmonson, University of Washington; Thomas Eisner, Cornell University; Michael Esler, Southern Illinois University; David E. Fairbrothers, Rutgers University; Paul P. Feeny, Cornell University; Richard S. Feldman, Marist College; Nancy Field, Bellevue Community College; Allan Fitzsimmons, University of Kentucky; Andrew J. Friedland, Dartmouth College; Kenneth O. Fulgham, Humboldt State University; Lowell L. Getz, University of Illinois at Urbana–Champaign; Frederick F. Gilbert, Washington State University; Jay Glassman, Los Angeles Valley College; Harold Goetz, North Dakota State University; Jeffery J. Gordon, Bowling Green State University; Eville Gorham, University of Minnesota; Michael Gough, Resources for the Future; Ernest M. Gould, Jr., Harvard University; Peter Green, Golden West College; Katharine B. Gregg, West Virginia Wesleyan College; Paul K. Grogger, University of Colorado at Colorado Springs; L. Guernsey, Indiana State University; Ralph Guzman, University of California, Santa Cruz; Raymond Hames, University of Nebraska, Lincoln; Raymond E. Hampton, Central Michigan University; Ted L. Hanes, California State University, Fullerton; William S. Hardenbergh, Southern Illinois University at Carbondale; John P. Harley, Eastern Kentucky University; Neil A. Harriman, University of Wisconsin, Oshkosh; Grant A. Harris, Washington State University; Harry S. Hass, San Jose City College; Arthur N. Haupt, Population Reference Bureau; Denis A. Hayes, environmental consultant; Stephen Heard, University of Iowa; Gene Heinze-Fry, Department of Utilities, State of Massachusetts; Jane Heinze-Fry, environmental educator; John G. Hewston, Humboldt State University; David L. Hicks, Whitworth College; Kenneth M. Hinkel, University of Cincinnati; Eric Hirst,

Oak Ridge National Laboratory; Doug Hix, University of Hartford; S. Holling, University of British Columbia; Donald Holtgrieve, California State University, Hayward; Michael H. Horn, California State University, Fullerton; Mark A. Hornberger, Bloomsberg University; Marilyn Houck, Pennsylvania State University; Richard D. Houk, Winthrop College; Robert J. Huggett, College of William and Mary; Donald Huisingh, North Carolina State University; Marlene K. Hutt, IBM; David R. Inglis, University of Massachusetts; Robert Janiskee, University of South Carolina; Hugo H. John, University of Connecticut; Brian A. Johnson, University of Pennsylvania, Bloomsburg; David I. Johnson, Michigan State University; Mark Jonasson, Crafton Hills College; Agnes Kadar, Nassau Community College; Thomas L. Keefe, Eastern Kentucky University; Nathan Keyfitz, Harvard University; David Kidd, University of New Mexico; Jesse Klingebiel, Kent School; Edward J. Kormondy, University of Hawaii–Hilo/West Oahu College; John V. Krutilla, Resources for the Future, Inc.; Judith Kunofsky, Sierra Club; E. Kurtz; Theodore Kury, State University of New York, Buffalo; Steve Ladochy, University of Winnipeg; Mark B. Lapping, Kansas State University; Tom Leege, Idaho Department of Fish and Game; William S. Lindsay, Monterey Peninsula College; E. S. Lindstrom, Pennsylvania State University; M. Lippiman, New York University Medical Center; Valerie A. Liston, University of Minnesota; Dennis Livingston, Rensselaer Polytechnic Institute; James P. Lodge, air pollution consultant; Raymond C. Loehr, University of Texas at Austin; Ruth Logan, Santa Monica City College; Robert D. Loring, DePauw University; Paul F. Love, Angelo State University; Thomas Lovering, University of California, Santa Barbara; Amory B. Lovins, Rocky Mountain Institute; Hunter Lovins, Rocky Mountain Institute; Gene A. Lucas, Drake University; Claudia Luke; David Lynn; Timothy F. Lyon, Ball State University; Stephen Malcolm, Western Michigan University; Melvin G. Marcus, Arizona State University; Gordon E. Matzke, Oregon State University; Parker Mauldin, Rockefeller Foundation; Marie McClune, The Agnes Irwin School (Rosemont, Pennsylvania); Theodore R. McDowell, California State University; Vincent E. McKelvey, U.S. Geological Survey; Robert T. McMaster, Smith College; John G. Merriam, Bowling Green State University; A. Steven Messenger, Northern Illinois University; John Meyers, Middlesex Community College; Raymond W. Miller, Utah State University; Arthur B. Millman, University of Massachusetts, Boston; Fred Montague, University of Utah; Rolf Monteen, California Polytechnic State University; Ralph Morris, Brock University, St. Catherine's, Ontario, Canada; William W. Murdoch, University of California, Santa Barbara; Norman Myers, environmental consultant; Brian C. Myres, Cypress College; A. Neale, Illinois State University; Duane Nellis, Kansas State University; Jan Newhouse, University of Hawaii, Manoa; Jim Norwine, Texas A&M University, Kingsville; John E. Oliver, Indiana State University; Eric Pallant, Allegheny College; Charles F. Park, Stanford University; Richard J. Pedersen, U.S. Department of Agriculture, Forest Service; David Pelliam, Bureau of Land Management, U.S. Department of Interior; Rodney Peterson, Colorado State University; Julie Phillips, De Anza College; William S. Pierce, Case Western Reserve University; David Pimentel, Cornell University; Peter Pizor, Northwest Community College; Mark D. Plunkett, Bellevue Community College; Grace L. Powell, University of Akron; James H. Price, Oklahoma College; Marian E. Reeve, Merritt College; Carl H. Reidel, University of Vermont; Charles C. Reith, Tulane University; Roger Revelle, California State University, San Diego; L. Reynolds, University of Central Arkansas; Ronald R. Rhein, Kutztown University of Pennsylvania; Charles Rhyne, Jackson State University; Robert A. Richardson, University of Wisconsin; Benjamin F. Richason III, St. Cloud State University; Ronald Robberecht, University of Idaho; William Van B. Robertson, School of Medicine, Stanford University; C. Lee Rockett, Bowling Green State University; Terry D. Roelofs, Humboldt State University; Christopher Rose, California Polytechnic State University; Richard G. Rose, West Valley College; Stephen T. Ross, University of Southern Mississippi; Robert E. Roth, The Ohio State University; Arthur N. Samel, Bowling Green State University; Floyd Sanford, Coe College; David Satterthwaite, I.E.E.D., London; Stephen W. Sawyer, University of Maryland; Arnold Schecter, State University of New York, Syracuse; Frank Schiavo, San Jose State University; William H. Schlesinger, Ecological Society of America; Stephen H. Schneider, National Center for Atmospheric Research; Clarence A. Schoenfeld, University of Wisconsin, Madison; Henry A. Schroeder, Dartmouth Medical School; Lauren A. Schroeder, Youngstown State University; Norman B. Schwartz, University of Delaware; George Sessions, Sierra College; David J. Severn, Clement Associates; Paul Shepard, Pitzer College and Claremont Graduate School; Michael P. Shields, Southern Illinois University at Carbondale; Kenneth Shiovitz; F. Siewert, Ball State University; E. K. Silbergold, Environmental Defense Fund; Joseph L. Simon, University of South Florida; William E. Sloey, University of Wisconsin, Oshkosh; Robert L. Smith, West Virginia University; Howard M. Smolkin, U.S. Environmental Protection Agency; Patricia M. Sparks, Glassboro State College; John E. Stanley, University of Virginia; Mel Stanley, California State Polytechnic University, Pomona; Norman R. Stewart, University of Wisconsin, Milwaukee; Frank E. Studnicka, University of Wisconsin, Platteville; Chris Tarp, Contra Costa College; Roger E. Thibault, Bowling

Green State University; William L. Thomas, California State University, Hayward; John D. Usis, Youngstown State University; Tinco E. A. van Hylckama, Texas Tech University; Robert R. Van Kirk, Humboldt State University; Donald E. Van Meter, Ball State University; Gary Varner, Texas A&M University; John D. Vitek, Oklahoma State University; Harry A. Wagner, Victoria College; Lee B. Waian, Saddleback College; Warren C. Walker, Stephen F. Austin State University; Thomas D. Warner, South Dakota State University; Kenneth E. F. Watt, University of California, Davis; Alvin M. Weinberg, Institute of Energy Analysis, Oak Ridge Associated Universities; Brian Weiss; Margery Weitkamp, James Monroe High School (Granada Hills, California); Anthony Weston, SUNY at Stony Brook; Raymond White, San Francisco City College; Douglas Wickum, University of Wisconsin, Stout; Charles G. Wilber, Colorado State University; Nancy Lee Wilkinson, San Francisco State University; John C. Williams, College of San Mateo; Ray Williams, Rio Hondo College; Roberta Williams, University of Nevada Las Vegas; Samuel J. Williamson, New York University; Ted L. Willrich, Oregon State University; James Winsor, Pennsylvania State University; Fred Witzig, University of Minnesota at Duluth; George M. Woodwell, Woods Hole Research Center; Robert Yoerg, Belmont Hills Hospital; Hideo Yonenaka, San Francisco State University; Malcolm J. Zwolinski, University of Arizona.

Brief Contents

Detailed Contents

Part I
HUMANS AND SUSTAINABILITY: AN OVERVIEW 2

Saguaro cacti, Arizona

Temperate deciduous forest, fall, Rhode Island

Temperate deciduous forest, winter, Rhode Island

Tree farm, North Carolina

Mangrove swamp, Colombia

Monoculture cropland, Blythe, California

Jack Carey

Crater Lake, Oregon

© Michael Grecco

Sulfur dioxide emissions from coal-burning power plant

Water hyacinth, Florida

Endangered cotton-top tamarin, tropical forest, South America

Invertebrate sea star

Warning coloration of poison-dart frog, South America

Environmental Science

Working with the Earth

EIGHTH EDITION

PART I

HUMANS AND SUSTAINABILITY: AN OVERVIEW

The environmental crisis is an outward manifestation of a crisis of mind and spirit. There could be no greater misconception of its meaning than to believe it is concerned only with endangered wildlife, human-made ugliness,

and pollution. These are part of it, but more importantly, the crisis is concerned with the kind of creatures we are and what we must become in order to survive.

LYNTON K. CALDWELL

1 ENVIRONMENTAL PROBLEMS, THEIR CAUSES, AND SUSTAINABILITY

Living in an Exponential Age

Once there were two kings from Babylon who enjoyed playing chess, with the winner claiming a prize from the loser. After one match, the winning king asked the loser to pay him by placing one grain of wheat on the first square of the chessboard, two on the second, four on the third, and so on. The number of grains was to double each time until all 64 squares were filled.

The losing king, thinking he was getting off easy, agreed with delight. It was the biggest mistake he ever made. He bankrupted his kingdom and still could not produce the 2^{63} grains of wheat he had promised. In fact, it's probably more than all the wheat that has ever been harvested!

This is an example of **exponential growth**, in which a quantity increases by a fixed percentage of the whole in a given time. As the losing king learned, exponential growth is deceptive. It starts off slowly, but after only a few doublings it grows to enormous numbers because each doubling is more than the total of all earlier growth.

Here is another example. Fold a piece of paper in half to double its thickness. If you could do this 42 times, the stack would reach from the earth to the moon, 386,400 kilometers (240,000 miles) away. If you could double it 50 times, the folded paper would almost reach the sun, 149 million kilometers (93 million miles) away!

The environmental problems we face—population growth, wasteful use of resources, destruction and degradation of wildlife habitats, extinction of plants and animals, poverty, and pollution—are interconnected and are growing exponentially. For example, world population has more than doubled in 49 years, from 2.5 billion in 1950 to 6 billion in 1999. Unless death rates rise sharply, it may reach 8 billion by 2028, 9 billion by 2054, and 10–15 billion by 2100 (Figure 1-1). Global economic output, much of it environmentally damaging, has increased almost eightfold since 1950.

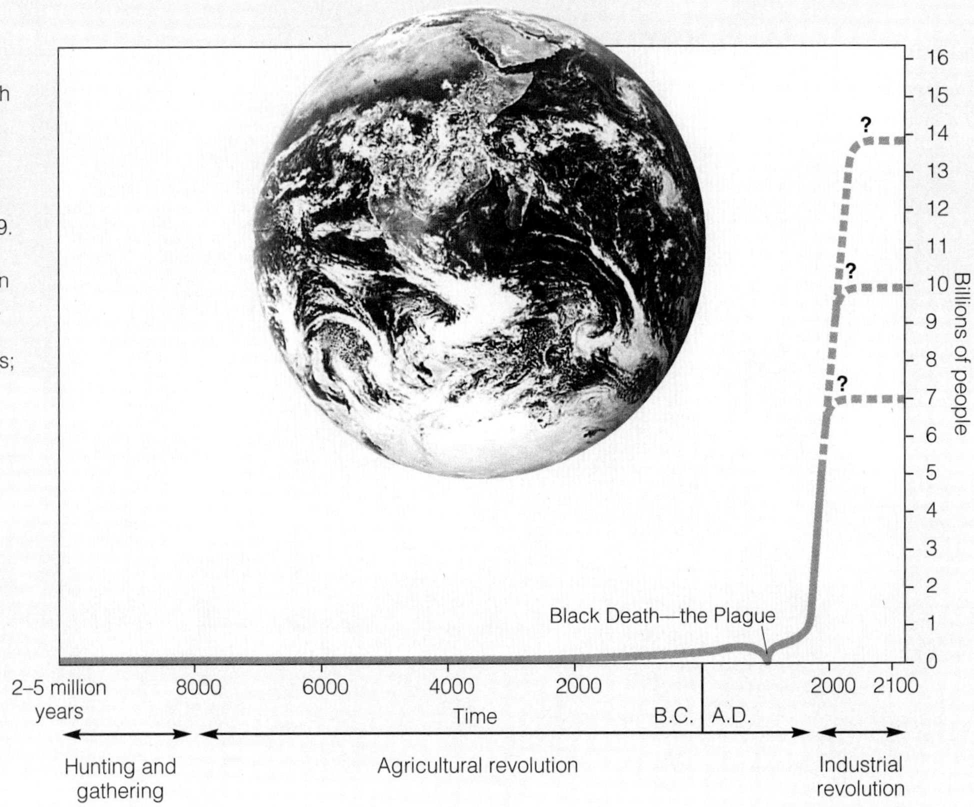

Figure 1-1 The J-shaped curve of past exponential world population growth, with projections beyond 2100. Notice that exponential growth starts off slowly, but as time passes the curve becomes increasingly steep. World population has more than doubled in only 49 years, from 2.5 billion in 1950 to 6 billion in 1999. Unless death rates rise sharply, it may reach 8 billion by 2028, 9 billion by 2054, and 10–15 billion by 2100. (This figure is not to scale.) (Data from World Bank and United Nations; photo courtesy of NASA)

Alone in space, alone in its life-supporting systems, powered by inconceivable energies, mediating them to us through the most delicate adjustments, wayward, unlikely, unpredictable, but nourishing, enlivening, and enriching in the largest degree—is this not a precious home for all of us? Is it not worth our love?

BARBARA WARD AND RENÉ DUBOS

This chapter is an overview of environmental problems, their root causes, the controversy over their seriousness, and ways we can live more sustainably. It discusses these questions:

- What are natural resources? What is a sustainable society?
- How fast is the human population increasing?
- What are the earth's main types of resources? How can they be depleted or degraded?
- What are the principal types of pollution? How can pollution be reduced and prevented?
- What are the root causes of today's environmental problems?
- What major effects have hunter–gatherer societies, agricultural societies, and industrialized societies had on the environment?
- How serious are environmental problems, and is our current course sustainable?

1-1 LIVING SUSTAINABLY

What Are Natural Resources and Sustainability? Our existence, lifestyles, and economies depend completely on the sun and the earth, a blue and white island in the black void of space. To economists *capital* is wealth used to sustain a business and to generate more wealth. We can think of energy from the sun as **solar capital** and the planet's air, water, soil, wildlife, minerals, and natural purification, recycling, and pest control processes as **natural resources** or **natural capital** (Figure 1-2).

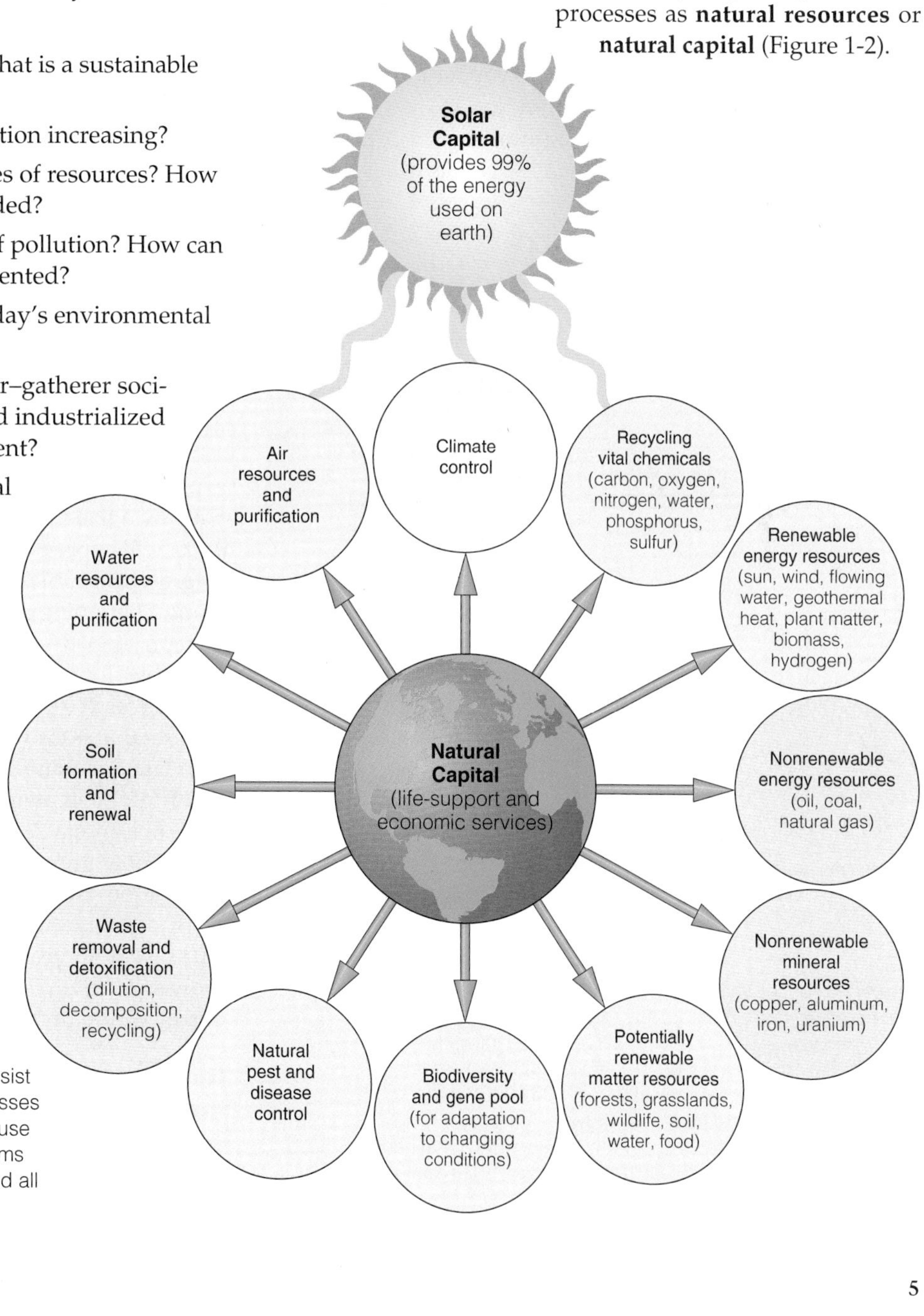

Figure 1-2 Solar and natural capital consist of the life-support resources and processes provided by the sun and the planet for use by us and other species. These two forms of capital support and sustain all life and all economies on the earth.

Sustainability is the ability of a specified system to survive and function over time. During this period, the system satisfies the needs of its inhabitants without depleting natural capital and thereby jeopardizing the prospects of current and future generations of humans and other species.

Living sustainably means living off income and not depleting the capital that supplies the income. Imagine that you inherit $1 million. Invest this capital at 10% interest per year and you will have a sustainable annual income of $100,000; that is, you can spend up to $100,000 a year without touching the capital that sustains your lifestyle.

Suppose you develop a taste for diamonds or a yacht, or all your relatives move in with you. If you spend $200,000 a year, your $1 million will be gone early in the 7th year; even if you spend just $110,000 a year, you will be bankrupt early in the 18th year. The lesson here is a very old one: *Don't kill the goose that lays the golden egg.* Deplete your capital, and you move from a sustainable to an unsustainable lifestyle.

The same lesson applies to natural capital (Figure 1-2). Environmentalists and many scientists believe that we are living unsustainably by depleting and degrading the earth's natural capital at an accelerating rate as our population (Figure 1-1) and demands on the earth's resources and natural processes increase exponentially (Guest Essay, p. 26).

The global ecological demand on the earth's resources from agriculture, mining, and industry is growing at about 5% per year, which doubles the demand on the earth's resources every 14 years. According to the World Wildlife Fund, the world lost about one-third of its natural capital between 1970 and 1999 because of a combination of exponential growth in population and in the use of the earth's natural resources.

Others, mostly economists and business leaders, disagree. They contend that there are no limits to human population growth and economic growth that can't be overcome by human ingenuity and technology. What do you think?

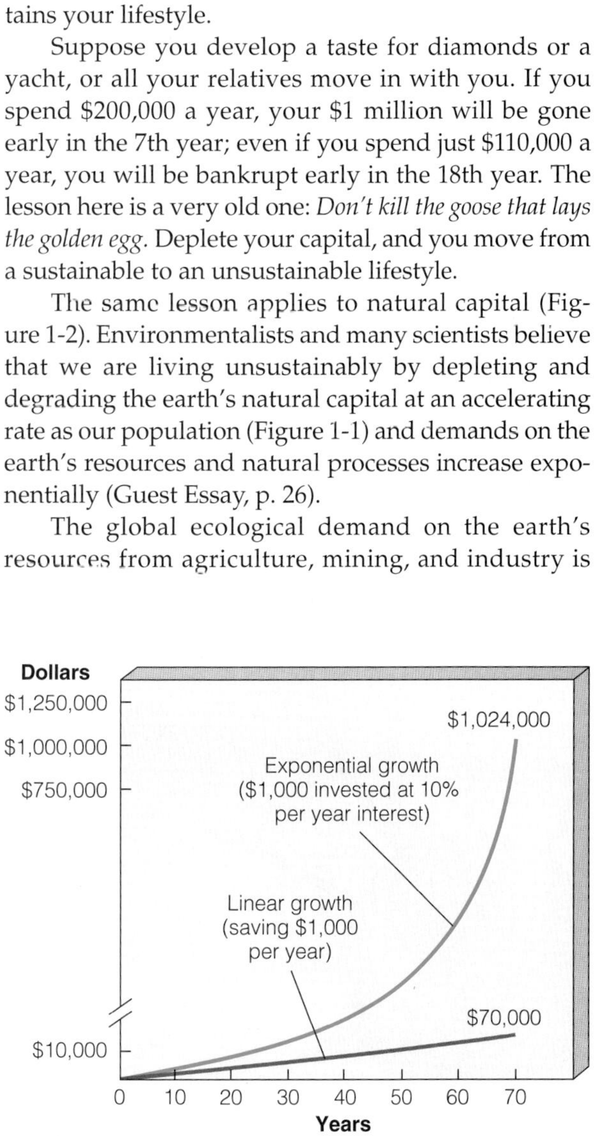

Figure 1-3 Linear and exponential growth. If you save $1,000 a year for a lifetime of 70 years, the resulting linear growth will allow you to save $70,000 (lower curve). If you invest $1,000 each year at 10% interest for 70 years and reinvest the interest, your money will grow exponentially to $1,024,000 (upper curve). If resource use, economic growth, or money in a savings account grows exponentially for 70 years (a typical human lifetime) at a rate of 10% a year, it will increase 1,024-fold.

1-2 GROWTH AND THE WEALTH GAP

What Is the Difference Between Linear Growth and Exponential Growth? Suppose you hop on a train that accelerates by 1 kilometer (0.6 mile) an hour every second. After 30 seconds, your speed would be 30 kilometers (19 miles) per hour. This is an example of **linear growth**, in which a quantity increases by a constant amount per unit of time, as in 1, 2, 3, 4, 5—or 1, 3, 5, 7, 9—and so on. If plotted on a graph, such growth in speed or growth of money in a savings account yields a straight line that slopes upward (Figure 1-3).

But suppose the train has a magic motor strong enough to double its speed every second. After 30 seconds, you would be traveling a billion kilometers (620 million miles) per hour!

This is an example of the astounding power of exponential growth. Any quantity growing exponentially by a fixed percentage, even as small as 0.001% or 0.1%, will experience extraordinary growth as its base of growth doubles again and again. If plotted on a graph, continuing exponential growth eventually yields a graph shaped somewhat like the letter J (Figure 1-3).

How long does it take to double the world's resource use or population size or money in a savings account that is growing exponentially? A quick way to calculate this **doubling time** in years is to use the **rule of 70**: 70/percentage growth rate = doubling time in years (a formula derived from the basic mathematics of exponential growth). For example, in 1999 the world's population grew by 1.33%. If that rate continues, the earth's population will double in about 53 years (70/1.33 = 53 years). In the example of exponential growth of savings by 10% a year (Figure 1-3), your money would double roughly every 7 years (70/10 = 7).

How Rapidly Is the Human Population Growing? The increasing size of the human population is an example of exponential growth (Figures 1-1 and 1-4 and Spotlight, right).

Infotrac Excercise: Soule, M. E. and D. Press. 1998. "What Is Environmental Studies?" *BioScience*, vol. 48, no. 5, p. 397(9).

World Population Reached

1 billion in 1804
2 billion in 1927 (123 years later)
3 billion in 1960 (33 years later)
4 billion in 1974 (14 years later)
5 billion in 1987 (13 years later)
6 billion in 1999 (12 years later)

World Population May Reach

7 billion in 2013 (14 years later)
8 billion in 2028 (15 years later)
9 billion in 2054 (26 years later)

Figure 1-4 World population milestones. (Data from United Nations Population Division, *World Population Prospects, 1998*)

SPOTLIGHT

Current Exponential Growth of the Human Population

The world's population is growing exponentially at a rate of about 1.33% per year. The ticking of this population clock means that in 1999 the world's population of 6.0 billion grew by 80 million people (6.0 billion × 0.0133 = 80 million), an average increase of 219,000 people a day, or 9,100 an hour.

At this 1.33% annual rate of exponential growth, it takes only about

- 5 days to add the number of Americans killed in all U.S. wars
- 2 months to add as many people as live in the Los Angeles basin
- 1.6 years to add the 129 million people killed in all wars fought in the past 200 years
- 3.4 years to add 273 million people (the population of the United States in 1999)
- 16 years to add 1.25 billion people (the population of China, the world's most populous country, in 1999)

Critical Thinking

Some economists argue that population growth is good because it provides more workers, consumers, and problem solvers to keep the global economy growing. Environmentalists argue that population growth threatens economies and the earth's life-support systems through increased pollution and environmental degradation. What is your position? Why?

Recent studies by researchers at Conservation International suggest that roughly 48% of the earth's total land area has been partially or totally disturbed by human activities (Figure 1-5). If uninhabitable areas of rock and ice are excluded, *73% of the habitable area of the planet has been disturbed by human activities.* What will happen to the earth's remaining wildlife habitat and wildlife species if the human population increases from 6 billion to 8 billion between 1999 and 2028 and perhaps to 9 billion by 2054?

What Is Economic Growth? Virtually all countries seek **economic growth**: an increase in their capacity to provide goods and services for people's final use. Such growth is normally achieved by increasing the flow or **throughput** of matter and energy resources used to produce goods and services. This increase is accomplished by population growth (more consumers and producers), more consumption per person, or both.

Economic growth usually is measured by an increase in a country's **gross national product (GNP)**: the market value in current dollars of all goods and services produced *within* and *outside* a country by the country's businesses for final use during a year. **Gross domestic product (GDP)** is the market value in current dollars of all goods and services produced *within* a country for final use during a year. To show one person's slice of the economic pie, economists often calculate the **per capita GNP**: the GNP divided by the total population.

The United Nations classifies the world's countries as economically developed or developing. The **developed countries** are highly industrialized. Most (except the countries of the former Soviet Union) have high average per capita GNPs (above $4,000). These countries, with 1.2 billion people (20% of the world's population in 1999), **(1)** have about 85% of the world's wealth and income, **(2)** use about 88% of its natural resources, and **(3)** generate about 75% of its pollution and waste.

All other nations are classified as **developing countries**, with low to moderate industrialization and per capita GNPs. Most are in Africa, Asia, and Latin America. Their 4.8 billion people (80% of the world's population in 1999) **(1)** have only about 15% of the wealth and income and **(2)** use only about 12% of the world's natural resources.

Figure 1-6 compares various characteristics of developed countries and developing countries. More than 95% of the projected increase in world population

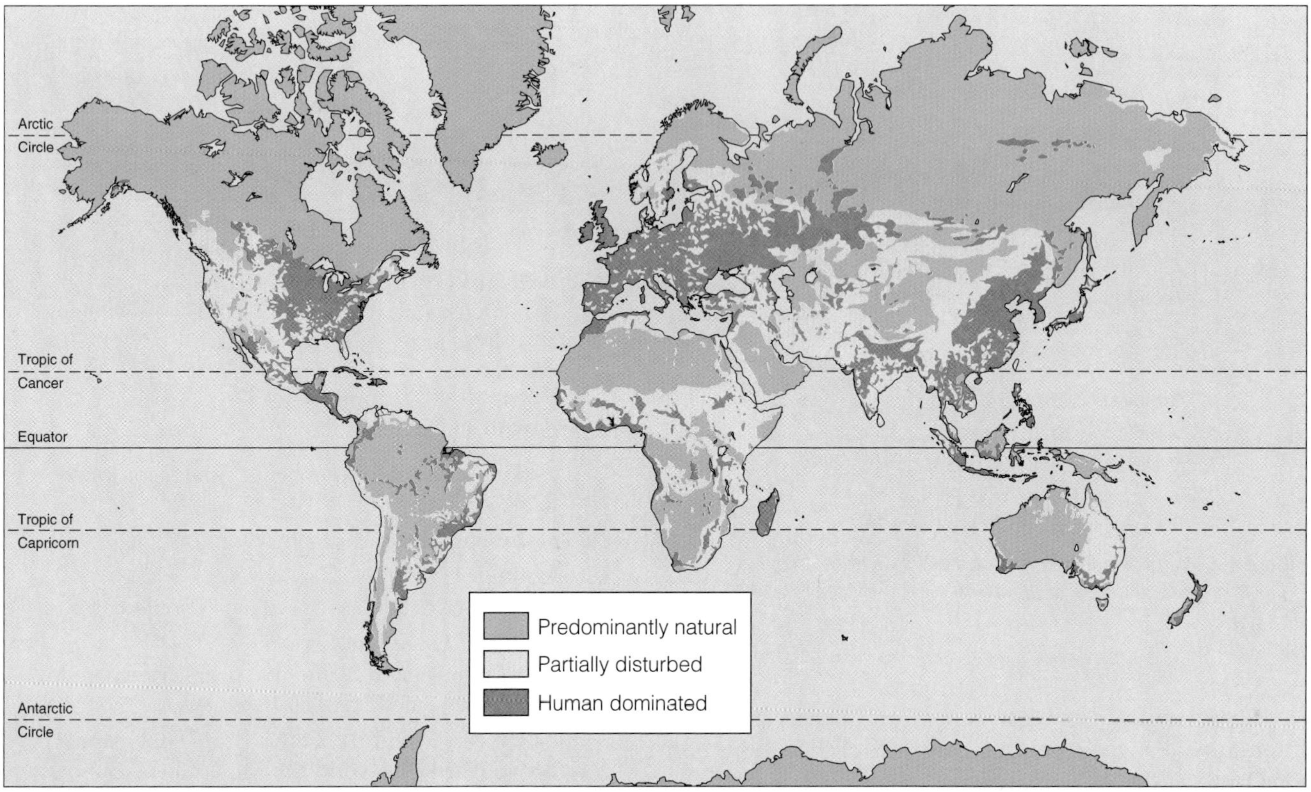

Figure 1-5 Human disturbance of the earth's land area. Green indicates undisturbed areas. Yellow indicates partially disturbed areas. Orange indicates seriously disturbed land areas, including deforestation, farmland, overgrazed grasslands, and urban areas. Excluding uninhabitable areas of rock, ice, desert, and steep mountain terrain, *only about 27% of the planet's land area remains undisturbed by human activities.* (Data from Conservation International)

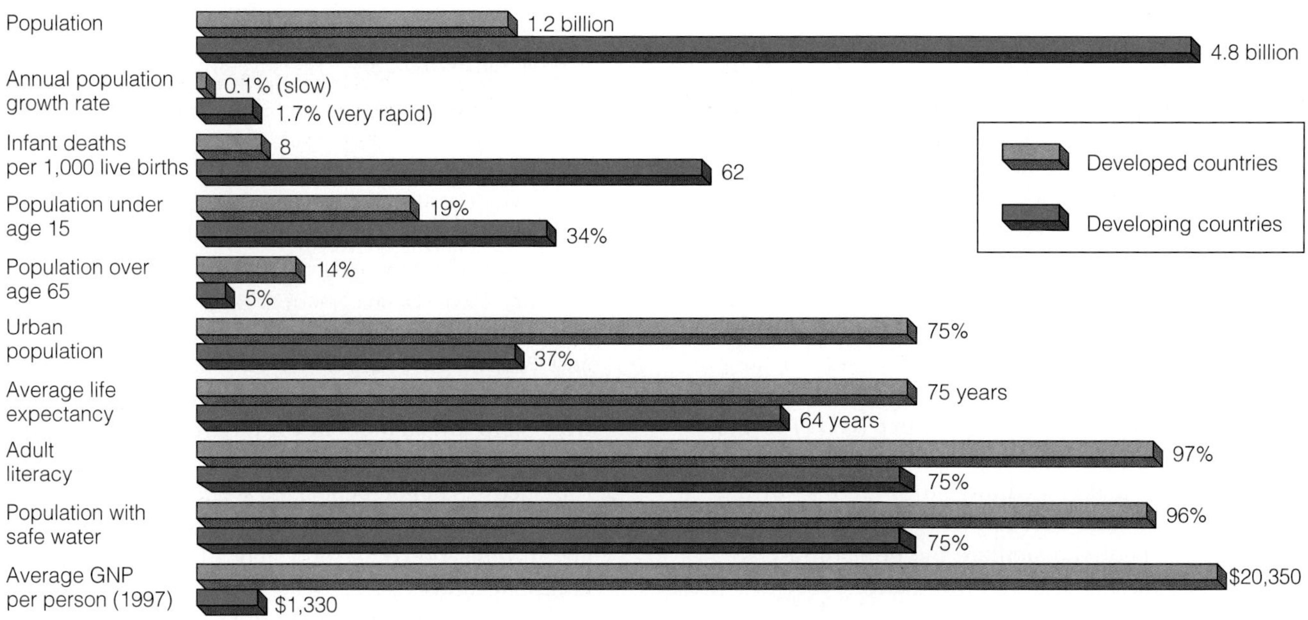

Figure 1-6 Some characteristics of developed countries and developing countries in 1999. (Data from United Nations and Population Reference Bureau)

is expected to take place in developing countries (Figure 1-7), *where 1 million people are added every 4 days*. The primary reason for such rapid population growth in developing countries (1.7% compared with 0.1% in developed countries) is the *large percentage of people who are under age 15* (34% compared to 19% in developed countries in 1999). As these young people move into their prime reproductive years over the next several decades, they will fuel rapid population growth.

Is Current Economic Growth Sustainable? Between 1950 and 1999 the global output of goods and services increased almost eightfold. Is such exponential economic growth sustainable? Some economists and business leaders say it is, but most environmental scientists and a growing number of economists disagree. They contend that many of the current forms of economic growth are not sustainable in the long run because of **(1)** the limits imposed by the earth's finite supplies of resources and **(2)** the environment's capacity to absorb, detoxify, and recycle the waste products of human societies (Guest Essay, p. 26).

Some of these analysts have called for emphasis on **sustainable development** in which we meet present human needs without preventing future generations of humans and other species from meeting their needs. Economist Herman Daly defines **sustainable development** as development without growth in the human throughput of matter and energy resources beyond the ability of the environment to supply and renew resources and to absorb or break down the wastes we produce.

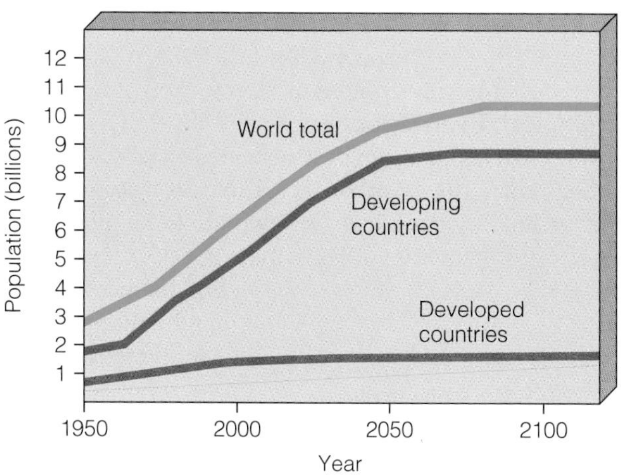

Figure 1-7 Past and projected population size for developed countries, developing countries, and the world, 1950–2120. More than 95% of the addition of 3.6 billion people between 1990 and 2030 is projected to occur in developing countries. (Data from United Nations)

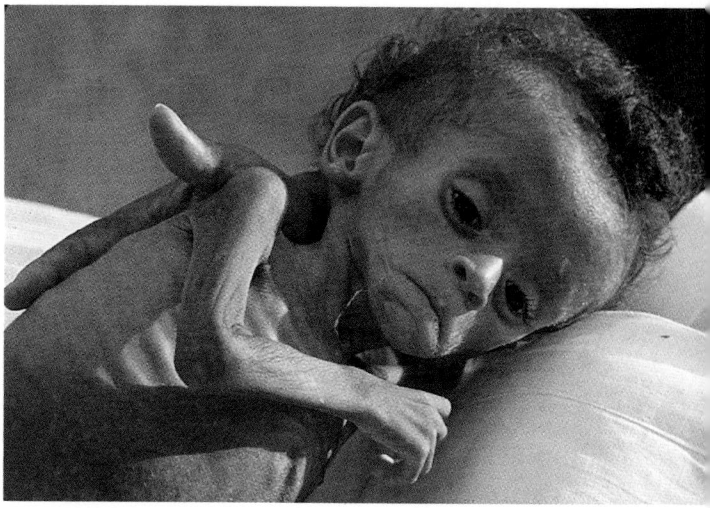

Figure 1-8 One in every three children under age 5, such as this Brazilian child, suffers from malnutrition. According to the World Health Organization, each day, at least 13,700 children die prematurely from malnutrition, or diseases related to malnutrition, and diseases from drinking contaminated water—an average of 10 preventable deaths each minute. Some analysts put the estimated death toll at twice this number. (John Bryson/Photo Researchers)

What Is the Wealth Gap? Since 1960, and especially since 1980, the gap between the per capita GNP of the rich, middle income, and poor has widened. Today, one person in five lives in luxury, and the next three get by. The fifth person (1.4 billion people worldwide) struggles to survive on less than $1 a day. One person in five is hungry or malnourished (Figure 1-8) and lacks decent housing and adequate health care. One of every four people lacks clean drinking water. One of every three people lacks enough fuel to keep warm and to cook food and does not have access to electricity. About two-thirds of humanity lacks sanitary toilets, and one of every four adults is illiterate.

Daily life for the estimated 1.4 billion desperately poor people in developing countries is a harsh struggle for survival. Parents, some with six or more children, are struggling to live on a cash income of about $1 a day or less. Having many children makes good sense to most poor parents. They see their children as a form of economic security to help them grow food, gather fuel (mostly wood and dung), haul drinking water, tend livestock, work, or beg in the streets. The desperately poor also tend to have many offspring because many of their children die at an early age. The two or three who live to adulthood can help their parents survive in old age (their 50s or 60s). Another 1.8 billion poor people struggle to survive on a cash income of about $3 per day.

However, poor families with several children may deplete and degrade local forests, soil, grasslands, wildlife, and water supplies. They do this for short-term survival even though they know it may lead to disaster in the long run.

The poor often have to live in areas with the highest levels of air and water pollution and with the greatest risk of natural disasters such as floods, earthquakes, hurricanes, and volcanic eruptions. They are also the ones who must take jobs (if they can find them) that subject them to unhealthy and unsafe working conditions at very low pay.

According to the World Health Organization, each year, at least 10 million of the desperately poor die from malnutrition (lack of protein and other nutrients needed for good health), diseases related to malnutrition and diseases from contaminated drinking water. *This premature death of about 27,400 human beings per day is equivalent to 69 jumbo jet planes, each carrying 400 passengers, crashing every day with no survivors.* Half of those dying are children under age 5 (Figure 1-8). A 1997 study by Johns Hopkins University researchers put this annual death toll acute from poverty at about 18 million a year, or an average of 49,300 premature deaths per day.

1-3 RESOURCES

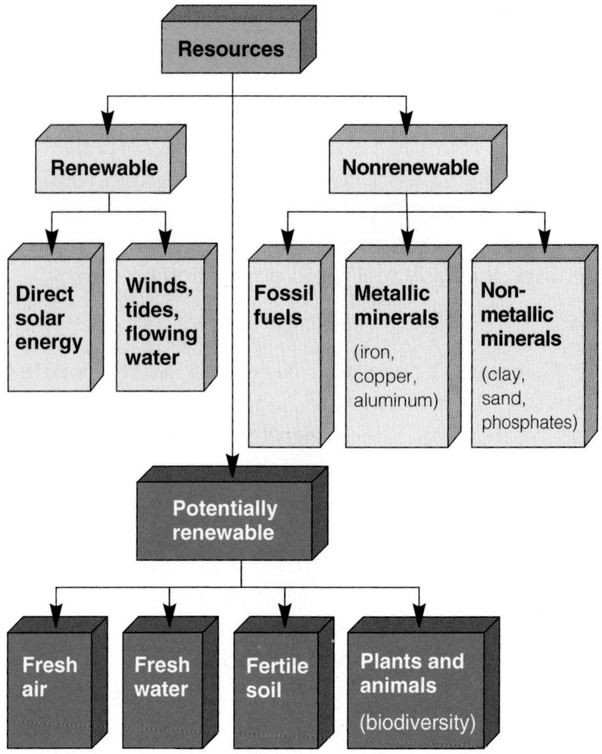

Figure 1-9 Major types of material resources. This scheme isn't fixed; potentially renewable resources can become nonrenewable resources if used for a prolonged period at a faster rate than they are renewed by natural processes.

What Is a Resource? Ecological Versus Economic Resources An **ecological resource** is anything required by an organism for normal maintenance, growth, and reproduction. Examples include habitat, food, water, and shelter.

An **economic resource** is anything obtained from the environment to meet human needs and wants. Examples include food, water, shelter, manufactured goods, transportation, communication, and recreation. On our short human time scale, we classify the material resources we get from the environment as *renewable, potentially renewable,* or *nonrenewable* (Figure 1-9; also see the orange boxes in the "Concepts and Connections" diagram inside the back cover).

Some resources, such as solar energy, fresh air, wind, fresh surface water, fertile soil, and wild edible plants, are directly available for use. Other resources, such as petroleum (oil), iron, groundwater (water found underground), and modern crops, aren't directly available. They become useful to us only with some effort and technological ingenuity. Petroleum, for example, was a mysterious fluid until we learned how to find, extract, and convert (refine) it into gasoline, heating oil, and other products that could be sold at affordable prices.

What Are Renewable Resources? Solar energy is called a **renewable** or **perpetual resource** because on a human time scale this solar capital (Figure 1-2) is continually renewed. It is expected to last at least 6 billion years as the sun completes its life cycle.

A **potentially renewable resource*** can be replenished fairly rapidly (hours to several decades) through natural processes. Examples are forest trees, grassland grasses, wild animals, fresh lake and stream water, groundwater, fresh air, and fertile soil.

One important potentially renewable resource is **biological diversity** or **biodiversity**: the different life forms (species) that can best survive the variety of conditions currently found on the earth. Kinds of biodiversity include **(1) genetic diversity** (variety in the genetic makeup among individuals within a species), **(2) species diversity** (variety among the species or distinct types of living organisms found in different habitats of the planet; Figure 1-10), and **(3) ecological diversity** (variety of forests, deserts, grasslands, streams, lakes, oceans, wetlands, and other biological communities).

This rich variety of genes, species, and biological communities gives us food, wood, fibers, energy, raw

*Most sources use the term *renewable resource.* I have added the word *potentially* to emphasize that these resources can be depleted if we use them faster than natural processes renew them.

Figure 1-10 Two species found in tropical forests are part of the earth's biodiversity. On the left is the world's largest flower, the flesh flower (*Rafflesia arnoldi*), growing in a tropical rain forest in Sumatra. The flower of this leafless plant can be as large as 1 meter (3.3 feet) in diameter and weigh 7 kilograms (15 pounds). The plant gives off a smell like rotting meat, presumably to attract flies and beetles that pollinate its flower. After blossoming once a year for a few weeks, the flower dissolves into a slimy black mass. On the right is a cotton top tamarin. (Left, Mitschuhiko Imanori/Nature Production; right, Gary Milburn/Tom Stack & Associates)

materials, industrial chemicals, and medicines, all of which pour hundreds of billions of dollars into the world economy each year. This biodiversity also provides us with free recycling, purification, and natural pest control services (Figure 1-2).

However, potentially renewable resources can be depleted. The highest rate at which a potentially renewable resource can be used *indefinitely* without reducing its available supply is called its **sustainable yield**. If we exceed a resource's natural replacement rate, the available supply begins to shrink, a process known as **environmental degradation**. Examples of such degradation include **(1)** urbanization of productive land, **(2)** waterlogging and salt buildup in soil, **(3)** excessive erosion of topsoil, **(4)** deforestation, **(5)** depletion of groundwater, **(6)** overgrazing of grasslands by livestock, **(7)** reduction in biodiversity by elimination of habitats and species, and **(8)** pollution. Such forms of environmental degradation can change potentially renewable resources into nonrenewable or unusable resources (Figure 1-11). A major cause of environmental degradation of potentially renewable resources is a phenomenon known as the *tragedy of the commons* (Connections, p. 14).

What Are Nonrenewable Resources? Resources that exist in a fixed quantity in the earth's crust are called **nonrenewable**, or **exhaustible**, **resources**. On a time scale of millions to billions of years, geological processes can renew such resources. However, on the much shorter human time scale of hundreds to thousands of years, these resources can be depleted much faster than they are formed.

These exhaustible resources include **(1)** *energy resources* (coal, oil, natural gas, and uranium, which cannot be recycled), **(2)** *metallic mineral resources* (iron, copper, and aluminum, which can be recycled), and **(3)** *nonmetallic mineral resources* (salt, clay, sand, and phosphates, which usually are difficult or too costly to recycle). A **mineral** is any hard, usually crystalline material that is formed naturally.

Figure 1-12 shows the production and depletion cycle of a nonrenewable energy or mineral resource. We never completely exhaust a nonrenewable mineral

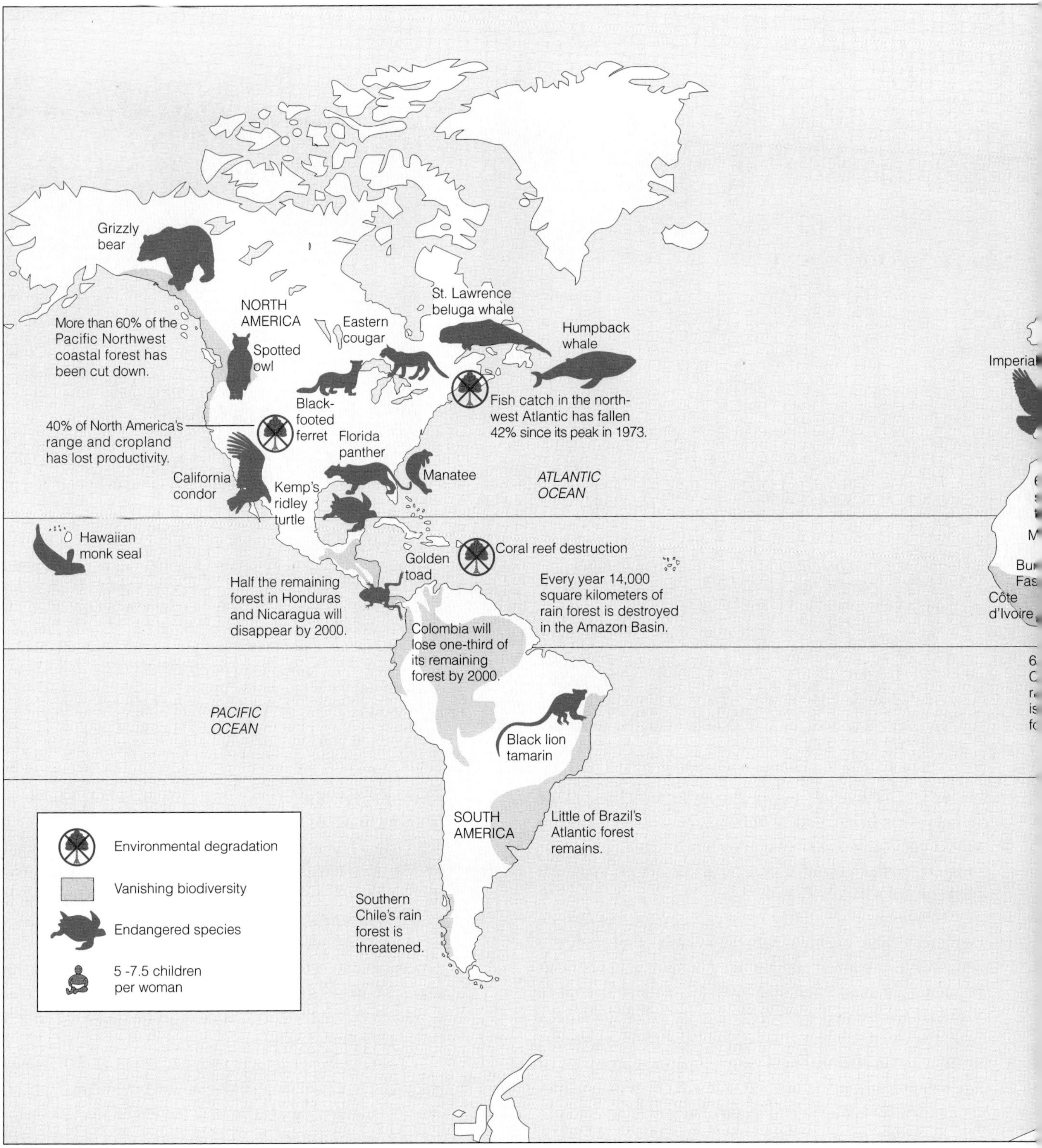

Figure 1-11 Examples of how some of the earth's natural systems are being assaulted at an accelerating rate as a result of the exponential growth of population and resource use and the resulting environmental degradation and loss of biodiversity. (Data from the World Conservation Union, World Wildlife Fund, Conservation International, United Nations, Population Reference Bureau, U.S. Fish and Wildlife Service, and Daniel Boivin)

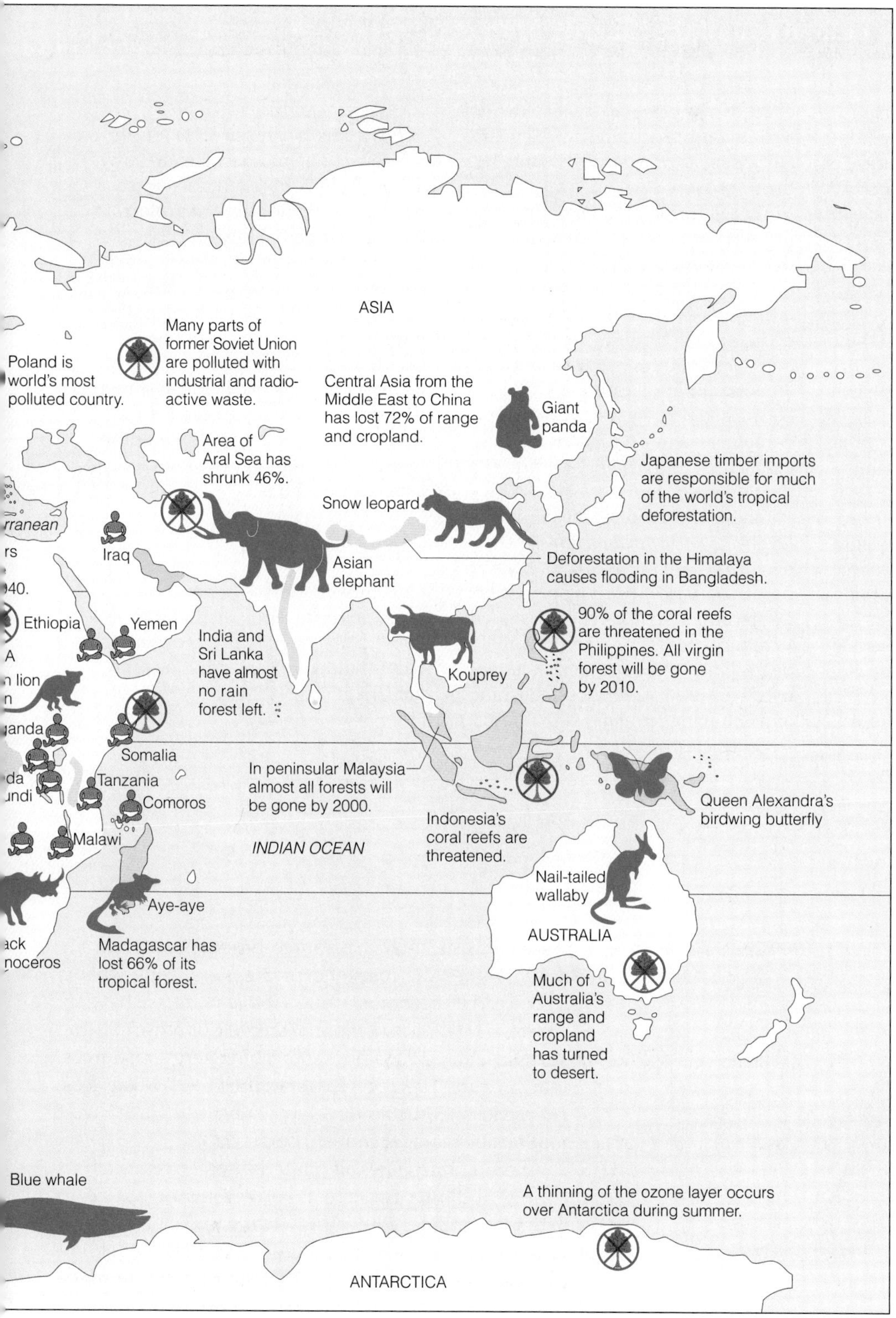
ASIA
Many parts of former Soviet Union are polluted with industrial and radio-active waste.
Poland is world's most polluted country.
Central Asia from the Middle East to China has lost 72% of range and cropland.
Giant panda
Area of Aral Sea has shrunk 46%.
Japanese timber imports are responsible for much of the world's tropical deforestation.
Snow leopard
rranean
rs
Iraq
Asian elephant
Deforestation in the Himalaya causes flooding in Bangladesh.
)40.
Ethiopia
Yemen
India and Sri Lanka have almost no rain forest left.
90% of the coral reefs are threatened in the Philippines. All virgin forest will be gone by 2010.
A
n lion
n
Kouprey
ganda
Somalia
da
Tanzania
undi
Comoros
In peninsular Malaysia almost all forests will be gone by 2000.
Queen Alexandra's birdwing butterfly
Malawi
INDIAN OCEAN
Indonesia's coral reefs are threatened.
Nail-tailed wallaby
Aye-aye
AUSTRALIA
ack
noceros
Madagascar has lost 66% of its tropical forest.
Much of Australia's range and cropland has turned to desert.
Blue whale
A thinning of the ozone layer occurs over Antarctica during summer.
ANTARCTICA

Renewable Resources and the Tragedy of the Commons

CONNECTIONS

One cause of environmental degradation is the overuse of **common-property resources**, which are owned by no one (or jointly by everyone in a country or area) but are available to all users at little or no charge. Most are potentially renewable. Examples include **(1)** clean air, **(2)** the open ocean and its fish, **(3)** migratory birds, **(4)** publicly owned lands (such as national forests, national parks, and wildlife refuges), **(5)** gases of the lower atmosphere, and **(6)** grazing lands, and **(7)** space.

In 1968, biologist Garrett Hardin (Guest Essay, p. 214) called the degradation of common-property resources the **tragedy of the commons**. It happens because each user reasons, "If I don't use this resource, someone else will. The little bit I use or pollute is not enough to matter."

With only a few users, this logic works. However, the cumulative effect of many people trying to exploit a common-property resource eventually exhausts or ruins it. Therein lies the tragedy.

One solution is to use common-property resources at rates below their sustainable yields or overload limits by reducing population, regulating access, or both. However, it is difficult to determine the sustainable yield of a forest, grassland, or an animal population, partly because yields vary with weather, climate, and unpredictable biological factors, and because getting such data is expensive.

These uncertainties mean that *it is best to use a potentially renewable resource at a rate well below its estimated sustainable yield*. This is a prevention or precautionary approach designed to reduce the risk of environmental degradation. This approach is rarely used because it requires establishing and enforcing regulations that restrict resource use.

Another approach is to convert common-property resources to private ownership. The reasoning is that owners of land or some other resource have a strong incentive to protect their investment. However, this approach is not practical for global common resources, such as the atmosphere, the open ocean, and migratory birds that cannot be divided up and converted to private property. Experience has also shown that private ownership can lead to short-term exploitation and environmental degradation instead of long-term sustainability.

Critical Thinking

Give three examples of how you cause environmental degradation as a result of the tragedy of the commons. How should we deal with this problem? Explain.

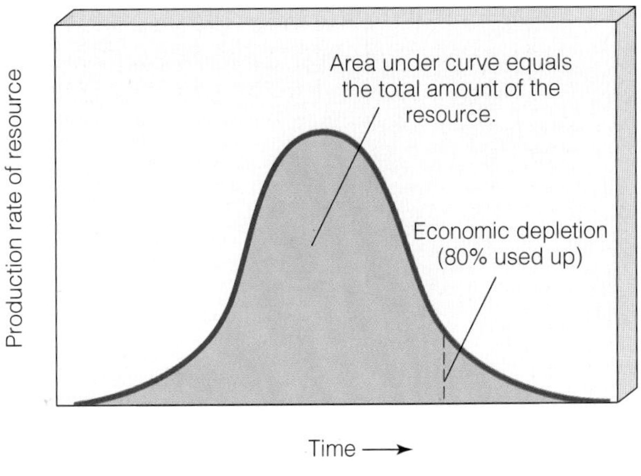

Figure 1-12 Full production and exhaustion cycle of a nonrenewable resource such as copper, iron, oil, or coal. Usually, a nonrenewable resource is considered *economically depleted* when 80% of its total supply has been extracted and used. Normally, it costs too much to extract and process the remaining 20%.

resource, but such a resource becomes *economically depleted* when the costs of extracting and using what is left exceed its economic value. At that point, we have six choices: **(1)** try to find more, **(2)** recycle or reuse existing supplies (except for nonrenewable energy resources, which cannot be recycled or reused), **(3)** waste less, **(4)** use less, **(5)** try to develop a substitute, or **(6)** wait millions of years for more to be produced.

Some nonrenewable material resources, such as copper and aluminum, can be recycled or reused to extend supplies. **Recycling** involves collecting and reprocessing a resource into new products. For example, glass bottles can be crushed and melted to make new bottles or other glass items. **Reuse** involves using a resource over and over in the same form. For example, glass bottles can be collected, washed, and refilled many times.

Recycling nonrenewable metallic resources requires much less energy, water, and other resources and produces much less pollution and environmental degradation than exploiting virgin metallic resources. Reuse of such resources requires even less energy and other resources and produces less pollution and environmental degradation than recycling.

Nonrenewable energy resources, such as coal, oil, and natural gas, can't be recycled or reused. Once burned, the useful energy in these fossil fuels is gone, leaving behind waste heat and polluting exhaust gases.

Most published estimates of the supply of a given nonrenewable resource refer to **reserves**: known deposits from which a usable mineral can be extracted profitably at current prices. Reserves increase when new deposits are found or when price increases make it profitable to extract deposits that previously were considered too expensive to exploit.

1-4 POLLUTION

What Is Pollution, and Where Does It Come From? Any addition to air, water, soil, or food that threatens the health, survival, or activities of humans or other living organisms is called **pollution**. Pollutants can enter the environment naturally (for example, from volcanic eruptions) or through human (anthropogenic) activities (for example, from burning coal). Most pollution from human activities occurs in or near urban and industrial areas, where pollutants are concentrated. Industrialized agriculture is also a major source of pollution.

Some pollutants contaminate the areas where they are produced; others are carried by wind or flowing water to other areas. Pollution does not respect local, state, or national boundaries.

Some pollutants come from single, identifiable sources, such as **(1)** the smokestack of a power plant, **(2)** the drainpipe of a meat-packing plant, or **(3)** the exhaust pipe of an automobile. These are called **point sources**. Other pollutants come from dispersed (and often difficult to identify) **nonpoint sources**. Examples are **(1)** the runoff of fertilizers and pesticides (from farmlands, golf courses, and suburban lawns and gardens) into streams and lakes and **(2)** pesticides sprayed into the air or blown by the wind into the atmosphere. It is much easier and cheaper to identify and control pollution from point sources than from widely dispersed nonpoint sources.

What Types of Harm Do Pollutants Cause? Unwanted effects of pollutants include **(1)** disruption of life-support systems for humans and other species, **(2)** damage to wildlife, human health, and property, and **(3)** nuisances such as noise and unpleasant smells, tastes, and sights.

Three factors determine how severe the harmful effects of a pollutant are. One is its *chemical nature*: how active and harmful it is to living organisms. Another is its **concentration**: the amount per unit of volume or weight of air, water, soil, or body weight. Concentration is sometimes expressed in *parts per million (ppm)*, with 1 ppm corresponding to 1 part pollutant per 1 million parts of the gas, liquid, or solid mixture in which the pollutant is found. Smaller concentration units are parts per billion (ppb) and parts per trillion (ppt).

One way to lower the concentration of a pollutant is to dilute it in a large volume of air or water. Until we started overwhelming the air and waterways with pollutants, dilution was *the* solution to pollution. Now it is only a partial solution.

The third factor is a pollutant's **persistence**: how long it stays in the air, water, soil, or body. **Degradable**, or **nonpersistent, pollutants** are broken down completely or reduced to acceptable levels by natural physical, chemical, and biological processes. Complex chemical pollutants broken down (metabolized) into simpler chemicals by living organisms (usually by specialized bacteria) are called **biodegradable pollutants**. Human sewage in a river, for example, is biodegraded fairly quickly by bacteria if the sewage is not added faster than it can be broken down.

Slowly degradable, or **persistent, pollutants** take decades or longer to degrade. Examples include the insecticide DDT and most plastics. **Nondegradable pollutants** cannot be broken down by natural processes. Examples include the toxic elements lead and mercury. The best ways to deal with slowly degradable and nondegradable pollutants are to avoid releasing them into the environment or to recycle or reuse them. Removing them from contaminated air, water, or soil is expensive and sometimes impossible.

We know little about the possible harmful effects of 90% of the 72,000 synthetic chemicals now in commercial use and the roughly 1,000 new ones added each year. Our knowledge about the effects of the other 10% of these chemicals is limited, mostly because it is quite difficult, time-consuming, and expensive to get this information. Even if we determine the main health and other environmental risks associated with a particular chemical, we know little about its possible interactions with other chemicals or the effects of such interactions on human health, other organisms, and life-support processes.

Solutions: What Can We Do About Pollution? There are two basic approaches to dealing with pollution: **(1)** prevent it from reaching the environment or

(2) clean it up if it does (Figure 1-13). **Pollution prevention** or **input pollution control** slows or eliminates the production of pollutants, often by using less harmful chemicals or processes. We can prevent (or at least reduce) pollution by following the four *R*s of resource use: *refuse (don't use), reduce, reuse,* and *recycle.*

Pollution cleanup or **output pollution control** involves cleaning up pollutants after they have been produced. Environmentalists have identified three problems with relying primarily on pollution cleanup:

- *It is only a temporary bandage as long as population and consumption levels grow without corresponding improvements in pollution control technology.* For example, adding catalytic converters to cars has reduced air pollution, but increases in the number of cars and in the total distance each travels have reduced the effectiveness of this cleanup approach.
- *It often removes a pollutant from one part of the environment only to cause pollution in another.* We can collect garbage, but the garbage is then **(1)** burned (perhaps causing air pollution and leaving a toxic ash that must be put somewhere), **(2)** dumped into streams, lakes, and oceans (perhaps causing water pollution), or **(3)** buried (perhaps causing soil and groundwater pollution).
- *Once pollutants have entered and become dispersed into the environment at harmful levels, it usually costs too much to reduce them to acceptable concentrations.*

Both pollution prevention and pollution cleanup are needed, but environmentalists and some economists urge us to emphasize prevention because it works better and is cheaper than cleanup. For nonpoint pollutants, hazardous wastes, and slowly degradable and nondegradable pollutants, pollution prevention is the most effective (perhaps the only) approach. As Benjamin Franklin reminded us long ago, "An ounce of prevention is worth a pound of cure." An increasing number of businesses have found that *pollution prevention pays.*

Governments can encourage both pollution prevention and pollution cleanup by **(1)** the *carrot approach* of using incentives such as various subsidies and tax write-offs or **(2)** the *stick approach* of regulations and

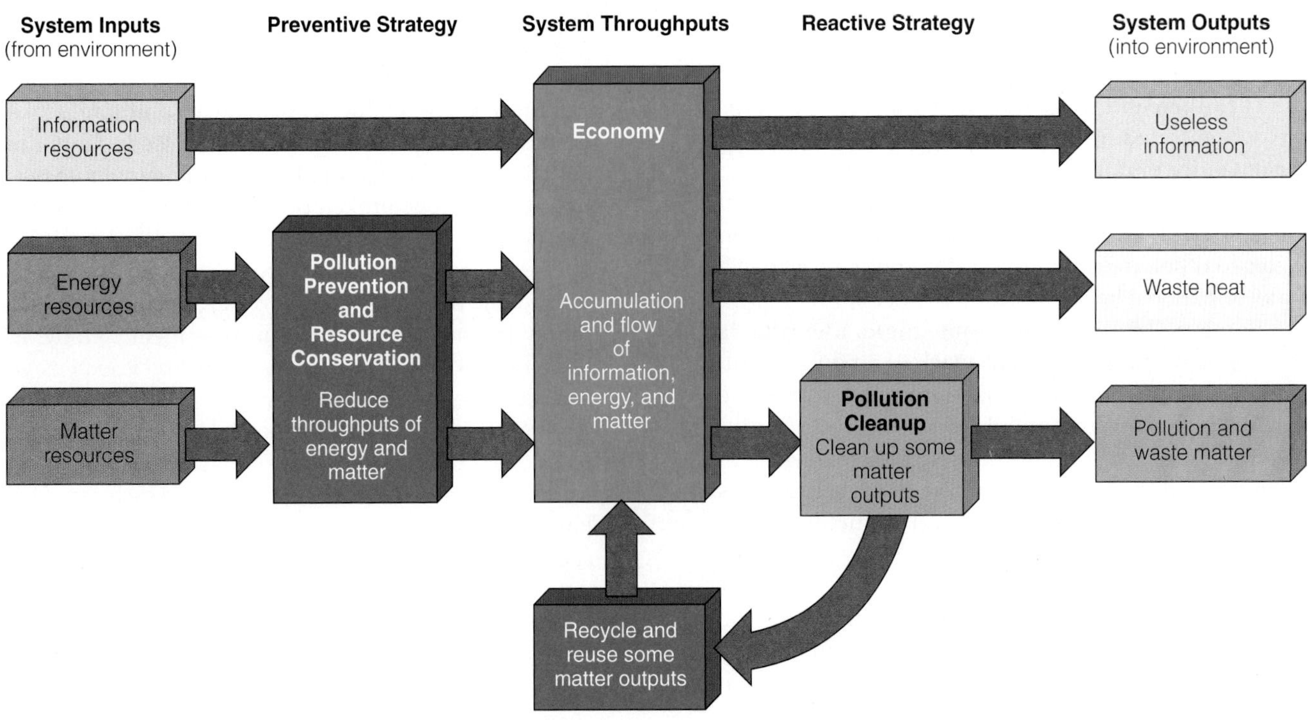

Figure 1-13 Inputs, throughputs, and outputs of a system (an economy) with two strategies for reducing pollution. Pollution prevention, or input pollution control, is based on reducing or eliminating pollution and conserving resources by reducing the flows of matter and energy resources through an economy. We can also reduce throughput by recycling or reusing some or most of the output of pollution or waste matter. Pollution cleanup, or output pollution control, involves trying to reduce pollutants to acceptable levels after they have been produced. Both approaches are needed, but environmentalists and a growing number of economists believe that we should place much greater emphasis on pollution prevention.

taxes. Most analysts believe that a mix of both approaches is best because excessive regulation and too much taxation can cause a political backlash. Achieving the right balance is difficult.

1-5 ENVIRONMENTAL AND RESOURCE PROBLEMS: CAUSES AND CONNECTIONS

What Are Key Environmental Problems and Their Root Causes? We face a number of interconnected environmental and resource problems (Figure 1-14). The first step in dealing with these problems is to identify their underlying causes. According to environmentalists, these include the following:

- Rapid population growth
- Use of resources with too little emphasis on pollution prevention and waste reduction
- Simplification and degradation of parts of the earth's life-support systems
- Poverty, which can drive poor people to use potentially renewable resources unsustainably for short-term survival and often exposes the poor to health risks and other environmental risks
- Failure of economic and political systems to encourage earth-sustaining forms of economic development and to discourage earth-degrading forms of economic growth
- Failure of economic and political systems to include the overall environmental cost of an economic good or service in its market price
- Attempts to dominate and manage nature for our use with far too little knowledge about how nature works

How Are Environmental Problems and Their Root Causes Connected? Once we have identified environmental problems and their root causes, the next step is to understand how they are connected to one another. The three-factor model in Figure 1-15 is a good starting point.

According to this simple model, the environmental impact of population on a given area depends on three factors: **(1)** the number of people (population size, P), **(2)** the average number of units of resources each person uses (per capita consumption or affluence, A), and **(3)** the amount of environmental degradation and pollution produced for each unit of resource used (the environmental destructiveness of the technologies used to provide and consume resources, T). This model, developed in the early 1970s by biologist Paul Ehrlich and physicist John Holdren, can be summarized as

$$\text{Population} \times \text{Affluence} \times \text{Technology} = \text{Environmental impact}$$

or

$$P \times A \times T = I$$

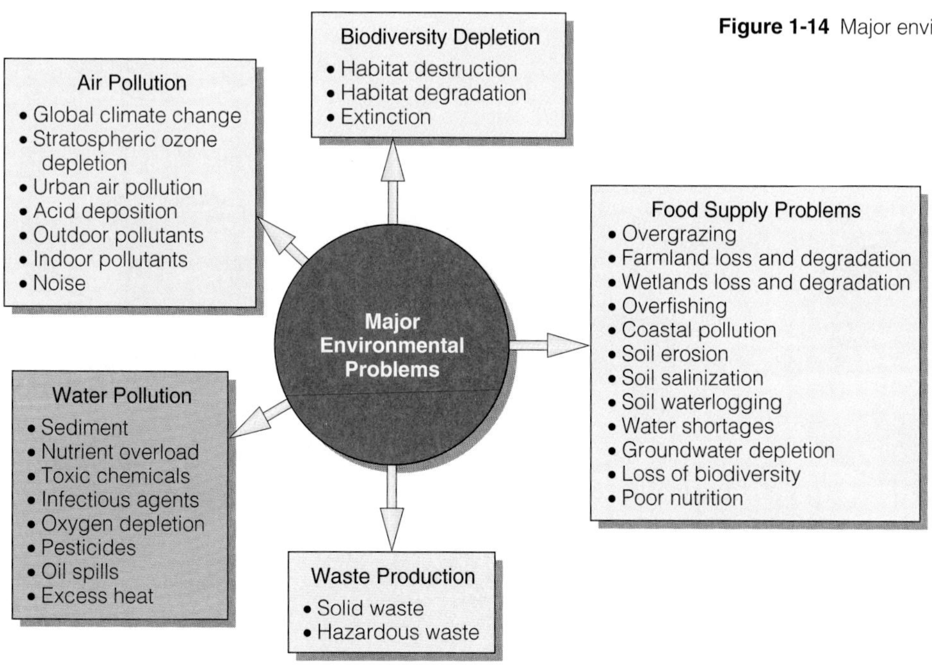

Figure 1-14 Major environmental and resource problems.

Figure 1-16 shows how the three factors depicted in Figure 1-15 can interact in developing countries and developed countries. In developing countries, population size and the resulting degradation of potentially renewable resources (as the poor struggle to stay alive) tend to be the key factors in total environmental impact.

In developed countries, high rates of per capita resource use and the resulting high levels of pollution and environmental degradation per person are usually the key factors determining overall environmental impact. For example, it is estimated that the average U.S. citizen consumes 35 times as much as the average citizen of India and 100 times as much as the average person in the world's poorest countries. *Thus, poor parents in a developing country would need 70–200 children to have the same lifetime environmental impact as 2 children in a typical U.S. family.*

Some forms of technology, such as polluting factories and motor vehicles and energy-wasting devices, increase environmental impact by raising the *T* factor

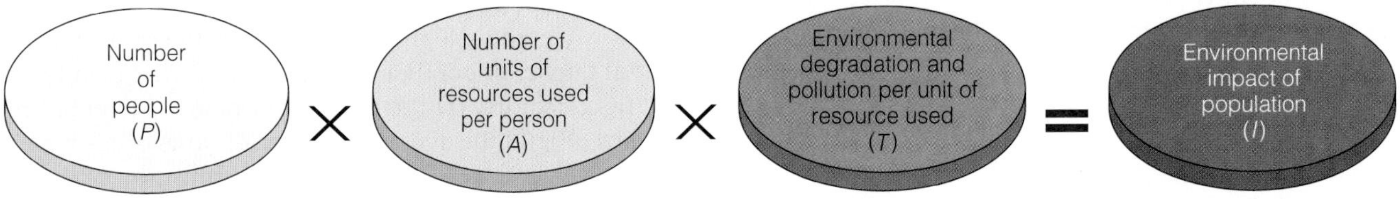

Figure 1-15 Simplified model of how three factors—population, affluence, and technology—affect the environmental impact of population. According to this model, the damage we do to the earth is equal to the number of people there are, multiplied by the amount of resources each person uses, multiplied by the amount of pollution, resource waste, and environmental degradation involved in extracting, making, and using each unit of resource.

Figure 1-16 Environmental impact of developing countries (top) and developed countries (bottom) based on the relative importance of the factors in the model shown in Figure 1-15. Circle size shows the relative importance of each factor. We can reduce the size of the *T* factor by improving technology for controlling and preventing pollution, resource waste, and environmental degradation.

in the equation. Other technologies, such as pollution control, solar cells, and more energy-efficient devices, can lower environmental impact by decreasing the *T* factor in the equation.

The three-factor model in Figure 1-15 can help us understand how key environmental problems and some of their causes are connected. However, these problems involve a number of poorly understood interactions among many more factors than those in the three-factor model (Figure 1-17).

1-6 CULTURAL CHANGES AND SUSTAINABILITY

What Major Human Cultural Changes Have Taken Place? Evidence from fossils and studies of ancient cultures suggests that the current form of our species, *Homo sapiens sapiens*, has walked the earth for only about 60,000 years (some recent evidence suggests 90,000 to 176,000 years), an instant in the planet's estimated 4.6-billion-year existence. Until about 12,000

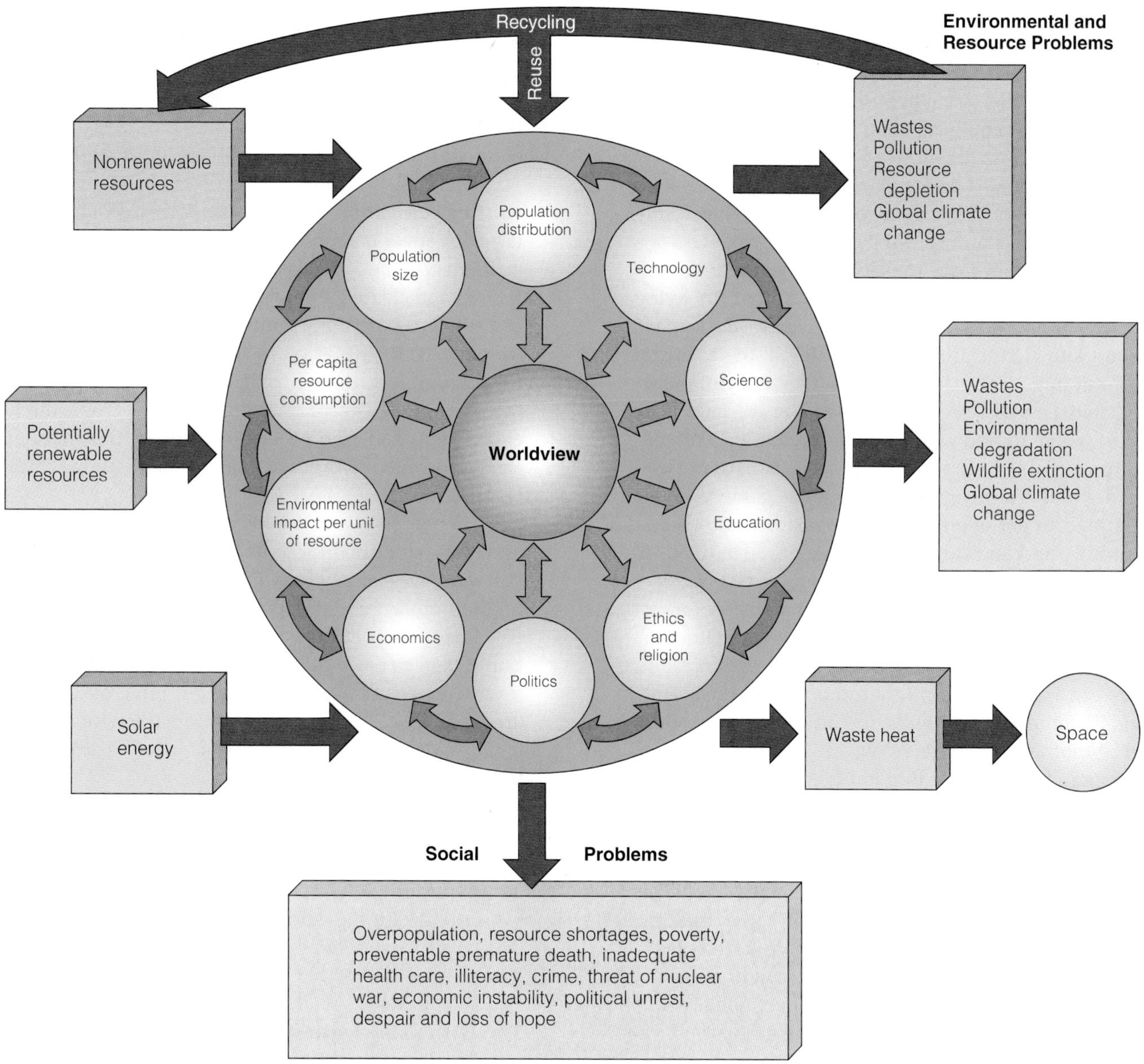

Figure 1-17 Environmental, resource, and social problems are caused by a complex, poorly understood mix of interacting factors, as illustrated by this simplified model.

years ago, we were mostly hunter–gatherers who moved as needed to find enough food for survival. Since then, there have been two major cultural shifts: **(1)** the **agricultural revolution,** which began 10,000–12,000 years ago, and **(2)** the **industrial revolution,** which began about 275 years ago.

These cultural revolutions have given us much more energy and new technologies with which to alter and control more of the planet to meet our basic needs and increasing desires. By expanding food supplies, lengthening life spans, and raising living standards for many people, each cultural shift contributed to the expansion of the human population and increased resource use, pollution, and environmental degradation.

How Did Ancient Hunting-and-Gathering Societies Affect the Environment? During most of our 60,000-year existence, we were **hunter–gatherers** who survived by collecting edible wild plant parts, hunting, fishing, and scavenging meat from animals killed by other predators. Our hunter–gatherer ancestors typically lived in small bands (of fewer than 50 people) who worked together to get enough food to survive. Many groups were nomadic, picking up their few possessions and moving from place to place to find enough food.

The earliest hunter–gatherers (and those still living this way today) survived through *environmental wisdom*: expert knowledge and understanding of their natural surroundings. They discovered that a variety of plants and animals could be eaten and used as medicines. They knew where to find water, how to predict the weather with reasonable accuracy, how plant availability changed throughout the year, and how some game animals migrated to get enough food. Because of high infant mortality and an estimated average life span of 30–40 years, hunter–gatherer populations grew very slowly.

Advanced hunter–gatherers had a greater impact on their environment than did early hunter–gatherers. Their use of fire converted forests into grasslands and they contributed to (perhaps even caused) the extinction of some large animals, including the mastodon, saber-toothed tiger, giant sloth, cave bear, mammoth, and giant bison. According to other researchers, such extinctions were caused mostly by rapid changes in climate, especially ice ages.

As these early humans moved to new areas, many probably carried plant seeds and roots with them. This altered the distribution of plants and, in some cases, the types of animals feeding on the newly introduced plants.

Hunter–gatherers exploited their environment to survive, but their environmental impact usually was limited and local. Because of their small groups and frequent migrations, natural processes repaired most of the damage they caused.

Most early and advanced hunter–gatherers were *dwellers in nature* who lived sustainably by using few resources per person and by learning to work with nature and with one another in small groups.

How Has the Agricultural Revolution Affected the Environment? Some 10,000–12,000 years ago, a cultural shift known as the **agricultural revolution** began in several regions of the world. It involved a gradual move from nomadic hunting-and-gathering groups to settled agricultural communities in which people domesticated wild animals and cultivated wild plants.

Plant cultivation probably developed in many areas, especially in the tropical forests of southeast Asia, northeast Africa, and Mexico. People discovered how to grow various wild food plants from roots or tubers (fleshy underground stems). To prepare the land for planting, they cleared small patches of tropical forests by cutting down trees and other vegetation and then burning the underbrush (Figure 1-18). The ashes fertilized the nutrient-poor soils in this **slash-and-burn cultivation**.

These early growers also used various forms of **shifting cultivation** (Figure 1-18). After a plot had been used for several years, the soil would be depleted of nutrients or reinvaded by the forest. Then the growers moved and cleared a new plot. They learned that each abandoned patch normally had to be left fallow (unplanted) for 10–30 years before the soil became fertile enough to grow crops again. While patches were regenerating, growers used them for tree crops, medicines, fuelwood, and other purposes. In this manner, early growers practiced sustainable cultivation within tropical forests.

Their dependence mostly on human muscle power and crude stone or stick tools meant that growers could cultivate only small plots; thus, they had little impact on their environment. However, the main reason that this form of agriculture was sustainable was that the population size and density of these early farmers were low. This meant that people could move on to other areas and leave abandoned plots unplanted for the several decades needed to restore soil fertility. The gradual shift from hunting and gathering to farming had several significant effects (Connections, p. 22).

The spread of agriculture meant that most of the world's human population gradually shifted from being hunter–gatherers, working with nature to survive, to becoming shepherds, farmers, and urban dwellers trying to tame and manage nature to survive and prosper.

How Has the Industrial Revolution Affected the Environment? The next great cultural shift, the **industrial revolution**, began in England in the mid-

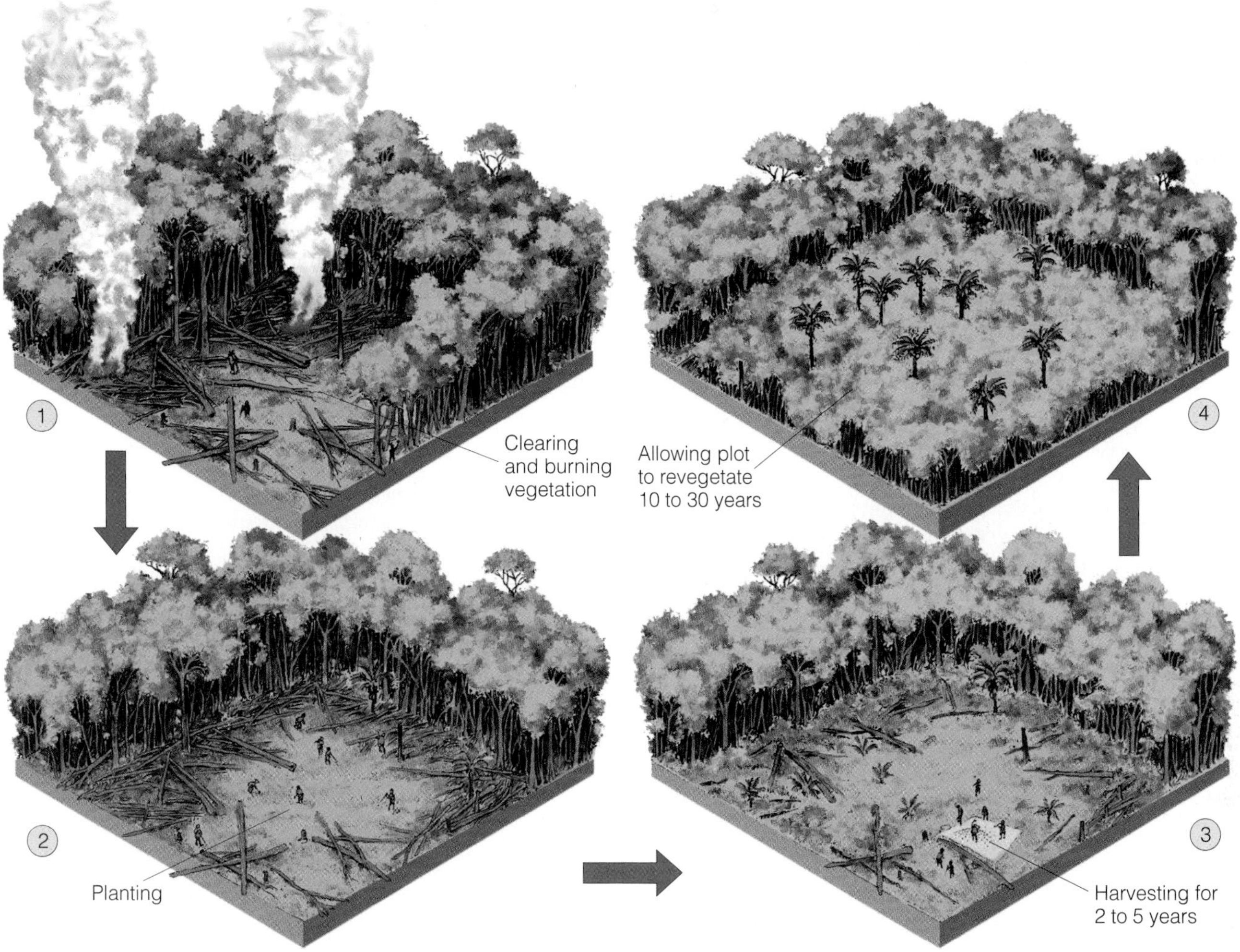

Figure 1-18 The first crop-growing technique may have been a combination of slash-and-burn and shifting cultivation in tropical forests. This method is sustainable only if small plots of the forest are cleared, cultivated for no more than 5 years, and then allowed to regenerate for 10–30 years to renew soil fertility. Indigenous cultures have developed many variations of this technique and have found ways to make some nondestructive uses of former plots while they are being regenerated.

1700s and spread to the United States in the 1800s. Production, commerce, trade, and distribution of goods all expanded rapidly.

With the industrial revolution came a shift from dependence on potentially renewable wood (with supplies dwindling in some areas because of unsustainable cutting) and flowing water to dependence on machines running on nonrenewable fossil fuels (first coal and later oil and natural gas). This led to a switch from small-scale, localized production of handmade goods to large-scale production of machine-made goods in centralized factories within rapidly growing industrial cities.

Factory towns grew into cities as rural people came to the factories for work. There they worked long hours under noisy, dirty, and hazardous conditions. Other workers toiled in dangerous coal mines. In these early industrial cities, coal smoke belching out of chimneys was so heavy that many people died of lung ailments. Ash and soot covered everything, and on some days the smoke was so thick that it blotted out the sun.

Fossil-fuel–powered farm machinery, commercial fertilizers, and new plant-breeding techniques increased per acre crop yields. This helped protect biodiversity by reducing the need to expand the area of cropland to grow food. Because fewer farmers were needed, more people migrated to cities. With a larger and more reliable food supply, the size of the human population began the sharp increase that continues today (Figures 1-1 and 1-4).

After World War I (1914–18), more efficient machines and mass-production techniques were developed.

Consequences of the Agricultural Revolution

CONNECTIONS

Here are some of the effects of the agricultural revolution:

- *Using domesticated animals to plow fields, haul loads, and perform other tasks increased the ability to expand agriculture and support more people.*
- *People* **(1)** *cut down vast forests to supply wood for fuel and building materials,* **(2)** *plowed up large expanses of grassland to grow crops, and* **(3)** *built irrigation systems to transfer water from one place to another.* Such extensive land clearing degraded or destroyed the habitats of many wild plants and animals, causing or hastening their extinction.
- *Soil erosion, salt buildup in irrigated soils, and overgrazing of grasslands by huge herds of livestock helped turn fertile land into desert; topsoil washed into streams, lakes, and irrigation canals.* This environmental degradation was a factor in the downfall of many great civilizations in the Middle East, North Africa, and the Mediterranean.

Historian Henry Kissinger reminds us, "As a historian, you have to be conscious of the fact that every civilization that has ever existed has ultimately collapsed."

- *People began accumulating material goods.* Nomadic hunter–gatherers could not carry many possessions in their travels, but farmers living in one place could acquire as much as they could afford.
- *Farmers could grow more than enough food for their families.* They could store the excess for a "rainy day" or use it to barter with craftspeople.
- *Urbanization—the formation of villages, towns, and cities—became practical.* Some villages grew into towns and cities, which served as centers for trade, government, and religion. Towns and cities concentrated sewage and other wastes, polluted the air and water, and greatly increased the spread of diseases.
- *Increased production and use of material goods created growing volumes of wastes.*
- *Conflict between societies became more common as ownership of land and water rights became crucial economic issues.* Armies and their leaders rose to power and conquered large areas of land and water supplies. These rulers forced powerless people (slaves and landless peasants) to do the hard, disagreeable work of producing food and constructing things such as irrigation systems, temples, and walled fortresses.
- *The survival of wild plants and animals, once vital to humanity, became less important.* Wild animals, which competed with livestock for grass and fed on crops, became enemies to be killed or driven from their habitats. Wild plants invading cropfields became weeds to be eliminated.

Critical Thinking

Would we be better off if agricultural practices had never been developed and we were still hunters and gatherers? Explain.

These technologies became the basis of today's advanced industrial societies in places such as the United States, Canada, Japan, Australia, and western Europe. Advanced industrial societies have provided numerous benefits along with environmental problems (Connections, right).

How Might the Information Revolution Affect the Environment? We are in the midst of a new cultural shift, the **information revolution**, in which new technologies such as the telephone, radio, television, and computers are enabling people to deal with more information more rapidly.

What the information revolution means for the environment is not yet clear. On the positive side, we can understand and respond to environmental problems more effectively. Today's information technologies also illustrate how we are globally interconnected with one another and with nature.

On the negative side, we are increasingly faced with an overload of information (infoglut). Much of it is untrue, misleading, or useless (infojunk). This can lead to confusion, distraction, and a sense of hopelessness as we try to identify useful environmental information and ideas in a rapidly growing sea of information.

A growing number of analysts believe that what we need is a *wisdom revolution* based on learning how to discern what information is important. Such analysts see an urgent need to revise the education system to train people in using *critical thinking* to evaluate and filter out incorrect, misleading, or useless information.

1-7 IS OUR PRESENT COURSE SUSTAINABLE?

Are Things Getting Better or Worse? Experts have conflicting views about how serious our population and environmental problems are and what we should do about them (Table 1-1, pp. 24–5). Some analysts, mostly economists, believe that human ingenuity and technological advances will allow us to **(1)** clean up pollution to acceptable levels, **(2)** find substitutes for any resources that become scarce, and **(3)** keep expanding the earth's ability to support more humans, as we have done in the past. They accuse most scientists and environmentalists of exaggerating the seriousness of

Suter, K. 1999. "The Club of Rome: The Global Conscience." *Contemporary Review*, vol. 275, i1602, p. 1(5).

Consequences of Advanced Industrial Societies

CONNECTIONS

Advanced industrial societies provide a variety of benefits to most people living in them, including

- Mass production of many useful and economically affordable products
- A sharp increase in agricultural productivity
- Lower infant mortality and longer life expectancy because of better sanitation, hygiene, nutrition, and medical care
- A decrease in the rate of population growth
- Better health, birth control methods, and education
- Methods for controlling pollution
- Greater average income and old-age security

However, the bad news is the resource and environmental problems we face today (Figure 1-14), mostly because of the rise of advanced industrial societies. Industrialization also isolates more people from nature and reduces understanding of the important ecological and economic services nature provides (Figure 1-2).

Critical Thinking

1. On balance, do you believe that the advantages of the industrial revolution have outweighed its disadvantages? Explain.

2. What three major things would you do to reduce the harmful environmental impacts of advanced industrial societies?

the problems we face and of failing to appreciate the progress (Table 1-1, left) we have made in improving quality of life and protecting the environment.

On the other hand, environmentalists and many scientists, and some economists contend that we are depleting and degrading the earth's natural capital at an accelerating rate and that this is leading to serious environmental and economic harm (Table 1-1, p. 24, and Guest Essay, p. 26). They are encouraged by the progress we have made. However, they point out how much more we must do to help make the earth more sustainable for present and future human generations and for other species that support us and other forms of life.

On November 18, 1992, some 1,680 of the world's senior scientists from 70 countries, including 102 of the 196 living scientists who are Nobel laureates, signed and sent an urgent warning to government leaders of all nations. According to this warning,

> *The environment is suffering critical stress. . . . Our massive tampering with the world's interdependent web of life—coupled with the environmental damage inflicted by deforestation, species loss, and climate change—could trigger widespread adverse effects, including unpredictable collapses of critical biological systems whose interactions and dynamics we only imperfectly understand. Uncertainty over the extent of these effects cannot excuse complacency or delay in facing the threats. . . . No more than one or a few decades remain before the chance to avert the threats we now confront will be lost and the prospects for humanity immeasurably diminished. . . . Whether industrialized or not, we all have but one lifeboat. No nation can escape injury when global biological systems are damaged. . . . We must recognize the earth's limited capacity to provide for us.*

Also in 1992, the prestigious U.S. National Academy of Sciences and the Royal Society of London issued a joint report, their first ever, which began,

> *If current predictions of population growth prove accurate and patterns of human activity on the planet remain unchanged, science and technology may not be able to prevent either irreversible degradation of the environment or continued poverty for much of the world.*

These two major warnings are not the views of a small number of scientists but the consensus of the mainstream scientific community, consisting of most of the world's key researchers on environmental problems.

Leading environmentalists call for us to launch an *environmental or sustainability revolution* to take place over the next 50 years (Guest Essay, p. 26). This would involve shifting our efforts from

- Pollution cleanup to pollution prevention (clean production)
- Waste disposal (mostly burial and burning) to waste prevention and reduction
- Protecting species to protecting the places where they live (habitats)
- Environmental degradation to environmental restoration
- Increased resource use to increased resource conservation

Do you agree or disagree with this assessment?

This chapter has presented an overview of the problems most environmentalists and many of the world's most prominent scientists believe we face and

Hint: Enter the search term *environmental problems* using the Subject Guide.

Table 1-1 Good and Bad News About Environmental Problems

Some Good News	Some Bad but Challenging News
Annual world population growth slowed from 2.2% to 1.33% between 1963 and 1999.	The world's population is still growing rapidly and is projected to increase from 6.0 billion to 8 billion between 1999 and 2028.
Global life expectancy rose from 33 to 66 years between 1900 and 1999.	Life expectancy in developing countries is 11 years less than in developed countries.
Annual global infant mortality dropped by 65% between 1900 and 1999.	Infant mortality in developing countries is eight times higher than in developed countries and could be improved in many developed countries.
Global grain production has outpaced population growth since 1978.	Since 1980 population growth has exceeded grain production in 69 developing countries, and the global per capita fish catch has declined since 1988. Future food production may be limited by its harmful enviromental impacts.
The percentage of the world's population that are hungry has decreased since 1960.	Because of population growth there are now more hungry people (841 million) than ever, and at least 600 million are seriously malnourished.
Biodiversity loss has been reduced by higher crop yields on less land.	The challenge is to increase high yields while sharply reducing the growing environmental impacts of such intensive production.
Biodiversity loss has been reduced by increased urbanization, with more people living on less land.	This effect is misleading because the higher resource use per person in cities must be supplied by croplands, forests, grasslands, fisheries, and mines elsewhere. This often diminishes biodiversity and increases environmental degradation in these areas that sustain otherwise unsustainable cities.
At least 5% of the world's land area has been set aside to protect wildlife and wildlife habitats.	This is half of the estimated minimum needed. Many wildlife sanctuaries exist on paper only and receive little protection, and many important types of biological communities are not protected. Every hour an estimated two to eight wildlife species become extinct because of human activities; this rate is expected to increase over the next few decades.
Total forest area in most temperate industrial countries increased during the 1980s.	Much of this increase came from replacing sustainable, biologically diverse old-growth forests with simplified and vulnerable tree farms and less diverse second growth forests. This has led to reduced forest biodiversity.
Tropical deforestation has been exaggerated, and enough wood, paper, and food can be produced by converting much of such diverse forests to tree farms, cropland, and grazing land.	If current deforestation rates continue, within 30 to 50 years little of these forests will remain. This will lead to a massive loss of biodiversity because such forests contain at least 50% of the world's terrestrial species.
Conservation tillage and no-till cultivation can cut soil erosion by about 75%.	These techniques are widely used in the United States but are rarely used in other countries. Topsoil is eroding faster that it forms on about one-third of the world's cropland. Deforestation, crop growing, and construction are increasing soil erosion in many countries.
Food industry scientists contend that the benefits of pesticides outweigh their harmful effects.	Most environmental and health scientists contend that the harmful effects of pesticides outweigh their beneficial effects. Reliance mostly on chemical pesticides (all of which eventually fail because of genetic resistance by pests) should be replaced with an ecological approach called integrated pest management (IPM).
A handful of scientists contend that global warming is not a serious problem.	The vast majority of scientists in this field contend that global warming is one of the world's most serious environmental and economic problems.

Table 1-1 ***(continued)***

Some Good News	Some Bad but Challenging News
In the long run, the ecological and economic benefits of reducing the threat of global warming by shifting from nonrenewable fossil fuels to renewable solar-based energy sources and reducing deforestation (especially in the tropics) will outweigh the costs. Such changes will also reduce pollution and environmental degradation and protect bioversity.	Some economists say that such changes will be too costly.
A handful of scientists say that depletion of ozone in the stratosphere (the atmosphere's second layer) is not a serious problem.	The vast majority of scientists in this field contend that ozone depletion in the stratosphere is a very serious problem.
Water is recycled by a natural cycle, and the earth has much more water that it needs. Dams and water transfer projects can meet projected water shortages.	Earth's water is not distributed equally, and much of it is inaccessible, unnecessarily wasted, or polluted. Currently, about 1.2 billion people lack access to clean drinking water. By 2025 at least 3 billion people in 90 countries are expected to face chronic water shortages. Large dams and water transfer projects lead to serious environmental degradation.
Since 1950 proven supplies (reserves) of virtually all nonrenewable fossil fuel and key mineral resources have increased significantly.	The exponential increase in the use of fossil fuel and mineral resources is causing massive land degradation, water pollution, and air pollution. These harmful environmental effects—not supplies or costs—may limit future resource use.
Adjusted for inflation, most nonrenewable fossil fuel and key mineral resources cost less today than in 1950.	These resource prices are low because most of the harmful environmental and health costs associated with their production are not included in their market prices. Such artificially low resource prices encourage waste, pollution, and environmental degradation and discourage waste reduction and pollution prevention.
In the United States, the recycling and composting of solid municipal solid waste (produced by households and businesses) increased from 7% to 30% between 1970 and 1999.	At least 60% of all municipal solid waste could be reused, recycled, or composted. Only about 1.5% of the estimated solid wasted produced in the United States is municipal solid waste. The other 98.5% comes mostly from mining, fossil fuel production, agriculture, and industrial activities, with very little of it being recycled, reused, or composted.
Some scientists contend that the health risks from exposure to toxic and hazardous chemicals are exaggerated.	The harmful effects of most chemicals are unknown, and a growing body of evidence suggests that their harmful effects (especially to the nervous, immune, and endocrine systems) have been underestimated.
Since 1970, air and water pollution levels in most industrialized countries have dropped significantly for most pollutants.	More people using more resources and a lack of emphasis on pollution prevention are threatening gains in pollution control in industrialized countries from laws enacted in the 1970s. Little progress has been made in reducing air and water pollution in most developing countries.

their root causes. It has also summarized the controversy over how serious environmental problems are and the cultural changes that have influenced how we treat the earth.

The rest of this book presents a more detailed analysis of these problems, the controversies they have created, and solutions proposed by analysts. To most environmentalists there are two important keys to dealing with the environmental problems and challenges we face:

- Learning as much as we can about how the earth sustains itself and adapts to ever-changing environmental conditions and integrating such lessons from nature into the ways we think and act.
- Recognizing that individuals matter. Anthropologist Margaret Mead has summarized our potential for change: "Never doubt that a small group of thoughtful, committed citizens can change the world. Indeed, it is the only thing that ever has."

What's the use of a house if you don't have a decent planet to put it on?

HENRY DAVID THOREAU

GUEST ESSAY

Launching the Environmental Revolution

Lester R. Brown

Lester R. Brown is president of the Worldwatch Institute, a private nonprofit research institute he founded in 1974 that is devoted to analysis of global environmental issues. Under his leadership, the institute publishes the annual State of the World Report, considered by many environmentalists and world leaders to be the best source of information about key environmental issues. It also publishes monographs on specific topics, World Watch *magazine, and a series of Environmental Alert books. He is author of more than a dozen books, recipient of the McArthur Foundation Genius Award, and winner of the United Nations' 1989 environment prize. He has been described by the* Washington Post *as "one of the world's most influential thinkers."*

The world faces potentially convulsive change. Two challenges facing us at the dawn of a new century are that the human population is 4 times as large as it was a century ago and the world economy is 17 times as large. This growth has allowed advances in living standards that our ancestors could not have imagined, but it has also undermined natural systems in ways they could not have feared.

The question is, "In what direction will change take us? Will the change come from strong worldwide initiatives that reverse the degradation of the planet and restore hope for the future, or will it come from continuing environmental deterioration that leads to economic decline and social instability?"

Muddling through will not work. Either we will turn things around quickly or the self-reinforcing internal dynamic of the deterioration-and-decline scenario will take over. The policy decisions we make in the years immediately ahead will determine whether our children live in a world of development or decline.

There is no precedent for the rapid and substantial change we need to make. Building an environmentally sustainable future depends on a restructuring of the global economy, major shifts in human reproductive behavior, and dramatic changes in values and lifestyles. Doing all this quickly adds up to a revolution that is driven and defined by the need to restore and preserve the earth's environmental systems. If this *environmental revolution* succeeds, it will rank with the agricultural and industrial revolutions as one of the great economic and social transformations in human history.

Like the agricultural revolution, it will dramatically alter population trends. Although the former set the stage for enormous increases in human numbers, this revolution will succeed only if it stabilizes human population size, reestablishing a balance between people and natural systems on which they depend. In contrast to the industrial revolution, which was based on a shift to fossil fuels, this new transformation will be based on a shift away from nonrenewable fossil fuels to a mix of renewable energy resources based mostly on the earth's sun, wind, and water flows.

The two earlier revolutions were driven by technological advances—the first by the discovery of farming and the second by the invention of the steam engine, which converted the energy in coal into mechanical power. The environmental revolution will be driven primarily by restructuring the global economy so that it does not destroy or degrade its natural support systems.

The pace of the environmental revolution must be far faster than that of its predecessors. The agricultural revolution began some 10,000 years ago, and the industrial revolution has been under way for about two centuries. But if the environmental revolution is to succeed, it must be compressed into a few decades.

Progress in the agricultural revolution was measured almost exclusively in the growth in food output that eventually enabled farmers to produce a surplus that could feed city dwellers. Similarly, industrial progress was gained by expanding the output of raw materials and manufactured goods. The environmental revolution will be judged by whether it can shift the world economy into an environmentally sustainable development path, one that leads to greater economic security, healthier lifestyles, and a worldwide improvement in the human condition.

Many still do not see the need for such an economic and social transformation. They see the earth's deteriorating physical condition as a peripheral matter that can

REVIEW QUESTIONS

1. Distinguish between *exponential growth* and *linear growth.*

2. Describe the connection between exponential growth and environmental problems.

3. Distinguish between *solar capital* and *natural capital.*

4. Explain the importance of natural capital to the environment, the economy of the country where you live, and your lifestyle.

5. Define *sustainability.* Distinguish between living on principal and living on interest and relate this to the sustainability of **(a)** the earth's life-support system and **(b)** your lifestyle.

be dealt with by minor policy adjustments. But 35 years of effort have failed to stem the tide of environmental degradation. There is now too much evidence on too many fronts to take these issues lightly.

While economic indicators such as investment, production, and trade are consistently positive, the key environmental indicators are increasingly negative. Forests are shrinking, water tables are falling, soils are eroding, wetlands are disappearing, fisheries are collapsing, rangelands are deteriorating, rivers are running dry, temperatures are rising, and plant and animal species are disappearing. The global economy as now structured cannot continue to expand much longer if the natural systems on which it depends continue to deteriorate at the current rate.

This planetary degradation is damaging human health. By the age of 10, thousands of children living in southern California's Los Angeles basin have respiratory systems that are permanently impaired by polluted air. The accelerated depletion of ozone in the stratosphere in the northern hemisphere will lead to about 20,000 additional skin cancer fatalities over the next half century in the United States alone. Worldwide, millions of lives are at stake. These examples, and countless others, show that our health is closely linked to that of the planet.

A scarcity of new cropland and fresh water and the effects of soil erosion, air pollution, and hotter summers on crop yields are slowing the growth of the world grain harvest. Combined with continuing rapid population growth, this has reversed the steady rise in grain output per person that the world had become accustomed to. Between 1950 and 1984, the historical peak year, world grain production per person climbed by nearly 40%. Since then, it has fallen roughly 1% a year, with the drop concentrated in poor countries. With food imports in these nations restricted by rising external debt, there are far more hungry people today than ever before.

On the economic front, the signs are equally ominous: Soil erosion, deforestation, and overgrazing are adversely affecting farming, forestry, and livestock productivity, slowing overall economic growth in agriculturally based economies. The decline in living standards that was once predicted by some ecologists from the combination of continuing rapid population growth, spreading environmental degradation, and rising external debt has become a reality for one-sixth of humanity.

Suppose we used a more comprehensive system of national economic accounting that incorporated losses of natural capital, such as topsoil and forests, the destruction of productive grasslands, the extinction of plant and animal species, and the health costs of air and water pollution, nuclear radiation, and increased ultraviolet radiation. Such an analysis might well show that most of humanity suffered a decline in living conditions in the 1980s and 1990s.

The key limits as we enter the 21st century are fresh water, forests, rangelands, oceanic fisheries, biological diversity, and the global atmosphere. Will we recognize the world's natural limits and adjust our economies accordingly, or will we expand our ecological footprint until it is too late?

Some good news is that there is a growing worldwide recognition outside the environmental community that the economy we now have cannot take us where we want to go. Three decades ago, only environmental activists were speaking out on the need for change. Now the ranks of activists have broadened to include CEOs of major corporations, government ministers, prominent scientists, and even intelligence agencies.

The goal is to develop a new type of economy. It is a solar-powered, bicycle- and rail-centered, reuse and recycle economy that uses energy, water, land, and materials much more efficiently and wisely than we do today. This challenge to humanity rivals any in our history.

Critical Thinking

1. Do you agree with the author that we need to bring about an environmental revolution within a few decades? Explain.

2. Do you believe that this can be done by making minor adjustments in the global economy or by restructuring the global economy to put less strain on the earth's natural systems? Explain.

6. What are *doubling time* and the *rule of 70*? Use the rule of 70 to calculate how many years it would take for the population of a country to double if it was growing at 2% per year.

7. Define *economic growth*, *throughput*, *gross national product*, *gross domestic product*, *per capita GNP*, *developed country*, and *developing country*.

8. What is *sustainable development*? Are all forms of economic growth sustainable? Explain.

9. Describe changes in the wealth gap since 1960 and connect them to environmental degradation.

10. Why does it make sense for a poor family to have a large number of children?

11. Define *ecological resource, economic resource, potentially renewable resource, nonrenewable resource, physically depleted resource, economically depleted resource, mineral reserves, reuse,* and *recycle.* Draw a depletion curve for a nonrenewable resource and explain how recycling and reuse affect depletion time.

12. What are the three components of biodiversity? Why is biodiversity important to **(a)** the earth's life-support systems, **(b)** the economy of the country where you live, and **(c)** your lifestyle?

13. What are *sustainable yield* and *environmental degradation*? Give four examples of environmental degradation. What is the *tragedy of the commons*? Give three examples of this tragedy on a global scale and explain how your lifestyle contributes to these examples. List some ways to deal with the tragedy of the commons.

14. Define *pollution, point sources, nonpoint sources, concentration, persistence, degradable pollutant, biodegradable pollutant, slowly degradable pollutant, nondegradable pollutant, pollution prevention (input pollution control),* and *pollution cleanup (output pollution control).* List three reasons that pollution prevention is better than pollution control.

15. Summarize six underlying causes of environmental problems.

16. Describe a simple model of relationships among population, resource use, environmental degradation, and overall environmental impact. How do these factors differ in developed and developing countries?

17. Distinguish between *hunter–gatherer societies, agricultural societies, industrial societies,* and *information societies.* Summarize the major beneficial and harmful environmental effects of each of these societies.

18. Define *slash-and-burn cultivation* and *shifting cultivation* and list the benefits and limitations of these types of cultivation.

19. Summarize the 1992 position statement of about 1,700 leading scientists about the state of the environment. Explain why you agree or disagree with this position.

20. List five major changes that environmentalists believe should take place over the next 50 years as part of an *environmental* or *sustainability revolution.* What are two important keys to bringing about such changes?

CRITICAL THINKING

1. How would you define *overpopulation*? Is the world overpopulated? Explain. Is the United States overpopulated? Explain.

2. **(a)** Do you believe that the society you live in is on an unsustainable path? Explain. What about the world as a whole? **(b)** Do you believe that it is possible for the society you live in to become a sustainable society within the next 50 years? Explain. **(c)** Would you classify yourself as a technological optimist, environmental pessimist, or hopeful environmental realist? Explain.

3. Do you favor instituting policies designed to reduce population growth and stabilize **(a)** the size of the world's population as soon as possible and **(b)** the size of the U.S. population (or the population of the country where you live) as soon as possible? Explain. If you agree that population stabilization is desirable, what three major policies do you believe should be implemented to accomplish this goal?

4. Explain why you agree or disagree with the following propositions:

- **a.** High levels of resource use by the United States and other developed countries are more beneficial than harmful.
- **b.** The economic growth from high levels of resource use in developed countries provides money for more financial aid to developing countries for reducing pollution, environmental degradation, and poverty.
- **c.** Stabilizing population is not desirable, because without more consumers, economic growth would stop.

5. Explain why you agree or disagree with the following proposition: The world will never run out of potentially renewable resources and most currently used nonrenewable resources because technological innovations will produce substitutes, reduce resource waste, or allow use of lower grades of scarce nonrenewable resources.

6. Do you believe that your current lifestyle is sustainable? If the answer is yes, explain why and include the impact of the world's other 6 billion people on your ability to sustain your current lifestyle. If your answer is no, explain why and list five things you could do now to make your lifestyle more sustainable. Which of these things do you actually plan to do?

***7.** What are the major resource and environmental problems in **(a)** the city, town, or rural area where you live and **(b)** the state where you live? Which of these problems affect you directly?

***8.** Write two-page scenarios describing what your life and that of any child you choose to have might be like 50 years from now if **(a)** we continue on our present path and **(b)** we shift to sustainable societies throughout most of the world.

***9.** Make a concept map of this chapter's major ideas using the section heads and subheads and the key terms (in boldface type). Look at the inside back cover and in Appendix 5 for information about concept maps.

FURTHER INFORMATION

For more information see Internet Sources and Further Readings at the back of the book. Online resources for this chapter can be found at **www.brookscole.com/biology**

---

*These are laboratory exercises or individual or class projects. Additional exercises and projects are found in the laboratory manual available with this textbook.

2 ECONOMICS, POLITICS, ETHICS, AND SUSTAINABILITY

A.D. 2060: Green Times on Planet Earth*

Mary Wilkins sat in the living room of the solar-powered earth-sheltered house she shared with her daughter Jane and her family. It was July 4, 2060: Independence Day.

She walked into the greenhouse and gazed out at the large blue and green earth flag (Figure 2-1) they flew with the American flag. Her eyes shifted to her grandchildren, Lynn and Jeffrey, running and hollering as they played tag on the neighborhood commons. She heard the hum of solar-powered pumps trickling water to rows of organically grown vegetables and glanced at the fish in the aquaculture and waste treatment tanks.

Mary returned to the coolness of her earth-sheltered house and began putting the finishing touches on the children's costumes for this afternoon's pageant in Rachel Carson Park. It would honor earth heroes who began the Age of Ecology in the 20th century, as well as those who continued this tradition in the 21st century. She smiled when she thought of Jeffrey's delight at being chosen to play Aldo Leopold, who in the late 1940s began urging people to work with the earth. Her pride swelled when Lynn was chosen to play Rachel Carson, who in the 1960s alerted us to threats from increasing exposure to pesticides and other harmful chemicals.

Even in her most idealistic dreams, she had never guessed that she would see the hemorrhaging loss of global biodiversity slowed to a trickle. There were no more wars over oil, and most air pollution gradually disappeared when energy from the sun, wind, and hydrogen (produced by using solar energy to decompose water) replaced fossil and nuclear fuels. Now there was greatly increased emphasis on pollution prevention and waste reduction. Walking and bicycling had increased in cities and towns designed as vibrant communities for people instead of cars. Low-polluting and safe ecocars got 128 kilometers per liter (300 miles per gallon), and there was efficient mass transportation.

World population had stabilized at 8 billion in 2028 and then had begun a slow decline. The rate of global warming had slowed significantly, and international treaties enacted in the 1990s effectively banned the chemicals that had begun depleting ozone in the upper atmosphere (stratosphere).

Two hours later, she, her daughter Jane, and her son-in-law Gene watched with pride as 40 beautiful children honored the leaders of the Age of Ecology. At the end, Lynn stepped forward and said, "Today we have honored many earth heroes, but the real heroes are the ordinary people in this audience and around the world who worked to help sustain the earth for us and other species. Thank you, Grandma, Mom, Dad, and everyone here for giving us such a wonderful gift. We promise to leave the earth even better for our children and grandchildren and all living creatures."

Environmentalists and many of the world's prominent scientists believe that we are at a critical turning point. One option is to start on a path that could lead to a world like the fictional (but possible) scenario just described; another is to stay on our present path, which could result in a world like the one described in the worst-case fictional scenario found on page 259.

Figure 2-1 Some people display an earth flag as a symbol of their commitment to working with the earth at the individual, local, national, and international levels. (Courtesy of Earth Flag Co., 33 Roberts Road, Cambridge, MA 02138)

*Compare this hopeful scenario with the worst-case scenario that opens Chapter 11.

A continent ages quickly once we come.
Ernest Hemingway

This chapter addresses the following questions:

- What are economic goods and resources, and how are they provided?
- How should we measure economic growth?
- How can economics help improve environmental quality?
- How can we sharply reduce poverty?
- How can we shift to an earth-sustaining economy?
- How is environmental policy made in the United States?
- How can people affect environmental policy?
- What major human-centered environmental worldviews guide most industrial societies?
- What are some life-centered and earth-centered environmental worldviews?
- What ethical guidelines might be used to help us work with the earth?
- How can we live more sustainably?

2-1 ECONOMIC SYSTEMS AND ENVIRONMENTAL PROBLEMS

What Supports and Drives Economies? An **economy** is a system of production, distribution, and consumption of *economic goods*: any material items or services that satisfy people's wants or needs. In an economy, individuals, businesses, and societies make **economic decisions** about what goods and services to produce, how to produce them, how much to produce, how to distribute them, and what to buy and sell.

The kinds of capital that produce material goods and services in an economy are called **economic resources**. They fall into three groups:

- **Natural capital**: goods and services produced by the earth's natural processes (Figure 1-2). These include **(1)** the planet's air, water, and land; **(2)** nutrients and minerals in the soil and deeper in the earth's crust; **(3)** wild and domesticated plants and animals (biodiversity); and **(4)** nature's dilution, waste disposal, pest control, and recycling services. There are no other sources for these natural resources, which support all economies and lifestyles.

- **Manufactured capital**: items made from natural capital with the help of human capital. This type of capital includes tools, machinery, equipment, factory buildings, and transportation and distribution facilities.

- **Human capital**: people's physical and mental talents. Workers sell their time and talents for wages. Managers take responsibility for combining natural capital, manufactured capital, and human capital to produce economic goods. In market-based systems, entrepreneurs and investors put up the money needed to produce an economic good with the intention of making a profit on their investment.

What Are the Major Types of Economic Systems? There are two major types of economic systems: centrally planned and market based. In a **pure command economic system**, or **centrally planned economy**, the government makes all economic decisions.

In a **pure market economic system**, also known as **pure capitalism**, all economic decisions are made in *markets*, in which buyers (demanders) and sellers (suppliers) of economic goods freely interact without government or other interference. All economic resources are owned by private individuals and institutions rather than by the government. All buying and selling are based on *pure competition*, in which no seller or buyer is powerful enough to control the supply, demand, or price of a good. All sellers and buyers have full access to the market and enough information about the beneficial and harmful aspects of economic goods to make informed decisions.

Economists often depict pure capitalism as a circular flow of economic goods and money between households and businesses operating essentially independently of the earth's life-support system or natural capital (Figure 2-2). By contrast, environmentalists and a growing number of economists emphasize the fundamental dependence of this or any economic system on the earth's natural capital (Figure 2-3).

In a pure capitalist system, a business has **(1)** no legal allegiance to a particular nation, **(2)** no obligation to supply any particular good or service, and **(3)** no obligation to provide jobs, safe workplaces, or environmental protection. A company's only obligation is to produce the highest profit for the owners or stockholders whose financial capital the company is using to do business. Business leaders may go beyond this obligation, but the pure capitalist system does not require such actions.

In reality, all countries have **mixed economic systems** that fall somewhere between the pure market and pure command systems. The economic systems of countries such as China and North Korea fall toward the command-and-control end of the economic spectrum, whereas those of countries such as the United States and Canada fall toward the market-based end of the spectrum. The economic systems of most other countries fall somewhere between these ends of the economic spectrum.

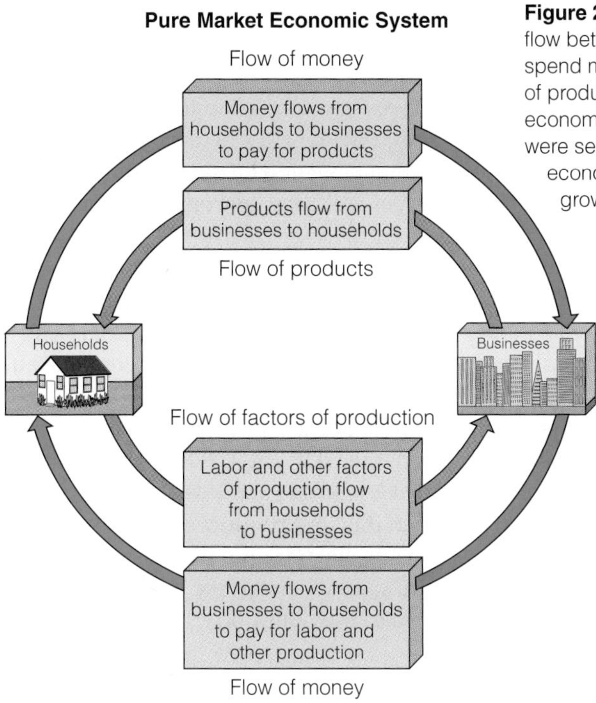

Figure 2-2 In a pure market economic system, economic goods and money would flow between households and businesses in a closed loop. People in households spend money to buy goods that firms produce, and firms spend money to buy factors of production (natural capital, manufactured capital, and human capital). In many economics textbooks, this and other economic systems are shown, as here, as if they were self-contained and thus independent of the natural capital that supports all economies and all life. This model reinforces the idea that unlimited economic growth of any kind is sustainable.

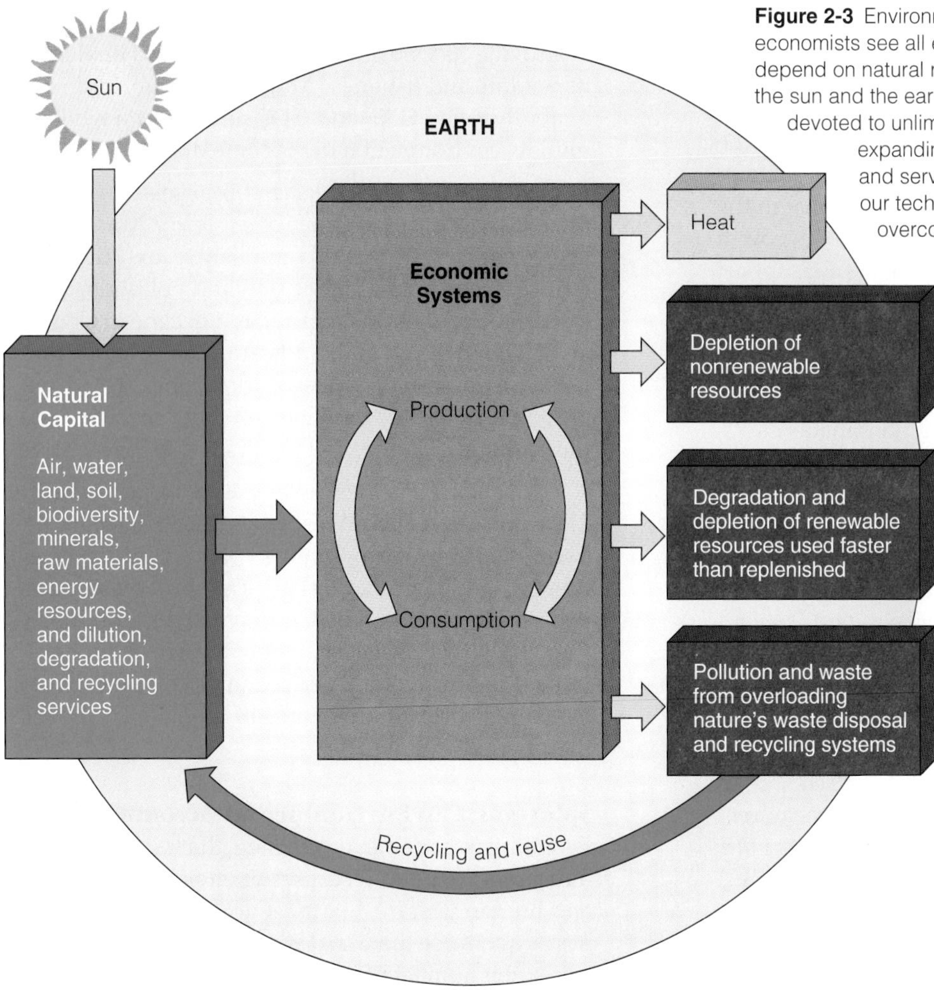

Figure 2-3 Environmentalists and a growing number of economists see all economies as artificial subsystems that depend on natural resources and services provided by the sun and the earth's natural capital. A consumer society devoted to unlimited economic growth to satisfy ever-expanding wants assumes that natural resources and services are essentially infinite or that our technological cleverness will allow us to overcome any natural limits. To environmentalists such a society is ultimately unsustainable.

2-2 ECONOMIC GROWTH AND EXTERNAL COSTS

How Do Economic Growth and Sustainability Differ? Virtually all economies seek **economic growth**: an increase in the capacity of the economy to provide goods and services for people's final use. Economies usually accomplish such growth by maximizing the flow of matter and energy resources (throughput) by means of population growth (more consumers), more consumption per person, or both (Figure 1-13). Economic growth usually is measured by an increase in a country's *gross national product (GNP)*: the market value in current dollars of all goods and services produced within and outside a country by the country's businesses for final use during a year.

Most economists, investors, and business leaders argue that we must have unlimited economic growth to create jobs, satisfy peoples' economic needs and desires, clean up the environment, and help eliminate poverty. Most of these analysts believe that the capacity for economic growth is virtually unlimited because of the earth's vast amount of resources and the ability of human ingenuity to overcome resource shortages and environmental problems through science and technology.

On the other hand, environmentalists and a growing number of economists and business leaders (Guest Essay, p. 34) contend that the notion of *sustainable economic growth* is nonsense because nothing based on the consumption of natural capital (Figure 1-2) is sustainable in the long run (Figure 2-3). They warn that in the long run, continuing unsustainable forms of growth will threaten both business and the planet's life-support systems for humans and other species.

To people with this view, the question is not so much, "To grow or not to grow?" but rather, "How can we grow without plundering the planet?" or "How can we grow as if the earth and future generations matter?"

According to system analysis expert and environmental scientist Donella Meadows,

> *What we need is smart development, not dumb growth. . . . When something grows it gets quantitatively bigger. When something develops it gets qualitatively better. . . . Smart development invests in insulation, efficient cars, and ever-renewed sources of energy. It ensures that forests and fields continue to produce wood, paper, and food, recharge wells, harbor wildlife, and attract tourists. Dumb growth crashes around looking for more oil. It clearcuts forests to keep loggers and sawmills going just a few more years until the trees run out. . . . It covers the landscape with the same kind of honkytonk ugliness tourists leave home to escape. . . .We need to meet dumb growth with smart questions. What really needs to grow? Who will benefit? Who will pay? What will last?*

Are GNP and GDP Useful Measures of Quality of Life and Environmental Degradation? We are urged to buy and consume more and more so that the GNP and GDP (p. 7) will rise, making the country where one lives and the world a better place for everyone. The truth is that GNP and GDP indicators are poor measures of human welfare, environmental health, or even economic health.

- *GNP and GDP hide the negative environmental effects of producing goods and services.* Pollution, crime, sickness, death, and depletion of natural resources are all counted as positive gains in the GDP or GNP. Every time an irreplaceable old-growth forest is cut down or a wetland is filled, the GDP and GNP go up. Pollution is counted as a *triple positive gain*. It is counted as a gain in GDP when it is first produced, counted again when society pays to clean it up partially, and counted as a third gain when people become sick or die from exposure to the pollution.
- *GNP and GDP don't include the depletion and degradation of natural resources or natural capital on which all economies depend.* A country can be headed toward ecological bankruptcy, exhausting its mineral resources, eroding its soils, cutting down its forests, destroying its wetlands and estuaries, and depleting its wildlife and fisheries. At the same time, it can have a rapidly rising GNP and GDP, at least for a while, until its environmental debts come due.
- *GNP and GDP hide or underestimate some of the positive effects of responsible behavior on society.* More energy-efficient light bulbs, appliances, and cars reduce electric and gasoline bills and pollution, but these beneficial effects register as a decline in GNP and GDP. GDP and GNP indicators also exclude the labor we put into volunteer work, the health care we give loved ones, the food we grow for ourselves, and the cooking, cleaning, and repairs we do for ourselves.
- *GNP and GDP tell us nothing about economic justice.* They don't reveal how resources, income, or the harmful effects of economic growth (pollution, waste dumps, land degradation) are distributed among the people in a country. The United Nations Children's Fund (UNICEF) suggests that countries should be ranked not by average per capita GNP or GDP but by average or median income of the poorest 40% of their people.

Solutions: Can Environmental Accounting Help? Economists have never claimed that GNP and GDP indicators are good measures of environmental health and human welfare, but most governments and business leaders use them that way. Environmentalists and a growing number of economists believe that GNP and GDP indicators should be replaced or supplemented with *environmental indicators* that give a more realistic picture by subtracting from the GDP and GNP things that lead to a lower quality of life and depletion of natural capital.

In 1972, economists William Nordhaus and James Tobin developed an indicator called *net economic welfare (NEW)* to estimate the annual change in a country's quality of life. They calculate a price tag for pollution and other "negative" goods and services included in the GNP and GDP and then subtract them to give the NEW. Recent studies tally harmful environmental costs (excluding those from global warming and fossil-fuel depletion) at 2% of the GDP in Japan and Australia, 12–15% in China, 23% in Germany, 40% in Sweden, and 45% in the United States.

Economist Robert Repetto and other researchers at the World Resources Institute have developed a *net national product (NNP)* that includes the depletion and destruction of natural resources as a factor in GNP. They have applied this indicator to Indonesia and Costa Rica.

Economists Herman E. Daly, John B. Cobb, Jr., and Clifford W. Cobb have developed an *index of sustainable economic welfare (ISEW)* and applied it to the United States. This comprehensive indicator of human welfare measures **(1)** per capita GNP adjusted for inequalities in income distribution, **(2)** depletion of nonrenewable resources, **(3)** loss of wetlands, **(4)** loss of farmland from soil erosion and urbanization, **(5)** the cost of air and water pollution, and **(6)** estimates of long-term environmental damage from ozone depletion and possible global warming. After rising by 42% between 1950 and 1976, this indicator fell 14% between 1977 and 1990.

A newer indicator is called the *genuine progress indicator (GPI)*. When this indicator is applied to the United States, the GPI per person has steadily declined since 1973. Indeed, between 1973 and 1994 the GDP rose from $12,500 to $17,000 per person while the GPI fell from $6,500 to $4,000 per person (Figure 2-4).

Such environmental indicators are far from perfect and include many crude estimates. However, many economists agree that they are more accurate than GNP and GDP as measures of life quality and environmental quality. Without such indicators, we don't know much about what is happening to people, the environment, and the planet's natural resource base and we don't have a way to measure what policies work. In effect, we are trying to guide national and global economies through treacherous economic and environmental waters at ever-increasing speeds using faulty radar.

The *good news* is that such indicators exist. The *bad news* is that so far they are not widely used.

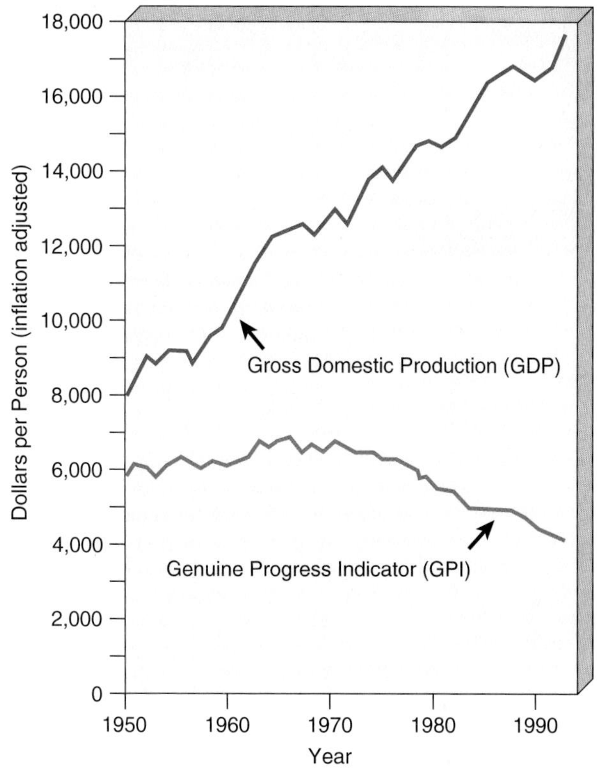

Figure 2-4 Comparison of per capita gross domestic product (GDP) and per capita genuine progress indicator (GPI) in the United States, 1950–94. Units of per capita GDP and GPI are inflation adjusted using 1982 dollars. Principal factors leading to the decline in the GPI were **(1)** long-term environmental damage, **(2)** depletion of domestic supplies of some nonrenewable resources, and **(3)** the growing gap between the incomes of the wealthy and the poor. (Data from Clifford Cobb, Ted Halstead, and Jonathan Rowe)

What Are Internal and External Costs? All economic goods and services have both internal and external costs. For example, the price a consumer pays for a car reflects the costs of the factory, raw materials, labor, marketing, and shipping, as well as a markup to allow the car company and its dealers some profits. After a car is purchased, the buyer must pay for gasoline, maintenance, and repair. All these direct costs, which are paid for by the seller and the buyer of an economic good, are called **internal costs**.

Making, distributing, and using any economic good or service also involve **externalities**: social costs or benefits not included in the market price. For example, if a car dealer builds an aesthetically pleasing showroom, it is an **external benefit** to people who enjoy the sight at no cost.

On the other hand, extracting and processing raw materials to make and propel cars **(1)** depletes nonrenewable energy and mineral resources, **(2)** produces solid and hazardous wastes, **(3)** disturbs land, **(4)** pollutes the air and water, **(5)** contributes to global climate change, and **(6)** reduces biodiversity. These harmful

GUEST ESSAY

Natural Capital

Paul G. Hawken

Paul G. Hawken is a practical visionary who understands both business and ecology and can communicate what is needed to make the transition to an earth-sustaining or restorative economy. In addition to founding Smith & Hawken, a retail company known for its environmental initiatives, he has written seven widely acclaimed books, including Growing a Business *(1987),* The Ecology of Commerce *(1993), and* Factor 10, The Next Industrial Revolution *(1998, with Amory and Hunter Lovins). He produced and hosted* Growing a Business, *a series for public television shown nationwide on 210 stations and now shown in 115 countries.* The Ecology of Commerce *was hailed as the best business book of 1993 and one of the most important books of the 20th century. In 1987,* Inc. *magazine named him one of the 12 best entrepreneurs of the 1980s, and in 1995 he was named by the* Utne Reader *as one of the 100 visionaries who could change our lives. He has also been described as the poet laureate of American capitalism.*

Great ideas, in hindsight, seem obvious. The concept of natural capital, [Figure 1-2] is such an idea. Natural capital is the myriad necessary and valuable resources and ecological processes that we rely on to produce our food, products, and services.

The concept of natural capital is not a new one. Economists have long noted that natural capital is a factor in industrial production, but a marginal factor. A new view is emerging: that our economic systems cannot long endure without taking the flow of renewable and nonrenewable resources through economies into account. The value of natural capital is becoming paramount to the success of all business.

This revision of neoclassical economics, yet to be accepted by most mainstream academicians, provides business and public policy with a powerful new tool for growth, profitability, and the public good. It is as if we had been sitting on a two-legged stool for the past century, wondering why our economies have become increasingly unbalanced and unsteady.

The concept of natural capital, when intelligently linked to human and manufactured capital, provides the critical connection between the satisfaction of human needs, the continued prosperity of business, and the preservation (some might even suggest the restoration) of the earth's living natural systems.

Most Americans are filled with cornucopian fantasies of technological prowess, where human ingenuity bypasses natural limits and creates unimagined abundance. Optimism easily intertwines with the belief that nanotechnology, biotechnology, computers, and technologies yet to developed will eliminate hunger, disease, and want.

Dreams of alleviating human suffering are worthy. However, they usually overlook the absolute necessity for fertile soil, ocean fisheries, a stable climate, biological diversity, and pure water, all of which we are losing and none of which can be created by any human-made technology known or imagined.

In our pursuit of dominance over the natural world, we have not taken into account the basic principle that industrialism, for all its sophistication, is enormously inefficient with respect to resources, energy, and waste. It is difficult for neoclassical economists, whose hypotheses and theories originated in a time of resource abundance, to understand that the very success of linear industrial systems has laid the groundwork for the next stage in economic evolution. This next stage, whatever it may be called, is being brought about by powerful and much-delayed feedback of information from high levels of inefficiency and waste. As that happens, the foundation of industrialism is giving way while the basis for the next industrial revolution is being established.

This shift is profoundly biological. It's not about the celebration of nature, although that is certainly part of it; it's about the incorporation of cyclical natural systems into our industrial life, our way of making things, and our way of processing things and deprocessing things. The reason this shift is going to happen is because cyclical industrial systems work better than linear ones. They close the loop and reincorporate wastes as part of the

effects are **external costs** passed on to workers, the public, and in some cases future generations. Car owners add to these external costs when they **(1)** throw trash out of a car, **(2)** drive a gas-guzzler (which produces more air pollution than a more efficient car), **(3)** disconnect or don't maintain a car's air-pollution control devices, or **(4)** don't keep the engine tuned.

Because these harmful costs aren't included in the market price, people don't connect them with car ownership. Still, everyone pays these hidden costs sooner or later, in the form of higher costs for health care and health insurance and higher taxes for pollution control. For example, in the United States gasoline costs about $1.10 to $1.60 per gallon (29¢ to 42¢ per liter). However, when you add all of the external costs of pollution, health, insurance and other social costs, American consumers are paying $5 to $7 per gallon ($1.30 to $1.90 per liter).

To some economists, external costs are minor defects in the flow of production and consumption in a self-contained economy not significantly dependent on natural capital (Figure 2-2). They assume that

Gallopin, G. C. and P. Raskin. 1998. "Windows on the Future: Global Scenarios and Sustainability." *Environment,* vol. 40(no. 3):6(17).

production cycle [Figure 2-3]. There are no landfills in a cyclical society.

If there is so much inefficiency in our system, why isn't it more apparent? The inefficiencies are masked by a financial system in which money, prices, and markets give us improper information. Markets are not giving us proper information about how much our suburbs, cars, and plastic drinking water bottles truly cost. Instead, we are getting warning signals from the beleaguered airsheds and watersheds, the overworked and eroded soils, the life-degrading inner cities and rural counties, the breakdown of stability worldwide, and the conflicts based on resource shortages. These feedbacks from nature are providing the information that our prices should give us but don't.

Prices don't give us good information for a simple reason: improper accounting. Natural capital has never been placed on the balance sheets of companies or the countries of the world. To paraphrase G. K. Chesterton, it could fairly be said that capitalism might be a good idea except that we have not tried it yet. And try we must and will, for capitalism cannot be fully attained or practiced until we have an accurate balance sheet, as any accounting student will tell us.

As it stands, our economic system is based on accounting principles that would bankrupt a company. Not surprisingly, it is posing problems for the world as a whole. When natural capital is placed on the balance sheet, not as a free resource of infinite supply but as an integral and valuable part of the production process, everything changes. The near obsessive pursuit of improvement in human productivity becomes balanced by the need for improved resource productivity. Using more and more resources to make fewer people more productive flies in the face of what we now need to improve our society and the environment. After all, it is people we have more of, not natural resources, so it is people we must use in order to reduce the flow of matter and energy resources through economies.

And that is what can happen when we move from linear extractive systems to cyclical ones. Instead of the huge capital investments required to find, extract, and process nonrenewable oil, when we shift to renewable wind energy we spend our money on windmill maintenance instead of hydrocarbons. We get more jobs, reduce capital requirements, and produce almost no pollution.

Many people sincerely believe that an economic system based on the integrity of natural systems is unworkable. To answer that concern, we may well want to reverse the question and ask, "How have we created an economic system that tells us it is cheaper to destroy natural capital than to maintain it?" We know this is not the way to take care of our cars, houses, and bridges, but somehow we have managed to overlook a pricing system that discounts the future and sells off the past. Or to put it another way, "How did we create an economic system that confuses capital with income?"

Can we devise and implement a more rational economic system? I think so. It is right before us. It requires no new theories, only common sense. It is based on the simple but powerful proposition that *all capital must be valued.*

There may be no *right* way to value a forest or a river, but there is a *wrong* way, which is to give it no value at all. If we have doubts about how to value a 500-year-old tree, we need only ask how much it would cost to make a new one from scratch. Or a new river. Or a new atmosphere.

The work of the future is the absorption and integration of the worth of living systems into every aspect of our culture and commerce, so that human systems mimic natural systems. Only by doing this can our cultures reflect growth and harmony rather than damage and discord.

Critical Thinking

1. If you were placed in charge of the world's economy, what are the three most important things you would do? Compare your answers with those of other members of your class.

2. Do you agree with the author of this essay that we must absorb and integrate the worth of living systems into every aspect of our culture and commerce? How would you go about doing this?

these defects can be rectified using the profits made from additional economic growth in a free-market economy.

To environmentalists and an increasing number of economists and business leaders, harmful externalities are a warning sign that our economic systems are stressing the environment and depleting natural capital (Figures 1-2 and 2-3). They believe that these harmful external costs should be included in the market prices of goods and services, a full-cost pricing process economists call *internalizing the external costs*.

2-3 SOLUTIONS: USING ECONOMICS TO IMPROVE ENVIRONMENTAL QUALITY

Should We Shift to Full-Cost Pricing? As long as businesses receive subsidies and tax breaks for extracting and using virgin resources and are not taxed for the pollutants they produce, few will volunteer to reduce short-term profits by becoming more environmentally responsible. Assume you own a company and believe it's wrong to subject your workers to hazardous

Hint: Enter the search terms *sustainable development, environmental aspects* using the Subject Guide.

conditions and pollute the environment beyond what natural processes can handle. Suppose you voluntarily improve safety conditions for your workers and install pollution controls, but your competitors don't. Then your product will cost more than theirs and you will be at a competitive disadvantage. Your profits will decline; you may eventually go bankrupt and have to lay off your employees.

One way of dealing with the problem of harmful external costs is for the government to levy taxes, pass laws, provide subsidies, or use other strategies that encourage or force producers to include all or most of these costs in the market prices of economic goods and services. Then that price would be the **full cost** of these goods and services: internal costs plus short- and long-term external costs. If this *user-pays* approach is used, people and businesses pay the full costs of the harm they do to others and the environment.

Full-cost pricing involves *internalizing the external costs*. This approach requires government action because few companies will intentionally increase their cost of doing business unless their competitors must do so as well. Internalizing the external costs of pollution and environmental degradation would make preventing pollution more profitable than cleaning it up, and waste reduction, recycling, and reuse more profitable than burying or burning most of the waste we produce.

Because external costs would be internalized, the market prices for most goods and services would rise. But the total price people would pay would be about the same because the hidden external costs related to each product would be included in its market price. Using full-cost pricing to internalize external costs provides consumers with information needed to make informed economic decisions about the effects of their lifestyles on the planet's life-support systems and on human health.

However, as external costs are internalized, economists and environmentalists warn that governments must reduce income, payroll, and other taxes and must withdraw subsidies formerly used to hide and pay for these external costs. Otherwise, consumers will face higher market prices without tax relief—a politically unacceptable policy guaranteed to fail.

The good news is that some goods and services would cost less because internalizing external costs encourages producers to **(1)** find ways to cut costs (by inventing more resource-efficient methods of production) and **(2)** offer more earth-sustaining (or *green*) products. Jobs would be lost in earth-degrading businesses, but at least as many (some analysts say more) jobs could be created in earth-sustaining businesses. If a shift to full-cost pricing took place over several decades, most current earth-degrading businesses would have time to transform themselves into profitable earth-sustaining businesses.

Full-cost pricing seems to make a lot of sense. Why isn't it more widely done? There are several reasons:

- Many producers of harmful and wasteful goods would have to charge so much that they couldn't stay in business.
- Producers of such goods would also have to give up government subsidies and tax breaks that have helped hide the external costs of their goods and services.
- It is hard to put a price tag on many harmful environmental and health costs.

Studies estimate that governments around the globe spend about $600 billion per year to subsidize **(1)** deforestation, **(2)** overfishing, **(3)** overgrazing, **(4)** unsustainable agriculture, **(5)** use of nonrenewable fossil fuels and nuclear energy, **(6)** groundwater depletion, and other environmentally destructive activities—what environmental expert Norman Myers calls *perverse subsidies*. According to some economists, eliminating these earth-degrading subsidies would allow a 7% cut in the global tax burden of $7.5 trillion per year and encourage job creation and investment.

Despite the difficulties, proponents believe that full-cost pricing for harmful environmental and health effects deserves a serious try. They argue that doing the best we can to estimate and internalize current external costs is far better than continuing the current pricing system, which gives too little or misleading information about the environmental and health effects of goods and services. The key question is whether the problems with the *tell-the-truth full-cost* pricing system are worse than those with the current *hide-the-true-cost* pricing system. What do you think?

How Useful Is Cost–Benefit Analysis? One of the chief tools corporations and governments use in making economic decisions is **cost–benefit analysis**. It involves comparing the estimated short-term and long-term costs (losses) with the estimated benefits (gains) for various courses of action. Cost–benefit analyses can be useful guides and can indicate the cheapest way to go, but they can also be misused (Spotlight, right).

To minimize possible abuses, environmentalists and economists advocate the following guidelines for all cost–benefit analyses: **(1)** Use uniform standards, **(2)** state all assumptions, **(3)** evaluate the reliability of all data inputs as high, medium, or low, **(4)** make projections using low, medium, and high discount rates, **(5)** show the estimated range of costs and benefits based on various sets of assumptions, **(6)** estimate the short- and long-term benefits and costs to all affected population groups, **(7)** estimate the effectiveness of the project or form of regulation instead of assuming (as is often done) that all projects and regulations will be executed with 100%

Some Problems with Cost–Benefit Analysis

SPOTLIGHT

There are several controversies about cost–benefit analysis. One involves the **discount rate**, an estimate of a resource's future economic value compared with its present value. *The size of the discount rate (usually given as a percentage) is a primary factor affecting the outcome of any cost–benefit analysis.*

At a zero discount rate, a stand of redwood trees worth $1 million today will still be worth $1 million 50 years from now; thus there is no need to cut them down for short-term economic gain. However, at a 10% annual discount rate (normally used by most businesses and by the U.S. Office of Management and Budget), the same trees will be worth only $10,000 in 50 years. As a result, it makes short-term economic sense to cut them down now and invest the profits in something else.

Proponents of high (5–10%) discount rates argue that inflation will reduce the value of their future earnings. They also fear that innovation or changes in consumer preferences will make a product or service obsolete.

Critics question this belief, pointing out that current economic systems are based on depletion of the natural capital that supports them (Guest Essay, p. 34). High discount rates worsen this situation by encouraging such rapid exploitation of resources for immediate payoffs that sustainable use of most potentially renewable natural resources is virtually impossible. These critics believe that unique and scarce resources should be protected by having a 0% or even a negative discount rate and that discount rates of 1–3% would make it profitable to use other resources sustainably or slowly.

Another problem with cost–benefit analysis is determining *who benefits and who is harmed.* In the United States, an estimated 100,000 people die each year because of exposure to hazardous chemicals and other safety hazards at work, and an additional 400,000 are seriously injured by such exposure. In many other countries (especially developing countries), the situation is much worse. Is this a necessary or an unnecessary cost of doing business?

Another limitation of cost–benefit analysis is that *many things we value cannot easily be reduced to dollars and cents.* We can put estimated price tags on human life, good health, clean air and water, pollution and accidents that are prevented, wilderness, the northern spotted owl, and various forms of natural capital (Figure 1-2). However, the dollar values different people assign to such things vary widely because of different assumptions, discount rates, and value judgments, leading to a wide range of projected costs and benefits.

Because these and other estimates of costs and benefits are so variable, *figures can easily be weighted to achieve the outcome desired by proponents or opponents of a proposed project or action.* For example, one industry-sponsored cost–benefit study estimated that compliance with a standard to protect U.S. workers from vinyl chloride would cost $65–90 billion; in fact, less than $1 billion was actually needed to comply with the standard.

Critics contend that analysis of various cost–benefit analyses indicates that using this method as the primary way to make decisions is very imprecise. It is somewhat like trying to detect a car speeding at 160 kilometers per hour (100 miles per hour) with a radar device so unreliable that at best it can tell us only that the car's speed is somewhere between 80 kph (50 mph) and 8,000 mph (5,000 mph).

Critical Thinking

Do you believe that cost–benefit analysis should be used as the primary way to evaluate whether any new environmental law or regulation should be put into effect and whether any existing environmental law or regulation should be weakened or strengthened? Explain. What should be the role of cost–benefit analysis? Why?

efficiency and effectiveness, and **(8)** open the evaluations to public review and discussion.

Should We Rely Mostly on Regulations or Market Forces? Most economists agree that controlling or preventing pollution and reducing resource waste require government intervention in the marketplace. Such government action can take the form of regulation, the use of market forces, or some combination of these approaches.

Regulation is a *command-and-control* approach. It involves enacting and enforcing laws that **(1)** set pollution standards, **(2)** regulate harmful activities, **(3)** ban the release of toxic chemicals into the environment, and **(4)** require that certain irreplaceable or slowly replenished resources be protected from unsustainable use.

Numerous studies of the effects of environmental regulations in the United States have found that they do very little harm to businesses. Indeed, in many cases they have led to improvements in resource use efficiency (which reduce costs) and to innovative products and industrial processes (which increase profits). A recent study by economist Robert Repetto at the World Resources Institute showed that between 1970 and 1990, the U.S. industries that spent the most on pollution control did significantly better than average in the global marketplace.

Business leaders and many environmentalists in the United States agree that some pollution control regulations discourage innovation by being too prescriptive and costly, and should be modified. They propose

Shifting the Tax Burden from Wages and Profits to Pollution and Waste

SOLUTIONS

Fully taxing pollution would raise about $1 trillion per year worldwide, enough to allow a 15% cut in taxes on wages and profits. A growing number of analysts favor taxing pollution instead of payrolls, which in the United States is a highly regressive tax for middle-class taxpayers. For example, U.S. workers making $30,000 a year pay 15.3% of their income in payroll taxes to support Social Security and Medicare, whereas their bosses making $300,000 a year pay only 5.7% of their income in payroll taxes.

Shifting more of the tax burden from wages and profits to pollution and waste **(1)** improves environmental quality by making polluters and resource depleters and degraders pay, **(2)** encourages pollution prevention and waste reduction and creativity in solving environmental problems (to avoid paying pollution taxes), **(3)** rewards recycling and reuse instead of use of nonrenewable virgin resources, **(4)** uses the marketplace, rather than regulations, to help protect the environment, **(5)** reduces unemployment, and **(6)** allows for cuts in highly resented and often regressive income, payroll, and sales taxes.

Critical Thinking

Do you favor making such a shift in what is taxed? Explain. How would you go about making such a shift politically feasible?

using regulations to set goals but freeing industries to meet such goals in any way that works.

Market forces can help improve environmental quality and reduce resource waste, mostly by encouraging the internalization of external costs. This is based on a fundamental principle of the marketplace in today's mixed economic systems: *What we reward (mostly by subsidies and tax breaks) we tend to get more of, and what we discourage (mostly by regulations and taxes) we tend to get less of.*

One way to put this principle into practice would be *to phase in government subsidies that encourage earth-sustaining behavior and phase out subsidies that encourage earth-degrading behavior*. The difficulty is that removing or adding such subsidies involves political decisions that can often be successfully opposed by powerful economic interests.

Another market approach is for the government to *grant tradable pollution and resource-use rights*. For example, a total limit on emissions of a pollutant or use of a resource could be set. Then permits would be used to allocate the total among manufacturers or users. Permit holders not using their entire allocation could **(1)** use it as a credit against future expansion, **(2)** use it in another part of their operation, or **(3)** sell it to other companies. Tradable rights could also be established among countries to help preserve biodiversity and reduce emissions of greenhouse gases and other pollutants with harmful regional (or global) effects.

Some environmentalists support tradable pollution rights as an improvement over the current regulatory approach. Other environmentalists believe that allowing companies to buy and trade rights to pollute is wrong because it **(1)** allows the wealthiest companies to continue polluting (thereby excluding smaller companies from the market), **(2)** tends to concentrate pollutants at the dirtiest plants (thereby jeopardizing the health people downwind of the plant), and **(3)** creates an incentive for fraud because most pollution control regulations are based on self-reporting of pollution outputs and government monitoring of such outputs is inadequate. What do you think?

Another market-based method is to *enact green taxes or effluent fees* that would help internalize many of the harmful external costs of production and consumption. This method could include taxes on each unit of pollution discharged into the air or water, each unit of hazardous or nuclear waste produced, each unit of virgin resources used, and each unit of fossil fuel used.

There are two major problems with this tax punishment, or stick, approach. *First*, because the taxes or effluent fees are set politically rather than by markets, elected officials find it easier to aim for popular approval rather then economic and ecological efficiency. As a result, the taxes often are too low to be effective and thus undermine the entire concept.

Second, elected officials are likely to see such taxes as ways of raising revenue instead of improving economic and ecological efficiency. However, economists point out that *green taxes* on pollution output, resource depletion, and environmental degradation would work if **(1)** they reduced or replaced income, payroll, or other taxes and **(2)** the poor and middle class were given a safety net to reduce the regressive nature of consumption taxes on essentials such as food, fuel, and housing. In other words, environmental taxes must be seen as a *tax-shifting* instead of a *tax-burden* approach (Solutions, above). This approach is now being tried in Sweden, Denmark, Spain, the Netherlands, the United Kingdom, and Germany.

Table 2-1 Economic Solutions to Pollution and Resource Waste					
Solution	**Internalizes External Costs**	**Innovation**	**International Competitiveness**	**Administrative Costs**	**Increases Government Revenue**
Regulation	Partially	Can encourage	Decreased*	High	No
Subsidies	No	Can encourage	Increased	Low	No
Withdrawing harmful subsidies	Yes	Can encourage	Decreased*	Low	Yes
Tradable rights	Yes	Encourages	Decreased*	Low	Yes
Green taxes	Yes	Encourages	Decreased*	Low	Yes
User fees	Yes	Can encourage	Decreased*	Low	Yes
Pollution-prevention bonds	Yes	Encourages	Decreased*	Low	No

*Unless more cost-effective and productive technologies are developed.

Charging user fees is another market-based method. For example, users would pay fees to cover all or most costs for **(1)** grazing livestock, extracting lumber and minerals from public lands, **(2)** using water provided by government-financed projects, and **(3)** using public lands for recreation.

Another market approach would *require businesses to post a pollution prevention or assurance bond* when they plan to develop a new mine, plant, incinerator, landfill, or development and before they introduce a new chemical or new technology. After a set length of time, the deposit (with interest) would be returned, minus actual or estimated environmental costs. This approach is similar to the performance bonds contractors are now required to post for major construction projects.

Each of these approaches has advantages and disadvantages (Table 2-1). Currently, in the United States, private industry and local, state, and federal government spend about $130 billion a year to comply with federal environmental regulations. Studies estimate that greater reliance on a variety of market-based policies could cut these expenditures by one-third to one-half.

Should We Emphasize Pollution Control or Pollution Prevention? Shouldn't our goal be zero pollution? Ideally, yes; in the real world, not necessarily. First, natural processes can handle some of our wastes, as long as we don't destroy, degrade, or overload these processes. However, environmentalists argue that harmful chemicals that cannot be degraded by natural processes or that break down very slowly should **(1)** not be released into the environment, or **(2)** be released only in small amounts and regulated by special permit.

Second, as long as we continue to rely on pollution control, we can't afford zero pollution. After we've removed a certain proportion of the pollutants in air, water, or soil, the cleanup cost per additional unit of pollutant rises sharply (Figure 2-5). Beyond a certain point, the cleanup costs exceed the harmful costs of pollution. Some businesses could then go bankrupt, and some people could lose jobs, homes, and savings. On the other hand, if we don't go far enough, dealing with the harmful external effects of pollution can cost more than pollution reduction.

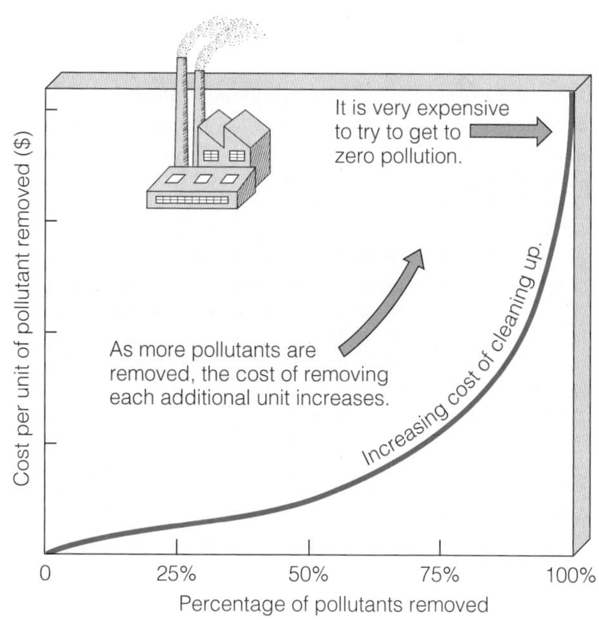

Figure 2-5 The cost of removing each additional unit of pollution rises exponentially, which explains why it is usually cheaper to prevent pollution than to clean it up.

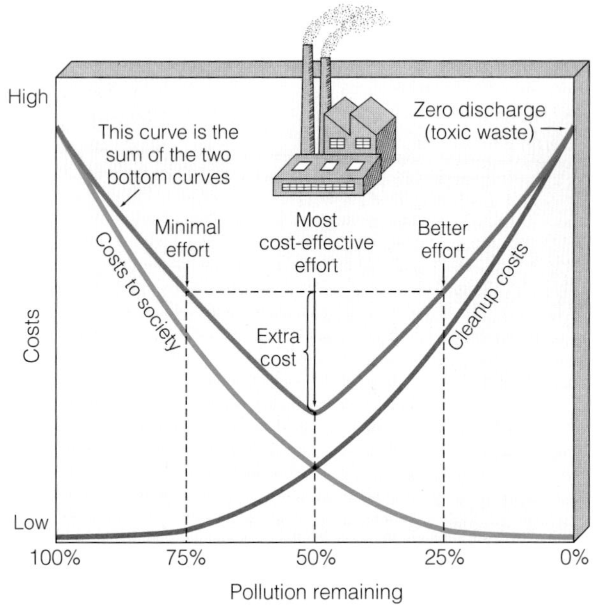

Figure 2-6 Finding the optimal level of pollution. This graph shows the optimal level at 50%, but the actual level varies depending on the pollutant.

To find the breakeven point, economists plot two curves: **(1)** a curve of the estimated economic costs of cleaning up pollution and **(2)** a curve of the estimated social (external) costs of pollution. They then add the two curves together to get a third curve showing the total costs. The lowest point on this third curve is the point of the *optimal level of pollution* (Figure 2-6).

On a graph, this looks neat and simple, but environmentalists and business leaders often disagree in their estimates of the harmful costs of pollution. This approach assumes that we know which substances are harmful and how much each part of the environment can handle without serious environmental harm—things we probably will never know precisely.

Most environmentalists and a growing number of economists and business leaders believe that environmental laws should emphasize pollution prevention, which avoids most of the regulatory problems and excessive costs of end-of-pipe pollution control. What do you think?

Is Encouraging Global Free Trade Environmentally Helpful or Harmful? On April 15, 1994, representatives of 120 nations signed the Uruguay Round of the General Agreement on Tariffs and Trade (GATT). This is a revised version of the 1948 GATT convention, which attempted to lower tariff barriers to world trade among member nations. The new GATT establishes a World Trade Organization (WTO), giving it the status of a major international organization (similar to the United Nations and the World Bank) and the power to oversee and enforce the agreement.

Proponents argue that agreements to reduce global trade barriers **(1)** will benefit developing countries, whose products are often at a competitive disadvantage in the global marketplace because of trade barriers erected by developed countries, and **(2)** can stimulate economic growth in all countries by allowing consumers to buy more things at lower prices.

Most environmental groups, and those concerned with consumer protection and worker health and safety, oppose some parts of the new version of GATT. They contend that the current version of GATT

- *Will not provide ample economic benefits for everyone.* Currently, 84% of the benefits of world trade flow to the richest one-fifth of the world's population, but only 0.9% to the poorest one-fifth. According to these critics, the new version of GATT would not change this situation, and it could worsen it.
- *Will increase the economic and political power of multinational corporations and decrease the power of small businesses, citizens, and democratically elected governments.*
- *Will weaken environmental and health and safety standards in developed countries.* Under the new GATT, any country could be fined (or be subjected to tariff fees) if any portion of it adopted stricter environmental, health, worker safety, resource use, or other standards than the uniform global standards established by the WTO (dominated by multinational companies based in developed countries).

Critics of the latest version of GATT call for it to be improved (Solutions, right). What do you think?

2-4 SOLUTIONS: REDUCING POVERTY

Does the Trickle-Down Approach to Reducing Poverty Work? Poverty usually is defined as the inability to meet one's basic economic needs. Currently, an estimated 1.4 billion people in developing countries—almost one of every four people on the planet—have an annual income of less than $370 per year. This income of roughly $1 per day is the World Bank's definition of poverty.

Poverty causes premature deaths and preventable health problems. It also tends to increase birth rates and often pushes people to use potentially renewable resources unsustainably in order to survive.

Most economists believe that a growing economy is the best way to help the poor. This is the so-called

Improving Trade Agreements

SOLUTIONS

Critics of current trade agreements would rewrite and correct what they believe are serious weaknesses in GATT and turn it into GAST: the *General Agreement for Sustainable Trade*. They offer the following suggestions for doing this:

- Having low or no trade tariffs on imports from nations that promote ecologically sustainable development by using environmental accounting and full-cost pricing
- Judging GATT or any trade agreement primarily on how it benefits the environment, workers, and the poorest 40% of humanity
- Setting minimum environmental, consumer protection, and worker health and safety standards for all participating countries
- Allowing any country to set more stringent versions of such standards without being penalized by fines or other means
- Requiring all panels or bodies setting and enforcing GATT standards to have environmental, labor, consumer, and health representatives from developed countries and developing countries
- Opening all discussions and findings of any GATT panel or other World Trade Organization body to global public scrutiny
- Allowing international environmental agreements to prevail when they conflict with GATT or any other trade agreement

Unless citizens exert intense pressure on legislators, critics warn that such safeguards will not be incorporated into international trade agreements.

Critical Thinking

1. Do you believe that the current GATT treaty will help or hinder environmental protection **(a)** globally, **(b)** in the country where you live, and **(c)** in your community? Explain.

2. Explain why you agree or disagree with each of the suggestions in this box. If you agree with all or most of these proposals, how could they be implemented? What role could you play in making such changes?

trickle-down hypothesis: Economic growth **(1)** creates more jobs, **(2)** enables more of the increased wealth to reach workers, and **(3)** provides greater tax revenues that can be used to help the poor help themselves.

However, instead of *trickling down*, most of the benefits of economic growth as measured by income have *flowed up* since 1960. This has made the top one-fifth of the world's people much richer and the bottom one-fifth poorer. Most people in between have lost, or gained only slightly, in real per capita GNP. This trend has accelerated in the 1980s and 1990s (Figure 2-7). According to a 1999 report by the United Nations Development Program, the combined wealth of the world's 225 richest people equaled the combined annual income of the world's poorest 2.5 billion people.

How Can Poverty Be Reduced? Analysts point out that reducing poverty requires the governments of most developing countries to make policy changes. Two examples are **(1)** shifting more of the national budget to help the rural and urban poor work their way out of poverty and **(2)** giving villages, villagers, and the urban poor title to common lands and to crops and trees they plant on them.

Analysts also urge developed countries and the wealthy in developing countries to help reduce poverty. Several controversial ways to do this have been suggested.

- *Forgive at least 60% of the almost $2 trillion that developing countries owe to developed countries and international lending agencies.* Some of this debt can be forgiven in exchange for carefully monitored agreements by the governments of developing countries to increase expenditures for family planning, health care, education, land redistribution, protection and restoration of biodiversity, and more sustainable use of renewable resources.
- *Increase nonmilitary aid to developing countries from developed countries, with the aid going directly to the poor to help them become more self-reliant.*
- *Shift most international aid from large-scale to small-scale projects intended to benefit local communities of the poor.*
- *Encourage banks and other organizations to make small loans to poor people wanting to increase their income.*
- *Require international lending agencies to use standard environmental and social impact analysis to evaluate any proposed development project.* No project should be supported unless **(1)** its net environmental impact (using full-cost accounting) is favorable, **(2)** most of its benefits go to the poorest 40% of the people affected, and

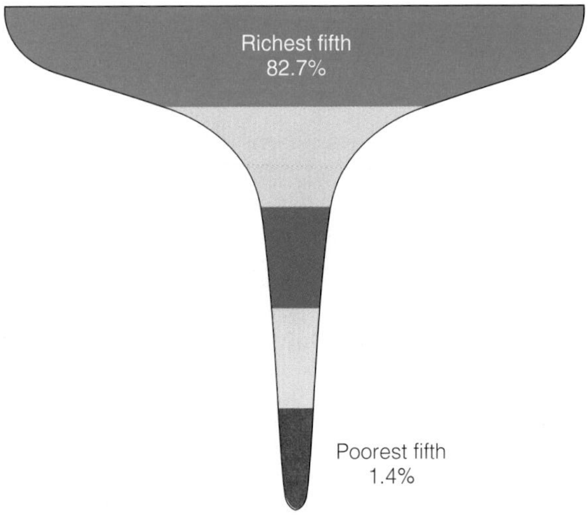

Figure 2-7 Data on the global distribution of income show that instead of trickling down, most of the world's income has flowed up, with the richest 20% of the world's population receiving more of the world's income than all of the remaining 80% in 1991. Each horizontal band in this diagram represents one-fifth of the world's population. This upward flow of global income has increased since 1960 and accelerated in the 1980s and 1990s. Do you favor or oppose this trend? Explain. If you oppose this trend, what do you believe are the most important ways to correct it? (Data from UN Development Programme)

(3) the local people it affects are involved in planning and executing the project.

- *Carefully monitor all projects and halt funding if environmental safeguards are not followed.*
- *Establish policies that encourage both developed countries and developing countries to slow population growth and stabilize their populations.*

2-5 SOLUTIONS: CONVERTING TO EARTH-SUSTAINING ECONOMIES

How Can We Make Working with the Earth Profitable? Greening Business In 1994, entrepreneur and business leader Paul Hawken (Guest Essay, p. 34) wrote *Ecology and Commerce*, a widely acclaimed book describing what's wrong with business and how it can be transformed to work with the earth. Hawken and several other business leaders and economists have laid out the following principles for transforming the planet's current earth-degrading economic systems into earth-sustaining, or restorative, economies over the next several decades:

- *Reward (subsidize) earth-sustaining behavior.*
- *Discourage (tax and don't subsidize) earth-degrading behavior.*
- *Use full-cost accounting to include the ecological value of natural resources in their market prices.*
- *Use environmental and social indicators to measure progress toward environmental and economic sustainability and human well-being* (Figure 2-4).
- *Use full-cost pricing to include the external costs of goods and services in their market prices.*
- *Replace taxes on wages and profits with taxes on throughput of matter and energy* (Solutions, p. 38).
- *Use low discount rates for evaluating future worth of irreplaceable or potentially nonrenewable resources* (Spotlight, p. 37).
- *Revoke the government-granted charters of environmentally and socially irresponsible businesses.*
- *Make environmental concerns a key part of all trade agreements and of all loans made by international lending agencies* (Solutions, p. 41).
- *Don't use potentially renewable resources (soil, water, forests, grasslands, and wildlife) faster than they are replenished.*
- *Don't waste resources.*
- *Don't release pollutants into the environment faster than the earth's natural processes can dilute or assimilate them.*
- *Emphasize pollution prevention and waste reduction.*
- *Help maintain the earth's capacity for self-repair.*
- *Preserve biodiversity.*
- *Reduce future ecological damage and repair past ecological damage.*
- *Reduce poverty.*
- *Slow population growth.*

Paul Hawken's simple golden rule for such an economy is, *"Leave the world better than you found it, take no more than you need, try not to harm life or the environment, and make amends if you do."*

Can We Make the Transition to an Earth-Sustaining Economy? Even if people believe that an earth-sustaining economy is desirable, is it possible to make such a drastic change in the way we think and act? Some environmentalists, economists, and business leaders say that it's not only possible but imperative and that it can be done over the next 40–50 years.

According to Paul Hawken, this new approach to economic thinking and actions recognizes that most business leaders are not evil, earth-degrading ogres.

INDIVIDUALS MATTER

Ray Anderson

Ray Anderson is CEO of Interface, an Atlanta-based company that makes carpet tiles. The company has 26 factories in 6 countries, customers in 110 countries, and more than $1 billion in annual sales.

Ray changed the way he viewed the world and his business after reading Paul Hawken's book *The Ecology of Commerce* (Guest Essay, p. 34). In 1994, he announced plans to develop the nation's first totally sustainable green corporation.

He has instituted hundreds of projects with the goals of zero waste, greatly reduced energy use, and eventually zero use of fossil fuels by relying on renewable solar energy. By 1998 the company had reduced resource waste by almost 25% and reduced energy waste enough to save $77 million. One of Interface's factories in California is being run on solar cells to produce the world's first solar-made carpet.

To achieve the goal of zero waste, Anderson plans to stop selling carpet and lease it as a way to control recycling. The company will install, clean, and repair the carpet, and when it wears out recycle it into new carpeting. As Anderson puts it, "We want to harvest yesterday's carpets and recycle them with zero scrap going to the landfill and zero emissions into the ecosystem—and run the whole thing on sunlight."

Some of his business friends think he's crazy. But Anderson is one of a growing number of business leaders committed to a finding a more economically and ecologically sustainable way of doing business while still making a profit for stockholders. Since he instituted his earth-friendly policies, the company's share price and earnings have increased and he says he's having a blast.

CONNECTIONS

Jobs and the Environment

Critics of environmental laws and regulations contend that they cause the loss of large numbers of jobs in the United States and in other developed countries. In fact, studies show that the opposite is true. For example, U.S. Department of Labor statistics reveal that only 0.1% of the jobs lost in the United States between 1987 to 1990 were lost as a result of environmental regulations.

Environmental protection is a major growth industry that creates new jobs, with about 1 million jobs expected to be added to the U.S. environmental protection industry between 1995 and 2005. A congressional study concluded that investing $115 billion per year in solar energy and in improving energy efficiency in the United States would eliminate about 1 million jobs in oil, gas, coal, and electricity production but would create 2 million other new jobs. Investment of the money saved by reducing energy waste could create another 2 million jobs.

Critical Thinking

What major things (if any) has the federal government in the United States (or the country where you live) done to stimulate the growth of environmental jobs? What major things (if any) has the government done to discourage the growth of environmental jobs?

Instead, they are trapped in a system that by design rewards them (with the highest profits and salaries and best chances for promotion) for maximizing short-term profits for owners and investors, regardless of the harmful short- and long-term environmental and social impacts.

Hawken argues that an earth-sustaining economy would free business leaders, workers, and investors from this ethical dilemma and allow them to be financially compensated and respected for doing socially and ecologically responsible work, improving environmental quality, and still making hefty profits for owners and stockholders. Making this shift should also create jobs (Connections, above).

The problem in making this shift is not economics, but politics. It involves the difficult task of convincing business leaders and elected officials to begin changing the current system of rewards and penalties, the profits from which have given them economic and political power.

The good news is that forward-looking investors, corporate executives (Individuals Matter, above), and the governments of countries such as Japan, Germany (Case Study, p. 44), the Netherlands, and the United States recognize that earth-sustaining businesses with good environmental management will prosper as the environmental revolution proceeds. *They recognize that the environmental revolution is also an economic revolution.* According to Lester R. Brown (Guest Essay, p. 26) and Christopher Flavin, "Converting the economy of the twentieth century into one that is environmentally

CASE STUDY

How Is Germany Investing in the Future and the Earth?

German political and business leaders see sales of environmental protection goods and services—already more than $500 billion per year—as a major source of new markets and income in the next century because environmental standards and concerns are expected to rise everywhere.

Stricter environmental standards and regulations in Germany have paid off in a cleaner environment and the development of innovative green technologies that can be sold at home and abroad in a rapidly growing market. Mostly because of stricter air pollution regulations, German companies have developed some of the world's cleanest and most efficient gas turbines, and they have invented the world's first steel mill that uses no coal to make steel. Germany sells these and other improved environmental technologies globally.

On the other hand, Germany continues to heavily subsidize its coal-mining industry, which increases environmental degradation and pollution and adds large amounts of carbon dioxide to the atmosphere. According to environmental expert Norman Myers (Guest Essay, p. 282), these perverse subsidies are now so high that the German government would save money by closing all of its coal mines and sending the miners home on full pay for the rest of their lives.

In 1977 the German government started the Blue Angel product-labeling program to inform consumers about products that cause the least environmental harm. Most international companies now use the German market to test and evaluate green products.

Germany has also revolutionized the recycling business. German car companies are required to pick up and recycle all domestic cars they make. Bar-coded parts enable disassembly plants to dismantle an auto for recycling in 20 minutes. Such *take-back* requirements are being extended to almost all products to reduce use of energy and virgin raw materials. Germany plans to sell its newly developed recycling technologies to other countries.

The German government has also supported research and development aimed at making Germany the world's leader in solar-cell technology (Section 20-3) and hydrogen fuel (Section 20-6), which it expects will provide a rapidly increasing share of the world's energy. Finally, Germany provides about $1 billion per year in green foreign aid to developing countries. Much of the aid is designed to stimulate demand for German technologies such as solar-powered lights, solar cells, and wind-powered water pumps.

Critical Thinking

1. What major steps (if any) are the government and businesses in the country where you live taking in making the transition to a more earth-sustaining economy?

2. What three major things could you do to help promote a shift to such an economy in your country and in your local community?

sustainable represents the greatest investment opportunity in history."

2-6 POLITICS AND ENVIRONMENTAL POLICY

How Does Social Change Occur in Representative Democracies? **Politics** is the process by which individuals and groups try to influence or control the policies and actions of governments at the local, state, national, or international levels. Politics is concerned with who has power over the distribution of resources and benefits: who gets what, when, and how.

Democracy is government by the people through elected officials and representatives. In a *constitutional democracy*, a constitution provides the basis of government authority, limits government power by mandating free elections, and guarantees freely expressed public opinion.

Political institutions in constitutional democracies are designed to allow gradual change to ensure economic and political stability. Rapid and destabilizing change in the United States, for example, is curbed by the system of checks and balances that distributes power among the three branches of government—executive, legislative, and judicial—and among federal, state, and local governments.

In passing laws, developing budgets, and formulating regulations, government officials must deal with pressure from many competing *special-interest groups*. Each group advocates passing laws favorable to its cause and weakening or repealing laws and regulations unfavorable to its position. Each group tries to obtain cheaper and better access to resources, subsidies, and tax relief, and to enact regulations that foster the group's goals.

Some special-interest groups are *profit-making organizations* such as corporations, and others are *nonprofit, nongovernment organizations (NGOs)*. Examples of NGOs

Pouyat, R. V. 1999. "Science and Environmental Policy—Making Them Compatible." *BioScience*, vol. 49, no. 4, p. 281(6).

Types of Environmental Laws in the United States

SOLUTIONS

Environmentalists and their supporters have persuaded the U.S. Congress to enact a number of important federal environmental and resource protection laws, as discussed throughout this text and listed in Appendix 3. These laws seek to protect the environment using the following approaches:

- *Setting standards for pollution levels or limiting emissions or effluents for various classes of pollutants* (Federal Water Pollution Control Act and Clean Air Acts)
- *Screening new substances for safety before they are widely used* (Toxic Substances Control Act)
- *Requiring comprehensive evaluation of the environmental impact of an activity before it is undertaken by a federal agency* (National Environmental Policy Act)
- *Setting aside or protecting various ecosystems, resources, and species from harm* (Wilderness Act and Endangered Species Act)
- *Encouraging resource conservation* (Resource Conservation and Recovery Act and National Energy Act)

Some environmental laws contain glowing rhetoric about goals but little guidance about how to meet them, leaving this task to regulatory agencies and the courts. In other cases, the laws or presidential executive orders specify one or more of the following general principles for setting regulations:

- *No unreasonable risk*: food regulations in the Food, Drug, and Cosmetic Act
- *No risk*: the zero-discharge goals of the Safe Drinking Water and Clean Water Acts
- *Standards based on best available technology*: the Clean Air, Clean Water, and Safe Drinking Water Acts
- *Risk–benefit balancing* (Section 8-6): pesticide regulations
- *Cost–benefit balancing* (Spotlight, p. 37): the Toxic Substances Control Act

Critical Thinking

Pick one of the U.S. environmental laws listed in Appendix 3 (or a law in the country where you live). Use the library or the Internet to evaluate the law's major strengths and weakness. Decide whether the law should be weakened, strengthened, or abolished and explain why. List the three most important ways you believe the law should be changed.

are educational institutions, labor unions, and mainstream and grassroots environmental organizations.

Most political decisions made in democracies result from bargaining, accommodation, and compromise among leaders of competing *elites*, or power brokers. The primary goal of government by competing elites is to maintain the overall economic and political stability of the system (status quo) by making only gradual change.

The net result of this fragmentation, lack of funding, and influence of the regulated on the regulators is *incremental decision making*. As a result, only small changes usually are made in existing policies and programs. A disadvantage of this design for stability is that democratic governments tend to *react* to crises instead of acting to prevent them.

Case Study: How Is Environmental Policy Made in the United States? The major function of the federal government in the United States is to develop and implement *policy* for dealing with various issues.

There are several steps in establishing federal environmental policy (or any other policy):

- Persuade lawmakers that a problem exists and that the government has a responsibility to address it.
- Try to pass laws to deal with the problem. Most environmental bills are evaluated by as many as 10 committees in the House of Representatives and the Senate. Effective proposals often are weakened by this fragmentation and by lobbying from groups opposing the law. Nonetheless, since the 1970s a number of environmental laws have been passed in the United States (Appendix 3 and Solutions, above).
- Appropriate enough funds to implement and enforce each law. Indeed, developing and adopting a budget is the most important and controversial thing the executive and legislative branches do. Developing a budget involves answering two key questions: What resource use and distribution problems will be addressed, and how much of the limited tax revenues will be used to address each problem?
- Have the appropriate government department or agency draw up regulations for implementing each law.
- Implement and enforce the approved regulations.

Solutions: How Can Individuals Affect Environmental Policy? A major theme of this book is that individuals matter. History shows that significant change usually comes from the *bottom up*. Without grassroots

Hint: Enter the search terms *policy, environmental science* using the Subject Guide.

How Can Environmental Laws and Regulations Be Improved?

SOLUTIONS

Environmentalists agree that some government laws and regulations go too far and that bureaucrats sometimes develop and impose ridiculous and excessively costly regulations. But they argue that the solution is to stop regulatory abuse, not throw out or seriously weaken the body of laws and regulations that help protect the public good.

According to environmentalist William Ashworth,

> *If we wish to make progress, it will do us no good to replace one failed system with another that failed just as badly. Government regulation, after all, didn't fall out of the sky; it was erected, piece by piece, as an attempt to deal with the damage caused by unrestrained property rights and the unregulated free-market system.... We do not need to deconstruct regulation, but to reconstruct it.*

To accomplish this, a growing number of analysts urge environmentalists to take a hard look at existing environmental laws and regulations. Which laws (or parts of laws) have worked and why? Which ones have failed, and why? Which government bureaucracies concerned with developing and enforcing environmental and resource regulations have abused their power or have not been responsive enough to the needs of ordinary people? How can such abuses be corrected? What existing environmental laws (or parts of laws) and regulations should be repealed or modified?

How can a balanced program of both regulation and market-based approaches (Table 2-1) be used to achieve environmental goals? What environmental problems lend themselves to market-based approaches (free-market environmentalism), and which ones do not? How can the principles of pollution prevention and waste reduction become the guiding principles for environmental legislation and regulation? What is the minimum amount of environmental legislation and regulation that is needed?

These are important issues that environmentalists, business leaders, elected officials, and government regulators need to address with a cooperative, problem-solving spirit.

Critical Thinking

Analyze a particular environmental law in the United States (Appendix 3) or in the country where you live to come up with ways in which it could be improved. Develop a strategy for bringing about such changes.

political action by millions of individual citizens and organized groups, the air you breathe and the water you drink today would be much more polluted, and much more of the earth's biodiversity would have disappeared.

Individuals can influence and change government policies in constitutional democracies by **(1)** voting for candidates and ballot measures, **(2)** contributing money and time to candidates seeking office, **(3)** lobbying, writing, faxing, e-mailing, or calling elected representatives, asking them to pass or oppose certain laws, establish certain policies, and fund various programs (Appendix 6), **(4)** using education and persuasion, **(5)** exposing fraud, waste, and illegal activities in government (whistleblowing), **(6)** filing lawsuits, and **(7)** participating in grassroots activities to bring about change or enforce existing laws and regulations.

Solutions: What Is Environmental Leadership? Four types of environmental leadership are

- *Leading by example* by using one's lifestyle to show others that change is possible and beneficial.
- *Working within existing economic and political systems to bring about environmental improvement, often in new, creative ways.* People can influence political elites by campaigning and voting for candidates and by communicating with elected officials (Appendix 6). They can also work within the system by choosing an environmental career (Individuals Matter, right).
- *Challenging the system and basic societal values.*
- *Proposing and working for better solutions to environmental problems.* Leadership is more than being against something; it also involves coming up with better ways to accomplish various goals.

All four types of leadership are needed.

Solutions: How Can We Make Government More Responsive to Ordinary Citizens? According to most political analysts, *campaign financing* is the biggest problem that keeps elected officials from being more responsive to the environmental and other needs and problems of ordinary citizens. The enormous amounts of money needed to run for office or to get reelected in the United States come mostly from wealthy and powerful individuals and corporate interests instead of ordinary citizens.

Most analysts and about 80% of citizens polled agree that the U.S. political system is based more on money than on citizens' votes—one reason so many Americans don't bother to vote. Even those who do vote often feel they're simply choosing the lesser of two

INDIVIDUALS MATTER

Environmental Careers

Besides committed earth citizens, the environmental movement needs dedicated professionals. In the United States (and in other developed countries), the *green job market* is one of the fastest-growing segments of the economy.

Many employers are actively seeking environmentally educated graduates. They are especially interested in people with scientific and engineering backgrounds and those with double majors (business and ecology, for example) or double minors.

Environmental career opportunities exist in a large number of fields: environmental engineering (currently the fastest-growing job market), sustainable forestry and range management, parks and recreation, air and water quality control, solid-waste and hazardous-waste management, recycling, urban and rural land-use planning, computer modeling, ecological restoration, and soil, water, fishery, and wildlife conservation and management.

Environmental careers can also be found in education, environmental planning, environmental management, environmental health, toxicology, geology, ecology, conservation biology, chemistry, climatology, population dynamics and regulation (demography), law, risk analysis, risk management, accounting, environmental journalism, design and architecture, energy conservation and analysis, renewable-energy technologies, hydrology, consulting, public relations, activism and lobbying, economics, diplomacy, development and marketing, publishing (environmental magazines and books), and law enforcement (pollution detection and enforcement teams).

Critical Thinking

1. Is the green job market really one of the fastest-growing segments of the economy? Use the library or the Internet to help answer this question.

2. Have you considered an environmental career? Why or why not?

evils. Many analysts see drastic reform in the way elections are financed as the key to making the government more responsive to ordinary citizens (Solutions, p. 48).

What Are the Roles of Mainstream and Grassroots Environmental Groups? Many types of environmental groups work at the local, state, national, and international levels (Appendix 1). Environmental organizations range from multimillion-dollar *mainstream* groups, led by chief executive officers and staffed by experts, to *grassroots* neighborhood groups formed to do battle on local environmental issues. More than 8 million U.S. citizens belong to environmental organizations.

Mainstream environmental groups are active primarily at the national level and to a lesser extent at the state level; often they form coalitions to work together on issues. Some mainstream organizations funnel substantial funds to local activists and projects.

Mainstream groups work within the political system; many have been major forces in persuading Congress to pass and strengthen environmental laws (Appendix 3) as well as fighting off attempts to weaken or repeal such laws. However, these groups must continually guard against having their efforts subverted by the political system they work to change and against losing touch with ordinary people and nature in the insulated atmosphere of national and state capitals.

The base of the environmental movement in the United States consists of at least 6,000 grassroots citizens' groups organized to protect themselves against local pollution and environmental damage. According to political analyst Konrad von Moltke, "There isn't a government in the world that would have done anything for the environment if it weren't for the citizen groups." Environmental grassroots groups are also active in colleges and public schools across the United States (Individuals Matter, p. 49).

What Are the Goals of the Anti-Environmental Movement in the United States? Since 1980 anti-environmentalists in the United States have mounted a massive campaign to weaken or repeal existing environmental laws and to destroy the credibility and effectiveness of the environmental movement. Appendix 4 lists some of the tactics used by the anti-environmental movement.

Since 1988, several hundred local and regional grassroots groups in the United States have formed a national anti-environmental coalition called the *Wise-Use movement*. Much of their money comes from real estate developers and from timber, mining, oil, coal, and ranching interests.

According to Ron Arnold, who played a key role in setting up this movement, the specific goals of the Wise-Use movement are as follows:

- *Cut all old-growth forests in the national forests and replace them with tree plantations.*
- *Modify the Endangered Species Act so that economic factors override preservation of endangered and threatened species.*

Election Finance Reform*

SOLUTIONS

A growing number of analysts of all political persuasions see *election finance reform* as the single most important way to reduce the influence of so-called money votes. They urge us to focus our efforts on this crucial issue as the key to making government more responsive to ordinary citizens.

One suggestion for reducing undue influence by powerful special interests would be to let the people (taxpayers) *alone* finance all federal, state, and local election campaigns, with low spending limits. Candidates could not accept direct or indirect donations from any other individuals, groups, or parties, with *absolutely no exceptions*.

Once elected, officials could not **(1)** use free mailing, staff employees, or other privileges to aid their election campaigns, **(2)** accept donations or any kind of direct or indirect financial aid from any individual, corporation, political parties, or interest group for their future election campaigns—or for any other reason that could even remotely influence their votes on legislation. Violators of this new *Public Funding Elections Act* would be barred from the campaign or removed from office. Anyone making illegal donations would be subject to large fines and possible jail sentences.

Having all elections financed entirely by public funds would cost each U.S. taxpayer only about $5–10 per year (as part of their income taxes) for all federal elections and a much smaller amount for state and local elections. Proponents see this as a small investment for making democracy more responsive to ordinary people.

With such a reform, elected officials could spend their time governing instead of raising money and catering to powerful special interests. Office seekers would not need to be wealthy. Public service could become a way to help the country and the earth. Special-interest groups would be heard because of the validity of their ideas, not the size of their pocketbooks.

Proponents of this reform contend that this is an issue that ordinary people of all political persuasions could work together on. From this fundamental political reform, other democratic reforms could flow.

The problem is getting members of Congress (and state legislators) to pass a virtually foolproof and constitutionally acceptable plan that would put them on equal financial footing with their challengers and that might decrease their chances of getting reelected.

Supporters of public financing of all elections argue that the way out of this dilemma is to band together to find and elect candidates who pledge to bring about this fundamental political reform, and then vote them out of office if they don't.

Critical Thinking

1. Do you agree or disagree with this approach to election reform? Explain. What role would you take in bringing about or opposing such a reform?

2. Why might the mainstream environmental groups in the United States not support such a reform? Should they?

*For an insightful in-depth analysis of this issue, see Moti Nissani, "Brass-Tacks Ecology," *The Trumpeter*, vol. 14, no. 3 (Summer 1997), pp. 143–148, and Chapter 10 in Moti Nissani, *Lives in the Balance: The Cold War and American Politics, 1945–1991* (Carson City, Nev.: Dowser Publishing Group, 1992).

- *Eliminate government restrictions on development of wetlands.*
- *Open all national parks, national wildlife refuges, and wilderness areas to oil drilling, mining, off-road vehicles (ORVs), and commercial development.*
- *Do away with the National Park Service and launch a 20-year construction program of new concessions run by private firms in the national parks.*
- *Continue mining on public lands under the provisions of the 1872 Mining Law (p. 326), which allows mining interests to pay no royalties to taxpayers for hard-rock minerals they remove and to buy public lands for a pittance.*
- *Recognize private property rights to mining claims, water, grazing permits, and timber contracts on public lands; do not raise fees for these activities. Ideally, sell resource-rich public lands to private enterprise.*
- *Provide civil penalties against anyone who legally challenges economic action or development on federal lands.*
- *Allow pro-industry (Wise-Use) groups or individuals to sue as "harmed parties" on behalf of industries threatened by environmentalists.*

Most anti-environmental groups either affiliated with or generally supportive of the Wise-Use movement and its offshoot, the *Alliance for America* (formed in 1991), use environmentally friendly–sounding names.*

Should We Expand the Concept of Security? Countries have legitimate *national security* interests, but

Environmental Action on Campuses

INDIVIDUALS MATTER

Since 1988 there has been a boom in environmental awareness on college campuses and some public schools across the United States. Much of this momentum began in 1989, when the Student Environmental Action Coalition (SEAC) then at the University of North Carolina at Chapel Hill held the first national student environmental conference on the UNC campus.

SEAC groups are active on 700 campuses, and the National Wildlife Federation's Campus Ecology Program (launched in 1989) has groups on about 600 campuses.* Most student environmental groups work with members of the faculty and administration to bring about environmental improvements on their own campuses and in their local communities.

Many of these groups focus on making an environmental audit of their own campuses or schools; they then use the data gathered to propose changes that will make their campuses or schools more ecologically sustainable, usually saving them money in the process.** In addition, students who have learned to do research on and develop solutions for environmental problems will be able to use these skills the rest of their lives.

Such audits have resulted in numerous improvements. For example, Morris A. Pierce, a graduate student at the University of Rochester in New York, developed an energy management plan adopted by that school's board of trustees. Under this plan, a capital investment of $33 million is projected to save the university $60 million over 20 years. Students have induced almost 80% of universities and colleges in the United States to develop recycling programs.

Students at Oberlin College in Ohio helped design a new sustainable environmental studies building (Figure 20-24, p. 537) that **(1)** uses energy efficiently, **(2)** is powered by solar cells that generate more electricity than they use, **(3)** includes living machines (Figure 12-29, p. 322) to discharge wastewater at least as pure as the water taken in, **(4)** includes materials grown or manufactured sustainably, **(5)** uses no materials known to cause human health problems, and **(6)** is landscaped in ways that promote biodiversity.

A 1997 report by the National Wildlife Foundation's Campus Ecology Program found that 23 student-researched and motivated projects had together saved the participating universities and colleges $16.3 million. According to this study, implementing similar programs in the nation's 3,700 universities and colleges could not only help improve environmental quality and environmental education but also lead to a savings of more than $2.6 billion.

*These efforts are described in the book *Ecodemia: Campus Environmental Stewardship at the Turn of the 21st Century* (Washington, D.C.: National Wildlife Federation, 1995) and in the *Campus Environmental Yearbook*, published annually by the National Wildlife Federation.

**Details for conducting such audits are found in April Smith and the Student Environmental Action Coalition, *Campus Ecology: A Guide to Assessing Environmental Quality and Creating Strategies for Change* (Los Angeles, Calif.: Living Planet Press, 1993).

without adequate soil, water, clean air, and biodiversity, no nation can be secure. To a growing number of policy analysts, national security ultimately depends on *global environmental security* based on sustainable use of the earth's life support system (Figure 2-3).

Countries are legitimately concerned with *economic security*. However, because all economies are supported by natural capital (Figure 2-3 and Guest Essay, p. 34), economic security also depends on environmental security.

Proponents of emphasizing environmental security propose that national governments have a council of advisers made up of highly qualified experts in environmental, economic, and military security. Any major decision would require integrating all three security concerns. These analysts also call for all countries to make environmental security a major focus of diplomacy and government policy at all levels. What do you think?

2-7 ENVIRONMENTAL WORLDVIEWS: CLASHING VALUES AND CULTURES

How Shall We Live? A Clash of Cultures and Values There are conflicting views about how serious our environmental problems are and what we should do

*For lists and information about these organizations, see *The Greenpeace Guide to Anti-Environmental Organizations* (1993, Odonian Press, Box 7776, Berkeley, CA 94707), *Masks of Deception: Corporate Front Groups in America* by Mark Megalli and Andy Friedman (1993, Essential Information, P.O. Box 19367, Washington, DC 20036), *Let the People Judge: A Reader on the Wise-Use Movement* (Washington, D.C.: Island Press, 1993), and *The War Against the Greens* (San Francisco: Sierra Club Books, 1994). Also see the Anti-Environmental Myth Web site at http://members.aol.com/jimn469897/myths.htm.

about them. These conflicts arise mostly out of differing **environmental worldviews**: how people think the world works, what they think their role in the world should be, and what they believe is right and wrong environmental behavior (**environmental ethics**).

People with widely differing environmental worldviews can take the same data, be logically consistent, and arrive at quite different conclusions (Table 1-1) because they start with different assumptions.

There are many different types of environmental worldviews, as summarized in Figure 2-8. Most can be divided into two groups according to whether they are *individual centered* (atomistic) or *earth centered* (holistic). Atomistic environmental worldviews tend to be *human centered* (anthropocentric) or *life centered* (biocentric, with the primary focus on either individual species or individual organisms). Holistic or ecocentric environmental worldviews are either *ecosystem centered* or *ecosphere* (life-support system) *centered.*

What Are the Major Human-Centered Environmental Worldviews? Most people in today's industrial consumer societies have a **planetary management worldview**, which has become increasingly common during the past 50 years. According to this human-centered environmental worldview, human beings, as the planet's most important and dominant species, can

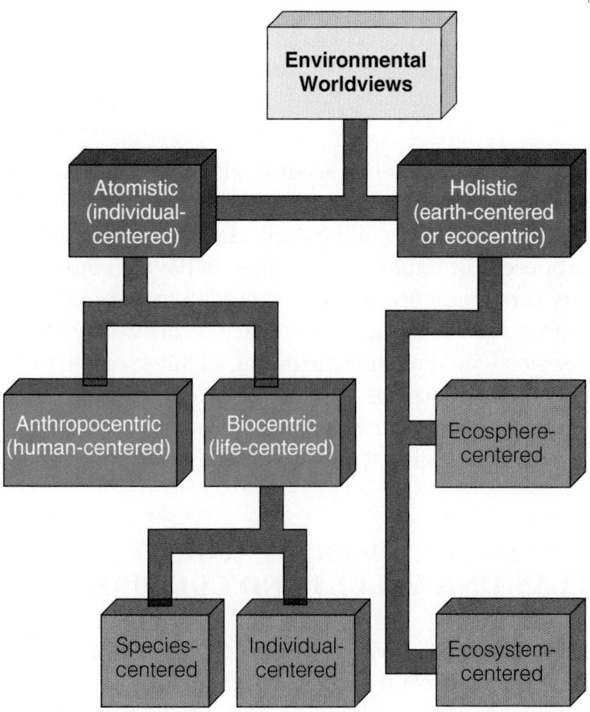

Figure 2-8 General types of environmental worldviews. (Diagram developed by Jane Heinze-Fry)

and should manage the planet mostly for their own benefit. Other species are seen as having only *instrumental value*; that is, their value depends on whether they are useful to us.

The basic environmental beliefs of this worldview include the following:

- *We are the planet's most important species and we are in charge of the rest of nature.* This idea crops up when people talk about "our" planet, "our" earth or "saving the earth."
- *There is always more.* The earth has an essentially unlimited supply of resources, to which we gain access via science and technology. If we deplete a resource, we will find substitutes. To deal with pollutants, we can invent technology to clean them up, dump them into space, or move into space ourselves. If we extinguish other species, we can use genetic engineering to create new and better ones.
- *All economic growth is good, more economic growth is better, and the potential for economic growth is essentially limitless.*
- *Our success depends on how well we can understand, control, and manage the earth's life-support systems for our benefit.*

This worldview is widely supported because it is said to be the primary driving force behind the major improvements in the human condition since the beginning of the industrial revolution (Table 1-1, left).

There are several variations of this environmental worldview. Some people belong to what might be called the *no-problem school*: There are no environmental, population, or resource problems that cannot be solved by more economic growth, better management, and better technology.

Another group, *the free-market school*, believes that the best way to manage the planet for human benefit is through a free-market global economy with minimal government interference and regulations. Free-market advocates would convert all public property resources to private property resources and let the global marketplace, governed by free-market competition (pure capitalism), decide essentially everything.

Another variation is that we do have serious environmental problems but that we can deal with them by engaging in *responsible planetary management*. People holding this view follow the pragmatic principle of *enlightened self-interest*: Better earth care is better self-care. They believe that we can sustain our species with a mixture of market-based competition, better technology, and some government intervention that **(1)** promotes ecologically sustainable forms of economic development, **(2)** protects environmental quality and

Biosphere 2: A Lesson in Humility

SPOTLIGHT

The failure of the first Biosphere 2 (the earth is Biosphere 1) project illustrated that technology cannot provide a sustainable substitute for natural life-support systems. In 1991 eight people were sealed into Biosphere 2, a $200-million facility designed to be a self-sustaining life-support system. The 1.3-hectare (3.2-acre) closed system, built in the desert near Tucson, Arizona, had miniature forests, lakes, a desert, streams, and a mini-ocean with coral reefs.

The Biospherians were supposed to be isolated for 2 years and to raise their own food, breathe air recirculated by plants, and drink water cleansed by natural processes. From the beginning they encountered numerous unexpected problems.

The life-support system began unraveling. So much oxygen disappeared mysteriously that more had to be pumped in from the outside to keep the Biospherians from suffocating.

Levels of nitrous oxide rose high enough to threaten the occupants with brain damage. Carbon dioxide skyrocketed to levels that threatened to poison the humans and spurred the growth of weedy vines that choked out food crops. Nutrients leached from the soil and polluted the water systems.

Tropical birds disappeared after the first freeze. An Arizona ant species got into the enclosure, proliferated, and killed off all other soft-bodied insects, and creatures as large as snakes had to hide to keep from being eaten alive. After the majority of the introduced insect species became extinct, the facility was overrun with cockroaches and katydids.

All together, 19 of the Biosphere's 25 small-animal species became extinct. Before the 2-year period was up, all plant-pollinating insects plants became extinct, thereby dooming to extinction most of the plant species. After this $200-million failure, the system was turned over to Columbia University for use as an environmental laboratory.

This humbling experience illustrated how little we know about the components, interactions, and processes of natural capital that sustain life on the earth. Scientists Joel Cohen and David Tilman, who evaluated the project, concluded, "No one yet knows how to engineer systems that provide humans with life-supporting services that natural ecosystems provide for free."

Critical Thinking

1. List three changes that you would consider making in your life as a result of the failure of the Biosphere 2 experiment.

2. Were the lessons learned from this experiment worth its $200-million cost? Explain.

private property rights, and **(3)** protects and manages public and common property resources.

Still another variation of the responsible planetary management environmental worldview is the *spaceship-earth worldview*. Earth is seen as a spaceship: a complex machine that we can understand, dominate, change, and manage in order to prevent environmental overload and provide a good life for everyone.

Other people advocate that the principle of *stewardship* should be used to guide our management of the earth. According to this view, we have an ethical responsibility to be caring and responsible managers or stewards of the earth. According to this view, we can and should make the world a better place for our species and other species through love, care, knowledge, and technology.

Can We Manage the Planet? Some people believe that any human-centered worldview will eventually fail because it wrongly assumes that we now have (or can gain) enough knowledge to become effective managers or stewards of the earth.

These people argue that the unregulated free-market approach won't work because it is based on mushrooming losses of natural capital (Figures 1-2, 1-5, and 1-11) and because it focuses on short-term economic benefits regardless of the harmful long-term consequences. They also contend that the spaceship-earth version of planetary management will not work because it is an oversimplified and misleading way to view an incredibly complex and ever-changing planet.

For example, these critics point out that we do not even know how many species live on the earth, much less what their roles are and how they interact with one another and their nonliving environment (Spotlight, above). We have only an inkling of what goes on in a handful of soil, a meadow, a patch of forest, a pond, or any other part of the earth. As biologist David Ehrenfeld puts it, "In no important instance have we been able to demonstrate comprehensive successful management of the world, nor do we understand it well enough to manage it even in theory."

Even if we had enough knowledge and wisdom to manage spaceship earth, some critics see this approach as requiring us to give up individual freedom in order to survive. Life on spaceship earth under a comprehensive system of planetary management or world government might be very much like the regimented life of astronauts in their capsule. The astronauts have virtually no individual freedom; essentially all of their actions are dictated by a central command (ground control).

Theologian Thomas Berry calls the industrial consumer society built on the human-centered, planetary management environmental worldview the "supreme pathology of all history":

> *We can break the mountains apart; we can drain the rivers and flood the valleys. We can turn the most luxuriant forests into throwaway paper products. We can tear apart the great grass cover of the western plains, and pour toxic chemicals into the soil and pesticides onto the fields, until the soil is dead and blows away in the wind. We can pollute the air with acids, the rivers with sewage, the seas with oil—all this in a kind of intoxication with our power for devastation. . . . We can invent computers capable of processing ten million calculations per second. And why? To increase the volume and speed with which we move natural resources through the consumer economy to the junk pile or the waste heap. . . . If, in these activities, the topography of the planet is damaged, if the environment is made inhospitable for a multitude of living species, then so be it. We are, supposedly, creating a technological wonderworld. . . . But our supposed progress . . . is bringing us to a wasteworld instead of a wonderworld.*

Do you agree with this assessment? Explain.

What Are Some Major Biocentric and Ecocentric Worldviews? Critics of human-centered environmental worldviews believe that such worldviews should be expanded to recognize the *inherent value* of all forms of life (that is, value that exists regardless of these life-forms' potential or actual use to us). They contend that recognizing the inherent value of all life and not degrading the earth's life-support systems are ways to benefit all people in this and future generations (Connections, right).

Most people with a life-centered (biocentric) worldview argue that our actions should not lead to the *premature* extinction of species. Some proponents go further and believe that each individual organism, not just each species, has an inherent right to survive.

Others believe that we must go beyond this biocentric worldview, which focuses on species and individual organisms. They see our primary role as limiting our actions to those that do not degrade or destroy the earth's life-support systems (and thus threaten the existence of our own species). In other words, they have an *earth-centered*, or *ecocentric*, environmental worldview, devoted to preserving the earth's biodiversity and the functioning of its life-support systems (ecological integrity) (Figure 2-8).

CONNECTIONS

Why Should We Care About Future Generations?

According to biologist David W. Ehrenfeld, caring about future generations enough not to degrade the earth's life-support systems is important because it gives future generations options for dealing with the problems they will face. He points out that if our ancestors had left for us the ecological devastation we are leaving our descendants, our options for enjoyment—perhaps even for survival—would be quite limited.

And in response to the question, "What can future generations do for us?" Ehrenfeld gives the following answer: "They give us a reason for treating our ecological home respectfully, so that our lives as well as theirs will be enriched."

In thinking about our responsibility toward future generations, some analysts believe that we should consider the wisdom given to us in the 18th century by the Iroquois Confederation of Native Americans: *In our every deliberation, we must consider the impact of our decisions on the next seven generations.*

Critical Thinking

What obligations, if any, concerning the environment do you have to future generations? Be honest about your feelings. To how many future generations do you have responsibilities? List the five most important environmental benefits and the five most important harmful conditions passed on to you by the previous two generations.

Their view is that we are part of, not apart from, the community of life and the ecological processes that sustain all life. Aldo Leopold summed up this idea in 1948: "All ethics rest upon a single premise: that the individual is a member of a community of interdependent parts."

There are many life-centered and earth-centered environmental worldviews, and several of them overlap in some of their beliefs. One ecocentric environmental worldview is the **earth-wisdom worldview**. It is based on the following major beliefs, which are the opposite of those making up the planetary management worldview:

- *Nature exists for all of the earth's species, not just for us.* We need the earth, but the earth does not need us.
- *There is not always more.* The earth's limited resources should not be wasted, but instead used efficiently and sustainably for us and all other species.

Deep Ecology

SPOTLIGHT

Deep ecology consists of the following ecocentric beliefs developed in 1972 by Norwegian philosopher Arnie Naess, in conjunction with philosopher George Sessions and sociologist Bill Devall:

- The well-being and flourishing of human and nonhuman life on earth have inherent value. These values are independent of the usefulness of the nonhuman world for human purposes.
- The fundamental interdependence, richness, and diversity of life-forms contribute to the flourishing of human and nonhuman life on earth.
- Humans have no right to reduce this interdependence, richness, and diversity except to satisfy vital needs.
- Present human interference with the nonhuman world is excessive, and the situation is worsening rapidly.
- Because of the damage caused by human interference in the nonhuman world, it would be better for humans, and much better for nonhumans, if there were a substantial decrease in the human population.
- Basic economic, technological, and ideological policies must therefore be changed.
- The ideological change is mainly that of appreciating *life quality* (involving situations of inherent value) rather than adhering to an ever-higher material standard of living.
- Those who subscribe to these points have an obligation directly or indirectly to try to implement the necessary changes.

Naess has also described some lifestyle guidelines compatible with the basic beliefs of deep ecology. They include **(1)** appreciating all forms of life, **(2)** protecting or restoring local ecosystems, **(3)** using simple means, **(4)** consuming less, **(5)** emphasizing satisfying vital needs rather than desires, **(6)** attempting to live *in* nature and promote community, **(7)** appreciating ethnic and cultural differences, **(8)** working to improve the standard of living for the world's poor, **(9)** working to eliminate injustice toward fellow humans or other species, and **(10)** acting nonviolently.

Deep ecology is not an ecoreligion, nor is it antireligious or antihuman, as some of its critics have claimed. Instead, it is a set of beliefs designed to have us think more deeply about the inherent value of all life on the earth and about our obligations toward both human and nonhuman life.

Critical Thinking

1. Which, if any, of the eight basic beliefs of the deep ecology environmental worldview do you agree with? Explain.

2. List five major changes that would occur in your life if most people lived by the beliefs of deep ecology.

- *Some forms of economic growth are environmentally beneficial and should be encouraged, but some are environmentally harmful and should be discouraged.*
- *Our success depends on learning to cooperate with one another and with the rest of nature by learning how to work with the earth.*

A related ecocentric environmental worldview is the *deep ecology worldview* (Spotlight, above).

Others say we do not need to be biocentrists or ecocentrists to value life or the earth because human-centered stewardship and planetary management environmental worldviews also call for us to value individuals, species, and the earth's life-support systems as part of our responsibility as earth's caretakers. What do you think?

2-8 SOLUTIONS: LIVING SUSTAINABLY

How Should We Evaluate Sustainability Proposals? *Sustainability* has become a buzzword that means different things to different people. Lester Brown (Guest Essay, p. 26) has this simple test for any sustainability proposal: "Does this policy or action lower carbon emissions? Does it reduce the generation of toxic wastes? Does it slow population growth? Does it increase the earth's tree cover? Does it cut ozone-depleting CFC emissions? Does it reduce air pollution? Does it reduce radioactive waste generation? Does it lead to less soil erosion? Does it protect the planet's biodiversity?"

Additional questions also help us evaluate sustainability proposals. Does it deplete natural capital? Does it diminish cultural diversity? Does it reduce poverty, hunger, and disease? Does it promote individual and community self-reliance? Does it prevent pollution? Does it reduce resource waste? Does it save energy? Does it transfer the most resource-efficient and environmentally benign technologies to developing countries? Does it enhance ecological and economic justice for all?

Solutions: What Are Some Ethical Guidelines for Working with the Earth? According to Robert Cahn, "The main ingredients of an environmental ethic are caring about the planet and all of its inhabitants, allowing unselfishness to control the immediate self-interest that harms others, and living each day so as to leave the lightest possible footprints on the planet." Ethicists and philosophers have developed a variety of ethical guidelines for living more sustainably or lightly on the earth. Such guidelines can be used by anyone, whether they have a human-centered stewardship environmental worldview or a biocentric or ecocentric environmental worldview.

Ecosphere and Ecosystems

- We should try to understand and work with the rest of nature to help sustain the ecological integrity, biodiversity, and adaptability of the earth's life-support systems.
- When we alter nature to meet our needs or wants, we should carefully evaluate our proposed actions and choose methods that do the least possible short- and long-term environmental harm.

Species and Cultures

- We should work to preserve as much of the earth's genetic variety as possible because it is the raw material for all future evolution and genetic engineering.
- We have the right to defend ourselves against individuals of species that do us harm and to use individuals of species to meet our vital needs, but we should strive not to cause the premature extinction of any wild species.
- The best ways to protect species and individuals of species are to protect the places where they live and to help restore places we have degraded.
- No human culture should become extinct because of our actions.

Individual Responsibility

- We should not inflict unnecessary suffering or pain on any animal we raise or hunt for food or use for scientific or other purposes.
- We should leave the earth as good as or better than we found it.
- We should use no more of the earth's resources than we need.
- We should work with the earth to help heal ecological wounds we have inflicted.

Do you agree or disagree with these guidelines?

What Is Earth Education? Most environmentalists believe that learning how to live sustainably requires a foundation of environmental or earth education that relies heavily on an interdisciplinary and holistic approach to learning (Guest Essay, p. 56). Achieving such ecological literacy means having the ability to synthesize and connect knowledge from a variety of disciplines in order to see the big picture (holistic thinking).

According to its proponents, the most important goals of such an education are

- *Develop respect or reverence for all life.*
- *Understand as much as we can about how the earth works and sustains itself, and use such earth wisdom to guide our lives, communities, and societies.*
- *Understand as much as we can about connections and interactions.* This includes those **(1)** within nature, **(2)** between people and the rest of nature, **(3)** among people with different cultures and beliefs, **(4)** among generations, **(5)** among the problems we face, and **(6)** among the solutions to these problems.

INDIVIDUALS MATTER

Mindquake: Evaluating One's Environmental Worldview

Questioning and perhaps changing one's environmental worldview can be difficult and threatening. It can set off a cultural *mindquake* that involves examining many of one's most basic beliefs. However, once people change their worldviews, it no longer makes sense for them to do things in the old ways. If enough people do this, then tremendous cultural change, once considered impossible, can take place rapidly.

Most environmentalists urge us to think about what our basic environmental beliefs are and why we have them. They believe that evaluating our beliefs, and being open to the possibility of changing them, should be one of our most important lifelong activities.

As this book emphasizes, most environmental issues are filled with controversy and uncertainty. A clearly right or wrong path is not easy to discover and usually is strongly influenced by one's environmental worldview. As philosopher Georg Hegel pointed out nearly two centuries ago, tragedy is not the conflict between right and wrong, but the conflict between right and right.

Critical Thinking

What are the basic beliefs of your current environmental worldview? At the end of this course, evaluate your answer to learn which (if any) of your basic environmental beliefs have changed.

CONNECTIONS

Learning from the Earth

Formal earth education is important, but many earth thinkers believe that it is not enough. They urge us to take the time to escape the cultural and technological body armor we use to insulate ourselves from nature and to experience nature directly.

They suggest that we reenchant our senses and kindle a sense of awe, wonder, and humility by standing under the stars, sitting in a forest, taking in the majesty and power of an ocean, or experiencing a stream, lake, or other part of untamed nature.

We might pick up a handful of soil and try to sense the teeming microscopic life in it that keeps us alive. We might look at a tree, mountain, rock, or bee and try to sense how they are a part of us, and we a part of them as interdependent participants in the earth's life-sustaining recycling processes.

Earth thinker Michael J. Cohen suggests that we recognize who we really are by saying,

I am a desire for water, air, food, love, warmth, beauty, freedom, sensations, life, community, place, and spirit in the natural world. . . . I have two mothers: my human mother and my planet mother, Earth. The planet is my womb of life.

Many psychologists believe that consciously or unconsciously we spend much of our lives in a search for roots: something to anchor us in a bewildering and frightening sea of change. As philosopher Simone Weil observed, "To be rooted is perhaps the most important and least recognized need of the human soul."

Earth philosophers say that to be rooted, each of us needs to find a *sense of place*: a stream, a mountain, a yard, a neighborhood lot, or any piece of the earth we feel at one with as a place we know, experience emotionally, and love. It can be a place where we live or a place we occasionally visit and experience in our inner being. When we become part of a place, it becomes a part of us. Then we are driven to defend it from harm and to help heal its wounds.

To many earth thinkers, emotionally experiencing our connectedness with the earth leads us to recognize that the healing of the earth and the healing of the human spirit are one and the same. They call for us to discover and tap into what Aldo Leopold calls "the green fire that burns in our hearts" and use this as a force for respecting and working with the earth and with one another.

Critical Thinking

Some analysts believe that learning earth wisdom by experiencing the earth and forming an emotional bond with its life forms and processes is unscientific, mystical poppycock based on a romanticized view of nature. They believe that better scientific understanding of how the earth works and improved technology are the only ways to achieve sustainability. Do you agree or disagree? Explain.

- *Use critical thinking skills to become wisdom seekers instead of vessels of information* (infoglut and infojunk).
- *Understand and evaluate one's worldview and see this as a lifelong process* (Individuals Matter, left).
- *Learn how to evaluate the beneficial and harmful consequences of one's lifestyle and profession on the earth, today and in the future.*
- *Use critical thinking skills to evaluate and resist advertising.* As humorist Will Rogers put it, "Too many people spend money they haven't earned to buy things they don't want, to impress people they don't like."
- *Foster a desire to make the world a better place and act on this desire.* As David Orr puts it, education should help students "make the leap from 'I know' to 'I care' to 'I'll do something.'"

According to environmental educator Mitchell Thomashow, four basic questions should be at the heart of environmental education:

- Where do the things I consume come from?
- What do I know about the place where I live?
- How am I connected to the earth and other living things?
- What are my purpose and responsibility as a human being?

How we answer these questions determines our *ecological identity*.

How Can Direct Experiences Help Us Learn How to Work with the Earth? In addition to formal education, some analysts believe that we need to experience nature directly to help us learn to walk more lightly on the earth (Connections, above).

A related goal is to *learn how to live more simply*. Although seeking happiness through the pursuit of material things is considered folly by virtually every major religion and philosophy, it is preached incessantly by modern advertising.

Some affluent people in developed countries are adopting a lifestyle of *voluntary simplicity*, doing and enjoying more with less by learning to live more simply. Voluntary simplicity is based on Mahatma Gandhi's *principle of enoughness*: "The earth provides enough to

Envisioning a Sustainable Society

GUEST ESSAY

Lester W. Milbrath

Lester W. Milbrath is director of the Research Program in Environment and Society and professor emeritus of political science and sociology at the State University of New York at Buffalo. During his distinguished career he has served as director of the Environmental Studies Center at SUNY/Buffalo (1976–87) and taught at Northwestern University, Duke University, the University of Tennessee, National Taiwan University in Taipei, and Aarhus University in Denmark. He has also been a visiting research scholar at the Australian National University and at Mannheim University in Germany. His research has focused on the relationships among science, society, and citizen participation in environmental policy decisions, with emphasis on environmental perceptions, beliefs, attitudes, and values. He has written numerous articles and books. His book Envisioning a Sustainable Society: Learning Our Way Out *(1989) summarizes a lifetime of studying our environmental predicament. It is considered one of the best analyses of what we can do to learn how to work with the earth. His most recent book is* Learning to Think Environmentally While There Is Still Time *(1995).*

The 1992 Earth Summit at Rio popularized the goal of sustainable development. Most of the heads of state meeting there believed that goal could be achieved by developing better technology and by writing better laws, agreements, and treaties, and by enforcing them. Unfortunately, their approach was flawed and will not achieve sustainability because they do not understand the nature of the crisis in our earthly home.

Try this thought experiment: Imagine that, suddenly, all the humans disappeared, but all the buildings, roads, shopping malls, factories, automobiles, and other artifacts of modern civilization were left behind. What then? After three or four centuries, buildings would have crumbled, vehicles would have rusted and fallen apart, and plants would have recolonized fields, roads, parking lots, even buildings. Water, air, and soil would gradually clear up; some endangered species would flourish. Nature would thrive splendidly without us.

That mental experiment makes it clear that we do not have an environmental crisis; we have a crisis of civilization. Heads of state meeting at the Earth Summit neither understood nor dealt with civilization's most crucial problems: Humans are reproducing at such epidemic rates that world population could rise to 9 billion by 2054; resource depletion and waste generation could easily triple or even quadruple over that period; waste discharges are already beginning to change the way the biosphere works; climate change and ozone loss will reduce the productivity of ecosystems just when hordes of new humans will be looking for sustenance, and will destroy the confidence people need in order to invest in the future.

Without intending to, we have created a civilization that is headed for destruction. Either we learn to control our growth in population and in economic activity, or nature will use death to control it for us.

Present-day society is not capable of producing a solution because it is disabled by the values our leaders constantly trumpet: economic growth, jobs, consumption, competitiveness, power, and domination. Societies pursuing these goals cannot avoid depleting their resources, degrading nature, poisoning life with wastes, and upsetting biospheric systems. *We have no choice but to change; resisting change will make us victims of change.*

But how do we transform to a sustainable society? My answer, which I believe is the only answer, is that *we must learn our way.* Nature, and the imperatives of its laws, will be our most powerful teacher as we learn our way to a new society. Most crucially, we must learn how to think about values.

Life in a viable ecosystem must become the core value of a sustainable society; that means all life, not just human life. Ecosystems function splendidly without humans, but human society would die without viable ecosystems. Individuals seeking life quality require a well-functioning society living in well-functioning ecosystems. We must give top priority to the ecosystems that support us and second priority to human societies.

A sustainable society would affirm love as a primary value and extend it not only to those near and dear, but to people in other lands, to future generations, and to other species. A sustainable society emphasizes partnership rather than domination, cooperation over competition, love over power. A sustainable society affirms justice and security as primary values.

A sustainable society would encourage self-realization—helping people to become all they are capable of being, rather than spending and consuming—as the key to a fulfilling life. A sustainable society would make

satisfy every person's need but not every person's greed. . . . When we take more than we need, we are simply taking from each other, borrowing from the future, or destroying the environment and other species." It means asking oneself, "How much is enough?" This is not an easy thing to do because people in affluent societies are conditioned to want more and more, and they often think of such wants as vital needs.

Voluntary simplicity by those who have more than they need should not be confused with the *forced simplicity* of the poor, who do not have enough to meet their basic needs for food, clothing, shelter, clean water long-lasting products to be cherished and conserved. People would learn a love of beauty and simplicity.

A sustainable society would use both planning and markets as basic and supplementary information systems. Markets fail us because they can neither anticipate the future nor make moral choices between objects and between policies. Markets also cannot provide public goods such as schools, parks, and environmental protection, which are just as important for life quality as private goods.

A sustainable society would continue further development of science and technology because we need practical creative solutions that are both environmentally sound and economically feasible. However, we should recognize that those who control science and technology could use them to dominate all other creatures; we must learn to develop social controls of science and technology to make our society more sustainable. We should not allow the deployment of powerful new technologies that can induce sweeping changes in economic patterns, lifestyles, governance, and social values without careful forethought regarding their long-term impacts.

Conscious social learning would become the dynamic of social change in a sustainable society, not only to deal with pressing problems, but also to realize a vision of a good society. Meaningful and lasting social change occurs when nearly everyone learns the necessity of change and the value of working toward it.

Ecological thinking is different from most thinking that guides modern society. For example, the following key maxims derived from the law of conservation and matter [Section 3-5], the laws of energy or thermodynamics [Section 3-7], and the workings of ecosystems [Chapter 4] are routinely violated in contemporary thinking and discourse: **(1)** *Everything must go somewhere (there is no away),* **(2)** *energy should not be wasted because all use of energy produces disorder in the environment,* **(3)** *we can never do just one thing (everything is connected),* and **(4)** *we must constantly ask, "And then what?"*

Every schoolchild and every adult should learn these simple truths; we need to reaffirm the tradition that knowledge of nature's workings and a respect for all life are basic to a true education. We should require such environmental education of all students, just as we now require every student to study history.

Ecological thinking recognizes that a proper understanding of the world requires people to learn how to think holistically, systematically, and futuristically. Because everything is connected to everything else, we must learn to anticipate second-, third-, and higher-order consequences for any contemplated major societal action.

A society learning to be sustainable would redesign government to maximize its ability to learn. It would use the government learning process to promote social learning. It would require that people who govern listen to citizens, not only to keep the process open for public participation, but also to cultivate mutual learning between officials and citizens.

A sustainable society would strive for an effective system of planetary politics because our health and welfare are vitally affected by how people, businesses, and governments in other lands behave. It would nurture planet-wide social learning.

Learning our way to a new society cannot occur, however, until enough people become aware of the need for major societal change. As long as contemporary society is working reasonably well and leaders keep telling us that society is on the right track, most people will not listen to a message urging significant change. For that reason, urgently needed change probably will be delayed, and conditions on our planet are likely to get worse before they can get better. Nature will be our most powerful teacher, especially when biospheric systems no longer work the way they used to. In times of great system turbulence, social learning can be extraordinarily swift.

Our species has a special gift: the ability to recall the past and foresee the future. Once we have a vision of the future, every decision becomes a moral decision. Even the decision not to act becomes a moral judgment. Those who understand what is happening to the only home for us and other species are not free to shrink from the responsibility to help make the transition to a more sustainable society.

Critical Thinking

1. Do you agree or disagree that we can only learn our way to a sustainable society? Explain.

2. Do you think we will learn our way to a sustainable society? Explain. What role, if any, do you intend to play in this process?

and air, and good health. Would you or do you use the principle of voluntary simplicity in your life? How?

Earth care also means *not using guilt and fear to motivate other people to work with the earth and other people.* We need to nurture, reassure, understand, and care for one another. Finally, we need to *have fun and take time to enjoy life.* Every day we should laugh and enjoy wild nature, beauty, friendship, and love.

How Can We Move Beyond Blame, Guilt, and Denial to Responsibility? According to many psychologists, when we first encounter an environmental problem, our initial response often is to find someone or something to blame: greedy industrialists, uncaring politicians, environmentalists, and misguided worldviews. It is the fault of such villains, and we are the victims.

This can lead to despair, denial, and inaction because we feel powerless to stop or influence these forces. There are also so many complex and interconnected environmental problems and conflicting views about their seriousness and possible solutions that we feel overwhelmed and wonder whether there is any way out—another emotion leading to denial and inaction.

Upon closer examination we may realize that we all make some direct or indirect contributions to the environmental problems we face. As Pogo said, "We have met the enemy and it is us." We don't want to feel guilty or bad about all of the things we are not doing, so we avoid thinking about them—another path to denial and inaction.

How do we move beyond immobilizing blame, fear, and guilt to engaging in more responsible environmental actions in our daily lives? Analysts have suggested several ways to do this.

First, recognize and avoid common mental traps that lead to denial, indifference, and inaction. These traps include **(1)** *gloom-and-doom pessimism* (it's hopeless), **(2)** *blind technological optimism* (science and technofixes will always save us), **(3)** *fatalism* (we have no control over our actions and the future), **(4)** *extrapolation to infinity* (if I can't change the entire world quickly, I won't try to change any of it), **(5)** *paralysis by analysis* (searching for the perfect worldview, philosophy, solutions, and scientific information before doing anything), and **(6)** *faith in simple, easy answers*.

Second, understand that no one can even come close to doing all of the things people suggest (or that we know we should be doing) to work with the earth. Try to determine what things you do have the greatest environmental impact and concentrate on them. Focus on things that you feel strongly about and that you can do something about. Rejoice in the good things you have done, then jump in and do more to make the earth a better place (Appendix 6).

Third, base your actions on a sense of *hope*, which history has shown to be the major energizing force for bringing about change. The secret is to keep your empowering feelings of hope and joy slightly ahead of your immobilizing feelings of despair.

Fourth, it recognize that there is no single correct or best solution to the environmental problems we face. Indeed, one of nature's most important lessons (from evolution) is that preserving diversity or a rainbow of flexible and adaptable solutions to the problems we face is the best way to adapt to earth's largely unpredictable, ever-changing conditions.

What Are the Major Components of an Environmental Revolution? The *environmental revolution* (Guest Essay, p. 26) that many environmentalists call for us to bring about would have several components:

- An *efficiency revolution* that involves not wasting matter and energy resources, using a combination of technological advances and lifestyle changes.
- A *pollution prevention revolution* that reduces pollution and environmental degradation by **(1)** reducing the waste of matter and energy resources, **(2)** keeping highly toxic substances from being released into the environment by recycling or reusing them within industrial processes, and **(3)** trying to find less harmful or easily biodegradable substitutes for such substances, or not producing them at all.
- A *sufficiency revolution*. This involves trying to meet the basic needs of all people on the planet and asking how many material things we really need to have a decent and meaningful life.
- A *demographic revolution* based on bringing the size and growth rate of the human population into balance with the earth's ability to support humans and other species.
- An *economic revolution* in which we use economic systems to reward earth-sustaining behavior and discourage earth-degrading behavior (Sections 2-3, 2-4, and 2-5).

Opponents of such a cultural change like to paint environmentalists as messengers of gloom, doom, and hopelessness. However, *the real message of environmentalism is not gloom and doom, fear, and catastrophe but hope and a positive vision of the future*. This is an exciting message of challenge and adventure as we struggle to find better and more responsible ways to live on this planet.

Envision the earth's life-sustaining processes as a beautiful and diverse web of interrelationships—a kaleidoscope of patterns, rhythms, and connections whose very complexity and multitude of possibilities remind us that cooperation, sharing, honesty, humility, and love should be the guidelines for our behavior toward one another and the earth.

When there is no dream, the people perish.
PROVERBS 29:18

REVIEW QUESTIONS

1. Distinguish **(a)** between *natural capital, manufactured capital,* and *human capital* and **(b)** between a *centrally planned economy, pure capitalism,* and a *mixed economic system.*

2. What is *economic growth,* and what are the pros and cons of such growth?

3. List four reasons that GNP and GDP are not useful ways to measure quality of life and environmental quality.

4. Distinguish between *internal costs* and *external costs* and give an example of each. What is *full-cost pricing,* and what are the pros and cons of using this approach to internalize external environmental costs?

5. What is *cost–benefit analysis,* and what are the pros and cons of using this tool to evaluate alternative courses of action?

6. List the pros and cons of using the following market forces to improve environmental quality: **(a)** phasing out environmentally harmful economic subsidies, **(b)** phasing in environmentally beneficial economic subsidies, **(c)** granting tradable pollution and resource use rights, **(d)** taxing environmentally harmful activities, **(e)** charging fees for resource use or resource extraction on public lands and water systems, and **(f)** requiring developers to post pollution prevention or assurance bonds. What are the pros and cons of decreasing taxes on wages and profits and increasing taxes on pollution and waste?

7. How do economists determine the optimal level of pollution for a particular chemical, and what are the pros and cons of using this approach?

8. List the pros and cons of encouraging global free trade as a way to improve environmental quality.

9. What is *poverty,* and what are its harmful consequences? What is the *trickle-down hypothesis* for dealing with poverty, and what are the pros and cons of using this approach?

10. List two ways in which the governments of developing countries can reduce poverty. List seven ways in which governments of developed countries can reduce poverty.

11. Explain how we could convert current economic systems to more sustainable ones. Explain why the environmental revolution is also an economic revolution.

12. What is *politics,* and why is political decision making so fragmented and incremental? What are the four main steps used by the federal government in the United States to establish environmental policy? What are four major types of environmental laws in the United States?

13. List seven ways individuals can influence environmental policy. List four types of environmental leadership.

14. Why is reforming election financing an important way to improve environmental quality?

15. Describe the key roles of mainstream and grassroots environmental groups. What are the goals of the anti-environmental movement? What is the Wise-Use movement?

16. Distinguish between environmental, economic, and military security and explain the importance of making environmental security a key priority of governments.

17. Distinguish between the *planetary management* and *earth-wisdom environmental worldviews* and list the pros and cons of each. What is *deep ecology*? Give two reasons for caring about future generations.

18. What are the main components of earth or environmental education?

19. What is *voluntary simplicity,* and what are the pros and cons of living this way?

20. What are four ways to move beyond blame, guilt, and denial and become more environmentally responsible?

21. What are five major components of an environmental revolution that most environmentalists believe we should bring about over the next 50 years?

CRITICAL THINKING

1. The primary goal of all current economic systems is to maximize economic growth by producing and consuming more and more economic goods. Do you agree with that goal? Explain. What are the alternatives?

2. Do you believe that cost–benefit analysis should be the *primary method* for making all decisions about how limited federal, state, and local government funds are used? Explain. If not, what are the alternatives?

3. Do you favor internalizing the external costs of pollution and unnecessary resource waste? Explain. How might it affect your lifestyle? Wildlife? Any children you might have?

4. (a) Do you believe that we should establish optimal levels or zero-discharge levels for toxic chemicals we release into the environment? Explain. What effects would adopting zero-discharge levels have on your life and lifestyle? **(b)** Should we assume that all chemicals we release or propose to release into the environment are potentially harmful until proven otherwise? Explain. Why hasn't this been done? What effects would adopting this principle have on your life? On the economy?

5. Do you agree or disagree with the proposals various analysts have made for sharply reducing poverty as discussed on pp. 41–42? Explain.

6. Do you agree or disagree with the guidelines for an earth-sustaining economy discussed on page 42? Explain.

7. What are the greatest strengths and weaknesses of the system of government in your country with respect to protecting the environment, working with the earth, and ensuring environmental justice for all? What three major changes, if any, would you make in this system?

8. Suppose that a presidential candidate ran on a platform calling for the federal government to phase in a tax on gasoline so that, over 5–10 years, the price of gasoline would rise to $5–7 a gallon (as is the case in Japan and most western European nations). The candidate argues that this tax increase is necessary to encourage conservation of oil and gasoline, reduce air pollution, and enhance future economic, environmental, and military security. The candidate also says that the tax revenue should be used to **(1)** reduce income taxes on the poor and middle class by an amount roughly equal to the increase in gasoline taxes, and **(2)** reduce taxes on wages and profits (Solutions, p. 38). Would you vote for this candidate who wants to triple the price of gasoline? Explain.

9. Do you agree or disagree with each of the major goals of the Wise-Use movement listed on p. 47–48? Explain. What positive effects has the Wise-Use movement had on the environmental movement?

10. What are the basic beliefs of your environmental worldview?

11. Which (if any) of the ethical guidelines on p. 54 do you disagree with? Explain. Can you suggest any additional ethical guidelines for working with the earth?

***12.** List all the economic goods you use, and then identify those that meet your basic needs and those that satisfy your wants. Identify any economic wants **(1)** you would be willing to give up, **(2)** those you believe you should give up but are unwilling to give up, and **(3)** those you hope to give up in the future. List the items that are likely to make you happy and improve the quality of your life. Relate the results of this analysis to your personal impact on the environment. Compare your results with those of your classmates.

***13.** If you knew you were going to die and had an opportunity to address everyone in the world for 5 minutes, what would you say? Write out your 5-minute speech and compare it with those of other members of your class.

***14.** Make a concept map of this chapter's major ideas, using the section heads and subheads and the key terms (in boldface type). Look at the inside back cover and in Appendix 5 for information about concept maps.

FURTHER INFORMATION

For more information see Internet Sources and Further Readings at the back of the book. Online resources for this chapter can be found at **www.brookscole.com/biology**

Part II

SCIENTIFIC PRINCIPLES AND CONCEPTS

Animal and vegetable life is too complicated a problem for human intelligence to solve, and we can never know how wide a circle of disturbance we produce in the harmonies of nature when we throw the smallest pebble into the ocean of organic life.

George Perkins Marsh

3 SCIENCE, SYSTEMS, MATTER, AND ENERGY

Saving Energy, Money, and Jobs in Osage, Iowa

Osage, Iowa (population about 4,000), has become the energy-efficiency capital of the United States. Its transformation began in 1974 when Wes Birdsall, general manager of Osage Municipal Gas and Electric Company, started urging the townspeople to save energy and reduce their natural gas and electric bills. The utility would also save money by not having to add new electrical generating facilities

Birdsall started his crusade by telling homeowners about the importance of insulating walls and ceilings and of plugging leaky windows and doors. He also advised people to replace their incandescent light bulbs with more efficient fluorescent bulbs and to turn down the temperature on water heaters and wrap them with insulation—economic boons to the local hardware and lighting stores. The utility company even gave away free water heater blankets. Birdsall also suggested saving water and fuel by installing low-flow shower heads.

Birdsall then stepped up his efforts by offering to give every building in town a free thermogram, an infrared scan that shows where heat escapes (Figure 3-1). When people could see the energy (and their money) hemorrhaging out of their buildings, they took action to plug the leaks, again helping the local economy. Birdsall then stepped up his campaign even more, announcing that no new houses could be hooked up to the company's natural gas line unless they met minimum energy-efficiency standards.

Since 1974, the town has reduced its natural gas consumption by 45%, no mean feat in a locale with frigid winter temperatures. In addition, the utility company saved enough money to accumulate a cash surplus and cut inflation-adjusted electricity rates by a third (which attracted two new factories to the area).

Each household saves more than $1,000 per year; this money supports jobs, and most of it circulates in the local economy. Before the town's energy-efficiency revolution, about $1.2 million a year left town to buy energy. What are your local utility companies and community doing to improve energy efficiency and stimulate the local economy?

Figure 3-1 An infrared photo showing heat loss (red, white, and yellow colors) around the windows, doors, roofs, and foundations of houses and stores in Plymouth, Michigan. Wes Birdsall provided similar thermograms of houses in Osage, Iowa. The average house in the United States has heat leaks and air infiltration equivalent to leaving a window wide open during the heating season. Because of poor design, most office buildings and houses in this country waste about half the energy used to heat and cool them. Americans pay about $300 billion a year for this wasted heat—more than the entire annual military budget. (VANSCAN® Continuous Mobile Thermogram by Daedalus Enterprises, Inc.)

The laws of thermodynamics control the rise and fall of political systems, the freedom or bondage of nations, the movements of commerce and industry, the origins of wealth and poverty, and the general physical welfare of the human race.

FREDERICK SODDY (NOBEL LAUREATE, CHEMISTRY)

This chapter addresses the following questions:

- What are science and technology? What is environmental science, and what are some of its limitations?
- What are major components and behaviors of complex systems?
- What are the basic forms of matter? What is matter made of? What makes matter useful to us as a resource?
- What are the major forms of energy? What makes energy useful to us as a resource?
- What are physical and chemical changes? What scientific law governs changes of matter from one physical or chemical form to another?
- What three main types of nuclear changes can matter undergo?
- What two scientific laws govern changes of energy from one form to another?
- How are the scientific laws governing changes of matter and energy from one form to another related to resource use and environmental disruption?

3-1 SCIENCE, TECHNOLOGY, ENVIRONMENTAL SCIENCE, AND CRITICAL THINKING

What Is Science and What Do Scientists Do?

Science is a pursuit of knowledge about how the world works. It is an attempt to discover order in nature and use that knowledge to make predictions about what is likely to happen in nature. As Albert Einstein once said, "The whole of science is nothing more than a refinement of everyday thinking." Figure 3-2 and the Guest Essay on p. 66 summarize the more systematic version of the everyday critical thinking process used by scientists.

The first thing scientists must do is ask a question or identify a problem to be investigated. Then scientists working on this problem collect **scientific data**, or facts, by making observations and taking measurements. The resulting scientific data or facts must be confirmed by repeated observations and measurements, ideally by several different investigators.

The primary goal of science is not facts themselves, but a new idea, principle, or model that connects and explains certain facts and leads to useful predictions about what is likely to happen in nature. Scientists working on a particular problem try to come up with a variety of possible explanations, or **scientific hypotheses**, of what they (or other scientists) observe in nature.

To be accepted, a scientific hypothesis not only must explain scientific data and phenomena but also should make predictions that can be used to test the validity of the hypothesis. Once a scientific hypothesis is invented, experiments are conducted (and repeated to be sure they are reproducible) to test the deductions or predictions.

One method scientists use to test a hypothesis is to develop a **model**, an approximate representation or simulation of a system being studied. There are many types of models: mental, conceptual, graphic, physical, and mathematical.

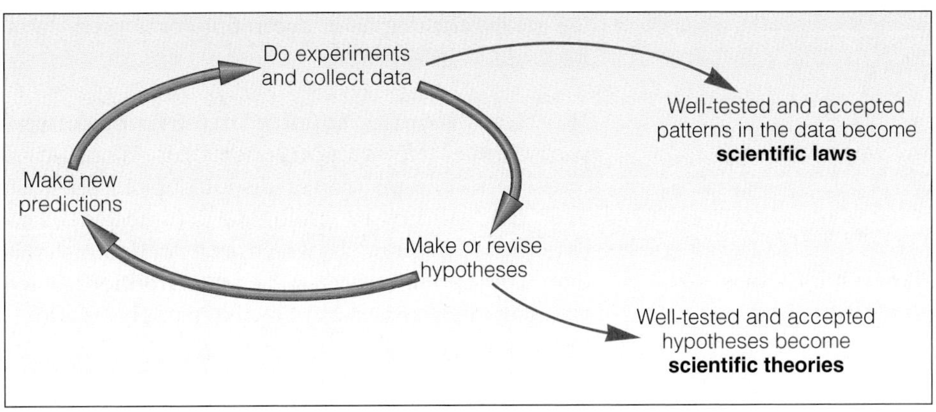

Figure 3-2 What scientists do. In the scientific process, a form of critical thinking, facts (data) are **(1)** gathered and verified by repeated experiments, **(2)** data are analyzed to see whether there is a consistent pattern of behavior that can be summarized as a scientific law, **(3)** hypotheses are proposed to explain the data, and **(4)** deductions or predictions are made and tested to evaluate each hypothesis. A hypothesis that is supported by a great deal of evidence and is widely accepted by the scientific community becomes a scientific theory.

If many experiments by different scientists support a particular hypothesis, it becomes a **scientific theory**: an idea, principle, or model that usually ties together and explains many facts that previously appeared to be unrelated and is supported by a great deal of evidence. To scientists, theories are not to be taken lightly. They are ideas or principles stated with a high degree of certainty because they are supported by extensive evidence.

Nonscientists often use the word *theory* incorrectly when they mean to refer to a *scientific hypothesis*, a tentative explanation that needs further evaluation. The statement, "Oh, that's just a theory," made in everyday conversation, implies a lack of knowledge and careful testing—the opposite of the scientific meaning of the word.

Another important result of science is a **scientific law**: a description of what we find happening in nature over and over in the same way. For example, after making thousands of observations and measurements over many decades, scientists discovered what is called the *second law of energy* or *thermodynamics*. One simple way of stating this law is that heat always flows spontaneously from hot to cold—something you learned the first time you touched a hot object. Scientific laws describe what we find happening in nature in the same way, whereas theories are widely accepted explanations of data and laws.

How Do Scientists Learn About Nature? We often hear about *the* scientific method. In reality, there are many **scientific methods**: ways scientists gather data and formulate and test scientific hypotheses, models, theories, and laws (Figure 3-2). Instead of being a recipe, each scientific method involves trying to answer a set of questions, with no particular guidelines for answering them:

- What is the question about nature to be answered?
- What relevant facts and data are already known?
- What new data (observations and measurements) should be collected, and how should this be done?
- After the data are collected, can they be used to formulate a scientific law?
- How can a hypothesis be invented that explains the data and predicts new facts? Is this the simplest and only reasonable hypothesis?
- What new experiments can be done to test the hypothesis (and modify it if necessary) so it can become a scientific theory?

New discoveries happen in many ways. Some follow this sequence:

data → law → hypothesis → theory.

At other times, scientists simply follow a hunch, bias, or belief and then do experiments to test the idea or hypothesis.

According to physicist Albert Einstein, "There is no completely logical way to a new scientific idea." Intuition, imagination, and creativity are as important in science as they are in poetry, art, music, and other great adventures of the human spirit.

Most processes or parts of nature that scientists seek to understand are influenced by a number of *variables* or *factors*. One way scientists test a hypothesis about the effects of a particular variable is to conduct a *controlled experiment* by setting up two groups. One is an *experimental group*, in which the chosen variable is changed in a known way. The second is a *control group*, in which the chosen variable is not changed.

The experiment is designed so that all components of each group are as identical as possible and experience the same conditions, except for the single factor being varied in the experimental group. If the experiment is designed properly, any difference between the two groups should result from a variable that was changed in the experimental group (Connections, right).

A basic problem is that many of nature's components and processes, especially those investigated by environmental scientists, involve a huge number of variables interacting in often poorly understood ways. In such cases, it is difficult or impossible to carry out meaningful controlled experiments.

How Valid Are the Results of Science? Scientists can do two major things: **(1)** disprove things and **(2)** establish that a particular model, theory, or law has a very high degree of validity and is extremely useful in explaining how nature works and in predicting what is likely to happen in nature. However, like scholars in virtually any field, scientists cannot prove that their ideas are *absolutely* true.

When people say that something has or has not been "scientifically proven," they can mislead us by falsely implying that science yields absolute proof or certainty. Although it may be extremely low, there is always some degree of uncertainty involved in any scientific model, theory, or law.

The goal of the rigorous scientific process is to reduce the degree of uncertainty as much as possible. However, the more complex the system being studied, the greater the degree of uncertainty or unpredictability about its behavior.

How Does Frontier Science Differ from Consensus Science? News reports often focus on new scientific "breakthroughs" and on disputes among scientists over the validity of preliminary (untested) data, hypotheses, and models (which are by definition tentative). These preliminary results, called **frontier science**, are controversial because they have not been widely tested and accepted.

CONNECTIONS

What's Harming the Robins?

Suppose a scientist observes an abnormality in the growth of robin embryos in a certain area. She knows that the area has been sprayed with a pesticide and suspects that the chemical may be causing the abnormalities she has observed.

To test this hypothesis, the scientist carries out a *controlled experiment*. She maintains two groups of robin embryos of the same age in the laboratory. Each group is exposed to exactly the same conditions of light, temperature, food supply, and so on, except that the embryos in the experimental group are exposed to a known amount of the pesticide in question.

The embryos in both groups are then examined over an identical period of time for the abnormality. If there is a statistically significantly larger number of the abnormalities in the experimental group than in the control group, then the results support the idea that the pesticide is the culprit.

To be sure there were no errors in the procedure, the experiment should be repeated several times by the original researcher, and ideally by one or more other scientists.

Critical Thinking

Can you find flaws in this experiment that might lead you to question the scientist's conclusions? (*Hint*: What other factors in nature—not the laboratory—and in the embryos themselves could possibly explain the results?)

By contrast, **consensus science** consists of data, theories, and laws that are widely accepted by scientists considered experts in the field involved. This aspect of science is very reliable but is rarely considered newsworthy. One way to find out what scientists generally agree on is to seek out reports by scientific bodies such as the U.S. National Academy of Sciences and the British Royal Society that attempt to summarize consensus among experts in key areas of science.

To give a sense of balance and fairness, the media often quote one or more of a handful of scientists who criticize the consensus view of the vast majority of scientists in a particular field. Sometimes these critics are not experts in the field they are criticizing. Instead of balance, this usually leads to a biased presentation because a minority view is given about the same weight as the consensus view held by most scientists.

What Is Technology? **Technology** is the creation of new products and processes intended to improve our efficiency, chances for survival, comfort level, and quality of life. Science and technology usually differ in the way the information and ideas they produce are shared. Many of the results of scientific research are published and distributed freely to be tested, challenged, verified, or modified. In contrast, many technological discoveries are kept secret until the new process or product is patented. However, the basis of some technology is published in journals and enjoys the same kind of public distribution and peer review as science.

What Is Environmental Science and What Are Its Limitations? **Environmental science** is the study of how we and other species interact with one another and with the nonliving environment (matter and energy). It is a *physical and social science* that integrates knowledge from a wide range of disciplines including physics, chemistry, biology (especially ecology), geology, meteorology, geography, engineering, resource conservation and management, demography (the study of population dynamics), economics, politics, sociology, psychology, and ethics. In other words, it is an interdisciplinary study of how the parts of nature and human societies operate and interact—a study of *connections* and *interactions* in nature (see inside back cover of this book).

There is controversy over some of the knowledge provided by environmental science, for much of it falls into the realm of frontier science. One problem involves *arguments over the validity of data.* There is no way to measure accurately how many metric tons of soil are eroded worldwide, how many hectares of tropical forest are cut, how many species become extinct, or how many metric tons of certain pollutants are emitted into the atmosphere or bodies of water each year.

The point environmental scientists want to make is that the trends in these phenomena are significant enough to be evaluated and addressed. Such environmental data should not be dismissed because they are "only estimates" (which are all we can ever have). However, this does not relieve investigators of the responsibility of getting the best estimates possible and pointing out that they *are* estimates.

Another limitation is that *most environmental problems involve so many variables and such complex interactions that we don't have enough information or sufficiently sophisticated models to aid in understanding them very well.* However, at some point we must use critical thinking skills (Guest Essay, p. 66) to evaluate the available (but always inadequate) information about the environmental problems we face and make political and economic decisions about what should be done.

GUEST ESSAY

Critical Thinking and Environmental Studies

Jane Heinze-Fry

Jane Heinze-Fry has a Ph.D. in environmental education and is the author of several supplements that accompany this textbook. She is author of several publications, including Critical Thinking and Environmental Studies: A Beginner's Guide *(with G. Tyler Miller as coauthor). Previously, she taught and directed environmental studies at Sweet Briar College. She also taught biology to students at the junior high, high school, and college levels, and has taught environmental science and biology at Emerson College and Bunker Hill Community College in Boston. Her interdisciplinary orientation is reflected in her concept maps, including the ones inside the back cover and in Appendix 5 of this textbook.*

Learners engaged in critical thinking try to

- Clarify their understanding of new concepts
- Connect new knowledge to prior knowledge and experience
- Make judgments about claims made by people in scientific articles, books, and advertisements
- Exercise both right- and left- brain capabilities
- Relate what they have learned to their own life experiences
- Understand and evaluate their environmental worldviews
- Take positions on issues
- Engage in problem-solving processes

Whenever we are faced with new information, we need to evaluate it by using critical thinking. Do we believe the information or not, and why? Do the claims seem reasonable or exaggerated? Here are some rules for evaluating evidence and claims:

1. Gather all the information you can.
2. Be sure that all key terms and concepts are defined and that you understand these definitions.
3. Question how the information (data) was obtained.
   - Were the studies well designed and carried out?
   - Was there an experimental group and a control group? Were the control and experimental groups treated identically except for the variable changed in the experimental group?
   - Did the investigators repeat their experiments several times and get essentially the same results?
   - Were the results verified by one or more other investigators?
4. Question the conclusions derived from the data.
   - Do the data support the claims, conclusions, and predictions?
   - Are there other possible interpretations? Are there more reasonable interpretations?
   - Do the conclusions involve a correlation or *apparent* connection between two or more variables, or do they imply a strong *cause-and-effect* relationship between such variables?
   - Are the conclusions based on the results of original research by experts in the field involved, or are they conclusions drawn by reporters or scientists in other fields?
5. Try to determine the assumptions and biases of the investigators and then question them.
   - Do the investigators have a monetary or political advantage in the outcome of the investigation or issue involved?
   - What are the underlying basic assumptions of the investigators? Would investigators with different basic assumptions or worldviews take the same data and come to different conclusions?
6. Expect and tolerate uncertainty. Recognize that science is a dynamic process that provides only a certain

3-2 MODELS AND BEHAVIOR OF SYSTEMS

Why Are Models of Complex Systems Useful? A **system** is set of components that function and interact in some regular and theoretically predictable manner. The environment consists of a vast number of interacting systems involving living and nonliving things.

Over time, people have learned the value of using models as approximate representations or simulations of real systems to find out how systems work and to evaluate which ideas or hypotheses work.

Some of the most powerful technologies invented by humans are mathematical models, which are used to supplement our mental models. *Mathematical models* consist of one or more equations used to describe the behavior of a system and to make predictions about the behavior of a system. Like all models, mathematical models are imperfect approximations of reality. They tend to make predictions ranging from fairly accurate to very accurate, depending on the model and the accuracy of the data used to formulate the model.

The basic process for developing mathematical models is essentially trial and error, similar to the degree of probability or certainty and that the more complex the system or process being investigated, the greater the degree of uncertainty.

- Are the data, claims, and conclusions based on the tentative results of *frontier science* or the more reliable and widely accepted results of *consensus science?*

7. Look at the big picture (think holistically).

- How do the results and conclusions fit into the whole system (the earth, ecosystem, economy, political system) involved?
- What additional data and experiments are needed to relate the results to the whole system?

8. Based on these steps, take a position by either rejecting or conditionally accepting the claims.

- Reject claims based on no evidence, insufficient evidence, or evidence coming from questionable sources.
- If evidence does not support a claim, reject it and state the conclusion you would draw.
- If the evidence supports the claims, conditionally accept the claims with the understanding that your support may change if new evidence disproves the claim.

In addition to these rules to enhance making reasoned judgments, there are a number of other strategies to improve your critical thinking skills. Some strategies focus on improving thinking, others on attitudes and values, others on actions.

Thinking strategies focus on learning how to use your left brain (which is good at logic and analysis) and your right brain (which is good at visualizing and creating). They include **(1)** constructing models, **(2)** clarifying concepts, **(3)** brainstorming, **(4)** defining problems, **(5)** creating alternative solutions, **(6)** visualizing future possibilities, and **(7)** exercising your creative brain paths.

Attitudes and values focus on **(1)** establishing an ecological identity, **(2)** reflecting on the interaction of lifestyle and the environment, **(3)** understanding and evaluating your worldview, and **(4)** using values analysis for evaluating proposed environmental policies and making personal environmental decisions. *Action strategies* focus on problem solving. They include techniques for **(1)** visualizing problems, **(2)** redefining problems, **(3)** considering alternative solutions, **(4)** creating plans of action, and **(5)** developing strategies for implementing action plans.

In the environmental course you are taking, there are many opportunities to develop your critical thinking skills. Your textbook offers Critical Thinking questions at the end of chapters and in most boxes. If your course uses the supplement *Critical Thinking and Environmental Studies: A Beginner's Guide*, you will learn the critical thinking strategies mentioned here.

Learning how to think critically is essential in helping you evaluate the validity and usefulness of what you read in newspapers, magazines, and books (such as this textbook), what you hear in lectures and speeches, and what you see and hear on the news and in advertisements.

Critical Thinking

1. Can you come up with an example in which critical thinking has helped you make a major change in one or more of your beliefs or helped you make an important personal decision? Can you think of a decision that may have come out better if you had used critical thinking skills such as those discussed in this essay?

2. Rote learning often involves the "memorize and spit back" strategy. Meaningful learning (including critical thinking) goes far beyond memorization and requires us to evaluate the validity of what we learn. Currently, about what percentage of your learning involves rote learning and what percentage involves critical thinking as defined at the beginning of this essay?

processes depicted in Figure 3-2. Making a mathematical model usually requires going many times through three familiar steps: **(1)** Make a guess and write down some equations, **(2)** compute the predictions implied by the equations, and **(3)** compare the predictions with observations, the predictions of mental models, existing experimental data, and scientific hypotheses, laws, and theories.

Mathematical models are important because they can give us improved perceptions and predictions, especially concerning matters for which our mental models are weak. Research has shown that mental models tend to be especially unreliable when **(1)** there are many interacting variables, **(2)** we attempt to extrapolate from too few experiences to a general case, **(3)** consequences follow actions only after long delays, **(4)** the consequences of actions lead to other consequences, **(5)** responses vary from one time to the next, and **(6)** controlled experiments (Connections, p. 65) are impossible, too slow, or too expensive to conduct. Under such conditions, a good mathematical model can do much better than most mental models.

After building and testing a model, scientists use it to predict what is likely to happen under a variety of

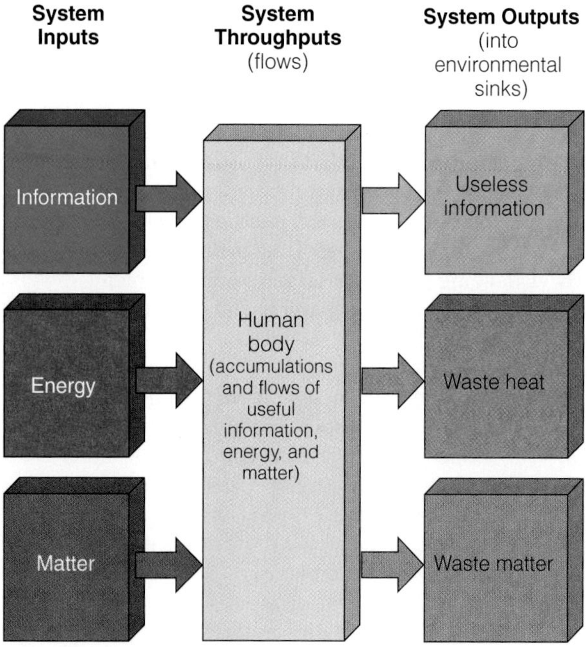

Figure 3-3 Generalized diagram of inputs, throughputs (flows), and outputs of information, matter, and energy in a system.

conditions. In effect, they use mathematical models to answer *if–then* questions: "*If* we do such and such, *then* what is likely to happen now and in the future?"

What Are Some Basic Components and Behaviors of System Models? Any system being studied or modeled has one or more **inputs** of things such as matter, energy, or information.

Inputs flow through a system at a certain rate. Such **flows** or **throughputs** of matter, energy, or information through a system are represented by arrows. Forms of matter, energy, or information flowing out of a system are called **outputs** and end up in *sinks* in the environment. Examples of such sinks are the atmosphere, bodies of water, underground water, soil, and land surfaces. Figure 3-3 is a generalized diagram of inputs, throughputs, and outputs for a system.

A **feedback loop** occurs when one change in a system leads to some other change, which eventually either reinforces or slows the original change. Feedback loops determine how things happen in a system over time.

Feedback loops occur when an output of matter, energy, or information is fed back into the system as an input. For example, recycling aluminum cans involves melting aluminum and feeding it back into an economic system to make new aluminum products. This feedback loop of matter reduces the need to find, extract, and process virgin aluminum ore. It also reduces the flow of waste matter (discarded aluminum cans) into the environment.

Feedback loops can be either positive or negative. A **positive feedback loop** is a runaway cycle in which a change in a certain direction provides information that causes a system to change further in the same direction. The environment is full of positive feedback loops, some desirable and some undesirable from a human standpoint. For example, as long as there are more human births than deaths on this planet, the human population will grow exponentially. This form of positive feedback can be viewed as desirable or undesirable depending on whether one sees population growth as an economic benefit (more workers and consumers) or an environmental drawback (more crowding, pollution, and resource depletion and degradation).

In a **negative feedback loop**, one change leads to a lessening of that change. To survive, you must maintain your body temperature within a certain range, regardless of whether the temperature outside is steamy or freezing. This phenomenon is called **homeostasis**: the maintenance of favorable internal conditions despite fluctuations in external conditions. Homeostatic systems consist of one or more negative feedback loops that help maintain constant internal conditions when changes occur.

Most systems, such as the temperature-regulating system of your body, contain one or a series of *coupled positive and negative feedback loops* (Figure 3-4). Normally a negative feedback regulates your body temperature. However, if your body temperature exceeds 42°C (108°F), your built-in negative feedback temperature control system breaks down as your body produces more heat than your sweat-dampened skin can get rid of. Then a positive feedback loop caused by overloading the system (Figure 3-4) overwhelms the negative or corrective feedback loop. These conditions produce a net gain in body heat, which produces even more body heat, and so on, until you die from heatstroke.

3-3 MATTER: FORMS, STRUCTURE, AND QUALITY

What Are Nature's Building Blocks? **Matter** is anything that has mass (the amount of material in an object) and takes up space. Matter includes the solids, liquids, and gases around us and within us. Matter is found in two *chemical forms*: **elements** (the distinctive building blocks of matter that make up every material substance) and **compounds** (two or more different elements held together in fixed proportions by attractive forces called *chemical bonds*). Various elements, compounds, or both can be found together in **mixtures**.

Hayden, T. 1998. "The Questions That Stump the Scientists." *Newsweek*, vol. 131, no. 3, p. 12(1).

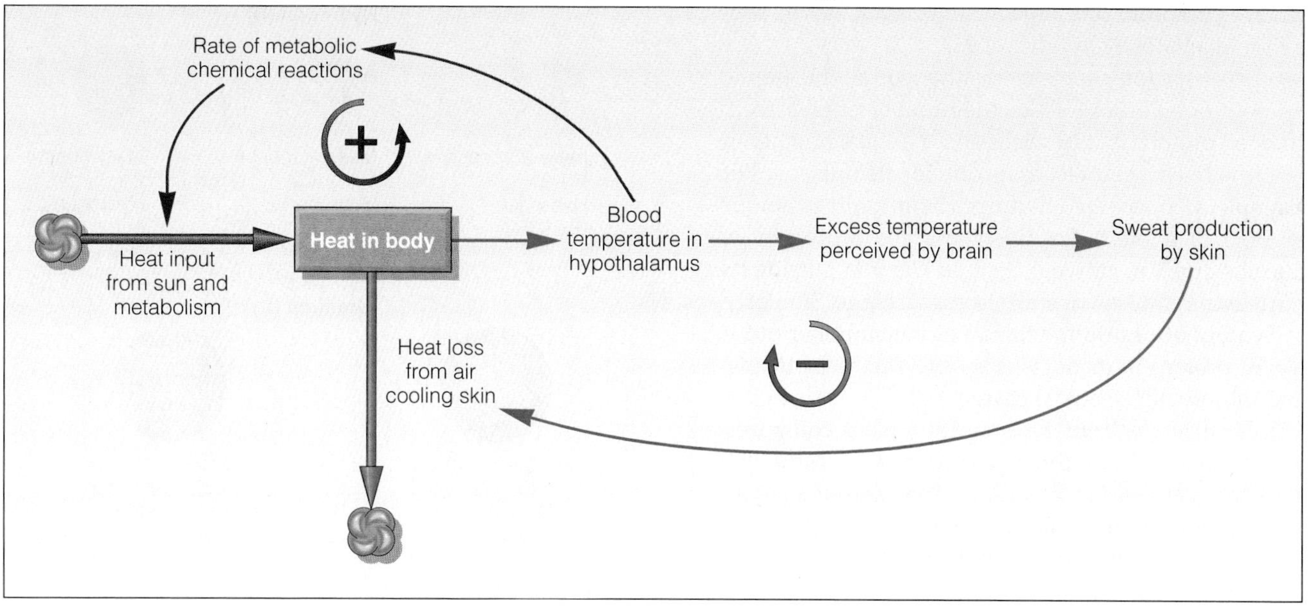

Figure 3-4 Coupled negative and positive feedback loops involved in temperature control of the human body. Homeostasis works in a limited range only: Above a certain body temperature, body metabolic rates get out of control and generate large amounts of heat. This positive (runaway) feedback loop generates more heat than the negative feedback loop can get rid of and body temperature increases out of control, resulting in death.

All matter is built from the 118 known chemical elements. (Ninety-two of them occur naturally and the other 26 have been synthesized in laboratories.) To simplify things, chemists represent each element by a one- or two-letter symbol; hydrogen (H), carbon (C), oxygen (O), nitrogen (N), phosphorus (P), sulfur (S), chlorine (Cl), fluorine (F), bromine (Br), sodium (Na), calcium (Ca), lead (Pb), mercury (Hg), and uranium (U) are but a few. These are the elements discussed in this book.

If you had a supermicroscope capable of looking at individual elements and compounds, you could see that they are made up of three types of building blocks: **(1) atoms** (the smallest unit of matter that is unique to a particular element), **(2) ions** (electrically charged atoms or combinations of atoms), and **(3) molecules** (combinations of two or more atoms of the same or different elements held together by chemical bonds). Because ions and molecules are formed from atoms, *atoms are the ultimate building blocks for all matter.*

Some elements are found in nature as molecules. Examples are nitrogen and oxygen, which together make up about 99% of the volume of the air you breathe. Two atoms of nitrogen (N) combine to form a nitrogen gas molecule, with the shorthand formula N_2 (read as "N-two"). The subscript after the element's symbol indicates the number of atoms of that element in a molecule. Similarly, most of the oxygen gas in the atmosphere exists as O_2 (read as "O-two") molecules. A small amount of oxygen, found mostly in the second layer of the atmosphere (stratosphere), exists as O_3 (read as "O-three") molecules, a gaseous form of oxygen called *ozone.*

What Are Atoms and Ions? If you increased the magnification of your supermicroscope, you would find that each different type of atom contains a certain number of *subatomic particles*. The main building blocks of an atom are **(1)** positively charged **protons** (*p*), **(2)** uncharged **neutrons** (*n*), and **(3)** negatively charged **electrons** (*e*).

Each atom consists of an extremely small center, or **nucleus**, containing protons and neutrons, and one or more electrons in rapid motion somewhere outside the nucleus. The location of an electron in an atom can be described only in terms of its probability of being at any given place outside the nucleus. This is analogous to saying that a certain number of tiny gnats are found somewhere in a cloud, without being able to identify their exact positions.

Each atom has an equal number of positively charged protons (inside its nucleus) and negatively charged electrons (outside its nucleus). Because these electrical charges cancel one another, *the atom as a whole has no net electrical charge.*

Each element has its own specific **atomic number**, equal to the number of protons in the nucleus of each of its atoms. The simplest element, hydrogen (H), has only 1 proton in its nucleus, so its atomic number is 1. Carbon (C), with 6 protons, has an atomic number of 6,

Hint: Enter the search term *matter* using the Subject Guide.

whereas uranium (U), a much larger atom, has 92 protons and an atomic number of 92.

Because atoms are electrically neutral, the atomic number of an atom tells us the number of positively charged protons in its nucleus and the equal number of negatively charged electrons outside its nucleus. For example, an uncharged hydrogen atom, with an atomic number of 1, has 1 positively charged proton in its nucleus and 1 negatively charged electron outside its nucleus and thus no overall electrical charge. Similarly, each atom of uranium, with an atomic number of 92, has 92 protons in its nucleus and 92 electrons outside, and thus no net electrical charge.

Because electrons have so little mass compared with the mass of a proton or a neutron, *most of an atom's mass is concentrated in its nucleus*. The mass of an atom is described in terms of its **mass number**: the total number of neutrons and protons in its nucleus.

Although all atoms of an element have the same number of protons in their nuclei, they may have different numbers of uncharged neutrons in their nuclei, and thus may have different mass numbers. Various forms of an element having the same atomic number but a different mass number are called **isotopes** of that element. Isotopes are identified by attaching their mass numbers to the name or symbol of the element. For example, hydrogen has three isotopes: hydrogen-1 (H-1), hydrogen-2 (H-2, common name deuterium), and hydrogen-3 (H-3, common name tritium). A natural sample of an element contains a mixture of its isotopes in a fixed proportion or percentage abundance by weight (Figure 3-5).

Atoms of some elements can lose or gain one or more electrons to form **ions:** atoms or groups of atoms with one or more net positive (+) or negative (–) electrical charges. For example, an atom of sodium (Na) (atomic number 11) with 11 positively charged protons and 11 negatively charged electrons can lose one of its electrons. It then becomes a sodium ion with a positive charge of 1 (Na^+) because it now has 11 positive charges (protons) but only 10 negative charges (electrons). An atom of chlorine (Cl) (with an atomic number of 17) can gain an electron and become a chlorine ion with a negative charge of 1 (Cl^-) because it then has 17 positively charged protons and 18 negatively charged electrons.

The number of positive or negative charges on an ion is shown as a superscript after the symbol for an atom or a group of atoms. Examples of other positive ions are calcium ions (Ca^{2+}) and ammonium ions (NH_4^+); other common negative ions are nitrate ions (NO_3^-), sulfate ions (SO_4^{2-}), and phosphate ions (PO_4^{3-}). These are the ions discussed in this book.

What Holds the Atoms and Ions in Compounds Together? Most matter exists as compounds. Chemists use a shorthand **chemical formula** to show the number of atoms (or ions) of each type in a compound. The formula contains the symbols for each of the elements present and uses subscripts to represent the number of atoms or ions of each element in the compound's basic structural unit. Compounds made up of oppositely charged ions are called *ionic compounds*, and those made up of molecules of uncharged atoms are called *covalent* or *molecular compounds*.

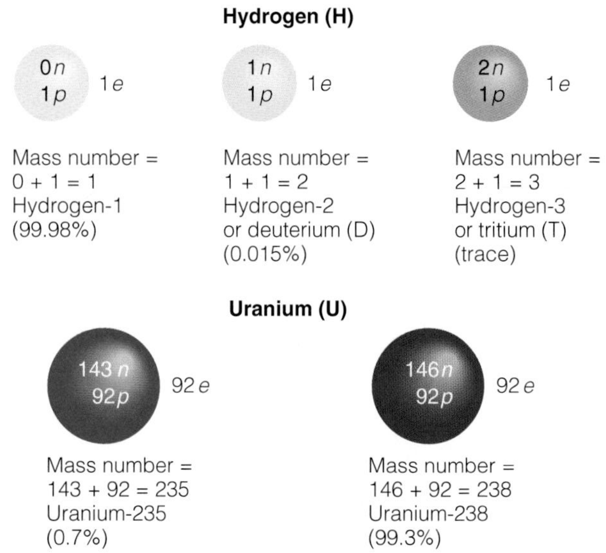

Figure 3-5 Isotopes of hydrogen and uranium. All isotopes of hydrogen have an atomic number of 1 because each has one proton in its nucleus; similarly, all uranium isotopes have an atomic number of 92. However, each isotope of these elements has a different mass number because its nucleus contains a different number of neutrons. Figures in parentheses indicate the percentage abundance by weight of each isotope in a natural sample of the element.

Sodium chloride (table salt), an *ionic compound*, is represented by NaCl. It consists of a network of oppositely charged *ions* (Na^+ and Cl^-) held together by the forces of attraction between opposite charges. The strong forces of attraction between such oppositely charged ions are called *ionic bonds*.

Water, a *covalent* or *molecular compound*, consists of molecules made up of uncharged atoms of hydrogen (H) and oxygen (O). Each water molecule consists of two hydrogen atoms chemically bonded to an oxygen atom, yielding H_2O (read as "H-two-O") molecules. The bonds between the atoms in such molecules are called *covalent bonds* and are formed when the atoms in the molecule share one or more pairs of their electrons.

What Are Organic and Inorganic Compounds? Table sugar, vitamins, plastics, aspirin, penicillin, and many other important materials have one thing in common: They are *organic compounds*, containing carbon atoms combined with each other and with atoms of one or more other elements such as hydrogen, oxygen, nitrogen, sulfur, phosphorus, chlorine, and fluorine. Virtually all organic compounds are molecular compounds held together by covalent bonds. Organic compounds can be either natural or synthetic (such as plastics and many drugs made by humans).

The millions of known organic (carbon-based) compounds include

- *Hydrocarbons*: compounds of carbon and hydrogen atoms. An example is methane (CH_4), the main component of natural gas.
- *Chlorinated hydrocarbons*: compounds of carbon, hydrogen, and chlorine atoms. Examples are DDT ($C_{14}H_9Cl_5$, an insecticide) and toxic PCBs (such as $C_{12}H_5Cl_5$), oily compounds used as insulating materials in electric transformers.
- *Chlorofluorocarbons* (CFCs): compounds of carbon, chlorine, and fluorine atoms. An example is Freon-12 (CCl_2F_2), until recently widely used as a coolant in refrigerators and air conditioners, as an aerosol propellant, and as a foaming agent for making some plastics.
- *Simple carbohydrates* (simple sugars): certain types of compounds of carbon, hydrogen, and oxygen atoms. An example is glucose ($C_6H_{12}O_6$), which most plants and animals break down in their cells to obtain energy.

Larger and more complex organic compounds, called *polymers*, consist of a number of basic structural or molecular units (*monomers*) linked by chemical bonds, somewhat like cars linked in a freight train. The three major types of organic polymers are complex carbohydrates, proteins, and nucleic acids.

Complex carbohydrates are made by linking a number of simple carbohydrate molecules such as glucose ($C_6H_{12}O_6$). Examples are the complex starches in rice and potatoes and cellulose found in the walls around plant cells.

Proteins are produced in cells by the linking of different sequences of about 20 different monomers known as *alpha-amino acids*, whose number and sequence in each protein are specified by the genetic code found in DNA molecules in an organism's cells. From only about 20 alpha-amino acid molecules, earth's life-forms can make tens of millions of different protein molecules.

Nucleic acids are made by linking hundreds to thousands of different types of monomers, called *nucleotides*. Each nucleotide consists of **(1)** a phosphate group, **(2)** a sugar molecule containing five carbon atoms (deoxyribose in DNA molecules and ribose in RNA molecules), and **(3)** one of four different nucleotide bases (represented by A, G, C, and T, the first letter in each of their names). In the cells of living organisms, these nucleotide units combine in different numbers and sequences to form nucleic acids such as various types of DNA and RNA. Bonds formed between parts of the four nucleotides in DNA hold two DNA strands together like a spiral staircase, forming a double helix. DNA molecules can unwind and replicate themselves. They contain the hereditary instructions for assembling new cells and for assembling the proteins each cell needs to survive and reproduce.

Genes consist of specific sequences of nucleotides in a DNA molecule. Each gene carries the codes (each consisting of three nucleotides) required to make various proteins. These coded units of genetic information about specific traits are passed on from parents to offspring during reproduction. Occasionally one or more of the nucleotide bases in a gene sequence are deleted, added, or replaced. Such changes, called **gene mutations**, can be helpful or harmful to an organism and its offspring, or not affect them at all.

Chromosomes are combinations of genes that make up a single DNA molecule, together with a number of proteins. Genetic information coded in your chromosomal DNA is what makes you different from an oak leaf, an alligator, or a flea, and from your parents. The relationships of genetic material to cells are depicted in Figure 3-6.

All other compounds are called *inorganic compounds*. Some of the inorganic compounds discussed in this book are sodium chloride (NaCl), water (H_2O), nitrous oxide (N_2O), nitric oxide (NO), carbon monoxide (CO), carbon dioxide (CO_2),* nitrogen dioxide (NO_2), sulfur dioxide (SO_2), ammonia (NH_3), hydrogen sulfide (H_2S), sulfuric acid (H_2SO_4), and nitric acid (HNO_3). These are the compounds discussed in this book.

What Is Matter Quality? **Matter quality** is a measure of how useful a matter resource is to us, based on its availability and concentration. **High-quality matter** is organized, concentrated, and usually found near the earth's surface, and has great potential for use as a matter resource. **Low-quality matter** is disorganized, dilute, and often deep underground or dispersed in the ocean or the atmosphere, and usually has little potential for use as a matter resource (Figure 3-7).

An aluminum can is a more concentrated, higher-quality form of aluminum than aluminum ore containing the same amount of aluminum. That's why it takes less energy, water, and money to recycle an aluminum can than to make a new can from aluminum ore.

*Classifying compounds as organic or inorganic is somewhat arbitrary. All organic compounds contain one or more carbon atoms, but CO and CO_2 are classified as inorganic compounds.

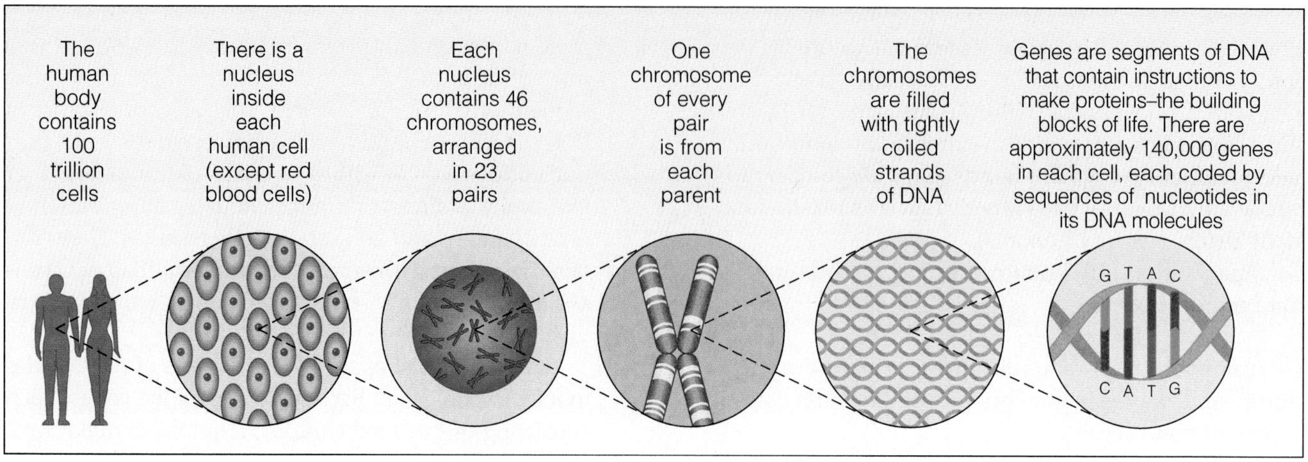

Figure 3-6 Relationships among cells, nuclei, chromosomes, DNA, and genes.

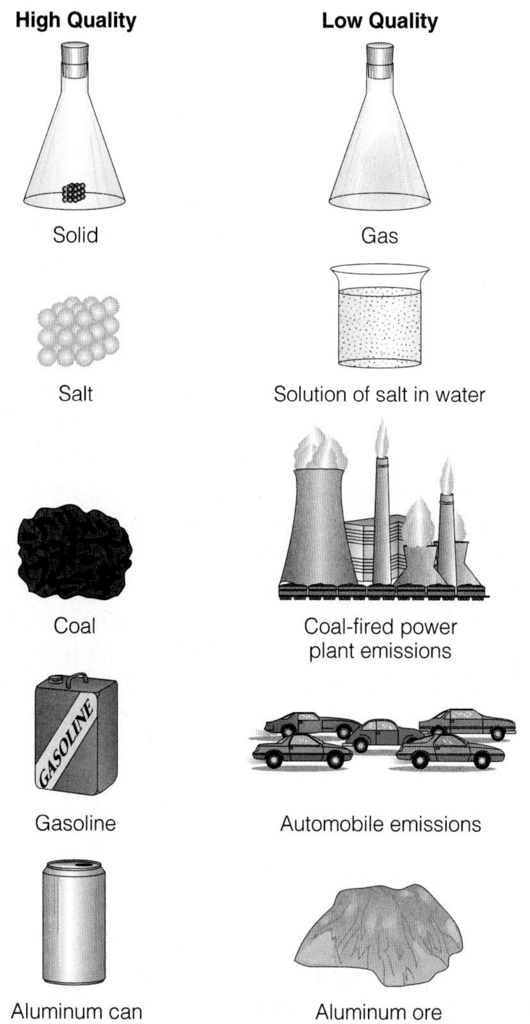

Figure 3-7 Examples of differences in matter quality. High-quality matter (left-hand column) is fairly easy to extract and is concentrated; low-quality matter (right-hand column) is more difficult to extract and is more dispersed than high-quality matter.

3-4 ENERGY: FORMS AND QUALITY

What Different Forms of Energy Do We Encounter? **Energy** is the capacity to do work and transfer heat. Work is performed when an object—be it a grain of sand, this book, or a giant boulder—is moved over some distance. Work, or matter movement, also is needed to boil water (to change it into the more dispersed and faster-moving water molecules in steam) or to burn natural gas to heat a house or cook food. Energy is also the heat that flows automatically from a hot object to a cold object when they come in contact.

Energy comes in many forms, including **(1)** light, **(2)** heat, **(3)** electricity, **(4)** chemical energy stored in the chemical bonds in coal, sugar, and other materials, **(5)** the mechanical energy of moving matter such as flowing water, wind (air masses), and joggers, and **(6)** nuclear energy emitted from the nuclei of certain isotopes.

Scientists classify energy as either kinetic or potential. **Kinetic energy** is the energy that matter has because of its mass and its speed or velocity. It is energy in action or motion. Wind (a moving mass of air), flowing streams, falling rocks, heat flowing from a body at a high temperature to one at a lower temperature, electricity (flowing electrons), moving cars—all have kinetic energy.

Radio waves, TV waves, microwaves, infrared radiation, visible light, ultraviolet radiation, X rays, gamma rays, and cosmic rays are all forms of kinetic energy known as **electromagnetic radiation** (Figure 3-8). Infrared radiation is being used to show us where energy is leaking out of our houses and buildings (Figure 3-1). Satellite scans use infrared radiation, ultraviolet radiation, and microwaves to show where plant life is found on land and at sea and to identify temperature changes in the upper layers of the oceans.

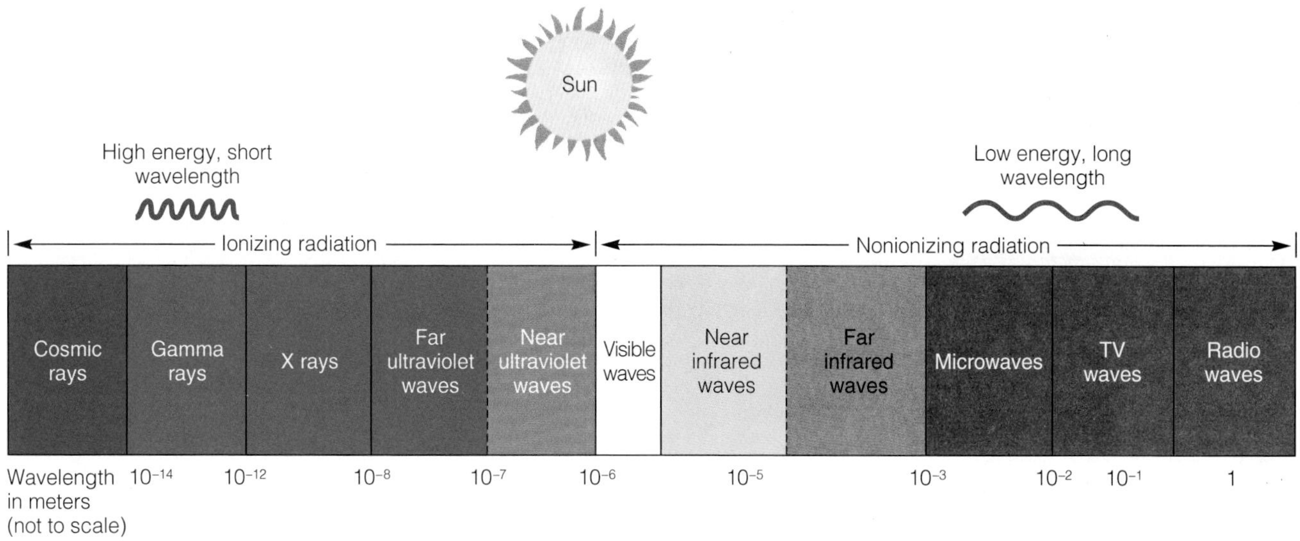

Figure 3-8 The electromagnetic spectrum: the range of electromagnetic waves, which differ in wavelength (distance between successive peaks or troughs) and energy content.

Cosmic rays, gamma rays, X rays, and ultraviolet radiation have enough energy to knock electrons from atoms and change them to positively charged ions. The resulting highly reactive electrons and ions can disrupt living cells, interfere with body processes, and cause many types of sickness, including various cancers. These potentially harmful forms of electromagnetic radiation are called **ionizing radiation**. The other forms of electromagnetic radiation do not contain enough energy to form ions and are called **nonionizing radiation**.

Heat is the total kinetic energy of all the moving atoms, ions, or molecules within a given substance, excluding the overall motion of the whole object. **Temperature** is a measure of the average speed of motion of the atoms, ions, or molecules in a sample of matter at a given moment. A substance can have a high heat content (much mass and many moving atoms, ions, or molecules) but a low temperature (low average molecular speed). For example, the total heat content of a lake is enormous, but its average temperature is low. Another substance can have a low heat content and a high temperature. A cup of hot coffee, for example, has a much lower heat content than a lake, but its temperature is much higher.

Potential energy is stored energy that is potentially available for use. A rock held in your hand, an unlit stick of dynamite, still water behind a dam, the gasoline in a car tank, and the nuclear energy stored in the nuclei of atoms all have potential energy because of their position or the position of their parts. Potential energy can be changed to kinetic energy. When you drop a rock, its potential energy changes into kinetic energy. When you burn gasoline in a car engine, the potential energy stored in the chemical bonds of its molecules changes into heat, light, and mechanical (kinetic) energy that propels the car.

What Is Energy Quality? From a human standpoint, the measure of an energy source's ability to do useful work is called its **energy quality** (Figure 3-9). **High-quality energy** is organized or concentrated and can perform much useful work. Examples are electricity, coal, gasoline, concentrated sunlight, nuclei of uranium-235 used as fuel in nuclear power plants, and heat concentrated in small amounts of matter so that its temperature is high.

By contrast, **low-quality energy** is disorganized or dispersed and has little ability to do useful work. An example is heat dispersed in the moving molecules of a large amount of matter (such as the atmosphere or a large body of water) so that its temperature is low. For example, the total amount of heat stored in the Atlantic Ocean is greater than the amount of high-quality chemical energy stored in all the oil deposits of Saudi Arabia. Yet the ocean's heat is so widely dispersed that it can't be used to move things or to heat things to high temperatures. It makes sense to match the quality of an energy source with the quality of energy needed to perform a particular task (Figure 3-9) because doing so saves energy and usually money.

3-5 PHYSICAL AND CHEMICAL CHANGES AND THE LAW OF CONSERVATION OF MATTER

What Is the Difference Between a Physical and a Chemical Change? A **physical change** involves no change in chemical composition. Cutting a piece of

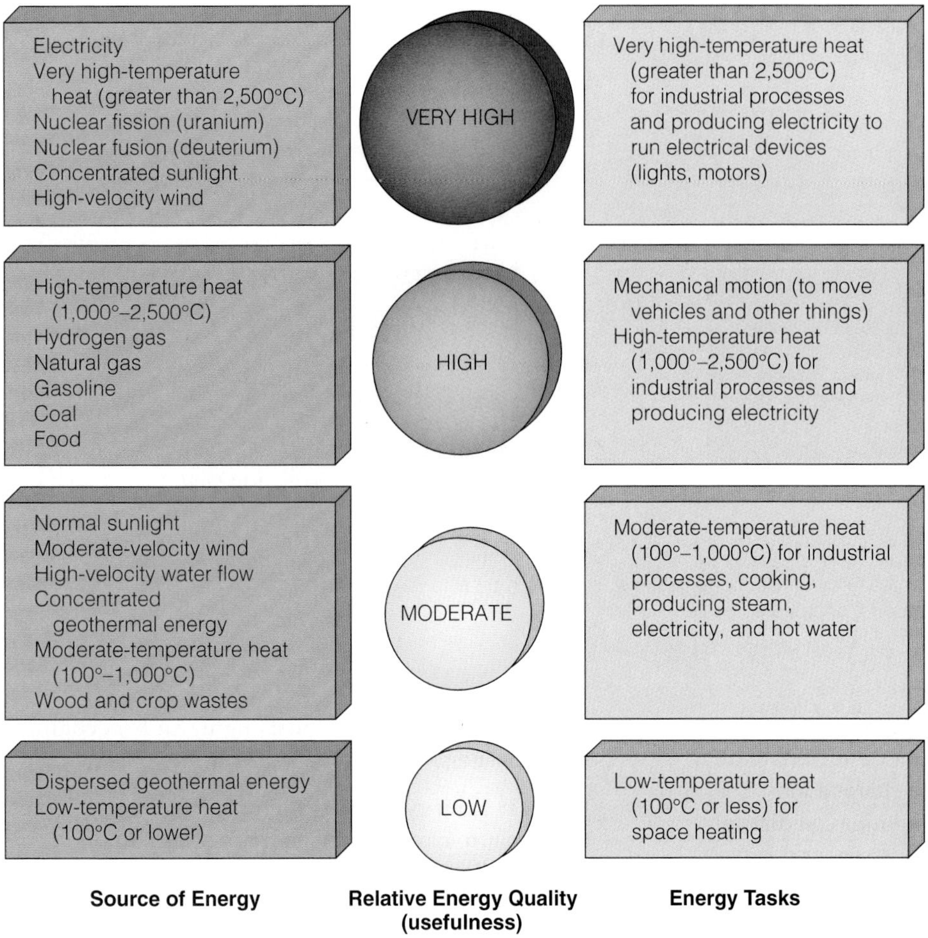

Figure 3-9 Categories of the quality (usefulness for performing various energy tasks) of different sources of energy. *High-quality energy* is concentrated and has great ability to perform useful work; *low-quality energy* is dispersed and has little ability to do useful work. To avoid unnecessary energy waste, it is best to match the quality of an energy source with the quality of energy needed to perform a task.

aluminum foil into small pieces is one example. Changing a substance from one physical state to another is a second example. When solid water (ice) is melted or liquid water is boiled, none of the H_2O molecules involved are altered; instead, the molecules are organized in different spatial (physical) patterns.

In a **chemical change** or **chemical reaction**, on the other hand, the chemical compositions of the elements or compounds are altered. Chemists use shorthand chemical equations to represent what happens in a chemical reaction. For example, when coal burns completely, the solid carbon (C) it contains combines with oxygen gas (O_2) from the atmosphere to form the gaseous compound carbon dioxide (CO_2):

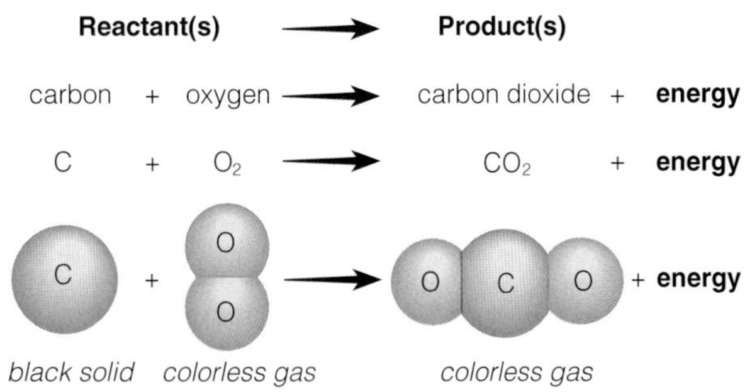

Energy is given off in this reaction, making coal a useful fuel. The reaction also shows how the complete burning of coal (or any of the carbon-containing compounds in wood, natural gas, oil, and gasoline) gives off carbon dioxide gas, which is a key gas that can lead to increased warming of the lower atmosphere (troposphere).

What Is the Law of Conservation of Matter: Why There Is No "Away" People commonly talk about consuming or using up material resources, but the truth is that *we don't consume matter*—we only use some of the earth's resources for a while. We take materials from the earth, carry them to another part of the globe, and process them into products that are used and then discarded, burned, buried, reused, or recycled.

In so doing, *we may change various elements and compounds from one physical or chemical form to another, but in no physical and chemical change can we create or destroy any of the atoms involved.* All we can do is rearrange them into different spatial patterns (physical changes) or different combinations (chemical changes). The italicized statement, based on many thousands of measurements, is known as the **law of conservation of matter**. In

Keeping Track of Atoms

SPOTLIGHT

In keeping with the law of conservation of matter, each side of a chemical equation must have the same number of *atoms* of each element involved. When this is the case, the equation is said to be *balanced.* The equation for the burning of carbon ($C + O_2 \longrightarrow CO_2$) is balanced because there are one atom of carbon and two atoms of oxygen on both sides of the equation.

Now consider the following chemical reaction: When electricity is passed through water (H_2O), the latter can be broken down into hydrogen (H_2) and oxygen (O_2), as represented by the following equation:

$$H_2O \longrightarrow H_2 + O_2$$

H_2O		H_2	+	O_2
2 H atoms 1 O atom		2 H atoms		2 O atoms

This equation is unbalanced because there is one atom of oxygen on the left but two atoms on the right.

We can't change the subscripts of any of the formulas to balance this equation because then we would be changing the arrangements of the atoms involved. Instead, we could use different numbers of the *molecules* involved to balance the equation. For example, we could use two water molecules:

$$2\,H_2O \longrightarrow H_2 + O_2$$

$2\,H_2O$		H_2	+	O_2
4 H atoms 2 O atoms		2 H atoms		2 O atoms

This equation is still unbalanced because even though the numbers of oxygen atoms on both sides are now equal, the numbers of hydrogen atoms are not.

We can correct this by having the reaction produce two hydrogen molecules:

$$2\,H_2O \longrightarrow 2\,H_2 + O_2$$

$2\,H_2O$		$2\,H_2$	+	O_2
4 H atoms 2 O atoms		4 H atoms		2 O atoms

Now the equation is balanced, and the law of conservation of matter has not been violated. We see that for every two molecules of water through which we pass electricity, two hydrogen molecules and one oxygen molecule are produced.

See if you can balance the chemical equation for the reaction of nitrogen gas (N_2) with hydrogen gas (H_2) to form ammonia gas (NH_3).

Critical Thinking

1. Balancing equations is based on the law of conservation of matter. Do you believe that this is an ironclad law of nature or one that through new scientific discoveries could be overthrown? Explain.

2. Imagine that you have the power to revoke the law of conservation of matter. List the three major ways this would affect your life.

describing chemical reactions, chemists use a shorthand system to make sure atoms are neither created nor destroyed, as required by the law of conservation of matter (Spotlight, above).

The law of conservation of matter means that there really is no "away" in "to throw away." *Everything we think we have thrown away is still here with us in one form or another.* We can collect dust and soot from the smokestacks of industrial plants, but these solid wastes must then be put somewhere. We can remove substances from polluted water at a sewage treatment plant, but the gooey sludge must be **(1)** burned (producing some air pollution), **(2)** buried (possibly contaminating underground water supplies), or **(3)** cleaned up and applied to the land as fertilizer (dangerous if the sludge contains nondegradable toxic metals such as lead and mercury). Banning use of the pesticide DDT in the United States but still selling it abroad means that it can return to the United States as DDT residues in imported coffee, fruit, and other foods, or as fallout from air masses moved long distances by winds—something environmentalists call the *circle of poison.*

We can make the environment cleaner and convert some potentially harmful chemicals into less harmful physical or chemical forms. However, the law of conservation of matter means that we will always face the problem of what to do with some quantity of wastes. By placing much greater emphasis on pollution prevention, waste reduction, and more efficient use of resources, we can greatly reduce the amount of wastes we add to the environment (Guest Essay, p. 78).

3-6 NUCLEAR CHANGES

What Is Natural Radioactivity? In addition to physical and chemical changes, matter can undergo a third type of change known as a **nuclear change**. This occurs when nuclei of certain isotopes spontaneously change or are made to change into one or more different isotopes. Three types of nuclear change are natural radioactive decay, nuclear fission, and nuclear fusion.

Natural radioactive decay is a nuclear change in which unstable isotopes spontaneously emit fast-moving particles, high-energy radiation, or both at a fixed rate. The unstable isotopes are called **radioactive isotopes** or **radioisotopes**. Radioactive decay into various isotopes continues until the original isotope is changed into a stable isotope that is not radioactive.

Radiation emitted by radioisotopes is damaging ionizing radiation. The most common form of ionizing energy released from radioisotopes is **gamma rays**, a form of high-energy electromagnetic radiation (Figure 3-8). High-speed ionizing particles emitted from the nuclei of radioactive isotopes are most commonly of two types: **alpha particles** (fast-moving, positively charged chunks of matter that consist of two protons and two neutrons) and **beta particles** (high-speed electrons).

Each type of radioisotope spontaneously decays at a characteristic rate into a different isotope. This rate of decay can be expressed in terms of **half-life**: the time needed for *one-half* of the nuclei in a radioisotope to decay and emit their radiation to form a different isotope. The decay continues, often producing a series of different radioisotopes, until a nonradioactive isotope is formed. Each radioisotope has a characteristic half-life, which may range from a few millionths of a second to several billion years (Table 3-1).

An isotope's half-life cannot be changed by temperature, pressure, chemical reactions, or any other known factor. Half-life can be used to estimate how long a sample of a radioisotope must be stored in a safe container before it decays to what is considered a safe level. A general rule is that such decay takes about 10 half-lives. Thus, people must be protected from radioactive waste containing iodine-131 (which concentrates in the thyroid gland and has a half-life of 8 days) for 80 days (10 × 8 days). Plutonium-239, which has a half-life of 24,000 years, is produced in nuclear reactors and used as the explosive in some nuclear weapons. This isotope can cause lung cancer when its particles are inhaled in minute amounts. Thus, it must be stored safely for 240,000 years (10 × 24,000 years)—about four times longer than the latest version of our species has existed.

Table 3-1 Half-Lives of Selected Radioisotopes

Isotope	Half-Life	Radiation Emitted
Potassium-42	12.4 hours	Alpha, beta
Iodine-131	8 days	Beta, gamma
Cobalt-60	5.27 years	Beta, gamma
Hydrogen-3 (tritium)	12.5 years	Beta
Strontium-90	28 years	Beta
Carbon-14	5,370 years	Beta
Plutonium-239	24,000 years	Alpha, gamma
Uranium-235	710 million years	Alpha, gamma
Uranium-238	4.5 billion years	Alpha, gamma

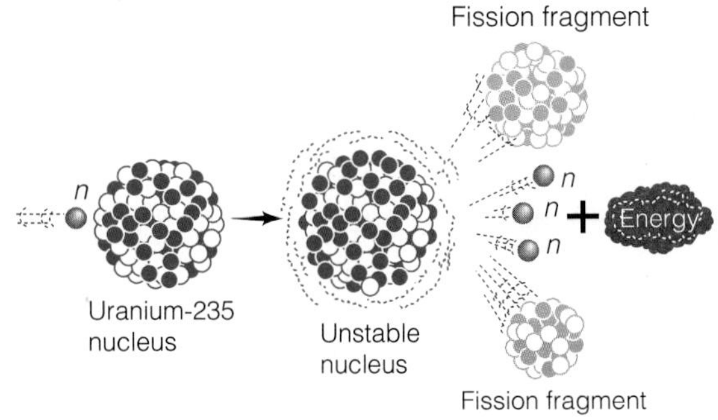

Figure 3-10 Fission of a uranium-235 nucleus by a neutron (*n*).

What Is Nuclear Fission? Splitting Nuclei **Nuclear fission** is a nuclear change in which nuclei of certain isotopes with large mass numbers (such as uranium-235) are split apart into lighter nuclei when struck by neutrons; each fission releases two or three more neutrons and energy (Figure 3-10). Each of these neutrons, in turn, can cause an additional fission. For these multiple fissions to take place, enough fissionable nuclei must be present to provide the **critical mass** needed for efficient capture of these neutrons.

Multiple fissions within a critical mass form a **chain reaction**, which releases an enormous amount of energy (Figure 3-11). Living cells can be damaged by the ionizing radiation released by the radioactive lighter nuclei and by high-speed neutrons produced by nuclear fission.

In an atomic bomb, an enormous amount of energy is released in a fraction of a second in an uncontrolled nuclear fission chain reaction. This reaction is initiated by an explosive charge, which suddenly pushes two masses of fissionable fuel together, causing the fuel to reach the critical mass needed for a chain reaction.

In the reactor of a nuclear power plant, the rate at which the nuclear fission chain reaction takes place is controlled so that under normal operation only one of every two or three neutrons released is used to split another nucleus. In conventional nuclear fission reactors, the splitting of uranium-235 nuclei releases heat, which produces high-pressure steam to spin turbines and thus generates electricity.

What Is Nuclear Fusion? Forcing Nuclei to Combine **Nuclear fusion** is a nuclear change in which two isotopes of light elements, such as hydrogen, are forced together at extremely high temperatures until they fuse to form a heavier nucleus, releasing energy in the process. Temperatures of at least 100 million degrees

Weiss, P. 1998. "Another Face of Entropy." *Science News*, vol. 154, no. 7, p. 108(2).

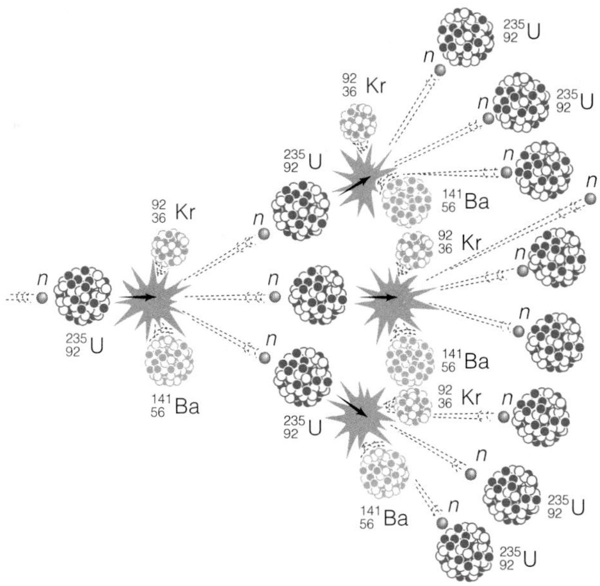

Figure 3-11 A nuclear chain reaction initiated by one neutron triggering fission in a single uranium-235 nucleus. This figure illustrates only a few of the trillions of fissions caused when a single uranium-235 nucleus is split within a critical mass of uranium-235 nuclei. The elements krypton (Kr) and barium (Ba), shown here as fission fragments, are only two of many possibilities.

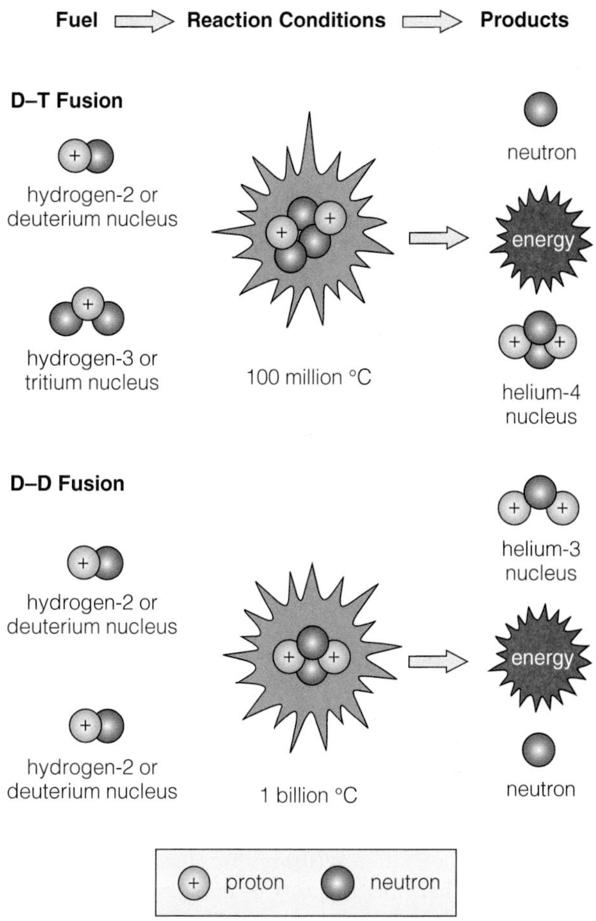

Figure 3-12 The deuterium–tritium (D–T) and deuterium–deuterium (D–D) nuclear fusion reactions, which take place at extremely high temperatures.

Celsius are needed to force the positively charged nuclei (which strongly repel one another) to fuse.

Nuclear fusion is much more difficult to initiate than nuclear fission, but once started it releases far more energy per unit of fuel than does fission. Fusion of hydrogen nuclei to form helium nuclei is the source of energy in the sun and other stars.

After World War II, the principle of *uncontrolled nuclear fusion* was used to develop extremely powerful hydrogen, or thermonuclear, weapons. These weapons use the D–T fusion reaction, in which a hydrogen-2, or deuterium (D), nucleus and a hydrogen-3 (tritium, T) nucleus are fused to form a larger, helium-4 nucleus, a neutron, and energy (Figure 3-12).

Scientists have also tried to develop *controlled nuclear fusion*, in which the D–T reaction is used to produce heat that can be converted into electricity. Despite more than 50 years of research, this process is still in the laboratory stage. Even if it becomes technologically and economically feasible, it probably won't be a practical source of energy until 2166 if then.

3-7 THE TWO IRONCLAD LAWS OF ENERGY

What Is the First Law of Energy? You Can't Get Something for Nothing Scientists have observed energy being changed from one form to another in millions of physical and chemical changes, but they have never been able to detect either the creation or destruction of any energy (except in nuclear changes). The results of their experiments have been summarized in the **law of conservation of energy**, also known as the **first law of energy** or the **first law of thermodynamics**: *In all physical and chemical changes, energy is neither created nor destroyed, but it may be converted from one form to another.*

This scientific law tells us that when one form of energy is converted to another form in any physical or chemical change *energy input always equals energy output.* No matter how hard we try or how clever we are, we can't get more energy out of a system than we put in; in other words, *we can't get something for nothing in terms of energy quantity.*

What Is the Second Law of Energy? You Can't Even Break Even Because the first law of energy states that energy can be neither created nor destroyed, it's tempting to think that there will always be enough energy, yet if we fill a car's tank with gasoline and drive around, or use a flashlight battery until it is dead,

Hint: Enter the search term *entropy* using the Subject Guide.

GUEST ESSAY

A New Environmentalism for the 21st Century

Peter Montague

Peter Montague is director of the Environmental Research Foundation in Washington, D.C., which studies and informs the public about environmental problems and the technologies and policies that might help solve them. He has served as project administrator of a hazardous waste research program at Princeton University and has taught courses in environmental impact analysis at the University of New Mexico. He is the coauthor of two books on toxic heavy metals in the natural environment and is editor of Rachel's Environment and Health Weekly, *an informative and readable newsletter on environmental problems and solutions.*

Environmentalism as we have known it for the last 30 years is dead. The environmentalism of the 1970s advocated strict numerical controls on releases into the environment of *dangerous wastes* (any unwanted or uncontrolled materials that can harm living things or disrupt ecosystems). But industry's ability to create new hazards quickly outstripped government's ability to establish adequate controls and enforcement programs.

After so many years of effort by government and by concerned citizens (the environmental movement), the overwhelming majority of dangerous chemicals is still not regulated in any way. Even those few that are covered by regulations have not been adequately controlled.

In short, the *pollution management* approach to environmental protection has failed and stands discredited; *pollution prevention* is our only hope. An ounce of prevention really is worth a pound of cure.

Here, in list form, is the situation facing environmentalists today:

- *All waste disposal—landfilling, incineration, deep-well injection—is polluting because "disposal" means dispersal into the environment.* Once wastes are created, they cannot be contained or controlled because of the scientific laws of matter and energy. The old environmentalism failed to recognize this important truth and thus squandered enormous resources trying to achieve the impossible. While we in the United States currently spend about $130 billion per year on pollution control, the global environment is increasingly threatened by a buildup of heat because of certain gases (such as carbon dioxide) that we emit into the atmosphere. At least half the surface of the planet is being subjected to damaging ultraviolet radiation from the sun as a result of ozone-depleting chemicals we have discharged into the atmosphere. Vast regions of the United States, Canada, and Europe are suffering from loss of forests, crop productivity, and fish as a result of a mixture of air pollutants, produced mostly by burning fossil fuels in power plants and automobiles. Soil and water are dangerously polluted at thousands of locales where municipal garbage and industrial wastes have been (and continue to be) dumped or incinerated; thousands of such sites remain to be discovered, according to U.S. government estimates.
- *The inevitable result of our reliance upon waste treatment and disposal systems has been an unrelenting buildup of toxic synthetic materials in humans and other forms of life worldwide.* For example, breast milk of women in industrialized countries is so contaminated with pesticides and industrial hydrocarbons that, if human milk were bottled and sold commercially, it could be banned by

something has been lost. If it isn't energy, what is it? The answer is *energy quality* (Figure 3-9), the amount of energy available that can perform useful work.

Countless experiments have shown that when energy is changed from one form to another, a decrease in energy quality always occurs. The results of these experiments have been summarized in what is called the **second law of energy** or the **second law of thermodynamics**: *When energy is changed from one form to another, some of the useful energy is always degraded to lower-quality, more dispersed, less useful energy.* This degraded energy usually takes the form of heat given off at a low temperature to the surroundings (environment). There it is dispersed by the random motion of air or water molecules and becomes even more disorderly and less useful. Another way to state the second law of energy is that *heat always flows spontaneously from hot (high-quality energy) to cold (low-quality energy).*

Basically, this law says that in any energy conversion, we always end up with *less* usable energy than we started with. So not only can we not get something for nothing in terms of energy quantity, *we can't even break even in terms of energy quality because energy always goes from a more useful to a less useful form.* The more energy we use, the more low-grade energy (heat) we add to the environment. No one has ever found a violation of this fundamental scientific law (see quotes at the beginning and end of this chapter).

Consider three examples of the second energy law in action. First, when a car is driven, only about 10% of the high-quality chemical energy available in its gasoline fuel is converted into mechanical energy (to propel the vehicle) and electrical energy (to run its electrical the U.S. Food and Drug Administration as unsafe for human consumption. If a whale today beaches itself on U.S. shores and dies, its body must be treated as "hazardous waste" because whales contain concentrations of PCBs (polychlorinated biphenyls, a class of industrial toxins) legally defined as hazardous.

- *The ability of humans and other life forms to adapt to changes in their environment is strictly limited by the genetic code each form of life inherits.* Continued contamination at a rate hundreds of times faster than we can adapt will subject humans to increasingly widespread sickness and degradation of the species, and could ultimately lead to extinction.
- *Damage to humans (and to other life forms) is abundantly documented.* Birds, fish, and humans in industrialized countries are enduring steadily rising levels of cancer, genetic mutations, and damage to their nervous, immune, and hormonal systems as a result of pollution.

If we will but look, the handwriting is on the wall everywhere.

To deal with these problems, industrial societies must abandon their reliance on waste treatment and disposal and on the regulatory system of numerical standards created to manage the damage that results from relying on waste disposal instead of waste prevention. We must quickly move the industrialized and industrializing countries to new technical approaches accompanied by new industrial goals: *clean production* or zero-discharge systems.

The concept of clean production involves industrial systems that avoid or eliminate dangerous wastes and dangerous products and minimize the use and waste of raw materials, water, and energy. Goods manufactured in a clean production process must not damage natural ecosystems throughout their entire life cycle, including **(1)** raw material selection, extraction, and processing, **(2)** product conceptualization, design, manufacture, and assembly, **(3)** material transport during all phases, **(4)** industrial and household usage, and **(5)** reintroduction of the product into industrial systems or into the environment when it no longer serves a useful function.

Clean production does not rely on *end-of-pipe* pollution controls such as filters or scrubbers or chemical, physical, or biological treatment. Measures that **(1)** pretend to reduce the volume of waste by incineration or concentration, **(2)** mask the hazard by dilution, or **(3)** transfer pollutants from one environmental medium to another are also excluded from the concept of clean production.

A new industrial pattern, and thus a new environmentalism, is emerging. It insists that the long-term well-being of humans and other species must be factored into our production and consumption plans. These new requirements are not optional; human survival and life quality depend on our willingness to make, and pay for, the necessary changes.

Critical Thinking

1. Do you agree with the author that the *pollution management* approach to environmental protection practiced during the past 30 years has failed and must be replaced with a *pollution prevention* approach? Explain.

2. List three beneficial and three undesirable economic, health, consumption, and lifestyle changes you might experience as a consequence of putting much greater emphasis on pollution prevention.

systems); the remaining 90% is degraded to low-quality heat that is released into the environment and eventually lost into space. Second, when electrical energy flows through filament wires in an incandescent light bulb, it is changed into about 5% useful light and 95% low-quality heat that flows into the environment; this so-called *light bulb* is really a *heat bulb*. Third, in living systems, solar energy is converted into chemical energy (food molecules) and then into mechanical energy (moving, thinking, and living); high-quality energy is degraded during this change of forms (Figure 3-13).

The second law of energy also means that *we can never recycle or reuse high-quality energy to perform useful work*. Once the concentrated energy in a serving of food, a liter of gasoline, a lump of coal, or a chunk of uranium is released, it is degraded to low-quality heat that is dispersed into the environment. We can heat air or water at a low temperature and upgrade it to high-quality energy, but the second law of energy tells us that it will take more high-quality energy to do this than we get in return.

Connections: How Does the Second Energy Law Affect Life? To form and maintain the highly ordered arrangement of molecules and the organized biochemical processes in your body, you must continually get and use high-quality matter and energy resources from your surroundings. As you use these resources, you add low-quality heat and low-quality waste matter to your surroundings. Your body continuously gives off heat equal to that of a 100-watt incandescent light bulb; this is why a closed room full of people gets warm. You also continuously break down solid, large molecules (such as glucose) into smaller

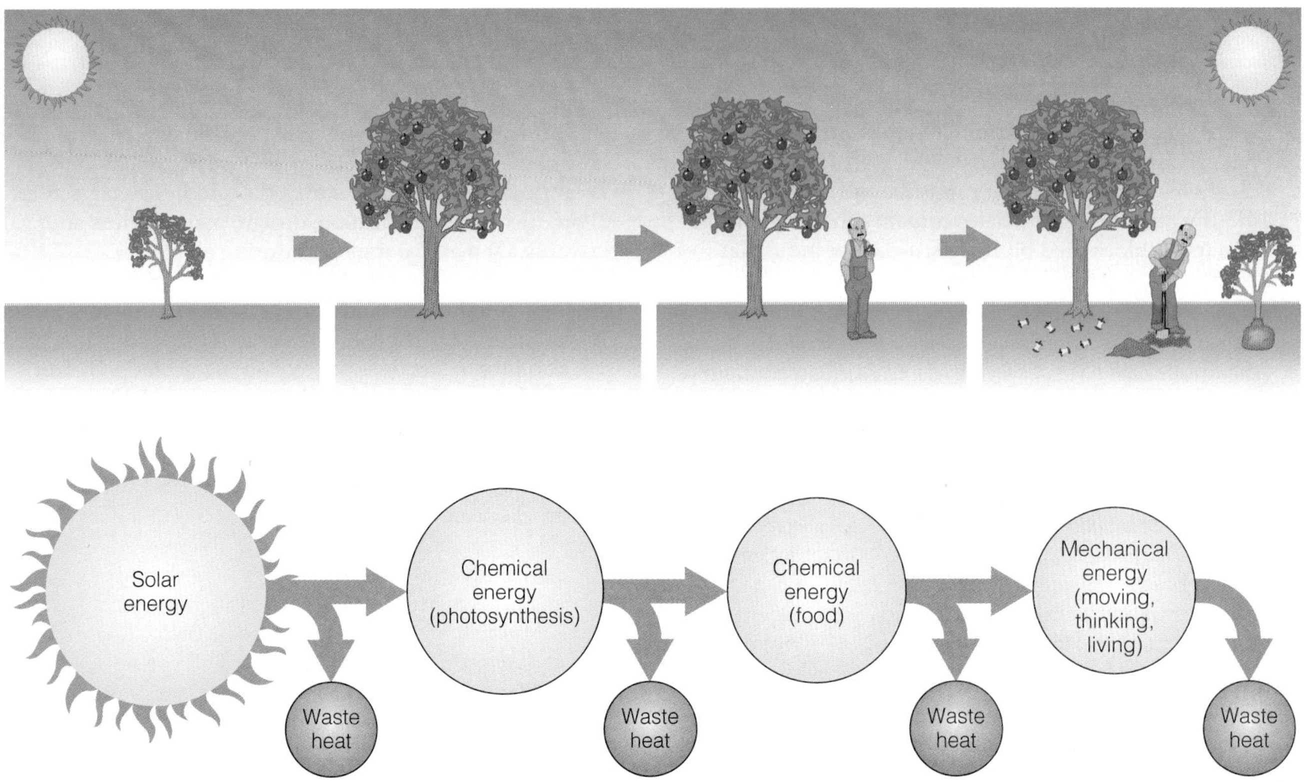

Figure 3-13 The second energy law in action in living systems. Each time energy is changed from one form to another, some of the initial input of high-quality energy is degraded, usually to low-quality heat that disperses into the environment.

molecules of carbon dioxide gas and water vapor, which are dispersed in the atmosphere.

Planting, growing, processing, and cooking food all require high-quality energy and matter resources that add low-quality heat and waste materials to the environment. In addition, enormous amounts of low-quality heat and waste matter are added to the environment when concentrated deposits of minerals and fuels are extracted from the earth's crust, processed, and used.

3-8 CONNECTIONS: MATTER AND ENERGY LAWS AND ENVIRONMENTAL PROBLEMS

What Are High-Throughput Economies? As a result of the law of conservation of matter and the second law of energy, individual resource use automatically adds some waste heat and waste matter to the environment. Most of today's advanced industrialized countries have **high-throughput economies** that attempt to sustain ever-increasing economic growth by increasing the *throughput* of matter and energy resources in their economic systems (Figure 3-14). These resources flow through the economies of such societies to planetary *sinks* (air, water, soil, organisms), where pollutants and wastes end up and can accumulate to harmful levels.

The scientific laws of matter and energy discussed in this chapter tell us that if more and more people continue to use and waste more and more energy and matter resources at an increasing rate, eventually the capacity of the environment to dilute and degrade waste matter and absorb waste heat will be exceeded. Thus, *at some point high-throughput societies become unsustainable.*

What Are Matter-Recycling Economies? A stopgap solution to this problem is to convert an unsustainable high-throughput economy to a **matter-recycling economy**. The goal of such a conversion is to allow economic growth to continue without depleting matter resources or producing excessive pollution and environmental degradation.

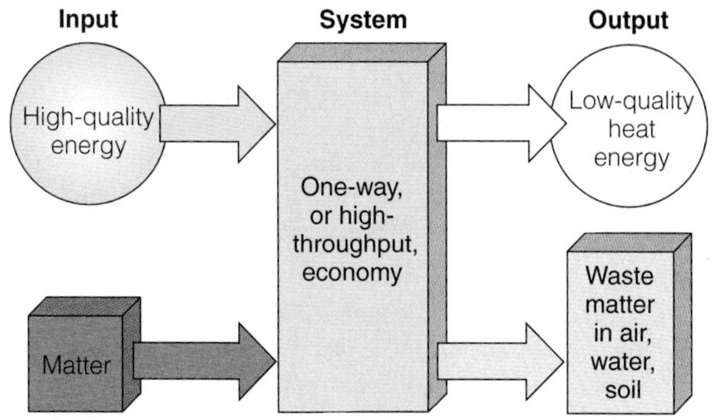

Figure 3-14 The eventually unsustainable high-throughput economies of most developed countries are based on maximizing the rates of energy and matter flow. This process rapidly converts high-quality matter and energy resources into waste, pollution, and low-quality heat.

Even though recycling matter saves energy, the two laws of energy tell us that *recycling matter resources always requires expenditure of high-quality energy (which cannot be recycled) and adds waste heat to the environment*. For the long run, a matter-recycling economy based on continuing population growth and per capita resource consumption must **(1)** have an inexhaustible supply of affordable high-quality energy, and **(2)** its environment must have an infinite capacity to absorb and disperse waste heat and to dilute and degrade waste matter.

There is also a limit to the number of times some materials, such as paper fiber, can be recycled before they become unusable. Changing to a matter-recycling economy is an important way to buy some time, but it does not allow more and more people to use more and more resources indefinitely, even if all of them were somehow perfectly recycled.

What Are Low-Throughput Economies? Learning from Nature The three scientific laws governing matter and energy changes suggest that the best long-term solution to our environmental and resource problems is to shift from an economy based on maximizing matter and energy flow (throughput) to a more sustainable or **low-throughput economy**. The major features of such a society are summarized in Figure 3-15.

The next four chapters are devoted to applying the three basic scientific laws of matter and energy to living systems and looking at some *biological principles* that can teach us how to work with nature.

The second law of thermodynamics holds, I think, the supreme position among laws of nature. . . . If your theory is found to be against the second law of thermodynamics, I can give you no hope.

ARTHUR S. EDDINGTON

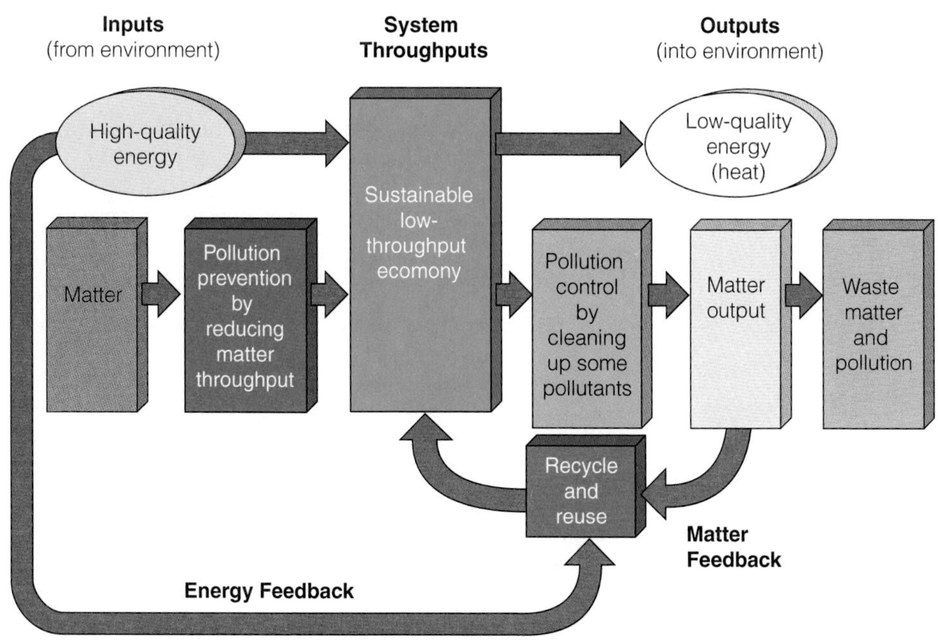

Figure 3-15 Lessons from nature. A sustainable low-throughput economy, based on energy flow and matter recycling, works with nature to reduce throughput. This is done by **(1)** reusing and recycling most nonrenewable matter resources, **(2)** using potentially renewable resources no faster than they are replenished, **(3)** using matter and energy resources efficiently, **(4)** reducing unnecessary consumption, **(5)** emphasizing pollution prevention and waste reduction, and **(6)** controlling population growth.

REVIEW QUESTIONS

1. Define *science* and explain how it works. Distinguish between *scientific data, scientific hypothesis, scientific model, scientific theory, scientific law,* and *scientific methods.* Explain why a scientific theory should be taken seriously. What is a *controlled experiment*?

2. Distinguish between *frontier science* and *consensus science* and between *science* and *technology.*

3. What is *environmental science* and what are some of its limitations?

4. What is a *system*? What are *mathematical models* and why are they important in understanding complex systems?

5. Distinguish between the *inputs, flows* (throughputs), and *outputs* of a system. Distinguish between a *negative feedback loop* and a *positive feedback loop* in a system and give an example of each. What is *homeostasis* and how is it maintained in a system?

6. Distinguish between *matter, elements, compounds,* and *mixtures.* Distinguish between *atoms, ions,* and *molecules.* What three major types of subatomic particles are found in atoms? Which two of these particles are found in the nucleus and which is found outside the nucleus? Distinguish between *atomic number* and *mass number.* What is an *isotope* of an atom? Distinguish between *organic compounds* and *inorganic compounds* and give an example of each.

7. Distinguish between *genes, gene mutations,* and *chromosomes.*

8. Distinguish between *high-quality matter* and *low-quality matter* and give an example of each.

9. What is *energy*? Distinguish between *kinetic energy* and *potential energy* and give an example of each. Give three types of *electromagnetic radiation.* Distinguish between *ionizing radiation* and *nonionizing radiation.* Distinguish between *heat* and *temperature.*

10. Distinguish between *high-quality energy* and *low-quality energy* and give an example of each.

11. Distinguish between a *physical change* and a *chemical change* and give an example of each.

12. What is the *law of conservation of matter*? What is its environmental significance? Explain why there is no "away" as a repository for pollution.

13. Distinguish between a *chemical change* and a *nuclear change.* Distinguish between *natural radioactive decay, radioisotopes, gamma rays, alpha particles,* and *beta particles.* What is the *half-life* of a radioactive isotope? For how many half-lives should radioactive material be stored safely before it decays to an acceptable level of radioactivity? Distinguish between *nuclear fission* and *nuclear fusion.* Distinguish between *critical mass* and a *nuclear chain reaction.*

14. Distinguish between the *first law of energy (thermodynamics)* and the *second law of energy (thermodynamics)* and give an example of each law in action. Use the second law of thermodynamics to explain why energy cannot be recycled.

15. Distinguish between a *high-throughput economy,* a *low-throughput economy,* and a *matter-recycling economy.* Use the law of conservation of matter and the first and second laws of thermodynamics to explain the need to shift from a high-throughput economy to a matter-recycling economy and eventually to a low-throughput economy.

CRITICAL THINKING

1. To what extent are scientists responsible for the applications of knowledge they discover? Should scientists abandon research because of its possible harmful uses? Explain.

2. Respond to the following statements:

a. It has never been scientifically proven that anyone has ever died from smoking cigarettes.

b. The greenhouse theory—that certain gases (such as water vapor and carbon dioxide) trap heat in the atmosphere and thus influence the average temperature and climate of the earth—is not a reliable idea because it is only a scientific theory.

3. Explain why we don't really consume anything and why we can never really throw matter away.

4. A tree grows and increases its mass. Explain why this isn't a violation of the law of conservation of matter.

5. If there is no "away," why isn't the world filled with waste matter?

6. Someone wants you to invest money in an automobile engine that will produce more energy than the energy in the fuel (such as gasoline or electricity) you use to run the motor. What is your response? Explain.

7. Use the second energy law to explain why a barrel of oil can be used only once as a fuel.

8. (a) Use the law of conservation of matter to explain why a matter-recycling economy will sooner or later be necessary. **(b)** Use the first and second laws of energy to explain why, in the long run, we will need a or low-throughput economy, not just a matter-recycling economy.

9. (a) Imagine that you have the power to violate the law of conservation of energy (the first energy law) for 1 day. What are the three most important things you would do with this power? **(b)** Repeat this process, imagining that you have the power to violate the second law of energy for 1 day.

*10. See whether you can find an advertisement or an article describing or using some aspect of science in which **(a)** the concept of scientific proof is misused, **(b)** the term *theory* is used when it should have been *hypothesis,* and **(c)** a consensus scientific finding is dismissed or downplayed because it is "only a theory."

*11. Use the library or Internet to find examples of various perpetual motion machines and inventions that allegedly violate the two laws of energy (thermodynamics) by producing more high-quality energy than the high-quality energy needed to make them run. What has happened to these schemes and machines (many of them developed by scam artists to attract money from investors)?

*12. Make a concept map of this chapter's major ideas using the section heads and subheads and the key terms (in boldface). Look inside the back cover and in Appendix 5 for information about concept maps.

FURTHER INFORMATION

For more information see Internet Sources and Further Readings at the back of the book. Online resources for this chapter can be found at **www.brookscole.com/biology**

4 ECOSYSTEMS AND HOW THEY WORK: CONNECTIONS IN NATURE

Have You Thanked the Insects Today?

Insects have a bad reputation. We classify many insect species as *pests* because they compete with us for food, spread human diseases (such as malaria), and invade our lawns, gardens, and houses. Some people have "bugitis," fear all insects, and think that the only good bug is a dead bug. However, this view fails to recognize the vital roles insects play in helping sustain life on earth.

A large proportion of the earth's plant species (including many trees) depend on insects to pollinate their flowers (Figure 4-1, right). In turn, we and other land-dwelling animals depend on plants for food, either by eating them or by consuming animals that eat them. If there were no pollinating insects, there would be very few fruits and vegetables for us and plant-eating animals to eat.

Insects, such as praying mantis (Figure 4-1, left), that eat other insects help control the populations of at least half the species of insects we call pests. This free pest control service is an important part of the natural capital (Figure 1-2) that helps sustain us.

Suppose all insects disappeared today. Within a year most of the earth's amphibians, reptiles, birds, and mammals would become extinct because of the disappearance of so much plant life. The earth would be covered with rotting vegetation and animal carcasses being decomposed by unimaginably huge hordes of bacteria and fungi.

Fortunately, this is not a realistic scenario because insects, which have been around for at least 400 million years, are phenomenally successful forms of life. They were the first animals to invade the land and, later, the air. Today they are by far the planet's most diverse, abundant, and successful animals.

Insects can rapidly evolve new genetic traits, such as resistance to pesticides. They also have an exceptional ability to evolve into new species when faced with new environmental conditions, and they are extremely resistant to extinction. Because of their ability to quickly develop genetic resistance to pesticides, using such chemical warfare to control or eradicate pests will always fail eventually. Moreover, many of the pesticides we use reduce the populations of insect species that help keep pest insect species under control.

Some people have wondered whether insects will take over the world if the human race extinguishes itself. This is the wrong question. Insects are already in charge. Insects can thrive without newcomers such as us, but we and most other land organisms would quickly perish without them.

Learning about the roles insects play in nature requires us to understand how insects and other organisms living in a biological *community* (such as a forest or pond) interact with one another and with the nonliving environment. *Ecology* is the science that studies such relationships and interactions in nature, as discussed in this chapter and the three chapters that follow.

Figure 4-1 Insects play important roles in helping sustain life on earth. The bright green caterpillar moth feeding on pollen in a crocus flower, right, and other insects pollinate flowering plants that serve as food for many plant eaters. The praying mantis eating a grasshopper (left) and many other insect species help control the populations of at least half of the insect species we classify as pests. (Left: M. H. Sharp/Photo Researchers, Inc. Right: Stephen Hopkin/Planet Earth Pictures)

Students who can begin early in their lives to think of things as connected, even if they revise their views every year, have begun the life of learning.

MARK VAN DOREN

This chapter addresses the following questions:

- What is ecology and what is life?
- What basic processes keep us and other organisms alive?
- What are the major living and nonliving parts of an ecosystem?
- What happens to energy in an ecosystem?
- What happens to matter in an ecosystem?
- How do scientists study ecosystems?

4-1 ECOLOGY AND LIFE

What Is Ecology? **Ecology** (from the Greek words *oikos*, "house" or "place to live," and *logos*, "study of") is the study of how organisms interact with one another and with their nonliving environment (including such factors as sunlight, temperature, moisture, and vital nutrients). Ecologists focus on trying to understand the interactions among organisms, populations, communities, ecosystems, and the ecosphere (Figure 4-2).

An **organism** is any form of life. Organisms can be classified into **species**, or groups of organisms that resemble one another in appearance, behavior, chemistry, and genetic endowment. Organisms that reproduce sexually are classified in the same species if, under natural conditions, they can actually or potentially breed with one another and produce live, fertile offspring.

We don't know how many species exist on the earth. Estimates range from 5 million to 100 million, most of which are insects and microorganisms. So far biologists have identified and named only about 1.8 million species. Biologists know a fair amount about roughly one-third of the known species, but understand the detailed roles and interactions of only a few.

A **population** consists of a group of interacting individuals of the same species that occupy a specific area at the same time (Figure 4-3). Examples are all sunfish in a pond, all white oak trees in a forest, and all people in a country. In most natural populations, individuals vary slightly in their genetic makeup, which is why they don't all look or behave exactly alike—a phenomenon called **genetic diversity** (Figure 4-4). Populations are dynamic groups that change in size, age distribution, density, and genetic composition as a result of changes in environmental conditions.

The place where a population (or an individual organism) normally lives is its **habitat**. Populations of all the different species occupying a particular place make up a **community**, or **biological community**: a complex interacting network of plants, animals, and microorganisms.

An **ecosystem** is a community of different species interacting with one another and with their nonliving environment of matter and energy. An ecosystem may be small, such as a particular stream or field or a patch of woods, desert, or marsh. Or the units may be large, generalized types of terrestrial (land) ecosystems such as a particular type of grassland, forest, or desert. Ecosystems can be natural or artificial (human-created). Examples of human-created ecosystems are cropfields, farm ponds, and reservoirs. All of the earth's ecosystems together make up what we call the **biosphere**, or **ecosphere**.

What Is Life? All life shares a set of basic characteristics that enable growth, survival, and reproduction:

- *Living organisms are made of cells that have highly organized internal structures and functions.* The **cell** is the basic unit of life. Each cell is bounded by an outer membrane or wall and contains genetic material (DNA) and other structures it needs to perform its life functions. Organisms may consist of a single cell (bacteria, for instance) or many cells.
- *Living organisms have characteristic types of deoxyribonucleic acid (DNA) molecules in each cell.* DNA is the stuff of which genes, the basic units of heredity, are made (Figure 3-6). These self-replicating molecules contain the instructions for making new cells and for assembling proteins and other molecules each cell needs for survival and reproduction.
- *Living organisms capture and transform matter and energy from their environment to supply their needs for survival, growth, and reproduction.* The complete set of chemical reactions that carries out this role in cells and organisms is called **metabolism**.
- *Living organisms maintain favorable internal conditions, despite changes in their external environment, if not overstressed.*
- *Living organisms perpetuate themselves through* ***reproduction:*** *the production of offspring by one or more parents.* **Asexual reproduction** generally occurs by simple cell division and is common in single-cell organisms. The mother cell divides to produce two identical cells that are clones or replicas of the original cell. **Sexual reproduction** occurs in organisms that produce offspring by combining sex cells or gametes (such as ovum and sperm) from both parents. This

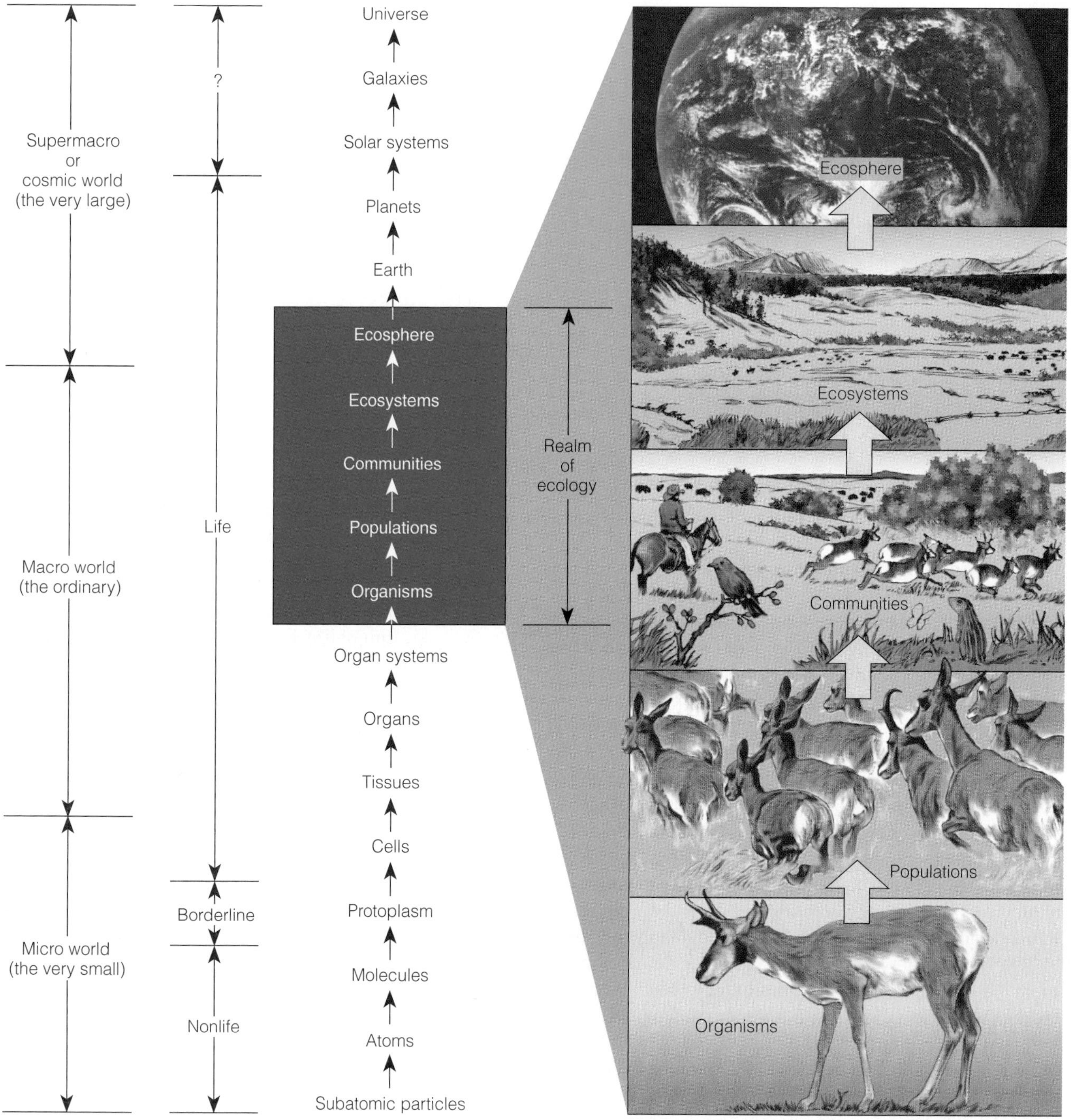

Figure 4-2 Model of levels of organization of matter in nature. Note that ecology focuses on five levels of this hierarchical model.

produces offspring that have combinations of traits from each parent. Sexual reproduction usually gives the offspring a greater chance of survival under changing environmental conditions than the genetic clones produced by asexual reproduction.

- *Living organisms adapt to changes in environmental conditions through the process of* **evolution**. Evolution involves changes in the genetic composition of organisms through time. When evolution continues long enough, it can result in new species. Ongoing processes of extinction and evolution, in response to environmental changes over millions of years, have led to the diversity of life-forms found on the earth today. This *biodiversity*, a vital part of natural capital, sustains life and provides the genetic raw material for adaptation to future environmental changes.

Figure 4-3 A population of monarch butterflies wintering in Michoacán, Mexico. The geographic distribution of this butterfly coincides with that of the milkweed plant, on which monarch larvae and caterpillars feed. (Frans Lanting/Bruce Coleman Ltd.)

Figure 4-4 The genetic diversity among individuals of one species of Caribbean snail is reflected in the variations in shell color and banding patterns. (Alan Solem)

4-2 EARTH'S LIFE-SUPPORT SYSTEMS

What Are the Major Parts of Earth's Life-Support Systems? We can think of the earth as being made up of several layers or concentric spheres (Figure 4-5). The **atmosphere** is a thin envelope of air around the planet. Its inner layer, the **troposphere**, extends only about 17 kilometers (11 miles) above sea level, but contains most of the planet's air, mostly nitrogen (78%) and oxygen (21%). The next layer, stretching 17–48 kilometers (11–30 miles) above the earth's surface, is the **stratosphere**. Its lower portion contains enough ozone (O_3) to filter out most of the sun's harmful ultraviolet radiation, thus allowing life to exist on land and in the surface layers of bodies of water.

The **hydrosphere** consists of the earth's liquid water (both surface and underground), ice (polar ice, icebergs, ice in permafrost), and water vapor in the atmosphere. The **lithosphere** is the earth's crust and upper mantle; the crust contains nonrenewable fossil fuels and minerals we use as well as potentially renewable soil chemicals (nutrients) required for plant life.

The **ecosphere**, or **biosphere**, is the portion of the earth in which living (biotic) organisms exist and interact with one another and with their nonliving (abiotic) environment. The ecosphere includes most of the hydrosphere and parts of the lower atmosphere and upper lithosphere, reaching from the deepest ocean floor, 20 kilometers (12 miles) below sea level, to the tops of the highest mountains. If the earth were an apple, the ecosphere would be no thicker than the apple's skin. *The goal of ecology is to understand the interactions in this thin, life-supporting global skin of air, water, soil, and organisms.*

What Sustains Life on Earth? Life on the earth depends on three interconnected factors (Figure 4-6):

- The *one-way flow of high-quality (usable) energy* from the sun, **(1)** through materials and living things in their feeding interactions, **(2)** then into the environment as low-quality energy (mostly heat dispersed into air or water molecules at a low temperature), and **(3)** eventually back into space as infrared radiation.

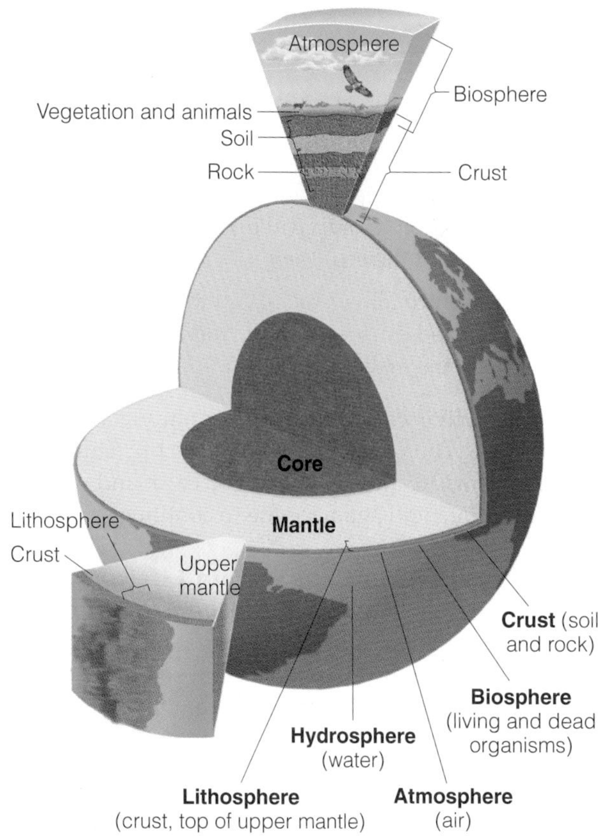

Figure 4-5 The general structure of the earth. The atmosphere consists of several layers, including the troposphere (innermost layer) and the stratosphere (second layer).

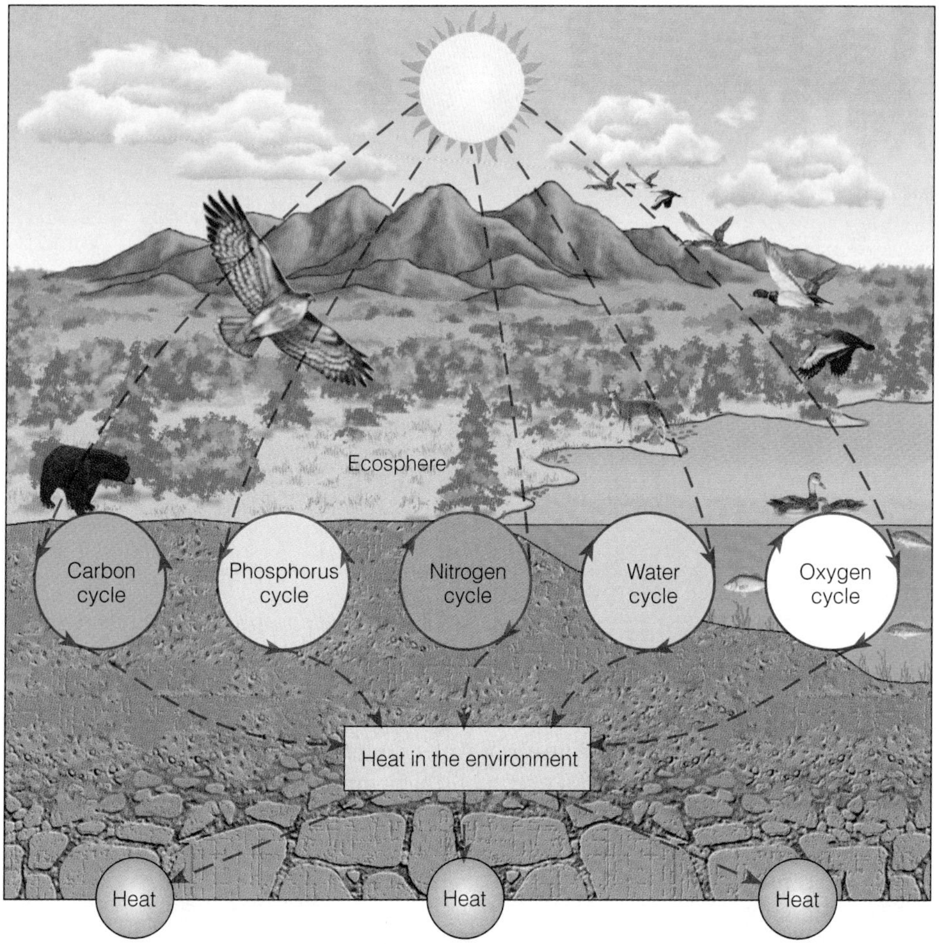

Figure 4-6 Life on the earth depends on **(1)** the *one-way flow of energy* (dashed lines) from the sun through the ecosphere, **(2)** the *cycling of crucial elements* (solid lines around circles), and **(3)** *gravity*, which keeps atmospheric gases from escaping into space and draws chemicals downward in the matter cycles. This simplified model depicts only a few of the many cycling elements.

- The *cycling of matter or nutrients* (all atoms, ions, or molecules needed for survival by living organisms) through parts of the ecosphere.
- *Gravity*, which allows the planet to hold onto its atmosphere and causes the downward movement of chemicals in the matter cycles

The earth is closed to significant inputs of matter from space. Thus, essentially all of the nutrients used by organisms are already present on earth and must be recycled again and again for life to continue.

How Does the Sun Help Sustain Life on Earth? The sun is a middle-aged star that lights and warms the planet. It also supplies the energy for *photosynthesis*, the process used by green plants and some bacteria to make the compounds that keep them alive and feed most other organisms. Solar energy also powers the cycling of matter and drives the climate and weather systems that distribute heat and fresh water over the earth's surface.

The sun is a gigantic fireball of hydrogen (72%) and helium (28%) gases. Temperatures and pressures in its inner core are so high that hydrogen nuclei fuse to form helium nuclei (Figure 3-12), releasing enormous amounts of energy. The sun, then, is really a gigantic nuclear fusion (thermonuclear) reactor running on hydrogen fuel. This enormous reactor radiates energy in all directions as electromagnetic radiation (Figure 3-8).

Because the earth is a tiny sphere in the vastness of space, it receives only about one-billionth of this output of energy. Much of this energy is either reflected away or absorbed by chemicals in its atmosphere. Most of what reaches the troposphere is visible light, infrared radiation (heat), and the small amount of ultraviolet radiation that is not absorbed by ozone in the stratosphere. About 28% of the solar energy reaching the troposphere is reflected back into space by clouds, chemicals, dust, and the earth's surface land and water (Figure 4-7).

Most of the remaining 72% of solar energy warms the troposphere and land, evaporates water and cycles it through the ecosphere, and generates winds. A tiny fraction (about 0.023%) is captured by green plants and bacteria, fueling photosynthesis to make the organic compounds most life-forms need to survive.

Most of this unreflected solar radiation is degraded into infrared radiation (which we experience as heat) as

 Chaffin, T. 1998. "Whole-Earth Mentor: A Conversation with Eugene P. Odum." *Natural History*, October, vol. 107, no. 8, p. 8(4).

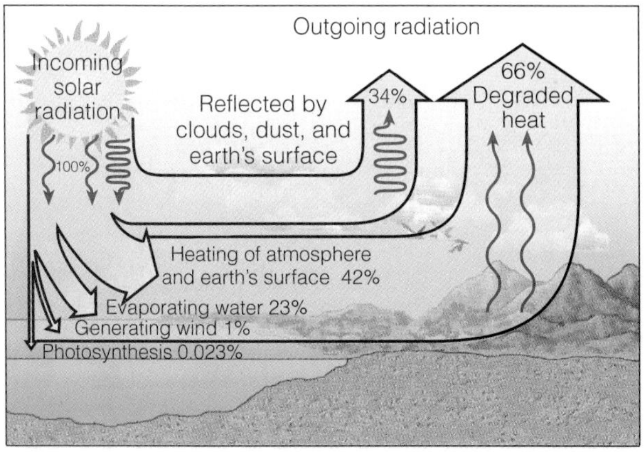

Figure 4-7 The flow of energy to and from the earth. The ultimate source of energy in most ecosystems is sunlight.

it interacts with the earth. Greenhouse gases (such as water vapor, carbon dioxide, methane, nitrous oxide, and ozone) reduce this flow of heat back into space. This helps warm the earth by acting somewhat like the glass in a greenhouse or the windows in a closed car, which allow a buildup of heat. Without this **natural greenhouse effect**, the earth would be nearly as cold as Mars, and life as we know it could not exist.

4-3 ECOSYSTEM CONCEPTS AND COMPONENTS

What Are the Major Types of Ecosystems? Viewed from outer space, the earth resembles an enormous jigsaw puzzle consisting of large masses of land and vast expanses of ocean (pp. 2–3 and Figure 1-1). Biologists have classified the terrestrial (land) portion of the ecosphere into **biomes**, large regions (such as forests, deserts, and grasslands) each characterized by a distinct climate and the specific life-forms, especially vegetation, adapted to it (Figure 4-8). **Climate**—long-term weather—is the main factor determining what type of life, especially what plants, will thrive in a given land area. Each biome consists of many ecosystems whose communities have adapted to differences in climate, soil, and other factors throughout the biome.

Marine and freshwater portions of the ecosphere can be divided into **aquatic life zones**, each containing

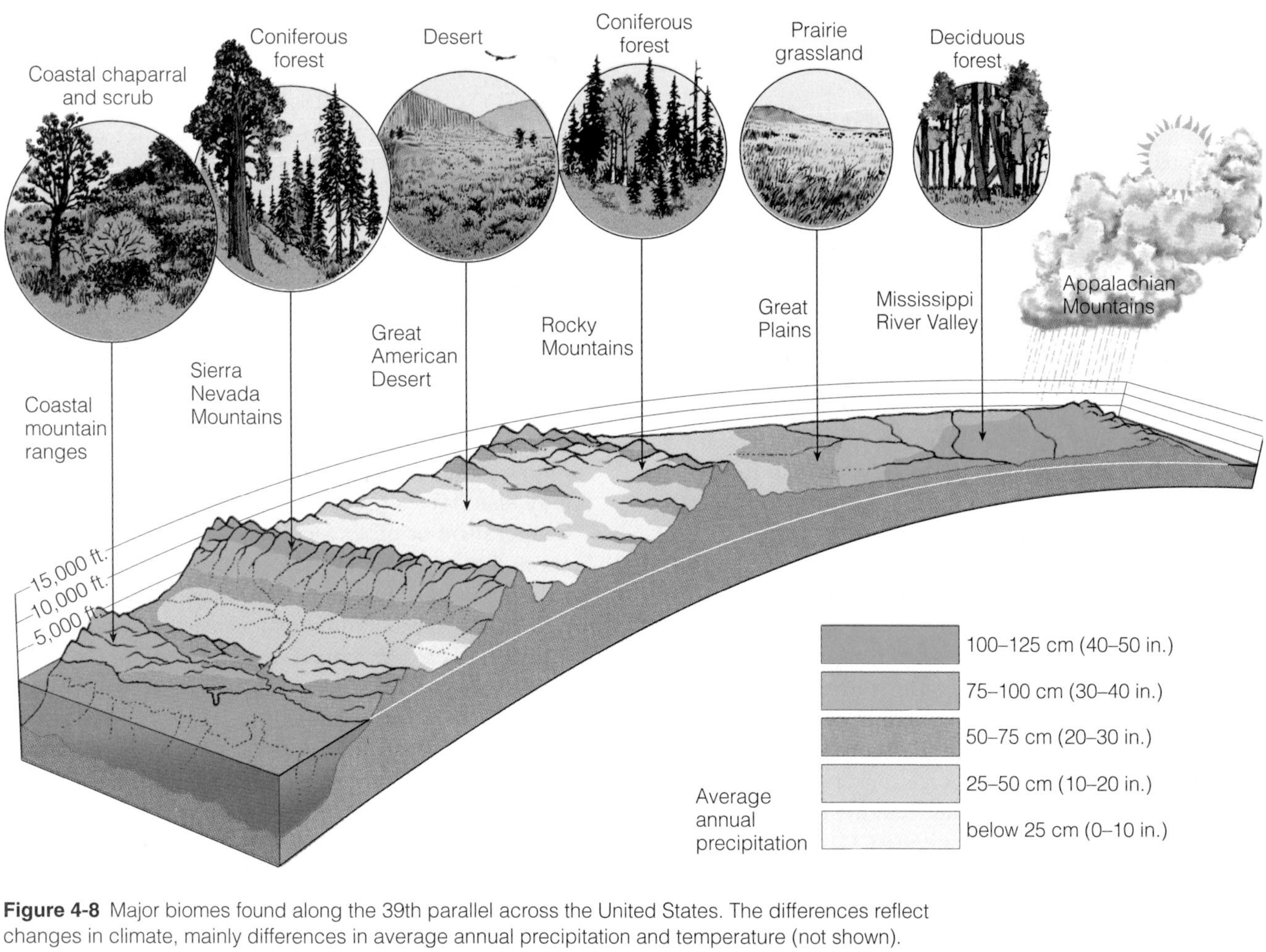

Figure 4-8 Major biomes found along the 39th parallel across the United States. The differences reflect changes in climate, mainly differences in average annual precipitation and temperature (not shown).

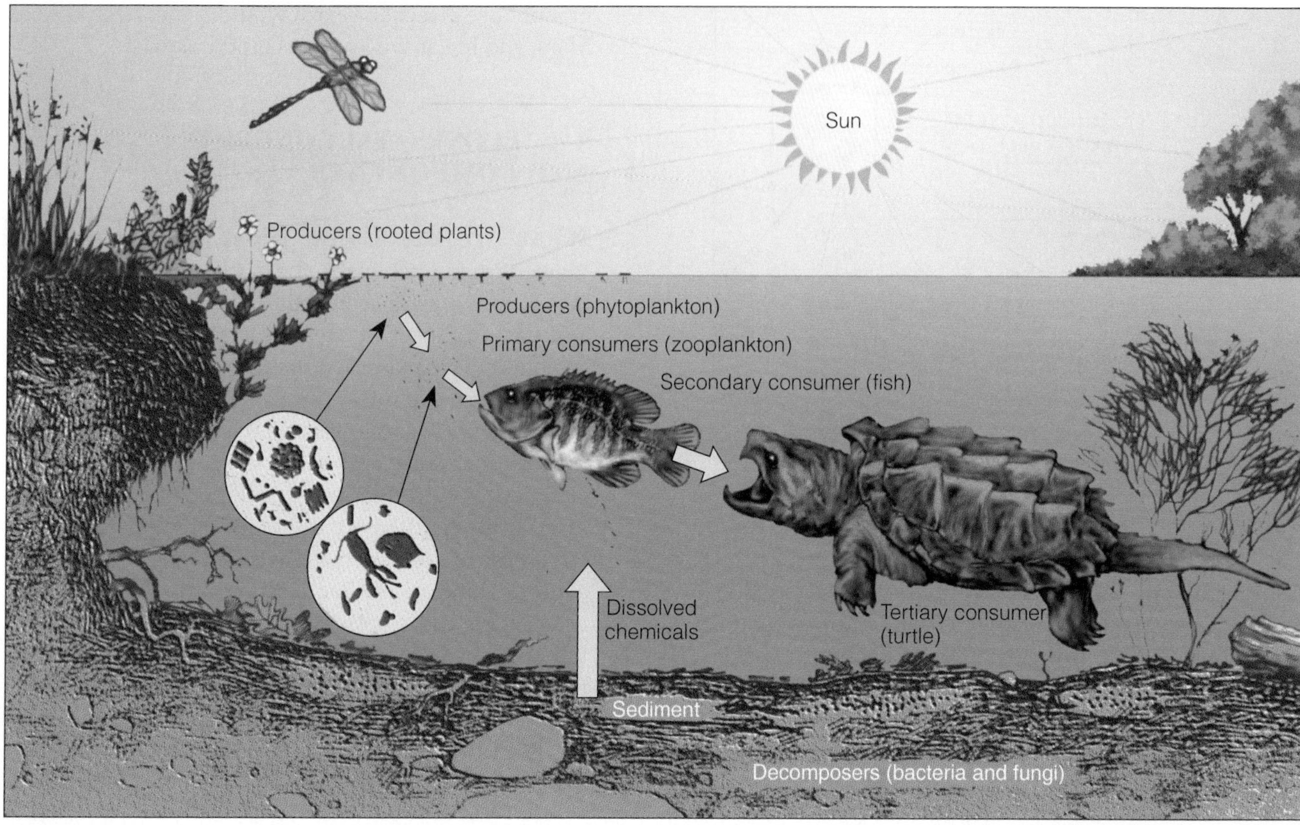

Figure 4-9 Major components of a freshwater ecosystem.

numerous ecosystems. Aquatic life zones are the aquatic equivalent of biomes. Examples include **(1)** *freshwater life zones* (such as lakes and streams) and **(2)** *ocean or marine life zones* (such as estuaries, coastlines, coral reefs, and the deep ocean). The earth's major land biomes and aquatic life zones are discussed in more detail in Chapter 6.

What Are the Major Components of Ecosystems? The ecosphere and its ecosystems can be separated into two parts: **(1) abiotic**, or nonliving, components (water, air, nutrients, and solar energy) and **(2) biotic**, or living, components (plants, animals, and microorganisms, sometimes called *biota*). Figures 4-9 and 4-10 are greatly simplified diagrams of some of the biotic and abiotic components in a freshwater aquatic ecosystem and a terrestrial ecosystem.

What Are the Major Nonliving Components of Ecosystems? The nonliving, or abiotic, components of an ecosystem are the physical and chemical factors that influence living organisms. Some important physical factors affecting land (terrestrial) ecosystems are **(1)** sunlight, **(2)** temperature, **(3)** precipitation, **(4)** wind, **(5)** latitude (distance from the equator), **(6)** altitude (distance above sea level), **(7)** frequency of fire, and **(8)** nature of the soil. For aquatic ecosystems, major physical factors include **(1)** water currents, **(2)** concentrations of dissolved nutrients (such as nitrogen and phosphorus), and **(3)** the amount of suspended solid material.

Different species thrive under different physical conditions. Some need bright sunlight, others thrive better in shade. Some require a hot environment, others a cool or cold one. Some do best under wet conditions, others under dry conditions.

Each population in an ecosystem has a **range of tolerance** to variations in its physical and chemical environment (Figure 4-11). Individuals within a population may also have slightly different tolerance ranges for temperature or other factors because of small differences in genetic makeup, health, and age. Thus, although a trout population may do best within a narrow band of temperatures (*optimum level or range*), a few individuals can survive above and below that band. As Figure 4-11 shows, tolerance has its limits, beyond which none of the trout can survive.

These observations are summarized in the **law of tolerance**: *The existence, abundance, and distribution of a species in an ecosystem are determined by whether the levels of one or more physical or chemical factors fall within the range tolerated by that species.* In other words, there are minimum and maximum limits for physical conditions (such as temperature) and concentrations of chemical

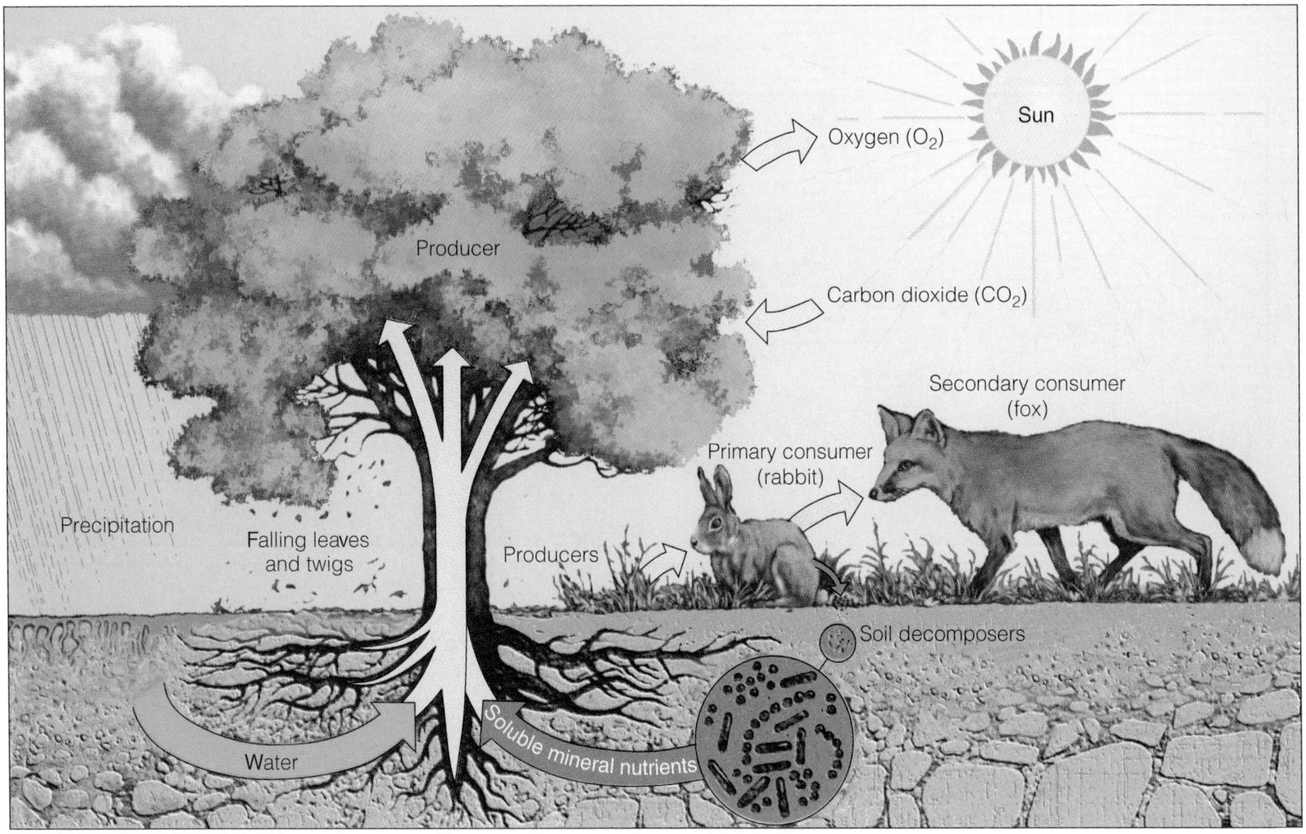

Figure 4-10 Major components of an ecosystem in a field.

substances, called **tolerance limits**, beyond which no members of a particular species can survive.

A species may have a wide range of tolerance to some factors and a narrow range of tolerance to others. Most organisms are least tolerant during juvenile or reproductive stages of their life cycles. Highly tolerant species can live in a variety of habitats with widely different conditions.

A variety of factors affect the number of organisms in a population. However, one factor, known as a **limiting factor**, often turns out to be more important than others in regulating population growth. This ecological principle, related to the law of tolerance, is called the **limiting factor principle**: *Too much or too little of any abiotic factor can limit or prevent growth of a population, even if all other factors are at or near the optimum range of tolerance.*

On land, precipitation often is the limiting factor. Lack of water in a desert limits the growth of plants. Soil nutrients can also act as a limiting factor on land. Suppose a farmer plants corn in phosphorus-poor soil. Even if water, nitrogen, potassium, and other nutrients are at optimum levels, the corn will stop growing when it uses up the available phosphorus.

Too much of an abiotic factor can also be limiting. For example, too much water or too much fertilizer can kill plants, a common mistake of many beginning gardeners.

Important limiting factors for aquatic ecosystems include **(1)** temperature, **(2)** sunlight, **(3)** **dissolved oxygen content** (the amount of oxygen gas dissolved in a given volume of water at a particular temperature and pressure), and **(4)** availability of nutrients. Another limiting factor in aquatic ecosystems is **salinity** (the amounts of various salts dissolved in a given volume of water).

What Are the Major Living Components of Ecosystems? Living organisms in ecosystems are usually classified as either *producers* or *consumers*, based on how they get food. **Producers**, sometimes called **autotrophs** (self-feeders), make their own food from compounds obtained from their environment. On land, most producers are green plants. In freshwater and marine ecosystems, algae and plants are the major producers near shorelines; in open water the dominant producers floating and drifting are *phytoplankton*, most of them microscopic. Only producers make their own food; all other organisms are consumers, which depend directly or indirectly on food provided by producers.

Most producers capture sunlight to make sugars (such as glucose, $C_6H_{12}O_6$) and other complex organic compounds from inorganic (abiotic) nutrients in the environment. This process is called **photosynthesis**. Although hundreds of chemical changes take place

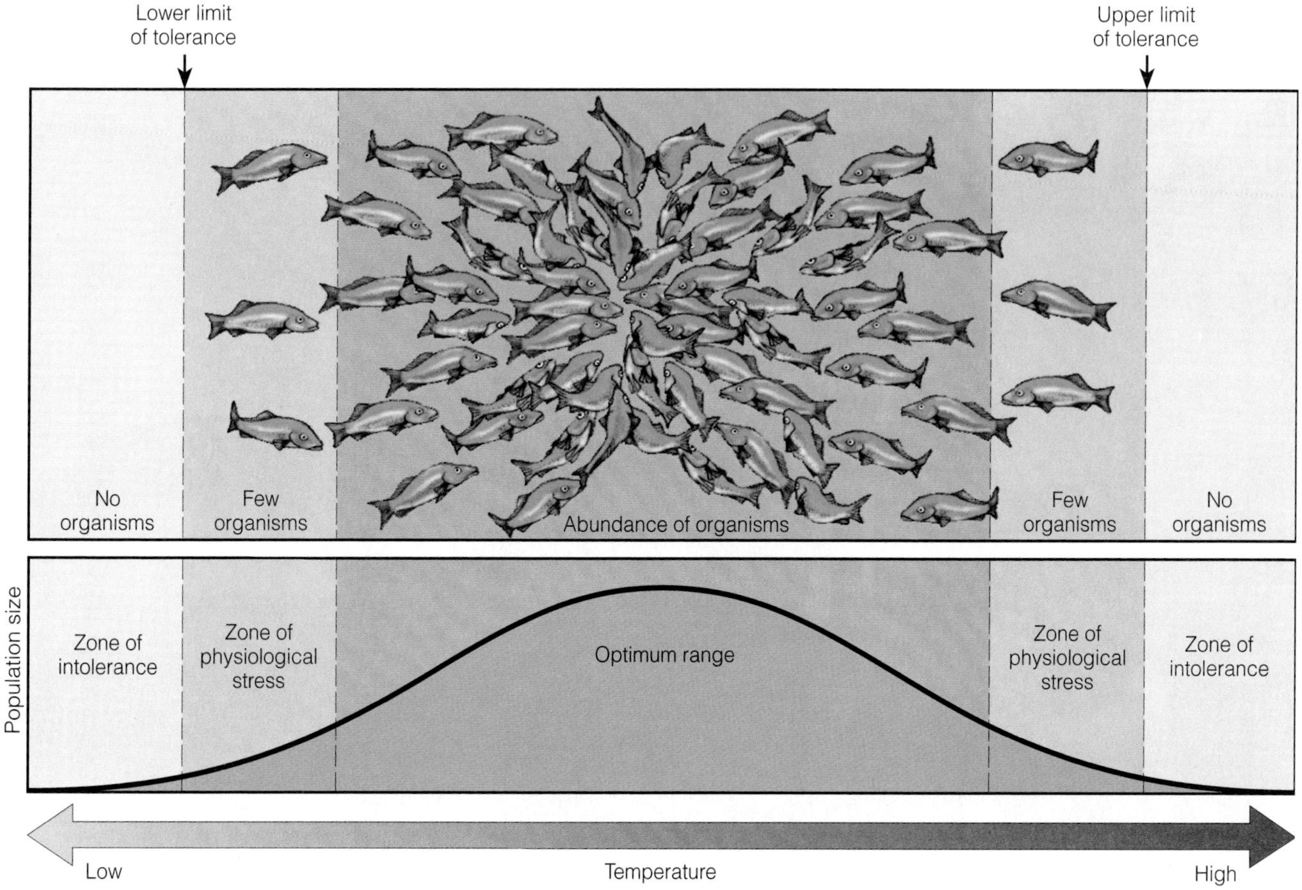

Figure 4-11 Range of tolerance for a population of organisms to an abiotic environmental factor—in this case, temperature.

during photosynthesis, the overall reaction can be summarized as follows:

Carbon dioxide + Water + **Solar energy** → Glucose + Oxygen

$$6\,CO_2 + 6\,H_2O + \textbf{Solar energy} \longrightarrow C_6H_{12}O_6 + 6\,O_2$$

A few producers, mostly specialized bacteria, can convert simple compounds from their environment into more complex nutrient compounds without sunlight, a process called **chemosynthesis**.

All other organisms in an ecosystem are **consumers** or **heterotrophs** ("other-feeders"), which get their energy and nutrients by feeding on other organisms or their remains. There are several classes of consumers, depending on their primary source of food. **Herbivores** (plant eaters) are called **primary consumers** because they feed directly on producers. **Carnivores** (meat eaters) feed on other consumers; those called **secondary consumers** feed only on primary consumers (herbivores). Most secondary consumers are animals, but a few (such as the Venus flytrap plant) trap and digest insects. **Tertiary (higher-level) consumers** feed only on other carnivores. **Omnivores** are consumers that eat both plants and animals; examples are pigs, rats, foxes, bears, cockroaches, and humans.

Other consumers, called **scavengers**, feed on dead organisms that were killed by other organisms or died naturally. Vultures, flies, crows, hyenas, and some species of sharks, beetles, and ants are examples of scavengers. **Detritivores** (detritus feeders and decomposers) live off **detritus** (pronounced "di-TRI-tus"), or parts of dead organisms and cast-off fragments and wastes of living organisms (Figure 4-12). **Detritus feeders**, such as crabs, carpenter ants, termites, earthworms, and wood beetles, extract nutrients from partly decomposed organic matter in leaf litter, plant debris, and animal dung.

Decomposers, mostly certain types of bacteria and fungi, are consumers that complete the breakdown and recycling of organic materials from the remains or wastes of all organisms. They recycle organic matter in ecosystems by breaking down dead organic material (detritus) to get nutrients and releasing the resulting simpler inorganic compounds into

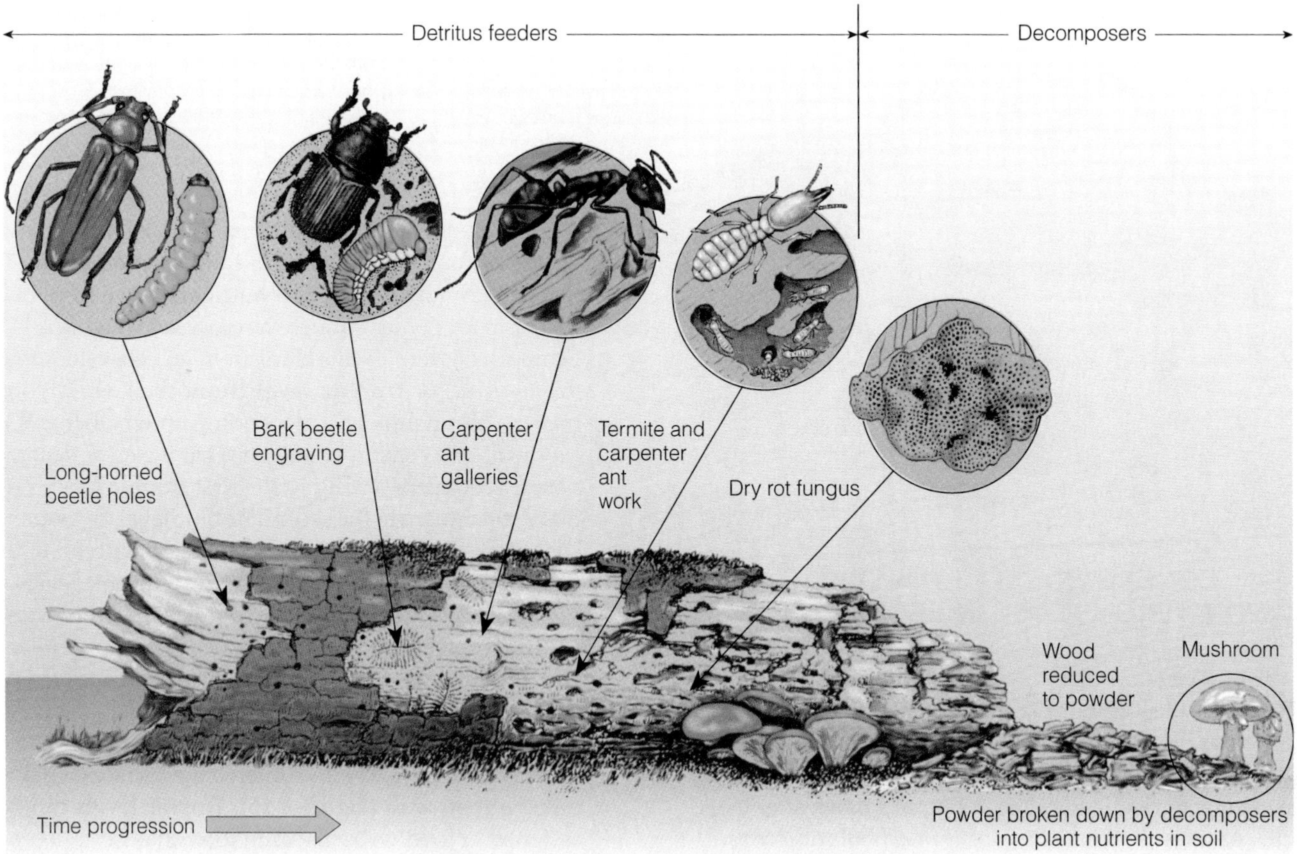

Figure 4-12 Some detritivores, called *detritus feeders*, directly consume tiny fragments of this log. Other detritivores, called *decomposers* (mostly fungi and bacteria), digest complex organic chemicals in fragments of the log into simpler inorganic nutrients. These nutrients can be used again by producers if they are not washed away or otherwise removed from the system.

the soil and water, where they can be taken up as nutrients by producers. In turn, decomposers are important food sources for worms and insects living in soil and water. Something that can be broken down by decomposers is said to be **biodegradable**. Figures 4-9 and 4-10 show various types of producers and consumers.

Both producers and consumers use the chemical energy stored in glucose and other organic compounds to fuel their life processes. In most cells, this energy is released by **aerobic respiration**, which uses oxygen to convert organic nutrients back into carbon dioxide and water. The net effect of the hundreds of steps in this complex process is represented by the following reaction:

Glucose + Oxygen → Carbon dioxide + Water + **Energy**

$$C_6H_{12}O_6 + 6\,O_2 \longrightarrow 6\,CO_2 + 6\,H_2O + \textbf{energy}$$

Although the detailed steps differ, the net chemical change for aerobic respiration is the opposite of that for photosynthesis.

Some decomposers get the energy they need through the breakdown of glucose (or other organic compounds) in the absence of oxygen. This form of cellular respiration is called **anaerobic respiration** or **fermentation**. Instead of carbon dioxide and water, the end products of this process are compounds such as methane gas (CH_4), ethyl alcohol (C_2H_6O), acetic acid ($C_2H_4O_2$, the main component of vinegar), and hydrogen sulfide (H_2S, when sulfur compounds are broken down).

The survival of any individual organism depends on the *flow of matter and energy* through its body. However, an ecosystem as a whole survives primarily through a combination of *matter recycling* (rather than one-way flow) and *one-way energy flow* (Figure 4-13). Decomposers complete the cycle of matter by breaking down detritus into inorganic nutrients that are used by producers. Without decomposers, the entire world would soon be knee-deep in plant litter, dead animal bodies, animal wastes, and garbage. Most life as we know it would no longer exist.

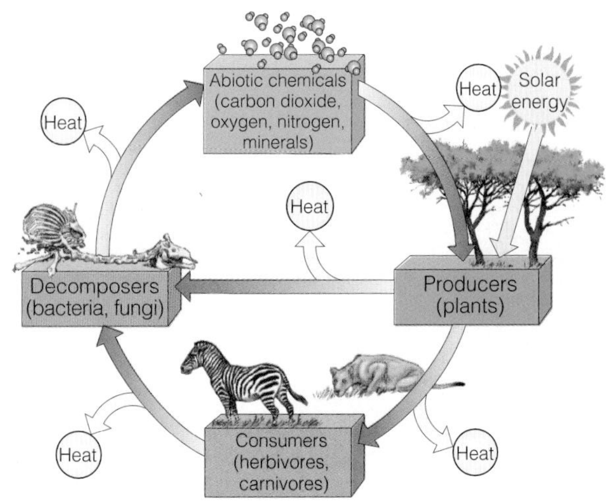

Figure 4-13 Model showing how an ecosystem's main structural components (energy, chemicals, and organisms) are linked by matter recycling and the one-way flow of energy from the sun, through organisms, and back to the environment as low-quality heat. Each type of organism in an ecosystem plays a unique role in the processes of energy flow and matter cycling.

4-4 CONNECTIONS: FOOD WEBS AND ENERGY FLOW IN ECOSYSTEMS

What Are Food Chains and Food Webs? All organisms, whether dead or alive, are potential sources of food for other organisms. A caterpillar eats a leaf, a robin eats the caterpillar, and a hawk eats the robin. Decomposers consume the leaf, caterpillar, robin, and hawk after they die. As a result, *there is little waste in natural ecosystems.*

The sequence of organisms, each of which is a source of food for the next, is called a **food chain**. It determines how energy and nutrients move from one organism to another through the ecosystem (Figure 4-14). Ecologists assign each organism in an ecosystem to a *feeding level*, or **trophic level** (from the Greek word *trophos*, "nourishment"), depending on whether it is a producer or a consumer and on what it eats or decomposes. Producers belong to the first trophic level, primary consumers to the second trophic level, secondary consumers to the third, and so on. Detritivores, or decomposers, process detritus from all trophic levels.

Real ecosystems are more complex than this. Most consumers feed on more than one type of organism, and most organisms are eaten by more than one type of consumer. Because most species participate in several different food chains, the organisms in most ecosystems form a complex network of interconnected food chains called a **food web** (Figure 4-15). Trophic levels can be assigned in food webs just as in food chains.

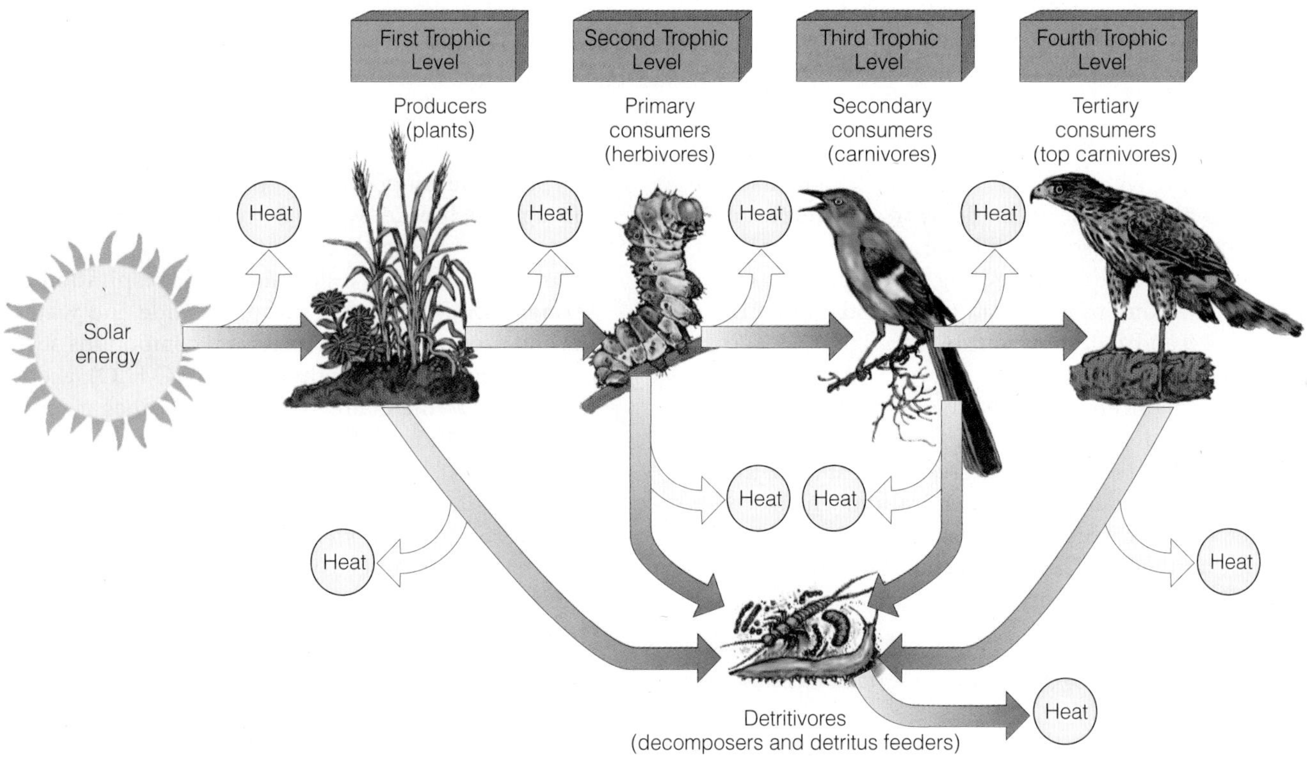

Figure 4-14 Model of a food chain. The arrows show how chemical energy in food flows through various *trophic levels* or energy transfers; most of the energy is degraded to heat, in accordance with the second law of energy. Food chains rarely have more than four trophic levels. Can you explain why?

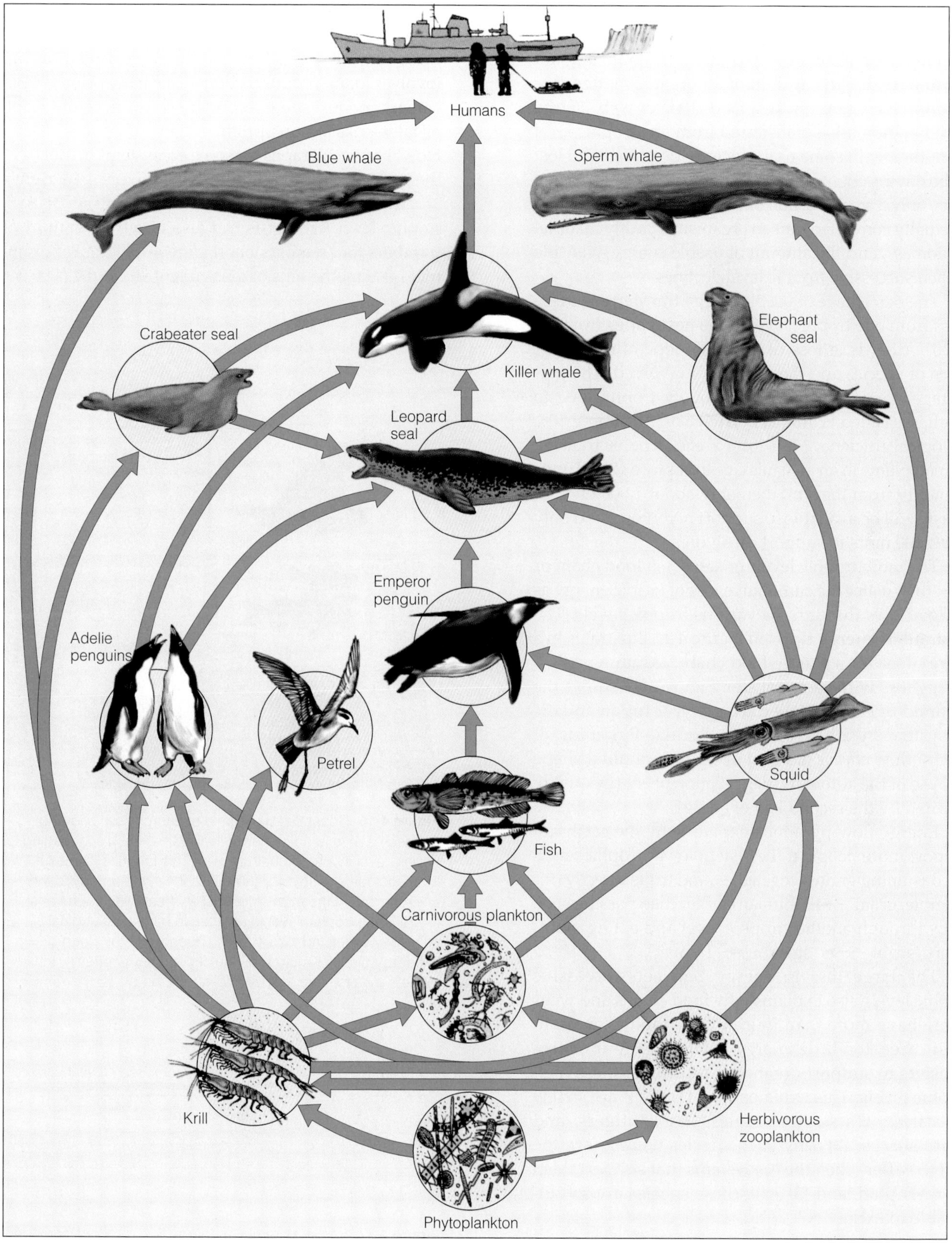

Figure 4-15 Model of a greatly simplified food web in the Antarctic. Many more participants in the web, including an array of decomposer organisms, are not depicted here.

How Can We Represent the Energy Flow and Storage of Biomass in an Ecosystem? Each trophic level in a food chain or web contains a certain amount of **biomass**, the dry weight of all organic matter contained in its organisms. In a food chain or web, energy stored in biomass is transferred from one trophic level to another, with some usable energy degraded and lost to the environment as low-quality heat in each transfer. Thus, only a small portion of what is eaten and digested is actually converted into an organism's bodily material or biomass, and the amount of usable energy available to each successive trophic level declines.

The percentage of usable energy transferred as biomass from one trophic level to the next varies from 5% to 20% (that is, a loss of 80–95%) depending on the types of species and the ecosystem involved. The percentage of energy transferred from one trophic level to another is called **ecological efficiency**. Assuming 10% ecological efficiency (90% loss) at each trophic transfer, if green plants in an area manage to capture 10,000 units of energy from the sun, then only about 1,000 units of energy will be available to support herbivores and only about 100 units to support carnivores.

The more trophic levels or steps in a food chain or web, the greater the cumulative loss of usable energy as energy flows through the various trophic levels. The **pyramid of energy flow*** in Figure 4-16 illustrates this energy loss for a simple food chain, assuming a 90% energy loss with each transfer. Figure 4-17 shows the pyramid of energy flow during 1 year for an aquatic ecosystem in Silver Springs, Florida. Pyramids of energy flow *always* have an upright pyramidal shape because of the automatic degradation of energy quality required by the second law of energy.

Energy flow pyramids explain why the earth can support more people if they eat at lower trophic levels by consuming grains, vegetables, and fruits directly (for example, grain ⟶ human) rather than passing such crops through another trophic level and eating grain-eaters (grain ⟶ steer ⟶ human).

The large loss in energy between successive trophic levels also explains why food chains and webs rarely have more than four or five trophic levels. In most cases, too little energy is left after four or five transfers to support organisms feeding at these high trophic levels. This explains why **(1)** there are so few top carnivores such as eagles, hawks, tigers, and white sharks, **(2)** why such species usually are the first to suffer when the ecosystems that support them are disrupted, and **(3)** why these species are so vulnerable to extinction.

*Because such pyramids represent energy flows, not energy storage, they should not be called pyramids of energy (a common error in many biology and environmental science textbooks).

The storage of biomass at various trophic levels in an ecosystem can be represented by a **pyramid of biomass** (Figure 4-18). Because the typical relationship involves many producers but not so many primary consumers and just a few secondary consumers, the graph usually looks like an upright pyramid (Figure 4-18, left). However, such graphs for some ecosystems do not have the typical upright pyramid shape (Figure 4-18, right).

By estimating the number of organisms at each trophic level, ecologists can also create a **pyramid of numbers** for an ecosystem (Figure 4-19, left). For example, 1,000 metric tons of grass might support 27,000,000

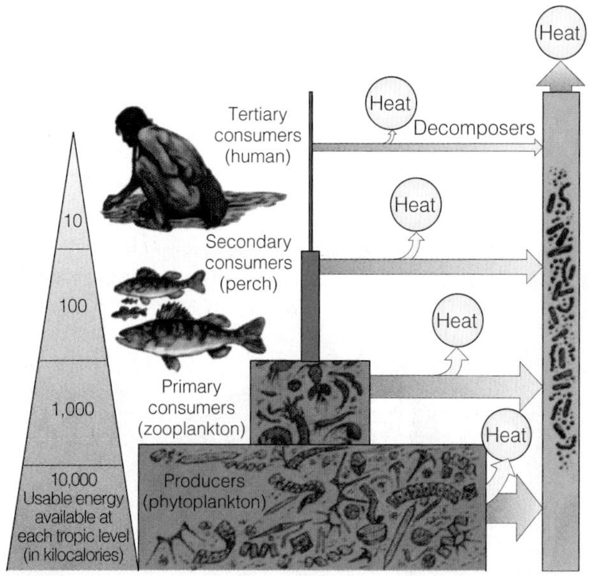

Figure 4-16 Generalized pyramid of energy flow, showing the decrease in usable energy available at each succeeding trophic level in a food chain or web. This model assumes a 10% ecological efficiency (90% loss in usable energy to the environment, in the form of low-quality heat) with each transfer from one trophic level to another. In nature, ecological efficiency varies from 5% to 20%. Because of the degradation of energy quality required by the second law of energy, these models always have a pyramidal shape.

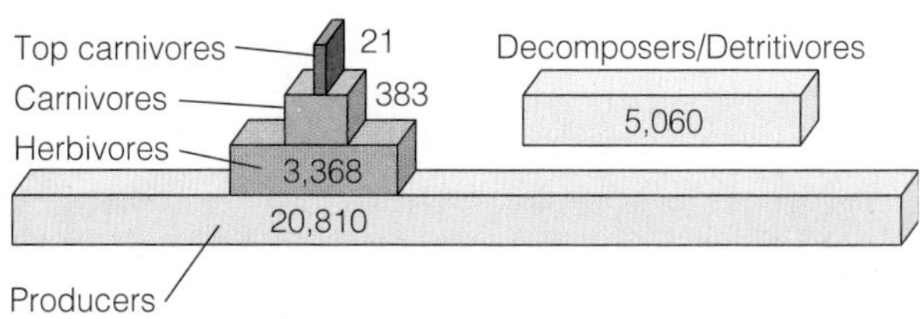

Figure 4-17 Annual pyramid of energy flow (in kilocalories per square meter per year) for an aquatic ecosystem in Silver Springs, Florida. (Used by permission from Cecie Starr, *Biology: Concepts and Applications*, 4th ed., Pacific Grove, Calif.: Brooks/Cole, 2000)

grasshoppers, which might support 90,000 frogs, which might support 300 trout, which in turn could feed 1 person for about 30 days. By eliminating trout from their diet, 30 people could survive for 30 days by consuming 10 frogs a day. If the food situation were more desperate, frogs could be eliminated from the diet and 900 people could survive for 30 days by eating about 1,000 grasshoppers a day. (Fried grasshoppers are a delicacy in some countries and contain more protein per unit of weight than most conventional forms of meat and meat products.) For some ecosystems these graphs do not have the typical upright pyramid shape (Figure 4-19, right).

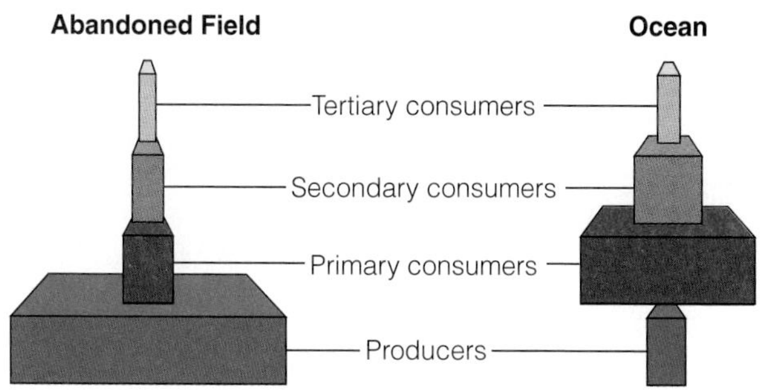

Figure 4-18 Generalized graphs of biomass of organisms in the various trophic levels for two ecosystems. The size of each tier in this conceptual model represents the dry weight per square meter of all organisms at that trophic level. For most land ecosystems, the total biomass at each successive trophic level decreases, yielding a pyramid of biomass with a large base of producers, topped by a series of increasingly smaller biomasses at higher trophic levels (left). In the open waters of aquatic ecosystems (right), the biomass of primary consumers (zooplankton) can actually exceed that of producers (which are microscopic phytoplankton that grow and reproduce rapidly, not large plants that grow and reproduce slowly). The zooplankton eat the phytoplankton almost as fast as they are produced. As a result, the producer population is never very large, and the graph is not an upright pyramid.

How Rapidly Do Producers in Different Ecosystems Produce Biomass? The *rate* at which an ecosystem's producers convert solar energy into chemical energy as biomass is the ecosystem's **gross primary productivity (GPP)**. In effect, it is the rate at which plants or other producers use photosynthesis to make more plant material (biomass). Figure 4-20 shows how this productivity varies across the earth.

To stay alive, grow, and reproduce, an ecosystem's producers must use some of the total biomass they produce for their own respiration. Only what is left, called **net primary productivity (NPP)**, is available for use as food by other organisms (consumers) in an ecosystem:

Net primary productivity	=	Rate at which producers store chemical energy as biomass (produced by photosynthesis)	−	Rate at which producers use chemical energy stored as biomass (through aerobic respiration)

Net primary productivity is the *rate* at which energy for use by consumers is stored in new biomass (cells, leaves, roots, and stems). It is measured in units of the energy or biomass available to consumers in a specified area over a given time. It is typically measured in kilocalories per square meter per year (kcal/m^2/yr) or in grams of biomass created per square meter per year (g/m^2/yr). Ultimately, the planet's total net primary productivity limits the number of consumers (including humans) that can survive on the earth. In other words, *the earth's total net primary productivity is the upper limit determining the planet's carrying capacity for all species.*

Various ecosystems and life zones differ in their net primary productivity (Figure 4-21). The most productive are estuaries, swamps and marshes, and tropical rain forests, and the least productive are open ocean, tundra (arctic and alpine grasslands), and desert.

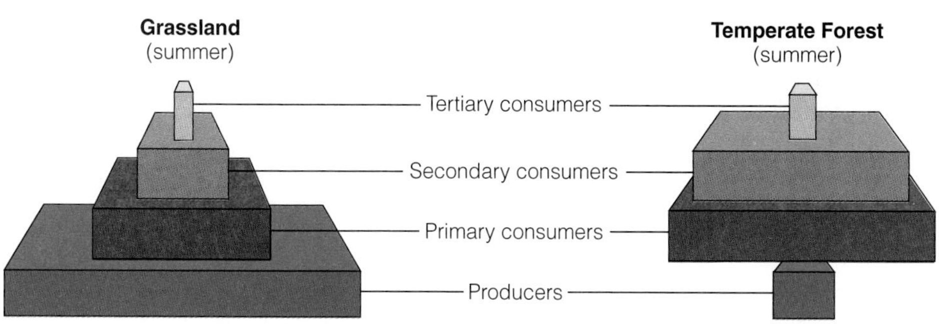

Figure 4-19 Generalized graphs of numbers of organisms in the various trophic levels for two ecosystems. Numbers of organisms for grasslands and many other ecosystems taper off from the producer level to the higher trophic levels, forming an upright pyramid (left). For other ecosystems, however, the graph can take a different shape. For example, a temperate forest (right) has a few large producers (the trees) that support a much larger number of small primary consumers (insects) that feed on the trees.

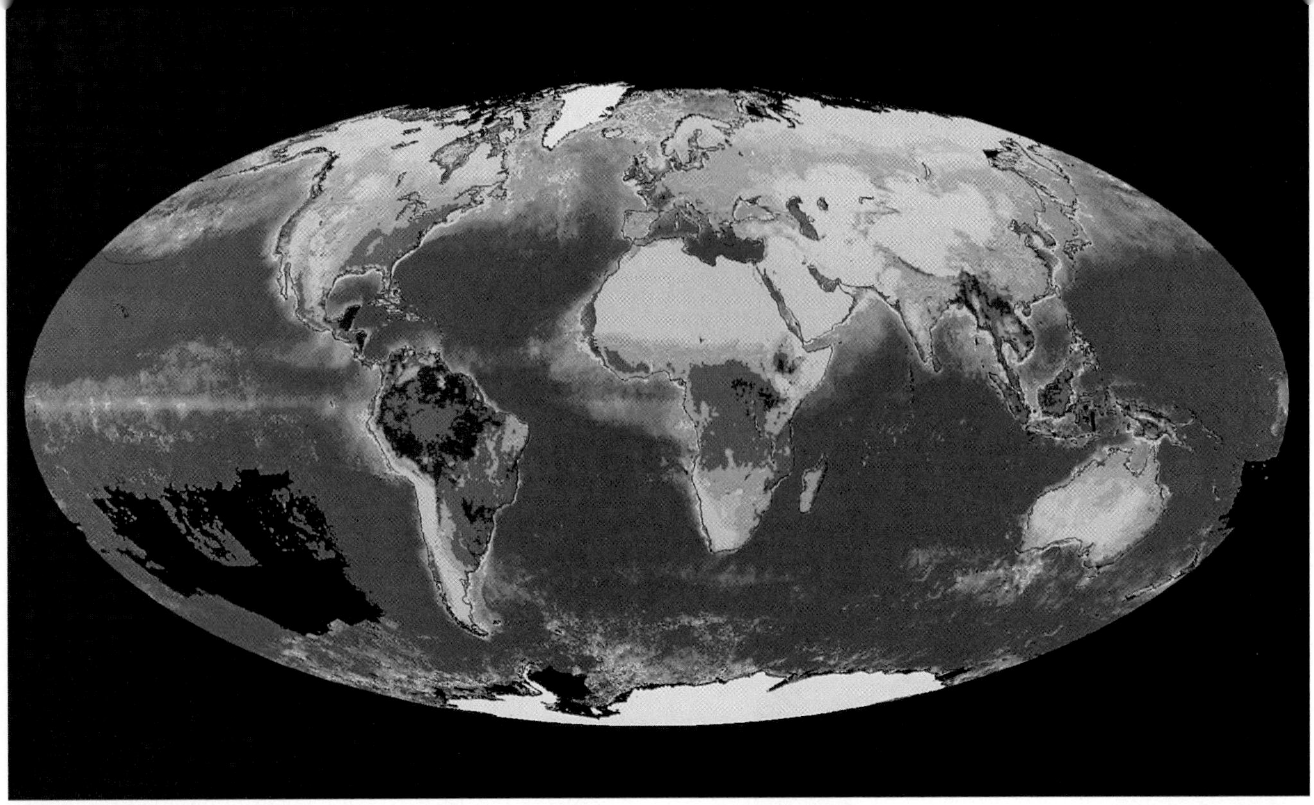

Figure 4-20 Three years of satellite data on the earth's gross primary productivity. Rain forests and other highly productive areas appear as dark green, deserts as yellow. The concentration of phytoplankton, a primary indicator of ocean productivity, ranges from red (highest) to orange, yellow, green, and blue (lowest). (Gene Carl Feldman, Compton J. Tucker–NASA/Goddard Space Flight Center)

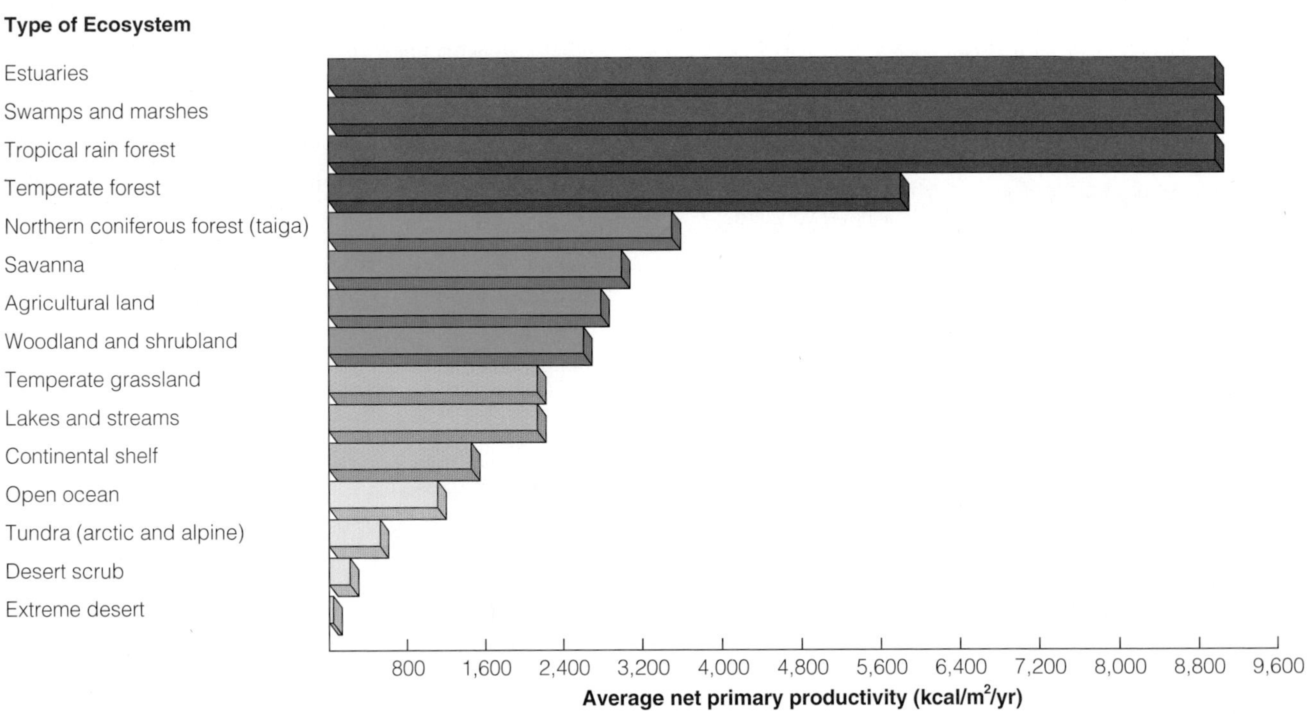

Figure 4-21 Estimated annual average net primary productivity per unit of area in major life zones and ecosystems, expressed as kilocalories of energy produced per square meter per year (kcal/m^2/yr). (Data from R. H. Whittaker, *Communities and Ecosystems*, 2d ed., New York: Macmillan, 1975)

It is tempting to conclude that to feed the world's hungry millions we should harvest plants in estuaries, swamps, and marshes, or clear tropical forests and plant crops. Most plants in estuaries, swamps, and marshes cannot be eaten by people, however, and they are vital food sources (and spawning areas) for fish, shrimp, and other aquatic life-forms that provide us and other consumers with protein.

In tropical forests, most nutrients are stored in the vegetation rather than in the soil. When the trees are removed, the nutrient-poor soils are rapidly depleted of their nutrients by frequent rains and growing crops. Crops can be grown only for a short time without massive and expensive applications of commercial fertilizers. This explains why many ecologists urge us not to clear large areas of tropical forest to supply food.

Agricultural land is a highly modified ecosystem in which we try to increase the net primary productivity and biomass of selected crop plants by adding water (irrigation) and nutrients (fertilizers). Nitrogen as nitrate (NO_3^-) and phosphorus as phosphate (PO_4^{3-}) are the most common nutrients in fertilizers because they are most often the nutrients limiting crop growth. Despite such inputs, the net primary productivity of agricultural land is not particularly high compared with that of other ecosystems (Figure 4-21).

Excluding uninhabitable areas of rock, ice, desert, and steep mountain terrain, humans have taken over, disturbed, or degraded about 73% of the earth's land surface (Figure 1-5). Ecologists have estimated that humans now use, waste, or destroy about 27% of the earth's total potential net primary productivity and 40% of the net primary productivity of the planet's terrestrial ecosystems.

This is the main reason why we are crowding out or eliminating the habitats and food supplies of a growing number of other species. If current estimates of our use of the earth's annual net primary productivity are reasonably correct, what will happen to us and to other species if the human population doubles over the next 40–50 years and per capita consumption of resources rises sharply? After one more doubling of the population, humans will be using 80% of all the products of sunlight, and shortly after that 100%, with virtually nothing left to support other forms of life.

4-5 CONNECTIONS: MATTER CYCLING IN ECOSYSTEMS

How Do Nutrient Cycles Sustain Life? A **nutrient** is any atom, ion, or molecule an organism needs to live, grow, or reproduce. Some elements (such as carbon, oxygen, hydrogen, nitrogen, phosphorus, sulfur, and calcium) are needed in fairly large amounts, whereas others (such as sodium, zinc, copper, and iodine) are needed in small or even trace amounts.

These nutrient atoms, ions, and molecules are continuously cycled from the nonliving environment (air, water, soil, and rock) to living organisms (biota) and then back again in what are called **nutrient cycles**, or **biogeochemical cycles** (literally, life–earth–chemical cycles). These cycles, driven directly or indirectly by incoming solar energy and gravity, include the carbon, oxygen, nitrogen, phosphorus, and hydrologic (water) cycles (Figure 4-6).

The earth's chemical cycles also connect past, present, and future forms of life. Some of the carbon atoms in your skin may once have been part of a leaf, a dinosaur's skin, or a layer of limestone rock. Some of the oxygen molecules you just inhaled may have been inhaled by your grandmother, Plato, or a hunter–gatherer who lived 25,000 years ago.

How Is Water Cycled in the Ecosphere? The **hydrologic cycle**, or **water cycle**, which collects, purifies, and distributes the earth's fixed supply of water, is shown in simplified form in Figure 4-22. The main processes in this water recycling and purifying cycle are **(1)** *evaporation* (conversion of water into water vapor), **(2)** *transpiration* (evaporation from leaves of water extracted from soil by roots and transported throughout the plant), **(3)** *condensation* (conversion of water vapor into droplets of liquid water), **(4)** *precipitation* (rain, sleet, hail, and snow), **(5)** *infiltration* (movement of water into soil), **(6)** *percolation* (downward flow of water through soil and permeable rock formations to groundwater storage areas called aquifers), and **(7)** *runoff* (downslope surface movement back to the sea to resume the cycle).

The water cycle is powered by energy from the sun and by gravity. Incoming solar energy evaporates water from oceans, streams, lakes, soil, and vegetation. About 84% of water vapor in the atmosphere comes from the oceans; the rest comes from land.

The amount of water vapor air can hold depends on its temperature; warm air can hold more water vapor than cold air. **Absolute humidity** is the amount of water vapor found in a certain mass of air; it is usually expressed as grams of water per kilogram of air. **Relative humidity** is the amount of water vapor in a certain mass of air, expressed as a percentage of the maximum amount it could hold at that temperature. For example, a relative humidity of 60% at 27°C (80°F) means that each kilogram (or other unit of mass) of air contains 60% of the maximum amount of water vapor it could hold at that temperature.

Winds and air masses transport water vapor over various parts of the earth's surface, often over long distances. Falling temperatures cause the water vapor to

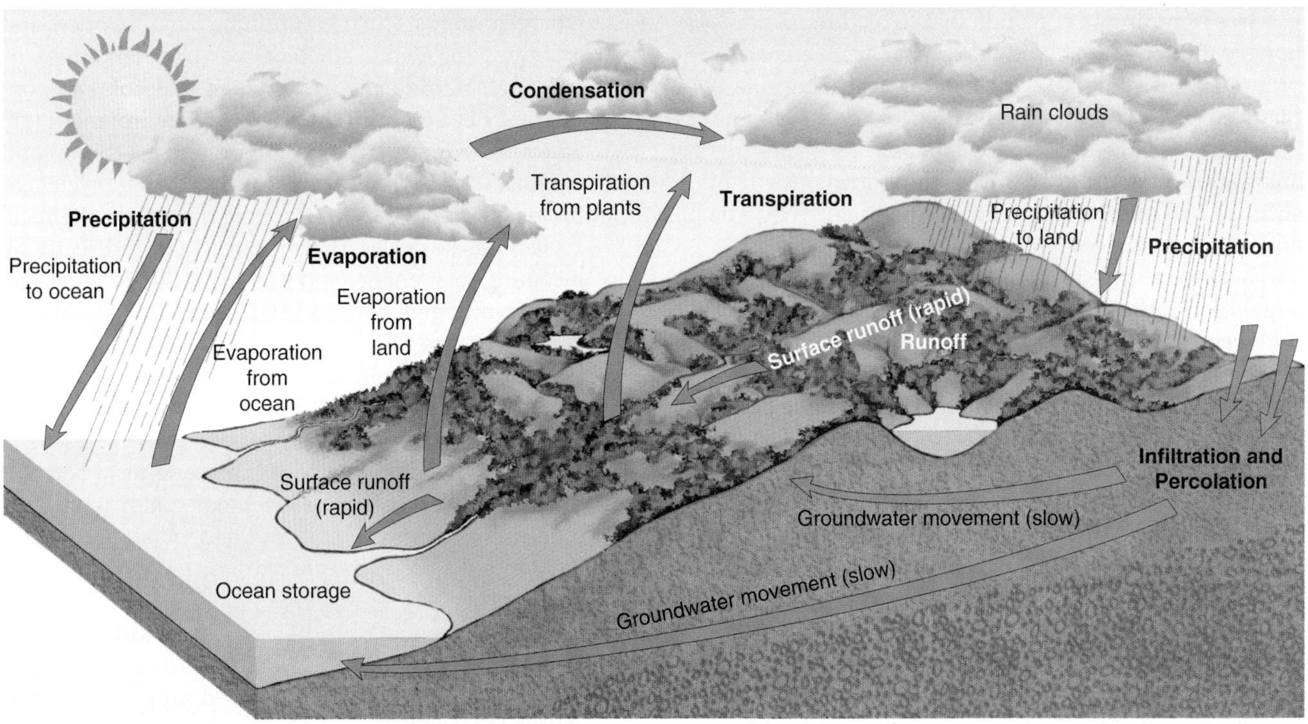

Figure 4-22 Simplified conceptual model of the hydrologic cycle.

condense into tiny droplets that form clouds or fog. For precipitation to occur, air must contain **condensation nuclei**: tiny particles on which droplets of water vapor can collect. Volcanic ash, soil dust, smoke, sea salts, and particulate matter emitted by factories, coal-burning power plants, and vehicles are sources of such particles. The temperature at which condensation occurs is called the **dew point**.

Some of the fresh water returning to the earth's surface as precipitation becomes locked in glaciers. Most of the precipitation falling on terrestrial ecosystems becomes *surface runoff* flowing into streams and lakes, which eventually carry water back to the oceans, where it can be evaporated to cycle again.

Besides replenishing streams and lakes, surface runoff also causes soil erosion, which moves soil and weathered rock fragments from one place to another. Water is thus the primary sculptor of the earth's landscape. Because water dissolves many nutrient compounds, it is a major medium for transporting nutrients within and among ecosystems.

Some of the water returning to the land soaks into (infiltrates) the soil and porous rock and then percolates downward, dissolving minerals from porous rocks on the way. This water is stored as groundwater in the pores and cracks of rocks. This underground water flows slowly downhill through rock pores and seeps out into streams and lakes or comes out in springs. Eventually, this water evaporates or reaches the sea to continue the cycle.

Humans intervene in the water cycle in several ways:

- Withdrawing large quantities of fresh water from streams, lakes, and underground sources. In heavily populated or heavily irrigated areas, withdrawals have led to groundwater depletion or intrusion of ocean salt water into underground water supplies.
- Clearing vegetation from land for agriculture, mining, road and building construction, and other activities. This increases runoff and reduces infiltration that recharges groundwater supplies; it also increases the risk of flooding and accelerates soil erosion and landslides.
- Modifying water quality by adding nutrients (such as phosphates) and other pollutants and by changing ecological processes that purify water naturally. Chapter 12 examines issues of water pollution and water quality.

How Is Carbon Cycled in the Ecosphere? The **carbon cycle** is based on carbon dioxide gas, which makes up 0.036% of the volume of the troposphere and is also dissolved in water. Carbon dioxide is a key component of nature's thermostat. If the carbon cycle removes too much CO_2 from the atmosphere, the earth will cool; if the cycle generates too much, the earth will get warmer. Thus, even slight changes in the carbon cycle can affect climate and ultimately the types of life that can exist on various parts of the planet.

Terrestrial producers remove CO_2 from the atmosphere and aquatic producers remove it from the water. They then use photosynthesis to convert CO_2 into complex carbohydrates such as glucose ($C_6H_{12}O_6$).

The cells in oxygen-consuming producers, consumers, and decomposers then carry out aerobic respiration, which breaks down glucose and other complex organic compounds and converts the carbon back to CO_2 in the atmosphere or water, for reuse by producers. This linkage between photosynthesis in producers and aerobic respiration in producers, consumers, and decomposers circulates carbon in the ecosphere and is a major part of the global carbon cycle (Figure 4-23). Oxygen and hydrogen, the other elements in carbohydrates, cycle almost in step with carbon.

Over millions of years, buried deposits of dead plant matter and bacteria are compressed between layers of sediment, where they form carbon-containing fossil fuels such as coal and oil (Figure 4-23). This carbon is not released to the atmosphere as CO_2 for recycling until these fuels are extracted and burned or until long-term geological processes expose these deposits to air. In only a few hundred years, we have extracted and burned fossil fuels that took millions of year to form. This explains why fossil fuels are nonrenewable resources on a human time scale.

The largest storage reservoir for the earth's carbon is sedimentary rocks such as limestone ($CaCO_3$) on the ocean floor and on continents. This carbon reenters the cycle very slowly, when some of the sediments dissolve and form dissolved CO_2 gas that can enter the atmosphere. Geologic processes can also bring bottom sediments to the surface, exposing the carbonate rock to chemical attack by oxygen and converting them to carbon dioxide gas. Carbon dioxide is also released into the atmosphere when acidic rain falls on and dissolves exposed limestone rock.

The oceans are the second largest storage reservoir in the carbon cycle. Oceans also play a major role in regulating the level of carbon dioxide in the atmosphere. Some carbon dioxide gas, which is readily soluble in water, stays dissolved in the sea, some is removed by photosynthesizing producers, and some reacts with seawater to form carbonate ions (CO_3^{2-}) and bicarbonate ions (HCO_3^-). As water warms, more dissolved CO_2 returns to the atmosphere, just as more carbon dioxide fizzes out of a carbonated beverage when it warms.

In marine ecosystems, some organisms take up dissolved CO_2 molecules, carbonate ions, or bicarbonate ions from ocean water. These ions can then react with calcium ions (Ca^{2+}) in seawater to form slightly soluble carbonate compounds such as calcium carbonate ($CaCO_3$) to build the shells and skeletons of marine organisms. When these organisms die, tiny particles of their shells and bone drift slowly to the ocean depths and are buried for eons (as long as 400 million years) in deep bottom sediments (Figure 4-23), where under immense pressure they are converted into limestone rock.

Since 1800 and especially since 1950, as world population and resource use have soared, we have disturbed the carbon cycle in two ways that add carbon dioxide to the atmosphere by

- Clearing trees and other plants that absorb CO_2 through photosynthesis
- Adding large amounts of CO_2 by burning fossil fuels and wood

Computer models of the earth's climate systems suggest that increased concentration of CO_2 and of other gases we're adding to the atmosphere could enhance the planet's natural greenhouse effect. Measurements show that we are adding more CO_2 to the atmosphere than oceans and plants have been able to remove. The resulting global warming could disrupt global food production and wildlife habitats and raise the average sea level, as discussed in more detail in Chapter 11.

How Is Nitrogen Cycled in the Ecosphere? Bacteria in Action Although chemically unreactive nitrogen gas (N_2) makes up 78% of the volume of the troposphere, it cannot be absorbed and used directly as a nutrient by multicellular plants or animals. Fortunately, lightning and certain bacteria convert nitrogen gas into compounds that can enter food webs as part of the **nitrogen cycle** (Figure 4-24, p. 104), a global gaseous cycle that is the most complex of the earth's chemical cycles.

In the first step in the nitrogen cycle, called *nitrogen fixation*, specialized bacteria convert gaseous nitrogen (N_2) to ammonia (NH_3) that can be used by plants. This is done mostly by cyanobacteria in soil and water and by *Rhizobium* bacteria living in small nodules (swellings) on the root systems of a wide variety of plant species, including soybeans, alfalfa, and clover. In a two-step process called *nitrification*, most of the ammonia in soil is converted by specialized aerobic bacteria to nitrite ions (NO_2^-), which are toxic to plants, and then to nitrate ions (NO_3^-), which are easily taken up by plants as a nutrient.

In a process called *assimilation*, plant roots then absorb inorganic ammonia, ammonium ions, and nitrate ions formed by nitrogen fixation and nitrification in soil

Figure 4-23 Simplified model of the global carbon cycle. The left portion shows the movement of carbon through marine ecosystems, and the right portion shows its movement through terrestrial ecosystems. Carbon reservoirs are shown as boxes; processes that change one form of carbon to another are shown in unboxed print. (Used by permission from Cecie Starr, *Biology: Concepts and Applications*, 4th ed., Pacific Grove, Calif.: Brooks/Cole, 2000)

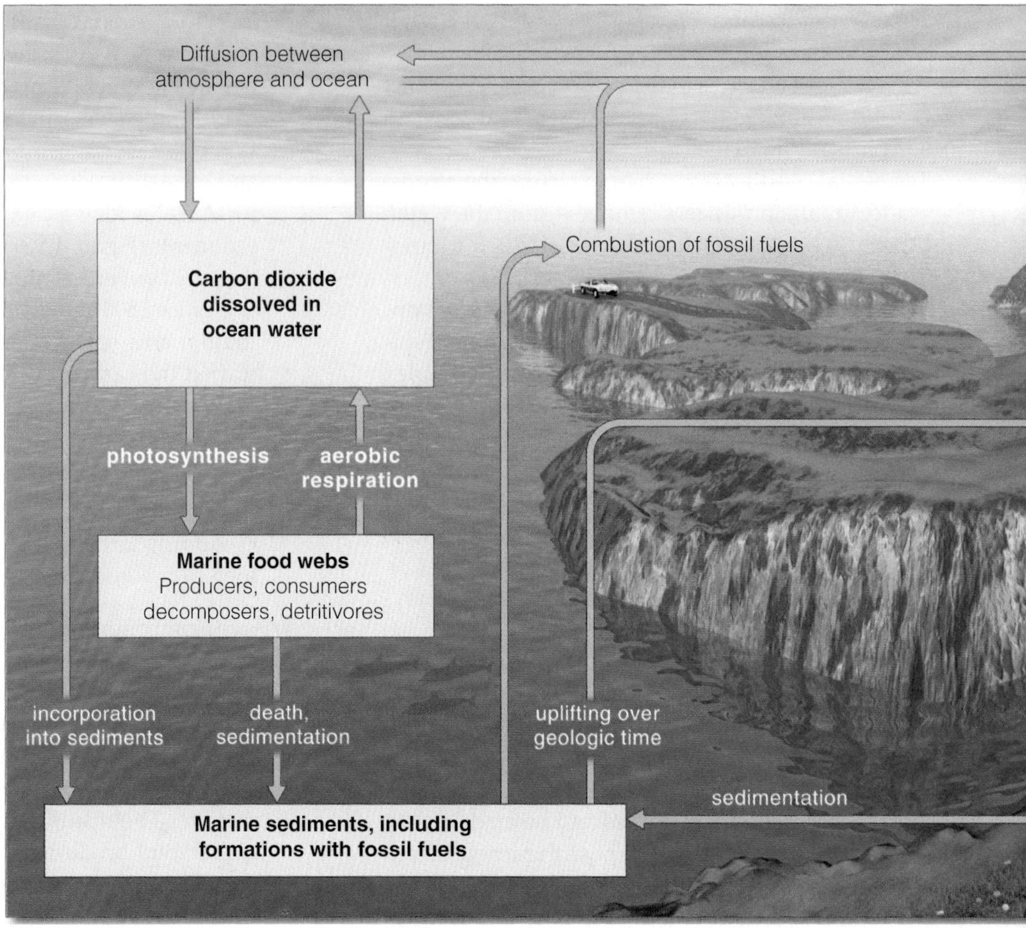

water. They use these ions to make nitrogen-containing organic molecules such as DNA, amino acids, and proteins. Animals in turn get their nitrogen by eating plants or plant-eating animals.

After nitrogen has served its purpose in living organisms, vast armies of specialized decomposer bacteria convert the nitrogen-rich organic compounds, wastes, cast-off particles, and dead bodies of organisms into simpler nitrogen-containing inorganic compounds such as ammonia (NH_3) and water-soluble salts containing ammonium ions (NH_4^+). This process is known as *ammonification*.

In a process called *denitrification*, other specialized bacteria (mostly anaerobic bacteria in waterlogged soil or in the bottom sediments of lakes, oceans, swamps, and bogs) then convert NH_3 and NH_4^+ back into nitrite (NO_2^-) and nitrate (NO_3^-) ions and then into nitrogen gas (N_2) and nitrous oxide gas (N_2O). These are then released to the atmosphere to begin the cycle again.

Humans intervene in the nitrogen cycle by

- Adding large amounts of nitric oxide (NO) into the atmosphere when we burn any fuel ($N_2 + O_2 \longrightarrow 2NO$). In the atmosphere, this nitric oxide combines with oxygen to form nitrogen dioxide gas (NO_2), which can react with water vapor to form nitric acid (HNO_3). Droplets of HNO_3 dissolved in rain or snow are components of *acid deposition*, commonly called *acid rain*. Nitric acid, along with other air pollutants, can **(1)** damage and weaken trees, **(2)** upset aquatic ecosystems, **(3)** corrode metals, and **(4)** damage marble, stone, and other building materials.
- Adding nitrous oxide (N_2O), a greenhouse gas, to the atmosphere through the action of anaerobic bacteria on livestock wastes and commercial inorganic

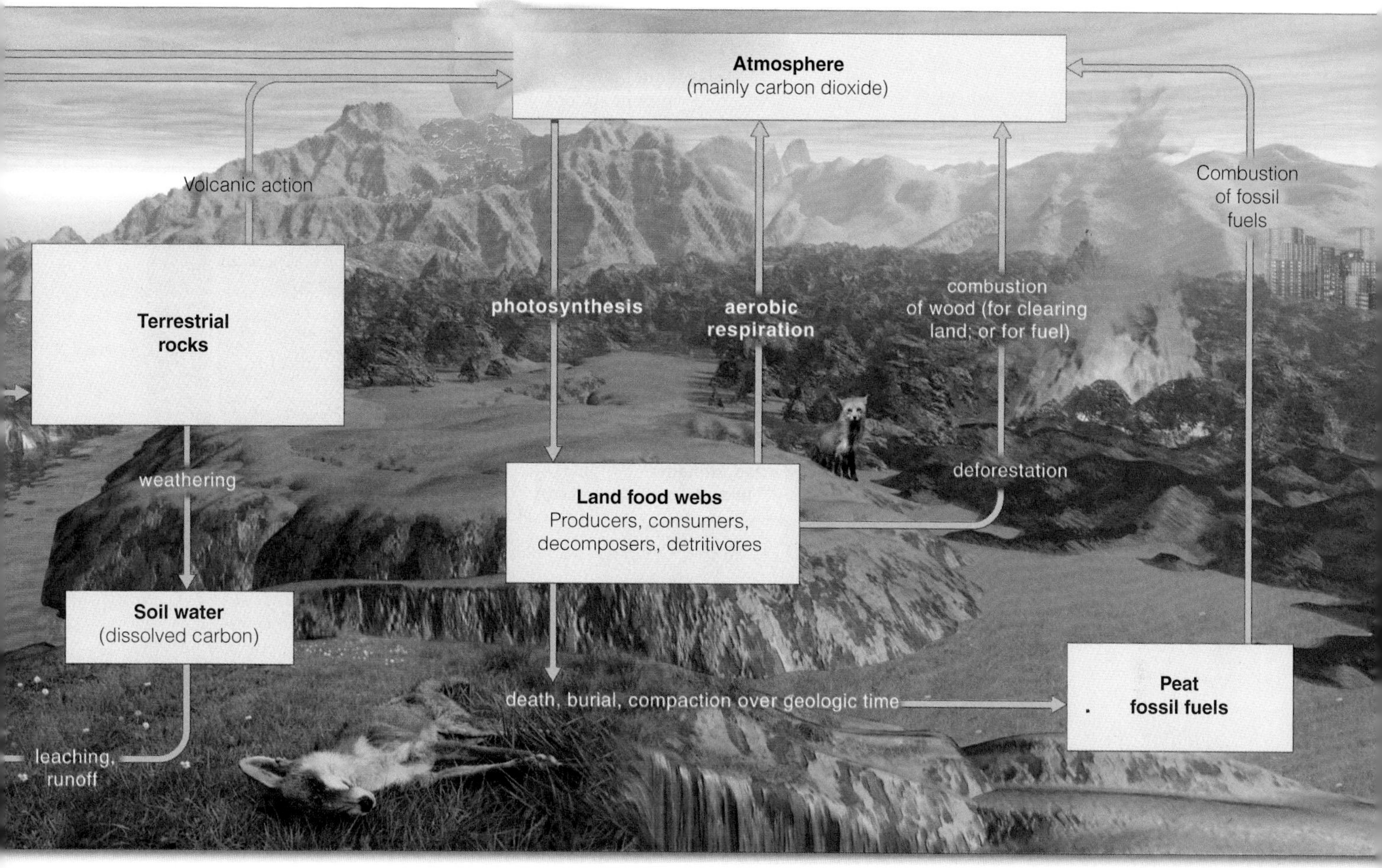

fertilizers applied to the soil. When N_2O reaches the stratosphere, it also contributes to depletion of the earth's ozone shield that filters out harmful ultraviolet radiation from the sun.

- Removing nitrogen from the earth's crust when we mine nitrogen-containing mineral deposits. This depletes nitrogen from topsoil by harvesting nitrogen-rich crops, and leaches water-soluble nitrate ions from soil through irrigation.

- Removing nitrogen from topsoil when we burn grasslands and clear forests before planting crops.

- Adding nitrogen compounds to aquatic ecosystems in agricultural runoff and discharge of municipal sewage. This excess of plant nutrients stimulates rapid growth of photosynthesizing algae and other aquatic plants. The subsequent breakdown of dead algae by aerobic decomposers can deplete the water of dissolved oxygen and can disrupt aquatic ecosystems by killing some types of fish and other oxygen-requiring (aerobic) organisms.

- Accelerating the deposition of acidic nitrogen compounds (such as NO_2 and HNO_3) from the atmosphere onto terrestrial ecosystems. This excessive input of nitrogen can stimulate the growth of weedy plant species, which can outgrow and perhaps eliminate other plant species that cannot use nitrogen as well.

How Is Phosphorus Cycled in the Ecosphere? Phosphorus circulates through water, the earth's crust, and living organisms in the **phosphorus cycle** (Figure 4-25). In this cycle, phosphorus moves slowly from phosphate deposits on land and in shallow ocean sediments to living organisms, and then much more slowly back to the land and ocean.

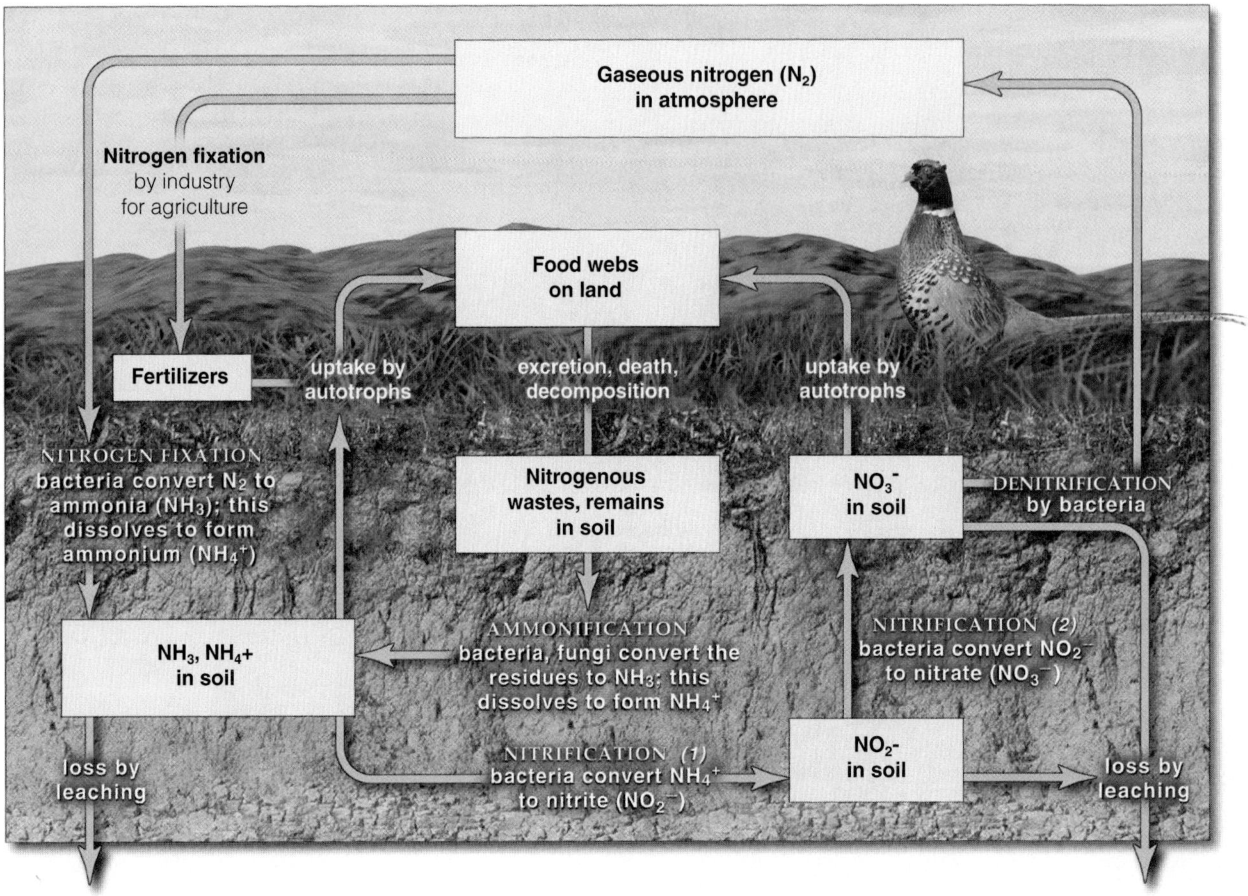

Figure 4-24 Greatly simplified model of the nitrogen cycle in a terrestrial ecosystem. Nitrogen reservoirs are shown as boxes and processes changing one form of nitrogen to another are shown in unboxed print. (Used by permission from Cecie Starr, *Biology: Concepts and Applications*, 4th ed., Pacific Grove, Calif.: Brooks/Cole, 2000)

Bacteria are less important here than in the nitrogen cycle. Unlike carbon and nitrogen, very little phosphorus circulates in the atmosphere because at the earth's normal temperatures and pressures, phosphorus and its compounds are not gases. Phosphorus is found in the atmosphere only as small particles of dust. In contrast to the carbon cycle, the phosphorus cycle is slow, and on a short human time scale much phosphorus flows one way from the land to the oceans.

Phosphorous typically is found as phosphate salts containing phosphate ions (PO_4^{3-}) in terrestrial rock formations and ocean bottom sediments. Because most soils contain little phosphate, it is often the limiting factor for plant growth on land unless phosphorus (as phosphate salts mined from the earth) is applied to the soil as a fertilizer. Phosphorus also limits the growth of producer populations in many freshwater streams and lakes because phosphate salts are only slightly soluble in water.

Humans intervene in the phosphorus cycle by

- Mining large quantities of phosphate rock for use in commercial inorganic fertilizers and detergents.
- Reducing the available phosphate in tropical forests by removing trees. When such forests are cut and burned, most remaining phosphorus and other soil nutrients are washed away by heavy rains, and the land becomes unproductive.
- Adding excess phosphate to aquatic ecosystems in **(1)** runoff of animal wastes from livestock feedlots, **(2)** runoff of commercial phosphate fertilizers from cropland, and **(3)** discharge of municipal sewage. Too much of this nutrient causes explosive growth of cyanobacteria, algae, and aquatic plants, which when they die and are decomposed use up dissolved oxygen and disrupt aquatic ecosystems.

How Is Sulfur Cycled in the Ecosphere? Sulfur circulates through the ecosphere in the **sulfur cycle**

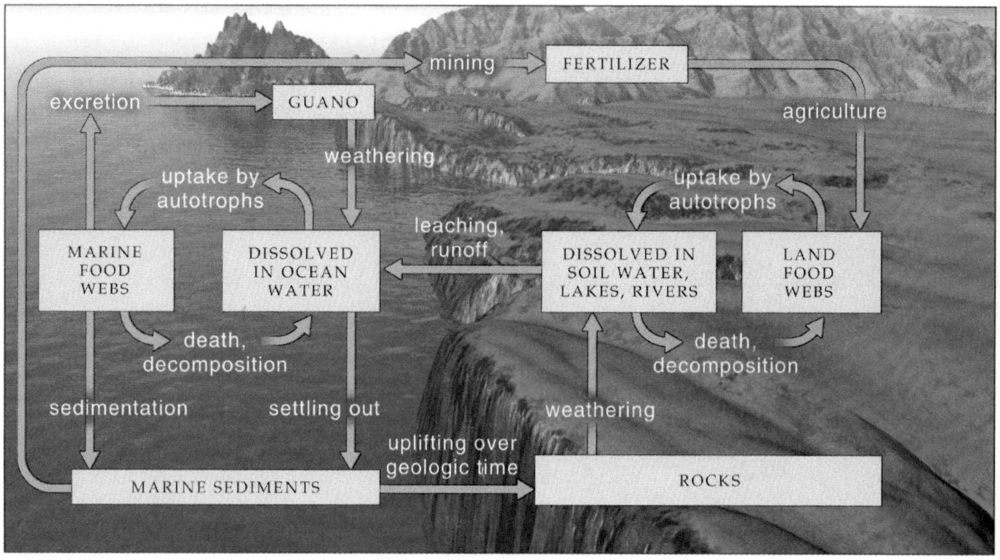

Figure 4-25 Simplified model of the phosphorus cycle. Phosphorus reservoirs are shown as boxes; processes that change one form of phosphorus to another are shown in unboxed print. (Used by permission from Cecie Starr, *Biology: Concepts and Applications*, 4th ed., Pacific Grove, Calif.: Brooks/Cole, 2000)

(Figure 4-26). Much of the earth's sulfur is stored underground in rocks and minerals, including sulfate (SO_4^{2-}) salts buried deep under ocean sediments.

Sulfur also enters the atmosphere from several natural sources. Hydrogen sulfide (H_2S), a colorless, highly poisonous gas with a rotten-egg smell. It is released from active volcanoes and by the breakdown of organic matter in swamps, bogs, and tidal flats caused by decomposers that don't use oxygen (anaerobic decomposers). Sulfur dioxide (SO_2), a colorless, suffocating gas, also comes from volcanoes. Particles of sulfate (SO_4^{2-}) salts, such as ammonium sulfate, enter the atmosphere from sea spray.

In the atmosphere, sulfur dioxide reacts with oxygen to produce sulfur trioxide gas (SO_3). Some of the sulfur dioxide then reacts with water droplets in the atmosphere to produce tiny droplets of sulfuric acid (H_2SO_4). Sulfur dioxide also reacts with other chemicals in the atmosphere such as ammonia to produce tiny particles of sulfate salts. These droplets and particles fall to the earth as components of acid deposition, which along with other air pollutants can harm trees and aquatic life.

About a third of all sulfur (including 99% of the sulfur dioxide) that reaches the atmosphere comes from human activities. We intervene in the atmospheric phase of the sulfur cycle by

- Burning sulfur-containing coal and oil to produce electric power, producing about two-thirds of the human inputs of sulfur dioxide
- Refining petroleum
- Using smelting to convert sulfur compounds of metallic minerals into free metals such as copper, lead, and zinc

4-6 HOW DO ECOLOGISTS LEARN ABOUT ECOSYSTEMS?

What Is Field Research? Ecologists try to unravel some of the workings of ecosystems in three general ways: **(1)** field research, **(2)** laboratory research, and **(3)** systems analysis. Each approach has its advantages and disadvantages.

Field research, sometimes called muddy-boots biology, involves going into nature and observing and measuring the structure of ecosystems and what happens in them. Most of what we know about the structure and functioning of ecosystems and how ecosystems change with changing environmental conditions has come from such research.

Increasingly, ecologists are using new technologies to collect field data. These include *remote sensing* from aircraft and satellites and *geographic information systems (GIS)* in which information gathered from broad geographic regions is stored in a spatial database and analyzed.

However, because even the most simple ecosystems are so complex, it is expensive, time-consuming, and difficult to get such information by carrying out field experiments.

What Is Laboratory Research? In the past 50 years, ecologists have increasingly supplemented field research

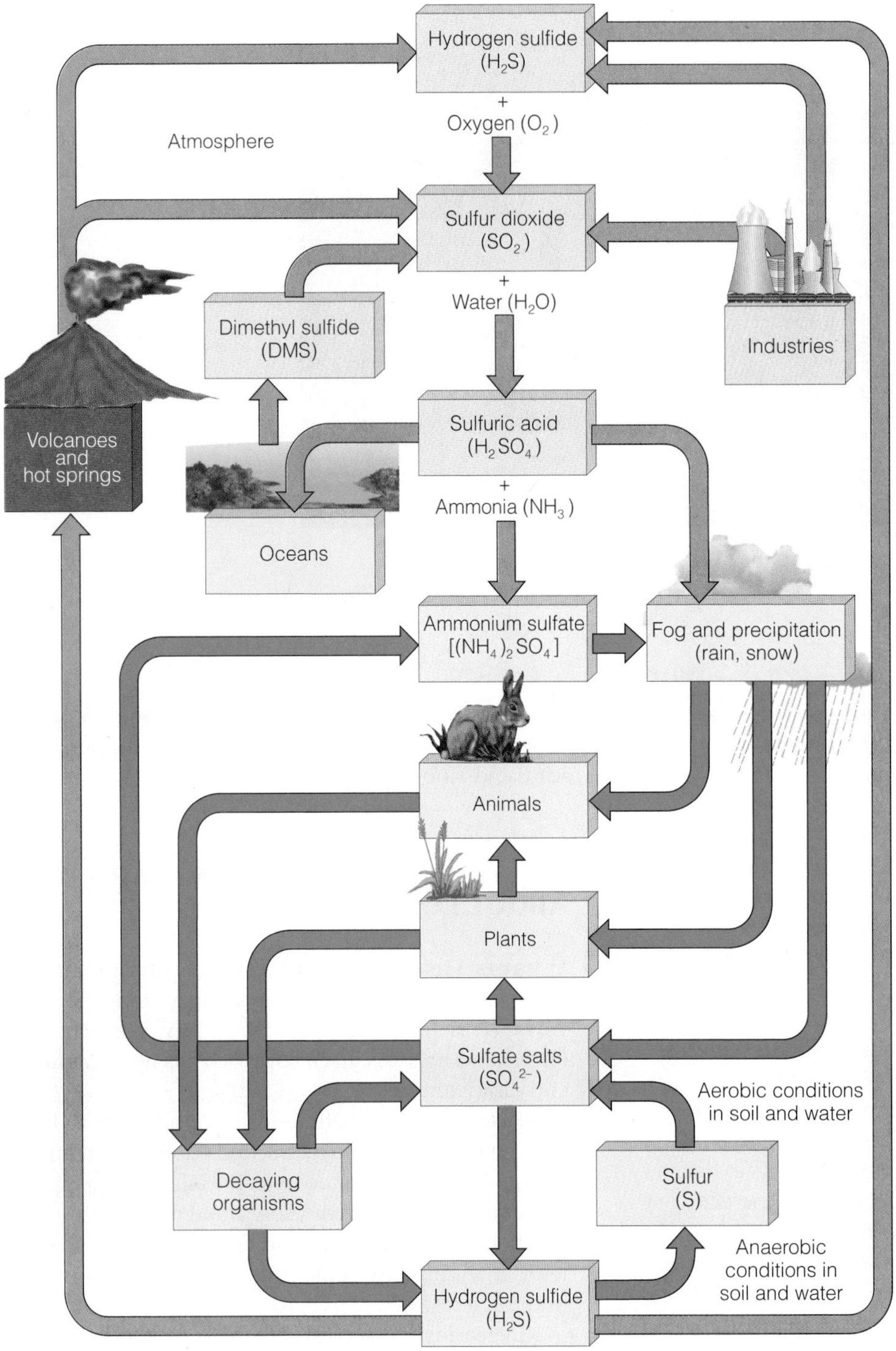

Figure 4-26 Simplified model of the sulfur cycle.

by using *laboratory research* to set up, observe, and make measurements of model ecosystems and populations under laboratory conditions. In such systems, it is easier for scientists to carry out controlled experiments. Such laboratory experiments are often quicker and cheaper than similar experiments in the field.

Often laboratory research can isolate cause-and-effect relationships that can be tested further in the field or by systems analysis using computer simulations. Indoor and outdoor chambers can also be used to study the effects of increasing levels of atmospheric CO_2 and ultraviolet light on populations, biodiversity, and the productivity of producers (including important agricultural crops). This information can help us project the consequences of ozone depletion in the stratosphere and global warming.

However, it is important to consider whether what scientists observe and measure in a simplified, con-

Chapin, F. S. et al. 1998. "Ecosystem Consequences of Changing Biodiversity." *BioScience,* vol. 48, no. 1, p. 45(8).

trolled system under laboratory conditions takes place in the same way in the more complex and dynamic conditions found in nature. Thus, the results of laboratory research must be coupled with and supported by field research.

What Is Systems Analysis? Since the late 1960s ecologists have made increasing use of *systems analysis* to develop models that simulate ecosystems. Computer simulation of such models can help us understand large and very complex systems (such as rivers, oceans, forests, grasslands, or cities) that cannot be adequately studied and modeled in field and laboratory research.

However, simulations and predictions made using ecosystem models are no better than the data and assumptions used to develop the models. Thus, careful field and laboratory ecological research must be used to provide the baseline data and the causal relationships among key variables needed to develop and test ecosystem models.

Sometimes systems analysis reveals surprising results and new questions that must be verified by a combination of field and laboratory research. Thus, meaningful ecological research involves a combination of field research, laboratory research, and systems analysis.

4-7 ECOSYSTEM SERVICES AND SUSTAINABILITY

What Are Ecosystem Services? We depend on nature for food, air, water, and virtually everything we use. Ecosystems provide humans and other species with a number of **ecosystem services**, which constitute the earth's natural capital (Figure 1-2). Without these services performed by diverse communities of species, we would be starving, gasping for breath, and drowning in our own wastes.

One of these ecosystem services is **biological diversity**, or **biodiversity**: the many forms of life in the conditions currently found on the earth. As you learned in Chapter 1, biodiversity includes **(1) genetic diversity** (variability in the genetic makeup among individuals in a single species, Figure 4-4), **(2) species diversity** (the variety of species in different habitats on the earth), and **(3) ecological diversity** (the variety of biological communities that interact with one another and with their nonliving environments).

This rich variety of genes, species, and ecosystems gives us food, wood, fibers, energy, raw materials, industrial chemicals, and medicines, and it pours hundreds of billions of dollars yearly into the global economy. The earth's life-forms and ecosystems also provide recycling, purification, and natural pest control. Every species here today contains genetic information that **(1)** represents thousands to millions of years of adaptation to the earth's changing environmental conditions and **(2)** is the raw material for future adaptations. Biodiversity is nature's insurance policy against disasters.

What Are the Two Basic Principles of Ecosystem Sustainability? In this chapter we have seen that almost all natural ecosystems and the ecosphere itself achieve *sustainability* by:

- Using renewable solar energy as their energy source
- Recycling the nutrients its organisms need for survival, growth, and reproduction

These two principles for sustainability arise from the structure and function of natural ecosystems (Figures 4-6 and 4-13) and from the law of conservation of matter (Section 3-5) and the two laws of energy (Section 3-7). Thus, the results of basic research in both the physical and biological sciences provide us with the same guidelines or lessons from nature on how we can live more sustainably on the earth (Figure 3-15).

All things come from earth, and to earth they all return.
MENANDER (342–290 B.C.)

REVIEW QUESTIONS

1. Define all boldfaced terms in this chapter.
2. What are the six basic characteristics of living things?
3. What are the five main components of the ecosphere (biosphere)?
4. What three processes sustain life on earth? How does the sun help sustain life on the earth? How is this related to the earth's natural greenhouse effect?
5. What are *biomes* and *climates* and how are they related? What are aquatic life zones?
6. What are the two major components of ecosystems? Give three examples of each component.
7. Distinguish between *range of tolerance* for a population in an ecosystem, the *law of tolerance*, and *tolerance limits*. How does each of these factors affect the composition (structure) of ecosystems? What is a *limiting factor* and how do such factors affect the composition of ecosystems? What are two important limiting factors for terrestrial ecosystems and for aquatic ecosystems?
8. What are the major biotic components of ecosystems?
9. Distinguish between *aerobic respiration* and *anerobic respiration*.

Hint: Enter the search terms *ecology, analysis* using the Subject Guide.

10. How do individual organisms and ecosystems differ in their dependence on energy and matter resources? Why are decomposers important, and what would happen without them?

11. Distinguish between a *food chain* and a *food web*. What is the *pyramid of energy flow* for an ecosystem? What is the effect of the second law of energy (thermodynamics) on the flow of energy through an ecosystem and on the amount of food energy available to top carnivores and humans? Distinguish between the *pyramid of biomass* and the *pyramid of numbers* and explain their significance to the structure of ecosystems.

12. Distinguish between *gross primary productivity* and *net primary productivity* and explain how net primary productivity affects the number of consumers in an ecosystem and on the earth. List two of the most productive ecosystems and two of the least productive ecosystems. Use the concept of net primary productivity to explain why harvesting plants from estuaries and clearing tropical forests to grow crops to feed the human population are not good ideas.

13. Explain how an ecosystem's main structural components are linked by matter recycling and energy flow (see Figure 4-13).

14. Define *nutrient* and give five examples of nutrients needed by living things. What are *nutrient cycles* (*biogeochemical cycles*), and why are they important to life on the earth?

15. Briefly describe the major components and functioning and the major human impacts on the **(a)** hydrologic cycle, **(b)** carbon cycle, **(c)** nitrogen cycle, **(d)** phosphorus cycle, and **(e)** sulfur cycle. Define *absolute humidity*, *relative humidity*, *condensation nuclei*, and *dew point*.

16. Explain how ecologists use field research, laboratory research, and systems analysis to learn about ecosystems and describe the advantages and disadvantages of each approach.

17. Define *ecosystem services* and give four examples of such services. Explain why biodiversity is such an important ecosystem service and distinguish between *genetic diversity*, *species diversity*, and *ecological diversity*.

18. What are two basic principles of ecosystem sustainability?

CRITICAL THINKING

1. Imagine that you are having a discussion with a maker of pesticides who believes that the only good insect is a dead insect, or with a child who has been culturally conditioned to have "bugitis" (a fear of all insects). How would you explain to these people the value of insects? Should we never kill insects? Under what circumstances should they be killed? Which kinds? Should a species of insects we have classified as a pest be wiped from the face of the earth? Is it likely that we could do this? Explain.

2. (a) A bumper sticker asks, "Have you thanked a green plant today?" Give two reasons for appreciating a green plant. **(b)** Trace the sources of the materials that make up the bumper sticker and decide whether the sticker itself is a sound application of the slogan. **(c)** Explain how decomposers help keep you alive.

3. (a) How would you set up a self-sustaining aquarium for tropical fish? **(b)** Suppose you have a balanced aquarium sealed with a clear glass top. Can life continue in the aquarium indefinitely as long as the sun shines regularly on it? **(c)** A friend cleans out your aquarium and removes all the soil and plants, leaving only the fish and water. What will happen?

4. Using the second law of energy, explain why there is such a sharp decrease in usable energy as energy flows through a food chain or web. Doesn't an energy loss at each step violate the first law of energy? Explain.

5. Using the second law of energy, explain why many poor people in developing countries live on a mostly vegetarian diet.

6. Carbon, nitrogen, and phosphorus are cycled in the ecosphere. Why do farmers not need to apply carbon to grow their crops but often need to add fertilizer containing nitrogen and phosphorus?

7. Carbon dioxide (CO_2) in the atmosphere fluctuates significantly on a daily and seasonal basis. Why are CO_2 levels higher during the day than at night?

8. Explain in ecological terms how the human practice of embalming a corpse and then entombing it in an airtight metal container could be viewed as contrary to the earth's natural processes of decay and recycling. Even though such practices are done for important cultural, religious, and health reasons, a growing number of people prefer to be cremated and then have their ashes spread on the land or at sea; they want their atoms back in action in the earth's vital chemical cycles as soon as possible after death. What is your preference? Why?

***9.** Write a brief scenario describing the sequence of consequences to us and to other forms of life (identify some of these organisms) if **(a)** all decomposers and detritus feeders were somehow eliminated and **(b)** all producers on land and in the upper zone of aquatic ecosystems were eliminated by drastic increases in ultraviolet radiation because of loss of the protective stratospheric ozone layer.

***10.** Make a concept map of this chapter's major ideas using the section heads and subheads and the key terms (in boldface). Look at the inside back cover and in Appendix 5 for information about concept maps.

FURTHER INFORMATION

For more information see Internet Sources and Further Readings at the back of the book. Online resources for this chapter can be found at **www.brookscole.com/biology**

5 EVOLUTION, BIODIVERSITY, AND COMMUNITY PROCESSES

Why Should We Care About Alligators?

The American alligator (Figure 5-1), North America's largest reptile, has no natural predators except humans. Hunters once killed large numbers of these animals for their exotic meat and their supple belly skin, used to make shoes, belts, and pocketbooks.

Other people considered alligators to be useless, dangerous vermin and hunted them for sport or out of hatred. Between 1950 and 1960, hunters wiped out 90% of the alligators in Louisiana, and by the 1960s the alligator population in the Florida Everglades was also near extinction.

People who say "So what?" are overlooking the alligator's important ecological roles in subtropical wetland ecosystems. Alligators dig deep depressions, or gator holes, that **(1)** collect fresh water during dry spells, **(2)** serve as refuges for aquatic life, and **(3)** supply fresh water and food for many animals. Large alligator nesting mounds provide nesting and feeding sites for species of herons and egrets. Alligators also eat large numbers of predatory gar fish and thus help maintain populations of game fish such as bass and bream.

As alligators move from gator holes to nesting mounds, they help keep areas of open water free of invading vegetation. Without these ecosystem services, freshwater ponds and coastal wetlands found in the alligator's habitat would be filled in by shrubs and trees, and dozens of species would disappear. Some ecologists classify the North American alligator as a *keystone species* because of its important ecological roles in helping maintain the structure, function, and sustainability of its natural ecosystems.

In 1967 the U.S. government placed the American alligator on the endangered species list. Protected from hunters, the alligator population made a strong comeback in many areas by 1975—too strong, according to those who find alligators in their backyards and swimming pools and to duck hunters, whose retriever dogs sometimes are eaten by alligators.

In 1977 the U.S. Fish and Wildlife Service reclassified the American alligator from an *endangered* to a *threatened* species in Florida, Louisiana, and Texas, where 90% of the animals live. In 1987 this reclassification was extended to seven other states. Alligators now number perhaps 3 million, most in Florida and Louisiana. It is generally illegal to kill members of a threatened species, but limited kills by licensed hunters are allowed in some areas of Florida, Louisiana, and South Carolina to control the population. The comeback of the American alligator is an important success story in wildlife conservation.

The increased demand for alligator meat and hides has created a booming business in alligator farms, especially in Florida. Such success reduces the need for illegal hunting of wild alligators.

Each species represents a long chain of evolution, and each species plays a unique ecological role in communities and ecosystems. Understanding such roles is important for **(1)** protecting species from premature extinction, **(2)** returning captive populations of endangered species to the wild, and **(3)** helping predict the effects of human actions on wild species.

Figure 5-1 The American alligator plays important ecological roles in its marsh and swamp habitats in the southeastern United States. After being classified as an endangered species in 1967, it has recovered enough to be changed from an endangered to a threatened species—an important success story in wildlife conservation. (M. H. Sharp/Photo Researchers, Inc.)

There is a grandeur to this view of life . . . that, whilst this planet has gone cycling on . . . endless forms most beautiful and most wonderful have been, and are being, evolved.

CHARLES DARWIN

This chapter addresses the following questions:

- What major types of life are found on the earth and how do scientists account for the emergence of life on the earth?
- What is evolution and how has it led to the current diversity of organisms on the earth?
- How do extinction of species and formation of new species affect biodiversity?
- What roles (ecological niches) do different types of organisms play in ecosystems?
- How do communities and ecosystems change as environmental conditions change?
- Does species diversity increase the stability of ecosystems?

5-1 LIFE AND ITS ORIGINS

What Types of Organisms Are Found on the Earth? Before we learn about how life evolved on the earth, it is useful to have a general knowledge of the earth's *biodiversity*, the many types of life we find on this planet.

On the basis of their cell structure, biologists classify all organisms as either eukaryotic or prokaryotic. All organisms except bacteria are **eukaryotic**: Their cells are surrounded by a membrane and have a *nucleus* (a membrane-bounded structure containing genetic material in the form of DNA) and several other internal parts.

Bacterial cells are **prokaryotic** (which means "before nucleus"): They are surrounded by a membrane but have no distinct nucleus or other internal parts enclosed by membranes. Although most familiar organisms are eukaryotic, they could not exist without hordes of microscopic prokaryotic organisms (bacteria).

Scientists group organisms into various categories based on their common characteristics, a process called *taxonomic classification*. The largest category is the *kingdom*, which includes all organisms that have one or perhaps several common features. In this book, the earth's organisms are classified into five kingdoms: *monera*, *protists*, *fungi*, *plants*, and *animals* (Figure 5-2). Most bacteria, fungi, and protists are **microorganisms**: organisms so small that they can be seen only through a microscope.

Monera (**bacteria** and **cyanobacteria**) are single-celled, microscopic prokaryotic organisms. Most bacteria play a vital role as decomposers. A few bacteria—*Streptococcus* (which causes strep throat) and *Salmonella* (which causes food poisoning), for example—can cause diseases in humans. However, your body is inhabited by billions of beneficial bacteria that help keep you healthy by aiding in food digestion and by crowding out disease-causing bacteria.

Protists (protista) are mostly single-celled eukaryotic organisms such as diatoms, dinoflagellates, amoebas, golden brown and yellow-green algae, and protozoans. Some protists cause human diseases such as malaria and sleeping sickness. **Fungi** are mostly many-celled (sometimes microscopic) eukaryotic organisms such as mushrooms, molds, mildews, and yeasts.

Plants (plantae) are mostly many-celled eukaryotic organisms such as red, brown, and green algae and mosses, ferns, conifers, and flowering plants (whose flowers produce seeds that perpetuate the species). Some plants such as corn and marigolds are **annuals**, which complete their life cycles in one growing season; others are **perennials**, which can live for more than 2 years, such as roses, grapes, elms, and magnolias.

Animals (animalia) are also many-celled, eukaryotic organisms. Most, called **invertebrates**, have no backbones. They include sponges, jellyfish, worms, arthropods (insects, shrimp, and spiders), mollusks (snails, clams, and octopuses), and echinoderms (sea urchins and sea stars). Insects play roles that are vital to our existence (p. 84). **Vertebrates** (animals with backbones and a brain protected by skull bones) include fishes (sharks and tuna), amphibians (frogs and salamanders), reptiles (crocodiles and snakes), birds (eagles and robins), and mammals (bats, elephants, whales, and humans).

How Did Life Emerge on the Earth? How did life on the earth evolve to its present system of diverse kingdoms and species (Figure 5-2), living in an interlocking network of matter cycles, energy flows, and species interactions? We don't know the full answer to this question, but a growing body of evidence suggests what might have happened.

Most of what we know of the earth's life history comes from **fossils**: mineralized or petrified replicas of

Figure 5-2 (right) Five kingdoms of the living world, a scheme for classifying the earth's diverse species into major groups. Most biologists believe that protocells gave rise to single-celled prokaryotes, and that these in turn evolved into the more complex protists, fungi, plants, and animals that make up the earth's stunning biodiversity. This is also a greatly simplified overview of *macroevolution*: changes in the genetic makeup of populations through successive generations over hundreds of millions of years, in which one species leads to the appearance of many other species and groups of species.

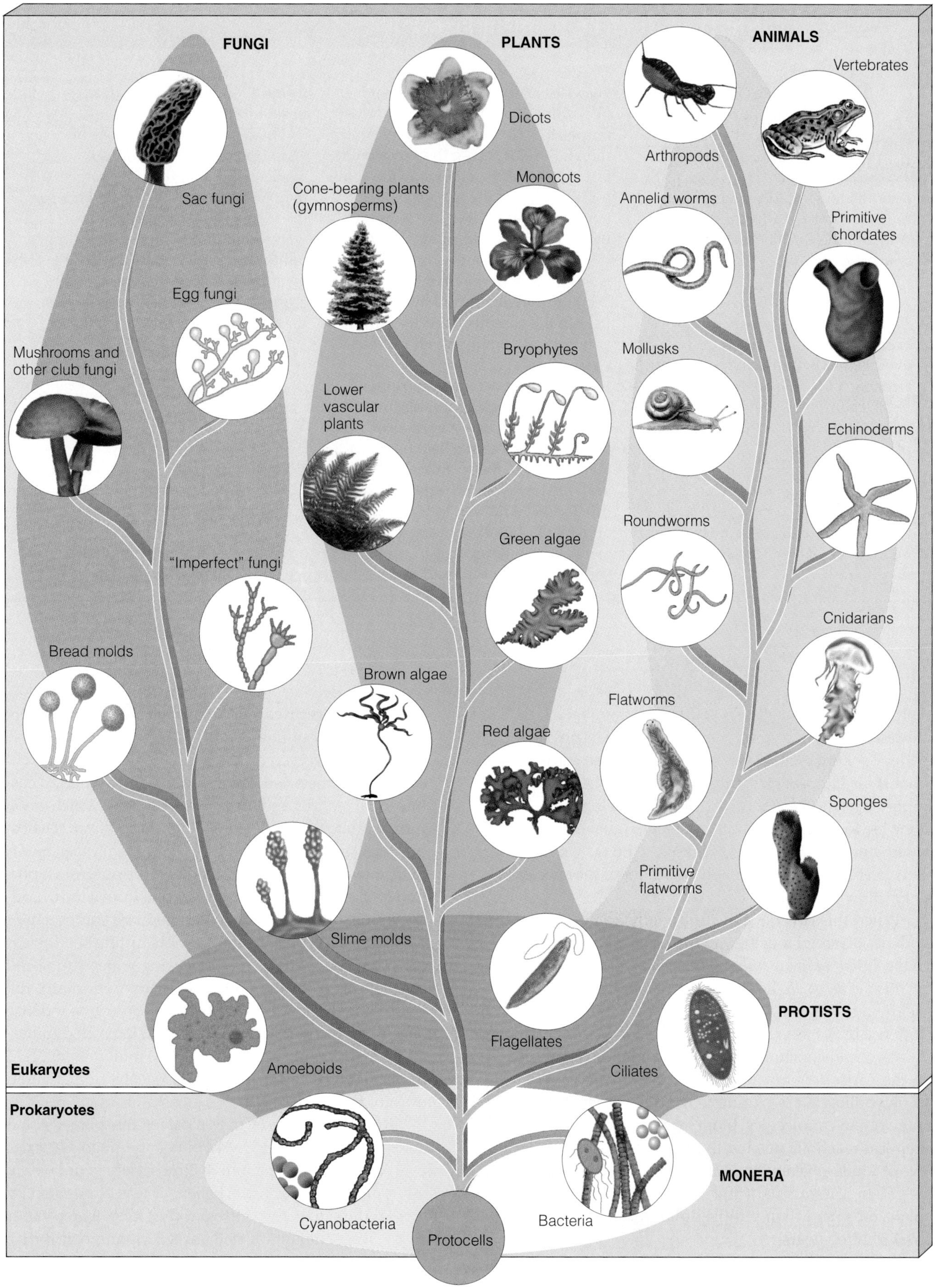
FUNGI
PLANTS
ANIMALS
Vertebrates
Dicots
Arthropods
Sac fungi
Cone-bearing plants (gymnosperms)
Monocots
Annelid worms
Primitive chordates
Egg fungi
Mushrooms and other club fungi
Bryophytes
Mollusks
Lower vascular plants
Echinoderms
Roundworms
Green algae
"Imperfect" fungi
Cnidarians
Bread molds
Brown algae
Flatworms
Red algae
Sponges
Primitive flatworms
Slime molds
PROTISTS
Flagellates
Eukaryotes
Amoeboids
Ciliates
Prokaryotes
MONERA
Cyanobacteria
Bacteria
Protocells

Earth: The Just-Right, Resilient Planet

CONNECTIONS

Like Goldilocks tasting porridge at the Three Bears' house, life on the earth as we know it requires a certain temperature range: Venus is much too hot and Mars is much too cold, but the earth is *just right*. (Otherwise, you wouldn't be reading these words.)

Life as we know it depends on liquid water. Again, temperature is crucial; life on the earth requires average temperatures between the freezing and boiling points of water, between 0°C and 100°C (32°F and 212°F) at the earth's range of atmospheric pressures.

The earth's orbit is the right distance from the sun to provide these conditions. If the earth were much closer, it would be too hot—like Venus—for water vapor to condense to form rain. If it were much farther away, its surface would be so cold—like Mars—that its water would exist only as ice. The earth also spins; if it didn't, the side facing the sun would be too hot and the other side too cold for water-based life to exist. So far, the temperature has been, like Baby Bear's porridge, just right.

The earth is also the right size; that is, it has enough gravitational mass to keep its iron–nickel core molten and to keep the gaseous molecules in its atmosphere from flying off into space. (A much smaller earth would be unable to hold onto an atmosphere consisting of such light molecules as N_2, O_2, CO_2, and H_2O.) The slow transfer of its internal heat (geothermal energy) to the surface also helps keep the planet at the right temperature for life. And thanks to the development of photosynthesizing bacteria over 2 billion years ago, an ozone sunscreen protects us and many other forms of life from an overdose of ultraviolet radiation.

On a time scale of millions of years, the earth is enormously resilient and adaptive. Its average temperatures have remained between the freezing and boiling points of water even though the sun's energy output has increased by about 30% over the 3.6 billion years since life arose. In short, the earth is just right for life as we know it.

Critical Thinking

1. Which do you believe is in greater danger: the earth or the human species? Explain.

2. Do you believe that humans can learn enough about the earth to manage it (mostly for the human species) on a global scale? Explain. If not, what are the alternatives?

skeletons, bones, teeth, shells, leaves, and seeds, or impressions of such items. Such fossils give us physical evidence of organisms that lived long ago and show us what their internal structures looked like. Despite its importance, the fossil record is uneven and incomplete. Some life forms left no fossils, some fossils have decomposed, and others are yet to be found. So far we have found fossils representing only about 1% of the species believed to have ever lived.

When this diverse and continually accumulating evidence is pieced together, an important idea emerges: *The evolution of life is linked to the physical and chemical evolution of the earth.* It becomes apparent that the earth has the right physical and chemical conditions for the existence of life as we know it (Connections, above).

A second conclusion from this evidence is that life on the earth developed in two phases over the past 4.7–4.8 billion years (Figure 5-3): **(1)** *chemical evolution* of the organic molecules, biopolymers, and systems of chemical reactions needed to form the first protocells (about 1 billion years) and **(2)** *biological evolution* of single-celled organisms (first prokaryotes and then eukaryotes) and then multicellular organisms (about 3.7–3.8 billion years).

How Did Chemical Evolution Take Place? Here is an overview of how scientists believe life may have formed and evolved on the earth, based on the current evidence. Some 4.6–4.7 billion years ago a cloud of cosmic dust condensed into planet earth, which soon turned molten from meteorite impacts and from the heat produced by the radioactive decay of chemical elements in its interior. As cooling took place, the outermost portion of the molten sphere solidified to form a thin, hardened crust of rock, devoid of atmosphere and oceans.

Volcanic eruptions and comets hitting the lifeless earth pierced its thin crust, releasing water vapor and other gases from the molten interior. Eventually the crust cooled enough for the water vapor to condense and fall to the surface as rain. This rain eroded minerals from rocks, and solutions of these minerals collected in depressions to form the early oceans that covered most of the globe.

Research suggests that most of the planet's first atmosphere had formed by 4.4 billion years ago. The exact composition of this primitive atmosphere is unknown, but atmospheric scientists believe it was dominated by carbon dioxide (CO_2), nitrogen (N_2), and water vapor (H_2O). Trace amounts of methane (CH_4), ammonia (NH_3),

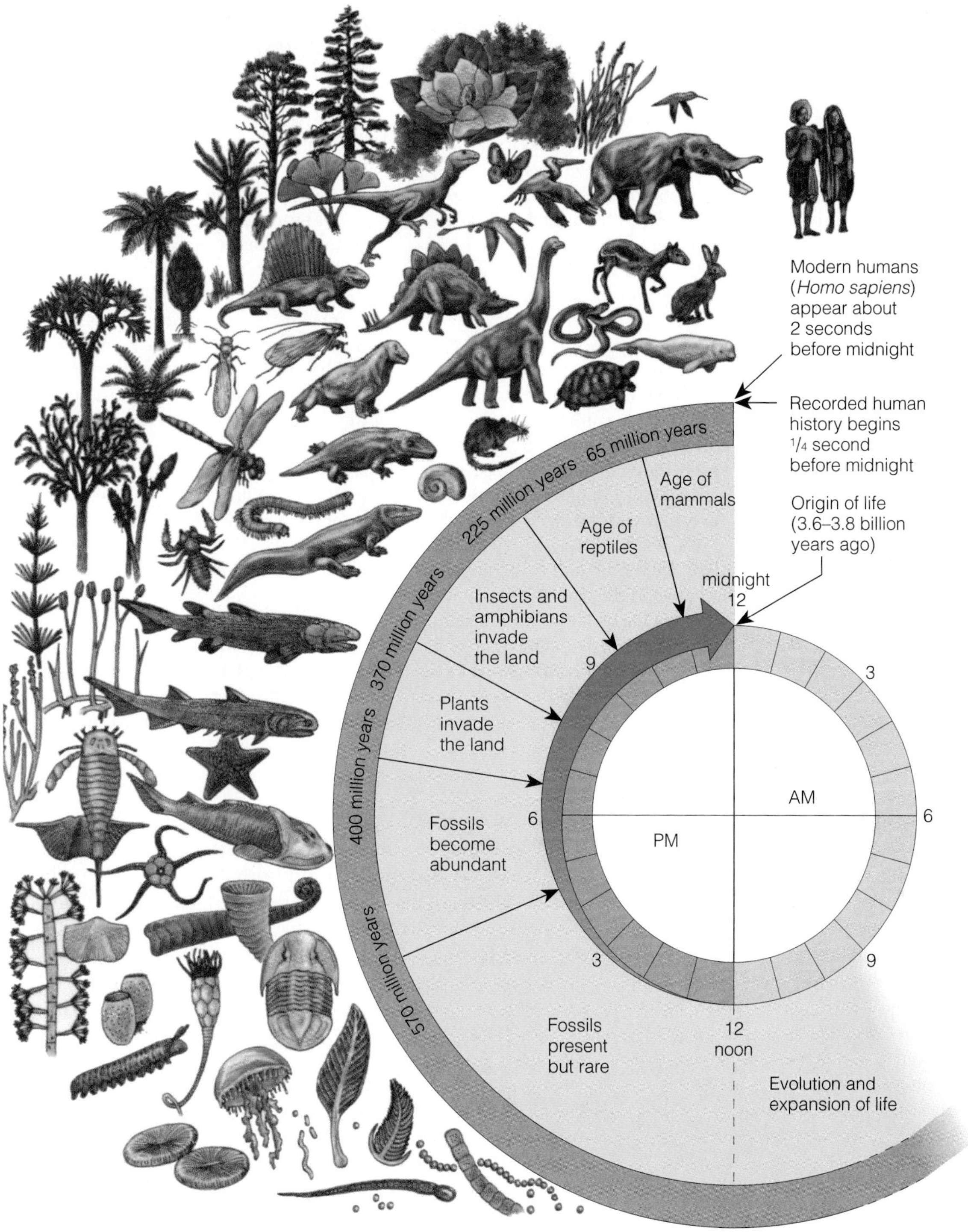

Figure 5-3 Greatly simplified overview of the biological evolution of life on the earth, which was preceded by about 0.5–1 billion years of chemical evolution. The early span of biological evolution on the earth, between about 3.7 billion and about 570 million years ago, was dominated by microorganisms (mostly bacteria and, later, protists) that lived in water. Plants and animals evolved first in the seas and moved onto land about 400 million years ago. Humans arrived on the scene only a very short time ago. If we compress the earth's roughly 3.7-billion-year history of biological evolution to a 24-hour time scale, the first human species (*Homo habilis*) appeared about 47–94 seconds before midnight, and our species (*Homo sapiens sapiens*) appeared about 2 seconds before midnight. Agriculture began only 0.25 second before midnight, and the industrial revolution has been around for only 0.007 second. (Adapted from George Gaylord Simpson and William S. Beck, *Life: An Introduction to Biology*, 2d ed., New York: Harcourt Brace Jovanovich, 1965)

How Did Life First Evolve?

SPOTLIGHT

Over time, it is believed that the earth's protocells evolved into single-celled, bacterialike prokaryotes having the properties we describe as life (although there is much debate over the details of how this might have happened). These anaerobic cells probably evolved either in the muddy sediments of tidal flats or at least 10 meters (30 feet) below the ocean's surface, protected from the intense UV radiation that bathed the earth at that time.

Scientists believe that these single-celled anaerobic bacteria multiplied and underwent genetic changes (mutated) for about a billion years in the earth's warm, shallow seas. The result was a variety of new types of prokaryotic cells that began forming about 2.7 billion years ago.

Scientists contend that during this long early period, life could not have survived on land—a hypothesis supported by a lack of land fossils. There was no ozone layer to shield the DNA and other molecules of early life from bombardment by intense ultraviolet radiation.

About 2.3–2.7 billion years ago the evolution of photosynthetic prokaryotes called *cyanobacteria* in the ocean drastically changed the earth. These cells could remove carbon dioxide from the water and (powered by sunlight) combine it with water to make the carbohydrates they needed. In the process, they released oxygen (O_2) into both the ocean and the atmosphere.

The resulting *oxygen revolution* took place over about half a billion years and opened the way for the evolution of a great variety of oxygen-using (aerobic) bacteria. Later came more complex organisms, first in the seas and then on land (after a protective ozone layer formed in the stratosphere).

As oxygen accumulated in the atmosphere, some was converted by incoming solar energy into ozone (O_3), which began forming in the lower stratosphere. This shield protected life-forms from the sun's deadly UV radiation, allowing green plants to live closer to the ocean's surface. Fossil evidence suggests that about 400–500 million years ago UV levels were low enough for the first plants to exist on the land. Over the next several hundred million years a variety of land plants and animals arose, followed by mammals (and eventually) the first humans (Figure 5-3).

Critical Thinking

If photosynthesizing bacteria had not developed, what kinds of life would you expect to find on the earth today? Where would most of it be found?

hydrogen sulfide (H_2S), and hydrogen chloride (HCl) probably were also present (although scientists disagree about the relative amounts of these gases).

Whatever the composition of this primitive atmosphere, scientists agree that it had no oxygen gas (O_2) because this element is so chemically reactive that it would have combined into compounds. (The only reason today's atmosphere has so much O_2 is that plants and some aerobic bacteria produce it in vast quantities through photosynthesis. But this is getting ahead of the story.)

Energy from electrical discharges (lightning), heat from volcanoes, and intense ultraviolet (UV) light and other forms of solar radiation was readily available for the synthesis of biologically important organic molecules from the inorganic chemicals found in the earth's primitive atmosphere. This idea was first proposed in 1923 by Russian biochemist Alexander Oparin.

In a number of experiments conducted since 1953, various mixtures of gases believed to have been in the earth's early atmosphere have been put in closed, sterilized glass containers. Then they were subjected to spark discharges to simulate lightning and heat. In these experiments, compounds necessary for life—various amino acids (the building-block molecules of proteins), simple carbohydrates, nucleic acids (the building-block molecules of DNA and RNA), and other small organic compounds—formed from the inorganic gaseous molecules. These experiments supported Oparin's hypothesis, although many details are missing or hotly debated.

Another possibility is that simple organic molecules necessary for life formed on dust particles in space and reached the earth on meteorites or comets (or on countless interplanetary dust particles floating around in space when the earth was formed). Another hypothesis is that life could have arisen around the mineral-rich and very hot *hydrothermal vents*, which sit atop cracks in the ocean floor leading to subterranean chambers of molten rock. We don't know which of these processes might have produced the organic molecules necessary for life, but all of these hypotheses are reasonable explanations of how this could have happened.

After these building-block organic molecules formed in the early atmosphere, they were removed by rain. Then they accumulated and underwent countless chemical reactions in the earth's warm, shallow waters. After several hundred million years of different chemical combinations in this hot organic soup, conglomerates of proteins, RNA, and other biopolymers may have combined to form membrane-bound *protocells*: small globules that could take up materials from their environment and grow and divide (much like living cells).

Eldredge, N. 1998. "Evolution and Environment: The Two Faces of Diversity." *Natural History,* vol. 107, no. 5, p. 54(2).

Again, there are several hypotheses explaining how this could have happened. With these forerunners of living cells, the stage was set for the drama of biological evolution (Figure 5-3 and Spotlight, left).

5-2 EVOLUTION AND ADAPTATION

What Is Evolution? The major driving force of adaptation to changes in environmental conditions is **biological evolution**, or **evolution**: the change in a population's genetic makeup through successive generations. Note that *populations, not individuals, evolve by becoming genetically different.*

According to the **theory of evolution**, all species descended from earlier, ancestral species. This theory explains the way life has changed over the past 3.7 billion years and why life is so diverse today.

Biologists use the term ***microevolution*** to describe the small genetic changes that occur in a population. The term ***macroevolution*** is used to describe long-term, large-scale evolutionary changes among groups of species, through which new species are formed from ancestral species and other species are lost through extinction (Figure 5-2).

How Does Microevolution Work? The first step in evolution is the development of *genetic variability* in a population. Genetic information in *chromosomes* is contained in various sequences of chemical units (called *nucleotides*) in DNA molecules. **Genes** found in chromosomes are segments of DNA that are coded for certain traits that can be passed on to offspring. A population's **gene pool** is the set of all genes in the individuals of the population of a species and *microevolution* is a change in a population's gene pool over time.

Although members of a population generally have the same number and kinds of genes, a particular gene may have two or more different molecular forms, called **alleles**. Sexual reproduction leads to a random shuffling or recombination of alleles. As a result, each individual in a population has a different combination of alleles. Without such genetic variability, evolution as we know it could not occur.

What Is the Role of Mutation in Microevolution? Genetic variability in a population originates through **mutations**: random changes in the structure or number of DNA molecules in a cell. One way mutations occur is by exposure to external agents such as radioactivity, X rays, and natural and human-made chemicals (called *mutagens*). Another source of mutations is random mistakes that are sometimes made in coded genetic instructions when DNA molecules are copied (each time a cell divides and whenever an organism reproduces). Mutations can occur in any cells, but only those in reproductive cells are passed on to offspring.

Some mutations are harmless, but many are harmful and alter traits so that an individual cannot survive (lethal mutations). Every so often a mutation is beneficial. The result is new genetic traits that give their bearer and its offspring better chances for survival and reproduction, either under existing environmental conditions or when such conditions change.

It is important to understand that mutations are **(1)** random and unpredictable, **(2)** the only source of totally new genetic raw material (alleles), and **(3)** very rare events.

What Role Does Natural Selection Play in Microevolution? The process of **natural selection** occurs when some individuals of a population have genetically based traits that increase their chances of survival and their ability to produce offspring. Three conditions are necessary for evolution of a population by natural selection to occur:

- There must be natural *variability*.
- The trait must be *heritable*, meaning that it must have a genetic basis such that it can be passed from one generation to another.
- The trait must somehow lead to **differential reproduction**, meaning that it must enable individuals with the trait to leave more offspring than other members of the population.

Natural selection causes any allele or set of alleles that result in a beneficial trait to become more common in succeeding generations and other alleles to become less common. A heritable trait that enables organisms to better survive and reproduce under a given set of environmental conditions is called an **adaptation**, or **adaptive trait**.

The process of microevolution can be summarized in three sentences: *Genes mutate, individuals are selected, and populations evolve.*

What Is an Example of Microevolution by Natural Selection? One of the best-documented examples of evolution by natural selection involves camouflage coloration in the peppered moth, which is found in England (Figure 5-4).

This example illustrates the important points of evolution by natural selection. Natural selection occurred because **(1)** there were two color forms (*variability*), **(2)** color form was genetically based (*heritability*), and **(3)** there was greater survival and reproduction by one of the color forms (*differential reproduction*). In this case, first an environmental change in the form of soot caused a change in the background color of tree trunks. This environmental change then led to a change

Hint: Enter the search terms *evolution, environment* using the Subject Guide.

Figure 5-4 Two varieties of peppered moths found in England illustrate one kind of adaptation: camouflage. Before the industrial revolution in the mid-1800s, the speckled light-gray form of this moth was prevalent. When these night-flying peppered moths rested on light-gray lichens on tree trunks during the day, their color camouflaged them from their predators (left). A dark-gray form also existed but was quite rare. During the industrial revolution, soot and other pollutants from factory smokestacks began killing lichens and darkening tree trunks. As a result, the dark form became the common one, especially near industrial cities. In this new environment, the dark form blended in with the blackened trees, whereas the light form was highly visible to predators. Through natural selection, the dark form began to survive and reproduce at a greater rate than its light-colored kin. (Both varieties appear in each photo. Can you spot them?) (Right, Michael Tweedie/Natural History Photographic Agency; left, Kim Taylor/Bruce Coleman Ltd.)

in selective force: Predators were able to find and eat the moths with the coloration that no longer blended in with the background.

What Is Coevolution? Some biologists have proposed that interactions between *species* can also result in microevolution in each of their populations. According to this hypothesis, when populations of two different species interact over a long time, changes in the gene pool of one species can lead to changes in the gene pool of the other species. This process is called **coevolution**.

Suppose, for example, that certain individuals in a population of carnivores (such as owls) become better at hunting prey (such as mice). Because of genetic variation, certain individuals of the prey have traits that allow them to escape or hide from their predators, and they pass these adaptive traits on to some of their offspring. But a few individuals in the predator population may also have traits (such as better eyesight or quicker reflexes) that allow them to successfully hunt the better-adapted prey. They would then pass these traits on to some of their offspring.

Similarly, individual plants in a population may evolve defenses such as camouflage, thorns, or poisons, against efficient herbivores. In turn, some herbivores in the population may have genetic characteristics that enable them to overcome these defenses and produce more offspring than those without such traits.

In coevolution, adaptation follows adaptation in something like an ongoing, long-term arms race among individuals in interacting populations of different species.

What Limits Adaptation? Shouldn't evolution lead to perfectly adapted organisms? Shouldn't adaptations to new environmental conditions allow our skin to become more resistant to the harmful effects of ultraviolet radiation, our lungs to cope with air pollutants, and our livers to become better at detoxifying pollutants? The answer to these questions is *no* because there are limits to adaptations in nature. First, *a change in environmental conditions can lead to adaptation only for traits already present in the gene pool of a population.*

Second, *even if a beneficial heritable trait is present in a population, that population's ability to adapt can be limited by its reproductive capacity.* If members of a population can't reproduce quickly enough to adapt to a particular environmental change, all of them can die. Populations of genetically diverse species that reproduce quickly—such as weeds, mosquitoes, rats, bacteria, or cockroaches—often can adapt to a change in environmental conditions in a short time. In contrast, populations of species such as elephants, tigers, sharks, and humans, which cannot produce large numbers of offspring rapidly, take a long time (typically thousands or even millions of years) to adapt through natural selection.

Finally, *even if a favorable genetic trait is present in a population, most of its members would have to die or become sterile so that individuals with the trait could predominate and pass the trait on.* This is hardly a desirable solution to the environmental problems humans face.

What Are Three Common Misconceptions About Evolution? One misconception about evolution arises from the interpretation of the expression "survival of the fittest" (which biologists almost never use), sometimes used to describe how natural selection works. This often is misinterpreted as "survival of the strongest." To biologists, however, *fitness* is a measure of reproductive success, so that the fittest individuals are those that leave the most descendants. Instead of "tooth and claw" competition, natural selection favors populations of species that *avoid* direct competition by producing offspring that can occupy ecological roles (niches) different from those of other species.

Some people have misinterpreted the theory of evolution as asserting that "humans evolved from apes." The theory of evolution makes no such claim. Instead, it states that apes and humans are descended from a common ancestor. In other words, at some time in the distant past a particular population of organisms had descendants, with some of them evolving into apes and others evolving into the human species.

A third misconception is that evolution involves some grand plan of nature in which species become progressively more perfect. This ignores the fact that the mutations and other processes that drive microevolution occur as a result of random, unpredictable events. From a scientific standpoint, there is no plan or goal of perfection in the evolutionary process.

5-3 SPECIATION, EXTINCTION, AND BIODIVERSITY

How Do New Species Evolve? Under certain circumstances natural selection can lead to an entirely new species. In this process, called **speciation**, two species arise from one.

The most common mechanism of speciation (especially among animals) takes place in two phases: geographic isolation and reproductive isolation. **Geographic isolation** occurs when two populations of a species or two groups of the same population become physically separated for fairly long periods into areas with different environmental conditions. For example, part of a population may migrate in search of food and then begin living in another area with different environmental conditions (Figure 5-5). Populations may also become separated **(1)** by a physical barrier (such as a mountain range, stream, lake, or road), **(2)** by a change such as a volcanic eruption or earthquake, or **(3)** when a few individuals are carried to a new area by wind or water.

The second phase of speciation is **reproductive isolation**. It occurs as mutation and natural selection operate independently in two geographically isolated populations and change the allele frequencies in different ways—a process called *divergence*. If divergence continues long enough, members of the geographically and reproductively isolated populations may become so different in genetic makeup that they can't interbreed—or if they do, they can't produce live, fertile offspring. Then one species has become two, and *speciation* has occurred through *divergent evolution*.

In a few rapidly reproducing organisms this type of speciation may occur within hundreds of years. However, with most species speciation takes from tens of thousands to millions of years. Given this time scale, it is difficult to observe and document the appearance of a new species. As a result, there are many controversial hypotheses about the details of speciation.

How Do Species Become Extinct? After evolution, the second process affecting the number and types of species on the earth is **extinction**. When environmental conditions change, a species may either evolve (become better adapted) or cease to exist (become extinct).

The earth's long-term patterns of speciation and extinction have been affected by several major factors: **(1)** large-scale movements of the continents (continental drift) over millions of years (Figure 5-6), **(2)** gradual climate changes caused by continental drift and slight

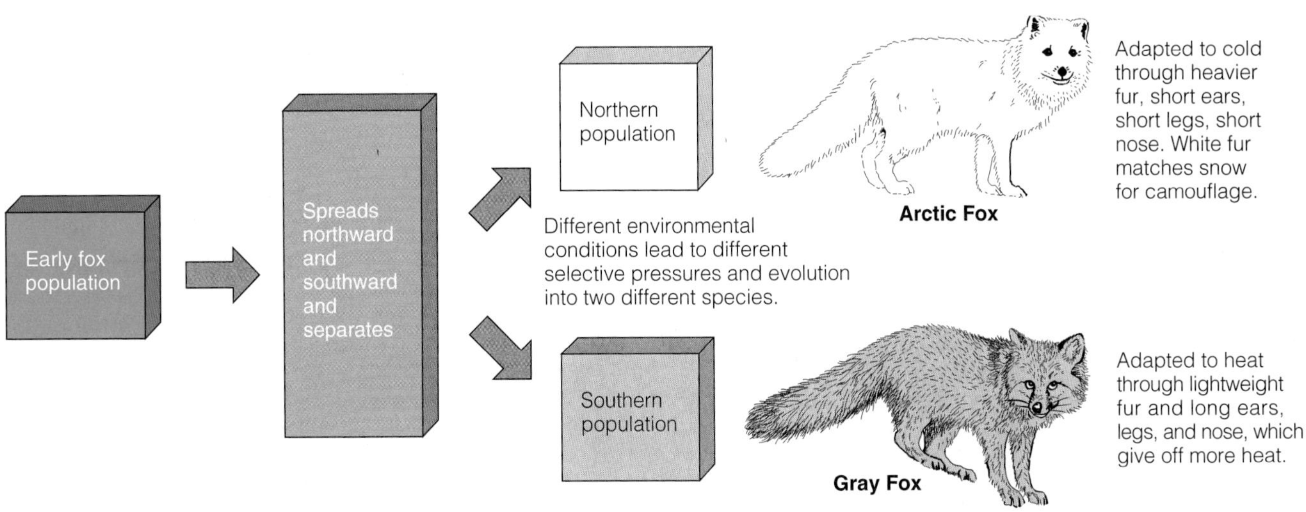

Figure 5-5 How geographic isolation can lead to reproductive isolation, divergence, and speciation.

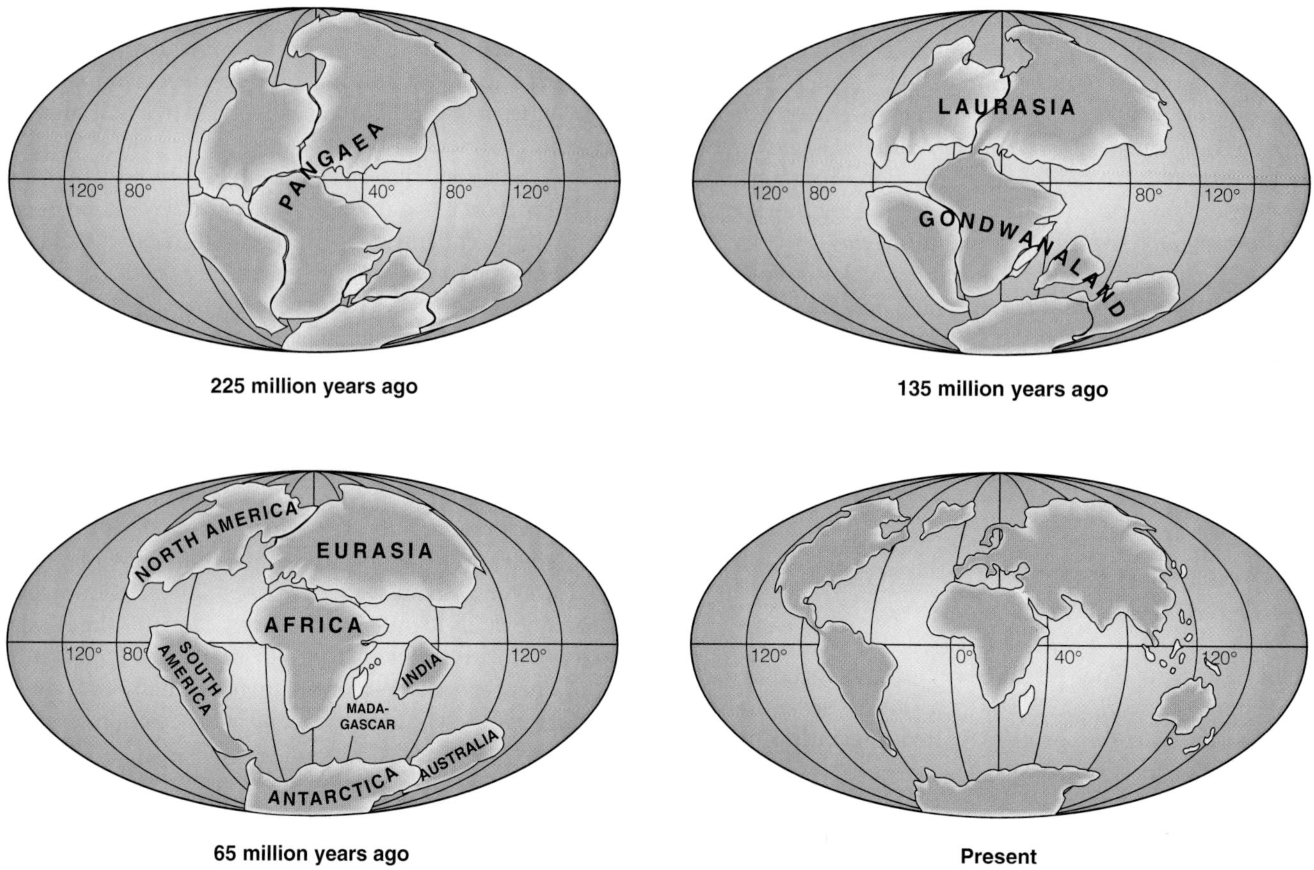

Figure 5-6 Continental drift, the extremely slow movement of continents over millions of years on several gigantic plates (discussed in more detail in Section 13-1). This process plays a role in the extinction of species and the rise of new species. Populations are geographically and eventually reproductively isolated as land masses float apart and new coastal regions are created. Rock and fossil evidence indicates that about 200–250 million years ago all of the earth's present-day continents were locked together in a supercontinent called Pangaea. About 180 million years ago, Pangaea began splitting apart as the earth's huge plates separated and eventually resulted in today's locations of the continents.

shifts in the earth's orbit around the sun, and **(3)** rapid climate change caused by catastrophic events (such as large volcanic eruptions and huge meteorites and asteroids crashing into the earth). Such events create dust clouds that shut down or sharply reduce photosynthesis long enough to eliminate huge numbers of producers and, soon thereafter, the consumers that fed on them.

Extinction is the ultimate fate of all species, just as death is for all individual organisms. Biologists estimate that 99.9% of all the species that have ever existed are now extinct.

As local conditions change, a certain number of species disappear at a low rate, called **background extinction**, as local conditions change. In contrast, **mass extinction** is an abrupt rise in extinction rates above the background level. It is a catastrophic, widespread (often global) event in which large groups of existing species (perhaps 25–90%) are wiped out.

Most mass extinctions are believed to result from global climate changes that kill many species and leave behind those able to adapt to the new conditions. Fossil and geological evidence indicates that the earth's species have experienced five great mass extinctions (20–60 million years apart) during the past 500 million years (Figure 5-7).

A crisis for one species is an opportunity for another. The existence of millions of species today means that speciation, on average, has kept ahead of extinction. Evidence shows that the earth's mass extinctions have been followed by periods of recovery called **adaptive radiations**, in which numerous new species have evolved over several million years to fill new or vacated ecological roles or niches in changed environments (Figure 5-8).

How Do Speciation and Extinction Affect Biodiversity? Speciation minus extinction equals *biodiversity*, the planet's genetic raw material for future

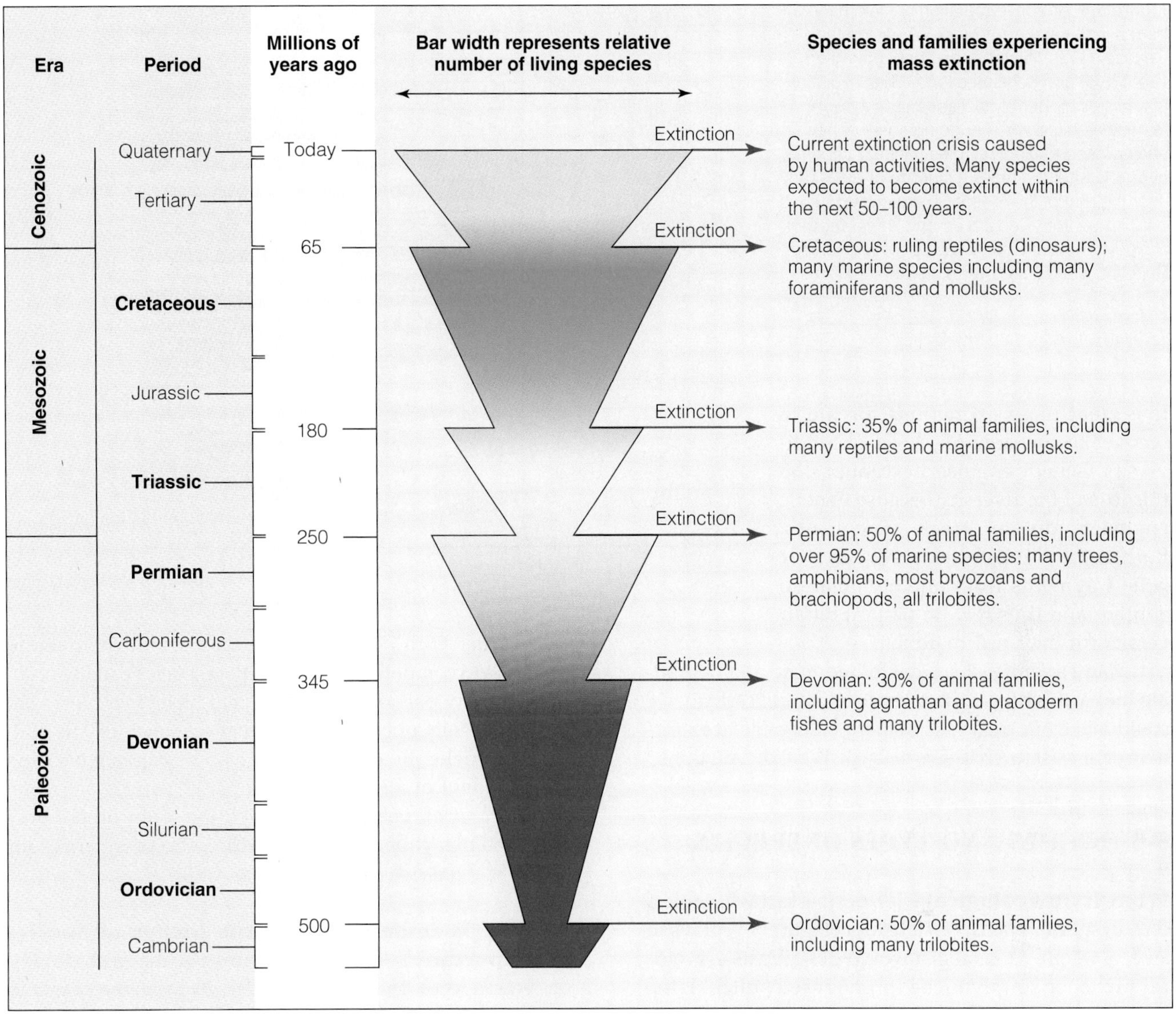

Figure 5-7 Over millions to hundreds of millions of years, macroevolution has consisted of dramatic exits (extinctions) and grand entrances (speciation and radiations) of large groups of species. Fossils and radioactive dating indicate that five major mass extinctions (indicated by arrows) have taken place over the past 500 million years. Mass extinctions leave large numbers of organism roles (niches) unoccupied and create new ones. As a result, each mass extinction has been followed by periods of recovery (represented by the wedge shapes) called *adaptive radiations* in which (over 10 million years or more) new species evolve to fill new or vacated ecological roles (niches). Many scientists say that we are now in the midst of a sixth mass extinction, caused primarily by overhunting and the increasing elimination, degradation, and fragmentation of wildlife habitats as a result of human activities.

evolution in response to changing environmental conditions. In this long-term give-and-take between extinction and speciation, mass extinctions temporarily reduce biodiversity. However, they also create evolutionary opportunities for surviving species to undergo adaptive radiations to fill unoccupied and new biological roles or niches (Figure 5-8).

Although extinction is a natural process, humans have become a major force in the premature extinction of species. As the human population and resource consumption increase over the next 50 years and we take over more and more of the planet's surface (Figure 1-5), we may cause the extinction of up to a quarter of the earth's current species. If this happens it will constitute a sixth mass extinction, caused by us.

On our short time scale, such catastrophic losses cannot be recouped by formation of new species; it took tens of millions of years after each of the earth's

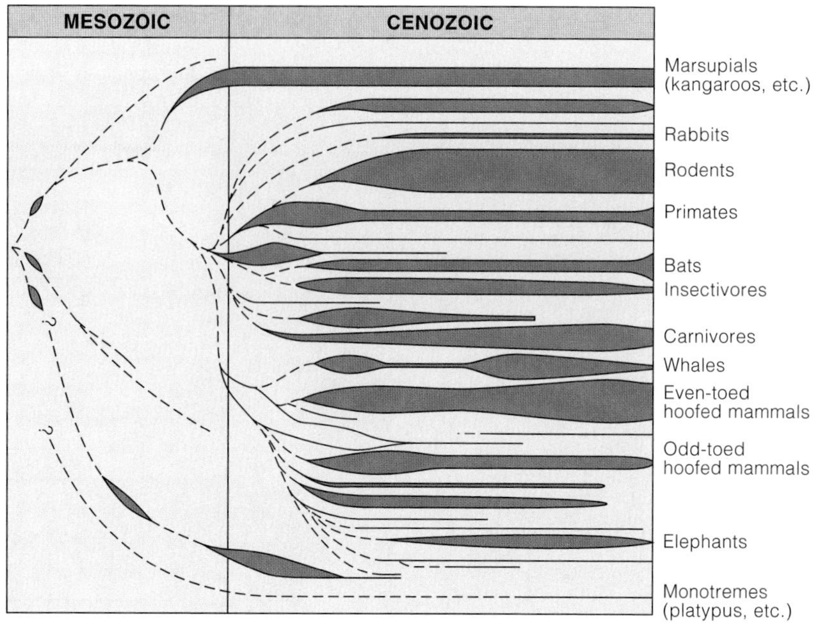

Figure 5-8 Adaptive radiation of mammals began in the first 10–12 million years of the Cenozoic era (which began about 65 million years ago) and continues today. This evolution of a large number of new species is thought to have resulted when huge numbers of new and vacated ecological roles (niches) became available after the mass extinction of dinosaurs near the end of the Mesozoic era. (Used by permission from Cecie Starr and Ralph Taggart, *Biology: The Unity and Diversity of Life*, 8th ed., Belmont, Calif.: Wadsworth, 1998)

five great mass extinctions for life to recover to the previous level of biodiversity. Genetic engineering cannot stop this loss of biodiversity because genetic engineers do not create new genes. Rather, they transfer existing genes or gene fragments from one organism to another and thus rely on natural biodiversity for their raw material.

Many scientists fear that by reducing and degrading the earth's life-support systems (natural capital), we could make our own species more vulnerable to extinction, or at least to a massive dieback. What do you think?

5-4 NICHES AND TYPES OF SPECIES

What Is an Ecological Niche? If asked what role a certain species such as an alligator (Figure 5-1) plays in an ecosystem, an ecologist would describe its **ecological niche**, or simply **niche** (pronounced "nitch"), the species' way of life or functional role in an ecosystem. A species' niche involves everything that affects its survival and reproduction. This includes **(1)** the range of tolerance for various physical and chemical conditions, such as temperature or water availability (Figure 4-11), **(2)** the types of resources it uses, such as food or nutrient requirements, **(3)** how it interacts with other living and nonliving components of the ecosystems in which it is found, such as what it eats or is eaten by, and **(4)** the role it plays in the flow of energy and cycling of matter in an ecosystem.

The ecological niche of a species is different from the **habitat** of the species. Habitat is the actual location where a species lives. Ecologists often say that a niche is like a species' occupation, whereas habitat is like its address.

A species' **fundamental niche** is the full potential range of physical, chemical, and biological conditions and resources it could theoretically use if there were no direct competition from other species. However, the niche of a species in a particular ecosystem tends to overlap with niches of other species. Niche overlap leads to competition. As a result, a species usually occupies only part of its fundamental niche in a particular community or ecosystem—what ecologists call its **realized niche**. By analogy, you may be capable of being president of a particular company (your *fundamental professional niche*), but competition from others may mean that you may become only a vice president (your *realized professional niche*).

Why Is Understanding the Niches of Species Important? As we seek to preserve the earth's biodiversity, ecologists call for greatly expanded research in understanding the niches of species. Understanding a species' niche is important because it can help us prevent it from becoming prematurely extinct and be useful in helping us assess the environmental of changes we make in terrestrial and aquatic systems. For example, how will the niches of various species be changed by clearing a forest, plowing up a grassland, filling in a wetland, or dumping pollutants into a lake or stream?

Unfortunately, we have little knowledge about the ecological roles of most of the earth's species because determining the niche of a species is a difficult, expensive, and time-consuming process.

Is It Better to Be a Generalist or a Specialist Species? Broad and Narrow Niches The niches of species can be used to broadly classify them as generalists or specialists. **Generalist species** have broad niches: They can live in many different places, eat a variety of foods, and tolerate a wide range of environmental conditions. Flies, cockroaches (Spotlight, right), mice, rats, white-tailed deer, raccoons, coyotes, copperheads, channel catfish, and humans are all generalist species.

Specialist species have narrow niches: They may be able to live in only one type of habitat, tolerate only a narrow range of climatic and other environmental conditions, or use only one or a few types of food. This makes them more prone to becoming endangered when environmental conditions change. Examples of specialists are **(1)** *tiger salamanders*, which can breed only in fishless ponds so their larvae won't be eaten, **(2)** *red-cockaded woodpeckers*, which carve nest-holes almost exclusively in old (at least 75 years) longleaf pines, **(3)** *spotted owls*, which require old-growth forests in the Pacific Northwest for food and shelter, and **(4)** China's highly endangered giant pandas, which feed almost exclusively on various types of bamboo.

Is it better then to be a generalist than a specialist? It depends. When environmental conditions are fairly constant, as in a tropical rain forest, specialists have an advantage because they have fewer competitors. However, under rapidly changing environmental conditions, the generalist usually is better off than the specialist.

What Different Roles Do Various Species Play in Ecosystems? When examining ecosystems, ecologists often apply particular labels—such as *native*, *nonnative*, *indicator*, or *keystone*—to various species to clarify their ecological roles or niches. Any given species may function as more than one of these four types in a particular ecosystem.

How Can Nonnative Species Cause Problems? Species that normally live and thrive in a particular ecosystem are known as **native species**. Others that migrate into an ecosystem or are deliberately or accidentally introduced into an ecosystem by humans are called **nonnative species**, **exotic species**, or **alien species**. Some of these introduced species (such as crop species and game for sport hunting) are beneficial to humans, but some thrive and crowd out many native species.

From a human standpoint, introduction of nonnative species can become a nightmare. In 1957, wild African bees were imported to Brazil to help increase honey production. Instead, these bees have displaced domestic honeybees and reduced the honey supply. Since then these nonnative bee species, popularly known as "killer bees," have moved northward into Central America (killing 150 people in Mexico since 1986); by the summer of 1994 they had become established in Texas (one death in 1994), Arizona (one death in 1993), New Mexico, Puerto Rico, and California. They are now heading north at 240 kilometers (150 miles) per year, although they will be stopped eventually by cold winters in the central United States (unless they can adapt genetically to cold weather).

SPOTLIGHT

Cockroaches: Nature's Ultimate Survivors

Cockroaches, the bugs many people love to hate, have been around for about 350 million years and are one of the great success stories of evolution. The major reason they are so successful is that they are *generalists*, able to eat almost anything (including algae, dead insects, salts in tennis shoes, electrical cords, glue, paper, soap, and—when times are bad—other, weaker cockroaches). The 4,000 known cockroach species can live and breed almost anywhere except in polar regions.

Some species can go for months without food, last a month without water, and withstand massive doses of radiation. One species can survive being frozen for 48 hours. The antennae of most cockroach species (which can detect minute movements of air), the vibration sensors in their knee joints, and their lightning-fast response times (faster than you can blink) allow them to evade predators and a human foot in hot pursuit. Some even have wings.

High reproductive rates also aid the survival of cockroaches. In only a year, a single Asian cockroach (especially prevalent in Florida) and its young can add about 10 million new cockroaches to the world. Their high reproductive rate also helps them quickly develop genetic resistance to almost any poison we throw at them. Most cockroaches also sample food before it enters their mouths and learn to shun foul-tasting poisons.

Only about 25 species of cockroach live in homes, but such species can carry viruses and bacteria that cause such diseases as hepatitis, polio, typhoid fever, plague, and salmonella. They can cause people to have allergic reactions ranging from watery eyes to severe wheezing. Indeed, about 60% of the 12 million Americans suffering from asthma are allergic to dead or live cockroaches.

Critical Thinking

How do you feel about cockroaches? If you could, would you exterminate them? What might be some ecological consequences of doing this?

Although they are not the killer bees portrayed in some horror movies, these bees are aggressive and unpredictable. They have killed thousands of domesticated animals and an estimated 1,000 people in the western hemisphere. Fortunately, most people not allergic to bee stings can run away. Most people killed by these honeybees have died because they fell down or became trapped and could not flee.

Hint: Enter the search terms *evolution, environmental* using the Subject Guide.

CONNECTIONS

Why Are Amphibians Vanishing?

Amphibians (frogs, toads, and salamanders) first appeared about 350 million years ago. These cold-blooded creatures range in size from a frog that can sit on your thumb to a Japanese salamander that is about 1.5 meters (5 feet) long.

Within the last decade or two, populations of hundreds of the world's estimated 5,100 amphibian species (including 2,700 species of frogs and toads) have been vanishing or declining in diverse locations, even in protected wildlife reserves and parks. In some locales, frog deformities are also occurring in unusually high numbers. According to the World Conservation Union, 25% of all known amphibians are extinct, endangered, or vulnerable.

Frogs are especially vulnerable to environmental disruption. As tadpoles, they live in water and eat plants and as adults they live mostly on land and eat insects. Their eggs have no protective shells. As adults, they take in water and air through their thin skins. Because each type of frog lives in a small, specialized habitat, they are sensitive indicators of environmental conditions in a variety of places.

Scientists have identified a variety of possible causes for such declines. They include

- Prolonged drought, which dries up breeding pools so that few tadpoles survive. Dehydration can also weaken amphibians, making them more susceptible to fatal viruses, bacteria, and fungi.
- Immigration or introduction of nonnative predators and competitors (such as fish).
- Pollution. Because amphibians live part of their lives in water and part on land, they are exposed to water, soil, and air pollutants. Their insect diet guarantees them abundant food, but in farming areas it also exposes them to pesticides. The soft, permeable skin that allows them to absorb oxygen from water also makes them extremely sensitive to waterborne pollutants.
- Increased acidity of the water in lakes and ponds from acid deposition (acid rain), which can harm their immune and endocrine systems and make them more vulnerable to bacterial infections and perhaps to a newly discovered type of skin fungus.
- Increases in ultraviolet radiation caused by reductions in stratospheric ozone.
- Overhunting, especially in Asia and France, where frog legs are a delicacy.
- Loss of habitat or its fragmentation into pieces too small to support populations of some amphibians.

Scientists are concerned about amphibians' decline for three reasons:

- It suggests that the world's environmental health is deteriorating rapidly, because amphibians are generally tough survivors and are sensitive sentinels of changes in environmental conditions.
- Adult amphibians play important roles in the world's ecosystems. For example, amphibians eat more insects (including mosquitoes) than do birds. In some habitats, extinction of certain amphibian species could also result in extinction of other species, such as reptiles, birds, aquatic insects, fish, mammals, and other amphibians that feed on them or their larvae.
- From a human perspective amphibians represent a genetic storehouse of pharmaceutical products waiting to be discovered. Hundreds of secretions from amphibian skin have been isolated, and some of these compounds are being used as painkillers and in treating burns and heart attacks. Others are being evaluated for their use as antiviral and antibacterial medicines.

As possible indicator species, amphibians may be sending us an important message. They don't need us, but we and other species need them.

Critical Thinking

On an evolutionary time scale, all species eventually become extinct. Some suggest that the widespread disappearance of amphibians is the result of natural responses to changing environmental conditions. Others contend that these losses are caused mostly by human actions and that such declines are a warning of possible danger for our own species and other species. What is your position? Why?

What Are Indicator Species? Species that serve as early warnings that a community or an ecosystem is being damaged are called **indicator species**. Birds are excellent biological indicators because they are found almost everywhere and respond quickly to environmental change. Research indicates that a major factor in the current decline of migratory, insect-eating songbirds in North America is loss or fragmentation of habitat. The tropical forests of Latin America and the Caribbean that are the birds' winter habitats are disappearing rapidly. Their summer habitats in North America are also disappearing or are being fragmented into patches that make the birds more vulnerable to attack by predators and parasites. Some amphibians (frogs, toads, and salamanders), which live part of their lives in water and part on land, are also believed to be indicator species (Connections, above).

What Are Keystone Species? The roles of some species in an ecosystem are much more important than their abundance or biomass suggests. Ecologists call such species **keystone species**, although this designa-

Why Are Sharks Important Species?

CASE STUDY

The world's 350 shark species vary widely in size. The smallest is the dwarf dog shark, about the size of a large goldfish, and the largest is the whale shark, the world's largest fish at 18 meters (60 feet) long. Various shark species, feeding at the top of food webs, cull injured and sick animals from the ocean and thus play an important ecological role. Without such shark species the oceans would be overcrowded with dead and dying fish.

Many people, influenced by movies and popular novels, think of sharks as people-eating monsters. But the two largest species—the whale shark and the basking shark—sustain their enormous bulk by filtering out and swallowing huge quantities of *plankton* (small free-floating sea creatures).

Every year, members of a few species of shark—mostly great white, bull, tiger, gray reef, lemon, and blue—injure about 50 to 100 people worldwide and kill between 5 and 12. Most attacks are by great white sharks, which feed on sea lions and other marine mammals and sometimes mistake divers in wet suits and surfers for their usual prey. A typical oceangoer is 150 times more likely to be killed by lightning than by a shark.

For every shark that injures a person, we kill 500,000 to 1 million sharks, for a total of more 50 to 100 million sharks each year. Sharks are killed mostly for their fins, widely used in Asia as a soup ingredient and as a pharmaceutical cure-all, and worth as much as $563 per kilogram ($256 per pound). In Hong Kong, a single bowl of shark fin soup can sell for as much as $90.

Sharks are also killed for their livers, meat (especially mako and thresher), hides (a source of exotic, high-quality leather), and jaws (especially great whites, whose jaws are worth thousands of dollars to collectors), or just because we fear them. Some sharks (especially blue, mako, and oceanic whitetip) die when they are trapped as bycatch in nets or lines deployed to catch swordfish, tuna, shrimp, and other commercially important species.

Sharks also help save human lives. In addition to providing people with food, they are helping us learn how to fight cancer (which sharks almost never get), bacteria, and viruses. Their highly effective immune system is being studied because it allows wounds to heal without becoming infected. This research might help prolong your life or the life of a loved one someday.

Sharks have several natural traits that make them prone to population declines from overfishing. They have only a few offspring (between 2 and 10) once every year or two. Depending on the species, sharks require 10 to 24 years to reach sexual maturity and begin reproducing. Sharks also have long gestation (pregnancy) periods—up to 24 months for some species.

Sharks are among the most vulnerable and least protected animals on the earth. Eight of the world's shark species, including great whites, sandtigers, and kitefins, are now considered critically endangered, endangered, or vulnerable to extinction.

With more than 400 million years of evolution behind them, sharks have had a long time to get things right. Preserving their evolutionary genetic development begins with the knowledge that sharks don't need us, but we and other species need them.

Critical Thinking

Do you fear sharks? Why? How do you feel about killing sharks in large numbers? Why?

tion is controversial.* In tropical forests, various species of bees, bats, ants, and hummingbirds play keystone roles by pollinating flowering plants, dispersing seed, or both. Some keystone species—including the wolf, leopard, lion, giant anteater, and great white shark (Case Study, above)—are top predators that exert a stabilizing effect on their ecosystems by feeding on and regulating the populations of certain species. The American alligator is also considered a keystone species in Florida and several other southern states (p. 109).

*All species play some role in their ecosystems and thus are important. Whereas some scientists consider all species equally important, others consider certain species to be more important than others, at least in helping maintain the ecosystems they are a part of.

Sea otters are keystone species that keep sea urchins from depleting kelp beds in offshore waters from Alaska to southern California. If the sea otter is removed, too little kelp is left to support the diverse community of crustaceans, mollusks, fish, and marine mammals such as fish-eating harbor seals. Along with sea urchins, sea otters also feed on abalone, crabs, and mollusks. This makes them unpopular with people who make a living harvesting crabs and abalone.

The loss of a keystone species can lead to population crashes and extinctions of other species that depend on it for certain services, a ripple or domino effect that spreads throughout an ecosystem. According to biologist Edward O. Wilson, "The loss of a keystone species is like a drill accidentally striking a power line. It causes lights to go out all over."

Figure 5-9 Specialized feeding niches of various species of birds in a coastal wetland. Such resource partitioning reduces competition and allows sharing of limited resources.

5-5 SPECIES INTERACTIONS

How Do Species Interact? An Overview When different species in an ecosystem have activities or resource requirements in common, they may interact with one another. Members of these species may be harmed by, benefit from, or be unaffected by the interaction. There are five basic types of interactions among species: *interspecific competition, predation, parasitism, mutualism,* and *commensalism.*

These interactions tend to regulate the populations of species and can help them to survive changes in environmental conditions, as discussed in more detail in Section 7-2.

How Do Species Compete with One Another? As long as commonly used resources are abundant, different species can share them. This allows each species to come closer to occupying the *fundamental niche* it could theoretically use if there were no competition from other species.

However, most species face competition from other species for one or more limited resources (such as food, sunlight, water, soil nutrients, space, nest sites, and good places to hide). Because of such **interspecific competition** (competition among species), parts of the fundamental niches of different species overlap. The more the niches of two species overlap, the more they compete with one another. With significant niche overlap, one of the competing species must **(1)** migrate to another area (if possible), **(2)** shift its feeding habits or behavior through natural selection and evolution, **(3)** suffer a sharp population decline, or **(4)** become extinct in that area. Competition among species can be an agent of natural selection that may lead to the evolution of new adaptations and even new species.

How Have Some Species Reduced or Avoided Competition? Over a time scale long enough for evolution to occur, species that compete for the same resources may evolve adaptations that reduce or avoid competition or overlap of their fundamental niches. One way this happens is through **resource partitioning**, the dividing up of scarce resources so that species with similar requirements use them at different times, in different ways, or in different places (Figure 5-9). In effect, they evolve traits that allow them to share the wealth, with each competing species occupying a realized niche that makes up only part of its fundamental niche.

For example, where lions and leopards live in the same area, lions take mostly larger animals as prey; leopards take smaller ones. Hawks and owls feed on similar prey, but hawks hunt during the day; owls hunt at night. Some bird species feed on the ground, whereas others seek food in trees and shrubs. Ecologist Robert H. MacArthur studied the feeding habits of five species of warblers (small insect-eating birds) that coexist in the forests of the northeastern United States and in the adjacent area of Canada. Although they appear to be competing for the same food resources, MacArthur found that the bird species reduce competition by spending at least half their time hunting for insects in different parts of trees (Figure 5-10).

How Do Predator and Prey Species Interact? In **predation**, members of one species (the *predator*) feed

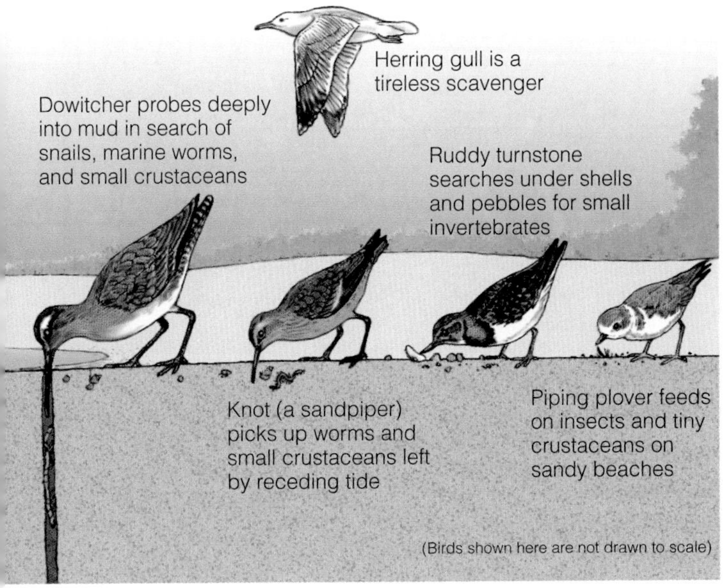

directly on all or part of a living organism of another species (the *prey*). However, they do not live on or in the prey. The prey may or may not die from the interaction. In this interaction the predator benefits, and the individual prey is clearly harmed. Together, the two kinds of organisms, such as lions (the predator or hunter) and zebras (the prey or hunted), are said to have a **predator–prey relationship**, as depicted in Figures 4-9, 4-10, 4-14, and 4-15.

Some people tend to view predators with contempt. When a hawk tries to capture and feed on a rabbit, some tend to root for the rabbit. Yet the hawk (like all predators) is merely trying to get enough food to feed itself and its young; in the process, it is playing an important role in controlling rabbit populations.

What Are Parasites and Why Are They Important? **Parasitism** occurs when one species (the *parasite*) feeds on part of another organism (the *host*) by living on or in the host for a significant portion of the host's life. In this relationship the parasite benefits and the host is harmed. Parasitism can be viewed as a special form of predation, but unlike a conventional predator, a parasite **(1)** usually is smaller than its host (prey), **(2)** remains closely associated with, draws nourishment from, and may gradually weaken its host over time, and **(3)** rarely kills its host.

Tapeworms, disease-causing microorganisms (pathogens), and other parasites live *inside* their hosts. Other parasites, such as ticks, fleas, mosquitoes, mistletoe plants, and fungi (that cause diseases such as athlete's foot) attach themselves to the *outside* of their hosts. Some parasites move from one host to another,

Figure 5-10 Resource partitioning of five species of common insect-eating warblers in spruce forests of Maine. Each species minimizes competition with the others for food by spending at least half its feeding time in a distinct portion (shaded areas) of the spruce trees; each also consumes somewhat different insect species from the others. (After R. H. MacArthur, "Population Ecology of Some Warblers in Northeastern Coniferous Forests," *Ecology*, vol. 36, 533–36, 1958)

as fleas and ticks do; others, such as tapeworms, spend their adult lives with a single host.

From the host's point of view parasites are harmful, but parasites play important ecological roles. Collectively, the incredibly complex matrix of parasitic relationships in an ecosystem acts somewhat like glue that helps hold the species in an ecosystem together. Parasites living within their hosts also help dampen drastic swings in population sizes and parasites promote biodiversity by helping prevent some organisms from becoming too plentiful.

How Do Species Interact So That Both Species Benefit? In **mutualism** two species interact in ways that benefit both. The honeybee and certain flowers have a mutualistic relationship: The honeybee feeds on a flower's nectar and in the process picks up pollen, pollinating female flowers when it feeds on them.

Other important mutualistic relationships exist between animals and the vast armies of bacteria in their stomachs or intestines that break down (digest) their food. The bacteria gain a safe home with a steady food supply; the animal gains access to a large source of energy. In addition to helping digest your food, bacteria in your intestines synthesize vitamin K and the B-complex vitamins, which you can't make.

Many mutualistic relationships provide a combination of nutrition and protection. One involves birds that ride on the backs of large animals such as African buffalo, elephants, and rhinoceroses (Figure 5-11). The birds remove and eat parasites from the animal's body and often make noises warning the animal when predators approach.

In another example of mutualism, various species of clownfish gain protection from sea anemones, marine animals with stinging tentacles that paralyze most fish that touch them (Figure 5-12). The clownfish are not harmed by sea anemones. They gain protection

Figure 5-11 Mutualism. Oxpeckers (or tickbirds) feed on the parasitic ticks that infest large, thick-skinned animals such as this endangered black rhinoceros in Kenya. The rhino benefits by having parasites removed from its body, and oxpeckers benefit by having a dependable source of food. The oxpecker also serves as a sentinel for the rhino by making a fierce hissing sound when alarmed. (Joe McDonald/Tom Stack & Associates)

Figure 5-12 This clownfish in the Coral Sea, Australia, has a mutualistic relationship with deadly sea anemones, whose tentacles quickly paralyze most other fish that touch them. The clownfish gains protection and food by feeding on scraps left over from fish killed by the sea anemone, which, in turn, gains protection by the clownfish against various fish that feed on sea anemones. (Carl Roessler/FPG International)

by living among the deadly tentacles, and they feed on the detritus left from the meals of the anemone. The sea anemones benefit because the clownfish protect them from some predators that eat anemones.

Mutualism is more common when resources are scarce. It is tempting to think of mutualism as an example of cooperation between species, but actually it involves each species benefiting by exploiting the other.

How Do Species Interact So That One Benefits but the Other Is Not Harmed? **Commensalism** is an interaction that benefits one species but neither harms nor helps the other species much, if at all. An example is the commensalistic relationship between various trees and other plants called *epiphytes* (such as various types of orchids and bromeliads) that attach themselves to the trunks or branches of large trees (Figure 5-13) in tropical and subtropical forests. These so-called air plants benefit by having a solid base on which to grow and by living in an elevated spot that gives them better access to sunlight. Their position in the tree allows them to get most of their water from the humid air and from rain collecting in their usually cupped leaves. They also absorb nutrient salts falling from the tree's upper leaves and limbs and from the dust in rainwater. Commensalism exists in cases in which an epiphyte does not harm the tree in which it grows. However, in some cases epiphytes become very abundant and can block light, which becomes an example of interspecific competition.

5-6 ECOLOGICAL SUCCESSION

How Do Ecosystems Respond to Change? One characteristic of all ecosystems is that their structures constantly change in response to changing environmental conditions. The gradual and fairly predictable change in species composition of a given area is called **ecological succession**. Succession involves a complex set of species interactions over time. During succession some species colonize an area and their populations become more numerous, whereas populations of other species decline and even disappear.

Ecologists recognize two types of ecological succession: primary and secondary, depending on the conditions present at the beginning of the process. **Primary succession** involves the gradual establishment of biotic communities in an area that has not been occupied by life before. In contrast, **secondary succession**, the more common type of succession, involves the reestablishment of a biotic community in an area where a biotic community was previously present.

Figure 5-13 Commensalism. This white orchid (an epiphyte or air plant from the tropical forests of Latin America) roots in the fork of a tree rather than the soil without penetrating or harming the tree. In this interaction, the epiphytes gain access to water, nutrient debris, and sunlight; the tree apparently remains unharmed. (Kenneth W. Fink/Ardea London)

What Is Primary Succession? Primary succession begins with an essentially lifeless area, where there is no soil in a terrestrial ecosystem, or no bottom sediment in an aquatic ecosystem (Figure 5-14). Examples of such barren areas include bare rock exposed by a retreating glacier or severe soil erosion, newly cooled lava, an abandoned highway or parking lot, or a newly created shallow pond or reservoir.

Establishment of a new biological community generally is slow. Before a community of plants (producers), consumers, and decomposers can become established on land, there must be *soil*: a complex mixture of rock particles, decaying organic matter, air, water, and living organisms. Depending mostly on the climate, it takes natural processes several hundred to several thousand years to produce fertile soil.

Soil formation begins when a rock or other lifeless area is colonized by a few hardy **pioneer species** (microbes, mosses, and lichens). They are usually species with the ability to establish large populations quickly in a new area, species that biologist Edward O. Wilson calls "nature's sprinters."

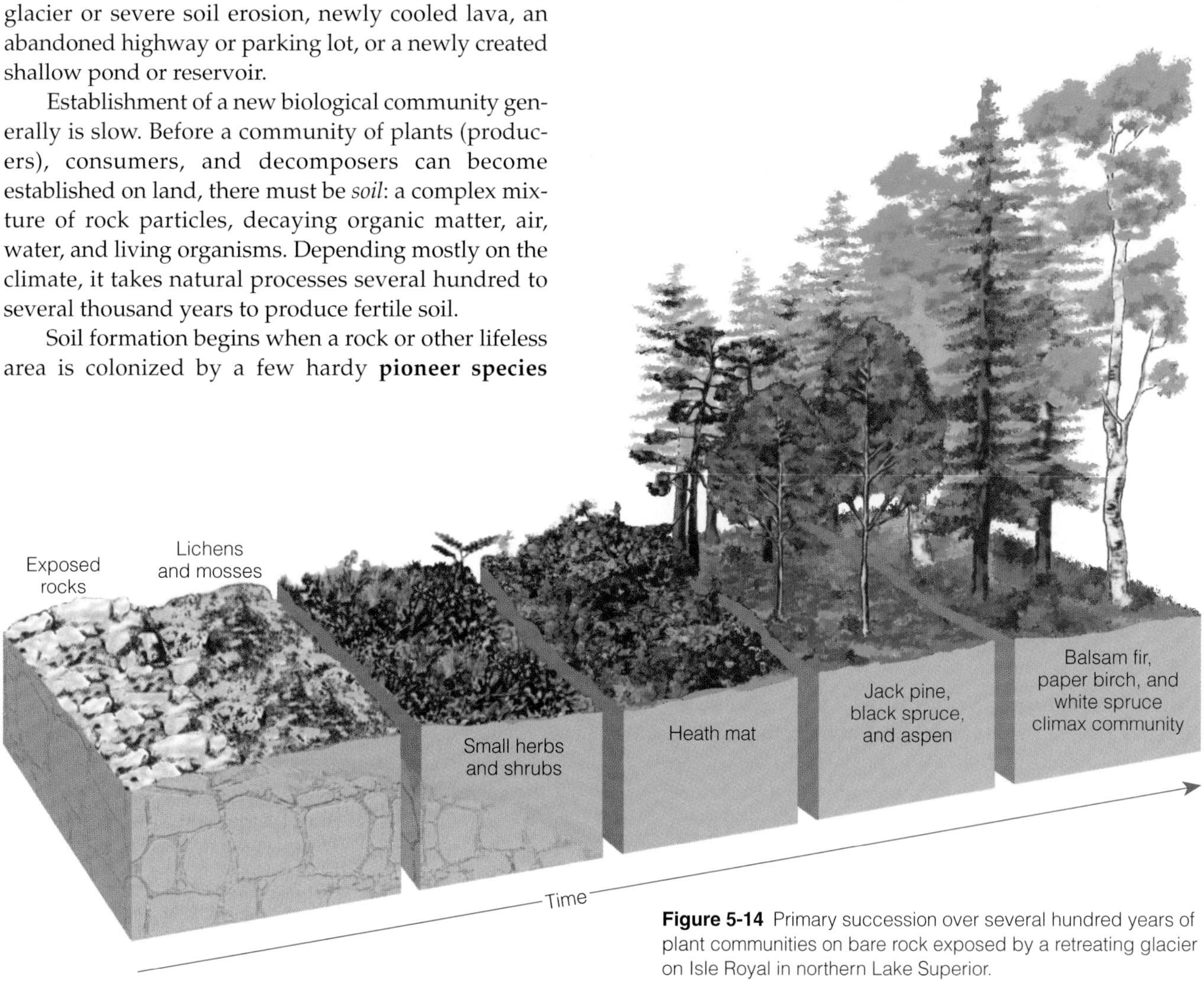

Figure 5-14 Primary succession over several hundred years of plant communities on bare rock exposed by a retreating glacier on Isle Royal in northern Lake Superior.

Figure 5-15 Secondary ecological succession of plant communities on an abandoned farm field in North Carolina. It took about 150–200 years after the farmland was abandoned for the area to be covered with a mature oak and hickory forest. A new disturbance such as deforestation or fire would create conditions favoring pioneer species, and in the absence of new disturbances, secondary succession would again occur over time, although not necessarily in the same sequence shown here.

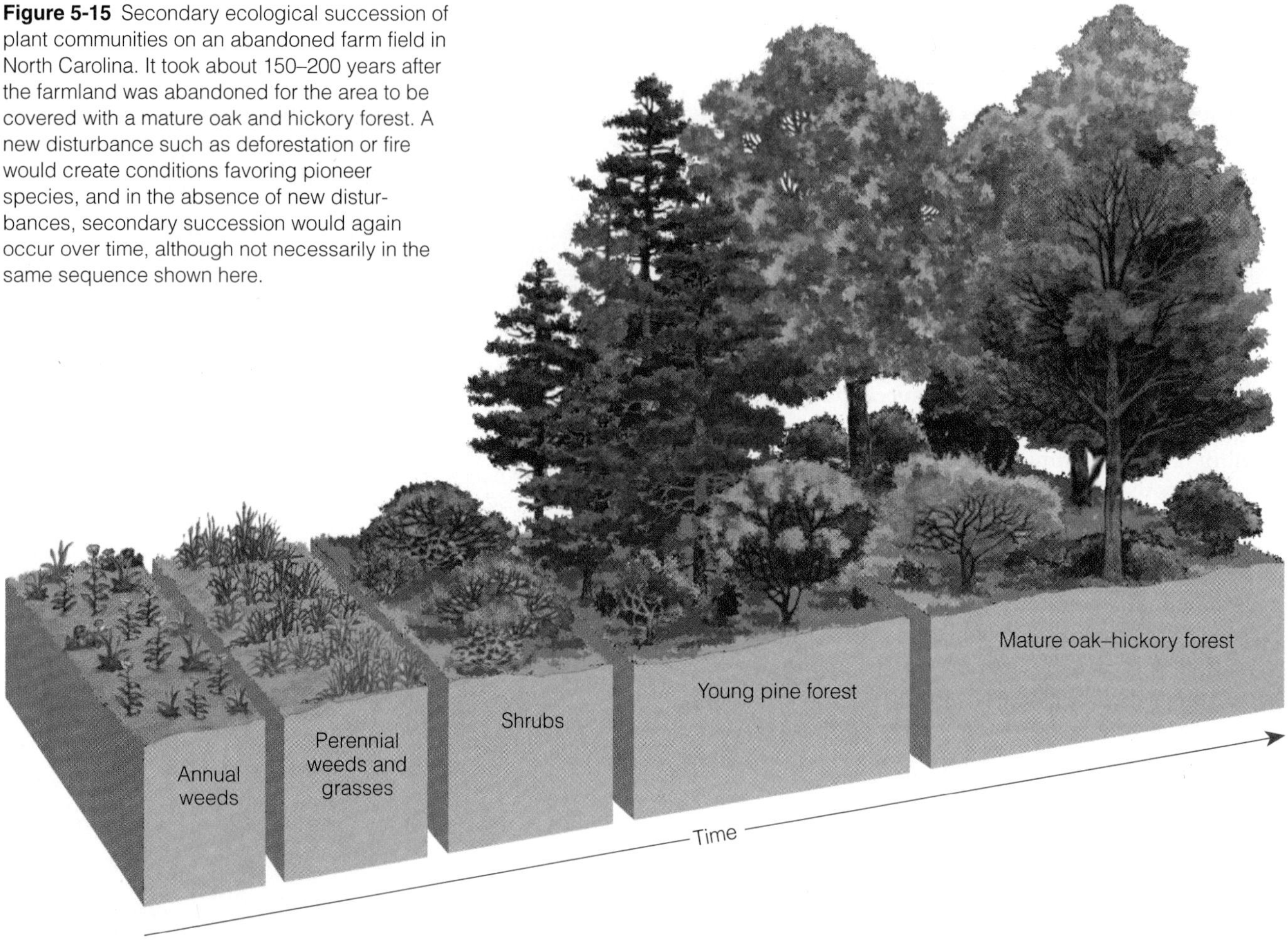

After hundreds of years the soil may be deep and fertile enough to store enough moisture and nutrients to support the growth of less hardy **midsuccessional plant species** of herbs, grasses, and low shrubs. Trees that require lots of sunlight and are adapted to the area's climate and soil replace these species.

As these tree species grow and create shade, they are replaced by **late successional plant species** (mostly trees) that can tolerate shade. Unless fire, flooding, severe erosion, tree cutting, climate change, or other natural or human processes disturb the area, what was once bare rock becomes a fairly stable and complex forest community (Figure 5-14).

What Is Secondary Succession? Secondary succession begins in an area where the natural community of organisms has been disturbed, removed, or destroyed, but some soil or bottom sediment remains. Candidates for secondary succession include abandoned farmlands, burned or cut forests, heavily polluted streams, and land that has been dammed or flooded. Because some soil or sediment is present, new vegetation usually can begin to germinate within a few weeks.

In the central (Piedmont) region of North Carolina, European settlers cleared the mature native oak and hickory forests and replanted the land with crops. Some of the land was subsequently abandoned because of erosion and loss of soil nutrients. Figure 5-15 shows how such abandoned farmland has undergone secondary succession.

Descriptions of ecological succession usually focus on changes in vegetation. However, these changes in turn affect food and shelter for various types of animals. Thus, as succession proceeds the numbers and types of animals and decomposers also change. Figure 5-16 shows some of the wildlife species likely to be found at various stages of secondary ecological succession in areas with a temperate climate.

Because primary and secondary succession involve changes in community structure, it is not surprising that the various stages of succession have different patterns of species diversity, trophic structure, niches, nutrient cycling, and energy flow and efficiency (Table 5-1).

At any time during primary or secondary succession, disturbances such as natural or human-caused fires or deforestation can convert a particular stage of

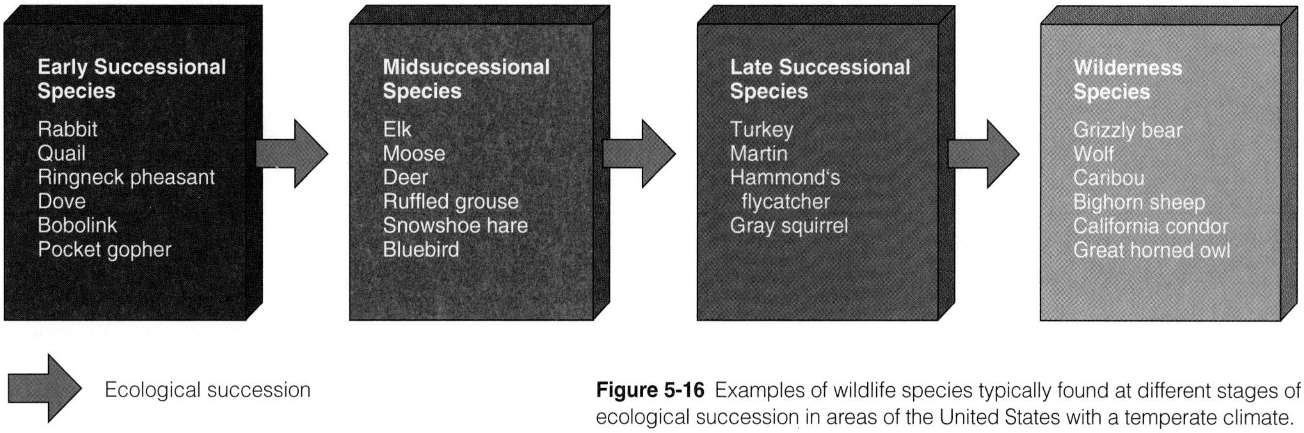

Figure 5-16 Examples of wildlife species typically found at different stages of ecological succession in areas of the United States with a temperate climate.

succession to an earlier stage. Such disturbances create new conditions that encourage some species and discourage or eliminate others.

How Do Agriculture and Plantation Forestry Relate to Succession? The primary goal of agriculture is high productivity by selected plant species such as wheat, rice, and corn. In modern industrialized agriculture, this is achieved by replacing species-rich late successional communities (such as mature grasslands and forests) with an early successional community. Such a community often consists of a single crop (monoculture) of opportunistic species that puts most of its primary productivity into edible parts instead of tall grass, trunks, thick limbs, and roots. Herbicides keep out other species (such as weeds) that would compete with the food crops for soil nutrients.

Similarly, timber companies attempt to increase timber productivity by replacing diverse forests with farms or plantations of a single, fast-growing species (see photo inside table of contents); they sometimes use herbicides to kill competing plants. In tropical areas, people have cut down large areas of species-rich rain forests and replaced them with single-species

Table 5-1 Ecosystem Characteristics at Immature and Mature Stages of Ecological Succession

Characteristic	Immature Ecosystem (Early Successional Stage)	Mature Ecosystem (Late Successional Stage)
Ecosystem Structure		
Plant size	Small	Large
Species diversity	Low	High
Trophic structure	Mostly producers, few decomposers	Mixture of producers, consumers, and decomposers
Ecological niches	Few, mostly generalized	Many, mostly specialized
Community organization (number of interconnecting links)	Low	High
Ecosystem Function		
Biomass	Low	High
Net primary productivity	High	Low
Food chains and webs	Simple, mostly plant → herbivore with few decomposers	Complex, dominated by decomposers
Efficiency of nutrient recycling	Low	High
Efficiency of energy use	Low	High

plantations of banana, rubber, and cocoa plants. Most homeowners spend lots of time and money establishing and keeping monoculture grass lawns at an early stage of secondary succession (and protecting them from invasions by opportunistic species such as crabgrass and other weeds).

How Predictable Is Succession? It is tempting to conclude that ecological succession is an orderly sequence in which each stage leads predictably to the next, more stable stage. According to this classic view, succession proceeds until an area is occupied by a predictable type of *climax community* dominated by a few long-lived plant species (which Edward O. Wilson calls "nature's long-distance runners").

However, research has shown that the sequences of species and community types that appear during primary or secondary succession can be highly variable and unpredictable. We cannot predict the course of a given succession or view it as some preordained progress toward an ideally adapted climax community. Rather, succession reflects the ongoing struggle by different species for enough light, nutrients, food, and space to survive and to gain a reproductive advantage over other species. This explains why a growing number of ecologists prefer terms such as *biotic change* instead of *succession* (which implies an ordered and predictable sequence of changes). Many ecologists have also replaced the term *climax community* with terms such as *mature community*.

5-7 ECOLOGICAL STABILITY AND SUSTAINABILITY

What Is Stability? All living systems, from single-celled organisms to the ecosphere, contain complex networks of negative and positive feedback loops (Section 3-2) that interact to provide some degree of stability or sustainability over each system's expected life span. This stability is maintained only by constant dynamic change in response to changing environmental conditions. For example, in a mature tropical rain forest, some trees die and others take their place. Unless the forest is cut or burned, however, you will still recognize it as a tropical rain forest 50 or 100 years from now.

It is useful to distinguish among three aspects of stability in living systems. **Inertia**, or **persistence**, is the ability of a living system to resist being disturbed or altered. **Constancy** is the ability of a living system such as a population to maintain a certain size or keep its numbers within the limits imposed by available resources. **Resilience** is the ability of a living system to bounce back after an external disturbance that is not too drastic.

Populations, communities, and ecosystems are so complex and variable that ecologists have little understanding of how they maintain inertia, constancy, and resilience while continually responding to changes in environmental conditions. Ecologists also find it difficult to predict which one or combination of environmental factors (Table 5-2) will stress populations or ecosystems beyond their range of tolerance (Figure 4-11).

However, scientists have learned that the signs of ill health in stressed ecosystems include **(1)** a drop in primary productivity, **(2)** increased nutrient losses, **(3)** decline or extinction of indicator species, **(4)** increased populations of insect pests or disease organisms, **(5)** decline in species diversity, and **(6)** the presence of contaminants.

How Do Time Delays Affect Complex Systems? Ecosystems and other complex systems often show **time delays** between the input of a stimulus and the response to it. A long a delay can mean that corrective negative feedback action comes too late. For example, a smoker exposed to cancer-causing chemicals in cigarette smoke may not get lung cancer for 20–30 years; by then it's too late for a negative feedback action (not smoking) to be effective.

Examples in which prolonged delays dampen the negative feedback mechanisms that might slow, prevent, or halt environmental problems are **(1)** population growth, **(2)** toxic dump leaks, **(3)** depletion of ozone in the stratosphere, **(4)** global warming and climate change from carbon dioxide and other greenhouse gases we add to the atmosphere, **(5)** destruction of forests from prolonged exposure to air pollutants, and **(6)** reduction of the population of a species to near extinction.

What Is Synergy and How Can It Affect Complex Systems? In arithmetic, 1 plus 1 always equals 2. But in some of the complex systems found in nature, 1 plus 1 may add up to more than 2 because of synergistic interactions. A **synergistic interaction** occurs when two or more processes interact so that the combined effect is greater than the sum of their separate effects.

Synergy can result when two people work together in accomplishing a task. For example, suppose you and I need to move a 140-kilogram (300-pound) tree that has fallen across the road. By ourselves, each of us can lift only, say, 45 kilograms (100 pounds). However, if we cooperate and use our muscles properly, together we can move the tree out of the way. That's using synergy to solve a problem.

Synergy amplifies the action of positive feedback loops and thus can promote change we believe is favorable. But synergy can also amplify harmful changes.

By identifying potentially harmful synergistic interactions and the leverage points that can activate

Table 5-2 Changes Affecting Ecosystems

Catastrophic*	
Natural	Drought
	Flood
	Fire
	Volcanic eruption
	Earthquake
	Hurricane/tornado
	Landslide
	Change in stream course
	Disease
Human-Caused	Deforestation
	Overgrazing
	Plowing
	Erosion
	Pesticide application
	Fire
	Mining
	Toxic contamination
	Urbanization
	Water and air pollution
	Loss and degradation of wildlife habitat
Gradual*	
Natural	Climatic changes
	Immigration
	Adaption/evolution
	Ecological succesion
	Disease
Human-Caused	Salinization and waterlogging of soils from irrigation
	Soil compaction
	Depletion of groundwater
	Water and air pollution
	Loss and degradation of wildlife habitat
	Elimination of "pests" and predators
	Introduction of nonnative species
	Overhunting/overfishing
	Toxic contamination
	Urbanization
	Excessive tourism

*Many changes can be either catastrophic or gradual.

them, we can anticipate, counteract, and even prevent some environmental problems. Thus, we may be able to counter harmful synergisms, promote beneficial ones, and improve the quality of life on the earth.

Does Species Diversity Increase Ecosystem Stability? In the 1960s, most ecologists believed that the greater the species diversity and the accompanying web of feeding and biotic interactions in an ecosystem, the greater its stability. According to this hypothesis, an ecosystem with a diversity of species and feeding paths has more ways to respond to most environmental stresses because it does not have "all its eggs in one basket." However, most recent research indicates that there are many exceptions to this intuitively appealing idea.

Of course, there is a minimum threshold of species diversity below which ecosystems cannot function; no ecosystem can function without some plants and decomposers. (Note that people and other animal consumers are not absolutely necessary for life to continue on the earth.) Beyond this, it is difficult to know whether simple ecosystems are less stable than complex ones, or to identify the threshold below which complex ecosystems fail. In part because some species play redundant roles (niches) in ecosystems, we don't know how many or which species can be eliminated before the entire ecosystem begins to lose stability or collapse.

Research indicates that ecosystems with more species tend to have higher net primary productivities than simpler ecosystems and can also be more resilient. For example, a recent study of grasslands showed that species-rich fields had a smaller decline in net primary productivity during prolonged drought and rebounded more quickly than species-poor fields in the same area.

This supports the idea that some level of biodiversity provides insurance against catastrophe. But there is uncertainty over how much biodiversity is needed in various ecosystems. For example, some recent research suggests that average annual net primary productivity reaches a peak at 10–40 producer species. Many ecosystems contain more producer species than this, but it is difficult to distinguish between those that are essential and those that aren't.

Part of the problem is that ecologists disagree on how to define *stability* and *diversity*. Does an ecosystem need both high inertia and high resilience to be considered stable? Evidence suggests that some ecosystems have one of these properties but not the other. For example, tropical rain forests have high species diversity and high inertia; that is, they are resistant to significant alteration or destruction. However, once a large tract of tropical forest is severely degraded, the ecosystem's resilience is so low that the forest may not be restored. Nutrients (which are stored primarily in the vegetation, not in the soil) and other factors needed for recovery may no longer be present. Such a large-scale loss of forest cover may so change the local or regional climate that forests can no longer be supported.

Grasslands, by contrast, are much less diverse than most forests, and because they burn easily they have low inertia. However, because most of their plant matter is stored in underground roots, these ecosystems have high resilience and recover quickly. A grassland can be destroyed only if its roots are plowed up and something else is planted in its place, or if it is severely overgrazed by livestock or other herbivores.

Another difficulty is that populations, communities, and ecosystems are rarely, if ever, at equilibrium. Instead, nature is in a continuing state of disturbance, fluctuation, and change; this means that an ecologist studying an ecosystem is trying to investigate a moving target. Indeed, ecologists recognize that disturbances of ecosystems are integral parts of the way nature works. Clearly, we have a long way to go in understanding how the factors in natural communities and ecosystems respond to changes in environmental conditions.

What Determines the Number of Species in an Ecosystem? There are no simple answers to this question. Two factors affecting the species diversity (or richness) of an ecosystem are its size and degree of isolation. In the 1960s Robert MacArthur and Edward O. Wilson began studying communities on islands to discover why large islands tend to have more species of a certain category (such as insects, birds, or ferns) than do small islands.

To explain these differences in species diversity with island size, MacArthur and Wilson proposed what is called the **species equilibrium model** or the **theory of island biogeography**. According to this model, the number of species found on an island is determined by a balance between two factors: the *immigration rate* of species to the island from other inhabited areas and the *extinction rate* of species established on the island. The model predicts that at some point the rates of immigration and extinction will reach an equilibrium point (Figure 5-17a) that determines the island's average number of different species (species diversity).

The model also predicts that immigration and extinction rates (and thus species diversity) are, in turn, affected by two important variables: the *size of the island* (Figure 5-17b) and its *distance from a mainland source of immigrant species* (Figure 5-17c). According to the model, a small island tends to have a lower species diversity than a large one for two reasons: First, a small island is a smaller target for potential colonizers, which results in lower immigration rates than for a larger island. Second, because a small island normally has fewer resources and less diverse habitats, it should have a higher extinction rate (because of increased competitive exclusion) than a larger island.

The model also predicts that an island's distance from a mainland source of new species is important in determining its species diversity. Assume that we have two islands of about equal size and that all other factors are roughly the same. According to the model, the

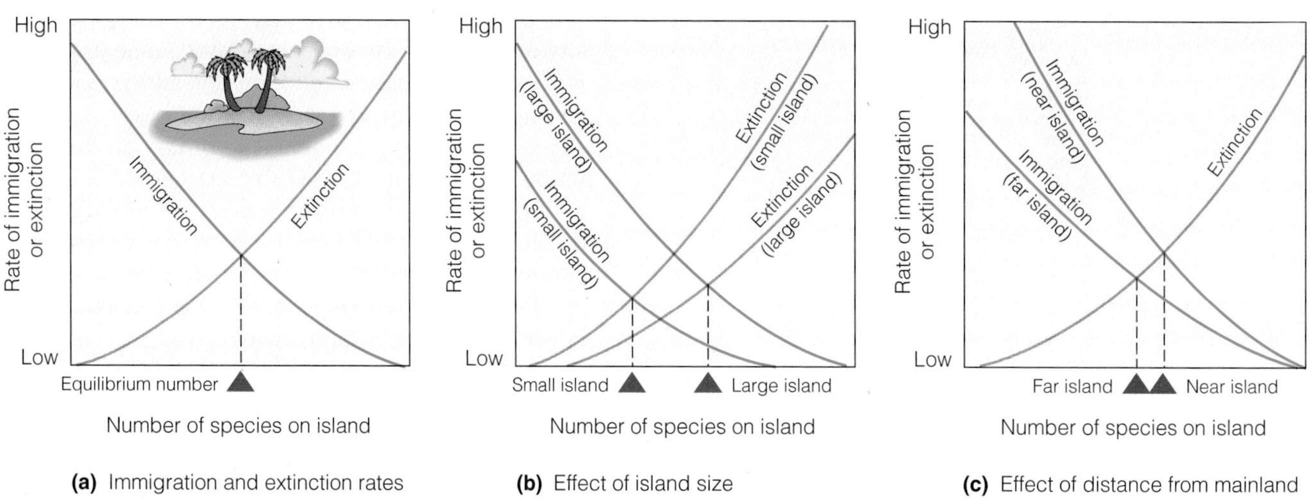

Figure 5-17 The species equilibrium model or theory of island biogeography, developed by Robert MacArthur and Edward O. Wilson. **(a)** The equilibrium number of species (blue triangle) on an island is determined by a balance between the immigration rate of new species and the extinction rate of species already on the island. **(b)** With time, large islands will have a larger equilibrium number of species than smaller islands because of higher immigration rates and lower extinction rates on large islands. **(c)** Assuming equal extinction rates, an island near a mainland will have a larger equilibrium number of species than a more distant island because the immigration rate to a near island is higher than that to a more distant one.

CONNECTIONS

Using Island Biogeography Theory to Protect Mainland Communities and Species

Ecologists and conservation biologists are now applying the species equilibrium model or theory of island biogeography (Figure 5-17) to communities on the mainland. Most remaining wildlife sanctuaries on land are *habitat islands* surrounded by an inhospitable "sea" of disturbed and unsuitable habitat. According to the species equilibrium model, the species diversity in these terrestrial patches or islands should be determined by their size and their distance from other patches that serve as sources of colonists.

Conservation biologists are using the model to help them locate (and try to protect) areas in the most danger of losing species diversity. Researchers are also using the model to estimate how large a particular nature preserve should be to prevent it from losing species.

The size of a preserve also depends on which species we hope to protect. For example, endangered species such as tigers and grizzly bears need much larger areas of undisturbed habitat than many other species to preserve a viable population.

Conservation biologists are also using the species equilibrium model to estimate how closely a series of small wildlife reserves should be spaced to allow the possibility of immigration from one preserve to another if a species in one reserve becomes locally extinct.

In addition, they are using the model to estimate the size and number of protected corridors needed to connect various reserves and encourage the spread of protected species among them. Much more research must be done to answer such questions, but progress is being made through this application of ecological theory to wildlife conservation.

Critical Thinking

How would you respond to the statement that the theory of island biogeography should not be taken too seriously because it is only a theory?

island closest to a mainland source of immigrant species will have a higher immigration rate and thus a higher species diversity (assuming that extinction rates on both islands are about the same).

MacArthur and Wilson's original model or scientific hypothesis has been tested and supported by a series of field experiments. As a result, biologists have accepted it well enough to elevate it to the status of an important and useful scientific theory. In recent years, it has even been applied to conservation efforts to protect wildlife on land (Connections, above).

In this chapter we have seen that interdependence and connectedness are the essential features of the living and nonliving parts of individual ecosystems, and of the ecosphere. The next two chapters show how this interdependence is the key to understanding the earth's major life zones and ecosystems and how populations can change when environmental conditions change.

We need to appreciate that a species, of whatever form, is a unique manifestation of life, the final product of some evolutionary development.

IRSTON R. BARNES

REVIEW QUESTIONS

1. Define all of the boldfaced terms in this chapter.

2. Why should you care about alligators? Explain their ecological importance.

3. Distinguish between *eukaryotic* and *prokaryotic* organisms and give an example of each. Describe the five major kingdoms of organisms and give three examples of each type.

4. Distinguish between *plants* and *animals* and between *vertebrates* and *invertebrates.*

5. What are *fossils* and how do they help us formulate ideas about how life developed on earth? Describe the conditions that make life on the earth just right for life as we know it. Distinguish between *chemical evolution* and *biological evolution*. Give a summary of how scientists think chemical evolution and biological evolution took place on the earth. What was the *oxygen revolution* and what is its importance to life as we know it?

6. Distinguish between *biological evolution*, *the theory of evolution*, *microevolution*, and *macroevolution*. Use the concepts of genes, gene pool, alleles, and mutations to explain how microevolution takes place. Define *natural selection*, *differential reproduction*, and *adaptation* and explain their roles in microevolution. Give an example of microevolution by natural selection. What is *coevolution* and what is its importance? List three factors that limit adaptation. What are three common misconceptions about evolution?

7. Explain *geographic isolation* and *reproductive isolation* and explain how they can lead to speciation through divergent evolution.

8. List three factors that have affected the earth's long-term patterns of speciation and extinction. Distinguish between *background extinction* and *mass extinction*. What is believed to have caused the five great mass extinctions in the past?

9. What is an *adaptive radiation?* How can such radiation lead to recovery after mass extinctions? Explain how speciation and extinction result in the planet's biodiversity. What evidence suggests that humans may be bringing about a sixth mass extinction?

10. What is the *ecological niche* of a species and why is it important to understand the niches of species? What is the difference between a species' *habitat* and *niche*? What is the difference between a species' *fundamental* niche and its *realized niche*?

11. List the major characteristics of *generalist* and *specialist* species. Give two examples of each type of species. Is it better to be a generalist or specialist? Explain.

12. Distinguish between *native, nonnative, indicator*, and *keystone* species and give an example of each. What can happen in an ecosystem that loses a keystone species?

13. Distinguish between *interspecific competition, predation, parasitism, mutualism,* and *commensalism* and give an example of each. What are the four options when there is significant overlap in the niches of two species in the same area?

14. Explain how *resource partitioning* allows species to avoid overlap of their fundamental niches and give two examples of this phenomenon.

15. Explain the *predator–prey relationship* and give two examples of this type of species interaction.

16. Explain the importance of *mutualism* and give an example of this type of species interaction.

17. Explain the importance of *commensalism* and give an example of this type of species interaction.

18. Distinguish between *primary succession* and *secondary succession*. Distinguish between *pioneer species, midsuccessional plant species,* and *late successional plant species.*

19. Distinguish between *inertia, constancy,* and *resilience* and explain how they help maintain stability in an ecosystem. What are six signs of ill health in a stressed ecosystem? How can time delays affect an ecosystem's ability to respond to stress? What is a *synergistic interaction* and how can it harm or benefit an ecosystem? Give examples of a harmful and a beneficial synergistic interaction.

20. Does species diversity increase ecosystem stability? List the pros and cons of species diversity.

21. Use the *theory of island biodiversity* to explain why ecosystems differ in the number of species they contain.

CRITICAL THINKING

1. In what ways are humans generalists? In what ways are we specialists? Do you think humans should become more dependent on a greater variety of foods and energy resources to promote long-term survival and sustainability? Explain and discuss the short- and long-term economic and political consequences of such a shift.

2. How would you respond to someone who says that they don't believe in evolution because it is "just a theory"?

3. Someone tells you not to worry about air pollution because through natural selection the human species will develop lungs that can detoxify pollutants. How would you reply?

4. What would happen if extinction had never occurred during evolutionary history? Do you think we would have more species than we do now? Why? What do you think would happen to biodiversity if extinction rates occurred throughout evolutionary history at the same rate they have been occurring during the past 50 years?

5. How would you respond to someone who says that extinction is a natural process and that we shouldn't worry about the loss of biodiversity.

6. Most domesticated animals and plants that we use for food benefit from a mutualistic relationship with us. In exchange for food, we care for them and ensure their continued reproduction. However, through crossbreeding and genetic engineering (to increase food production) we have eliminated many of the adaptive traits they need for survival in the wild. How do you feel about this form of mutualism we have imposed on other species? How would your life be changed if we decided to end this mutualistic relationship? Do you believe that domesticated animals we raise should be treated as humanely as possible during the time we are fattening them up for slaughter? Explain your answer. What effect might this have on the price and availability of meat produced from such animals?

***7.** An important adaptation of humans is a strong opposable thumb, which allows us to grip and manipulate things with our hands. As a demonstration of the importance of this trait, fold each of your thumbs into the palm of its hand and then tape them securely in that position for an entire day. After the demonstration, make a list of the things you could not do without the use of your thumbs.

***8.** Conduct field studies and try to find the research results of others to help you identify the major native, nonnative, indicator, and keystone species in the natural areas around where you live.

***9.** Visit a nearby land area or aquatic system (such as a wetland) and look for and record signs of ecological succession. Study the area carefully to see whether you can find patches that are at different stages of succession because of various disturbances.

***10.** Make a concept map of this chapter's major ideas, using the section heads and subheads and the key terms (in boldface). Look at the inside back cover and in Appendix 5 for information about concept maps.

FURTHER INFORMATION

For more information see Internet Sources and Further Readings at the back of the book. Online resources for this chapter can be found at **www.brookscole.com/biology**

6 CLIMATE, WEATHER, AND BIODIVERSITY

Connections: Blowing in the Wind

Environmental science is a study of connections. One of the things that connects all life on the earth is wind, a vital part of the planet's circulatory system. Without wind, most of the earth would be uninhabitable: The tropics would be unbearably hot, and most of the rest of the planet would freeze.

Winds also transport nutrients from one place to another. Dust rich in phosphates blows across the Atlantic from the Sahara Desert in Africa (Figure 6-1), helping to replenish rain forest soils in Brazil and build up agricultural soils in the Bahamas. Iron-rich dust blowing from China's Gobi Desert falls into the Pacific Ocean between Hawaii and Alaska and stimulates the growth of phytoplankton, the minute producers that support ocean food webs. That's the *good news.*

The *bad news* is that wind also transports harmful substances. Particles of reddish-brown soil and pesticides banned in the United States are blown from Africa's deserts and eroding farmlands into the sky over Florida. This makes it difficult for the state to meet federal air pollution standards and is suspected to be been a factor in killing coral reefs in the Florida Keys and the Caribbean.

Industrial pollution and dust from rapidly industrializing China and central Asia blow across the Pacific Ocean and degrade air quality over the western United States, especially Washington and Oregon. Studies show that Asian pollution contributes as much as 10% to West Coast smog.

There's *mixed news* as well. Particles from volcanic eruptions ride the winds, circle the globe, and change the earth's climate for a while. Emissions from the 1991 eruption of Mount Pinatubo in the Philippines—the largest eruption this century—cooled the earth slightly for 3 years, temporarily masking signs of global warming. On the other hand, volcanic ash, like the blowing desert dust, adds valuable trace minerals to the soil where it settles.

The lesson, once again, is that *there is no "away,"* and wind—acting as part of the planet's circulatory system for heat, moisture, and plant nutrients—is one reason. Movement of soil particles from one place to another by wind and water is a natural phenomenon, but when we disturb the soil and leave it unprotected we hasten the process.

Wind is also an important factor in climate through its influence on global air circulation patterns. Climate, in turn, is crucial for determining what kinds of plant and animal life are found in the major biomes of the different geographic regions of the ecosphere, as we shall see in this chapter.

Figure 6-1 Some of the dust shown here blowing from Africa's Sahara Desert can end up as soil nutrients in Amazonian rain forests. (NOAA. USGS/NMD EROS Data Center)

When we try to pick out anything by itself, we find it hitched to everything else in the universe.

John Muir

This chapter addresses the following questions:

- What key factors determine variations in the earth's weather and climate?
- How does climate determine where the major biomes are found?
- What are the major types of desert, grassland, and forest biomes, and how do human activities affect them?
- What are the basic types of aquatic life zones, and what factors influence the kinds of life they contain?
- What are the major types of saltwater life zones, and how do human activities affect them?
- What are the major types of freshwater life zones, and how do human activities affect them?

6-1 WEATHER AND CLIMATE: A BRIEF INTRODUCTION

How Does Weather Differ from Climate? At every moment at any spot on the earth, the *troposphere* (the inner layer of the atmosphere containing most of the earth's air) has a particular set of physical properties. Examples are temperature, pressure, humidity, precipitation, sunshine, cloud cover, and wind direction and speed. These short-term properties of the troposphere at a given place and time are what we call **weather**.

Climate is the average long-term weather of an area; it is a region's general pattern of atmospheric or weather conditions, including seasonal variations and weather extremes (such as hurricanes or prolonged drought or rain) averaged over a long period (at least 30 years). Figure 6-2 is a generalized map of the earth's major climate zones.

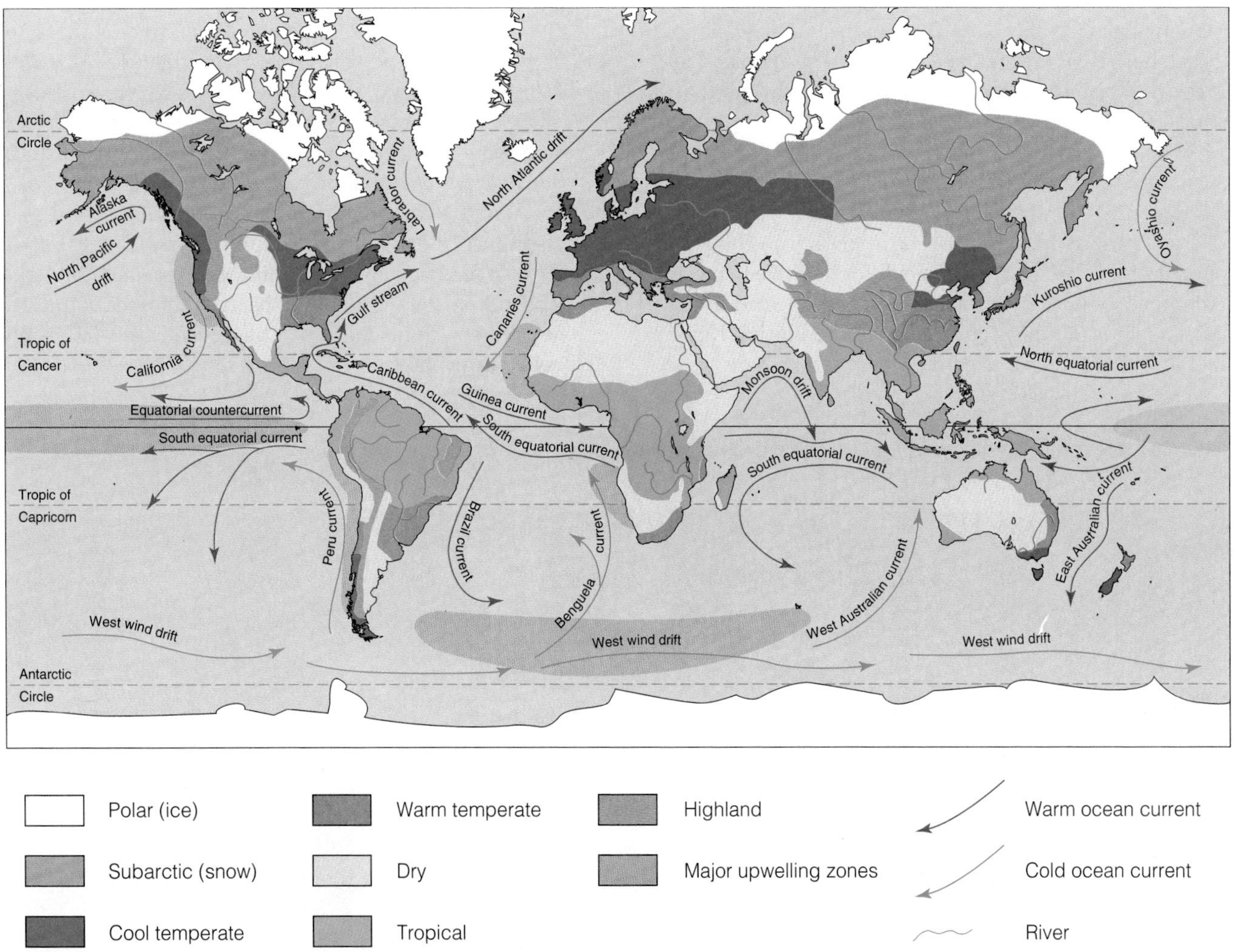

Figure 6-2 Generalized map of global climate zones, showing the major contributing ocean currents and drifts. Large variations in climate are dictated mainly by temperature (with its seasonal variations) and by the quantity and distribution of precipitation.

Galtie, A. 1999. "Is El Nino Now a Man-Made Phenomenon?" *The Ecologist,* vol. 29, no. 2, p. 64(4).

How Does the Global Circulation of Air Affect Regional Climates? *Average temperature* and *average precipitation* are the two main factors determining a region's climate. The temperature and precipitation patterns that lead to different climates (Figure 6-2) are caused primarily by the way air circulates over the earth's surface.

The following factors determine global air circulation patterns:

- *Uneven heating of the earth's surface* because air is heated much more at the equator (where the sun's rays strike directly throughout the year) than at the poles (where sunlight strikes at an angle and thus is spread out over a much greater area). These differences help explain why tropical regions near the equator are hot, polar regions are cold, and temperate regions in between generally have intermediate average temperatures (Figure 6-2).
- *Seasonal changes because the earth's axis* (an imaginary line connecting the north and south poles) is *tilted*. As a result, various regions are tipped toward or away from the sun as the earth makes its annual revolution (Figure 6-3). This creates opposite seasons in the northern and southern hemispheres.
- *Rotation of the earth on its axis*, which prevents air currents from moving due north and south from the equator. Forces in the atmosphere created by this rotation deflect winds (moving air masses). The result is six huge convection cells of swirling air masses—three north and three south of the equator—that transfer heat and water from one area to another (Figure 6-4).
- *Long-term variations in the amount of solar energy striking the earth.* These are caused by occasional changes in solar output and slight planetary shifts in which the earth's axis wobbles (22,000-year cycle) and tilts (44,000-year cycle) as it revolves around the sun.
- *Properties of air and water*. Heat from the sun evaporates ocean water and transfers heat from the oceans to the atmosphere, especially near the hot equator. As this moisture-laden air rises, it cools and releases moisture as condensation (because cold air can hold less water vapor than warm air). The resulting cooler, drier air becomes denser, sinks (subsides), and creates an area of high pressure. As these air masses flow across the earth's surface, they pick up heat and moisture and begin to rise again. The resulting convection cells circulate air, heat, and moisture both vertically and from place to place in the troposphere, leading to different climates and patterns of vegetation (Figure 6-5).

How Do Ocean Currents Affect Regional Climates? The factors just listed, plus differences in water density, create warm and cold ocean currents (Figure 6-2). These currents, like air currents, redistribute heat received from the sun and thus influence climate and vegetation, especially near coastal areas.

For example, without the warm Gulf Stream, which transports 25 times more water than all the world's rivers, the climate of northwestern Europe would be subarctic. Currents also help mix ocean waters and distribute nutrients and dissolved oxygen needed by aquatic organisms.

Along some steep western coasts of continents, almost constant trade winds blow offshore, pushing

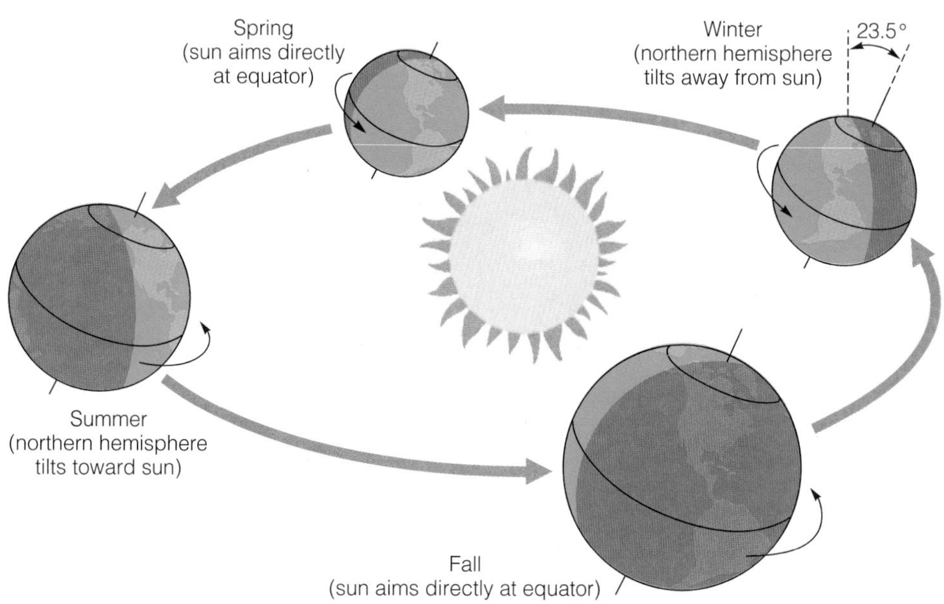

Figure 6-3 The effects of the earth's tilted axis on climate. As the planet makes its annual revolution around the sun on an axis tilted about 23.5°, various regions are tipped toward or away from the sun. The resulting variations in the amount of solar energy reaching the earth create the seasons.

Hint: Enter the search terms *El Nino, environmental aspects* using the Subject Guide.

Figure 6-4 Formation of prevailing surface winds, which disrupt the general flow of air from the equator to the poles and back to the equator. As the earth rotates, its surface turns faster beneath air masses at the equator and slower beneath those at the poles. This deflects air masses moving north and south to the west or east, creating six huge convection cells in which air swirls upward and then descends toward the earth's surface at different latitudes. The direction of air movement in these cells sets up belts of prevailing winds that distribute air and moisture over the earth's surface. These winds affect the general types of climate found in different areas and drive the circulation of ocean currents. (Used by permission from Cecie Starr and Ralph Taggart, *Biology: The Unity and Diversity of Life*, 8th ed., Belmont, Calif.: Wadsworth, 1998)

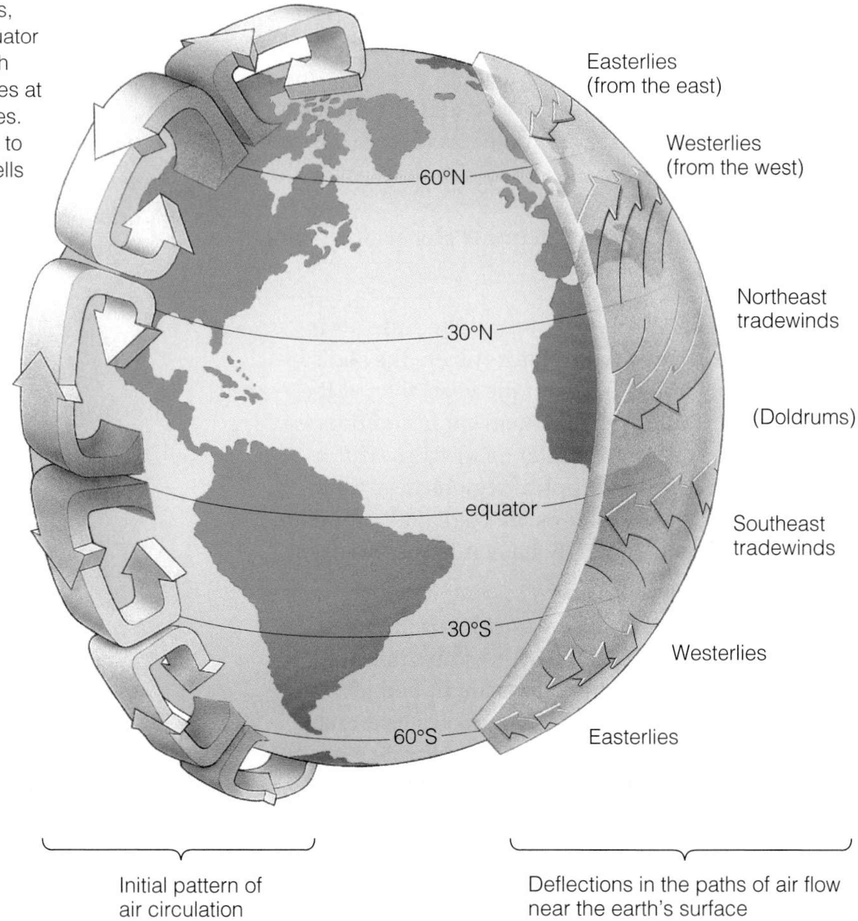

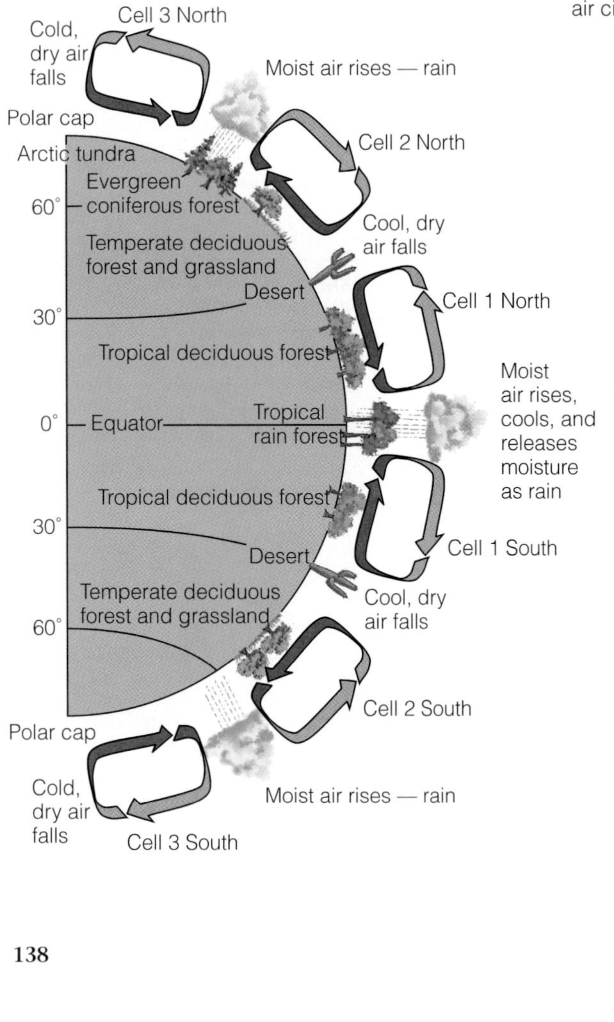

Figure 6-5 Model of global air circulation and biomes. Heat and moisture are distributed over the earth's surface by vertical currents that form six large convection cells (called Hadley cells) at different latitudes. The direction of air flow and the ascent and descent of air masses in these convection cells determine the earth's general climatic zones. The resulting uneven distribution of heat and moisture over the planet's surface leads to the forests, grasslands, and deserts that make up the earth's biomes.

surface water away from the land. This outgoing surface water is replaced by an **upwelling** of cold, nutrient-rich bottom water. Upwellings, whether far from shore or near shore (Figure 6-2), bring plant nutrients from the deeper parts of the ocean to the surface and support large populations of phytoplankton, zooplankton, fish, and fish-eating seabirds.

What Is the El Niño–Southern Oscillation? Every few years, in the Pacific Ocean, normal coastal upwelling (Figure 6-6, left) is affected by changes in climate patterns called the *El Niño–Southern Oscillation* or *ENSO* (Figure 6-6, right), often called *El Niño.* In an ENSO, the prevailing westerly winds weaken or cease (for reasons that are not clear) and surface water along the South and North American coasts becomes warmer. The normal upwelling of cold, nutrient-rich water is suppressed, which reduces primary productivity and causes a sharp decline in the populations of some fish species.

A strong ENSO can trigger extreme weather changes over at least two-thirds of the globe, especially in lands along the Pacific and Indian Oceans (Figure 6-7). In 1997 and 1998, the world experienced the strongest ENSO (El Niño) during the past 200 years. It came only 14 years after the second strongest ENSO in 1982–83 and only 2 years after a moderate El Niño between 1992 and 1997 (the longest ever documented). Some models project that if global warming continues, El Niño–like weather may become the norm and cause major ecological and socioeconomic problems.

Sometimes an *El Niño* is followed by its cooling counterpart *La Niña,* as occurred from July 1998 into 1999. Typically a *La Niña* **(1)** means more Atlantic Ocean hurricanes, **(2)** colder winters in Canada, **(3)** warmer and drier winters in the southeastern and southwestern U.S., **(4)** wetter winters in the Pacific Northwest, **(5)** torrential rains in Southeast Asia, **(6)** lower wheat yields in Argentina, and **(7)** more wildfires in Florida.

How Does the Chemical Makeup of the Atmosphere Lead to the Greenhouse Effect? Small amounts of gases such as water vapor (H_2O), carbon dioxide (CO_2), ozone (O_3), methane (CH_4), nitrous oxide (N_2O), and chlorofluorocarbons (CFCs) play a key role in determining the earth's average temperatures and thus its climates.

Together, these gases, known as **greenhouse gases**, act somewhat like the glass panes of a greenhouse: They allow light, infrared radiation, and some ultraviolet radiation from the sun (Figure 4-7) to pass through the troposphere. The earth's surface absorbs much of this solar energy and degrades it to longer-wavelength, infrared radiation (that is, heat), which then rises into the troposphere (Figure 4-7). Some of this heat escapes into space and some is absorbed by molecules of greenhouse gases and emitted into the

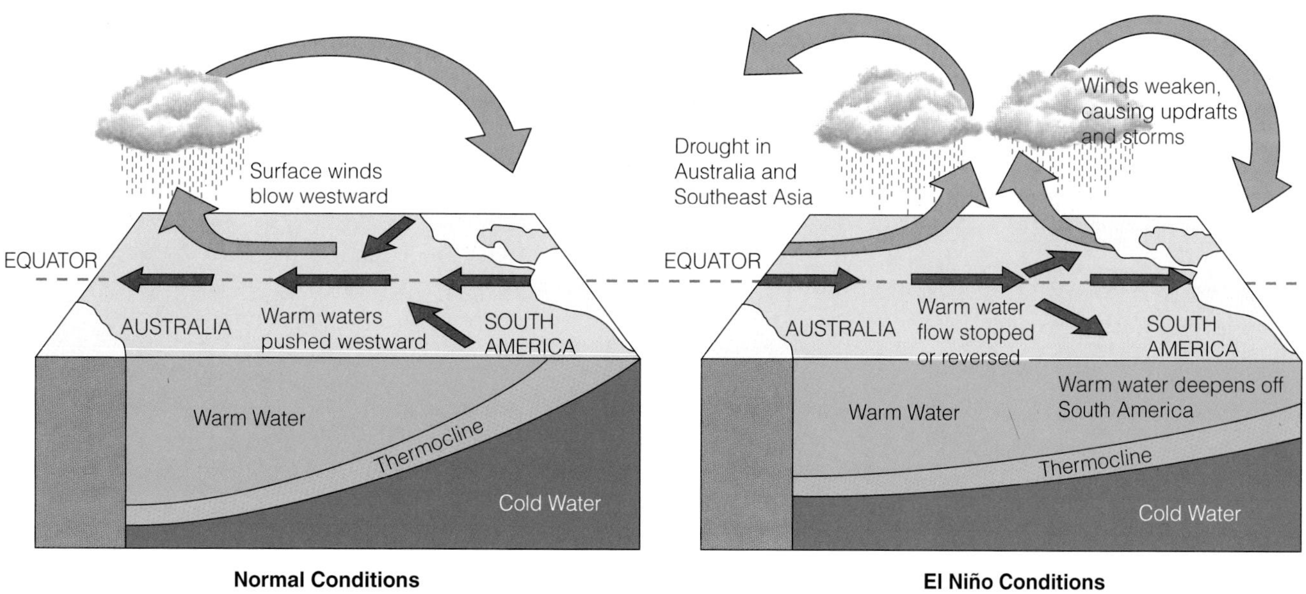

Figure 6-6 Normal surface winds blowing westward cause shore upwellings of cold, nutrient-rich bottom water in the tropical Pacific Ocean near the coast of Peru (left). The warm and cold water are separated by a zone of gradual temperature change called the thermocline. Every few years a climate shift known as the *El Niño–Southern Oscillation* (*ENSO*) disrupts this pattern. Westward surface winds weaken, which depresses the coastal upwellings and warms the surface waters off South America (right). When an ENSO lasts 12 months or longer, it severely disrupts populations of plankton, fish, and seabirds in upwelling areas and can trigger extreme weather changes over much of the globe (Figure 6-7).

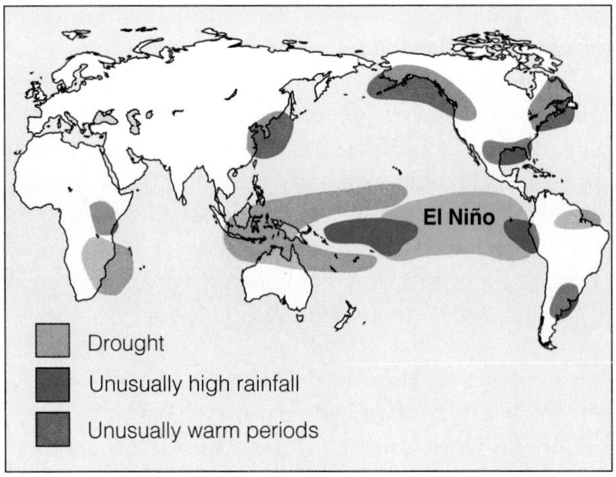

Figure 6-7 Typical global climatic effects of an El Niño–Southern Oscillation. During the 1996–98 ENSO, huge waves battered the California coast and torrential rains caused widespread flooding and mudslides. In Peru, hundreds of people were killed by floods and mudslides, about 250,000 people were left homeless, and harvests were ruined. Drought in Brazil, Indonesia, and Australia led to massive wildfires in tinder-dry forests. India and parts of Africa also experienced severe drought. A catastrophic ice storm hit Canada and the northeastern United States, but the southeastern U.S. had fewer hurricanes. (Data from United Nations Food and Agriculture Organization)

troposphere as infrared radiation, which warms the air. This natural warming effect of the troposphere is called the **greenhouse effect** (Figure 6-8).

The basic principle behind the greenhouse effect is well established and has been confirmed by satellite measurements. Indeed, without its current greenhouse gases (especially water vapor), the earth would be a cold and lifeless planet with an average surface temperature of −18°C (0°F) instead of its current 15°C (59°F).

We and other species currently benefit from a comfortable level of greenhouse gases that typically undergo only minor, slow fluctuations over hundreds to thousands of years. However, mathematical models of the earth's climate indicate that natural or human-induced global warming taking place over a few decades could be disastrous for human societies and many forms of life, as discussed in Chapter 11.

How Does the Chemical Makeup of the Atmosphere Lead to the Ozone Layer? In a band of the stratosphere 16–26 kilometers (11–16 miles) above the earth's surface, oxygen (O_2) is continuously converted to ozone (O_3) and back to oxygen by a sequence of reactions initiated by ultraviolet radiation from the sun ($3O_2 + UV \rightleftharpoons 2O_3$). The result is a thin veil of protective ozone at very low concentrations (up to 10 parts per million). Normally, the average levels of ozone in this life-saving layer don't change much, because the rate of ozone destruction is equal to its rate of formation.

Ultraviolet (UV) radiation reaching the stratosphere is composed of three bands: A, B, and C. The ozone layer blocks out nearly all the highest-energy, shortest-wavelength radiation (UV-C), approximately half of the next highest band (UV-B), and a small part of the lowest-energy radiation (UV-A). Stratospheric

a. Rays of sunlight penetrate the lower atmosphere and warm the earth's surface.

b. The earth's surface absorbs much of the incoming solar radiation and degrades it to longer-wavelength infrared radiation (heat), which rises into the lower atmosphere. Some of this heat escapes into space and some is absorbed by molecules of greenhouse gases and emitted as infrared radiation, which warms the air.

c. As concentrations of greenhouse gases rise, their molecules absorb and emit more infrared radiation, which adds more heat to the lower atmosphere.

Figure 6-8 The greenhouse effect. Without the atmospheric warming provided by this natural effect, the earth would be a cold and mostly lifeless planet. According to the widely accepted greenhouse theory, when concentrations of greenhouse gases in the atmosphere rise, the average temperature of the troposphere also rises. (Modified by permission from Cecie Starr, *Biology: Concepts and Principles*, 4th ed., Pacific Grove, Calif.: Brooks/Cole, 2000) ozone prevents at least 95% of the sun's harmful ultraviolet radiation from reaching the earth's surface. This ozone creates warm layers of air that prevent churning gases in the troposphere from entering the stratosphere. This *thermal cap* is important in determining the average temperature of the troposphere and thus the earth's current climates. There is considerable evidence that chemicals added to the atmosphere by human activities are decreasing levels of protective ozone in the stratosphere, as discussed in more detail in Chapter 11.

6-2 BIOMES: CLIMATE AND LIFE ON LAND

Why Are There Different Organisms in Different Places? Why is one area of the earth's land surface a desert, another a grassland, and another a forest? Why are there different types of deserts, grasslands, and forests?

The general answer to these questions is differences in *climate* (Figure 6-2), caused mostly by differences in average temperature and precipitation caused by global air circulation (Figure 6-5). Figure 6-9 and the photo on

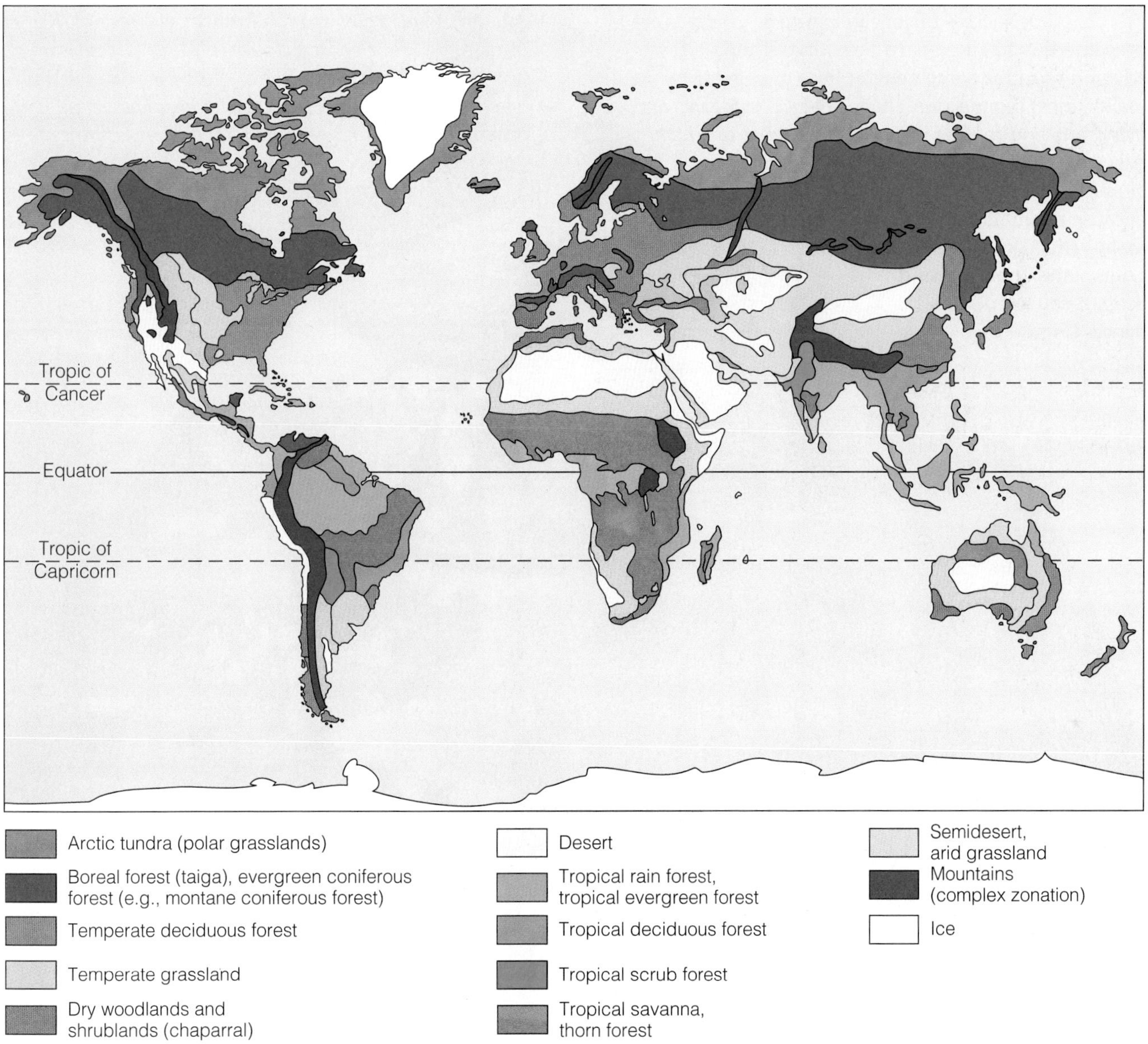

Figure 6-9 The earth's major biomes—the main types of natural vegetation in different undisturbed land areas—result primarily from differences in climate. Each biome contains many ecosystems whose communities have adapted to differences in climate, soil, and other environmental factors. In reality, people have removed or altered much of this natural vegetation for farming, livestock grazing, harvesting lumber and fuelwood, mining, and constructing villages and cities, thereby altering the biomes (Figure 1-5).

pp. 2–3 show global distributions of *biomes*: terrestrial regions with characteristic types of natural, undisturbed ecological communities adapted to the climate of the region. By comparing Figure 6-9 with Figure 6-2, you can see how the world's major biomes vary with climate. Figure 4-8 shows major biomes in the United States as one moves through different climates along the 39th parallel.

For plants, *precipitation generally is the limiting factor that determines whether a land area is desert, grassland, or forest.* Taken together, average annual precipitation and temperature (along with soil type) are the most important factors in producing tropical, temperate, or polar deserts, grasslands, and forests (Figure 6-10).

On maps such as the one in Figure 6-9, biomes are presented as having sharp boundaries and as being covered with the same general type of vegetation. In reality, most biomes don't have sharp boundaries and blend into one another in transitional zones or *ecotones*. Also, the types and numbers of plants in a biome vary from one location to another because of small variations in climate (microclimates), soil types, and natural and human-caused disturbances (Figure 1-5).

Climate and vegetation both vary with **latitude** (distance from the equator) and **altitude** (elevation above sea level). If you travel from the equator toward either pole, you will generally encounter colder climates and zones of vegetation adapted to those climates. Similarly, as elevation above sea level increases, climate becomes colder. Thus, if you climb a tall mountain from its base to its summit, you can observe changes in plant life similar to those you would encounter in traveling from the equator to the poles (Figure 6-11).

Why Do Plant Sizes, Shapes, and Survival Strategies Differ? Arctic soils are wet and nutrient rich. So why are there no trees in the Arctic, and why are the plants there so close to the ground? Why don't desert plants such as cacti have leaves? Why do trees in most forests found in both the warm tropics and in cold areas such as Canada and Sweden keep their leaves year-round, whereas most trees in temperate forests lose their leaves in winter?

Plants exposed to cold air year-round or during winter have traits that keep them from losing too much heat and water. For example, trees or tall plants can't survive in the cold, windy arctic grass-

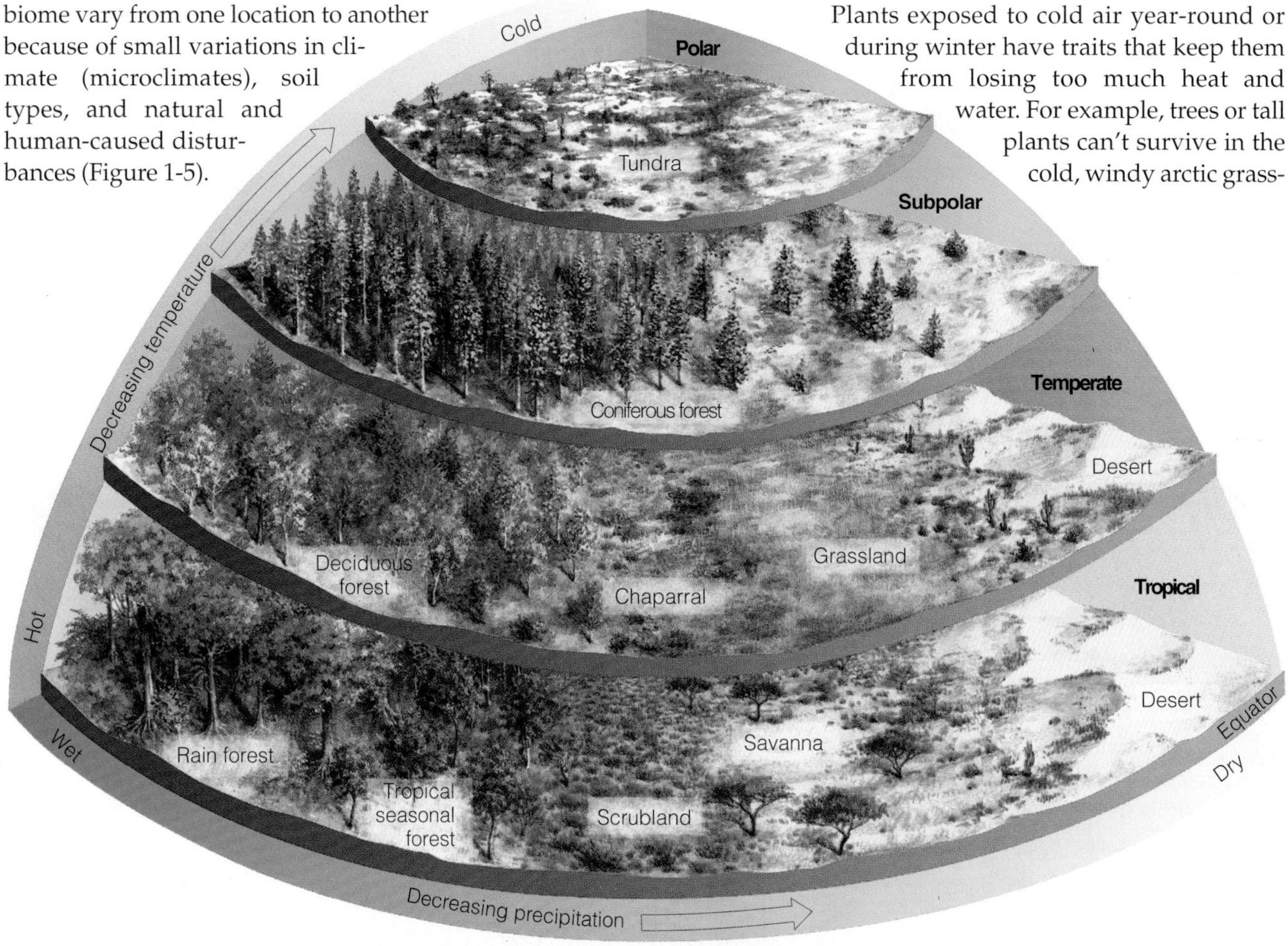

Figure 6-10 Average precipitation and average temperature, acting together as limiting factors over a period of 30 or more years, determine the type of desert, grassland, or forest biome in a particular area. Although the actual situation is much more complex, this simplified diagram explains how climate determines the types and amounts of natural vegetation found in an area left undisturbed by human activities. (Used by permission of Macmillan Publishing Company, from Derek Elsom, *The Earth*, New York: Macmillan, 1992. Copyright © 1992 by Marshall Editions Developments Limited)

lands (tundra) because they would lose too much of their heat for survival.

Desert plants exposed to the sun all day long must be able to lose enough heat so that they don't overheat and die. They must also conserve enough water for survival. **Succulent** (fleshy) **plants**, such as the saguaro (pronounced "sah-WAH-row") cactus (photo in table of contents, p. xii), survive in dry climates by having a vertical orientation of most surfaces, no leaves, and the ability to store water and synthesize food in their expandable, fleshy tissue. The plant's shape and lack of leaves give it a small surface area exposed to sunlight and a large area away from the sun for radiating heat out. Succulent plants also reduce water loss by opening their pores (stomata) to take up carbon dioxide (CO_2) only at night.

Trees of wet tropical rain forests tend to be **broadleaf evergreen plants**, which keep most of their broad leaves year-round. The large surface area of the leaves allows them to collect ample sunlight for photosynthesis and radiate heat during hot weather.

In a climate with a cold (and sometimes dry) winter, keeping such leaves would cause plants to lose too much heat and water for survival. In such climates, **broadleaf deciduous plants**, such as oak and maple trees, survive drought and cold by shedding their leaves and becoming dormant during such periods.

If we move further north to areas such as Canada and Sweden, where summers are cool and short, this strategy is less successful. Instead evolution has favored **coniferous** (cone-bearing) **evergreen plants** (such as spruces, pines, and firs). These plants keep some of their narrow pointed leaves (needles) all year. The waxy coating, shape, and clustering of conifer needles slow down heat loss and evaporation during the long, cold winter. Additionally, by keeping their leaves all winter, such trees are ready to take advantage of the brief summer without having to take time to grow new needles.

6-3 DESERT AND GRASSLAND BIOMES

What Are the Major Types of Deserts? A **desert** is an area where evaporation exceeds precipitation. Precipitation typically is less than 25 centimeters (10 inches) a year and often is scattered unevenly throughout the year. Deserts have sparse, widely spaced, mostly low vegetation.

A combination of low rainfall and different average temperatures creates tropical, temperate, and cold deserts (Figure 6-10). In *tropical deserts,* such as the southern Sahara in Africa, temperatures are usually high year-round and there is little rainfall. These driest places on earth typically have few plants and a hard, windblown surface strewn with rocks and some sand.

In *temperate deserts,* such as the Mojave in southern California, daytime temperatures are hot in summer and cool in winter, and there is more precipitation than in tropical deserts. The vegetation is sparse, consisting mostly of widely dispersed, drought-resistant shrubs and cacti or other succulents, and animals are adapted to the lack of water and temperature variations (Figure 6-12, p.146).

In *cold deserts,* such as the Gobi Desert in China, winters are cold, summers are warm or hot, and precipitation is low. In the semiarid zones between deserts and grasslands, we find *semidesert.* This biome is dominated by thorn trees and shrubs adapted to a long dry spells followed by brief, sometimes heavy rains.

How Do Desert Plants and Animals Survive? Adaptations for survival in the desert have two themes: "Beat the heat" and "Every drop of water counts." Some desert plants are evergreens with wax-coated leaves (creosote bush) that minimize transpiration. Perennial shrubs such as mesquite and creosote plants grow deep roots to tap into groundwater, and they drop their leaves to survive in a dormant state during long dry spells. Other perennials such as short (prickly pear; Figure 6-12, p. 146) and tall (saguaro) cacti use widely spread shallow roots to collect water after brief showers and store it in their spongy tissues.

Many desert plants are annual wildflowers and grasses that store much of their biomass in seeds during dry periods and remain inactive (sometimes for years) until they receive enough water to germinate. Shortly after a rain, in a frenzy of biological activity, the seeds germinate, grow, carpet the desert with a dazzling array of colorful flowers, produce new seed, and die—all in only a few weeks.

Most desert animals are small. They beat the heat and reduce water loss by evaporative cooling. They hide in cool burrows or rocky crevices by day and come out at night or in the early morning. Some desert animals have physical adaptations for conserving water (Spotlight, p. 145). Insects and reptiles have thick outer coverings to minimize water loss through evaporation. They also reduce water loss by having dry feces and by excreting a dried concentrate of urine. Some of the smallest desert animals, such as spiders and insects, get their water only from dew or from the food they eat. Some desert animals become dormant during periods of extreme heat or drought and are active only during the cooler months of the year.

What Impacts Do Humans Have on Desert Ecosystems? Deserts take a long time to recover

Figure 6-11 Generalized effects of latitude and altitude on climate and biomes. Parallel changes in vegetation type occur when we travel from the equator to the poles or from lowlands to mountaintops.

from disturbances because of their **(1)** slow plant growth, **(2)** low species diversity, **(3)** slow nutrient cycling (because of little bacterial activity in their soils), and **(4)** water shortages. Vegetation destroyed by livestock overgrazing and off-road vehicles may take decades to grow back. Vehicles can also collapse underground burrows where many desert animals live.

Some major human impacts on deserts are as follows:

- The rapid growth of large desert cities. Examples are in Saudi Arabia and Egypt and in the southwestern United States (such as Palm Springs, California; Las Vegas, Nevada; and Phoenix, Arizona). Increasingly, people in such cities are destroying fragile desert soil, plants, and animal burrows with four-wheel-drive vehicles and motorcycles.
- Irrigation of desert areas (such as parts of southern California) to grow crops. When this is done, salts may accumulate in the soil (salinization) as the water evaporates, and crop productivity is limited.
- Depletion of underground water (aquifers) as desert cities and irrigation expand.
- Disruption and pollution by the extraction of oil and a variety of minerals and building materials (such as road stone and sand).

Abundant sunlight is the largest and most untapped resource of deserts. Over the next 40–50 years many analysts expect industrialized societies to make a transition from reliance on nonrenewable fossil and nuclear fuels to greatly increased use of renewable solar energy. If such a shift takes place, large areas of many deserts near urban areas will be covered with arrays of solar collectors and solar cells to produce electricity and hydrogen gas (Section 20-7). Great care must be taken to do this in ways that don't seriously disrupt desert ecosystems.

What Are the Major Types of Grasslands? **Grasslands** are regions with enough average annual precipitation to allow grass (and in some areas, a few trees) to prosper, but with precipitation so erratic that drought and fire prevent large stands of trees from growing. If they are not overgrazed, grasses (many of them perennials) in these biomes are renewable resources because these plants grow out from the bottom. This allows their stems to grow again after being nibbled off by grazing animals.

The three main types of grasslands—tropical, temperate, and polar (tundra)—result from combinations of low average precipitation and various average temperatures (Figure 6-10). *Tropical grasslands* are found in areas with high average temperatures, low to moderate precipitation, and a prolonged dry season.

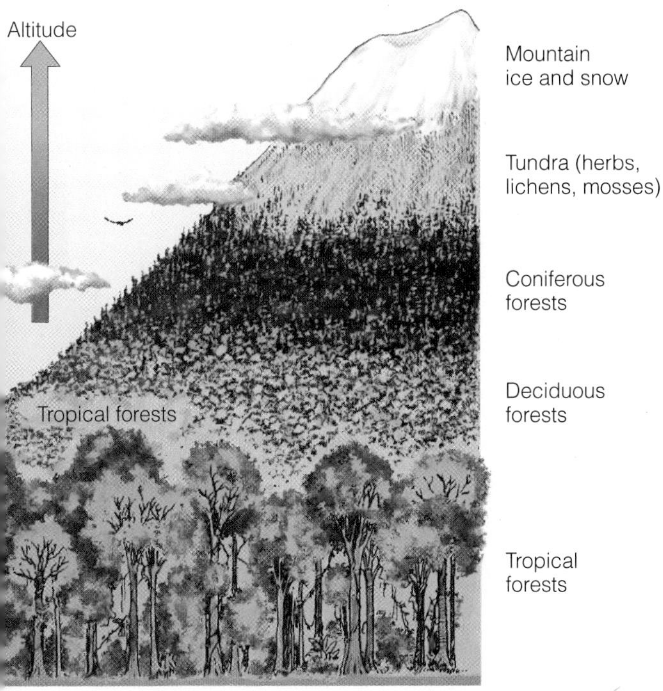

One type of tropical grassland, called a *savanna*, usually has scattered shrubs, and occasional deciduous trees, warm temperatures year-round, two prolonged dry seasons, and abundant rain the rest of the year. African tropical savannas contain enormous herds of *grazing* (grass- and herb-eating) and *browsing* (twig- and leaf-nibbling) hoofed animals, including wildebeests (Figure 6-13), gazelles, zebras, giraffes, and antelopes. These and other large herbivores have evolved specialized eating habits that minimize interspecific competition for resources (resource partitioning): Giraffes eat leaves and shoots from the tops of trees, elephants eat leaves and branches further down, Thompson's gazelles and wildebeests prefer short grass, and zebras graze on longer grass and stems.

In *temperate grassland*, winters are bitterly cold and summers are hot and dry. Annual precipitation is fairly sparse and falls unevenly through the year. Drought, occasional fires, and intense grazing inhibit the growth of trees and bushes, except along rivers. Deep and nutrient-rich soil in such grasslands is held in place by a thick network of intertwined roots of drought-tolerant grasses—unless the topsoil is plowed up and allowed to blow away by prolonged exposure to high winds found in these biomes.

Types of temperate grasslands are the *tall-grass prairies* (Figure 6-14, p.148) and *short-grass prairies* of the Midwestern and Western United States and Canada, the South American *pampas*, the African *veldt*, and the *steppes* of central Europe and Asia. Here winds blow almost continuously and evaporation is rapid, often leading to fires in the summer and fall.

Polar grasslands, or *arctic tundra*, occur just south of the Arctic polar ice cap (Figure 6-9). During most of the year these treeless plains are bitterly cold, swept by frigid winds, and covered with ice and snow. Winters are long and dark, and the scant precipitation falls mostly as snow. This biome is carpeted with a thick,

SPOTLIGHT

The Kangaroo Rat: Water Miser and Keystone Species

The kangaroo rat (Figure 6-12) is a remarkable mammal superbly adapted for conserving water in its desert environment.

As the desert's chief seed eater, it is also a keystone species that helps support other desert species and helps keep desert shrubland from becoming grassland.

This rodent comes out of its burrow only at night, when the air is cool and water evaporation has slowed. It seeks dry seeds that it quickly stuffs into its cheek pouches.

After a night of foraging it empties its cache of seeds into its cool burrow, where they soak up water exhaled in the rodent's breath. When the rodent eats these seeds, it gets this water back.

The kangaroo rat does not drink water; its water comes from the recycled moisture in the seeds and from water produced when sugars in the seeds undergo aerobic respiration during digestion.

Some of the water vapor in the rat's breath also condenses on the cool inside surface of its nose. This condensed water then diffuses back to its body.

Kangaroo rats have no sweat glands, so they don't lose water by perspiration. In addition, they save water by excreting hard, dry feces and thick, nearly solid urine produced by their extremely efficient kidneys.

Critical Thinking

Water is scarce in much of the southwestern United States, where the kangaroo rat lives. However, this area has one of the highest rates of human population growth. As this happens, what ecological lesson can we learn from the kangaroo rat about how to survive in this area (and other water-short areas throughout the world)?

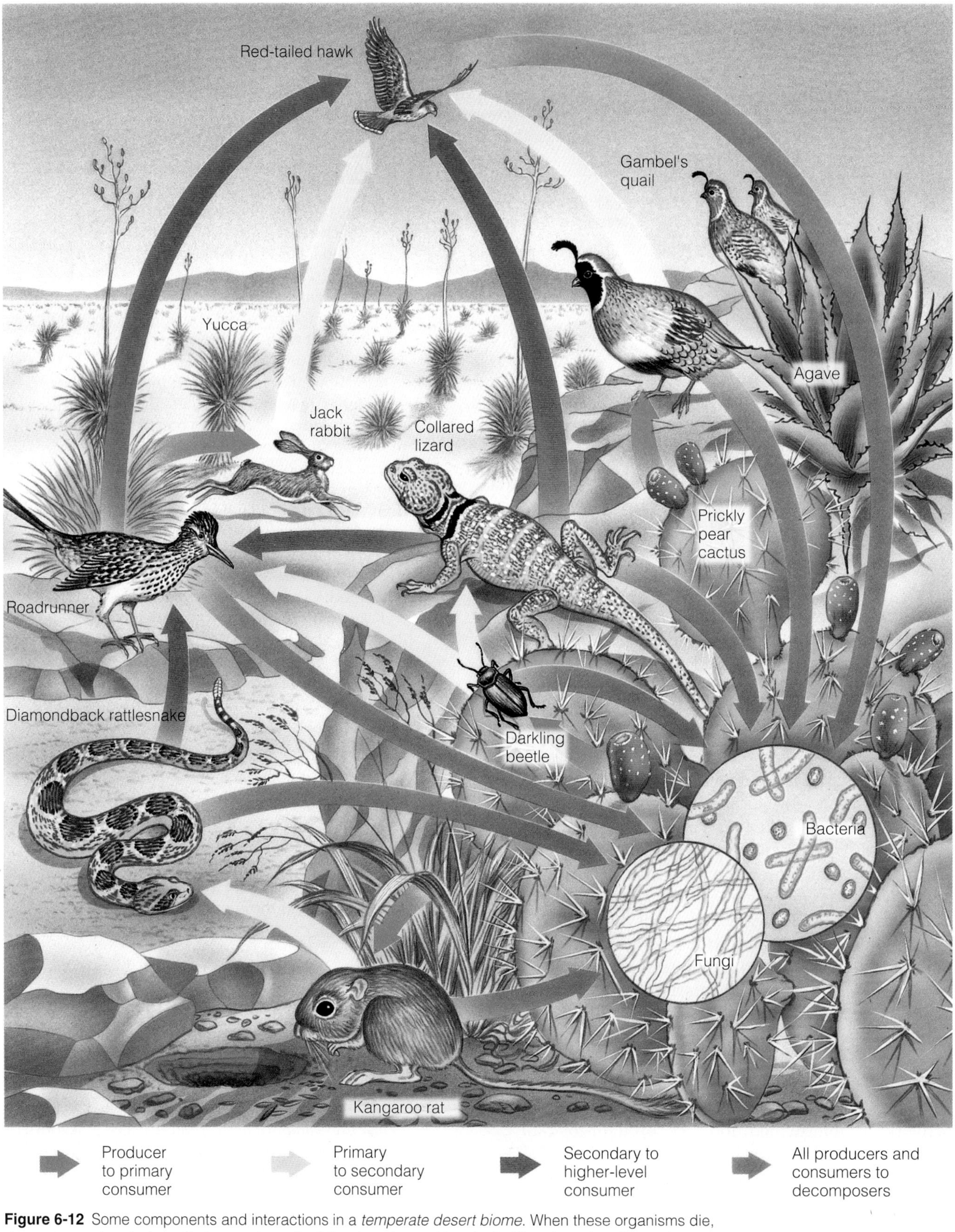

Figure 6-12 Some components and interactions in a *temperate desert biome.* When these organisms die, decomposers break down their organic matter into minerals used by plants. Transfers of matter and energy among producers, primary consumers (herbivores), and secondary (or higher-level) consumers (carnivores) are indicated by colored arrows. Organisms are not drawn to scale.

 Friedrich, R. L and R. V. Blystone. 1998. Internet Teaching Resources for Remote Sensing and GIS. *BioScience*, v48 n3 p187(6).

Figure 6-13 Serengeti tropical savanna in Tanzania, Africa, an example of one type of tropical grassland biome. Most savannas consist of grasslands punctuated by stands of deciduous shrubs and trees, which shed their leaves during the dry season and thus avoid excessive water loss. More large, hoofed, plant-eating mammals (ungulates), such as the wildebeest shown here, live in this biome than anywhere else. As the seasons change, wildebeest and many other large herbivores living in the Serengeti biome migrate to find enough water and high-quality grasses. (Jonathan Scott/Planet Earth Pictures)

spongy mat of low-growing plants, primarily grasses, mosses, and dwarf woody shrubs (Figure 6-15). Most of the annual growth of these plants occurs during the 6- to 8-week summer, when sunlight shines almost around the clock.

One effect of the extreme cold is **permafrost**, a perennially frozen layer of the soil that forms when the water there freezes. In summer, water near the surface thaws, but the permafrost soil layer below stays frozen and prevents liquid water at the surface from seeping into the ground. Thus, during the brief summer the soil above the permafrost layer remains waterlogged, forming a large number of shallow lakes, marshes, bogs, ponds, and other seasonal wetlands. Hordes of mosquitoes, blackflies, and other insects thrive in these shallow surface pools. They feed large colonies of migratory birds (especially waterfowl) that return from the south to nest and breed in the bogs and ponds.

What Impacts Do Humans Have on Grassland Ecosystems? Because grasslands have a low rate of decomposition, shallow soil, short growing season, and slow plant growth rate, vegetation destroyed by human activities can take decades to grow back. Some major human impacts on deserts are as follows:

- Burning, plowing up, and converting some areas of savanna into cropland. By releasing large quantities of carbon dioxide into the atmosphere, this may contribute to the greenhouse effect as much as (if not more than) the more publicized clearing and burning of tropical rain forests.

- Overgrazing by livestock in tropical and temperate grasslands as governments and aid agencies encourage the drilling of water wells. With more water, the livestock population increases, and pastures around wells are overgrazed and trampled by thousands of hooves. This can convert grassland into less productive desert and semidesert.

- Plowing large areas of fertile temperate grasslands in North America, western Europe, and Ukraine and converting them to highly productive cropland (Figure 6-16). As long as temperate grasslands keep their fertile soil and the climate does not change, they can continue producing much of the world's cereal grains. However, poor farming practices, overgrazing, mismanagement, and occasional prolonged droughts lead to severe wind erosion and loss of topsoil, which can convert temperate grasslands into desert or semidesert shrubland.

- Damaging the fragile arctic tundra in Alaska and Siberia. This is caused by oil exploration and drilling, air pollution, spills or leaks of oil and toxic wastes, and disruption of soil and vegetation by vehicles.

6-4 FOREST AND MOUNTAIN BIOMES

What Are the Major Types of Forests? Undisturbed areas with moderate to high average annual precipitation tend to be covered with **forest**, which contains various species of trees and smaller forms of vegetation. The three main types of forest—*tropical*, *temperate*, and *boreal* (polar)—result from combinations of

Figure 6-14 Some components and interactions in a temperate tall-grass prairie ecosystem in North America. When these organisms die, decomposers break down their organic matter into minerals used by plants. Transfers of matter and energy among producers, primary consumers (herbivores), and secondary (or higher-level) consumers (carnivores) are indicated by colored arrows. Organisms are not drawn to scale.

Figure 6-15 Polar grassland (arctic tundra) in Alaska in summer. During the long, dark, cold winter, this land is covered with snow and ice. Its low-growing plants are adapted to the lack of sunlight and water, freezing temperatures, and constant high winds. Below the surface there is a frozen layer of soil called permafrost. Most of this layer stays frozen year-round, except in the brief summer, when its top portion melts and forms seasonal ponds (wetlands). (Charlie Ott/Photo Researchers, Inc.)

Figure 6-16 Replacement of a temperate grassland with a monoculture crop near Blythe, California. When the tangled root network of natural grasses is removed, the fertile topsoil is subject to severe wind erosion unless it is covered with some type of vegetation. If global warming accelerates over the next 50 years, many of these grasslands may become too hot and dry for farming, thus threatening the world's food supply. (National Archives/EPA Documerica)

this precipitation level and various average temperatures (Figure 6-10).

Tropical rain forests are a type of broadleaf evergreen forest (Figure 6-17) found near the equator (Figure 6-9), where hot, moisture-laden air rises and dumps its moisture (Figure 6-5). These forests have a warm annual mean temperature (which varies little, daily or seasonally), high humidity, and heavy rainfall almost daily.

Tropical rain forests have incredible biological diversity. These diverse forms of life occupy a variety of specialized niches in distinct layers, based mostly on their need for sunlight (Figure 6-18). The stratification of specialized plant and animal niches in various layers of a tropical rain forest enables coexistence of a great variety of species (biodiversity). Although tropical rain forests cover only about 2% of the earth's land surface, they are habitats for 50–80% of the earth's terrestrial species.

Dropped leaves, fallen trees, and dead animals decompose quickly because of the warm, moist conditions and hordes of decomposers. This rapid recycling of scarce soil nutrients is why there is usually little litter on the ground. Instead of being stored in the soil, most minerals released by decomposition are quickly taken up by plants. Thus, most of a tropical rain forest's nutrients are stored in the biomass of its living organisms.

Because of the dense vegetation, little wind blows in tropical rain forests, eliminating the possibility of wind pollination. Many of the plants have evolved elaborate flowers (Figure 1-10, left) that attract particular insects, birds, or bats as pollinators.

Moving a little farther from the equator (Figure 6-9), we find *tropical deciduous forests* (sometimes called *tropical monsoon forests* or *tropical seasonal forests*). These forests are warm year-round and get most of their plentiful rainfall during a wet (monsoon) season that is followed by a long dry season. They contain a mixture of deciduous trees (which lose their leaves to survive the dry season) and drought-tolerant evergreen trees (which retain most of their leaves year-round). Where the dry season is especially long, we find *tropical scrub forests* (Figure 6-9) containing mostly small deciduous trees and shrubs.

Temperate deciduous forests (Figure 6-19, p.152) grow in areas with moderate average temperatures that change significantly with the season. These areas have **(1)** long, warm summers, **(2)** cold but not too severe winters, and, **(3)** abundant precipitation, often spread fairly evenly throughout the year.

This biome is dominated by a few species of broadleaf deciduous trees such as oak, hickory, maple, poplar, and sycamore. They survive cold winters by dropping their leaves in the fall and becoming dormant (photo in table of contents, p. xiii). Each spring they grow new leaves that change in the fall into an array of reds and golds (photo in table of contents, p. xii) before

Figure 6-17 Some components and interactions in a tropical rain forest ecosystem. When these organisms die, decomposers break down their organic matter into minerals used by plants. Transfers of matter and energy among producers, primary consumers (herbivores), and secondary (or higher-level) consumers (carnivores) are indicated by colored arrows. Organisms are not drawn to scale.

Figure 6-18 Stratification of specialized plant and animal niches in various layers of a tropical rain forest. The presence of these specialized niches enables species to avoid or minimize competition for resources and results in the coexistence of a great variety of species (biodiversity).

dropping. Because of the fairly low rate of decomposition, these forests accumulate a thick layer of slowly decaying leaf litter that is a storehouse of nutrients.

Evergreen coniferous forests, also called *boreal forests* and *taigas* (pronounced "TIE-guhs"), are found just south of the arctic tundra in northern regions across North America, Asia, and Europe (Figure 6-9). In this subarctic climate, winters are long, dry, and extremely cold; in the northernmost taiga, sunlight is available only 6–8 hours a day. Summers are short, with mild to warm temperatures, and the sun typically shines 19 hours a day.

Most boreal forests are dominated by a few species of evergreen conifer trees such as spruce, fir, cedar, hemlock, and pine. The small, needle-shaped, waxy-coated leaves of these trees can withstand the intense cold and drought of winter when snow blankets the ground. Plant diversity is low in these forests because few species can survive the winters when soil moisture is frozen.

Beneath the stands of trees, there is a deep layer of partially decomposed conifer needles and leaf litter. Decomposition is slow because of the low temperatures, the waxy coating of conifer needles, and the high acidity. As the conifer needles decompose, they make the thin, nutrient-poor soil acidic and prevent most other plants (except certain shrubs) from growing on the forest floor.

These biomes contain a variety of wildlife (Figure 6-20). During the brief summer the soil becomes waterlogged, forming acidic bogs, or *muskegs*, in low-lying areas of these forests. Warblers and other insect-eating birds feed on hordes of flies, mosquitoes, and caterpillars.

Figure 6-19 Some components and interactions in a temperate deciduous forest ecosystem. When these organisms die, decomposers break down their organic matter into minerals used by plants. Transfers of matter and energy among producers, primary consumers (herbivores), and secondary (or higher-level) consumers (carnivores) are indicated by colored arrows. Organisms are not drawn to scale.

Figure 6-20 Some components and interactions in an evergreen coniferous (boreal or taiga) ecosystem. When these organisms die, decomposers break down their organic matter into minerals used by plants. Transfers of matter and energy among producers, primary consumers (herbivores), and secondary (or higher-level) consumers (carnivores) are indicated by colored arrows. Organisms are not drawn to scale.

Coastal coniferous forests or *temperate rain forests* are found in scattered coastal temperate areas with ample rainfall or moisture from dense ocean fogs. Along the coast of North America, from Canada to northern California, undisturbed areas of biomes are dominated by dense stands of large conifers such as Sitka spruce, Douglas fir, and redwoods.

What Impacts Do Humans Have on Forest Ecosystems? Some major human impacts on forests are as follows:

- Clearing and degradation of tropical rain forests and tropical seasonal forests for timber, grazing land, and agriculture, leading to severe erosion of their already nutrient-poor soils. Within 50 years only scattered fragments of these diverse forests might remain.
- Clearing of *temperate deciduous forests* in Europe, Asia, and North America, mostly for timber, cropland, and urban development. In North America, about 99.9% of the original stands of temperate deciduous forests have been cleared for such purposes. Some have been converted to managed *tree farms* or *tree plantations,* where a single species is grown for timber, pulpwood, or Christmas trees (photo in table of contents, p. xiii). Through secondary succession (Figure 5-15), many of North America's original virgin forests have regrown as less diverse secondary forests.
- Clearing of large areas of *evergreen coniferous forests* by loggers in North America, Finland, Sweden, and Canada. Within a decade most of the vast boreal forests of Siberia and Russia may disappear because of logging, mining, and flooding by the reservoirs created for hydroelectric projects. Because trees grow slowly in the cold northern climate, these forests can take a long time to recover from disturbance.

Why Are Mountains Ecologically Important? Some of the world's most spectacular and important environments are mountains, which make up about 20% the earth's land surface. Mountains are places where dramatic changes in altitude, climate, soil, and vegetation take place over a very short distance (Figure 6-11). Because of the steep slopes, mountain soils are especially prone to erosion when the vegetation holding them in place is removed by natural disturbances (such as landslides and avalanches) or human activities.

Many freestanding mountains are *islands of biodiversity* surrounded by a sea of lower-elevation landscapes transformed by human activities. As a result, many mountain areas contain species found nowhere else on earth; they are also sanctuaries for animal species driven from lowland areas.

The ice and snow of mountaintops help regulate the earth's climate by reflecting solar radiation back into space. Sea levels depend on the melting of glacial ice, most of which is locked up in Antarctica, the most mountainous of all continents. Mountain regions also contain the majority of the world's forests, which contain much of the world's biodiversity. Mountains also play a critical role in the hydrologic cycle (Figure 4-22) by gradually releasing melting ice, snow, and water stored in the soils and vegetation of mountainsides to small streams.

What Impacts Do Humans Have on Mountain Ecosystems? Despite their ecological, economic, and cultural importance, the fate of mountain ecosystems has not been a high priority of governments or many environmental organizations. Mountain ecosystems are coming under increasing environmental pressure from several major trends:

- Rapidly increasing population, especially in developing countries. This is forcing landless poor people, refugees, and minority populations to migrate uphill and try to survive on less stable soils. These newcomers often use mountain soils and forests unsustainably in a desperate struggle to survive or because of a lack of knowledge about how to grow food, raise livestock, and harvest wood in these habitats.
- Increased commercial extraction of timber and mineral resources. People often do this with little regard for the resulting environmental consequences such as **(1)** excessive soil erosion, **(2)** flooding in the valleys below, **(3)** loss of wildlife habitat, and **(4)** air and water pollution from mining activities.
- A growing number of hydroelectric dams and reservoirs. Rivers in mountains are attractive sites for dams and reservoirs because the elevation and slope of mountains increase the force of flowing water. These reservoirs flood mountain slopes, and the dams alter the types and abundance of species in rivers.
- Environmental degradation from the global boom in skiing, trekking, and other forms of recreation and tourism in mountainous areas.
- Increased air pollution from growing urban and industrial centers and increased use of automobiles. Trees and other vegetation at high elevations (especially conifers such as spruce) are bathed year-round in air pollutants such as ozone and acidic compounds, carried there by prevailing winds from cities, factories, and coal-burning power plants.
- Changes in climate and levels of ultraviolet (UV) radiation brought about by human activities. If global warming occurs as projected during the next century, many species of mountain plants could be displaced

by lowland species that have higher rates of growth and reproduction. Similarly, any increase in UV radiation brought about by depletion of the ozone layer may have a pronounced effect on mountain life, which is already exposed to high levels of ultraviolet radiation because of altitude.

- Increased warfare in mountainous areas.

6-5 SALTWATER LIFE ZONES

What Are the Two Major Types of Aquatic Life Zones? The aquatic equivalents of biomes are called *aquatic life zones*. The major types of organisms found in aquatic environments are determined by the water's *salinity* (the amounts of various salts such as sodium chloride [NaCl] dissolved in a given volume of water). As a result, aquatic life zones are divided into two major types: **(1)** *saltwater* or *marine* (particularly estuaries, coastlines, coral reefs, coastal marshes, mangrove swamps, the ocean above the continental shelf, and the deep ocean) and, **(2)** *freshwater* (particularly lakes and ponds, streams and rivers, and inland wetlands).

Living in water has its advantages. Water's buoyancy provides physical support. Limited fluctuations in temperature greatly reduce the risks from drying out or becoming overheated. Required nutrients are dissolved and readily available. And potentially toxic metabolic wastes secreted by aquatic organisms are diluted and dispersed.

Most aquatic life zones can be divided into surface, middle, and bottom layers. Important environmental factors determining the types and numbers of organisms found in these layers are **(1)** *temperature*, **(2)** *access to sunlight for photosynthesis*, **(3)** *dissolved oxygen content*, and **(4)** *availability of nutrients* such as carbon (as dissolved CO_2 gas), nitrogen (as NO_3^-), and phosphorus (mostly as PO_4^{3-}) for producers.

Why Are the Oceans Important? A more accurate name for Earth would be *Ocean* because saltwater oceans cover about 71% of the planet's surface (Figure 6-21). The oceans play key roles in the survival of virtually all life on the earth. These roles include

- Regulating the earth's climate by **(1)** distributing solar heat through ocean currents (Figure 6-2), **(2)** evaporating ocean water as part of the global hydrologic cycle (Figure 4-22), and **(3)** serving as a giant reservoir for removing carbon dioxide (a greenhouse gas) from the atmosphere.
- Providing habitats for about 250,000 known species of marine plants and animals, which are sources of food for many other organisms (including humans).

Figure 6-21 We live on an ocean planet, with salty oceans covering about 71% of the earth's surface. About 97% of the earth's water is in the interconnected oceans, which cover 90% of the planet's mostly ocean southern hemisphere (left) and 50% of its land–ocean northern hemisphere (right). The average depth of the world's oceans is 3.8 kilometers (2.4 miles).

- Dispersing and diluting many human-produced wastes that flow into or are dumped into the ocean.

What Is the Coastal Zone? Oceans have two major life zones: the coastal zone and the open sea (Figure 6-22). The **coastal zone** is the warm, nutrient-rich, shallow water that extends from the high-tide mark on land to the gently sloping, shallow edge of the *continental shelf* (the submerged part of the continents).

Although it makes up less than 10% of the ocean's area, the coastal zone contains 90% of all marine species and is the site of most large commercial marine fisheries. Most ecosystems found in the coastal zone have a very high primary productivity (Figure 4-20) and net primary productivity per unit of area (Figure 4-21). This occurs because of the zone's ample supplies of sunlight and plant nutrients (deposited from land and stirred up by wind and ocean currents).

What Are Estuaries, Coastal Wetlands, and Mangrove Swamps? One highly productive area in the coastal zone is an **estuary**, a partially enclosed area of coastal water where seawater mixes with fresh water and nutrients from rivers, streams, and runoff from land (Figure 6-23). The constant water movement stirs up the nutrient-rich silt, making it available to producers.

According to an estimate by ecologists at the University of Georgia, just 0.4 hectares (1 acre) of tidal estuary provides an estimated $75,000 worth of free waste treatment and has a value of about $83,000 when recreation and fish for food are included. By comparison, 0.4 hectare (1 acre) of prime farmland in Kansas has a top value of about $1,200 and an annual production value of about $600.

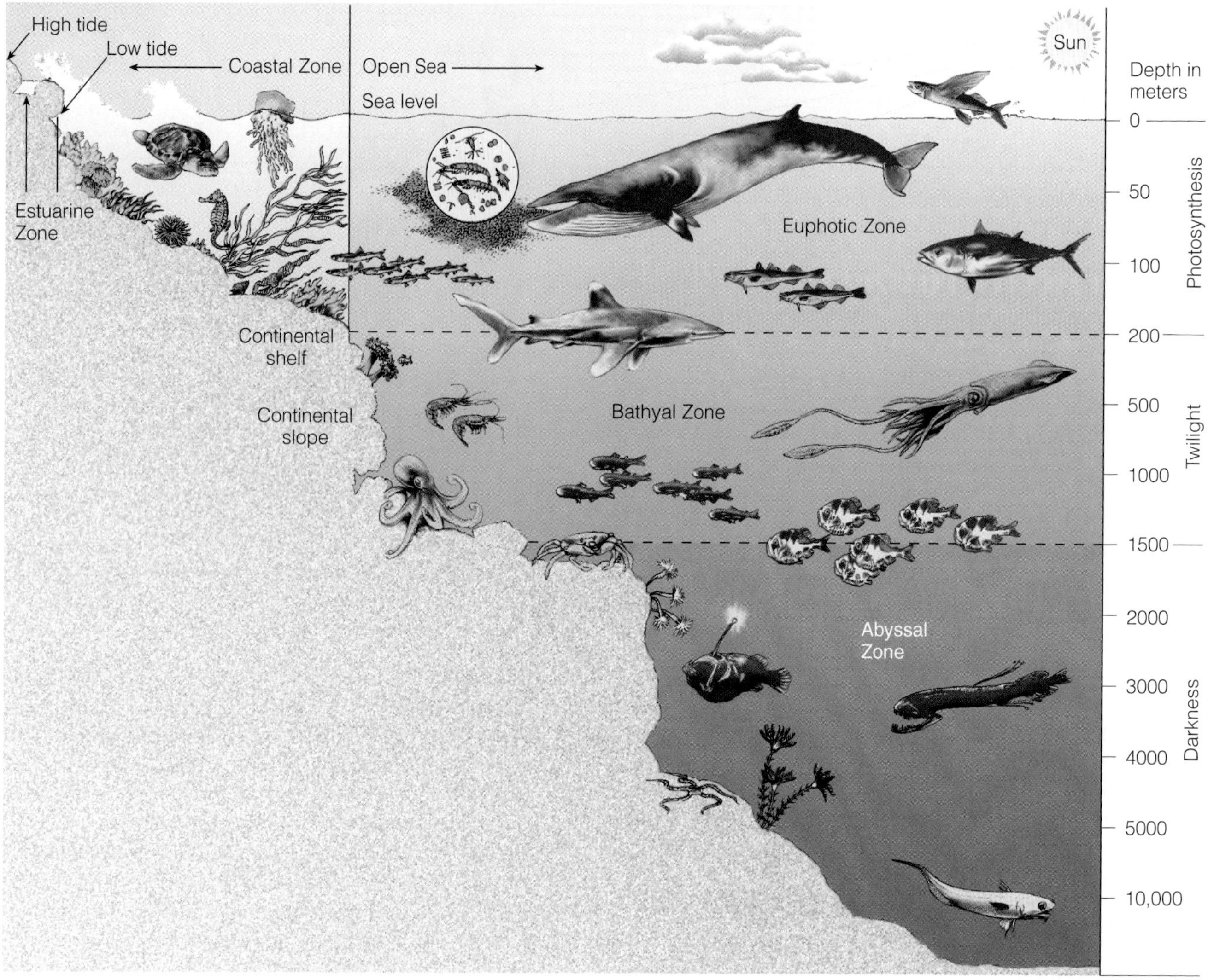

Figure 6-22 Major life zones in an ocean. (Not drawn to scale. Actual depths of zones may vary.)

Areas of coastal land that are covered all or part of the year with salt water are called **coastal wetlands**. They provide many services, such as

- Acting as breeding grounds and habitats for a variety of waterfowl and other wildlife
- Serving as popular areas for recreational activities such as boating, fishing, and hunting
- Helping to maintain the quality of coastal waters by diluting, filtering, and settling out sediments, excess nutrients, and pollutants
- Protecting lives and property during floods by absorbing and slowing the flow of water and buffering shores against damage and erosion during storms

In temperate areas, including the United States, coastal wetlands usually consist of a mixture of *bays, lagoons, salt flats, mud flats,* and *salt marshes* (Figure 6-24). Grasses are the dominant vegetation. About 95% of commercially harvested fish and shellfish depend on these highly productive coastal wetlands for some period of their lives.

Highly productive **mangrove swamps** (photo in table of contents, p. xiv) are found along warm tropical coasts where there is too much silt for coral reefs to grow. They are dominated by salt-tolerant trees or shrubs known as mangroves, of which there are about 55 species worldwide. These swamps help protect the coastline from erosion and reduce damage from typhoons and hurricanes. They also trap sediment washed off the land

and provide breeding, nursery, and feeding grounds for some 2,000 species of fish, invertebrates, and plants.

What Are Rocky and Sandy Shores? Some coasts have steep *rocky shores* pounded by waves. Visitors are often surprised at the variety of life found in the numerous pools and other niches in the rocks in the area of shoreline between low and high tides (the *intertidal zone*) of rocky shores (Figure 6-25, top).

Other coasts have gently sloping *barrier beaches*, or *sandy shores*, with niches for different marine organisms, including crabs, lugworms, clams, ghost shrimp, sand dollars, and flounder (Figure 6-25, bottom). These sandy beaches and their adjoining coastal wetlands are also home to a variety of shorebirds that feed in specialized niches on crustaceans, insects, and other organisms (Figure 5-9).

If not destroyed by human activities, one or more rows of natural sand dunes on barrier beaches (with the sand held in place by the roots of grasses) serve as the first line of defense against the ravages of the sea (Figure 6-26, p.160). However, such beaches are prime sites for development. When coastal developers remove the protective dunes or build behind the first set of dunes, storms can flood and even sweep away seaside buildings and severely erode the unprotected beaches.

Barrier islands are long, thin, low offshore islands of sediment that generally run parallel to the shore. They are found along some coasts such as most of North America's Atlantic and Gulf coasts. These islands help protect the mainland, estuaries, lagoons, and coastal wetlands by dispersing the energy of approaching storm waves.

Their low-lying beaches are constantly shifting: Gentle waves build them up, and storms flatten and erode them. Sooner or later many of the structures humans build on low-lying barrier islands, such as Atlantic City, Miami Beach, and Ocean City, Maryland (Figure 6-27, p.160) are damaged or destroyed by flooding, severe beach erosion, or major storms (including hurricanes).

Case Study: Why Are Coral Reefs Important? In the shallow coastal zones of warm tropical and subtropical oceans we often find **coral reefs** (Figure 6-28, p.160). These beautiful natural wonders are among the world's oldest, most diverse, and most productive ecosystems and are homes for about one-fourth of all marine species (Figure 6-29, p.161).

Coral reefs are formed by massive colonies of tiny animals called *polyps* that are close relatives of jellyfish. They slowly build reefs by secreting a protective crust of limestone (calcium carbonate) around their soft bodies. When they die, their empty crusts or outer skeletons remain as a platform for more reef growth.

Figure 6-23 View from the space shuttle of the sediment plume at the mouth of Madagascar's Betsiboka River as it flows through the estuary and into the Mozambique Channel, which separates this huge island from the African coast. Because of its topography, heavy rainfall, and the clearing of forests for agriculture, Madagascar is the world's most eroded country. (NASA)

Coral reefs are a joint venture between the polyps and tiny single-celled algae called *zooxanthellae* that live in the tissues of the polyps. In this *mutualistic relationship*, the microscopic photosynthesizing algae provide the polyp with food and oxygen. The polyps in turn provide a well-protected home for the algae. In turn, the algae make nutrients such as nitrogen and phosphorus available that usually are scarce in tropical waters. The algae and other producers give corals most of their bright colors and through photosynthesis provide plentiful food for a variety of marine life. Many species of reef fish maintain mutualistic (Figure 5-12) or commensalistic relationships with other reef species.

Coral reefs provide a number of important services, including

- Removing some of carbon dioxide from the atmosphere as part of the carbon cycle (when coral polyps form limestone shells)
- Acting as natural barriers that help protect 15% of the world's coastlines from battering waves and storms
- Building beaches, atolls, and islands
- Providing food, jobs, and building materials for some of the world's poorest countries

Figure 6-24 Some components and interactions in a salt marsh ecosystem. When these organisms die, decomposers break down their organic matter into minerals used by plants. Transfers of matter and energy among consumers (herbivores) and secondary (or higher-level) consumers (carnivores) are indicated by colored arrows. Organisms are not drawn to scale.

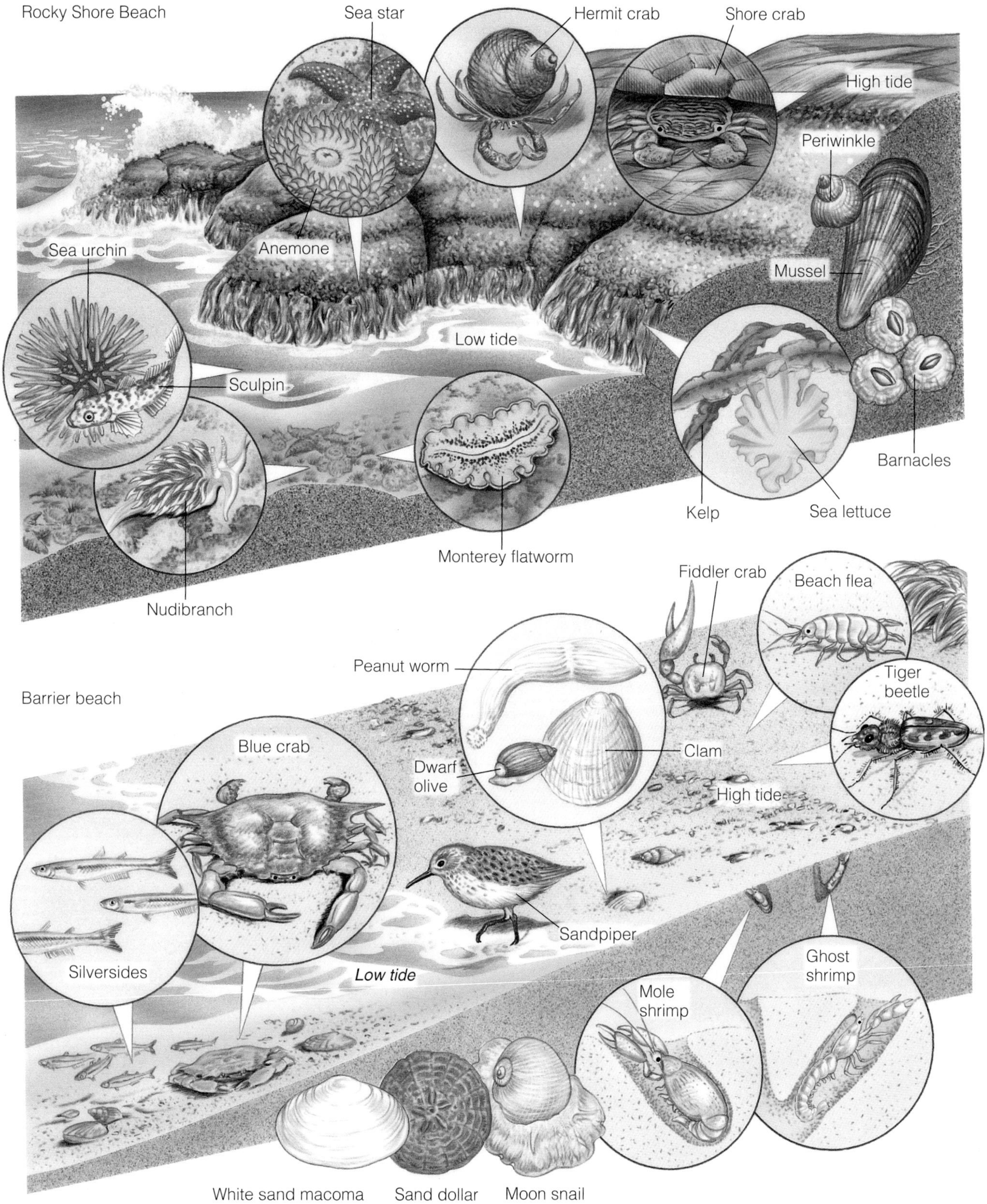

Figure 6-25 Living between the tides: some organisms with specialized niches found in various zones on rocky shore beaches (top) and barrier or sandy beaches (bottom).

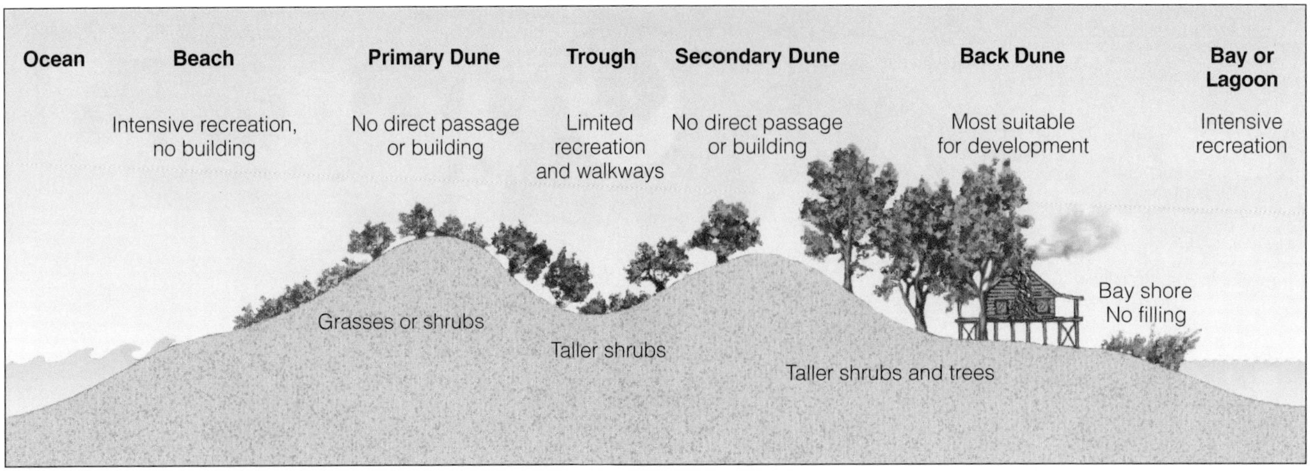

Figure 6-26 Primary and secondary dunes on gently sloping sandy beaches play an important role in protecting the land from erosion by the sea. The roots of various grasses that colonize the dunes help hold the sand in place. Ideally, construction and development should be allowed only behind the second strip of dunes; walkways to the beach should be built over the dunes to keep them intact. This not only helps preserve barrier beaches, but also protects structures from being damaged and washed away by wind, high tides, beach erosion, and flooding from storm surges. This type of protection is rare, however, because the short-term economic value of oceanfront land is considered to be much higher than its long-term ecological and economic values.

Figure 6-27 A developed barrier island, Ocean City, Maryland, has 8 million visitors a year. To keep up with shifting sands, taxpayers spend millions of dollars to pump sand onto the beaches and to rebuild natural sand dunes. They may end up spending millions more to keep buildings from sinking, because barrier islands lack effective protection against flooding and damage from severe storms. Within a few hours, barrier islands can be cut in two or severely damaged by a hurricane. If global warming raises average sea levels, as projected, most of these valuable pieces of real estate will be under water. (G. H. Demetrakas/O. C. Camera)

Figure 6-28 A healthy coral reef in the Philippines covered by colorful algae (bottom) and a bleached coral reef in the Bahamas that has lost most of its algae (top) because of changes in the environment (such as cloudy water or extreme temperatures). With the algae gone, the white limestone of the coral skeleton becomes visible. If the environmental stress is not removed, the corals die. These diverse and productive ecosystems are being damaged and destroyed at an alarming rate. (Top, Robert I. Wicklund; bottom, Karl & Jill Wallin/FPG International)

Figure 6-29 Some components and interactions in a coral reef ecosystem. When these organisms die, decomposers break down their organic matter into minerals used by plants. Transfers of matter and energy among producers, primary consumers (herbivores), and secondary (or higher-level) consumers (carnivores) are indicated by colored arrows. Organisms are not drawn to scale.

- Serving as nurseries for many species of fish
- Supporting almost 1 million marine species
- Supporting fishing and tourism industries worth billions of dollars each year
- Giving us an underwater world to study and enjoy

What Impacts Do Humans Have on Coastal Zones? Coastal zones are among the most densely populated and most intensely used and polluted ecosystems. Currently, about 40% of the world's population live along coasts or within 160 kilometers (100 miles) of a coast. By 2025, it's estimated that two-thirds of the world's people will reside on or near coastal areas.

Since 1900, the world has lost approximately half of its coastal wetlands, primarily through coastal development. Since 1780 about 53% of the area of estuaries and coastal wetlands in the contiguous United States has been destroyed or damaged, primarily because of dredging and filling and waste contamination. California alone has lost 91% of its original coastal wetlands, but Florida has lost the largest area of such wetlands in the United States. A 1994 study by researchers at the University of California at Berkeley estimated that the value of the long-term natural capital services provided by California's remaining wetlands is $124.5 billion.

Since the mid-1960s, some tropical coastal countries have lost half or more of their mangrove forests because of **(1)** industrial logging for timber and fuelwood, **(2)** conversion to ponds for raising fish and shellfish (aquaculture), **(3)** conversion to rice fields and other agricultural land, and **(4)** urban development.

Coastal ecosystems are particularly vulnerable to toxic contamination because they trap pesticides, heavy metals, and other pollutants and concentrate them to high levels. On any given day about 37% of U.S. coastal shellfish beds are closed to commercial or sport fishing, usually because of contamination from sewage treatment plants, septic tank systems, and urban runoff.

Despite their ecological importance, coral reefs are disappearing and being degraded at an alarming rate. They are vulnerable to damage because they **(1)** grow slowly, **(2)** are disrupted easily, **(3)** and thrive only in clear, warm, and fairly shallow water of constant high salinity.

The biggest threats to many of the world's coral reefs come from human activities, especially the deposition of eroded soil produced by deforestation, and construction, agriculture, mining, dredging, and poor land management along increasingly populated coastlines. The suspended soil sediment that washes downriver to the sea or erodes from coastal areas smothers coral polyps or blocks their sunlight.

Silting and increased ocean temperatures are two major causes of coral reef *bleaching*. It occurs when corals under stress expel their mutualistic algae and thus are unable to grow or repair themselves. This expulsion exposes the colorless coral animals and the underlying ghostly white skeleton of calcium carbonate (Figure 6-28, top). Robbed of nutrients and more vulnerable to pathogens (viruses, bacteria, and fungi), the corals eventually die unless the stress is removed in time.

In recent years record sea temperatures, caused by a combination of a strong ENSOs (Figures 6-6 and 6-7) and global warming, have been linked to mass bleaching of many of the world's coral reefs. Other threats to coral reefs include **(1)** increased ultraviolet radiation resulting from depletion of stratospheric ozone, **(2)** runoff of toxic pesticides, industrial chemicals, and fertilizers, **(3)** commercial fishing boats' use of cyanide or dynamite to stun fish so they can be harvested, **(4)** removal of coral for building material, aquariums, and jewelry, **(5)** oil spills, and **(6)** damage from tourists and recreational divers.

Marine biologists estimate that humans have directly or indirectly caused the death of 10% of the world's coral reefs, especially those in Southeast Asia and the Caribbean. Another 30% of the remaining reefs are in critical condition, and 30% more are threatened; only 30% are stable. If the current rates of destruction continue, another 60% of the reefs could be gone in the next 20–40 years. Like the canaries miners used in coal mines to warn of dangerous levels of explosive methane gas, coral reefs are oceanic sentinels warning us of the threats to their habitats.

The *good news* is that protected coral reefs can often recover. However, protecting reefs is difficult and expensive, and only half the countries with coral reefs have set aside reserves that receive some protection from human activities.

What Biological Zones Are Found in the Open Sea? The sharp increase in water depth at the edge of the continental shelf separates the coastal zone from the **open sea**, which is divided into three vertical zones—euphotic, bathyal, and abyssal—based mostly on the penetration of sunlight (Figure 6-22, p. 156). This vast volume of ocean contains only about 10% of all marine species.

Average primary productivity and net primary productivity per unit of area are quite low in the open sea (Figure 4-21), except at an occasional equatorial upwelling, where currents bring up nutrients from the ocean bottom. However, because the open sea covers so much of the earth's surface (Figure 6-21), it makes the largest contribution to the earth's overall net primary productivity. Without strong international agreements, overfishing and pollution can degrade the open ocean because of the tragedy of the commons (Connections, p. 14).

6-6 FRESHWATER LIFE ZONES

What Are Freshwater Life Zones? Freshwater life zones occur where water with a dissolved salt concentration of less than 1% by volume accumulates on or flows through the surfaces of terrestrial biomes. Examples are **(1)** *standing* (lentic) bodies of fresh water such as lakes, ponds, and inland wetlands and **(2)** *flowing* (lotic) systems such as streams and rivers.

Runoff from nearby land provides freshwater life zones with inputs of organic material, inorganic nutrients, and pollutants. Thus, these life zones are closely connected to their nearby terrestrial biomes.

What Are the Major Characteristics of Freshwater Lakes? **Lakes** are large natural bodies of standing fresh water formed when precipitation, runoff, or groundwater seepage fills depressions in the earth's surface. Causes of such depressions include **(1)** glaciation (the Great Lakes of North America), **(2)** crustal displacement (Lake Nyasa in East Africa), and **(3)** volcanic activity (Crater Lake in Oregon, photo on p. xv).

Lakes are fed by rainfall, melting snow, and the streams that drain the surrounding watershed. Lakes normally consist of distinct zones (Figure 6-30), providing habitats and niches for different species.

Ecologists classify lakes according to their nutrient content and their primary productivity. A newly formed lake generally has a small supply of plant nutrients and is called an **oligotrophic** (poorly nourished) **lake** (Figure 6-31, bottom). This type of lake is often deep, with steep banks. Because of its low net primary productivity, such a lake usually has crystal-clear blue or green water, with small populations of phytoplankton and fish, such as smallmouth bass and trout.

Over time, sediment washes into an oligotrophic lake, and plants grow and decompose to form bottom sediments. A lake with a large or excessive supply of nutrients (mostly nitrates and phosphates) needed by producers is called a **eutrophic** (well-nourished) **lake** (Figure 6-31, top). Such lakes typically are shallow, and their water is generally a murky brown or green with very poor visibility. Because of their high levels of nutrients, these lakes have a high net primary productivity. Human inputs of nutrients from the atmosphere and from nearby urban and agricultural areas can accelerate the eutrophication of lakes—a process called *cultural eutrophication*. Many lakes fall somewhere between the two extremes of nutrient enrichment and are called **mesotrophic lakes**.

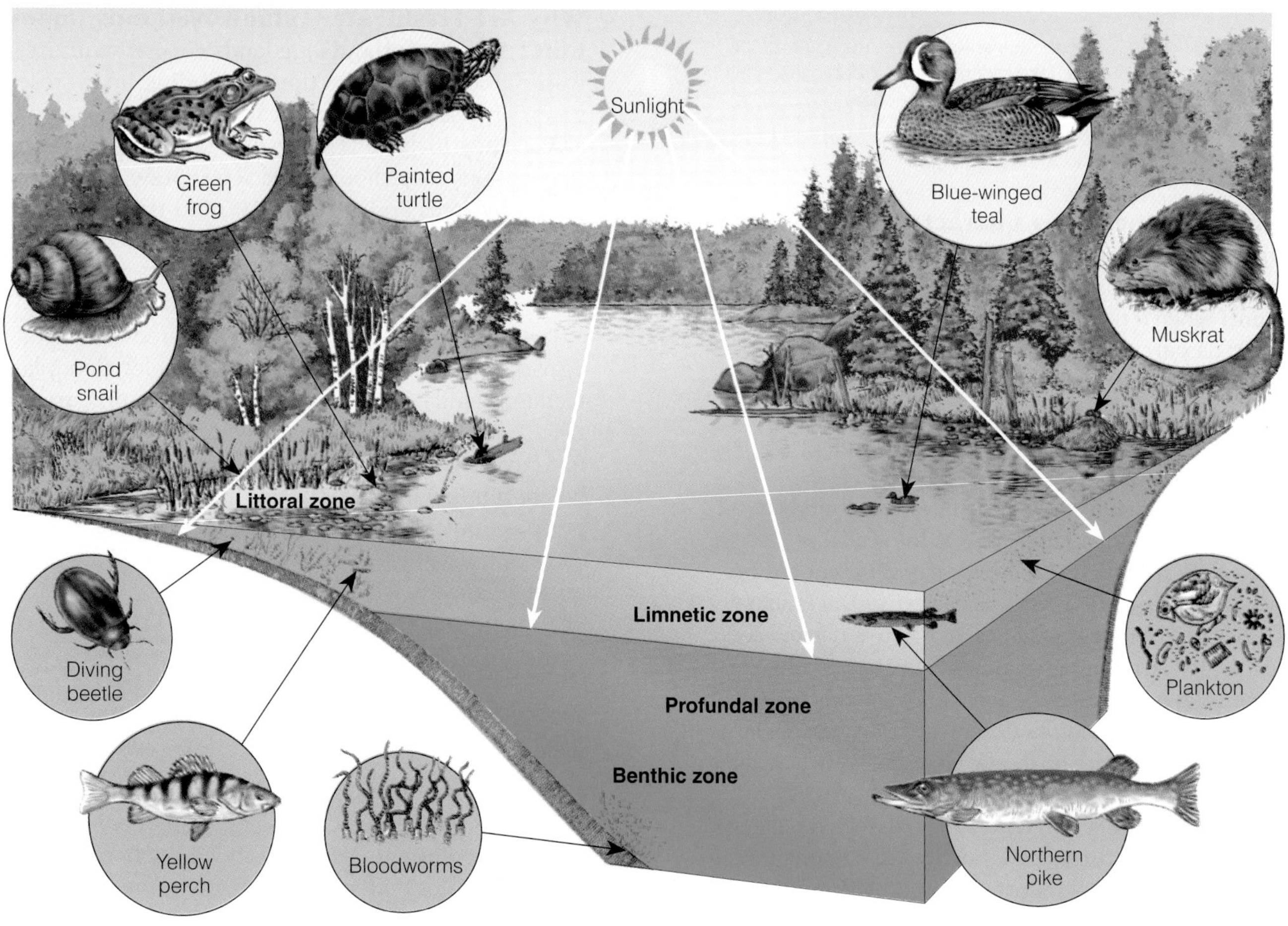

Figure 6-30 The distinct zones of life in a temperate-zone lake.

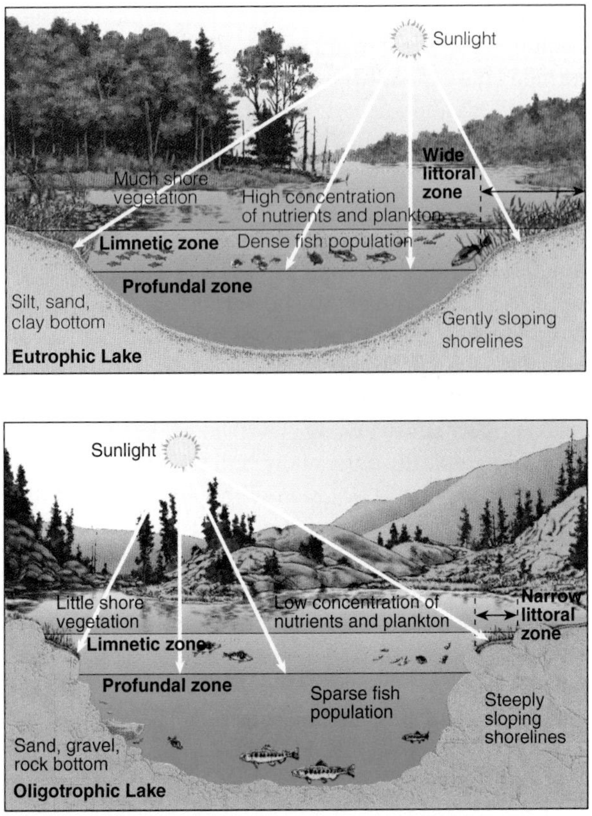

Figure 6-31 A eutrophic, or nutrient-rich, lake (top) and an oligotrophic, or nutrient-poor, lake (bottom). Mesotrophic lakes fall between these two extremes of nutrient enrichment.

What Are the Major Characteristics of Freshwater Streams? Precipitation that doesn't sink into the ground or evaporate is **surface water**. It becomes **runoff** when it flows into streams and eventually to the ocean as part of the hydrologic cycle (Figure 4-22). A **watershed**, or **drainage basin**, is the entire land area, which delivers water, sediment, and dissolved substances via small streams to a larger stream or river and ultimately to the sea.

The downward flow of surface water and groundwater from mountain highlands to the sea takes place in three zones (Figure 6-32). Because of different environmental conditions in each zone, a *river system* is actually a series of different ecosystems. In the first, narrow zone, headwater or mountain highland streams of cold, clear water rush over waterfalls and rapids. As this turbulent water flows and tumbles downward, it dissolves large amounts of oxygen from the air. Here plants such as algae and mosses are attached to rocks. The fish are cold-water fish such as trout, which need lots of dissolved oxygen.

In the second zone, the headwater streams merge to form wider, deeper streams that flow down gentler slopes with fewer obstacles. The warmer water and other conditions in this zone support more producers (phytoplankton and a variety of cool-water and warm-water fish species such as black bass) with slightly lower oxygen requirements.

In the third zone, streams join into wider and deeper rivers that meander across broad, flat valleys. Water in this zone usually has higher temperatures and less dissolved oxygen than water in the first two zones. The main channels of these slow-moving, wide, and murky rivers support distinctive varieties of fish (carp and catfish), whereas their backwaters support species similar to those present in lakes.

Humans sometimes straighten, deepen, and widen meandering streams to improve navigation and to help reduce flooding and bank erosion, but such *stream channelization* is controversial. At its mouth, a river may divide into many channels as it flows through coastal wetlands and estuaries, where the river water mixes with ocean water (Figure 6-23).

As streams flow downhill, they become powerful shapers of land. Over millions of years the friction of moving water levels mountains and cuts deep canyons; the rock and soil the water removes are deposited as sediment in low-lying areas. Human impacts on streams are discussed in Section 12-5.

Why Are Freshwater Inland Wetlands Important? Inland wetlands are lands covered with fresh water all or part of the time (excluding lakes, reservoirs, and streams) and located away from coastal areas. They include **(1)** marshes, **(2)** prairie potholes (depressions carved out by glaciers), **(3)** swamps (dominated by trees and shrubs), **(4)** mud flats, **(5)** floodplains, **(6)** bogs (rain-fed, peat-rich areas), **(7)** wet meadows, and **(8)** the wet arctic tundra in summer. Some inland wetlands are huge; others are small.

Some of these wetlands are covered with water year-round. Others, such as prairie potholes, floodplain wetlands, and bottomland hardwood swamps, are *seasonal wetlands,* which are usually underwater or soggy for only a short time each year. Some stay dry for years before filling with water again. In such cases, only the composition of the soil or the presence of plants such as cattails, bulrushes, or red maples may indicate that a given area really is a wetland.

Inland wetlands play important ecological and economic roles. They

- Provide food and habitats for fish, migratory waterfowl, and other wildlife.
- Improve water quality by filtering, diluting, and degrading toxic wastes, excess nutrients, sediments, and other pollutants. The Audubon Society conservatively estimates that inland wetlands in the United States provide water-quality protection worth at least $1.6 billion per year.

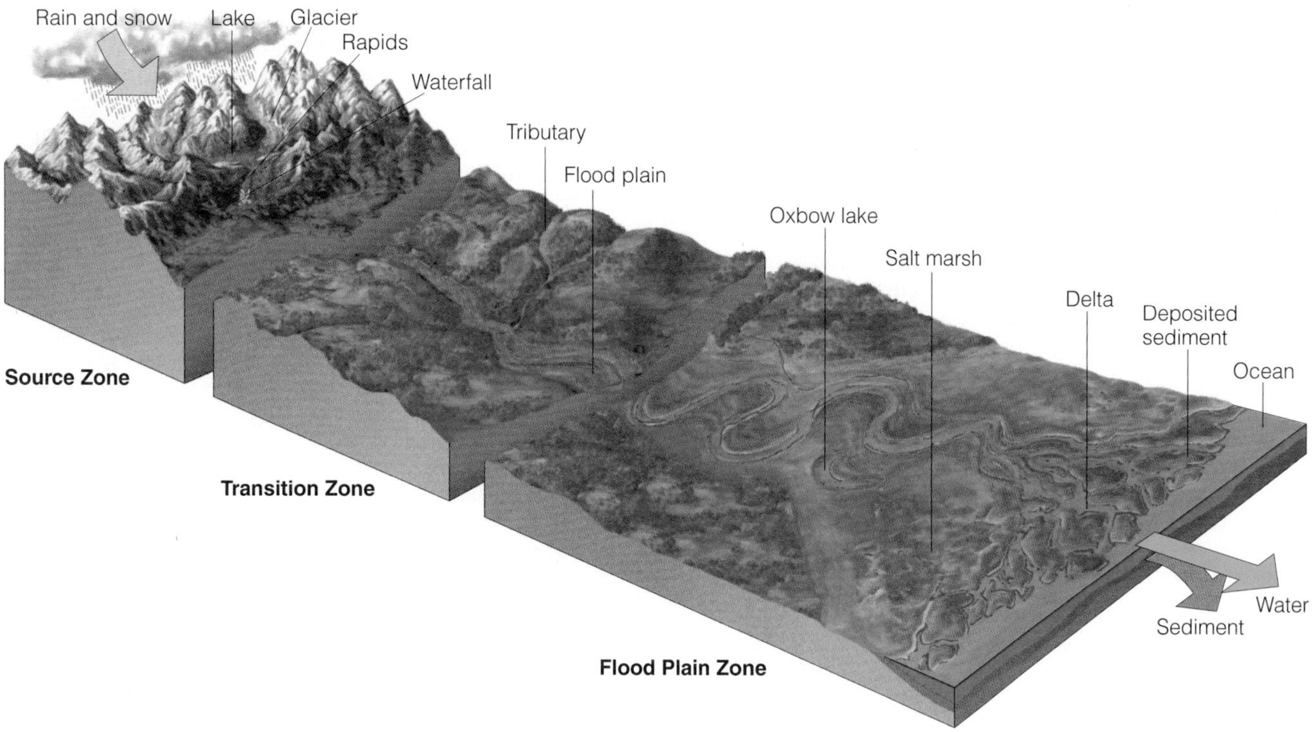

Figure 6-32 The three zones in the downhill flow of water: **(1)** Source Zone with mountain (headwater) streams; **(2)** Transition Zone with wider, lower-elevation streams; and **(3)** Flood Plain Zone with rivers, which empty into the ocean.

- Reduce flooding and erosion by absorbing stormwater and releasing it slowly and by absorbing overflows from streams and lakes. According to Audubon Society estimates, if the remaining wetlands in the United States were destroyed, additional flood-control costs would be $7.7 billion to $31 billion per year.
- Help replenish groundwater supplies, a primary water source for over 50% of the U.S. population.
- Play significant roles in the global carbon, nitrogen, sulfur, and water cycles.
- Provide recreation (especially waterfowl hunting) and are used to grow crops such as blueberries, cranberries, and rice.

What Impacts Do Humans Have on Inland Wetlands? Despite the ecological importance of year-round and seasonal inland wetlands, many are drained, dredged, filled in, or covered over. Each year some 475 square kilometers (183 square miles) of inland wetland in the United States are lost, about 80% to agriculture and the rest to mining, forestry, oil and gas extraction, highways, and urban development. Other countries have suffered similar losses.

In the United States, a federal permit is required to fill wetlands or deposit dredged or fill material in them. According to the U.S. Fish and Wildlife Service, this law has helped cut the average annual loss of wetlands by 75% since 1969. However, there are continuing attempts to weaken it by using unscientific criteria to classify areas as wetlands. Only about 8% of remaining inland wetlands are under federal protection, and federal, state, and local protection of wetlands is weak.

The stated goal of current U.S. federal policy is zero net loss in the function and value of wetlands. However, a policy known as *mitigation banking* allows destruction of existing wetlands as long as an equal area of the same type of wetland is created or restored. Experience has shown that **(1)** at least half of the attempts to create new wetlands fail to replace lost ones, **(2)** most of those that are created bear little resemblance to natural wetlands, and **(3)** restoring and creating wetlands is expensive.

In this chapter we have seen that climate is the key factor determining the types of plant and animal life we find at a particular spot on the earth's surface. Change the climate (as we may be doing; see Chapter 11) and we change the earth's patterns of biodiversity and disrupt economic activities. Human activities are also destroying and degrading the habitats and species found in a growing area of the earth's terrestrial and aquatic areas.

Ecological concepts may be regarded as the epitome of scientific thinking.

G. T. RENNER, JR.

REVIEW QUESTIONS

1. Define all of the boldfaced terms in this chapter.

2. How does wind affect climate, global distribution of nutrients, and pollution?

3. Distinguish between *climate* and *weather.* What two main factors determine a region's climate?

4. What five factors affect the global circulation of air? How does each factor affect global air circulation? What causes opposite seasons in the northern and southern hemispheres?

5. How do oceans affect regional climates? What would happen to the climate of northwestern Europe if the Gulf Stream didn't exist? What is an *upwelling,* and why are upwellings important to life?

6. What is the *El Niño–Southern Oscillation (ENSO)*? How does it affect ocean life and weather in various parts of the world? What is *La Niña,* and how does it affect weather in various parts of the world?

7. What is the *greenhouse effect,* and how does it affect the earth's climate? What would life on the earth be like without the natural greenhouse effect? Why is there concern over human activities enhancing the natural greenhouse effect?

8. What is the *ozone layer,* and how does it help protect life on the earth and affect the earth's climate?

9. What is a *biome*? What limiting factor usually determines whether a land area is a desert, grassland, or forest? What two factors determine whether an area of the earth's surface is a tropical, temperate, or polar desert, grassland, or forest? What is an *ecotone*? How do climate and vegetation vary with latitude and longitude?

10. Why are there no trees in the Arctic, and why are the plants there so close to the ground? Why don't desert plants such as cacti have leaves? Why do trees in most forests found in both the warm tropics and in cold areas such as Canada and Sweden keep their leaves year-round, whereas most trees in temperate forests lose their leaves in winter? What are *succulent plants,* where are they found, and how do they survive? What are *broadleaf evergreen plants,* where are they found, and how do they survive? What are *broadleaf deciduous plants,* where are they found, and how do they survive? What are *coniferous evergreen plants,* where are they found, and how do they survive?

11. What is a *desert*? What are the three major types of desert, and how do they differ in climate and biological makeup? How do desert plants and animals survive? Why are desert ecosystems vulnerable to disruption, and what four types of human activities have harmful impacts on deserts?

12. What is *grassland*? What are the three major types of grassland, and how do they differ in climate and biological makeup? Why are grasslands vulnerable to disruption, and what four types of human activities have harmful impacts on grasslands?

13. What is a *forest*? What are the three major types of forest, and how do they differ in climate and biological makeup? What three types of human activities have harmful impacts on forests?

14. Why are mountains ecologically important, and what factors make them vulnerable to ecological disruption? What seven types of human activities have harmful impacts on mountains?

15. What are the two major types of aquatic life zones? What two factors determine the major types of organisms found in aquatic systems? What four advantages do aquatic organisms have over terrestrial organisms? What are four major factors determining the types and numbers of organisms found in the surface, middle, and bottom layers of aquatic systems?

16. What are three reasons that oceans are important to life on the earth? What is the *coastal zone*? Distinguish among *estuaries, coastal wetlands,* and *mangrove swamps* and describe the important services each of these coastal systems provides.

17. Distinguish between *rocky shores* and *sandy shores* and describe the major types of aquatic life found in each. Why is it important to preserve the dunes on barrier beaches? What are *barrier beaches,* why are they so attractive for human development, and why are human structures built there so vulnerable to destruction?

18. What are *coral reefs,* how are they formed, and what eight major services do they provide? What six harmful impacts do human activities have on coral reefs?

19. What is the *open sea,* and what are its three major zones?

20. What is a *freshwater life zone*? What is a *lake*? What are three types of lakes, based on their nutrient content and primary productivity?

21. Describe the three zones of a river as it flows from mountain highlands to the sea.

22. What are *inland wetlands,* and what six important roles do they play? What is *mitigation banking,* and what are three problems with using this approach?

CRITICAL THINKING

1. Why do deserts and arctic tundra support a much smaller biomass of animals than do tropical forests?

2. What factors in your lifestyle contribute to the destruction and degradation of tropical forests?

3. How would you respond to someone who claims that it is not important to protect areas of temperate and polar biomes because most of the world's biodiversity is in the tropics?

4. How would you respond to someone who proposes that we use the deep portions of the world's oceans as depositories for our radioactive and other hazardous wastes because the deep oceans are vast and are located far away from human habitats? Give reasons for your response.

5. What factors in your lifestyle contribute to the destruction and degradation of coastal and inland wetlands?

6. Someone tries to sell you several brightly colored pieces of dry coral. Explain in biological terms why this transaction is probably a rip-off.

7. You are a defense attorney arguing in court for sparing an undeveloped old-growth tropical rain forest and a coral reef from severe degradation or destruction by development. Write out a brief closing statement for the defense of each of these ecosystems. If the judge decides you can save only one of the ecosystems, which one would you choose, and why?

***8.** How has the climate changed in the area where you live during the past 50 years? Investigate the beneficial and harmful effects of these changes. How have these changes benefited or harmed you personally?

***9.** What type of biome do you live in or near? What effects have human activities over the past 50 years had on the characteristic vegetation and animal life normally found in the biome you live in? How is your own lifestyle affecting this biome?

***10.** If possible, visit a nearby lake, pond, or reservoir. Would you classify it as oligotrophic, mesotrophic, or eutrophic? What are the primary factors contributing to its nutrient enrichment? Which of these factors are related to human activities?

***11.** Make a concept map of this chapter's major ideas, using the section heads and subheads and the key terms (in boldface). Look at the inside back cover and in Appendix 5 for information about concept maps.

FURTHER INFORMATION

For more information see Internet Sources and Further Readings at the back of the book. Online resources for this chapter can be found at **www.brookscole.com/biology**

7 POPULATION DYNAMICS, CARRYING CAPACITY, AND CONSERVATION BIOLOGY

Two Islands: Can We Treat This One Better?

Easter Island (Rapa Nui) is a small, isolated island in the great expanse of the South Pacific. It was first colonized by Polynesians about 2,500 years ago.

The civilization they developed was based on the island's towering palm trees, which were used for shelter, tools, boats, fuel, food, rope, and clothing. Using these resources, they developed an impressive civilization and a technology capable of making and moving large stone structures, including their famous statues (Figure 7-1).

The people flourished, the population peaking at about 10,000 (with estimates ranging from 7,000 to 20,000). But they were unable to stop themselves from using up the island's precious trees faster than they were regenerated—an example of the tragedy of the commons (p. 14). Each person who cut a tree reaped immediate personal benefits while helping to doom the civilization as a whole.

Once the trees were gone the islanders could not build canoes for hunting porpoises. Without the forest to absorb and slowly release water, springs and streams dried up, exposed soils eroded, crop yields plummeted, and famine struck.

The starving people turned to warfare and, possibly, cannibalism. Both the population and the civilization collapsed. When Dutch explorers first reached the island on Easter Day, 1722, they found only about 2,000 inhabitants, struggling under primitive conditions on a mostly barren island.

Like Easter Island at its peak, the earth is an isolated island (in the vastness of space) with a thriving technological civilization and no other suitable planet to migrate to. As on Easter Island, our population is growing and we are consuming exhaustible and potentially renewable resources at a rapid pace.

Will the humans on Earth Island recreate the tragedy of Easter Island on a grander scale, or will we learn how to live sustainably on this planet that is our only home? The choice is ours to make.

Figure 7-1 These massive stone figures on Easter Island are the remains of the technology created by an ancient civilization of Polynesians. This civilization collapsed because the people used up the trees (especially large palm trees) that were the basis of their livelihood. More than 200 of these stone statues once stood on huge stone platforms lining the coast. At least 700 additional statues were abandoned in rock quarries or on ancient roads between the quarries and the coast. No one knows how the early islanders (with no wheels, no draft animals, and no sources of energy except their own muscles) transported these gigantic structures for miles before erecting them. However, it is presumed that they did this by felling large trees and using them to roll and erect the statues. (George Holton/Photo Researchers, Inc.)

In looking at nature…never forget that every single organic being around us may be said to be striving to the utmost to increase its numbers.

CHARLES DARWIN, 1859

This chapter addresses the following questions:

- How do populations change in size, density, and makeup in response to environmental stress?
- What different reproductive strategies do species use to enhance their survival?
- What is conservation biology?
- What impacts do human activities have on populations, communities, and ecosystems?
- What efforts are we making to restore ecosystems damaged by human activities?

7-1 POPULATION DYNAMICS AND CARRYING CAPACITY

What Are the Major Characteristics of a Population? Populations are dynamic. They change in **(1)** *size* (number of individuals), **(2)** *density* (number of individuals in a certain space), **(3)** *dispersion* (spatial pattern such as clumping, uniform dispersion, or random dispersion, in which the members of a population are found in their habitat depending mostly on resource availability, Figure 7-2), and **(4)** *age distribution* (proportion of individuals of each age in a population) in response to environmental stress or changes in environmental conditions. These changes are called **population dynamics**.

What Limits Population Growth? Four variables—*births, deaths, immigration,* and *emigration*—govern changes in population size. A population gains individuals by birth and immigration and loses them by death and emigration:

Population change = (Births + Immigration) − (Deaths + Emigration)

These variables depend on changes in resource availability or on other environmental changes (Figure 7-3). If the number of individuals added from births and immigration equals the number lost to deaths and immigration, then there is **zero population growth**.

Populations vary in their capacity for growth, also known as the **biotic potential** of the population. The **intrinsic rate of increase** (r) is the rate at which a population would grow if it had unlimited resources. Generally, individuals in populations with a high intrinsic rate of increase **(1)** *reproduce early in life*, **(2)** *have short generation times* (the time between successive generations), **(3)** *can reproduce many times* (have a long reproductive life), and **(4)** *have many offspring each time they reproduce.*

No population can grow indefinitely. In the real world, a rapidly growing population reaches some size limit imposed by a shortage of one or more limiting factors, such as light, water, space, or nutrients. *There are always limits to population growth in nature.*

Environmental resistance consists of all the factors acting jointly to limit the growth of a population. The population size of a species in a given place and time is determined by the interplay between its biotic potential and environmental resistance (Figure 7-3).

Together biotic potential and environmental resistance determine the **carrying capacity (K)**, the number of individuals of a given species that can be sustained indefinitely in a given space (area or volume).

What Is the Difference Between Exponential and Logistic Growth? A population that does not have resource limitations grows exponentially. *Exponential growth* starts out slowly and then proceeds faster and faster as the population increases. If number of individuals is plotted against time, this sequence yields a J-shaped exponential growth curve (Figure 7-4a).

Logistic growth involves exponential population growth when the population is small and a steady

Clumped (elephants)

Uniform (creosote bush)

Random (dandelions)

Figure 7-2 Generalized dispersion patterns for individuals in a population throughout their habitat. The most common pattern is one in which members of a population exist in clumps throughout their habitat (left), mostly because resources usually are found in patches.

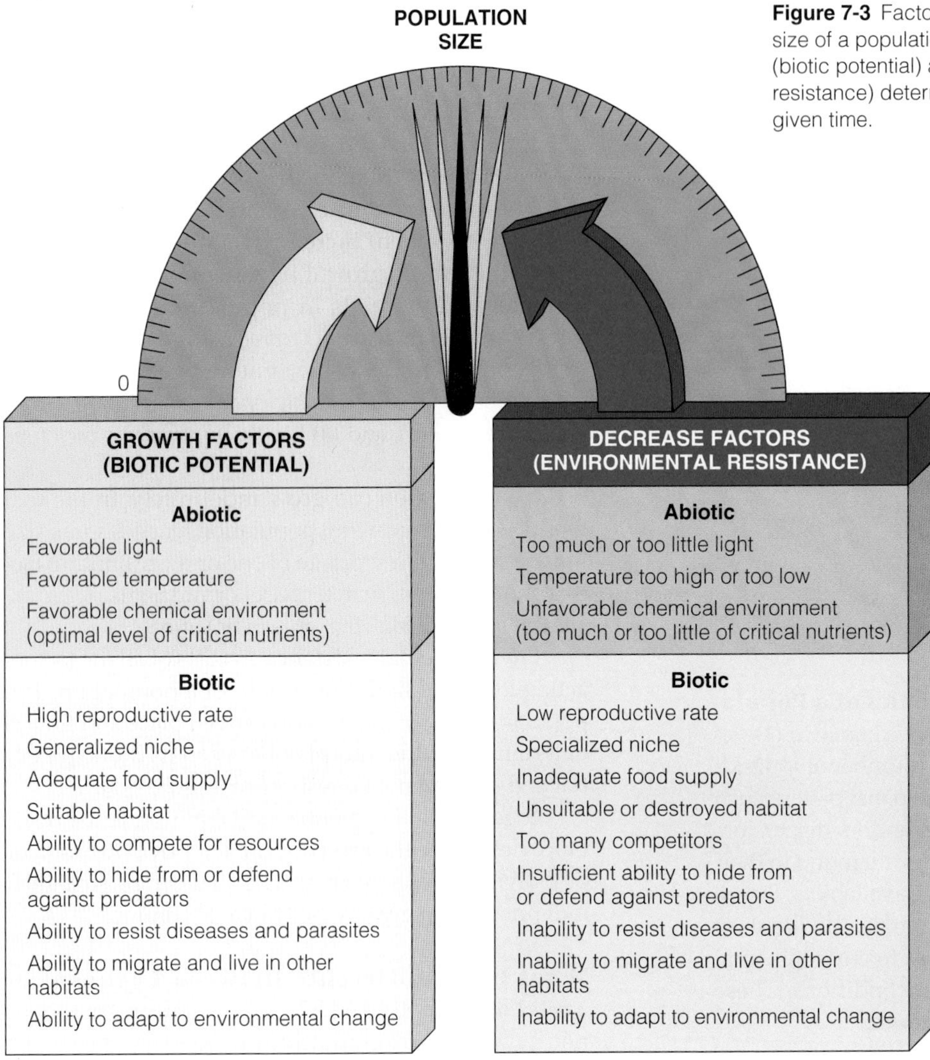

Figure 7-3 Factors that tend to increase or decrease the size of a population. The balance between growth factors (biotic potential) and decrease factors (environmental resistance) determines the size of a population at any given time.

decrease in population growth with time as the population encounters environmental resistance and approaches the carrying capacity of its environment. After leveling off, a population with this type of growth typically fluctuates slightly above and below the carrying capacity. A plot of number of individuals against time yields a sigmoid or S-shaped logistic growth curve (Figure 7-4b). A classic case of logistic growth involves the increase of the sheep population on the island of Tasmania, south of Australia, in the early 19th century (Figure 7-5).

What Happens if the Population Size Exceeds the Carrying Capacity? The populations of some species don't make such a smooth transition from exponential growth to logistic growth. Instead, such a population uses up its resource base and temporarily *overshoots* or exceeds the carrying capacity of its environment. This overshoot occurs because of a *reproductive time lag*, the period required for the birth rate to fall and the death rate to rise in response to resource overconsumption.

In such cases the population will suffer a *dieback* or *crash* unless the excess individuals switch to new resources or move to an area with more favorable conditions. A classic case of such a population crash occurred when reindeer were introduced in 1910 onto a small island off the southwest coast of Alaska (Figure 7-6). In nature we find that over time species have three general types of *population cycles*: stable, irruptive, and cyclic (Figure 7-7).

Humans are not exempt from overshoot and dieback, as shown by the tragedy on Easter Island (p. 168). Ireland also experienced a population crash after a fungus destroyed the potato crop in 1845. About 1 million people died and 3 million people emigrated to other countries.

The earth's carrying capacity for the human species has been extended by technological, social, and other

Thorson, B. 1998. "Boom and Bust." *Canadian Geographic*, vol. 118, no. 2, p. 68(7).

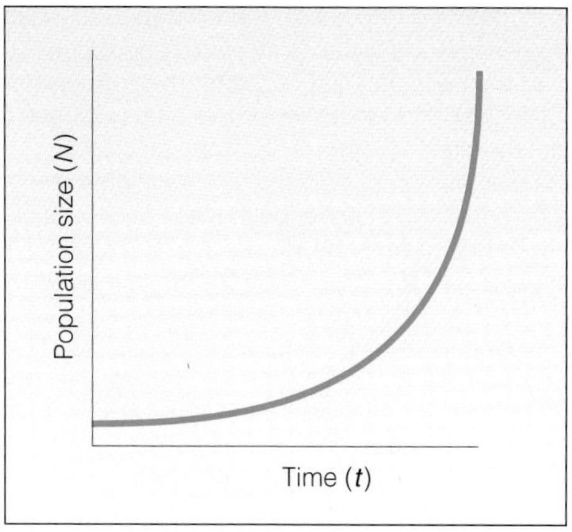

(a) Exponential growth

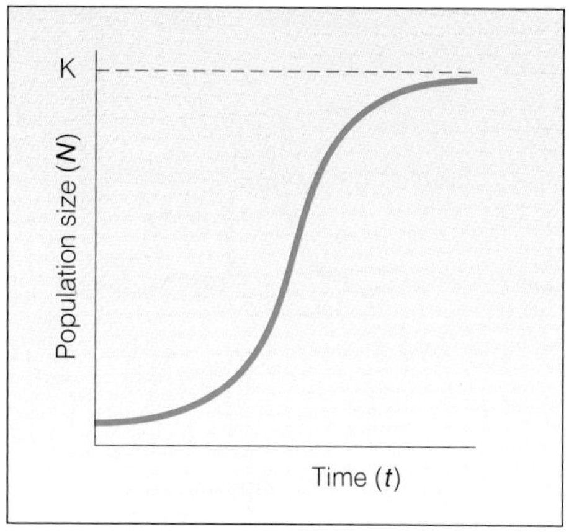

(b) Logistic growth

Figure 7-4 Theoretical population growth curves. **(a)** *Exponential growth,* in which the population's growth rate increases with time. Exponential growth occurs when resources are not limiting and a population can grow at its *intrinsic rate of increase* (r). Exponential growth of a population cannot continue forever because eventually some factor limits population growth. **(b)** *Logistic growth,* in which the growth rate decreases as the population gets larger. With time, the population size stabilizes at or near the *carrying capacity* (K) of its environment.

cultural changes. We have increased food production and used large amounts of energy and matter resources to make normally uninhabitable areas of the earth habitable. However, there is growing concern about how long we will be able to keep doing this on a planet of finite size and resources with an exponentially growing population and per capita resource use.

Carrying capacity is not a simple, fixed quantity and is affected by many factors. Examples are **(1)** competition within and among species, **(2)** immigration and emigration, **(3)** natural and human-caused catastrophic events, and **(4)** seasonal fluctuations in the supply of food, water, hiding places, and nesting sites.

How Does Population Density Affect Population Growth? *Density-independent population controls* affect a population's size regardless of its population density. Examples include floods, hurricanes, severe

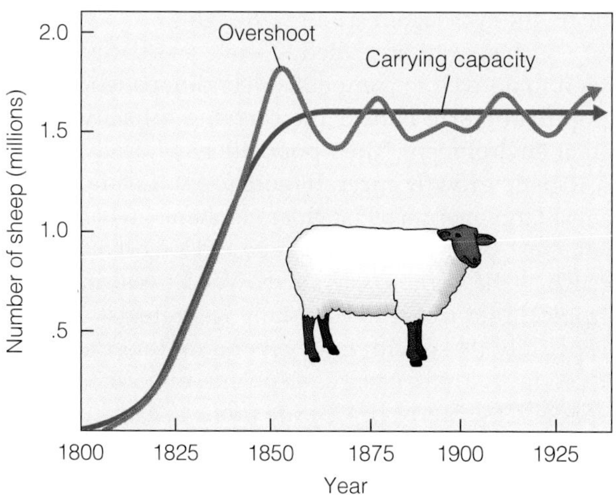

Figure 7-5 *Logistic growth* of a sheep population on the island of Tasmania between 1800 and 1925. After sheep were introduced in 1800 their population grew exponentially because of ample food. By 1855 they overshot the land's carrying capacity. Their numbers then stabilized and fluctuated around a carrying capacity of about 1.6 million sheep.

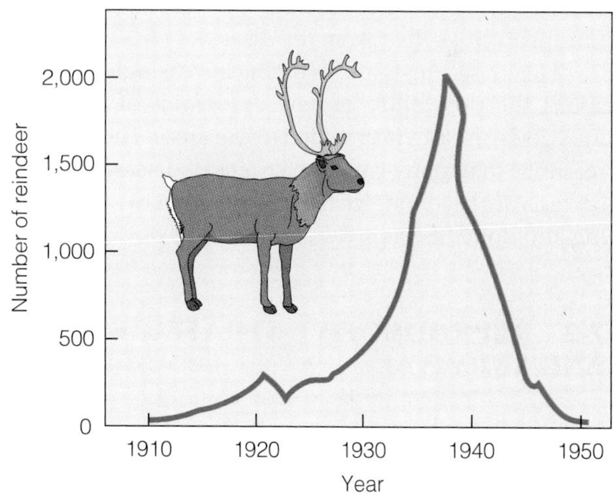

Figure 7-6 Exponential growth, overshoot, and population crash of reindeer introduced to a small island off the southwest coast of Alaska. When 26 reindeer (24 of them female) were introduced in 1910, lichens, mosses, and other food sources were plentiful. By 1935 the herd's population had soared to 2,000, overshooting the island's carrying capacity. This led to a population crash, with the herd plummeting to only 8 reindeer by 1950.

Hint: Enter the search term *lemmings* using the Subject Guide.

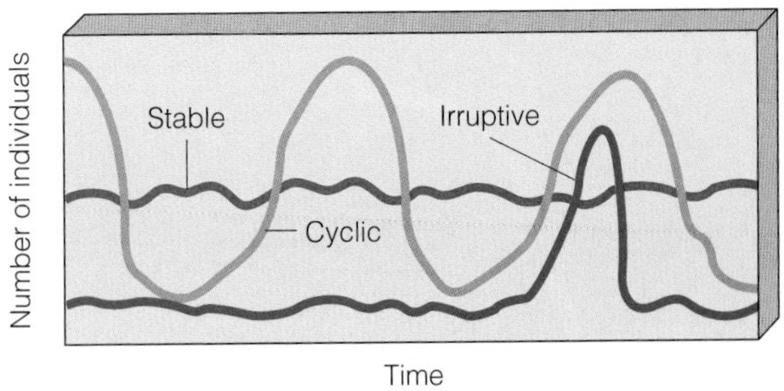

Figure 7-7 General types of idealized population change curves found in nature. The population size of a species with a fairly *stable* population fluctuates slightly above and below its carrying capacity. The populations of some species may occasionally explode, or *irrupt*, to a high peak and then crash to a more stable lower level (Figure 7-6). Other species undergo sharp increases in their numbers, followed by seemingly periodic crashes. Predators are sometimes blamed, but the actual causes of such boom–bust cycles are poorly understood.

drought, unseasonable weather, fire, destruction of habitat (such as clearing a forest of its trees or filling in a wetland), and pesticide spraying. For example, a severe freeze in late spring can kill many individuals in a plant population, regardless of its density.

Some limiting factors have a greater effect as a population's density increases. Examples of such *density-dependent population controls* are competition for resources, predation, parasitism, and disease. As prey populations become denser, their members compete more for limited resources. Individuals in a dense population are also more likely to be infected by parasites or contagious disease organisms.

Infectious disease is a classic example of density-dependent population control. An example is the *bubonic plague*, which swept through Europe during the 14th century. The bacterium that causes this disease normally lives in rodents. It was transferred to humans by fleas that fed on infected rodents and then bit humans. The disease spread like wildfire through crowded cities, where sanitary conditions were poor and rats were abundant. At least 25 million people in European cities died from the disease.

Health scientists are becoming increasingly alarmed about the possibility of new epidemics of common infectious diseases in crowded urban areas. The primary reason is that many common strains of disease-causing bacteria are becoming genetically resistant to most existing antibiotics (Section 8-5).

7-2 REPRODUCTIVE STRATEGIES AND SURVIVAL

What Are r-Strategist Species? Each species has a characteristic mode of reproduction. At one extreme are species that reproduce early. They have many (usually small) offspring each time they reproduce, reach reproductive age rapidly, and have short generation times. They give their offspring little or no parental care or protection to help them survive. Species with this reproductive strategy overcome the massive loss of their offspring by producing so many young that a few will survive to reproduce many offspring to begin the cycle again.

Species with such a capacity for a high intrinsic rate of increase (r) are called **r-strategists** (Figure 7-8, left). Algae, bacteria, rodents, annual plants (such as dandelions), and most insects (p. 84) are examples. Such species tend to be *opportunists*. They reproduce and disperse rapidly when conditions are favorable or when a new habitat or niche becomes available for invasion, as in the early stages of ecological succession (Figures 5-14 and 5-15). Changing or unfavorable environmental conditions can cause such populations to crash. Hence, most r-strategists go through irregular and unstable boom–bust cycles.

What Are K-Strategist Species? At the other extreme are **K-strategists**, species that tend to reproduce late and have few offspring with long generation times. Typically these offspring develop inside their mothers (where they are safe). They are fairly large and mature slowly. They are cared for and protected by one or both parents until they reach reproductive age. This strategy results in a few big and strong individuals that can compete for resources and reproduce a few young to begin the cycle again (Figure 7-8, right).

Such species are called K-strategists because they tend to do well in competitive conditions when their population size is near the carrying capacity (K) of their environment. Their populations typically follow a logistic growth curve (Figure 7-4b). Examples are most large mammals (such as elephants, whales, and humans), birds of prey, and large and long-lived plants (such as the saguaro cactus, redwood trees, and most tropical rain forest trees). Many K-strategist species, especially those with long generation times and low reproductive rates (such as elephants, rhinoceroses, and sharks, Case Study, p. 123), are prone to extinction.

Many organisms have reproductive strategies between the extremes of r-strategists and K-strategists, or they change from one extreme to the other under certain environmental conditions. In agriculture we raise both r-strategists (crops) and K-strategists (livestock).

The reproductive strategy of a species may give it a temporary advantage, but *the availability of suitable habitat for individuals of a population in a particular area is what determines its ultimate population size.* Regardless of how

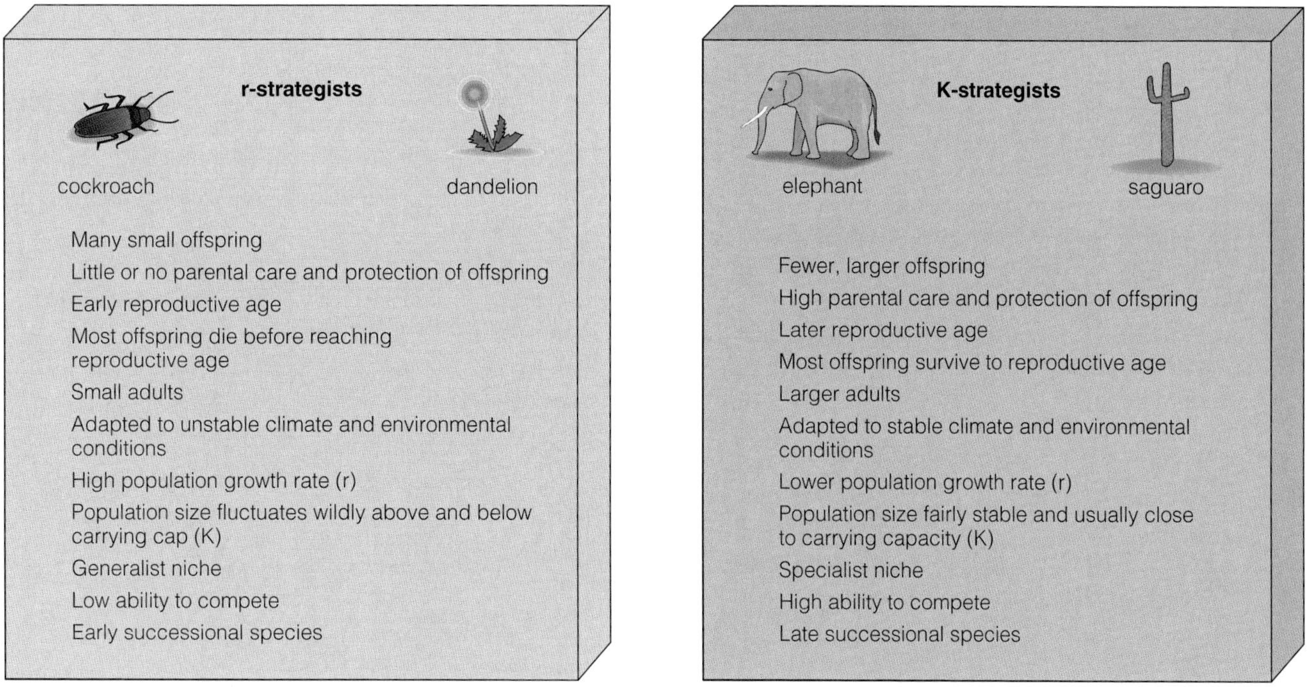

Figure 7-8 Generalized characteristics of r-strategists and K-strategists. Many species have characteristics between these two extremes.

fast a species can reproduce, there can be no more dandelions than there is dandelion habitat and no more zebras than there is zebra habitat in a particular area.

What Are Survivorship Curves? Individuals of species with different reproductive strategies tend to have different *life expectancies*. One way to represent the age structure of a population is with a **survivorship curve**, which shows the number of survivors of each age group for a particular species. There are three generalized types of survivorship curves: *late loss, early loss,* and *constant loss* (Figure 7-9).

A table of the numbers of individuals at each age from a survivorship curve is called a *life table*. It shows the projected life expectancy and probability of death for individuals at each age. Insurance companies use life tables of human populations in various countries or regions to determine policy costs for their customers. Because life tables show that women in the United States survive an average of 7 years longer than men, a 65-year-old man normally will pay more for life insurance than a 65-year-old woman.

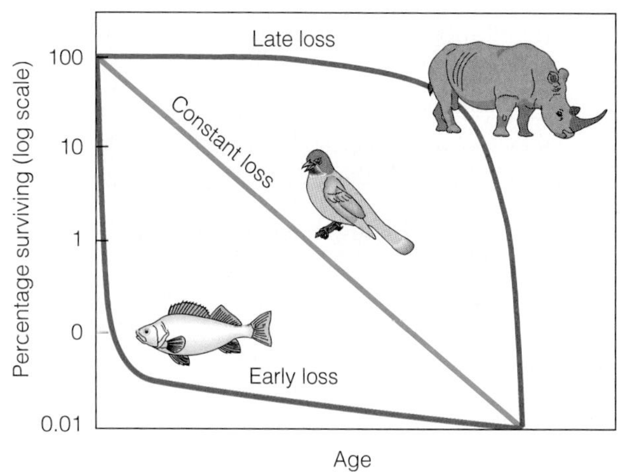

Figure 7-9 Three general types of survivorship curves for populations of different species, obtained by showing the percentages of the members of a population surviving at different ages. For a *late loss* population (such as elephants, rhinoceroses, and humans), there is typically high survivorship to a certain age, then high mortality. A *constant loss* population (such as many songbirds) shows a fairly constant death rate at all ages. For an *early loss* population (such as annual plants and many bony fish species), survivorship is low early in life. These generalized survivorship curves only approximate the behavior of species.

7-3 CONSERVATION BIOLOGY: SUSTAINING WILDLIFE POPULATIONS

What Is Conservation Biology? Conservation involves sensible and careful use of natural resources by humans. **Conservation biology** is an interdisciplinary science that originated in the 1970s. Its goals are to investigate human impacts on biodiversity and to develop practical approaches to maintaining biodiversity to ensure the continued existence of populations of wild species.

Conservation biology differs from *wildlife management*, which is devoted primarily to the manipulation of the population sizes of various animal species, especially game species prized by hunters and fishers (Sections 18-4 and 18-5).

Conservation biologists believe that we must preserve **ecological integrity**, the conditions and natural processes (such as energy flow and matter cycling in ecosystems, Figure 4-13) that generate and maintain biodiversity and allow evolutionary change as a key mechanism for adapting to changes in environmental conditions. An ecological integrity report card for an ecosystem would include an evaluation of

- Habitat quality
- Integrity of the biological community (biodiversity, community composition and structure, key species, alien species, and endangered and threatened species)
- Integrity of ecological processes (chemical cycling, energy flows, net primary productivity, succession)
- Water and soil characteristics and quality
- Degree of actual or potential disturbance from **(1)** natural variability (frequency and intensity of fires, flooding, drought and storms; **(2)** disease and pest outbreaks; **(3)** human interventions such as land clearing and pollution; and **(4)** outside influences such as climate change and sea-level rise)
- Identification of key ecological, economic, and other societal factors determining the current and long-term ecological integrity of the ecosystem

Conservation biology rests on three underlying principles:

- Biodiversity and ecological integrity are necessary to all life on earth and should not be reduced by human actions.
- Humans should not cause or hasten the premature extinction of wildlife populations and species or disrupt vital ecological processes.
- The best way to preserve earth's biodiversity and ecological integrity is to protect intact ecosystems that provide sufficient habitat for sustaining natural populations of species.

Conservation biology is based on Aldo Leopold's ethical principle that something is right when it tends to maintain the earth's life-support systems for us and other species and wrong when it doesn't.

Chapter 17 examines methods for sustaining forests, rangelands, parks, and wilderness ecosystems through land management practices, and Chapter 18 examines issues of premature extinction, with an emphasis on rare and endangered species.

How Can Conservation Biology Help Prevent Premature Extinction and Ensure Ecological Integrity? Conservation biology seeks answers to three questions:

- *What is the status of natural populations, and which species are in danger of extinction?*
- *What is the status of the integrity of ecosystems, and what ecosystem services of value to humans and other species are we in danger of losing?*
- *What measures can we take to ensure that we maintain habitat of the quality and size needed to ensure viable populations of wild species and that ecosystem integrity can be sustained?*

These are challenging questions, and the answers require extensive field research and a strong grounding in ecological theory. To understand the status of natural populations, we must **(1)** measure the current population size, **(2)** determine how population size is likely to change with time, and **(3)** determine whether existing populations are likely to be sustainable.

7-4 HUMAN IMPACTS ON ECOSYSTEMS: LEARNING FROM NATURE

How Have Humans Modified Natural Ecosystems? To survive and support growing numbers of people, we have greatly increased the number and area of the earth's natural systems that we have modified, cultivated, built on, or degraded (Figure 1-5). We have used technology to alter much of the rest of nature in several ways by

- *Fragmenting and degrading habitat.*
- *Simplifying natural ecosystems.* When we plow grasslands and clear forests, we often replace their thousands of interrelated plant and animal species with one crop or one kind of tree—called *monocultures*—or with buildings, highways, and parking lots. Then we spend a lot of time, energy, and money trying to protect such monocultures from invasion by opportunist species of plants (weeds), pests (mostly insects), and pathogens (fungi, viruses, or bacteria that harm the plants we want to grow).
- *Strengthening some populations of pest species and disease-causing bacteria by speeding up natural selection and causing genetic resistance through overuse of pesticides.*

Rapport, D. J. and W. G. Whitford. 1999. "How Ecosystems Respond to Stress." *BioScience*, vol. 49, no. 3, p. 193(1).

- *Eliminating some predators.* Some ranchers want to eradicate bison or prairie dogs that compete with their sheep or cattle for grass. They also want to eliminate wolves, coyotes, eagles, and other predators that occasionally kill sheep. Big game hunters also push for elimination of predators that prey on game species.
- *Deliberately or accidentally introducing new species,* some beneficial and some harmful to us and other species. In the late 1800s several Chinese chestnut trees brought to the United States were infected with a fungus that spread to the American chestnut, once found throughout much of the eastern United States. Between 1910 and 1940, this accidentally introduced fungus virtually eliminated the American chestnut.
- *Overharvesting potentially renewable resources.* Ranchers and nomadic herders sometimes allow livestock to overgraze grasslands until erosion converts these ecosystems to less productive semideserts or deserts. Farmers sometimes deplete the soil of nutrients by excessive crop growing. Species of fish are overharvested. Wildlife species with economically valuable parts (such as elephant tusks, rhinoceros horns, and tiger skins) are endangered by illegal hunting (poaching).
- *Interfering with the normal chemical cycling and energy flows in ecosystems.* Soil nutrients can erode from monoculture crop fields, tree farms, construction sites, and other simplified ecosystems and can overload and disrupt other ecosystems such as lakes and coastal ecosystems. Chemicals such as chlorofluorocarbons (CFCs) released into the atmosphere can increase the amount of harmful ultraviolet energy reaching the earth by reducing ozone levels in the stratosphere. Emissions of carbon dioxide and other greenhouse gases—from burning fossil fuels and from clearing and burning forests and grasslands—can trigger global climate change by disrupting energy flow through the atmosphere.

CONNECTIONS

Ecological Surprises

Malaria once infected 9 out of 10 people in North Borneo, now known as Brunei. In 1955 the World Health Organization (WHO) began spraying the island with dieldrin (a DDT relative) to kill malaria-carrying mosquitoes. The program was so successful that the dreaded disease was virtually eliminated.

However, unexpected things began to happen. The dieldrin also killed other insects, including flies and cockroaches living in houses. The islanders applauded this turn of events, but then small lizards that also lived in the houses died after gorging themselves on dieldrin-contaminated insects. Next, cats began dying after feeding on the lizards. Then, in the absence of cats, rats flourished and overran the villages. When the people became threatened by sylvatic plague carried by rat fleas, the World Health Organization (WHO) parachuted healthy cats onto the island to help control the rats.

Then the villagers' roofs began to fall in. The dieldrin had killed wasps and other insects that fed on a type of caterpillar that either avoided or was not affected by the insecticide. With most of its predators eliminated, the caterpillar population exploded, munching its way through its favorite food: the leaves used in thatched roofs.

Ultimately, this episode ended happily: Both malaria and the unexpected effects of the spraying program were brought under control. Nevertheless, the chain of unforeseen events emphasizes the unpredictability of interfering with an ecosystem. It reminds us that when we intervene into nature we need to ask the question, "And then what?"

Critical Thinking

Do you believe that the beneficial effects of spraying pesticides on North Borneo outweighed the resulting unexpected and harmful effects? Explain.

To survive we must exploit and modify parts of nature. However, we are beginning to understand that any human intrusion into nature has multiple effects, most of them unpredictable (Connections, right).

The challenge is to maintain a balance between simplified, human-altered ecosystems and the neighboring, more complex natural ecosystems on which we and other life-forms depend and to slow down the rates at which we are altering nature for our purposes. If we simplify and degrade too much of the planet to meet our needs and wants, what's at risk is not the earth but our own species.

According to biodiversity expert E. O. Wilson, "If this planet were under surveillance by biologists from another world, I think they would look at us and say, 'Here is a species in the mid-stages of self-destruction.'" The evolutionary lesson to be learned from nature is that no species can get "too big for its britches," at least not for long.

Solutions: What Can We Learn from Nature About Living Sustainably? Scientific research indicates that living systems have six key features: *interdependence, diversity, resilience, adaptability, unpredictability,* and *limits*. Many biologists believe that the best way for us to live more sustainably is to learn about the processes and adaptations by which nature sustains itself (Solutions, right) and then mimic these lessons from nature.

Biologists have used these lessons from nature (Solutions, right) to formulate several principles to guide us in our search for more sustainable lifestyles:

- *Our lives, lifestyles, and economies are totally dependent on the sun and the earth.*
- *We can never do merely one thing.* Any human intrusion into nature has mostly unpredictable side effects. When we alter nature, we should ask the question, "And then what?"
- *Everything is connected to everything else.* The primary goal of ecology is to discover which connections in nature are the strongest, most important, and most vulnerable to disruption.
- *We should reduce and minimize the damage we do to nature*—a prevention strategy—and help heal some of the ecological wounds we have inflicted.
- *We should use care, restraint, humility, and cooperation with nature as we alter the ecosphere to meet our needs and wants.*

Principles of Sustainability: Learning from Nature

SOLUTIONS

Here are five basic ecological lessons or principles of sustainability derived from observing how nature works:

- *Most ecosystems use renewable solar energy as their primary source of energy.*
- *Ecosystems replenish nutrients and dispose of wastes by recycling chemicals.* There is virtually no waste in nature and there is no "away." The waste outputs and decomposed remains of one organism are resource inputs for other organisms.
- *Ecosystems minimize energy waste because the energy they use cannot be recycled and is always degraded to less useful forms of energy.*
- *Biodiversity helps maintain sustainability and ecological integrity and serves as a source of adaptations to changing environmental conditions.*
- *In nature there are always limits to population growth.* The population size and growth rate of all species are controlled by their interactions with other species and with their nonliving environment.

There is considerable evidence that we are violating these five principles of sustainability. No one knows how long we can continue doing this.

Critical Thinking

List two ways that your lifestyle violates each of these five principles of sustainability. Would you be willing to change these practices? What beneficial and harmful effects would such changes have on your lifestyle?

7-5 SOLUTIONS: WORKING WITH NATURE TO HELP ECOSYSTEMS HEAL

How Can We Rehabilitate and Restore Damaged Ecosystems? Much of the environmental damage we have inflicted on nature is at least partially reversible. We can protect degraded ecosystems and allow them to replenish themselves through secondary ecological succession (Figure 5-15). We can replant forests, clean up streams, and at least partially restore wetlands (Individuals Matter, right). Researchers are creating a new discipline of *restoration ecology* devoted to renewing, repairing, or reconstructing damaged ecosystems.

Farmer and philosopher Wendell Berry says we should try to answer three questions in deciding whether and how to modify or rehabilitate natural ecosystems: **(1)** What is here? **(2)** What will nature permit us to do here? and **(3)** What will nature help us do here?

By studying how natural ecosystems recover, scientists are learning how to speed up repair operations. *Rehabilitation* involves making degraded land productive again (by stopping soil erosion, for example).

Active restoration is more ambitious. It involves working with nature to restore the biodiversity and ecological processes that occur in wild ecosystems. Often, it's not necessary to plant anything; we need only to find and protect the strongest types of natural growth and allow natural ecological succession to proceed. Another approach is *replacement* of a degraded ecosystem with another type of ecosystem, such as replacing a degraded area of forest with a productive pasture.

Repairing and protecting ecosystems require that a pool of suitable existing species be available to occupy a wide variety of habitats and niches. This is why it is so important to preserve biodiversity throughout the world and to protect sustainable patches of ecosystems that harbor species that recolonize ecosystems after a disturbance.

Restoration of a Wetland

INDIVIDUALS MATTER

Humans have drained, filled in, or covered over swamps, marshes (Figure 6-24), and other wetlands relentlessly for centuries. This has been done to **(1)** create rice fields and other land used to grow crops, **(2)** create land for urban development and highways, **(3)** reduce disease such as malaria caused by mosquitoes, **(4)** extract minerals, oil, and natural gas.

Some have begun to question such practices as we learn more about the ecological and economic importance of coastal wetlands (Section 6-5) and inland wetlands (Section 6-6). Can we turn back the clock to restore or rehabilitate lost marshes?

California rancher Jim Callender decided to try. In 1982 he bought 20 hectares (50 acres) of Sacramento Valley ricefield that had been a marsh until the early 1970s. To grow rice, the previous owner had destroyed it, bulldozing, draining, leveling, uprooting the native plants and spraying with chemicals to kill the snails and other food of the waterfowl.

Callender and his friends set out to restore the marshland. They hollowed out low areas, built up islands, replanted tules and bulrushes, reintroduced smartweed and other plants needed by birds, and planted fast-growing Peking willows. After 6 years of care, hand-planting, and annual seeding with a mixture of watergrass, smartweed, and rice, the marsh is once again a part of the Pacific Flyway used by migratory waterfowl.

Jim Callender and others have shown that at least part of the continent's degraded or destroyed wetlands can be partially reclaimed with scientific knowlege and hard work. Such restoration is useful, but to most ecologists the real challenge is to protect remaining wetlands (and other undisturbed ecosystems) from harm in the first place. What do you think?

What Are the Limits of Ecosystem Rehabilitation and Restoration? Some analysts worry that environmental restoration could encourage continuing environmental destruction by suggesting that any ecological harm we do can be undone. However, ecological restoration is imperfect at best and falls far short of fully restoring damaged ecosystems.

Another concern is government policies that allow developers to destroy one ecosystem or wetland if they protect, restore, or "create" a similar one of roughly the same size. This tradeoff or *mitigation* approach is preferable to wanton destruction of ecosystems. However, by legitimizing further destruction, it partially defeats the main purpose of ecosystem restoration: *repairing previous damage*. If we prevent ecosystem destruction and degradation, then expensive and only partially effective ecological restoration is not needed.

We cannot command nature except by obeying her.

SIR FRANCIS BACON

REVIEW QUESTIONS

1. Define the boldfaced terms found in this chapter.

2. Relate the fall of the civilization on Easter Island to the tragedy of the commons.

3. What are four factors that affect population change?

4. Write an equation showing how population change is related to births, deaths, immigration, and emigration.

5. What is the *biotic potential* of a population? What are four characteristics of a population with a high intrinsic rate of increase? What are *environmental resistance* and *carrying capacity*? How do biotic potential and environmental resistance interact to determine carrying capacity? What factors can alter an area's carrying capacity?

6. Distinguish between *exponential* and *logistic growth* of a population and give an example of each type.

7. How can a population overshoot its carrying capacity, and what are the consequences of doing this?

8. Distinguish between stable, irruptive, and cyclic forms of population change.

9. Distinguish between *density-dependent* and *density-independent factors* that affect a population's size and give an example of each.

10. List the characteristics of *r-strategist* and *K-strategist* species and give two examples of each type.

11. What is a *survivorship curve*, and how is it used?

12. Distinguish between *conservation, conservation biology, wildlife management*, and *ecological integrity*. List seven things that determine the ecological integrity of an ecosystem.

13. What are the three underlying principles of conservation biology? What three questions does conservation biology try to answer?

14. List seven potentially harmful ways in which humans modify natural ecosystems.

15. List five *principles of sustainability* derived from observing how natural systems are sustained.

16. List four principles we could use to help us live more sustainably.

CRITICAL THINKING

1. Why do you think the people on Easter Island (p. 168) didn't realize that they were consuming the resources that supported them and stop before it was too late? Do you believe that we are facing a similar situation on Earth Island? Explain.

2. Why are pest species likely to be extreme r-strategists? Why are many endangered species likely to be extreme K-strategists?

3. Given current environmental conditions, if you had a choice would you rather be an r-strategist or a K-strategist? Explain your answer. What implications does your decision have for your current lifestyle?

4. If after your death you could come back as a member of a particular type of species, what type of survivorship curve (Figure 7-9) would you like to have, and why?

5. Explain why a simplified ecosystem such as a cornfield is usually much more vulnerable to harm from insects and plant diseases than a more complex, natural ecosystem such as a grassland.

***6.** Use the library and Internet to choose one wild plant species and one animal species and analyze the factors that are likely to limit the populations of each species. How are the factors that limit animal and plant populations similar? How are they different?

***7.** Use the principles of sustainability derived from the scientific study of how nature sustains itself (Solutions, p. 176) to evaluate the sustainability of the following parts of human systems: **(a)** transportation, **(b)** cities, **(c)** agriculture, **(d)** manufacturing, **(e)** waste disposal, and **(f)** your own lifestyle. Compare your analysis with those made by other members of your class.

***8.** Use the library and Internet to find three examples of ecological restoration. For each example, explain how the restoration was achieved. What ecological principles were used in restoring each of the ecosystems? How successful was each of the restoration efforts? How could you improve on each of the efforts?

***9.** Make a concept map of this chapter's major ideas, using the section heads and subheads and the key terms (in boldface). Look inside the back cover and in Appendix 5 for information about concept maps.

FURTHER INFORMATION

For more information see Internet Sources and Further Readings at the back of the book. Online resources for this chapter can be found at **www.brookscole.com/biology**

8 RISK, TOXICOLOGY, AND HUMAN HEALTH

The Big Killer

What is roughly the diameter of a 30-caliber bullet, can be bought almost anywhere, is highly addictive, and kills about 8,200 people every day or 342 per hour? It's a cigarette. *Cigarette smoking is the single most preventable major cause of death and suffering among adults.*

The World Health Organization (WHO) estimates that each year tobacco contributes to the premature deaths of at least 3 million people from heart disease, lung cancer, other cancers, bronchitis, emphysema, and stroke. The annual death toll from smoking-related diseases is projected to reach 10 million by 2030 (70% of them in developing countries)—an average of about 27,400 preventable deaths per day.

Smoking kills about 419,000 Americans per year, an average of 1,150 deaths per day (Figure 8-1). This death toll is equivalent to three fully loaded jumbo jets crashing every day with no survivors. Smoking causes more deaths each year in the United States than do all illegal drugs, alcohol (the second most harmful drug after nicotine), automobile accidents, suicide, and homicide combined (Figure 8-1). According to a 1998 study, secondhand smoke (inhaled by nonsmokers) causes 30,000 to 60,000 premature deaths per year in the United States.

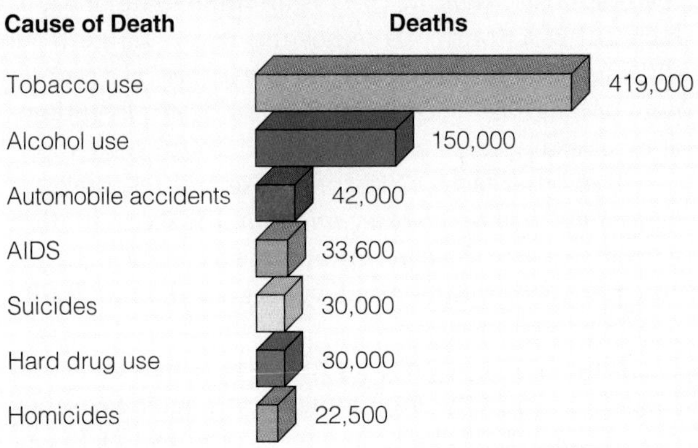

Figure 8-1 Annual deaths in the United States from tobacco use and other causes. Smoking is by far the nation's leading cause of preventable death, causing more premature deaths each year than all the other categories in this figure combined. (Data from National Center for Health Statistics)

The overwhelming consensus in the scientific community is that the nicotine (and probably the acetaldehyde) inhaled in tobacco smoke is highly addictive. Only 1 in 10 people who try to quit smoking succeed—about the same relapse rate as for recovering alcoholics and those addicted to heroin or crack cocaine. A British government study showed that adolescents who smoke more than one cigarette have an 85% chance of becoming smokers.

Government agencies and independent economists estimate that the country's 48 million smokers cost the United States $70–$100 billion a year in medical bills, increased insurance costs, disability, lost earnings and productivity because of illness, and property damage from smoking-caused fires. This is an average of $3–4 per pack of cigarettes sold.

Many health experts urge that a $2–4 federal tax be added to the price of a pack of cigarettes. Such a tax would mean that the users of cigarettes, not the rest of society, would pay a much greater share of the health, economic, and social costs associated with their smoking: a *user-pays* approach.

Other suggestions for reducing the death toll and health effects of smoking in the United States include **(1)** banning all cigarette advertising, **(2)** forbidding the sale of cigarettes and other tobacco products to anyone under 21 (with strict penalties for violators), **(3)** banning all cigarette vending machines, **(4)** classifying nicotine as an addictive and dangerous drug (and placing its use in tobacco or other products under the jurisdiction of the Food and Drug Administration), **(5)** eliminating all federal subsidies and tax breaks to U.S. tobacco farmers and tobacco companies, and **(6)** using cigarette tax income to finance a massive antitobacco advertising and education program.

For the first time in the history of the world, every human being is now subjected to dangerous chemicals from the moment of conception until death.

RACHEL CARSON

This chapter addresses the following questions:

- What types of hazards do people face?
- What is toxicology, and how do scientists measure toxicity?
- What chemical hazards do people face, and how can they be measured?
- What physical hazards do people face from earthquakes and volcanic eruptions?
- What types of disease (biological hazards) threaten people in developing countries and developed countries?
- How can risks be estimated, managed, and reduced?

8-1 RISK AND HAZARDS

What Is Risk? **Risk** is the possibility of suffering harm from a *hazard* that can cause injury, disease, economic loss, or environmental damage. Risk is expressed in terms of **probability**: a mathematical statement about how likely it is that some event or effect will occur.

The probability of a risk is expressed as a fraction ranging from 0 (absolute certainty that there is no risk, which can never be shown) to 1.0 (absolute certainty that there is a risk). For example, the lifetime cancer risk from exposure to a particular chemical may have a probability of 0.000001, or 1 in 1 million. This means that one of every 1 million people exposed to the chemical at a specified average daily dose will develop cancer over a typical lifetime (usually considered to be 70 years).

Risk assessment involves using data, hypotheses, and models to estimate the probability of harm to human health, society, or the environment that may result from exposure to specific hazards.

What Are the Major Types of Hazards? The various kinds of hazards we face can be categorized as follows:

- *Cultural hazards* such as unsafe working conditions, smoking (p. 179), poor diet, drugs, drinking, driving, criminal assault, unsafe sex, and poverty
- *Chemical hazards* from harmful chemicals in the air (Chapter 10), water (Chapter 12), soil (Chapter 13), and food (Chapter 15). The bodies of most human beings contain small amounts of about 500 synthetic organic chemicals—whose health effects are mostly unknown—that did not exist in 1920
- *Physical hazards* such as ionizing radiation, noise, fire, tornadoes, hurricanes, earthquakes, volcanic eruptions, and floods
- *Biological hazards* from pathogens (bacteria, viruses, and parasites), pollen and other allergens, and animals such as bees and poisonous snakes

According to a 1998 study by Cornell University scientist David Pimentel (Guest Essay, p. 352), environmental factors such as malnutrition, smoking, cooking fires, skin cancer, exposures to pesticides and other hazardous chemicals, and air and water pollution cause about 40% of the world's annual deaths.

8-2 TOXICOLOGY

What Determines Whether a Chemical Is Harmful? Dose and Response **Toxicity** is a measure of how harmful a substance is. Whether a chemical (or other agent such as ionizing radiation) is harmful depends on several factors. One is the **dose**, the amount of a potentially harmful substance that a person has ingested, inhaled, or absorbed through the skin. Whether a chemical is harmful depends on **(1)** the size of the dose over a certain period of time, **(2)** how often an exposure occurs, **(3)** who is exposed (adult or child, for example), and **(4)** how well the body's detoxification systems (liver, lungs, and kidneys) work.

The type and amount of health damage that result from exposure to a chemical or other agent are called the **response**. An *acute effect* is an immediate or rapid harmful reaction to an exposure; it can range from dizziness or a rash to death. A *chronic effect* is a permanent or long-lasting consequence (kidney or liver damage, for example) of exposure to a harmful substance.

Two other factors affecting dose and response are *bioaccumulation* and *biomagnification*. **Bioaccumulation** is an increase in the concentration of a chemical in specific organs or tissues at a level higher than normally would be expected.

Biomagnification is a process in which the levels of some toxins in the environment are magnified as they pass through food chains and webs (Figure 8-2). Examples of chemicals that can be biomagnified include long-lived, fat-soluble organic compounds such as the pesticide DDT and PCBs (oily chemicals used in electrical transformers) and some radioactive isotopes (such as strontium-90, Table 3-1). Stored in body fat, such chemicals can be passed along to offspring during gestation or egg laying and as mothers nurse their young.

The detection of trace amounts of a chemical in air, water, or food does not necessarily mean that it is there at a level harmful to most people. Practically any synthetic or natural chemical, even water, can be harmful

 1998. "New Global Health Report Warns About Health Risks of Environmental Degradation." *Journal of Environmental Health,* v61 n2 p31(3).

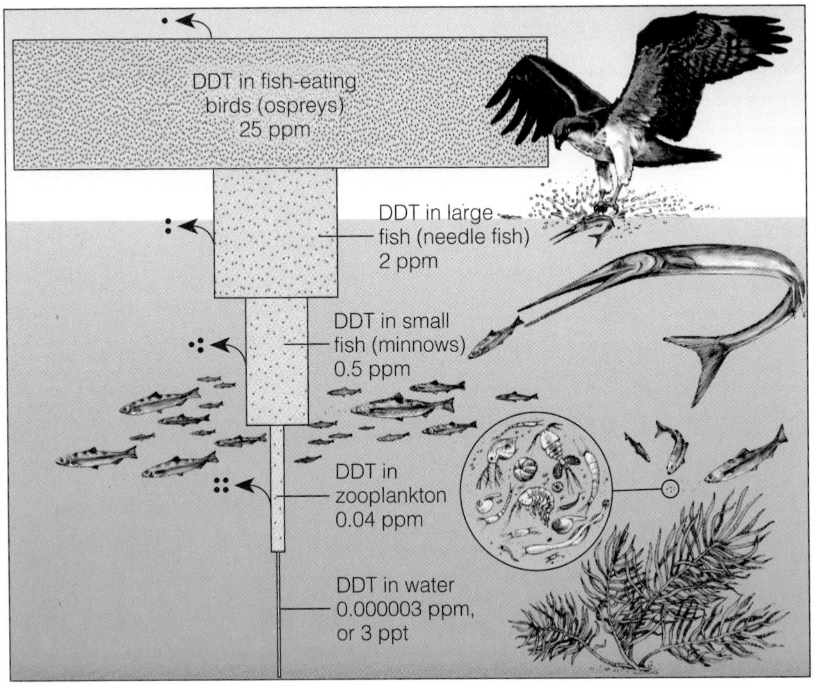

Figure 8-2 Bioaccumulation and biomagnification. DDT is a fat-soluble chemical that can bioaccumulate in the fatty tissues of animals. In a food chain or food web, the accumulated concentrations of DDT can be biologically magnified in the bodies of animals at each higher trophic level. This diagram shows that the concentration of DDT in the fatty tissues of organisms was biomagnified about 10 million times in this food chain in an estuary near Long Island Sound. If each phytoplankton organism in such a food chain takes up from the water and retains one unit of DDT, a small fish eating thousands of zooplankton (which feed on the phytoplankton) will store thousands of units of DDT in its fatty tissue. Then each large fish that eats 10 of the smaller fish will ingest and store tens of thousands of units, and each bird (or human) that eats several large fish will ingest hundreds of thousands of units. Dots represent DDT, and arrows show small losses of DDT through respiration and excretion.

if ingested in a large enough quantity. The critical question is how much exposure to a particular toxic chemical causes a harmful response.

Some people have the mistaken idea that all natural chemicals are safe and all synthetic chemicals are harmful. In fact, many synthetic chemicals are quite safe if used as intended, and many natural chemicals are deadly.

What Is a Poison? Legally, a **poison** is a chemical that has an LD_{50} of 50 milligrams or less per kilogram of body weight. An $\mathbf{LD_{50}}$ is the **median lethal dose**: the amount of a chemical received in one dose that kills exactly 50% of the animals (usually rats and mice) in a test population (usually 60–200 animals) within a 14-day period. *LD* stands for *lethal dose*, and the subscript *50* refers to the percentage of test organisms for which the dose was lethal.

Chemicals vary widely in their toxicity (Table 8-1). Some poisons can cause serious harm or death after a single acute exposure at extremely low doses. Others cause such harm only at such huge doses that it is nearly impossible to get enough into the body. Most chemicals fall between these two extremes.

How Do Scientists Determine Toxicity? Scientists use three methods to determine the level at which a substance poses a health threat. One is *case reports* (usually made by physicians) about people suffering some adverse health effect or death after exposure to a chemical. Such information often involves accidental poisonings, drug overdoses, homicides, or suicide attempts.

A second method relies on *laboratory investigations* (usually on test animals) to determine toxicity, residence time, what parts of the body are affected, and (sometimes) how the harm takes place. The third method is *epidemiology*, which involves studies of populations of humans exposed to certain chemicals or diseases.

Case reports usually are the least valuable source for determining toxicity because the actual dose and the exposed person's health status often are not known. However, such reports can provide clues about environmental hazards and suggest the need for laboratory investigations.

Acute toxicity and chronic toxicity usually are determined by tests on live laboratory animals (especially mice and rats, which are small and prolific and can be housed inexpensively in large numbers) and on bacteria. Such tests also are made on cell and tissue cultures and chicken egg membranes.

Acute toxicity tests are run to develop a **dose-response curve**, which shows the effects of various doses of a toxic agent on a group of test organisms (Figure 8-3). Such tests are *controlled experiments* in which the effects of the chemical on a *test group* are compared with the responses of a *control group* of organisms not exposed to the chemical. Care is taken to ensure that organisms in each group are as identical as possible in

Hint: Enter *environmental health* or *environmental* using Key Words.

Table 8-1 Toxicity Ratings and Average Lethal Doses for Humans

Toxicity Rating	LD_{50} (milligrams per kilogram of body weight)*	Average Lethal Dose†	Examples
Supertoxic	Less than 0.01	Less than 1 drop	Nerve gases, botulism toxin, mushroom toxins, dioxin (TCDD)
Extremely toxic	Less than 5	Less than 7 drops	Potassium cyanide, heroin, atropine, parathion, nicotine
Very toxic	5–50	7 drops to 1 teaspoon	Mercury salts, morphine, codeine
Toxic	50–500	1 teaspoon to 1 ounce	Lead salts, DDT, sodium hydroxide fluoride, sulfuric acid, caffeine, carbon tetrachloride
Moderately toxic	500–5,000	1 ounce to 1 pint	Methyl (wood) alcohol, ether, phenobarbital, amphetamines (speed), kerosene, aspirin
Slightly toxic	5,000–15,000	1 pint to 1 quart	Ethyl alcohol, Lysol, soaps
Essentially nontoxic	15,000 or greater	More than 1 quart	Water, glycerin, table sugar

*Dosage that kills 50% of individuals exposed.

†Amounts of substances that are liquids at room temperature when given to a 70.4-kilogram (155-pound) human.

age, health status, and genetic makeup and that they are exposed to the same environmental conditions.

Fairly high dose levels are used to reduce the number of test animals needed, obtain results quickly, and lower costs. Otherwise, tests would have to be run on millions of laboratory animals for many years, and manufacturers couldn't afford to test most chemicals. For the same reasons, the results of high-dose exposures are usually extrapolated to low-dose levels using mathematical models. Then the extrapolated low-dose results on the test organisms are extrapolated to humans to estimate LD_{50} values for acute toxicity (Table 8-1).

According to the *linear dose–response model*, any dose of a toxic chemical or ionizing radiation has a certain risk of causing harm (Figure 8-3, left). With the *threshold dose–response model* (Figure 8-3, right) there is a threshold dose below which no detectable harmful effects occur, presumably because the body can repair the damage caused by low doses of some substances. It is extremely difficult to establish which of these models applies at low doses. To be on the safe side, the linear or nonthreshold dose–response model often is assumed.

Some scientists challenge the validity of extrapolating data from test animals to humans because human physiology and metabolism are often different from those of the test animals. Others counter that such tests and models work fairly well (especially for revealing cancer risks) when the correct experimental animal is chosen or when a chemical is toxic to several different species of test animals.

Animal tests take 2–5 years and cost $200,000 to $2 million per substance tested. Animal rights groups want to limit or ban all use of test animals (or to ensure that experimental animals are treated in the most humane manner possible). Scientists are looking for substitute methods, but they point out that some animal testing is needed because the known alternatives cannot adequately mimic the complex biochemical interactions of a live animal. What do you think?

Another approach to testing for toxicity and identifying the agents causing diseases is **epidemiology**: the study of the patterns of disease or toxicity to find out why some people get sick and others do not. Typically, the health of people exposed to a particular toxic agent or disease organism (the experimental group) is compared with the health of another group of statistically similar people not exposed to these conditions (the control group).

Epidemiology has limitations. For many toxic agents, too few people have been exposed to sufficiently high levels to allow detection of statistically significant differences. It is usually impossible to conclusively link an observed epidemiological effect with exposure to a particular toxic agent because people are exposed to many different toxic agents and disease-causing factors throughout their lives. All an epidemiological study can do is to establish strong, moderate, weak, or no statistical associations between a hazard and a health problem. Epidemiology is rarely useful for predicting the effects of new technologies, substances, or diseases because it

1998. "The Five Worst Environmental Threats to Children's Health." *Journal of Environmental Health*, vol. 60, no. 9, p. 46(2).

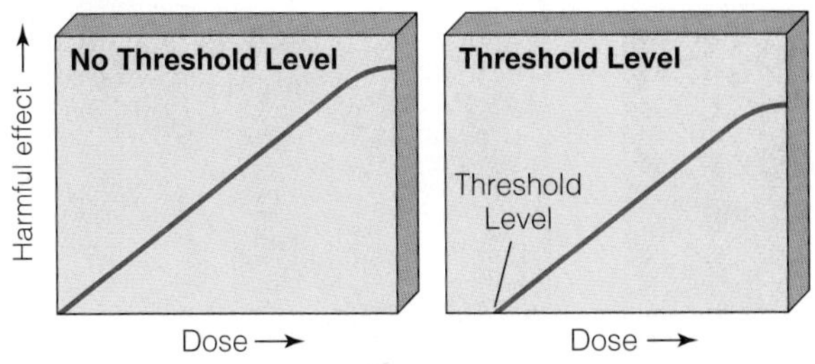

Figure 8-3 Two hypothetical dose–response curves. The curve on the left represents harmful effects that occur with increasing doses of a chemical or ionizing radiation; no dose is considered to be safe. The curve on the right shows the response of exposure to a chemical or ionizing radiation in which harmful effects appear only when the dose exceeds a certain threshold level. There is considerable uncertainty and controversy over which of these models applies to various harmful agents because of the difficulty in estimating the response to very low doses. Nonlinear threshold and nonthreshold dose–response curves, in which the graphs are curved instead of being straight lines, have also been observed.

cannot be used to evaluate hazards to which people have not been exposed.

Thus, all methods for estimating toxicity levels and risks have serious limitations, but they are all we have. To take this uncertainty into account and minimize harm, standards for allowed exposure to toxic substances and radiation typically are set at levels 1/100 or even 1/1,000 of the estimated harmful levels.

Despite their many limitations, carefully conducted and evaluated toxicity studies are important sources of information we can use to understand dose and response effects and to estimate and set exposure standards. But citizens, lawmakers, and regulatory officials must recognize the huge uncertainties and guesswork involved in all such studies.

8-3 CHEMICAL HAZARDS

What Are Toxic and Hazardous Chemicals? **Toxic chemicals** generally are defined as substances that are fatal to over 50% of test animals (LD_{50}) at given concentrations. **Hazardous chemicals** cause harm by **(1)** being flammable or explosive, **(2)** irritating or damaging the skin or lungs (strong acidic or alkaline substances such as oven cleaners), **(3)** interfering with or preventing oxygen uptake and distribution (asphyxiants such as carbon monoxide and hydrogen sulfide), or **(4)** inducing allergic reactions of the immune system (allergens).

What Are Mutagens, Teratogens, and Carcinogens? **Mutagens** are agents, such as chemicals and radiation, that cause random *mutations*, or changes in the DNA molecules found in cells. Mutations in a sperm or egg cell may be passed on to future generations and cause diseases such as manic depression, cystic fibrosis, hemophilia, sickle-cell anemia, Down's syndrome, and some types of cancer. Mutations in other cells are not inherited but may cause harmful effects. However, most mutations are harmless, probably because all organisms have biochemical repair mechanisms that can find and correct mistakes or changes in the DNA code.

Teratogens are chemicals, radiation, or viruses that cause birth defects while the human embryo is growing and developing during pregnancy, especially during the first 3 months. Chemicals known to cause birth defects in laboratory animals include PCBs, thalidomide, steroid hormones, and heavy metals such as arsenic, cadmium, lead (Section 14-7), and mercury.

Carcinogens are chemicals, radiation, or viruses that cause or promote the growth of a malignant (cancerous) tumor, in which certain cells multiply uncontrollably. Many cancerous tumors spread by **metastasis** when malignant cells break off from tumors and travel in body fluids to other parts of the body. There, they start new tumors, making treatment much more difficult.

Because there are more than 100 types of cancer (depending on the types of cells involved), there are many different causes. These include genetic predisposition, viral infections, and exposure to various mutagens and carcinogens.

According to WHO, environmental and lifestyle factors play a key role in causing or promoting up to 80% of all cancers. Major sources of carcinogens are cigarette smoke (30–40% of cancers), diet (20–30%), occupational exposure (5–15%), and environmental pollutants (1–10%). Inherited genetic factors and certain viruses cause about 10–20% of all cancers. However, there is no agreement on ways to test substances for genetic damage in humans.

Typically, 10–40 years may elapse between the initial exposure to a carcinogen and the appearance of detectable symptoms. Partly because of this time lag, many healthy teenagers and young adults have trouble believing that their smoking (p. 179), drinking, eating, and other lifestyle habits today could lead to some form of cancer before they reach age 50.

According to most scientists, there are no threshold levels for damage from mutagens, teratogens, and carcinogens. The only safe exposure is zero.

How Can Chemicals Harm the Immune, Nervous, and Endocrine Systems? Since the 1970s there has been a growing body of research on wildlife and laboratory animals and epidemiological studies of

Hint: Enter *pollution* or *health* using Key Words.

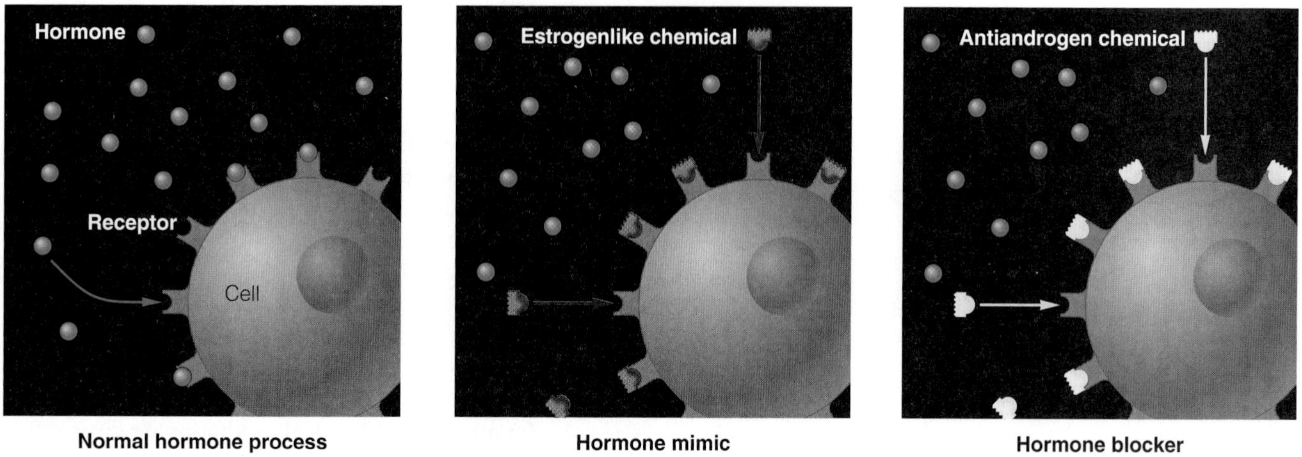

Figure 8-4 Hormones are molecules that act as messengers in the endocrine system to regulate various bodily processes, including reproduction, growth, and development. Each type of hormone has a unique molecular shape that allows it to attach to specially shaped receptors on the surface of or inside cells and transmit its chemical message (left). Molecules of certain pesticides and other molecules have shapes similar to those of natural hormones. Some of these hormone impostors, called *hormone mimics*, disrupt the endocrine system by attaching to estrogen receptor molecules (center). Others, called *hormone blockers*, prevent natural hormones such as androgens (male sex hormones) from attaching to their receptors (right). Some pollutants called *thyroid disrupters* may disrupt hormones released by thyroid glands and cause growth and weight disorders and brain and behavioral disorders.

humans. These studies indicate that long-term (often low-level) exposure to various toxic chemicals in the environment can disrupt the body's immune, nervous, and endocrine systems.

The *immune system* consists of numerous specialized cells and tissues that protect the body against disease and harmful substances by forming antibodies to invading agents and rendering them harmless. However, some synthetic chemicals, viruses such as the human immunodeficiency virus (HIV), ionizing radiation, and malnutrition can weaken the human immune system and leave the body wide open to attacks by allergens, infectious bacteria, viruses, and protozoans. Recent studies of laboratory animals and wildlife as well as epidemiological studies of humans (especially in developing countries) have linked suppression of the immune system to several widely used pesticides.

Synthetic chemicals in the environment threaten the human *nervous system* (brain, spinal cord, and peripheral nerves). Many poisons are *neurotoxins*, which attack nerve cells (neurons). Examples are **(1)** chlorinated hydrocarbons (DDT, PCBs, dioxins), **(2)** organophosphate pesticides (Section 16-3), **(3)** formaldehyde, **(4)** various compounds of arsenic, mercury, lead, and cadmium, and **(5)** widely used industrial solvents such as trichloroethylene (TCE), toluene, and xylene.

The *endocrine system* is a complex network of glands and hormones that regulates many of the body's functions. Each type of hormone has a specific molecular shape that allows it to attach only to certain cell receptors; they fit together somewhat like a molecular key fitting into a specially shaped keyhole on a receptor molecule (Figure 8-4, left). Once bonded together, the hormone and its receptor molecule move to the cell's nucleus to execute the chemical message carried by the hormone.

The endocrine glands release extremely small amounts of *hormones* into the bloodstream that act as natural chemical messengers to control body functions such as sexual reproduction, growth, development, and behavior in humans and other animals (Connections, right).

Why Do We Know So Little About the Harmful Effects of Chemicals? According to risk assessment expert Joseph V. Rodricks, "Toxicologists know a great deal about a few chemicals, a little about many, and next to nothing about most." The U.S. National Academy of Sciences estimates that only about 10% of the 72,000 chemicals in commercial use have been thoroughly screened for toxicity, and only 2% have been adequately tested to determine whether they are carcinogens, teratogens, or mutagens. Hardly any of the chemicals in commercial use have been screened for damage to the nervous, endocrine, and immune systems.

Each year we introduce into the marketplace about 1,000 new synthetic chemicals, with little knowledge about their potentially harmful effects. Currently, federal and state governments do not regulate about 99.5% of the commercially used chemicals in the United

Are Hormone Disrupters and Mimics a Health Threat?

CONNECTIONS

Over the last decade experts from a number of disciplines have been piecing together field studies on wildlife, studies on laboratory animals, and epidemiological studies of human populations. This analysis suggests that a variety of human-made chemicals can act as *hormone* or *endocrine disrupters.*

Some, called *hormone mimics,* are estrogenlike chemicals that disrupt the endocrine system by being able to attach to estrogen receptor molecules (Figure 8-4, center). Others, called *hormone blockers,* disrupt the endocrine system by preventing natural hormones such as androgens (male sex hormones) from attaching to their receptors (Figure 8-4, right). There is also growing concern about pollutants that can act as *thyroid disrupters* and cause growth, weight, brain, and behavioral disorders.

Most natural hormones are broken down or excreted. However, many synthetic hormone impostors are stable, fat-soluble compounds whose concentrations can be biomagnified as they move through food chains and webs (Figure 8-2). Thus, they can pose a special threat to humans and other carnivores dining at the top of food webs.

Numerous wildlife and laboratory studies reveal various possible effects of estrogen mimics and hormone blockers. Here are a few of many examples:

- Ranch minks fed Lake Michigan fish contaminated with endocrine disrupters such as DDT and PCBs failed to reproduce.
- Exposure to PCBs has reduced penis size in some test animals and in 118 boys born to women who were exposed to a PCB spill in Taiwan in 1979.
- In 1973, estrogen mimics called PBBs accidentally got into cattle feed in Michigan, and from there into beef. Pregnant women who ate the beef (and whose breast milk had high levels of PBBs) had sons with undersized penises and malformed testicles.
- During the past 50 years there have been dramatic increases in testicular and prostate cancer in humans almost everywhere.
- Average sperm counts among men in the United States and Europe have declined by 50% during the past 60 years.

Much more research is needed to verify such frontier science findings and to determine whether low levels of most hormone-disrupting chemicals in the environment pose a threat to the human population.

If such research shows that exposure to small amounts of hormone disrupters is harmful to humans and some forms of wildlife, the only reasonable choice may be to prevent such chemicals from reaching the environment. This will be a difficult and controversial economic and political decision.

Critical Thinking

Do you consider the possible threat from hormone disrupters a problem that could affect you? Any child you might have? List three important ways for dealing with this problem.

States. There are three major reasons for this lack of information and regulation.

- Under existing laws most chemicals are considered innocent until proven guilty. No one is required to investigate whether they are harmful.
- There are not enough funds, personnel, facilities, and test animals to provide such information for more than a small fraction of the many chemicals we encounter in our daily lives.
- It is too difficult and expensive to analyze the combined effects of multiple exposures to various chemicals and the possible interactions among such chemicals. For example, just to study the possible different three-chemical interactions among the top 500 most widely used industrial chemicals would require 20.7 million experiments—a physical and financial impossibility.

The difficulty and expense of getting information about the harmful effects of chemicals is one reason an increasing number of scientists and health officials are pushing for much greater emphasis on *pollution prevention.* This strategy greatly reduces the need for statistically uncertain and controversial toxicity studies and exposure standards. It also reduces the risk posed by potentially hazardous chemicals and products and their possible but poorly understood multiple interactions.

This approach is based on the **precautionary principle**. According to this concept, when there is considerable scientific uncertainty about potentially serious harm from chemicals or technologies, decision-makers should act to prevent harm to humans and the environment. It is based on familiar axioms: Look before you leap. Better safe than sorry. A ounce of prevention is worth a pound of cure.

Under this approach, those proposing to introduce a new chemical or technology would bear burden of establishing its safety. In other words, new chemicals and technologies would be assumed to be guilty until proven innocent. Critics contend that doing this would make it too expensive and almost impossible to introduce any new chemical or technology, Do you agree with this principle?

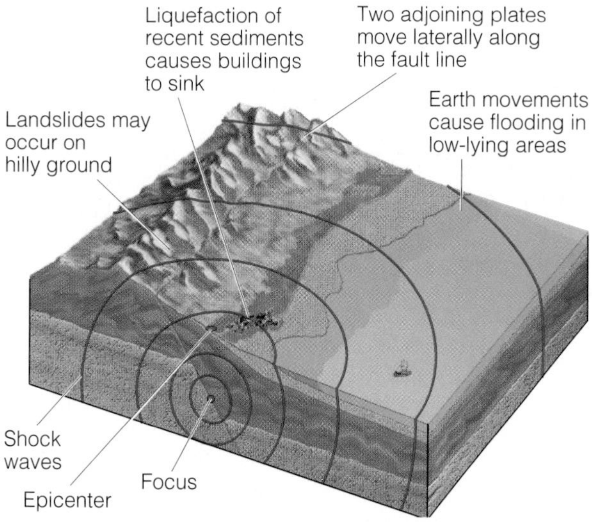

Figure 8-5 Major features and effects of an earthquake.

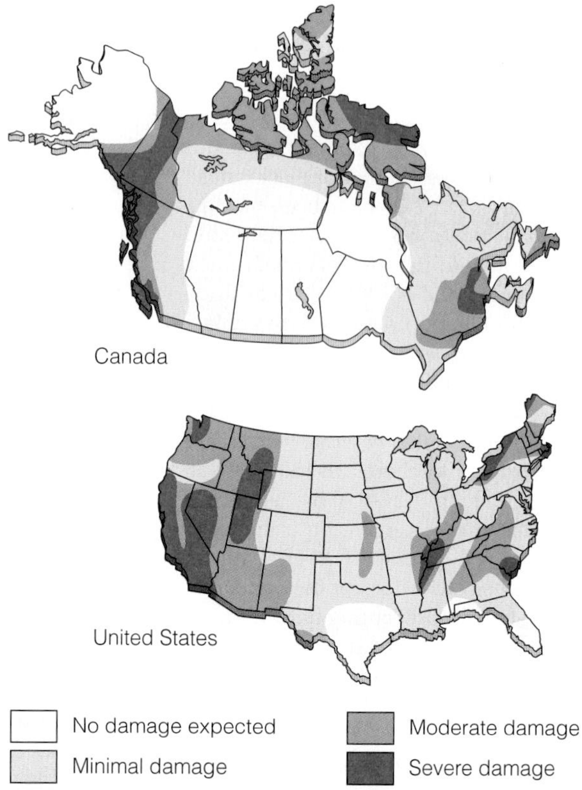

Figure 8-6 Expected damage from earthquakes in Canada and the contiguous United States. Except for a few regions along the Atlantic and the Gulf coasts, virtually every part of the continental United States is subject to some risk from earthquakes. Several areas have a risk of moderate to major damage. This map is based on earthquake records. (Data from U.S. Geological Survey)

8-4 PHYSICAL HAZARDS: EARTHQUAKES AND VOLCANIC ERUPTIONS*

What Are Earthquakes? Stress in the earth's crust can cause solid rock to deform until it suddenly fractures and shifts along the fracture, producing a *fault*. The faulting or a later abrupt movement on an existing fault causes an **earthquake**. An earthquake has certain features and effects (Figure 8-5).

The primary effects of earthquakes include shaking and sometimes a permanent vertical or horizontal displacement of the ground. These effects may have serious consequences for people and for buildings, bridges, freeway overpasses, dams, and pipelines. Secondary effects of earthquakes include rockslides, urban fires, and flooding caused by subsidence (sinking) of land. Coastal areas also can be severely damaged by large, earthquake-generated water waves, called *tsunamis* (misnamed "tidal waves," even though they have nothing to do with tides) that travel as fast as 950 kilometers (590 miles) per hour.

Solutions: How Can We Reduce Earthquake Hazards? We can reduce loss of life and property from earthquakes by **(1)** examining historical records and making geologic measurements to locate active fault zones, **(2)** making maps showing high-risk areas (Figure 8-6), **(3)** establishing building codes that regulate the placement and design of buildings in areas of high risk, and **(4)** trying to predict when and where earthquakes will occur.

***Kenneth J. Van Dellen**, professor emeritus of geology and environmental science, Macomb Community College, is the primary author of this section, with assistance from G. Tyler Miller, Jr.

What Are Volcanoes? An active **volcano** occurs where magma (molten rock) reaches the earth's surface through a central vent or a long crack (fissure). Volcanic activity can release *ejecta* (debris ranging from large chunks of lava rock to ash that may be glowing hot), liquid lava, and gases (such as water vapor, carbon dioxide, and sulfur dioxide) into the environment (Case Study, right).

We tend to think negatively of volcanic activity, but it also provides some benefits. One is outstanding scenery in the form of majestic mountains, some lakes (such as Crater Lake in Oregon; see photo in table of contents, p. xv), and other landforms. Perhaps the most important benefit of volcanism is the highly fertile soils produced by the weathering of lava.

Solutions: How Can We Reduce Volcano Hazards? We can reduce the loss of human life and sometimes property from volcanic eruptions by **(1)** land-use planning, **(2)** better prediction of volcanic eruptions, and **(3)** effective evacuation plans. The eruptive history of a

The Mount St. Helens Eruption

CASE STUDY

On May 18, 1980, Mount St. Helens, in the Cascade Range near the Washington–Oregon border, erupted in what has been described as the worst volcanic disaster in U.S. history. Fifty-seven people died in the eruption, and several hundred cabins and homes were destroyed or severely damaged.

Tens of thousand of hectares of forest were obliterated, along with campgrounds and bridges. An estimated 7,000 large animals (bear, deer, elk, and mountain lions) died, as did millions of smaller animals and birds and some 11 million fish. Salmon hatcheries were damaged and crops (including alfalfa, apples, potatoes, and wheat) were lost. Many people living in the area also lost their jobs.

On the plus side, trace elements from ash that were added to the soil may benefit agriculture in the long run. Furthermore, increased tourism to the area brought new jobs and income.

By 1990 many biologists were surprised at how fast various forms of life had begun colonizing many of the most devastated areas. This rapid recovery has taught biologists important and often surprising lessons about nature's ability to recover from what seems to be almost total devastation.

Critical Thinking

If vegetation can recover within 2 to 3 decades from the effects of a devastating volcanic eruption, why should we spend money and time on ecological restoration of highly degraded forests, grasslands, and wetlands (Section 7-5)?

volcano or volcanic center can provide some indication of where the risks are. Scientists are also studying phenomena that precede an eruption such as **(1)** tilting or swelling of the cone, **(2)** changes in magnetic and thermal properties of the volcano, **(3)** changes in gas composition, and **(4)** increased seismic activity.

8-5 BIOLOGICAL HAZARDS: DISEASE IN DEVELOPED AND DEVELOPING COUNTRIES

What Are Nontransmissible Diseases? A **nontransmissible disease** is not caused by living organisms and does not spread from one person to another. Examples are cardiovascular (heart and blood vessel) disorders, most cancers, diabetes, bronchitis, emphysema, and malnutrition. Such diseases typically have multiple (and often unknown) causes and tend to develop slowly and progressively over time.

What Are Transmissible Diseases? A **transmissible disease** is caused by a living organism (such as a bacterium, virus, protozoa, or parasite) and can be spread from one person to another. These infectious agents are called *pathogens* and are spread by air, water, food, body fluids, and some insects, and other nonhuman carriers called *vectors*.

Typically, a *bacterium* is a one-celled microorganism capable of replicating itself by simple cell division. A *virus* is a microscopic, noncellular infectious agent. Its DNA contains instructions for making more viruses, but it has no apparatus to do so. In order to replicate, a virus must invade a host cell and take over the cell's DNA to create a factory for producing more viruses.

According to the WHO and UNICEF, every year in developing countries at least 11 million children under the age of 5 die of mostly preventable infectious diseases—an average of at least 30,000 premature deaths per day (Spotlight, p. 188). About 80% of all illnesses in developing countries are caused by waterborne infectious diseases (such as diarrhea, hepatitis, typhoid fever, and cholera), mainly from unsafe drinking water and inadequate sanitation systems. Travel, migration, deforestation, loss of biodiversity, agriculture, urbanization, climate change, and natural disasters all contribute to the introduction and spread of infectious agents in new populations.

In 1998, the world's seven deadliest infectious diseases were **(1)** *acute respiratory infections*, mostly pneumonia and flu (caused by bacteria and viruses and killing about 3.7 million people per year), **(2)** *HIV/AIDS* (a viral disease, 2.5 million), **(3)** *diarrheal diseases* (caused by bacteria and viruses, 2.5 million), **(4)** *malaria* (caused by parasitic protozoa, 1.5–2.7 million), **(5)** *tuberculosis* (a bacterial disease, 2 million; Case Study, p. 189), **(6)** *measles* (a viral disease, 1 million), and **(7)** *hepatitis B* (a viral disease, 1 million)

How Rapidly Are Viral Diseases Spreading? Viral diseases include **(1)** *influenza* or *flu* (transmitted by the bodily fluids or airborne emissions of an infected person), **(2)** *Ebola* (transmitted by the blood or other body fluids of an infected person), **(3)** *rabies* (transmitted by dogs, coyotes, raccoons, skunks, and bats), and **(4)** *HIV/AIDS*. Viruses, like bacteria, can genetically adapt rapidly to different conditions.

Although health officials worry about the emergence of new viral diseases (such those caused by Ebola viruses), they recognize that the greatest virus health threat to humans is the emergence of new, very virulent

Are We Losing the War Against Infectious Bacteria?

SPOTLIGHT

There is growing and alarming evidence that we may be losing our war against infectious bacterial diseases because bacteria are among the earth's ultimate survivors. When a colony of bacteria is dosed with an antibiotic such as penicillin, most of the bacteria are killed. However, a few have mutant genes that make them immune to the drug. Through natural selection (Section 5-2), a single mutant can pass such traits on to most of its offspring, which can amount to 16,777,216 in only 24 hours!

Each time this strain of bacterium is exposed to penicillin or some other antibiotic, a larger proportion of its offspring are genetically resistant to the drug. The rapid multiplication of resistant bacteria in a victim is made easier because the antibiotics also wipe out their bacterial competitors.

Even worse, bacteria can become genetically resistant to antibiotics they have never been exposed to. When a resistant and a nonresistant bacterium touch one another (say, on a hospital bedsheet or in a human stomach), they can exchange a loop of DNA called a plasmid, thereby transferring genetic resistance from one organism to another.

The incredible genetic adaptability of bacteria is one reason the world now faces a potentially serious rise in the incidence of some infectious bacterial diseases once controlled by antibiotics. Other factors also play a key role, including **(1)** spread of bacteria (some beneficial and some harmful) around the globe by human travel and the trade of goods, **(2)** overuse of antibiotics by doctors, often at the insistence of their patients, **(3)** failure of many patients to take all of their prescribed antibiotics, which promotes bacterial resistance, **(4)** availability of antibiotics in many countries without prescriptions, **(5)** overuse of pesticides (Section 16-3), which increases populations of pesticide-resistant insects and other disease carriers, and **(6)** widespread use of antibiotics in the livestock and dairy industries.

The result of these factors acting together is that every major disease-causing bacterium now has strains that resist at least one of the roughly 160 antibiotics we use to treat bacterial infections. In 1998, health officials were alarmed to learn of the existence of a strain of bubonic plague in Madagascar that is resistant to multiple antibiotics.

In 1998, officials at the U.S. Centers for Disease Control and Prevention estimated that about 2 million patients (most with a weakened immunity system) develop a hospital-acquired infection in the United States each year, and about 90,000 of these patients die. I was a victim of a hospital-acquired infection in 1997.

There is growing controversy over the widespread use of antibiotics to increase the growth rate of livestock animals. Most of the roughly 30 million cows slaughtered each year in the United States are moved to a feedlot to be fattened up for several months before being killed. During this period, each cow gets 200–300 milligrams of antibiotics per day to fight infections and speed up growth. This use alone accounts for about half of all antibiotics used in the United States each year. Poultry and pigs being raised for slaughter also receive high doses of antibiotics for the same reasons.

In 1999, an epidemiologist with the U.S. Centers for Disease Control and Prevention urged that an antibiotic (virginiamycin) widely used since 1974 to make cattle, poultry, and pigs grow faster not be used anymore. He warned that this antibiotic could cause genetic resistance to a chemically similar new antibiotic (Synercid) developed for human use to wipe out bacteria that are resistant to all other antibiotics. In 1999 researchers found a ninefold rise in two types of drug-resistant bacteria in humans between 1992 and 1998. They attributed this to growing use of certain antibiotic drugs (quinolones) approved in 1995 and 1996 for use in chickens and cattle.

Critical Thinking

1. What role, if any, have you played in the increase in genetic resistance of bacteria to widely used antibiotics? List three ways to reduce this threat.

2. Do you believe that the use of the same antibiotics to treat human illness and to fatten livestock should be banned (as was done in Europe in 1999)? Explain. Would you favor using small amounts of such antibiotics to treat disease in livestock?

stains of influenza. Flu viruses move through the air and are highly contagious. In 1918–19 a flu epidemic infected more than half the world's population and killed 20–30 million people.

There is a growing threat from the spread of *acquired immune deficiency syndrome* or *AIDS*, which is caused by HIV. The virus itself isn't deadly, but it kills immune cells and leaves the body defenseless against infectious bacteria and other viruses. HIV can be transmitted **(1)** during unprotected sexual activity, **(2)** from one intravenous (IV) drug user to another through shared needles, **(3)** from an infected mother to an infant during birth, and **(4)** by exposure to infected blood.

According to the WHO, in 1998 at least 34 million people (22 million of them in sub-Saharan Africa alone) were infected with HIV. An average of 15,300 people are newly infected with the virus each day, 80% of them in Asia and Africa.

CASE STUDY

The Global Tuberculosis Epidemic

One of the world's most underreported stories during the last decade has been the rapid spread of tuberculosis (TB). According to the WHO, this highly infectious bacterial disease kills about 2 million people and infects about 8 million people per year (Figure 8-7). In India, where nearly half the population is infected with TB, the disease kills a half a million people each year.

The bacterium causing TB infection moves from person to person, mainly in airborne droplets produced by coughing, sneezing, singing, or even talking. Whereas TB kills about 2 million people per year, highly publicized Ebola viruses, which are hard to transmit, have killed about 650 people over the past 20 years—an average of 33 people per year.

About half of the infected people who spread highly contagious TB don't know they are infected. Because many of the infected people don't appear to be sick, this global health problem has been called a *silent global epidemic*.

Until this resurgence, the incidence of TB had fallen sharply (except among the poor). This occurred mostly because of prevention programs (such as X ray screening and detection of people with active TB) and treatment with antibiotics, which began in the 1940s. Before antibiotics, the spread of the disease was controlled by isolating patients with active TB in sanitariums until they either died or recovered.

Major reasons for the recent increase in TB are **(1)** poor TB screening and control programs (especially in developing countries, where about 95% of the new cases occur), **(2)** development of strains of the tuberculosis bacterium that are genetically resistant to virtually all effective antibiotics (typically leading to mortality rates of over 50%), **(3)** population growth and increased urbanization (which increase contacts among people), **(4)** poverty, and **(5)** the spread of AIDS, which greatly weakens the immune system and allows TB bacteria to multiply.

Slowing the spread of the disease involves early identification and treatment of people with active TB, usually those with a chronic cough. Treatment with a combination of four inexpensive drugs can cure 90% of those with active TB. However, to be successful the drugs must be taken *every* day for 6 to 8 months. Because the symptoms disappear after a few weeks, many patients think they are cured and stop taking the drugs. This allows the disease to recur in a hard-to-treat form. It then spreads to other people, and drug-resistant strains of TB bacteria develop.

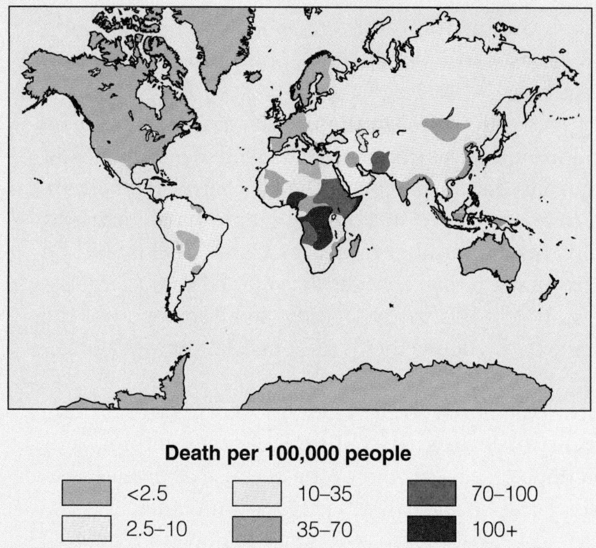

Figure 8-7 The current global tuberculosis (TB) epidemic. This easily transmitted disease is spreading rapidly and now kills about 2 million people a year. Without global effort to identify and fully treat people with active TB, the death toll is projected to rise to about 4 million a year by 2005. (Data from World Health Organization)

According to the World Health Organization, a worldwide campaign to help control TB would cost only about $360 million to help save at least 20 million lives during the next decade.

Critical Thinking

Before you read this report, were you aware of the serious global TB epidemic? Why do you think that this important story has gotten so little media attention compared to other diseases that cause many fewer deaths per year?

Within about 7–10 years, 95% of those with HIV develop AIDS. This long incubation period means that infected people often spread the virus for several years before they learn that they are infected. There is as yet no cure for AIDS, although drugs may help some infected people live longer (if they can afford the treatment).

By the end of 1998, about 16 million people had died of AIDS-related diseases. About 2.5 million of these deaths occurred in 1998 (1.8 million of them in Africa). In the 29 African countries hit hardest by AIDS, the average life expectancy at birth is 7 years less than it would have been without the presence of AIDS.

Once a viral infection starts, it is much harder to fight than infections by bacteria and protozoans. Only a few antiviral drugs exist because most drugs that will kill a virus also harm the cells of its host. Treating viral infections (such as colds, flu, and most mild coughs and sore throats) with antibiotics is useless and increases genetic resistance in disease-causing bacteria. Medicine's only effective weapon against viruses are preventive vaccines that stimulate the body's immune system to produce antibodies to ward off viral infections. Immunization with vaccines has helped tame viral diseases such as smallpox, polio, rabies, influenza, and measles.

Timing is crucial in the fight against viruses. The smallpox virus was declared eradicated the year medical authorities became aware of the HIV virus. If smallpox had not been wiped out then, it would have increased again and the number of deaths from AIDS would be much greater today. The reason is that HIV-infected people cannot be vaccinated against smallpox because the vaccine overwhelms their weakened immune systems.

Connections: What Factors Can Affect the Spread of Transmissible Diseases? Outbreaks of infectious diseases often occur because of a change in the physical, social, or biological environment of disease reservoirs, carrier vectors, or hosts to new populations.

For instance, the Asian tiger mosquito that is a vector for dengue fever (called "breakbone fever" by those who experience the excruciating pain it causes in joints), yellow fever, and other viruses. In 1985 it was brought accidentally to the United States inside used tires shipped from Asia. Today, this species has become established in Hawaii and the southeastern United States, and has begun extending its range northwards toward Chicago and Washington, D.C.

Migration to urban areas increases the probability of infection from such diseases as TB (Spotlight, p. 189), cholera, and sexually transmitted diseases. Conversely, migration to uninhabited rural areas and deforestation can expose people to new diseases and disease vectors.

In 1998, scientists warned that increased outbreaks of many tropical infectious diseases in developing countries (such as malaria, sleeping sickness, and river blindness) are related to reducing biodiversity by destroying forests and wiping out other species that help control disease vectors. For example, as more people clear and move into tropical forests, they come into more contact with monkeys that harbor mostly unknown viruses that humans are susceptible to.

Another example is increases in tick-transmitted Lyme disease in the United States. Deer mice carry the spirochete bacterium that causes Lyme disease which causes fever, lethargy, and (sometimes) long-lasting arthritis in humans. The mice are bitten by certain species of ticks that can then transfer the infection to people. As more people have moved into wooded suburbs (especially in the eastern United States), they have come into greater contact with ticks infected with Lyme disease bacteria.

The increased cultivation of rice in many parts of the world also increases populations of mosquitoes and other insects that transmit diseases to humans. When they plant rice, farmers often flood their fields or paddies, creating an ideal breeding ground for such insects.

Climate change can also affect the spread of infectious diseases. For example, scientists have established a link between the *El Niño–Southern Oscillation* (ENSO) (Figure 6-6, right) that occurs every few years and the spread of the bacterium that causes cholera. These bacteria can live inside tiny crustaceans called copepods that in turn live in microscopic aquatic animals (zooplankton) found in the coastal waters and the open sea. The warming of surface waters during an ENSO can produce plankton blooms and increase the populations of copepods containing cholera-causing bacteria. The bacteria can then infect populations of humans living along the Pacific coast of South America, as occurred during the 1991 ENSO. In 1991, a cholera epidemic struck a 2,000-kilometer (1,200-mile) stretch of Peruvian coast, making 500,000 people ill and killing 5,000 people. Scientists also warn that projected global warming can lead to the spread of tropical infectious diseases such as malaria, dengue fever, and yellow fever to temperate areas.

Natural disasters such as floods, earthquakes, landslides, and hurricanes can also spread disease-causing organisms. Flooding is an especially dangerous threat because it often contaminates water supplies with raw sewage and creates areas of standing water and moist soil that are ideal breeding grounds for mosquitoes and other insects that spread infectious diseases.

Earthquakes and landslides can release fungi and other soil-residing infectious organisms into the air, where they can be spread by winds. The high winds of hurricanes can transfer infectious organisms and carriers of disease (such as insects) from tropical to temperate areas.

Case Study: Malaria, a Protozoal Disease About 45% of the world's population live in tropical and subtropical regions in which malaria is present (Figure 8-8). Currently, an estimated 300–500 million people are infected with malaria parasites worldwide, and there are 270–480 million new cases each year.

Malaria's symptoms come and go and include fever and chills, anemia, an enlarged spleen, severe abdominal pain and headaches, extreme weakness, and greater susceptibility to other diseases. The disease kills 1.5–2.7 million people each year, more than half of them children under age 5.

Malaria is caused by four species of protozoa of the genus *Plasmodium*. Most cases of the disease are transmitted when an uninfected female of any one of 60

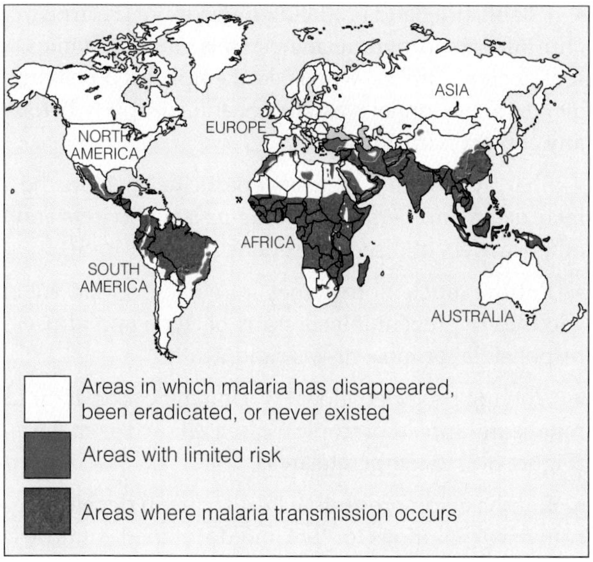

Figure 8-8 Worldwide distribution of malaria today (red) and its projected distribution in 2046 (orange). About 45% of the world's current population lives in areas in which malaria is present, with the disease killing 1.5–2.7 million people a year. If the world becomes warmer, as projected by current climate models, by 2046 malaria could affect 60% of the world's population. (Data from the World Health Organization)

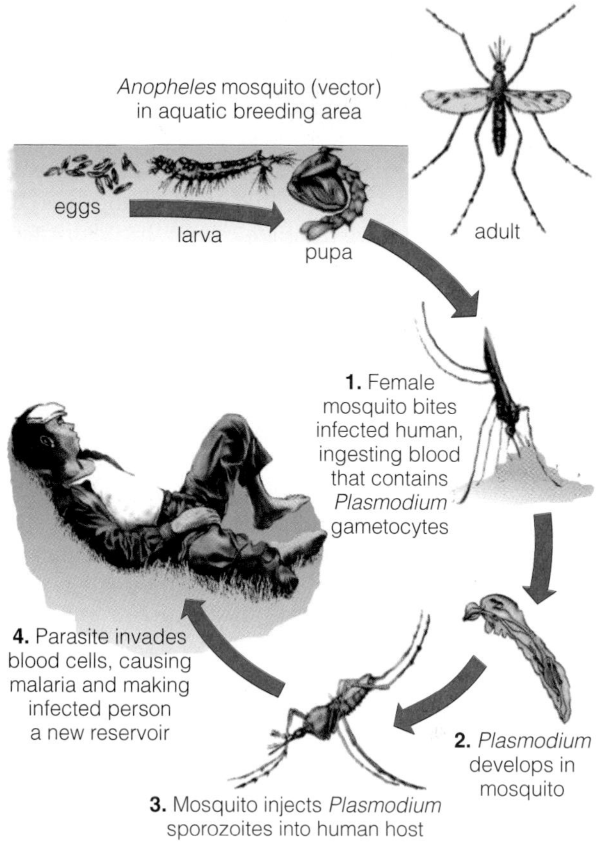

Figure 8-9 The life cycle of malaria. This life cycle of *Plasmodium* circles from mosquito to human and back to mosquito.

species of *Anopheles* mosquito bites an infected person, ingests blood that contains the parasite, and later bites an uninfected person (Figure 8-9). When this happens, *Plasmodium* parasites move out of the mosquito and into the human's bloodstream, multiply in the liver, and then enter blood cells to continue multiplying. Malaria can also be transmitted by blood transfusions or by sharing needles.

The malaria cycle repeats itself until immunity develops, treatment is given, or the victim dies. Over the course of human history, rapidly reproducing and highly adaptable malarial protozoa probably have killed more people than all the wars ever fought.

During the 1950s and 1960s, the spread of malaria was sharply curtailed by draining swamplands and marshes, spraying breeding areas with insecticides, and using drugs to kill the parasites in the bloodstream. Since 1970, however, malaria has come roaring back. Most species of the malaria-carrying *Anopheles* mosquitoes have become genetically resistant to most of the insecticides used. Worse, the *Plasmodium* parasites have become genetically resistant to the common antimalarial drugs.

Researchers are working to develop new antimalarial drugs, vaccines, and biological controls for *Anopheles* mosquitoes. Such approaches are underfunded and have proved more difficult than originally thought. Researchers also are studying the feasibility of altering the mosquitoes' genetic makeup so that they cannot carry and transmit the parasite to humans.

According to health experts, prevention is the best approach to slowing the spread of malaria. Methods include **(1)** increasing water flow in irrigation systems to prevent mosquito larvae from developing (an expensive and wasteful use of water), **(2)** using mosquito nets dipped in a nontoxic insecticide (permethrin) in windows and doors of homes, **(3)** cultivating fish that feed on mosquito larvae (biological control), **(4)** clearing vegetation around houses, and **(5)** planting trees that soak up water in low-lying marsh areas where mosquitoes thrive (a method that can degrade or destroy ecologically important wetlands).

What Are the Major Diseases in Developed Countries? As a country industrializes, it usually makes an *epidemiological transition*. The infectious diseases of childhood become less important, and the chronic diseases of adulthood (heart disease and stroke, cancer, and respiratory conditions) become more important in causing mortality.

In 1996, 39% of the deaths in the United States were from heart attacks and strokes, 24% from cancer, 5% from infectious diseases (mostly pneumonia, influenza, and AIDS), and 4% from accidents (half from automobile accidents). Almost two-thirds of these deaths result from chronic diseases (such as heart attack, stroke, and cancer) that take a long time to develop and that have multiple causes. In 1980, infectious diseases ranked as America's fifth leading killer. By 1996 they were the third leading cause of death, representing a partial reversal of the epidemiologic transition in the United States.

According to the Department of Health and Human Services, about 95% of the money spent on health care in the United States each year is used to *treat* rather than to *prevent* disease—one reason that health-care costs are so high and continue to climb. Health experts estimate that changing harmful lifestyle factors could prevent 40–70% of all premature deaths, one-third of all cases of acute disability, and two-thirds of all cases of chronic disability in the United States.

Solutions: How Can We Reduce Infectious and Other Diseases? According to health scientists and public health officials, the majority of infectious diseases that affect humanity could be prevented or their incidence greatly reduced by

- Greatly increasing research on tropical diseases. The WHO estimates that only 3% of the money spent worldwide each year for such research is devoted to malaria and other tropical diseases, even though more people suffer and die worldwide from these diseases than from all others combined.
- Mounting a global campaign to reduce overcrowding, unsafe drinking water, poor sanitation, inadequate health-care systems, malnutrition, and poverty.
- Increasing funding for monitoring, diagnosing, and responding to disease outbreaks to reduce the chances that a small outbreak can turn into an epidemic.
- Not using powerful antibiotics to treat minor infections or undiagnosed symptoms.
- Educating the public to understand the dangers of overuse of antibiotics and the need to take all of the antibiotics in any prescription.
- Not using antibiotics that have caused widespread genetic resistance and rotating from one antibiotic to another.
- Not selling antibiotics without a prescription (allowed in many countries).
- Sharply reducing the use of antibiotics to promote growth in livestock.
- Establishing and enforcing much more rigorous anti-infection programs in hospitals, nursing homes, and doctor's offices, with special emphasis on having all health personnel wash their hands properly before any contact with a patient.
- Sharply reducing the use of pesticides to slow the increase in numbers of pesticide-resistant insects and other carriers of disease (Sections 16-3 and 16-5).
- Putting much more money into the development of vaccines to prevent infections by bacteria and viruses responsible for most disease and death.
- Slowing projected global warming (Section 11-5) to reduce the spread of tropical diseases such as malaria (Figure 8-8) to temperate areas.
- Emphasizing preventive health care in developing countries (Solutions, p. 195) and developed countries.

8-6 RISK ANALYSIS

How Can We Estimate Risks? Risk analysis involves **(1)** identifying hazards and evaluating their associated risks (*risk assessment*), **(2)** ranking risks (*comparative risk analysis*), **(3)** determining options and making decisions about reducing or eliminating risks (*risk management*), and **(4)** informing decision makers and the public about risks (*risk communication*).

Risk assessment involves **(1)** determining the types of hazards involved, **(2)** estimating the probability that each hazard will occur, and **(3)** estimating how many people are likely to be exposed to it and how many may suffer serious harm. Statistical probabilities based on past experience, animal testing and other tests, and epidemiological studies are used to estimate risks from older technologies and products (Section 8-1). To evaluate new technologies and products, risk evaluators use more uncertain statistical probabilities, based on models rather than actual experience.

The left side of Figure 8-10 is an example of *comparative risk analysis*, summarizing the greatest ecological and health risks identified by a panel of scientists acting as advisers to the U.S. Environmental Protection Agency. Note the difference between the comparison of relative risk by scientists (Figure 8-10, left) and the general public (Figure 8-10, right). These differences result largely from failure of professional risk evaluators to communicate the nature of risks and their relative importance to the public, teachers, and members of the media. Some risk experts contend that much of our risk education is based on often misleading media reports on the latest risk scare of the week (based mainly on frontier science) that do not put such risks in perspective.

Scientists (Not in rank order in each category)	Citizens (In rank order)
High-Risk Health Problems • Indoor air pollution • Outdoor air pollution • Worker exposure to industrial or farm chemicals • Pollutants in drinking water • Pesticide residues on food • Toxic chemicals in consumer products **High-Risk Ecological Problems** • Global climate change • Stratospheric ozone depletion • Wildlife habitat alteration and destruction • Species extinction and loss of biodiversity	**High-Risk Problems** • Hazardous waste sites • Industrial water pollution • Occupational exposure to chemicals • Oil spills • Stratospheric ozone depletion • Nuclear power-plant accidents • Industrial accidents releasing pollutants • Radioactive wastes • Air pollution from factories • Leaking underground tanks
Medium-Risk Ecological Problems • Acid deposition • Pesticides • Airborne toxic chemicals • Toxic chemicals, nutrients, and sediment in surface waters	**Medium-Risk Problems** • Coastal water contamination • Solid waste and litter • Pesticide risks to farm workers • Water pollution from sewage plants
Low-Risk Ecological Problems • Oil spills • Groundwater pollution • Radioactive isotopes • Acid runoff to surface waters • Thermal pollution	**Low-Risk Problems** • Air pollution from vehicles • Pesticide residues in foods • Global climate change • Drinking water contamination

Figure 8-10 Comparative risk analysis of the most serious ecological and health problems according to scientists acting as advisers to the U.S. Environmental Protection Agency (left column). Risks in each of these categories are not listed in rank order. The right side of this figure represents polls showing how U.S. citizens rank the ecological and health risks they perceive as being the most serious. Why do you think there is such a great difference between the ranking by risk experts and by the general public? (Data from Science Advisory Board, *Reducing Risks*, Washington, D.C.: Environmental Protection Agency, 1990)

The key question is whether the estimated short- and long-term risks of using a particular technology or product outweigh the estimated short- and long-term benefits of other alternatives. One method for making such evaluations is **risk–benefit analysis**, which involves estimating such benefits and risks involved.

What Are the Greatest Risks People Face? The greatest risks most people face today are rarely dramatic enough to make the daily news. In terms of reduced life span from malnutrition, exposure to disease-causing organisms and dangerous chemicals, and lack of basic health care, *the greatest risk by far is poverty* (Figure 8-11).

After the health risks associated with poverty, the greatest risks of premature death are mostly the result of voluntary—and thus correctable—choices people make about their lifestyles (Figures 8-1 and 8-11). By far the best ways to reduce one's risk of premature death and serious health risks are to **(1)** not smoke, **(2)** avoid excess sunlight (which ages skin and causes skin cancer), **(3)** not drink alcohol or drink only in moderation (no more than two drinks in a single day), **(4)** reduce consumption of foods containing cholesterol and saturated fats, **(5)** eat a variety of fruits and vegetables, **(6)** exercise regularly, **(7)** lose excess weight, and **(8)** for those who can afford a car, drive as safely as possible in a vehicle with the best available safety equipment.

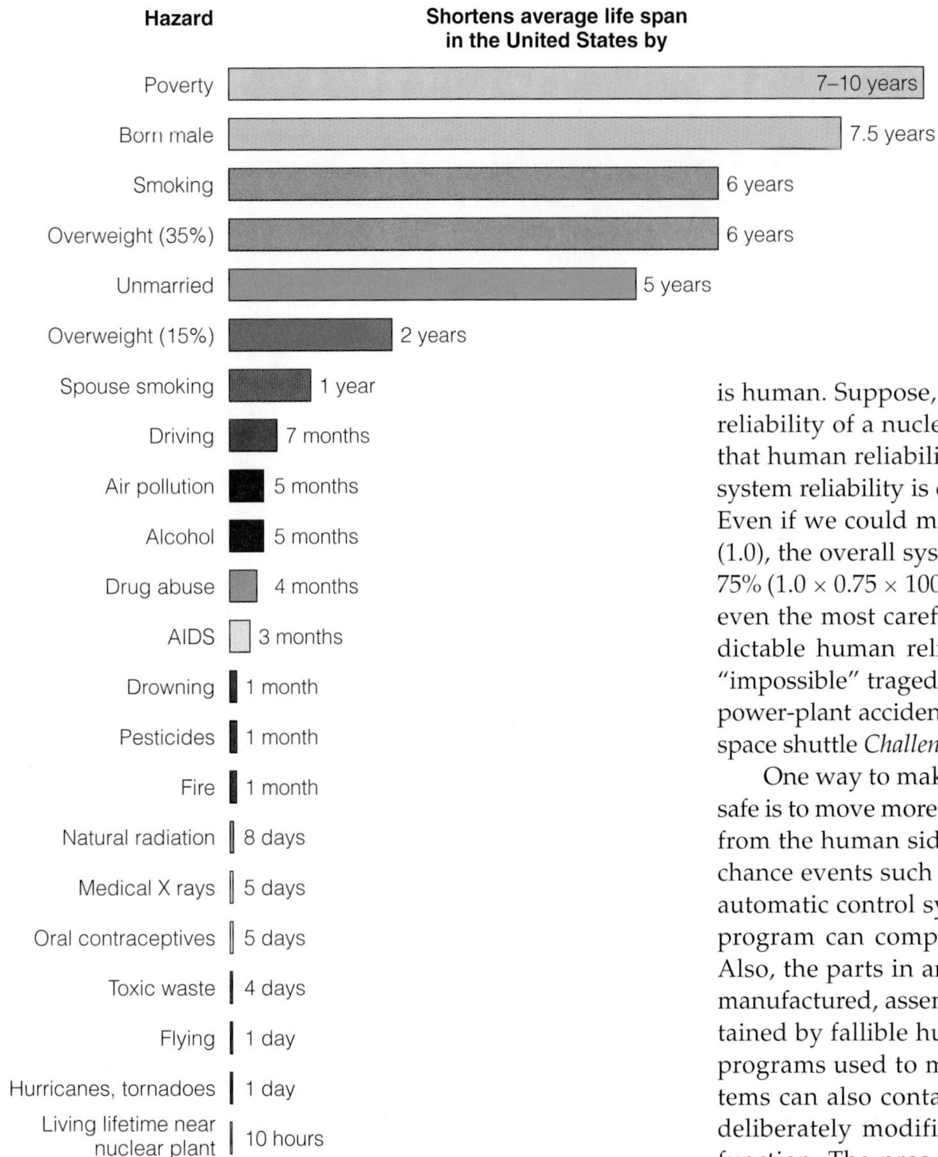

Figure 8-11 Comparison of risks that people face, expressed in terms of shorter average life span. After poverty, the greatest risks people face are mostly from voluntary choices they make about their lifestyles. (Data from Bernard L. Cohen)

How Can We Estimate Risks for Technological Systems? The more complex a technological system and the more people needed to design and run it, the more difficult it is to estimate the risks. The overall reliability of any technological system (expressed as a percentage) is the product of two factors (multiplied by 100):

$$\text{System reliability (\%)} = \text{Technology reliability} \times \text{Human reliability} \times 100$$

With careful design, quality control, maintenance, and monitoring, a highly complex system such as a nuclear power plant or a space shuttle can achieve a high degree of technology reliability. However, human reliability is usually much lower than technology reliability and is virtually impossible to predict: To err is human. Suppose, for example, that the technology reliability of a nuclear power plant is 95% (0.95) and that human reliability is 75% (0.75). Then the overall system reliability is only 71% ($0.95 \times 0.75 \times 100 = 71\%$). Even if we could make the technology 100% reliable (1.0), the overall system reliability would still be only 75% ($1.0 \times 0.75 \times 100 = 75\%$). The crucial dependence of even the most carefully designed systems on unpredictable human reliability helps explain essentially "impossible" tragedies such as the Chernobyl nuclear power-plant accident (p. 496) and the explosion of the space shuttle *Challenger*.

One way to make a system more foolproof or fail-safe is to move more of the potentially fallible elements from the human side to the technical side. However, chance events such as a lightning bolt can knock out automatic control systems. No machine or computer program can completely replace human judgment. Also, the parts in any automated control system are manufactured, assembled, tested, certified, and maintained by fallible human beings. Computer software programs used to monitor and control complex systems can also contain human errors, or they can be deliberately modified by computer viruses to malfunction. The pros and cons of genetic engineering reveal the difficulty in evaluating a new technology (Pro/Con, p. 196).

What Are the Limitations of Risk Assessment and Risk–Benefit Analysis? Risk assessment has many built-in uncertainties and limitations. It depends on toxicology assessments that have scientific and economic limitations (Section 8-2). Each additional step in risk assessment (and related risk–benefit analysis) also has uncertainties and economic limitations.

Here are some of the key questions involved in risk assessment:

- How reliable are risk assessment data and models?
- Who profits from allowing certain levels of harmful chemicals into the environment, and who suffers? Who decides this?

Improving Health Care in Developing Countries

SOLUTIONS

With adequate funding, the health of people in developing countries (and the poor in developed countries) can be improved dramatically, quickly, and cheaply by providing the following forms of mostly preventive health care:

- Better nutrition, prenatal care, and birth assistance for pregnant women. At least 600,000 women in developing countries die each year of mostly preventable pregnancy-related causes, compared with about 6,000 in developed countries.

- Better nutrition for children.

- Greatly improved postnatal care (including promotion of breast-feeding) to reduce infant mortality. Breast-fed babies get natural immunity to many diseases from antibodies in their mothers' milk.

- Immunization against the world's five largest preventable infectious diseases: tetanus, measles, diphtheria, typhoid fever, and polio. Between 1971 and 1992, the percentage of children in developing countries immunized against these diseases rose from 10% to 80%, saving about 9 million lives a year.

- Oral rehydration therapy for victims of diarrheal diseases, which cause about one-fourth of all deaths of children under the age of 5. A simple solution of boiled water, salt, and sugar or rice, at a cost of only a few cents per person, can prevent death from dehydration. According to the British medical journal *Lancet*, this simple treatment is "the most important medical advance of the 20th century."

- Careful and selective use of antibiotics for infections.

- Clean drinking water and sanitation facilities for the one-third of the world's population that lacks them.

According to the World Health Organization, extending such primary health care to all the world's people would cost an additional $10 billion per year, about 4% of what the world spends every year on cigarettes or devotes every 4 days to military spending. The cost of this program is about $1 per child.

Critical Thinking

1. Do you believe that developed countries should foot at least half of the bill implementing such proposals? What major economic and environmental advantages would this have for developed countries?

2. How many dollars per year of your taxes would you be willing to spend for such a preventive health program in developing countries?

- Should estimates emphasize short-term risks, or should more weight be put on long-term risks? Who should make this decision?

- Should the primary goal of risk analysis be to determine how much risk is acceptable (the current approach) or to figure out how to do the least damage (a prevention approach)?

- Who should do a particular risk–benefit analysis or risk assessment, and who should review the results? A government agency? Independent scientists? The public?

- Should cumulative effects of various risks be considered, or should risks be considered separately, as is usually done? Suppose a pesticide is found to have an annual risk of killing one person in a million from cancer, the current EPA limit. Cumulatively, however, effects from 40 such pesticides might kill 40, or 400, of every 1 million people. Is this acceptable?

- Should risk levels be higher for workers (as is almost always the case) than for the general public? What say should workers and their families have in this decision? According to government estimates, the exposure of workers to toxic chemicals in the United States causes 50,000–70,000 deaths (at least half from cancer) and 350,000 new cases of illness per year. The situation is much worse in developing countries with more than 1 million work-related deaths occurring worldwide each year. Is this a necessary cost of doing business?

If carried out carefully, risk analysis is a useful and much-needed tool (Spotlight, p. 198). However, according to the National Academy of Sciences, exposure to toxic chemicals is responsible for 2–4% of the 521,000 cancer deaths in the United States; this amounts to 10,400–20,800 premature cancer deaths per year. There is controversy over how much emphasis should be placed on using risk-assessment as the *primary* way to evaluate the risks from chemicals and technologies (Spotlight, p. 198).

Critics of risk-assessment understand that all dangerous chemicals can't be banned, but they argue that the emphasis should shift from determining acceptable risk levels to trying to reduce the risks as much as possible by pollution prevention, based on the precautionary principle. These critics also accuse industries of favoring risk analysis because so little is known about

Good and Bad News About Genetic Engineering

PRO/CON

Genetic engineers have learned how to splice genes and recombine sequences of existing DNA molecules in organisms to produce DNA with new genetic characteristics (recombinant DNA). These advances allow the transfer of genes with certain traits from one species to another (Figure 8-12) without waiting for new genetic combinations to evolve through the process of natural selection.

This rapidly developing technology excites many scientists and investors. They see it as a way to produce, patent, and sell high-yield crops and livestock with more protein and greater resistance to diseases, pests, frost, and drought. Genetic engineers also hope to create bacteria that can destroy oil spills, degrade toxic wastes, concentrate metals found in low-grade ores, and serve as biological factories for new vaccines, drugs, and therapeutic hormones. Gene therapy, its proponents argue, could also eliminate certain genetic diseases by removing, turning off, or blocking harmful genes.

Genetic engineering has produced a drug that reduces heart attack damage and agents that fight diabetes, hemophilia, and some forms of cancer; it has also been used to diagnose AIDS and cancer. Genetically altered bacteria have been used to manufacture more effective vaccines and hormones. In 1998, researchers bioengineered a potato to contain a protein that causes diarrhea (a major killer). When people eat the potato, they develop antibodies to the diarrhea bug.

In agriculture, gene transfer has been used to develop **(1)** strawberries and tomatoes that resist frost because they contain "antifreeze" genes from fish, **(2)** tomatoes that stay fresh and tasty longer, **(3)** potatoes that resist disease because they contain chicken genes, and **(4)** smaller cows that produce more milk.

Some people worry that biotechnology may run amok. They argue that it should be kept under strict control because we don't understand enough about how nature works to allow unregulated genetic alteration of humans and other species. They also point to serious and widespread ecological problems that have resulted from the accidental or deliberate introduction nonnative organisms into biological communities (Section 18-2).

Figure 8-12 An example of genetic engineering. The 6-month-old mouse on the left is normal; the other mouse of the same age contains a human growth hormone gene in the chromosomes of all its cells. Mice with the human growth hormone gene normally grow two to three times faster and reach a size twice that of mice without the gene. (R. L. Brinster and R. E. Hammer/School of Veterinary Medicine, University of Pennsylvania)

health risks from pollutants and because the data that do exist are controversial. The result is that risk assessment and risk–benefit analysis can be crafted to support almost any conclusion and then called scientific decision making. Critics charge that the huge uncertainties in risk assessment and risk–benefit analysis (Spotlight, p. 198) also allow industries to delay regulatory decisions for decades by challenging data in the courts. What do you think?

How Should Risks Be Managed? Once an assessment of risk is made, decisions must be made about what to do about the risk. **Risk management** includes the administrative, political, and economic actions taken to decide whether and how to reduce a particular societal risk to a certain level and at what cost.

Risk management involves deciding **(1)** which of the vast number of risks facing society should be evaluated and managed and in what order or priority with the limited funds available, **(2)** how reliable the risk–benefit analysis or risk assessment performed for each risk is, **(3)** how much risk is acceptable, **(4)** how much money it will take to reduce each risk to an acceptable level, **(5)** how much each risk will be reduced if available funds are limited (as is almost always the case), and **(6)** how the risk management plan will be communicated to the public, monitored, and enforced. Each step in this process involves making value judgments and weighing trade-offs to find some reasonable compromise among conflicting political, economic, health, and environmental interests.

How Well Do We Perceive Risks? Most of us do poorly in assessing the relative risks from the hazards that surround us (Figures 8-10 and 8-11). Many people deny or shrug off the high-risk chances of dying (or

To such critics, the resulting *genetic pollution* may eventually prove to be more harmful than chemical pollution. Unlike chemicals, genetically engineered organisms might mutate, change their form and behavior, migrate to other areas, and alter the genetic traits of existing wild species. Unlike defective cars and other products, they couldn't be recalled.

Critics are also concerned that one of the most serious effects of widespread use of biotechnology is reduction of the world's vital genetic diversity and thus biodiversity. Already the world's 20 major food crops have become 70% less genetically diverse because a wide range of wild strains have been replaced with only a few varieties developed by crossbreeding. Widespread use of bioengineered crop strains could hasten this loss of biodiversity.

Mixing certain traits found in wild varieties is used to develop all new varieties by crossbreeding or genetic engineering. This reducing of the natural genetic diversity of wild plants and animals undermines the genetic engineers' ability to produce new combinations in the laboratory and undermines the future success of the biotech industry.

Genetically engineered plants can also affect human health. Use of the genetically engineered bovine growth hormone (recently approved by the FDA to increase milk production in cows) raises the incidence of udder infections. This requires giving cows higher doses of antibiotics, which can accelerate the problem of genetic resistance in humans to widely used antibiotics.

In 1989, a committee of prominent ecologists appointed by the Ecological Society of America warned that the ecological effects of new combinations of genetic traits from different species would be difficult to predict. Their report called for a case-by-case review of any proposed environmental releases, as well as carefully regulated, small-scale field tests before any bioengineered organism is put into commercial use—practices that are not being followed adequately.

Proponents argue that the risk of biotech-caused ecological catastrophes is small. However, insurance companies have refused to insure the release of genetically engineered organisms into the environment against the possibility of catastrophic environmental damage because they have no way to evaluate the risks of any given introduction.

This controversy illustrates the difficulty of balancing the actual and potential benefits of a technology with its actual and potential risks of harm. Proponents of this new technology may be right in their belief that the benefits far exceed the potential and mostly unknown risks. But what if they are wrong?

Critical Thinking

1. What government controls, if any, do you believe should be applied to the development and use of genetic engineering? How would you enforce any restrictions?

2. Use the library or the Internet to find out what controls now exist on biotechnology in the United States (or the country where you live) and how well such controls are enforced.

injury) from voluntary activities they enjoy, such as motorcycling (1 death in 50 participants), smoking (1 in 300 participants by age 65 for a pack-a-day smoker), hang-gliding (1 in 1,250), and driving (1 in 2,500 without a seat belt and 1 death in 5,000 with a seat belt).

Yet some of these same people may be terrified about the possibility of dying from a commercial airplane crash (1 in 4.6 million), train crash (1 in 20 million), snakebite (1 in 36 million), shark attack (1 in 300 million), or exposure to trichloroethylene (TCE) in drinking water at the trace levels allowed by the EPA (1 in 2 billion).

Being bombarded with news about people killed or harmed by various hazards distorts our sense of risk. However, *the most important good news each year is that about 99.1% of the people on the earth didn't die*. But that's not what people see on TV, hear, or read about every day. Despite the greatly increased use of synthetic chemicals in food production and processing, the general health and average life expectancy of people in the United States (and most developed countries) have increased during the past 50 years.

Our perceptions of risk and our responses to perceived risks often have little to do with how risky the experts say something is (Figures 8-10 and 8-11). The public generally sees a technology or a product as being riskier than experts do when

- *It is new or complex rather than familiar*. Examples include genetic engineering or nuclear power, as opposed to large dams or coal-fired power plants.
- *It is perceived as being mostly involuntary*. Examples include nuclear power plants or food additives, as opposed to driving or smoking.
- *It is viewed as unnecessary rather than as beneficial or necessary*. Examples might include

How Useful Are Risk Assessment and Risk–Benefit Analysis?

SPOTLIGHT

When conducted properly, risk assessment and risk–benefit analysis are sincere attempts to find a rational basis for making regulatory decisions by analyzing available scientific evidence about the safety of a chemical or technology. Proponents contend that they are useful ways to **(1)** organize and analyze available scientific information, **(2)** identify significant hazards, **(3)** focus on areas that need more research, **(4)** help regulators decide how money for reducing risks should be allocated, and **(5)** stimulate people to make more informed decisions about health and environmental goals and priorities.

However, critics point out that when we add new chemicals or technologies we are mostly *flying blind* about their possible harmful effects. For example, a recent study has documented the significant uncertainties involved in even simple risk assessment. Eleven European governments established 11 different teams of their best scientists and engineers (including those from private companies) to assess the hazards and risks from a small plant storing only one hazardous chemical (ammonia). The 11 teams, consisting of world-class experts analyzing this very simple system, disagreed with one another on fundamental points and varied in their assessments of the hazards by a factor of 25,000!

The current built-in uncertainty in risk assessment and risk–benefit analysis (and cost–benefit analysis, Spotlight, p. 37) is analogous to a radar device that can detect a car speeding at 160 kilometers (100 miles) per hour, but can tell us only that the car is traveling somewhere between 16 kilometers (10 miles) per hour and 160,000 kilometers (100,000 miles) per hour. Such inherent uncertainty explains why regulators setting human exposure levels for toxic substances usually divide the best results by 100 to 1,000 to provide the public with a margin of safety.

Risk assessments and risk–benefit analyses yield only a range of probabilities and uncertainties based on different assumptions, not the precise numbers that decision makers and the public wish they had. Environmentalists and health officials believe that all such analyses should contain the wide range of assumptions, probabilities, and uncertainties involved and be open to full public review.

According to critics, risk assessment pretends to provide a rational scientific assessment of risk or safety but cannot do so because the required data are not available. Also, politics, economics, and value judgments are involved at every step of the risk analysis process. To its critics, risk assessment is mostly an art, not a science.

Critics argue that the main decision-making tools we should rely on are *looking at the available alternatives to a particular chemical or technology and having a full public discussion of the major advantages (benefits) and disadvantages (costs) of those alternatives.*

They argue that if the alternatives are examined fairly, emphasis will shift from trying to determine "acceptable" risk levels to trying to reduce the risks as much as possible by *pollution prevention* using the precautionary principle.

Critical Thinking

Do you believe that **(a)** risk assessment, risk–benefit analysis, and cost–benefit analysis (Spotlight, p. 37) should be used as the *primary* tools for establishing any federal health, safety, or environmental regulation or that **(b)** emphasis should be on fair evaluation of alternatives with the goal of finding the least harmful (and most affordable) alternative? Explain. Why is the second approach rarely used?

using chlorofluorocarbon (CFC) propellants in aerosol spray cans or using food additives that increase sales appeal, as opposed to cars or aspirin.

- *Its use involves a large, well-publicized death toll from a single catastrophic accident rather than the same or an even larger death toll spread out over a longer time.* Examples might include a severe nuclear power plant accident, an industrial explosion, or a plane crash, as opposed to coal-burning power plants, automobiles, or smoking.

- *Its use involves unfair distribution of the risks.* Citizens are outraged when government officials decide to put a hazardous-waste landfill or incinerator in or near their neighborhood, even when the decision is based on risk–benefit analysis. This is usually seen as politics, not science. Residents will not be satisfied by estimates that the lifetime risks of cancer death from the facility are no greater than, say, 1 in 100,000. Living near the facility means that they, not the 99,999 people living farther away, have a much higher risk of dying from cancer by having this risk involuntarily imposed on them.

- *The people affected are not involved in the decision-making process from start to finish.*

- *Its use does not involve a sincere search for and evaluation of alternatives.* People who believe that their lives and the lives of their families are being threatened want to know what the alternatives are and which alternative results in the least harm to them and the earth.

Better education and communication about the nature of risks will help bring the public's perceptions of various risks closer to those of professional risk evaluators. However, such education will not eliminate the emotional, cultural, and ethical factors that decision makers must take into account in determining the acceptability of a particular risk and in evaluating the possible alternatives.

Not all waste and pollution can be eliminated. . . . What is absolutely crucial, however, is to recognize that pollution prevention should be the first choice and the option against which all other options are judged. The burden of proof imposed on individuals, companies, and institutions should be to show that pollution prevention options have been thoroughly examined, evaluated, and used before lesser options are chosen.

JOEL HIRSCHORN

REVIEW QUESTIONS

1. Define the boldfaced terms found in this chapter.

2. What human activity kills the largest number of people each year?

3. What are *risk* and *risk assessment*? What is *probability*? List two **(a)** cultural hazards, **(b)** chemical hazards, **(c)** physical hazards, and **(d)** biological hazards.

4. What is *toxicity*? Distinguish between *dose* and *response* for a potentially harmful substance. List four factors that determine whether a chemical is harmful. Distinguish between *bioaccumulation* and *biomagnification.*

5. What is a *poison*? List three methods used to determine toxicity and list the limitations of each method. What is a *dose–response curve*? Distinguish between a *linear dose–response curve* and a *threshold dose–response curve.*

6. Distinguish between *toxic chemicals* and *hazardous chemicals.* Distinguish between *mutagens, teratogens,* and *carcinogens* and give one example of each. What are the two largest causes of cancers?

7. Distinguish between the *immune system, nervous system,* and *endocrine system* and give an example of something that causes harm to each system. What are *hormone disrupters* and *hormone mimics*? Give two examples of such chemicals.

8. About what percentage of the 72,000 chemicals in commercial use in the United States have been screened **(a)** to assess toxicity, **(b)** to determine whether they are carcinogens, teratogens, or mutagens, and **(c)** to determine whether they damage the nervous, endocrine, or immune systems? What percentage of the commercially used chemicals in the United States is regulated by federal and state governments? List three reasons for the lack of information about most chemicals in commercial use. Distinguish between regulation strategy and the pollution prevention strategy for protecting the public from potentially harmful chemicals. What is the *precautionary principle*? Why is it rarely used?

9. Distinguish between an earthquake and a volcanic eruption and list ways to reduce the hazards from each of these events. What are some of the benefits of volcanic eruptions?

10. Distinguish between *nontransmissible* and *transmissible diseases* and give two examples of each type. About how many children die each year in developing countries from mostly preventable infectious diseases? What are the seven deadliest infectious diseases in order of the number of deaths they cause each year?

11. What are the major types of diseases in **(a)** developing countries, **(b)** developed countries, and **(c)** the United States? What is an *epidemiologic transition*?

12. How do infectious bacteria become resistant to antibiotics? List seven factors that have led to an increase in infectious diseases that can't be controlled by most antibiotics.

13. What is the best way to treat **(a)** a bacterial disease and **(b)** a viral disease? Give two examples of each type of disease.

14. What causes tuberculosis and how is it transmitted? About how many people died from TB during the past year? Give five reasons for the increase in TB infections in recent years. How can the spread of this bacterial infectious disease be slowed?

15. Distinguish between HIV and AIDS. List four ways in which HIV can be transmitted. About how many people were infected with HIV and how many people died of AIDS during the past year? List ways to prevent the spread of this viral infectious disease.

16. What causes malaria? About how many people die from malaria each year? List five ways to prevent this protozoal infectious disease.

17. List ten ways to prevent or reduce the incidence of infectious diseases throughout the world. List seven major ways to improve health care in developing countries. About how much would implementing these measures cost per year?

18. Distinguish between *risk analysis, risk assessment,* and *risk–benefit analysis.* What are the major limitations of risk assessment and risk–benefit analysis?

19. What are the greatest risks people face in terms of reduced life span? List eight ways to reduce your risk of premature death and serious health problems.

20. What are the pros and cons of genetic engineering?

21. What is *risk management*? About what percentage of the people on the earth die each year? List seven reasons that lead people to perceive that certain risks are greater than experts say they are.

CRITICAL THINKING

1. Explain why you agree or disagree with the proposals made by health officials for reducing the death toll and other harmful effects of smoking given on p. 179.

2. What do you believe are the three major things we should do to reduce the development of genetically resistant infectious bacteria?

3. Do you think chemicals should be regulated based not only on their carcinogenic properties, but also on their effects on the nervous, immune, and endocrine systems? Do the potential costs of doing this outweigh the potential benefits? Explain.

4. Should we have zero pollution levels for all hazardous chemicals? Explain.

5. Do you believe that health and safety standards in the workplace should be strengthened and enforced more vigorously, even if this causes a loss of jobs when companies transfer operations to countries with weaker standards? Explain.

6. Evaluate the following statements:

- **a.** We shouldn't get so worked up about exposure to toxic chemicals because almost any chemical can cause some harm in a large enough dose.
- **b.** We shouldn't worry so much about exposure to toxic chemicals because through genetic adaptation we can develop immunity to such chemicals.
- **c.** We shouldn't worry so much about exposure to toxic chemicals because we can use genetic engineering to reduce or eliminate such problems.

7. Which type of natural disaster (flood, hurricane, earthquake, or volcanic eruption) are you most likely to experience? Why? What can you do to escape or reduce the harm if such a disaster strikes? What actions can you take when it occurs?

8. What are the five major risks you face from your lifestyle, where you live, and what you do for a living? Which of these risks are voluntary and which are involuntary? List the five most important things you can do to reduce these risks. Which of these things do you actually plan to do?

9. How would you answer each of the questions raised about risk assessment and risk management on pp. 194–195? Explain each of your answers.

10. Do you believe that we should shift the emphasis from risk assessment and risk–benefit analysis to pollution prevention and risk reduction? Explain. What beneficial and harmful effects might such a shift have on your life? On the life of any child you might have?

***11.** Conduct a survey to determine the major physical, chemical, and biological risks to human health at your school. Develop a plan for preventing or reducing these hazards. Present your findings to school officials.

***12.** Conduct a survey of officials, citizens, hospitals, and medical personnel to determine the effectiveness of efforts to control the transmission of infectious diseases in your community. List major improvements that could be made and submit them to the appropriate officials.

***13.** Make a concept map of this chapter's major ideas, using the section heads and subheads and the key terms (in boldface). Look at the inside back cover and in Appendix 5 for information about concept maps.

FURTHER INFORMATION

For more information see Internet Sources and Further Readings at the back of the book. Online resources for this chapter can be found at **www.brookscole.com/biology**

Part III

HUMAN POPULATION, RESOURCES, AND SUSTAINABILITY

I recognize the right and duty of this generation to develop and use our natural resources, but I do not recognize the right to waste them; or to rob by wasteful use, the generations that come after us.

THEODORE ROOSEVELT, 1900

9 THE HUMAN POPULATION: GROWTH AND DISTRIBUTION

Slowing Population Growth in Thailand

Can a country sharply reduce its population growth in only 15 years? Thailand did.

In 1971, Thailand adopted a national policy to control and reduce its population growth. When the program began the country's population was growing at a rate of 3.2% per year and the average Thai family had 6.4 children. Fifteen years later, the country's population growth rate had been cut in half to 1.6%. By 1999 the rate had fallen to 1.1%, and the average number of children per family was 2.0. Thailand's population is projected to grow from 62 million in 1999 to 73 million by 2025.

There are several reasons for this impressive feat: **(1)** the creativity of the government-supported family-planning program, **(2)** high literacy among women (90%), **(3)** an increasing economic role for women and advances in women's rights, **(4)** better health care for mothers and children, **(5)** the openness of the Thai people to new ideas, **(6)** the willingness of the government to encourage and financially support family planning and to work with the private, nonprofit Population and Community Development Association (PCDA), and **(7)** support of family planning by the country's religious leaders (95% of Thais are Buddhist).

This remarkable transition was catalyzed by the charismatic leadership of Mechai Viravidaiya, a public relations genius and former government economist who launched the PCDA in 1974 to help make family planning a national goal. PCDA workers handed out condoms anywhere there was a crowd: at festivals, movie theaters, even traffic jams. Between 1971 and 1999 the percentage of married women using modern birth control rose from 15% to 70%—higher than the 59% usage in developed countries and the 49% usage in developing countries.

Mechai helped establish a German-financed revolving loan plan to enable people participating in family-planning programs to install toilets and drinking water systems. Low-rate loans were offered to farmers practicing family planning. The government also offers loans to individuals from a fund that increases as their village's level of contraceptive use rises.

All is not completely rosy. Although Thailand has done well in slowing population growth and raising per capita income, it has been less successful in reducing pollution and improving public health, especially maternal health and control of AIDS and other sexually transmitted diseases.

Its capital, Bangkok, remains one of the world's most polluted and congested cities. It is plagued with notoriously high levels of traffic congestion and air pollution (Figure 9-1). The typical motorist in Bangkok spends 44 days per year sitting in traffic, costing $2.3 billion in lost work time.

Figure 9-1 This policeman in Bangkok, Thailand, is wearing a mask to reduce his intake of air polluted mainly by automobiles. Bangkok is one of the world's most car-clogged cities, with car commutes averaging 3 hours per day. Roughly one of every nine of its residents has a respiratory ailment. (Martin Harvey/Natural History Photographic Agency)

The problems to be faced are vast and complex, but come down to this: 6 billion people are breeding exponentially. The process of fulfilling their wants and needs is stripping earth of its biotic capacity to produce life; a climactic burst of consumption by a single species is overwhelming the skies, earth, waters, and fauna.

PAUL HAWKEN

This chapter addresses the following questions:

- How is population size affected by birth, death, fertility, and migration rates?
- How is population size affected by the percentage of males and females at each age level?
- How can population growth be slowed?
- What success have India and China had in slowing population growth?
- How is the world's population distributed between rural and urban areas?
- How do transportation systems shape urban areas and growth?
- How can cities be made more livable and sustainable?

9-1 FACTORS AFFECTING HUMAN POPULATION SIZE

How Is Population Size Affected by Birth Rates and Death Rates? Populations grow or decline through the interplay of three factors: births, deaths, and migration. **Population change** is calculated by subtracting the number of people leaving a population (through death and emigration) from the number entering it (through birth and immigration) during a specified period of time (usually a year):

$$\text{Population change} = (\text{Births} + \text{Immigration}) - (\text{Deaths} + \text{Emigration})$$

When births plus immigration exceed deaths plus emigration, population increases; when the reverse is true, population declines. When these factors balance out, population size remains stable, a condition known as **zero population growth (ZPG)**.

Instead of using the total numbers of births and deaths per year, demographers use **(1)** the **birth rate**, or **crude birth rate** (the number of live births per 1,000 people in a population in a given year), and **(2)** the **death rate**, or **crude death rate** (the number of deaths per 1,000 people in a population in a given year). Figure 9-2 shows the crude birth and death rates for various groupings of countries in 1999.

Birth rates and death rates are coming down worldwide, but death rates have fallen more sharply than birth rates. As a result, there are more births than deaths; every time your heart beats 2.5 more babies are added to the world's population. At this rate, each day we share the earth and its resources with about 219,000 more people than the day before, 95% of them in developing countries.

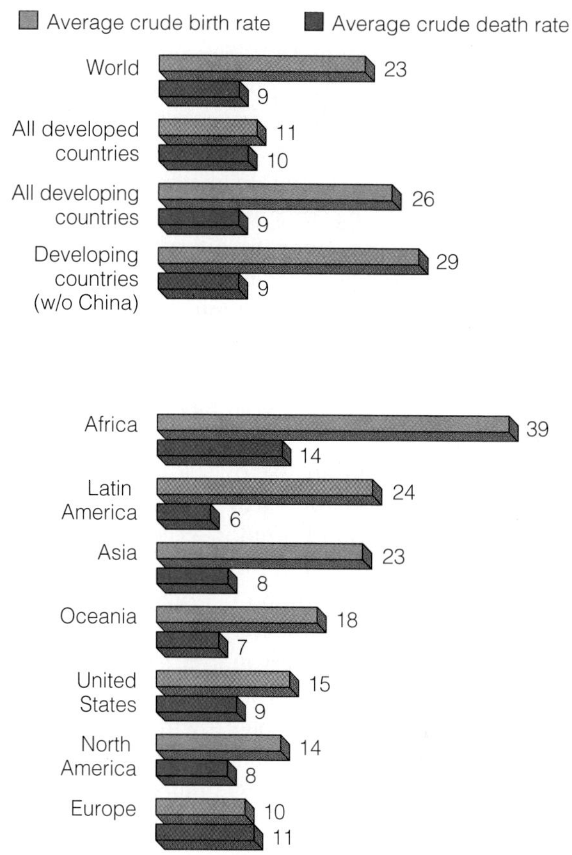

Figure 9-2 Average crude birth and death rates for various groupings of countries in 1999. (Data from Population Reference Bureau)

The rate of the world's annual population change (excluding migration) usually is expressed as a percentage:*

$$\text{Annual rate of natural population change (\%)} = \frac{\text{Birth rate} - \text{Death rate}}{1{,}000 \text{ persons}} \times 100 = \frac{\text{Birth rate} - \text{Death rate}}{10}$$

*Crude birth and death rates that have not been rounded off to the nearest whole number often are used to calculate natural change; the result is then rounded off to the nearest tenth of a percent. Consequently, use of the rounded-off crude birth and death rate figures shown in Figure 9-2 will not always produce the rounded-off percentage growth figures shown in Figure 9-3.

The rate of the world's annual population growth (natural increase) dropped 40% between 1963 and 1999, from 2.2% to 1.33%. This is good news, but during the same period the population base rose by about 88%, from 3.2 billion to 6 billion. This drop in the rate of population increase is roughly analogous to learning that the truck heading straight at you has slowed from 100 kilometers per hour to 60 while its weight increased by 88%. Exponential population growth has not disappeared; it's just occurring at a slower rate.

Figure 9-3 presents the annual rates of population change for major parts of the world in 1999. An annual natural increase rate of 1–3% may seem small, but such exponential growth rates lead to enormous increases in population size over a 100-year period. For example, a population growing by 3% a year will increase its population 19-fold in a century.

The current annual population increase rate of 1.33% adds about 80 million people per year (6 billion × 1.33%=80 million), roughly equal to adding another New York City every month, a Germany every year, and a United States every 3.5 years. Despite the drop in the rate of population growth, the larger base of population means that 80 million people were added in 1999, compared with only 69 million in 1963, when the world's population growth rate reached its peak.

In numbers of people, China (with 1.26 billion in 1999, about one of every five people in the world) and India (with 1 billion) dwarf all other countries. Together they make up 37% of the world's population. The United States, with 273 million people in 1999, has the world's third largest population but only 4.6% of the world's people. Figure 9-4 gives projected population growth in various regions between 1999 and 2025. More than 95% of this growth is projected to take place in developing countries, where acute poverty has become a way of life for about 1.5 billion people.

How Have Global Fertility Rates Changed? Two types of fertility rates affect a country's population size and growth rate. The first type, **replacement-level fertility**, is the number of children a couple must bear to replace themselves. It is slightly higher than two children per couple (2.1 in developed countries and as high as 2.5 in some developing countries in 1999), mostly because some female children die before reaching their reproductive years.

Does reaching replacement level fertility mean an immediate halt in population growth (zero population growth)? No, because there are so many future parents already alive that if each had an average of 2.1 children and their children also had 2.1 children, the population would continue to grow for 50 years or more (assuming that death rates don't rise).

The second type of fertility rate is the **total fertility rate (TFR)**: an estimate of the average number of children

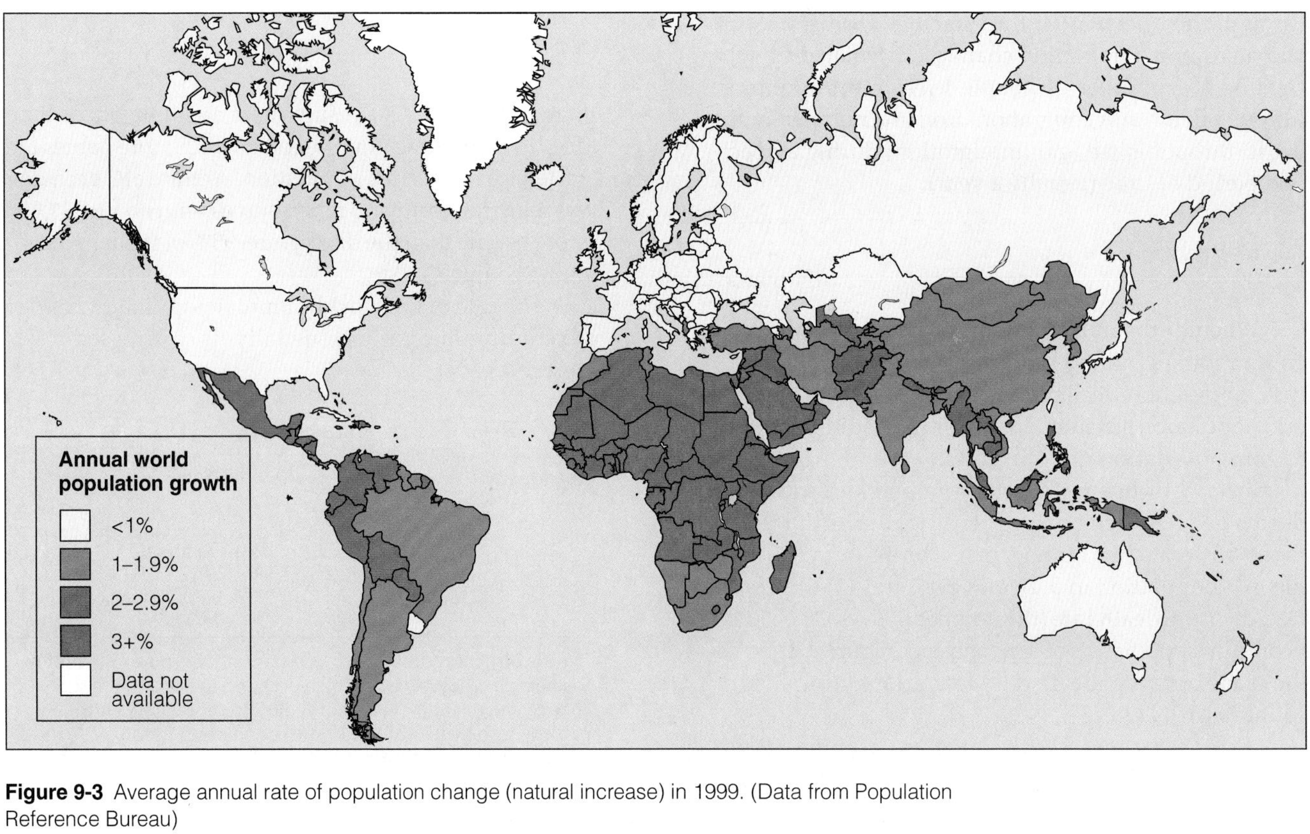

Figure 9-3 Average annual rate of population change (natural increase) in 1999. (Data from Population Reference Bureau)

Brown, L. R. et al. 1999. "Sixteen Impacts of Population Growth." *The Futurist*, vol. 33, no. 2, p. 36(6).

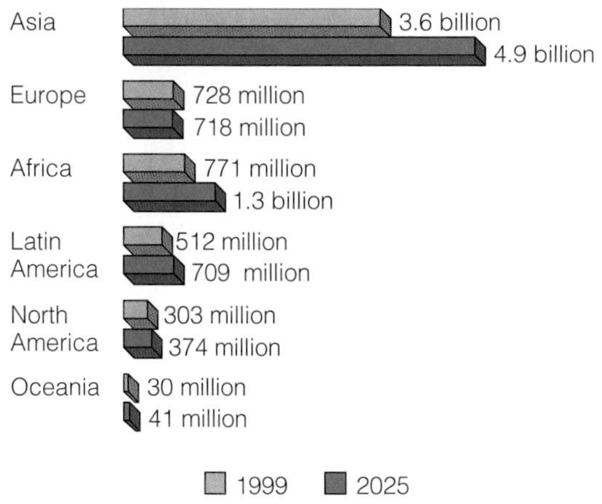

Figure 9-4 Population projections by region, 1999–2025. (Data from United Nations and Population Reference Bureau)

a woman will have during her childbearing years under current age-specific birth rates. The *good news* is that TFRs have dropped considerably since 1950. In 1999 the worldwide average TFR was 2.9 children per woman (down from 5 in 1950). It was 1.5 in developed countries (down from 2.5 in 1950) and 3.2 in developing countries (down from 6.5 in 1950). In 1999, 66 countries with 44% of the world's population had TFRs at or below the level of 2.1 children per woman.

The *bad news* is that the average global TFR of 2.9 (3.2 in developing countries) is still far above the replacement level of 2.1. TFRs vary considerably throughout the world (Figure 9-5), with the highest rate by far in Africa (5.4 children per woman). United Nations population projections to 2050 vary depending on the world's projected average TFR (Figure 9-6).

How Have Fertility Rates Changed in the United States? The population of the United States has grown from 76 million in 1900 to 273 million in 1999—a 2.6-fold increase—even though the country's TFR has oscillated wildly (Figure 9-7). In 1957, the peak of the post–World War II baby boom, the TFR reached 3.7 children per woman. Since then it has generally declined, remaining at or below replacement level since 1972.

The drop in the TFR has led to a decline in the rate of population growth in the United States. However, the country's population is still growing faster than that of most developed countries and is not even close to zero population growth. Including immigration, the U.S. population of 273 million grew by 1% in 1999—more than double the mean rate of the world's industrialized nations. This growth added about 2.74 million people: 1.64 million more births than deaths (accounting for about 60% of the growth), 800,000 legal immigrants and refugees, and an estimated 300,000 illegal immigrants.

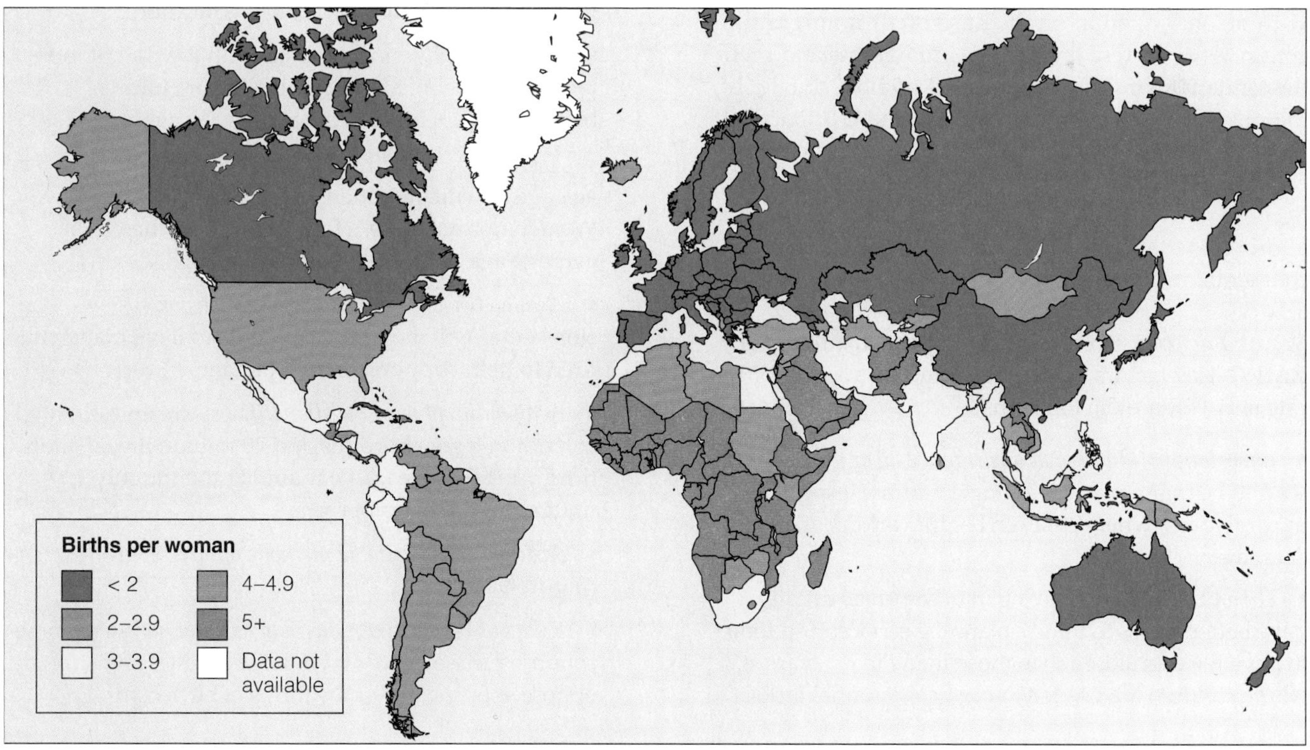

Figure 9-5 Total fertility rates in 1999. (Data from Population Reference Bureau)

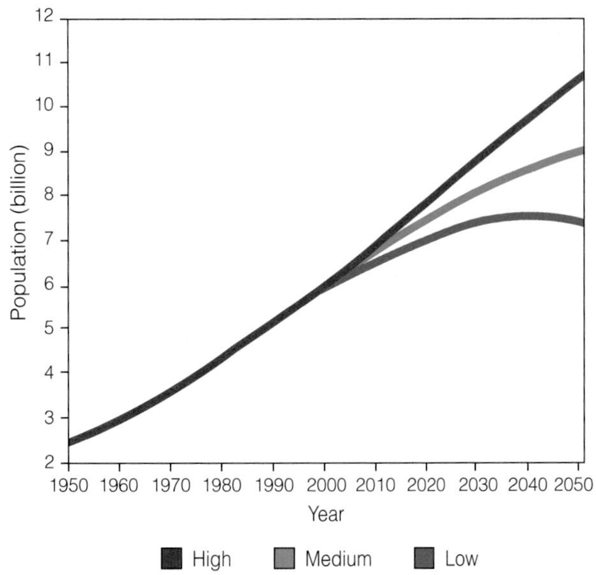

Figure 9-6 UN population projections to 2050, assuming that the world's total fertility rate will fall in a range between 2.5 (high), 2.0 (medium), and 1.6 (low) children per woman. (Data from United Nations)

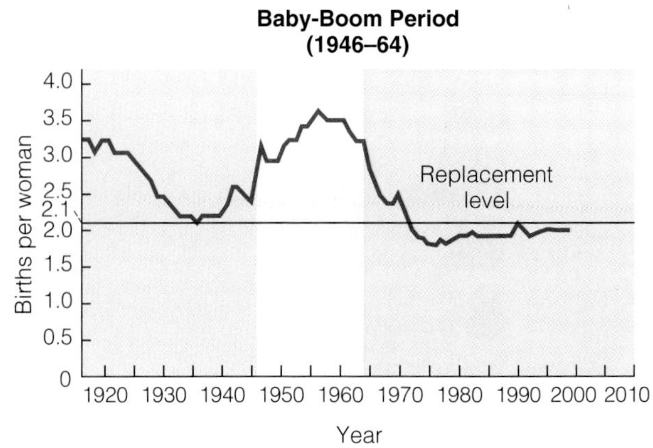

Figure 9-7 Total fertility rate for the United States between 1917 and 1999. (Data from Population Reference Bureau and U.S. Census Bureau)

This is roughly equivalent to adding another California (with 31 million people) every 13 years.

According to U.S. Bureau of Census medium projections, the U.S. population will increase from 273 million to 394 million by the year 2050—a 44% increase, with no stabilization on the horizon. If immigration continues at current levels, new immigrants and their descendants would account for 80 million, or 66% of the projected 121-million increase in the U.S. population between 1999 and 2050. The Census Bureau's high estimate is 507 million by 2050, an 86% increase. Because of a high per capita rate of resource use, each addition to the U.S. population has an enormous environmental impact.

What Factors Affect Birth Rates and Fertility Rates? Key factors affecting a country's average birth rate and TFR are the following:

- *Importance of children as a part of the labor force.* Rates tend to be higher in developing countries (especially in rural areas, where children begin working at an early age).
- *Urbanization.* People living in urban areas usually have better access to family-planning services and tend to have fewer children than those living in rural areas, where children are needed to perform essential tasks.
- *Cost of raising and educating children.* Rates tend to be lower in developed countries, where raising children is much more costly because children don't enter the labor force until their late teens or early 20s.
- *Educational and employment opportunities for women.* Rates tend to be low when women have access to education and paid employment outside the home. TFRs tend to decline as the female literacy rate increases. In developing countries, women with no education generally have two more children than women with a secondary school education.
- *Infant mortality rate.* In areas with low infant mortality rates, people tend to have fewer children because fewer children die at an early age.
- *Average age at marriage* (or, more precisely, the average age at which women have their first child). Women normally have fewer children when their average age at marriage is 25 or older.
- *Availability of private and public pension systems.* Pensions eliminate the need of parents to have many children to help support them in old age.
- *Availability of legal abortions.* There are an estimated 26 million legal abortions and 20 million illegal abortions worldwide each year among the roughly 190 million pregnancies per year.
- *Availability of reliable methods of birth control* (Figure 9-8).
- *Religious beliefs, traditions, and cultural norms.* In some countries, these factors favor large families and strongly oppose abortion and some forms of birth control.

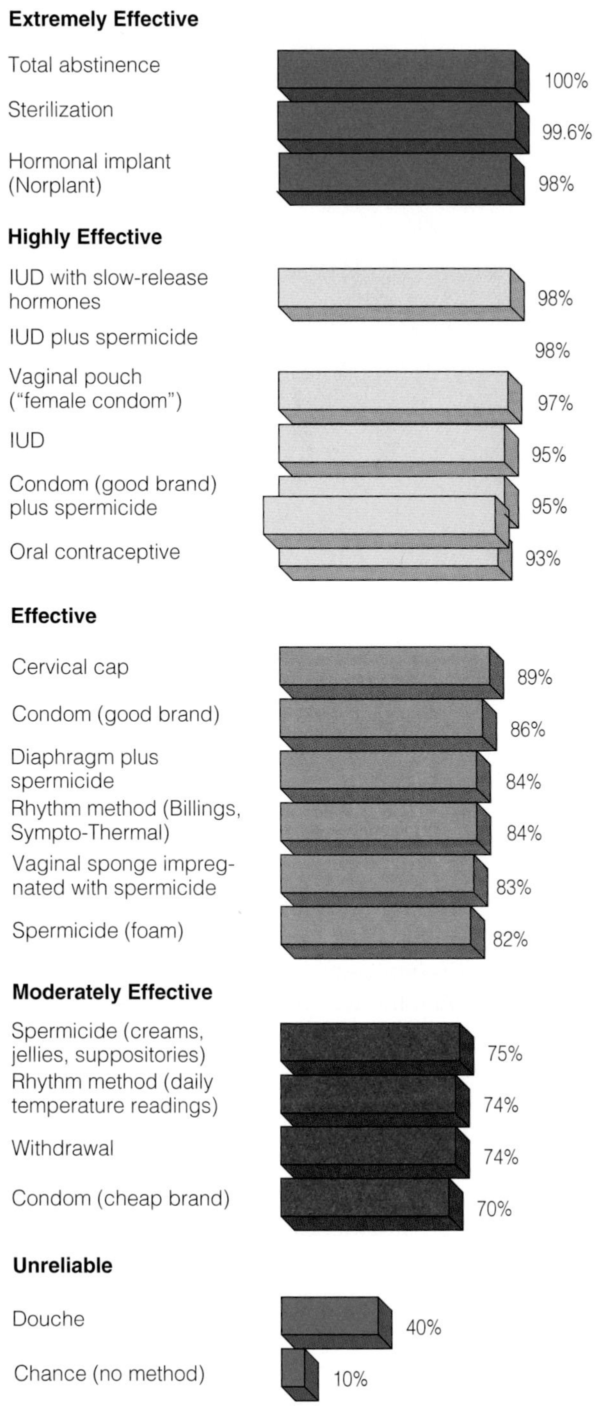

Figure 9-8 Typical effectiveness of birth control methods in the United States. Percentages are based on the number of undesired pregnancies per 100 couples using a specific method as their sole form of birth control for a year. For example, a 93% effectiveness rating for oral contraceptives means that for every 100 women using the pill regularly for 1 year, 7 will get pregnant. Effectiveness rates tend to be lower in developing countries, primarily because of lack of education. (Data from Alan Guttmacher Institute)

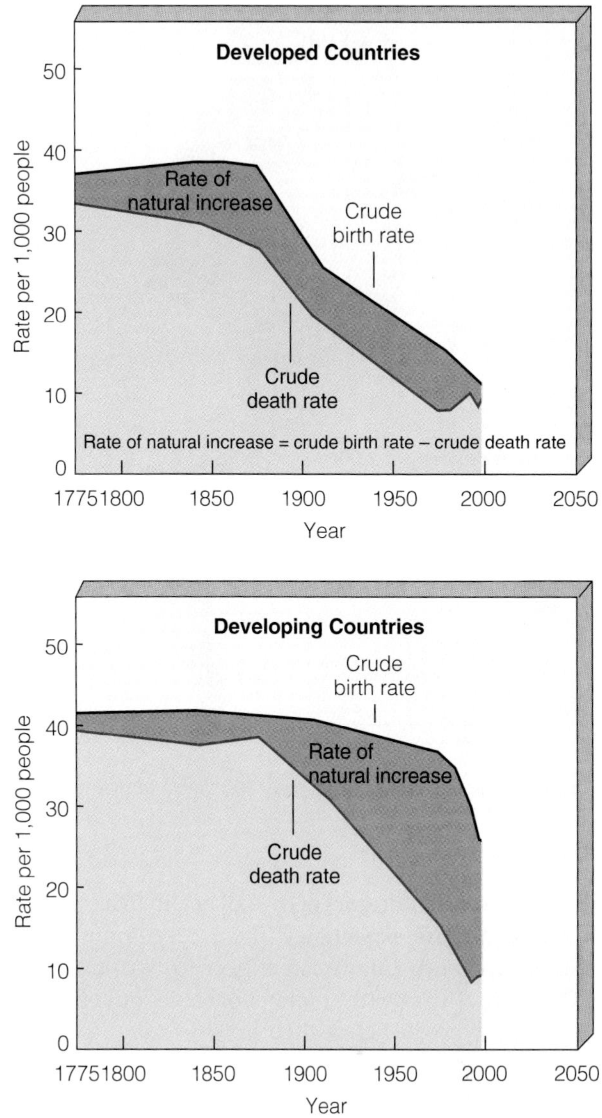

Figure 9-9 Changes in crude birth and death rates for developed and developing countries between 1775 and 1999. (Data from Population Reference Bureau and United Nations)

What Factors Affect Death Rates? The rapid growth of the world's population over the past 100 years is not the result of a rise in the crude birth rate; rather, it has been caused largely by a decline in crude death rates (especially in developing countries; Figure 9-9). More people started living longer (and fewer infants died) because of **(1)** increased food supplies and distribution, **(2)** better nutrition, **(3)** improvements in medical and public health technology (such as immunizations and antibiotics), **(4)** improvements in sanitation and personal hygiene, and **(5)** safer water supplies (which have curtailed the spread of many infectious diseases).

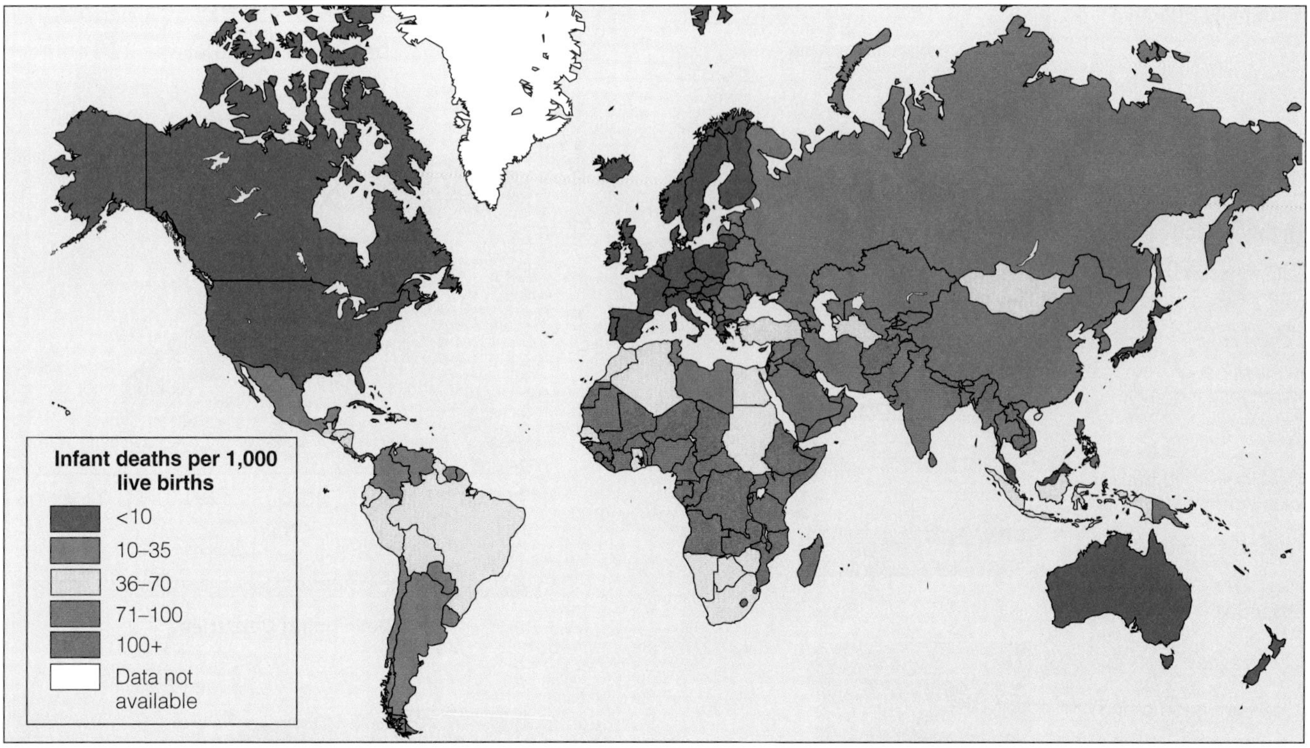

Figure 9-10 Infant mortality rates in 1999. (Data from Population Reference Bureau)

Two useful indicators of overall health in a country or region are **life expectancy** (the average number of years a newborn infant can expect to live) and the **infant mortality rate** (the number of babies out of every 1,000 born who die before their first birthday). In most cases, a low life expectancy in an area is the result of high infant mortality. In 1999 life expectancy averaged 75 years in developed countries and 63 years in developing countries. The *good news* is that globally life expectancy at birth increased from 48 in 1955 to 66 in 1999 and is projected to reach 73 by 2025. The *bad news* is that in the world's 41 poorest countries, mainly in Asia and Africa, life expectancy is only about 50 years.

Because it reflects the general level of nutrition and health care, infant mortality is probably the single most important measure of a society's quality of life (Figure 9-10). A high infant mortality rate usually indicates **(1)** insufficient food (undernutrition), **(2)** poor nutrition (malnutrition), and **(3)** a high incidence of infectious disease (usually from contaminated drinking water).

Between 1965 and 1999, the world's infant mortality rate dropped from 20 per 1,000 live births to 8 in developed countries and from 118 to 62 in developing countries. This is an impressive achievement, but it still means that at least 8 million infants die of preventable causes during their first year of life—an average of 22,000 mostly unnecessary infant deaths per day. About 99% of these deaths occur in developing countries.

The U.S. infant mortality rate of 7.0 per 1,000 in 1999 was down from 16.1 in 1975. Despite this improvement, 34 other countries had lower rates in 1999. Three factors that keep the U.S. infant mortality rate higher than it could be are **(1)** inadequate health care (for poor women during pregnancy and for their babies after birth), **(2)** drug addiction among pregnant women, and **(3)** the high birth rate among teenage women.

The United States has the highest teenage pregnancy rate of any industrialized country. Each year nearly 1 million teenage girls become pregnant in the United States and more than 300,000 of them have abortions. Babies born to teenagers are more likely to have low birth weights, the most important factor in infant deaths.

9-2 POPULATION AGE STRUCTURE

What Are Age Structure Diagrams? As mentioned earlier, even if the replacement-level fertility rate of 2.1 were magically achieved globally tomorrow, the world's population would keep growing for at least another 50 years, stabilizing at about 9 billion by 2054 (assuming no increase in death rates). The reason for this is a population's **age structure**: the proportion of the population (or of each sex) at each age level.

Demographers typically construct a population age structure diagram by plotting the percentages or

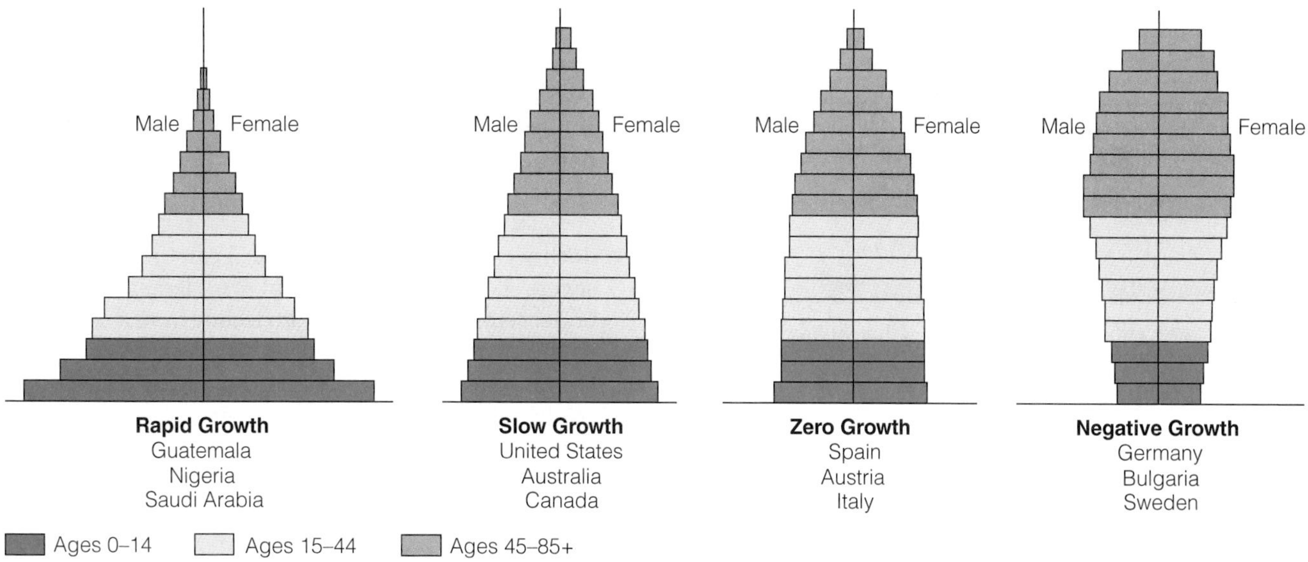

Figure 9-11 Generalized population age structure diagrams for countries with rapid, slow, zero, and negative population growth rates. (Data from Population Reference Bureau)

numbers of males and females in the total population in each of three age categories: *prereproductive* (ages 0–14), *reproductive* (ages 15–44), and *postreproductive* (ages 45 and up). Figure 9-11 presents generalized age structure diagrams for countries with rapid, slow, zero, and negative population growth rates.

How Does Age Structure Affect Population Growth? Any country with many people below age 15 (represented by a wide base in Figure 9-11, left) has a powerful built-in *momentum* to increase its population size unless death rates rise sharply. The number of births rises even if women have only one or two children because of the large number of women who will soon be moving into their reproductive years.

In 1999, 31% of the people on the planet were under 15 years old. These 1.9 billion young people are poised to move into their prime reproductive years. In developing countries the number is even higher: 34%, compared with 19% in developed countries. This powerful force for continued population growth, mostly in developing countries (Figure 9-12), will be slowed only by an effective program to reduce birth rates or by a catastrophic rise in death rates.

How Can Age Structure Diagrams Be Used to Make Population and Economic Projections? The 78-million-person increase that occurred in the U.S. population between 1946 and 1964, known as the *baby boom* (Figure 9-7), will continue to move up through the country's age structure as the members of this group grow older (Figure 9-13).

Baby boomers now make up nearly half of all adult Americans. As a result, they dominate the population's demand for goods and services and play an increasingly important role in deciding who gets elected and what laws are passed. Baby boomers who created the youth market in their teens and 20s are now creating the 50-something market.

The economic burden of helping support a large number of retired baby boomers will fall on the *baby-bust generation*: people born since 1965 (when TFRs fell sharply and have remained below 2.1 since 1970; Figure 9-7). Retired baby boomers may use their political clout to force the smaller number of people in the baby-bust generation to pay higher income, health-care, and Social Security taxes. This could lead to much resentment and conflicts between the two generations.

In other respects, the baby-bust generation should have an easier time than the baby-boom generation. Fewer people will be competing for educational opportunities, jobs, and services, and labor shortages may drive up their wages, at least for jobs requiring education or technical training beyond high school. On the other hand, members of the baby-bust group may find

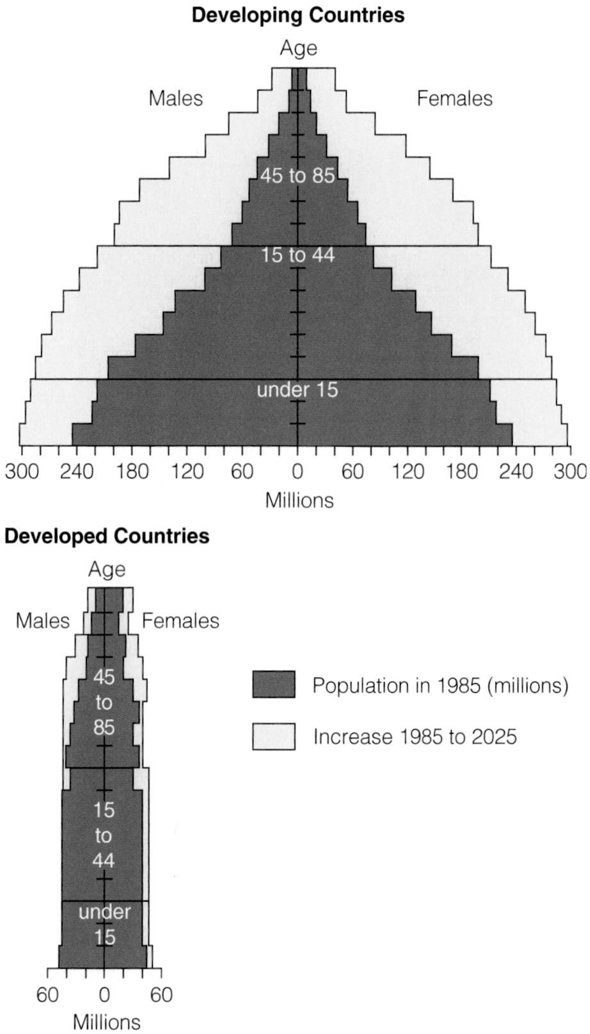

Figure 9-12 Age structure and projected population growth in developing countries and developed countries, 1985–2025. In 1999, there were 1 billion young people in their prime reproductive years of 15–24 and 1.9 billion people under age 15 who will soon be moving into their reproductive years. (Data from United Nations Population Division and Population Reference Bureau)

it difficult to get job promotions as they reach middle age because most upper-level positions will be occupied by members of the much larger baby-boom group. Many baby boomers may delay retirement because of improved health and the need to accumulate adequate retirement funds.

From these few projections we can see that any booms or busts in the age structure of a population create social and economic changes that ripple through a society for decades.

What Are Some Effects of Population Decline? The populations of most of the world's countries are projected to grow throughout most of the 21st century. By 1999, however, 28 countries with 754 million people—13% of humanity—had roughly stable populations (annual growth rates below 0.3%) or declining populations. In other words, about one-eighth of humanity has achieved a stable population. As the projected age structure of the world's population changes between 1990 and 2150 and the percentage of people age 60 or older increases (Figure 9-14), more and more countries will begin experiencing population declines.

If population decline is gradual, its negative effects usually can be managed. But rapid population decline, like rapid population growth, can lead to severe economic and social problems. Countries undergoing rapid population decline have a sharp rise in the proportion of older people, who consume a large share of medical care, Social Security, and other costly public services, with such services supported by working taxpayers. A country with a declining population can also face labor shortages unless it relies on greatly increased automation or immigration of foreign workers.

Case Study: The Graying of Japan In only 7 years, between 1949 and 1956, Japan (Figure 9-15) cut its birth, total fertility, and population growth rates in half. The main reason was widespread access to family planning implemented by the post–World War II U.S. occupation forces and the Japanese government.

Since 1956 these rates have declined further, mostly because of access to family-planning services and other factors: cramped housing, high land prices, late marriage ages, and high costs of education. Japan, the world's second largest economy, is now one of the world's least fertile and fastest-aging societies.

If this trend continues and the country's almost negligible immigration rate doesn't rise, Japan's population of 127 million in 1999 is projected to fall to 121 million by 2025, 100 million by 2050, and 67 million in 2100.

As Japan approaches zero population growth, it is beginning to face some of the problems of an aging population. Japan's universal health insurance and pension systems use about 45% of its national income. This economic burden is projected to rise to 60% or higher in 2020. Japanese economists worry that the steep taxes needed to fund these services could discourage economic growth.

Since 1980 Japan has been feeling the effects of a declining workforce, with a projected drop of about 650,000 workers a year between 2000 and 2050. The population of Japan is 99% Japanese. Fearing a breakdown in its social cohesiveness, the government has been unwilling to increase immigration to provide more workers. Despite this official policy, the country

 Pimentel, D. and M. Pimentel. 1999. Population Growth, Environmental Resources, and the Food. *Social Research,* v66 i1 p417.

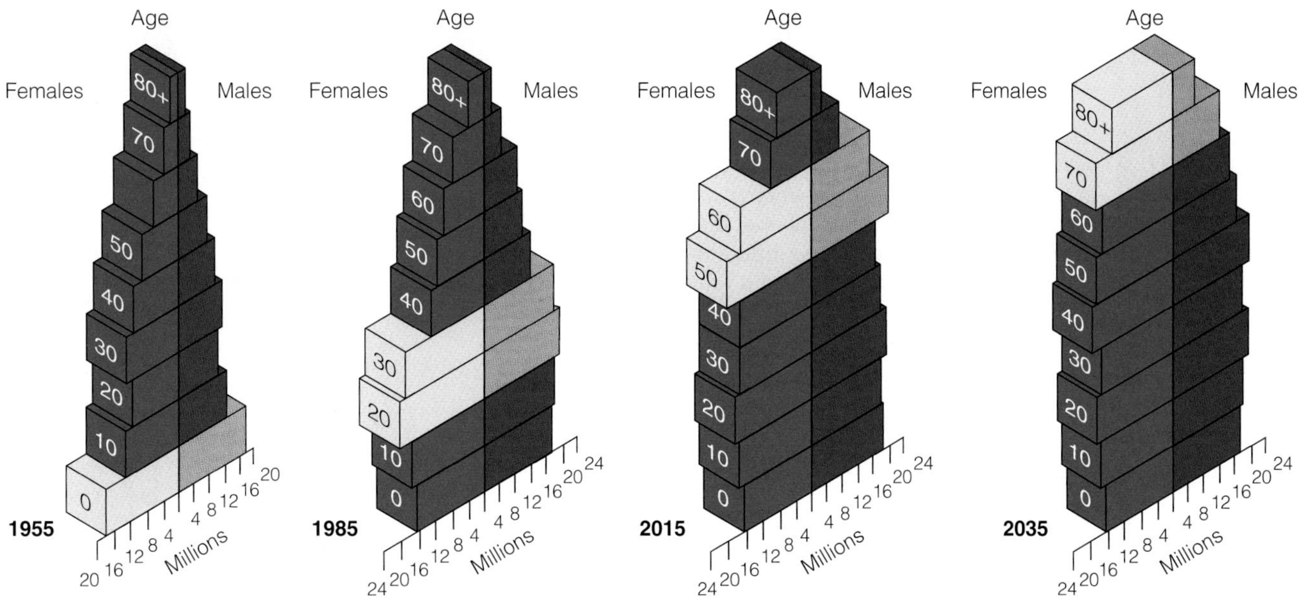

Figure 9-13 Tracking the baby-boom generation in the United States. (Data from Population Reference Bureau and U.S. Census Bureau)

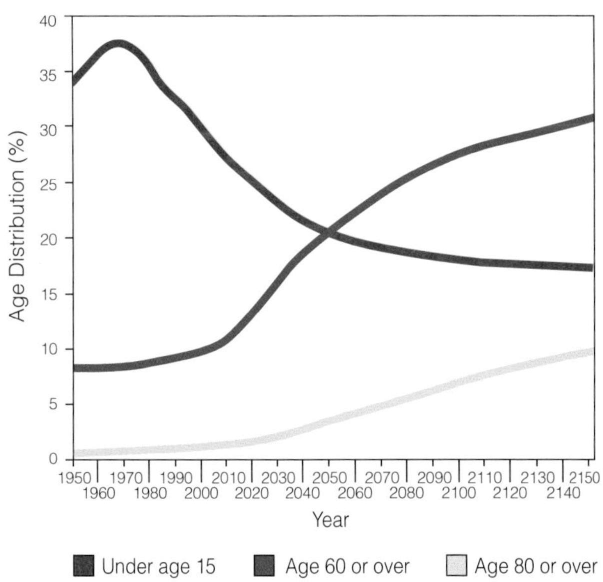

Figure 9-14 Projected percentage of world population under age 15, age 60 or over, and age 80 or over, 1950–2150, assuming the medium fertility projection shown in Figure 9-6. Between 1998 and 2050 the number of people over age 80 is projected to increase from 66 million to 370 million. The cost of supporting a much larger elderly population will place enormous strains on the world's economy. (Data from the United Nations)

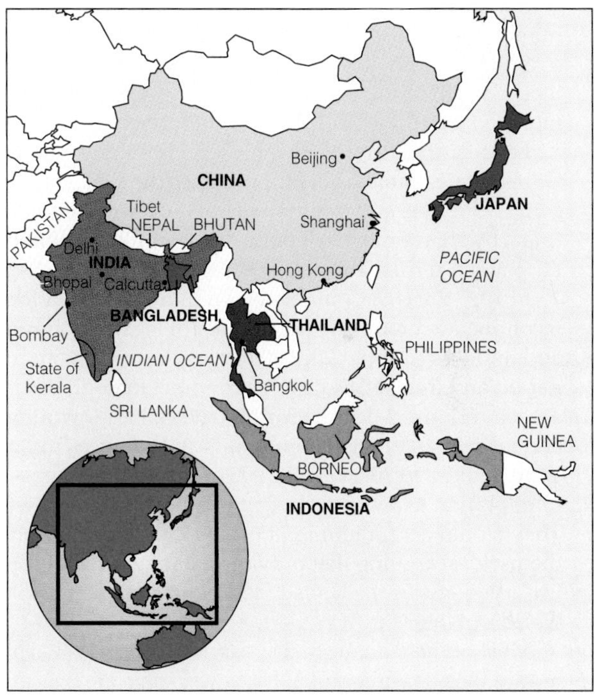

Figure 9-15 Where are Japan, Thailand, Indonesia, India, China, and Bangladesh? Some of the countries highlighted here are discussed in other chapters.

is becoming increasingly dependent on illegal immigrants to keep its economic engines running. How Japan deals with these problems will be watched closely by other countries as they make the transition to zero population growth and, eventually, to population decline.

9-3 SOLUTIONS: INFLUENCING POPULATION SIZE

How Is Population Size Affected by Migration?

The population of an area is also affected by movement of people into (immigration) and out of (emigration)

Immigration in the United States

CASE STUDY

Between 1820 and 1996, the United States admitted almost twice as many immigrants and refugees as all other countries combined. However, the number of legal immigrants has varied during different periods because of changes in immigration laws and rates of economic growth (Figure 9-16).

In 1999, the United States received about 800,000 legal immigrants and refugees (Figure 9-16) and 300,000 illegal immigrants, together accounting for 43% of the country's population growth. The Immigration and Naturalization Service estimates that there are 2–5 million illegal immigrants in the United States. Currently, more than 75% of all legal immigrants live in six states: California, Florida, Illinois, New York, New Jersey, and Texas. If illegal immigrants are included, this figure rises to about 90%.

Immigrants place a tax burden on residents of such states. In California, for example, the average household pays an extra $1,178 in taxes per year because of immigrants. However, according to a 1997 study by the National Academy of Sciences, the work and taxes paid by immigrants add as much as $10 billion per year to the overall U.S. economy. The study estimated that during a lifetime each immigrant will pay an average of $80,000 more in taxes than he or she costs in services.

Between 1820 and 1960, most legal immigrants to the United States came from Europe; since then, most have come from Asia and Latin America. If current trends continue, by 2050 almost half of the U.S. population will be Hispanic (up from 10% in 1993).

In 1995 the U.S. Commission on Immigration Reform recommended reducing the number of legal immigrants and refugees to about 700,000 per year for a transition period and then to 550,000 a year. Some demographers and environmentalists go further and call for lowering the annual ceiling for legal immigrants and refugees into the United States to 300,000–450,000 or for limiting legal immigration to about 20% of annual population growth. They would accept immigrants only if they can support themselves, arguing that providing immigrants with public services turns the United States into a magnet for the world's poor.

Most of these analysts also support efforts to sharply reduce illegal immigration. However, some are concerned that a crackdown on illegal immigrants can also lead to discrimination against legal immigrants.

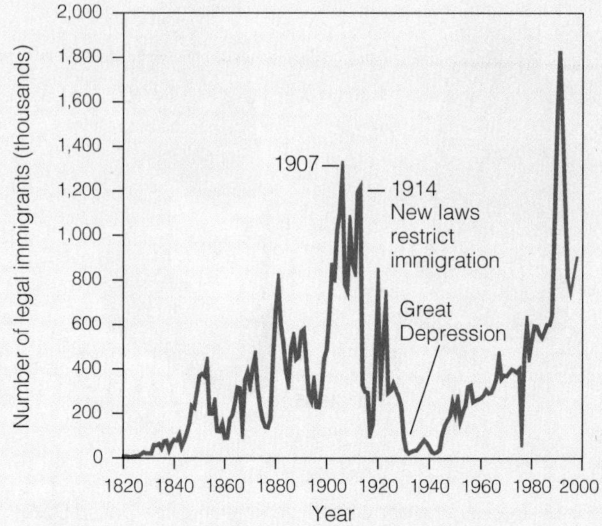

Figure 9-16 Legal immigration to the United States, 1820–1997. The large increase in immigration since 1989 resulted mostly from the Immigration Reform and Control Act of 1986, which granted legal status to illegal immigrants who could show that they had been living in the country for several years. (Data from U.S. Immigration and Naturalization Service)

Proponents argue that reducing immigration would allow the United States to stabilize its population sooner and help reduce the country's enormous environmental impact. Others oppose reducing current levels of legal immigration, arguing that **(1)** it would diminish the historical role of the United States as a place of opportunity for the world's poor and oppressed, **(2)** immigrants pay taxes and take many menial, low-paying jobs that Americans shun, **(3)** few immigrants receive public assistance, and **(4)** many immigrants open businesses and create jobs.

Critical Thinking

Should the United States reduce its current level of legal immigration and tighten up on illegal immigration? Explain.

that area. Most countries influence their rates of population growth to some extent by restricting immigration. Only a few countries—chiefly Canada, Australia, and the United States (Case Study, above)—allow large annual increases in population from immigration.

Some migration is involuntary and involves refugees displaced by armed conflict, environmental degradation, or natural disaster. According to the International Federation of the Red Cross, in 1998 an estimated 25 million international *environmental refugees* moved from one country to another because of problems such as drought, desertification, deforestation, soil erosion, and resource shortages. The problem is worsened by the fact that emergency aid to developing countries dropped by 40% between 1993 and 1998. The number of environmental refugees could reach 50–150 million in the next century if global warming projections are correct.

International migration to developed countries absorbs only about 1% of the annual population growth in developing countries. Thus, population change for most countries is determined mainly by the difference between their birth rates and death rates.

Migration within countries, especially from rural to urban areas, plays an important role in the population dynamics of cities, towns, and rural areas, as discussed in Section 9-6.

What Are the Pros and Cons of Reducing Births? Currently about 93% of the world's population (and 91% of the people in developing countries) live in countries with fertility reduction programs. However, most governments spend less than 1% of their national budgets on such programs.

The projected increase of the human population from 6 to 9 billion or more between 1999 and 2054 raises an important question: *Can the world provide an adequate standard of living for 3 billion more people without causing massive environmental damage?*

There is intense controversy over this question, whether the earth is overpopulated, and what measures, if any, should be taken to slow population growth. To some the planet is already overpopulated (Figure 9-17), but others disagree. Some analysts—mostly economists—argue that we should encourage population growth.

Others believe that asking how many people the world can support is the wrong question, equivalent to asking how many cigarettes one can smoke before getting lung cancer. Instead, they say, we should be asking what the *optimum sustainable population* of the earth might be, based on the planet's *cultural carrying capacity* (Guest Essay, p. 214). Such an optimum level would allow most people to live in reasonable comfort and freedom without impairing the ability of the planet to sustain future generations. No one knows what this optimum population might be. Some consider it a meaningless concept; some put it at 20 billion, others at 8 billion, and others at a level below today's population size.

Those who don't believe that the earth is overpopulated point out that the average life span of the world's 6 billion people is longer today than at any time in the past. They say that talk of a population crash is alarmist, that the world can support billions more people, and that people are the world's most valuable resource for solving the problems we face and stimulating economic growth by becoming consumers.

Some people view any form of population regulation as a violation of their religious beliefs, whereas others see it as an intrusion into their privacy and personal freedom. They believe that all people should be free to have as many children as they want. Some developing countries and some members of minorities in developed countries regard population control as a form of genocide to keep their numbers and power from rising.

Proponents of slowing and eventually stopping population growth point out that we fail to provide the basic necessities for one out of six people on the earth

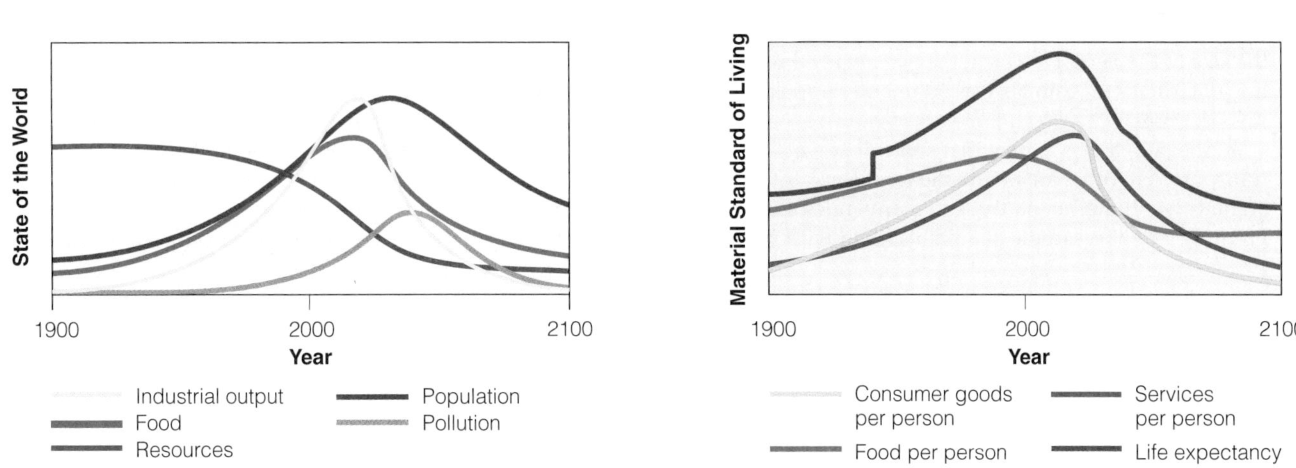

Figure 9-17 Plots of a computer model projecting what might happen if the world's population and economy continue growing exponentially at 1990 levels, assuming no major policy changes or technological innovations. This scenario projects that the world has already overshot some of its limits and that if current trends continue unchanged, we face global economic and environmental collapse sometime in the 21st century. (Used by permission from Donella Meadows et al., *Beyond the Limits: Confronting Global Collapse, Envisioning a Sustainable Future*, White River Junction, Vt.: Chelsea Green, 1992)

GUEST ESSAY

Moral Implications of Cultural Carrying Capacity

Garrett Hardin

As longtime professor of human ecology at the University of California at Santa Barbara, Garrett Hardin made important contributions to relating ethics to biology. He has raised hard ethical questions, sometimes taken unpopular stands, and forced people to think deeply about environmental problems and their possible solutions. He is best known for his 1968 essay "The Tragedy of the Commons," which has had a significant impact on the disciplines of economics and political science and on the management of potentially renewable resources. His many books include Filters Against Folly: How to Survive Despite Economists, Ecologists, and the Merely Eloquent; Living Within Limits; *and* The Ostrich Factor: Our Population Myopia *(see Further Readings).*

For many years, Angel Island in San Francisco Bay was plagued with too many deer. A few animals transplanted there in the early 1900s lacked predators and rapidly increased to nearly 300 deer—far beyond the carrying capacity of the island. Scrawny, underfed animals tugged at the heartstrings of Californians, who carried extra food for them from the mainland to the island.

Such well-meaning charity worsened the plight of the deer. Excess animals trampled the soil, stripped the bark from small trees, and destroyed seedlings of all kinds. The net effect was to lower the island's carrying capacity, year by year, as the deer continued to multiply in a deteriorating habitat.

State game managers proposed that the excess deer be shot by skilled hunters. "How cruel!" some people protested. Then the managers proposed that coyotes be introduced onto the island. Though not big enough to kill adult deer, coyotes can kill fawns, thereby reducing the size of the herd. However, the Society for the Prevention of Cruelty to Animals was adamantly opposed to this proposal.

In the end, it was agreed that some deer would be transported to other areas suitable for deer. A total of 203 animals were caught and trucked many miles away. From the fate of a sample of animals fitted with radio collars, it was estimated that 85% of the transported deer died within a year (most of them within 2 months) from various causes: predation by coyotes, bobcats, and domestic dogs, shooting by poachers and legal hunters, and being run over by cars.

The net cost (in 1982 dollars) for relocating each animal surviving for a year was $2,876. The state refused to continue financing the program, and no volunteers stepped forward to pay future bills.

Angel Island is a microcosm of the planet as a whole. Organisms reproduce exponentially, but the environment doesn't increase at all. The moral is a simple ecological commandment: *Thou shalt not transgress the carrying capacity.*

Now let's examine the situation for humans. A competent physicist has placed global human carrying capacity at 50 billion, about 8 times the current world population. Before you give in to the temptation to urge women to have more babies, consider what Robert Malthus said nearly 200 years ago: "There should be no more people in a country than could enjoy daily a glass of wine and piece of beef for dinner."

A diet of grain or bread and water is symbolic of minimum living standards; wine and beef are symbolic of higher living standards that make greater demands on the environment. When land that could produce plants for direct human consumption is used to grow grapes for wine or corn for cattle, more energy is expended to feed the human population. Because carrying capacity is defined as the *maximum* number of animals (humans) an area can support, using part of the area to support such cultural luxuries as wine and beef reduces the carrying capacity. This reduced capacity is called the *cultural carrying capacity,* and it is always smaller than simple carrying capacity.

Energy is the common "coin of the realm" for all competing demands on the environment. Energy saved by giving up a luxury can be used to produce more food staples and support more people. We could increase the simple carrying capacity of the earth by giving up any (or all) of the following "luxuries": street lighting, vacations, private cars, air conditioning, and

today. If we can't (or won't) do this now, how will we be able to do this for 3 billion more people by 2054?

Proponents of slowing population growth consider population growth in developed and developing countries as a threat to the earth's life-support systems and to the human population. They contend that if we don't sharply lower birth rates, we are deciding by default to raise death rates for humans and greatly increase environmental harm. In 1992, for example, the highly respected U.S. National Academy of Sciences and the Royal Society of London issued the following joint statement: "If current predictions of population growth and patterns of human activity on the planet remain unchanged, science and technology may not be able to prevent either irreversible degradation of the environment or continued poverty for much of the world."

Proponents of this view recognize that population growth is not the only cause of environmental and resource problems. However, they argue that adding several hundred million more people in developed artistic performances of all sorts. But what we consider "luxuries" depends on our values as individuals and societies, and values are largely matters of choice. At one extreme, we could maximize the number of human beings living at the lowest possible level of comfort. Or we could try to optimize the quality of life for a much smaller human population.

The carrying capacity of the earth is a scientific question. It may be possible to support 50 billion people at a bread-and-water level. Is that what we choose? The question, "What is the cultural carrying capacity?" requires that we debate questions of value, about which opinions differ.

An even greater difficulty must be faced. So far, we have been treating carrying capacity as a *global* issue, as if there were some global sovereignty capable of enforcing a solution on all people. But there is no global sovereignty ("one world"), nor is there any prospect of one in the foreseeable future. Thus, we must ask how some 200 nations are to coexist in a finite global environment if different sovereignties adopt different standards of living.

Consider a protected redwood forest that produces neither food for humans nor lumber for houses. Because people must travel many kilometers to visit it, the forest is a net loss in the national energy budget. However, for those fortunate enough to wander through the cathedral-like aisles beneath an evergreen vault, a redwood forest does something precious for the human spirit. But then intrudes an appeal from a distant land, where millions are starving because their population has overshot the carrying capacity; we are asked to save lives by sending food. So long as we have surpluses, we may safely indulge in the pleasures of philanthropy. But after we have run out of our surpluses, then what?

A spokesperson for the needy from that land makes a proposal: "If you would only cut down your redwood forests, you could use the lumber to build houses and then grow potatoes on the land, shipping the food to us. Since we are all passengers together on Spaceship Earth, are you not duty bound to do so? Which is more precious, trees or human beings?"

This last question may sound ethically compelling, but let's look at the consequences of assigning a preemptive and supreme value to human lives. At least 2 billion people in the world are poorer than the 35 million "legally poor" in America, and their numbers are increasing by about 1 million per year. Unless this increase is halted, sharing food and energy on the basis of need would require the sacrifice of one amenity after another in rich countries. The ultimate result of sharing would be complete poverty everywhere on the earth in order to maintain the earth's simple carrying capacity. Is that the best humanity can do?

To date, there has been overwhelmingly negative reaction to all proposals to make international philanthropy conditional on the cessation of population growth by poor, overpopulated recipient nations. Foreign aid is governed by two apparently inflexible assumptions:

- The right to produce children is a universal, irrevocable right of every nation, no matter how hard it presses against the carrying capacity of its territory.
- When lives are in danger, the moral obligation of rich countries to save human lives is absolute and undeniable.

Considered separately, each of these two well-meaning doctrines might be defensible; taken together, they constitute a fatal recipe. If humanity gives maximum carrying capacity precedence over problems of cultural carrying capacity, the result will be universal poverty and environmental ruin.

Or do you see an escape from this harsh dilemma?

Critical Thinking

1. What population size would allow the world's people to have good quality of life? What do you believe is the cultural carrying capacity of the United States? Should the United States have a national policy to establish this population size as soon as possible? Explain.

2. Do you support the two principles this essay lists as the basis of foreign aid to needy countries? If not, what changes would you make in the requirements for receiving such aid?

countries and several billion more in developing countries can only intensify existing environmental and social problems. They call for drastic changes to prevent accelerating environmental decline and a rise in death rates (Figure 9-18).

These analysts believe that people should have the freedom to produce as many children as they want. However, such freedom would apply only if it did not reduce the quality of other people's lives now and in the future, either by impairing the earth's ability to sustain life or by causing social disruption. They point out that limiting the freedom of individuals to do anything they want in order to protect the freedom of other individuals is the basis of most laws in modern societies. What is your opinion on this issue?

How Can Economic Development Help Reduce Birth Rates? Demographers have examined the birth and death rates of western European countries that industrialized during the 19th century. From these data

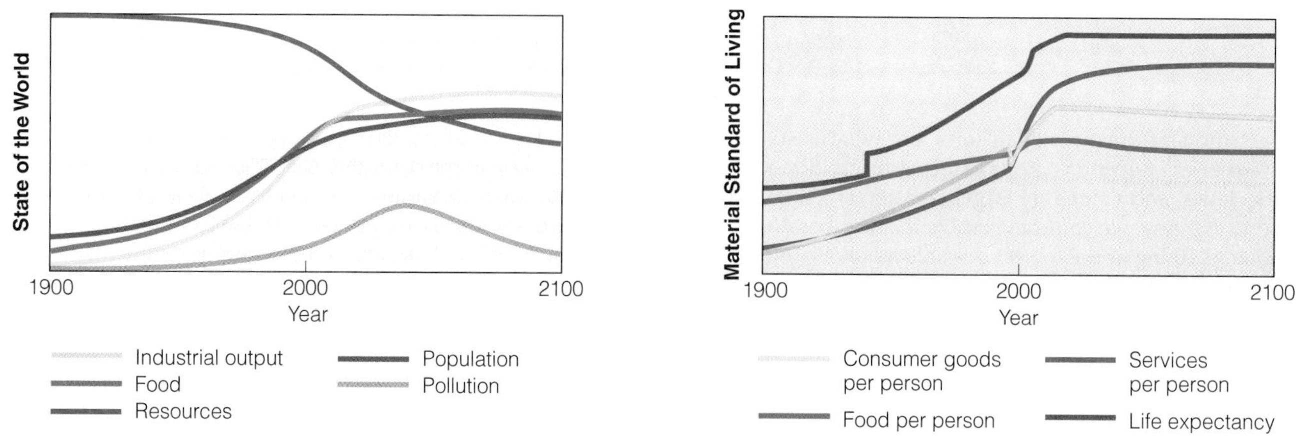

Figure 9-18 Computer-generated scenario projecting how we can avoid overshoot and collapse and make a fairly smooth transition to a sustainable future. It assumes that **(1)** technology allows us to double supplies of nonrenewable resources, double crop and timber yields, cut soil erosion in half, and double the efficiency of resource use within 20 years, **(2)** 100% effective birth control was made available to everyone by 1995, **(3)** no couple has more than two children, beginning in 1995, and **(4)** per capita industrial output is stabilized at 1990 levels. Another computer run projects that waiting until 2015 to implement these changes would lead to overshoot and collapse sometime around 2075, followed by a transition to sustainability by 2100. (Used by permission from Donella Meadows et al., *Beyond the Limits: Confronting Global Collapse, Envisioning a Sustainable Future,* White River Junction, Vt.: Chelsea Green, 1992)

they developed a hypothesis of population change known as the **demographic transition**: As countries become industrialized, first their death rates and then their birth rates decline.

According to this hypothesis, the transition takes place in four distinct stages (Figure 9-19). In the *preindustrial stage,* harsh living conditions lead to a high birth rate (to compensate for high infant mortality) and a high death rate. Thus there is little population growth.

In the *transitional stage,* industrialization begins, food production rises, and health care improves. Death rates drop and birth rates remain high, so the population grows rapidly (typically 2.5–3% a year).

In the *industrial stage,* industrialization is widespread. The birth rate drops and eventually approaches the death rate. Reasons for this convergence of rates include better access to birth control, decline in the infant mortality rate, increased job opportunities for women, and the high costs of raising children who don't enter the workforce until after high school or college. Population growth continues, but at a slower and perhaps fluctuating rate, depending on economic conditions. Most developed countries are now in this third stage, and a few developing countries are entering this stage.

In the *postindustrial stage,* the birth rate declines even further, equaling the death rate and thus reaching zero population growth. Then the birth rate falls below the death rate and total population size slowly decreases. Emphasis shifts from unsustainable to sustainable forms of economic development. Twenty-eight countries (most of them in Europe) containing about 13% of the world's population have entered this stage. To most population experts, the challenge is to help the remaining 87% of humanity reach this stage.

In most developing countries today, death rates have fallen much more than birth rates. In other words, these developing countries—mostly in Southeast Asia, Africa, and Latin America—are still in the transitional stage, halfway up the economic ladder, with high population growth rates. Some economists believe that developing countries will make the demographic transition over the next few decades without increased family-planning efforts.

However, despite encouraging declines in fertility, some population analysts fear that the still-rapid population growth in many developing countries will outstrip economic growth and overwhelm local life-support systems. This could cause many of these countries to be caught in a *demographic trap,* something that is currently happening in a number of developing countries, especially in Africa.

Analysts also point out that some of the conditions that allowed developed countries to develop are not available to many of today's developing countries. Even with large and growing populations, many developing countries do not have enough skilled workers to produce the high-tech products needed to compete in the global economy. Most low- and middle-income

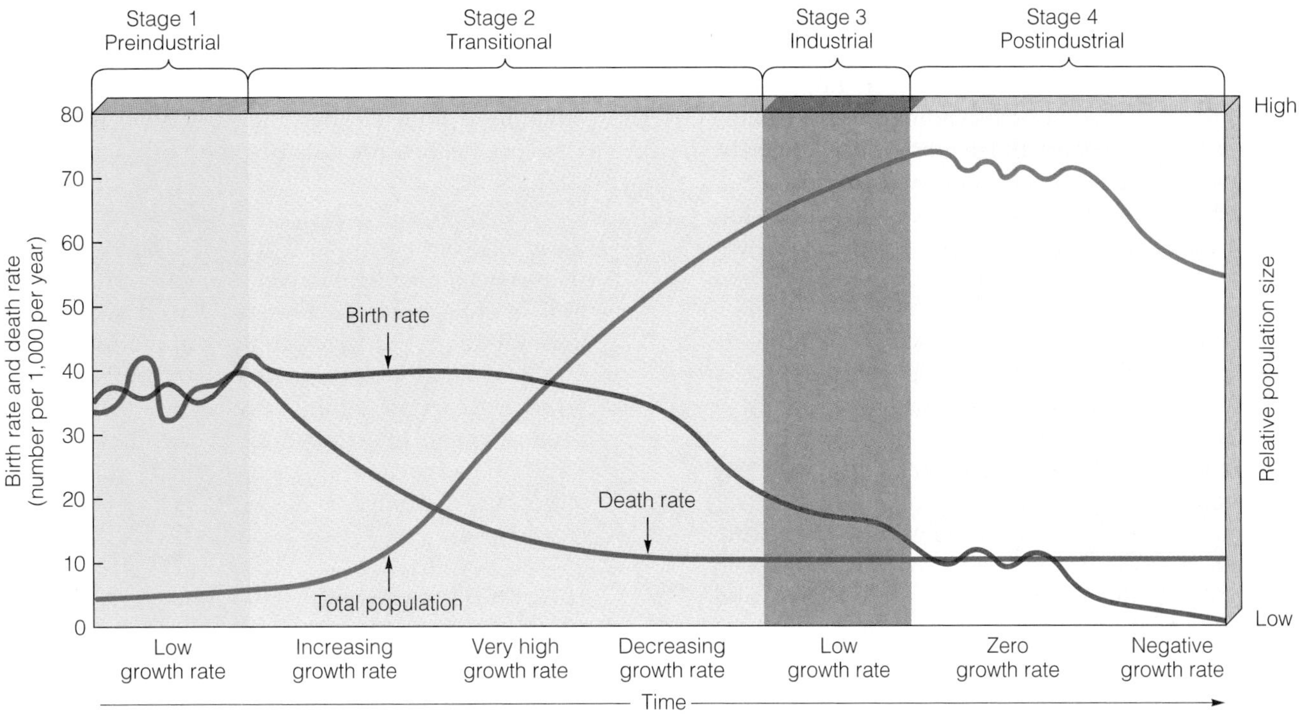

Figure 9-19 Generalized model of the demographic transition.

developing countries also lack the capital and resources needed for rapid economic development. Furthermore, the amount of economic assistance for developing countries, which are struggling under tremendous debts, has been decreasing since 1980, with a 25% drop between 1992 and 1997. Indeed, since the mid-1980s, developing countries have paid developed countries \$40–50 billion a year (mostly in debt interest) more than they have received from these countries.

How Can Family Planning Help Reduce Birth and Abortion Rates and Save Lives? **Family planning** provides educational and clinical services that help couples choose how many children to have and when to have them. Such programs vary from culture to culture, but most provide information on birth spacing, birth control, breast-feeding, and prenatal care.

Family planning has been an important factor in increasing the proportion of married women in developing countries who use modern contraception. It has gone from 10% in the 1960s to 49% in 1999 (37%, if China is excluded); this is close to the 59% usage by married women in developed countries. Research indicates that family planning is responsible for at least 55% of the drop in TFRs in developing countries, from 6 in 1960 to 2.9 in 1999 (3.8 if China is excluded).

Family-planning programs save a society money by reducing the need for children's social services. Proponents also argue that providing access to family planning throughout the world would bring about a sharp decline in the number of legal and illegal abortions per year.

Family planning, coupled with improved prenatal care and health services, can also help reduce the risk of death from childbearing. Each year at least 600,000 women die from pregnancy-related causes (an average of 1,600 deaths per day), with 99% of such deaths occurring in developing countries. In Africa as a whole, there is one doctor for every 10,000 people and a woman has a 1 in 48 chance of dying in pregnancy.

The effectiveness of family planning varies with program design and funding. It has been a significant factor in reducing birth and fertility rates in populous countries such as China (Section 9-4), Indonesia, Brazil, and Bangladesh. Family planning has also played a major role in reducing population growth in Japan, Thailand (p. 202), Mexico, South Korea, Taiwan, Iran, and several other countries with moderate to small populations. These successful programs, based on committed leadership, local implementation, education and publicity, and wide availability of contraceptives, demonstrate that population growth rates can be decreased significantly within 15–30 years.

Despite such successes, family planning has had moderate to poor results in more populous developing countries such as India, Egypt, Pakistan, and Nigeria. Results also have been poor in 79 less populous developing countries, especially in Africa and Latin America, with high or very high population growth rates.

According to UN studies, an estimated 300 million women in developing countries want to limit the number and determine the spacing of their children, but they lack access to services. Extending family-planning services to these women and to those who will soon be entering their reproductive years could prevent an estimated 5.8 million births a year and over 5 million abortions a year. Other analysts call for expanding existing family-planning programs to include teenagers and sexually active unmarried women, who now are often excluded. Some analysts urge pro-choice and pro-life groups to join forces in greatly reducing unplanned births and abortions, especially among teenagers. What do you think?

Some analysts urge that programs be broadened to educate men about the importance of having fewer children and taking more responsibility for raising them. Proponents also argue that much more research is needed on developing new, more effective, and acceptable methods of birth control for men.

Since 1998, Chinese men have been able to put a contraceptive device about the size of a pager in their underwear for 1 hour per month. This new male contraceptive emits sperm-killing electronic pulses that can render a man sterile for up to a month. Fertility returns 2 months after the man stops using the device.

Family planning could be provided in developing countries to all couples who want it for about $17 billion a year, the equivalent of less than a week's worth of worldwide military expenditures. If developed countries provided one-third of this $17 billion, each person in the developed countries would spend only $4.80 a year. This would help shrink average family size from 2.7 to 2.1 children and reduce the world's population by about 2.9 billion people.

How Can Economic Rewards and Penalties Be Used to Help Reduce Birth Rates? Some population experts argue that family planning, even coupled with economic development, cannot lower birth and fertility rates quickly enough to avoid a sharp rise in death rates in many developing countries. They point to studies showing that most couples in developing countries want three or four children. This is well above the replacement-level fertility required to bring about eventual population stabilization.

These analysts believe that we must go beyond family planning and offer economic rewards and penalties to help slow population growth. About 20 countries offer small payments to people who agree to use contraceptives or to be sterilized. However, such payments are most likely to attract people who already have all the children they want.

Some countries, including China, penalize couples who have more than one or two children by raising their taxes, charging other fees, or eliminating income tax deductions for a couple's third child (as in Singapore, Hong Kong, and Ghana). Families who have more children than the prescribed limit may also lose health-care benefits, food allotments, and job options.

Economic rewards and penalties designed to lower birth rates work best if they **(1)** encourage (rather than coerce) people to have fewer children, **(2)** reinforce existing customs and trends toward smaller families, **(3)** do not penalize people who produced large families before the programs were established, and **(4)** increase a poor family's economic status. Once a country's population growth is out of control it may be forced to use coercive methods to prevent mass starvation and hardship, as has been the case for China (Section 9-4).

How Can Empowering Women Help Reduce Birth Rates? Studies show that women tend to have fewer and healthier children and live longer when they have **(1)** access to education and to paying jobs outside the home and **(2)** when they live in societies in which their rights are not suppressed.

Women, roughly half of the world's population, do almost all of the world's domestic work and child care and provide more health care with little or no pay than all the world's organized health services combined. They also do more than half the work associated with growing food, gathering fuelwood, and hauling water. As one Brazilian woman put it, "For poor women the only holiday is when you are asleep." Women's unpaid work, at an estimated value of $11 trillion annually (almost half of the annual total global output), is not included in a country's GDP.

Women work two-thirds of all hours worked, but receive only one-tenth of the world's income and own a mere 0.01% of the world's property. In most developing countries, women don't have the legal right to own land or borrow money to increase agricultural productivity. Women also make up 70% of the world's poor and almost two-thirds of the more than 960 million adults who can neither read nor write.

Women are largely excluded from economic and political decision making. They hold only 14% of the world's administrative and managerial positions and occupy only 10% of parliamentary seats.

Most analysts believe that women everywhere should have full legal rights and the opportunity to become educated and earn income outside the home. This would not only slow population growth but also promote human rights and freedom. However, empowering women by seeking gender equality will require some major social changes that will be difficult to achieve in male-dominated societies.

9-4 CASE STUDIES: SLOWING POPULATION GROWTH IN INDIA AND CHINA

What Success Has India Had in Controlling Its Population Growth? The world's first national family-planning program began in India (Figure 9-15) in 1952, when its population was nearly 400 million. In 1999, after 47 years of population control efforts, India was the world's second most populous country, with a population of 1 billion.

In 1952, India added 5 million people to its population; in 1999 it added 18 million—49,300 more mouths to feed each day. With 36% of its people under age 15, India's population is projected to reach 1.5 billion by 2050 and possibly 1.9 billion before leveling off early in the 22nd century. In 1999, women in India averaged 3.4 children, down from 5.3 in 1970. Despite this decline in fertility, India's population is growing exponentially at about 1.9% a year.

India's people are among the poorest in the world, with an average per capita income of about $370 a year; for 30% of the population it is less than $100 a year, or 27¢ a day. Nearly half of India's labor force is unemployed or can find only occasional work. Although India currently is self-sufficient in food-grain production, about 40% of its population suffers from malnutrition, mostly because of poverty. Life expectancy is 60 years, and the infant mortality rate is 72 deaths per 1,000 live births.

Some analysts fear that India's already serious malnutrition and health problems will worsen as its population continues to grow rapidly. With 16% of the world's people, India has just 2.3% of the world's land resources and 1.7% of the world's forests. About half of India's cropland is degraded as a result of soil erosion, waterlogging, salinization, overgrazing, and deforestation. About 70% of India's water is seriously polluted, and sanitation services often are inadequate.

Without its long-standing family-planning program, India's population and environmental problems would be growing even faster. Still, to its supporters the results of the program have been disappointing because of **(1)** poor planning, **(2)** bureaucratic inefficiency, **(3)** the low status of women (despite constitutional guarantees of equality), **(4)** extreme poverty, and **(5)** a lack of administrative and financial support.

Even though the government has provided information about the advantages of small families for years, Indian women still have an average of 3.4 children because most couples believe they need many children to do work and care for them in old age. Many social and cultural norms favor large families, including the strong preference for male children; some couples keep having children until they produce one or more boys. These factors in part explain why even though 90% of Indian couples know of at least one modern birth control method, only 36% actually use one.

What Success Has China Had in Controlling Its Population Growth? Since 1970, China (Figure 9-15) has made impressive efforts to feed its people and bring its population growth under control. Between 1972 and 1999 China reduced its crude birth rate, from 32 to 16 per 1,000 people, and its TFR dropped from 5.7 to 1.8 children per woman. Since 1985 its infant mortality rate has been almost one-half the rate in India and its illiteracy is about almost one-third that of India. Life expectancy in China is 71 years, 11 years higher than in India. China's per capita income of $867 is more than twice that in India.

Despite these achievements, with the world's largest population (1.26 billion) and a growth rate of 1.0%, China had about 12 million more mouths to feed in 1999. Its population is projected to reach 1.5 billion by 2025.

To achieve its sharp drop in fertility, China has established the most extensive, intrusive, and strict population control program in the world. Couples are strongly urged to postpone the age at which they marry and to have no more than one child. Married couples have ready access to free sterilization, contraceptives, and abortion. Paramedics and mobile units ensure access even in rural areas.

Couples who pledge to have no more than one child receive extra food, larger pensions, better housing, free medical care, and salary bonuses; their child will be given free school tuition and preferential treatment in employment when he or she enters the job market. Couples who break their pledge lose all the benefits. The result is that 81% of married women in China are using modern contraception, compared with 60% in developed countries and only 36% in other developing countries.

Government officials realized in the 1960s that the only alternative to strict population control was mass starvation. China is a dictatorship, and thus, unlike India, it has been able to impose a consistent population policy throughout its society.

China's large and still growing population has an enormous environmental impact that could reduce its ability to produce enough food and threaten the health of many of its people. China has 21% of the world's population, but only 7% of its fresh water and cropland, 3% of its forests, and 2% of its oil. Soil erosion in China is serious and is apparently getting worse.

Most countries prefer to avoid the coercive elements of China's program. However, other parts of this program could be used in many developing countries. Especially useful is the practice of localizing the program, rather than asking people to go to distant family-planning centers. Perhaps the best lesson for other countries is to act to curb population growth before they must choose between mass starvation and coercive measures that severely restrict human freedom.

9-5 CUTTING GLOBAL POPULATION GROWTH

How Could Population Growth Be Reduced to More Sustainable Levels? In 1994 the United Nations held its third once-in-a-decade Conference on Population and Development in Cairo, Egypt. One of the conference's goals was to encourage action to stabilize the world's population at 7.8 billion by 2054, instead of the projected 9 billion (Figure 9-6). The major goals of the resulting population plan, endorsed by 180 governments, are to do the following by 2015:

- Provide universal access to family-planning services and reproductive health care
- Improve the health care of infants, children, and pregnant women
- Encourage development and implementation of national population policies as part of social and economic development policies
- Bring about more equitable relationships between men and women, with emphasis on improving the status of women and expanding education and job opportunities for young women
- Increase access to education, especially for girls
- Increase the involvement of men in child-rearing responsibilities and family planning
- Take steps to eradicate poverty
- Reduce and eliminate unsustainable patterns of production and consumption

Many analysts applaud these goals, but some call them wishful thinking. Even if they wanted to, most governments could not afford to implement many of these goals.

However, the experience of Japan, Thailand (p. 202), South Korea, Taiwan, and China indicates that a country can achieve replacement-level fertility within 15–30 years. Such experience also suggests that the best way to slow population growth is a combination of investing in family planning, reducing poverty, and elevating the status of women.

Because countries differ in population growth rates, use and availability of resources, and social structure, the combination of these factors must be tailored to each country's situation. Furthermore, most analysts believe that government policy makers should devise policies that minimize the environmental impact of population growth in their effort to achieve sustainability.

9-6 POPULATION DISTRIBUTION: URBANIZATION AND URBAN GROWTH

How Fast Are Urban Areas Growing? An **urban area** is often defined as a town or city plus its adjacent suburban fringes with a population of more than 2,500 people (although some countries set the minimum at 10,000–50,000 residents). A **rural area** is usually defined as an area with a population of less than 2,500 people.

A country's **degree of urbanization** is the percentage of its population living in an urban area. **Urban growth** is the rate of increase of urban populations.

Urban areas grow in two ways: by *natural increase* (more births than deaths) and by *immigration* (mostly from rural areas). Rural people are *pulled* to urban areas in search of jobs, a better life, and freedom from the constraints of village cultural life. Urban growth in developing countries is also fueled by government policies that distribute most income and social services to urban dwellers (especially in capital cities, where a country's leaders live) at the expense of rural dwellers. People may also be *pushed* from rural areas into urban areas by factors such as poverty, lack of land, declining agricultural jobs because of increased use of mechanized agriculture, famine, and war. The urban poor fortunate enough to find employment usually must work long hours for low wages at jobs that may expose them to dust, hazardous chemicals, excessive noise, and dangerous machinery.

Between 1950 and 1999 the number of people living in the world's urban areas increased from 200 million to 2.7 billion. By 2025 it is projected to reach 5.5 billion, with 90% of this urban growth occurring in developing countries.

Several trends are important in understanding the problems and challenges of urban growth on this rapidly urbanizing planet. *First*, the proportion of the global population living in urban areas increased between 1850 and 1999 from 2% to 45%. If UN projections are correct, by 2025 about 63% of the world's people will be living in urban areas. Although India is largely a rural country, its urban population of 256 million is about equal to the entire population of the United States.

Second, the number of large cities is mushrooming. In 1960 there were 111 cities with populations of more than 1 million. Today there are 293, and some analysts expect this number to increase to 650 by 2025. Currently, 1 person out of every 10 lives in a city with a million or more inhabitants, and many live in the world's 15 *megacities*, with 10 million or more people. The world's five largest cities are **(1)** Tokyo, Japan (28 million), **(2)** Mexico City, Mexico (18.1 million), **(3)** Bombay, India (18 million), **(4)** São Paulo, Brazil (17.7 million), and **(5)** New York in the United States (16.6 million).

According to the United Nations the world now has 23 megacities, 18 of them in developing countries.

As they grow outward, separate urban areas may merge to form a *megalopolis*. For example, the remaining open space between Boston, Massachusetts and Washington, D.C., is rapidly urbanizing and coalescing. This 800-kilometer-long (500-mile-long) urban area, sometimes called *Bowash* (Figure 9-20), contains almost 60 million people, about twice Canada's entire population.

Third, developing countries, with 37% urbanization, contain 1.8 billion urban dwellers and are projected to reach at least 57% urbanization by 2025. Most of this growth will occur in large cities, which already have trouble supplying their residents with water, food, housing, jobs, sanitation, and basic services. For example, during the 5 years between 1990 and 1995, cities in developing countries grew by about 263 million people—equivalent to another Los Angeles every 3 months. Some 73% of Latin Americans live in cities. The most explosive urban growth is expected to take place in Africa and Asia, which are currently only 30–35% urbanized. Many of these cities will be water short, waste filled, and choked with pollution.

Fourth, in developed countries (with 75% urbanization), urban growth is much slower than in developing countries. Still, developed countries should reach 84% urbanization by 2025.

Finally, poverty is becoming increasingly urbanized as more poor people migrate from rural to urban areas. The United Nations estimates that at least 1 billion people live in the crowded *slums* of inner cities or in vast illegal *squatter settlements* and *shantytowns*. In such settlements, people move onto undeveloped land (usually without the owner's permission) and build shacks made of packing crates, plastic sheets, corrugated metal pipes, or whatever they can find (Spotlight, p. 222).

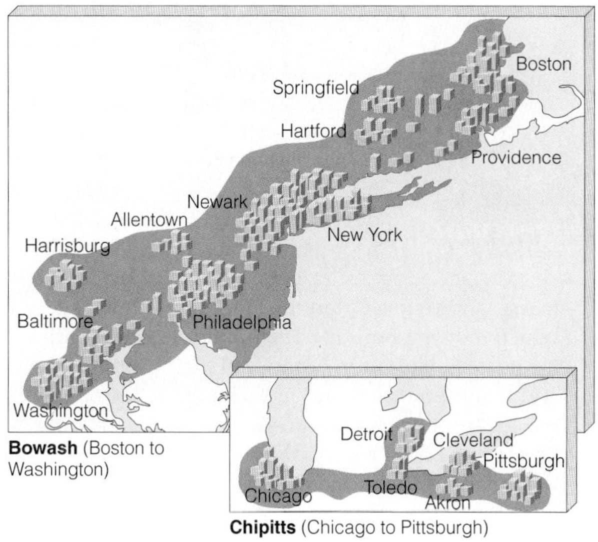

Figure 9-20 Two megalopolises: Bowash, consisting of urban sprawl and coalescence between Boston and Washington, D.C., and Chipitts, extending from Chicago to Pittsburgh.

Case Study: Mexico City About 18.1 million people—about one of every five Mexicans—live in Mexico City (Figure 9-21), the world's second most populous city. Every day an additional 2,000 poverty-stricken rural peasants pour into the city, hoping to find a better life. This adds about 750,000 new people per year.

Mexico City suffers from severe air pollution, high unemployment (close to 50%), deafening noise, congestion, and a soaring crime rate. More than one-third of its residents live in crowded slums (called *barrios*) or squatter settlements, without running water or electricity. At least 8 million people have no sewer facilities. This means that huge amounts of human waste are deposited in gutters and vacant lots every day, attracting armies of rats and swarms of flies. When the winds pick up dried excrement, a *fecal snow* often falls on parts of the city, leading to widespread salmonella and hepatitis infections, especially among children.

Some 4 million motor vehicles and 30,000 factories spew pollutants into the atmosphere. Air pollution is intensified because the city lies in a basin surrounded by mountains, and frequent thermal inversions trap pollutants at ground level. Since 1982 the amount of contamination in the city's smog-choked air has more than tripled. Indeed, breathing the city's air is said to be roughly equivalent to smoking two packs of cigarettes a day.

The city's air and water pollution cause an estimated 100,000 premature deaths per year. Writer Carlos Fuentes has nicknamed this megacity "Makesicko City." Because of the air pollution, many foreign companies and governments give imported workers additional hazard pay for working in Mexico City.

The Mexican government is industrializing other parts of the country in an attempt to slow migration to Mexico City. Cars have been banned from a 50-block central zone. Taxis built before 1985 have been taken off the streets, and trucks can run only on liquefied petroleum gas (LPG). The government began phasing in unleaded gasoline in 1991, but it will be years before millions of older vehicles, which burn leaded gas, are eliminated. The government has also planted 25 million trees to help clean the air and has purchased some land to provide green space for the city.

If the city's population continues to grow as projected, these problems, already at crisis levels, will become even worse. If you were in charge of Mexico City, what would you do?

How Urbanized Is the United States? In 1800 only 5% of Americans lived in cities. Since then, three major

The Urban Poor

SPOTLIGHT

Squatter settlements and shantytowns in developing countries usually lack clean water supplies, sewers, electricity, and roads. Often the land on which they are built is not suitable for human habitation because of air and water pollution and hazardous wastes from nearby factories, or because the land is especially subject to landslides, flooding, earthquakes, or volcanic eruptions.

An estimated 100 million people are homeless and sleep on the streets or wherever they can and the number exceeds 1 billion if squatters and others with temporary or insecure living areas are included. Half of all urban children under age 15 in developing countries live in conditions of extreme poverty, and about one-fifth of them are street children with little or no family support. In Cairo, Egypt, children of kindergarten age can be found digging through clods of ox dung, looking for undigested kernels of corn to eat.

Many cities do not provide squatter settlements and shantytowns with adequate drinking water, sanitation facilities, electricity, food, health care, housing, schools, or jobs. Not only do these cities lack the needed money, but their officials fear that improving services will attract even more of the rural poor. Many city governments regularly bulldoze squatter shacks and send police to drive the illegal settlers out. The people then move back in or develop another shantytown somewhere else.

Shantytowns and squatter settlements are also found in some developed countries. For example, in Texas along the southern Rio Grande River, living conditions in shantytowns are as bad as in similar settlements in the cities of developing countries. In the United States, most inner cities have concentrations of poor people.

Despite joblessness, squalor, overcrowding, environmental hazards, and rampant disease, most squatter and slum residents are better off than the rural poor. With better access to family-planning programs, they tend to have fewer children, who have better access to schools. Most residents are adaptable and resilient and have hope for a better future. Many squatter settlements provide a sense of community and a vital safety net of neighbors, friends, and relatives for the poor. A few squatter communities have organized to improve their living conditions.

Critical Thinking

1. What three things do you believe should be done to reduce the numbers and improve living conditions for the urban poor? For the rural poor?

2. Should squatters around cities of developing countries be given title to land they don't own? Explain. What are the alternatives?

internal population shifts have taken place in the United States. As a result of the first shift, *migration to large central cities*, 76% of Americans live in 350 *metropolitan areas* (cities and towns with at least 50,000 people). Nearly half of the country's population lives in consolidated metropolitan areas containing 1 million or more residents (Figure 9-22).

In the second shift, more people began *migrating from large central cities to suburbs and smaller cities.* Since 1970 this type of migration has followed new jobs to such areas. Today about 41% of the country's urban dwellers live in central cities and 59% live in suburbs.

The third shift, which has been taking place for several decades, is *migration from the North and East to the South and West*. Since 1980, about 80% of the U.S. population increase has occurred in the South and West, particularly near the coasts. This shift is expected to continue.

In a recent fourth shift, people have been *migrating from urban areas back to rural areas.* According to the Census Bureau, between 1900 and 1997 rural counties had a net influx of about 2 million people, virtually all of it from domestic migration.

What Are the Major Urban Problems in the United States? Here is some *good news.* Since 1920, many of the worst urban environmental problems in the United States (and other developed countries) have been reduced significantly.

Most people have better working and housing conditions and air and water quality have improved. Better sanitation, public water supplies, and medical care have slashed death rates and the prevalence of sickness from malnutrition and transmittable diseases such as measles, diphtheria, typhoid fever, pneumonia, and tuberculosis. Furthermore, concentrating most of the population in urban areas has helped protect the country's biodiversity by reducing the destruction and degradation of wildlife habitat.

The biggest problems facing numerous cities in the United States (especially older ones) are **(1)** deteriorating services, **(2)** aging infrastructures (streets, schools, bridges, housing, sewers), **(3)** budget crunches from lost tax revenues and rising costs as businesses and more affluent people move out, and **(4)** rising poverty in many central city areas. As a result, violence, drug traffic and abuse, crime, decay, and blight have increased

Figure 9-21 The locations of Mexico, Brazil, Belize, and Costa Rica. Many other countries highlighted here are discussed in other chapters.

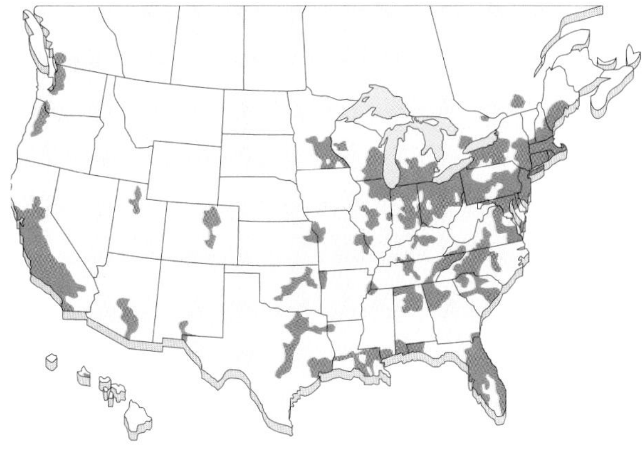

Figure 9-22 Major urban regions in the United States. Nearly half (48%) of Americans live in *consolidated metropolitan areas* with 1 million or more people. (Data from U.S. Census Bureau)

in parts of central cities. Unemployment rates in some inner-city areas are typically 50% or higher.

Another major problem in the United States is *urban sprawl*—the growth of low-density development on the edges of cities and towns. Some of the undesirable consequences of sprawl are

- Traffic congestion.
- Longer commutes that steal time from family and work.
- Worsening air and water pollution.
- Increased flooding.
- Loss of farmland, open fields, forests, and wetlands.
- Higher taxes to pay for expanded services.
- Destruction of downtown locally owned stores and restaurants by development of large suburban shopping malls.
- Erosion of a city's tax base as people and businesses move to the suburbs. This forces cities to raise taxes on remaining taxpayers, which causes more people and businesses to flee the city.
- Increased unemployment and poverty in decaying central cities.

9-7 URBAN RESOURCE AND ENVIRONMENTAL PROBLEMS

What Are the Environmental Pros and Cons of Urban Areas? Most of today's urban areas don't even come close to being self-sustaining; they survive only by importing food, water, energy, minerals, and other resources from farms, forests, mines, and watersheds. They also produce enormous quantities of wastes that can pollute air, water, and land within and outside their boundaries (Figure 9-23). Thus, today's cities affect not only the health of their inhabitants but also the environmental health of rural areas and the planet.

The 45% of the world's people currently living in urban areas occupy only about 2% of the planet's land area, but consume 75% of the earth's resources (including 60% of the planet's water). Supplying these urban dwellers with resources is a major reason that humans have disturbed 73% of the earth's land area, if uninhabitable ice and rock areas are excluded (Figure 1-5). According to Mathis Wackernagel and William Rees, 58 times the land area of London is needed to supply its residents with food and timber. Meeting the needs of all the world's people at the same rate of resource use as London would require at least three more Earths.

However, urbanization has some environmental benefits, including the following:

- Recycling is more economically feasible because of the large concentration of recyclable materials.
- Environmental pressures from population growth are reduced because birth rates in urban areas usually are one-fourth to one-third of those in rural areas.
- Per capita expenditures on environmental protection are higher in urban areas.
- Concentrating people in urban areas helps preserve biodiversity by reducing the stress on wildlife habitats. This effect is important but is not as great as it first appears because of the large areas of the earth's land area must be disturbed and degraded to provide

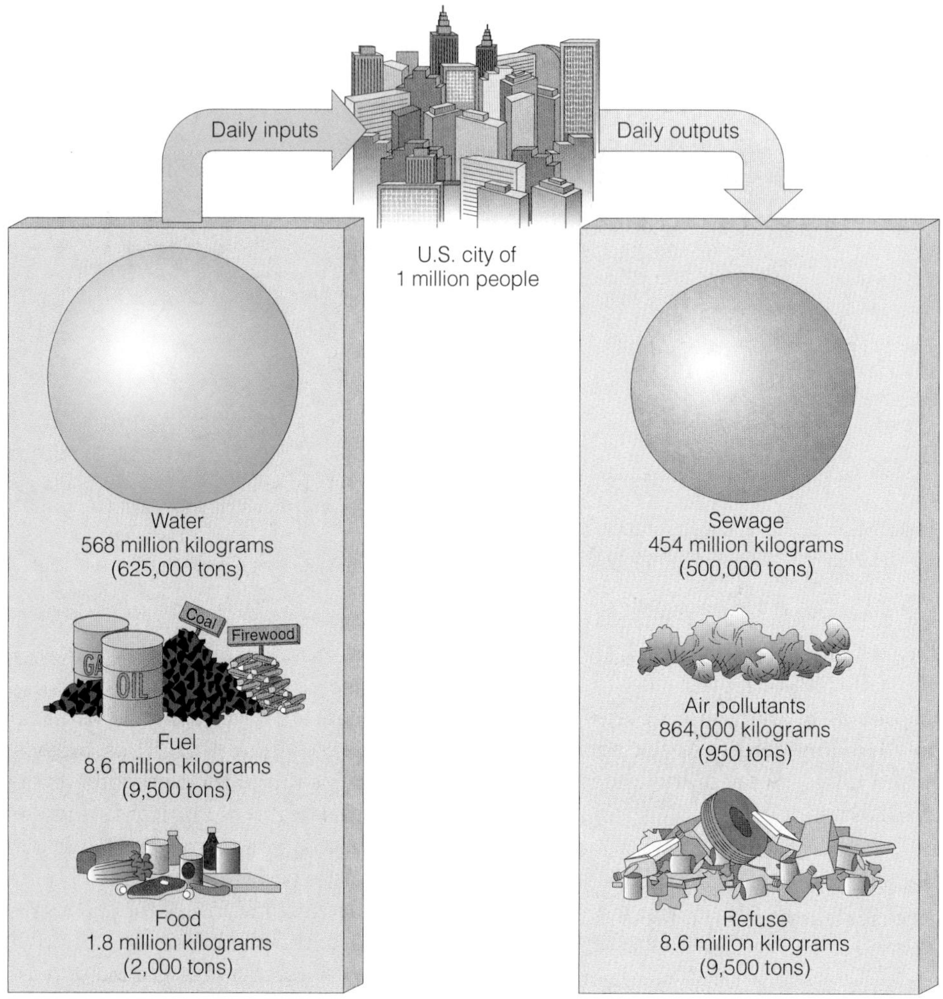

Figure 9-23 Typical daily inputs and outputs of matter and energy for a U.S. city of 1 million people.

urban dwellers with food, water, energy, minerals, and other resources (Figure 1-5). Furthermore, rural or wildlife areas downwind or downstream from urban areas receive much of the pollution produced in urban areas.

Some analysts call for developing a more sustainable relationship between cities and the living world. Doing this will require converting high-waste, unsustainable cities with a *linear metabolism* (based on an ever-increasing throughput of resources and output of wastes; Figure 3-14) to low-waste, more sustainable cities with a *circular metabolism* (based on efficient use of resources, reuse, recycling, pollution prevention, and waste reduction; Figure 3-15).

What Are the Major Resource and Environmental Problems of Urban Areas? Major urban areas face a number of resource and environmental problems. *Most cities have few trees, shrubs, or other plants* that absorb air pollutants, give off oxygen, help cool the air (as water transpires from their leaves), muffle noise, provide wildlife habitats, and give aesthetic pleasure. As one observer remarked, "Most cities are places where they cut down the trees and then name the streets after them." According to the American Forestry Association, one city tree provides over $57,000 worth of air conditioning, erosion and stormwater control, wildlife shelter, and air pollution control over a 50-year lifetime.

Population growth, urban sprawl, and poor land-use planning are the major causes of tree loss in urban areas. For example, between 1962 and 1998 the population of the Puget Sound watershed surrounding Seattle, Washington, doubled. A 1999 study comparing satellite surveys of this area between 1962 and 1998 found a loss of almost half of the area's tree cover. Peak storm runoff, which would have been absorbed by the trees and soil, increased by 29%, costing communities an estimated $2.4 billion. The lost trees would also have reduced air pollution and helped cool the city at no charge. Satellite surveys also show that Atlanta, Georgia, once famed for its tree cover, lost more than half of its heavy tree cover between 1974 and 1998.

Most cities produce little of their own food. However, people can grow their own food by **(1)** planting community gardens in unused lots, **(2)** using window-box and balcony planters, **(3)** creating gardens or greenhouses on the roofs of apartment buildings and on patios, and **(4)** raising fish in tanks and sewage lagoons. Rooftop gardens need little or no irrigation or fertilizer and are widely used in the Netherlands and Germany. According to the United Nations, about 800 million urban farmers provide about 15% of the world's food, and this proportion could be increased. The city of Seattle, Washington, provides community garden space for 4,500 urban gardeners in more than 1,600 garden plots throughout the city. Cities can also encourage farmers' markets, which lower food prices by allowing farmers to sell directly to customers. This also helps prevent nearby farmland from being swallowed up by urban sprawl.

Cities generally are warmer, rainier, foggier, and cloudier than suburbs and nearby rural areas. The enormous amounts of heat generated by cars, factories, furnaces, lights, air conditioners, heat-absorbing dark roofs and roads, and people in cities create an **urban heat island** (Figure 9-24) surrounded by cooler suburban and rural areas. The dome of heat also traps pollutants, especially tiny solid particles (suspended particulate matter), creating a **dust dome** above urban areas. If wind speeds increase, the dust dome elongates downwind to form a **dust plume**, which can spread the city's pollutants for hundreds of kilometers.

Cities can save money and counteract the heat island effect by **(1)** instituting tree-planting programs, **(2)** requiring lighter and more reflective paints and building materials, **(3)** adding light-colored sand to asphalt to increase reflectivity, **(4)** establishing high energy-efficiency standards for vehicles, buildings, and appliances, and **(5)** establishing gardens on the roofs of large buildings (as Chicago, Illinois, plans to do for most of the city's skyscrapers to help cool the city and reduce air pollution). In Malaysia, a 15-story building is covered with tropical plants to cool it naturally. This reduces energy use and the emission of heat by the building.

Many cities have water supply and flooding problems. As cities grow and their water demands increase, expensive reservoirs and canals must be built and deeper wells drilled. The transfer of water to urban areas deprives rural and wild areas of surface water and sometimes depletes groundwater faster than it is replenished. Covering land with buildings, asphalt, and concrete causes precipitation to run off quickly. This can overload sewers and storm drains and contribute to water pollution and flooding in cities and downstream areas. In many cities water is wasted and poorly

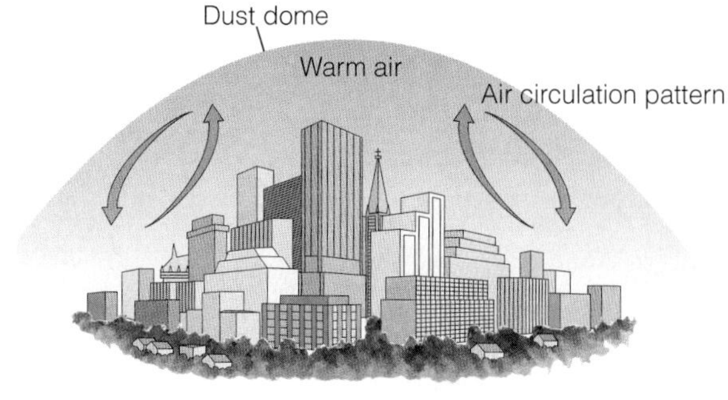

Figure 9-24 An urban heat island causes patterns of air circulation that create a dust dome over the city. Winds can elongate the dome toward downwind areas. A strong cold front can blow the dome away; this lowers local pollution levels but increases pollution in downwind areas.

managed. In urban areas of many developing countries, 50–70% of water is lost or wasted because of leaky pipes, lack of water meters, and artificially low prices for water. Cities in developing and developed countries could reduce demand on reservoirs and waste-treatment systems by **(1)** fixing leaks in water distribution systems, **(2)** installing water meters in all houses and buildings, **(3)** installing water-saving toilets and other fixtures, **(4)** separating household and industrial wastewater, **(5)** developing water reuse systems, and **(6)** educating schoolchildren and adults about how to save water.

About 58% of the water in Manila is lost because of leaks and poor management, compared with an 8% loss in Singapore, where pipes are better maintained. After Bogor, Indonesia, installed water meters and raised prices to encourage households to conserve water, water use dropped by one-third.

Many cities are built on floodplain areas subject to natural flooding. Floodplains are considered prime land for urbanization because they are flat, accessible, and near rivers. The poor often have little choice but to live on areas experiencing frequent flooding, landslides, and earthquakes. According to the International Red Cross, 40–50 of the world's fastest-growing cities are at risk of earthquakes.

Many of the world's largest cities are in coastal areas. If an enhanced greenhouse effect increases the average atmospheric temperature as projected (Sections 11-3 and 11-4), a rise in average sea level of even a meter could flood many of these cities, perhaps sometime during the 21st century.

Urban residents are generally subjected to much higher concentrations of pollutants than are rural residents. According to the World Health Organization, more than 1.1 billion people—about one-fifth of humanity—live in urban areas where air pollution levels exceed healthful levels. The World Bank estimates that almost two-thirds

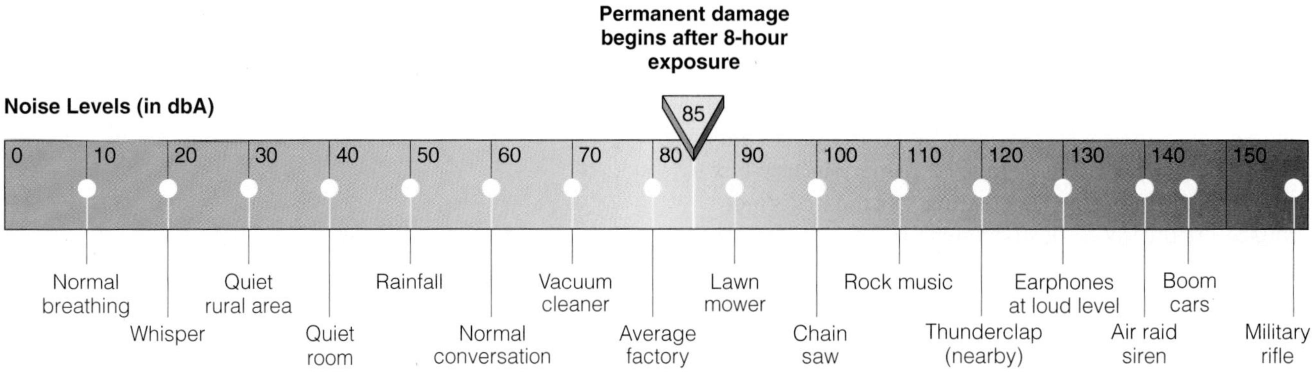

Figure 9-25 Noise levels (in decibel-A [dbA] sound pressure units) of some common sounds. You are being exposed to a sound level high enough to cause permanent hearing damage if you need to raise your voice to be heard above the racket, if a noise causes your ears to ring, or if nearby speech seems muffled. Prolonged exposure to lower noise levels and occasional loud sounds may not damage your hearing but can greatly increase internal stress. There are five major ways to control noise: **(1)** Modify noisy activities and devices to produce less noise, **(2)** shield noisy devices or processes, **(3)** shield workers or other receivers from the noise, **(4)** move noisy operations or things away from people, and **(5)** use antinoise, a new technology that cancels out one noise with another.

of urban residents in developing countries don't have adequate sanitation facilities. About 90% of all sewage in developing countries is discharged directly into rivers, lakes, and coastal waters without treatment of any kind.

Most urban dwellers also are subjected to excessive noise (Figure 9-25). Harmful effects from prolonged exposure to excessive noise include permanent hearing loss, high blood pressure (hypertension), muscle tension, migraine headaches, higher cholesterol levels, gastric ulcers, irritability, insomnia, and psychological disorders, including increased aggression.

Urban areas have beneficial and harmful effects on human health. Many aspects of urban life benefit human health, including better access to education, social services, and medical care. On the other hand, high-density city life increases **(1)** the spread of infectious diseases (especially if adequate drinking water and sewage systems are not available), **(2)** physical injuries (mostly from industrial and traffic accidents), and **(3)** health problems caused by increased exposure to pollution and noise (Figure 9-25). The WHO estimates that 600 million urban dwellers in developing countries live or work in life- and health-threatening environments. According to the World Bank, at least 220 million people in cities in developing countries don't have safe drinking water and an estimated 420 million people don't have access to even the simplest latrine.

Another problem is the *loss of rural cropland, fertile soil, forests, wetlands, and wildlife habitats as cities expand.* Once prime agricultural land or forestland is paved over or built on, it is lost for food production and habitat for most of its former wildlife. As a city expands, more energy is needed to transport food to its people, which causes more pollution. In coastal areas, urban growth destroys or pollutes ecologically valuable wetlands.

9-8 TRANSPORTATION AND URBAN DEVELOPMENT

How Do Transportation Systems Affect Urban Development? Transportation and land-use decisions are linked together and determine **(1)** where people live, **(2)** how far they must go to get to work and buy food and other necessities, **(3)** how much land is paved over, and **(4)** how much air pollution people are exposed to.

If a city cannot spread outward, it must grow vertically—upward and downward (below ground)—so that it occupies a small land area with a high population density. Most people living in such compact cities walk, ride bicycles, or use energy-efficient mass transit. Many European cities and urban areas such as Hong Kong and Tokyo are compact and tend to be more energy-efficient than the dispersed cities in the United States, Canada, and Australia, where ample land often is available for outward expansion.

A combination of cheap gasoline, plentiful land, and a network of highways produces dispersed, automobile-oriented cities with low population density, often called *urban sprawl.* Most people in such urban areas live in single-family houses with unshared walls that lose and gain heat rapidly unless they are well insulated and airtight. Urban sprawl also gobbles up unspoiled natural habitats, paves over fertile farmland, and promotes heavy dependence on the automobile.

Good and Bad News About Motor Vehicles

PRO/CON

The automobile provides convenience and unprecedented mobility. To many people, cars are also symbols of power, sex, excitement, social status, and success. Moreover, much of the world's economy is built on producing motor vehicles and supplying roads, services, and repairs for them. In the United States, $1 of every $4 spent and one of every six nonfarm jobs is connected to the automobile.

Despite their benefits, motor vehicles have many destructive effects on people and the environment. Since 1885, when Karl Benz built the first automobile, almost 18 million people have been killed in motor vehicle accidents. According to the World Health Organization, this global death toll increases by an estimated 885,000 people per year (an average of 2,400 deaths per day, equal to eight fatal jumbo jet crashes each day). In addition, each year about 15 million people are injured or permanently disabled in motor vehicle accidents.

In the United States alone, 16 million motor vehicle accidents (up from 7 million in 1970) kill about 42,000 people each year and injure about 4 million people, at least 300,000 of them severely. *More Americans have been killed by cars than by all the country's wars.*

Motor vehicles also are the largest source of air pollution (including 15% of global CO_2 emissions), laying a haze of smog over the world's cities. In the United States, motor vehicles produce at least 50% of the air pollution, even though emission standards are as strict as any in the world. Two-thirds of the oil used in the United States and one-third of the world's total oil consumption are devoted to transportation.

By making long commutes and shopping trips possible, automobiles and highways have helped create urban sprawl and reduced use of more efficient forms of transportation. Worldwide, at least a third of urban land is devoted to roads, parking lots, gasoline stations, and other automobile-related uses. In the United States, more land is devoted to cars than to housing. Half the land in an average U.S. city is used for cars, prompting urban expert Lewis Mumford to suggest that the U.S. national flower should be the concrete cloverleaf.

In 1907 the average speed of horse-drawn vehicles through the borough of Manhattan was 18.5 kilometers (11.5 miles) per hour; today cars and trucks creep along Manhattan streets at an average speed of 5 kilometers (3 miles) per hour. If current trends continue, U.S. motorists will spend an average of 2 years of their lifetimes in traffic jams, imprisoned in metal boxes that were supposed to provide speed, freedom, and mobility.

The U.S. economy loses at least $74 billion a year because of wasted fuel and work time lost in traffic delays. Even if the money is available, building more roads is not the answer because, as economist Robert Samuelson put it, "cars expand to fill available concrete."

According to a recent World Bank study, the high costs of dependence on the automobile hinder economic development. A 1998 survey of U.S. real estate investors and analysts also concluded that more compact cities that promote mass transit, bike paths, and other alternatives to the car offer better investment opportunities than sprawling urban areas.

Critical Thinking

If you own a car (or hope to own one), what conditions (if any) would encourage you to rely less on the automobile and travel to school or work by bicycle or motor scooter, on foot, by mass transit, or by a car pool or van pool?

Dispersed car-centered cities use up to 10 times more energy per person for transportation than more compact cities that allow better use of mass transit, bicycles, and walking. In addition, spread-out cities require more building materials, roads, power lines, and water and sewer lines.

Who Has Most of the World's Motor Vehicles? There are two main types of ground transportation: *individual* (such as cars, motor scooters, bicycles, and walking) and *mass* (mostly buses and rail systems). Only about 8% of the world's population own cars, and only 5–10% (1% in many developing countries) can afford to. Also, about one-third of the people in most countries are too young or too old to drive.

Between 1999 and 2020 the number of motor vehicles in the world is expected to increase from 501 million to almost 1 billion, with much of this increase in Asia, Latin America, and eastern Europe. This will greatly increase congestion, pollution, land disruption, and use of energy and matter resources.

Despite having only 4.6% of the world's people, the United States has 35% of the world's cars and trucks (one car for every two people). In the United States the car is used for 98% of all urban transportation and for 91% of travel to work (with 80% of Americans driving to work alone).

Americans drive 3 billion kilometers (2 billion miles) each year—as far as the rest of the world combined. No wonder British author J. B. Priestley remarked, "In America, the cars have become the people." Such "automania" is spreading to developing countries. Despite their many advantages, there are serious drawbacks to relying on motor vehicles as the major form of transportation (Pro/Con, above).

Are Motor Scooters the Answer? A growing number of people in developing countries who cannot afford cars are using motor scooters. Most burn a mixture of oil and kerosene in small, inefficient, and noisy engines that emit clouds of air pollutants.

One solution is to replace these smog machines with quiet electric scooters that produce very little pollution except at power plants supplying the electricity for battery recharging. Recently Taiwan and Indonesia introduced air pollution control legislation that may spur the use of electric scooters.

Are Riding Bicycles and Walking Alternatives to the Car? Globally, bicycles outsell cars by almost three to one because most people can afford a bicycle, whereas fewer than 10% can afford a car. Besides being inexpensive to buy and maintain, bicycles **(1)** produce no pollution, **(2)** are rarely a serious danger to pedestrians or cyclists, **(3)** take few resources to make, and **(4)** are the most energy-efficient form of transportation (including walking).

In 1996 a California firm developed a simple $15 bicycle that eliminates the chain by attaching the pedals to the front wheel. This inexpensive design could greatly increase bicycle use in developing countries.

In urban traffic, cars and bicycles move at about the same average speed. Using separate bike paths or lanes running along roads, cyclists can make most trips shorter than 8 kilometers (5 miles) faster than drivers.

In China, at least 50% of urban trips are made by bicycle and the government gives subsidies to those who bicycle to work. Many cities in western Europe and Japan have taken back the streets for pedestrians, cyclists, and children by banning cars or slowing motor traffic in residential and shopping areas. In the Netherlands (with more bicycle paths than any other country), bicycle travel makes up 30% of all urban trips, compared to 1% in the United States.

In Copenhagen, Denmark, 2,300 bicycles are available for public use at no charge. The system is financed by ads attached to the bicycle frames and wheels. The system is so popular that each bicycle is used on average once every 8 minutes.

For longer trips, secure bike parking spaces can be provided at mass transit stations, and buses and trains can be equipped to carry bicycles. Such *bike-and-ride* systems are widely used in Japan, Germany, the Netherlands, and Denmark. In Seattle, Washington, all city buses are equipped with bicycle racks.

Only about 2% of Americans bicycle to work (the "no-pollute commute"), even though half of all U.S. commutes are under 8 kilometers (5 miles). According to recent polls, 20% of Americans say they would bicycle to work if safe bike lanes were available and if their employers provided secure bike storage and showers at work. More than 300 U.S. cities now have part of their police force on bicycles.

However, as developing countries experience economic growth, some governments are discouraging bicycle use, viewing it as a sign of backwardness. In 1993, for example, Chinese officials banned bicycles from Shanghai's main street and prestigious shopping area so they wouldn't be seen by tourists and wealthy Chinese shoppers. Similarly, officials in Jakarta, Indonesia, have confiscated thousands of cycle rickshaws to project a more modern image for visitors.

One development that may increase bike use is the explosive growth in the use of electric bicycles. Battery-powered bikes can extend the range of bike trips, allow cyclists to travel over hilly terrain, and replace noisy and polluting mopeds and motor scooters.

What Are the Pros and Cons of Mass Transit? In the United States mass transit accounts for only 3% of all passenger travel, compared with 15% in Germany and 47% in Japan. In the United States 80% of federal gasoline tax revenues goes for highways and only 20% goes for mass transit. This encourages states and cities to invest in highways instead of mass transit.

Rapid-rail, suburban train, and trolley systems can transport large numbers of people at high speed. They are also more energy-efficient, produce less air pollution, cause fewer injuries and deaths, and take up less land than motor vehicles. However, they are efficient and cost-effective only where many people live along a narrow corridor and can easily reach properly spaced stations (as in Hong Kong).

In western Europe and Japan, a new generation of streamlined, comfortable, and low-polluting high-speed rail (HSR) lines is being used for medium-distance travel between cities. These *bullet* or supertrains travel on new or upgraded tracks at speeds up to 330 kilometers (200 miles) per hour.

A high-speed train network could replace airplanes, buses, and private cars for most medium-distance travel between major American cities (Figure 9-26). Critics say that such a system would cost too much in government subsidies. However, this ignores the fact that motor vehicle transportation receives subsidies of $300–600 billion per year in the United States. Some analysts believe that phasing out some of these automobile subsidies and applying them to developing a national rail system might be a far more efficient use of limited government funds. What do you think?

Bus systems are more flexible than rail systems because they can run throughout sprawling cities and be rerouted overnight if transportation patterns change. Bus systems also require less capital and have lower operating costs than heavy-rail systems. Curitiba, Brazil, has developed one of the world's best bus systems and has led the way in becoming a more sustainable city (Solutions, p. 230).

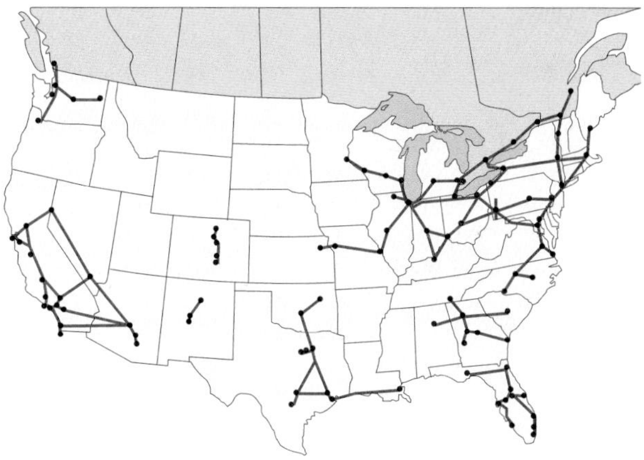

Figure 9-26 Potential routes for high-speed bullet trains in the United States and parts of Canada. Such a system would allow rapid, comfortable, safe, and affordable travel between major cities in a region. It would greatly reduce dependence on cars, buses, and airplanes for trips between these urban areas. (Data from High Speed Rail Association)

However, bus systems often cost more to operate than they bring in because they must offer low fares to attract riders. Furthermore, unless they operate in separate express lanes, buses often get caught in traffic. Sometimes buses are supplemented by car pools, van pools, and jitneys (small vans or minibuses traveling along regular and stop-on-demand routes).

Is It Feasible to Reduce Automobile Use? Environmentalists and a number of economists suggest that one way to reduce the harmful effects of automobile use is to make drivers pay directly for most of the true or external costs of automobile use.

According to a study by the World Resources Institute, federal, state, and local government automobile subsidies in the United States amount to $300–600 billion a year (depending on the costs included), an average subsidy of $1,600–3,200 per vehicle. Taxpayers (drivers and nondrivers alike) foot this bill mostly unknowingly.

If drivers had to pay these hidden costs directly in the form of a gasoline tax, the tax on each gallon would be about $5–6. Pollution emission fees, toll charges on roads (especially during peak traffic times), and higher parking fees are also ways to have motorists pay for the environmental and social costs of driving—a *user-pays* approach based on full-cost pricing (Section 2-3).

Making heavy trucks pay for the road damage they cause would shift more freight to energy-efficient rail systems. It's estimated that heavy trucks cause 95% of all damage to U.S. highways, with one heavy truck causing as much highway wear and tear as 9,600 cars. Current U.S. government subsidies for trucks not only drain public funds but also give trucking an unfair economic advantage over more efficient and less damaging rail freight.

Including the hidden costs in the market prices of cars, trucks, and gasoline up front makes economic and environmental sense. However, this approach faces massive political opposition from the public (mostly because they are unaware of the huge hidden costs they are already paying) and from the powerful transportation-related industries. In addition, tools for encouraging more use of mass transportation will not work unless fast, efficient, reliable, and affordable mass transit options are available. Furthermore, most people who can afford cars are virtually addicted to them and most people who can't afford them hope to buy one someday.

One way to reduce the incentive to drive is to use *car-sharing networks* to disconnect car use from car ownership. Such networks have been established in over 300 cities in Germany, Austria, Italy, Switzerland, and the Netherlands. Such networks give members easy access to car use without the hassles of renting or the costs of ownership. Each member pays for a card that opens lockers containing keys to cars parked around a city. Members call the network and are directed to the closest locker and car.

Analysts point out that developing countries now have a unique economic and environmental opportunity to avoid the "car trap" by not subsidizing motor vehicle transport. Instead, they can invest most of their money in developing modern, efficient, low-polluting public transit systems and providing citywide paths for bicycling and walking.

9-9 SOME SOLUTIONS TO URBAN PROBLEMS

What Is Conventional Land-Use Planning? Most urban areas and some rural areas use some form of **land-use planning** to determine the best present and future use of each parcel of land in the area.

Much land-use planning is based on the assumption that substantial future population growth and economic development should be encouraged, regardless of the environmental and other consequences. Typically this leads to uncontrolled or poorly controlled urban growth and sprawl.

A major reason for this often destructive process is that in the United States 90% of the revenue local governments use to provide schools, police and fire protection, public water and sewer systems, and other public services comes from *property taxes* levied on all buildings and property based on their economic value.

The new costs of urban growth and sprawl rarely are covered by increased revenue from property taxes. Local governments often try to raise more money by promoting economic growth because they usually cannot raise property taxes enough to meet expanding

Curitiba, Brazil

SOLUTIONS

One of the world's showcase ecocities is Curitiba, Brazil, with a population of 2.2 million. Trees are everywhere in Curitiba because city officials have given neighborhoods over 1.5 million trees to plant and care for. No tree in the city can be cut down without a permit; for every tree cut, two must be planted.

The air is clean because the city is not built around the car and officials have integrated land-use and transportation planning. There are 145 kilometers (90 miles) of bike paths, and more are being built. With the support of shopkeepers, many streets in the downtown shopping district have been converted to pedestrian zones in which no cars are allowed.

Curitiba probably has the world's best bus system. Each day a network of clean and efficient buses carries over 1.5 million passengers—75% of the city's commuters and shoppers—at a low cost (20–40¢ per ride with unlimited transfers) on express bus lanes. Only high-rise apartment buildings are allowed near major bus routes, and each building must devote the bottom two floors to stores, which reduces the need for residents to travel. Extra-large buses run on popular routes and the system's speed and convenience are enhanced by tube-shaped bus shelters where riders pay in advance.

As a result, Curitiba has one of Brazil's lowest outdoor air pollution rates. Between 1974 and 1995, car traffic dropped by 30%, while the city's population doubled.

The city recycles roughly 70% of its paper and 60% of its metal, glass, and plastic, which is sorted by households for collection. Recovered materials are sold mostly to the city's more than 340 major industries.

The city bought a plot of land downwind of downtown as an industrial park. The city put in streets, services, housing, and schools, ran a special worker's bus line to the area, and enacted stiff air and water pollution control laws. This development has attracted clean national and foreign corporations.

To equip the poor with basic technical training skills needed for jobs, the city set up old buses as roving technical training schools; courses cost the equivalent of two bus tokens. Each bus gives courses for 3 months in a particular area and then moves to another area.

The poor can swap sorted trash for food or bus tokens and receive free medical, dental, and child care; there are also 40 feeding centers for street children. As a result, the infant mortality rate has fallen by more than 60% since 1977.

All of these things have been accomplished despite Curitiba's enormous population growth from 300,000 in 1950 to 2.2 million in 1999, as rural poor people have flocked to the city.

Curitiba has slums, shantytowns, and most of the problems of other cities. But most of its citizens have a sense of vision, solidarity, pride, and hope, and they are committed to making their city even better. The entire program is the brainchild of architect and former college teacher Jaime Lerner, an energetic and charismatic leader who has served as the city's mayor three times since the 1970s.

Critical Thinking

1. Why do you think Curitiba has been so successful in its efforts to become an ecocity, compared with the generally unsustainable cities found in most developed and developing countries?

2. What is the city or area where you live doing to make itself more ecologically and economically sustainable?

needs. Typically the long-term result is a destructive positive feedback loop of more economic growth leading to more environmental degradation.

What Is Ecological Land-Use Planning? Environmentalists urge communities to use comprehensive, regional **ecological land-use planning**, in which additional variables are integrated into a model designed to anticipate a region's present and future needs and problems. It is a complex process that takes into account geological, ecological, economic, health, and social factors (Solutions, right).

Ecological land-use planning sounds good on paper, but it is not widely used for several reasons. First, local officials seeking reelection every few years usually focus on short-term rather than long-term problems and often can be influenced by economically powerful developers.

Second, officials (and often the majority of citizens) are often unwilling to pay for costly ecological land-use planning and implementation, even though a well-designed plan can prevent or ease many urban problems and save money in the long run. Finally, it's difficult to get municipalities within a region to cooperate in planning efforts. As a result, an ecologically sound development plan in one area may be undermined by unsound development in nearby areas.

In developing countries, most cities don't have the information or funding to carry out ecological land use planning. Urban maps often are 20–30 years old and lack descriptions of large areas of cities, especially those growing rapidly because of squatter settlements.

How Can Land Use Be Controlled? Once a plan is developed, governments control the uses of various

Ecological Land-Use Planning

SOLUTIONS

Six basic steps are involved in ecological land-use planning:

1. *Make an environmental and social inventory.* Experts survey geological factors (soil type, floodplains, water availability), ecological factors (wildlife habitats, stream quality, pollution), economic factors (housing, transportation, industrial development), and health and social factors (disease, crime rates, poverty). A top priority is to identify and protect areas that are critical for preserving water quality, supplying drinking water, and reducing erosion, and that are most likely to suffer from toxic wastes and natural hazards such as flooding.

2. *Identify and prioritize goals.* What are the primary goals? To encourage or discourage further economic development (at least some types) and population growth? To preserve prime cropland, forests, and wetlands from development? To reduce soil erosion?

3. *Develop individual and composite maps.* Data for each factor surveyed in the environmental and social inventory are plotted on separate transparent plastic maps. The transparencies are then superimposed or combined by computer into three composite maps, one each for geological, ecological, and socioeconomic factors.

4. *Develop a master composite.* The three composite maps are combined to form a master composite, which shows how the variables interact and indicates the suitability of various areas for different types of land use.

5. *Develop a master plan.* The master composite (or a series of alternative master composites) is evaluated by experts, public officials, and the public, and a final master plan is drawn up and approved.

6. *Implement the master plan.* The plan is set in motion, monitored, updated, and revised as needed by the appropriate government, legal, environmental, and social agencies.

Critical Thinking

Why do you think so few areas use ecological land-use planning? Would you be willing to pay slightly higher local taxes to support such planning where you live? Explain.

parcels of land by legal and economic methods. The most widely used approach is **zoning**, in which various parcels of land are designated for certain uses. Zoning can be used to control growth and protect areas from certain types of development. However, zoning can be influenced or modified by developers. Overly strict zoning can also discourage innovative approaches to solving urban problems.

Local governments can also control the rate of development by

- Limiting the number of building permits, sewer hookups, roads, and other services.
- Requiring an environmental impact analysis for proposed roads and development projects.
- Requiring developers to pay a fee that includes the full cost of additional services, such as roads and water and sewer lines, for development projects.
- Taxing land on the basis of its *actual* use as agricultural land or forestland, rather than on the basis of its higher and most profitable *potential* use as developed land. The lower tax bill would keep owners of farmland and forestland from having to sell such land to pay their tax bills.
- Promoting compact development by shifting property taxes from buildings to land, as is being done in parts of Denmark, Australia, New Zealand, South Africa, South Korea, and the state of Pennsylvania in the United States.
- Preserving open space.
- Reducing sprawl by offering tax breaks and funds for environmental cleanup to encourage redevelopment of vacant or abandoned parcels of land within cities (especially in inner cities and abandoned industrial areas).
- Establishing urban growth boundaries (lines surrounding a city beyond which new development is not allowed), as has been done in Portland, Oregon.
- Establishing green belts (rings of undeveloped land surrounding cities) like the one in Oslo, Norway.
- Giving tax breaks to landowners who agree in legally binding conservation easements to use their land only for specified purposes, such as agriculture, wildlife habitat, or nondestructive forms of recreation.
- Purchasing development rights that restrict land use and using land trusts to buy and protect ecologically valuable land. Such purchases can be made by private groups such as the Nature Conservancy or the Audubon Society, by local nonprofit, tax-exempt charitable organizations, and by government agencies.
- Changing zoning laws to encourage reintegration of homes, workplaces, and shopping areas and thus reduce urban sprawl, energy waste, and loss of community.
- Concentrating high-density development along bus lines or near other forms of mass transportation

GUEST ESSAY

The Ecological Design Arts

David W. Orr

David W. Orr is professor of environmental studies at Oberlin College and one of the nation's most respected environmental educators. He is author of numerous environmental articles and three books, including Ecological Literacy *and* Earth in Mind. *He is education editor for* Conservation Biology *and a member of the editorial advisory board of* Orion Nature Quarterly.

If *Homo sapiens sapiens* entered its industrial civilization in an intergalactic design competition, it would be tossed out in the qualifying round. It doesn't fit. It won't last. The scale is wrong. And even its defenders admit that it's not very pretty. The most glaring design failures of industrial and technologically driven societies are the loss of diversity of all kinds, impending climate change, pollution, and soil erosion.

Of course, industrial civilization wasn't designed at all; it was mostly imposed by single-minded individuals, armed with one doctrine of human progress or another, each requiring a homogenization of nature and society. These individuals for the most part had no knowledge of ecological design arts: the set of perceptual and analytic abilities, ecological wisdom, and practical wherewithal needed to make things that fit into a world of microbes, plants, animals, and energy laws.

Good ecological design incorporates understanding about how nature works into the ways we design, build, and live. It is required in our designs of farms, houses, neighborhoods, cities, transportation systems, technologies, economies, energy policies, and just about anything that directly or indirectly requires energy or materials or governs their use.

When human artifacts and systems are well designed, they are in harmony with the ecological patterns in which they are embedded. When poorly designed, they undermine those larger patterns, creating pollution, higher costs, and social stress. Bad design is not simply an engineering problem, although better engineering often would help. Its roots go deeper.

Good ecological design has certain common characteristics, including correct scale, simplicity, efficient use of resources, a close fit between means and ends, durability, redundancy, and resilience. These characteristics are often place-specific, or, in John Todd's words, "elegant solutions predicated on the uniqueness of place." Good design also solves more than one problem at a time and promotes human competence, efficient and frugal use of resources, and sound regional economies. Where good design becomes part of the social fabric at all levels, unanticipated positive side effects multiply. When people fail to design with ecological competence, unwanted side effects and disasters multiply.

The pollution, violence, social decay, and waste all around us indicate that we have designed things badly, for, I think, three primary reasons. First, as long as land and energy were cheap and the world was relatively empty, we did not need to master the discipline of good design. The result was sprawling cities, wasteful economies, waste dumped into the environment, bigger and less efficient automobiles and buildings, and conversion of entire forests into junk mail and Kleenex—all in the name of economic growth and convenience.

Second, design intelligence fails when greed, narrow self-interest, and individualism take over. Good design is a cooperative community process requiring people who share common values and goals that bring them together and hold them together. American cities, with their

to reduce car use and urban sprawl, as has been done in Portland, Oregon; Toronto, Canada; Stockholm, Sweden; and Curitiba, Brazil (Solutions, p. 230).

- Requiring comprehensive public review of all aspects of a new development, as is required by law in Vermont.
- Promoting integrated planning by electing a common government for a city and its surrounding suburbs and county, as is done in Portland, Oregon, and Curitiba, Brazil (Solutions, p. 230).
- Establishing national and state comprehensive land-use plans, as is done in Japan, much of western Europe, and Oregon in the United States.

How Can We Make Cities More Sustainable? An important goal in coming decades should be to make urban areas more self-reliant, sustainable, and enjoyable places to live through good ecological design (Guest Essay, above). At the same time, we need to reduce the demands cities make on the earth's resources by having urban systems mimic the circular metabolism of nature (Figure 3-15). The struggle to learn how to live more sustainably on the earth will largely be won or lost in the world's cities.

In a sustainable and ecologically healthy city, called an *ecocity* or *green city*, matter and energy resources are used efficiently. Emphasis is placed on pollution prevention, reuse, recycling, and efficient use of energy and matter resources. Per capita solid waste production is greatly reduced and at least 60% of what is produced is reused, recycled, or composted. An ecocity takes advantage of locally available energy sources and requires that all buildings, vehicles, and appliances meet high energy-efficiency standards.

Trees and plants adapted to the local climate and soils are planted throughout an ecocity to provide shade extremes of poverty and opulence, are products of people who believe they have little in common with one another. Greed, suspicion, and fear undermine good community and good design alike.

Third, poor design results from poorly equipped minds. Good design can be done only by people who understand harmony, patterns, and systems. Industrial cleverness, on the contrary, is evident mostly in the minutiae of things, not in their totality or in their overall harmony. Good design requires a breadth of view that causes people to ask how human artifacts and purposes fit within a particular culture and place. It also requires ecological intelligence, by which I mean an intimate familiarity with how nature works.

An example of good ecological design is found in John Todd's living machines, which are carefully orchestrated ensembles of plants, aquatic animals, technology, solar energy, and high-tech materials to purify wastewater, but without the expense, energy use, and chemical hazards of conventional sewage treatment technology. Todd's living machines resemble greenhouses filled with plants and aquatic animals [Figure 12-29]. Wastewater enters at one end and purified water leaves at the other. In between, an ensemble of organisms driven by sunlight use and remove nutrients, break down toxins, and incorporate heavy metals in plant tissues.

Ecological design standards also apply to the making of public policy. For example, the Clean Air Act of 1970 required car manufacturers to install catalytic converters to remove air pollutants. Two decades later, emissions per vehicle are down substantially, but because more cars are on the road, air quality is about the same—an example of inadequate ecological design. A sounder design approach to transportation would create better access among housing, schools, jobs, stores, and recreation areas, build better public transit systems, restore and improve railroads, and create bike trails and walkways.

An education in the ecological design arts would foster the ability to see things in their ecological context, integrating firsthand experience and practical competence with theoretical knowledge about how nature works. It would aim to equip people to build households, institutions, farms, communities, corporations, and economies that **(1)** do not emit carbon dioxide or other heat-trapping gases, **(2)** operate on renewable energy, **(3)** preserve biological diversity, **(4)** recycle material and organic wastes, and **(5)** promote sustainable local and regional economies.

The outline of a curriculum in ecological design arts can be found in recent work in ecological restoration, ecological engineering, solar design, landscape architecture, sustainable agriculture, sustainable forestry, energy efficiency, ecological economics, and least-cost, end-use analysis. A program in ecological design would weave these and similar elements together around actual design objectives that aim to make students smarter about systems and about how specific things and processes fit in their ecological context. With such an education we can develop the habits of mind, analytical skills, and practical competence needed to help sustain the earth for us and other species.

Critical Thinking

1. Does your school offer courses or a curriculum in ecological design? If not, suggest some reasons that it does not.

2. Use the principles of good ecological design to evaluate how well your campus is designed. Suggest ways to improve its design.

and beauty, to reduce pollution and noise, and to supply wildlife habitats. Abandoned inner-city and industrial sites are recycled to productive uses. Abandoned lots and polluted creeks are cleaned up and restored. Nearby forests, grasslands, wetlands, and farms are preserved instead of being devoured by urban sprawl. Much of an ecocity's food comes from nearby organic farms, solar greenhouses, community gardens, and small gardens on rooftops, in yards, and in window boxes.

An ecocity is people oriented, not car oriented. Its residents are able to walk or bike to most places, including work, and to take low-polluting mass transit. In such cities, land use planning is closely linked to transportation planning and policies. Political support for such integrated planning is growing as a backlash against sprawl is spreading across much of the United States. Efforts to institute "smart growth" planning are under way in states such as New Jersey, California, and Maryland. However, analysts warn that using "smart growth" policies to make cities more sustainable is merely a short-term band-aid solution unless the population growth that fuels urban growth is sharply reduced.

People in sustainable cities take seriously the advice Lewis Mumford gave over three decades ago: "Forget the damned motor car and build cities for lovers and friends." Examples of cities that have attempted to become more ecologically sustainable include Curitiba, Brazil (Solutions, p. 230), and Davis, California.

Case Study: Davis, California The ecocity is not a futuristic dream. The citizens and elected officials of Davis, California, a city of about 54,000 people northeast of San Francisco, committed themselves in the early 1970s to making their city ecologically sustainable.

City building codes encourage the use of solar energy for water and space heating. All new homes must meet high standards of energy efficiency, and when an existing home changes hands the buyer must bring it up to the energy conservation standards for new homes. Houses are heated by solar energy in Davis's Village Homes development, America's first solar neighborhood. They face into a common open space reserved for people and bicycles; cars are restricted to streets, which are located only on the periphery of the development.

Since 1975 Davis has cut its use of energy for heating and cooling in half. It has a solar power plant, and some of the electricity it produces is sold to the regional utility company. Eventually the city plans to generate all of its own electricity using renewable energy.

The city discourages the use of automobiles and encourages the use of bicycles by closing some streets to automobiles, building bike lanes on major streets, and building bicycle paths. Any new housing tract must have a separate bike lane, and some city employees are given bikes. As a result, bicycles account for 40% of all in-city transportation, and less land is needed for parking spaces. This heavy dependence on the bicycle is aided by the city's warm climate and flat terrain.

Davis limits the type and rate of its growth, and it maintains a mix of homes for people with low, medium, and high incomes. Development of the fertile farmland surrounding the city for residential or commercial use is restricted. The city also limits the size of shopping centers to encourage smaller neighborhood shopping centers, each easily reached by foot or bicycle.

The test of the quality of life in an advanced economic society is now largely in the quality of urban life. Romance may still belong to the countryside—but the present reality of life abides in the city.

JOHN KENNETH GALBRAITH

REVIEW QUESTIONS

1. Define the boldfaced terms found in this chapter.

2. How did Thailand reduce its birth rate?

3. How is population change calculated? What is *zero population growth*? What are the *crude birth rate* and the *crude death rate*? About how many people are added to the world's population each year? Each day? How is the annual rate of population change calculated? What three countries have the world's largest populations?

4. Distinguish between *replacement-level fertility* and *total fertility rate*. Explain why replacement-level fertility is higher than 2. Explain why reaching replacement-level fertility does not mean an immediate halt in population growth. List 10 factors that affect birth rates and fertility rates.

5. List five reasons that the world's death rate has declined over the past 100 years. Distinguish between *life expectancy* and *infant mortality rate*. Why is the infant mortality the best measure of a society's quality of life? About how many infants die each year during their first year of life? List three factors that keep the U.S. infant mortality higher than it could be.

6. What is the *age structure* of a population? Explain why the current age structure of the world's population means that the world's population will keep growing for at least another 50 years even if replacement-level rate of 2.1 is somehow reached globally tomorrow. Draw the general shape of an age structure diagram for a country undergoing **(a)** rapid population growth, **(b)** moderate population growth, and **(b)** slow or zero population growth. What percentage of population is under age 15 in **(a)** the world, **(b)** developed countries, and **(c)** developing countries?

7. What are the benefits and potentially harmful effects of population decline for a country such as Japan?

8. Give the major arguments for and against reducing births globally. What is your position on this issue?

9. What is the *demographic transition* and what are its four phases? What factors might keep many developing countries from making the demographic transition?

10. What is *family planning*, and what are the advantages of using this approach for reducing the birth rate? What economic rewards and penalties have some countries used to reduce birth rates? What four conditions enhance the success of using such economic rewards and penalties? What three factors lead women to have fewer and healthier children?

11. Briefly describe and compare the success China and India have had in reducing their birth rates. What are the major components of China's population control program?

12. What are the eight goals of the current United Nations plan to stabilize the world's population at 7.8 billion by 2054, instead of the projected 9 billion?

13. Distinguish between **(a)** an *urban area* and a *rural area*, and **(b)** *degree of urbanization* and *urban growth*. What are five major trends in urbanization and urban growth? Summarize the major urban problems of Mexico City.

14. What four major shifts in the population of the United States have taken place since 1800? What percentage of the U.S. population lives in urban areas? What are six major problems facing numerous cities in the United States? What is *urban sprawl* and what are nine of its undesirable consequences?

15. What are the environmental pros and cons of urban areas? Distinguish between unsustainable cities with a linear metabolism and more sustainable cities with a circular metabolism.

16. Describe the major resource and environmental problems of urban areas relating to **(a)** trees and plants, **(b)** food production, **(c)** microclimate, **(d)** water supply and natural hazard problems (such as flooding, landslides, and earthquakes), **(e)** noise, **(f)** pollution, **(g)** beneficial and harmful effects on human health, and **(h)** loss of rural cropland, fertile soil, forests, wetlands, and wildlife habitats as cities expand. What are four ways cities can grow more of their food? What are *urban heat islands, dust domes,* and *dust plumes*? What are five ways to save money and counteract the heat-island effect? List six ways cities can promote water conservation.

17. How do transportation systems affect urban development? What are two basic types of cities in terms of density and transportation systems? What are the pros and cons of **(a)** automobiles, **(b)** motor scooters, **(c)** bicycles, and **(d)** various forms of mass transit such as rail systems and buses? How could dependence on the automobile be reduced by **(a)** using the economic system to provide rewards and penalties and **(b)** car sharing networks?

18. Distinguish between *conventional land-use planning* and *ecological land-use planning*. Why isn't ecological land-use planning used more widely? Describe the cycle that leads to uncontrolled urban growth and sprawl. What is *zoning* and what are its pros and cons? What are 15 ways local governments can control the rate of urban development?

19. Describe the major characteristics of a sustainable and ecologically healthy *ecocity* or *green city*. Briefly describe the progress that Curitiba, Brazil, and Davis, California, have made in becoming ecocities.

CRITICAL THINKING

1. Why are falling birth rates not necessarily a reliable indicator of future population growth trends?

2. Why is it rational for a poor couple in India to have five or six children? What changes might induce such a couple to consider their behavior irrational?

3. Do you believe that the population of your own country is too high? Explain. What about the population of the area in which you live?

4. Evaluate the claims made by those opposing a reduction in births and those promoting a reduction in births, as discussed on pp. 213–215. Which position do you support, and why?

5. Explain why you agree or disagree with each of the following proposals:

- **a.** The number of legal immigrants and refugees allowed into the United States each year should be reduced sharply.
- **b.** Illegal immigration into the United States should be decreased sharply. If you agree, how would you go about achieving this?
- **c.** Families in the United States should be given financial incentives to have more children to prevent population decline.
- **d.** The United States should adopt an official policy to stabilize its population and reduce unnecessary resource waste and consumption as rapidly as possible.
- **e.** Everyone should have the right to have as many children as they want.

6. Some people have proposed that the earth could solve its population problem by shipping people off to space colonies, each containing about 10,000 people. Assuming that we could build such large-scale, self-sustaining space stations (which we can't do with existing technology; Spotlight, p. 5), how many people would have to be shipped off each day to provide living spaces for the 80 million people being added to the earth's population each year? Current space shuttles can handle about 6 to 8 passengers. Assuming that this capacity can be increased to 100 passengers per shuttle, how many shuttles would have to be launched per day to take care of the 80 million people being added each year? According to your calculations, determine whether this proposal is a logical solution to the earth's population problem.

7. Why has China been more successful than India in reducing its rate of population growth? Do you agree with China's current population control policies? Explain. What alternatives, if any, would you suggest?

8. Congratulations—you have just been put in charge of the world. List the five most important features of your population policy.

9. Do you believe that the United States (or the country where you live) should develop a comprehensive and integrated mass transit system over the next 20 years, including building an efficient rail network for travel within and among its major cities? How would you pay for such a system?

10. Congratulations—you have just been put in charge of the world. List the five most important features of your urban policy.

***11.** Assume that your entire class (or manageable groups of your class) is charged with coming up with a plan for halving the world's rate of population growth within the next 20 years. Develop a detailed plan that would achieve this goal, including any differences between policies in developing countries and developed countries. Justify each part of your plan. Predict what problems you might face in implementing the plan and devise strategies for dealing with these problems.

***12.** As a class project, evaluate land use and land-use planning by your school. Draw up an improved plan based on ecological principles, and submit the plan to school officials.

***13.** Make a concept map of this chapter's major ideas, using the section heads and subheads and the key terms (in boldface). Look at the inside back cover and in Appendix 5 for information about concept maps.

FURTHER INFORMATION

For more information see Internet Sources and Further Readings at the back of the book. Online resources for this chapter can be found at **www.brookscole.com/biology**

10 AIR AND AIR POLLUTION

When Is a Lichen Like a Canary?

Nineteenth-century coal miners took canaries with them into the mines—not for their songs, but for the moment when they stopped singing. Then the miners knew it was time to get out of the mine because the air contained methane, which could ignite and explode.

Today we use sophisticated equipment to monitor air quality, but living things such as lichens (Figure 10-1) still can warn us of bad air. A lichen consists of a fungus and an alga living together, usually in a mutually beneficial (mutualistic) partnership.

With more than 20,000 known species, lichens can live almost anywhere—on rocks, trees, bare soil, buildings, gravestones, and even sun-bleached bones. Some lichens survive for more than 4,000 years.

These hearty pioneer species are good air pollution detectors because they are always absorbing air as a source of nourishment. Certain lichen species are sensitive to specific air-polluting chemicals. Old man's beard (*Usnea trichodea*) (Figure 10-1, right) and yellow *Evernia* lichens, for example, sicken or die in the presence of too much sulfur dioxide.

Because lichens are widespread, long-lived, and reliably anchored in place, they can also help track pollution to its source. The scientist who discovered sulfur dioxide pollution on Isle Royale in Lake Superior, where no car or smokestack had ever intruded, used *Evernia* lichens to point the finger northward to coal-burning facilities at Thunder Bay, Canada.

Radioactive particles spewed into the atmosphere by the Chernobyl nuclear power-plant disaster (p. 496) fell to the ground over much of northern Scandinavia, spread, and were absorbed by lichens that carpet much of Lapland. The area's Saami people depend on reindeer meat for food, and the reindeer feed on lichens. After Chernobyl more than 70,000 reindeer had to be killed and the meat discarded because it was too radioactive to eat. However, scientists helped the Saami identify which of the remaining reindeer to move by analyzing lichens (which absorbed some of the radioactive fallout) to pinpoint the most contaminated areas.

Last but not least, lichens can replace electronic monitoring stations that would cost more than $100,000 each. This is not so much a triumph of nature over technology as a partnership between the two, for technicians use highly sophisticated methods to analyze lichens for pollution and measure their rates of photosynthesis.

Air pollution can harm human health, trees and other plants, animals, and lakes and has thinned the protective ozone layer above the earth (Section 11-6). It also damages buildings, monuments, statues, rubber, and other materials and reduces visibility.

Because we all must breathe air from a global atmospheric commons, air pollution anywhere is a potential threat elsewhere, and in some cases everywhere. Lichens can alert us to the danger, but as with all forms of pollution, the best solution is prevention.

Figure 10-1 Red and yellow crustose lichens growing on slate rock in the foothills of the Sierra Nevada near Merced, California (left), and *Usnea trichodea* lichen growing on a branch of a larch tree in Gifford Pinchot National Park, Washington (right). The vulnerability of various lichen species to specific air pollutants can help researchers detect levels of these pollutants and track down their sources. (Left, Kenneth W. Fink/Ardea London; right, Milton Rand/Tom Stack & Associates)

I thought I saw a blue jay this morning. But the smog was so bad that it turned out to be a cardinal holding its breath.

MICHAEL J. COHEN

This chapter addresses the following questions:

- What layers are found in the atmosphere?
- What are air pollutants, and where do they come from?
- What is smog?
- What is acid deposition?
- What are the harmful effects of air pollutants?
- How can we prevent and control air pollution?

10-1 THE ATMOSPHERE

What Is the Troposphere? Weather Breeder We live at the bottom of a "sea" of air called the **atmosphere**. This thin envelope of life-sustaining gases surrounding the earth is divided into several spherical layers (Figure 10-2). Each layer is characterized by abrupt changes in temperature, the result of differences in the absorption of incoming solar energy.

About 75% of the mass of the earth's air is found in the atmosphere's innermost layer, the **troposphere**, which extends only about 17 kilometers (11 miles) above sea level at the equator and about 8 kilometers (5 miles) over the poles. If the earth were an apple, this lower layer containing the air we breathe would be no thicker than the apple's skin. This thin and turbulent layer of rising and falling air currents and winds is the planet's weather breeder.

Throughout the earth's long history the composition of the troposphere has varied considerably (Section 5-1). Today, about 99% of the volume of clean, dry air in the troposphere consists of two gases: nitrogen (78%) and oxygen (21%). The remainder has slightly less than 1% argon (Ar), 0.036% carbon dioxide (CO_2), and trace amounts of several other gases. Air in the troposphere also holds water vapor in amounts varying from 0.01% by volume at the frigid poles to 5% in the humid tropics.

The average pressure exerted by the gases in the atmosphere decreases with altitude because the average density (mass of gases per unit volume) decreases with altitude. Temperature also declines with altitude in the troposphere, but abruptly begins to rise at the top of this zone, called the *tropopause*. This temperature change limits mixing between the troposphere and upper layers of the atmosphere.

What Is the Stratosphere? Earth's Global Sunscreen The atmosphere's second layer is the **stratosphere**, which extends from about 17 to 48 kilometers (11 to 30 miles) above the earth's surface (Figure 10-2). Unlike air in the troposphere, air in the stratosphere is calm, with little vertical mixing. Although the stratosphere contains less matter than the troposphere, its composition is similar, with two notable exceptions: **(1)** Its volume of water vapor is about 1/1,000 as much, and **(2)** its volume of ozone (O_3) is about 1,000 times greater.

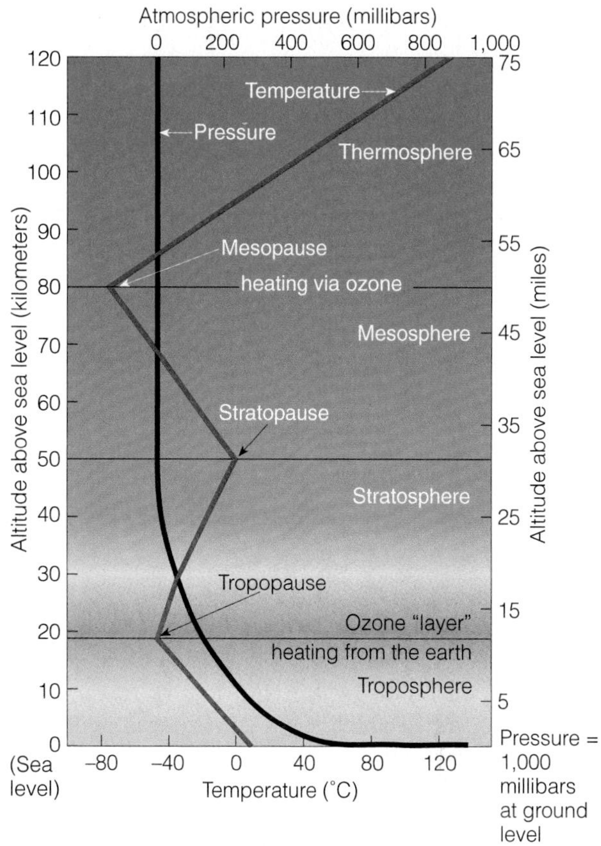

Figure 10-2 The earth's current atmosphere consists of several layers. Most UV radiation from the sun is absorbed by ozone (O_3) in the stratosphere, which is found primarily in the *ozone layer* between 17 and 26 kilometers (10 to 16 miles) above sea level.

Stratospheric ozone is produced when some of the oxygen molecules there interact with lightning and ultraviolet (UV) radiation emitted by the sun. This "global sunscreen" of ozone in the stratosphere keeps about 95% of the sun's harmful UV radiation from reaching the earth's surface (Section 11-6). This UV filter **(1)** allows humans and other forms of life to exist on land, **(2)** helps protect humans from sunburn, skin and eye cancer, cataracts, and damage to the immune system, and **(3)** prevents much of the oxygen in the troposphere from being converted to ozone, a harmful air pollutant. The trace amounts of ozone that do form in the troposphere are a component of urban smog and they damage plants, the respiratory systems of humans and other animals, and materials such as rubber.

Table 10-1 Characteristics of Two Important Global Processes		
	Greenhouse Effect	**Ozone Shield**
Where in the atmosphere does this occur?	Troposphere	Stratosphere
What process occurs?	Traps heat near the earth's surface	Filters ultraviolet (UV) radiation from the sun
What natural gases are involved?	Water (H_2O), carbon dioxide (CO_2), methane (CH_4)	Oxygen (O_2), ozone (O_3)
What are important human inputs?	Carbon dioxide (CO_2), methane (CH_4), chlorofluorocarbons (CFCs), nitrous oxide (N_2O)	Chlorofluorocarbons (CFCs), halons, carbon tetrachloride, methyl choloroform (stable), halogen-containing gases
What problem results?	Global warming	Ozone depletion

Thus, our good health, and that of many other species, depends on having enough ozone in the stratosphere and little ozone in the troposphere. There is considerable evidence that some human activities are increasing the amount of harmful ozone in the tropospheric air we must breathe and decreasing the amount of beneficial ozone in the stratosphere (Table 10-1 and Section 11-6).

What Are Two Important Global Processes Taking Place in the Atmosphere? Two natural processes that play crucial roles in the earth's climate and biodiversity are the *greenhouse effect* (Figure 6-8), which helps heat the troposphere, and the *ozone shield* in the stratosphere (Figure 10-2), which filters out most of the sun's UV radiation (Table 10-1). Both of these natural processes are necessary for life as we know it. However, there is considerable evidence that chemicals we add to the atmosphere can enhance the natural greenhouse effect and lead to *global warming* and reduce the concentration of ozone in the stratosphere, an effect called *ozone depletion* (Table 10-1).

10-2 URBAN OUTDOOR AIR POLLUTION FROM SMOG

What Are the Major Types and Sources of Air Pollution? Air pollution is the presence of one or more chemicals in the atmosphere in sufficient quantities and duration to cause harm to humans, other forms of life, and materials. As clean air in the troposphere moves across the earth's surface, it collects the products of natural events (dust storms and volcanic eruptions) and human activities (emissions from cars and smokestacks). These potential pollutants, called **primary pollutants**, are mixed vertically and horizontally and are dispersed and diluted by the churning air in the troposphere. While in the troposphere, some of these primary pollutants may react with one another or with the basic components of air to form new pollutants, called **secondary pollutants** (Figure 10-3).

Long-lived primary and secondary pollutants can travel great distances before they return to the earth's surface as solid particles, droplets, or chemicals dissolved in precipitation. Table 10-2 lists the major classes of pollutants commonly found in outdoor air.

Indoor pollutants come from infiltration of polluted outside air and from various chemicals used or produced inside buildings. Risk analysis experts rate indoor and outdoor air pollution as high-risk human health problems (Figure 8-10, left).

Most outdoor pollutants in urban areas enter the atmosphere from the burning of fossil fuels in power plants and factories (*stationary sources*) and motor vehicles (*mobile sources*). In car-clogged cities such as Los Angeles, California; São Paulo, Brazil; Bangkok, Thailand (Figure 9-1); Rome, Italy; and Mexico City, Mexico (Case Study, p. 221), motor vehicles are responsible for 80–88% of the air pollution. According to the World Health Organization (WHO), more than 1.1 billion people—one of every five—live in urban areas where the air is unhealthy to breathe.

Because they contain large concentrations of cars and factories, cities normally have higher air-pollution

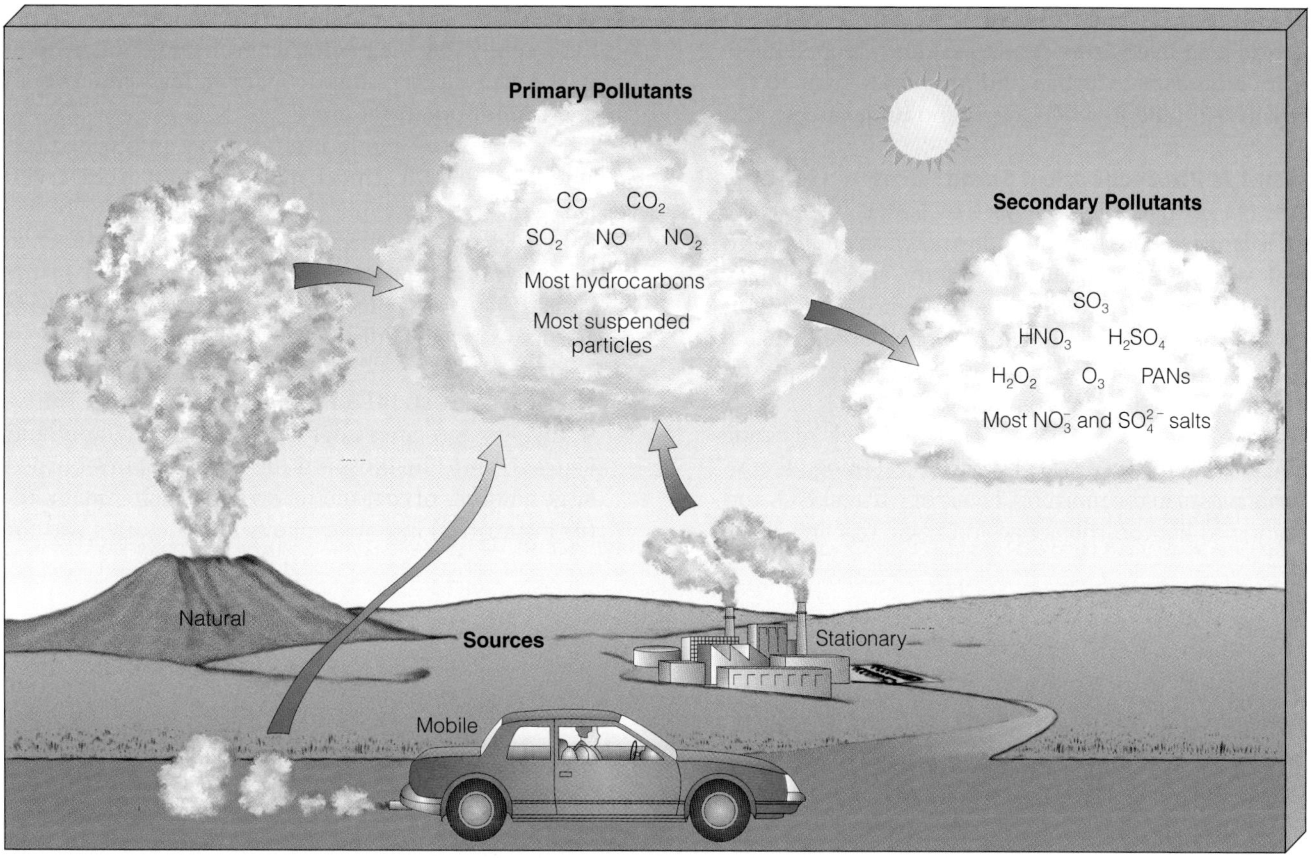

Figure 10-3 Sources and types of air pollutants. Human inputs of air pollutants may come from mobile sources (cars) and stationary sources (industrial and power plants). Some primary air pollutants may react with one another or with other chemicals in the air to form secondary air pollutants.

Table 10-2 Major Classes of Air Pollutants

Class	Examples
Carbon oxides	Carbon monoxide (CO) and carbon dioxide (CO_2)
Sulfur oxides	Sulfur dioxide (SO_2) and sulfur trioxide (SO_3)
Nitrogen oxides	Nitric oxide (NO), nitrogen dioxide (NO_2), nitrous oxide (N_2O) (NO and NO_2 are often lumped together and labeled NO_x)
Volatile organic compounds (VOCs)	Methane (CH_4), propane (C_3H_8), chlorofluorocarbons (CFCs)
Suspended particulate matter (SPM)	Solid particles (dust, soot, asbestos, lead, nitrate, and sulfate salts), liquid droplets (sulfuric acid, PCBs, dioxins, and pesticides)
Photochemical oxidants	Ozone (O_3), peroxyacyl nitrates (PANs), hydrogen peroxide (H_2O_2), aldehydes
Radioactive substances	Radon-222, iodine-131, strontium-90, plutonium-239 (Table 3-1)
Hazardous air pollutants (HAPs), which cause health effects such as cancer, birth defects, and nervous system problems	Carbon tetrachloride (CCl_4), methylchloride (CH_3Cl), chloroform ($CHCl_3$), benzene (C_6H_6), ethylene dibromide ($C_2H_2Br_2$), formaldehyde (CH_2O_2)

levels than rural areas. However, prevailing winds can spread long-lived primary and secondary air pollutants from emissions in urban and industrial areas to the countryside and to other downwind urban areas.

What Is Photochemical Smog? Brown-Air Smog Any chemical reaction activated by light is called a *photochemical reaction*. Air pollution known as **photochemical smog** is a mixture of primary and secondary pollutants formed under the influence of sunlight (Figure 10-4). The resulting mixture of more than 100 chemicals is dominated by ozone, a highly reactive gas that harms most living organisms.

The hotter the day, the higher the levels of ozone and other components of photochemical smog. As traffic increases in the morning, levels of NO and NO_2 and unburned hydrocarbons rise and begin reacting in the presence of sunlight to produce photochemical smog. On a sunny day the photochemical smog builds up to peak levels by early afternoon, irritating people's eyes and respiratory tracts.

Virtually all modern cities have photochemical smog. However, it is much more common in cities with sunny, warm, dry climates and lots of motor vehicles. Examples of such cities are Los Angeles, California; Denver, Colorado; and Salt Lake City, Utah in the United States, as well as Sydney, Australia; Mexico City, Mexico; and São Paulo and Buenos Aires in Brazil.

What Is Industrial Smog? Gray-Air Smog Thirty-five years ago cities such as London, England, and Chicago and Pittsburgh in the United States burned large amounts of coal and heavy oil (which contain sulfur impurities) in power plants and factories and for

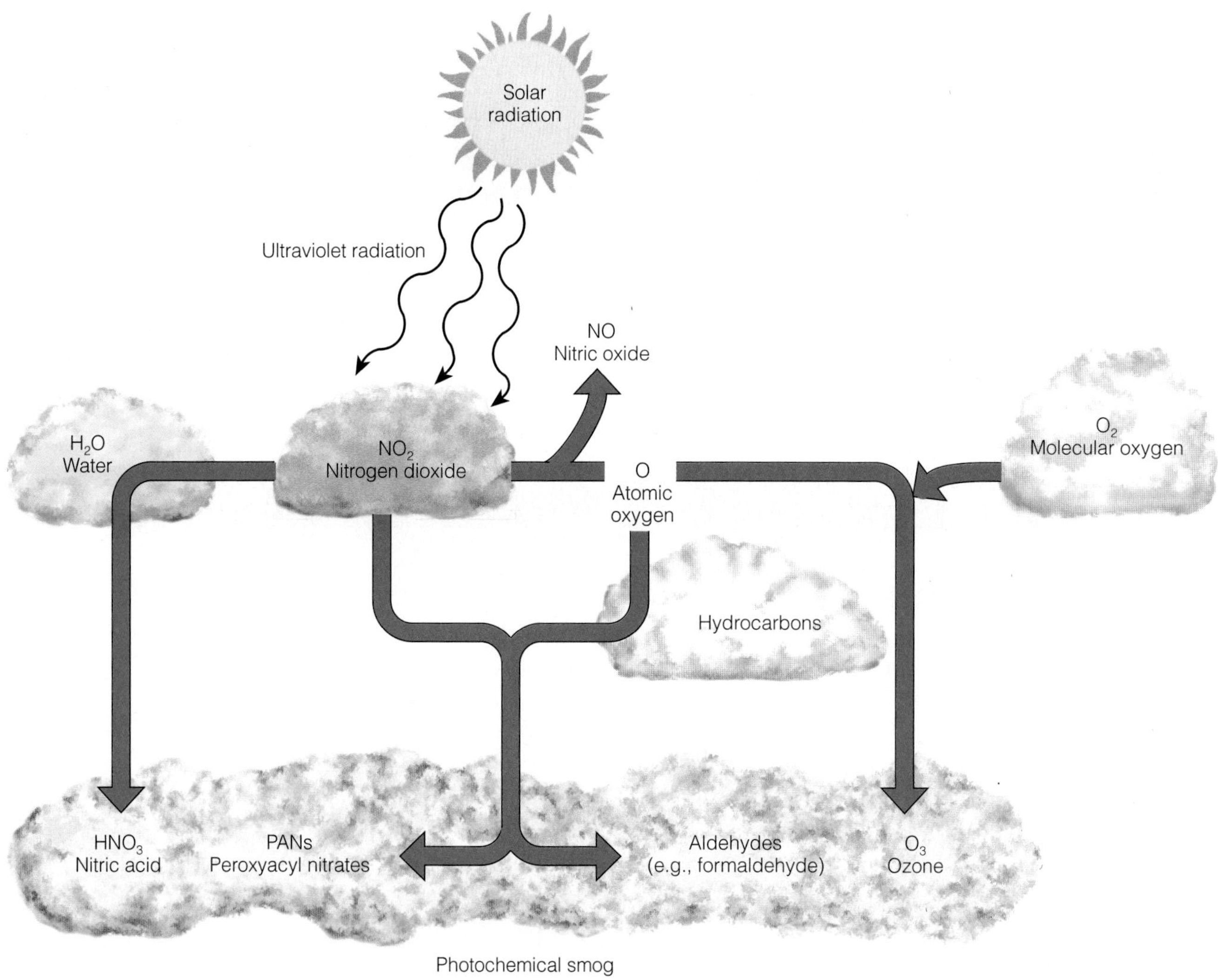

Figure 10-4 Simplified scheme of the formation of photochemical smog. The severity of smog generally is associated with atmospheric concentrations of ozone at ground level.

space heating. During winter, people in such cities were exposed to **industrial smog** consisting mostly of sulfur dioxide, suspended droplets of sulfuric acid (formed from some of the sulfur dioxide, Figure 10-3), and a variety of suspended solid particles and droplets (called aerosols).

Urban industrial smog is rarely a problem today in most developed countries because coal and heavy oil are burned only in large boilers with reasonably good pollution control or with tall smokestacks. However, industrial smog is a problem in industrialized urban areas of China, India, Ukraine, and some eastern European countries, where large quantities of coal are burned with inadequate pollution controls.

What Factors Influence the Formation of Photochemical and Industrial Smog? The frequency and severity of smog in an area depend on several things: **(1)** the local climate and topography, **(2)** the population density, **(3)** the amount of industry, and **(4)** the fuels used in industry, heating, and transportation. In areas with high average annual precipitation, rain and snow help cleanse the air of pollutants. Winds help sweep pollutants away and bring in fresh air, but they may also transfer some pollutants to downwind areas.

Hills and mountains tend to reduce the flow of air in valleys below them and allow pollutant levels to build up at ground level. Buildings in cities generally slow wind speed, thereby reducing dilution and removal of pollutants.

During the day the sun warms the air near the earth's surface. Normally this heated air expands and rises, carrying low-lying pollutants higher into the troposphere (Figure 10-5, left). Colder, denser air from surrounding high-pressure areas then sinks into the low-pressure area created when the hot air rises. This continual mixing of the air helps keep pollutants from reaching dangerous concentrations near the ground.

Sometimes, however, a lower layer of dense, cool air can be trapped beneath a layer of less dense, warm air in an urban basin or valley, causing a phenomenon known as a **temperature inversion** or **thermal inversion** (Figure 10-5, right). The changing temperature (temperature gradient) in the warm air above the pool of cool air prevents ascending air currents (that would disperse and dilute pollutants) from developing. These inversions usually last for only a few hours. However, when a high-pressure air mass stalls over an area they can last for several days. This allows air pollutants at ground level to build up to harmful and even lethal concentrations.

The United States suffered its first major air pollution disaster in Donora, Pennsylvania, a small industrial town southeast of Pittsburgh. The town is surrounded by mountains and nestled in the Monongahela Valley, which is subject to frequent thermal inversions (Figure 10-5, right). In 1948, Donora experienced a thermal inversion that trapped pollutants from steel mills, a zinc smelter, a sulfuric acid plant, and several other industrial plants for 5 days. About 6,000 of the town's 14,000 inhabitants fell ill, and 20 of them died.

A city with several million people and motor vehicles in an area with a sunny climate, light winds, mountains on three sides, and the ocean on the other has ideal conditions for photochemical smog worsened by frequent thermal inversions. This describes California's Los Angeles basin, which has 14 million people, 23 million motor vehicles, thousands of factories, and thermal inversions at least half of the year. Despite having the world's toughest air-pollution control program, Los Angeles has severe air pollution. Other cities with frequent thermal inversions are Denver, Colorado, in the United States; Mexico City, Mexico (Case Study, p. 221); Rio de Janeiro and São Paulo in Brazil; and Beijing and Shenyang in China.

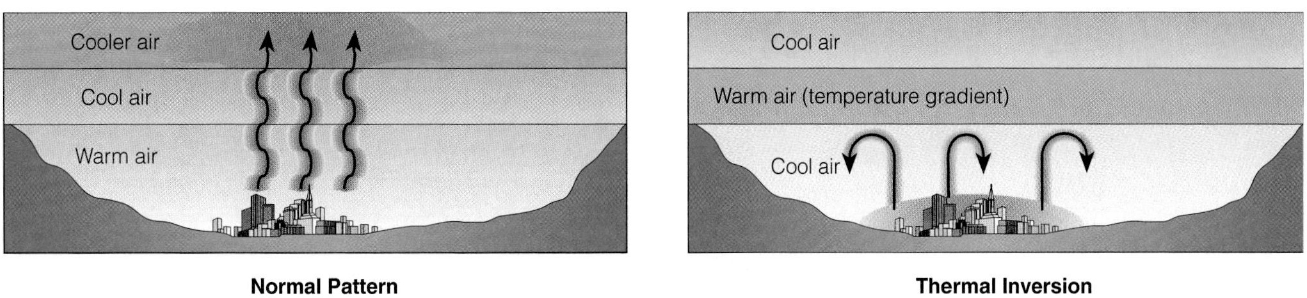

Figure 10-5 Thermal inversion. The layer of warm air (right) prevents ascending air currents from dispersing and diluting pollutants from the cool pool of air near the ground. Because of their topography, Los Angeles in the United States and Mexico City in Mexico have frequent thermal inversions, many of them prolonged during the summer.

10-3 REGIONAL OUTDOOR AIR POLLUTION FROM ACID DEPOSITION

What Is Acid Deposition? Most coal-burning power plants, ore smelters, and other industrial plants in developed countries use tall smokestacks to emit sulfur dioxide, suspended particles, and nitrogen oxides above the inversion layer. This "dilution solution" reduces local air pollution and allows companies to meet government standards without having to add expensive air-pollution control devices. However, it increases pollution downwind because what goes up must come down—another example of connections or unintended consequences.

As these primary pollutants, sulfur dioxide and nitrogen oxides, are transported as much as 1,000 kilometers (600 miles) by prevailing winds, they form secondary pollutants such as nitric acid vapor, droplets of sulfuric acid, and particles of acid-forming sulfate and nitrate salts (Figure 10-3). These chemicals descend to the earth's surface in two forms: **(1)** *wet deposition* (as acidic rain, snow, fog, and cloud vapor), and **(2)** *dry deposition* (as acidic particles). The resulting mixture is called **acid deposition** (Figure 10-6). This form of pollution is commonly called *acid rain.* However, *acid deposition* is a better term because the acidity can reach the earth's surface not only in rain but also as gases and as solid particles.

Acidity of substances in water is commonly expressed in terms of *pH*. Solutions with pH values less than 7 are *acidic*, and those with pH values greater than 7 are *alkaline* or *basic* (Figure 10-7). Natural precipitation is slightly acidic, with a pH of 5.0–5.6. However, primarily because of acid deposition, typical rain in the eastern United States is now about 10 times more acidic, with a pH of 4.3. In some areas it is 100 times more acidic, with a pH of 3—as acidic as vinegar. Some cities and mountaintops are bathed in a fog as acidic as lemon juice, with a pH of 2.3—about 1,000 times the acidity of normal precipitation.

What Areas Are Most Affected by Acid Deposition? Acid deposition occurs on a regional rather than a global basis because the acidic components remain in the atmosphere for only a few days. However, acid deposition is a serious regional problem (Figure 10-8) in many areas downwind from coal-burning power plants, smelters, factories, and large urban areas. How seriously vegetation and aquatic life in nearby lakes are affected by acid deposition depends mostly on whether the soil in the area is acidic or basic.

In some areas, soils are basic enough to neutralize or buffer some inputs of acids. The ecosystems most harmed by acid deposition are **(1)** those containing thin, acidic soils without such natural buffering (Figure 10-8, green areas) and **(2)** those in which the buffering capacity of soils has been depleted by decades of exposure to acid deposition.

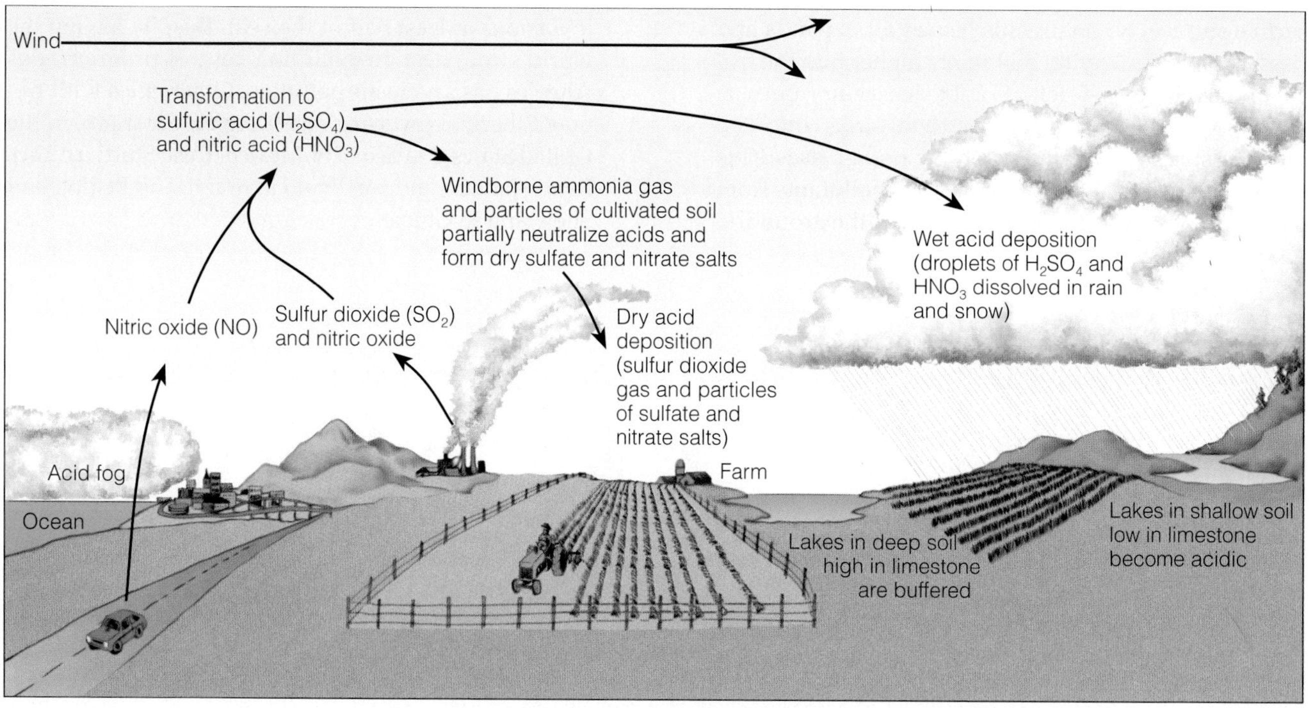

Figure 10-6 Acid deposition, which consists of rain, snow, dust, or gas with a pH lower than 5.6, is commonly called acid rain. Soils and lakes vary in their ability to buffer or remove excess acidity.

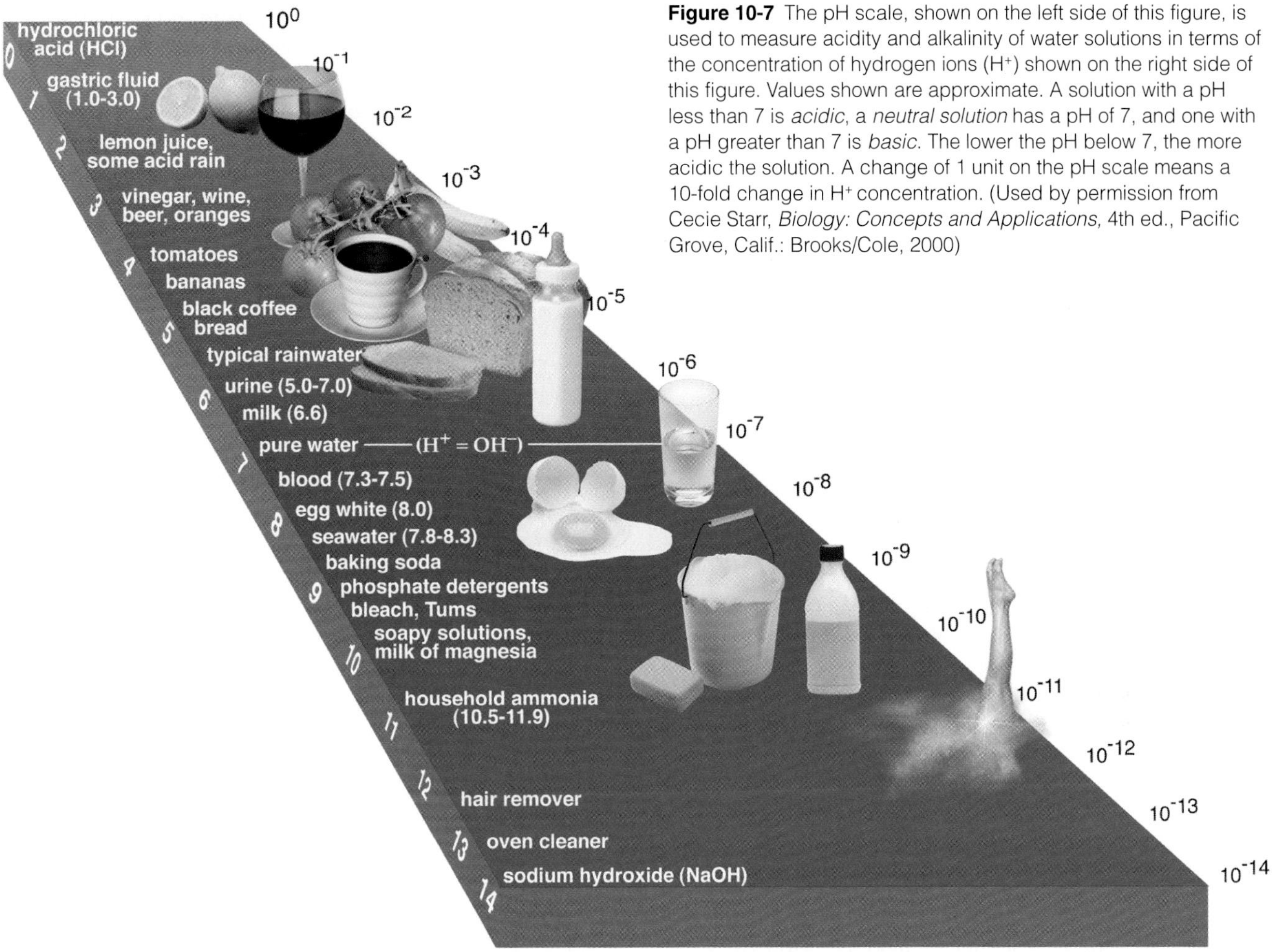

Figure 10-7 The pH scale, shown on the left side of this figure, is used to measure acidity and alkalinity of water solutions in terms of the concentration of hydrogen ions (H^+) shown on the right side of this figure. Values shown are approximate. A solution with a pH less than 7 is *acidic*, a *neutral solution* has a pH of 7, and one with a pH greater than 7 is *basic*. The lower the pH below 7, the more acidic the solution. A change of 1 unit on the pH scale means a 10-fold change in H^+ concentration. (Used by permission from Cecie Starr, *Biology: Concepts and Applications*, 4th ed., Pacific Grove, Calif.: Brooks/Cole, 2000)

Many acid-producing chemicals generated by power plants, factories, smelters, and cars in one country may be exported to other countries by prevailing winds. For example, more than three-fourths of the acid deposition in Norway, Switzerland, Austria, Sweden, the Netherlands, and Finland is blown to those countries from industrialized areas of western Europe (especially the United Kingdom and Germany) and eastern Europe. Acid deposition is also a growing problem in China (affecting 40% of its land area), parts of the former Soviet Union, India, Nigeria, Brazil, Venezuela, and Colombia.

Chemical detective work indicates that more than half the acid deposition in southeastern Canada and the eastern United States originates from coal- and oil-burning power plants and factories in Ohio, Indiana, Pennsylvania, Illinois, Missouri, West Virginia, and Tennessee. In areas near and downwind from large urban areas, emissions of NO and NO_2 (mostly from cars) leading to the formation of nitric acid may be the main culprit.

Within a few decades, NO_x and SO_2 emissions from developing countries are expected to outstrip those from developed countries, leading to greatly increased damage from acid deposition over a much wider area, especially where soils are sensitive to acidification.

In 1999 researchers at the Scripps Institution of Oceanography found a thick, brown haze of air pollution covering much of the Indian Ocean during winter. The haze is about the size of the United States and rises as high as 3,000 meters (10,000 feet)—five times higher than a similar pollution haze that rises over North America and Europe.

The researchers found that the air pollution comes from burning fossil fuels for transportation and industry in China, Southeast Asia, and India. These pollutants are blown out over the ocean during the winter monsoon season when prevailing winds flow down from the Himalayas and out to sea. In the late spring and summer, the winds reverse. This blows some of the haze back across the land, where it combines with monsoon rains and can fall as acid deposition.

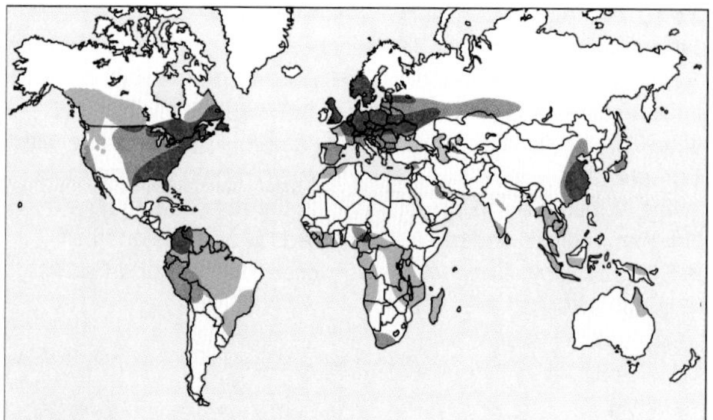

Figure 10-8 Regions where acid deposition is now a problem and regions with the potential to develop this problem, either because of increased air pollution (mostly from power plants, industrial plants, and ore smelters) or because of soils that cannot neutralize inputs of acidic compounds (green areas and most red areas). (Data from World Resource Institute and U.S. Environmental Protection Agency)

What Are the Effects of Acid Deposition? Acid deposition has many harmful ecological effects, especially when the pH falls below 5.1 for terrestrial systems and 5.5 for aquatic systems. It also contributes to human respiratory diseases such as bronchitis and asthma (which can cause premature death) and it damages statues, buildings, metals, and car finishes.

Acid deposition and other air pollutants such as ozone (O_3) can damage tree foliage directly, but the most serious effect is weakening trees so they become more susceptible to other types of damage (Figure 10-9). The areas hardest hit by acid deposition are mountaintop forests because **(1)** they tend to have thin soils without much buffering capacity and **(2)** trees on mountaintops (especially conifers such as red spruce that keep their leaves year-round) are bathed almost continuously in very acidic fog and clouds.

A combination of acid deposition and other air pollutants (especially ozone) can make trees more susceptible to stresses such as **(1)** cold temperatures, **(2)** diseases, **(3)** insects, **(4)** drought, **(5)** fungi (which thrive under acidic conditions), and **(6)** a drop in net primary productivity from loss of soil plant nutrients.

The final cause of tree damage or death may be mosses, insect attacks, diseases, and lack of nutrients. However, the underlying cause is often years of exposure to an atmospheric cocktail of air pollutants and soil overloaded with acids. Drops in forest productivity caused by depletion of soil nutrients and acid-buffering chemicals can reduce biodiversity and have significant economic implications for timber companies.

It can take decades to hundreds of years for soil to replenish nutrients leached out by acid deposition. Thus, losses in plant productivity could continue for decades even if emissions of sulfur dioxide and nitrogen oxides are reduced by air-pollution control programs. Until recently scientists expected some of the lost forest productivity to be offset by increased productivity from the larger input of nitric acid and nitrate salts from acid deposition in areas where this nutrient is the limiting factor. However, recent research shows that much of the nitrate raining down on forests in parts of Germany and Norway damaged by air pollution is not being taken up by the trees or by nitrogen-using microbes in the soil.

Acid deposition can also release aluminum ions (Al^{3+}) attached to soil minerals. Once released from soil particles, these water-soluble ions can damage tree roots. When washed into lakes, aluminum ions can also kill many kinds of fish by stimulating excessive mucus formation. This asphyxiates the fish by clogging their gills.

Excess acidity can contaminate fish in some lakes with highly toxic methylmercury. Increased acidity of lakes apparently converts moderately toxic inorganic mercury compounds in lake-bottom sediments into highly toxic methylmercury, which is more soluble in the fatty tissue of animals and can be biomagnified to higher concentrations in aquatic food chains and webs (Figure 8-2). However, acid runoff into lakes and streams is rated by risk analysis experts as a low-risk ecological problem (Figure 8-10, left).

How Serious Is Acid Deposition in the United States? A large-scale, government-sponsored research study on the *ecological effects* of acid deposition in the United States in the 1980s concluded that the problem was serious but not yet at a crisis stage. Some *good news* is that a 1993 study by the U.S. Geological Survey found that the concentration of sulfate ions, a key component of acid deposition, declined at 26 of 33 U.S. rainwater collection sites between 1980 and 1991. In addition, U.S. sulfur dioxide emissions dropped 40% between 1970 and 1999.

The *bad news* is that a 1999 study showed that acid deposition has spread from the northeast to the southeast, Sierra Nevada, the Cascade Range, and parts of the Rocky Mountains. To reduce these threats, scientists say that by 2005 the United States must **(1)** reduce SO_2 emissions by 50% more than the cuts enacted in 1990 and **(2)** reduce emissions of nitrogen oxides from power plants to 70% below 1990 levels.

Representatives of U.S. coal companies and industries that burn coal and oil claim that adding expensive

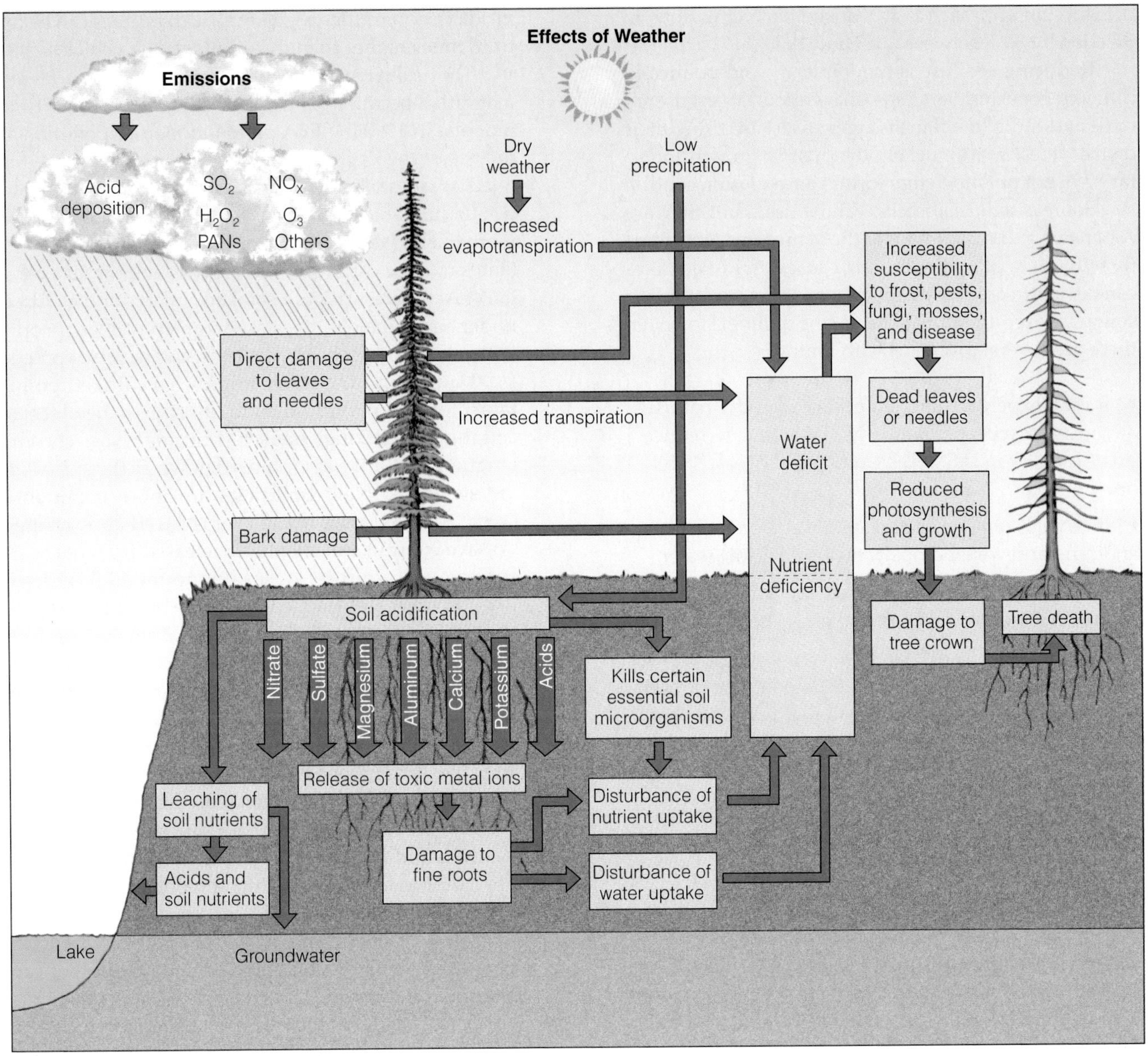

Figure 10-9 Possible or suspected harmful effects of prolonged exposure to an atmospheric cocktail of air pollutants on trees.

air-pollution control equipment or burning low-sulfur coal or oil costs more than the resulting health and environmental benefits are worth. According to the Environmental Protection Agency (EPA), however, the actual cleanup costs of SO_2 in 1994 were about one-tenth of the estimate given by industries when they opposed the new standards set by the Clean Air Act of 1990. A comprehensive 1997 study by Resources for the Future found that the environmental and public health benefits of reductions in SO_2 from 1995 to 2030 will generate more than $12 in benefits for every $1 in compliance costs.

Solutions: What Can Be Done to Reduce Acid Deposition? According to most scientists studying acid deposition, the best solutions are *prevention approaches*. They include **(1)** reducing energy use and thus air pollution by improving energy efficiency (Section 20-2), **(2)** switching from coal to cleaner-burning natural gas (Section 19-3) and renewable energy resources (Chapter 20), **(3)** removing sulfur from coal before it is burned, **(4)** burning low-sulfur coal, **(5)** removing SO_2, particulates, and nitrogen oxides from smokestack gases, **(6)** removing nitrogen oxides from motor vehicle exhaust, and **(7)** taxing emissions of sulfur dioxide (an approach that reduced such emissions in Sweden by 30% between 1989 and 1996).

Reducing coal use is economically and politically difficult. For example, China (the world's largest user of coal) and India (the fourth-largest user) are using their own coal reserves to fuel rapid industrial growth and so far have not put much money into air pollution control.

Some *cleanup approaches* can be used, but they are expensive and mask some of the symptoms temporarily without treating underlying causes. Acidified lakes can be neutralized by treating them or the surrounding soil with large amounts of limestone or lime. However, there are several problems with liming.

- It is an expensive and temporary remedy that usually must be repeated annually. Using lime to reduce excess acidity in U.S. lakes would cost at least $8 billion per year.
- It can kill some types of plankton and aquatic plants and can harm wetland plants that need acidic water.
- It's difficult to know how much of the lime to put where (in the water or at selected places on the ground).
- Some recent research suggests that liming can increase populations of microbes that deplete carbon stored in the slowly decaying soil matter (humus) and thus reduce forest productivity. Recently, however, researchers in England found that adding a small amount of phosphate fertilizer can neutralize excess acidity in a lake.

10-4 INDOOR AIR POLLUTION

What Are the Types and Sources of Indoor Air Pollution? If you are reading this book indoors, you may be inhaling more air pollutants with each breath than if you were outside (Figure 10-10). According to EPA studies, in the United States levels of 11 common pollutants are generally two to five times higher inside homes and commercial buildings than outdoors, and as much as 100 times higher in some cases. Levels of fine particles, which can contain toxins and metals such as lead and cadmium, can be as much as 60% higher indoors than outdoors.

Researchers found that concentrations of several pesticides (such as chlordane), approved for outdoor use only, were 10 times greater inside than outside homes. Most indoor pollution from these chemicals is believed to come from pesticides used to fumigate houses around their foundations, but some comes from pesticide dust tracked in on shoes. The pesticides used to fumigate house foundations seep into the soil and foundations and give off gases that can seep into houses for years. A 1993 study found that pollution levels inside cars in traffic-clogged U.S. urban areas can be up to 18 times higher than those outside the vehicles.

The health risks from exposure to such chemicals are magnified because people spend 70–98% of their time indoors. In 1990 the EPA placed indoor air pollution at the top of the list of 18 sources of cancer risk, and it is rated by risk analysis scientists as a high-risk health problem for humans (Figure 8-10, left). According to the EPA, more than 3,000 cases of cancer per year in the United States may be caused by exposure to indoor air pollutants. At greatest risk are smokers, infants and children under age 5, the old, the sick, pregnant women, people with respiratory or heart problems, and factory workers.

Danish and U.S. EPA studies have linked pollutants found in buildings to dizziness, headaches, coughing, sneezing, nausea, burning eyes, chronic fatigue, and flulike symptoms, known as the *sick building syndrome*. New buildings are more commonly "sick" than old ones because of reduced air exchange (to save energy) and chemicals released from new carpeting and furniture. According to the EPA, at least 17% of the 4 million commercial buildings in the United States are considered "sick" (including EPA headquarters). Indoor air pollution in the United States costs an estimated $100 billion per year in absenteeism, reduced productivity, and health-care costs.

According to the EPA and public health officials, the three most dangerous indoor air pollutants are **(1)** cigarette smoke (p. 179), **(2)** formaldehyde, and **(3)** radioactive radon-222 gas. Worker exposure to asbestos fibers in mines and in factories making asbestos material is also a serious indoor air pollution problem, especially in developing countries. A number of research studies on laboratory animals have also identified tiny fibers of *fiberglass* as a widespread and potentially potent carcinogen in indoor air.

The chemical that causes most people difficulty is *formaldehyde*, an extremely irritating gas. As many as 20 million Americans suffer from chronic breathing problems, dizziness, rash, headaches, sore throat, sinus and eye irritation, wheezing, and nausea caused by daily exposure to low levels of formaldehyde emitted (outgassed) from common household materials. These include **(1)** building materials (such as plywood, particleboard, paneling, and high-gloss wood used in floors and cabinets), **(2)** furniture, **(3)** drapes, **(4)** upholstery, **(5)** adhesives in carpeting and wallpaper, **(7)** fingernail hardener, and **(7)** wrinkle-free coating on permanent-press clothing (Figure 10-10). The EPA estimates that as many as 1 out of every 5,000 people who live in manufactured homes for more than 10 years will develop cancer from formaldehyde exposure.

In developing countries, the burning of wood, dung, and crop residues in open fires or in unvented or poorly vented stoves for cooking and heating

Oliver, L. C. and B. W. Shackleton. 1998. "The Indoor Air We Breathe." *Public Health Reports*, vol. 113, no. 5 p. 398(12).

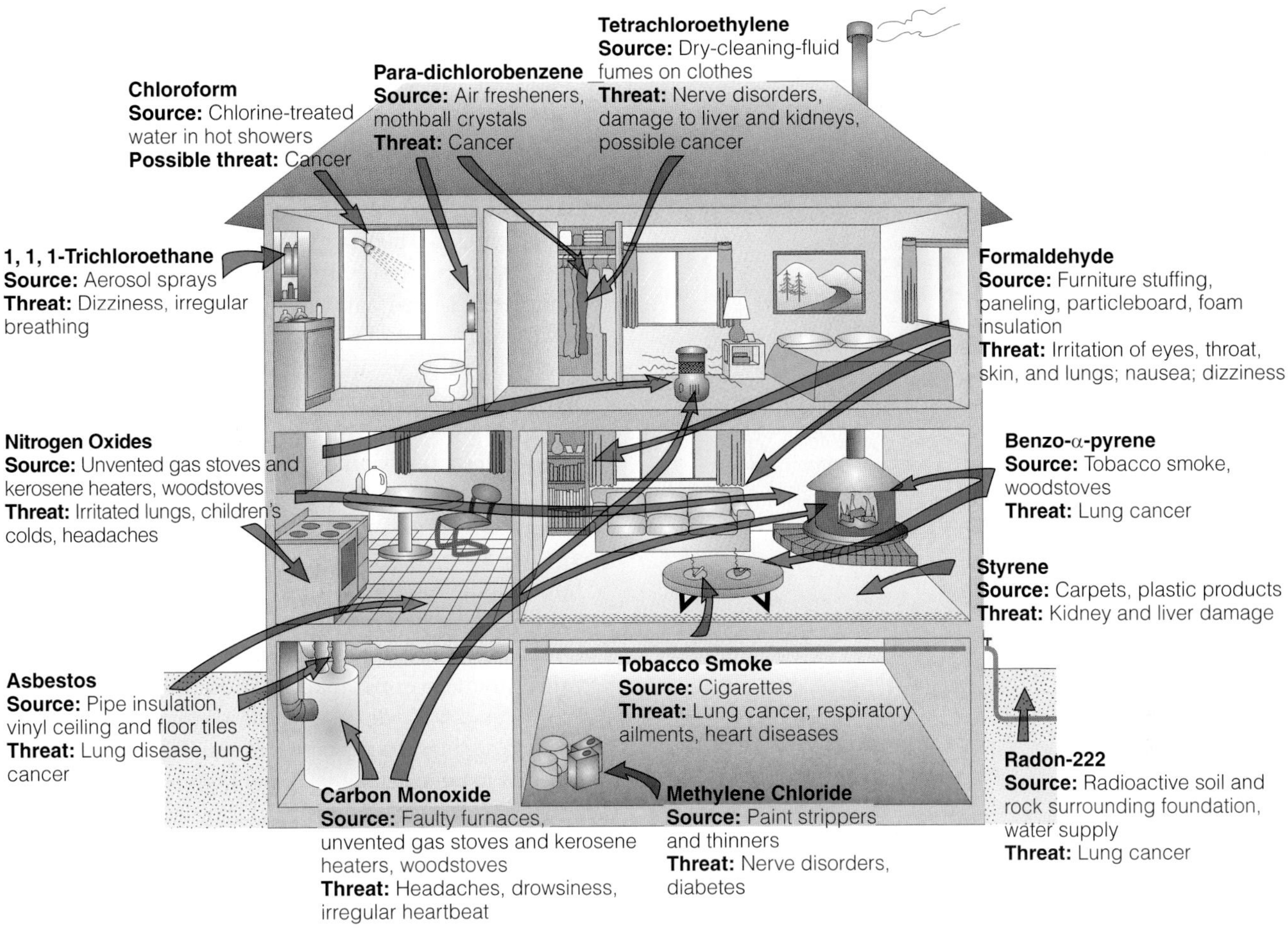

Figure 10-10 Some important indoor air pollutants. (Data from U.S. Environmental Protection Agency)

exposes inhabitants, especially women and young children, to very high levels of particulate air pollution. According to a World Bank study, each year fuelwood cooking smoke causes respiratory diseases leading to premature death in approximately 4 million children worldwide.

Case Study: Is Your Home Contaminated with Radon Gas? Radon-222 is a naturally occurring radioactive gas that you can't see, taste, or smell. It is produced by the radioactive decay of uranium-238. Small amounts of uranium-238 are found in most soil and rock, but this isotope is much more concentrated in underground deposits of minerals such as uranium, phosphate, granite, and shale.

When radon gas from such deposits seeps upward through the soil and is released outdoors, it disperses quickly in the atmosphere and decays to harmless levels. Radon gas can enter buildings above such deposits through **(1)** cracks in foundations and walls, **(2)** openings around sump pumps and drains, and **(3)** hollow concrete blocks (Figure 10-11) and build up to high levels, especially in unventilated lower levels of homes and buildings.

Radon-222 gas quickly decays into solid particles of other radioactive elements that, if inhaled, expose lung tissue to a large amount of ionizing radiation from alpha particles. This can damage lung tissue and lead to lung cancer over the course of a 70-year lifetime. Your chances of getting lung cancer from radon depend mostly on **(1)** how much radon is in your home, **(2)** the amount of time you spend in your home, and **(3)** whether you are a smoker or have ever smoked.

In 1998 the National Academy of Science estimated that prolonged exposure for a lifetime of 70 years to low levels of radon or radon acting together with smoking is responsible for 15,000 to 22,000 (or 12%) of the lung cancer deaths each year in the United States. This makes radon the second leading cause of lung cancer after smoking. Most of the deaths are among smokers or former smokers, with about 2,100 to 2,900 among nonsmokers. Scientists who believe there is a threshold dose (Figure 8-3, right) before radon is harmful say that these estimates are too high.

Hint: Enter the search terms *indoor air quality* using the Subject Guide.

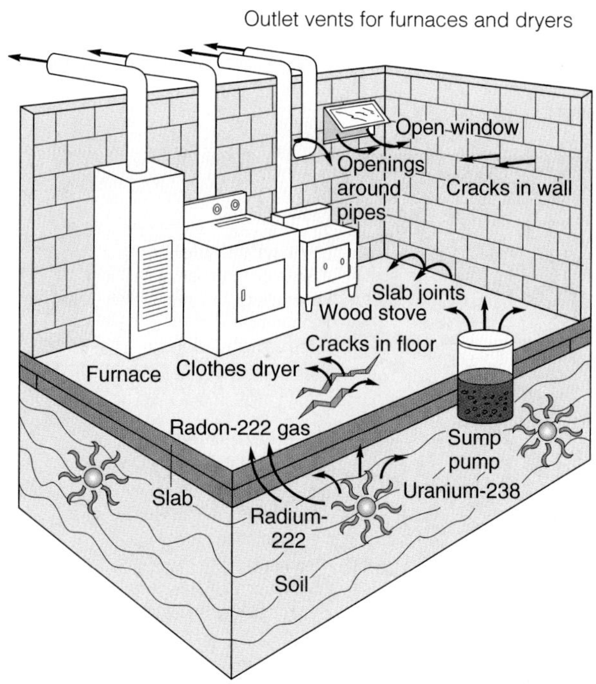

Figure 10-11 Sources and paths of entry for indoor radon-222 gas. (Data from U.S. Environmental Protection Agency)

EPA indoor radon surveys suggest that 4–5 million U.S. homes may have annual radon levels above 4 picocuries per liter of air* and that 50,000–100,000 homes may have levels above 20 picocuries per liter. If the 4 picocuries per liter standard is adopted (as proposed by the EPA), the cost of testing and correcting the problem could run about $50 billion, with a 15–20% reduction in radon-related deaths. Some researchers argue that it makes more sense to spend perhaps only $500 million to find and fix homes and buildings with radon levels above 20 picocuries per liter until more reliable data are available on the threat from exposure to lower levels of radon.

In 1988 the EPA and the U.S. Surgeon General's Office recommended that everyone living in a detached house, a town house, a mobile home, or on the first three floors of an apartment building test for radon. Ideally, radon levels should be continuously monitored in the main living areas (not basements or crawl spaces) for 2 months to a year. By 1999 only about 6% of U.S. households had conducted radon tests (most lasting only 2 to 7 days and costing $20–100 per home).

If testing reveals an unacceptable level, homeowners can consult the free EPA publication *Radon Reduction Methods* for ways to reduce radon levels and health risks. According to the EPA, radon control could add $350–500 to the cost of a new home and correcting a radon problem in an existing house could run $800–2,500.

*A *picocurie* is a trillionth of a curie, which is the amount of radioactivity emitted by a gram of radium.

Case Study: What Should Be Done About Asbestos? Asbestos is a name given to several different fibrous forms of silicate minerals. For decades it has been widely used as a building material and for large water pipelines because of its strength, flexibility, and low cost compared with competing materials. Unless completely sealed within a product, asbestos crumbles easily into a dust of fibers tiny enough to become suspended in the air and inhaled deep into the lungs, where they remain for many years.

Prolonged exposure to asbestos fibers can cause **(1)** *asbestosis* (a chronic, sometimes fatal disease that eventually makes breathing nearly impossible and was recognized as a hazard among asbestos workers as early as 1924),**(2)** *lung cancer*, and **(3)** *mesothelioma* (a fatal cancer of the lung's lining). Epidemiological studies have shown that smokers exposed to asbestos fibers have a much greater chance of dying from lung cancer than do nonsmokers exposed to such fibers.

Most of these diseases occur in workers exposed for years to high levels of asbestos fibers. This group includes asbestos miners, insulators, pipefitters, shipyard employees, and workers in asbestos-producing factories, who often spent years working in clouds of asbestos dust. Family members who breathe asbestos dust brought home in the clothes and hair of asbestos workers also have higher cancer rates, as do people who live near asbestos manufacturing plants with inadequate control of asbestos emissions.

According to health officials, asbestos fibers caused the premature cancer deaths of 171,500 asbestos workers in the United States between 1967 and 1998, the worst occupational health disaster of the 20th century. An additional 119,000 premature cancer deaths among U.S. asbestos workers are predicted before 2025, mostly from workers exposed to unsafe conditions before working conditions were improved.

After being swamped with health claims from workers, most U.S. asbestos manufacturing companies either declared bankruptcy or moved their operations to other countries (such as Mexico and Brazil) with weaker environmental laws and lax enforcement.

Since 1980 the focus in the United States has shifted from asbestos workers to concern over possible health effects of inhalation of low levels of asbestos fibers by the public in buildings. Between 1900 and 1984, asbestos was sprayed on ceilings and walls of schools and other public and private buildings in the United States for fireproofing, soundproofing, insulation of heaters and pipes, and wall and ceiling decoration. The EPA banned those uses in 1984.

In 1989 the EPA ordered a ban on almost all remaining uses of asbestos (such as brake linings, roofing shingles, and water pipes) in the United States by 1997. Representatives of the asbestos industry in the United States and Canada (which now produces most of the asbestos used in the United States) challenged the ban in court. They contended that with proper precautions these asbestos products can be used safely and that the costs of the ban outweigh the benefits. In 1991 a federal appeals court overturned the 1989 EPA ban.

In 1979, the EPA recommended removal of existing asbestos in one of every seven commercial and public buildings in the United States (including 30,000 schools), at a cost of at least $100 billion. However, risk analysis and scientific studies indicate that the risk from indoor exposure to asbestos fibers (even in buildings rich in asbestos materials) is extremely low and is about one-tenth the risk from breathing asbestos fibers found in outdoor air from natural sources (wind and erosion).

Critics argued that removing asbestos from buildings is a waste of money, with each life saved costing $100–500 million. They called for sealing, wrapping, and other forms of containment instead of removal asbestos, except where it has been damaged or disturbed. Indeed, improper or unnecessary removal can release more asbestos fibers than does sealing off asbestos that is not crumbling. In 1990, the EPA agreed with this position but has not widely publicized this reversal of their earlier policy. In 1998 chemists developed a foam that lets building owners treat asbestos-containing fireproofing material without removing it.

According to health experts, the major risk health risk from asbestos today is occurring among asbestos miners and workers in developing countries (especially Russia, China, Brazil, India, and Thailand), where the use of asbestos as an inexpensive building material is growing rapidly. Experts say that 90% of the deaths from asbestos exposure can be prevented by using a good-fitting face mask, wetting asbestos to control dust, and changing clothes before and after handling asbestos.

10-5 EFFECTS OF AIR POLLUTION ON LIVING ORGANISMS AND MATERIALS

How Is Human Health Harmed by Air Pollutants? Your respiratory system (Figure 10-12) has a number of mechanisms that help protect you from air pollution. Hairs in your nose filter out large particles. Sticky mucus in the lining of your upper respiratory tract captures smaller (but not the smallest) particles and dissolves some gaseous pollutants. Sneezing and

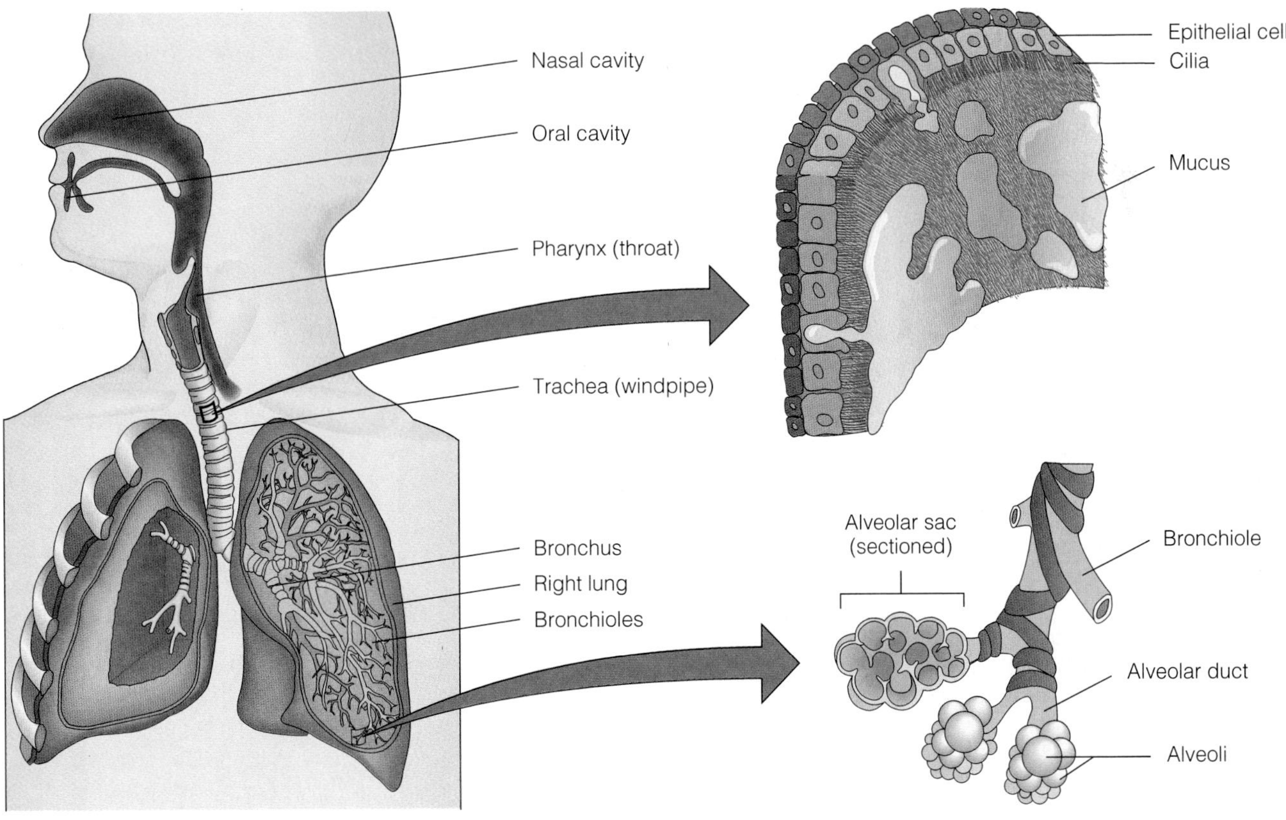

Figure 10-12 Major components of the human respiratory system.

coughing expel contaminated air and mucus when your respiratory system is irritated by pollutants. The cells of your upper respiratory tract are also lined with hundreds of thousands of tiny, mucus-coated hairlike structures called *cilia* that continually wave back and forth, transporting mucus and the pollutants they trap to your throat (where they are swallowed or expelled).

Years of smoking and exposure to air pollutants can overload or break down these natural defenses, causing or contributing to respiratory diseases such as **(1)** *lung cancer*, **(2)** *asthma* (typically an allergic reaction causing sudden episodes of muscle spasms in the bronchial walls, resulting in acute shortness of breath), **(3)** *chronic bronchitis* (persistent inflammation and damage to the cells lining the bronchi and bronchioles, causing mucus buildup, painful coughing, and shortness of breath), and **(4)** *emphysema* (irreversible damage to air sacs or alveoli leading to abnormal dilation of air spaces, loss of lung elasticity, and acute shortness of breath, Figure 10-13). Elderly people, infants, pregnant women, and people with heart disease, asthma, or other respiratory diseases are especially vulnerable to air pollution.

About 90% of the *carbon monoxide* (CO)—a colorless, odorless, poisonous gas—in the troposphere comes from natural sources. Most of this is produced by reaction in the upper troposphere between methane (emitted mostly by the anaerobic decay of organic matter in swamps, bogs, and marshes) and oxygen. Because this CO is diluted by the turbulent air flows in the troposphere, it does not build up to harmful levels.

However, the remaining 10% of the CO added to the atmosphere comes from the incomplete burning of carbon-containing chemicals (primarily fossil fuels). Cigarette smoking (p. 179) is responsible for the largest human exposure to CO, but this gas is also released by motor vehicles, kerosene heaters, woodstoves, fireplaces, and faulty heating systems.

CO reacts with hemoglobin in red blood cells and thus reduces the ability of blood to carry oxygen. This impairs perception and thinking, slows reflexes, and causes headaches, drowsiness, dizziness, and nausea. CO can also trigger heart attacks and angina attacks in people with heart disease. It can damage the development of fetuses and young children and aggravate chronic bronchitis, emphysema, and anemia. Exposure to high levels of CO causes collapse, coma, irreversible damage to brain cells, and even death. Carbon monoxide detectors (similar in size to smoke detectors) are now available for about $50 and are a good safety investment for homeowners.

Inhaling *suspended particulate matter* (microscopic solid particles and tiny drops of liquids that can remain suspended in air for extended periods) aggravates bronchitis and asthma, and long-term exposure can contribute to development of chronic respiratory disease and cancer (Spotlight, right).

Sulfur dioxide, produced mostly by burning of coal in power plants and in some industrial processes such as paper making and smelting of metals, causes some constriction of the airways in healthy people and severe restriction in people with asthma. Chronic exposure

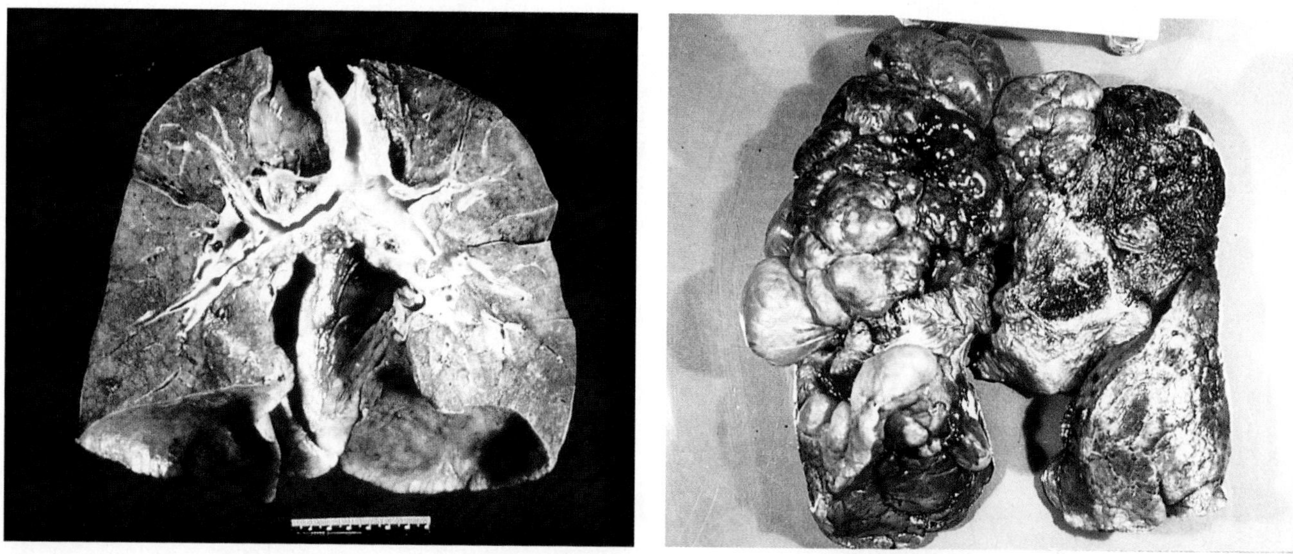

Figure 10-13 Normal human lungs (left) and the lungs of a person who died of emphysema (right). Prolonged smoking and exposure to air pollutants can cause emphysema in anyone, but about 2% of emphysema cases result from a defective gene that reduces the elasticity of the air sacs in the lungs. Anyone with this hereditary condition, for which testing is available, should not smoke and should not live or work in a highly polluted area. (O. Auerbach/Visuals Unlimited)

causes a condition similar to bronchitis. Sulfur dioxide and suspended particles react to form far more hazardous acid sulfate particles, which are inhaled more deeply into the lungs than SO_2 and remain there for long periods. According to the WHO, at least 625 million people are exposed to unhealthy levels of sulfur dioxide from fossil-fuel burning.

Nitrogen oxides (especially NO_2), released mostly by burning fossil fuels, can **(1)** irritate the lungs, **(2)** aggravate asthma or chronic bronchitis,**(3)** cause conditions similar to chronic bronchitis and emphysema, and **(4)** increase susceptibility to respiratory infections such as the flu and common colds (especially in young children and elderly people).

Research indicates that many *volatile organic compounds* (VOCs, such as benzene and formaldehyde) and *toxic particulates* (such as lead, cadmium, PCBs, and dioxins) can cause mutations, reproductive problems, or cancer. Most VOCs come from burning fossil fuels, solvents, paints, and other products.

Evidence also shows that inhaling *ozone*, a component of photochemical smog (Figure 10-4), causes coughing, chest pain, shortness of breath, and eye, nose, and throat irritation. It also aggravates chronic diseases such as asthma, bronchitis, emphysema, and heart disease and reduces resistance to colds and pneumonia. Many U.S. cities often exceed safe levels, especially during warm weather. A 1987 study also showed that after factoring out smoking, long-time Los Angeles residents exposed to ozone had twice the risk of cancer compared with residents of cleaner air cities.

According to a 1998 U.S. EPA report, all outdoor air in the United States is contaminated with eight cancer-causing industrial chemicals or HAPs (Table 10-2) at levels that exceed the EPA's safety standards.

How Many People Die Prematurely from Air Pollution? It is difficult to estimate by risk analysis how many people die prematurely from respiratory or cardiac problems caused or aggravated by air pollution because people are exposed to so many different pollutants over their lifetimes.

In the United States, estimates of annual deaths related to outdoor air pollution range from 65,000 to 200,000 (most from exposure to fine particles). If indoor air pollution is included, estimated annual deaths from air pollution in the United States range from 150,000 to 350,000 people—equivalent to 1–2 fully loaded 400-passenger jumbo jets crashing *each day* with no survivors.

Millions more become ill and lose work time. According to the EPA and the American Lung Association, air pollution in the United States costs at least \$150 billion annually in health care and lost work productivity, with \$100 billion of that caused by indoor air pollution.

Health Dangers from Fine Particles

SPOTLIGHT

Research indicates that invisible particles—especially *fine particles* with diameters less than 10 microns (PM-10) and *ultrafine particles* with diameters less than 2.5 microns (PM-2.5)—pose a significant health hazard. Such particles are emitted by incinerators, motor vehicles, radial tires, wind erosion, wood-burning fireplaces, and power and industrial plants.

Such tiny particles are not effectively captured by modern air-pollution control equipment, and they are small enough to penetrate the respiratory system's natural defenses against air pollution. They can also bring with them droplets or other particles of toxic or cancer-causing pollutants that become attached to their surfaces. Once they are lodged deep within the lungs, these fine particles can cause chronic irritation that can trigger asthma attacks, aggravate other lung diseases, cause lung cancer, and interfere with the blood's ability to take in oxygen and release CO_2. This strains the heart, increasing the risk of death from heart disease.

Several recent studies of air pollution in U.S. cities have indicated that fine and ultrafine particles prematurely kill 65,000–150,000 Americans each year. There is no known threshold level below which the harmful effects of fine particles disappear.

Exposure to particulate air pollution is much worse in most developing countries, where urban air quality has generally deteriorated. The World Bank estimates that if particulate levels were reduced globally to WHO guidelines, 300,00–700,000 premature deaths per year could be prevented.

Critical Thinking

In the United States, the EPA has proposed stricter national standards for emissions of ultrafine microscopic particles with diameters less than 2.5 microns (PM-2.5). Business leaders have vigorously opposed this proposal, arguing that implementing it would be too costly and that it is based on flimsy scientific evidence. EPA officials say that their review of the scientific evidence—one of the most exhaustive ones ever undertaken by the agency—supports the need for a standard for ultrafine particles. Do you believe that a stricter standard for emissions of ultrafine particles should be put into effect? Explain. Use the library or the Internet to determine what has happened to this proposed standard.

According to a 1997 study by the WHO and the World Bank, worldwide at least 3 million people (most of them in Asia) die prematurely each year from the effects of indoor (2.7 million deaths) and outdoor (0.3 million deaths) air pollution. China reported at least 3 million premature deaths from urban air pollution between 1994 and 1996. Most children who live in large cities in developing countries breathe air that is the equivalent of smoking two packs of cigarettes a day.

How Are Plants Damaged by Air Pollutants? Some gaseous pollutants (especially ozone) damage leaves of crop plants and trees directly when they enter leaf pores (Figure 10-9). Spruce, fir, and other conifers, especially at high elevations, are most vulnerable to air pollution because of their long life spans and the year-round exposure of their needles to a mixture of air pollutants.

However, the effects may not become visible for several decades, when large numbers of trees suddenly begin dying off because of depletion of soil nutrients and increased susceptibility to pests, diseases, fungi, and drought. This phenomenon, known as *Waldsterben* (forest death), has turned whole forests of spruce, fir, and beech into stump-studded meadows and mountainsides (see photo on p. 201). It is estimated that air pollution has been a key factor in reducing the overall productivity of European forests by about 16% and causing damage valued at roughly $30 billion per year.

Forest diebacks have also occurred in the United States, mostly on high-elevation slopes that face moving air masses and are dominated by red spruce. The most seriously affected areas are in the Appalachian Mountains. Ozone and other air pollutants are believed to be responsible for this forest degradation (Figure 10-9) and loss of biodiversity.

Air pollution, mostly by ozone, also threatens some crops—especially corn, wheat, and soybeans, the three most important U.S. crops—and is reducing U.S. food production by 5–10%. In the United States, estimates of economic losses to agriculture as a result of air pollution range from $1.9 to $5.4 billion per year.

How Can Air Pollutants Damage Aquatic Life? High acidity (low pH) can severely harm the aquatic life in freshwater lakes where the surrounding soils have little acid-neutralizing capacity or where there is significant winter snowfall. Much of the damage to aquatic life in such areas is a result of *acid shock* caused by the sudden runoff of large amounts of highly acidic water and aluminum ions into lakes and streams, when snow melts in the spring or after unusually heavy rains. The aluminum ions leached from the soil and lake sediment by this sudden input of acid can kill fish and inhibit their reproduction.

As the acidity of a lake increases and its food chain is disrupted, there is a decline in net primary productivity. This can turn a moderately eutrophic lake (Figure 6-31, top) into a clear blue oligotrophic lake (Figure 6-31, bottom).

Because of excess acidity, at least 16,000 lakes in Norway and Sweden contain no fish, and 52,000 more lakes have lost most of their acid-neutralizing capacity. In Canada some 14,000 acidified lakes are almost fish graveyards, and 150,000 more are in peril.

In the United States, about 9,000 lakes are threatened with excess acidity, one-third of them seriously. Most of them are concentrated in the northeast and the upper Midwest—especially Minnesota, Wisconsin, and the upper Great Lakes—where 80% of the lakes and streams are threatened by excess acidity. Over 10% of some 200 lakes in New York's Adirondack Mountains are too acidic to support fish.

What Are the Harmful Effects of Air Pollutants on Materials? Each year, air pollutants cause billions of dollars in damage to various materials we use (Table 10-3). The fallout of soot and grit on buildings, cars, and clothing requires costly cleaning. Air pollutants break down exterior paint on cars and houses, and they deteriorate roofing materials. Irreplaceable marble statues, historic buildings, and stained glass windows around the world have been pitted, gouged, and discolored by air pollutants. Damage to buildings in the United States from acid deposition alone is estimated by the EPA at $5 billion per year.

10-6 SOLUTIONS: PREVENTING AND REDUCING AIR POLLUTION

How Have Laws Been Used to Reduce Air Pollution in the United States? The U.S. Congress passed Clean Air Acts in 1970, 1977, and 1990, providing federal air-pollution regulations that are enforced by each state. These laws required the EPA to establish *national ambient air quality standards (NAAQS)* for seven outdoor pollutants: suspended particulate matter, sulfur oxides, carbon monoxide, nitrogen oxides, ozone, volatile organic compounds, and lead. Each standard specifies the maximum allowable level, averaged over a specific period, for a certain pollutant in outdoor (ambient) air.

Here is some *good news*. The Clean Air Acts have worked. Between 1970 and 1997, concentrations of major air pollutants decreased nationally by 32%, despite significant increases in population size and a doubling of economic growth and the average number of kilometers driven. According to EPA data, between 1986 and 1997 lead levels in U.S. air

Baker, B. 1999. "Court Rulings Question Air Pollution Standards at EPA." *BioScience*, vol. 49, no. 18, p. 618(1).

Table 10-3 Harmful Effects of Air Pollution on Materials		
Material	**Effects**	**Principal Air Pollutants**
Stone and concrete	Surface erosion, discoloration, soiling	Sulfur dioxide, sulfuric acid, nitric acid, particulate matter
Metals	Corrosion, tarnishing, loss of strength	Sulfur dioxide, sulfuric acid, nitric acid, particulate matter, hydrogen sulfide
Ceramics and glass	Surface erosion	Hydrogen fluoride, particulate matter
Paints	Surface erosion, discoloration, soiling	Sulfur dioxide, hydrogen sulfide, ozone, particulate matter
Paper	Embrittlement, discoloration	Sulfur dioxide
Rubber	Cracking, loss of strength	Ozone
Leather	Surface deterioration, loss of strength	Sulfur dioxide
Textiles	Deterioration, fading, soiling	Sulfur dioxide, nitrogen dioxide, ozone, particulate matter

decreased 98%, carbon monoxide 37%, sulfur dioxide 37%, suspended particulate matter (10 micrometers or less in diameter) 25%, ground level ozone 15%, and nitrogen dioxide 10%. Without the 1970 standards for emissions of pollutants, air-pollution levels would be much higher today.

Here is some *bad news*. In 1999, studies by the EPA and the Environmental Defense Fund found that 220 million Americans breathe air that is 100 times more toxic than the goal set by Congress in 1990. The studies showed that much of the toxic pollutants came from cars, trucks, and small businesses.

Between 1970 and 1990, the United States spent about $436 billion (in 1990 dollars) to comply with clean air regulations. Total human health and ecological benefits during the same 20-year period were estimated to range from $2.7 to $14.6 trillion (in 1990 dollars), 6 to 33 times the costs. A 1999 study by the EPA estimated that the health benefits of the Clean Air Act of 1990 between 1990 and 2010 should be about $110 billion (including preventing 23,000 premature deaths)—over four times the estimated $27 million in compliance costs to industries and consumers.

How Could U.S. Air-Pollution Laws Be Improved? The Clean Air Act of 1990 was an important step in the right direction, but most environmentalists point to the following major deficiencies in this law:

- *Continuing to rely almost entirely on pollution cleanup rather than pollution prevention.* In the United States, the air pollutant with the largest drop (98% between 1970 and 1997) in its atmospheric level was lead, which was virtually banned in gasoline. This shows the effectiveness of the pollution prevention approach.
- *Failing to sharply increase fuel efficiency standards for cars and light trucks.* According to environmental scientists, this would reduce oil imports and air pollution more quickly and effectively than any other method and would save consumers enormous amounts of money (Section 20-2).
- *Not requiring stricter emission standards for fine particulates.* In 1997 the EPA proposed stricter standards for fine particles that would prevent an estimated 20,000 premature deaths, 500,000 asthma attacks, and 9,000 hospital admissions per year and save $1 billion worth of crop losses. The EPA estimates the cost of implementing the standards at $7 billion per year, with the resulting health and other benefits estimated at $120 billion per year. Industries opposing these regulations put the cost of implementing the regulations at $200 billion per year.
- *Giving municipal trash incinerators 30-year permits,* which locks the nation into hazardous air pollution emissions and toxic waste from incinerators well into the 21st century. This also undermines pollution prevention, recycling, and reuse (Chapter 14).
- *Setting weak standards for air-pollution emissions from incinerators,* thus allowing unnecessary emissions of mercury, lead, dioxins, and other toxic pollutants.
- *Doing too little to reduce emissions of carbon dioxide and other greenhouse gases* (Section 11-5).

Should We Use the Marketplace to Reduce Pollution? To help reduce SO_2 emissions, the Clean Air Act of 1990 allows an *emissions trading policy*, which enables the 110 most polluting power plants in 21 states (primarily in the Midwest and East) to buy and sell SO_2 pollution rights.

Each year a power plant is given a certain number of pollution credits or rights that allow it to emit a certain amount of SO_2. A utility that emits less SO_2 than its limit receives more pollution credits. It can use these credits to avoid reductions in SO_2 emissions from some of its other facilities, bank them for future plant expansions, or sell them to other utilities, private citizens, or environmental groups. Proponents of this system argue that it allows the marketplace to determine the cheapest, most efficient way to get the job done instead of having the government dictate how to control pollution.

Some environmentalists see this market approach as an improvement over the current regulatory approach, as long as it achieves net reduction in SO_2 pollution. This would be done by limiting the total number of credits and gradually lowering the annual number of credits, something that is not required by the 1990 Clean Air Act.

Some environmentalists contend that marketing pollution rights allows utilities with older, dirtier power plants to buy their way out and keep on emitting unacceptable levels of SO_2. They also warn that this approach creates incentives to cheat. Air quality regulation is based largely on self-reporting of emissions, and pollution monitoring is incomplete and imprecise. Thus, sellers of permits will benefit by understating their reductions (to get more permits), and permit buyers will benefit by underreporting emissions (to reduce their permit purchases).

Because of a loophole in the 1970 Clean Air Acts, existing coal-burning power plants have been able to operate without further restrictions on their emissions. This loophole plus emissions trading have not resulted in any significant reduction of emissions of SO_2, particulates, and toxic mercury since 1970, even though the plants have been modified and in some cases expanded. In 1999, New York state officials and the EPA filed lawsuits claiming that these older plants have been expanded and modified enough to be classified as new plants and should meet modern air pollution emission standards.

Here is some *good news*. Between 1994 and 1997, the emission trading system helped reduce SO_2 emissions in the United States by 30%. The cost of doing this was less than one-tenth the cost projected by industry because this market-based system motivated companies to reduce emissions in more efficient ways. In 1997, the EPA proposed a voluntary emissions trading program involving smog-forming nitrogen oxides (NO_x) for 22 eastern states and the District of Columbia. Emissions trading programs may also be implemented for particulates emissions.

How Can We Reduce Outdoor Air Pollution? Figure 10-14 summarizes ways to reduce emissions of sulfur oxides, nitrogen oxides, and particulate matter from stationary sources (such as electric power plants and industrial plants that burn coal). Until recently, emphasis has been primarily on dispersing and diluting the pollutants by using tall smokestacks or adding equipment that removes some of the particulate pollutants after they are produced (Figure 10-15). Under the sulfur reduction requirements of the Clean Air Act of 1990, more utilities are switching to low-sulfur coal to reduce SO_2 emissions. Environmentalists call for taxes on air pollutant emissions and greater emphasis on prevention. What do you think?

Figure 10-16 lists ways to reduce emissions from motor vehicles, the primary culprits in producing photochemical smog. Use of alternative vehicle fuels to reduce air pollution is evaluated in Table 10-4, p. 257.

A University of Colorado professor has developed a 1-second highway test for auto emissions. A beam of infrared light is sent across a highway to an air pollution detector that measures levels of carbon monoxide, hydrocarbons, and nitrogen oxides in a car's exhaust as it passes by. The information is fed into a computer and a video camera captures the car's license plate number and stores it in the computer. Drivers can be sent notices directing them to correct faulty emissions and have the vehicle is rechecked at an official testing station or face large fines.

An important way to make significant reductions in air pollution is to get older, high-polluting vehicles off the road. According to EPA estimates, 10% of the vehicles on the road in the United States emit 50–70% of the pollutants. A problem is that many old cars are owned by people who can't afford to buy a newer car. One suggestion would be to pay people to take their old cars off the road, which would result in huge savings in health and air-pollution control costs.

California's South Coast Air Quality Management District Council developed a drastic program to pro-

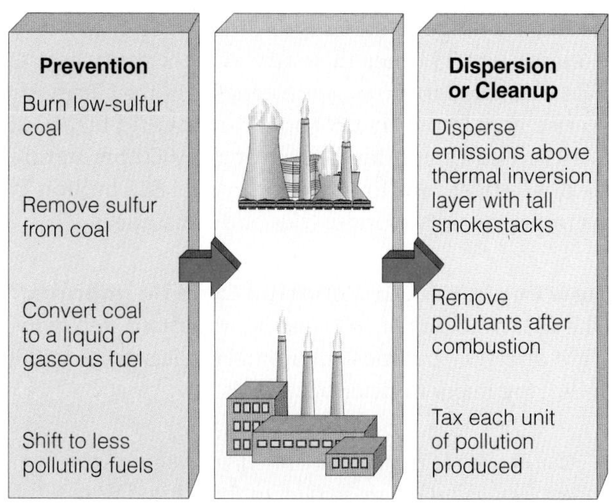

Figure 10-14 Solutions: methods for reducing emissions of sulfur oxides, nitrogen oxides, and particulate matter from stationary sources such as coal-burning electric power plants and industrial plants.

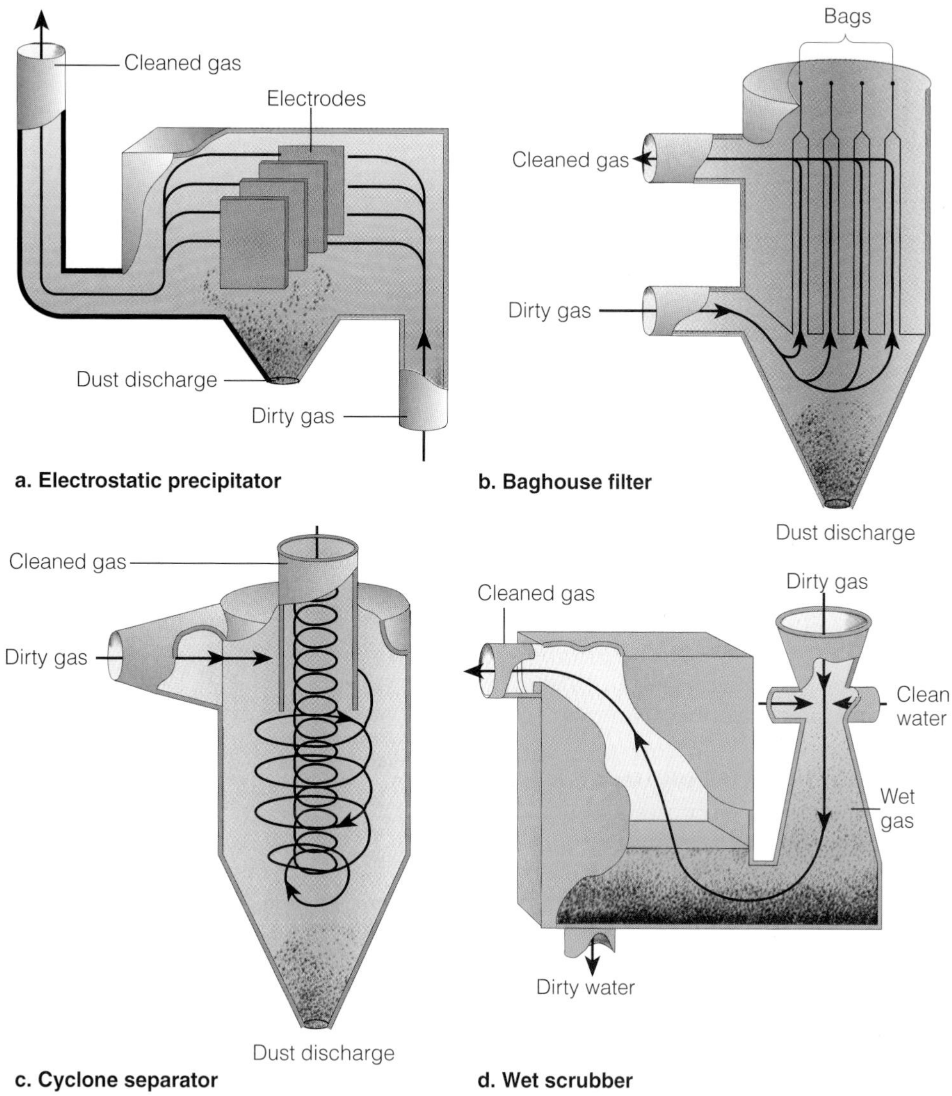

Figure 10-15 Solutions: four commonly used methods for removing particulates from the exhaust gases of electric power and industrial plants. Of these, only baghouse filters remove many of the more hazardous fine particles. All these methods produce hazardous materials that must be disposed of safely, and except for cyclone separators, all of them are expensive. The wet scrubber can also reduce sulfur dioxide emissions.

duce an 80% reduction in ozone, photochemical smog, and other major air pollutants in the Los Angeles area by 2009. This plan would **(1)** sharply reduce use of gasoline-burning engines over two decades by converting cars, trucks, buses, chain saws, outboard motors, and lawn mowers to run on electricity or alternative fuels, **(2)** outlaw drive-through facilities to keep vehicles from idling in lines, **(3)** substantially raise parking fees and assess high fees for families owning more than one car, **(4)** require gas stations to use a hydrocarbon-vapor recovery system on gas pumps and to sell alternative fuels (Table 10-4), **(5)** strictly control or relocate industrial plants and businesses that release large quantities of hydrocarbons and other pollutants, and **(6)** find substitutes for or ban consumer products that release hydrocarbons, including aerosol propellants, paints, household cleaners, and barbecue starter fluids. Such measures are a glimpse of what many other cities may have to do if people, cars, and industries continue to proliferate.

Here is some *good news*. Since the 1960s, Tokyo, Japan (with a current population of about 28 million), has implemented a strict air-pollution control program that has sharply reduced levels of sulfur dioxide, carbon monoxide, and ozone. During the past 25 years outdoor air quality in most western European cities has also improved. The *bad news* is that outdoor air quality has

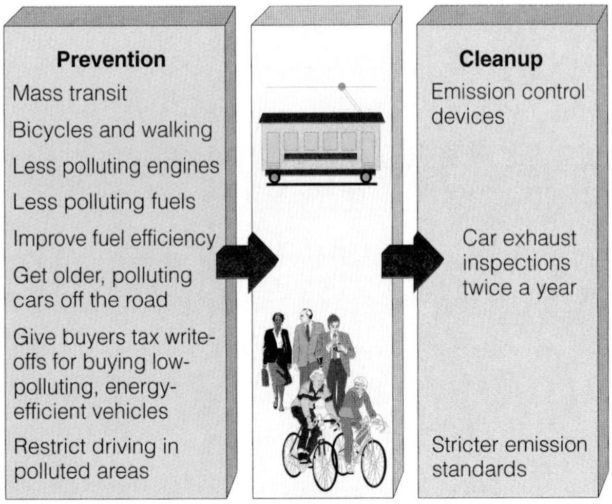

Figure 10-16 Solutions: methods for reducing emissions from motor vehicles.

remained about the same or has gotten worse in most rapidly growing urban areas in developing countries.

How Can We Reduce Indoor Air Pollution? In the United States indoor air pollution poses a much greater health risk for many people than outdoor air pollution. Yet the EPA spends about $500 million per year fighting outdoor air pollution and only about $13 million a year on indoor air pollution.

To reduce indoor air pollution, it's not necessary to impose indoor air quality standards and monitor the more than 100 million homes and buildings in the United States. Instead, air-pollution experts suggest that indoor air pollution reduction can be achieved by several means (Figure 10-17). Another possibility for cleaner indoor air in high-rise buildings is rooftop greenhouses through which building air can be circulated. Some actions you can take to reduce your exposure to indoor air pollutants are listed in Appendix 6.

In developing countries, indoor air pollution from open fires and leaky and inefficient wood- or charcoal-burning stoves (and the resulting high levels of respiratory illnesses) could be reduced if governments gave people simple stoves that burn biofuels more efficiently (which would also reduce deforestation) and that are vented outside, or provided them with simple solar cookers (Figure 20-21d).

How Can We Protect the Atmosphere? An Integrated Approach As population and consumption levels rise, we can generate new air pollution faster than we can clean up the old, even in developed countries with strict air-pollution control laws. As a result, environmentalists believe that protecting the atmosphere, and thus the health of people and many other organisms, will require a global approach that integrates many different strategies. Suggestions for doing this over the next 40–50 years include the following:

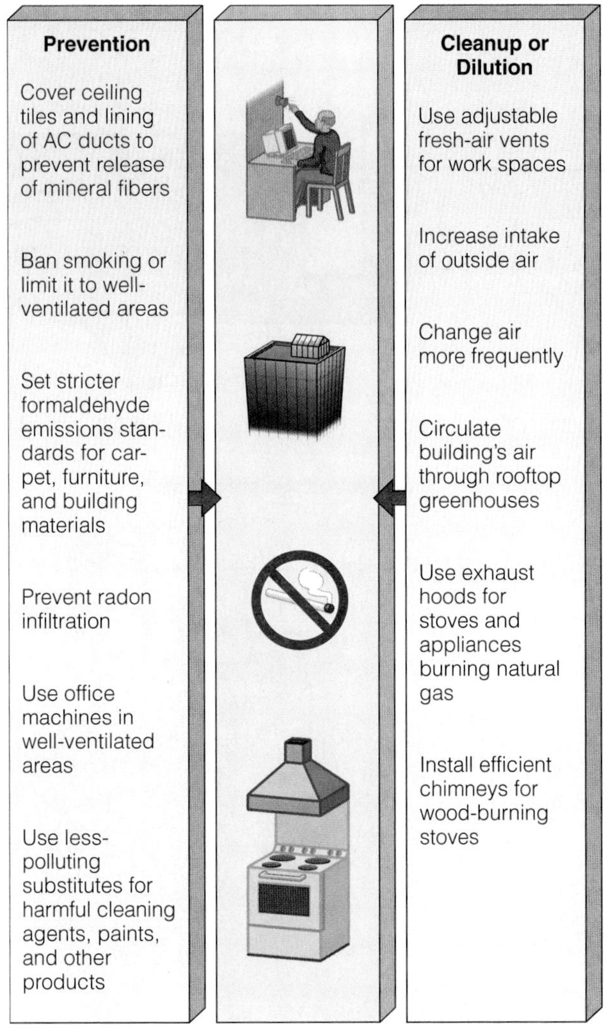

Figure 10-17 Solutions: ways to prevent and reduce indoor air pollution.

- *Putting more emphasis on pollution prevention*
- *Improving energy efficiency*
- *Reducing use of fossil fuels (especially coal and oil)*
- *Increasing use of renewable energy*
- *Slowing population growth*
- *Integrating air-pollution, water-pollution, energy, land-use, population, economic, and trade policies*
- *Regulating air quality for an entire region or airshed*

Table 10-4 Evaluation of Alternatives to Gasoline

Advantages	Disadvantages
Compressed Natural Gas	
Fairly abundant, inexpensive domestic and global supplies Low emissions of hydrocarbons, CO, and CO_2 Vehicle development advanced; well suited for fleet vehicles Reduced engine maintenance	Large fuel tank required; one-fourth the range Expensive engine modification required ($2,000) New filling stations required Nonrenewable resource
Electricity	
Renewable if not generated from fossil fuels or nuclear power Zero vehicle emissions Electric grid in place Efficient and quiet	Limited range and power Batteries expensive Slow refueling (6–8 hours) Power-plant emissions if generated from coal or oil
Reformulated Gasoline (Oxygenated Fuel)	
No new filling stations required Low to moderate emissions reduction of CO No engine modification required	Nonrenewable resource Dependence on imported oil perpetuated No reduction of CO_2 emissions Higher cost Water resources contaminated by leakage and spills
Methanol	
High octane Reduction of CO_2 emissions (total amount depends on method of production) Reduced total air pollution (30–40%)	Large fuel tank required; one-half the range Corrosive to metal, rubber, plastic Increased emissions of potentially carcinogenic formaldehyde High CO_2 emissions if generated by coal High capital cost to produce Hard to start in cold weather
Ethanol	
High octane Reduction of CO_2 emissions (total amount depends on distillation process and efficiency of crop growing) Reduction of CO emissions Potentially renewable	Large fuel tank required; lower range Much higher cost Corn supply limited Competition with food growing for cropland Smog formation possible Corrosive Hard to start in cold weather
Solar–Hydrogen	
Renewable if produced using solar energy Lower flammability Virtually emission-free No emissions of CO_2 Nontoxic	Nonrenewable if generated by fossil fuels or nuclear power Large fuel tank required No distribution system in place Engine redesign required Currently expensive

- *Phasing in full-cost pricing, mostly by taxing the production of air pollutants*
- *Distributing cheap and efficient cookstoves and solar cookstoves in developing countries*
- *Transferring the latest energy-efficiency, renewable-energy, pollution prevention, and pollution control technologies to developing countries*

Making such changes will be controversial and expensive. However, proponents argue that not implementing such an integrated approach will cost far more in money, poor human health, premature death, and ecological damage. What do you think?

Turning the corner on air pollution requires moving beyond patchwork, end-of-pipe approaches to confront pollution at its sources. This will mean reorienting energy, transportation, and industrial structures toward prevention.

HILARY F. FRENCH

REVIEW QUESTIONS

1. Define the boldfaced terms used in this chapter.

2. How can lichens be used to detect air pollutants?

3. Distinguish between *atmosphere*, *troposphere*, and *stratosphere*. What key role does the stratosphere play in maintaining life on the earth?

4. Distinguish between the *greenhouse effect* and the *ozone shield* and explain their importance in sustaining life on the earth.

5. Distinguish between *air pollution*, *primary air pollutants*, and *secondary air pollutants*. List the major classes of pollutants found in outdoor air. Distinguish between *stationary* and *mobile* sources of pollution for outdoor air.

6. What is *photochemical smog* and how does it form? What is *industrial smog* and how does it form? What major factors influence the formation of photochemical and industrial smog? What is a *temperature* or *thermal inversion* and what are its harmful effects? Why do Los Angeles, California, and Mexico City, Mexico, experience frequent thermal inversions?

7. Distinguish among *acid deposition*, *wet deposition*, and *dry deposition*. What areas tend to be affected by acid deposition? What are the major harmful effects of acid deposition on **(a)** trees and other plants, **(b)** soils, **(c)** aquatic life, **(d)** human health, and **(e)** materials? How serious is acid deposition in the United States? List seven ways to prevent acid deposition. What are the advantages and disadvantages of liming acidified lakes to reduce the effects of acid deposition?

8. How serious is indoor air pollution and what are some of its sources? What is the *sick-building syndrome*? According to the EPA, what are the three most dangerous indoor air pollutants in the United States? What is the most dangerous indoor air pollutant in most developing countries?

9. Summarize the problem of indoor pollution from radioactive radon gas. What can be done about it?

10. Summarize the problem of exposure to asbestos fibers. What can be done about it?

11. List four defenses your body has against air pollution. What are the major harmful health effects of **(a)** carbon monoxide, **(b)** suspended particulate matter, **(c)** sulfur dioxide, **(d)** nitrogen oxides, **(e)** volatile organic compounds, and **(f)** ozone?

12. About how many people die prematurely each year from exposure to air pollutants in **(a)** the United States and **(b)** the world? List ways that plants (including food crops) are harmed by air pollution. List ways that air pollutants damage aquatic life.

13. What is the Clean Air Act and how has it helped reduce outdoor air pollution in the United States? According to environmentalists, what are six weaknesses of the current Clean Air Act in the United States?

14. What is an *emission trading policy* and what are the pros and cons of using this approach to help reduce air pollution?

15. List the major prevention and cleanup methods for dealing with air pollution from **(a)** emissions of sulfur oxides, nitrogen oxides, and particulate matter from stationary sources, **(b)** automobile emissions, **(c)** indoor air pollution in developed countries, and **(d)** indoor air pollution in developing countries.

16. What are 10 components of an integrated approach for dealing with air pollution?

CRITICAL THINKING

1. Evaluate the pros and cons of the following statement: "Because we have not proven absolutely that anyone has died or suffered serious disease from nitrogen oxides, current federal emission standards for this pollutant should be relaxed."

2. What climate and topographical factors in your local community intensify the level of air pollution, and which of these factors help reduce air pollution?

3. Should all tall smokestacks be banned? Explain.

4. Should annual government-held auctions of marketable trading permits be used as a primary way of controlling and reducing air pollution? Explain. What conditions, if any, would you put on this approach?

5. Do you agree or disagree with the possible weaknesses of the U.S. Clean Air Act listed on pp. 253–254? Defend each of your choices. Can you identify other weaknesses?

***6.** Evaluate your exposure to some or all of the indoor air pollutants in Figure 10-10 in your school, workplace, and home. Come up with a plan for reducing your exposure to these pollutants.

***7.** Have buildings at your school been tested for radon? If so, what were the results? What has been done about areas with unacceptable levels? If this testing has not been done, talk with school officials about having it done.

***8.** Make a concept map of this chapter's major ideas, using the section heads and subheads and the key terms (in boldface). Look at the inside back cover and in Appendix 5 for information about concept maps.

FURTHER INFORMATION

For more information see Internet Sources and Further Readings at the back of the book. Online resources for this chapter can be found at **www.brookscole.com/biology**

11 GLOBAL WARMING AND OZONE LOSS

A.D. 2060: Hard Times on Planet Earth*

Mary Wilkins sat in the living room of her underground home in Illinois (Figure 11-1), which she shared with her daughter Jane and her two children. It was July 4, 2060: Independence Day. There would be no parade or barbecue or fireworks today; people didn't stay outside for long because of the searing heat from global warming—called the Warming by most people.

Millions of Americans had long ago migrated to Canada to find a cooler climate and more plentiful food supply after much of America's Midwestern breadbasket and central and southern California had been mostly abandoned because of a lack of water and food. By 2050 much of Louisiana's coastline and wetlands were under water, as were most of the low-lying areas along the East Coast of the United States.

A door opened and her daughter Jane came out, ready to go to work at the refugee center. Thousands of these overwhelmed centers had been opened up throughout the United States to help the estimated 60 million people displaced from their jobs and homes by the economic and social chaos caused by the Warming.

Her 10- and 12-year-old grandchildren, Lynn and Jeffrey, followed their mother out of the bedroom. Their father had died 5 years ago while fighting almost endless fires that had wiped out huge areas of forest and released carbon dioxide that accelerated the Warming.

*Compare this fictional worst-case scenario with the fictional hopeful scenario that opens Chapter 2.

She began telling them about the old days, before the Warming. "Almost everyone drove cars that ran on gasoline. But in 2040 use of gasoline and coal was banned because burning these fuels emitted a gas called carbon dioxide that helped make the climate warmer. Now most people get around by foot or bicycle or on buses running on less polluting natural gas, like the one your mother takes to work. There is no air conditioning because coal-burning electric power plants have been banned and the few remaining nuclear power plans are too expensive to run.

"Since the 1970s scientists urged us to get most of the energy we needed from the sun, wind, flowing water, underground heat, and hydrogen gas produced using sunlight. But we waited too long to do this."

"Why did people let things get so bad?" asked Lynn, the older child. Mary's eyes filled with tears as she took the child in her arms, "Because we didn't want to believe anything bad could happen with a problem whose effects we wouldn't see until 50 years later."

Many of the world's leading scientists believe that two of the greatest threats we face are **(1)** global warming from fossil fuel burning and deforestation that enhances the earth's natural greenhouse effect (Figure 6-8) and **(2)** depletion of stratospheric ozone, caused by our use of chlorofluorocarbons and other chemicals (Table 10-1).

Other analysts believe that these problems are exaggerated or that we need more research before acting to reduce such trends by phasing out fossil fuels and halting deforestation. Most environmental scientists argue that even without global warming we should be doing these things to save money, reduce pollution, and preserve biodiversity.

Figure 11-1 An earth-sheltered house in Will County, Illinois, in the United States. About 13,000 families across the United States have built such houses. Mary Wilkins's fictional house in 2060 could be similar to this one. (Pat Armstrong/Visuals Unlimited)

We are embarked on the most colossal ecological experiments of all time—doubling the concentration in the atmosphere of an entire planet of one of its most important gases—and we really have little idea of what might happen.

PAUL A. COLINVAUX

This chapter addresses the following questions:

- What is the difference between the greenhouse effect and global warming?
- What do we know about the earth's temperatures in the past and about possible temperatures in the future?
- Can we really make the earth warmer, and if so, will a few degrees matter?
- What are some possible effects of global warming?
- What can we do about global warming?
- Are we depleting ozone in the stratosphere, and why should we care?
- What can we do to slow ozone depletion?

11-1 THE GREENHOUSE EFFECT AND GLOBAL WARMING

What Is the Greenhouse Effect? The *greenhouse effect* (Figure 6-8 and Table 10-1) is the natural warming of the lower atmosphere (troposphere) because of the presence of certain atmospheric gases, called *greenhouse gases*. This natural effect, first proposed by Swedish chemist Svante Arrhenius in 1896, has been confirmed by numerous laboratory experiments and atmospheric measurements. It is one of the most widely accepted theories in the atmospheric sciences. Indeed, without the heating provided by the natural greenhouse effect, the earth would be a lifeless, ice-covered rock spinning through space.

The major greenhouse gases are water vapor (H_2O), carbon dioxide (CO_2), ozone (O_3), methane (CH_4), nitrous oxide (N_2O), chlorofluorocarbons (CFCs), and perfluorocarbons (PFCs, such as CF_4). The two predominant greenhouse gases in the troposphere are **(1)** water vapor, controlled by the hydrologic cycle (Figure 4-22), and **(2)** carbon dioxide, controlled by the global carbon cycle (Figure 4-23).

The primary greenhouse gas is water vapor. However, because its concentration in the atmosphere is fairly high (1–5%), inputs of water vapor from human activities have little effect on this chemical's greenhouse effects. By contrast, the concentration of carbon dioxide in the atmosphere is so small (0.036%) that a fairly large input of CO_2 from human activities can affect the average temperature of the troposphere.

What Is Global Warming? Measured atmospheric levels of certain greenhouse gases—CO_2, CFCs, methane, and nitrous oxide—have risen substantially in recent decades (Figure 11-2) and are projected to enhance the earth's natural greenhouse effect, a phenomenon called **global warming** (Table 10-1).

Most of the increases in levels of these greenhouse gases since 1958 have been caused by human activities, primarily burning fossil fuels, agriculture, deforestation, and use of CFCs. Although molecules of CFCs, methane, and nitrous oxide release much more heat per molecule than CO_2, the much larger input of CO_2 makes it the most important greenhouse gas produced by human activities.

Developed countries account for about 60% of current CO_2 emissions and developing countries for the remaining 40%. The United States alone accounts for about 23% of global CO_2 emissions from human activities (with 80% of these emissions produced by 122 large corporations, according to a 1999 report by the Natural Resources Defense Council), followed by China (14%), Russia (7%), and Japan (5%). However, CO_2 emissions are increasing rapidly in developing countries such as China and India that are industrializing rapidly.

Long cylinders of ice drilled out of antarctic ice sheets contain a long historical record of snowfall, dust, air bubbles, and chemicals (such as carbon dioxide and methane) found in the atmosphere. The clues, trapped in the ice layers like tree rings, enable scientists to reconstruct past climates. Such ice core analysis reveals that at the beginning of the industrial revolution the atmospheric concentration of CO_2 was about 280 parts per million (referred to as the *preindustrial level*). Between 1860 and 1998 the concentration of CO_2 in the atmosphere grew exponentially to 370 parts per million (Figure 11-2a), higher than any time in the past 160,000 years. The atmospheric concentrations of CO_2 and other greenhouse gases are projected to double from preindustrial (1860) levels sometime during the 21st century—probably by 2050—and then continue to rise.

11-2 PAST AND FUTURE CHANGES IN THE EARTH'S CLIMATE

What Is the Scientific Consensus About the Earth's Past Temperatures? The Intergovernmental Panel on Climate Change (IPCC) is a network of about 2,500 of the world's leading climate experts from 70 nations organized by the United Nations to study climate change. Since 1990 the IPCC has published several reports evaluating the best available evidence

Figure 11-2 Increases in average concentrations of major greenhouse gases in the troposphere, mostly as a result of human activities. (Data from Electric Power Research Institute. Adapted and updated by permission from Cecie Starr, *Biology: Concepts and Applications*, 4th ed., Pacific Grove, Calif.: Brooks/Cole, 2000)

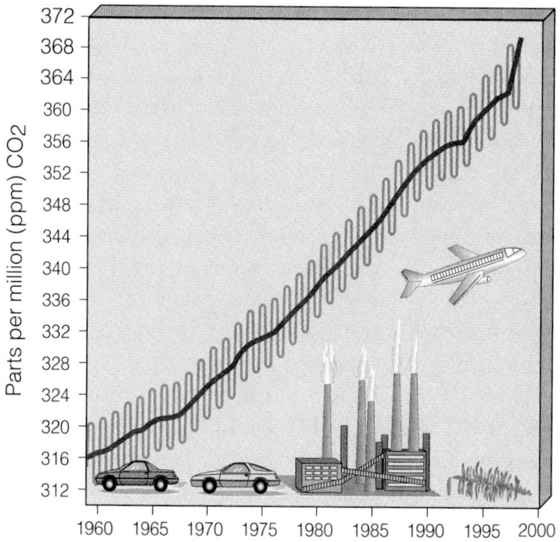

(a) Carbon dioxide (CO_2) is responsible for 50–60% of the global warming from greenhouse gases produced by human activities since preindustrial times. The main sources are **(1)** fossil-fuel burning (70–75%) and **(2)** land clearing and plant burning (20–25%). Most of the CO_2 comes from burning coal, but an increasing fraction is released from motor vehicle exhaust. CO_2 remains in the atmosphere for 50–200 years. The annual rise and fall of CO_2 levels shown in the graph result from less photosynthesis during winter and more during summer.The 3 parts per million increase in the atmospheric concentration of carbon dioxide in 1998 was the largest ever recorded.

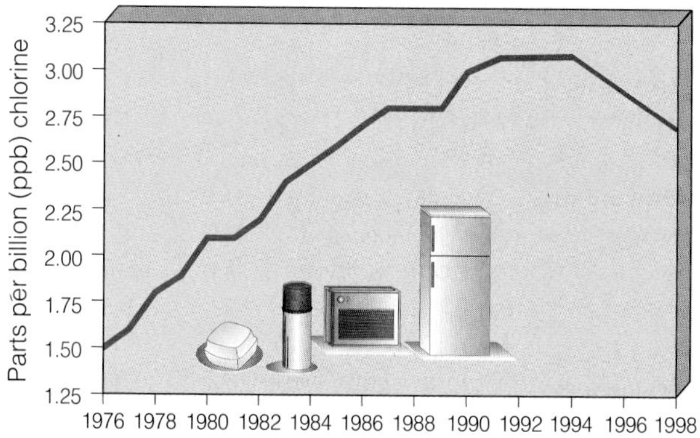

(b) Chlorofluorocarbons (CFCs) contribute to global warming in the troposphere and deplete ozone in the stratosphere. The main sources are **(1)** leaking air conditioners and refrigerators, **(2)** evaporation of industrial solvents, **(3)** production of plastic foams, and **(4)** aerosol propellants. CFCs take 10–20 years to reach the stratosphere and generally trap 1,500–7,000 times as much heat per molecule as CO_2 while they are in the troposphere. This heating effect in the troposphere may be partially offset by the cooling caused when CFCs deplete ozone during their 65- to 135-year stay in the stratosphere. Their use is being phased out.

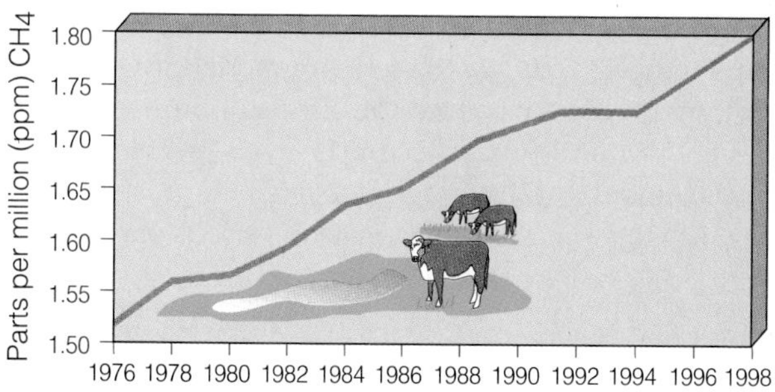

(c) Methane (CH_4) accounts for about 20% of the overall warming effect. It is produced when anaerobic bacteria break down dead organic matter in moist places that lack oxygen. These areas include **(1)** swamps and other natural wetlands, **(2)** rice paddies, **(3)** landfills, and **(4)** the intestinal tracts of cattle, sheep, and termites. Production and use of oil and natural gas (especially from leaks in natural gas pipelines) and incomplete burning of organic materials (including biomass burning in the tropics) also are significant sources. CH_4 stays in the troposphere for 9–15 years. Each CH_4 molecule traps about 20 times as much heat as a CO_2 molecule.

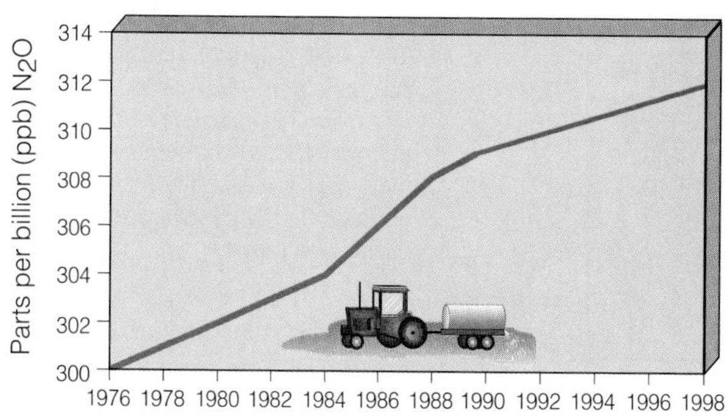

(d) Nitrous oxide (N_2O) contributes to global warming and can also deplete ozone in the stratosphere. It is released from **(1)** nylon production, **(2)** burning of biomass and nitrogen-rich fuels (especially coal), **(3)** smog-fighting catalytic converters on motor vehicles, and **(4)** the breakdown of nitrogen fertilizers in soil, livestock wastes, and nitrate-contaminated groundwater. Its life span in the troposphere is about 120 years, and it traps about 200 times as much heat per molecule as CO_2.

concerning **(1)** the greenhouse effect, **(2)** past changes in global temperatures, **(3)** climate models projecting future changes in global temperatures and climate, **(4)** what we can do to delay or reduce global warming, and **(5)** what we can do to adjust its effects.

Analysis of the content of ancient ice in antarctic glaciers and other data show that the earth's average surface temperature has fluctuated considerably over geologic time. These data show that during the past 800,000 years several ice ages have covered much of the planet with thick ice. Each glacial period lasted about 100,000 years and was followed by a warmer interglacial period of 10,000–12,500 years (Figure 11-3).

For the past 10,000 years we have enjoyed the warmth of the latest interglacial period (called the Holocene). This climatic stability has prevented drastic changes in the nature of soils and vegetation patterns throughout the world, allowing large increases in food production and thus in population. However, even small temperature changes during this period have led to hardship and premature deaths and to large migrations of peoples in response to changed agricultural and grazing conditions.

Analysis of gases in bubbles trapped in ancient ice shows that over the past 160,000 years tropospheric water vapor levels (the dominant greenhouse gas) have remained fairly constant. During most of this period levels of CO_2 have fluctuated between 190 and 290 parts per million. Estimated changes in the levels of tropospheric CO_2 correlate fairly closely with estimated variations in the earth's mean surface temperature during the past 160,000 years (Figure 11-4).

Since 1860 (when measurements began), mean global temperature (after correcting for urban heat island effects; see Figure 9-24) has risen almost 0.7°C (1.3°F), with most of this rise occurring since 1946 (Figure 11-5). The 15 warmest years on record since 1860 have occurred in the past two decades, with 1998 the hottest. The average global temperature in 1998 exceeded the previous record set in 1995 by 0.25°C (0.45°F), a huge jump. In 1999, researchers reported that a 3.2-kilometer-long ice core drilled out of the antarctic ice sheet shows that the levels of greenhouse gases are higher now than at any time in the past 420,000 years.

Some critics contend that the slight cooling trend between 1940 and 1975 (Figure 11-5) contradicts current climate models. However, research has shown that this cooling trend was caused by increased levels of light-obscuring fine particles (produced by burning fossil fuels) and dust in the troposphere during this period. Current climate models account for this effect.

Many uncertainties remain. Some or most of the roughly 0.7°C rise since 1860 could result from normal fluctuations in the mean global temperature. On balance, however, the IPCC concluded in its 1995 report that "the observed increase over the last century is unlikely to be entirely due to natural causes" and that "the balance of evidence suggests that there is a discernible human influence on global climate."

How Do Scientists Model Climate Changes? Computer Models as Crystal Balls To project the effects of increases in greenhouse gases on average global temperature and changes in the earth's climate, scientists develop *mathematical models* of such systems and run them on supercomputers. How well the results correspond to the real world depends on **(1)** the design and assumptions of each model, **(2)** the accuracy of the data used, **(3)** magnification of tiny errors over time, **(4)** factors in the earth's climate system that amplify (positive feedback) or dampen (negative feedback) changes in average global temperatures, and **(5)** the effects of totally unexpected or unpredictable events (chaos).

Figure 11-6 shows the series of models involved in making predictions about the effects of human activities and natural processes on **(1)** levels of greenhouse gases (model 1), **(2)** average global temperature (model 2), **(3)** changes in regional climate in various parts of the

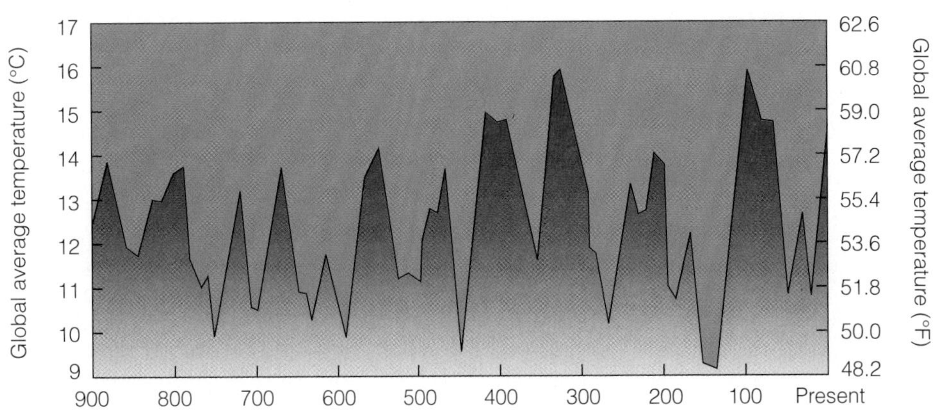

Figure 11-3 During the past 900,000 years the earth has experienced cycles of ice ages, each lasting about 100,000 years and then followed by warmer interglacial periods lasting 10,000–12,500 years. The warm interglacial period during the past 10,000 years has been a major factor in the development of agriculture, human civilizations, and population growth.

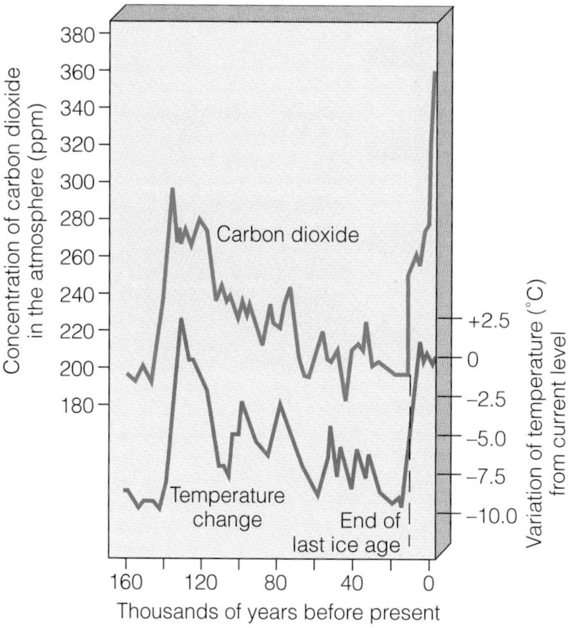

Figure 11-4 Estimated long-term variations in mean global surface temperature and average tropospheric carbon dioxide levels over the past 160,000 years. These CO_2 levels were obtained by inserting metal tubes deep into antarctic glaciers, removing the ice, and analyzing bubbles of ancient air trapped in ice at various depths throughout the past. Such analyses reveal that since the last great ice age ended about 10,000 years ago, we have enjoyed a warm interglacial period. The rough correlation between tropospheric CO_2 levels and temperature shown in these estimates based on ice core data suggests a connection.

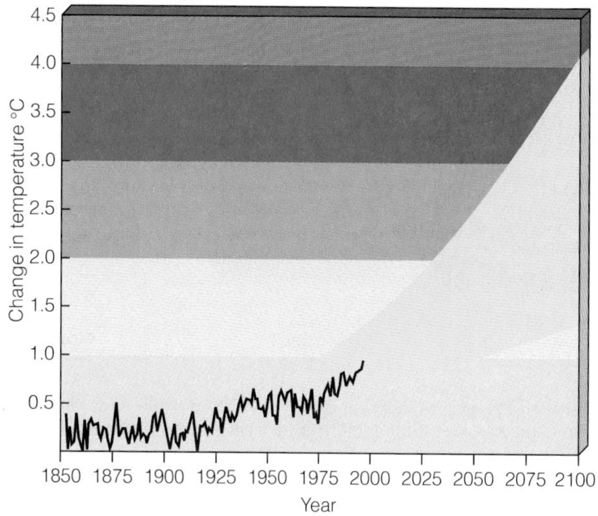

Figure 11-5 Recorded changes in the earth's mean surface temperature between 1860 and 1998 (dark line). The curved yellow region shows global warming projected by various computer models of the earth's climate systems. Note that the climate model projections match the almost 0.7°C (1.3°F) recorded temperature increase between 1860 and 1998. Current models indicate that the average global temperature will rise by 1.3–4.1°C (2.3–7.3°F) sometime during the next century. However, this projection assumes that air pollution from sulfate aerosols will continue to increase and exert a slight cooling effect, despite the fact that sulfur emissions have leveled off since 1960. The projected warming shown here could be *overestimated* or *underestimated* by a factor of two. (Data from U.S. National Academy of Sciences and National Center for Atmospheric Research)

world (model 3), and **(4)** possible effects of such changes in these different areas (models 4 through 7). Current improved versions of models 1 and 2 account for the temperature changes observed since 1860 and project that the earth's average global temperature will increase. However, they disagree on how such warming might affect the climate in different areas (model 3) and the effects of such changes in these areas (models 4 through 7).

Climate models are improving but are still cloudy crystal balls. However, without them we are groping around in darkness trying to understand and deal with serious issues.

What Is the Scientific Consensus About Future Global Warming and Its Effects? In 1995 the IPCC projected that according to the latest climate models, the earth's mean surface temperature should rise 1.3–4.1°C (2.3–7.3°F) between 1990 and 2100 (Figure 11-5). This temperature rise may not seem like much, but even at the lowest projected increase of 1.3°C (2.3°F), the earth would be warmer than it has been for 10,000 years. Current models project that climate change, once begun, will continue for hundreds of years.

According to the models, the northern hemisphere should warm more and faster than the southern hemisphere and there should be a more pronounced warming at the earth's poles. Both of these effects have occurred (Connections, p.265).

Other signs of global warming that have been observed include **(1)** increased retreat of some glaciers on the tops of mountains in the Alps, Andes, Himalayas, and Northern Cascades of Washington during the last 30 years, **(2)** northward migrations of some warm-climate fish and trees, **(3)** spread of some tropical diseases away from the equator, and **(4)** increased bleaching of coral reefs in tropical areas with warmer water.

Climate models also project that as the earth's atmosphere warms, the rate of water evaporation will increase and global average precipitation will rise. Much of it will come in heavy showers or thunderstorms (thus increasing flooding) rather than in gentler, longer-lasting rainfalls. Such changes are already being observed in many parts of the northern hemisphere.

With a warmer climate, global sea levels will rise, mainly because water expands slightly when heated. Current climate models of global warming project that global sea levels will rise by 15–99 centimeters (7–39 inches) between 1990 and 2100. However, the rise could be much greater if all or a large portion of Antarctica's western ice shelf breaks up (Connections, p.265). It is

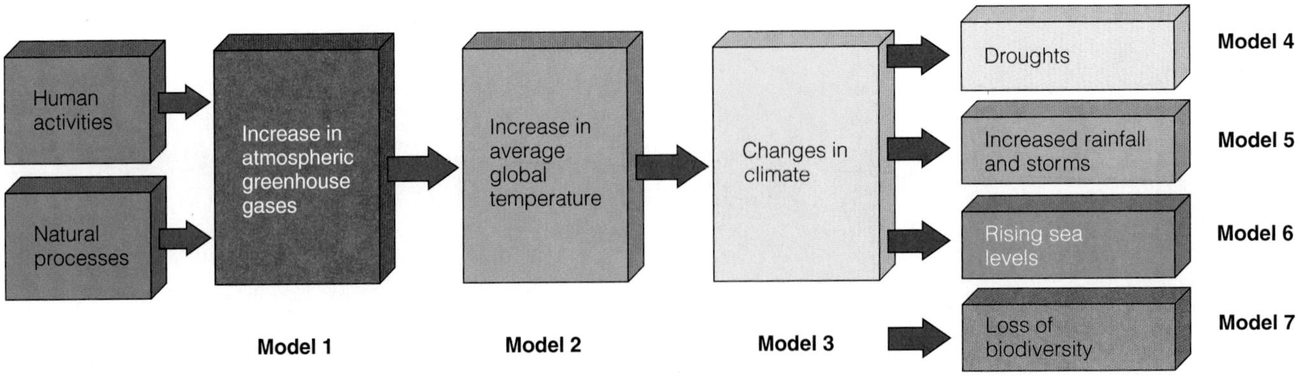

Figure 11-6 Generalized schematic of the greenhouse effect and models of its possible consequences. The greenhouse effect itself (model 1) is well established. The increase in average global temperature (Figure 11-5) resulting from higher concentrations of greenhouse gases (model 2) is projected by mathematical climate models. Current models cannot consistently project either the specific climate changes (model 3) or their consequences (models 4–7) in various parts of the world.

sobering to realize that the earth's sea level rose about 92 meters (300 feet) as the planet warmed at the end of the last ice age.

IPCC scientists warn that warming or cooling by more than 1°C (1.8°F) over a few decades (instead of over many centuries, as has been the pattern during the last 10,000 years) will cause serious disruptions of the earth's ecosystems and human societies.

11-3 GLOBAL WARMING OR A LOT OF HOT AIR? HOW SERIOUS IS THE THREAT?

Will the Earth Really Get Warmer? There is controversy over **(1)** whether we are already experiencing global warming, **(2)** how warm temperatures might be in the future, and **(3)** the effects of such temperature increases. One problem is that we have only about 100 years of fairly accurate data. This makes it difficult to separate the normal short-term ups and downs of global temperatures (called *climate noise*) from an overall rise in average global temperature.

Scientists have also identified a number of factors that might amplify (positive feedback) or dampen (negative feedback) a rise in average atmospheric temperature. These factors influence how fast temperatures might climb and what the effects might be on various areas. Let's look more closely at the possible effects of such factors.

How Do Changes in Solar Output Affect Earth's Temperatures? Solar output varies by about 0.1% over the 11-year and 22-year sunspot cycles—and over 80-year and other much longer cycles. These up-and-down changes in solar output can temporarily warm or cool the earth and thus affect the projections of climate models.

However, no known mechanism can explain how the small observed changes in solar output over the past hundred years could result in the warming trend observed during this period. In addition, increased inputs of greenhouse gases to the troposphere from human activities can worsen any temperature increase caused by increased solar output.

How Do the Oceans Affect Climate? Currently the oceans help moderate tropospheric temperature by removing about 29% of the excess CO_2 we pump into the atmosphere, but we don't know whether they can absorb more. If the oceans warm up enough, some of the dissolved CO_2 will bubble out into the atmosphere (just as in a glass of carbonated ginger ale left out in the sun), amplifying and accelerating global warming.

Global warming could be dampened if the oceans absorbed more heat, but this depends on how long the heat takes to reach deeper layers. Recent measurements indicate that deep vertical mixing in the ocean occurs extremely slowly (taking hundreds of years).

In 1999, researchers found that from 1991 to 1994, when a strong El Niño–Southern Oscillation (ENSO, Figure 6-6) warmed the Pacific Ocean, the ocean released 30–80% less CO_2. Thus, strong ENSOs could help slow global warming, but they also have their own harmful effects (Figure 6-7). Since 1970 there have been more and longer-lasting ENSOs.

There is also concern that deep ocean currents could be disrupted. At present, these currents (driven largely by differences in water density and winds) act like a gigantic conveyor belt, transferring heat from one place to another and storing carbon dioxide in the deep sea

Early Warnings from the Earth's Ice

CONNECTIONS

During the past 50 years scientists have observed dramatic changes at the earth's two frigid poles, which are regarded as early warning sentinels of global warming.

News from the Arctic is alarming. Between 1968 and 1997, surface temperatures at nine stations north of the arctic circle have risen by about 5.5°C (9.9°F)—over 10 times the average global temperature increase. Also, the area covered by sea ice around the Arctic decreased by about 5% between 1978 and 1997.

The bright color of the floating sea ice cover in the Arctic Ocean (a body of water half the size of the United States) helps cool the earth by reflecting sunlight. Because it is floating, large-scale melting of Arctic Sea ice wouldn't raise global sea levels. However, it would reroute some major ocean currents (Figure 6-2) and affect the climate of other parts of the globe.

Studies also show that as the frozen north warms and thaws, peat buried in the arctic tundra soil would decay and release large amounts of CO_2. This positive feedback mechanism would speed up the rate of increase of CO_2 in the atmosphere. According to climate researchers at San Diego State University (California), the arctic tundra has been removing carbon dioxide from the atmosphere for 9,000 years but since 1982 has become a net emitter of carbon dioxide to the atmosphere.

The news from the Antarctic is also disturbing. The huge antarctic ice cap contains 70% of the world's fresh water and 90% of the world's highly reflective ice. In addition to having a major influence on global climate, the antarctic ice cap can affect the global sea level.

Since 1947 the average temperature at Antarctica has risen by more than 2.5°C (4.5°F) in summer and by 5.6°C (10°F) in winter. As a result, 10 huge pieces (some as large as the state of Delaware) have broken off of the floating western antarctic ice shelf since 1950 (four between 1994 and 1998). During the past few years, this breakup of ice shelves has accelerated.

Until recently scientists had projected that partial melting or breakdown of the western antarctic ice sheet would raise global sea levels by as much as 0.9 meters (3 feet) by 2100. Now some are considering the possibility that virtually all of this sheet will be gone by 2100. If this happens, the global sea level would rise about 5 meters (16 feet) and flood many of the world's coastal cities, wetlands, and low-lying islands.

Critical Thinking

What difference might it make in your life and in that of any child you might have if human activities play an important role in continued warming of the earth's polar regions? What three important things could you do to help slow down such warming?

(Figure 11-7). Global warming could halt this thermal conveyor belt by reducing the density and salinity of water in the North Atlantic so that the water would not sink. If this loop stalls out, evidence from past climate changes indicates that this could trigger atmosphere temperature changes of more than 5°C (9°F) over periods as short as 40 years. Increased melting of Antarctica's western ice sheet could accelerate this change (Connections, above).

Evidence is also accumulating that a warmer planet will pour fresh water into the oceans in the form of additional rainfall and melting ice and snow. Jorge Sarmiento, an oceanographer at Princeton University, has run computer models showing that an upper layer of less dense fresh water could reduce the ability of the oceans to absorb excess carbon dioxide by 10–30% by 2050. This would accelerate global warming.

How Do Water Vapor Content and Clouds Affect Climate? Warmer temperatures would increase evaporation and the water-holding capacity of the air and create more clouds. These additional clouds could have a warming effect by trapping heat or a cooling effect by reflecting sunlight back into space. The net result of these two opposing effects depends on whether it is day or night and on the type (thin or thick) and altitude of clouds. Scientists don't know which of these factors might predominate or how cloud types and heights might vary in different parts of the world as a result of global warming.

In 1999, researchers at the University of Colorado reported that the wispy condensation trails left behind by jet planes may have a greater impact on the earth's climate than scientists had thought. Using infrared satellite images, they found that jet contrails expand and turn into large cirrus clouds that tend to trap heat in the upper atmosphere. If these preliminary results are confirmed, jet planes could be responsible for as much as half of the regional warming in the northern hemisphere.

How Might Changes in Polar Ice Affect Climate? The earth's light-colored ice sheets act like enormous mirrors, reflecting sunlight back into space. If warmer

Figure 11-7 This loop of ocean water stores carbon dioxide in the deep sea and brings warmth to Europe. It occurs when ocean water in the North Atlantic is dense enough (because of its salt content and cold temperature) to sink to the ocean bottom and well up in the warmer Pacific (helping cool that part of the world). Then a shallower return current aided by winds brings warmer and less salty—and thus less dense—water to the Atlantic, which can then cool and sink to begin the cycle again. If this heat conveyor belt or loop stalls out because of a drop in density of ocean water in the North Atlantic (possibly from changes caused by global warming), massive climate changes over much of the earth's surface could occur within only a few decades. Models indicate that this oceanic heat conveyor belt would return, but only after hundreds or thousands of years. (Used by permission from Cecie Starr, *Biology: Concepts and Applications*, 4th ed., Pacific Grove, Calif.: Brooks/Cole, 2000)

temperatures melted some of this ice and exposed darker ground or ocean, more sunlight would be absorbed and warming would be accelerated. Then more ice would melt, further accelerating the rise in atmospheric temperature (Connections, p. 265).

On the other hand, the early stages of global warming might increase the amount of the earth's water stored as ice. Warmer air would carry more water vapor, which could drop more snow on some polar glaciers, especially the gigantic antarctic ice sheets. If snow accumulated faster than ice was lost, the ice sheet would grow, reflect more sunlight, and help cool the atmosphere, perhaps leading to a new ice age within a thousand years.

How Might Air Pollution Affect Climate? Projected global warming might be partially offset by *aerosols* (tiny droplets and solid particles) of various air pollutants released or formed in the atmosphere by volcanic eruptions and human activities. It is hypothesized that higher levels of aerosols attract enough water molecules to form condensation nuclei, leading to increased cloud formation.

The resulting clouds would reflect more incoming sunlight back into space during daytime and help counteract the heating effects of increased greenhouse gases. However, nights would be warmer because the clouds would still be there and prevent some of the heat stored in the earth's surface (land and water) during the day from being radiated into space. These pollutants may explain why most of the recent warming in the northern hemisphere occurs at night.

However, these interactions are complex. Pollutants in the lower troposphere can either warm or cool the air, depending on the reflectivity of the underlying surface. These uncertainties, plus improved air pollution control, make it unlikely that air pollutants will counteract projected global warming very much in the next 50 years. Aerosols also fall back to the earth or are washed out of the atmosphere within weeks or months, whereas CO_2 and other greenhouse gases remain in the atmosphere for decades to several hundred years.

These aerosols are also major components of acid deposition (Section 10-3), which slows forest growth and weakens or kills many trees. This reduces the ability of growing trees to absorb some of the CO_2 we are putting into the atmosphere and could accelerate global warming.

We could maintain or increase levels of aerosol air pollutants to offset possible global warming. However, because these pollutants already kill hundreds of thousands of people a year and damage food crops, trees, and other forms of vegetation (Section 10-5), they must be reduced (Section 10-6).

How Might Increased CO_2 Levels Affect Photosynthesis and Methane Emissions? Some studies suggest that more CO_2 in the atmosphere could increase the rate of photosynthesis in areas with adequate amounts of water and other soil nutrients. This would remove more CO_2 from the atmosphere and help slow global warming.

Other studies suggest that this effect varies with different types of plants and in different climate zones. Also, much of any increased plant growth could be offset by plant-eating insects that breed more rapidly and year-round in warmer temperatures. Weeds may also grow more rapidly at the expense of food crops.

Global warming could be accelerated by increased release of methane (a potent greenhouse gas) from wetlands. A 1994 study indicated that increased uptake of CO_2 by wetland plants could boost emissions of methane by providing more organic matter for methane-producing anaerobic bacteria to decompose. Some scientists also speculate that in a warmer world huge amounts of methane now tied up in arctic tundra soils and in muds on the bottom of the Arctic Ocean might be released if the blanket of permafrost in tundra soils melts and the Arctic Ocean warms considerably. Conversely, some scientists believe that bacteria in tundra soils would rapidly oxidize the escaping methane to CO_2, a less potent but still important greenhouse gas.

Goldsmith, E. and C. Henderson. 1999. "The Economic Costs of Climate Change." *The Ecologist*, vol. 29, no. 12, p. 98(3).

How Might Forest Turnover Affect CO_2 Levels? Another factor that can affect CO_2 levels is *forest turnover*: the rate at which trees grow and die in a forest. A 1994 study indicated that the turnover in tropical forests worldwide is accelerating. Besides reducing the biodiversity of tree species in tropical forests, this change could enhance global warming by reducing removal of CO_2 from the atmosphere because less dense, faster-growing trees require less CO_2 for growth.

How Rapidly Could Climate Shift? If moderate change takes place gradually over several hundred years, people in areas with unfavorable climate changes may be able to adapt to the new conditions. However, if the projected global temperature change takes place over several decades, we may not be able to switch food-growing regions and relocate the large portion of the world's population living near coastal areas fast enough. The result would be large numbers of premature deaths from lack of food, as well as social and economic chaos. Such rapid changes would also reduce the earth's biodiversity because many species couldn't move or adapt.

Recent data from analyses of ice cores and deep-sea sediments suggest that average temperatures during the warm interglacial period that began about 125,000 years ago (Figure 11-4) varied as much as 10°C (18°F) in only a decade or two and that such warming and cooling periods each lasted 1,000 years or more. If these findings are correct and also apply to the current interglacial period, fairly small rises in greenhouse gas concentrations could trigger rapid up-and-down shifts in average global temperatures.

As a result of the factors discussed in this section, climate scientists estimate that *projections about global warming and rises in average sea levels during the next 50–100 years could be half (the best-case scenario) or twice (the worst-case scenario) the current projections.* In any event, possible climate change and its effects are likely to be erratic and mostly unpredictable.

11-4 SOME POSSIBLE EFFECTS OF A WARMER WORLD

Why Should We Worry if the Earth's Temperature Rises a Few Degrees? So what's the big deal? Why should we worry about a possible rise of only a few degrees in the earth's average temperature? We often have that much change between May and July, or even between yesterday and today.

This is a common critical thinking trap that many people fall into. The key point is that we are not talking about normal swings in *local weather*, but about a projected *global* change in *climate*—weather averaged over decades, centuries, and millennia. A warmer global climate could have a number of possible effects (Figure 11-8), some harmful and some beneficial, depending on where one lives.

How Might Global Warming Affect Food Production? A warmer global climate could increase food production in some areas and lower it in others. Food production in some areas is likely to be disrupted by a combination of more intense storms that damage crops and increased temperatures that dry out soils.

Archeological evidence and computer models indicate that climate belts would shift toward the earth's poles by 100–150 kilometers (60–90 miles) or upward 150 meters (500 feet) in altitude for each 1°C (1.8°F) rise in global temperature. Whether such poleward shifts lead to increased crop productivity in new crop-growing areas depends mainly on two factors: **(1)** the fertility of the soil in such regions and **(2)** the availability of enormous amounts of money to build a new agricultural infrastructure (for irrigation and food storage and distribution).

In parts of Asia, food production in more northern areas could increase because of favorable soils. However, in North America the northward expansion of crop-growing regions from the Midwestern United States into Canada would be limited by the thinner and less fertile soils there.

Because winter cold kills insects and their eggs, temperature determines how far from the equator crop-destroying pests can live. As temperatures rise in the United States and other currently temperate areas, food production could be lowered by exploding populations of insects, crop diseases (such as rusts and molds), and weeds. Rising sea levels could flood low-lying coastal areas where crops are grown and decrease seafood supplies by flooding many of the world's productive coastal wetlands, marshes, and mangrove swamps. Increases in the average temperature and depth of tropical ocean waters would reduce biodiversity by destroying highly productive coral reefs.

Laboratory experiments suggest that increased concentrations of carbon dioxide could stimulate photosynthesis and increase the productivity of some types of crops. This is particularly true for C3 plants (so called because the product of their first sequence of reactions during photosynthesis has three carbon atoms). C3 plants include food staples such as rice, wheat, and soybeans. However, experiments indicate that the response of C4 plants (such as maize, sorghum, sugarcane, millet, and many pasture and forage grasses) to increased carbon dioxide levels would not be as great.

Current climate models can't predict where changes in crop-growing capacity might occur or how long such

Hint: Enter the search terms *global warming, economic* using the Subject Guide.

Figure 11-8 Some possible effects of global warming.

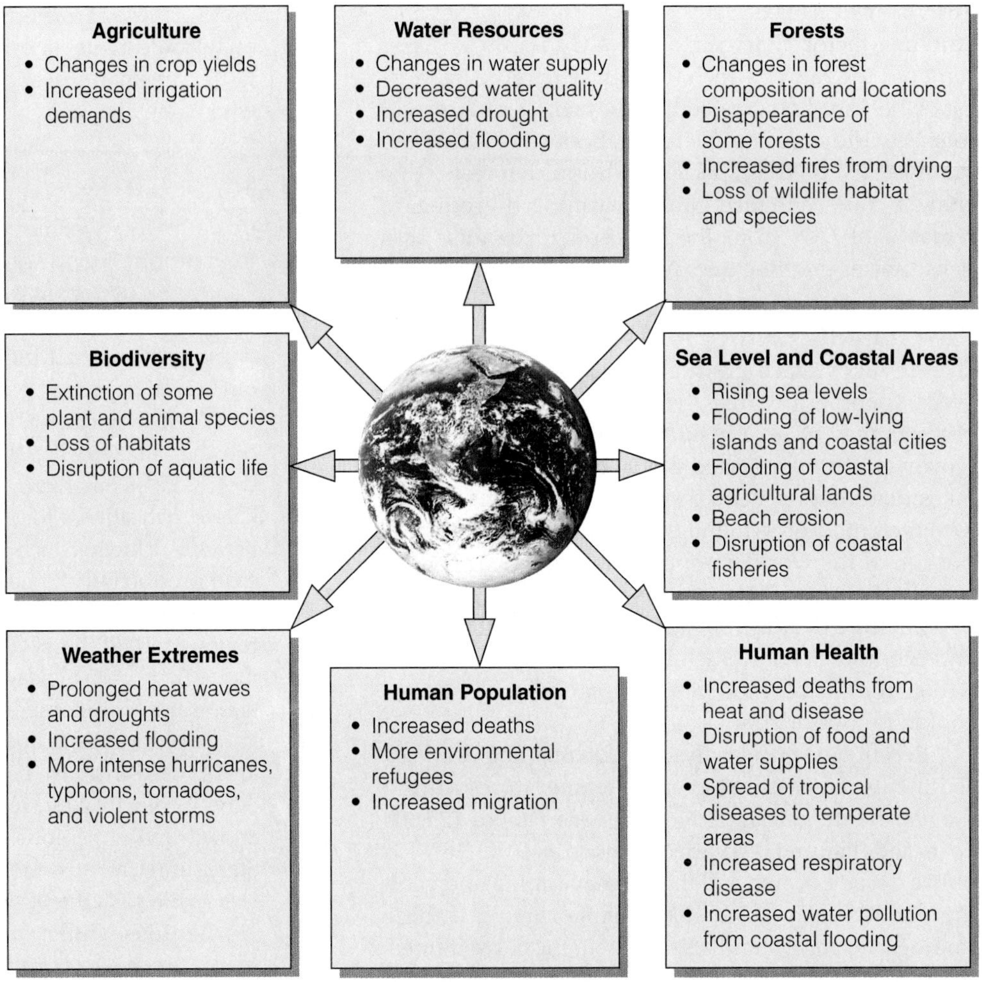

changes might last. However, we do know that just a 10% drop in global crop yields would cause large increases in hunger and starvation (especially in poor countries) and disrupt economic and social systems.

How Might Global Warming Affect Water Supplies? Global warming would also reduce water supplies in some areas. Indeed, lakes, streams, and aquifers in some populated areas could shrink or dry up altogether. This would force entire populations to migrate to areas with adequate water supplies—if they could. So far we can't say with much certainty where this might happen.

More intense rainfall in some areas and earlier melting of winter snow could increase the intensity and frequency of inland flooding, especially along rivers. Lower water levels in streams, bays, and lakes would increase pollution concentrations. Reduced precipitation and rising sea levels would increase the salinity of estuaries, disrupting the life cycles of some fish and shellfish species and polluting underground drinking water supplies in coastal areas.

How Might Global Warming Affect Forests? According to scientists, global warming will change the makeup and location of many of the world's forests. With a warmer climate, forests in temperate and subarctic regions would move toward the poles or to higher altitudes, leaving more grassland and shrubland in their wake. Models indicate that at least one-third of the world's forests and the wildlife that depends on them will be seriously affected by global warming.

According to the 1995 report of the IPCC, midlatitude climate zones are projected to shift northward by 550 kilometers (340 miles) over the next century. Trees whose seeds are spread by birds may be able to spread at this rate. But nut-bearing trees such as oaks and trees whose seeds are spread by wind may not be able to migrate fast enough and would die out (Figure 11-9). Poor soils may also limit the rate at which tree species can spread northward.

According to the IPCC, during the 21st century "entire forest types might disappear, including half of the world's dry tropical forests." Models also indicate that as much as 40% of the boreal forests in Alaska,

Canada, and Russia could disappear. Such forest diebacks would release carbon stored in their biomass and in surrounding soils and accelerate global warming.

Oregon State University scientists project that drying from global warming could cause massive *wildfires* in up to 90% of North American forests. If widespread fires occurred, large numbers of homes and large areas of wildlife habitats would be destroyed. Huge amounts of CO_2 would be added to the atmosphere, which would accelerate global warming.

How Might Global Warming Affect Biodiversity? Climate change would lead to reductions in biodiversity in many areas. Large-scale forest diebacks would cause mass extinction of plant and animal species that couldn't migrate to new areas. Fish would die as temperatures soared in streams and lakes, dissolved oxygen levels dropped, and lower water levels concentrated pesticides.

Species that live at high latitudes and on mountaintops would have nowhere to go, and many would become extinct as vegetation zones shifted to higher elevations. Any shifts in regional climate would threaten many parks, wildlife reserves, wilderness areas, wetlands, and coral reefs, wiping out many current efforts to stem the loss of biodiversity.

Projected global warming may sharply reduce populations of some species, especially those with specialized niches. For example, Kirkland's warbler, an endangered bird that nests exclusively under young jack pines in northern Michigan, would probably become extinct. The reason is that if northern Michigan ends up with less rain and more heat, jack pines there will probably die off in the next 30–90 years.

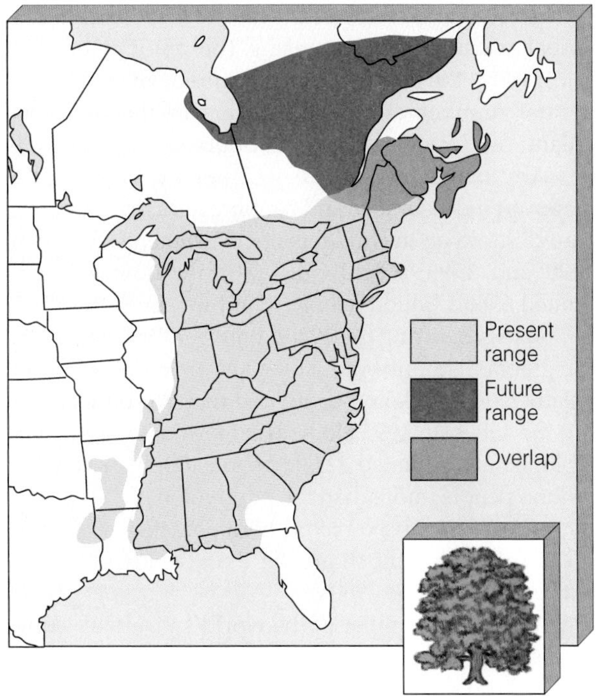

Figure 11-9 Possible effects of global warming on the geographic range of beech trees. According to one projection, if CO_2 emissions doubled between 1990 and 2050, beech trees (now common throughout the eastern United States) would survive only in a greatly reduced range in northern Maine and southeastern Canada. This is only one of a number of tree species whose geographic ranges would be changed drastically by global warming. (Data from Margaret B. Davis and Catherine Zabinski, University of Minnesota)

How Might Global Warming Affect Sea Levels? In a warmer world, sea levels would rise, primarily because ocean water expands when heated (and to a lesser degree because some land-based glacial ice would melt). Even the modest sea-level rise projected to occur by 2100 would

- Flood coastal regions, where about one-third of the world's people and economic infrastructure are concentrated
- Flood agricultural lowlands and deltas in parts of Bangladesh, India, and China, where much of the world's rice is grown
- Cause the disappearance of most existing beaches on the U.S. East Coast
- Cover low-lying islands such as the Cook and Marshall Islands in the Pacific, the Maldives, and some Caribbean islands with water and submerge large, densely populated coastal plains in places such as South Florida and Bangladesh
- Place the world's already dwindling fisheries at risk by flooding coastal estuaries, wetlands, and coral reefs
- Move barrier islands farther inland, accelerate coastal erosion, contaminate coastal aquifers with salt water, and flood tanks storing oil and other hazardous chemicals in coastal areas

If warming at the poles caused ice sheets and glaciers to melt even partially, the global sea level would rise far more. One comedian jokes that he plans to buy land in Kansas because it will probably become valuable beach-front property; another boasts that she isn't worried because she lives in a houseboat—the "Noah strategy."

How Might Global Warming Affect Weather Extremes? In a warmer world, *weather extremes* are expected to increase in number and severity. In many areas, heat waves will be more frequent and hotter and humidity will be higher. Prolonged heat waves and droughts could become the norm in many areas, taking a huge toll on many humans and ecosystems.

As the upper layers of seawater warm, the intensity of hurricanes, typhoons, tornadoes, and other violent storms is expected to increase. Hurricanes can enhance global warming by increasing the amount of CO_2 released by the ocean during summer. Since 1900 the annual number of intense storms in the northern Atlantic and Pacific oceans has doubled.

Any major increase in weather extremes could bankrupt the insurance and banking industries and put a huge strain on the budgets of governments. Between 1990 and 1998, worldwide natural disaster losses totaled $360.1 billion—more than four times the $77.5 billion losses during the 1980s. Between 1990 and 1998, the global insurance industry paid out $159 billion in weather-related claims, compared with $83 billion during the 1980s. In 1998 alone, weather-related disasters **(1)** killed an estimated 32,000 people, **(2)** displaced 300 million people (more than the population of the United States), and **(3)** caused record-high damages of $92 billion worldwide—much greater than the losses for the entire decade of the 1980s.

A growing number of the world's insurance companies are sharply raising their premiums for coverage of weather-related damages and eliminating coverage in high-risk areas. Executives of some insurance companies are pressuring government leaders to get more serious about slowing possible global warming and are spurring improvements in energy efficiency and increased use of renewable energy through their buying and investment policies.

How Might Global Warming Affect Human Health? According to the 1995 IPCC report, global warming would bring more heat waves. This would double or triple heat-related deaths, especially among the elderly, the young, and people with heart disease.

A warmer world would also disrupt supplies of food and fresh water, displacing millions of people and altering disease patterns in unpredictable ways. The spread of warmer and wetter tropical climates from the equator would bring malaria (Figure 8-8), encephalitis, yellow fever, dengue fever, and other insectborne diseases to formerly temperate zones.

Atmospheric warming also affects the respiratory tract by increasing air pollution in winter and increasing exposure to dusts, pollens, and smog in summer. Sea-level rise could spread infectious disease by flooding coastal sewage and sanitation systems.

How Might Global Warming Affect the Number of Environmental Refugees? According to Norman Myers (Guest Essay, p. 282), by 2050 global warming could produce as many as 50–150 million environmental refugees (compared with 7 million war refugees in Europe after World War II). Most of these refugees would migrate illegally to other countries, causing social disorder and political instability.

11-5 SOLUTIONS: DEALING WITH THE THREAT OF GLOBAL WARMING

Should We Do More Research or Act Now? There are three schools of thought concerning global warming. A dozen or so scientists (many of them not experts in climate research and many heavily funded by the oil and coal industries) contend that global warming is not a threat. Some popular press commentators and writers even claim that it is a hoax. Widespread reporting of this *no-problem* minority view in the media has clouded the issue, cooled public support for action, and slowed international negotiations to deal with this threat.

A second group of scientists and economists question whether we should spend hundreds of billions of dollars phasing out fossil fuels and replacing deforestation with reforestation (and in the process risk disrupting national and global economies) to help ward off something that might not happen. They call for more research before making such far-reaching decisions.

A third group of scientists and economists urge us to adopt a *precautionary strategy*. They believe that when dealing with risky and far-reaching environmental problems such as global warming, the safest course is to take informed preventive action *before* there is overwhelming scientific knowledge to justify acting. As Jonathan Overpeck, a paleontologist with the U.S. National Center for Atmospheric Research, puts it, "With the planet, we get just one experiment. And we can't turn things around if we don't like what happens."

Economists at a 1997 meeting of the American Economics Association, led by Nobel laureates Kenneth Arrow and Robert Solow, declared, "As economists, we believe that global climate change carries with it significant environmental, economic, social, and geopolitical risks and that preventive steps are justified." Since 1997, CEOs of several major oil companies (including British Petroleum, Royal Dutch Shell, and Sun Oil) and the CEO of Ford Motor have joined most insurance company executives in agreeing that there was enough evidence that human activities were contributing to global warming to begin taking precautionary action.

Those who favor doing nothing or waiting before acting point out that there is a 50% chance that we are *overestimating* the impact of rising greenhouse gases. However, those urging action point out that there is also a 50% chance that we are *underestimating* such effects. Most scientists no longer debate whether human-induced global warming is taking place. Instead, they are trying to project what might happen as such warming increases during this century and evaluate strategies for slowing such warming and adapting to its effects.

In the early stages of projected global warming it will be easy for people to deny that anything serious is happening because the projected harmful effects won't occur for several decades. Psychologist Robert Ornstein calls this denial the *boiled frog syndrome*. He describes it as like trying to alert a frog to danger as it sits in a pan of water very slowly being heated on the stove. If the frog could talk it would say, "I'm a little warmer, but I'm doing fine." As the water gets hotter, we would warn the frog that it will die, but it might reply, "The temperature has been increasing for a long time, and I'm still alive."

Eventually the frog dies because it has no experience of the lethal effects of boiling water and thus cannot perceive its situation as dangerous. Like the frog, humans face a possible future without precedent, and our senses are unable to pick up warnings of impending danger.

Some say that we should take the actions needed to slow global warming even if there were no threat because such actions lead to other important environmental and economic benefits (Solutions, p. 272). For example, a reduction in the combustion of fossil fuels, especially coal, will lead to sharp reductions in air pollution that harms and prematurely kills large numbers of people, lowers food and timber productivity, and decreases biodiversity. This *no-regrets strategy* is an important part of the precautionary strategy.

How Can We Slow Global Warming? According to the 1995 IPCC report, stabilizing CO_2 levels at the current level would require reducing current global CO_2 emissions by 66–83%. This is a highly unlikely and politically charged change.

Figure 11-10 presents a variety of solutions analysts have suggested to slow possible global warming; none of these solutions is being pursued vigorously. The quickest, cheapest, and most effective way to reduce emissions of CO_2 and other air pollutants over the next two to three decades is to use energy more efficiently (Section 20-2 and Solutions, p. 272).

Some analysts call for increased use of nuclear power (Section 19-5) because it produces only about one-sixth as much CO_2 per unit of electricity as coal. Other analysts argue that the danger of large-scale releases of highly radioactive materials from nuclear power-plant accidents and the very high cost of nuclear power make it a much less desirable option than improving energy efficiency and relying more on renewable energy resources (Chapter 20).

Using natural gas (Section 19-3) could help us make the 40- to 50-year transition to an age of energy efficiency and renewable energy. When burned, natural gas emits only half as much CO_2 per unit of energy as coal, and it emits far smaller amounts of most other air pollutants. However, without effective maintenance, more reliance on natural gas can increase inputs of methane (a potent greenhouse gas) from leaking tanks and pipelines and thus increase global warming.

One method for reducing CO_2 emissions would be to **(1)** phase out government subsidies for fossil fuels (about $300 billion per year globally and $21 billion annually in the United States) over a decade and **(2)** gradually phase in *carbon taxes* on fossil fuels (especially coal and gasoline) based on their emissions of CO_2 and other air pollutants. To be politically feasible, analysts argue that these consumption tax increases should be matched by declines in taxes on income, labor, or capital. As fossil fuel subsidies are removed, subsidies for increased energy efficiency and renewable energy would be phased in to help speed up the switch to these alternatives.

Another approach is to agree to global and national limits on greenhouse gas emissions and allow industries and countries to sell and trade emission permits in the marketplace. For example, a coal-burning power plant in Ohio could earn offset CO_2 credits for financing a tree-planting project in Oregon or Costa Rica. Some believe that this approach is more politically feasible than carbon taxes.

Another suggestion is for developed countries to finance the transfer of renewable energy technologies to developing countries to reduce their need to burn fossil fuels. Proponents contend that such a global public-works program would create millions of jobs and foster economic growth in both developing and developed countries. One way to finance this technology transfer might involve an expanded version of the Tobin tax already used on all international currency transactions. An assessment of 0.25% on the almost $1.4

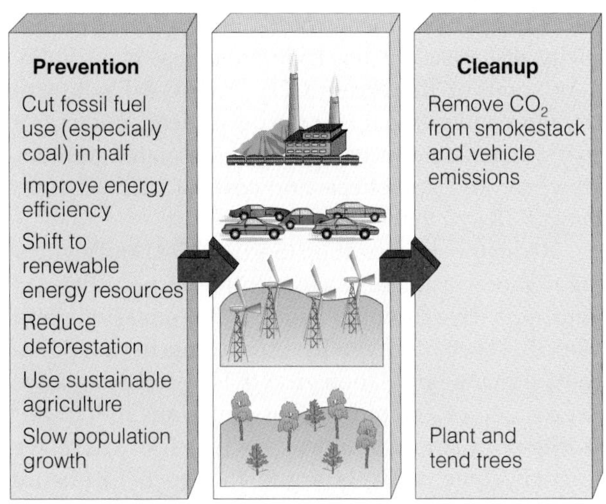

Figure 11-10 Solutions: methods for slowing global warming.

Energy Efficiency to the Rescue

SOLUTIONS

According to energy expert Amory Lovins (Guest Essay, p. 528), *the major remedies for slowing possible global warming are things we should be doing already even if there were no threat of global warming.* He argues that we should **(1)** waste less energy, **(2)** reduce air pollution by cutting down on our use of fossil fuels and switching to renewable forms of energy, and **(3)** harvest trees more sustainably. He contends that if we do these things, we and other forms of life—in this and in future generations—would be better off even if these actions had nothing to do with global climate.

According to Lovins, improving energy efficiency (Sections 20-1 and 20-2) would be the fastest, cheapest, and surest way to slash emissions of CO_2 and most other air pollutants within two decades, using existing technology. He estimates that increased energy efficiency would also save the world up to $1 trillion per year in reduced energy costs—as much as the annual global military budget.

Using energy more efficiently would also **(1)** reduce pollution, **(2)** help protect biodiversity, and **(3)** deter arguments among governments about how CO_2 reductions should be divided up and enforced, **(4)** make the world's supplies of fossil fuel last longer, **(5)** reduce international tensions over who gets the dwindling oil supplies, and **(6)** allow more time to phase in renewable energy.

According to a 1990 government report, controlling emissions of greenhouse gases in the United States will cost about $10 billion per year over the next 100 years, for a total of $1 trillion. Cutting annual U.S. oil imports by 20% by wasting less energy would cover the annual $10 billion projected costs of reducing greenhouse gas emissions.

To Lovins and most environmentalists, greatly improving worldwide energy efficiency is a money-saving, life-saving, biodiversity-saving, win–win, no-regrets solution to many of our environmental problems, even if climate change were not an issue. A small but growing number of companies have gotten the message and are saving money and energy while reducing carbon dioxide emissions.

Critical Thinking

1. Do you agree that improving energy efficiency **(a)** is an important way to reduce the input of CO_2 into the atmosphere and **(b)** should be done regardless of its impact on the threat of global warming?

2. Why do you think there has been little emphasis on improving energy efficiency? Explain. (See Sections 20-1 and 20-2.)

trillion daily commerce in foreign exchange would generate an estimated $200 billion per year—more than enough to finance the transfer.

It has also been suggested that we remove CO_2 from the exhaust gases of fossil fuel–burning vehicles, furnaces, power plants, and industrial boilers. However, available methods can remove only about 30% of the CO_2, and using them would double the cost of electricity. Scientists are also trying to devise ways to liquefy some of the excess CO_2 produced by human activities and pump it into the deep ocean, unminable coal seams, abandoned oil fields, and isolated aquifers. However, the side effects and costs of such schemes may be unacceptable.

Reducing deforestation (Section 17-4) and switching to more sustainable agriculture (Section 15-4 and Solutions, right) would reduce CO_2 emissions and help preserve biodiversity. According to most analysts, slowing population growth is also crucial. If we cut per capita greenhouse gas emissions in half but world population doubles, we're back where we started. Some analysts argue that to help ensure global environmental security (Guest Essay, p. 282), developed countries should transfer energy efficiency, renewable energy, pollution prevention, and waste reduction technologies to developing countries as soon as possible.

Some call for a massive global reforestation program as a strategy for slowing global warming. A 1998 paper by U.S. government scientists suggested that North American forests and soils were absorbing all of the carbon dioxide being emitted by the burning of fossil fuels in North America. However, other studies have challenged these preliminary results.

Under ideal conditions, forests might be able to absorb as much as 50% of excess carbon dioxide from the atmosphere. But to do that, all trees would have to be young and as efficient in removing carbon dioxide as the most responsive species: the loblolly pine. In addition, when the trees die they release carbon dioxide back into the atmosphere. Studies suggest that a global reforestation program (requiring each person in the world to plant and tend an average of 1,000 trees every year) would offset only about 3 years of our current CO_2 emissions from burning fossil fuels. However, such a program would help restore deforested and degraded land and reduce soil erosion and loss of biodiversity.

Some scientists have suggested various technofixes for dealing with possible global warming, including **(1)** adding iron to the oceans to stimulate the growth of marine algae (which could remove more CO_2 through photosynthesis), **(2)** unfurling gigantic foil-surfaced sun mirrors in space to reduce solar input, and **(3)** injecting sunlight-reflecting sulfate particulates into the stratosphere to cool the earth's surface.

Many of these costly schemes might not work, and most would probably produce unpredictable short- and long-term harmful environmental effects. Moreover, once started, those that work could never be stopped without a renewed rise in CO_2 levels. Instead of spending huge sums of money on such schemes, many scientists believe it would be more effective and less expensive to improve energy efficiency (Solutions, left) and shift to renewable forms of energy that don't produce carbon dioxide (Chapter 20).

What Has Been Done to Reduce Greenhouse Gas Emissions? At the 1992 Earth Summit in Rio de Janeiro, Brazil, 106 nations approved a Convention on Climate Change in which developed countries committed themselves to reducing their emissions of CO_2 and other greenhouse gases to 1990 levels by the year 2000. However, the convention did not *require* countries to reach this goal and most will not achieve this goal.

In December 1997 more than 2,200 delegates from 161 nations met in Kyoto, Japan, to negotiate a new treaty to help slow global warming. The resulting treaty would **(1)** require 38 developed countries to cut greenhouse emissions to an average of 5.2% below 1990 levels between 2008 and 2012, **(2)** not require developing countries to make any cuts in their greenhouse gas emissions, and **(3)** allow emissions trading, in which a country that beats its goal for reducing greenhouse gas emissions can sell its excess reductions to countries that failed to meet their reduction goals.

Some analysts praise the agreement as a small but important step in dealing with the problem of potential global warming and hope that the conditions of the treaty will be strengthened in future negotiating sessions. However, climate scientists estimate that it would take at least a 60% reduction in global emissions of greenhouse gases below 1990 levels to slow projected global warming to an acceptable level. According to computer models, the 5.2% reduction goal of the Kyoto Protocol would shave only about 0.06°C (0.1°F) off the roughly 1.4°C (2.5°F) temperature rise projected by 2050.

There is also controversy over what role developing countries should take in reducing their CO_2 emissions. By early 2000 the U.S. Congress had not ratified the treaty, mostly because of its failure to require emission reductions from developing countries.

Organic Farming to the Rescue

SOLUTIONS

Recent studies by the Rodale Institute in Pennsylvania and other scientists suggest that a wholesale switch to organic farming could help slow global warming. They have shown that crops grown with organic fertilizer produce equivalent yields but much less CO_2 through respiration than crops grown with commercial organic fertilizer.

CO_2 outputs are also reduced because organic farming uses 50% less energy than conventional farming methods. Farmers can also reduce greenhouse emissions by adopting well-known soil-conservation methods such as using cultivation tillage to reduce or eliminate plowing, using cover crops in winter, and preserving buffer strips of trees along river banks (Section 13-8).

Critical Thinking

List three things that government could do to encourage farmers to switch to more sustainable organic farming and soil-conservation methods.

According to a 1999 study by Tellus Institute and the World Wildlife Fund, aggressive policies to curb global warming would spur substantial job and economic growth throughout the United States. By using economic incentives and regulatory changes to improve energy efficiency and land use and increased use of renewable energy, the United States could save as much as $43 billion per year and create more than 870,000 new jobs by 2010. Together, by 2010 these policies would **(1)** reduce U.S. energy use by 18% and electricity use by 30%, **(2)** cut SO_2 emissions by 50% and emissions of nitrogen oxides by 25%, and **(3)** reduce U.S. CO_2 emissions by 14% below 1990 levels, twice the amount specified in the 1997 Kyoto climate treaty.

How Can We Prepare for Possible Global Warming? It seems clear that many (perhaps most) of the things climate experts have recommended (Figure 11-10) either will not be done or will be done too slowly. As a result, a growing number of analysts suggest that we should also begin to prepare for the effects of long-term global warming. Figure 11-11 shows some ways to do this.

Implementing the key measures for slowing or responding to climate change listed in Figures 11-10 and 11-11 will cost a lot of money. However, studies indicate that in the long run the savings would greatly exceed the costs. Some actions you can take to reduce the threat of global warming are given in Appendix 6.

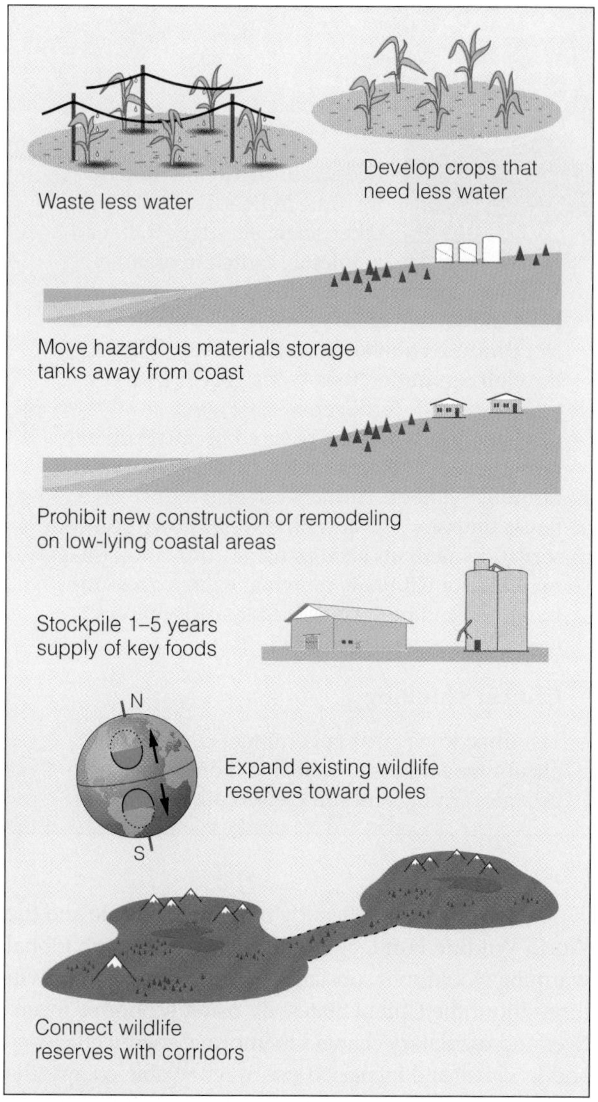

Figure 11-11 Solutions: ways to prepare for the possible long-term effects of global warming.

11-6 OZONE DEPLETION: IS IT A SERIOUS THREAT?

What Is the Threat from Ozone Depletion? Evolution of photosynthetic, oxygen-producing bacteria has produced a stratospheric global sunscreen: the ozone layer (Figure 10-2). Its presence for the past 450 million years has allowed life to develop and expand on land and in the surface layers of aquatic systems.

A handful of scientists dismiss the threat of ozone depletion from human-produced chemicals. Based on measurements and models (Figure 11-12), however, the overwhelming consensus of researchers in this field is that ozone depletion by certain chlorine- and bromine-containing chemicals emitted into the atmosphere by human activities is a serious long-term threat to **(1)** humans, **(2)** many other animals, and **(3)** the sunlight-driven primary producers (mostly plants) that support the earth's food chains and webs.

Measurements of the concentrations of ozone in the stratosphere at numerous sites around the world show that during the 1980s normal ozone levels dropped 5–15% in winter above the temperate and tropical zones of both hemispheres—three times the losses measured in the 1970s. Globally, the earth lost an average of about 4% of its stratospheric ozone between 1979 and 1994. According to a 1995 report by prominent atmospheric scientists, average global ozone levels are projected to drop 7–13% during the 1990s.

What Causes Ozone Depletion? From Dream Chemicals to Nightmare Chemicals This situation started when Thomas Midgley, Jr., a General Motors chemist, discovered the first **chlorofluorocarbon (CFC)** in 1930, and chemists made similar compounds to create a family of highly useful CFCs. The two most widely used are CFC-11 (trichlorofluoromethane, CCl_3F) and CFC-12 (dichlorofluoromethane, CCl_2F_2), known by their trade name as Freons.

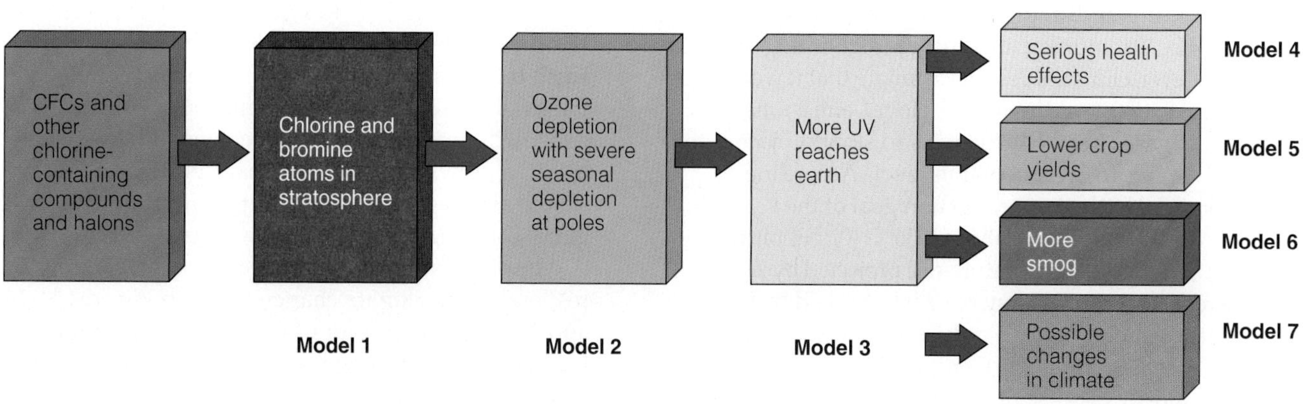

Figure 11-12 Generalized schematic of ozone depletion and models of its possible effects. Models 1–3 are widely accepted. The nature and extent of the effects in models 4–7 are still being studied.

Voynick, S. 1999. Living with Ozone. *World and I*, vol. 14, no. 7, p192.

These chemically stable, odorless, nonflammable, nontoxic, and noncorrosive compounds seemed to be dream chemicals. Cheap to make, they became popular as **(1)** coolants in air conditioners and refrigerators (replacing toxic sulfur dioxide and ammonia), **(2)** propellants in aerosol spray cans, **(3)** cleaners for electronic parts such as computer chips, **(4)** sterilants for hospital instruments, **(5)** fumigants for granaries and ship cargo holds, and **(6)** bubbles in plastic foam used for insulation and packaging.

But CFCs were too good to be true. In 1974 calculations by chemists Sherwood Rowland and Mario Molina (building on earlier work by Paul Crutzen) indicated that CFCs were creating a global chemical time bomb by lowering the average concentration of ozone in the stratosphere. They shocked both the scientific community and the $28-billion-per-year CFC industry by calling for an immediate ban of CFCs in spray cans (for which substitutes were readily available).

Here's what Rowland and Molina found: Spray cans, discarded or leaky refrigeration and air conditioning equipment, and the production and burning of plastic foam products release CFCs into the atmosphere. Because these molecules are insoluble in water and are chemically unreactive, they are not removed from the troposphere. As result, mostly through convection, random drift, and the turbulent mixing of air in the troposphere, they rise slowly into the stratosphere, taking 11–20 years to make the journey.

In the stratosphere, under the influence of high-energy UV radiation, these molecules break down and release highly reactive chlorine atoms, which speed up the breakdown of highly reactive ozone (O_3) into O_2 and O in a cyclic chain of chemical reactions (Figure 11-13). This causes ozone in the stratosphere to be destroyed faster than it is formed.

Each CFC molecule can last in the stratosphere for 65–110 years (depending on its type). During that time, each chlorine atom released from these molecules can convert up to 100,000 molecules of O_3 to O_2. If Rowland and Molina's calculations and later models and atmospheric measurements of CFCs in the stratosphere are correct (as almost all scientists in this field believe), these dream molecules have turned into a nightmare of global ozone destroyers.

Rowland and Molina warned us of this problem in 1974. However, it took 15 years of interaction between the scientific and political communities before countries agreed to begin phasing out CFCs. The CFC industry (led by the DuPont Company), a powerful, well-funded adversary with a lot of profits and jobs at stake, attacked Rowland and Molina. However, they held their ground, expanded their research, and explained the meaning of their calculations to other scientists, elected officials, and the media. It was not until 1988—14 years after Rowland and Molina's study—that DuPont officials acknowledged that CFCs were depleting the ozone layer and agreed to stop producing them

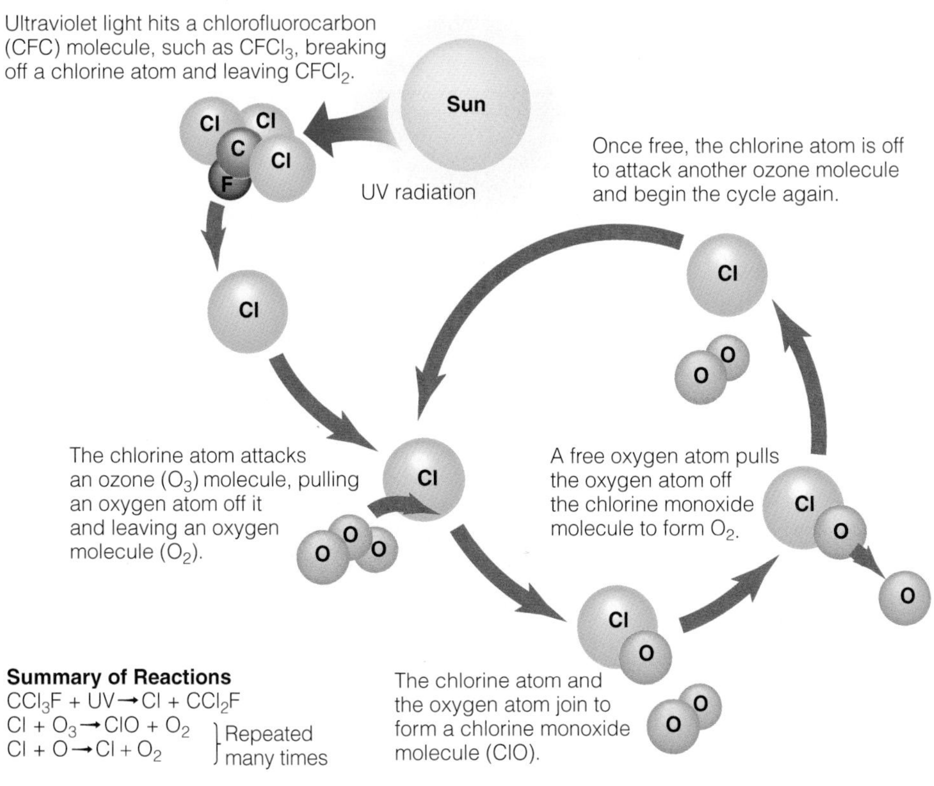

Figure 11-13 A simplified summary of how chlorofluorocarbons (CFCs) and other chlorine-containing compounds destroy ozone in the stratosphere. Note that chlorine atoms are continuously regenerated as they react with ozone. Thus, they act as *catalysts*, chemicals that speed up chemical reactions without being used up by the reaction. Bromine atoms released from bromine-containing compounds that reach the stratosphere also destroy ozone by a similar mechanism.

Hint: Enter the search term *ozone layer depletion* using the Subject Guide.

once they found substitutes.* In 1995, Rowland and Molina (along with Paul Crutzen) received the Nobel prize in chemistry for their work.

What Other Chemicals Deplete Stratospheric Ozone? Ozone-depleting compounds, collectively called *ODCs*, include

- *Halons* and *HBFCs*, both used in fire extinguishers
- *Methyl bromide* (CH_3Br), a widely used fumigant
- *Carbon tetrachloride* (CCl_4), a cheap, highly toxic solvent
- *Methyl chloroform*, or 1,1,1-trichloroethane ($C_2H_3Cl_3$), used as a cleaning solvent for clothes and metals and as a propellant in more than 160 consumer products such as correction fluid, dry-cleaning sprays, spray adhesives, and other aerosols
- *Hydrogen chloride (HCl),* emitted into the stratosphere by U.S. space shuttles

Why Is There Seasonal Thinning of Ozone over the Poles? At times the news about ozone loss has taken scientists by surprise. The first surprise came in 1984, when researchers analyzing satellite data discovered that 40–50% of the ozone in the upper stratosphere over Antarctica was being destroyed during the antarctic spring and early summer (September–December), when sunlight returned after the dark antarctic winter. Since then, this seasonal antarctic *ozone thinning* (incorrectly called an ozone hole) has expanded in most years, with a typical seasonal loss of about 50% (but as high as 100% in some spots). In 1998, seasonal ozone thinning above Antarctica was the largest ever, and covered an area larger than the size of North America. Ozone loss in 1999 was slightly less because of some easing of polar winds (Figure 11-14). However, the thinning still covered an area about the size of Europe.

Measurements indicate that CFCs are the primary culprits. Each sunless winter, steady winds blow in a circular pattern over the earth's poles. This creates a *polar vortex*: a huge swirling mass of very cold air that is isolated from the rest of the atmosphere until the sun returns a few months later. When water droplets in clouds enter this circling stream of extremely frigid air, they form tiny ice crystals. The surfaces of these ice crystals collect CFCs and other ozone-depleting chemicals in the stratosphere and speed up (catalyze) the chemical reactions that release Cl atoms and ClO. Instead of entering a chain reaction of ozone destruction (Figure 11-13), the ClO atoms combine with one another to form Cl_2O_2 molecules. In the dark of winter the Cl_2O_2 molecules cannot react with ozone, so they accumulate in the polar vortex.

When sunlight and the antarctic spring return 2–3 months later (in October), the light breaks up the stored Cl_2O_2 molecules, releasing large numbers of Cl atoms and initiating the catalyzed chlorine cycle (Figure 11-13). Within weeks, this destroys 40–50% of the ozone above Antarctica (and 100% in some places). The returning sunlight gradually melts the ice crystals, breaks up the vortex of trapped polar air, and allows it to begin mixing again with the rest of the atmosphere. Then new ozone forms over Antarctica until the next dark winter.

When the vortex breaks up, huge masses of ozone-depleted air above Antarctica flow northward and linger for a few weeks over parts of Australia, New Zealand, South America, and South Africa. This raises biologically damaging UV-B levels in these areas by 3–10%, and in some years by as much as 20%.

In 1988 scientists discovered that similar but less severe ozone thinning occurs over the north pole during the arctic spring and early summer (February–June), with a seasonal ozone loss of 11–38% (compared with a typical 50% loss above Antarctica). When this mass of air above the Arctic breaks up each spring, large masses of ozone-depleted air flow south to linger over parts of Europe, North America, and Asia.

So far, seasonal ozone loss over the north pole is much lower than that over the south pole. However, scientists warn that the problems of ozone depletion and global warming can interact synergistically to enhance both effects and could lead to more serious ozone depletion above the Arctic and Antarctic.

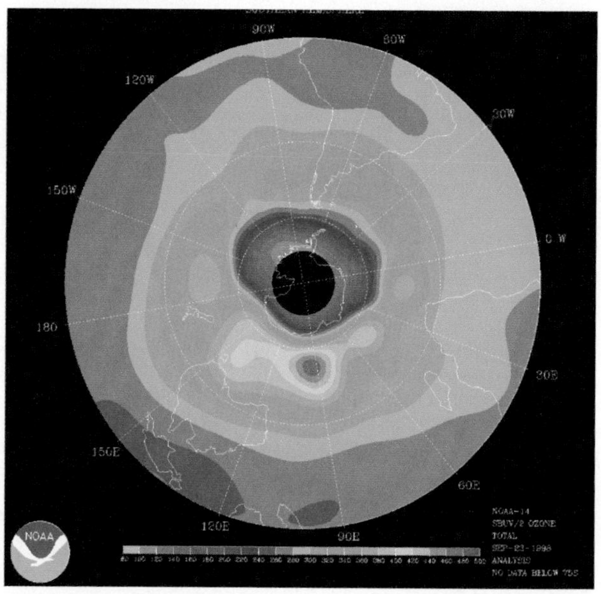

Figure 11-14 Seasonal ozone thinning or loss (shown by shades of pink) in the upper stratosphere over Antarctica as measured by satellite on September 23, 1999. (NOAA/NESDIS)

*For a fascinating account of how corporate stalling, politics, economics, and science can interact, see Sharon Roan's *Ozone Crisis: The 15-Year Evolution of a Sudden Global Emergency* (New York: Wiley, 1989).

PRO/CON

Is Ozone Depletion a Hoax?

Charge: There is no ozone hole, and the whole idea is merely a scientific theory.

Response: Technically, it is not correct to call this phenomenon an ozone *hole*. Instead, it is a thinning or partial depletion of normal ozone levels in the stratosphere. A scientific theory is a widely accepted idea or principle that has a very high degree of certainty. Nothing in science can be proven absolutely, but there is overwhelming scientific evidence of ozone thinning in the stratosphere.

Charge: There is no scientific proof that an ozone hole exists.

Response: Ground-based and satellite measurements clearly show seasonal ozone thinning above Antarctica and the Arctic, and other measurements reveal a lower overall thinning everywhere except over the tropics.

Charge: CFC molecules can't reach the stratosphere because they are heavier than air, and no measurements have detected them in the stratosphere.

Response: The atmosphere is like a turbulent fluid in which gases, both lighter and heavier than air, are mixed thoroughly by the churning movements of large air masses. Since 1975 measuring instruments on balloons and satellites have shown that CFCs are transported high into the stratosphere.

Charge: Sodium chloride (NaCl) from the evaporation of sea spray contributes more chlorine to the stratosphere than do CFCs.

Response: Particles of NaCl from sea spray (unlike CFCs) are soluble in water and are washed out of the lower atmosphere. Measurements have detected no sodium in the lower stratosphere.

Charge: Volcanic eruptions and biomass burning have added much more chlorine (Cl) to the stratosphere than have human-caused CFC emissions.

Response: Most of the water-soluble HCl injected into the troposphere from occasional large-scale volcanic eruptions is washed out by rain before it reaches the stratosphere. Even Fred Singer (who is highly skeptical about some aspects of ozone depletion models) stated in 1993 that "CFCs make the major contribution to stratospheric chlorine." However, recent research indicates that aerosol particles of sulfate salts produced from sulfur dioxide emitted by volcanic eruptions can destroy stratospheric ozone for several years after an eruption. These volcanic aerosols, plus the human-related inputs of ozone-destroying chemicals, can seriously deplete ozone in the stratosphere.

Charge: Seasonal ozone thinning over Antarctica is a natural phenomenon because it appeared long before CFCs were in wide use.

Response: In the late 1950s, British scientist Gordon Dobson discovered that the antarctic polar vortex leads to some seasonal ozone loss through natural causes. Measurements indicate that between 1956 and 1976 the natural pattern of slight ozone loss above Antarctica did not change significantly. Since 1976, however, seasonal losses of ozone above Antarctica have increased dramatically and have been linked by measurements and models to rising levels of CFCs in the stratosphere. A 1994 review of Dobson's 1958 data concluded that in 1958 there was no credible evidence for a significant ozone thinning (such as that observed since 1976) over Antarctica.

Charge: The best models of ozone production and destruction estimate that global ozone levels should be 10% less than we actually observe.

Response: Scientists recognize that their models must be improved. However, they point out that their models of ozone depletion by chemicals produced by human activities still account for about 90% of the observed ozone depletion.

Charge: The expected increase in UV-B radiation caused by stratospheric ozone loss has not been detected in urban areas in the United States and most other developed countries.

Response: Increased UV levels have been detected in urban areas in countries in the southern hemisphere such as Australia, New Zealand, South Africa, Argentina, and Chile (as well as in Toronto, Canada) that are exposed to ozone-depleted air drifting away from Antarctica and the Arctic after the seasonal polar vortex breaks up each year. Some ozone depletion experts hypothesize that significant ground-level increases in UV-B have not been observed in many urban areas (especially in industrialized countries in the northern hemisphere) because ozone-laden smog over most cities may filter out some of the UV-B. They argue that polluting our way out of increased UV-B levels in such cities is unacceptable because of the premature deaths, health problems, and damage to trees, crops, and materials caused by increasing air pollution (Section 11-4).

Critical Thinking

Do you believe that ozone depletion in the stratosphere is a serious problem or one that has been greatly exaggerated? Explain.

Is Ozone Depletion Really a Serious Problem? Political talk show commentator Rush Limbaugh, zoologist and former head of the Atomic Energy Commission Dixy Lee Ray (now deceased), physicist and climate scientist S. Fred Singer, and articles and books in the popular press have claimed that ozone depletion by CFCs is a hoax, or at least a vastly exaggerated problem. The Pro/Con box above summarizes some of their charges and the answers by prominent research scientists studying ozone depletion.

The Cancer You Are Most Likely to Get

CONNECTIONS

Considerable research indicates that years of exposure to UV-B ionizing radiation in sunlight is the primary cause of *squamous cell* (Figure 11-15, left) and *basal cell* (Figure 11-15, center) *skin cancers*. Together these two types make up 95% of all skin cancers. Typically there is a 15- to 40-year lag between excessive exposure to UV-B and development of these cancers.

Caucasian children and adolescents who get only a single severe sunburn double their chances of getting these two types of cancers. Some 90–95% of these types of skin cancer can be cured if detected early enough, although their removal may leave disfiguring scars. These cancers kill only 1–2% of their victims, but this still amounts to about 2,300 deaths in the United States each year.*

A third type of skin cancer, *malignant melanoma* (Figure 11-15, right), occurs in pigmented areas such as moles anywhere on the body's surface. This type of cancer can spread rapidly (within a few months) to other organs. It kills about one-fourth of its victims (most under age 40) within 5 years, despite surgery, chemotherapy, and radiation treatments. Each year it kills about 100,000 people (including 7,300 Americans in 1998), mostly Caucasians. It can often be cured if detected early enough, but recent studies show that some melanoma survivors have a recurrence more than 15 years later.

Recent evidence suggests that about 90% of sunlight's melanoma-causing effect may come from exposure to UV-A and 10% from UV-B. Some sunscreens don't protect from UV-A, and tanning booth lights emit mostly UV-A.

Evidence indicates that people (especially Caucasians) who get three or more blistering sunburns before age 20 are five times more likely to develop malignant melanoma than those who have never had severe sunburns. About 10% of those who get malignant melanoma have an inherited gene that makes them especially susceptible to the disease.

To protect yourself, the safest course is to stay out of the sun (especially between 10 A.M. and 3 P.M., when UV levels are highest) and avoid tanning parlors. When you are in the sun, wear tightly woven protective clothing, a wide-brimmed hat, and sunglasses that protect against UV-A and UV-B radiation (ordinary sunglasses may actually harm your eyes by dilating your pupils so that more UV radiation strikes the retina).

Because UV rays can penetrate clouds, overcast skies do not protect you; neither does shade, because UV rays can reflect off sand, snow, water, or patio floors. People who take antibiotics and women who take birth control pills are more susceptible to UV damage.

Use a sunscreen that offers protection against both UV-A and UV-B and has a protection factor of 15 or more (25 if you have light skin). Apply to all exposed skin and reapply it after swimming or excessive perspiration. Most people don't realize that the protection factors for sunscreens are based on using one full ounce of the product—far more than most people apply. Some people are also increasing their risk of skin cancer by falsely assuming that sunscreens allow them to spend more time in the sun.

Children who use a sunscreen with a protection factor of 15 every time they are in the sun from age 1 to age 18 decrease their chance of getting skin cancer by 80%. Babies under a year old should not be exposed to the sun at all and should not have sunscreens applied until they are at least 6 months old.

Become familiar with your moles and examine your skin at least once a month. The warning signs of skin cancer **(1)** are a change in the size, shape, or color of a mole or wart (the major sign of malignant melanoma, which must be treated quickly), **(2)** sudden appearance of dark spots on the skin, or **(3)** a sore that keeps oozing, bleeding, and crusting over but does not heal. Be alert for precancerous growths (reddish-brown spots with a scaly crust). If you observe any of these signs, consult a doctor immediately.

Critical Thinking

What precautions, if any, do you take to reduce your chances of getting skin cancer from exposure to sunlight? Explain why you do or don't take such precautions.

*I have had six basal cell cancers on my face and neck because of "catching too many rays" in my younger years. I wish I had known then what I know now.

Why Should We Be Worried About Ozone Depletion? Life in the Ultraviolet Zone Why should we care about ozone loss? From a human standpoint the answer is that with less ozone in the stratosphere, more biologically damaging UV-A and UV-B radiation will reach the earth's surface and give humans **(1)** worse sunburns, **(2)** more cataracts (a clouding of the eye's lens that reduces vision and can cause blindness if not corrected), and **(3)** more skin cancers (Connections, above). The number of new cases of skin cancer in the United States has quadrupled over the past 20 years, resulting in nearly 10,000 deaths per year.

According to UN Environment Programme estimates, the additional UV-B radiation reaching the earth's surface resulting from an annual 10% loss of global ozone (already a strong possibility within a few years) could lead to **(1)** 300,000 additional cases of squamous cell cancer (Figure 11-15, left) and basal cell cancer (Figure 11-15, center) worldwide each year, **(2)** 4,500–9,000 additional cases of potentially fatal malignant melanoma

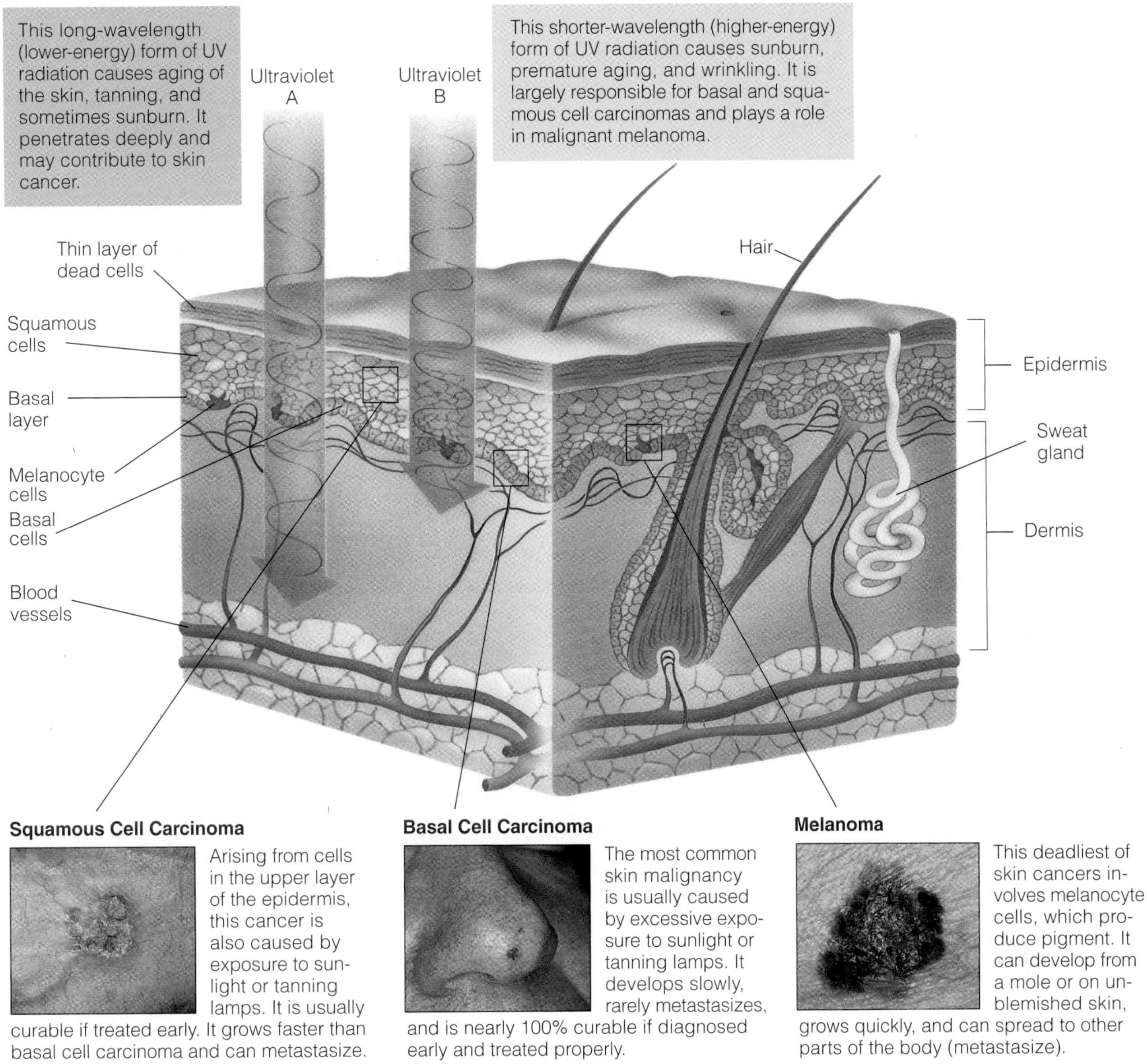

Figure 11-15 Structure of the human skin and the relationships between ultraviolet (UV-A and UV-B) radiation and the three types of skin cancer. The incidence of these types of cancer is rising, mostly because more fair-skinned people have increased their exposure to sunlight by moving to areas with sunnier climates and spending more of their leisure time exposed to sunlight. If ozone-destroying chemicals continue to reduce stratospheric ozone levels, incidence of these types of cancers is expected to rise. (The Skin Cancer Foundation)

(Figure 11-15, right) each year, and **(3)** 1.5 million new cases of cataracts (which account for over half of the world's 25–35 million cases of blindness) each year.

Assuming that we phase out *all* ozone-destroying chemicals over the next three decades, the U.S. Environmental Protection Agency estimates that ozone thinning during the 1980s and 1990s will lead to 12 million new cases of skin cancer and 200,000 additional skin cancer deaths in the United States alone over the next 50 years.

Other effects of increased UV exposure are

- Suppression of the immune system. This makes the body more susceptible to infectious diseases and some forms of cancer.
- An increase in eye-burning, highly damaging acid deposition (Figure 10-6) and photochemical smog in the troposphere (Figure 10-4).
- Lower yields of key crops such as corn, rice, cotton, soybeans, beans, peas, sorghum, and wheat,

with estimated losses totaling $2.5 billion per year in the United States before the middle of the 21st century.

- Decline in forest productivity of the many tree species sensitive to UV-B radiation. This could reduce CO_2 uptake and enhance global warming.
- Increased breakdown and degradation of materials such as various types of paints, plastics, and outdoor materials. Such damage could cost billions of dollars per year.
- Reduction in the productivity of surface-dwelling phytoplankton, which could **(1)** upset aquatic food webs, **(2)** decrease yields of seafood eaten by humans, and **(3)** possibly accelerate global warming by decreasing the oceanic uptake of CO_2 by phytoplankton.

Some critics who believe that the threat of ozone depletion has been exaggerated argue that we should not worry about a 10%, 20%, or even 50% increase in UV-B radiation reaching the earth's surface. According to Fred Singer, moving just 97 kilometers (60 miles) closer to the equator involves potential exposure to 10% more UV-B radiation.

Many scientists contend that this argument ignores two crucial points. First, the southward movement of the U.S. population (and people in many other countries) is one reason skin cancer rates have risen. Any increased exposure to UV-B radiation because of ozone depletion would add to this already serious health threat. Second, we are talking about the likely possibility of rapid (within decades), widespread, long-lasting, and essentially unpredictable ecological

Table 11-1 CFC Substitutes

Types	Pros	Cons
HCFCs (hydrochlorofluorocarbons)	Break down faster (2–20 years). Pose about 90% less danger to ozone layer. Can be used in aerosol sprays, refrigeration, air conditioning, foam, and cleaning agents.	Are greenhouse gases. Will still deplete ozone, especially if used in large quantities. Health effects largely unknown. Expensive. May lower energy efficiency of appliances. Can be degraded to trifluoroacetate (TFA), which can inhibit plant growth in wetlands.
HFCs (hydrofluorocarbons)	Break down faster (2–20 years). Do not contain ozone-destroying chlorine. Can be used in aerosol sprays, refrigeration, air conditioning, and insulating foam.	Are potent greenhouse gases. Expensive. Safety questions about flammability and toxicity still unresolved. Can be degraded to trifluoroacetate (TFA), which can inhibit plant growth in wetlands. May lower energy efficiency of appliances. Production of HFC-134a, a refrigerant substitute, yields an equal amount of methyl chloroform, a serious ozone depleter.
Hydrocarbons (HCs) (such as propane and butane)	Cheap and readily available. Can be used in aerosol sprays, refrigeration, foam, and cleaning agents. Not patentable. Can be made locally in developing countries.	Can be flammable and poisonous if released. Some increase in ground-level air pollution.
Ammonia	Simple alternative for refrigerators; widely used before CFCs.	Toxic if inhaled. Must be handled carefully.
Water and steam	Effective for some cleaning operations and for sterilizing medical instruments.	Creates polluted water that must be treated. Wastes water unless the used water is cleaned up and reused.
Terpenes (from the rinds of lemons and other citrus fruits)	Effective for cleaning electronic parts.	None.
Helium	Effective coolant for refrigerators, freezers, and air conditioners.	This rare gas may become scarce if use is widespread, but very little coolant is needed per appliance.

disruptions in species adapted to existing levels of background UV-B radiation.

Humans can quickly make cultural adaptations to increased UV-B radiation by staying out of the sun, protecting their skin with clothing, and applying sunscreens. However, plants and other animals that help support us and other forms of life can't make such changes except through the long process of biological evolution.

11-7 SOLUTIONS: PROTECTING THE OZONE LAYER

How Can We Protect the Ozone Layer? The scientific consensus of researchers in this field is that we should immediately stop producing all ozone-depleting chemicals. Even with immediate action, the models indicate that it will take 50–60 years for the ozone layer to return to 1975 levels and another 100–200 years for full recovery to pre-1950 levels.

Substitutes are already available for most uses of CFCs (Table 11-1, left), and others are being developed (Individuals Matter, right). One substitute for CFCs is *hydrochlorofluorocarbons* or *HCFCs* (such as $CHClF_2$, containing fewer chlorine atoms per molecule than CFCs). If used in massive quantities, however, HCFCs would still cause ozone depletion and act as potent greenhouse gases.

Another substitute for CFCs is *hydrofluorocarbons* or *HFCs* (such as CF_3, containing fluorine but no chlorine or bromine). Recent research indicates that HFCs have a negligible effect on ozone depletion. However, they may need to be restricted or phased out because they are powerful greenhouse gases that remain in the atmosphere much longer than CFCs or CO_2.

To a growing number of scientists, hydrocarbons (HCs) such as propane and butane are a better way to reduce ozone depletion while doing little to increase global warming. Developing countries could use HC refrigerator technology to leapfrog ahead of industrialized countries without having to invest in costly HFC and HCFC technologies that will have to be phased out within a few decades. This approach is also less costly because HCs cannot be patented and can be manufactured locally, reducing the need to import expensive HFCs and HCFCs.

Can Technofixes Save Us? What about a quick fix from technology, so that we can keep on using CFCs? One proposed scheme for removing ozone from the atmosphere is to launch a fleet of 20–30 football-field-long, radio-controlled blimps into the stratosphere above Antarctica. Hanging from each blimp would be a huge curtain of electrical wires that would inject negatively charged electrons into the stratosphere. The electrons would convert ozone-destroying chlorine atoms (Cl) in the stratosphere to chloride ions ($Cl + e \longrightarrow Cl^-$) that would not react with ozone. A second suspended sheet of positively charged wires could be used to attract the negatively charged ions. However, atmospheric chemist Ralph Ciecerone believes that this plan won't work because other chemical species in the stratosphere snatch electrons more readily than does chlorine. This scheme could also have unpredictable side effects on atmospheric chemistry.

Others have suggested using tens of thousands of lasers to blast CFCs out of the atmosphere before they can reach the stratosphere. However, the energy required to do this would be enormous and expensive, and decades of research would be needed to perfect the types of lasers needed. Moreover, we can't predict the possible effects of such powerful laser blasts on climate, birds, or planes.

What Is Being Done to Reduce Ozone Depletion? Some Hopeful Progress In 1987, 36 nations meeting in Montreal developed a treaty, commonly known as the *Montreal Protocol*, to cut emissions of CFCs (but not other ozone depleters) into the atmosphere by

INDIVIDUALS MATTER

Ray Turner and His Refrigerator

Ray Turner, an aerospace manager at Hughes Aircraft in California, made an important low-tech, ozone-saving discovery by using his head—and his refrigerator. His concern for the environment led him to look for a cheap and simple substitute for the CFCs used as cleaning agents for removing films of oxidation from the electronic circuit boards manufactured at his plant.

He started by looking in his refrigerator. He decided to put drops of various substances on a corroded penny to see whether any of them removed the film of oxidation. Then he used his soldering gun to see whether solder would stick to the surface of the penny, indicating that the film had been cleaned off.

First, he tried vinegar. No luck. Then he tried some ground-up lemon peel, also a failure. Next he tried a drop of lemon juice and watched as the solder took hold. The rest, as they say, is history.

Today, Hughes Aircraft uses inexpensive citrus-based solvents that are CFC-free to clean circuit boards. This new cleaning technique has reduced circuit board defects by about 75% at Hughes. And Turner got a hefty bonus. Now other companies, such as AT&T, clean computer boards and chips using acidic chemicals extracted from cantaloupes, peaches, and plums. Maybe you can find a solution to an environmental problem in your refrigerator, grocery or drugstore, or backyard.

GUEST ESSAY

Environmental Security

Norman Myers

Norman Myers is an international consultant in environment and development, with emphasis on conservation of wildlife species and tropical forests, and is one of the world's leading environmental experts. His research and consulting have taken him to 80 countries. He has served as a consultant for many development agencies and research organizations, including the U.S. National Academy of Sciences, the World Bank, the Organization for Economic Cooperation and Development, various UN agencies, and the World Resources Institute. Among his many publications (see Further Readings) are The Primary Source: Tropical Forests and Our Future *(1992),* The Gaia Atlas of Planet Management *(1993),* Scarcity or Abundance: A Debate on Environment *(1994),* Ultimate Security: The Environmental Basis of Political Security *(1996), and* Perverse Subsidies *(1998).*

There is a new and different threat to our national security emerging—the destruction of our environment. I believe that one of our key national security objectives must be to reverse the accelerating pace of environmental destruction around the globe.

—Senator Sam Nunn, former chairman of the U.S. Senate Armed Service Committee

We are already engaged in World War III. It is a war against nature, and it is simply no contest. As a result, the threat from the skies is no longer missiles but ozone-layer depletion and global warming; we also face water shortages and excessive soil erosion.

These and other environmental threats, including overpopulation and widespread poverty in developing countries, have been described by a growing number of political leaders and military planners as the greatest threat we face short of nuclear war. So a new concept is emerging in councils of foreign policy makers and military strategists: the environmental dimension to security issues.

According to this idea, we should move beyond traditional thinking about security concepts and consider a series of environmental factors underpinning our material welfare. These factors include such natural resources as a nation's soil, water, forests, grasslands, and fisheries, and the climatic patterns and biogeochemical cycles that maintain the life-support systems of all nations [Figures 1-2 and 2-3].

If a nation's environmental foundations are degraded or depleted, its economy may well decline, its social fabric deteriorate, and its political structure become destabilized as growing numbers of people seek to sustain themselves from declining resource stocks. The likely outcome is tension and conflict within a nation and possibly with other nations.

Thus, national security is no longer about fighting forces and weaponry alone. It relates increasingly to watersheds, croplands, forests, genetic resources, climate, and other factors that, taken together, are as crucial to a nation's security as are military factors.

In many developing countries, decline of the natural resource base that underpins agriculture has led to increased imports, rising prices, and ultimately food shortages. In turn, these shortages have helped trigger civil disorders, military outbreaks, and downfalls of governments.

Food supply disputes have also occurred among developed countries. For example, in the North Atlantic, Great Britain and Iceland have come to the edge of hostilities over declining cod stocks. At least 16 similar major clashes over declining fishery stocks have occurred in other parts of the world. Such dis-

about 35% between 1989 and 2000. After hearing more bad news about ozone depletion, representatives of 93 countries met in London in 1990 and in Copenhagen in 1992 and adopted a protocol accelerating the phaseout of key ozone-depleting chemicals.

The agreements reached so far and signed by 165 countries are important examples of global cooperation in response to serious threats to global environmental security (Guest Essay, above). Because of these agreements, CFC production fell by 85% between 1988 (its peak production year) and 1998 and CFC levels in the troposphere have declined slightly (Figure 11-26). Global production of carbon tetrachloride and methyl chloroform has also dropped sharply, but that of methyl bromide, halons, and HCFCs continues to rise.

Methyl bromide is widely used as a fumigant to control certain soil-dwelling pathogens for more than 100 crops worldwide and will be banned in the United States by 2005. In 1999, U.S. Department of Agriculture scientists announced the discovery of a naturally occurring, nontoxic, non–ozone-depleting chemical (benzaldehyde) in peaches and several other fruits that could be used as an alternative for methyl bromide.

There is also the problem of how to destroy CFCs removed from old refrigerators, air conditioners, and other devices. In 1998, researchers at Yale University putes are likely to increase in the future in light of the failure of fisheries around the world to maintain sustainable yields.

Another source of conflict is water supplies. In the Middle East, competition for scarce water is a major factor in political confrontations, and shortages are projected to grow even more acute [p. 286 and Section 12-3]. Israel went to war in 1967 in part because the Arabs were trying to divert the headwaters of the River Jordan. Currently, Israel—the world's most water-efficient country—is beginning to suffer from critical water deficits.

Of 200 major river systems, almost 150 are shared by two nations, and more than 50 by three to ten nations; all in all, these rivers support almost 40% of the world's population. As many as 80 countries, with 40% of the world's population, already experience water shortages.

Other conflicts occur because of deforestation. The Ganges River system is dependent on tree cover in its catchment zone in the Himalayas. Because of deforestation, flooding during the annual monsoon has become so widespread that it regularly imposes damages to crops, livestock, and property worth $1 billion a year in downstream communities in India, even though most of the deforestation occurs in Nepal.

In the instances cited here, the environmental issues are readily apparent. In other cases, the impact is more deferred and diffuse. Examples are species extinctions and gene depletion, with all this ultimately implies for future genetic contributions to agriculture, medicine, industry, and energy.

Probably the most deferred, diffuse, and significant impact of all will prove to be climatic change. Buildup of carbon dioxide and other greenhouse gases in the global atmosphere, if it persists as projected, will cause far-reaching disruptions in temperature and rainfall patterns, as discussed in this chapter.

We cannot dispatch battalions to turn back the deserts, we cannot launch a flotilla to resist the rising seas, we cannot send fighter planes to counter the greenhouse effect. Instead, we can achieve more enduring, widespread, and true security by increasing environmental security. Increasingly, it is becoming a case of trees now or tanks later.

Major environmental problems recognize no geographic boundaries; the winds carry no passports. The new security will be security for us all, or for none. Peace with the earth means peace with each other, and vice versa. Hopefully, the prospect of catastrophic breakdown can motivate us to achieve ultimate peace on the earth, by learning how to cooperate with one another and with the rest of nature.

Critical Thinking

1. Do you agree or disagree with the position that we need to place much more emphasis on environmental security? Should we treat it with the same degree of seriousness, analysis, and funding as we do economic security and military security, or should environmental security have higher or lower priority? Defend your answers.

2. Some 300 million couples in developing countries want to reduce their family size but lack the family planning services to do so. If we were to meet this need, we would reduce the ultimate global population by at least 2 billion. The cost would be $6 billion per year, equivalent to two and one-half days of military spending worldwide. Why do you think it is not being done? What might be the environmental and economic security benefits?

extracted a chemical from rhubarb leaves and heated it to a high temperature in a tube to transform CFCs into table salt, carbon, and sodium fluoride, which is used in some toothpastes.

Even if the 1992 agreements are upheld, scientists estimate that the ozone layer will continue to be depleted until around 2080 and will cause ozone losses of 10–30% over the northern hemisphere (where most of the world's people live). However, without the 1992 international agreement, ozone depletion would be a much more serious threat.

The ozone treaty has set an important precedent for global cooperation and action to avert potential global disaster. However, international cooperation in dealing with projected global warming is much more difficult because the evidence of global warming is less clear-cut. Moreover, lowering our inputs of greenhouse gases by greatly reducing use of fossil fuels and greatly slowing deforestation is economically and politically difficult.

Some people have wondered whether there is intelligent life in other parts of the universe; others wonder whether there is intelligent life on the earth. To them, if we can seriously deal with the interconnected global problems of loss of biodiversity, possible climate change, and depletion of stratospheric ozone from human activities, then the answer is a hopeful *yes*. Otherwise, they believe the answer is a tragic *no*.

The atmosphere is the key symbol of global interdependence. If we can't solve some of our problems in the face of threats to this global commons, then I can't be very optimistic about the future of the world.

Margaret Mead

REVIEW QUESTIONS

1. Define the boldfaced terms used in this chapter.

2. What is the *greenhouse effect*? How widely is this theory accepted? What are the two major greenhouse gases? Why is there more concern about the carbon dioxide as a greenhouse gas than about water vapor? What is *global warming*? What human activities contribute to global warming?

3. What is the scientific consensus about how the earth's temperatures have varied in the past and how past temperature changes have varied with atmospheric concentrations of carbon dioxide? According to the latest models, between 1990 and 2100 about how much increase is projected for **(a)** the global average temperature and **(b)** global sea levels?

4. Explain how each of the following factors might enhance or dampen global warming: **(a)** changes in solar output, **(b)** the oceans, **(c)** water vapor content and clouds, **(d)** changes in polar ice, **(e)** air pollution, **(f)** effects of increased carbon dioxide levels on photosynthesis, and **(g)** forest turnover. How rapidly might climate shift? What is the range of error in current climate change projections?

5. Why should we worry about a possible rise of only a few degrees in the earth's average temperature?

6. Explain how global warming might affect each of the following: **(a)** food production, **(b)** water supplies, **(c)** forests, **(d)** biodiversity, **(e)** sea levels, **(f)** weather extremes, **(g)** human health, and **(h)** the number of environmental refugees.

7. What are the three schools of thought about what should be done about global warming? Which position do you support? Why? What is the boiled frog syndrome and how does it apply to our perception of the dangers of global warming?

8. According to climate scientists, by how much will we need to reduce the current level of fossil fuel use to stabilize carbon dioxide levels at their current level?

9. List six prevention methods and two cleanup methods for dealing with global warming.

10. Explain how improving energy efficiency and relying more on organic farming could help reduce emissions of greenhouse gases.

11. Summarize progress made at the international level toward reducing emissions of greenhouse gases.

12. List seven ways that we might prepare for and adjust to the effects of global warming.

13. What is *ozone depletion* and how serious is this problem? What types of chemicals cause ozone depletion? How do these chemicals cause such depletion?

14. Explain how seasonal ozone thinning occurs each year over the earth's poles.

15. Summarize the arguments supporting the idea that ozone depletion is a serious problem.

16. What are the major harmful effects of ozone depletion on **(a)** human health, **(b)** crop yields, **(c)** forest productivity, **(d)** materials such as plastics and paints, and **(e)** productivity of plankton?

17. Distinguish between *squamous cell skin cancer*, *basal cell skin cancer*, and *malignant melanoma*. List ways that you can reduce your chances of getting skin cancer. Which of these things do you do?

18. If all ozone-depleting chemicals were banned now, about how long would it take for the ozone layer to recover from past damage caused by these chemicals?

19. Summarize the progress that has been made in reducing the threat of ozone depletion.

CRITICAL THINKING

1. Do you believe that global warming from an enhanced natural greenhouse effect caused at least partly by human activities is a serious problem or one that has been greatly exaggerated? Explain.

2. In preparation for the 1992 UN Conference on the Human Environment in Rio de Janeiro, President George Bush's top White House economic adviser gave an address in Williamsburg, Virginia, to representatives of governments from a number of countries. He told his audience not to worry about global warming because the average temperature increases scientists are predicting were much less than the temperature increase he experienced in coming from Washington, D.C., to Williamsburg. What is the fundamental flaw in this reasoning?

3. Explain why you agree or disagree with each of the proposals listed in Figure 11-10 (for slowing down emissions of greenhouse gases into the atmosphere) and those given in Figure 11-11 (for preparing for the effects of global warming). What effects would carrying out these proposals have on your own lifestyle and on those of any children and grandchildren you might have? What might be the effects of *not* carrying out these actions?

4. What consumption patterns and other features of your lifestyle directly add greenhouse gases to the atmosphere? Which, if any, of these things would you be willing to give up to slow projected global warming and reduce other forms of air pollution?

5. Explain how ozone can be beneficial or harmful to people depending on where it is located in the atmosphere.

6. Use the information in Table 11-1 to determine which chemicals should be given priority as substitutes for CFCs. Explain your choices.

7. In 1995 and 1996, some members of the U.S. House of Representatives introduced legislation that would cancel U.S. participation in the international ozone treaty on the basis that there is no scientific proof of ozone thinning. Criticize this position on the basis of the nature of science (Section 3-1) and the information in the Pro/Con box on p. 277.

***8.** How has the climate changed in the area where you live during the past 50 years? Investigate the beneficial and harmful effects of these changes. How have these changes benefited or harmed you personally?

***9.** As a class, conduct a poll of students at your school to determine whether they believe that global warming from an enhanced greenhouse effect is a very serious problem, a moderately serious problem, or of little concern. Tally the results to see whether there are differences related to year in school, political leaning (liberal, conservative, independent), or sex of poll participants.

***10.** As a class, conduct a poll of students at your school to determine whether they believe that stratospheric ozone depletion is a very serious problem, a moderately serious problem, or of little concern. Tally the results to see whether there are differences related to year in school, political leaning (liberal, conservative, independent), or sex of poll participants.

***11.** Make a concept map of this chapter's major ideas, using the section heads and subheads and the key terms (in boldface). Look at the inside back cover and in Appendix 5 for information about concept maps.

FURTHER INFORMATION

For more information see Internet Sources and Further Readings at the back of the book. Online resources for this chapter can be found at **www.brookscole.com/biology**

12 WATER RESOURCES AND WATER POLLUTION

Water Wars in the Middle East

The next war in the Middle East may well be fought over water, not oil. Most water in this dry region comes from three shared river basins: the Jordan, the Tigris–Euphrates, and the Nile (Figure 12-1). Water in much of this arid region is already in short supply, and the human population in this region is projected to double in 25 years.

Ethiopia, which controls the headwaters that feed 86% of the Nile's flow, plans to divert more of this water; so does Sudan. This could reduce the amount of water available to water-short Egypt, whose terrain is desert except for a thin green strip of irrigated cropland running down its middle along the Nile and its delta. Between 1999 and 2025, Egypt's population is expected to increase from 67 million to 96 million, greatly increasing the demand for already scarce water.

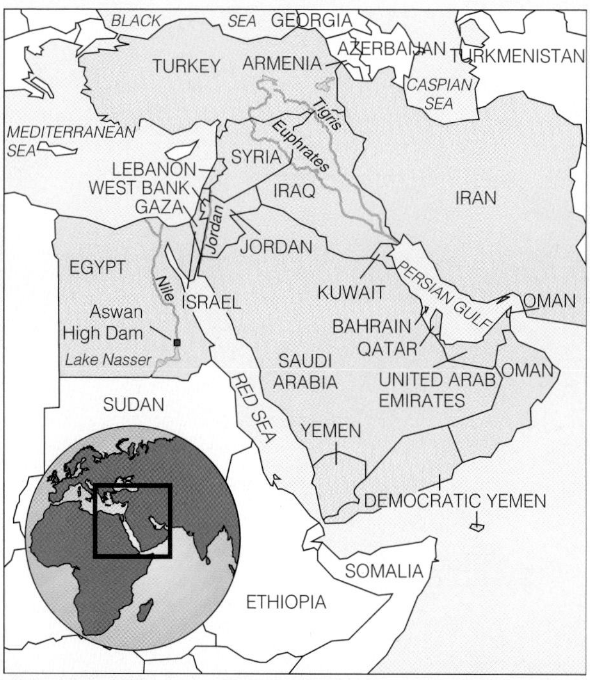

Figure 12-1 The Middle East, whose countries have some of the highest population growth rates in the world. Because of the dry climate, food production depends heavily on irrigation. Existing conflicts among countries in this region over access to water may soon overshadow both long-standing religious and ethnic clashes and attempts to take over valuable oil supplies.

Egypt's options are to **(1)** go to war with Sudan and Ethiopia to obtain more water, **(2)** slash population growth, **(3)** improve irrigation efficiency, **(4)** spend $2 billion to build the world's longest concrete canal and a massive pumping station to pump water out of Lake Nasser (the reservoir created from the Nile by the Aswan High Dam) and create a lush new valley of irrigated farmland in the middle of the desert, **(5)** import more grain to reduce the need for irrigation water, **(6)** work out water-sharing agreements with other countries, or **(7)** suffer the harsh human and economic consequences.

There is also fierce competition for water among Jordan, Syria, and Israel, which get most of their water from the Jordan River basin (Figure 12-1). Jordan, which gets about 75% of its water from the Jordan River system, must double its supply over the next 20 years to keep up with projected population growth.

Some 92 million people currently live in the water-short basins of the Tigris and Euphrates rivers (Figure 12-1). By 2020 the population there is projected to almost double to 170 million. Turkey, located at the headwaters of these two rivers, plans to build 22 dams along the upper Tigris and Euphrates to generate huge quantities of electricity and irrigate a large area of land. These dams will reduce the flow of water to downstream Syria and Iraq.

Clearly, distribution of water will be a key issue in any future peace talks in this region. Resolving these problems will require a combination of **(1)** regional cooperation in allocating water supplies, **(2)** slowed population growth, **(3)** improved efficiency in the use of water, and **(4)** increased water prices to encourage conservation and improve irrigation efficiency.

By 2050, almost 3 billion more people will be trying to share the same amount of fresh water the earth has today. Already, 1.4 billion people lack access to clean drinking water and two-thirds of the world's households don't have running water.

As fresh water becomes scarcer, access to water resources will be a major factor in determining the economic, environmental, and military security of a growing number of countries around the globe (Guest Essay, p. 282). According to Lester Brown (Guest Essay, p. 26) and Christopher Flavin of the Worldwatch Institute, "The spreading scarcity of fresh water may be the most underestimated resource issue facing the world as it enters the new millennium."

Our liquid planet glows like a soft blue sapphire in the hard-edged darkness of space. There is nothing else like it in the solar system. It is because of water.

JOHN TODD

This chapter addresses the following questions:

- What are water's unique physical properties?
- How much fresh water is available to us, and how much of it are we using?
- What areas face water shortages, and what can be done about this problem?
- What areas suffer from flooding, and what can be done to reduce the risk of flooding and flood damage?
- What pollutes water, where do the pollutants come from, and what effects do they have?
- What are the water pollution problems in streams, lakes, oceans, and groundwater?
- How can we prevent and reduce water pollution?

12-1 WATER'S IMPORTANCE AND UNIQUE PROPERTIES

Why Is Water So Important? We live on the water planet, with a precious film of water—most of it salt water—covering about 71% of the earth's surface (Figure 6-21). The earth's organisms are made up mostly of water; a tree is about 60% water by weight, and most animals (including humans) are about 50–65% water.

Each of us needs only a dozen or so cupfuls of water per day to survive, but huge amounts of water are needed to supply us with food, shelter, and our other needs and wants. Water also plays a key role in sculpting the earth's surface, moderating climate, and diluting pollutants.

What Are Some Important Properties of Water? Water is a remarkable substance with a unique combination of properties:

- *There are strong forces of attraction (called hydrogen bonds) between molecules of water.* These attractive forces between water molecules are the major factor determining water's unique properties.

- *Water exists as a liquid over a wide temperature range because of the strong forces of attraction between water molecules.* Its high boiling point of 100°C (212°F) and low freezing point of 0°C (32°F) mean that water remains a liquid in most climates on the earth.

- *Liquid water changes temperature very slowly because it can store a large amount of heat without a large change in temperature.* This high heat capacity helps protect living organisms from the shock of abrupt temperature changes; it also moderates the earth's climate and makes water an excellent coolant for car engines, power plants, and heat-producing industrial processes.

- *It takes a lot of heat to evaporate liquid water because of the strong forces of attraction between its molecules.* Water absorbs large amounts of heat as it changes into water vapor and releases this heat as the vapor condenses back to liquid water. This is a primary factor in distributing heat throughout the world as part of the world's global air circulation system (Figure 6-5). This property also makes evaporation of water an effective cooling process, which is why you feel cooler when perspiration or bathwater evaporates from your skin.

- *Liquid water can dissolve a variety of compounds.* This enables it to **(1)** carry dissolved nutrients into the tissues of living organisms, **(2)** flush waste products out of those tissues, **(3)** serve as an all-purpose cleanser, and **(4)** help remove and dilute the water-soluble wastes of civilization. Water's superiority as a solvent also means that it is easily polluted by water-soluble wastes.

- *The strong attractive forces between the molecules of liquid water cause its surface to contract (high surface tension) and also to adhere to and coat a solid (high wetting ability).* Together these properties allow water to rise through a plant from the roots to the leaves (capillary action).

- *Unlike most liquids, water expands when it freezes.* This means that ice has a lower density (mass per unit of volume) than liquid water. Thus ice floats on water and bodies of water freeze from the top down instead of from the bottom up. Without this property, lakes and streams in cold climates would freeze solid and eliminate most of their current forms of aquatic life. Because water expands upon freezing, it can also break pipes, crack engine blocks (which is why we use antifreeze), and break up streets and fracture rocks (thus forming soil).

Water is the lifeblood of the ecosphere. It is truly a wondrous substance that connects us to one another, to other forms of life, and to the entire planet. Despite its importance, water is one of our most poorly managed resources. We waste it and pollute it. We also charge too little for making it available, thus encouraging still greater waste and pollution of this potentially renewable resource, for which there is no substitute.

12-2 SUPPLY, RENEWAL, AND USE OF WATER RESOURCES

How Much Fresh Water Is Available? Only a tiny fraction of the planet's abundant water is available to us as fresh water. About 97% by volume is found in the oceans and is too salty for drinking, irrigation, or industry (except as a coolant).

The remaining 3% is fresh water. About 2.997% of the earth's water is locked up in ice caps or glaciers or is buried so deep that it costs too much to extract. Only about 0.003% of the earth's total volume of water is easily available to us as soil moisture, usable groundwater, water vapor, and lakes and streams. If the world's water supply were only 100 liters (26 gallons), our usable supply of fresh water would be only about 0.003 liter (one-half teaspoon) (Figure 12-2).

Fortunately, the available fresh water amounts to a generous supply. And this water is continuously collected, purified, recycled, and distributed in the solar-powered *hydrologic cycle* (Figure 4-22) as long as we don't **(1)** overload it with slowly degradable and nondegradable wastes or **(2)** withdraw it from underground supplies faster than it is replenished. Unfortunately, we are doing both.

Differences in average annual precipitation divide the world's countries and people into water haves and have-nots. For example, Canada, with only 0.5% of the world's population, has 20% of the world's fresh water supply. By contrast, China, with 21% of the world's people, has only 7% of the world's fresh water supply. Already more than 300 of China's 640 largest cities are short of water, and 100 of them are very short.

As population, irrigation, and industrialization increase, water shortages in already water-short regions will intensify and wars over water may erupt (p. 286). Projected global warming also might cause changes in rainfall patterns and disrupt water supplies in unpredictable ways (Section 11-4).

What Is Surface Water? The fresh water we use first arrives as the result of precipitation. Precipitation that does not infiltrate the ground or return to the atmosphere by evaporation (including transpiration) is called **surface runoff** that flows into streams, lakes, wetlands, and reservoirs.

A **watershed**, also called a **drainage basin**, is a region from which water drains into a stream, stream system, lake, reservoir, or other water body.

What Is Groundwater? Some precipitation infiltrates the ground and percolates downward through voids (pores, fractures, crevices, and other spaces) in soil and rock (Figure 12-3). The water in these voids is called **groundwater**.

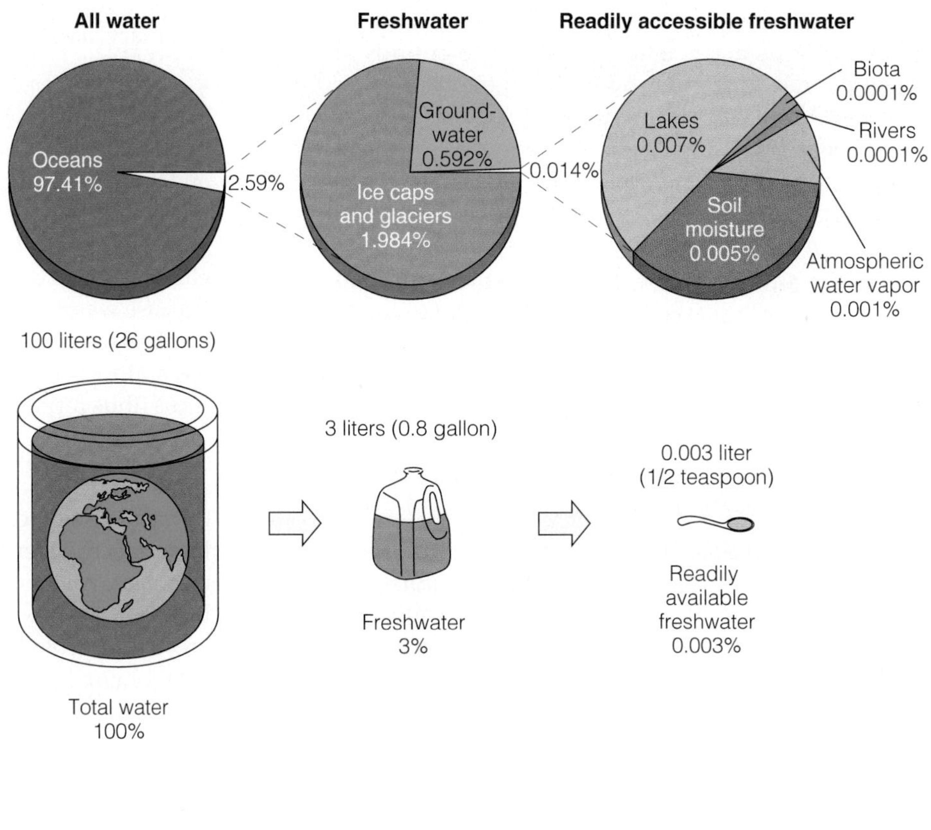

Figure 12-2 The planet's water budget. Only a tiny fraction by volume of the world's water supply is fresh water available for human use.

Close to the surface, the voids have little moisture in them. However, below some depth, in what is called the **zone of saturation**, they are filled with water except for occasional air bubbles. The boundary between the surface of the zone of saturation and the unsaturated zone above is called the **water table**. The water table falls in dry weather and rises in wet weather.

Porous, water-saturated layers of sand, gravel, or bedrock through which groundwater flows are called **aquifers** (Figure 12-3). Any area of land through which water passes downward or laterally into an aquifer is called a **recharge area**. Aquifers are replenished naturally by precipitation that percolates downward through soil and rock in what is called **natural recharge**, but some are recharged from the side by *lateral recharge*.

Groundwater moves from the recharge area through an aquifer and out to a discharge area (well, spring, lake, geyser, stream, or ocean) as part of the hydrologic cycle. Groundwater normally moves from points of high elevation and pressure to points of lower elevation and pressure. This movement is quite slow, typically only a meter or so (about 3 feet) per year and rarely more than 0.3 meter (1 foot) per day.

Some aquifers get very little (if any) recharge. Often found fairly deep underground and formed tens of thousands of years ago, they are (on a human time scale) nonrenewable resources. Withdrawals from such aquifers amount to *water mining* that, if kept up, will deplete these ancient deposits of water.

How Do We Use the World's Fresh Water? Since 1950, the global rate of water withdrawal from surface and groundwater sources has increased almost fivefold and per capita use has tripled. According to a 1996 study, humans currently use about 54% of the global surface runoff that is realistically available from the hydrologic cycle. Because of increased population growth and economic development, global withdrawal rates of surface water are projected to at least double in the next two decades and exceed the available surface runoff in a growing number of areas.

Uses of withdrawn water vary from one region to another and from one country to another (Figure 12-4). Worldwide, about 70% of all water withdrawn each year from rivers, lakes, and aquifers is used to irrigate 17% of the world's cropland. Some 60–80% of this

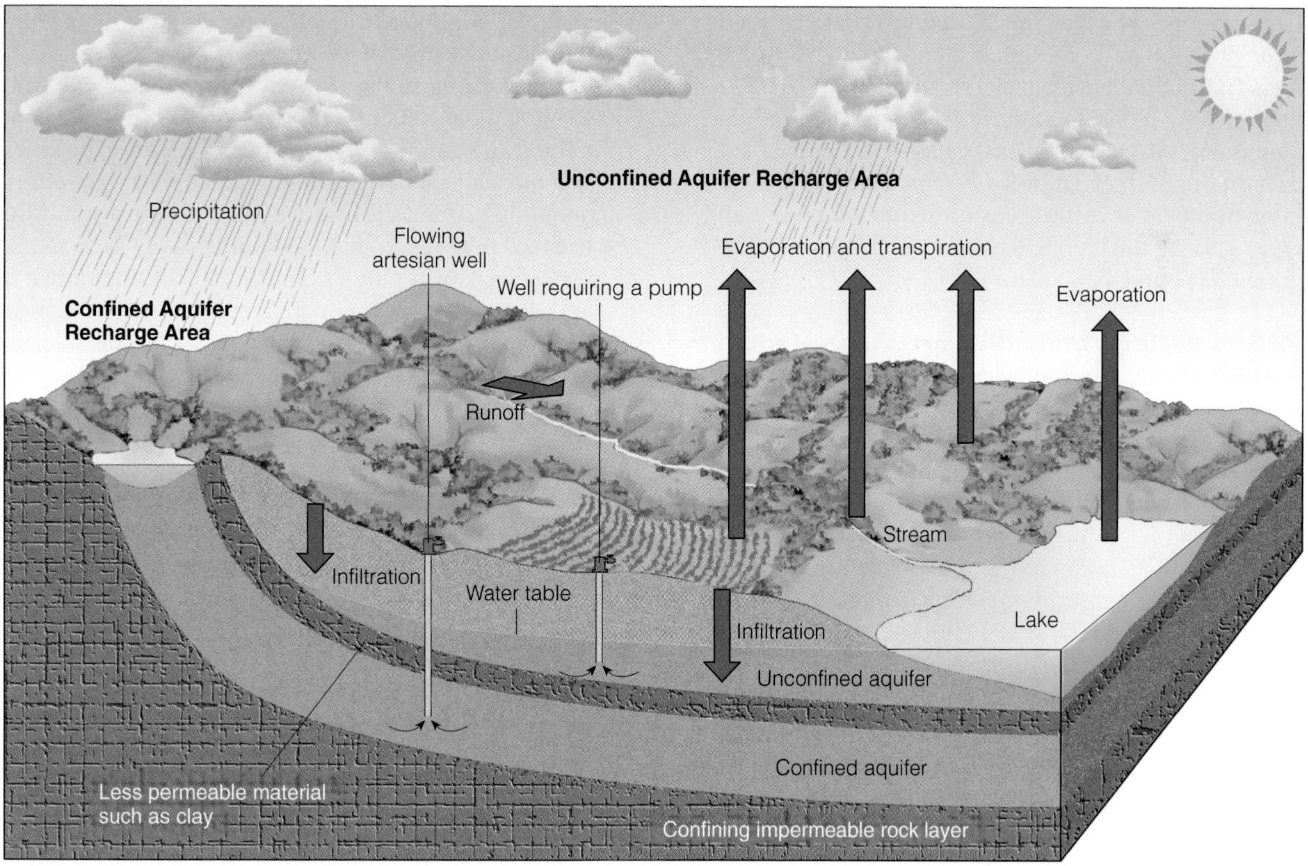

Figure 12-3 The groundwater system. An *unconfined aquifer* is an aquifer with a water table. A *confined aquifer* is bounded above and below by less permeable beds of rock. Groundwater in this type of aquifer is confined under pressure.

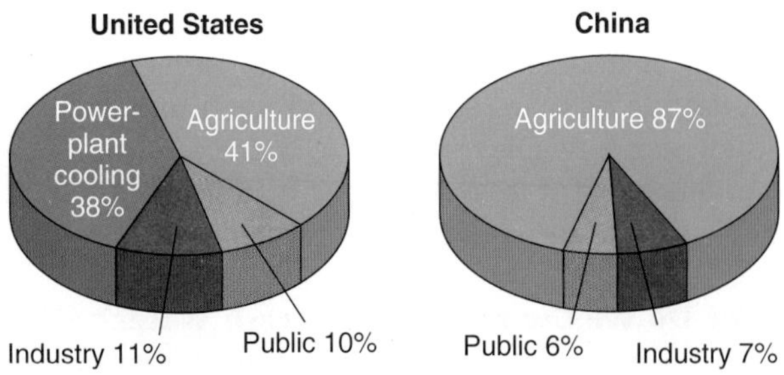

Figure 12-4 Use of water in the United States and China. The United States has the world's highest per capita use of water, amounting to an average of 4,800 liters (1,280 gallons) per person every day in 1995 (down by 4% since 1990). About half of the water used in the United States is unnecessarily wasted. However, between 1980 and 1995 total water use in the United States decreased by 10% despite a 16% increase in population, mostly because of more efficient irrigation. (Data from Worldwatch Institute and World Resources Institute)

water either evaporates or seeps into the ground before reaching crops. About 20% of the water withdrawn each year is used for industry and 10% is used by residences and cities.

Agricultural and manufactured products require large amounts of water, much of which could be used more efficiently and reused. It takes about **(1)** 380,000 liters (100,000 gallons) to make an automobile, **(2)** 3,800 liters (1,000 gallons) to produce 454 grams (1 pound) of aluminum, **(3)** 3,000 liters (800 gallons) to produce 454 grams (1 pound) of grain-fed beef in a feedlot (where large numbers of cattle are confined to a fairly small area), and **(4)** 100 liters (26 gallons) to produce 1 kilogram (2.2 pounds) of paper.

Case Study: Fresh Water Resources in the United States Although the United States has plenty of fresh water, much of it is in the wrong place at the wrong time or is contaminated by agricultural and industrial practices. The Eastern states usually have ample precipitation, whereas many of the Western states have too little. In the East, the largest uses for water are for energy production, cooling, and manufacturing; in the West, the largest use by far is for irrigation (which accounts for about 85% of all water use).

In many parts of the Eastern United States the most serious water problems are flooding, occasional urban shortages, and pollution. For example, the 3 million residents of Long Island, New York, get most of their water from an aquifer that is becoming severely contaminated.

The major water problem in the arid and semiarid areas of the Western half of the country is a shortage of runoff caused by low precipitation, high evaporation, and recurring prolonged drought. Water tables in many areas are dropping rapidly as farmers and cities deplete groundwater aquifers faster than they are recharged.

12-3 TOO LITTLE WATER

What Causes Freshwater Shortages? According to water expert Malin Falkenmark, there are four causes of water scarcity: **(1)** a *dry climate* (Figure 6-2), **(2)** *drought* (a period in which precipitation is much lower and evaporation is higher than normal), **(3)** *desiccation* (drying of the soil because of such activities as deforestation and overgrazing by livestock), and **(4)** *water stress* (low per capita availability of water caused by increasing numbers of people relying on limited levels of runoff).

Since the 1970s water scarcity intensified by prolonged drought (mostly in areas with dry climates, where 40% of the world's people live) has killed more than 24,000 people per year and created many environmental refugees. In water-short areas, many women and children must walk long distances each day, carrying heavy jars or cans, to get a meager supply of sometimes contaminated water. Millions of poor people in developing countries have no choice but to try to survive on drought-prone land. If global warming occurs as projected, severe droughts may become more common in some areas of the world.

According to the United Nations and the World Bank, almost 500 million people in 31 countries containing 40% of the world's population now experience chronic water shortages that threaten their agriculture and industry and the health of their people. By 2025, at least 2.8 billion people in 48 countries are expected to face severe water stress, and by 2050 the number is expected to be at least 4 billion and perhaps as high as 7 billion. In most of these countries the problem is not a shortage of water but the wasteful and unsustainable use of normally available supplies.

In the United States, many major urban centers (especially those in the West and Midwest) are located in areas that don't have enough water (Figure 12-5). Experts project that current shortages and conflicts over water supplies will get much worse as more industries and people migrate west and compete with farmers for scarce water. These shortages could worsen if climate warms as a result of an enhanced greenhouse effect.

Evidence of water stress is seen in the draining of rivers and falling water tables. The Colorado River in the southwestern United States rarely reaches the sea anymore. In northern China, the Yellow River has run dry part of every year since 1985. The Nile, the largest river in the Middle East (Figure 12-1), has little water in it when it reaches the sea. Water tables are falling on every continent, including major food-producing areas

Barlow, M. 1999. We Are Running out of Water. *The Ecologist,* vol. 29, no.3, p. 182(1).

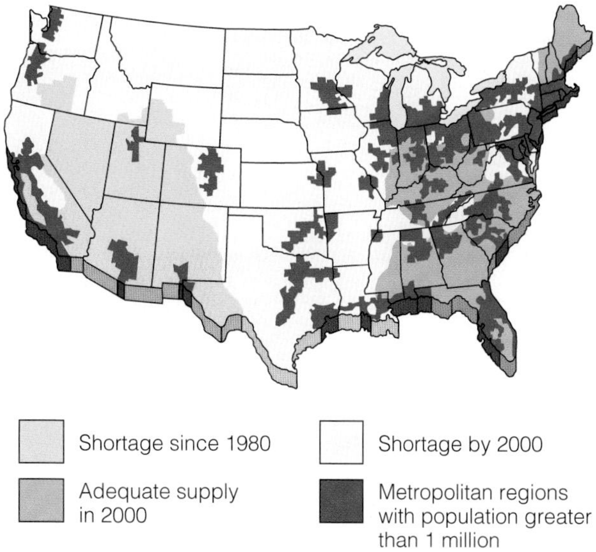

Figure 12-5 Current and projected water deficit regions in the continental United States and their proximity to metropolitan areas having populations greater than 1 million. (Data from U.S. Water Resources Council and U.S. Geological Survey)

such as the North China Plain, the U.S. southern Great Plains, and most of India.

A number of analysts believe that *access to water resources, already a key foreign policy and environmental security issue for water-short countries, will become even more important over the next 10–20 years* (Guest Essay, p. 282). Almost 150 of the world's 214 major river systems (57 of them in Africa) are shared by 2 countries; another 50 are shared by 3 to 10 nations. Some 40% of the world's population already clashes over water, especially in the Middle East (p. 285).

Some areas have lots of water, but the largest rivers (which carry most of the runoff) are far from agricultural and population centers where the water is needed. For example, South America has the largest annual water runoff of any continent, but 60% of the runoff flows through the Amazon River in remote areas where few people live.

How Can We Increase Water Supplies? There are five ways to increase the supply of fresh water in a particular area: **(1)** Build dams and reservoirs to store runoff, **(2)** bring in surface water from another area, **(3)** withdraw groundwater, **(4)** convert salt water to fresh water (desalination), and **(5)** improve the efficiency of water use.

In developed countries, people tend to live where the climate is favorable and then bring in water from another watershed. In developing countries, most people (especially the rural poor) must settle where the water is and try to capture and use as much precipitation as they can.

What Are the Pros and Cons of Large Dams and Reservoirs? Large dams and reservoirs have benefits and drawbacks (Figure 12-6). Large reservoirs created by damming streams can capture and store water from rain and melting snow. This water can then be released as desired to **(1)** produce hydroelectric power at the dam site, **(2)** irrigate land below the dam, **(3)** control flooding of land below the reservoir, and **(4)** provide water carried to towns and cities by aqueducts. Reservoirs also provide recreational activities such as swimming, fishing, and boating.

Dams also have a number of drawbacks (Figure 12-6). A series of dams on a river, especially in arid areas, can reduce downstream flow to a trickle and prevent it from reaching the sea as a part of the hydrologic cycle (Figure 4-22). In the United States, the heavily dammed Colorado River rarely makes it to the Gulf of California. Instead it fizzles into a trickle that disappears into the Mexican desert or (in drought years) the Arizona desert. This **(1)** threatens the survival of species that spawn in such rivers, **(2)** destroys estuaries that serve as breeding grounds for numerous aquatic species, and **(3)** increases saltwater contamination of aquifers near coasts.

When completed, China's Three Gorges project on the mountainous upper reaches of the Yangtze River will be the world's largest hydroelectric dam and reservoir. This superdam, with the electric output of 18 large nuclear power plants, will supply power to industries and to 150 million Chinese. The project will also help China reduce its dependence on coal and help hold back the Yangtze River's floodwaters (which have claimed at least 500,000 lives during the past 100 years). The project is supposed to be completed by 2009 and cost $17–30 billion. However, estimates by independent experts put the completion date at around 2019 and the cost at $70 billion or more.

This project will flood large areas of some of the best farmland in the region and 800 existing factories. It will also displace 1.9 million people, including 100 towns and 2 cities (each with about 100,000 people), and flood some of China's most scenic land to create a reservoir 596 kilometers (370 miles) long.

Critics charge that the dam will transform the Yangtze into a sewer for industrial wastes and the immense pressure from the huge volume of water in the reservoir could trigger landslides and earthquakes. Furthermore, as the reservoir fills up with sediment, the dam could overflow and expose half a million people to severe flooding (especially if the reservoir is kept filled at a high level, as planned, to provide maximum hydroelectric power).

Hint: Enter the search terms *water, shortage* using the Subject Guide.

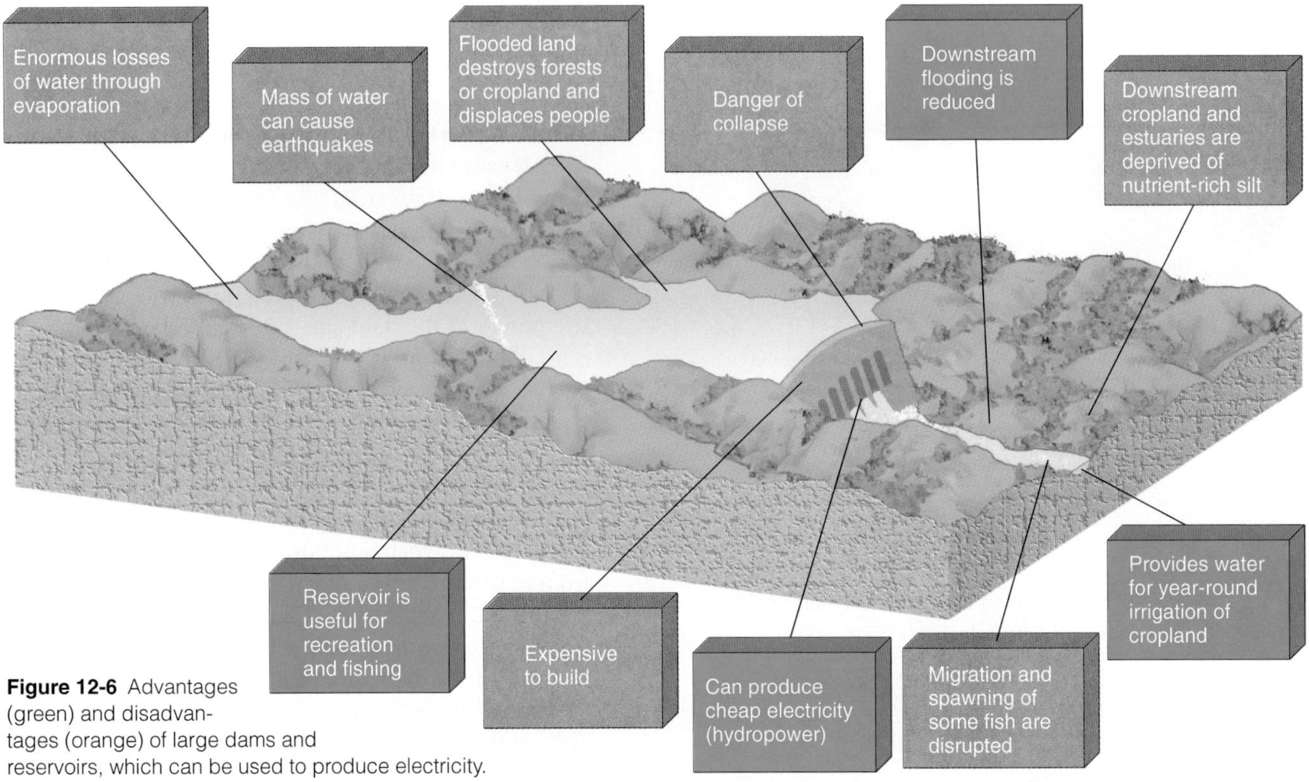

Figure 12-6 Advantages (green) and disadvantages (orange) of large dams and reservoirs, which can be used to produce electricity.

What Are the Pros and Cons of Watershed Transfers? The California Experience Tunnels, aqueducts, and underground pipes can transfer stream runoff collected by dams and reservoirs from water-rich watersheds to water-poor areas. Although such transfers have benefits, they also create environmental problems. Indeed, most of the world's large-scale water transfers illustrate the important ecological principle that *you can't do just one thing*.

Some cities are finding that protecting watersheds that provide them with clean water is cheaper than trying to clean up polluted water for drinking. For example, New York City plans to spend $1.4 billion on a project to buy and protect upstate watersheds that provide the city with drinking water. This means that the city won't have to spend $3–8 billion for a new water filtration system.

One of the world's largest watershed transfer projects is the California Water Project. In California, the basic water problem is that 75% of the population lives south of Sacramento but 75% of the rain falls north of it. The California Water Project uses a maze of giant dams, pumps, and aqueducts to transport water from water-rich northern California to heavily populated areas and to arid and semiarid agricultural regions, mostly in southern California (Figure 12-7).

For decades, northern and southern Californians have been feuding over how the state's water should be allocated under this project. Southern Californians say they need more water from the north to support Los Angeles, San Diego, and other growing urban areas and to grow more crops. Although agriculture uses 74% of the water withdrawn in California, it accounts for only 2.5% of the state's economic output. Irrigation for just two crops, alfalfa (to feed cattle) and cotton, uses as much water as the homes of all 33 million Californians.

Opponents in the north say that sending more water south would degrade the Sacramento River, threaten fisheries, and reduce the flushing action that helps clean San Francisco Bay of pollutants. They also argue that **(1)** much of the water sent south is wasted unnecessarily and **(2)** making irrigation just 10% more efficient would provide enough water for domestic and industrial uses in southern California.

If water supplies in northern California were to drop sharply because of global warming, the amount of water delivered by the huge distribution system would plummet. Most irrigated agriculture in California would have to be abandoned and much of the population of southern California might have to move to areas with more water. The 6-year drought that California experienced between 1986 and 1992 was perhaps just a small taste of the future.

Pumping out more groundwater is not the answer. Throughout much of California, groundwater is already being withdrawn faster than it is replenished.

Figure 12-7 The California Water Project and the Central Arizona Project involve large-scale transfers of water from one watershed to another. Arrows show the general direction of water flow.

Figure 12-8 If completed, the James Bay project in northern Quebec will alter or reverse the flow of 19 major rivers and flood an area the size of Washington State to produce hydropower for consumers in Quebec and the United States, especially in New York State. Phase I of this 50-year project is completed; phase II was postponed indefinitely in 1994 because of a surplus of electricity and opposition by environmentalists and the indigenous Cree, whose ancestral hunting grounds would have been flooded.

To most analysts, quicker and cheaper solutions are **(1)** improving irrigation efficiency and **(2)** allowing farmers to sell their legal rights to withdraw certain amounts of water from rivers.

Case Study: The James Bay Watershed Transfer Project Another major watershed transfer project is the James Bay project. It is a $60-billion, 50-year scheme to harness the wild rivers that flow into Quebec's James and Hudson Bays to produce electric power for Canadian and U.S. consumers (Figure 12-8). If completed, this megaproject would **(1)** construct 600 dams and dikes that will reverse or alter the flow of 19 giant rivers covering a watershed three times the size of New York State, **(2)** flood an area of boreal forest and tundra equal in area to Washington State or Germany, and **(3)** displace thousands of indigenous Cree and Inuit, who for 5,000 years have lived off James Bay by subsistence hunting, fishing, and trapping.

After 20 years and $16 billion, Phase I has been completed. The second and much larger phase was postponed indefinitely in 1994 because of **(1)** an excess of power generated, **(2)** opposition by the Cree (whose ancestral hunting grounds would have been flooded) and Canadian and U.S. environmentalists, and **(3)** New York State's cancellation of two contacts to buy electricity produced by Phase II.

Case Study: The Aral Sea Watershed Transfer Disaster The shrinking of the Aral Sea (Figure 12-9) is a result of a large-scale water transfer project in an area of the former Soviet Union with the driest climate in

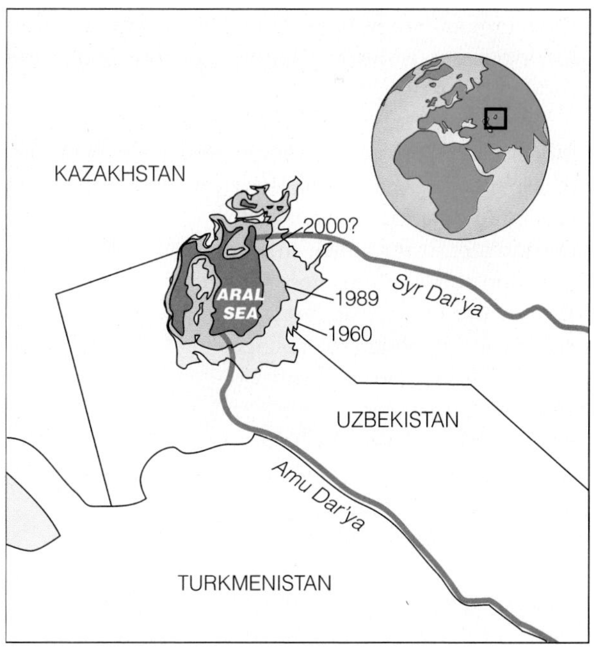

Figure 12-9 Once the world's fourth-largest freshwater lake, the Aral Sea has been shrinking and getting saltier since 1960 because most of the water from the rivers that replenish it has been diverted to grow cotton and food crops. As the lake shrinks, it leaves behind a salty desert, economic ruin, increasing health problems, and severe ecological disruption.

Central Asia. Since 1960, enormous amounts of irrigation water have been diverted from the inland Aral Sea and its two feeder rivers to irrigate cotton, vegetable, fruit, and rice crops. The irrigation canal, the world's longest, stretches over 1,300 kilometers (800 miles).

This water diversion (coupled with droughts) has caused a regional ecological, economic, and health disaster, described by one former Soviet official as "ten times worse than the 1986 Chernobyl nuclear power-plant accident." Since 1960 **(1)** the sea's salinity has tripled, **(2)** its surface area has shrunk by 54%, **(3)** its volume has decreased by 80%, and **(4)** its two supply rivers are now mere trickles. About 36,000 square kilometers (14,000 square miles) of former lake bottom has become a human-made salt desert. The process continues, and within a decade the once enormous Aral Sea may break up into three small brine lakes.

All 24 of the area's native fish species have become extinct, devastating the area's fishing industry, which once provided work for more than 60,000 people. Roughly half of the area's bird and mammal species have also disappeared.

Winds pick up the salty dust that encrusts the lake's now-exposed bed and blow it onto fields as far away as 300 kilometers (190 miles) away. As the salt spreads, it kills crops, trees, and wildlife and destroys pastureland.

These changes have also affected the area's already semiarid climate. The once-huge Aral Sea acted as a thermal buffer, moderating the heat of summer and the extreme cold of winter. Now there is less rain, summers are hotter, winters are colder, and the growing season is shorter. Cotton and crop yields have dropped dramatically.

To raise yields, farmers have increased inputs of herbicides, insecticides, and fertilizers on some crops. Many of these chemicals have percolated downward and accumulated to dangerous levels in the groundwater, from which most of the region's drinking water comes. Rising water tables from excessive irrigation have brought these chemicals near the surface and contaminated drinking water supplies. The low river flows have also concentrated salts, pesticides, and other toxic chemicals, making surface water supplies hazardous to drink.

Winds whip up fertilizer and pesticide residues from the poisoned agricultural land and from the bare floor of the shriveled Aral Sea. The combination of toxic dust and contaminated water have caused serious health problems for a growing number of the 35 million people living in the Aral Sea's watershed. Such problems include abnormally high rates of infant mortality, tuberculosis, anemia, respiratory illness (one of the world's highest), kidney and liver diseases (especially cancers), arthritic diseases, typhoid fever, and hepatitis.

Between 1999 and 2002 the United Nations and the World Bank plan to spend $600 million to address some of the area's serious ecological, economic, and health problems. Emphasis will be on projects to purify drinking water and upgrade irrigation and drainage systems. More efficient irrigation will lower the region's water table, flush salt from croplands, and boost crop productivity. However, this process will take decades and will not prevent the shrinkage of the Aral sea into a few brine lakes.

What Are the Pros and Cons of Withdrawing Groundwater? Pumping groundwater from aquifers has several advantages over tapping more erratic flows from streams: **(1)** it can be removed as needed year-round, **(2)** it is not lost by evaporation, and **(3)** it is usually less expensive to develop than surface water systems. In the United States, about half of the drinking water (96% in rural areas and 20% in urban areas) and 40% of irrigation water is pumped from aquifers. In Florida, Hawaii, Idaho, Mississippi, Nebraska, and New Mexico, more than 90% of the population depends on groundwater for drinking water.

However, overuse of groundwater can cause or intensify several problems: **(1)** *aquifer depletion*, **(2)** *aquifer subsidence* (sinking of land when groundwater is withdrawn), and **(3)** *intrusion of salt water into aquifers* (Fig-

Figure 12-10 Areas of greatest aquifer depletion, subsidence, saltwater intrusion, and groundwater contamination in the United States. (Data from U.S. Water Resources Council and U.S. Geological Survey)

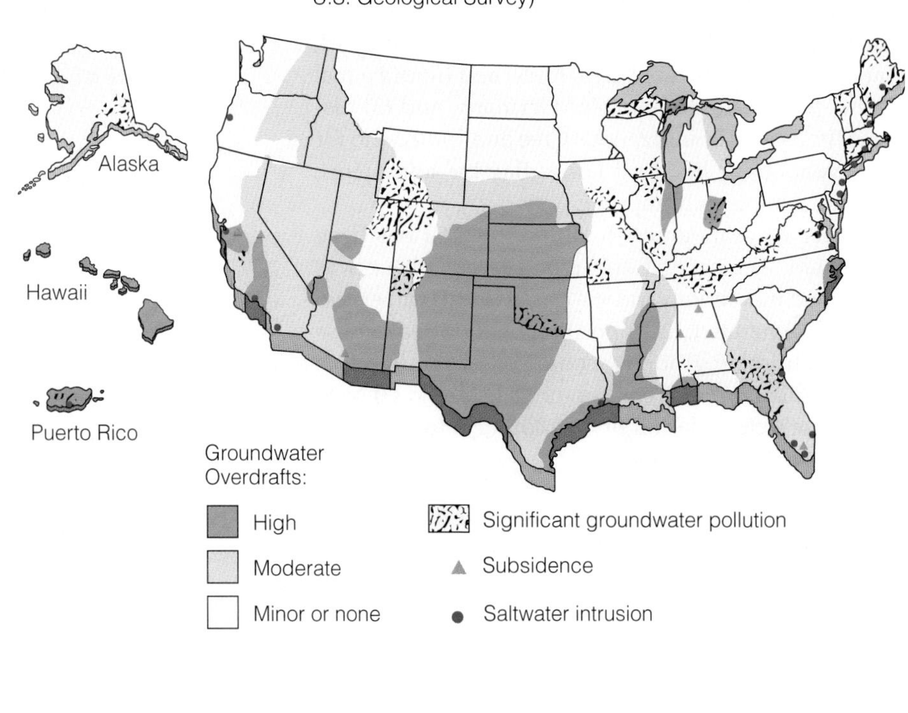

ure 12-10). Groundwater can also become contaminated by industrial and agricultural activities, septic tanks, and other sources, as discussed in Section 12-5. Because groundwater is the source of about 40% of the stream flow in the United States, groundwater depletion also robs streams of water.

Currently, *groundwater in the United States is being withdrawn at four times its replacement rate.* The most serious overdrafts are occurring **(1)** in parts of the huge Ogallala Aquifer, extending from southern South Dakota to central Texas (Figures 12-10, 12-11, and Case Study, p. 297) and **(2)** in parts of the arid southwestern United States (Figure 12-5), especially California's Central Valley, which supplies about half the country's vegetables and fruits. Aquifer depletion is also a problem in Saudi Arabia, northern China, northern Africa (especially Libya and Tunisia), southern Europe, the Middle East, and parts of Mexico, Thailand, and India.

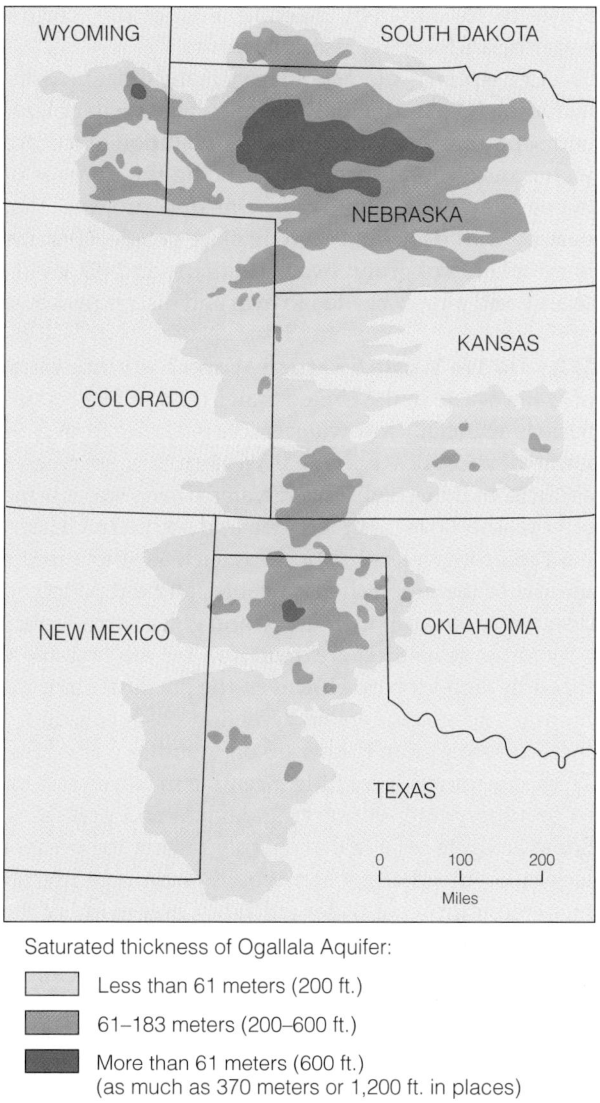

Figure 12-11 The Ogallala, the world's largest known aquifer. If the water in this aquifer were above ground, it could cover all 50 states with 0.5 meter (1.5 feet) of water. This fossil aquifer, which is renewed very slowly, is being depleted (especially at its thin southern end in parts of Texas, New Mexico, Oklahoma, and Kansas) to grow crops, raise cattle, and provide urban dwellers and industries with water. (Data from U.S. Geological Survey)

In China, water tables are falling almost everywhere there is flat land, especially in the North China Plain, which supplies nearly 40% of the country's grain. In India, water is being withdrawn from aquifers two or more times faster than the recharge rate. As the country's aquifers are depleted, cutbacks in irrigation could reduce grain harvests by 25%. This will make it very difficult to feed the 600 million more people projected to be added between 1998 and 2050.

When fresh water from an aquifer near a coast is withdrawn faster than it is recharged, salt water intrudes into the aquifer (Figure 12-12). Such intrusion threatens to contaminate the drinking water of many towns and cities along the Atlantic and Gulf coasts (Figure 12-10) and in the coastal areas of Israel, Syria, and the Arabian Gulf states.

Ways to prevent or slow groundwater depletion include **(1)** controlling population growth, **(2)** not planting water-thirsty crops in dry areas, **(3)** developing crop strains that require less water and are more resistant to heat stress, and **(4)** wasting less irrigation water.

How Useful Is Desalination? The removal of dissolved salts from ocean water or from brackish (slightly salty) groundwater, called **desalination**, is another way to increase freshwater supplies. Distillation and reverse osmosis are the two most widely used methods. *Distillation* involves heating salt water until it evaporates and condenses as fresh water, leaving salts behind in solid form. In *reverse osmosis*, salt water is pumped at high pressure through a thin membrane whose pores allow water molecules, but not dissolved salts, to pass through.

About 11,100 desalination plants in 120 countries (especially in the arid Middle East and parts of North Africa) meet less than 0.2% of the world's water needs. However, desalination has a downside. Because it uses vast amounts of electricity, water produced in this way costs three to five times more than water from conventional sources. Distributing the water from coastal desalination plants costs even more in terms of the energy needed to pump the water uphill and inland.

Desalination also produces large quantities of brine containing high levels of salt and other minerals. Dumping the concentrated brine into the ocean near the plants might seem to be the logical solution, but this would increase the local salt concentration and threaten food resources in estuary waters. If these wastes were dumped on the land, they could contaminate groundwater and surface water.

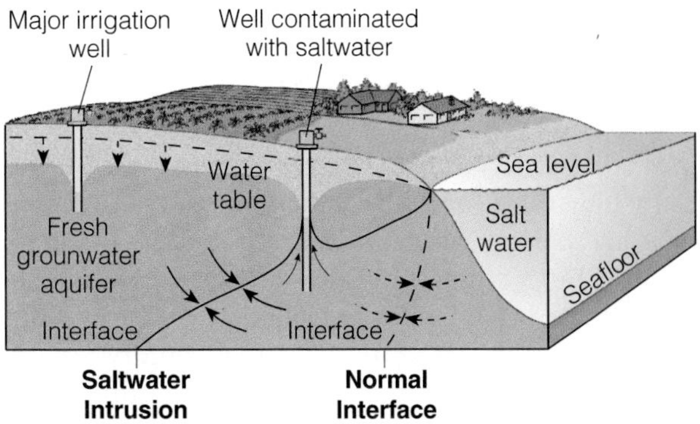

Figure 12-12 Saltwater intrusion along a coastal region. When the water table is lowered, the normal interface (dotted line) between fresh and saline groundwater moves inland (solid line).

Desalination can provide fresh water for coastal cities in arid countries (such as sparsely populated Saudi Arabia), where the cost of getting fresh water by any method is high. However, desalinated water probably will never be cheap enough to irrigate conventional crops or to meet much of the world's demand for fresh water unless affordable, efficient solar-powered distillation plants can be developed and someone can figure out what to do with the resulting mountains of salt.

Can Cloud Seeding and Towing Icebergs Improve Water Supplies? For years several countries, particularly the United States, have been experimenting with seeding clouds with tiny particles of chemicals (such as silver iodide). The particles form water condensation nuclei and thus produce more rain over dry regions and more snow over mountains. However, cloud seeding is not useful in very dry areas, where it is most needed, because rain clouds rarely are available there. Furthermore, widespread cloud seeding would introduce large amounts of the cloud-seeding chemicals into soil and water systems, possibly harming people, wildlife, and agricultural productivity.

Another obstacle to cloud seeding is legal disputes over the ownership of water in clouds. During the 1977 drought in the Western United States, the attorney general of Idaho accused officials in neighboring Washington of "cloud rustling" and threatened to file suit in federal court.

There also have been proposals to tow massive icebergs to arid coastal areas (such as Saudi Arabia and southern California) and then to pump the fresh water from the melting bergs ashore. However, the technology for doing this is not available and the costs may be too high, especially for water-short developing countries.

Solutions: Why Is Reducing Water Waste So Important? Increasing the water supply in some areas is important, but rising population, food needs, industrialization, and unpredictable shifts in water supplies from global warming eventually will outstrip this approach. It makes much more sense, economically and environmentally, to use water more efficiently.

Mohamed El-Ashry of the World Resources Institute estimates that *65–70% of the water people use throughout the world is wasted through evaporation, leaks, and other losses*. The United States, the world's largest user of water, does slightly better but still loses about 50% of the water it withdraws. El-Ashry believes that it is economically and technically feasible to reduce water waste to 15%, thereby meeting most of the world's water needs for the foreseeable future.

Conserving water would have many other benefits, including **(1)** reducing the burden on wastewater plants and septic systems, **(2)** decreasing pollution of surface water and groundwater, **(3)** reducing the number of expensive dams and water-transfer projects that destroy wildlife habitats and displace people, **(4)** slowing depletion of groundwater aquifers, and **(5)** saving energy and money needed to treat and distribute water.

Why Do We Waste So Much Water? A prime cause of water waste in the United States (and in most countries) is artificially low water prices resulting from government subsidies. Each year Western farmers receive about $2 billion in subsidized water. Cheap water is the only reason farmers in Arizona and southern California can grow water-thirsty crops such as alfalfa in the middle of the desert. It also enables affluent people in desert areas to keep their lawns and golf courses green.

Water subsidies are paid for by all taxpayers through higher taxes. However, the harmful environmental and groundwater depletion costs caused by these subsidies don't show up on monthly water bills. Thus, consumers have little incentive to use less water or to adopt water-conserving devices and processes. Some analysts believe that sharply raising the price of federally subsidized water would encourage investments in improving water efficiency, and many of the West's water supply problems could be eased.

Another reason for the water waste in the United States (and many other countries) is that the responsibility for water resource management in a particular watershed may be divided among many state and local governments rather than being handled by one authority. The Chicago, Illinois, metropolitan area, for example, has 349 water-supply systems divided among some 2,000 local units of government over a six-county area.

In sharp contrast is the regional approach to water management used in England and Wales. The British

Mining Groundwater: The Shrinking Ogallala Aquifer

CASE STUDY

Large amounts of water pumped from the Ogallala, the world's largest known aquifer (Figure 12-11), have helped transform vast areas of arid high plains prairie land into one of the largest and most productive agricultural regions in the United States.

With yields two to three times those of dryland farming, irrigated farming is a profitable venture. As a result, this region now produces 20% of U.S. agricultural output (including 40% of its feedlot beef), valued at $32 billion per year. This production has brought prosperity to many farmers and merchants in this region, but the hidden environmental (and economic) cost has been increasing depletion of the aquifer in some areas.

Although this aquifer is gigantic, it is essentially a nonrenewable aquifer (stored during the retreat of the last ice age about 15,000–30,000 years ago) with an extremely slow recharge rate. In some areas, water is being pumped out of the aquifer 8–10 times faster than the natural recharge rate. The northernmost states (Wyoming, North Dakota, South Dakota, and parts of Colorado) still have ample supplies. However, supplies in parts of the southern states, where the aquifer is thinner (Figure 12-11) are being depleted rapidly.

Water experts project that at the current rate of withdrawal, one-fourth of the aquifer's original supply will be depleted by 2020, and much sooner in areas where it is shallow. It will take thousands of years to replenish the aquifer.

Government farm policies encourage depletion of the aquifer by subsidizing water-thirsty crops that require irrigation and by providing crop disaster payments. The greater the use of groundwater, the greater this tax break—an example of positive feedback in action.

Depletion of this essentially nonrenewable water resource can be delayed. Farmers can use more efficient forms of irrigation (Figure 12-13); in some areas they may have to switch to crops that require less water, or to dryland farming. Total irrigated area is already declining in five of the seven states using this aquifer because of rising drilling and pumping costs as the water table drops. Many farmers in northwest Texas have reduced their water use by 20–25% by switching to more efficient irrigation technologies.

Cities using this groundwater can improve their water management by wasting less water and using less for lawns and golf courses. People enjoying the benefits of this aquifer can pitch in by installing water-saving toilets and showerheads and converting their lawns to plants that can survive in an arid climate with little watering.

Critical Thinking

Do you believe that farmers and cattle raisers using water withdrawn from the Ogallala should receive federal subsidies to grow crops and raise livestock that require large amounts of irrigation water? Explain. What are the alternatives?

Water Act of 1973 replaced more than 1,600 agencies with 10 regional water authorities based on natural watershed boundaries. Each water authority owns, finances, and manages all water supply, water pollution control, and waste treatment facilities in its region. Each is managed by a group of elected local officials and a smaller number of officials appointed by the national government.

Solutions: How Can We Waste Less Water in Irrigation? Worldwide, about 65% of the water that is diverted from rivers or pumped out of aquifers is used for irrigating about 17% of the world's cropland to produce almost 40% of the world's food. Because only about 40% of this water reaches crops, more efficient use of even a small amount of irrigation water would free large amounts of water for other uses.

Most irrigation systems distribute water from a groundwater well or a surface canal by downslope or gravity flow through unlined ditches in cropfields so that the water can be absorbed by crops (Figure 12-13, left). However, this flood irrigation method delivers far more water than needed for crop growth. Moreover, only 50–60% of the water reaches crops because of evaporation, deep percolation (seepage), and runoff.

Seepage can be reduced by placing plastic, concrete, or tile liners in the large irrigation canals that distribute water to unlined irrigation ditches. Lasers can be used to make sure that fields are level so that water flowing into unlined irrigation ditches is distributed more evenly. Small check dams of earth and stone can capture runoff from hillsides and channel it to fields. Holding ponds can store rainfall or capture irrigation water for recycling to crops.

Many U.S. farmers served by the dwindling Ogallala Aquifer now use center-pivot sprinkler systems (Figure 12-13, right), with which 70–80% of the water reaches crops. Some of these farmers are switching to

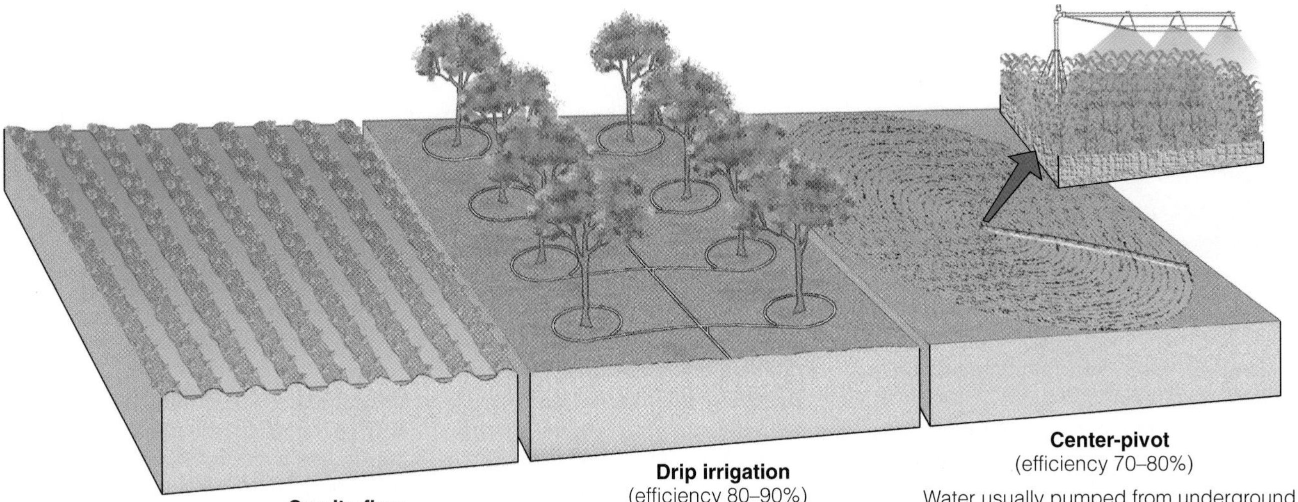

Figure 12-13 Major irrigation systems.

low-energy precision application (LEPA) sprinklers. These systems bring 75–85% (some claim 85–95%) of the water to crops by spraying it closer to the ground and in larger droplets than the center-pivot system; they also reduce energy use and costs by 20–30%. However, because of the high initial costs, such sprinklers are used on only about 1% of the world's irrigated cropland.

In the 1960s, highly efficient trickle or drip irrigation systems were developed in arid Israel. A network of perforated piping, installed at or below the ground surface, releases a trickle of water close to plant roots (Figure 12-13, center), minimizing evaporation and seepage and bringing 80–90% of the water to crops. Drip systems tend to be too expensive for most poor farmers and for use on low-value row crops. However, they are economically feasible for high-profit fruit, vegetable, and orchard crops and for home gardens. Researchers are testing drip irrigation systems that cost one-third to one-tenth as much as current drip systems.

Irrigation efficiency can also be improved by computer-controlled systems that monitor soil moisture and provide water only when necessary. Farmers can also switch to more water-efficient, drought-resistant, and salt-tolerant crop varieties and use nighttime irrigation to reduce loss to evaporation. Farmers can also use organic farming techniques, which **(1)** produce higher crop yields per hectare **(2)** require only one-fourth of the water and commercial fertilizer of conventional farming and **(3)** reduce CO_2 emissions (Solutions, p. 273).

Since 1950 Israel has used many of these techniques to slash irrigation water waste by about 84% while irrigating 44% more land. The government also gradually removed most government subsidies to raise the price of irrigation water to one of the highest rates in the world. Israel also imports most of its water-intensive wheat and meat, concentrating on fruits, vegetables, and flowers that require less water. However, in other countries where government-funded water supply schemes and other subsidies provide artificially cheap irrigation water, farmers have little incentive to invest in water-saving techniques.

As fresh water becomes scarce and cities consume water once used for irrigation, carefully treated urban wastewater (which is rich in nitrate and phosphate plant nutrients) could be used for irrigation. Israel now treats and reuses 55% of its municipal sewage water, mostly for irrigating nonedible crops such as cotton and flax, and plans to reuse 80% of this flow within the next few years.

Solutions: How Can We Waste Less Water in Industry, Homes, and Businesses? Manufacturing processes can use recycled water or be redesigned to save water. Japan and Israel lead the world in conserving and recycling water in industry. One paper mill in Hadera, Israel, uses one-tenth as much water as most of the world's other paper mills. Manufacturing aluminum from recycled scrap rather than from virgin ores can reduce water needs by 97%.

Nearly half of the water supplied by municipal water systems in the United States is used to flush toilets and water lawns, and another 15% is lost through leaky pipes. In a typical U.S. home, flushing toilets, washing hands, and bathing account for about 78% of the water used. In the arid Western United States and in dry Australia, lawn and garden watering can take 80% of a household's daily usage, and much of this water is unnecessarily wasted.

Green lawns in an arid or semiarid area can be replaced with vegetation adapted to a dry climate—a form of landscaping called *xeriscaping*, from the Greek word *xeros*, meaning "dry." A xeriscaped yard typically uses 30–80% less water than a conventional one. Drip irrigation can water gardens and other vegetation around homes.

More than half the water supply in Cairo, Egypt; Lima, Peru; Mexico City, Mexico; and Jakarta, Indonesia, disappears before it can be used, mostly from leaks. In Cairo people often have to wade ankle deep across streets because of leaky water pipes. Leaky pipes, water mains, toilets, bathtubs, and faucets waste 20–35% of water withdrawn from public supplies in the United States and the United Kingdom.

Many cities offer no incentive to reduce leaks and waste. About one-fifth of all U.S. public water systems don't have water meters and charge a single low rate (often less than $100 a year for an average family) for virtually unlimited use of high-quality water. Many apartment dwellers have little incentive to conserve water because their water use is included in their rent.

In Boulder, Colorado, the introduction of water meters reduced water use by more than one-third. Tucson, Arizona, is a desert city with ordinances that require conserving and reusing water. As a result, it consumes half as much water per person as Las Vegas, a desert city with much less emphasis on water conservation. Las Vegas, one of the fastest-growing cities in the United States, uses more water per person than any other city in the world. Water experts project that even if current water conservation efforts are successful, the city will run short of water by 2007.

In 1988 Bogor, Indonesia, reduced water demand by one-third by installing water meters and raising the price of water. You might think that raising the price of water hurts the poor. However, artificially low water prices discourage cities from extending water services to poor areas. As a result, many of the poor in developing countries end up paying 4–100 times the price of city water to private water vendors.

A California water utility that gave rebates for water-saving toilets and distributed some 35,000 water-saving showerheads cut per capita water use 40% in only 1 year. On January 1, 1994, the Comprehensive Energy Act of 1992 required that all new toilets sold in the United States use no more than 6 liters (1.6 gallons) per flush. Similar laws have been passed in Mexico and in Ontario, Canada.

A low-flow showerhead costing about $20 saves about $34–56 per year in water heating costs—a one-time investment that can give you a 50% return on your money with no financial risk. Audits conducted by students in Brown University's environmental studies program showed that the school could save $44,000 a year by using low-flow showerheads in dormitories.

Between 1987 and 1998 Boston, Massachusetts, reduced total water demand 24% by **(1)** fixing leaky pipes, **(2)** installing water-saving fixtures, and **(3)** educating the public about how to save water. This allowed the city to avoid diverting two large rivers to supply more water, which would have cost 2–3 times more than its water conservation program.

Two decades of experience with droughts in northern California has shown that water demand can be cut by more than 50% for homes, 60% for parks, and 20% for businesses without economic hardships. Savings in water use—called *negaliters* or *negagallons*—can also save money. For instance, a 1994 New York City study showed that providing a $240 rebate to customers who convert to water-saving toilets and showerheads would cost only one-third of the $8- to $12-billion price tag for developing new water supplies for the city's residents.

Used water (gray water) from bathtubs, showers, bathroom sinks, and clothes washers can be collected, stored, treated, and reused for irrigation, raising fish, and other purposes. Some 70% of the wastewater in Israel is used in this way.

In the United States, California has become the first state to legalize reuse of gray water to irrigate landscapes. An estimated 50–75% of the water used by a typical house could be reused as gray water. Treatment of wastewater for reuse would be easier and cheaper if wastewater from industries were kept separate from household wastewater. Cities in developing countries that are putting in or expanding water systems could save money and water by separating flows from industry and households and using low-quality water flows for flushing toilets.

In some parts of the United States, people can lease systems that purify and completely recycle wastewater from houses, apartments, or office buildings. Such a system can be installed in a small outside shed and serviced for a monthly fee roughly equal to that charged by most city water-and-sewer systems. In Tokyo, Japan, all the water used in Mitsubishi's 60-story office building is purified for reuse by an automated recycling system.

Another way to conserve water and reduce flooding is to collect and use rainwater for nondrinking purposes such as flushing toilets, irrigating gardens, watering lawns, and putting out fires. In Tokyo, Japan, large tanks on top of 579 city buildings capture and use rainwater.

Another approach is to coordinate the water policies of national and local governments. When a national government subsidizes the cost of water, this can undermine a local government's efforts to conserve water. Some actions you can take to waste less water are listed in Appendix 6.

12-4 TOO MUCH WATER

What Are the Causes and Effects of Flooding?

Natural flooding by streams, the most common type of flooding, is caused primarily by heavy rain or rapid melting of snow. This causes water in the stream to overflow its normal channel and to cover the adjacent area, called a **floodplain** (Figure 12-14). Floodplains, which include highly productive wetlands, **(1)** provide natural flood and erosion control, **(2)** help maintain high water quality, and **(3)** contribute to the recharging of groundwater.

People have settled on floodplains since the beginnings of agriculture. The soil is fertile, and ample water is available for irrigation. Communities can use the water for transportation of people or goods, and floodplains (being flat) are suitable for cropland, buildings, highways, and railroads. People may decide that these benefits outweigh the risk of flooding (if they are aware of the risk).

On marine coasts, flooding results most often from wind-driven storm surges and rain-swollen streams associated with tropical cyclones (typhoons and hurricanes). Flooding can also occur on the shorelines of large inland lakes.

Floods are a natural phenomenon and have several benefits. They **(1)** provide the world's most productive farmland because they are regularly covered with nutrient-rich silt left after flood waters recede, **(2)** recharge groundwater under plains, and **(3)** refill wetlands, which help keep rivers flowing during droughts and provide important breeding and feeding grounds for fish and waterfowl.

Each year, flooding (Figure 12-15) kills thousands of people and causes tens of billions of dollars in property damage. Floods, like droughts, usually are considered natural disasters, but since the 1960s human activities have contributed to the sharp rise in flood deaths and damages. Indeed, floods account for about 39% of all deaths from natural disasters—more than any other type.

Two ways humans increase the likelihood of flooding and the resulting damage are by **(1)** removing water-absorbing vegetation, especially on hillsides (Figure 12-16) and **(2)** living on floodplains. In the United States 10 million households and businesses with property valued at $1 trillion exist in flood-prone areas. In many developing countries, the poor have little choice but to try to survive in

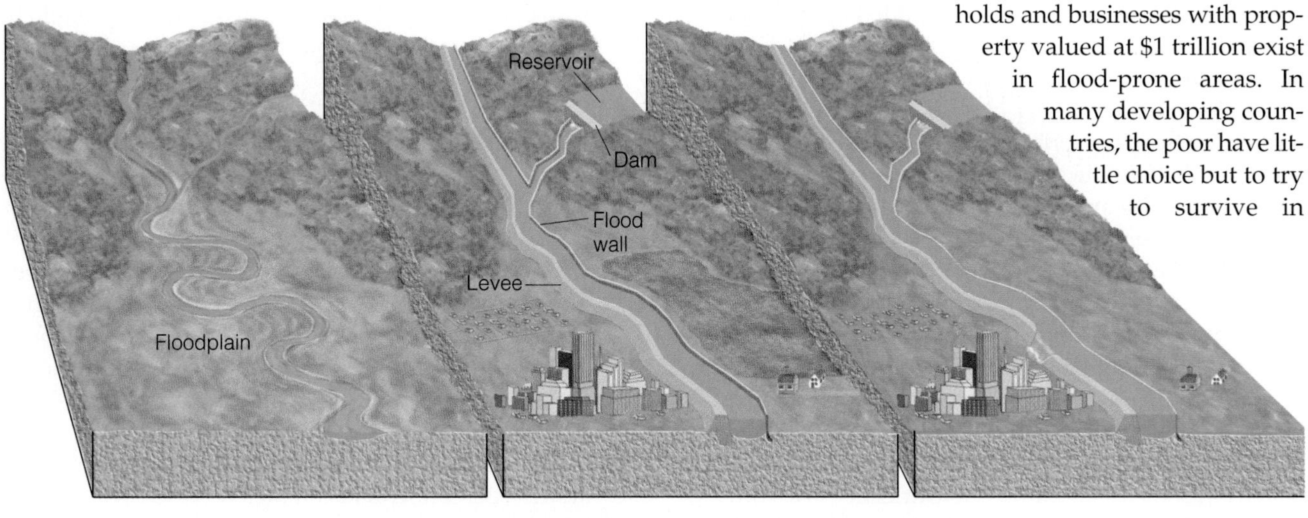

Figure 12-14 Land in a natural floodplain (left) often is flooded after prolonged rains. When the floodwaters recede, alluvial deposits of silt are left behind, creating a nutrient-rich soil. To reduce the threat of flooding (and thus allow people to live in floodplains), rivers have been dammed to create reservoirs that store and release water as needed. They have also been narrowed, straightened, and equipped with protective levees and walls (middle). However, these alterations can give a false sense of security to floodplain dwellers, who actually live in high-risk areas. In the long run, such measures can greatly increase flood damage. Although dams, levees, and walls prevent flooding in most years, they can be overwhelmed by prolonged rains (right), as happened in the Midwestern United States during the summer of 1993 (Figure 12-15).

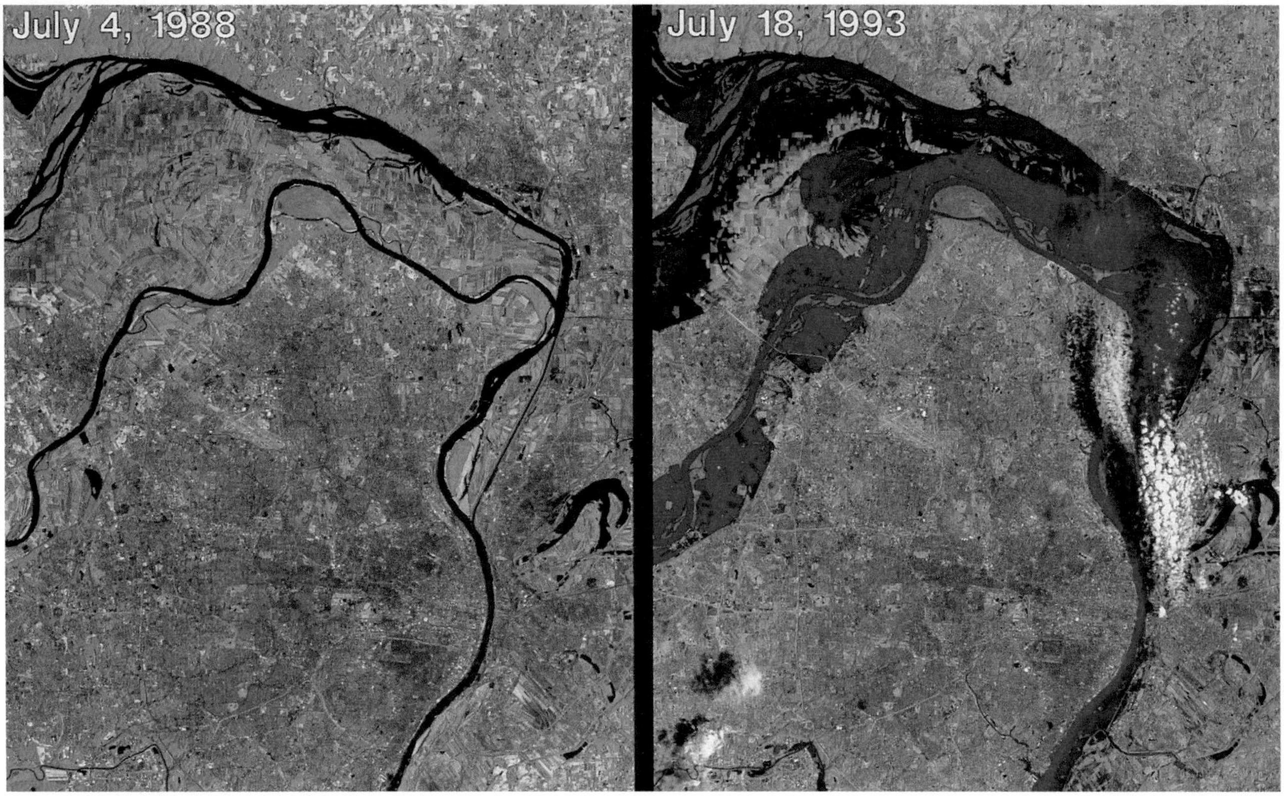

Figure 12-15 Satellite images of the area around St. Louis before flooding on July 4, 1988 (left), and on July 18, 1993, after severe flooding from prolonged rains. (Earth Satellite Corporation)

flood-prone areas. In developed countries, however, people deliberately settle on floodplains and then expect dams, levees, and other devices to protect them from flood waters. Such solutions don't work when heavier-than-normal rains come.

Urbanization also increases flooding (even with only moderate rainfall) by replacing vegetation and soil with highways, parking lots, and buildings, which leads to rapid runoff of rainwater. If sea levels rise during the next century, as projected, many low-lying coastal cities, wetlands, and croplands will be under water.

Case Study: Living Dangerously in Bangladesh Bangladesh (Figure 9-15) is one of the world's most densely populated countries, with 123 million people packed into an area roughly the size of Wisconsin. Its population is projected to reach 166 million by 2025. Bangladesh is also one of the world's poorest countries, with an average per capita GNP of about $260, or 71¢ per day.

More than 80% of the country consists of floodplains and shifting islands of silt formed by a delta at the mouths of three major rivers. Runoff from annual monsoon rains in the Himalaya Mountains of India, Nepal, Bhutan, and China flows down the rivers through Bangladesh into the Bay of Bengal.

The people of Bangladesh are used to moderate annual flooding during the summer monsoon season, and they depend on the floodwaters to grow rice, their primary source of food. The annual deposit of eroded Himalayan soil in the delta basin also helps maintain soil fertility.

In the past, great floods occurred every 50 years or so, but since the 1970s they have come about every 4 years. Bangladesh's flood problems begin in the Himalayan watershed. A combination of rapid population growth, deforestation, overgrazing, and unsustainable farming on steep, easily erodible mountain slopes has greatly diminished the ability of the soil to absorb water. Instead of being absorbed and released slowly, water from the monsoon rains runs off the denuded Himalayan foothills, carrying vital topsoil with it (Figure 12-16). This runoff, combined with heavier-than-normal monsoon rains, has increased the severity of flooding along Himalayan rivers and in Bangladesh—another example of connections.

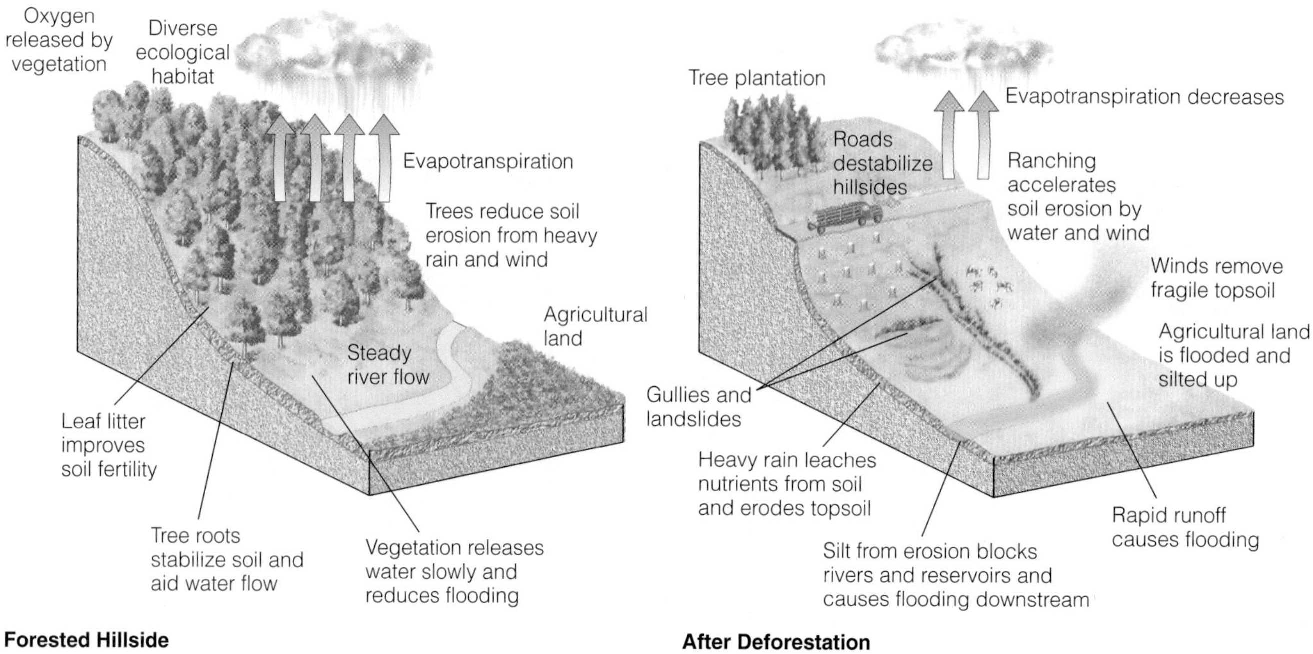

Figure 12-16 A hillside before and after deforestation. Once a hillside has been deforested—for timber and fuelwood, grazing livestock, or unsustainable farming—water from precipitation rushes down the denuded slopes, eroding precious topsoil and flooding downstream areas. A 3,000-year-old Chinese proverb says, "To protect your rivers, protect your mountains."

In 1988 a disastrous flood covered two-thirds of Bangladesh's land area for 3 weeks and leveled 2 million homes after the heaviest monsoon rains in 70 years. At least 2,000 people drowned, and 30 million people—one of every four—were left homeless. More than a quarter of the country's crops were destroyed, costing at least $1.5 billion and causing thousands of people to die of starvation.

In their struggle to survive, the poor in Bangladesh have cleared many of the country's coastal mangrove forests (see photo, p. XIV)for fuelwood, cultivation of crops, and aquaculture ponds for raising shrimp. This has led to more severe flooding because these coastal wetlands shelter the low-lying coastal areas from storm surges and cyclones.

Sixteen devastating cyclones have slammed into Bangladesh since 1961; in 1970, as many as 1 million people drowned in one storm. Another surge killed an estimated 140,000 people in 1991. Damages and deaths from cyclones in areas still protected by mangrove forests are much lower than in areas where the forests have been cleared.

The severity of this problem can be reduced if Bangladesh, Bhutan, China, India, and Nepal cooperate in reforestation and flood control measures and reduce their population growth.

Solutions: How Can We Reduce Flood Risks? One controversial way of reducing flooding is *channelization*, in which a section of a stream is deepened, widened, or straightened to allow more rapid runoff (Figure 12-14, middle). Although channelization can reduce upstream flooding, the increased flow of water can also increase upstream bank erosion and downstream flooding and deposition of sediment.

Artificial levees and embankments along stream banks can reduce the chances of water overflowing into nearby floodplains. They may be permanent or temporary (such as sandbags placed when a flood is imminent). Levees, like channelization, contain and speed up stream flow and increase the water's capacity for doing damage downstream. If a levee breaks or the water spills over it, floodwater may persist long after the stream discharge has decreased (Figure 12-14, right).

Many people in the Midwestern United States learned this harsh lesson during massive flooding from heavy rains during the summer of 1993 (Figure 12-15). After the floodwaters receded, a toxic sludge laced with pesticides, industrial wastes, and raw sewage was left on large areas of land along the floodplains of the Mississippi, Missouri, and Illinois rivers. Severe erosion from the 1993 flooding reduced crop yields on already heavily eroded fields.

Watershed Management Is Key to Improving America's Water Resources 1999. *Journal of Environmental Health,* v61 i9 p43(2).

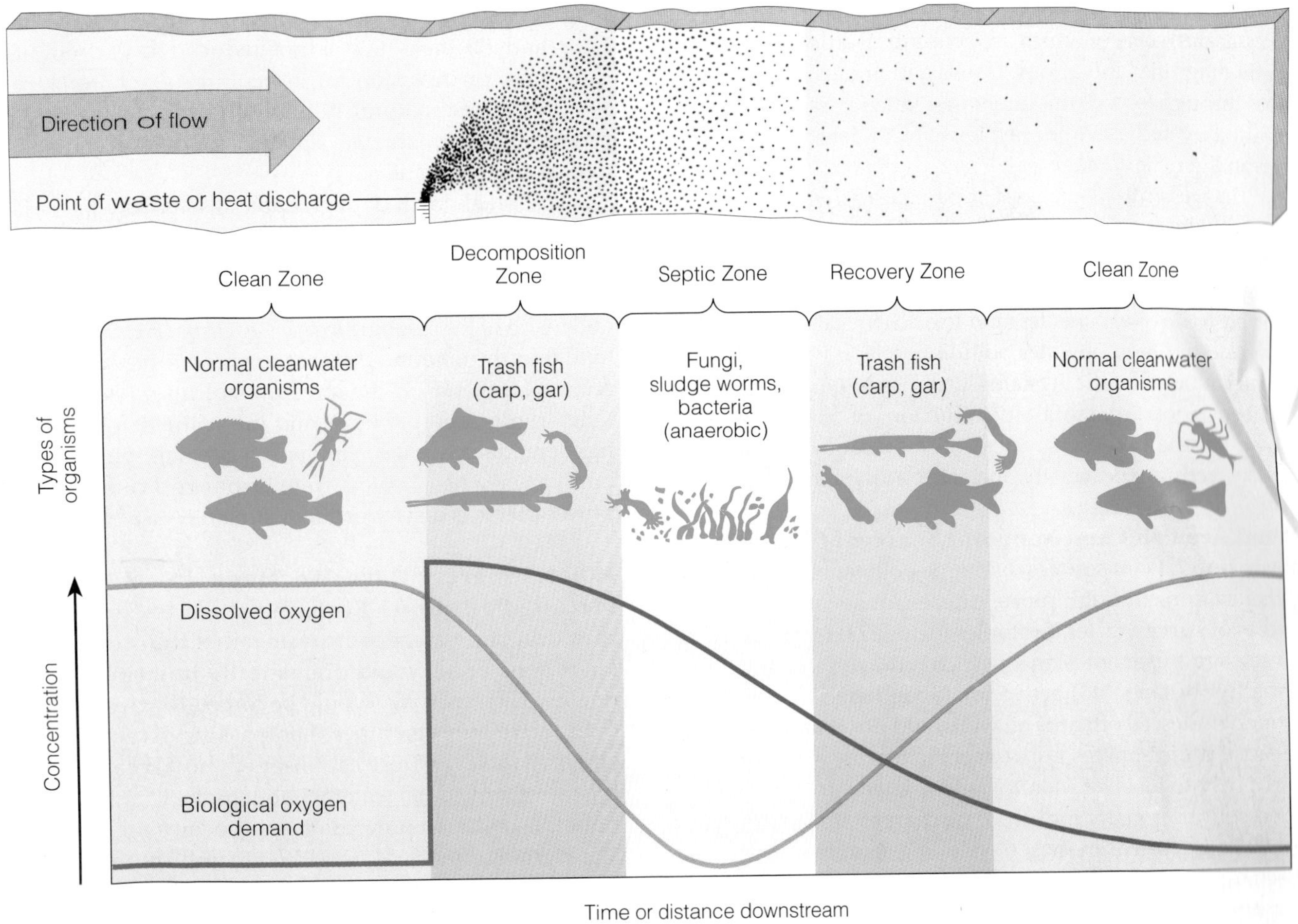

Figure 12-18 Dilution and decay of degradable, oxygen-demanding wastes and heat, showing the oxygen sag curve (orange) and the curve of oxygen demand (blue). Depending on flow rates and the amount of pollutants, streams recover from oxygen-demanding wastes and heat if they are given enough time and are not overloaded.

than 30 years of effort, $250 million of British taxpayers' money, and millions more spent by industry, the Thames has made a remarkable recovery. Commercial fishing is thriving, and many species of waterfowl and wading birds have returned to their former feeding grounds.

What Is the Bad News About Stream Pollution? Despite progress in improving stream quality in most developed countries, large fish kills and contamination of drinking water still occur. Most of these disasters are caused by **(1)** accidental or deliberate releases of toxic inorganic and organic chemicals by industries, **(2)** malfunctioning sewage treatment plants, and **(3)** nonpoint runoff of pesticides and nutrients (eroded soil, fertilizer, and animal waste) from cropland or animal feedlots (Individuals Matter, p. 311).

Available data indicate that stream pollution from huge discharges of sewage and industrial wastes is a serious and growing problem in most developing countries, where waste treatment is practically nonexistent. Numerous streams in the former Soviet Union and in eastern European countries are severely polluted. Currently, more than two-thirds of India's water resources are polluted with industrial wastes and sewage. Of the 78 streams monitored in China, 54 are seriously polluted with untreated sewage and industrial wastes. About 20% of China's rivers are too polluted to use for irrigation and 80% of them are so degraded they no longer support fish. In Latin America and Africa, most streams passing through urban or industrial areas are severely polluted.

What Are the Pollution Problems of Lakes? In lakes, reservoirs, and ponds, dilution is often less effective than in streams. Lakes and reservoirs often contain stratified layers that undergo little vertical mixing, and ponds contain small volumes of water. Stratification

A levee can contribute to increased destruction when major floods occur, but the destruction happens *downriver* from each levee. Thus, downriver landowners need levees to protect against upriver levees, which are needed to protect against flood crests caused by levees further upriver. The result is a spiraling *levee race* that can eventually cause destruction for almost everyone when the incredible force of a hugely flood-swollen river inevitably spills over or takes out levees.

A *flood control dam* built across a stream can hold back, store, and release water more gradually (Figure 12-14, middle). The dam and its reservoir may also provide such secondary benefits as hydroelectric power, water for irrigation, and recreational facilities. A reservoir can reduce floods only if its water level is kept low. However, it is much more profitable for dam operators to keep water levels high (for producing electricity and supplying irrigation water). As a result, after prolonged rains the reservoir can overflow, or operators may release large volumes of water to prevent overflow, thereby worsening the severity of flooding downstream. The reservoir gradually fills with sediment (that once fertilized downstream floodplains) until it is useless, and it also has other drawbacks (Figure 12-6).

Some flood control dams have failed, causing sudden, catastrophic flooding and threatening lives, property, and wildlife. According to the Federal Emergency Management Agency, the United States has about 1,300 unsafe dams in populated areas, and the dam safety programs of most states are inadequate because of weak laws and budget cuts.

Since 1934 the U.S. Army Corps of Engineers has spent almost $250 billion on channelization, dams, and levees, saving at least $380 billion in damages. Even so, the average financial losses from flood damage (adjusted for inflation) in the United States more than doubled between 1934 and 1999. The massive floods in the Midwestern United States in 1993 caused an estimated $6.5 billion in damage to property and crops; billions more will have to be spent to repair the more than 800 levees and embankments that were topped or breached.

This disaster also demonstrated that dams, levees, and channelization can give people a false sense of security and encourage them to settle on floodplains, thus worsening the severity of flood damage. Some countries are realizing that the risks of managing natural water flow can outweigh the benefits; Germany, for example, plans to bulldoze through dikes and allow parts of floodplains to flood regularly.

From an environmental viewpoint, *floodplain management* is the best approach. The first step is to construct a *flood frequency curve* (based on historical records and an examination of vegetation) to determine how often *on average* a flood of a certain size occurs in a particular area. This doesn't tell us exactly when floods will occur, but it provides a general idea of how often they might occur, based on the area's history.

Using these data, a plan is developed **(1)** to prohibit certain types of buildings or activities in high-risk zones, **(2)** to elevate or otherwise floodproof buildings that are allowed on the legally defined floodplain, and **(3)** to construct a floodway that allows floodwater to flow through the community with minimal damage. Floodplain management based on thousands of years of experience can be summed up in one idea: *Sooner or later the river (or the ocean) always wins.*

In the United States, the Federal Flood Disaster Protection Act of 1973 requires local governments to adopt floodplain development regulations to be eligible for federal flood insurance. It also denies federal funding to proposed construction projects in designated flood hazard areas. The federal flood insurance program underwrites $185 billion in policies because private insurance companies are unwilling to fully insure people who live in flood-prone areas. This federal program thus can encourage people to build on floodplains and low-lying coastal areas.

Some economists argue that it would make more economic sense for the government to buy the 2% of the country's land that repeatedly floods instead of continuing to make disaster payments. Others believe that people should be free to live in flood-prone areas, but that those choosing to do so should accept the risk without the assistance of federal flood insurance. Attempts to reduce or eliminate federal flood insurance coverage (or to restrict development on floodplains) usually are defeated politically by intense protests from property owners who already live in these risky areas.

12-5 POLLUTION OF STREAMS, LAKES, AND GROUNDWATER

What Are the Major Water Pollutants? Water pollution is any chemical, biological, or physical change in water quality that has a harmful effect on living organisms or makes water unsuitable for desired uses. There are several classes of water pollutants. One is *disease-causing agents (pathogens)*, which include bacteria, viruses, protozoa, and parasitic worms that enter water from domestic sewage and untreated human and animal wastes (Table 12-1).

A good indicator of the quality of water for drinking or swimming is the number of colonies of *coliform bacteria* present in a 100-milliliter (0.1-quart) sample of water. The World Health Organization recommends a coliform bacteria count of 0 colonies per

Hint: Enter the search terms *water quality, standards* using the Subject Guide.

Table 12-1 Common Diseases Transmitted to Humans Through Contaminated Drinking Water

Type of Organism	Disease	Effects
Bacteria	Typhoid fever	Diarrhea, severe vomiting, enlarged spleen, inflamed intestine; often fatal if untreated
	Cholera	Diarrhea, sever vomiting, dehydration; often fatal if untreated
	Bacterial dysentery	Diarrhea; rarely fatal except in infants without proper treatment
	Enteritis	Severe stomach pain, nausea, vomiting; rarely fatal
Viruses	Infectious hepatitis	Fever, severe headache, loss of appetite, abdominal pain, jaundice, enlarged liver; rarely fatal but may cause permanent liver damage
Parasitic protozoa	Amoebic dysentery	Severe diarrhea, headache, abdominal pain, chills, fever; if not treated can cause liver abscess, bowel perforation, and death
	Giardiasis	Diarrhea, abdominal cramps, flatulence, belching, fatigue
Parasitic worms	Schistosomiasis	Abdominal pain, skin rash, anemia, chronic fatigue, and chronic general ill health

100 milliliters for drinking water, and the EPA recommends a maximum level for swimming water of 200 colonies per 100 milliliters.

According to the U.S. Centers for Disease Control and Prevention, as many as 900,000 people get sick and 900 die each year in the United States because of waterborne microbial infections. In developing countries hundreds of millions of people get sick and at least 11 million die annually from such diseases.

A second category of water pollutants is *oxygen-demanding wastes*, organic wastes that can be decomposed by aerobic (oxygen-requiring) bacteria. Large populations of bacteria decomposing these wastes can degrade water quality by depleting water of dissolved oxygen (Figure 12-17), causing fish and other forms of oxygen-consuming aquatic life to die. The quantity of oxygen-demanding wastes in water can be determined by measuring the **biological oxygen demand (BOD)**: the amount of dissolved oxygen needed by aerobic decomposers to break down the organic materials in a certain volume of water over a 5-day incubation period at 20°C (68°F).

A third class of water pollutants is *water-soluble inorganic chemicals*, which include acids, salts, and compounds of toxic metals such as mercury and lead. High levels of these chemicals can **(1)** make water unfit to drink, **(2)** harm fish and other aquatic life, **(3)** lower crop yields, and **(4)** accelerate corrosion of metals exposed to such water.

Inorganic plant nutrients are another class of water pollutants. They are water-soluble nitrates and phosphates that can cause excessive growth of algae and other aquatic plants. When the excess populations of these plants die and decay, they deplete water of dissolved oxygen and kill fish. Drinking water with excessive levels of nitrates lowers the oxygen-carrying capacity of the blood; this can kill unborn children and infants, especially those under 1 year old.

Water can also be polluted by a variety of *organic chemicals*, which include oil, gasoline, plastics, pesticides, cleaning solvents, detergents, and many other chemicals. They threaten human health and harm fish and other aquatic life.

By far the biggest class of water pollutants by weight is *sediment*, or *suspended matter*—insoluble particles of soil and other solids that become suspended in water, mostly when soil is eroded from the land. Sediment **(1)** clouds water and reduces photosynthesis, **(2)** disrupts aquatic food webs, **(3)** carries pesticides, bacteria, and other harmful substances, **(4)** settles out and destroys feeding and spawning grounds of fish, and **(5)** clogs and fills lakes, artificial reservoirs, stream channels, and harbors.

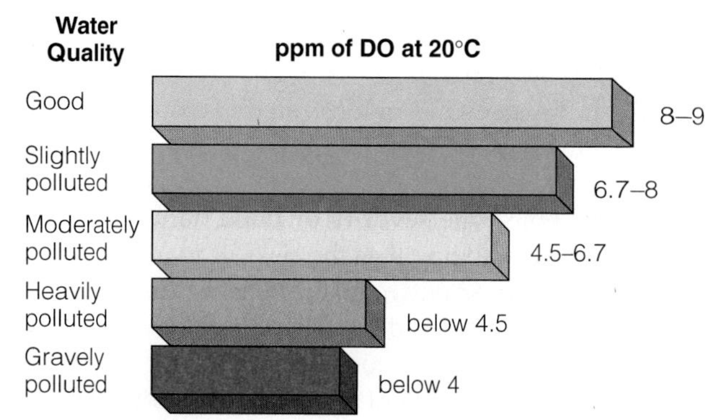

Figure 12-17 Water quality and dissolved oxygen (DO) content in parts per million (ppm) at 20°C (68°F). The solubility of oxygen decreases as the water temperature increases. Only a few species of fish can survive in water with fewer than 4 ppm of dissolved oxygen.

Water can also be polluted by *water-soluble radioactive isotopes*, some of which are concentrated or biologically magnified in various tissues and organs as they pass through food chains and webs. Ionizing radiation emitted by such isotopes can cause birth defects, cancer, and genetic damage.

Heat absorbed by water used to cool industrial and power plants can lower water quality. The resulting rise in water temperature, called *thermal pollution*, lowers dissolved oxygen levels and makes aquatic organisms more vulnerable to disease, parasites, and toxic chemicals.

Another form of water pollution, *genetic pollution*, occurs when aquatic systems are disrupted by the deliberate or accidental introduction of *nonnative species*. Some of these species can crowd out native species, reduce biodiversity, and cause economic losses.

What Are Point and Nonpoint Sources of Water Pollution? **Point sources** discharge pollutants at specific locations through pipes, ditches, or sewers into bodies of surface water. Examples include **(1)** factories, **(2)** sewage treatment plants (which remove some but not all pollutants), **(3)** active and abandoned underground mines, **(4)** offshore oil wells, and **(5)** oil tankers. Because point sources are at specific places, they are fairly easy to identify, monitor, and regulate. In developed countries many industrial discharges are strictly controlled, whereas in most developing countries such discharges are largely uncontrolled.

Nonpoint sources are sources that cannot be traced to any single site of discharge. They are usually large land areas or airsheds that pollute water by runoff, subsurface flow, or deposition from the atmosphere. Examples include **(1)** acid deposition (Figure 10-6), **(2)** runoff of chemicals into surface water (including stormwater), and **(3)** seepage into the ground from croplands, livestock feedlots, logged forests, streets, lawns, and parking lots.

In the United States, nonpoint pollution from agriculture—mostly in the form of sediment, inorganic fertilizers, manure, salts dissolved in irrigation water, and pesticides—is responsible for an estimated 64% of the total mass of pollutants entering streams and 57% of those entering lakes. Each year livestock in the United States produce enough manure to fill an imaginary train of 6.7 million boxcars circling the earth 12.5 times. Little progress has been made in controlling nonpoint water pollution because of the difficulty and expense of identifying and controlling discharges from so many diffuse sources.

What Are the Pollution Problems of Streams? Flowing streams, including large ones called *rivers*, can recover rapidly from degradable, oxygen-demanding wastes and excess heat by a combination of dilution and bacterial decay. This natural recovery process works as long as **(1)** streams are not overloaded with these pollutants and **(2)** their flow is not reduced by drought, damming, or diversion for agriculture and industry. However, these natural dilution and biodegradation processes do not eliminate slowly degradable and nondegradable pollutants.

The breakdown of degradable wastes by bacteria depletes dissolved oxygen, which reduces or eliminates populations of organisms with high oxygen requirements until the stream is cleansed of wastes. The depth and width of the resulting *oxygen sag curve* (Figure 12-18) (and thus the time and distance required for a stream to recover) depend on the stream's volume, flow rate, temperature, and pH level, and the volume of incoming degradable wastes. Similar oxygen sag curves can be plotted when heated water from industrial and power plants is discharged into streams.

What Progress Have We Made in Reducing Stream Pollution? Requiring cities to withdraw their drinking water downstream rather than upstream (as is done now) would dramatically improve water quality. Then each city would be forced to clean up its own waste outputs rather than passing them downstream. However, upstream users, who already have the use of fairly clean water without high cleanup costs, fight this pollution prevention approach.

Here is some *good news*. Water pollution control laws enacted in the 1970s have greatly increased the number and quality of wastewater treatment plants in the United States and many other developed countries. Laws have also required industries to reduce or eliminate point-source discharges into surface waters. These efforts, spurred by individuals (Individuals Matter, p. 307), have enabled the United States to hold the line against increased pollution of most of its streams by disease-causing agents and oxygen-demanding wastes. This is an impressive accomplishment considering the rise in economic activity and population since the laws were passed.

One success story is the cleanup of Ohio's Cuyahoga River, which was so polluted that in 1959 and again in 1969 it caught fire and burned for several days as it flowed through Cleveland, Ohio. The highly publicized image of this burning river prompted city, state, and federal officials to enact laws limiting the discharge of industrial wastes into the river and sewage systems and to appropriate funds to upgrade sewage treatment facilities. Today the river has made a comeback and is widely used by boaters and anglers.

Pollution control laws passed since 1970 have also led to improvements in dissolved oxygen content in many streams in Canada, Japan, and most western European countries. A spectacular cleanup has occurred in Great Britain. In the 1950s the Thames River was little more than a flowing anaerobic sewer. But after more

A levee can contribute to increased destruction when major floods occur, but the destruction happens *downriver* from each levee. Thus, downriver landowners need levees to protect against upriver levees, which are needed to protect against flood crests caused by levees further upriver. The result is a spiraling *levee race* that can eventually cause destruction for almost everyone when the incredible force of a hugely flood-swollen river inevitably spills over or takes out levees.

A *flood control dam* built across a stream can hold back, store, and release water more gradually (Figure 12-14, middle). The dam and its reservoir may also provide such secondary benefits as hydroelectric power, water for irrigation, and recreational facilities. A reservoir can reduce floods only if its water level is kept low. However, it is much more profitable for dam operators to keep water levels high (for producing electricity and supplying irrigation water). As a result, after prolonged rains the reservoir can overflow, or operators may release large volumes of water to prevent overflow, thereby worsening the severity of flooding downstream. The reservoir gradually fills with sediment (that once fertilized downstream floodplains) until it is useless, and it also has other drawbacks (Figure 12-6).

Some flood control dams have failed, causing sudden, catastrophic flooding and threatening lives, property, and wildlife. According to the Federal Emergency Management Agency, the United States has about 1,300 unsafe dams in populated areas, and the dam safety programs of most states are inadequate because of weak laws and budget cuts.

Since 1934 the U.S. Army Corps of Engineers has spent almost $250 billion on channelization, dams, and levees, saving at least $380 billion in damages. Even so, the average financial losses from flood damage (adjusted for inflation) in the United States more than doubled between 1934 and 1999. The massive floods in the Midwestern United States in 1993 caused an estimated $6.5 billion in damage to property and crops; billions more will have to be spent to repair the more than 800 levees and embankments that were topped or breached.

This disaster also demonstrated that dams, levees, and channelization can give people a false sense of security and encourage them to settle on floodplains, thus worsening the severity of flood damage. Some countries are realizing that the risks of managing natural water flow can outweigh the benefits; Germany, for example, plans to bulldoze through dikes and allow parts of floodplains to flood regularly.

From an environmental viewpoint, *floodplain management* is the best approach. The first step is to construct a *flood frequency curve* (based on historical records and an examination of vegetation) to determine how often *on average* a flood of a certain size occurs in a particular area. This doesn't tell us exactly when floods will occur, but it provides a general idea of how often they might occur, based on the area's history.

Using these data, a plan is developed **(1)** to prohibit certain types of buildings or activities in high-risk zones, **(2)** to elevate or otherwise floodproof buildings that are allowed on the legally defined floodplain, and **(3)** to construct a floodway that allows floodwater to flow through the community with minimal damage. Floodplain management based on thousands of years of experience can be summed up in one idea: *Sooner or later the river (or the ocean) always wins.*

In the United States, the Federal Flood Disaster Protection Act of 1973 requires local governments to adopt floodplain development regulations to be eligible for federal flood insurance. It also denies federal funding to proposed construction projects in designated flood hazard areas. The federal flood insurance program underwrites $185 billion in policies because private insurance companies are unwilling to fully insure people who live in flood-prone areas. This federal program thus can encourage people to build on floodplains and low-lying coastal areas.

Some economists argue that it would make more economic sense for the government to buy the 2% of the country's land that repeatedly floods instead of continuing to make disaster payments. Others believe that people should be free to live in flood-prone areas, but that those choosing to do so should accept the risk without the assistance of federal flood insurance. Attempts to reduce or eliminate federal flood insurance coverage (or to restrict development on floodplains) usually are defeated politically by intense protests from property owners who already live in these risky areas.

12-5 POLLUTION OF STREAMS, LAKES, AND GROUNDWATER

What Are the Major Water Pollutants? **Water pollution** is any chemical, biological, or physical change in water quality that has a harmful effect on living organisms or makes water unsuitable for desired uses. There are several classes of water pollutants. One is *disease-causing agents (pathogens)*, which include bacteria, viruses, protozoa, and parasitic worms that enter water from domestic sewage and untreated human and animal wastes (Table 12-1).

A good indicator of the quality of water for drinking or swimming is the number of colonies of *coliform bacteria* present in a 100-milliliter (0.1-quart) sample of water. The World Health Organization recommends a coliform bacteria count of 0 colonies per

Type of Organism	Disease	Effects
Bacteria	Typhoid fever	Diarrhea, severe vomiting, enlarged spleen, inflamed intestine; often fatal if untreated
	Cholera	Diarrhea, sever vomiting, dehydration; often fatal if untreated
	Bacterial dysentery	Diarrhea; rarely fatal except in infants without proper treatment
	Enteritis	Severe stomach pain, nausea, vomiting; rarely fatal
Viruses	Infectious hepatitis	Fever, severe headache, loss of appetite, abdominal pain, jaundice, enlarged liver; rarely fatal but may cause permanent liver damage
Parasitic protozoa	Amoebic dysentery	Severe diarrhea, headache, abdominal pain, chills, fever; if not treated can cause liver abscess, bowel perforation, and death
	Giardiasis	Diarrhea, abdominal cramps, flatulence, belching, fatigue
Parasitic worms	Schistosomiasis	Abdominal pain, skin rash, anemia, chronic fatigue, and chronic general ill health

Table 12-1 Common Diseases Transmitted to Humans Through Contaminated Drinking Water

100 milliliters for drinking water, and the EPA recommends a maximum level for swimming water of 200 colonies per 100 milliliters.

According to the U.S. Centers for Disease Control and Prevention, as many as 900,000 people get sick and 900 die each year in the United States because of waterborne microbial infections. In developing countries hundreds of millions of people get sick and at least 11 million die annually from such diseases.

A second category of water pollutants is *oxygen-demanding wastes*, organic wastes that can be decomposed by aerobic (oxygen-requiring) bacteria. Large populations of bacteria decomposing these wastes can degrade water quality by depleting water of dissolved oxygen (Figure 12-17), causing fish and other forms of oxygen-consuming aquatic life to die. The quantity of oxygen-demanding wastes in water can be determined by measuring the **biological oxygen demand (BOD)**: the amount of dissolved oxygen needed by aerobic decomposers to break down the organic materials in a certain volume of water over a 5-day incubation period at 20°C (68°F).

A third class of water pollutants is *water-soluble inorganic chemicals*, which include acids, salts, and compounds of toxic metals such as mercury and lead. High levels of these chemicals can **(1)** make water unfit to drink, **(2)** harm fish and other aquatic life, **(3)** lower crop yields, and **(4)** accelerate corrosion of metals exposed to such water.

Inorganic plant nutrients are another class of water pollutants. They are water-soluble nitrates and phosphates that can cause excessive growth of algae and other aquatic plants. When the excess populations of these plants die and decay, they deplete water of dissolved oxygen and kill fish. Drinking water with excessive levels of nitrates lowers the oxygen-carrying capacity of the blood; this can kill unborn children and infants, especially those under 1 year old.

Water can also be polluted by a variety of *organic chemicals*, which include oil, gasoline, plastics, pesticides, cleaning solvents, detergents, and many other chemicals. They threaten human health and harm fish and other aquatic life.

By far the biggest class of water pollutants by weight is *sediment*, or *suspended matter*—insoluble particles of soil and other solids that become suspended in water, mostly when soil is eroded from the land. Sediment **(1)** clouds water and reduces photosynthesis, **(2)** disrupts aquatic food webs, **(3)** carries pesticides, bacteria, and other harmful substances, **(4)** settles out and destroys feeding and spawning grounds of fish, and **(5)** clogs and fills lakes, artificial reservoirs, stream channels, and harbors.

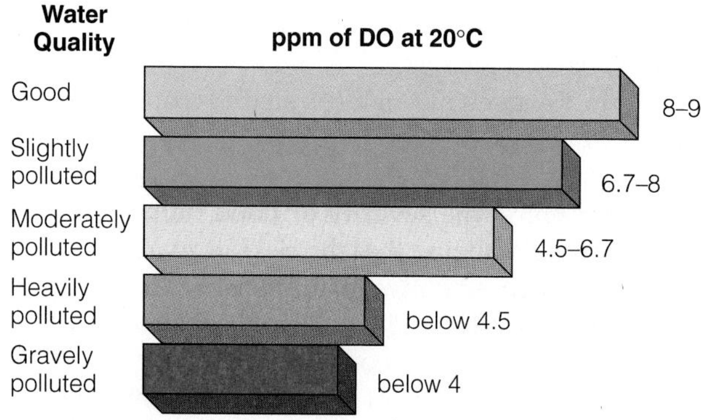

Figure 12-17 Water quality and dissolved oxygen (DO) content in parts per million (ppm) at 20°C (68°F). The solubility of oxygen decreases as the water temperature increases. Only a few species of fish can survive in water with fewer than 4 ppm of dissolved oxygen.

Water can also be polluted by *water-soluble radioactive isotopes*, some of which are concentrated or biologically magnified in various tissues and organs as they pass through food chains and webs. Ionizing radiation emitted by such isotopes can cause birth defects, cancer, and genetic damage.

Heat absorbed by water used to cool industrial and power plants can lower water quality. The resulting rise in water temperature, called *thermal pollution*, lowers dissolved oxygen levels and makes aquatic organisms more vulnerable to disease, parasites, and toxic chemicals.

Another form of water pollution, *genetic pollution*, occurs when aquatic systems are disrupted by the deliberate or accidental introduction of *nonnative species*. Some of these species can crowd out native species, reduce biodiversity, and cause economic losses.

What Are Point and Nonpoint Sources of Water Pollution? **Point sources** discharge pollutants at specific locations through pipes, ditches, or sewers into bodies of surface water. Examples include **(1)** factories, **(2)** sewage treatment plants (which remove some but not all pollutants), **(3)** active and abandoned underground mines, **(4)** offshore oil wells, and **(5)** oil tankers. Because point sources are at specific places, they are fairly easy to identify, monitor, and regulate. In developed countries many industrial discharges are strictly controlled, whereas in most developing countries such discharges are largely uncontrolled.

Nonpoint sources are sources that cannot be traced to any single site of discharge. They are usually large land areas or airsheds that pollute water by runoff, subsurface flow, or deposition from the atmosphere. Examples include **(1)** acid deposition (Figure 10-6), **(2)** runoff of chemicals into surface water (including stormwater), and **(3)** seepage into the ground from croplands, livestock feedlots, logged forests, streets, lawns, and parking lots.

In the United States, nonpoint pollution from agriculture—mostly in the form of sediment, inorganic fertilizers, manure, salts dissolved in irrigation water, and pesticides—is responsible for an estimated 64% of the total mass of pollutants entering streams and 57% of those entering lakes. Each year livestock in the United States produce enough manure to fill an imaginary train of 6.7 million boxcars circling the earth 12.5 times. Little progress has been made in controlling nonpoint water pollution because of the difficulty and expense of identifying and controlling discharges from so many diffuse sources.

What Are the Pollution Problems of Streams? Flowing streams, including large ones called *rivers*, can recover rapidly from degradable, oxygen-demanding wastes and excess heat by a combination of dilution and bacterial decay. This natural recovery process works as long as **(1)** streams are not overloaded with these pollutants and **(2)** their flow is not reduced by drought, damming, or diversion for agriculture and industry. However, these natural dilution and biodegradation processes do not eliminate slowly degradable and nondegradable pollutants.

The breakdown of degradable wastes by bacteria depletes dissolved oxygen, which reduces or eliminates populations of organisms with high oxygen requirements until the stream is cleansed of wastes. The depth and width of the resulting *oxygen sag curve* (Figure 12-18) (and thus the time and distance required for a stream to recover) depend on the stream's volume, flow rate, temperature, and pH level, and the volume of incoming degradable wastes. Similar oxygen sag curves can be plotted when heated water from industrial and power plants is discharged into streams.

What Progress Have We Made in Reducing Stream Pollution? Requiring cities to withdraw their drinking water downstream rather than upstream (as is done now) would dramatically improve water quality. Then each city would be forced to clean up its own waste outputs rather than passing them downstream. However, upstream users, who already have the use of fairly clean water without high cleanup costs, fight this pollution prevention approach.

Here is some *good news*. Water pollution control laws enacted in the 1970s have greatly increased the number and quality of wastewater treatment plants in the United States and many other developed countries. Laws have also required industries to reduce or eliminate point-source discharges into surface waters. These efforts, spurred by individuals (Individuals Matter, p. 307), have enabled the United States to hold the line against increased pollution of most of its streams by disease-causing agents and oxygen-demanding wastes. This is an impressive accomplishment considering the rise in economic activity and population since the laws were passed.

One success story is the cleanup of Ohio's Cuyahoga River, which was so polluted that in 1959 and again in 1969 it caught fire and burned for several days as it flowed through Cleveland, Ohio. The highly publicized image of this burning river prompted city, state, and federal officials to enact laws limiting the discharge of industrial wastes into the river and sewage systems and to appropriate funds to upgrade sewage treatment facilities. Today the river has made a comeback and is widely used by boaters and anglers.

Pollution control laws passed since 1970 have also led to improvements in dissolved oxygen content in many streams in Canada, Japan, and most western European countries. A spectacular cleanup has occurred in Great Britain. In the 1950s the Thames River was little more than a flowing anaerobic sewer. But after more

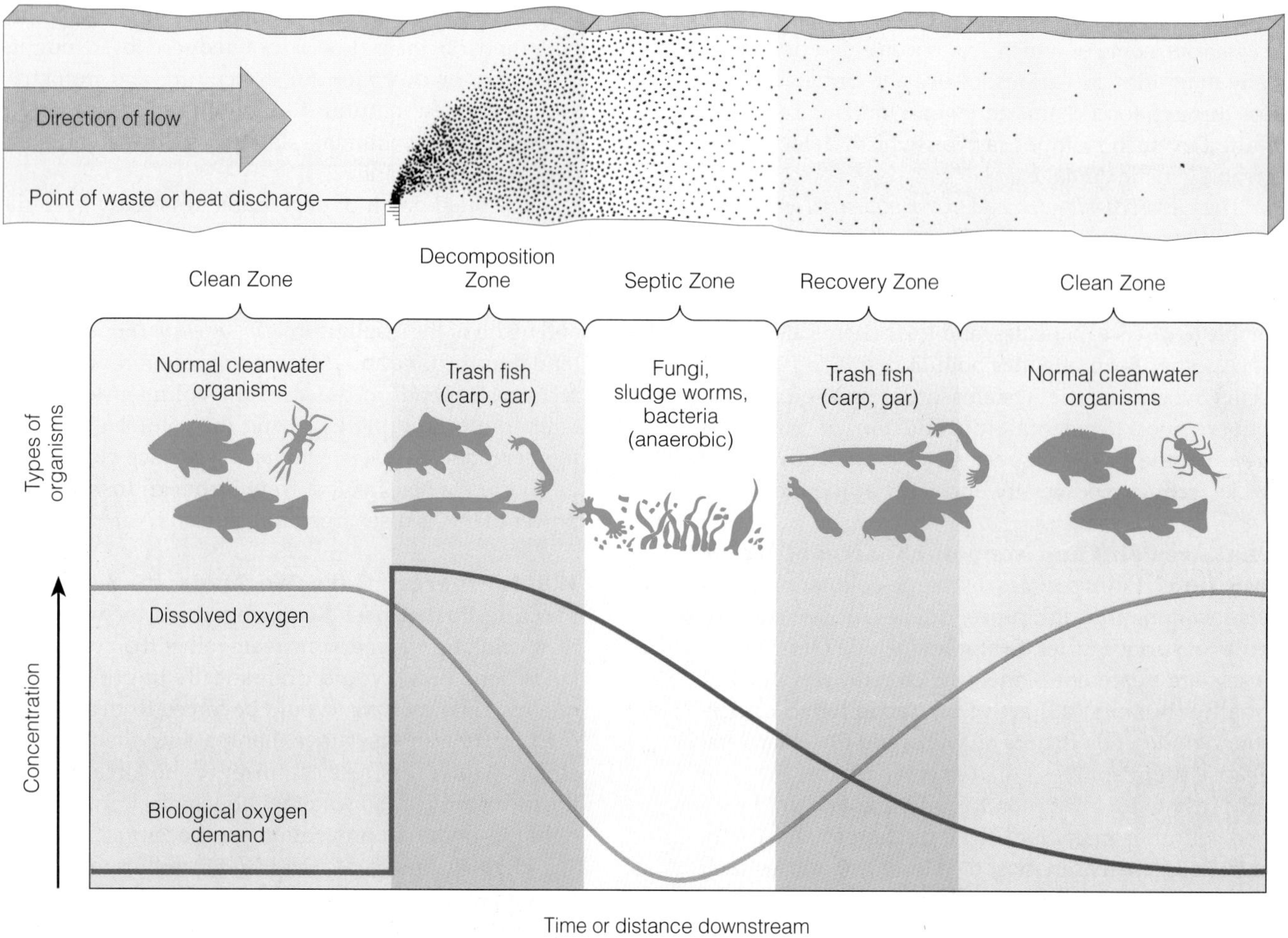

Figure 12-18 Dilution and decay of degradable, oxygen-demanding wastes and heat, showing the oxygen sag curve (orange) and the curve of oxygen demand (blue). Depending on flow rates and the amount of pollutants, streams recover from oxygen-demanding wastes and heat if they are given enough time and are not overloaded.

than 30 years of effort, $250 million of British taxpayers' money, and millions more spent by industry, the Thames has made a remarkable recovery. Commercial fishing is thriving, and many species of waterfowl and wading birds have returned to their former feeding grounds.

What Is the Bad News About Stream Pollution? Despite progress in improving stream quality in most developed countries, large fish kills and contamination of drinking water still occur. Most of these disasters are caused by **(1)** accidental or deliberate releases of toxic inorganic and organic chemicals by industries, **(2)** malfunctioning sewage treatment plants, and **(3)** nonpoint runoff of pesticides and nutrients (eroded soil, fertilizer, and animal waste) from cropland or animal feedlots (Individuals Matter, p. 311).

Available data indicate that stream pollution from huge discharges of sewage and industrial wastes is a serious and growing problem in most developing countries, where waste treatment is practically nonexistent. Numerous streams in the former Soviet Union and in eastern European countries are severely polluted. Currently, more than two-thirds of India's water resources are polluted with industrial wastes and sewage. Of the 78 streams monitored in China, 54 are seriously polluted with untreated sewage and industrial wastes. About 20% of China's rivers are too polluted to use for irrigation and 80% of them are so degraded they no longer support fish. In Latin America and Africa, most streams passing through urban or industrial areas are severely polluted.

What Are the Pollution Problems of Lakes? In lakes, reservoirs, and ponds, dilution is often less effective than in streams. Lakes and reservoirs often contain stratified layers that undergo little vertical mixing, and ponds contain small volumes of water. Stratification

Rescuing a River

INDIVIDUALS MATTER

In the 1960s, Marion Stoddart (Figure 12-19) moved to Groton, Massachusetts, on the Nashua River, then considered one of the nation's filthiest rivers. Dead fish bobbed on its waves, and at times the water was red, green, or blue from pigments discharged by paper mills.

Instead of thinking nothing could be done, she committed herself to restoring the Nashua and establishing public parklands along its banks. She didn't start by filing lawsuits or organizing demonstrations; instead she created a careful cleanup plan and approached state officials with it in 1962. They laughed, but she was not deterred and began practicing the most time-honored skill of politics: one-on-one persuasion. She identified the power brokers in the riverside communities and began to educate them, to win them over, and she got them to cooperate in cleaning up the river.

She got the state to ban open dumping in the river. When promised federal matching funds for building the treatment plant failed to materialize, Stoddart gathered 13,000 signatures on a petition sent to President Richard Nixon. The funds arrived in a hurry.

Stoddart's next success was getting a federal grant to beautify the river. She hired high school dropouts to clear away mounds of debris. When the river cleanup was completed, she persuaded communities along the river to create some 2,400 hectares (6,000 acres) of riverside park and woodlands along both banks.

Now, almost four decades later, the Nashua is still clean. Several new water treatment plants have been built, and a citizens' group founded by Stoddart keeps watch on water quality. The river supports many kinds of fish and other wildlife, and its waters are used for canoeing and other kinds of recreation.

For her efforts, Stoddart has been named by the UN Environment Programme as an outstanding worldwide worker for the environment. However, she might say that the blue and canoeable Nashua itself is her best reward.

Figure 12-19 Earth citizen Marion Stoddart canoeing down the Nashua River near Groton, Massachusetts. She spent over two decades spearheading successful efforts to have this river cleaned up. (Seth Resnick)

also reduces levels of dissolved oxygen, especially in the bottom layer. In addition, lakes, reservoirs, and ponds have little flow, further reducing dilution and replenishment of dissolved oxygen. The flushing and changing of water in lakes and large artificial reservoirs can take from 1 to 100 years, compared with several days to several weeks for streams.

Thus, lakes, reservoirs, and ponds are more vulnerable than streams to contamination by plant nutrients, oil, pesticides, and toxic substances such as lead, mercury, and selenium. These contaminants can destroy both bottom life and fish and birds that feed on contaminated aquatic organisms. Atmospheric fallout and runoff of acids is a serious problem in lakes vulnerable to acid deposition (Figure 10-8).

Concentrations of some chemicals, such as DDT (Figure 8-2), PCBs (Figure 12-20), some radioactive isotopes, and some mercury compounds can be biologically magnified as they pass through food webs in lakes. Many toxic chemicals also enter lakes and reservoirs from the atmosphere.

Lakes receive inputs of nutrients and silt from the surrounding land basin as a result of natural erosion and runoff. This natural nutrient enrichment of lakes is called **eutrophication**. Over time, some lakes become more eutrophic (Figure 6-31, top), but others don't because of differences in the surrounding drainage basin. Near urban or agricultural areas, human activities can greatly accelerate the input of nutrients to a lake, which results in a process known as **cultural eutrophication**. Such a change is caused mostly by nitrate- and phosphate-containing effluents from sewage treatment plants, runoff of fertilizers and animal wastes, and accelerated erosion of nutrient-rich topsoil (Figure 12-21).

During hot weather or drought, this nutrient overload produces dense growths of organisms such as algae, cyanobacteria, water hyacinths, and duckweed. Dissolved oxygen (in both the surface layer of water near the shore and in the bottom layer) is depleted when large masses of algae die, fall to the bottom, and are decomposed by aerobic bacteria. This oxygen depletion can kill fish and other aquatic animals. If

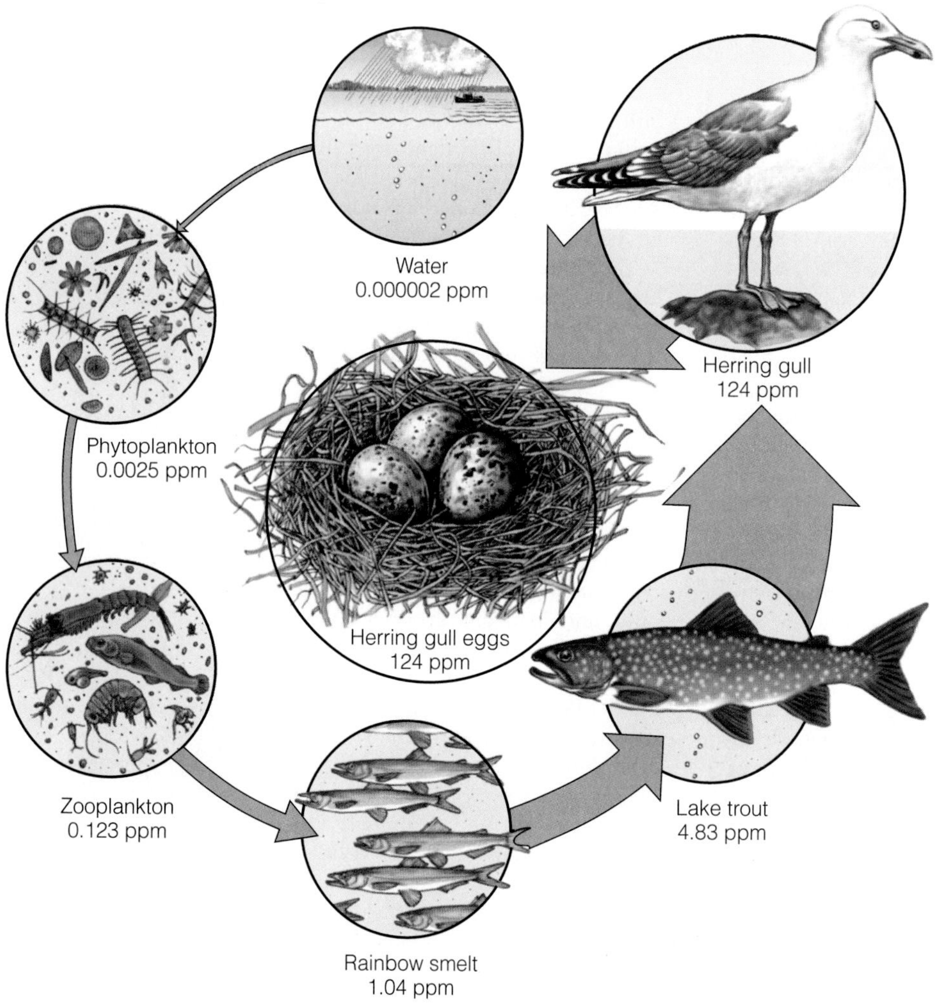

Figure 12-20 Biological magnification of PCBs (polychlorinated biphenyls) in an aquatic food chain in the Great Lakes. Most of the 209 different PCBs are insoluble in water, soluble in fats, and resistant to biological and chemical degradation—properties that result in their accumulation in the tissues of organisms and their biological amplification in food chains and webs. Although the long-term health effects on people exposed to low levels of PCBs are unknown, high doses of PCBs in laboratory animals produce **(1)** liver and kidney damage, **(2)** gastric disorders, **(3)** birth defects, **(4)** skin lesions, **(5)** hormonal changes, **(6)** smaller penis size, and **(7)** tumors. Boys in Taiwan exposed to PCBs while in their mothers' wombs developed smaller penises than other Taiwanese boys. In the United States, manufacture and use of PCBs have been banned since 1976; before then, millions of metric tons of these long-lived chemicals were released into the environment, many of them ending up in bottom sediments of lakes, streams, and oceans.

excess nutrients continue to flow into a lake, anaerobic bacteria take over and produce gaseous decomposition products such as smelly, highly toxic hydrogen sulfide and flammable methane.

About one-third of the 100,000 medium to large lakes and about 85% of the large lakes near major population centers in the United States suffer from some degree of cultural eutrophication. One-fourth of China's lakes are classified as eutrophic.

Ways to *prevent* or reduce cultural eutrophication include **(1)** advanced waste treatment (Section 12-7), **(2)** bans or limits on phosphates in household detergents and other cleaning agents, and **(3)** soil conservation and land-use control to reduce nutrient runoff.

Major *cleanup methods* are **(1)** dredging bottom sediments to remove excess nutrient buildup, **(2)** removing excess weeds, **(3)** controlling undesirable plant growth with herbicides and algicides, and **(4)** pumping air through lakes and reservoirs to avoid oxygen depletion (an expensive and energy-intensive method).

As usual, pollution prevention is more effective and usually cheaper in the long run than pollution con-

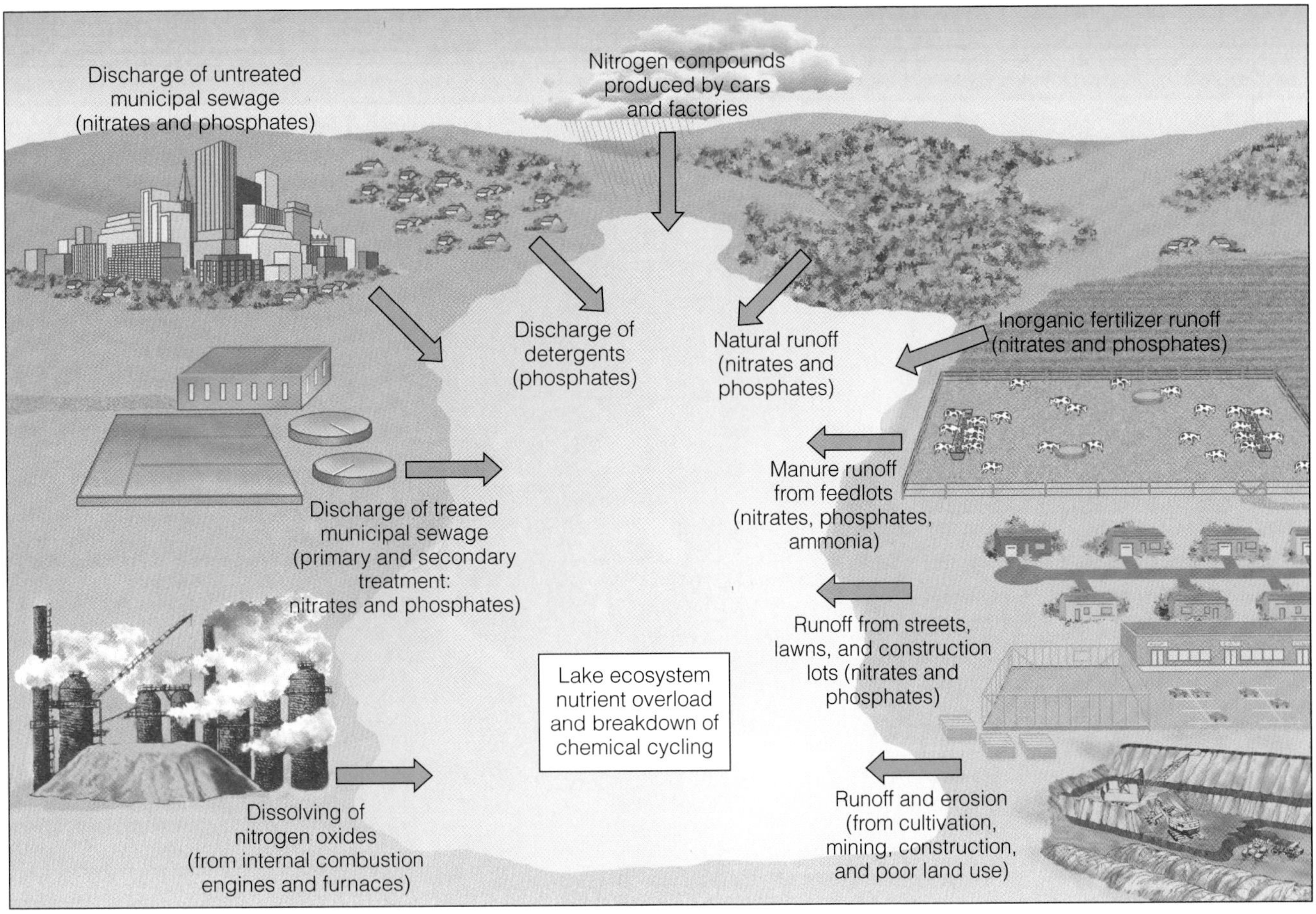

Figure 12-21 Principal sources of nutrient overload causing *cultural eutrophication* in lakes. The amount of nutrients from each source varies according to the types and amounts of human activities occurring in each airshed and watershed. Levels of dissolved oxygen (Figure 12-17) drop when enlarged populations of algae and plants (stimulated by increased nutrient input) die and are decomposed by aerobic bacteria. Lowered oxygen levels can kill fish and other aquatic life and reduce the aesthetic and recreational value of the lake.

trol. If excessive inputs of limiting plant nutrients stop, a lake usually can return to its previous state.

Seattle's Lake Washington is a success story of recovery from severe eutrophication caused by decades of sewage inputs. The recovery took place after the sewage was diverted into Puget Sound. This worked for three reasons: **(1)** A large body of water (Puget Sound) was available to receive the sewage wastes, **(2)** the lake had not yet filled with weeds and sediment because of its large size and depth, and **(3)** preventive corrective action was taken before the lake had become a shallow, highly eutrophic lake (Figure 6-31, top).

Case Study: Chemical Pollution in the Great Lakes

The five interconnected Great Lakes (Figure 12-22) contain at least 95% of the fresh surface water in the United States and 20% of the world's fresh surface water. The Great Lakes basin is home for about 38 million people, about 30% of the Canadian population and 14% of the U.S. population.

Despite their enormous size, these lakes are vulnerable to pollution from point and nonpoint sources because less than 1% of the water entering the Great Lakes flows out to the St. Lawrence River each year. In addition to land runoff, these lakes receive large quantities of acids, pesticides, and other toxic chemicals by deposition from the atmosphere (often blown in from hundreds or thousands of kilometers away).

By the 1960s many areas of the Great Lakes were suffering from severe cultural eutrophication, huge fish kills, and contamination from bacteria and other wastes. The impact on Lake Erie was particularly intense because it is the shallowest of the Great Lakes. Many bathing beaches had to be closed, and by 1970 the lake had lost nearly all its native fish.

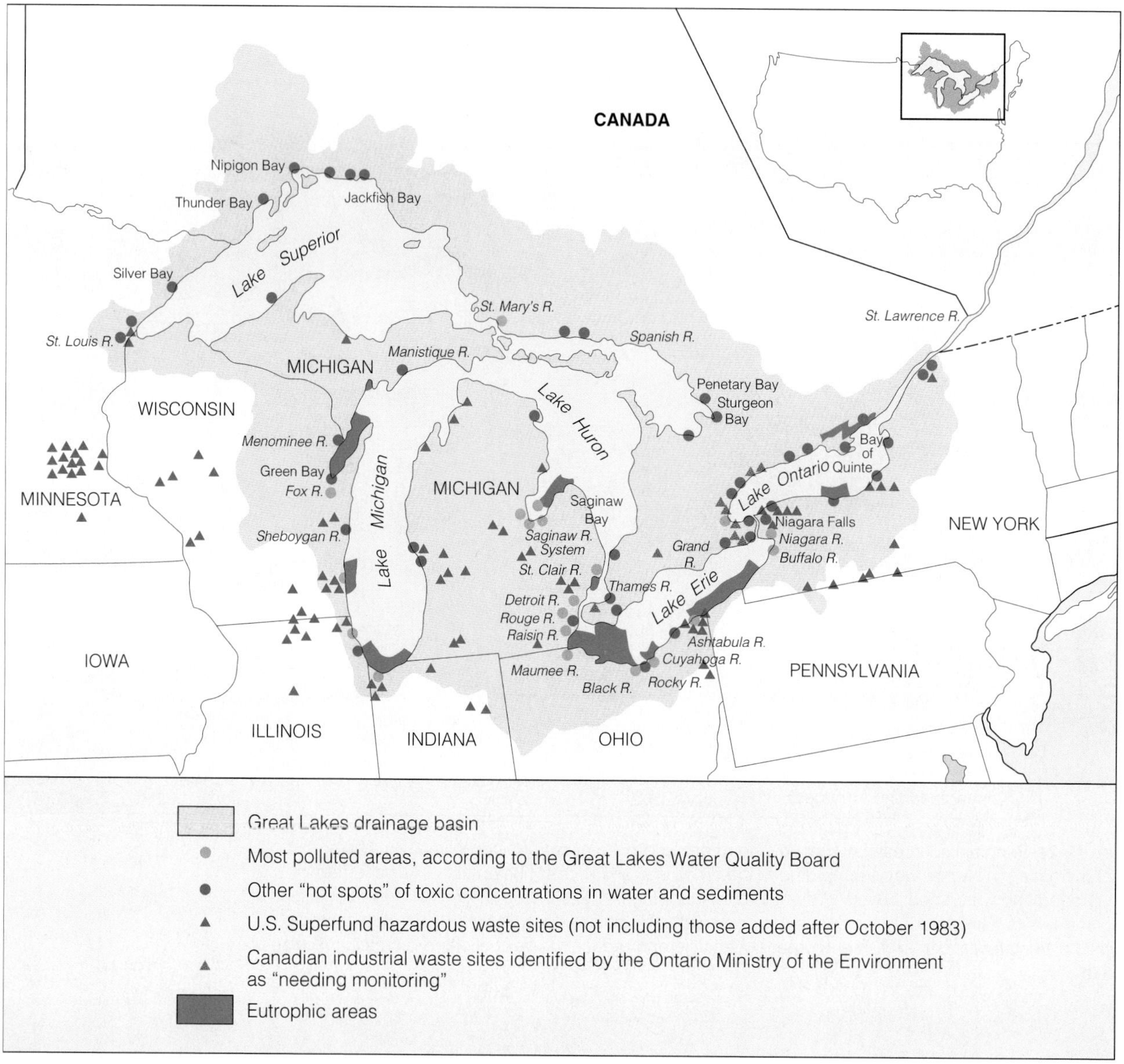

Figure 12-22 The Great Lakes basin and the locations of some of its water quality problems. (Data from Environmental Protection Agency)

Here is some *good news.* Since 1972, a $20-billion Great Lakes pollution control program has been carried out jointly by Canada and the United States. This joint program has **(1)** significantly decreased levels of phosphates, coliform bacteria, and many toxic industrial chemicals, **(2)** decreased algae blooms, **(3)** increased dissolved oxygen levels and sport and commercial fishing catches, and **(4)** allowed most swimming beaches to reopen.

These improvements were brought about mainly by **(1)** new or upgraded sewage treatment plants, **(2)** better treatment of industrial wastes, and **(3)** banning of phosphate detergents, household cleaners, and water conditioners.

Here is some *bad news:*

- Less than 3% of the lakes' shoreline is clean enough for swimming or for supplying drinking water.
- Nonpoint land runoff of pesticides and fertilizers from urban sprawl has surpassed industrial pollution as the greatest threat to the lakes.
- Atmospheric deposition of pesticides, mercury from coal-burning plants, and other toxic chemicals from as far as Mexico and Russia account for an estimated 50% of the input of toxic compounds.
- A survey by Wisconsin biologists revealed that one fish in four taken from the Great Lakes is unsafe for human consumption.

Tracking Down the Pfiesteria Cell from Hell

INDIVIDUALS MATTER

Dr. JoAnn M. Burkholder is an aquatic botanist at North Carolina State University. She knows what it is to be sickened by a newly identified fish-killing microbe and to experience the political heat when you go public with your research to alert people about a potentially serious health threat.

In 1986 she investigated why a colleague's laboratory research fish were dying mysteriously and discovered that the culprit was a new microbe so tiny that dozens could fit on the head of a pin. She named it *Pfiesteria piscida* (pronounced "fee-STEER-e-ah pis-kuh-SEED-uh"), and some biologists call it the cell from hell.

She discovered that this complex microscopic organism can behave as both a plant and an animal and assume at least 24 guises in its lifetime. Without suitable prey, the microbe can masquerade as a plant or lie dormant for years. Then under certain conditions these microbes can change from algae eaters into fish-killing dinoflagellates that release neurotoxins that stun fish in rivers and coastal estuaries, and usually kills them within 10 minutes to several hours.

The neurotoxin can also form an aerosol above the water. In 1993, Burkholder and her chief research aid experienced nausea, burning eyes and cramps, weakness, slow-healing sores, difficulty breathing, and severe loss of memory and mental powers from breathing toxic fumes released in tanks of fish dying from *Pfiesteria* attacks. They eventually recovered, but still cannot exercise strenuously without severe shortness of breath and the onset of respiratory illness.

Since then more than 100 researchers, fishermen, and water-skiers in North Carolina, Virginia, and Maryland have experienced one or more of these symptoms when exposed to water or air contaminated by *Pfiesteria* toxins. With the proper nutrient environment and other conditions, the organism can live in waters as far south as the Gulf of Mexico and thrive throughout much of the world.

Through lab and field research, Burkholder developed evidence that connected outbreaks or blooms of *Pfiesteria* with excessive levels of nitrogen (as nitrates) and phosphorus (as phosphates) in rivers and estuaries. High levels of such nutrients are found in runoff from fertilized croplands, industrial development, and feedlots (especially those used to raise hogs and chickens) into rivers flowing into coastal estuaries. In 1997, *Pfiesteria* outbreaks cost $60 million in losses to U.S. fisheries and tourism.

In 1991, Burkholder went public with her findings and urged North Carolina state legislators to put curbs on hog farming and enact much tougher laws to reduce the flow of nutrients and other pollutants into the state's rivers. Hog farmers, developers, farming interests, fishing industry officials (worried about whether it is safe to eat fish and shellfish from affected rivers and estuaries), tourist industry officials (alarmed about a negative image of the state's huge coastal recreational industry), and some state officials reacted negatively to her political activism. Some challenged her character and competence and accused her of using the results of preliminary research to push for questionable policies. She also received some anonymous death threats.

Burkholder has not backed down and continues to criticize state health officials and legislators for not taking her concerns about public health seriously enough. Under the glare of state and national publicity,* the state now supports research on the problem, and in 1997 the North Carolina legislature enacted a 2-year moratorium on new hog farms.

Research by other scientists has confirmed the link between *Pfiesteria* outbreaks and nutrient overloading of rivers. By 1997, a number of federal and state environmental, health, and agricultural agencies had set up a coordinated research effort to learn more about what triggers outbreaks of the organism and how they affect humans and other organisms.

*For a popularized description of her research and political battle to alert the public and elected officials to the dangers posed by this microbe, see Rodney Barker's *And the Waters Turned to Blood: The Ultimate Biological Threat* (New York: Simon & Schuster, 1997).

- In what is called the *grasshopper effect*, some of the toxic chemicals (such as DDT and other banned pesticides) that rode air currents to the Great Lakes are evaporating from the lakes and being transmitted by winds to the Canadian and Scandinavian arctic and other parts of the globe.
- Water levels in the lakes are at their lowest in three decades, and according to climate models they should drop much more by 2050 because of global warming. This will reduce production of hydroelectric power, cause freighters to run closer to the bottom, disturb bottom sediments, release toxic chemicals such as PCBs and dioxins, and contaminate food webs.

In 1991 the U.S. government passed a law requiring accelerated cleanup of the lakes, especially 42 toxic hot spots, and an immediate reduction in air pollutant emissions in the region. However, meeting these goals may be delayed by a lack of federal and state funds.

Some environmentalists call for a ban on **(1)** the use of chlorine as a bleach in the pulp and paper industry around the Great Lakes, **(2)** all new incinerators in the area, and **(3)** discharge into the lakes of

70 toxic chemicals that threaten human health and wildlife. Officials of these industries strongly oppose such bans.

Case Study: Genetic Pollution in the Great Lakes There is more *bad news.* Fisheries in the Great Lakes face threats from *genetic pollution.* In 1986, larvae of a nonnative species, the *zebra mussel,* arrived in water discharged from a European ship near Detroit. With no known natural enemies, these tiny mussels have run amok; they **(1)** deplete the food supply for other lake species, **(2)** clog irrigation pipes, **(3)** shut down water intake systems for power plants and city water supplies, **(4)** foul beaches, **(5)** grow in huge masses on boat hulls, piers, and other surfaces and **(6)** cost the Great Lakes basin at least $700 million per year (annual costs could reach $5 billion within a few years).

Another problem is that zebra mussels pick up and store toxic pollutants from the water. Mussel-eating gobies can then pass these pollutants on to fish that prey on the gobies. This can transfer the poisons further up the food web and contaminate fish that people eat.

Within a few years, the zebra mussel is expected to spread and dramatically alter most freshwater communities in parts of the United States and southern Canada. This could lead to damages costing tens of billions of dollars.

However, zebra mussels may be *good news* for a number of aquatic plants. By consuming algae and other microorganisms, the mussels increase water clarity. Clearer waters permit deeper penetration of sunlight and more photosynthesis, allowing some native plants to thrive and return the plant composition of Lake Erie (and presumably other lakes) closer to what it was 100 years ago. Because the plants provide food and increase dissolved oxygen, their comeback may benefit certain aquatic animals (including the mussels).

There is more *bad news.* In 1991 a larger and potentially more destructive species, the *quagga mussel,* invaded the Great Lakes, probably brought in by a Russian freighter. It can survive at greater depths and tolerate more extreme temperatures than the zebra mussel. There is concern that it may eventually colonize areas such as the Chesapeake Bay and waterways in parts of Florida.

By 1999 all of the Great Lakes had been invaded by a European fish, the round goby. The *good news* is that these tiny predators have a voracious appetite for zebra mussels. The *bad news* is that these bottom-dwelling fish also devour eggs and fry of any fish sharing their habitat. This includes prized recreational fish such as perch, walleye, and smallmouth bass.

Why Is Groundwater Pollution Such a Serious Problem? Highly visible oil spills get a lot of media attention, but a much greater threat to human health is the out-of-sight pollution of groundwater (Figure 12-23), a prime source of water for drinking and irrigation. Groundwater in aquifers (Figure 12-3) is easy to deplete and pollute because much of it is renewed very slowly. Although experts rate groundwater pollution as a low-risk ecological problem, they consider pollutants in drinking water (much of it from groundwater) a high-risk health problem (Figure 8-10, left). According to environmentalists, laws protecting groundwater are weak in the United States and nonexistent in most countries.

When groundwater becomes contaminated, it cannot cleanse itself of degradable wastes, as surface water can if it is not overloaded. Because groundwater flows are slow and not turbulent, contaminants are not diluted and dispersed effectively. Groundwater also has much smaller populations of decomposing bacteria than surface-water systems, and its cold temperature slows down decomposition reactions. Thus, it can take hundreds to thousands of years for contaminated groundwater to cleanse itself of degradable wastes; nondegradable wastes are there permanently on a human time scale.

Crude estimates indicate that up to 25% of usable groundwater in the United States is contaminated (and in some areas as much as 75%). In New Jersey, for example, every major aquifer is contaminated. In California, pesticides contaminate the drinking water of more than 1 million people. In Florida, where 92% of the residents rely on groundwater for drinking, over 1,000 wells have been closed. The EPA has documented groundwater contamination by 74 pesticides in 38 states.

Groundwater can be contaminated from a number of sources, including **(1)** underground storage tanks, **(2)** landfills, **(3)** abandoned hazardous waste dumps, **(4)** deep wells used to dispose of liquid hazardous wastes, and industrial waste and **(5)** livestock waste storage lagoons located above or near aquifers (Figure 12-23). An EPA survey found that one-third of 26,000 industrial waste ponds and lagoons have no liners to prevent toxic liquid wastes from seeping into aquifers, and one-third of those sites are within 1.6 kilometers (1 mile) of a drinking water well.

The EPA estimates that at least 1 million underground tanks storing gasoline, diesel fuel, and toxic solvents are leaking their contents into groundwater. A slow gasoline leak of just 4 liters (1 gallon) per day can seriously contaminate the water supply for 50,000 people. Such slow leaks usually remain undetected until someone discovers that a well is contaminated.

Determining the extent of a leak can cost $25,000–250,000. Cleanup costs range from $10,000 for a small spill to $250,000 or more if the chemical reaches

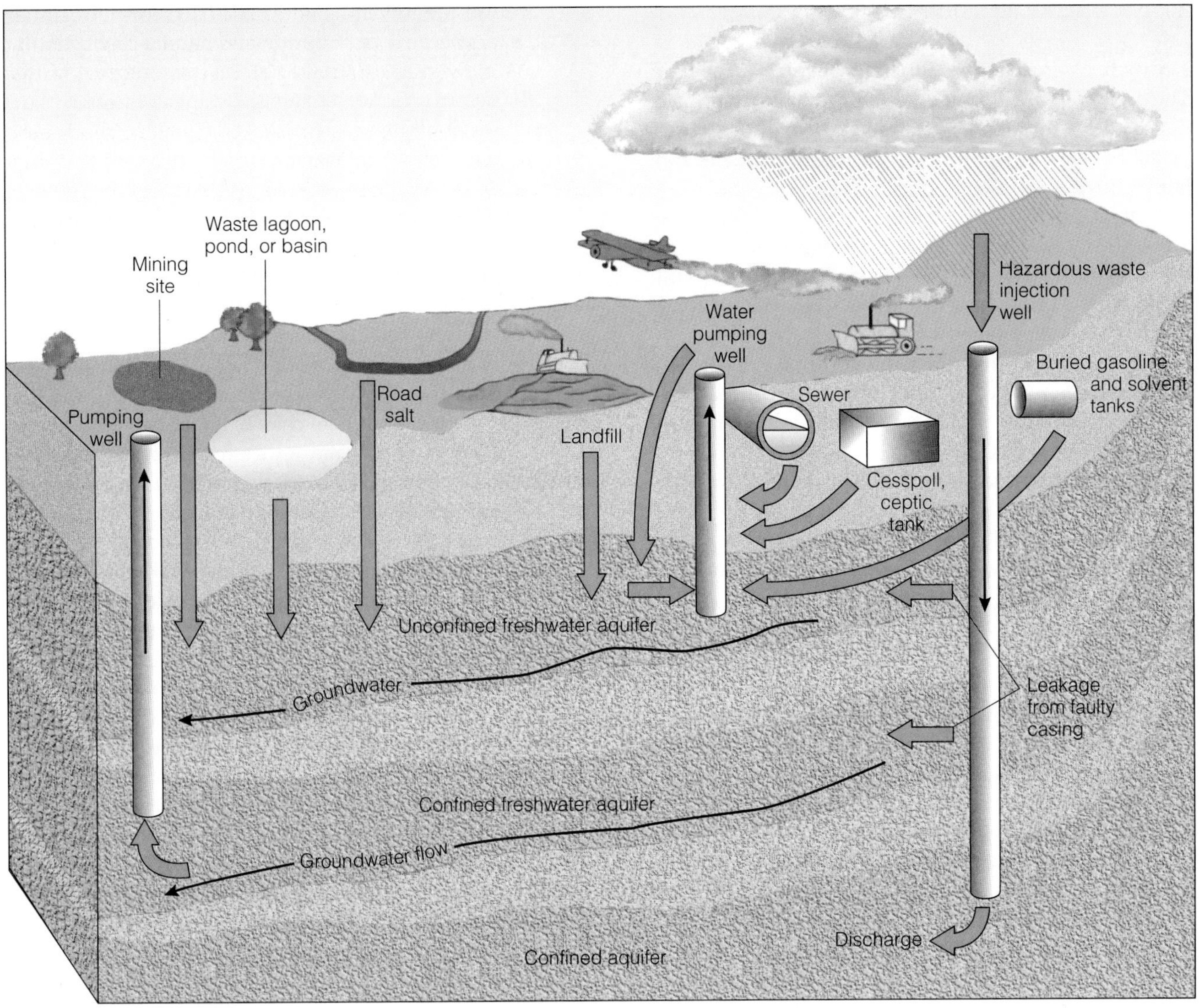

Figure 12-23 Principal sources of groundwater contamination in the United States.

an aquifer, and complete cleanup is rarely possible. Replacing a leaking tank adds an additional $10,000–60,000. Legal fees and damages to injured parties can run into the millions. Current regulations should reduce leakage from new tanks in the United States but do little about the millions of older tanks that the EPA calls toxic time bombs.

Solutions: How Can We Protect Groundwater? Pumping polluted groundwater to the surface, cleaning it up, and returning it to the aquifer is usually too expensive (for a single aquifer, $5 million or more). Recent attempts to pump and treat contaminated aquifers indicate that it may take 50–1,000 years of continuous pumping before all the contamination is forced to the surface and drinking water quality is achieved. Thus, *preventing contamination is considered the only effective way to protect groundwater resources.* Ways to do this include

- *Monitoring aquifers near landfills and underground tanks*
- *Requiring leak detection systems for existing and new underground tanks used to store hazardous liquids*
- *Requiring liability insurance for old and new underground tanks used to store hazardous liquids*
- *Banning or more strictly regulating disposal of hazardous wastes in deep injection wells and landfills*
- *Storing hazardous liquids above ground in tanks with systems that detect and collect leaking liquids*

12-6 OCEAN POLLUTION

How Much Pollution Can the Oceans Tolerate? The oceans are the ultimate sink for much of the waste matter we produce, as summarized in the African proverb, "Water may flow in a thousand channels, but it all returns to the sea."

Oceans can dilute, disperse, and degrade large amounts of raw sewage, sewage sludge, oil, and some types of industrial waste, especially in deep-water areas. Some forms of marine life have also proved to be much more resilient than some scientists had expected. This has led some analysts to suggest that it is safer to dump sewage sludge and most other hazardous wastes into the deep ocean than to bury them on land or burn them in incinerators.

Other scientists dispute this idea, pointing out that we know less about the deep ocean than we do about outer space. They add that dumping waste in the ocean would delay urgently needed pollution prevention and promote further degradation of this vital part of the earth's life-support system. What do you think?

How Do Pollutants Affect Coastal Areas? In 1999, ecologist Jane Lubchenco pointed to new data indicating that human activities are dramatically altering the earth's oceans and impairing their ability to provide a wide range of ecological services for humans and other species. Coastal areas—especially wetlands and estuaries, coral reefs, and mangrove swamps—bear the brunt of our enormous inputs of wastes into the ocean. This is not surprising because about 40% the world's population lives on or within 100 kilometers (160 miles) of the coast, and coastal populations are growing more rapidly than global population. Fourteen of the world's 15 largest metropolitan areas, each with 10 million people or more, are near coastal waters.

In most coastal developing countries (and in some coastal developed countries), municipal sewage and industrial wastes are dumped into the sea without treatment. The most polluted seas lie off the densely populated coasts of Bangladesh, India, Pakistan, Indonesia, Malaysia, Thailand, and the Philippines. About 85% of the sewage from large cities along the Mediterranean Sea, which has a coastal population of 200 million people during tourist season, is discharged into the sea untreated. This causes widespread beach pollution and shellfish contamination.

In the United States about 35% of all municipal sewage ends up virtually untreated in marine waters. Most U.S. harbors and bays are badly polluted by municipal sewage, industrial wastes, and oil. Each year at least one-third of the area of U.S. coastal waters around the lower 48 states is closed to shellfish harvesters because of pollution and habitat disruption.

New studies show that vast colonies of human viruses from raw sewage, sewage treatment plants (which don't remove viruses), and leaking septic tanks migrate into coastal waters. One study found that about one-fourth of the people using coastal beaches develop ear infections, sore throats and eyes, respiratory disease, or gastrointestinal disease.

California's Santa Monica Bay has been a filming site for the widely watched television show *Baywatch*, which gives an illusion of a clean California beach lifestyle. What viewers don't know is that the bay is so polluted that the actors get extra pay each time they enter the water and are chemically cleaned afterward.

Runoff of sewage and agricultural wastes into coastal waters and acid deposition from the atmosphere (Figure 9-6) introduce large quantities of nitrogen and phosphorus, which can cause explosive growth of harmful algae. These algal blooms—called red, brown, or green tides, depending on their color—can release waterborne and airborne toxins that damage fisheries, kill some fish-eating birds, reduce tourism, and poison seafood. When the algae die and decompose, coastal waters are depleted of oxygen and a variety of marine species die. A Harvard study showed that between 1979 and 1996 harmful algae blooms in North American coastal waters increased more than fourfold.

According to ecologist Jane Lubchenco, there are 50 "dead zones" where much aquatic life is destroyed because of excessive inputs of fertilizers that deplete dissolved oxygen. The largest area of such coastal cultural eutrophication in the western hemisphere is in the Gulf of Mexico.

For half the year, this zone in the Gulf of Mexico is rich with shrimp, fish, and many other forms of aquatic life. However, the spring snowmelt and rains flush nitrogen and phosphorus fertilizers from farms and cities throughout the Mississippi River's giant watershed, which covers more than 40% of the continental United States from Montana to western New York. As a result, this zone becomes overfertilized, depleted of oxygen, and devoid of fish and other creatures for about 6 months a year. By summer the "dead zone" covers about 20,700 square kilometers (8,000 square miles).

Case Study: The Chesapeake Bay The Chesapeake Bay, the largest estuary in the United States, is in trouble because of human activities. Between 1940 and 1998, the number of people living in the Chesapeake Bay area grew from 3.7 million to 15.7 million, and within a few years its population may reach 18 million.

The estuary receives wastes from point and nonpoint sources scattered throughout a huge drainage basin that includes 9 large rivers and 141 smaller streams and creeks in parts of six states (Figure 12-24). The bay has become a huge pollution sink because **(1)** it is quite shallow and **(2)** only 1% of the waste entering it is flushed into the Atlantic Ocean.

Levels of phosphates and nitrates have risen sharply in many parts of the bay, causing algae blooms and oxygen depletion (Figure 12-24). Studies have shown that point sources, primarily sewage treatment plants, contribute about 60% by weight of the phosphates. Nonpoint sources—mostly runoff from urban, suburban, and agricultural land and deposition from the atmosphere—are the origins of about 60% by weight of the nitrates.

Air pollutants account for nearly 40% of the nitrogen, 90% of the mercury (mostly from older coal-burning power plants), and 95% of the lead entering the estuary. Large quantities of pesticides run off cropland and urban lawns and industries discharge large amounts of toxic wastes, often in violation of their discharge permits. Commercial harvests of oysters, crabs, and several important fish have fallen sharply since 1960 because of a combination of overfishing, pollution, and disease.

In the 1980s the Chesapeake Bay Program, the country's most ambitious attempt at integrated coastal management, was implemented. Results have been impressive. Between 1985 and 1993, phosphorus levels declined 16% and nitrogen levels dropped 7%—a significant achievement given the increasing population in the watershed and the fact that nearly 40% of the nitrogen inputs come from the atmosphere.

Reaching the goals of a 40% reduction in nutrient levels and a significant improvement in habitat water quality throughout the bay will be especially difficult because the area's population is expected to grow by 25% between 1995 and 2020. Moreover, the bay will soon be invaded by zebra and quagga mussels. So far, however, the Chesapeake Bay Program shows what can be done when diverse interested parties work together to achieve goals that benefit both wildlife and people.

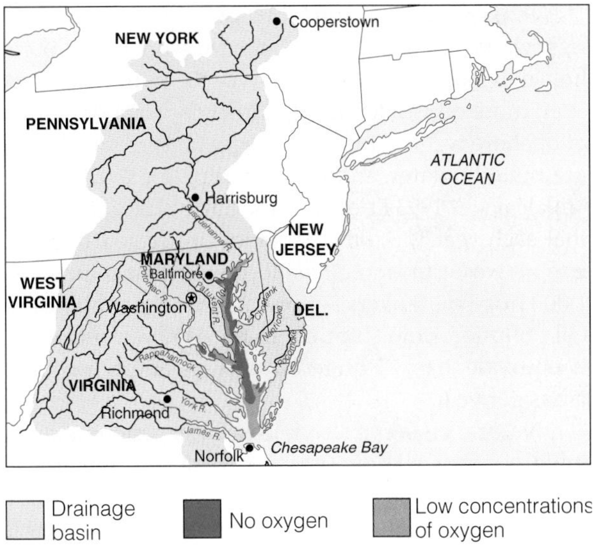

Figure 12-24 Chesapeake Bay, the largest estuary in the United States, is severely degraded as a result of water pollution from point and nonpoint sources in six states and from deposition of air pollutants.

What Pollutants Do We Dump into the Ocean? Dumping of industrial waste off U.S. coasts has stopped, although it still occurs in a number of other developed countries and some developing countries. However, barges and ships still legally dump large quantities of **dredge spoils** (materials, often laden with toxic metals, scraped from the bottoms of harbors and rivers to maintain shipping channels) at 110 sites off the Atlantic, Pacific, and Gulf coasts.

In addition, many countries dump into the ocean large quantities of sewage **sludge**: a gooey mixture of toxic chemicals, infectious agents, and settled solids removed from wastewater at sewage treatment plants. Since 1992 this practice has been banned in the United States.

Fifty countries with at least 80% of the world's merchant fleet have agreed not to dump sewage and garbage at sea, but this agreement is difficult to enforce and often is violated. Most ship owners save money by dumping wastes at sea and risk only small fines if they are caught. Each year as many as 2 million seabirds and more than 100,000 marine mammals (including whales, seals, dolphins, and sea lions) die when they ingest or become entangled in fishing nets, ropes, and other debris dumped into the sea and discarded on beaches.

Under the London Dumping Convention of 1972, 100 countries agreed not to dump highly toxic pollutants and high-level radioactive wastes in the open sea beyond the boundaries of their national jurisdictions. Since 1983 these same nations have observed a moratorium on the dumping of low-level radioactive wastes at sea, which in 1994 became a permanent ban. However, France, Great Britain, Russia, China, and Belgium may legally exempt themselves from this ban. In 1992 it was learned that for decades the former Soviet Union had been dumping large quantities of high- and low-level radioactive wastes into the Arctic Ocean and its tributaries.

What Are the Effects of Oil on Ocean Ecosystems? *Crude petroleum* (oil as it comes out of the ground) and *refined petroleum* (fuel oil, gasoline, and other processed petroleum products) are accidentally or deliberately released into the environment from a number of sources.

Tanker accidents and blowouts at offshore drilling rigs (when oil escapes under high pressure from a borehole in the ocean floor) get most of the publicity. However, more oil is released **(1)** during normal operation of offshore wells, **(2)** from washing tankers and releasing the oily water, and **(3)** from pipeline and storage tank leaks. A 1993 Friends of the Earth study estimated that each year U.S. oil companies unnecessarily spill, leak, or waste an amount of oil equal to that shipped by 1,000 huge oil tankers—more oil than Australia uses. Oil pollution from shipping in the Mediterranean Sea is equivalent to 17 huge oil tankers emptying their tanks per year.

Natural oil seeps also release large amounts of oil into the ocean at some sites, but most ocean oil pollution comes from activities on land. Almost half (some experts estimate 90%) of the oil reaching the oceans is waste oil dumped, spilled, or leaked onto the land or into sewers by cities, individuals, and industries. Each year, a volume of oil equal to 33 times the amount spilled by the *Exxon Valdez* tanker (Case Study, below) is disposed of improperly by U.S. motorists changing their own motor oil. Worldwide, about 10% of the oil that reaches the ocean comes from the atmosphere, mostly from smoke emitted by oil fires.

The effects of oil on ocean ecosystems depend on a number of factors: **(1)** type of oil (crude or refined), **(2)** amount released, **(3)** distance of release from shore, **(4)** time of year, **(5)** weather conditions, **(6)** average water temperature, and **(7)** ocean currents. Research shows that most (but not all) forms of marine life recover from exposure to large amounts of crude oil within 3 years. However, recovery from exposure to refined oil, especially in estuaries, may take 10 years or longer. The effects of spills in cold waters and in shallow enclosed gulfs and bays generally last longer.

Oil slicks that wash onto beaches can have a serious economic impact on coastal residents, who lose income from fishing and tourist activities. Oil-polluted beaches washed by strong waves or currents become clean after about a year, but beaches in sheltered areas remain contaminated for several years. Estuaries and salt marshes suffer the most and longest-lasting damage. Despite their localized harmful effects, oil spills are rated by experts as a low-risk ecological problem (Figure 8-10, left).

Case Study: The *Exxon Valdez* Oil Spill Crude oil from Alaska's North Slope fields near Prudhoe Bay is carried by pipeline to the port of Valdez and then shipped by tanker to the West Coast. On March 24, 1989, the *Exxon Valdez*, a tanker more than three football fields long, went off course in a 16-kilometer-wide (12-mile-wide) channel in Prince William Sound near Valdez, Alaska. It hit submerged rocks, creating the worst oil spill ever in U.S. waters (Figure 12-25).

The rapidly spreading oil slick coated more than 1,600 kilometers (1,000 miles) of shoreline, almost the length of the U.S. shoreline between New Jersey and South Carolina. The full loss of wildlife will never be known because most of the dead animals sank and decomposed without being counted. By 1999, 10 years after the accident, only 2 of 23 damaged populations (the bald eagle and river otters) had recovered, 12 were gradually recovering, and the other 8 (including orcas and harbor seals) had not recovered.

Exxon spent $2.2 billion directly on the cleanup, but some aspects of the cleanup effort did more harm than good. For example, the use of high-pressure jets of hot water to clean beaches killed coastal plants and animals that had survived the spill. As a result, a year after the spill the oiled sites had recovered more rapidly than the washed sites.

In 1991, Exxon pleaded guilty to federal felony and misdemeanor charges and agreed to pay the federal government and the state of Alaska $1 billion in fines and civil damages. In 1994 a jury awarded members of the fishing industry, landowners, and other Alaskan residents $5 billion in punitive damages and penalties. However, Exxon has appealed this decision and it may be tied up in the courts for decades.

Studies show that others must share the blame for this tragedy. State officials had been lax in monitoring Alyeska's preparedness for dealing with oil spills, and the Coast Guard did not effectively monitor tanker traffic because of inadequate radar equipment and personnel. Americans who drive fuel-inefficient cars and don't insist that we use and waste less oil by requiring greatly increased fuel efficiency standards also share part of the blame.

This roughly $8.5-billion accident (one-fourth of it paid for by U.S. taxpayers because of tax write-offs by Exxon) might have been prevented if Exxon had spent only $22.5 million to fit the tanker with a double hull (which it still has not done). In the early 1970s interior secretary Rogers Morton told Congress that all oil tankers using Alaskan waters would have double hulls, but under pressure from oil companies the requirement was dropped.

Today, virtually all merchant ships have double hulls, but only 15% of oil tankers have such hulls. Legislation passed since the spill requires all new tankers to have double hulls and all existing large single-hulled oil tankers to be phased out by 2015. However, the oil industry is working to weaken these and other stricter requirements.

This spill highlighted the importance of pollution prevention and the advantages of shifting to improving

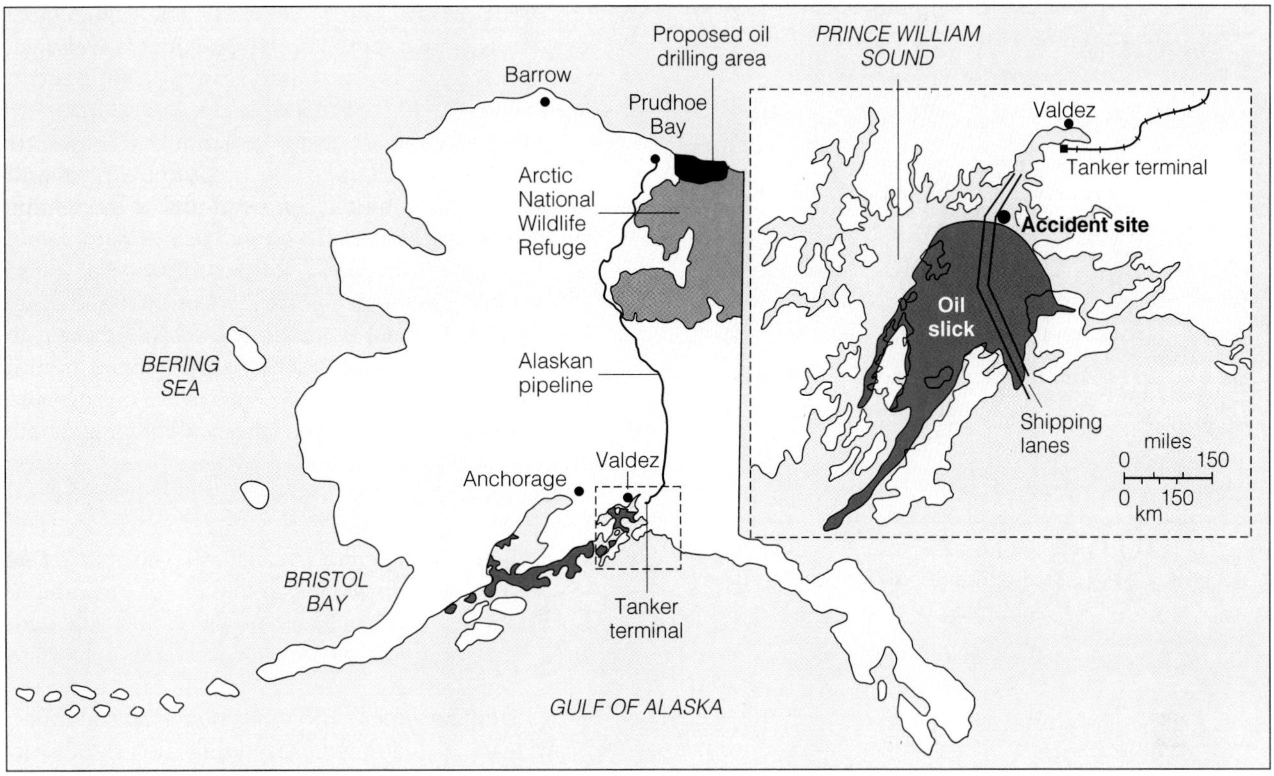

Figure 12-25 Alaska's Prince William Sound, the site of the oil spill from the tanker *Exxon Valdez* on March 24, 1989.

energy efficiency and renewable energy (Chapter 20) to reduce dependence on oil. Even with the best technology and a fast response by well-trained people, scientists estimate that no more than 12–15% of the oil from a major spill can be recovered.

Solutions: How Can We Protect Coastal Waters? The key to protecting oceans is to reduce the flow of pollution from the land and from streams emptying into the ocean. Such efforts must be integrated with efforts to prevent and control air pollution because an estimated 33% of all pollutants entering the ocean worldwide comes from air emissions from land-based sources.

Analysts have suggested the following measures to prevent and reduce excessive pollution of coastal waters:

Prevention

- *Encourage or require separate sewage and storm runoff lines in coastal urban areas*
- *Discourage ocean dumping of sludge and hazardous dredged materials*
- *Protect sensitive and ecologically valuable coastal areas from development, oil drilling, and oil shipping*
- *Use ecological land-use planning (Solutions, p. 231) to control and regulate coastal development*
- *Require double hulls for all oil tankers by 2003*
- *Recycle used oil*
- *Reduce genetic pollution from nonnative aquatic species by using heat to kill organisms in ballast water or developing filters to trap the organisms when ballast water is taken into or discharged from a ship*

Cleanup

- *Improve oil spill cleanup capabilities*
- *Require at least secondary treatment of coastal sewage, or use wetlands, solar aquatic, or other environmentally acceptable methods*

12-7 SOLUTIONS: PREVENTING AND REDUCING WATER POLLUTION

What Can We Do About Water Pollution from Nonpoint Sources? The leading nonpoint source of water pollution is agriculture. Ways to help control nonpoint water pollution include the following:

- Reducing fertilizer runoff into surface waters and leaching into aquifers by using moderate amounts of slow-release fertilizer and using none on steeply sloped land.

- Reducing the need for fertilizer by alternating plantings between row crops and soybeans or other nitrogen-fixing plants.
- Planting buffer zones of vegetation between cultivated fields and nearby surface water.
- Reducing pesticide runoff by applying pesticides only when needed and using biological control or integrated pest management (Section 16-5).
- Sharply reducing nonfarm uses of inorganic fertilizers and pesticides on golf courses, lawns, and public lands.
- Controlling runoff and infiltration of manure from animal feedlots and barnyards by managing animal density, planting buffers, and not locating feedlots on land near surface water when that land slopes steeply toward the water and in flood zones.
- Reducing soil erosion and flooding by reforesting critical watersheds. This would also help slow global warming and loss of wildlife habitat.

What Can We Do About Water Pollution from Point Sources? The Legal Approach In many developing countries and in some developed countries, sewage and waterborne industrial wastes are discharged without treatment into the nearest waterway or into wastewater lagoons.

In developed countries, most wastes from point sources are purified to varying degrees. The Federal Water Pollution Control Act of 1972 (renamed the Clean Water Act when it was amended in 1977) and the 1987 Water Quality Act form the basis of U.S. efforts to control pollution of the country's surface waters. The main goals of the Clean Water Act were to make all U.S. surface waters safe for fishing and swimming by 1983 and to restore and maintain the chemical, physical, and biological integrity of the nation's waters. The United States has made progress but has not met these goals.

In 1995 the EPA developed a *discharge trading policy* designed to use market forces to reduce water pollution (as has been done with sulfur dioxide for air pollution control). The policy would allow a water pollution source, such as an industrial plant or a sewage treatment plant, to sell credits for its excess reductions to another facility that can't reduce its discharges as cheaply.

Here is some *good news.* The Clean Water Act of 1972 led to the following improvements in U.S. water quality between 1972 and 1998: **(1)** The percentage of U.S. rivers and lakes tested that are fishable and swimmable increased from 36% to 62%, **(2)** the amount of topsoil lost through agricultural runoff was cut by about 1.1 billion metric tons (1 billion tons) annually, **(3)** the number of people served by sewage treatment plants increased from 85 million to 173 million, and **(4)** annual wetland losses decreased by 83%.

Here is some *bad news:* **(1)** About 44% of lakes, 38% of rivers (up from 26% in 1984), and 32% of estuaries in the United States are still unsafe for fishing, swimming, and other recreational uses, **(2)** hog, poultry, and cattle farm runoff pollutes 70% of U.S. rivers, **(3)** despite aggressive monitoring nearly 110,000 metric tons (100,000 tons) of toxic industrial wastes are dumped into U.S. rivers each year, **(4)** fish caught in more than 1,400 different waterways are unsafe to eat because of high levels of pesticides and other toxic substances, and **(5)** antiquated sewage systems in 1,100 cities still dump poorly treated sewage into streams, lakes, and coastal waters.

Some environmentalists call for the Clean Water Act to be strengthened by **(1)** increasing funding and authority to control nonpoint sources of pollution, **(2)** strengthening programs to prevent and control toxic water pollution, including phasing out certain toxic discharges (such as many organic chemicals containing chlorine), **(3)** providing more funding and authority for integrated watershed and airshed planning to protect groundwater and surface water from contamination, **(4)** permitting states with good environmental records to take over parts of the clean water program, under looser federal control, and **(5)** expanding citizens' ability to bring lawsuits to ensure that water pollution laws are enforced.

Many people oppose these proposals, contending that the Clean Water Act's regulations are already too restrictive and costly. Farmers and developers also see the law as a curb on their rights as property owners to fill in wetlands. They also believe that they should be compensated for property value losses because of federal wetland protection regulations. State and local officials want more discretion in testing for and meeting water quality standards. They argue that in many communities it is too expensive and unnecessary to test for all the water pollutants required by federal law.

What Can We Do About Water Pollution from Point Sources? The Technological Approach As population, urbanization, and industrialization grow, the volume of wastewater needing treatment will increase enormously. In rural and suburban areas with suitable soils, sewage from each house usually is discharged into a **septic tank** (Figure 12-26). About 25% of all homes in the United States are served by septic tanks, which should be cleaned out every 3–5 years by a reputable contractor so that they won't contribute to groundwater pollution.

In urban areas, most waterborne wastes from homes, businesses, factories, and storm runoff flow through a network of sewer pipes to wastewater treatment plants. Some cities have separate lines for stormwater runoff, but in 1,200 U.S. cities the lines for these two systems are combined because it is cheaper. When rains cause combined sewer systems to overflow, they discharge untreated sewage directly into surface waters.

When sewage reaches a treatment plant, it can undergo up to three levels of purification. **Primary sewage treatment** is a mechanical process that uses screens to filter out debris such as sticks, stones, and rags; suspended solids settle out as sludge in a settling tank (Figure 12-27). Improved primary treatment uses chemically treated polymers to remove suspended solids more thoroughly.

Secondary sewage treatment is a biological process in which aerobic bacteria are used to remove up to 90% of biodegradable, oxygen-demanding organic wastes (Figure 12-27). Some plants use *trickling filters*, in which aerobic bacteria degrade sewage as it seeps through a bed of crushed stones covered with bacteria and protozoa. Others use an *activated sludge process*, in which the sewage is pumped into a large tank and mixed for several hours with bacteria-rich sludge and air bubbles to facilitate degradation by microorganisms. The water then goes to a sedimentation tank, where most of the suspended solids and microorganisms settle out as sludge. The sludge produced from primary or secondary treatment is broken down in an anaerobic digester and then incinerated, dumped into the ocean or a landfill, or applied to land as fertilizer.

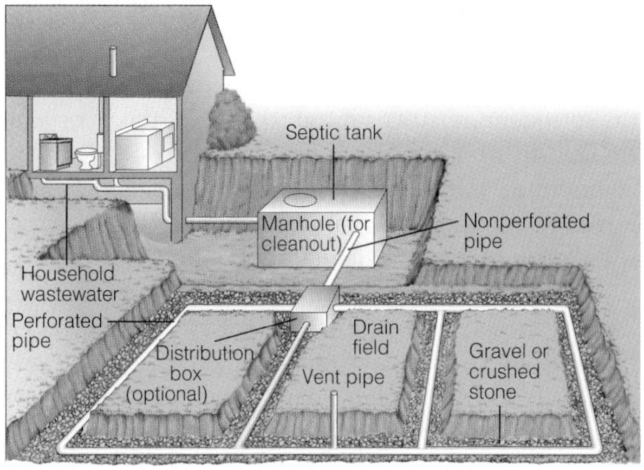

Figure 12-26 Septic tank system used for disposal of domestic sewage and wastewater in rural and suburban areas. This system traps greases and large solids and discharges the remaining wastes over a large drainage field. As these wastes percolate downward, the soil filters out some potential pollutants, and soil bacteria decompose biodegradable materials. To be effective, septic tank systems must be **(1)** properly installed in soils with adequate drainage, **(2)** not placed too close together or too near well sites, **(3)** and pumped out when the settling tank becomes full.

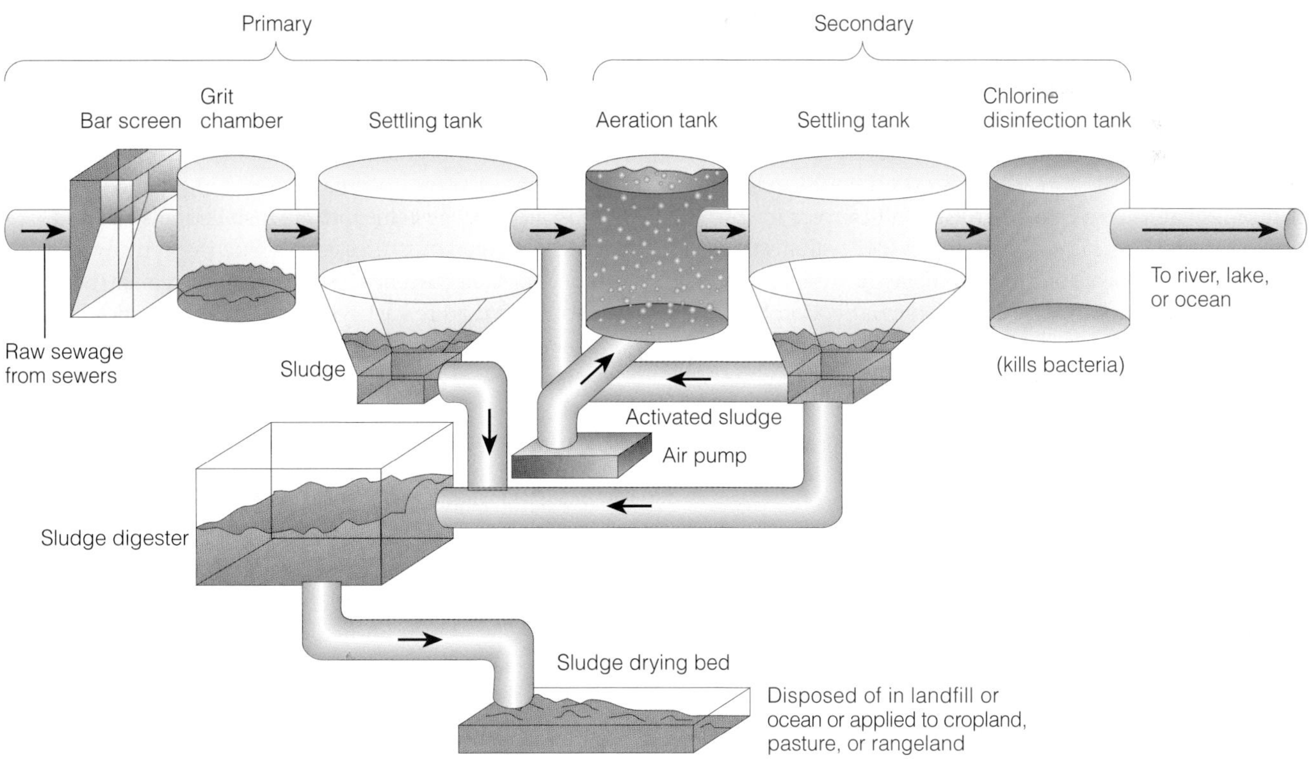

Figure 12-27 Primary and secondary sewage treatment.

Even after secondary treatment, wastewater still contains about 3–5% by weight of the original oxygen-demanding wastes, 3% of the suspended solids, 50% of the nitrogen (mostly as nitrates), 70% of the phosphorus (mostly as phosphates), and 30% of most toxic metal compounds and synthetic organic chemicals. Virtually no long-lived radioactive isotopes or persistent organic substances such as pesticides are removed.

As a result of the Clean Water Act, most U.S. cities have secondary sewage treatment plants. However, government studies have found that at least two-thirds of sewage treatment plants have either water quality or public health problems and most industries have violated water pollution regulations. Moreover, 500 cities have failed to meet federal standards for sewage treatment plants, and 34 East Coast cities simply screen out large floating objects from their sewage before discharging it into coastal waters. Another problem is that primary and secondary treatment do not remove most hazardous wastes.

Advanced sewage treatment is a series of specialized chemical and physical processes that remove specific pollutants left in the water after primary and secondary treatment (Figure 12-28). Advanced treatment is rarely used because such plants typically cost twice as much to build and four times as much to operate as secondary plants.

Before water is discharged after primary, secondary, or advanced treatment, it is bleached (to remove water coloration) and disinfected (to kill disease-carrying bacteria and some but not all viruses). The usual method for doing this is *chlorination*. However, chlorine can react with organic materials in water to form small amounts of chlorinated hydrocarbons, some of which cause cancers in test animals. According to some preliminary research in 1992, chlorinated drinking water may cause 7–10% of all cancers in the United States. There is also growing evidence that some chlorinated hydrocarbons may damage the human nervous, immune, and endocrine systems (Section 8-3). Other disinfectants such as ozone and ultraviolet light are used in some places, but they cost more than chlorination and are not as long-lasting.

Sewage treatment produces a toxic, gooey sludge that must be disposed of or recycled as fertilizer. About 36% by weight of all municipal sludge produced in the United States is applied to farmland, forests, highway medians, and degraded land as fertilizer, and 9% is composted. The rest is **(1)** dumped in conventional landfills (where it can contaminate groundwater) or **(2)** incinerated (which can pollute the air with traces of toxic chemicals. The resulting toxic incinerator ash usually is buried in landfills that EPA experts say will leak eventually.

Before it is applied to land, sewage sludge can be heated to kill harmful bacteria, as is done in Switzerland and parts of Germany. It can also be treated to remove toxic metals and organic chemicals before application. However, such treatment can be expensive and is not used for sludge applied to cropland in the United States. Untreated sludge can be applied to land not used for crops or livestock, such as forests, surface-mined land, golf courses, lawns, cemeteries, and highway medians.

Since 1992, when the U.S. Congress banned ocean dumping of sludge, its use as a fertilizer has increased sharply. However, most sewage sludge used as fertilizer in the United States is not adequately treated to kill harmful bacteria and remove toxic metals and organic chemicals, which may end up in food products or leach into groundwater. A growing number of health problems and lawsuits have resulted from use of sludge to fertilize crops.

Environmental scientist Peter Montague (Guest Essay, p. 78) believes that the entire sewage treatment approach should be redesigned with the goal of preventing toxic and hazardous waste from reaching sewage treatment plants. This would be accomplished by

- Requiring industries and businesses to remove all toxic and hazardous wastes from water sent to municipal sewage treatment plants.
- Encouraging or requiring industries to reduce or eliminate toxic chemical use and waste through clean production (Guest Essay, p. 78 and Solutions, p. 360).
- Encouraging the use of less harmful household chemicals (Table 14-2, p. 361) and having harmful chemicals picked up and disposed of safely on a weekly basis as part of garbage collection services.
- Having households, apartment buildings, and offices eliminate sewage outputs by switching to modern composting toilet systems that are installed, maintained, and managed by professionals. Such systems would be cheaper than current sewage systems because they wouldn't require vast systems of underground pipes connected to centralized sewage treatment plants and would conserve huge amounts of water.

How Can We Treat Sewage by Working with Nature? Some communities and individuals are seeking better ways to purify contaminated water by working with nature. A low-tech, low-cost alternative to expensive waste treatment plants is to create an artificial wetland, as the residents of Arcata, California, did, led by Humboldt State University professors Robert Gearheart and George Allen.

In this coastal town of 17,000, some 63 hectares (155 acres) of wetlands has been created between the town and the adjacent Humboldt Bay. The marshes, devel-

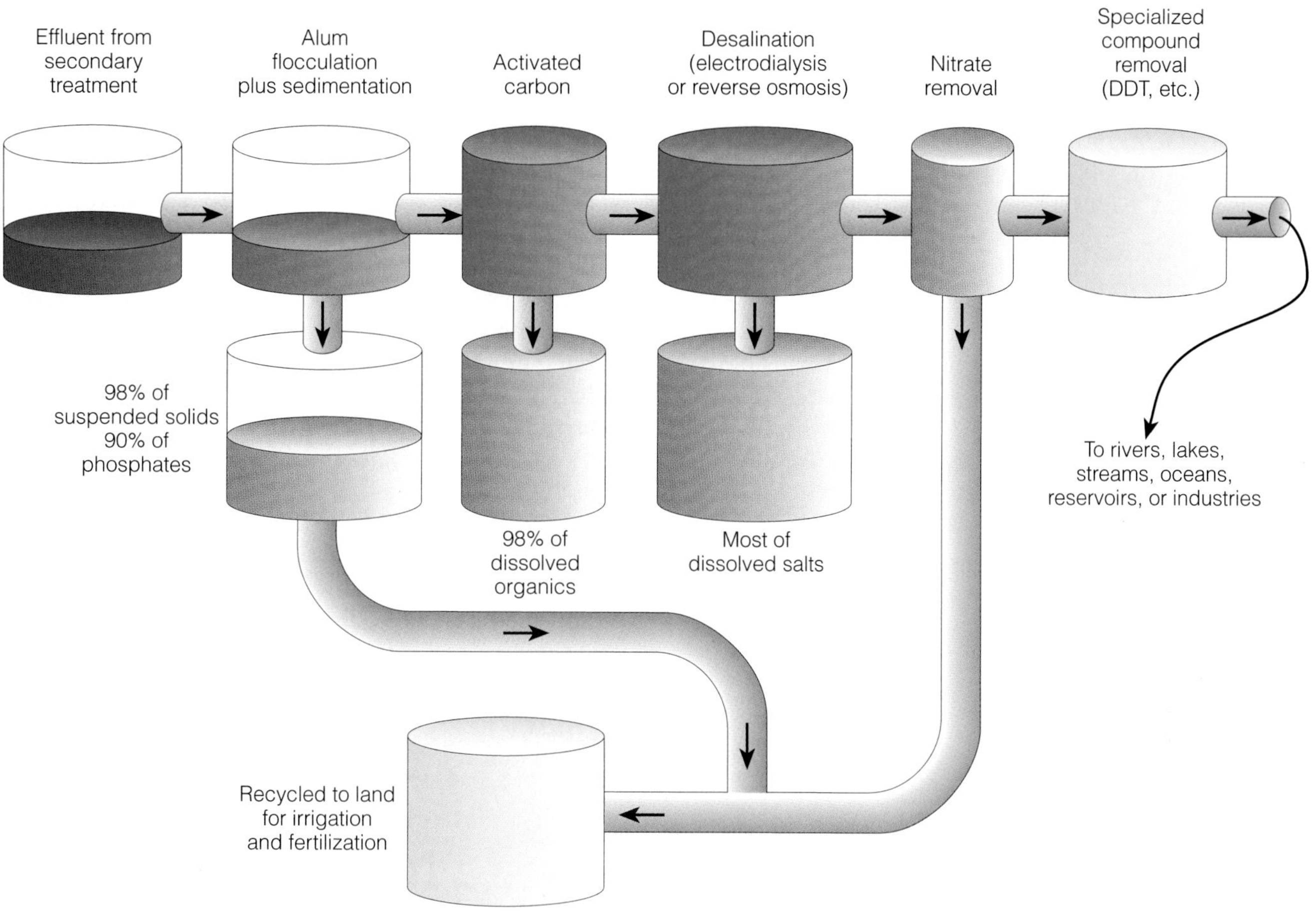

Figure 12-28 Advanced sewage treatment. Often only one (or two) of these processes is used to remove specific pollutants in a particular area. This expensive method is not widely used.

oped on land that was once a dump, act as an inexpensive, natural waste treatment plant. The project was completed in 1974 for less than half the estimated cost of a conventional treatment plant.

Here's how it works: First, sewage is held in sedimentation tanks, where the solids settle out as sludge that is removed and processed for use as fertilizer. The liquid is pumped into oxidation ponds, where remaining wastes are broken down by bacteria. After a month or so, the water is released into the artificial marshes, where it is further filtered and cleansed by plants and bacteria. Although the water is clean enough to be discharged directly into the bay, state law requires that it first be chlorinated. So the town chlorinates the water and then dechlorinates it before sending it into the bay, where oyster beds thrive.

The marshes and lagoons are an Audubon Society bird sanctuary and provide habitats for thousands of otters, seabirds, and marine animals; the treatment center is a city park and attracts more than 150,000 visitors a year. The town even celebrates its natural sewage treatment system with an annual "Flush with Pride" festival. Over 150 cities and towns in the United States now use natural and artificial wetlands for treating sewage. For example, Phoenix, Arizona, is creating wetlands to clean up a portion of its sewage. This is much cheaper than spending $625 million to upgrade its conventional wastewater treatment system.

How can natural processes be used for treating municipal sewage and industrial wastewater if there isn't a wetland available or enough land on which to create one? According to ecologist John Todd, you set up a greenhouse lagoon and use natural food chains and sunshine the way nature does (Figure 12-29). The purification process in such *living machines* begins when sewage flows into a greenhouse containing rows of large tanks full of aquatic plants such as water hyacinths, cattails, and bulrushes. In these tanks, algae and microorganisms decompose wastes into nutrients absorbed by the plants.

The decomposition is speeded up by sunlight streaming into the greenhouse. Then the water passes through an artificial marsh of sand, gravel, and bulrush plants, which filters out algae and organic waste. Next the water flows into aquarium tanks, where snails and zooplankton consume microorganisms and are in turn consumed by crayfish, tilapia, and other fish that can be eaten or sold as bait.

After 10 days, the now-clear water flows into a second artificial marsh for final filtering and cleansing. Todd's living machine purification systems now operate in 13 states around the United States and in seven other countries.

Another approach is to use wastewater to grow forests. The idea of planting forests over wastewater leach fields originated from Ed Burton about 25 years ago. This approach can remove carbon dioxide from the atmosphere while the trees are growing and thus help slow projected global warming.

Is the Water Safe to Drink? Many rivers in eastern Europe, Latin America, and Asia are used as sources of drinking water but are severely polluted, as are some rivers in more developed countries. Aquifers used as sources of drinking water in many developed countries and developing countries are becoming contaminated with pesticides, fertilizers, and hazardous organic chemicals.

About one-fourth of people in developing countries don't have access to clean drinking water. In China, an estimated 700 million people drink contaminated water. Of China's 27 largest cities, only 6 provide drinking water that meets government standards. Contaminated drinking water is considered a key factor in the doubling of liver and cancer deaths since 1970. In Russia, half of all tap water is unfit to drink, and a third of the aquifers are too contaminated for drinking purposes.

The United Nations estimates that it would cost about $25 billion a year to bring low-cost safe water and sanitation to the 1.4 billion people—one of every four—who don't have access to clean drinking water. These expenditures could prevent many of the 11 million deaths and 3.4 billion cases of illness caused each year by unsafe water. Currently, the world is spending only about $8 billion a year on clean water efforts. The $17-billion shortfall is about equal to what people in Europe and the United States spend each year on pet food or about what the world spends every 8 days for military purposes.

Figure 12-29 At the Providence, Rhode Island, Solar Sewage Plant, biologist John Todd demonstrates how ecological waste engineering in a greenhouse can be used to purify wastewater. Todd and others are conducting research to perfect solar–aquatic systems based on working with nature. (Ocean Arks International)

How Is Drinking Water Purified? Treatment of water for drinking by city dwellers is much like wastewater treatment. Areas that depend on surface water usually store it in a reservoir for several days to improve clarity and taste by allowing the dissolved oxygen content to increase and suspended matter to settle out. The water is then pumped to a purification plant, where it is treated to meet government drinking water standards. Usually the water is run through sand filters and activated charcoal before it is disinfected. In areas with very pure sources of groundwater, little treatment is necessary.

How Is the Quality of Drinking Water Protected? About 54 countries, most of them in North America and Europe, have safe drinking water standards. The U.S. Safe Drinking Water Act of 1974 requires the EPA to establish national drinking water standards, called *maximum contaminant levels*, for any pollutants that *may* have adverse effects on human health. This act has helped improve drinking water in much of the United States, but attempts to weaken this law continue. In addition, maximum contaminant levels have not been set for many potentially dangerous water pollutants such as certain synthetic organic compounds, radioactive materials, toxic metals, and pathogens.

Privately owned wells are not required to meet federal drinking water standards, primarily because of the costs of testing each well regularly (at least $1,000) and opposition to mandatory testing and compliance by some homeowners.

According to a 1994 study by the Natural Resources Defense Council (NRDC), in 1992 the drinking water of 50 million people—one in five Americans—violated one or more EPA pollutant standards. In most cases people were not notified (as required by the 1974 law) when their drinking water was contaminated. This study also estimated that contaminated drinking water is responsible for 7 million illnesses and 1,200 deaths per year in the United States.

Such dangers were revealed when residents of Milwaukee, Wisconsin, were told on April 7, 1993, that water from their taps was unsafe to drink. Before this health crisis was over, 104 people had died and 400,000 people had become sick as a result of exposure to *Cryptosporidium*, a parasitic organism. Milwaukee spent an estimated $54 million to deal with this outbreak. In May 1994, another outbreak of this parasite killed 19 and sickened more than 100 people in Las Vegas, Nevada.

According to the NRDC, U.S. drinking water supplies could be made safer at a cost of only about $30 a year per household. However, Congress is being pressured by water-polluting industries to weaken the Safe Drinking Water Act by **(1)** eliminating national tests of drinking water, **(2)** eliminating the requirement that the media be advised of emergency water health violations and that water system officials notify their customers of such violations, **(3)** allowing states to give drinking water systems a permanent right to violate the standard for a given contaminant if the provider claims it can't afford to comply, and **(4)** eliminating the requirement that water systems use affordable, feasible technology to remove cancer-causing contaminants.

Environmentalists call for the U.S. Safe Drinking Water Act to be strengthened by **(1)** improving treatment by combining at least half of the 50,000 water systems that serve fewer than 3,300 people each with larger ones nearby, **(2)** strengthening and enforcing public notification requirements about violations of drinking water standards, and **(3)** banning all lead in new plumbing pipes, faucets, and fixtures (current law allows fixtures with up to 10% lead to be sold as lead-free).

Is Bottled Water the Answer? The United States has one of the world's best drinking water supply systems. Yet about half of all Americans worry about getting sick from tap water contaminants and many either drink bottled water or install expensive water-purification systems. Studies indicate that many of these consumers are being ripped off and in some cases may end up drinking water that is dirtier than they can get from their taps. An estimated one-third of the bottled water purchased in the United States is contaminated with bacteria.

To be safe, consumers purchasing bottled water should determine whether the bottler belongs to the International Bottled Water Association (IBWA) and adheres to its testing requirements.* Some companies pay $2,500 annually to obtain more stringent certification by the National Sanitation Foundation, an independent agency that tests for 200 chemical and biological contaminants.

Before drinking expensive bottled water and buying costly home water purifiers, health officials suggest that consumers have their water tested by local health authorities or private labs to identify what contaminants, if any, must be removed; then they should buy a unit that does the required job. Independent experts contend that unless tests show otherwise, for most urban and suburban Americans served by large municipal drinking water systems, home water treatment systems aren't worth the expense and maintenance hassles.

Buyers should be suspicious of door-to-door salespeople, telephone appeals, scare tactics, and companies offering free water tests (which often are neither accurate nor carried out by certified labs). They should carefully check out companies selling such equipment and demand a copy of purifying claims by EPA-certified laboratories. Buyers should also be wary of claims that a treatment device has been *approved* by the EPA. Although the EPA does *register* such devices, it neither tests nor approves them.

Solutions: How Can We Use Water Resources More Sustainably? According to water resource expert Sandra Postel, none of the technological solutions for supplying more water—dams, watershed transfers, tapping groundwater, and using water more efficiently—deal with the three underlying forces that can lead humans to use such a potentially renewable resource in unsustainable ways: **(1)** depletion or degradation of a shared resource, which shrinks the resource pie to be shared (the tragedy of the commons), **(2)** population growth, which forces the resource pie to be divided into smaller slices, and **(3)** unequal distribution or access (primarily because of poverty), which means that some countries and people get larger slices than others.

Sustainable water use is based on the commonsense principle stated in an old Inca proverb: "The frog does not drink up the pond in which it lives."

Sustainable use of potentially renewable groundwater means that the rate of extraction should not exceed the rate of recharge. Sustainable water use also requires an integrated plan governing water use, sewage treatment, and water pollution among all users of a water basin (as is done in Great Britain).

Many analysts believe that a key to reducing water waste and providing more equitable access to water resources is for governments to phase out subsidies

*Check for the IBWA seal of approval on the bottle or contact the International Bottled Water Association (113 North Henry Street, Alexandria, VA 22314; phone: 703-683-5213) for a member list.

that reduce the price of water for those benefiting from such subsidies.

According to environmentalists, a sustainable approach to dealing with water pollution requires that we shift our emphasis from pollution cleanup to pollution prevention. This involves **(1)** *reducing* the toxicity or volume of pollutants (for example, replacing organic solvent-based inks and paints with water-based materials), **(2)** *reusing* wastewater instead of discharging it (for example, reusing treated wastewater for irrigation), and **(3)** *recycling* pollutants (for example, cleaning up and recycling contaminated solvents for reuse) instead of discharging them.

To make such a shift, we need to accept that the environment—air, water, soil, life—is an interconnected whole. Without an integrated approach to all forms of pollution, environmentalists argue that we will continue to shift environmental problems from one part of the environment to another. Some actions you can take to help reduce water pollution are listed in Appendix 6.

It is not until the well runs dry that we know the worth of water.

BENJAMIN FRANKLIN

REVIEW QUESTIONS

1. Define the boldfaced terms used in this chapter.

2. Explain why there is a danger of water wars in the Middle East.

3. List seven unique properties of water and explain the importance of each property.

4. What percentage of the earth's total volume of water is available for use by us?

5. Distinguish between *surface runoff, watershed, groundwater, zone of saturation, water table, aquifer, recharge area,* and *natural recharge.* Explain how the water in some aquifers can be depleted. What percentage of global surface runoff do humans use? About what percentage of the water we withdraw each year is used for **(a)** irrigation, **(b)** industry, and **(c)** residences and cities? What are the major water uses and problems of **(a)** the Eastern United States and **(b)** the Western United States?

6. List four causes of water scarcity. About how many people in the world face chronic water shortages? How many people are expected to face such shortages by 2050?

7. List five ways to increase the supply of fresh water in a particular area. List the major pros and cons of building large dams and reservoirs to supply fresh water.

8. List the major pros and cons of supplying water by transferring it from one watershed to another. Summarize the problems with this approach for **(a)** California, **(b)** the James Bay project in Canada, and **(c)** the Aral Sea in Central Asia.

9. List the pros and cons of supplying more water by withdrawing groundwater. Summarize the problems with this approach in the Ogallala Aquifer in the United States. List four ways for preventing or slowing groundwater depletion.

10. List the pros and cons of increasing supplies of freshwater by **(a)** desalination of salt water, **(b)** cloud seeding, and **(c)** towing icebergs to water-short areas.

11. What percentage of the water used by people throughout the world is wasted? List six benefits of conserving water. What are two major reasons why we waste so much water?

12. How can we waste less water in **(a)** irrigation and **(b)** industry, homes, and businesses?

13. What is a *floodplain* and why do so many people live on floodplains? List three advantages of floods. What are the major disadvantages of floods? Explain three ways that humans increase the damages from floods. Describe the nature and causes of the flooding problems in Bangladesh. List the pros and cons of trying to reduce flood risks by **(a)** stream channelization, **(b)** building artificial levees and embankments, **(c)** building dams, and **(d)** floodplain management.

14. What is *water pollution*? What are nine types of water pollutants and what are the major effects of each type? Distinguish between *point* and *nonpoint sources of water pollution* and give examples of each type. Which type is easier to control? Why?

15. What are the major water pollution problems of streams? Summarize the good news and the bad news about attempts to prevent or control stream pollution.

16. What are the major water pollution problems of lakes? Summarize the good news and the bad news about attempts to prevent or control lake pollution. Distinguish between *eutrophication* and *cultural eutrophication.* What are the major causes of cultural eutrophication? List three methods for **(a)** preventing cultural eutrophication and **(b)** cleaning up cultural eutrophication. List three reasons why Seattle's Lake Washington was able to recover from cultural eutro-phication. Give the good news and the bad news about attempts to reduce water pollution in the Great Lakes. Describe three types of genetic pollution affecting the Great Lakes.

17. List four reasons why groundwater pollution is such a serious problem. List five major sources of groundwater contamination. List five ways of preventing groundwater contamination.

18. List the major pollution problems of the oceans. Why are most of these problems found in coastal areas? Summarize the major pollution problems of the Chesapeake Bay in the United States. Distinguish between ocean pollution from dredge spoils and from sewage sludge. Distinguish between *crude petroleum* and *refined petroleum* and summarize the major water pollution problems caused by oil. What are the three major sources of oil pollution in the world's oceans? Summarize the causes and effects of the oil spill from the *Exxon Valdez* tanker in Alaska's Prince William Sound. List seven ways to help prevent oil pollution of the oceans and two ways to help clean up such pollution.

19. List seven ways to help prevent water pollution from nonpoint sources. Why has there been so little emphasis on dealing with this problem? Explain how the United States (and most developed countries) have reduced water pollution from point sources by enacting laws, and summarize the good and bad news about such efforts. List five ways in which environmentalists believe water pollution control laws in the United States should be strengthened.

20. Distinguish between *septic tanks, primary sewage treatment, secondary sewage treatment,* and *advanced sewage treatment* as ways to reduce water pollution. List ways to deal with the sludge produced by waste treatment methods. What are the pros and cons of each approach? List three ways to treat sewage based on working with nature. Describe the approach used to treat sewage **(a)** in Arcata, California, and **(b)** by using John Todd's living machines.

21. What percentage of the world's people don't have access to clean drinking water? How is drinking water purified? How is the quality of drinking water protected in the United States? How successful have these efforts been? List three ways in which environmentalists believe the U.S. Safe Drinking Water Act should be strengthened. List the pros and cons of drinking bottled water.

22. List five ways of using the world's water more sustainably.

CRITICAL THINKING

1. How do human activities increase the harmful effects of prolonged drought? How can we reduce these effects?

2. Explain how dams and reservoirs can cause more flood damage than they prevent. Should all proposed large dam and reservoir projects be scrapped? Explain.

3. Should federal subsidies of irrigation projects in the Western United States (or in the country where you live) be phased out? Explain.

4. Should the prices of water for all uses be raised sharply to include more of its environmental costs and to encourage water conservation? Explain. What harmful and beneficial effects might this have on **(a)** business and jobs, **(b)** your lifestyle and the lifestyles of any children or grandchildren you might have, **(c)** the poor, and the **(d)** environment?

5. List five major ways to conserve water on a personal level (see Appendix 6). Which, if any, of these practices do you now use or intend to use?

6. How do human activities contribute to flooding and flood damage? How can these effects be reduced?

7. Some analysts argue that the U.S. Federal Flood Disaster Protection Act should be repealed because it encourages people to build on areas with a high risk of flooding by providing them with government-backed insurance (paid for mostly by people who not do live in high flood-risk areas) to rebuild after a flood. Instead, they would replace it with an act that provides emergency relief to victims of a disastrous flood but does not help flood victims rebuild in such risky areas. Do you agree or disagree with this proposal? Explain. Would your position be the same if you lived in an area prone to flooding?

8. Why is dilution not always the solution to water pollution? Give examples and conditions for which this solution is, and is not, applicable.

9. How can a stream cleanse itself of oxygen-demanding wastes? Under what conditions will this natural cleansing system fail?

10. Should all dumping of wastes in the ocean be banned? Explain. If so, where would you put the wastes instead? What exceptions would you permit, and why?

11. Your town (Town B) is located on a river between towns A and C. What are the rights and responsibilities of upstream communities to downstream communities? Should sewage and industrial wastes be dumped at the upstream end of a community that generates them?

***12.** In your community,

- **a.** What are the major sources of the water supply?
- **b.** How is water use divided among agricultural, industrial, power-plant cooling, and public uses? Who are the biggest consumers of water?
- **c.** What has happened to water prices during the past 20 years? Are they too low to encourage water conservation and reuse?
- **d.** What water supply problems are projected?
- **e.** How is water being wasted?

***13.** In your community,

- **a.** What are the principal nonpoint sources of contamination of surface water and groundwater?
- **b.** What is the source of drinking water?
- **c.** How is drinking water treated?
- **d.** Is groundwater contamination a problem? If so, where, and what has been done about the problem?

***14.** Develop a water conservation plan for your school and submit it to school officials.

***15.** Make a concept map of this chapter's major ideas, using the section heads and subheads and the key terms (in boldface). Look at the inside back cover and in Appendix 5 for information about concept maps.

FURTHER INFORMATION

For more information see Internet Sources and Further Readings at the back of the book. Online resources for this chapter can be found at **www.brookscole.com/biology**

13 MINERAL AND SOIL RESOURCES

The Great Terrain Robbery

Want to get rich at the taxpayers' expense? You can if you know how to make use of a little known mining law passed in 1872 to encourage mining of gold, silver, lead, copper, uranium, and other hard-rock minerals on U.S. public lands.

Under this 1872 law, a person or corporation can assume legal ownership of any public land not classified as wilderness or park simply by *patenting* it. This involves **(1)** declaring their belief that the land contains valuable hard-rock minerals, **(2)** spending $500 to improve the land for mineral development, **(3)** filing a claim, and **(4)** paying the federal government $6–12 per hectare ($2.50–5.00 an acre) for the land. So far public lands containing at least $385 billion of the public's mineral resources have been transferred to private interests at these bargain-basement prices.

Figure 13-1 Bear Trap Creek in Montana is one example of how gold mining can contaminate water with highly toxic cyanide or mercury used to extract gold from its ore. Air and water also convert the sulfur in gold ore to sulfuric acid, which releases toxic metals such as cadmium and copper into streams and groundwater. (Bryan Peterson)

In 1993 the Manville Corporation paid $10,000 for federal land in Montana that contains an estimated $32 billion of platinum and palladium. Secretary of Interior Bruce Babbitt was forced by the courts in 1994 to sell to a Canadian corporation for only $10,000 federal land in Nevada containing an estimated $10–15 billion worth of gold. In 1995 he had to sign over federal land in Idaho containing $1 billion worth of minerals to a Danish company that paid just $275 for the bonanza.

Domestic and foreign mining companies operating under this law remove mineral resources worth about $2–3 billion on once public land they have bought at absurdly low prices. They pay no royalties to the U.S. Treasury.

The 1872 law does not even require that patented property be mined. Land speculators have often purchased such property at 1872 prices and then sold it or leased it for thousands of times what they paid. In 1986 a mining company paid the government $42,500 for oil shale land in Colorado and a month later sold it to Shell Oil for $37 million.

Other owners have developed sites as casinos, ski resorts, golf courses, and vacation home developments. According to former senator Dale Bumpers, "The 1872 mining law is a license to steal and the biggest scam in America." This government subsidy to mining companies costs U.S. taxpayers about $200 million a year.

There is also no provision in the 1872 law requiring reclamation of damaged land. Acids and toxic metal residues continue to leach from thousands of the almost 558,000 abandoned hard-rock mines and open pits, mostly in the West. One result is some 19,000 kilometers (12,000 miles) of polluted streams (Figure 13-1); another result is that 56 abandoned mine sites are now listed on the EPA's Superfund list of the nation's worst hazardous waste dump sites.

It's common for a company to mine a site, abandon it, file for bankruptcy, and leave the public with the cleanup bill. It is estimated that the taxpayers' cleanup costs for all land damaged by hard-rock mining on existing or sold public lands will be $33–72 billion, depending on whether groundwater and toxic waste cleanup are included. Environmentalists have been trying, without success, to have this law revised to protect taxpayers and the environment.

Below that thin layer comprising the delicate organism known as the soil is a planet as lifeless as the moon.

G. Y. JACKS AND R. O. WHYTE

This chapter addresses the following questions:

- What are the major geologic processes occurring on and in the earth?
- How does the rock cycle recycle earth materials and concentrate mineral resources?
- How fast are nonfuel mineral supplies being used up, and how can we increase supplies of key minerals?
- What are the environmental effects of extracting and using mineral resources?
- What is soil, and what types are best for growing crops?
- Why should we worry about soil erosion?
- How can we control erosion and reduce the loss of nutrients from topsoil?

13-1 GEOLOGIC PROCESSES*

What Is the Earth's Structure? As the primitive earth cooled over eons, its interior separated into three major concentric zones, which geologists identify as the core, the mantle, and the crust (Figure 4-5). Various indirect measurements indicate that the earth's innermost zone, the **core**, is made mostly of iron (with perhaps some nickel). The core has a solid inner part, surrounded by a liquid core of molten material.

The earth's core is surrounded by a thick, solid zone called the **mantle**. This largest zone of the earth's interior is rich in iron (its major constituent), silicon, oxygen, and magnesium. Most of the mantle is solid rock, but under its rigid outermost part there is a zone of very hot, partly melted rock that flows like soft plastic. This plastic region of the mantle is called the *asthenosphere*.

The outermost and thinnest zone of the earth is called the **crust**. It consists of **(1)** the *continental crust*, which underlies the continents (including the continental shelves extending into the oceans), and **(2)** the *oceanic crust*, which underlies the ocean basins and covers 71% of the earth's surface (Figure 13-2).

What Is Plate Tectonics? Internal Geologic Processes We tend to think of the earth's crust, mantle, and core as fairly static and unchanging. However, these parts of the earth are constantly changing because of geologic processes taking place within the earth and on the earth's surface, most over thousands to millions of years.

A map of the earth's earthquakes and volcanoes shows that most of these phenomena occur along certain lines or belts on the earth's surface (Figure 13-3a). The areas of the earth outlined by these major belts are called **plates** (Figure 13-3b). They are about 100 kilometers

*__Kenneth J. Van Dellen__, professor emeritus of geology and environmental science, Macomb Community College, is the primary author of this and the next section, with assistance from G. Tyler Miller, Jr.

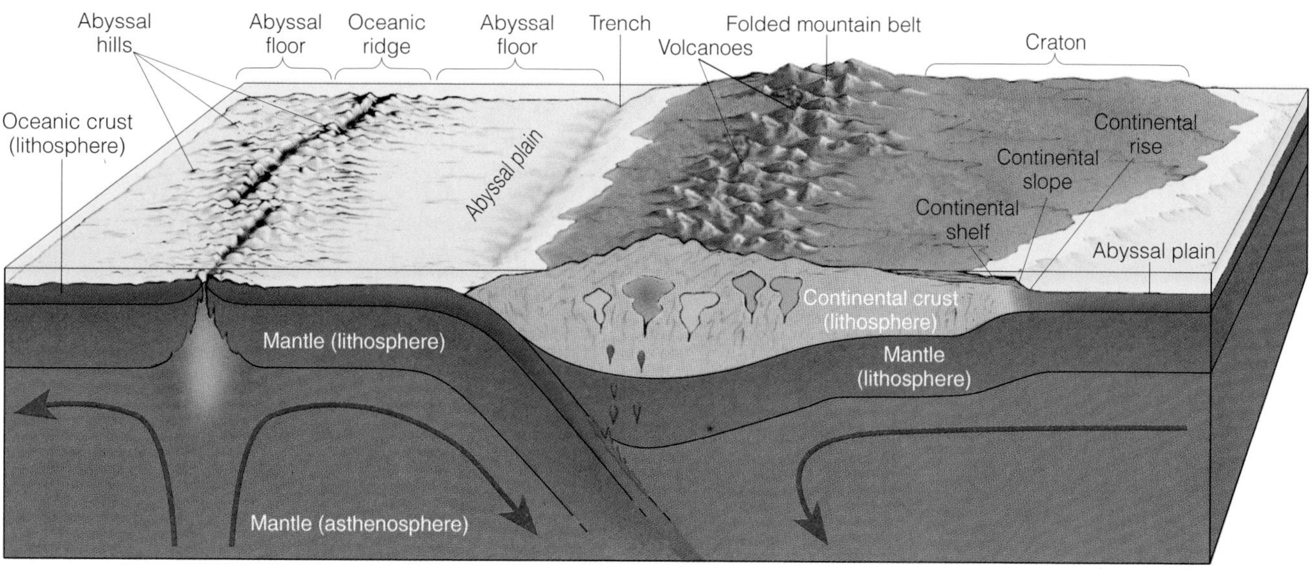

Figure 13-2 Major features of the earth's crust and upper mantle. The lithosphere, composed of the crust and outermost mantle, is rigid and brittle. The asthenosphere, a zone in the mantle, can be deformed by heat and pressure (that is, it is plastic).

(60 miles) thick and are composed of the crust and the rigid, outermost part of the mantle (above the asthenosphere)—a combination called the **lithosphere**. These plates move constantly, supported by the slowly flowing asthenosphere like large pieces of ice floating on the surface of a lake during the spring breakup. Some plates move faster than others, but a typical speed is about the rate at which fingernails grow.

The theory explaining the movements of the plates and the processes that occur at their boundaries is called **plate tectonics**. The concept, which became widely accepted by geologists in the 1960s, was developed from an earlier idea called *continental drift*. Throughout the earth's history, continents have split and joined as plates have drifted thousands of kilometers back and forth across the planet's surface (Figure 5-6).

Plate motion produces mountains (including volcanoes), the oceanic ridge system, trenches, and other features of the earth's surface (Figure 13-2). Natural hazards such as volcanoes and earthquakes are likely to be found at plate boundaries (Figure 13-3a), and plate movements and interactions concentrate many of the minerals we extract and use.

The theory of plate tectonics also helps explain how certain patterns of biological evolution occurred. By reconstructing the course of continental drift over millions of years (Figure 5-6), we can trace how life-forms migrated from one area to another when continents that are now far apart were still joined together. As the continents separated, speciation occurred as populations became geographically and reproductively isolated (Figure 5-5).

Lithospheric plates have three types of boundaries (Figure 13-4). At a **divergent plate boundary** the plates move apart in opposite directions (Figure 13-4, top), and at a **convergent plate boundary** they are pushed together by internal forces (Figures 13-3b and 13-4, middle). At most convergent plate boundaries, oceanic lithosphere is carried downward (subducted) under the island arc or the continent at a **subduction zone**. A trench ordinarily forms at the boundary between the two converging plates (Figure 13-4, middle). Stresses in the plate undergoing subduction cause earthquakes at convergent plate boundaries.

The third type of plate boundary, called a **transform fault**, occurs where plates move in opposite but parallel directions along a fracture (fault) in the lithosphere (Figure 13-4, bottom). In other words, the plates slide past one another. Like the other types of plate boundaries, most transform faults are on the ocean floor.

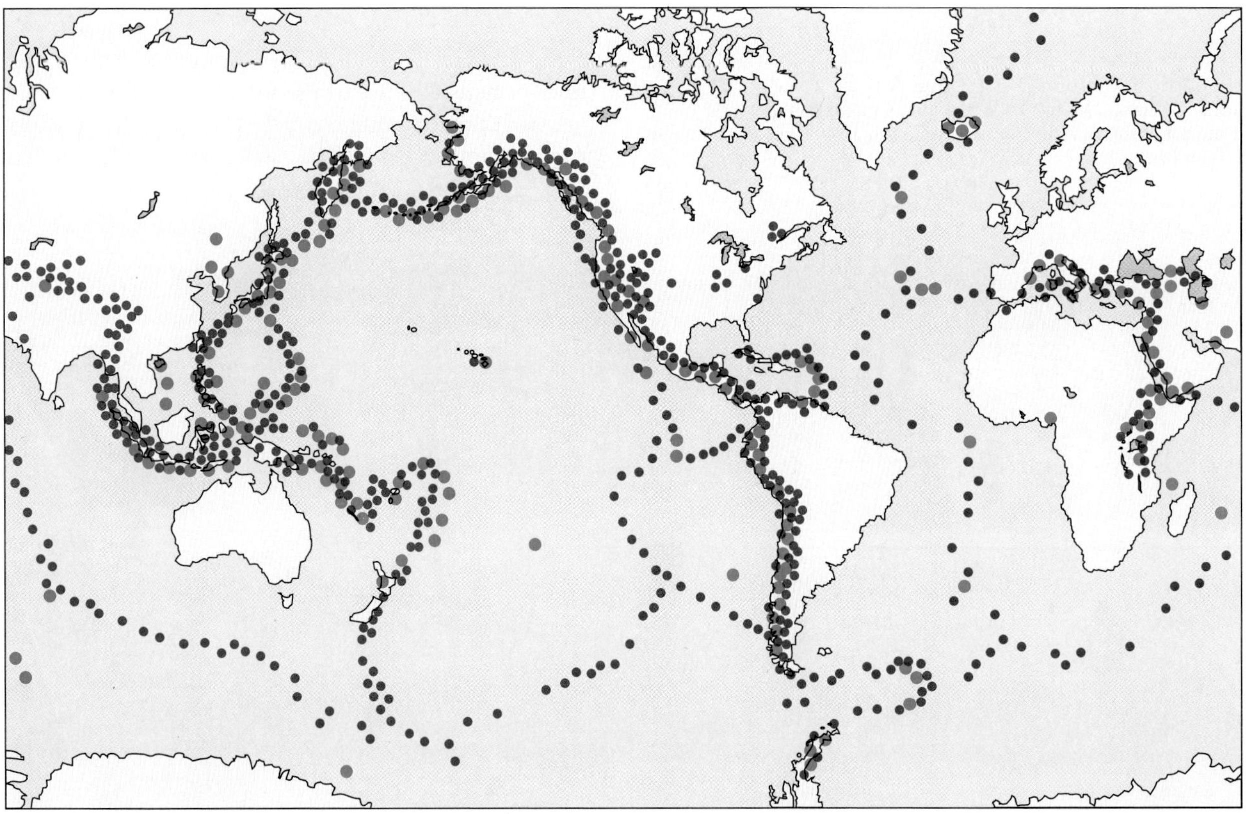

Figure 13-3 **(a)** Earthquake and volcano sites are distributed mostly in bands along the planet's surface **(b)** These bands correspond to the patterns for the types of lithospheric plate boundaries shown in Figure 13-4.

What Geologic Processes Occur on the Earth's Surface? Geological changes based directly or indirectly on energy from the sun and on gravity (rather than on heat in the earth's interior) are called *external processes*. Whereas internal processes generally build up the earth's surface, external processes tend to wear it down.

A major external process is **erosion.** It is the process by which material is dissolved, loosened, or worn away from one part of the earth's surface and deposited in other places. Streams, the most important agent of erosion, operate everywhere on the earth except in the polar regions (Figure 13-5). They produce ordinary valleys and canyons, and they may form deltas where streams flow into lakes and oceans (Figure 6-23). Some erosion is also caused when wind blows particles of soil from one area to another (Figure 6-1). Human activities, particularly those that destroy vegetation, accelerate erosion, as discussed in Section 13-6.

13-2 MINERAL RESOURCES AND THE ROCK CYCLE

How Are the Earth's Rocks Recycled? A **mineral** is an element or inorganic compound that occurs naturally and is solid. It usually has a crystalline internal structure made up of an orderly, three-dimensional arrangement of atoms or ions. **Rock** is any material that makes up a large, natural, continuous part of the earth's crust. Geologic processes constantly redistribute the chemical elements within and at the earth's surface. Based on the way it forms, rock is placed in three broad classes: igneous, sedimentary, and metamorphic.

Igneous rock can form below or on the earth's surface when molten rock material (magma) wells up from the earth's upper mantle or deep crust, cools, and hardens into rock. Examples are granite (formed underground) and lava rock (formed above ground when molten lava cools and hardens). Although often covered by sedimentary rocks or soil, igneous rocks form the bulk of the earth's crust. They also are the main source of many nonfuel mineral resources.

Sedimentary rock forms from sediment. Most such rocks are formed when preexisting rocks are weathered and eroded into small pieces, transported from their sources, and deposited in a body of surface water. As these deposited layers become buried and compacted, the resulting pressure causes their particles to bond together to form sedimentary rocks such as

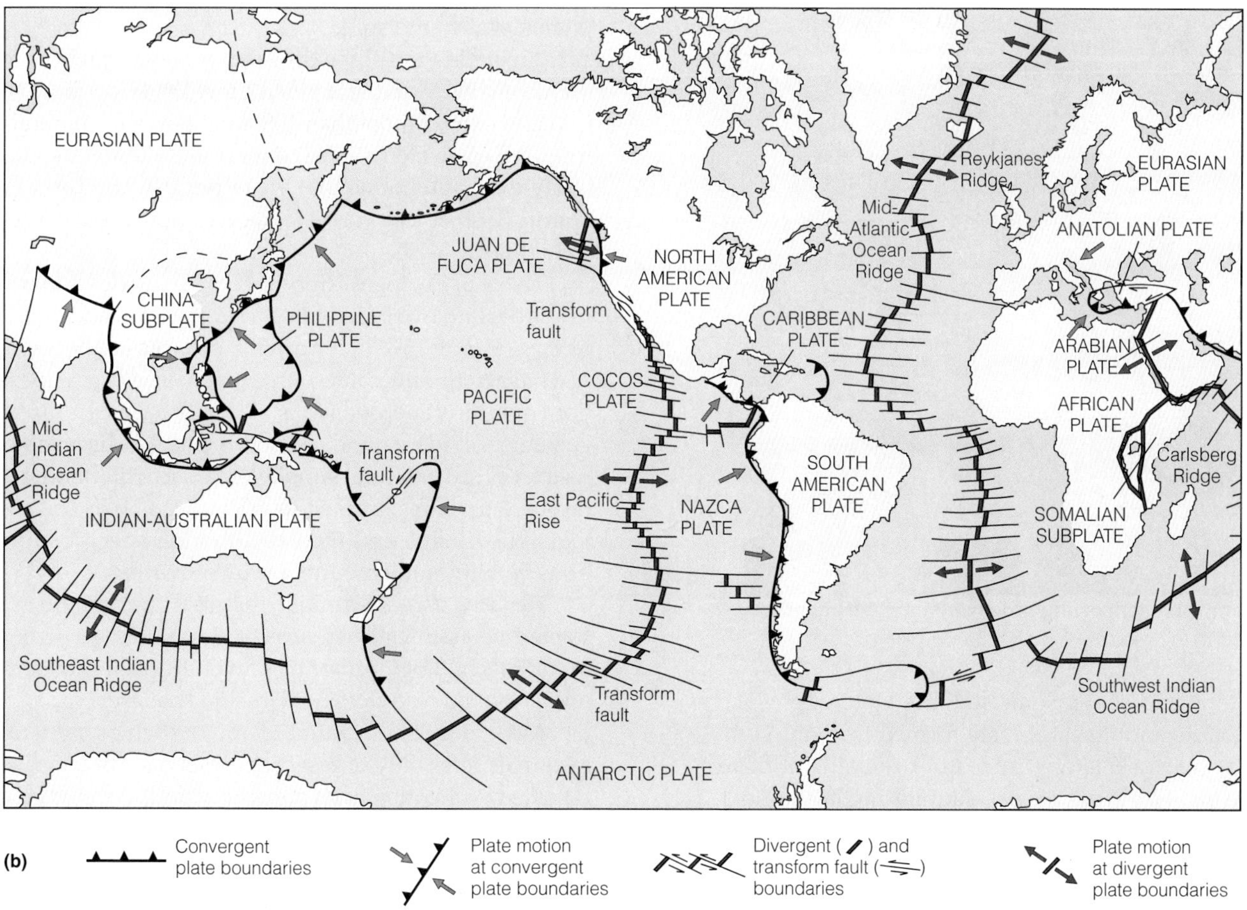

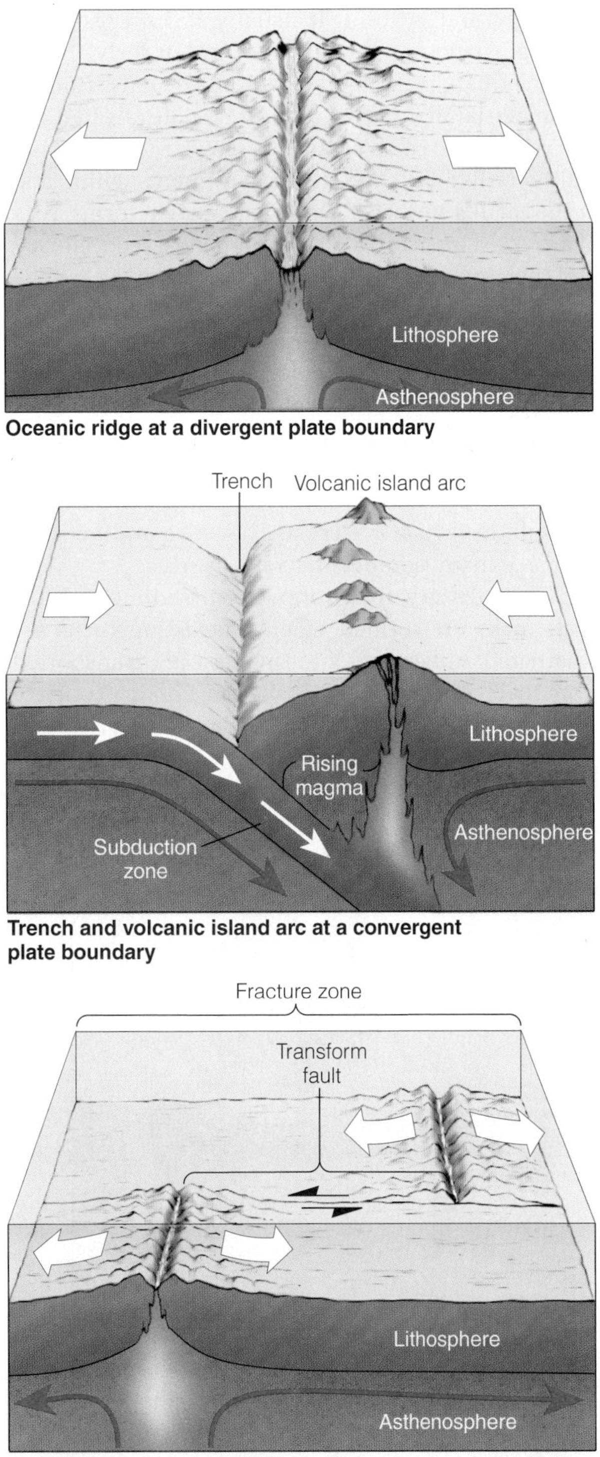

Oceanic ridge at a divergent plate boundary

Trench and volcanic island arc at a convergent plate boundary

Transform fault connecting two divergent plate boundaries

Figure 13-4 Types of boundaries between the earth's lithospheric plates. All three boundaries occur both in oceans and on continents.

sandstone and shale. Some sedimentary rocks, such as dolomite and limestone, are formed from the compacted shells, skeletons, and other remains of dead organisms. Two types of coal—lignite and bituminous coal—are sedimentary rocks derived from plant remains.

Metamorphic rock is produced when a preexisting rock is subjected to high temperatures (which may cause it to melt partially), high pressures, chemically active fluids, or a combination of these agents. Examples are anthracite (a form of coal), slate, and marble.

Rocks are constantly exposed to various physical and chemical conditions that can change them over time. The interaction of processes that change rocks from one type to another is called the **rock cycle** (Figure 13-6). The slowest of the earth's cyclic processes recycles material over millions of years and is responsible for concentrating the planet's nonrenewable mineral resources on which humans depend.

What Are Mineral Resources? A **mineral resource** is a concentration of naturally occurring solid, liquid, or gaseous material in or on the earth's crust that can be extracted and processed into useful materials at an affordable cost. The earth's internal and external processes have produced numerous mineral resources, which on a human time scale are essentially nonrenewable because of the slowness of the rock cycle. Mineral resources include **(1)** *energy resources* (coal, oil, natural gas, uranium, geothermal energy), **(2)** *metallic mineral resources* (iron, copper, aluminum), and **(3)** *nonmetallic mineral resources* (salt, gypsum, clay, sand, phosphates, water, and soil).

An **ore** is a metal-yielding material that can be economically extracted at a given time. We know how to find and extract more than 100 nonrenewable minerals from the earth's crust. We convert these minerals into many everyday items that we either use and discard (Figure 3-14) or learn to reuse, recycle, or use less wastefully (Figure 3-15).

The U.S. Geological Survey (USGS) divides mineral resources into two broad categories: *identified* and *undiscovered* (Figure 13-7). **Identified resources** are deposits of a particular mineral resource that have a *known* location, quantity, and quality, or deposits based on direct geological evidence and measurements. **Undiscovered resources** are potential supplies of a particular mineral resource that are assumed to exist on the basis of geologic knowledge and theory (although specific locations, quality, and amounts are unknown).

Reserves are identified resources that can be extracted economically at current prices using current mining technology. **Other resources** are identified and undiscovered resources not classified as reserves.

Most published estimates of particular mineral resources refer only to reserves. Reserves can increase **(1)** when exploration finds previously undiscovered economic-grade mineral resources or **(2)** when identified subeconomic-grade mineral resources become economically viable because of new technology or higher prices. Figure 13-7 shows a region labeled *potential*

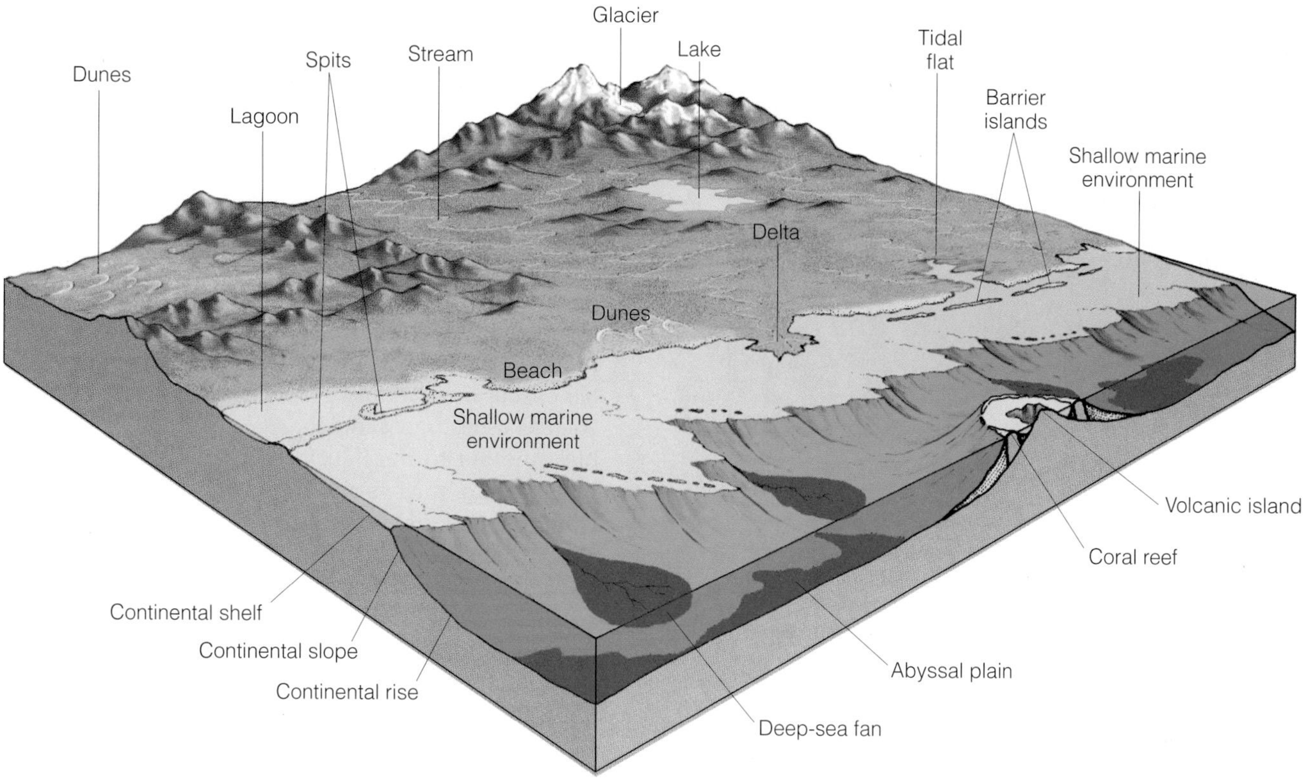

Figure 13-5 The variety of landforms and sedimentary environments depicted here is mainly the result of *external processes*, powered primarily by solar energy (as it drives the hydrologic cycle and wind) and gravity, with some assistance from organisms such as reef-building corals.

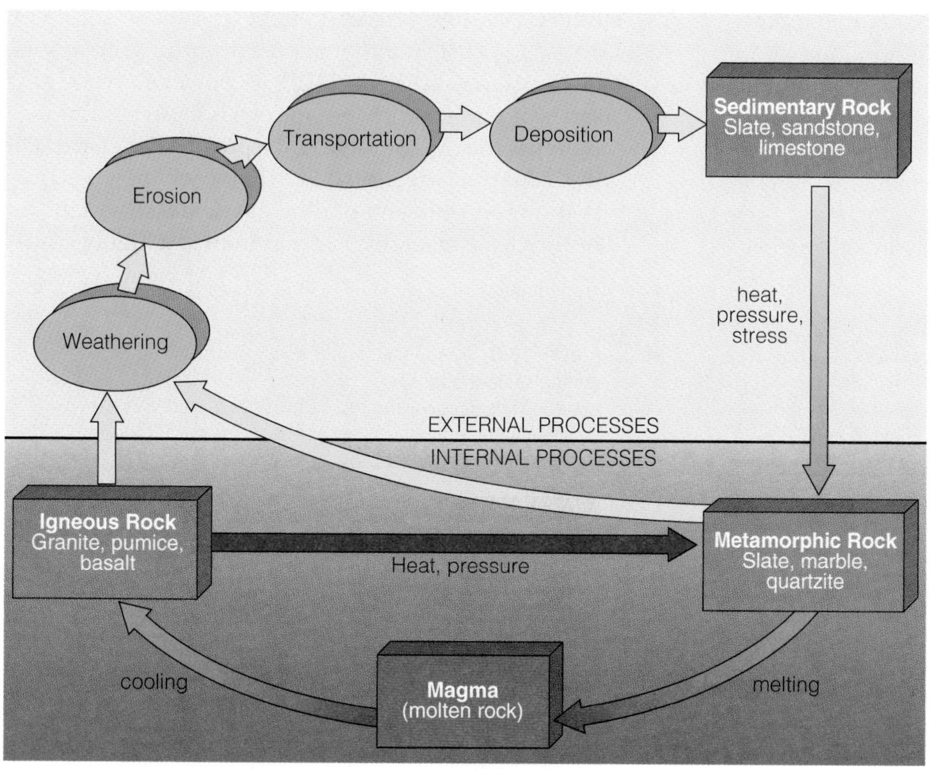

Figure 13-6 The rock cycle, the slowest of the earth's cyclic processes. The earth's materials are recycled over millions of years by three processes: melting, erosion, and metamorphism, which produce igneous, sedimentary, and metamorphic rocks. Rock of any of the three classes can be converted to rock of either of the other two classes (or can even be recycled within its own class).

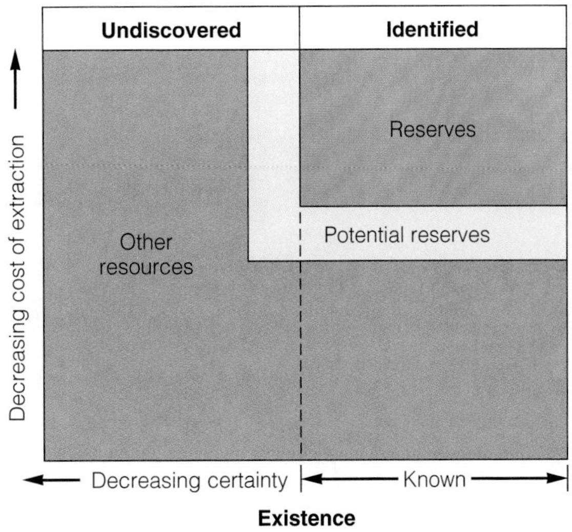

Figure 13-7 General classification of mineral resources. (The area shown for each class does not represent its relative abundance.) In theory, all mineral resources classified as *other resources* could become reserves because of rising mineral prices or improved mineral location and extraction technology. In practice, geologists expect only a fraction of other resources to become reserves. The area labeled *potential reserves* shows the way reserves normally increase.

reserves, indicating how resources can be expanded. Theoretically, all of the *other resources* could eventually be converted to reserves, but this is highly unlikely.

How Do We Find and Remove Mineral Deposits?

Mining companies use several methods to find promising mineral deposits. Geological information about plate tectonics and mineral formation suggests areas worthy of closer study. Aerial photos and satellite images sometimes reveal rock formations associated with certain minerals. Other instruments on aircraft and satellites can detect mineral deposits by their effects on the earth's magnetic or gravitational fields.

After profitable deposits of minerals are located, deep deposits are removed by **subsurface mining** and shallow deposits by **surface mining**. In surface mining, mechanized equipment strips away the **overburden** of soil and rock and usually discards it as waste material called **spoil**. Surface mining extracts about 90% of the mineral and rock resources and more than 60% of the coal by weight in the United States.

The type of surface mining used depends on the resource being sought and on local topography. In **open-pit mining** (Figure 13-8), machines dig holes and remove ores such as iron and copper. This method is also used for sand and gravel and for building stone such as limestone, sandstone, slate, granite, and marble. Another form of surface mining is **dredging**, in which chain buckets and draglines scrape up underwater mineral deposits.

Strip mining is surface mining in which bulldozers, power shovels, or stripping wheels remove the overburden in strips. It is used mostly for extracting coal and some phosphate rock. Most surface-mined coal is removed by area strip mining or contour strip mining, depending on the terrain.

Area strip mining is used where the terrain is fairly flat. An earthmover strips away the overburden and a power shovel digs a cut to remove the mineral deposit, such as coal. After the mineral is removed, the trench is filled with overburden and a new cut is made parallel to the previous one. The process is repeated over the entire site. If the land is not restored, area strip mining leaves a wavy series of highly erodible hills of rubble called *spoil banks* (Figure 13-9).

Contour strip mining is used on hilly or mountainous terrain (Figure 13-10). A power shovel cuts a series of terraces into the side of a hill. An earthmover removes the overburden, and a power shovel extracts the coal; the overburden from each new terrace is dumped onto the one below. Unless the land is restored, a wall of dirt is left in front of a highly erodi-

Figure 13-8 This open-pit copper mine in Bingham, Utah, is the largest human-made hole in the world, 4.0 kilometers (2.5 miles) in diameter and 0.8 kilometer (0.5 mile) deep. The amount of material removed from this mine is seven times the amount moved to build the Panama Canal. (Don Green/Kennecott Copper Corporation, now owned by British Petroleum)

Figure 13-9 Spoil banks, the results of unrestored area strip mining of coal near Mulla, Colorado. Restoration of newly strip-mined areas is now required in the United States, but many previously mined areas have not been restored. (National Archives/EPA Documerica)

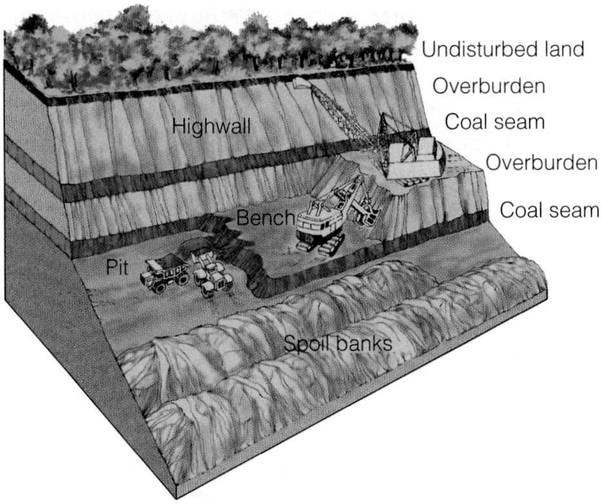

Figure 13-10 Contour strip mining of coal.

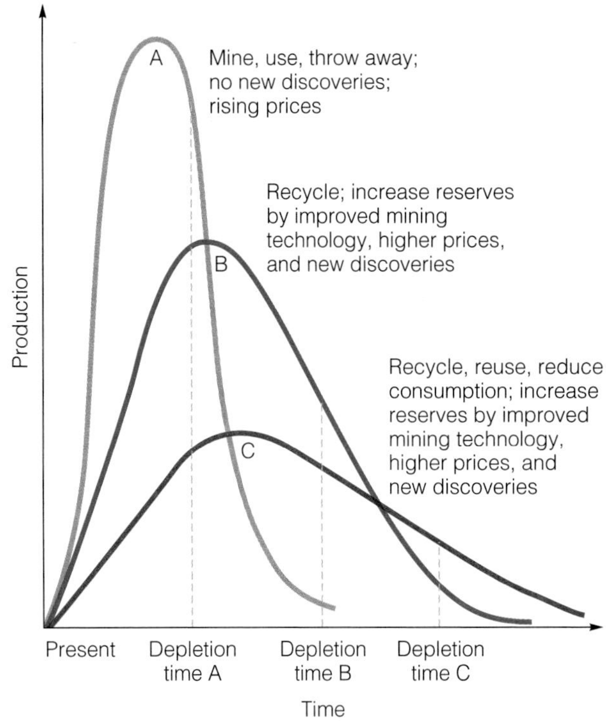

Figure 13-11 Depletion curves for a nonrenewable resource (such as aluminum or copper) using three sets of assumptions. Dashed vertical lines represent times when 80% depletion occurs.

ble bank of soil and rock called a *highwall*. Sometimes giant augers are used to drill horizontally into a hillside to extract underground coal.

In the United States, contour strip mining for coal is used mostly in the mountainous Appalachian region. Restoring land on surface mining sites can reverse mining's devastating effects. However, much of the land previously strip-mined for coal in the Appalachian region has not been restored. In addition, about three-fourths of the coal that can be surface mined in the United States is in arid and semiarid regions in the West, where the soil and climate usually prevent full restoration.

Subsurface mining is used to remove coal and various metal ores that are too deep to be extracted by surface mining. Miners dig a deep vertical shaft, blast subsurface tunnels and chambers to get to the deposit, and haul the coal or ore to the surface. Subsurface mining disturbs less than one-tenth as much land as surface mining and usually produces less waste material. However, it leaves much of the resource in the ground and is more dangerous and expensive than surface mining. Roofs and walls of underground mines collapse, trapping and killing miners, explosions of dust and natural gas injure or kill them, and prolonged inhalation of mining dust causes lung diseases.

13-3 ESTIMATING SUPPLIES OF NONRENEWABLE MINERAL RESOURCES

Will There Be Enough Mineral Resources? The future supply of nonrenewable minerals depends on two factors: **(1)** the actual or potential supply and **(2)** the rate at which that supply is used. We never completely run out of any mineral. However, a mineral becomes *economically depleted* when the costs of finding, extracting, transporting, and processing the remaining deposits exceed the returns. At that point we have five choices: recycle or reuse existing supplies, waste less, use less, find a substitute, or do without.

As mentioned earlier, most published estimates of the supplies of a given resource refer to *reserves*: known deposits from which a usable mineral can be extracted profitably at current prices (Figure 13-7). **Depletion time** is the time it takes to use up a certain proportion (usually 80%) of the reserves of a mineral at a given rate of use. When experts disagree about depletion times, they are often using different assumptions about supply and rate of use (Figure 13-11). A traditional measure of the projected availability of nonrenewable resources is the **reserve-to-production ratio**: the number of years that proven reserves of a particular nonrenewable mineral will last at current annual production rates.

Estimates of reserves are continually changing because new deposits are often discovered, and new mining and processing can allow some of the minerals classified as other resources (Figure 13-7) to be converted to reserves. Under these circumstances, the reserve-to-production ratio is the best available projection of the current estimated supply and its estimated depletion time.

The shortest depletion time assumes no recycling or reuse and no increase in reserves (curve A, Figure 13-11).

A longer depletion time assumes that recycling will stretch existing reserves and that better mining technology, higher prices, and new discoveries will increase reserves (curve B, Figure 13-11). An even longer depletion time assumes that new discoveries will further expand reserves and that recycling, reuse, and reduced consumption will extend supplies (curve C, Figure 13-11). Finding a substitute for a resource leads to a new set of depletion curves for the new resource.

While world population doubled between 1950 and 1993, global production of six key metals (aluminum, copper, lead, nickel, tin, and zinc) increased more than eightfold. During this same period, world reserves of copper increased almost fivefold, lead almost threefold, zinc fourfold, and aluminum almost ninefold. Furthermore, the prices of most metals today have changed little in constant dollars over the last 150 years, mostly because of government subsidies and failure to include the harmful environmental effects of metal mining and processing in their market prices.

Case Study: Mineral Resources in the United States Currently Americans use about 30% of the world's resources produced each year, including 33% of the paper, 25% of the fossil fuel, and 20% of the metals consumed globally each year. Excluding food, the average American uses about 101 kilograms (222 pounds) of wood, paper, chemicals, metals, plastics, fossil fuels, and rock and other mineral resources each day, or about 37 metric tons (41 tons) per year! Researchers at the University of British Columbia estimate that trying to sustain the entire world at an American (or Canadian) level of resource use would require the land area of three earths.

According to resource experts, the United States will never again be self-sufficient in oil or many key metal resources. Currently, the United States imports 50% or more of 24 of its 42 most important nonfuel minerals. Some are imported because they are used faster than they can be produced from domestic supplies; others are imported because foreign ore deposits are of a higher grade and cheaper to extract than remaining U.S. reserves.

Most U.S. mineral imports come from reliable and politically stable countries. However, experts are concerned about the availability of four minerals—manganese, cobalt, platinum, and chromium—of which the United States has little or no reserves and which it imports from potentially unstable countries in the former Soviet Union and Africa (South Africa, Zambia, Zaire).

13-4 INCREASING MINERAL RESOURCE SUPPLIES

How Does Economics Affect Resource Supplies? Geologic processes determine the quantity and location of a mineral resource in the earth's crust, but economics determines what part of the known supply is used.

According to standard economic theory, in a competitive free market a plentiful resource is cheap because supply exceeds demand; when a resource becomes scarce its price rises, stimulating exploration and development of better mining technology. Rising prices also make it profitable to mine ores of lower grades and encourage the search for substitutes. However, this theory may no longer apply to most developed countries because in the economic systems of such countries, industry and government control supply, demand, and prices of minerals to such a large extent that a truly competitive free market does not exist.

Most mineral prices are low because countries subsidize development of their domestic mineral resources to help promote economic growth and national security. In the United States, for instance, mining companies get depletion allowances amounting to 5–22% of their gross income, depending on the mineral. Some mining companies recoup more money through this tax deduction than they invest in the mine. In addition, the companies can deduct much of their costs for finding and developing mineral deposits. Moreover, hard-rock mining companies operating in the United States get public land and the minerals they extract essentially for free (p. 326).

Between 1982 and 1998, government mining subsidies cost U.S. taxpayers almost $6 billion. Critics charge that such subsidies **(1)** encourage environmentally destructive mining that would otherwise not be economically feasible and **(2)** discourage metal recycling industries (which receive far fewer government subsidies).

Another problem is that the cost of nonfuel mineral resources is only a small part of the final cost of goods. Thus, because scarcity of minerals does not raise the market prices of products very much, industries and consumers have no incentive to reduce demand for products in time to avoid economic depletion of the minerals.

Yet another problem is that most of the harmful environmental costs of mining and processing are not included in market prices. As a result, mining companies and manufacturers have little incentive to reduce resource waste and pollution because they can pass many of the harmful environmental costs of their production on to society and to future generations.

Environmentalists and some economists argue that taxing rather than subsidizing the extraction of nonfuel mineral resources would **(1)** provide governments with revenue, **(2)** create incentives for more efficient resource use, **(3)** promote waste reduction and pollution prevention, and **(4)** encourage recycling and reuse. Mining company representatives say they need subsidies and low taxes to keep the prices of minerals low for consumers. What do you think?

Can We Find Enough New Land-Based Mineral Deposits? Geologic exploration guided by better knowledge, satellite surveys, and other new techniques will increase current reserves of most minerals. Although most easily accessible, high-grade deposits are already known, new deposits will be found, mostly in unexplored areas of developing countries.

However, exploring for new resources takes lots of capital and is a risky financial venture. Typically, if geologists identify 10,000 possible deposits of a given resource, only 1,000 sites are worth exploring; only 100 justify drilling, trenching, or tunneling; and only 1 becomes a producing mine or well. Even if large new supplies are found, no nonrenewable mineral supply can stand up to continued exponential growth in its use.

One factor limiting production of nonfuel minerals is lack of investment capital. With today's fluctuating mineral markets and rising costs, investors are wary of tying up large sums for long periods with no assurance of a reasonable return.

Can We Get Enough Minerals by Mining Lower-Grade Ores? Some analysts contend that all we need to do to increase supplies of any mineral is to extract lower grades of ore. They point to new earth-moving equipment, improved techniques for removing impurities, and other technological advances in mineral extraction and processing during the past few decades.

In 1900 the average copper ore mined in the United States was about 5% copper by weight; today it is 0.5%, and copper costs less (adjusted for inflation). New methods of mineral extraction may allow even lower-grade ores of some metals to be used (Solutions, right).

Several factors limit the mining of lower-grade ores, however. As poorer grade ores are mined, a point is reached at which it costs more to mine and process such resources than they are worth (unless we have a virtually inexhaustible source of cheap energy). Availability of fresh water also may limit the supply of some mineral resources. Finally, exploitation of lower-grade ores may be limited by the environmental impact of waste material produced during mining and processing. At some point, the costs of land restoration and pollution control exceed the current value of the minerals, unless we continue to pass these harmful costs on to society and to future generations.

Can We Get Enough Minerals by Mining the Oceans? Ocean mineral resources are found in **(1)** seawater, **(2)** sediments and deposits on the shallow continental shelf (Figure 13-5), and **(3)** sediments and nodules on the deep-ocean floor. Most of the chemical elements found in seawater occur in such low concentrations that recovering them takes more energy and money than they are worth. Only magnesium, bromine, and sodium chloride are abundant enough to be extracted profitably at current prices with existing technology.

Deposits of minerals (mostly sediments) along the continental shelf and near shorelines are already significant sources of sand, gravel, phosphates, sulfur, tin, copper, iron, tungsten, silver, titanium, platinum, and diamonds.

The deep ocean floor at various sites may be a future source of manganese and other metals. Manganese-rich nodules may cover 20% of the world's ocean floors and have been found in large quantities at a few sites. These cherry- to potato-sized rocks are 30–40% manganese by weight; they also contain small amounts of other important metals such as nickel, copper, and

Mining with Microbes

SOLUTIONS

One way to improve mining technology is the use of microorganisms for in-place (*in situ*) mining. This biological approach to mining would **(1)** remove desired metals from ores while leaving the surrounding environment undisturbed, **(2)** reduce air pollution associated with the smelting of metal ores, and **(3)** reduce water pollution associated with using hazardous chemicals such as cyanides to extract gold.

Once an ore deposit has been identified and deemed economically viable, wells are drilled into it and the ore is fractured. Then the ore is inoculated with either natural or genetically engineered bacteria to extract the desired metal. Next the ore is flooded with water, which is then pumped to the surface, where the desired metal is removed.

This technique permits economical extraction from low-grade ores, which are increasingly being used as high-grade ores are depleted. Since 1958, the copper industry has been using natural strains of the bacterium *Thiobacillus ferroxidans* to remove copper from low-grade copper ore. Currently, at least 25% of all copper produced worldwide, worth more than $1 billion a year, comes from such biomining.

Microbiological processing of ores is slow, however. It can take decades to remove the same amount of material that conventional methods can remove within months or years. So far, biological methods are economically feasible only with low-grade ore (such as gold and copper) for which conventional techniques are too expensive.

Critical Thinking

If you had a large sum of money to invest, would you invest it in the microbiological processing of aluminum ore? Explain.

cobalt. They might be sucked up from the ocean floor by giant vacuum pipes or scooped up by buckets on a continuous cable operated by a mining ship. Rich deposits of gold, silver, zinc, and copper are also produced as sulfide deposits around very hot volcanic springs and vents found at various areas in the deep ocean.

Nodule beds in international waters are not being developed today because of squabbles over who owns them and how any profits should be distributed among the world's nations. An international Law of the Sea Treaty (signed by the United States in 1994) may resolve some of these issues.

Some environmentalists believe that seabed mining probably would cause less harm than mining on land. However, they are concerned that removing seabed mineral deposits and dumping back unwanted material will stir up ocean sediments. This could destroy seafloor organisms and have unknown effects on poorly understood ocean food webs. Surface waters might also be polluted by the discharge of sediments from mining ships and rigs.

Some biologists warn that deep-sea mining could have a devastating effect on mostly unknown marine biodiversity. Scientists now believe there may be 10 million—and perhaps as many as 100 million—previously unknown species of microbes and small animals residing in the waters and sediments of the deep. These scientists contend that we should understand more about their nature and functions before disrupting such marine life.

Can We Find Substitutes for Scarce Nonrenewable Mineral Resources? The Materials Revolution Some analysts believe that even if supplies of key minerals become very expensive or scarce, human ingenuity will find substitutes. They point to the current materials revolution in which silicon and new materials, particularly ceramics and plastics, are being developed and used as replacements for metals.

Ceramics have many advantages over conventional metals. They are harder, stronger, lighter, and longer-lasting than many metals and they withstand intense heat and do not corrode. Within a few decades we may have high-temperature ceramic superconductors in which electricity flows without resistance. Such a development may lead to faster computers, more efficient power transmission, and affordable electromagnets for propelling magnetic levitation trains.

Plastics also have advantages over many metals. High-strength plastics and composite materials strengthened by lightweight carbon and glass fibers are likely to transform the automobile and aerospace industries. They cost less to produce than metals because they require less energy, don't need painting, and can be molded into any shape. New plastics and gels are also being developed to provide superinsulation without taking up much space. One new plastic can withstand extremely high temperatures and is not even affected by exposure to the most intense laser beams.

Substitutes can undoubtedly be found for many scarce mineral resources. However, the search is costly and phasing a substitute into a complex manufacturing process takes time. While a vanishing mineral is being replaced, people and businesses dependent on it may suffer economic hardships. Moreover, finding substitutes for some key materials may be difficult or impossible. Examples are helium, phosphorus for phosphate fertilizers, manganese for making steel, and copper for wiring motors and generators. Finally, some substitutes are inferior to the minerals they replace. For example, even though aluminum could replace copper in electrical wiring, producing aluminum takes much more energy than producing copper, and aluminum wiring presents a greater fire hazard than copper wiring.

13-5 ENVIRONMENTAL EFFECTS OF EXTRACTING AND USING MINERAL RESOURCES

What Are the Environmental Impacts of Using Mineral Resources? The mining, processing, and use of crustal resources requires enormous amounts of energy and often causes land disturbance, erosion, and air and water pollution (Figure 13-12).

Mining can affect the environment in several ways, including

- Scarring and disruption of the land surface (Figures 13-8, 13-9, and 13-10)
- Underground fires in coal mines that cannot always be put out
- Collapse or subsidence of land above underground mines, which can cause houses to tilt, sewer lines to crack, gas mains to break, and groundwater systems to be disrupted
- Wind- or water-caused erosion of toxin-laced mining wastes such as spoil heaps and tailings
- Acid mine drainage, when rainwater seeping through a mine or mine wastes **(1)** carries sulfuric acid (H_2SO_4, produced when aerobic bacteria act on iron sulfide minerals in spoil) to nearby streams and groundwater (Figure 13-13), **(2)** contaminates water supplies, and **(3)** destroys aquatic life

After extraction from the ground, many resources must be separated from other matter, a process that can pollute the air and water. Ore, for example, typically contains two parts: **(1)** the ore mineral, which contains the desired metal, and **(2) gangue**, which is the waste mineral material. **Beneficiation**, or separation in a mill

Ziemer, L. S. 1998. The 1872 Mining Law and the 20th Century Collide. *Environmental Law,* vol. 28 no. 1 p. 145(13).

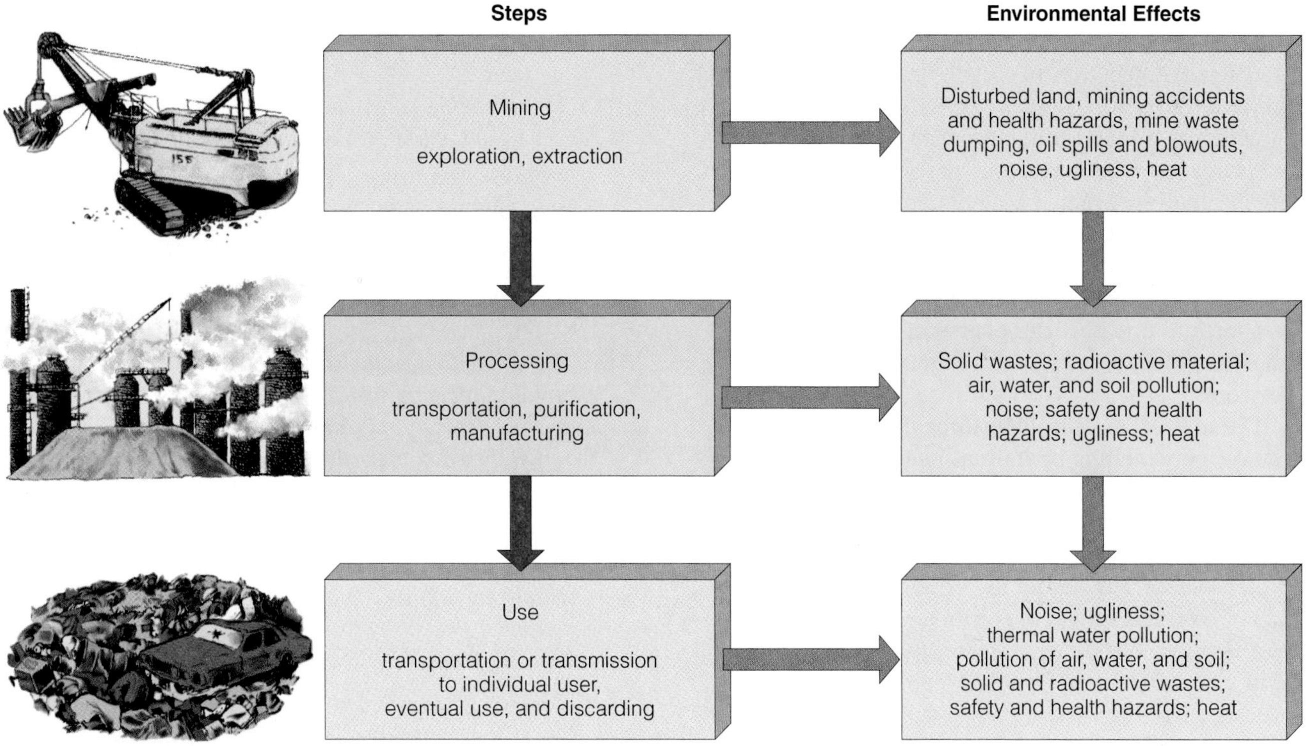

Figure 13-12 Some harmful environmental effects of resource extraction, processing, and use. The energy used to carry out each step causes additional pollution and environmental degradation. Harm could be minimized by requiring mining, processing, and manufacturing companies to include the full costs of the pollution and environmental degradation in the prices of their products. Many of these *external* costs are now passed on to society in the form of poorer health, increased health and insurance costs, and increased taxes to deal with pollution and environmental degradation.

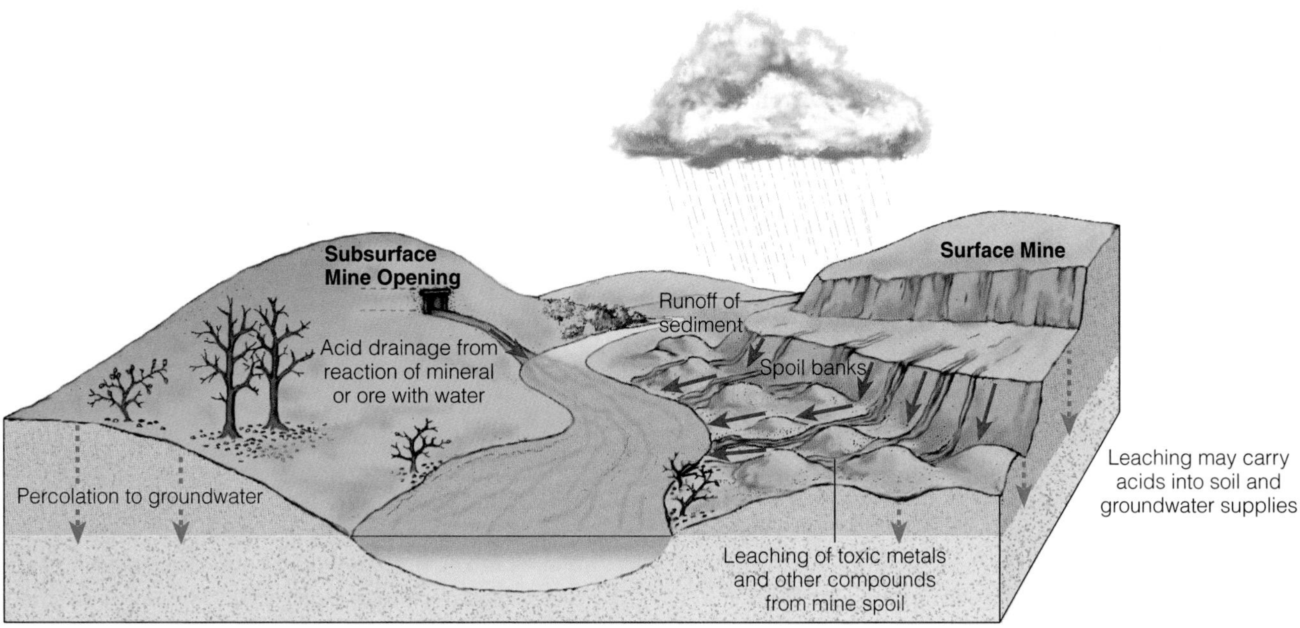

Figure 13-13 Pollution and degradation of a stream and groundwater by runoff of acids—called *acid mine drainage*—and by toxic chemicals from surface and subsurface mining. These substances can kill fish and other aquatic life. Acid mine drainage has damaged over 19,000 kilometers (12,000 miles) of streams in the United States, mostly in Appalachia and in the West.

Hint: Enter the search term *mining act* using the Subject Guide.

of the ore mineral from the gangue, produces waste called **tailings.** Piles of tailings are ugly, and toxic metals blown or leached from them by rainfall can contaminate surface and groundwater.

Most ore minerals do not consist of pure metal, so **smelting** is done to separate the metal from the other elements in the ore mineral. Without effective pollution control equipment, smelters emit enormous quantities of air pollutants, which damage vegetation and soils in the surrounding area. Pollutants include sulfur dioxide, soot, and tiny particles of arsenic, cadmium, lead, and other toxic elements and compounds found in many ores.

Decades of uncontrolled sulfur dioxide emissions from copper-smelting operations near Copperhill and Ducktown, Tennessee, killed all the vegetation over a wide area around the smelter; another dead vegetation zone was created around the Sudbury, Ontario, nickel smelter. New dead vegetation zones have formed in parts of eastern Europe, the former Soviet Union, and Chile. Smelters also cause water pollution and produce liquid and solid hazardous wastes that must be disposed of safely. Some companies are using improved technology to reduce pollution from smelting, reduce production costs, and save costly cleanup bills and liability for damages.

Are There Environmental Limits to Resource Extraction and Use? Some environmentalists and resource experts believe that the greatest danger from high levels of resource consumption may not be the exhaustion of resources but the damage that their extraction, processing, and conversion to products do to the environment (Figure 13-12).

The mineral industry accounts for 5–10% of world energy use, making it a major contributor to air and water pollution and to emissions of greenhouse gases. The *grade* of an ore—its percentage of metal content—largely determines the environmental impact of mining it (Case Study, right); more accessible and higher-grade ores generally are exploited first. As they are depleted, it takes more money, energy, water, and other materials to exploit lower-grade ores, and environmental effects increase accordingly.

Should the U.S. 1872 Mining Law Be Reformed to Reduce Environmental Harm and Save Taxpayers Money? Environmentalists point out that mining is the only natural resource industry in the United States that by law can buy public lands; it is also the only resource industry that pays no rents or royalties for resource extraction. Environmentalists and a growing number of citizens support a drastic reform of the 1872 mining law (p. 326), including the following changes:

- *Prohibiting buying of public land for mining but allowing such land to be leased for mining for up to 20 years.*
- *Requiring a full environmental impact assessment of the proposed mining activities before a mining lease is approved.*
- *Setting strict environmental standards for preventing and controlling pollution and environmental degradation resulting from mining activities during the period of the lease.*
- *Requiring companies leasing public land to post an environmental performance bond to cover estimated environmental damage and ecological restoration costs based on the environmental impact study of the project.* After such costs are deducted as needed and restoration is complete, remaining funds plus interest would be returned to the mining company.
- *Requiring mining companies to pay rent to cover all government (taxpayer) costs in evaluating and monitoring any leased mining site.*
- *Requiring mining companies to pay a 12.5% royalty on the gross (not net) values of all minerals removed.*
- *Making mining companies legally and financially responsible for environmental cleanup and restoration of each site.*

Mining companies claim that charging royalties for minerals taken from public lands and requiring them to pay for cleanup will force them to do their mining in other countries, which would cost American jobs and reduce tax revenues. They also argue that their average cost for patenting public land under the 1872 law is about $42,000 per hectare ($17,000 per acre) when their mining development costs are included.

Environmentalists counter that mining companies would still make a reasonable profit on the high-value minerals such as gold and platinum they get from public lands and that threats to move operations elsewhere are a rarely implemented scare tactic (greenmail). For example, gold costs miners about $30 per ounce to extract, but in recent years it has been sold for $320–395 per ounce. Even with a 12.5% royalty and responsibility for cleanup costs—as required for oil, gas, and coal companies—hard-rock mining companies can turn a hefty profit on high-price minerals such as gold and platinum.

Environmentalists also point out that Canada, Australia, South Africa, and other countries that are major extractors of hard-rock minerals don't sell public lands to mining companies, and they require the companies to pay rent on any public land they lease and royalties on the minerals they extract. What do you think?

The Environment and the New Gold Rush

CASE STUDY

Miners must extract and process massive quantities of soil and rock to end up with small quantities of gold. Gold miners typically remove ore equal to the weight of 50 automobiles to extract an amount of gold that would fit inside your clenched fist. Most newlyweds would be surprised to know that about 5.5 metric tons (6 tons) of mining waste was created to make their two gold wedding rings.

The mountains of solid waste remaining after gold is extracted from its ore are left piled near the mine sites and can pollute the air, surface water (Figure 13-1), and groundwater. Gold-bearing rock tends to contain large quantities of sulfur, which forms sulfuric acid when exposed to air and water (Figure 13-13). In addition to killing aquatic life, the acid puts highly toxic metals such as cadmium and copper into solution.

In Australia and North America, a new mining technology, called *cyanide heap leaching*, is cheap enough to allow mining companies to level entire mountains containing very low-grade gold ore. To extract the gold, miners spray a cyanide solution (which reacts with gold) onto huge open-air piles of crushed ore. They then collect the solution in leach beds and overflow ponds, recirculate it a number of times, and extract gold from it. Unfortunately, cyanide is extremely toxic and can be harmful or lethal to people, plants, and wildlife, especially to birds and mammals drawn to cyanide collection ponds as a source of water.

Cyanide leach pads and collection ponds can also leak or overflow, posing threats to underground drinking water supplies and to wildlife (especially fish) in lakes and streams. Special liners beneath the ore heaps and in the collection ponds are supposed to prevent leaks, but some have failed; according to the EPA, all such liners will eventually leak.

In the United States, companies have used the 1872 mining law to buy public land for practically nothing, mine a site, abandon it, file for bankruptcy, and leave the public with the cleanup bill (p. 326). A glaring example is the Summitville gold mine site in the San Juan Mountains of southern Colorado.

A Canadian company bought the land from the federal government for a pittance, spent $1 million developing the site, and then removed $98 million worth of gold. Shoddy construction allowed acids and toxic metals to leak from the site and poison a 27-kilometer (17-mile) stretch of the Alamosa River, the source of irrigation water for farms and ranches in the San Luis Valley.

The company then declared bankruptcy and abandoned the property, but only after being allowed to retrieve $2.5 million of the $7.5 million reclamation bond it had posted with the state. Summitville is now a Superfund site; the EPA spends $40,000 a day just to contain the site's toxic wastes. Ultimately, the EPA expects to spend about $120 million to finish the cleanup.

The gold rush of the 1980s and 1990s has also caused millions of miners—many of them landless poor—in various Latin American, Asian, and African developing countries to stream into tropical forests and other areas in search of gold. These small-scale miners use destructive mining techniques such as **(1)** digging large pits by hand, **(2)** river dredging, and **(3)** hydraulic mining (a technique, outlawed in the United States, in which water jets wash entire hillsides into sluice boxes).

Highly toxic mercury usually is used to extract the gold from the other materials. In the process, much of the mercury ends up contaminating water supplies and fish consumed by people.

Critical Thinking

Do you believe that the harmful environmental impacts of gold mining should be much more strictly regulated in **(a)** the United States and **(b)** other countries? If so, what major regulations would you like to see? How would you see that such controls are implemented and enforced?

13-6 SOIL: THE BASE OF LIFE

What Are the Major Layers Found in Mature Soils? The material we call **soil** is a complex mixture of eroded rock, mineral nutrients, decaying organic matter, water, air, and billions of living organisms, most of them microscopic decomposers (Figure 13-14). Although soil is a potentially renewable resource, it is produced very slowly by the weathering of rock, deposit of sediments by erosion, and decomposition of organic matter in dead organisms.

Mature soils are arranged in a series of zones called **soil horizons**, each with a distinct texture and composition that varies with different types of soils. A cross-sectional view of the horizons in a soil is called a **soil profile.** Most mature soils have at least three of the possible horizons (Figure 13-14).

The top layer, the *surface-litter layer*, or *O horizon*, consists mostly of freshly fallen and partially decomposed leaves, twigs, animal waste, fungi, and other organic materials. Normally, it is brown or black. The *topsoil layer*, or *A horizon*, is a porous mixture of partially

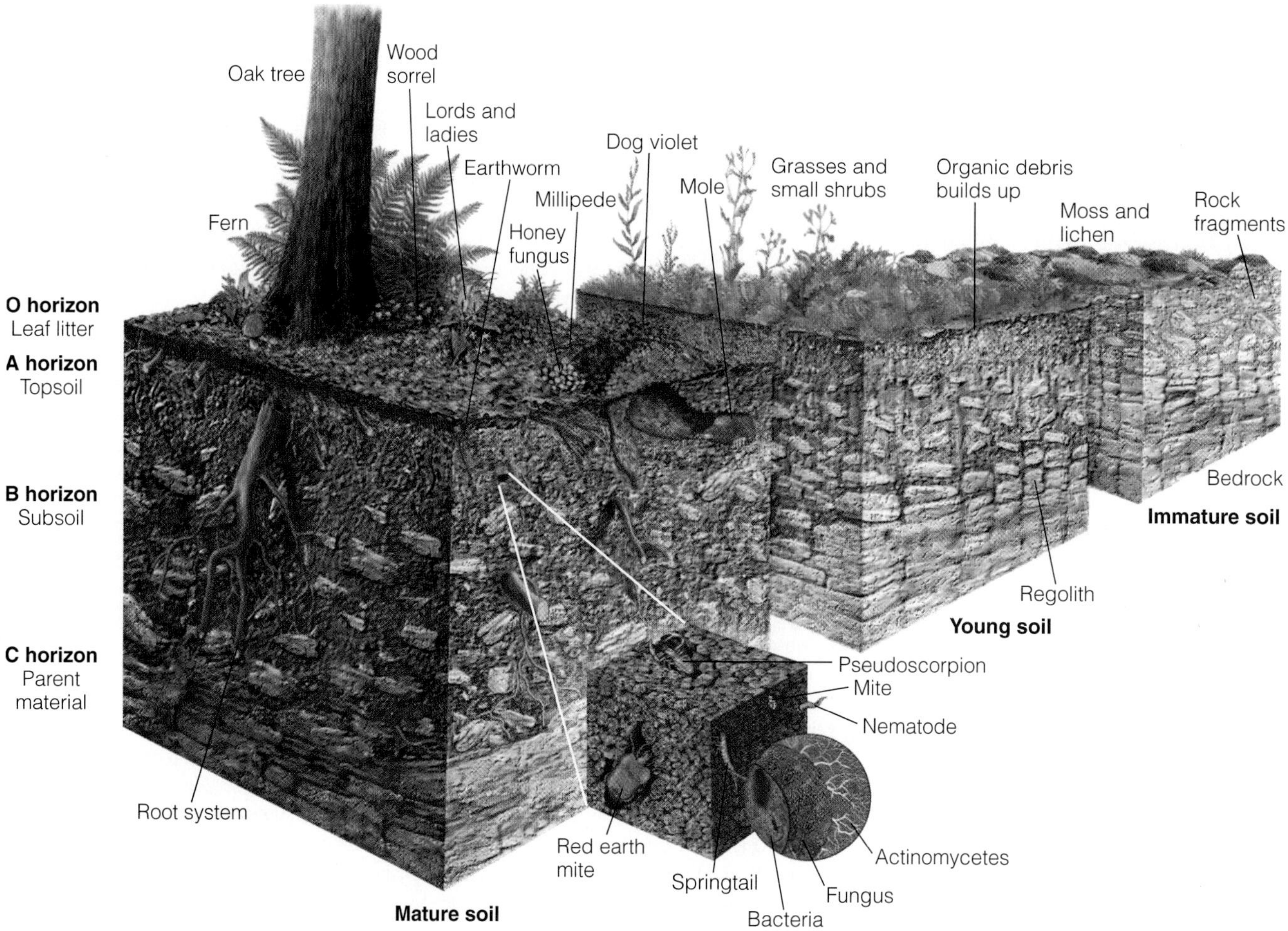

Figure 13-14 Formation and generalized profile of soils. Horizons, or layers, vary in number, composition, and thickness, depending on the type of soil. (Used by permission of Macmillan Publishing Company from Derek Elsom, *Earth*, New York: Macmillan, 1992. Copyright ©1992 by Marshall Editions Developments Limited)

decomposed organic matter, called **humus**, and some inorganic mineral particles. It is usually darker and looser than deeper layers. The roots of most plants and most of a soil's organic matter are concentrated in these two upper layers. As long as these layers are anchored by vegetation, soil stores water and releases it in a nourishing trickle instead of a devastating flood.

The two top layers of most well-developed soils teem with bacteria, fungi, earthworms, and small insects that interact in complex food webs (Figure 13-15). Bacteria and other decomposer microorganisms found by the billions in every handful of topsoil recycle the nutrients we and other land organisms need (Figure 13-16). They break down some complex organic compounds into simpler inorganic compounds soluble in water. Soil moisture carrying these dissolved nutrients is drawn up by the roots of plants and transported through stems and into leaves.

Some organic litter in the two top layers is broken down into a sticky, brown residue of partially decomposed organic material (humus). Because this humus is only slightly soluble in water, most of it stays in the topsoil layer. A fertile soil that produces high crop yields has a thick topsoil layer with lots of humus. This helps topsoil hold water and nutrients taken up by plant roots.

The color of its topsoil tells us a lot about how useful a soil is for growing crops. For example, dark-brown or black topsoil is nitrogen-rich and high in organic matter. Gray, bright yellow, or red topsoils are low in organic matter and will need nitrogen enrichment to support most crops.

The *B horizon (subsoil)* and the *C horizon (parent material)* contain most of a soil's inorganic matter, mostly broken-down rock consisting of varying mixtures of sand, silt, clay, and gravel. The C horizon lies on a base of unweathered parent rock called *bedrock*.

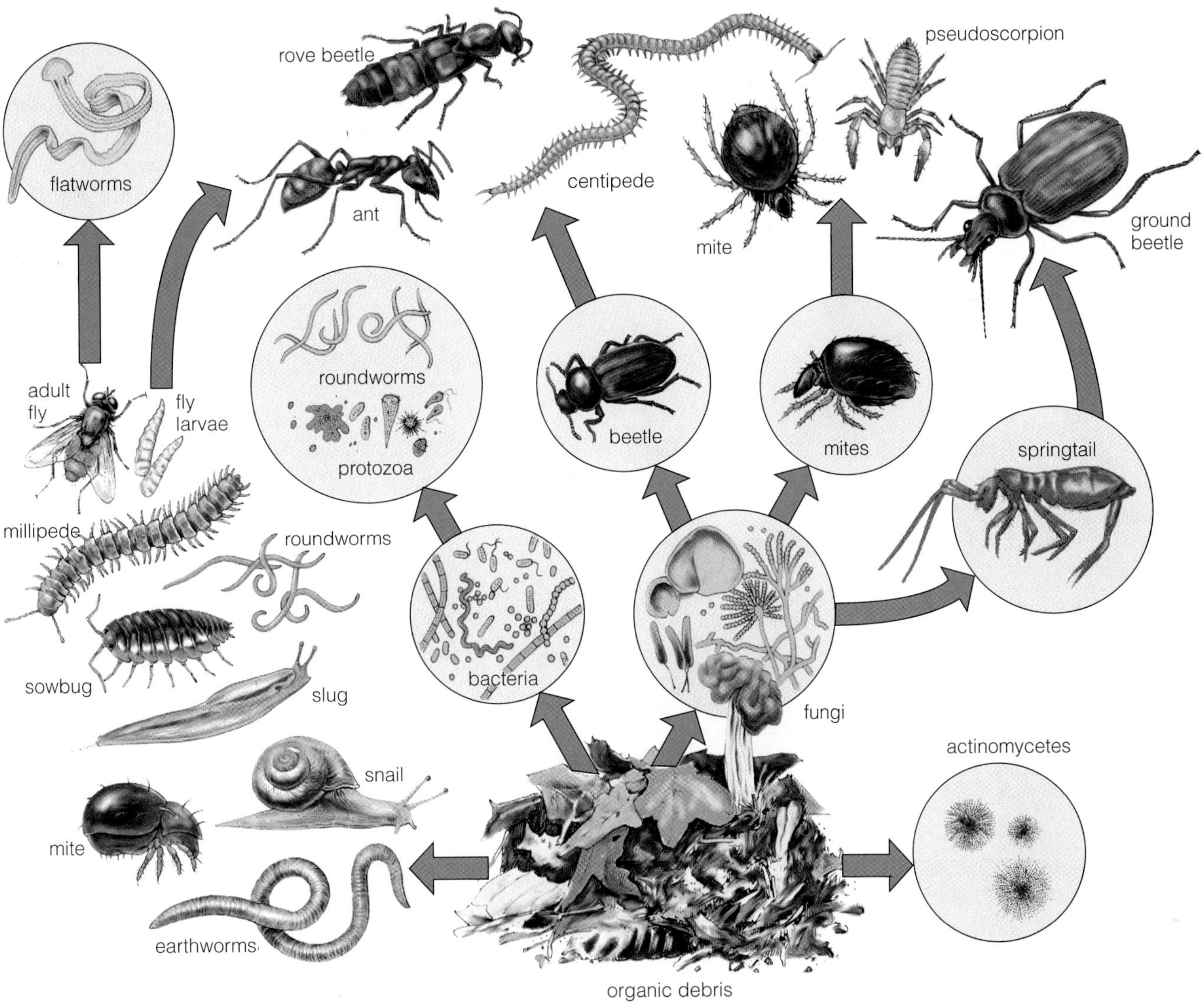

Figure 13-15 Greatly simplified food web of living organisms found in soil.

The spaces, or pores, between the solid organic and inorganic particles in the upper and lower soil layers contain varying amounts of air (mostly nitrogen and oxygen gas) and water. Plant roots need oxygen for cellular respiration.

Some of the precipitation that reaches the soil percolates through the soil layers and occupies many of the soil's open spaces or pores. This downward movement of water through soil is called **infiltration**. As the water seeps down, it dissolves various soil components in upper layers and carries them to lower layers in a process called **leaching**.

Soils develop and mature slowly. The earth's current mature soils vary widely from biome to biome in color, content, pore space, acidity, and depth. Five important soil types, each with a distinct profile, are shown in Figure 13-17. Most of the world's crops are grown on soils exposed when grasslands and deciduous forests are cleared.

How Do Soils Differ in Texture, Porosity, and Acidity? Soils vary in their content of *clay* (very fine particles), *silt* (fine particles), *sand* (medium-size particles), and *gravel* (coarse to very coarse particles). The relative amounts of the different sizes and types of mineral particles determine **soil texture**, as depicted in Figure 13-18. Soils with roughly equal mixtures of clay, sand, silt, and humus are called **loams**.

To get an idea of a soil's texture, take a small amount of topsoil, moisten it, and rub it between your fingers and thumb. A gritty feel means that it contains a lot of sand. A sticky feel means a high clay content, and you should be able to roll it into a clump. Silt-laden soil feels smooth, like flour. A loam topsoil, which is best suited for plant growth, has a texture between these extremes—a crumbly, spongy feeling—with many of its particles clumped loosely together.

Soil texture helps determine **soil porosity**, a measure of the volume of pores or spaces per volume of soil and of the average distances between those spaces. A porous soil has many pores and can hold more water and air than a less porous soil. The average size of the spaces or pores in a soil determines **soil permeability**: the rate at which water and air move from upper to lower soil layers. Soil porosity is also influenced by **soil structure**: how soil particles are organized and clumped together.

Loams are the best soils for growing most crops because they hold lots of water, but not too tightly for plant roots to absorb. Sandy soils are easy to work, but water flows rapidly through them. They are useful for growing irrigated crops or those with low water requirements, such as peanuts and strawberries.

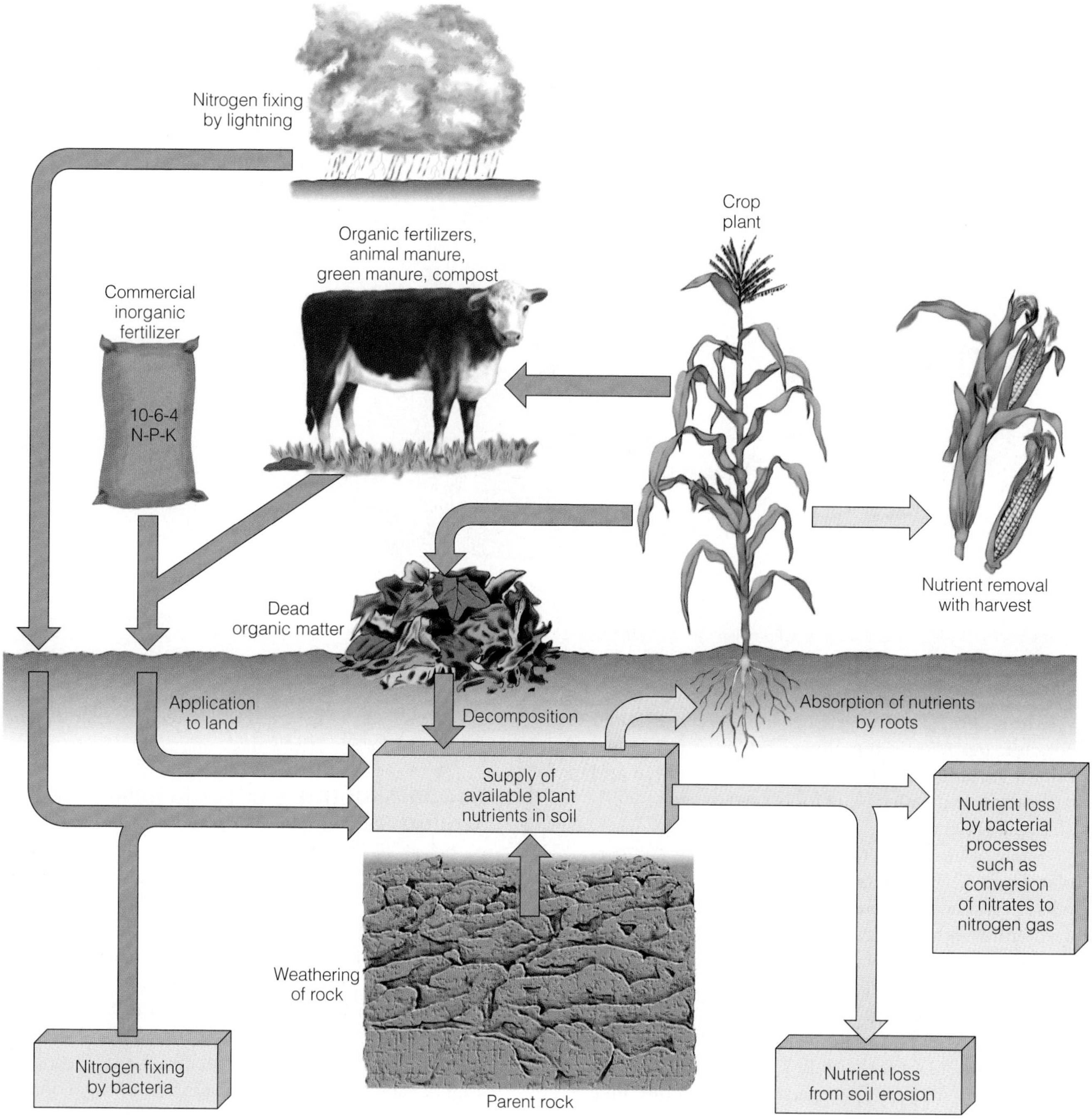

Figure 13-16 Pathways of plant nutrients in soils.

The particles in clay soils are very small and easily compacted. When these soils get wet, they form large, dense clumps, which is why wet clay can be molded into bricks and pottery. Clay soils are more porous and have a greater water-holding capacity than sandy soils, but the pore spaces are so small that these soils have a low permeability. Because little water can infiltrate to lower levels, the upper layers can easily become too waterlogged for most crops.

A numerical scale of *pH* values is used to compare the acidity and alkalinity in water solutions (Figure 10-7). The pH of a soil influences the uptake of soil nutrients by plants, which vary in the pH ranges they can tolerate.

When soils are too acidic, the acids can be partially neutralized by an alkaline substance such as lime. Because lime speeds up the decomposition of organic matter in the soil, however, manure or another organic fertilizer should also be added to maintain soil fertility.

In dry regions such as much of the western and southwestern United States, rain does not leach away calcium and other alkaline compounds, so soils in such areas may be too alkaline (pH above 7.5) for some crops. Adding sulfur, which is gradually converted into sulfuric acid by soil bacteria, reduces soil alkalinity.

13-7 SOIL EROSION

What Causes Soil Erosion? **Soil erosion** is the movement of soil components, especially surface litter and topsoil, from one place to another. The two main agents of erosion are flowing water and wind. Some soil erosion is natural: the long-term wearing down of mountains and building up of plains and deltas by the combined action of physical, chemical, and biological forces (Figure 13-5). In undisturbed vegetated ecosystems, the roots of plants help anchor the soil, and usually soil is not lost faster than it forms.

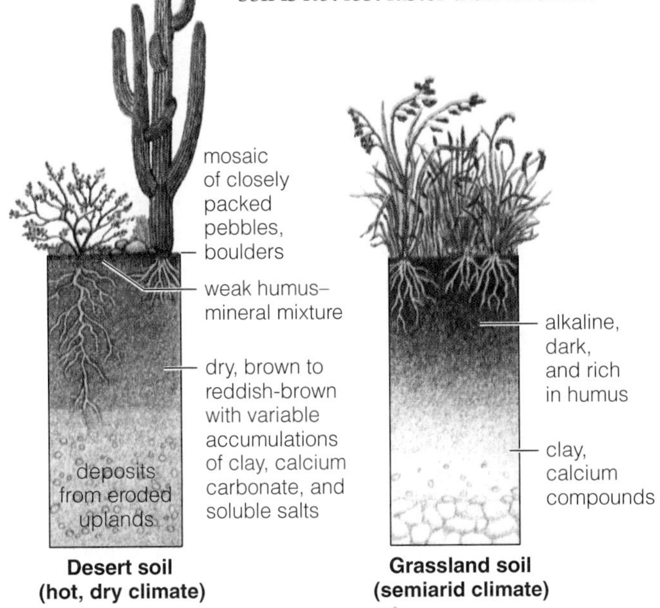

Figure 13-17 Soil profiles of the principal soil types typically found in five different biomes.

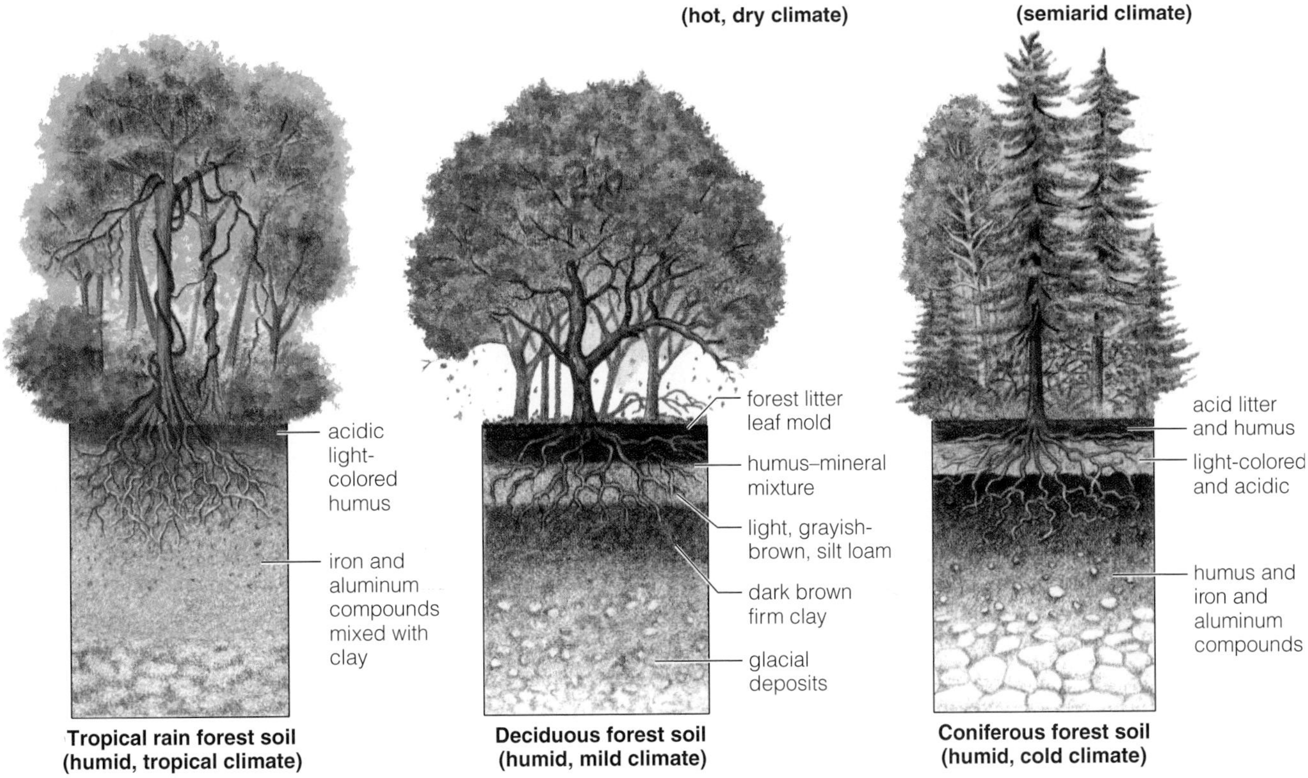

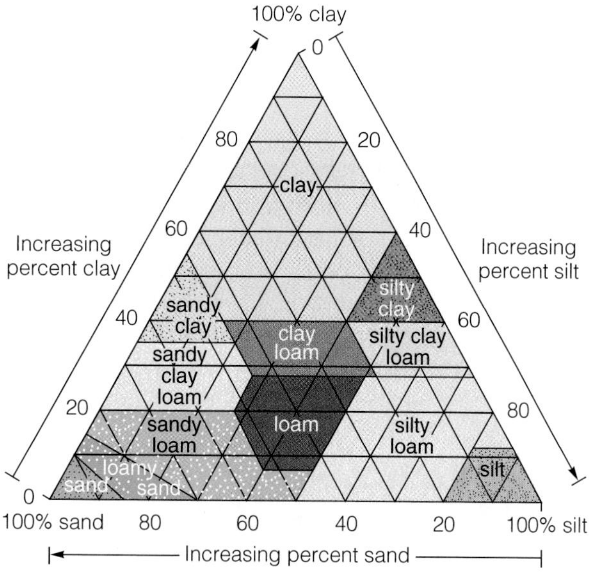

Figure 13-18 Soil texture depends on the proportions of clay, silt, and sand particles in the soil. Soil texture affects soil porosity, the average number and spacing of pores in a given volume of soil. Loams—roughly equal mixtures of clay, sand, silt, and humus—are the best soils for growing most crops. (Data from National Resources Conservation Service)

Figure 13-19 Rill and gully erosion of vital topsoil from irrigated cropland in Arizona. (USDA Soil Conservation Service)

However, farming, logging, construction, overgrazing by livestock, off-road vehicles, deliberate burning of vegetation, and other activities that destroy plant cover leave soil vulnerable to erosion. Such human activities can speed up erosion and destroy in a few decades what nature took hundreds to thousands of years to produce. In 1937, U.S. President Franklin D. Roosevelt sent a letter to the governors of the states in which he said, *"The nation that destroys its soil destroys itself."*

Most soil erosion is caused by moving water. Soil scientists distinguish among three types of water erosion. *Sheet erosion* occurs when surface water moves down a slope or across a field in a wide flow and peels off fairly uniform sheets or layers of soil. Because the topsoil disappears evenly, sheet erosion may not be noticeable until much damage has been done. In *rill erosion* the surface water forms fast-flowing rivulets that cut small channels in the soil (Figure 13-19). In *gully erosion*, rivulets of fast-flowing water join together and with each succeeding rain cut the channels wider and deeper until they become ditches or gullies (Figure 13-19). Gully erosion usually happens on steep slopes where all or most vegetation has been removed.

Losing topsoil makes a soil less fertile and less able to hold water. The resulting sediment—the largest source of water pollution—clogs irrigation ditches, boat channels, reservoirs, and lakes. The sediment-laden water is cloudy and tastes bad, fish die, and flood risk increases. Rivers running brown with silt from human-accelerated soil erosion contain nutrients that are hemorrhaging from the land (Figure 6-23).

Soil, especially topsoil, is classified as a potentially renewable resource because it is continuously regenerated by natural processes. However, in tropical and temperate areas it takes 200–1,000 years (depending on climate and soil type) for 2.54 centimeters (1 inch) of new topsoil to form. If topsoil erodes faster than it forms on a piece of land, the soil becomes a nonrenewable resource. Annual erosion rates for farmlands throughout the world are 7–100 times the natural renewal rate (Guest Essay, p. 352). Soil erosion is milder on forestland and rangeland than on cropland, but forest soil takes two to three times longer to restore itself than does cropland. Construction sites usually have the highest erosion rates by far.

How Serious Is Global Soil Erosion? According to a UN Environment Programme survey, topsoil is eroding faster than it forms on about one-third of the world's cropland, causing an estimated 85% of the world's land degradation from human activities.

A 1992 study by the World Resources Institute and the UN Environment Programme found that soil on an area equal to the size of China and India combined had been seriously eroded since 1945 (Figure 13-20). The

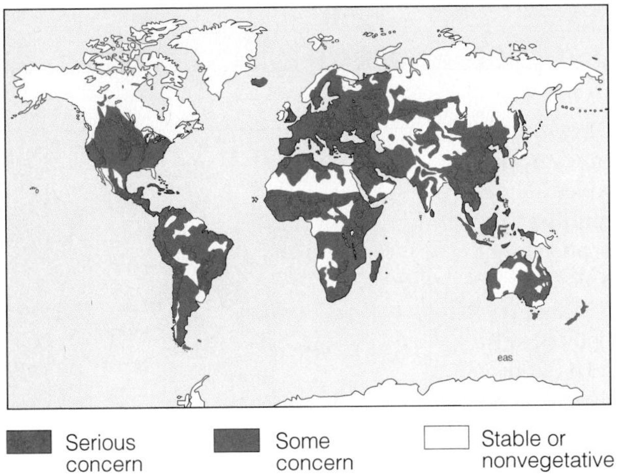

Figure 13-20 Global soil erosion. (Data from UN Environment Programme and the World Resources Institute)

study also found that about 15% of land scattered across the globe was too eroded to grow crops anymore because of a combination of overgrazing (35%), deforestation (30%), and unsustainable farming (28%). Two-thirds of these seriously degraded lands are in Asia and Africa.

Each year we must feed about 80 million more people with an estimated 24 billion metric tons (26 billion tons) less topsoil, eroded as a result of human activities. The topsoil that washes and blows into the world's streams, lakes, and oceans each year would fill a train of freight cars long enough to encircle the planet 150 times. At that rate, the world is losing about 7% of its topsoil from actual or potential cropland each decade.

The situation is worsening as many poor farmers in some developing countries plow up marginal (easily erodible) lands to survive. Soil expert David Pimentel (Guest Essay, p. 352) estimates that soil erosion causes at least $375 billion per year worldwide in direct damage to agricultural lands and indirect damage to waterways, infrastructure, and human health, an average of $42 million in damages per hour.

How Serious Is Soil Erosion in the United States? According to the Soil Conservation Service (now the National Resources Conservation Service), about one-third of the nation's original prime topsoil has been washed or blown into streams, lakes, and oceans, mostly as a result of overcultivation, overgrazing, and deforestation (Case Study, p. 346).

Today, it is estimated that soil on cultivated land in the United States is eroding about 16 times faster than it can form. Erosion rates are even higher in heavily farmed regions, including the Great Plains, which has lost one-third or more of its topsoil in the 150 years since it was first plowed. Some of the country's most productive agricultural lands, such as those in Iowa, have lost about half their topsoil. California's soil is eroding about 80 times faster than it can be formed.

In 1999 researcher Stanley Trimble reported results indicating that in some areas of the United States, soil erosion rates may be much lower than official government estimates. His analysis of 60 years of data on the amount of sediment from soil erosion flowing into a creek from an area of steep slopes and intensive agriculture in Wisconsin revealed that the current rate of sedimentation is only about 5% of what it was in the 1930s.

Other researchers caution about applying these results to much of the rest of the United States. Soil expert David Pimentel (Guest Essay, p. 352) also points out that Trimble's study looked specifically at sediments in a watershed, and not at what was happening on cropfields.

Because of soil conservation efforts, the U.S. Department of Agriculture (USDA) estimates that soil erosion in the United States decreased by about 40% between 1985 and 1997. Using these data, USDA researchers estimate that soil erosion cost the United States about $30 billion in 1997, an average loss of $3.4 million per hour.

Critics say that even these lowered estimates of soil erosion and damages from such erosion are exaggerated. They point to studies by several soil scientists concluding that if current rates of cropland erosion in the United States continue for 100 years, crop yields will be only 3–10% less than they would be without such erosion.

However, David Pimentel and others point out that this estimate does not include all of the ecological effects caused by soil erosion. Such effects include reductions in soil depth, availability of soil water for crops, and soil organic matter and nutrients. When such effects are included, soil scientists and ecologists estimate that soil erosion causes a 15–30% reduction in crop productivity. This requires costly use of inorganic fertilizers to help replace lost soil nutrients. However, because fertilizers are not a substitute for fertile soil, there is a limit to the amount of fertilizer that can be applied before crop yields level off and then begin to decline.

What Is Desertification, and How Serious Is This Problem? **Desertification** is a process whereby the productive potential of arid or semiarid land falls by 10% or more; this phenomenon results mostly from human

The Dust Bowl

CASE STUDY

In the 1930s, Americans learned a harsh environmental lesson when much of the topsoil in several dry and windy Midwestern states was lost through a combination of poor cultivation practices and prolonged drought.

Before settlers began grazing livestock and planting crops there in the 1870s, the deep and tangled root systems of native prairie grasses anchored the fertile topsoil firmly in place (Figure 13-17). Plowing the prairie tore up these roots, and the agricultural crops the settlers planted annually in their place had less extensive root systems.

After each harvest, the land was plowed and left bare for several months, exposing it to the plains winds. Overgrazing also destroyed large expanses of grass, denuding the ground. The stage was set for severe wind erosion and crop failures; all that was needed was a long drought.

Such a drought occurred between 1926 and 1934. In the 1930s, dust clouds created by hot, dry windstorms darkened the sky at midday in some areas; rabbits and birds choked to death on the dust. During May 1934, the entire eastern United States was blanketed by a cloud of topsoil blown off the Great Plains, some 2,400 kilometers (1,500 miles) away. Journalists gave the Great Plains a new name: the *Dust Bowl* (Figure 13-21).

During the so-called Dirty Thirties, large areas of cropland were stripped of topsoil and severely eroded. Thousands of displaced farm families from Oklahoma, Texas, Kansas, and Colorado migrated to California or to the industrial cities of the Midwest and East. Most found no jobs because the country was in the midst of the Great Depression.

In May 1934, Hugh Bennett of the U.S. Department of Agriculture (USDA) went before a congressional hearing in Washington to plead for new programs to protect the country's topsoil. Lawmakers took action when Great Plains dust began seeping into the hearing room.

In 1935, the United States passed the Soil Erosion Act, which established the Soil Conservation Service (SCS) as part of the USDA. With Bennett as its first head, the SCS (now called the Natural Resources Conservation Service) began promoting sound conservation practices, first in the Great Plains states and later elsewhere. Soil conservation districts were formed throughout the country, and farmers and ranchers were given technical assistance in setting up soil conservation programs.

Unfortunately, these heroic efforts have not yet stopped human-accelerated erosion in the Great Plains. The basic problem is that much of the region is better suited for moderate grazing than for farming. If the earth warms as projected, the region could become even drier and farming might have to be abandoned.

Figure 13-21 The Dust Bowl of the Great Plains, where a combination of extreme drought and poor soil conservation practices led to severe wind erosion of topsoil in the 1930s. Note the connection of this area with the Ogallala aquifer in Figure 12-11.

Climate researchers see signs of a returning Dust Bowl period because of a megadrought that lasts 2 to 4 decades. By examining tree rings, archeological finds, lake sediments, and sand dunes scientists have found that prolonged megadroughts generally hit twice a century as part of a complex drought cycle. They also found that smaller 2-year droughts strike every 20 years or so.

Critical Thinking

Do you think Americans learned a lesson about protecting soil as a result of the Dust Bowl in the 1930s? Explain.

activities. *Moderate desertification* is a 10–25% drop in productivity, *severe desertification* is a 25–50% drop, and *very severe desertification* is a drop of 50% or more, usually creating huge gullies and sand dunes. Desertification is considered to be a serious and growing problem in many parts of the world (Figure 13-22).

Practices that leave topsoil vulnerable to desertification include **(1)** overgrazing on fragile arid and semiarid rangelands, **(2)** deforestation without reforestation, **(3)** surface mining without land reclamation, **(4)** irrigation techniques that lead to increased erosion, **(5)** salt buildup and waterlogged soil, **(6)** farming on land with unsuitable terrain or soils, and **(7)** soil compaction by farm machinery and cattle. The consequences of desertification include **(1)** worsening drought, **(2)** famine, **(3)** declining living standards, and **(4)** swelling numbers of environmental refugees whose land is too eroded to grow crops or feed livestock.

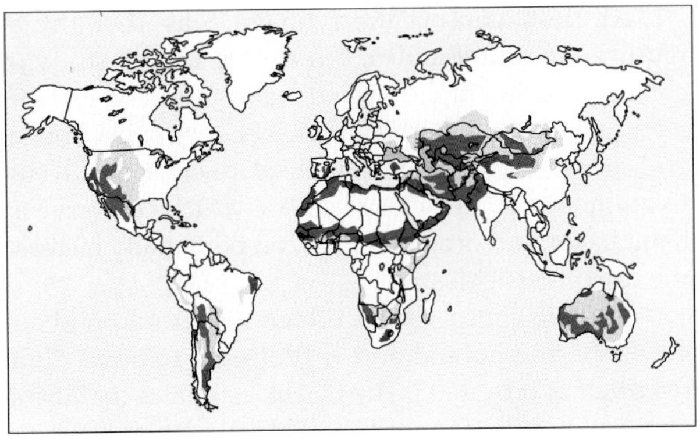

Figure 13-22 Desertification of arid and semiarid lands. (Data from UN Environmental Programme and Harold E. Drengue)

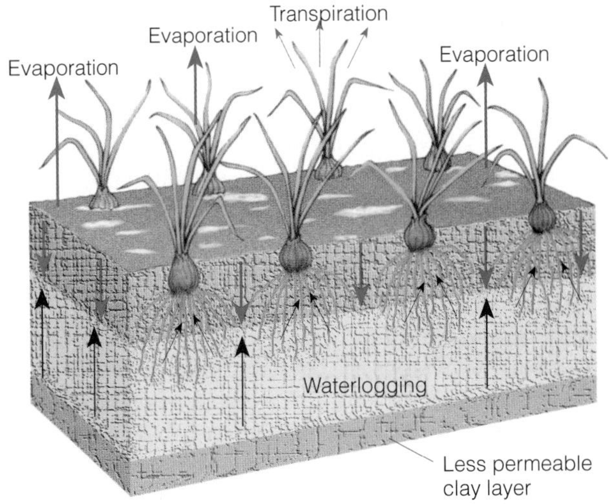

Salinization

1. Irrigation water contains small amounts of dissolved salts.
2. Evaporation and transpiration leave salts behind.
3. Salt builds up in soil.

Waterlogging

1. Precipitation and irrigation water percolate downward.
2. Water table rises.

Figure 13-23 Salinization and waterlogging of soil on irrigated land without adequate drainage lead to decreased crop yields.

An estimated 8.1 million square kilometers (3.1 million square miles)—an area the size of Brazil and 12 times the size of Texas—have become desertified in the past 50 years. According to a 1999 United Nations conference on desertification, **(1)** about 40% of the world's land is suffering from the effects of desertification and **(2)** each year about 150,000 square kilometers (58,000 square miles) of land—an area larger than Greece—becomes desertified. This affects the livelihoods of at least 1 billion people in 100 countries.

Solutions: How Can We Slow Desertification?

The most effective way to slow desertification is to drastically reduce overgrazing, deforestation, and the destructive forms of planting, irrigation, and mining that are to blame. In addition, planting trees and grasses will anchor soil and hold water while slowing desertification and reducing the threat of global warming (Section 11-5).

Such prevention and rehabilitation would cost $10–22 billion annually for the next 20 years. This expenditure is considerably less than the estimated $42 billion annual loss in agricultural productivity from desertified land. Once this potential productivity is restored, the cost of the program could be recouped quickly. So far, less than $1 billion per year is spent globally to halt this form of land degradation.

How Do Excess Salts and Water Degrade Soils?

The approximately 17% of the world's cropland that is now irrigated by various methods (Figure 12-13) produces almost 40% of the world's food. Irrigated land can produce crop yields that are two to three times greater than those from rain watering, but irrigation has its downside. Most irrigation water is a dilute solution of various salts, picked up as the water flows over or through soil and rocks. Small quantities of these salts are essential nutrients for plants, but they are toxic in large amounts.

Irrigation water not absorbed into the soil evaporates, leaving behind a thin crust of dissolved salts (such as sodium chloride) in the topsoil. The accumulation of these salts, called **salinization** (Figure 13-23), stunts crop growth, lowers yields, and eventually kills plants and ruins the land. According to a 1995 study, severe salinization has reduced yields on 20% of the world's irrigated cropland, and another 30% has been moderately salinized. The most severe salinization occurs in Asia, especially in China, India, and Pakistan.

Precipitation can desalinate soil, but this takes thousands of years in arid and semiarid areas where irrigation is used. Salts can be flushed out of soil by applying much more irrigation water than is needed for crop growth, but this practice increases pumping and crop production costs, wastes enormous amounts of water, and waterlogs plants if the water table rises close to the surface.

Heavily salinized soil can also be renewed by taking the land out of production for 2 to 5 years, installing an underground network of perforated drainage pipes, and flushing the soil with large quantities of low-salt water. However, this costly scheme only slows the salt buildup; it does not stop the process. Flushing salts from the soil also makes downstream irrigation water saltier unless the saline water can be drained into evaporation ponds rather than returned to the stream or canal.

Another problem with irrigation is **waterlogging** (Figure 13-23). Farmers often apply large amounts of irrigation water to leach salts deeper into the soil. Without adequate drainage, however, water accumulates underground, gradually raising the water table. Saline water then envelops the deep roots of plants, lowering their productivity and killing them after prolonged exposure. At least one-tenth of all irrigated land worldwide suffers from waterlogging, and the problem is getting worse.

13-8 SOLUTIONS: SOIL CONSERVATION

How Can Conservation Tillage Reduce Soil Erosion? **Soil conservation** involves reducing soil erosion and restoring soil fertility. For hundreds of years, farmers have used various methods to reduce soil erosion, most of which involve keeping the soil covered with vegetation.

In **conventional-tillage farming** the land is plowed and then the soil is broken up and smoothed to make a planting surface. In areas such as the Midwestern United States, harsh winters prevent plowing just before the spring growing season. Thus, cropfields often are plowed in the fall, baring the soil during the winter and early spring and leaving it vulnerable to erosion.

To reduce erosion, many U.S. farmers are using **conservation-tillage farming** (either *minimum-tillage* or *no-till farming*). The idea is to disturb the soil as little as possible while planting crops. With *minimum-tillage farming,* special tillers break up and loosen the subsurface soil without turning over the topsoil, previous crop residues, and any cover vegetation. In *no-till farming,* special planting machines inject seeds, fertilizers, and weed-killers (herbicides) into slits made in the unplowed soil.

Besides reducing soil erosion, conservation tillage **(1)** saves fuel, **(2)** cuts costs, **(3)** holds more water in the soil, **(4)** keeps the soil from getting packed down, and **(5)** allows more crops to be grown during a season (multiple cropping). Yields are at least as high as those from conventional tillage. It also reduces the release of carbon dioxide from the soil to the air, helping to ease the threat of global warming.

At first, conservation tillage was thought to require more herbicides, but a 1990 USDA study of corn production in the United States found no real difference in levels of herbicide use between conventional and conservation tillage systems. However, no-till cultivation of corn does leave stalks, which can serve as habitats for the corn borer; this can potentially increase the use of pesticides.

By 1998 conservation tillage was used on about 40% of U.S. cropland and is projected to be used on over half of it by 2005. The USDA estimates that using conservation tillage on 80% of U.S. cropland would reduce soil erosion by at least half. So far, the practice is not widely used in other parts of the world.

How Can Terracing, Contour Farming, Strip Cropping, and Alley Cropping Reduce Soil Erosion? **Terracing** can reduce soil erosion on steep slopes, each of which is converted into a series of broad, nearly level terraces that run across the land contour (Figure 13-24a). Terracing retains water for crops at each level and reduces soil erosion by controlling runoff.

In mountainous areas such as the Himalayas and the Andes, farmers traditionally built elaborate systems of terraces to grow crops. Today, however, some of these slopes are being farmed without terraces, leaving the land too nutrient poor to grow crops or generate new forest after only 10–40 years. Although most poor farmers know the risk of not terracing, many have too little time and too few workers to build terraces; they must plant crops or starve. The resultant loss of protective vegetation and topsoil (Figure 12-16) also greatly intensifies flooding below these watersheds.

Soil erosion can be reduced by 30–50% on gently sloping land by means of **contour farming**: plowing and planting crops in rows across, rather than up and down, the sloped contour of the land (Figure 13-24b). Each row planted along the contour of the land acts as a small dam to help hold soil and slow water runoff.

In **strip cropping**, a row crop such as corn alternates in strips with another crop (such as a grass or a grass–legume mixture) that completely covers the soil and thus reduces erosion (Figure 13-24b). The strips of the cover crop trap soil that erodes from the row crop. They also catch and reduce water runoff and help prevent the spread of pests and plant diseases. Nitrogen-fixing legumes such as soybeans or alfalfa planted in some of the strips help restore soil fertility.

Erosion can also be reduced by **alley cropping**, or **agroforestry**, a form of *intercropping* in which several crops are planted together in strips or alleys between trees and shrubs that can provide fruit or fuelwood

Toy, T. J., et al. 1999. RUSLE for Mining, Construction and Reclamation Lands. *Journal of Soil & Water Conservation*, v54 i2 p462(6).

(a)

(b)

(c)

(d)

Figure 13-24 Soil conservation methods: **(a)** terracing in Bali, **(b)** contour planting and strip cropping in Illinois, **(c)** alley cropping in Peru, and **(d)** windbreaks in South Dakota. (Clockwise from top left, Prato/Bruce Coleman Ltd.; USDA Soil Conservation Service; P. A. Sanchez/North Carolina State University; USDA Soil Conservation Service)

(Figure 13-24c). The trees provide shade (which reduces water loss by evaporation) and help to retain and slowly release soil moisture. The tree and shrub trimmings can be used as mulch (green manure) for the crops and as fodder for livestock.

How Can Gully Reclamation, Windbreaks, Land Classification, and PAM Reduce Soil Erosion? **Gully reclamation** can restore sloping bare land on which water runoff quickly creates gullies. Small gullies can be seeded with quick-growing plants such as oats, barley, and wheat for the first season, whereas deeper gullies can be dammed to collect silt and gradually fill in the channels. Fast-growing shrubs, vines, and trees can also be planted to stabilize the soil, and channels can be built to divert water from the gully and prevent further erosion.

Windbreaks, or **shelterbelts**, can reduce wind erosion. Long rows of trees planted to partially block the wind (Figure 13-24d). Windbreaks, which are especially effective if uncultivated land is kept covered with vegetation, also help retain soil moisture, supply some wood for fuel, and provide habitats for birds, pest-eating and pollinating insects, and other animals. Unfortunately, many of the windbreaks planted in the upper Great Plains of the United States after the 1930s Dust Bowl disaster have been cut down to make way for large irrigation systems and modern farm machinery.

Hint: Enter the search term *soil degradation* using the Subject Guide.

Land can be evaluated with the goal of identifying easily erodible (marginal) land that should be neither planted in crops nor cleared of vegetation. In the United States, the National Resources Conservation Service has set up a classification system to identify types of land that are suitable or unsuitable for cultivation. Such efforts and recent farm legislation have helped reduce soil erosion in the United States (Case Study, below).

A chemical called polyacrylamide (PAM) has been used recently to sharply reduce erosion of some irrigated fields, at moderate cost in most soils (except desert soils). Tests show that adding just 10 parts per million (ppm) of this white crystal to water during the first hour of irrigation can reduce soil erosion by 70–99%. After that, irrigation can continue for 19–24 hours without further treatment. The negatively charged PAM particles may work by binding to the positively charged clay particles in soils, reducing erosion by increasing the cohesiveness of surface soil particles.

How Can We Maintain and Restore Soil Fertility? Fertilizers partially restore plant nutrients lost by erosion, crop harvesting, and leaching. Farmers can use either **organic fertilizer** from plant and animal materials or **commercial inorganic fertilizer** produced from various minerals.

Three basic types of *organic fertilizer* are animal manure, green manure, and compost. **Animal manure** includes the dung and urine of cattle, horses, poultry, and other farm animals. It improves soil structure, adds organic nitrogen, and stimulates beneficial soil bacteria and fungi. Despite its effectiveness, the use of animal manure in the United States has decreased. One reason is that separate farms for growing crops and raising animals have replaced most mixed animal-raising and crop-farming operations. Animal manure is available at feedlots near urban areas, but transporting it to distant rural crop-growing areas usually costs too much. Thus, much of this valuable resource is wasted and can end up polluting nearby bodies of water. In addition, tractors and other motorized farm machinery have replaced horses and other draft animals that naturally added manure to the soil.

Green manure is fresh or growing green vegetation plowed into the soil to increase the organic matter and humus available to the next crop. **Compost** is a rich natural fertilizer and soil conditioner that aerates soil, improves its ability to retain water and nutrients, helps prevent erosion, and prevents nutrients from being wasted in landfills. Farmers, homeowners, and communities produce compost by piling up alternating layers of nitrogen-rich wastes (such as grass clippings,

Slowing Soil Erosion in the United States

CASE STUDY

Of the world's major food-producing countries, only the United States is sharply reducing some of its soil losses through conservation tillage and government-sponsored soil conservation programs.

The 1985 Farm Act established a strategy for reducing soil erosion in the United States. In the first phase of this program, farmers are given a subsidy for highly erodible land they take out of production and replant with soil-saving grass or trees for 10 years. The land in such a *conservation reserve* cannot be farmed, grazed, or cut for hay. Farmers who violate their contracts must pay back all subsidies plus interest.

Between 1985 and 1995, this program cut soil losses on cropland in the United States by about 60%—a shining example of good news—and could eventually cut such losses as much as 80%. In 1996 Congress reauthorized the Conservation Reserve Program until 2002.

The second phase of the program required all farmers with highly erodible land to develop government-approved 5-year soil conservation plans for their entire farms by the end of 1990. A third provision of the Farm Act authorizes the government to forgive all or part of farmers' debts to the Farmers Home Administration if they agree not to farm highly erodible cropland or wetlands for 50 years. The farmers must plant trees or grass on this land or restore it to wetland.

Despite some weaknesses, the 1985 Farm Act makes the United States the first major food-producing country to make soil conservation a national priority. Even though these efforts to slow soil erosion are an important step, effective soil conservation is practiced on only about half of all U.S. agricultural land and on less than half of the country's most erodible cropland.

Critical Thinking

Do you believe that U.S. tax dollars should be used to pay farmers for taking highly erodible land out of production? Explain. What are the alternatives?

weeds, animal manure, and vegetable kitchen scraps), carbon-rich plant wastes (dead leaves, hay, straw, sawdust), and topsoil (Individuals Matter, right). This mixture provides a home for microorganisms that aid the decomposition of the plant and manure layers. Composting also reduces the amount of waste taken to landfills and incinerators.

Another method for conserving soil nutrients is **crop rotation**. Corn, tobacco, and cotton can deplete the topsoil of nutrients (especially nitrogen) if planted on the same land several years in a row. Farmers using crop rotation may plant areas or strips with such nutrient-depleting crops one year; the next year, however, they plant the same areas with legumes, whose root nodules add nitrogen to the soil, or with crops such as oats, barley, rye, or sorghum. This method helps restore soil nutrients and reduces erosion by keeping the soil covered with vegetation. It also helps reduce crop losses to insects by presenting them with a changing target.

INDIVIDUALS MATTER

Fast-Track Composting

Compost piles must be turned over every few days for aeration to speed up decomposition. Ruth Beckner has invented a special drill bit that attaches to a cordless electric drill to inject air into a compost pile with very little physical exertion. Using her COMPOST AIR™* device for about 3 minutes per day, you can create good compost in about 3 weeks, instead of waiting 12–18 months for the pile to decompose naturally.

Worms, the planet's champion recyclers, can also help create compost efficiently and without any odors. Recycling coordinators in Seattle, Washington, and Toronto, Canada, have given out thousands of worm bins to homeowners.

*For information, contact Beckner & Beckner, 15 Portola Avenue, San Rafael, CA 94903 or call 1-800-58COMPOST.

Will Inorganic Fertilizers Save the Soil? Today, many farmers (especially in developed countries) rely on *commercial inorganic fertilizers* containing nitrogen (as ammonium ions, nitrate ions, or urea), phosphorus (as phosphate ions), and potassium (as potassium ions). Other plant nutrients may also be present in low or trace amounts.

Inorganic commercial fertilizers are easily transported, stored, and applied. Worldwide, their use increased about 10-fold between 1950 and 1989, but declined by 12% between 1990 and 1998. Today, the additional food they help produce feeds one of every three people in the world; without them, world food output would drop an estimated 40%.

Commercial inorganic fertilizers have some disadvantages, however. They do not add humus to the soil. Unless animal manure and green manure are also added, the soil's content of organic matter, and thus its ability to hold water, will decrease, and the soil will become compacted and less suitable for crop growth. By decreasing the soil's porosity, inorganic fertilizers also lower its oxygen content and keep added fertilizer from being taken up as efficiently. In addition, most commercial fertilizers supply only 2 or 3 of the 20 or so nutrients needed by plants. Moreover, producing, transporting, and applying inorganic fertilizers require large amounts of energy and release nitrous oxide (N_2O), a greenhouse gas (Figure 11-2d).

The widespread use of commercial inorganic fertilizers, especially on sloped land near streams and lakes, also causes water pollution as some fertilizer nutrients are washed into nearby bodies of water. The resulting plant nutrient enrichment (cultural eutrophication; Figure 12-21) causes algae blooms that use up oxygen dissolved in the water, thereby killing fish. Rainwater seeping through the soil can also leach nitrates in commercial fertilizers into groundwater. Drinking water drawn from wells containing high levels of nitrate ions can be toxic, especially for infants.

Environmental historian Donald Worster reminds us that fertilizers are not a substitute for fertile soil:

> *We can no more manufacture a soil with a tank of chemicals than we can invent a rain forest or produce a single bird. We may enhance the soil by helping its processes along, but we can never recreate what we destroy. The soil is a resource for which there is no substitute.*

According to soil scientists, responsibility for reducing soil erosion should not be limited to farmers. Timber cutting, overgrazing, mining, and urban development carried out without proper regard for soil conservation cause at least 40% of soil erosion in the United States. Some things you can do to reduce soil erosion are listed in Appendix 6.

Land Degradation and Environmental Resources

GUEST ESSAY

David Pimentel

David Pimentel is professor of insect ecology and agricultural sciences in the College of Agriculture and Life Sciences at Cornell University. He has published over 440 scientific papers and 22 books on environmental topics, including land degradation, agricultural pollution and energy use, biomass energy, and pesticides. He was one of the first ecologists to use an interdisciplinary, holistic approach in investigating complex environmental problems.

At a time when the world's human population is rapidly expanding and its need for more land to produce food, fiber, and fuelwood is also escalating, valuable land is being degraded through erosion and other means at an alarming rate. Soil degradation is of great concern because soil reformation is extremely slow. Globally it takes an average of 500 years (with a range of 220 to 1,000 years) to renew 2.5 centimeters (1 inch) of soil in tropical and temperate areas—a renewal rate of about 1 metric ton of topsoil per hectare of land per year. Worldwide annual erosion rates for agricultural land are about 20–100 times this average.

Erosion rates vary in different regions because of topography, rainfall, wind intensity, and the type of agricultural practices used. In China, for example, the average annual soil loss is reported to be about 40 metric tons per hectare (18 tons per acre), whereas the U.S. average is 18 metric tons per hectare (8 tons per acre). In states such as Iowa and Missouri, however, annual soil erosion averages more than 35 metric tons per hectare (16 tons per acre).

Worldwide, about 10 million hectares (25 million acres) of land—an area about the size of Virginia—are abandoned for crop production each year because of high erosion rates plus waterlogging, salinization, and other forms of soil degradation. In addition, according to the UN Environment Programme, crop production becomes uneconomical on about 20 million hectares (49 million acres) each year because soil quality has been severely degraded.

Soil erosion also occurs in forestlands, but it is not as severe as that in the more exposed soil of agricultural land. Soil erosion in managed forests is a primary concern because the soil reformation rate in forests is about one-third to one-half as fast as that in agricultural land. To compound this erosion problem, at least 24 million hectares (59 million acres) of forest are being cleared each year throughout the world; most of this land is used to grow food and graze cattle.

The effects of agriculture and forestry are interrelated in other ways. Deforestation reduces fuelwood supplies and forces the poor in developing countries to substitute crop residue and manure for fuelwood. When these plant and animal wastes are burned instead of being returned to the land as ground cover and organic fertilizer, erosion is intensified and productivity of the land is decreased. These factors, in turn, increase pressure to convert more forestland into agricultural land, further intensifying soil erosion.

One reason that soil erosion is not a high priority among many governments and farmers is that it usually occurs so slowly that its cumulative effects may take decades to become apparent. For example, the removal of 1 millimeter (0.04 inch) of soil is so small that it goes undetected. But over a 25-year period the loss would be 25 millimeters (1 inch), which would take about 500 years to replace by natural processes.

Besides reduced soil depth, soil erosion leads to reduced crop productivity because of losses of water, organic matter, and soil nutrients. A 50% reduction of soil organic matter on a plot of land has been found to reduce corn yields as much as 25%.

At some point, either the loss of topsoil from the world's croplands will have to be checked by effective soil conservation practices, or the growth in the world's population will be checked by hunger and malnutrition.

LESTER R. BROWN

REVIEW QUESTIONS

1. Define the boldfaced terms used in this chapter.

2. Describe the basic features of the 1872 Mining Law in the United States.

3. Distinguish between the earth's *core, mantle,* and *crust.*

4. What are *tectonic plates*? What is the *lithosphere*? What is the *theory of plate tectonics* and what is its importance to physical and biological processes on the earth? What are the three different types of boundaries between the earth's lithospheric plates?

5. What is *erosion* and what are its two major causes?

6. Distinguish between a *mineral* and a *rock.* Distinguish between *igneous, sedimentary,* and *metamorphic rock* and give two examples of each type.

7. Describe the *rock cycle* and explain its importance.

When soil erodes, vital plant nutrients such as nitrogen, phosphorus, potassium, and calcium are also lost. With U.S. annual cropland erosion rates of about 18 metric tons per hectare (8 tons per acre), an estimated $18 billion of plant nutrients is lost annually. Using fertilizers to replace these nutrients substantially adds to the cost of crop production.

Some analysts who are unaware of the numerous and complex effects of soil erosion have falsely concluded that the damages are minor. For example, they report that soil loss causes an annual reduction in crop productivity of only 0.1–0.5% in the United States. However, we must consider all the ecological effects caused by erosion, including reductions in soil depth, availability of water for crops, and soil organic matter and nutrients. When this is done, agronomists and ecologists report a 15–30% reduction in crop productivity, leading to increased use of costly fertilizer. Because fertilizers are not a substitute for fertile soil, they can be applied only up to certain levels before crop yields begin to decline.

Reduced agricultural productivity is only one of the effects of soil erosion. In the United States, water runoff is responsible for transporting about 3 billion metric tons (3.3 billion tons) of sediment (about 60% from agricultural land) each year to waterways in the lower 48 states. Off-site damages to U.S. water storage capacity, wildlife, and navigable waterways from these sediments cost an estimated $6 billion each year. About 25% of new water storage capacity in U.S. reservoirs is built solely to compensate for sediment buildup.

When soil sediments that include pesticides and other agricultural chemicals are carried into streams, lakes, and reservoirs, fish production is adversely affected. These contaminated sediments interfere with fish spawning, increase predation on fish, and destroy fisheries in estuarine and coastal areas.

Increased erosion and water runoff on mountain slopes flood agricultural land in the valleys below, further decreasing agricultural productivity. Eroded land also does not hold water very well, again decreasing crop productivity. This effect is magnified in the 80 countries (with nearly 40% of the world's population) that experience frequent droughts.

Thus, soil erosion is one of the world's critical problems, and if not slowed, it will seriously reduce agricultural and forestry production and degrade the quality of aquatic ecosystems. Solutions that are not particularly difficult often are not implemented because erosion occurs so gradually that we fail to acknowledge its cumulative effects until damage is irreversible. Many farmers have been conditioned to believe that losses in soil fertility can be remedied by applying more fertilizer or by using more fossil-fuel energy.

The principal way to control soil erosion and its accompanying runoff of sediment is to maintain adequate vegetative coverage on soils [by using various methods discussed in Section 13-8]. These methods are also cost-effective, especially when off-site costs of erosion are included. Scientists, policy makers, and agriculturists need to work together to implement soil and water conservation practices before world soils lose most of their productivity.

Critical Thinking

1. Some analysts contend that average soil erosion rates around the world are low and that the soil erosion problem can be solved easily with improved agricultural technology such as no-till cultivation and increased use of commercial inorganic fertilizers. Do you agree or disagree with this position? Explain.

2. What three major things do you believe elected officials should do to decrease soil erosion and the resulting water pollution by sediment in the United States or in the country where you live?

8. Distinguish between a *mineral resource* and an *ore*. Distinguish between *identified resources, undiscovered resources, reserves,* and *other resources*. What are two ways in which the reserves of a mineral resource can increase?

9. Distinguish between *subsurface mining* and *surface mining*. Distinguish between *open-pit mining, dredging,* and *strip mining*. Distinguish between *area strip mining* and *contour strip mining*.

10. What is *economic depletion* of a mineral resource? When such depletion occurs, what five choices are available? Distinguish between *depletion time* and the *reserve-to-production ratio* for a mineral resource. How can the estimated depletion time for a resource be extended? Explain why the United States will never again be self-sufficient in oil and many key metal resources.

11. Explain how a competitive free market should increase supplies of a scarce mineral resource and explain why this idea may not be effective under today's economic conditions. Explain how mining lower-grade ores can increase the supply of a mineral resource and list three factors that might limit the usefulness of this approach.

12. Evaluate the potential of **(1)** using the oceans as a source of additional supplies of key minerals and **(2)** finding substitutes for scarce nonrenewable mineral resources.

13. List five major environmental impacts of mining, processing, and using mineral resources. What is *smelting,* and what are its major environmental impacts? What are three environmental limits associated with the extraction and use of mineral resources?

14. How can microbes be used to mine some ores, and how useful is this approach? What are some of the problems associated with the increased mining of gold in developed and developing countries?

15. What is *soil?* Distinguish between a *soil horizon* and a *soil profile.* What is *humus,* and what is its importance? What does the color of topsoil tell you about how useful a soil is for growing crops? Distinguish between *soil infiltration* and *leaching.* Distinguish between *soil texture, soil porosity,* and *soil permeability.* What are *loams* and why are they the best soils for growing most crops?

16. What are the major natural and human-related causes of soil erosion? Describe three types of soil erosion. What are the major harmful effects of soil erosion?

17. How serious is soil erosion **(a)** on a global scale and **(b)** in the United States?

18. What is *desertification?* How serious is this problem, and what are its major causes? How can desertification be slowed?

19. Distinguish between *salinization* and *waterlogging* of soils. How serious are these problems? What can be done about them?

20. What is *soil conservation?* Distinguish between *conventional-tillage farming* and *conservation-tillage farming.* What are the advantages of conservation-tillage farming?

21. Distinguish between *terracing, contour farming, strip cropping, alley cropping, gully reclamation,* and *windbreaks* as methods for reducing soil erosion.

22. Distinguish between *organic fertilizer* and *commercial inorganic fertilizer* and list the advantages of each approach for maintaining or restoring soil fertility. Distinguish between *animal manure, green manure,* and *compost* as methods for fertilizing soil. What is *crop rotation,* and why is it useful in helping maintain soil fertility? List the advantages and disadvantages of using commercial inorganic fertilizers to maintain and restore soil fertility.

CRITICAL THINKING

1. (a) Explain what would happen if plate tectonics stopped. **(b)** Explain what would happen if erosion stopped.

2. Use the second law of energy (Section 3-7) to analyze the feasibility of each of the following processes. Which, if any, could be profitable without subsidies?

a. Extracting most minerals dissolved in seawater

b. Recycling minerals that are widely dispersed

c. Mining increasingly lower-grade deposits of minerals

d. Using inexhaustible solar energy to mine minerals

e. Continuing to mine, use, and recycle minerals at increasing rates

3. Explain why you support or oppose each of the proposals listed on p. 338 for reforming the U.S. 1872 Mining Law.

4. Why should everyone, not just farmers, be concerned about soil conservation?

5. What are the main advantages and disadvantages of commercial inorganic fertilizers? Why should both inorganic and organic fertilizers be used?

***6.** Use the library or Internet to find out what key mineral resources are found in the country where you live. How long could the estimated reserves of each of these meet the current needs of your country? How long will the estimated reserves last if the use of each of these resources increases by 2% a year?

***7.** What mineral resources are extracted in your local area? What mining methods are used? Do local, state, or federal laws require restoration of the landscape after mining is completed? If so, how stringently are those laws enforced?

***8.** As a class project, evaluate soil erosion on your school grounds. Use this information to develop a soil conservation plan for your school and then present it to school officials.

***9.** Make a concept map of this chapter's major ideas, using the section heads and subheads and the key terms (in boldface type). Look at the inside back cover and in Appendix 5 for information about concept maps.

FURTHER INFORMATION

For more information see Internet Sources and Further Readings at the back of the book. Online resources for this chapter can be found at **www.brookscole.com/biology**

14 SOLID AND HAZARDOUS WASTE

There Is No "Away": The Love Canal Tragedy

Between 1942 and 1953, Hooker Chemicals and Plastics (owned by OxyChem since 1968) sealed chemical wastes containing at least 200 different chemicals into steel drums and dumped them into an old canal excavation (called Love Canal after its builder, William Love) near Niagara Falls, New York.

In 1953 Hooker Chemicals filled the canal, covered it with clay and topsoil, and sold it to the Niagara Falls school board for $1. The company inserted in the deed a disclaimer denying legal liability for any injury caused by the wastes. In 1957 Hooker warned the school board not to disturb the clay cap because of the possible danger from toxic wastes.

By 1959 an elementary school, playing fields, and 949 homes had been built in the 10-square-block Love Canal area (Figure 14-1). Roads and sewer lines crisscrossed the dump site, some of them disrupting the clay cap covering the wastes. An expressway built at one end of the dump in the 1960s blocked groundwater from migrating to the Niagara River. This created a "bathtub effect" that allowed contaminated groundwater and rainwater to build up and overflow the disrupted cap.

Residents began complaining to city officials in 1976 about chemical smells and chemical burns their children received playing in the canal area, but these complaints were ignored. In 1977 chemicals began leaking from the badly corroded steel drums into storm sewers, gardens, basements of homes next to the canal, and the school playground.

In 1978, after considerable media publicity and pressure from residents led by Lois Gibbs (a mother galvanized into action as she watched her children come down with one illness after another; Guest Essay, p. 362), the state acted. It closed the school and arranged for the 239 homes closest to the dump to be evacuated, purchased, and destroyed.

Two years later, after protests from families still living fairly close to the landfill, President Jimmy Carter declared Love Canal a federal disaster area, had the remaining families relocated, and offered federal funds to buy 564 more homes. Residents of all but 72 of the homes moved out.

The dump site was covered with a new clay cap and surrounded by a drainage system that pumps leaking wastes to a new treatment plant. Some of the residents who remained claim that the entire problem was exaggerated by other residents, environmentalists, and the media.

After more than 15 years of court cases, OxyChem agreed in 1994 to a $98-million settlement with New York State and agreed to be responsible for all future treatment of wastes and wastewater at the Love Canal site. In 1999 the company agreed to reimburse the federal government and New York state $7.1 million for the Love Canal cleanup.

Because of the difficulty in linking exposure to a variety of chemicals to specific health effects (Section 8-2), the long-term health effects of exposure to hazardous chemicals on Love Canal residents remain unknown and controversial. However, for the rest of their lives the evacuated families will worry about the possible effects of the chemicals on themselves and their children and grandchildren.

In June 1990 the EPA declared the area (renamed Black Creek Village) safe and allowed state officials to begin selling 234 of the remaining houses at 10–20% below market value. The dump has not been cleaned up but only fitted with a drainage system. The EPA acknowledges that the dump site will leak again, sooner or later. Buyers must sign an agreement stating that New York State and the federal government make no guarantees or representations about the safety of living in these homes.

The Love Canal incident is a vivid reminder that we can never really throw anything away, that wastes don't stay put, and that preventing pollution is much safer and cheaper than trying to clean it up.

Figure 14-1 The Love Canal housing development near Niagara Falls, New York, was built near a hazardous-waste dump site. The photo shows the area when it was abandoned in 1980. In 1990 the EPA allowed people to buy some of the remaining houses and move back into the area, despite protests from environmentalists. (NY State Department of Environmental Conservation)

Solid wastes are only raw materials we're too stupid to use.
ARTHUR C. CLARKE

This chapter addresses the following questions:

- What are solid waste and hazardous waste, and how much of each type do we produce?
- What can we do to reduce, reuse, and recycle solid waste and hazardous waste?
- What are we doing to recycle aluminum, paper, and plastics?
- What are the advantages and disadvantages of burning or burying wastes?
- What can we do to reduce exposure to lead, dioxins, and hazardous chlorine compounds?
- How can we make the transition to a more sustainable low-waste society?

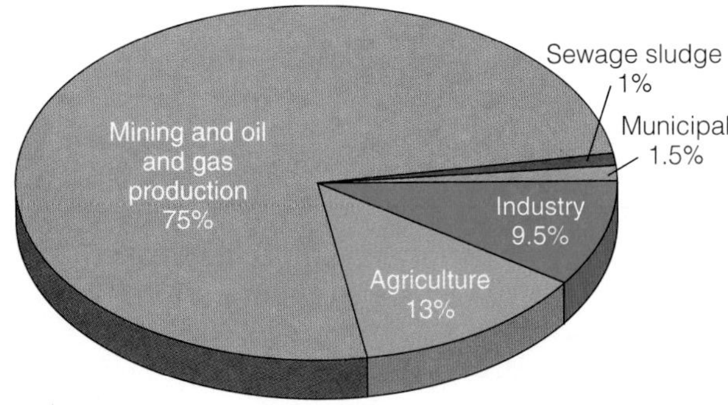

Figure 14-2 Sources of the estimated 11 billion metric tons (12 billion tons) of solid waste produced each year in the United States. Mining and industrial activities produce 65 times as much solid waste as household activities. (Data from U.S. Environmental Protection Agency and U.S. Bureau of Mines)

14-1 WASTING RESOURCES: THE HIGH-WASTE APPROACH

What Is Solid Waste, and How Much Is Produced The United States, with only 4.6% of the world's population, produces about 33% of the world's **solid waste**: any unwanted or discarded material that is not a liquid or a gas. The United States generates about 11 billion metric tons (12 billion tons) of solid waste per year—an average of 40 metric tons (44 tons) per person. Although garbage produced directly by households and businesses is a significant problem, about 98.5% of the solid waste in the United States comes from mining, oil and natural gas production, agriculture, and industrial activities used to produce goods and services for consumers (Figure 14-2).

The remaining 1.5% of solid waste produced in the United States is **municipal solid waste** (MSW) from homes and businesses in or near urban areas. The amount of MSW, often called *garbage*, produced in the United States in 1998 was enough to fill a bumper-to-bumper convoy of garbage trucks encircling the globe almost eight times. This amounted to an average of 860 kilograms (1,900 pounds) per person in the United States—two to three times that in most other developed countries and many times that in developing countries.

About 30% of the resources in MSW produced in the United States in 1998 was recycled or composted (up from 7% in 1960); the other 70% was hauled away and either dumped in landfills (54%) or burned in incinerators and waste-to-energy plants (16%) at a cost of about $40 billion (projected to rise to $75 billion by 2005).

What Does It Mean to Live in a High-Waste Society? Here are a few of the solid wastes U.S. consumers throw away:

- Enough aluminum to rebuild the country's entire commercial airline fleet every 3 months
- Enough tires each year to encircle the planet almost three times
- About 18 billion disposable diapers per year, which if linked end-to-end would reach to the moon and back seven times
- About 2 billion disposable razors, 10 million computers, and 8 million television sets each year
- About 2.5 million nonreturnable plastic bottles every hour
- Some 14 billion catalogs (an average of 54 per American) and 38 billion pieces of junk mail each year

This is only part of the 1.5% of all solid waste labeled "municipal" in Figure 14-2.

What Is Hazardous Waste, and How Much Is Produced? In the United States, **hazardous waste** is legally defined as any discarded solid or liquid material that **(1)** contains one or more of 39 toxic, carcinogenic, mutagenic, or teratogenic compounds at levels that exceed established limits (including many solvents, pesticides, and paint strippers), **(2)** catches fire easily (gasoline, paints, and solvents), **(3)** is reactive or unstable enough to explode or release toxic fumes (acids, bases, ammonia, chlorine bleach), or **(4)** is capable of corroding metal containers such as tanks, drums, and barrels (industrial cleaning agents and oven and drain cleaners).

This narrow official definition of hazardous wastes (mandated by Congress) does *not* include the following materials: **(1)** radioactive wastes (Section 19-5), **(2)** hazardous and toxic materials discarded by households (Table 14-1), **(3)** mining wastes, **(4)** oil- and gas-drilling wastes (routinely discharged into surface waters or dumped into unlined pits and landfills), **(5)** liquid waste containing organic hydrocarbon compounds (80% of all liquid hazardous waste), **(6)** cement kiln dust, produced when liquid hazardous wastes are burned in a cement, and **(7)** wastes from the thousands of small businesses and factories that generate less than 100 kilograms (220 pounds) of hazardous waste per month.

The EPA estimates that at least 5.5 billion metric tons (6 billion tons) of hazardous waste are produced each year in the United States—an average of 2 metric tons (22 tons) per person. This amounts to about 75% of the world's hazardous waste. However, because only about 5% of the total is *legally* defined as hazardous waste, *95% of the country's hazardous waste is not regulated by hazardous-waste laws.* In most other countries, especially developing countries, even less of the hazardous waste is regulated.

Table 14-1 Common Household Toxic and Hazardous Materials

Cleaning Products
Disinfectants
Drain, toilet, and window cleaners
Oven cleaners
Bleach and ammonia
Cleaning solvents and spot removers
Septic tank cleaners
Paint and Building Products
Latex and oil-based paints
Paint thinners, solvents, and strippers
Stains, varnishes, and lacquers
Wood preservatives
Acids for etching and rust removal
Asphalt and roof tar
Gardening and Pest Control Products
Pesticide sprays and dusts
Weed killers
Ant and rodent killers
Flea powder
Automotive Products
Gasoline
Used motor oil
Antifreeze
Battery acid
Solvents
Brake and transmission fluid
Rust inhibitor and rust remover
General Products
Dry-cell batteries (mercury and cadmium)
Artists' paints and inks
Glues and cements

14-2 PRODUCING LESS WASTE AND POLLUTION: REDUCING THROUGHPUT

What Are Our Options? There are two ways to deal with the solid and hazardous waste we create: **(1)** *waste management* and **(2)** *pollution (waste) prevention.* Waste management is a *high-waste approach* that views waste production as an unavoidable product of economic growth. It attempts to manage the resulting wastes in ways that reduce environmental harm, mostly by burying them, burning them, or shipping them off to another state or country. The goal is to move increasing amounts of matter and energy resources through the economy to enhance economic growth (Figure 3-14).

Preventing pollution and waste is a *low-waste approach* that views most solid and hazardous waste either **(1)** as potential resources (that we should be recycling, composting, or reusing) or **(2)** as harmful substances that we should not be using in the first place (Figures 3-15, 14-3, and 14-4). With this approach, taxes and subsidies (Table 2-1) are used to discourage waste production and encourage waste prevention (Guest Essays, p. 78 and 362).

According to the U.S. National Academy of Sciences (Figures 14-3 and 14-4), the low-waste approach should have the following hierarchy of goals: **(1)** *reduce* waste and pollution, **(2)** *reuse* as many things as possible, **(3)** *recycle and compost* as much waste as possible, **(4)** *chemically or biologically treat or incinerate* waste that can't be reduced, reused, recycled, or composted, and **(5)** *bury* what is left in state-of-the-art landfills or aboveground vaults after the first four goals have been met.

Scientists estimate that in a low-waste society 60–80% of the solid and hazardous waste produced could be eliminated through reduction, reuse, and recycling (including composting). The remaining 20–40% of

Figure 14-3 Solutions: priorities suggested by prominent scientists for dealing with material use and solid waste. To date, these priorities have not been followed in the United States (and in most other countries); most efforts are devoted to waste management (bury it or burn it). (U.S. Environmental Protection Agency and U.S. National Academy of Sciences)

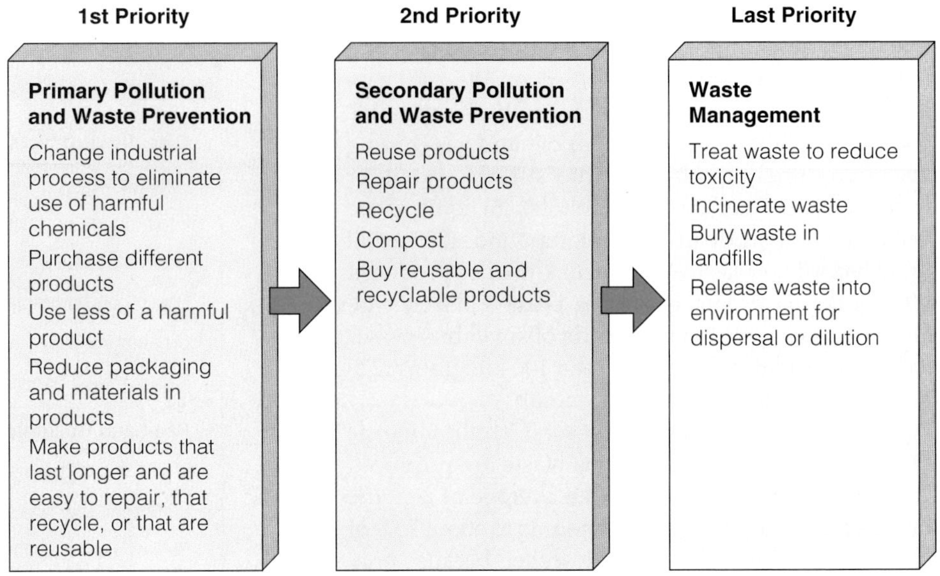

Figure 14-4 Solutions: priorities suggested by prominent scientists for dealing with hazardous waste. To date, these priorities have not been followed in the United States (and in most other countries). (U.S. National Academy of Sciences)

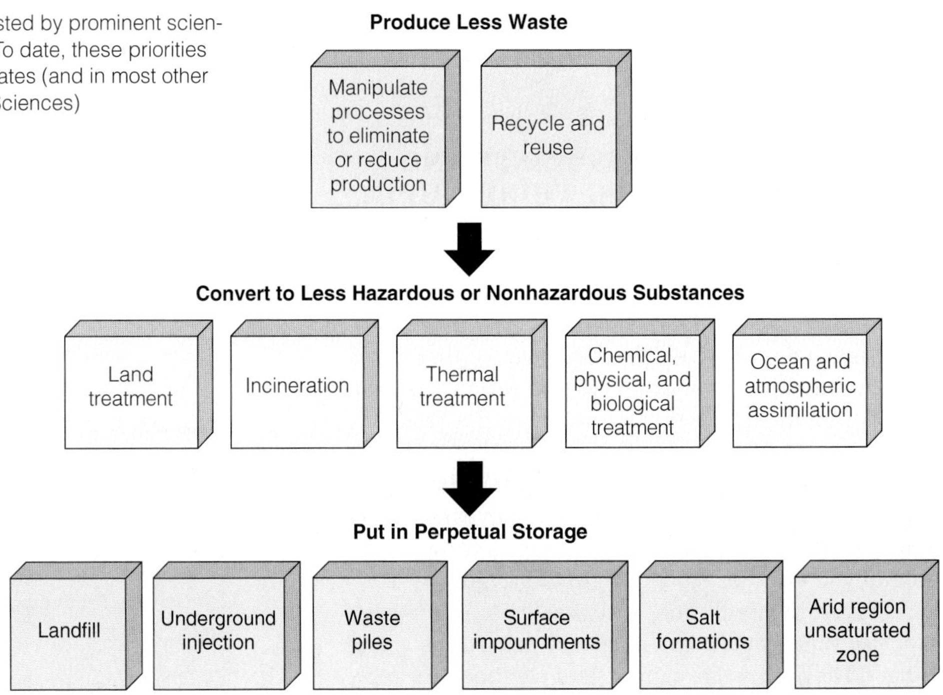

such wastes would then be treated to reduce their toxicity, and what's left would be burned or buried under carefully regulated conditions. Currently, the order of priorities shown in Figures 14-3 and 14-4 for dealing with solid and hazardous wastes is reversed in the United States (and in most other countries). The major reason for this is that the costs of producing and dealing with these wastes are not included in the market prices of products.

Why Is Producing Less Waste and Pollution the Best Choice? A small but increasing number of companies are learning that reducing waste and pollution can be good for corporate profits, worker health and safety, the local community, consumers, and the environment as a whole. Such methods

- Save energy and virgin resources by keeping the quality of matter resources high (Figure 3-7) with a lower input of high-quality energy
- Reduce the environmental effects of extracting, processing, and using resources (Figure 13-12)
- Improve worker health and safety by reducing exposure to toxic and hazardous materials

- Decrease pollution control and waste management costs and future liability for toxic and hazardous materials
- Usually are less costly on a life cycle basis than trying to clean up pollutants and manage wastes once they are produced

A 1992 study of 181 waste reduction initiatives in 27 U.S. firms found that two-thirds of the initiatives took 6 months or less to implement. One-fourth required no capital investment, two-thirds paid back their capital investments in 6 months or less, and 93% got them back within 3 years.

In 1975 the Minnesota Mining and Manufacturing Company (3M), which makes 60,000 different products in 100 manufacturing plants, began a Pollution Prevention Pays (3P) program. It **(1)** redesigned equipment and processes, **(2)** used fewer hazardous raw materials, **(3)** identified hazardous chemical outputs (and recycled or sold them as raw materials to other companies), and **(4)** began making more nonpolluting products. By 1995 3M's overall waste production was down by one-third, emissions of air pollutants were reduced by 70%, and the company had saved over $750 million in waste disposal costs.

Solutions: How Can We Reduce Waste and Pollution? There are several ways to reduce waste and pollution or resource throughput (Figure 3-15). One is to *decrease consumption*. Before any purchase this involves asking questions such as: Do I really need this? Can I buy it secondhand? Can I borrow or lease it?

Another way to reduce throughput is to *redesign manufacturing processes and products to use less material* (Figure 14-5). For example, cars made today are lighter because an increasing proportion of their steel parts are being replaced with aluminum (80% from recycled metal) and plastic parts.

Another strategy is to *design products that produce less pollution and waste fewer resources when used* (Figure 14-5). Examples include paints that use fewer volatile solvents and more energy-efficient cars, lights, and appliances (Section 20-2). Biodegradable enzymes have replaced water-polluting phosphates (Figure 12-21) in 90% of all detergents used in Japan and Europe and in 50% of those used in the United States. In 1998, Israeli scientists developed an edible, clear, tasteless, odorless coating (derived from seaweed) for food that could make synthetic plastic wrapping obsolete and make foods last longer.

Manufacturing processes can also be redesigned to produce less waste and pollution (Figure 14-5). Most toxic organic solvents can be recycled within plants or replaced with water-based or citrus-based solvents (Individuals Matter, p. 281). Some processes can be redesigned to eliminate washing altogether. Hydrogen peroxide can be used instead of toxic chlorine to bleach paper and other materials. For clothes, wet cleaning (with water and steam or with detergents dissolved in liquid CO_2) and microwave drying can replace dry cleaning with toxic organic solvents such as perchloroethylene (PERC). In effect, over the next two decades we could change the ways we make most things (Solutions, p. 360).

Individuals can use less hazardous (and usually cheaper) cleaning products (Table 14-2). Three inexpensive chemicals—*baking soda* (which can also be used as a deodorant and a toothpaste), *vinegar*, and *borax*—can be used for most cleaning and clothes bleaching. People can also use pesticides and other hazardous chemicals only when absolutely necessary and in the smallest amount possible.

Green design and life cycle assessment can help develop products that are easy to repair, reuse, remanufacture, compost, or recycle. Several European auto manufacturers now design their cars for easy disassembly and for reuse and recycling of up to 80% of their parts (75% in the United States). They also are trying to minimize the use of nonrecyclable or hazardous materials.

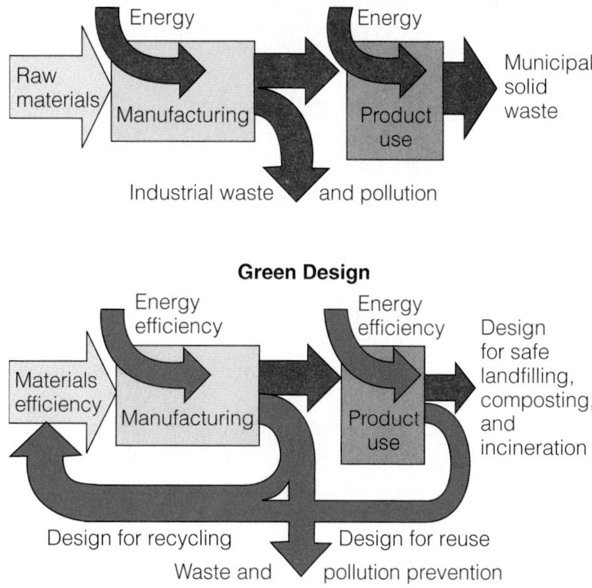

Figure 14-5 How product design affects throughputs of matter and energy resources and outputs of pollution and solid waste. Most conventional manufacturing processes involve a one-way flow of materials from raw materials to waste and pollution. Green design reduces the overall impact by emphasizing efficient use of matter and energy resources, reduction of outputs of solid and hazardous waste, and reuse and recycling of materials. (Adapted from U.S. Congress Office of Technology Assessment, *Green Products by Design: Choices for a Cleaner U.S. Environment*, Washington, D.C.: Government Printing Office, 1992)

An Ecoindustrial Revolution: Cleaner Production

SOLUTIONS

Some analysts urge us to bring about an *ecoindustrial revolution* over the next 50 years as a way to help achieve industrial, economic, and environmental sustainability. All industrial products and processes would be redesigned and integrated into an essentially closed system of cyclical material flows.

The goals of this emerging concept of *cleaner production* (Guest Essay, p. 78), or *industrial ecology*, are to reduce resource throughput, waste, and pollution (Figure 3-15) by **(1)** minimizing the input of energy and matter resources and the output of wastes, **(2)** using less material and energy per unit of output, **(3)** substituting less toxic or nontoxic chemicals in manufacturing processes or products, **(4)** recycling or reusing toxic chemicals used or produced in manufacturing processes, and **(5)** developing materials exchange systems in which one company's wastes become another company's resources and companies take back packaging and used products from consumers for reuse, recycling, repair, or remanufacturing.

In effect, companies would mimic natural chemical cycles (Section 4-5) and interact in complex *resource exchange webs* similar to food webs in natural ecosystems. With such an exchange and chemical cycling network, producing a large amount of easily reusable waste might sometimes be preferable to designing a more efficient process that produces a small amount of unusable waste.

A prototype of this concept exists in Kalundborg, Denmark, where a coal-fired power plant, an oil refinery, a municipal heating authority, a producer of sulfuric acid, a sheet rock plant, a biotechnology company, local farms, some greenhouses, a fish farm, and several other businesses are working together to exchange and convert their wastes into resources. A similar arrangement in Fiji links together a chicken-raising operation, a mushroom farm, a brewery, fish farms, hydroponic gardens, and a methane gas production unit. Ecoindustrial parks are in the planning stages in areas such as Baltimore (Maryland), Rochester (New York), Chattanooga (Tennessee), Halifax (Nova Scotia), and the Brownsville/Matamoros region along the Texas–Mexico border.

Such an ecoindustrial revolution, carried out over the next 50 years, will challenge designers, engineers, and scientists to revolutionize the way we design and make things (Guest Essay, p. 232). This will reduce pollution, improve human health, and help preserve biodiversity and ecological integrity.

This important form of learning to work with nature will also provide economic benefits to businesses by **(1)** reducing the costs of controlling pollution and complying with pollution regulations, **(2)** improving the health of workers (thus reducing company health-care insurance expenses), **(3)** reducing legal liability for toxic and hazardous wastes, **(4)** stimulating companies to come up with new, environmentally friendly processes and products that can be sold worldwide, and **(5)** giving companies a better image among consumers, based on results rather than on public relations campaigns.

Critical Thinking

What short- and long-term disadvantages (if any) might there be with an ecoindustrial revolution? Do you believe that it will be possible to phase in such a revolution in the country where you live over the next 2–3 decades? Explain. What are the three most important strategies for doing this?

Products can be designed to last longer. For example, although tires are now being produced with an average life of 97,000 kilometers (60,000 miles), researchers believe this could be extended to at least 160,000 kilometers (100,000 miles).

Eliminating or reducing unnecessary packaging is another important strategy. Here are some key questions that environmentalists believe manufacturers and consumers should ask about packaging: Is it necessary? Can it use fewer materials? Can it be reused? Are the resources that went into it nonrenewable or potentially renewable? Does it contain the highest feasible amount of consumer-discarded (postconsumer) recycled material? Is it designed to be recycled easily? Can it be incinerated without producing harmful air pollutants or a toxic ash? Can it be buried and decomposed in a landfill without producing chemicals that can contaminate groundwater?

Trash taxes can also be used to reduce waste. In 1986 Denmark imposed a tax on many types of solid waste to promote reuse and recycling and reduce solid waste. By 1996 the country had reduced the amount of waste brought to its municipal landfills and incinerators by 26% and achieved an overall recycling rate of 61%. Several other European countries also have taxed solid waste. A related *pay-as-you-throw* system reduces solid waste and encourages recycling by

Table 14-2 Alternatives to Some Common Household Chemicals

Chemical	Alternative	Chemical	Alternative
Deodorant	Sprinkle baking soda on a damp wash cloth and wipe skin; dab on vanilla extract.	General surface cleaner	Mixture of vinegar, salt, and water, or mixture of borax, soap, lemon juice, and water.
Oven cleaner	Baking soda and water paste, scouring pad.	Bleach	Baking soda or borax.
Toothpaste	Baking soda.	Mildew remover	Mix ½ cup vinegar, ½ cup borax, and warm water.
Drain cleaner	Pour ½ cup salt down drain, followed by boiling water; or pour 1 handful baking soda and ½ cup white vinegar and cover lightly for one minute.	Disinfectant and general cleaner	Mix ½ cup borax in 1 gallon hot water.
Window cleaner	Add 2 teaspoons white vinegar to 1 quart warm water.	Furniture or floor polish	Mix ½ cup lemon juice and 1 cup vegetable or olive oil.
Toilet bowl, tub, and tile cleaner	Mix a paste of borax and water; rub on and let set one hour before scrubbing. Can also scrub with baking soda and a brush.	Carpet and rug shampoos	Sprinkle on cornstarch, baking soda, or borax and vacuum.
Floor cleaner	Add ½ cup vinegar to a bucket of hot water; sprinkle a sponge with borax for tough spots.	Detergents and detergent boosters	Washing soda or borax and soap powder.
Shoe polish	Polish with inside of a banana peel, then buff.	Spray starch	In a spray bottle, mix 1 tablespoon cornstarch in a pint of water.
Silver polish	Clean with baking soda and warm water.	Fabric softener	Add 1 cup white vinegar or ¼ cup baking soda to final rinse.
Air freshener	Set vinegar out in an open dish. Use an opened box of baking soda in closed areas such as refrigerators and closets. To scent the air, use pine boughs or make sachets of herbs and flowers.	Dishwasher soap	1 part borax and 1 part washing soda.
		Pesticides (indoor and outdoor)	Use natural biological controls (Section 16-5).

basing garbage collection charges on the amount of waste a household generates for disposal.

A more revolutionary step in reducing materials use and waste production would be a *shift to a true service economy*. Instead of selling goods (such as washing machines, cars, and carpets), companies would become lessors and make their profits by supplying the services products provide (cleaning clothes and leasing transportation and carpets). For a fee, companies would provide all of the materials and products needed to supply and maintain their services and would take back all materials at the end of their useful lives. This would give such companies a strong incentive to provide products that last and that are easy to repair, upgrade, dismantle, and reuse or recycle.

The Xerox corporation has taken this approach since 1992. It now leases most of its copy machines as part of its mission to provide document services instead of selling photocopiers. The company replaces or upgrades parts in its modularly designed copiers at no extra cost to the customer. When the service contract expires, Xerox takes the machine back for reuse or remanufacture. By 1997, 28% of its copiers were remanufactured from older copiers, saving $100 million per year. The company expects to raise this percentage to 87% and recycle 97% of its material. Ray Anderson, CEO of a large carpet tile company, plans to lease rather than sell carpet (Individuals Matter, p. 43).

Cities such as Berkeley, California, and Takoma Park, Maryland, have pioneered the idea of *tool libraries*. People who use tools only occasionally can check out a variety of power and hand tools.

Making this shift to a low-throughput economy will **(1)** require removing subsidies that make virgin materials seem artificially cheap and **(2)** taxing emissions of pollutants, as discussed in Section 2-3. Some firms and cities are leading the way by eliminating almost all of their waste. The city of Canberra, Australia, has a goal of achieving no waste by 2010 and the Netherlands has set a goal of reducing its waste by 70–90%. What is your community and country doing?

GUEST ESSAY

We Have Been Asking the Wrong Questions About Wastes

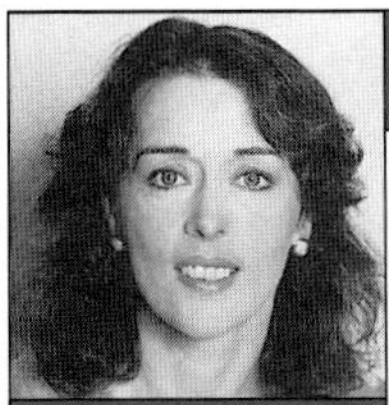

Lois Marie Gibbs

In 1977, Lois Marie Gibbs was a young housewife with two children living near the Love Canal toxic dump site. She had never engaged in any sort of political action until her children began experiencing unexplained illnesses and she learned that toxic chemicals were oozing from the dump site into many of the area's yards and basements. Then she organized her neighborhood and became the president and major strategist for the Love Canal Homeowners Association. This grassroots political action brought hazardous-waste issues to national prominence and spurred passage of the federal Superfund legislation to help clean up abandoned hazardous-waste sites. Lois Gibbs then moved to Washington, D.C., and formed Citizens' Clearinghouse for Hazardous Wastes (renamed the Center for Health, Environment, and Justice), an organization that has helped over 7,000 community organizations protect themselves from hazardous wastes. Her story is told in her autobiography, Love Canal: My Story *(State University of New York Press, 1982); she was also the subject of a CBS movie,* Lois Gibbs: The Love Canal, *which aired in 1982. Her latest book is* Dying from Dioxin *(Boston: South End Press, 1995). She is an inspiring example of what a dedicated citizen can do to change the world.*

Just about everyone knows our environment is in danger. One of the most serious threats is the massive amount of waste we put into the air, water, and ground every year. All across the United States and around the world are thousands of places that have been, and continue to be, polluted by toxic chemicals, radioactive waste, and just plain garbage.

For generations, the main question people have asked is, "Where do we put all this waste? It's got to go somewhere." That is the wrong question, as has been shown by a series of experiments in waste disposal and by the simple fact that there is no "away" in "throwaway."

We tried dumping our waste in the oceans. That was wrong. We tried injecting it into deep, underground wells. That was wrong. We've been trying to build landfills that don't leak, but that doesn't work. We've been trying to get rid of waste by burning it in high-tech incinerators; that only produces different types of pollution, such as air pollution and toxic ash. We've tried a broad range of "pollution" controls, but all that does is allow legalized, high-tech pollution. Even recycling, which is a very good thing to do, suffers from the same problem as all the other methods: It addresses waste *after* it has been produced.

For many years, people have been assuming that "it's got to go somewhere," but now many people, especially young people, are starting to ask why. Why do we produce so much waste? Why do we need products and services that have so many toxic by-products? Why can't industry change the way it makes things so that it stops producing so much waste?

These are the *right* questions. When you start asking them, you start getting answers that lead to *pollution prevention* and *waste reduction* instead of *pollution control* and *waste management*. People, young and old, who care about pollution prevention are challenging companies to stop making products with gases that reduce ozone in the ozone layer [Section 11-6] and contribute to the threatening possibility of global warming [Section 11-3]. They are asking why so many goods are wrapped in excessive, throwaway packaging. They are challenging companies that sell pesticides, cleaning fluids, batteries, and other hazardous products to either remove the toxins from those products or take them back for recovery or recycling rather than disposing of them in the environment. They are demanding alternatives to throwaway materials in general.

14-3 REUSE

What Are the Advantages of Refillable Containers? *Reuse* is a form of waste reduction that **(1)** extends resource supplies, **(2)** keeps high-quality matter resources from being reduced to low-matter-quality waste (Figure 3-7), and **(3)** reduces energy use and pollution even more than recycling. Two examples of reuse are refillable glass beverage bottles and refillable soft drink bottles made of polyethylene terephthelate (PET) plastic. Unlike throwaway and recyclable cans and bottles, refillable beverage bottles create local jobs related to their collection and refilling. Moreover, studies by Coca-Cola and PepsiCo of Canada show that 0.5-liter (16-ounce) bottles of their soft drinks cost one-third less in refillable bottles.

In 1964, 89% of all soft drinks and 50% of all beer in the United States were sold in refillable glass bottles. Today such bottles made up only about 7% of the beer and soft drink market, and only 10 states even have refillable glass bottles. The disappearance of most local bottling companies has led to a loss of local jobs, income, and tax revenues. Some call for reinstatement of this bottling reuse system in the United States; others

Since 1988, hundreds of student groups have contacted my organization to get help and advice in taking these effective types of actions. Many of these groups begin by working to get polystyrene food packaging out of their school cafeterias and local fast-food restaurants.

Oregon students even took legal action to get rid of cups and plates made from bleached paper because the paper contains the deadly poison dioxin. They were asking the right questions, and they got the right answer when the school systems switched to reusable cups, plates, and utensils.

Dozens of student groups have joined with local grassroots organizations to get toxic-waste sites cleaned up or to stop construction of new toxic-waste sites, radioactive-waste sites, or waste incinerators.

Waste issues are not simply environmental issues; they are also tied up with our economy, which is geared to producing and then disposing of waste. *Somebody* is making money from every scrap of waste and has a vested interest in keeping things the way they are. Environmentalists and industry officials constantly argue about "cost–benefit analysis," which, simply stated, poses the question of whether the benefit of controlling pollution or waste will be greater than the cost. But this is another example of asking the wrong question. The right questions are, "Who will benefit, and who will pay the cost?"

Waste issues are also issues of *justice* and *fairness*. Again, there's a lot of debate between industry officials and environmentalists, especially those in federal and state environmental agencies, about so-called acceptable risk [Section 8-6]. These industry officials and environmentalists decide the degree of people's exposure to toxic chemicals, but don't ask the people who will actually be exposed how they feel about it. Again, asking what is an "acceptable risk" of exposure is the wrong question. It's simply not just to expose people to chemical poisons without their consent.

Risk analysts often say, "But there's only a one in a million chance of increased death from this toxic chemical." That may be true. But suppose I took a pistol and went to the edge of your neighborhood and began shooting into it. There's probably only a one in a million chance that I'd hit somebody, but would you issue me a license to do that? As long as we don't stand up for our rights and demand that "bullets" in the form of hazardous chemicals not be "fired" in our neighborhoods, we are giving environmental regulators and waste producers a license to kill a certain number of us without our even being consulted.

From my personal experience, I know that decisions made to dump wastes at Love Canal and in thousands of other places were not made purely on the basis of the best available scientific knowledge. The same holds true for decisions about how to manage the wastes we produce today and how to produce less waste.

Instead, the world we live in is shaped by decisions based on money and power. If you really want to understand what's behind any given environmental issue, the first question you should ask is, "Who stands to profit from this?" Then ask, "Who is going to pay the price?" You can then identify both sides of the issue and decide whether you want to be part of the problem or part of the solution.

Critical Thinking

1. Do you believe that we should put primary emphasis on pollution prevention and waste reduction? Explain.

2. What changes would you be willing to make in your own lifestyle to prevent pollution and reduce waste?

say it isn't practical because the system of collections and returns has been dismantled.

Denmark has led the way by banning all beverage containers that can't be reused. However, in 1999 the government of Denmark was brought to trial because its law prohibiting sale of canned beer and soft drinks and nonrefillable glass and plastic bottles violates the European Union's packaging waste law. To encourage use of refillable glass bottles, Ecuador has a refundable beverage container deposit fee that is 50% of the cost of the drink. In Finland, 95% of the soft drink, beer, wine, and spirits containers are refillable, and in Germany, 73% are refillable.

Another reusable container is the metal or plastic lunchbox. Sandwiches and refrigerator leftovers can be put in small reusable plastic containers instead of plastic wrap and aluminum foil (most of which is not recycled). This practice and most forms of reuse save money and reduce resource waste. Another way to reduce resource waste and pollution is by the use of reusable grocery and shopping bags (Solutions, p. 364).

In 1991 Toyota shifted entirely to reusable shipping containers. A similar move by the Xerox corporation saves the company $2–5 million per year.

What Kind of Grocery Bags Should We Use?

SOLUTIONS

When you're offered a choice between plastic or paper bags for your groceries, which should you choose? The answer is *neither*. Both are environmentally harmful, and the question of which is the more damaging has no clear-cut answer.

On one hand, plastic bags degrade slowly in landfills and can harm wildlife if swallowed, and producing them pollutes the environment. On the other hand, producing the brown paper bags used in most supermarkets uses trees and pollutes the air and water. Overall, white or clear polyethylene plastic bags require less energy for manufacture and cause less damage to the environment than do paper bags not made mostly from recycled paper.

Instead of having to choose between paper and plastic bags, you can bring your own *reusable* canvas or string containers to the store, and save and reuse any paper or plastic bags you get. Using a reusable bag just five times displaces the pollution caused by the manufacture of the bag. To encourage people to bring their own reusable bags, stores in the Netherlands charge for paper or plastic bags.

Critical Thinking

1. Apply similar reasoning to determine what kind of cup (plastic, paper, or reusable) you should use whenever possible. How could you solve the problem of getting coffee or other beverages at fast-food places and at workplaces?

2. Do you believe that grocery stores should charge for paper or plastic bags and sell reusable bags to encourage reuse? Explain. How would you go about implementing such a policy in all major grocery stores to provide an even economic playing field for all consumers?

14-4 RECYCLING

How Can We Recycle Organic Solid Wastes? Community Composting **Compost** is a sweet-smelling, dark-brown, humuslike material that is rich in organic matter and soil nutrients. It is produced when microorganisms (mostly fungi and aerobic bacteria) in soil break down organic matter such as leaves, food wastes, paper, and wood in the presence of oxygen.

Biodegradable wastes make up about 35% by weight of the MSW output in the United States. Only about 5% of the MSW in the United States is composted (compared with 17% in France and 10% in Switzerland). Such wastes can be composted by consumers in backyard bins (Individual's Matter, p. 351) or indoor containers or collected and composted in centralized community facilities (as in many western European countries).

The resulting compost can be used as an organic soil fertilizer or conditioner, as topsoil, or as a landfill cover. Compost also can be used to help restore eroded soil on hillsides and along highways, strip-mined land, overgrazed areas, and eroded cropland.

To be successful, a large-scale composting program must **(1)** overcome siting problems (few people want to live near a giant compost pile or plant), **(2)** control odors, and **(3)** exclude toxic materials that can contaminate the compost and make it unsuitable for use as a fertilizer on crops and lawns. Three ways to control or reduce odors for large-scale composting operations are **(1)** enclosing the facilities and filtering the air inside (but residents near large composting plants still complain of unacceptable odors), **(2)** creating municipal compost operations near existing landfills or at other isolated sites, and **(3)** decomposing biodegradable wastes in a closed metal container in which air is recirculated to give precise control of available oxygen and temperature (a technique that has been used successfully in the Netherlands for 20 years).

What Are the Two Types of Recycling? There are two types of recycling for materials such as glass, metals, paper, and plastics: primary and secondary. The most desirable type is *primary* or *closed-loop recycling*, in which wastes discarded by consumers (*postconsumer wastes*) are recycled to produce new products of the same type (such as newspaper into newspaper and aluminum cans into aluminum cans).

A still useful but less desirable type is *secondary*, or *open-loop, recycling*, in which waste materials are converted into different products. Primary recycling reduces the amount of virgin materials in a product by 20–90%, whereas secondary recycling reduces virgin material by 25% at most.

Environmentalists urge us not to be misled by labels claiming that paper and plastic bags or other items are recyclable. Just about anything is in theory recyclable. What counts is whether an item is actually recycled, and whether we complete the recycling loop by buying products using the maximum feasible content of postconsumer recycled materials.

Case Study: Recycling Municipal Solid Waste in the United States In 1998, about 30% of U.S. MSW was recycled or composted—the highest rate of any industrialized country. This included recycling of 96% of car batteries, 43% of wastepaper, 42% of aluminum

 Gardner, G. and P. Sampat. 1999. "Making Things Last: Reinventing Our Material Culture." *The Futurist,* vol. 33, no. 5, p. 24(5).

and ferrous metals, 40% of yard waste, and 27% of glass containers.

By 1998 the United States had more than 8,800 municipal curbside recycling programs serving 51% of the population. Recent pilot studies in several U.S. communities show that a 60–80% recycling and composting rate is possible.

One of the best ways to encourage recycling is a *pay-as-you-throw* program that bases garbage collection charges on the amount of waste a household generates for disposal; materials sorted out for recycling are hauled away free. Currently, more than 2,800 communities in North America have such systems.

Recycling also benefits communities by creating more jobs than burying or burning wastes. Recycling 1 million tons of solid waste in the United States requires about 1,800 workers, compared with 600 workers for landfilling these wastes and only 80 for incinerating them.

Is Centralized Recycling of Mixed Solid Waste the Answer? Large-scale recycling can be accomplished by collecting mixed urban waste and transporting it to centralized *materials-recovery facilities (MRFs)*. There, machines shred and automatically separate the mixed waste to recover glass (which can be melted and converted to new bottles or fiberglass insulation), iron, aluminum, and other valuable materials, which are sold to manufacturers as raw materials (Figure 14-6). The remaining paper, plastics, and other combustible wastes are recycled or burned. The resulting heat produces steam or electricity to run the recovery plant or to sell to nearby industries or homes.

By 1998 the United States had more than 220 MRFs, and at least 60 more were in the planning stages. However, such plants are expensive to build and maintain, and once trash is mixed it takes a lot of money and energy to separate it. These facilities also can emit toxic air pollutants and produce a toxic ash that must be disposed of safely. Because of high operating costs, some MRFs have shut down. MRFs must also have a large input of garbage to make them financially successful. Thus, their owners have a vested interest in increased

Figure 14-6 Schematic of a generalized materials-recovery facility used to sort mixed wastes for recycling and burning to produce energy. Because such plants require high volumes of trash to be economical, they discourage reuse and waste reduction.

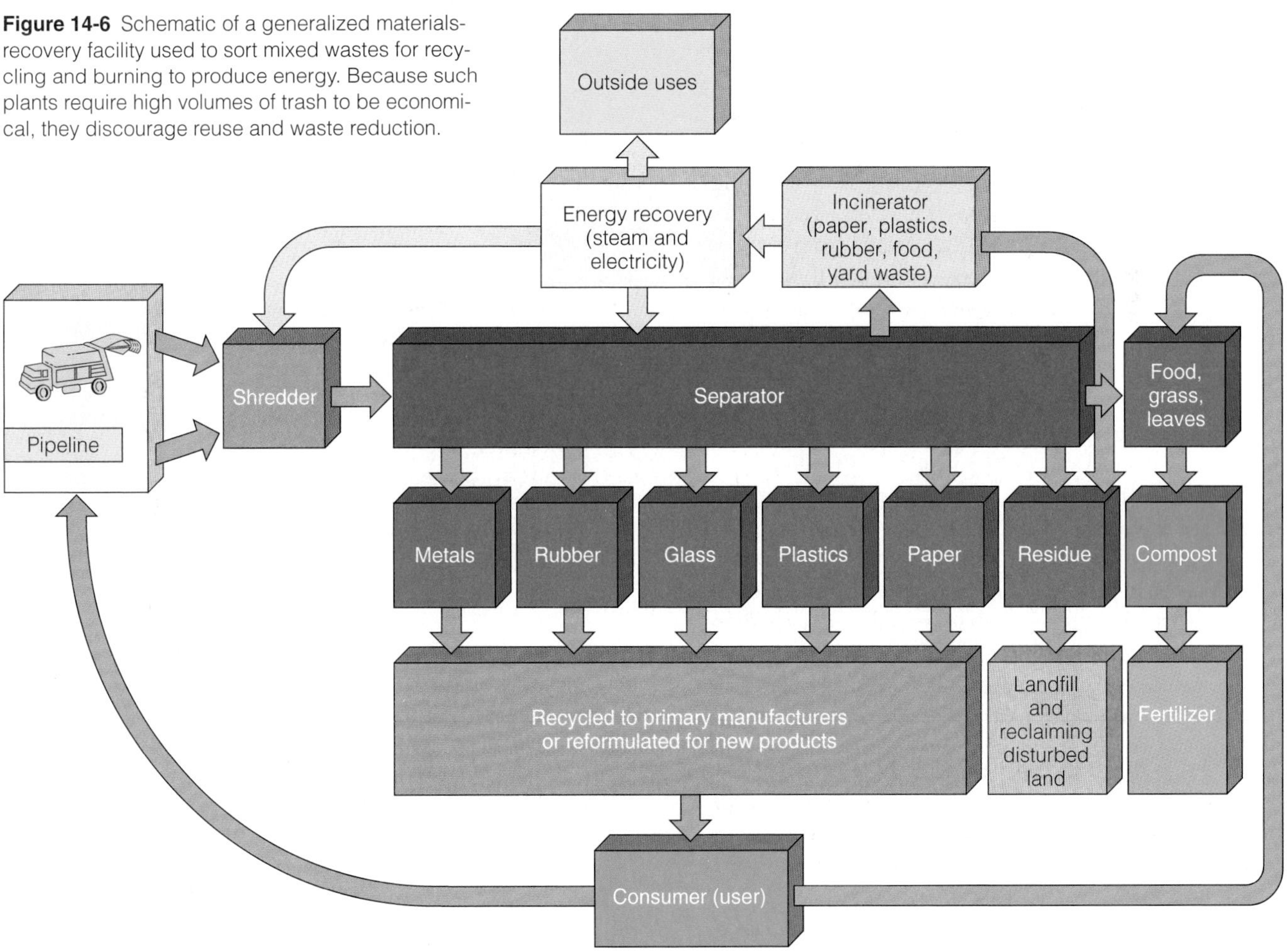

Hint: Enter *recycling* or *economics* using Key Words.

throughput of matter and energy resource to produce more trash—the reverse of what prominent scientists believe we should be doing (Figure 14-3).

Is Separating Solid Wastes for Recycling the Answer? Many solid-waste experts argue that it makes more sense economically and environmentally for households and businesses to separate trash into recyclable and reusable categories (such as glass, paper, metals, certain types of plastics, and compostable materials) before it is picked up. Compartmentalized city collection trucks, private haulers, or volunteer recycling organizations pick up the segregated wastes and sell them to scrap dealers, compost plants, and manufacturers. Another alternative (especially in less populated areas) is the establishment of a network of drop-off centers, buyback centers, and deposit-refund programs in which people deliver and either sell or donate their separated recyclable materials.

The source-separation approach produces little air and water pollution, reduces litter, and has low start-up costs and moderate operating costs. It also saves more energy and provides more jobs for unskilled workers than centralized MRFs, and it creates three to six times more jobs per unit of material than landfilling or incineration. In addition, separated recyclables are cleaner and usually can be sold for a higher price. Source separation also educates people about the need for waste reduction, reuse, and recycling (Individuals Matter, right).

INDIVIDUALS MATTER

Source Separation Recycling in Some Georgia Schools

In Rome, Georgia, environmental educator Steve Cordle has set up a project that combines the important concept of source-separation recycling with environmental education.

He has designed and implemented a source-separation recycling program for 17 Floyd County schools in Georgia. Before his program, which began in 1992, the schools were a major contributor to the county landfill and hardly any of their waste output was recycled.

During a 7-month period in 1996, approximately 114 metric tons (250,000 pounds) of paper, corrugated cardboard, and aluminum and steel cans were collected in separate containers. The school system stored and sold the items for recycling. During this period the project added $10,139.69 to school funds, although school officials consider the money incidental to the educational value of the program.

This important experiment in *environmental education* teaches students (and teachers) from kindergarten through high school about the need for recycling and engages them in hands-on participation in source separation.

Cordle is working hard to expand this idea to a larger area and ideally to Georgia's entire school system. With proper backing and support, this program could become a model for use throughout the United States and other parts of the world. In 1998, Cordle's project received a $250,000 grant from federal and state sources.

Critical Thinking

Use the second law of energy (thermodynamics) to explain why a *properly designed* source-separation recycling program takes less energy and produces less pollution than a centralized program that collects mixed waste over a large area and hauls it to a centralized facility where workers or machinery separate the wastes for recycling.

Aluminum and paper separated out for recycling are worth a lot of money. As a result, in a growing number of cities people steal these materials—from curbside containers set out by residents and from unprotected recycling drop-off centers—and sell them. This undermines municipal recycling programs by lowering the income available from selling these high-value materials.

Does Recycling Make Economic Sense? The answer is yes and no, depending on different ways of looking at the economic and environmental costs and benefits of recycling. Critics contend that recycling **(1)** has become almost a religion that is above criticism regardless of how much it costs communities, **(2)** does not make sense if it costs more to recycle materials than to send them to a landfill or incinerator (as is the case in some areas), **(3)** is often not needed to save landfill space because most areas in the United States are not running out of landfill space, and **(4)** may make economic sense for valuable and easy-to-recycle materials (such as aluminum, paper, and steel) but not for cheap or plentiful resources (such as glass from silica) and most plastics (which are expensive to recycle).

Many communities established recycling programs with the idea that they should pay for themselves. But recycling proponents argue that *recycling programs should not be judged on whether they pay for themselves any more than are conventional garbage disposal systems based on land burial or incineration.*

Moreover, recycling proponents contend that with full-cost accounting (Section 2-3), the net economic, health, and environmental benefits of recycling far outweigh the costs. An analysis of various recycling programs revealed that cities and private waste collection firms tend to make money if they have high recycling rates and a single pickup system (for both materials to be recycled and garbage that can't be recycled). Cities cited by critics as losing money on recycling often have expensive dual collection systems and have not designed programs to encourage more cost-effective, high rates of recycling.

Finally, environmentalists point out that the primary benefit from recycling is *not* a reduction in the use of landfills and incinerators. Instead, the major reasons for recycling are

- Reduced use of virgin resources
- Protection of biodiversity and wildlife habitat because of less conversion of diverse natural forests into monoculture tree farms to produce paper products
- Less use of toxic chemicals
- Less pollution and environmental degradation
- Reduced throughput of matter and energy resources (Figure 3-15)

Why Don't We Have More Reuse and Recycling? Three factors that hinder recycling (and reuse) are **(1)** failure to include the environmental and health costs of raw materials in the market prices of consumer items, **(2)** more tax breaks and subsidies for resource-extracting industries ($2.6 billion per year in the United States) than for recycling and reuse industries, and **(3)** lack of large, steady markets for recycled materials.

Analysts have suggested various ways to overcome the obstacles to recycling, including the following:

- Taxing virgin resources and phasing out subsidies for extraction of virgin resources
- Lowering or eliminating taxes on recycled materials based on postconsumer waste content
- Providing subsidies for reuse and postconsumer waste recycling
- Requiring households and businesses to pay directly for garbage collection based on how much they throw away, with lower charges or no charges for materials separated for recycling and composting (a *pay-as-you throw* system)
- Encouraging or requiring government purchases of recycled products to help increase demand and lower prices
- Viewing landfilling and incineration of solid wastes as last resorts to be used only for wastes that can't be reused, composted, or recycled (Figure 14-3)
- Requiring ecolabels on all products, evaluating life cycle environmental costs and listing preconsumer and postconsumer recycled content

CASE STUDY

Recycling, Reuse, and Waste Reduction in Germany

In 1991, Germany enacted the world's toughest packaging law, which was designed to **(1)** reduce the amount of waste being landfilled or incinerated, **(2)** reduce waste production, and **(3)** recycle or reuse 65% of the nation's packaging by 1995. Product distributors must take back their boxes and other containers for reuse or recycling, and incineration of packaging is not allowed.

To implement the system, over 600 German manufacturers and distributors pay fees to a nonprofit company they formed to collect, sort, and reprocess the packaging discards (coded with a green dot) of member firms.

The system has been so successful that in 1997, 86% of all packaging in Germany was recycled (up from 12% in 1992). Plastic packaging has lost one-third of its market share to glass and cardboard, and four out of five German manufacturers have reduced their use of packaging. About 75% of all produce shipped through Germany is boxed in leased, reusable shipping crates.

By 1998, 28 countries had enacted *take-back* laws for packaging materials, 16 have such laws for batteries, and 12 are considering such legislation for electronic items. In the Netherlands all packaging waste is banned from landfills.

Critical Thinking

What might be some long-term disadvantages of Germany's program? Do you believe that a similar program should be put into place in the country where you live? Explain.

Countries such as Germany are leading the way in recycling, reuse, and waste reduction (Case Study, above).

14-5 CASE STUDIES: RECYCLING ALUMINUM, WASTEPAPER, AND PLASTICS

How Much Aluminum Is Being Recycled? Recycling aluminum produces 95% less air pollution and 97% less water pollution, and it requires 95% less energy than mining and processing aluminum ore. In 1998, 74% (compared with 15% in 1973) of aluminum beverage cans in the United States were recycled. Despite this progress, about 26% of the 105 billion aluminum cans produced in 1998 in the United States were thrown away. Laid end to end, these cans would wrap around the planet more than 120 times.

Recycling aluminum cans is great, but many environmentalists believe that these cans are an example of an unnecessary item that could be replaced by more energy-efficient and less polluting refillable glass or PET plastic bottles—a switch from recycling to reuse that also creates local jobs. One way to encourage this change would be to place a heavy tax on nonrefillable containers and no tax on reusable beverage containers, as is done in at least nine other countries. What do you think?

How Much Wastepaper Is Being Recycled? Paper (especially newspaper and cardboard) is one of the easiest materials to recycle. To make paper, a chemical such as caustic soda (sodium hydroxide) is used to convert wood chips into a soft mush (pulp) that is pressed into a thin sheet and then dried. Recycling paper involves removing its ink, glue, and coating and reconverting it to pulp that is pressed again into new paper. This process breaks down some of the paper fibers, requiring addition of some new pulp to maintain paper strength.

In 1998 the United States recycled about 43% of its wastepaper (up from 25% in 1989). At least 10 other countries recycle 50–98% of their wastepaper and paperboard, with a global recycling rate of 41%.

In addition to saving trees, recycling paper **(1)** saves energy because it takes 30–64% less energy to produce the same weight of recycled paper as to make the paper from trees, **(2)** reduces air pollution from pulp mills by 74–95%, **(3)** lowers water pollution by 35%, **(4)** helps prevent groundwater contamination by toxic ink left after paper rots in landfills over a 30- to 60-year period, **(5)** conserves large quantities of water, **(6)** can save landfill space, **(7)** creates five times more jobs than harvesting trees for pulp, and **(8)** can save money.

Buying recycled paper products can save trees and energy and reduce pollution, but it does not necessarily reduce solid waste. Only products made from *postconsumer waste*—waste intercepted on its way from consumer to the landfill or incinerator—do that.

Most recycled paper is actually made from *preconsumer waste*: scraps and cuttings recovered from paper and printing plants. Because paper manufacturers have always recycled this waste, it has never contributed to landfill problems. Now this paper is labeled "recycled" as a marketing ploy, giving the false impression that people who buy such products (often at higher prices) are helping the solid-waste problem. Most "recycled" paper has no more than 50% recycled fibers, with only 10% from postconsumer waste. Environmentalists propose that the government require companies to report the amount of postconsumer recycled materials in paper and other products and reserve the term *recycled* only for items using postconsumer recycled materials.

Is It Feasible to Recycle Plastics? Plastics are various types of polymer molecules made by chemically linking monomer molecules (called *petrochemicals*) produced mostly from oil and natural gas. The plastics industry is among the leading producers of hazardous waste.

Plastics now account for about 9% by weight and 22% by volume of MSW in the United States and about 60% of the debris found on U.S. beaches. In landfills, toxic cadmium and lead compounds used as binders, colorants, and heat stabilizers can leach out of plastics and ooze into groundwater and surface water. Most plastics used today are nondegradable or take 200–400 years to degrade. Even biodegradable plastics take decades to partially decompose in landfills because they lack sufficient oxygen and moisture.

Currently, only about 7% by weight of all plastic wastes in the United States are recycled, mostly because there are so many different types of plastic resin. Before they can be recycled, plastics in trash must be separated into different types of resins by consumers or separated from mixed trash (a costly, labor-intensive procedure unless affordable automated separating technologies can be developed).

Another problem is that the current price of oil is so low that the price of virgin plastic resins (except for PET, used mostly in plastic drink bottles) is about 40% lower than that of recycled resins. However, in 1998 Chrysler Corporation built an experimental automobile with a body made from plastic derived entirely from recycled PET bottles and expects to sell such vehicles within a decade. Chrysler says the plastic cars will be cheaper and lighter and get better gas mileage.

Environmentalists recognize the beneficial qualities of plastics: **(1)** durability (in products such as car and machine parts, carpeting, toys, furniture, reusable tubs and containers, and refillable bottles), **(2)** light weight, **(3)** unbreakability (compared with glass), and in some cases **(4)** reusability as containers. But many environmentalists believe that some widespread uses of plastics—especially excessive and often unnecessary single-use packaging and throwaway beverage and food containers—should be sharply reduced and replaced with less harmful and wasteful alternatives. What do you think?

14-6 DETOXIFYING, BURNING, BURYING, AND EXPORTING WASTES

How Can Hazardous Waste Be Detoxified? Denmark has the most comprehensive and effective hazardous-waste detoxification program. Each Danish municipality has at least one facility that accepts paints,

 Piasecki, B., D. Rainey, and K. Fletcher. 1998. "Is Combustion of Plastics Desirable?" *American Scientist*, July–August v86 n4 p364(10).

solvents, and other hazardous wastes from households. Hazardous and toxic waste from industries is delivered to 21 transfer stations throughout the country. All waste is then transferred to a large treatment facility, where about 75% of it is detoxified; the rest is buried in a carefully designed and monitored landfill.

Some consider biological treatment of hazardous waste, or *bioremediation*, to be the wave of the future for cleaning up some types of toxic and hazardous waste. In this process, microorganisms, usually bacteria, are used to destroy toxic or hazardous substances or convert them to harmless forms. This approach mimics nature by using decomposers to recycle matter.

If toxin-degrading bacteria or fungi can be found or engineered for specific hazardous chemicals, they could clean up contaminated sites at less than half the cost of disposal in landfills, and at only about one-third the cost of on-site incineration. Bioremediation might also clean up contaminated groundwater at an affordable cost by pumping it to the surface, treating it with microorganisms, and then returning it to its aquifer.

Preliminary testing reveals that bioremediation is effective for a number of specific organic wastes, but it does not appear to work very well for toxic metals, highly concentrated chemical wastes, or complete digestion of some complex mixtures of toxic chemicals.

Another biological way to treat hazardous wastes is *phytoremediation*, which involves using natural or genetically engineered plants to remove contaminants—again mimicking nature. For example, fast-growing poplar trees can be planted at contaminated industrial or other sites. The trees act like straws to suck up contamination from soil and groundwater and either safely store the chemicals in their tissues or metabolize them into compounds released into the atmosphere. This method is inexpensive, but more research is needed to determine whether the by-products released into the air by the poplars are harmful.

Is Burning Solid and Hazardous Waste the Answer? In 1998 about 16% of the MSW and 7% of the officially regulated hazardous waste in the United States was burned in incinerators. Most of this mixed MSW was burned in 170 *mass-burn incinerators* (Figure 14-7), which burn mixed trash without separating out hazardous materials (such as batteries and polyvinyl

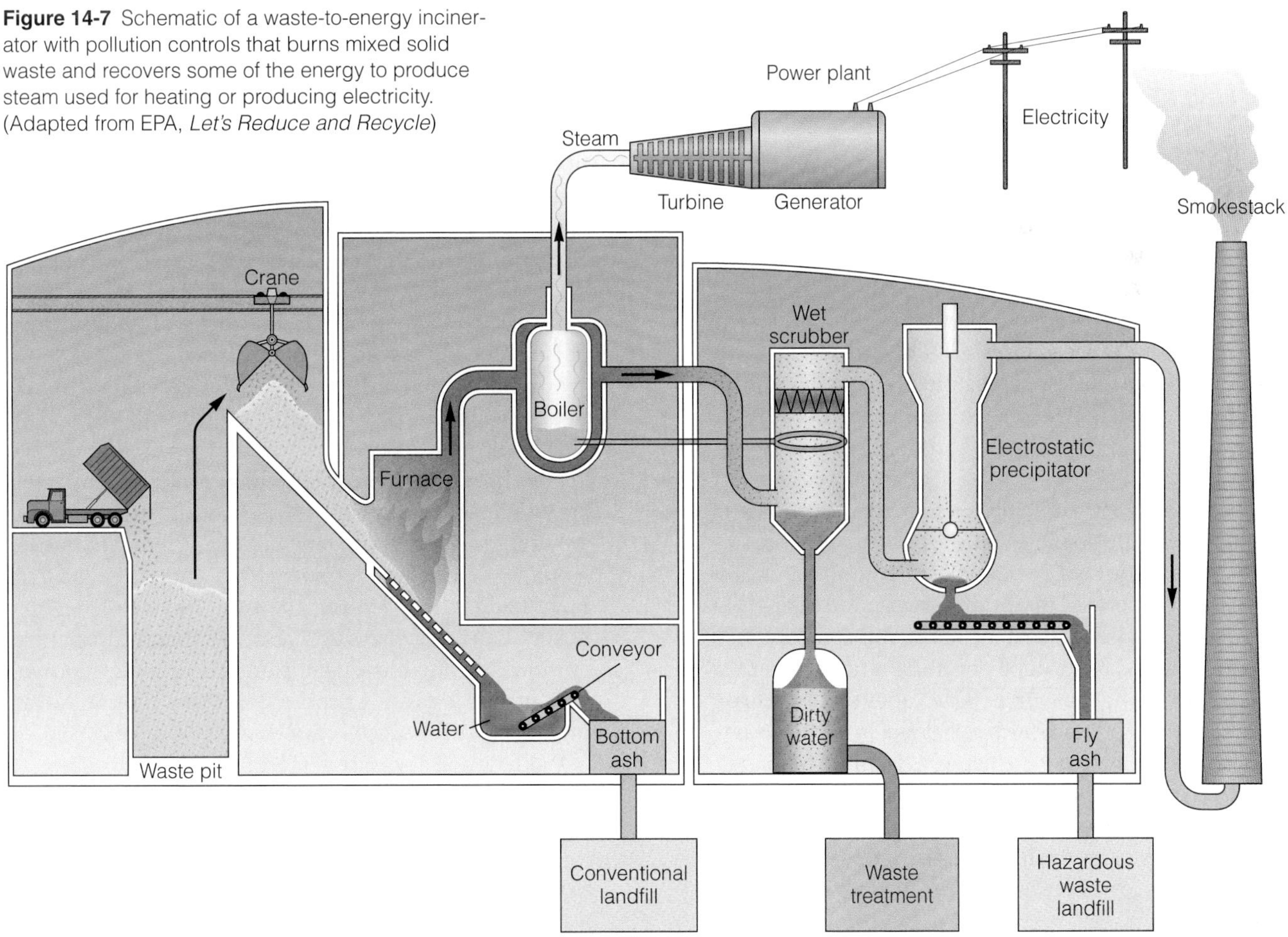

Figure 14-7 Schematic of a waste-to-energy incinerator with pollution controls that burns mixed solid waste and recovers some of the energy to produce steam used for heating or producing electricity. (Adapted from EPA, *Let's Reduce and Recycle*)

Hint: Enter *incineration* using Key Words.

chloride [PVC] plastic materials) or noncombustible materials that can interfere with combustion and pollute the air. Hazardous waste in the United States is also burned in cement-factory kilns.

Incinerators are costly to build, operate, and maintain, and they create very few long-term jobs. Without continual maintenance and good operator training and supervision, the air-pollution control equipment on incinerators often fails and emission standards are exceeded.

According to some EPA scientists and other hazardous waste experts, even with advanced air-pollution control devices, *all incinerators burning hazardous or solid waste release toxic air pollutants.* They also leave a highly toxic ash to be disposed of in landfills that the EPA concedes will eventually leak.

According to EPA hazardous-waste expert William Sanjour, EPA incinerator regulations don't work because

> *the regulations require no monitoring of the outside air in the vicinity of the incinerator. Because operators maintain the records, they can easily cheat. . . . Government inspectors are poorly trained and have low morale and high turnover. . . . Government inspectors typically work from nine to five Monday through Friday. So if there is anything particularly nasty to burn, it will be done at night or on weekends. When complaints come in, . . . the inspector may visit the plant but rarely finds anything. The enforcement officials tend to view the incinerator operator as their client and the public as a nuisance. . . . There is no reward to inspectors for finding serious violations.*

Sweden banned the construction of new incinerators in 1985. In 1993 the *Wall Street Journal* warned that burning municipal trash spells financial disaster for local governments in the United States. Since 1992 Rhode Island and West Virginia have banned solid-waste incineration because of its health threats and high cost. Since 1985 several solid-waste incinerators in the United States have been shut down because of excessive costs and pollution, and over 280 new incinerator projects have been blocked, delayed, or canceled because of intense public opposition and high costs.

Japan depends on incinerators more than any country, burning about 65% of its MSW in more than 1,850 government-operated incinerators and more than 3,300 privately owned industrial incinerators. Many of these incinerators are built near populous areas. Recently there has growing concern in Japan about reported rising infant deaths and health problems in areas downwind of many Japanese incinerators from emissions of toxic dioxins (Section 14-7). The head of a government advisory committee on dioxins found that the concentration of dioxins in the air in Japan is three times that in the United States and some European countries. In 1999, the government of Japan announced a goal of eliminating dioxin emissions by 2005.

Is Land Disposal of Solid Waste the Answer? Currently, about 54% by weight of the MSW in the United States is buried in sanitary landfills (compared with 90% in the United Kingdom, 80% in Canada, 15% in Japan, and 12% in Switzerland). A **sanitary landfill** is a garbage graveyard in which solid wastes are spread out in thin layers, compacted, and covered daily with a fresh layer of clay or plastic foam.

Modern state-of-the-art landfills on geologically suitable sites are lined with clay and plastic before being filled with garbage (Figure 14-8). The bottom is covered with a second impermeable liner, usually made of several layers of clay, thick plastic, and sand. This liner collects *leachate* (rainwater contaminated as it percolates through the solid waste) and is intended to prevent its leakage into groundwater. Collected leachate ("garbage juice") is pumped from the bottom of the landfill, stored in tanks, and sent to a regular sewage treatment plant or an on-site treatment plant. When full, the landfill is covered with clay, sand, gravel, and topsoil to prevent water from seeping in. Several wells are drilled around the landfill to monitor any leakage of leachate into nearby groundwater.

Sanitary landfills for solid wastes offer certain benefits. Air-polluting open burning is avoided. Odor is seldom a problem, and rodents and insects cannot thrive. If located properly they can reduce water pollution from leaching, but proper siting is not always achieved. A sanitary landfill can be put into operation fairly quickly, has low operating costs, and can handle a huge amount of solid waste. After a landfill has been filled, the land can be graded, planted, and used as a park, golf course, ski hill, athletic field, or wildlife area, or for some other recreational purpose.

Solid-waste landfills also have drawbacks. While in operation, they cause traffic, noise, and dust; most also emit toxic gases. Paper and other biodegradable wastes break down very slowly in today's compacted and water- and oxygen-deficient landfills. Newspapers dug up from some landfills are still readable after 30 or 40 years; hot dogs, carrots, and chickens that have been dug up after 10 years have not decomposed. Landfills also deprive present and future generations of valuable reusable and recyclable resources and encourage waste production instead of pollution prevention and waste reduction.

The underground anaerobic decomposition of organic wastes at landfills produces explosive methane (a greenhouse gas), carbon dioxide (a greenhouse gas), toxic hydrogen sulfide gas, and smog-forming volatile organic compounds that escape into the air. Large

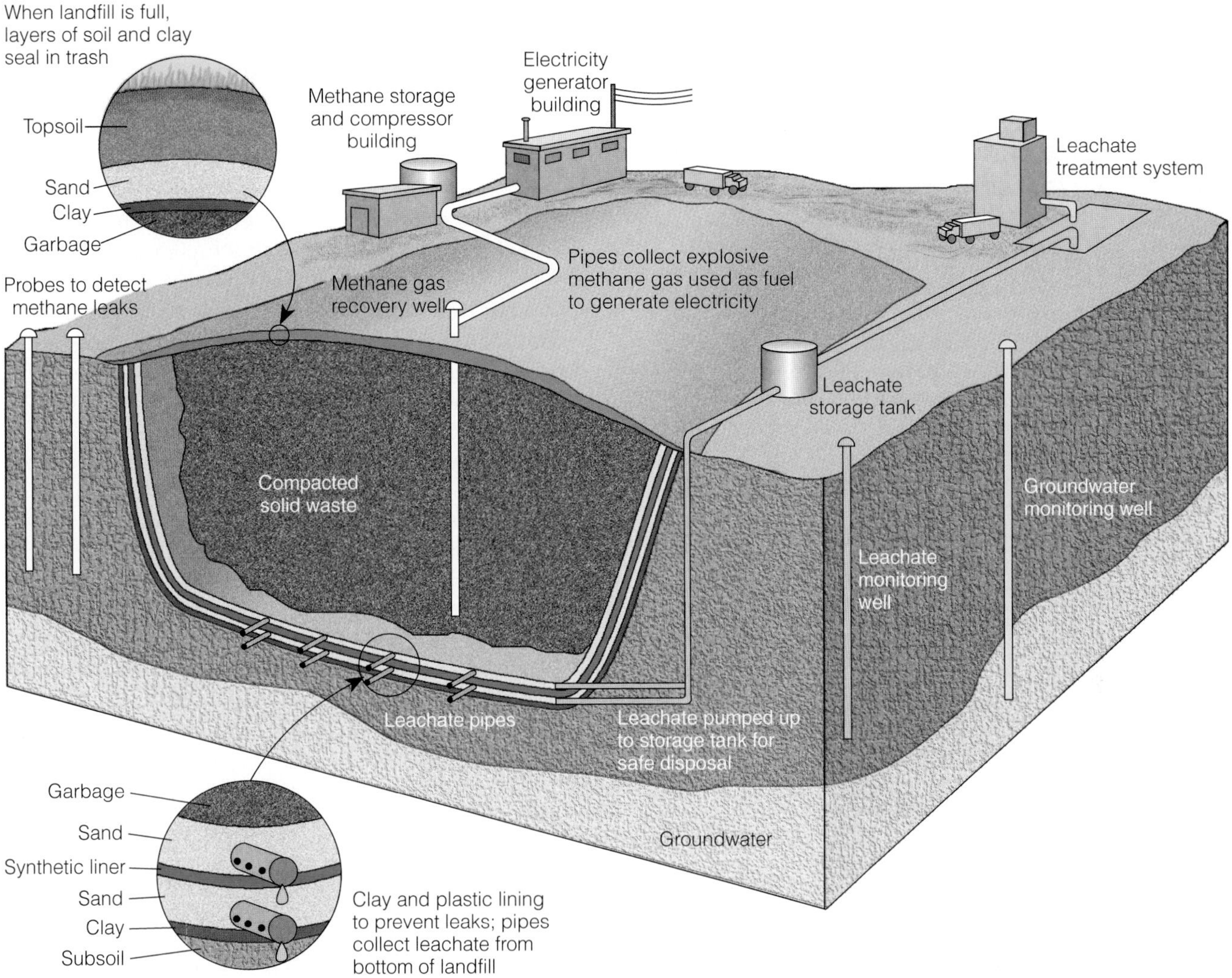

Figure 14-8 State-of-the-art sanitary landfills are designed to eliminate or minimize environmental problems that plague older landfills. Only a few of the 3,000 municipal and industrial landfills in the United States have such a state-of-the-art design, and 85% of U.S. landfills are unlined. Even state-of-the-art landfills will eventually leak, passing both the effects of contamination and cleanup costs on to future generations.

landfills can be equipped with vent pipes to collect these gases, and the collected methane can be burned in small power plants or in fuel cells to produce steam or electricity (Figure 14-8, and Individuals Matter, p. 372). Methane must be collected in all new landfills in the United States, but thousands of older and abandoned landfills don't have such systems and will emit methane for decades. Currently, U.S. landfills are responsible for about 36% of U.S. methane emissions and 10% of global methane emissions.

A 1998 study by the New York State Department of Health found that women living near solid-waste landfills where gases are escaping have a 400% greater risk of bladder cancer or leukemia (cancer of blood-forming cells). Several other studies of landfills in 339 U.S. counties, Canada, and Germany found increased incidence of leukemias and cancers of the bladder, liver, prostate, lung, and cervix.

Contamination of groundwater and nearby surface water by leachate from both unlined and lined landfills is another serious problem. Some 86% of the U.S. landfills studied have contaminated groundwater, and a fifth of all Superfund hazardous-waste sites are former municipal landfills that will cost billions of dollars to clean up. Once groundwater is contaminated, it is extremely expensive and difficult—often impossible—to clean up.

Modern double-lined landfills (Figure 14-8), required in the United States since 1996, delay the

Using Landfill Methane to Heat a School

INDIVIDUALS MATTER

In January 1997, Pattonville High School in Maryland Heights, Missouri (near St. Louis), became the first public school to use methane gas produced by a nearby landfill as a source of heat for the school's 117 classrooms and two gymnasiums.

Most students at the high school looked at the nearby landfill and saw garbage. In 1993, however, members of the school's Ecology Club saw it as an opportunity to reduce resource waste and pollution.

After doing their homework, they convinced school officials to build a pipeline to transfer waste methane gas from the landfill to the school and to rebuild two basement heating system boilers to burn the methane to heat the school. The school is expected to recoup its investment in about 5 years and save money thereafter.

This project reduces resource waste and pollution because before the landfill was burning off the methane gas, which emitted carbon dioxide, a greenhouse gas (Figure 11-2a), into the atmosphere. In addition, students at the school are getting a lesson in applied environmental science by studying the operation and design of the system.

Critical Thinking

This project may not have wide application because most schools are not near landfills. Are there any renewable energy resources (Chapter 20) available at your school site that could be used to provide all or part of the school's heat or electricity? If so, consider evaluating such resources and presenting a plan to school officials.

release of toxic leachate into groundwater below landfills, but they do not prevent it. These landfills are designed to accept waste for 10–40 years, and current EPA regulations require owners to maintain and monitor landfills for at least 30 years after they are closed. However, they could begin to leak after this period, passing the health risks and costs of contamination to future generations.

According to G. Fred Lee, an experienced landfill consultant, the best solution to the leachate problem is to apply clean water to landfills continuously and then collect and treat the resulting leachate in carefully designed and monitored facilities. He contends that after 10–20 years of such washing, little potential for groundwater pollution should remain. This wetting would also hasten the breakdown of wastes and thus allow old landfills to be dug out and used again.

Since 1979 there has been a sharp drop in the number of solid-waste landfills in the United States as existing landfills have reached their capacity or closed; since 1997 only more expensive, state-of-the art landfills (Figure 14-8) are allowed to operate. Although a few U.S. cities, notably Philadelphia and New York, are having trouble finding nearby landfills, there is no national shortage of landfill volume because most small local landfills are being replaced by larger local and regional landfills.

Is Land Disposal of Hazardous Waste the Answer? Most hazardous waste in the United States is disposed of by deep-well injections, surface impoundments, and state-of-the-art landfills. In *deep-well disposal*, liquid hazardous wastes are pumped under pressure through a pipe into dry, porous geologic formations or into fracture zones of rock far beneath aquifers tapped for drinking and irrigation water (Figure 13-3). In theory, these liquids soak into the porous rock material and are isolated from overlying groundwater by essentially impermeable layers of rock.

Because this method is simple and cheap, its use is increasing rapidly as other methods are legally restricted or become too expensive. By 1998, 60% of the liquid hazardous wastes in the United States were injected underground. If sites are chosen according to the best geological and seismic data, a number of scientists believe deep wells may be a reasonably safe way to dispose of fairly dilute solutions of organic and inorganic waste; with proper site selection and care, they may be safer than incineration. Furthermore, if some use eventually is found for the waste, it could be pumped back to the surface.

However, many scientists believe that current regulations for geologic evaluation, long-term monitoring, and long-term liability if wells contaminate groundwater are inadequate. Until deep-well disposal is more carefully evaluated and regulated, many scientists and environmentalists believe that it should not be allowed to proliferate. What do you think?

Surface impoundments such as ponds, pits, or lagoons (Figure 12-23) used to store hazardous waste are supposed to be sealed with a plastic liner on the bottom. Solid wastes settle to the bottom and accumulate, whereas water and other volatile compounds evaporate into the atmosphere. According to the EPA, however, 70% of these storage basins have no liners, and as many as 90% may threaten groundwater. Eventually all liners will leak, and waste will percolate into groundwater. Moreover, volatile compounds (such as hazardous organic solvents) can evaporate into the atmosphere, promote smog formation, and return to the ground to contaminate surface water and groundwater in other locations.

About 5% by weight of the legally regulated hazardous waste produced in the United States is concentrated, put into drums, and buried in specially designed and monitored commercial hazardous-waste landfills. Sweden goes further and buries its concentrated hazardous wastes in underground vaults (Figure 14-9). By contrast, in the United Kingdom most hazardous wastes are mixed with household garbage and stored in hundreds of conventional landfills all over the country.

When current and future commercial hazardous-waste landfills in the United States eventually leak and threaten water supplies, many of their operators will declare bankruptcy. Then the EPA may have to put the landfills on the Superfund list and taxpayers will pick up the tab for cleaning them up. As EPA hazardous-waste expert William Sanjour points out,

> *The real cost of dumping is not borne by the producer of the waste or the disposer, but by the people whose health and property values are destroyed when the wastes migrate onto their property and by the taxpayers who pay to clean it up. . . . It is better for liners to leak sooner rather than later, because then there will be responsible parties that they can get to clean it up. Liners don't protect communities. They protect the people who put the waste there and the politicians who let them put the waste there, because they are long since gone when the problem comes up.*

Some engineers and scientists have proposed storing hazardous wastes above ground in large, two-story, reinforced-concrete buildings until better technologies are developed (Figure 14-10). The first floor would contain no wastes but would have inspection walkways so people could easily check for leaks from the upper story. Any leakage would be collected, treated, solidified, and returned to the storage building. The buildings would have negative air pressure—caused by fans blowing in air and then filtering it—to prevent the

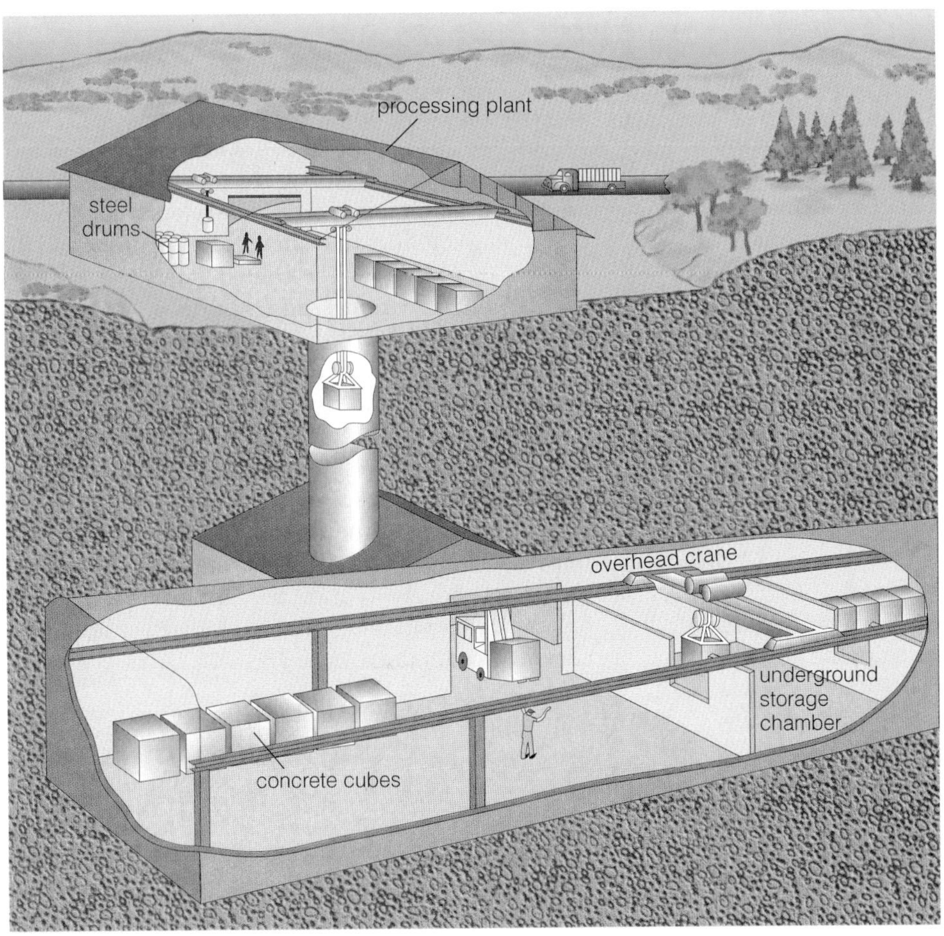

Figure 14-9 Swedish method for handling hazardous waste. Hazardous materials are placed in drums, which are embedded in concrete cubes and stored in an underground vault.

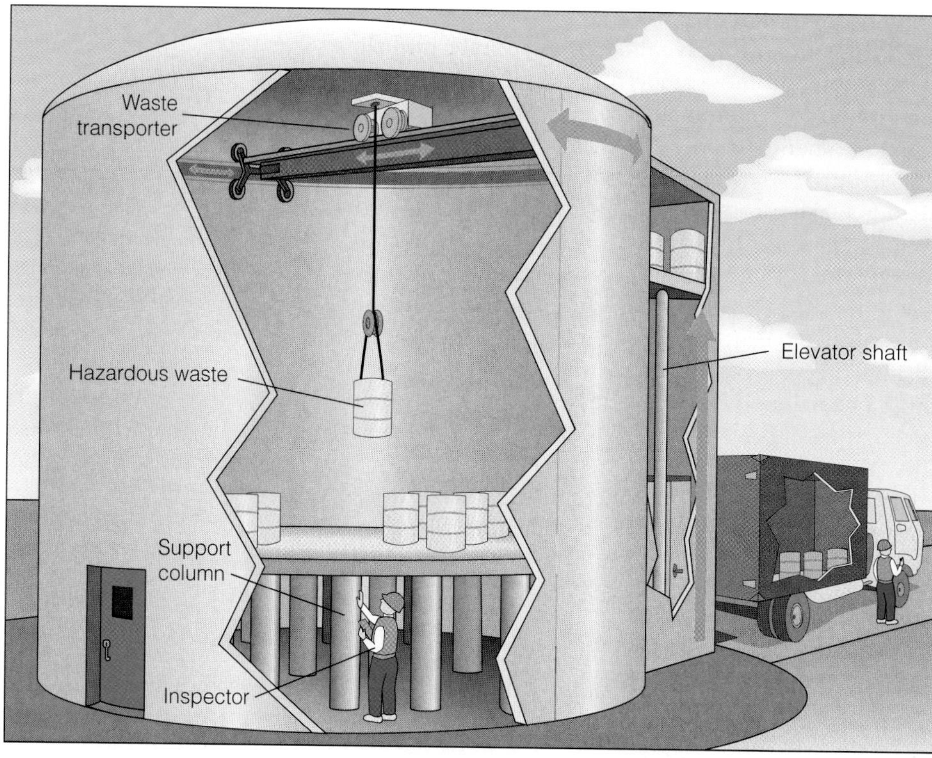

Figure 14-10 An aboveground hazardous-waste storage facility would help keep wastes from contaminating groundwater.

release of toxic gases. Proponents believe that such an *in-sight* approach would be cheaper and safer than *out-of-sight* landfills, deep wells, or incinerators for many hazardous wastes. What do you think?

There is also growing concern about accidents during some of the more than 500,000 shipments of hazardous wastes (mostly to landfills and incinerators in trucks or by train) in the United States each year. Between 1988 and 1992, approximately 34,500 toxic chemical accidents caused 100 deaths, over 11,000 injuries, and evacuation of over 500,000 people. Typically, trucks and trains carrying hazardous materials are involved in about 13,000 accidents per year. Few communities have the equipment and trained personnel to deal with hazardous-waste spills.

Is Exporting Waste the Answer? Some U.S. cities that lack sufficient landfill space are shipping some of their solid waste and hazardous waste to other states or other countries, especially developing countries. However, some developing countries are refusing to accept delivery.

Waste disposal firms can charge high prices for picking up hazardous wastes. If they can then dispose of them (legally or illegally) at low costs, they pocket huge profits. Often officials of developing countries find it difficult to resist the income (often bribes) derived from receiving these wastes.

In 1994, 64 nations met in Geneva, Switzerland, and approved an immediate ban on all exports of hazardous wastes for disposal from developed countries to developing countries (including eastern Europe and the former Soviet Union). The United States was the only participating nation that did not support the ban.

An effective worldwide ban on all hazardous-waste exports from one country to another would help. But this would not end illegal trade in these wastes because the potential profits are much too great. To most environmentalists, the only real solution to the hazardous-waste problem is to produce as little as possible in the first place (pollution prevention). In 1998, 34 industrial nations adopted agreements to sharply reduce the use of 16 persistent organic pollutants and toxic metals (such as lead, mercury, and cadmium).

14-7 CASE STUDIES: LEAD, DIOXINS, AND CHLORINE

How Can We Reduce Exposure to Lead? Each year 12,000–16,000 American children under age 9 are treated for acute lead poisoning, and about 200 die.

About 30% of the survivors suffer from palsy, partial paralysis, blindness, and mental retardation.

Lead can also cause damage at levels far below those that cause acute lead poisoning, especially in children and unborn fetuses. Research indicates that children under age 6 and unborn fetuses with fairly low blood levels of lead are especially vulnerable to **(1)** nervous system impairment, **(2)** a lowered IQ (by 4–6 points), **(3)** a shortened attention span, **(4)** hyperactivity, **(5)** hearing damage, and **(6)** various behavior disorders.

It is great news that between 1976 and 1995 the percentage of U.S. children ages 1–5 with lead levels above the current blood level standard dropped from 85% to 4.4%, preventing at least 9 million childhood lead poisonings. The primary reason is government regulations that phased out leaded gasoline and lead-based solder, a pollution prevention approach instituted primarily because of the pioneering research on lead toxicity in children published in 1979 by Dr. Herbert Needleman.

Even with the encouraging drop in average blood levels of lead, the U.S. Centers for Disease Control and Prevention estimate that at least 890,000 U.S. children still have unsafe levels of lead from exposure to a number of sources (Figure 14-11). Numerous studies have also shown that there is no safe level of lead in children's blood. A 1999 study by the U.S. General Accounting Office found that most states were not testing young children on Medicaid for lead poisoning, as required by a 1989 law.

Health scientists have proposed a number of ways to help protect children from lead poisoning:

- *Testing all children for lead by age 1.*
- *Banning incineration of solid and hazardous waste or greatly increasing current pollution control standards for old and new incinerators.*
- *Phasing out leaded gasoline worldwide over the next decade.*
- *Testing older housing and buildings for leaded paint and lead dust and removing this hazard.* According to government estimates, 83% of U.S. homes (and 86% of public housing units) built before 1980 contain lead-based paint.
- *Banning all lead solder in plumbing pipes and fixtures and in food cans.*
- *Removing lead from municipal drinking water systems within 10 years.*
- *Washing fresh fruits and vegetables and hands thoroughly.*
- *Testing ceramicware used to serve food for lead glazing.*
- *Banning imports of lead-laden PVC miniblinds.*

Doing most of these things will cost an estimated $50 billion in the United States. But health officials say the alternative is to keep poisoning and mentally handicapping large numbers of children.

Although the threat from lead has been reduced in the United States, this is not the case in developing countries. The World Health Organization (WHO) estimates that 15–18 million children in developing countries suffer permanent brain damage because of lead poisoning. Blood levels of lead are especially high in India, Bangladesh, China, and Russia.

How Dangerous Are Dioxins? Dioxins are a family of 219 different chlorinated hydrocarbon compounds formed as unwanted by-products in chemical reactions (usually at high temperatures, especially in incinerators) involving chlorine and hydrocarbons. Basically, any time you expose chlorine-containing compounds to high temperature, you have conditions that can produce dioxins.

One of these compounds, TCDD—sometimes simply called dioxin—is the most harmful and the most widely studied. Dioxins such as TCDD are persistent chemicals that linger in the environment for decades, especially in soil and human fat tissue.

In 1990 representatives of the paper and chlorine industries claimed that the inconsistent results of health studies exonerated TCDD and other dioxins, and they pushed the EPA for a reassessment of the health risks of dioxins. However, the EPA's 1994 reevaluation indicated that dioxins are an even greater health threat than previously thought.

This new review concluded that dioxin is "likely to present a cancer threat to humans." According to WHO studies, TCDD is a known carcinogen and many of the less potent dioxins are probable human carcinogens.

In 1998 a group of German scientists concluded that dioxins may be responsible for 12% of human cancers in industrialized countries. If this estimate is correct, dioxins are responsible for about 120,000 cancers each year in the United States.

Because of high levels of dioxin in breast milk, newborns in the United States get 80 times their lifetime "safe" dose during their first 6 months of life. In 1998, *Consumer Reports* published test results showing that all major baby food brands had alarming levels of dioxin in meat-based products.

The EPA's comprehensive, 3-year, $4-million review by more than 100 scientists around the world also revealed that the most powerful effects of exposure to various dioxins at or near levels already occurring in the U.S. population are seen in the reproductive system, endocrine (hormone) system, and immune system. These effects may pose an even greater threat to human

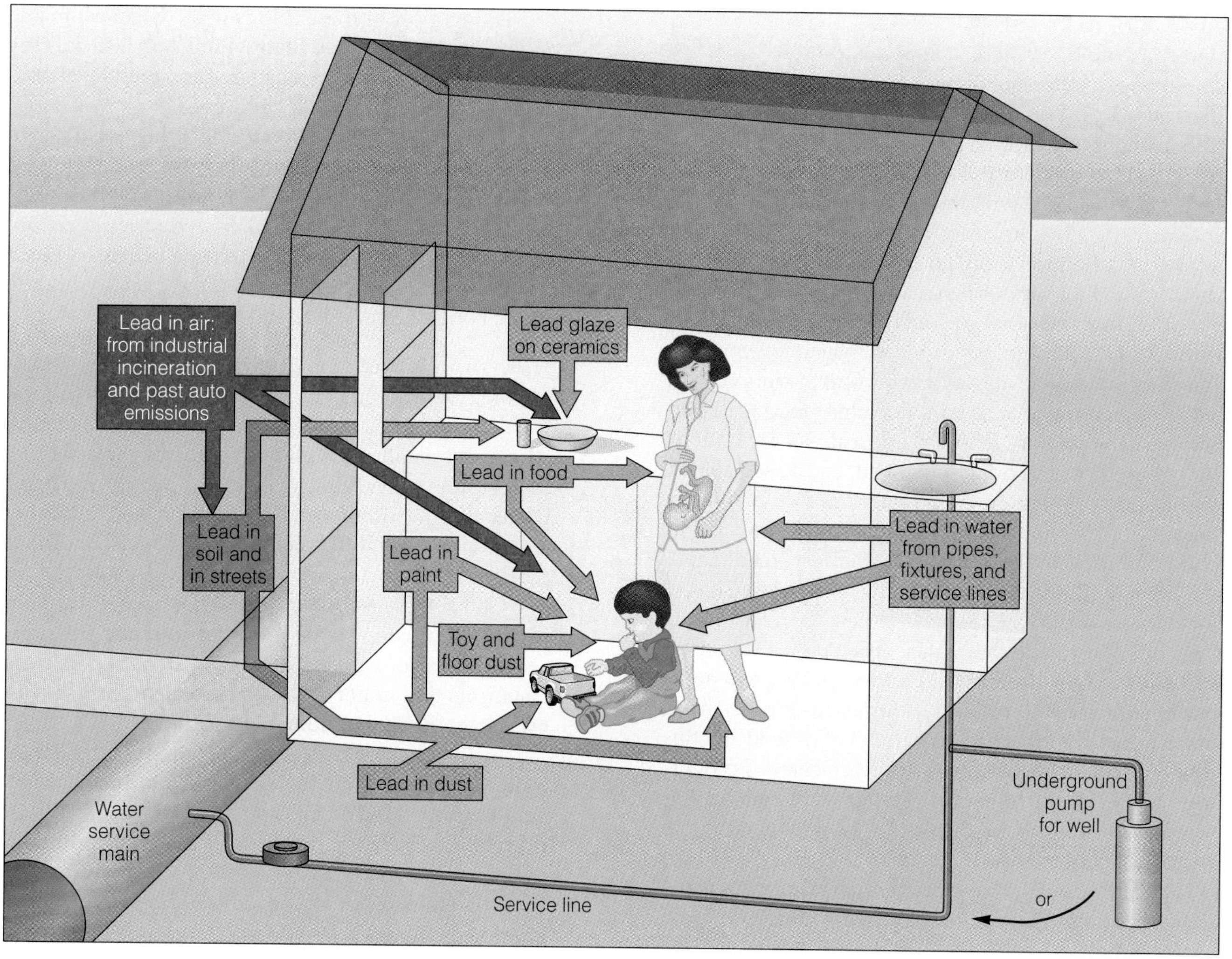

Figure 14-11 Sources of lead exposure for children and fetuses.

health (especially for fetuses and newborns) than the cancer-promoting ability of various dioxins (Section 8-2).

In 1996, biologist Barry Commoner and other researchers released a study showing that about 86% of the dioxins produced in the United States could be eliminated without economic sacrifice and with important environmental gains. This would be accomplished mostly by eliminating the sources of dioxin from medical waste incinerators, MSW generators, paper mills, iron ore sintering plants, and cement kilns burning hazardous wastes.

What Should We Do About Chlorine? Modern society depends heavily on chlorine (Cl_2) and chlorine-containing compounds. The problem is that some of the chlorine-containing chemicals we produce and use in fairly large amounts **(1)** are persistent, **(2)** accumulate in body fat, and **(3)** are harmful to human health (according to animal and other toxicity studies; Section 8-2).

In 1993, the Governing Council of the American Public Health Association (APHA), a major U.S. scientific and medical association, unanimously approved a statement urging the American chemical industry to phase out use of chlorine except for producing some pharmaceuticals and for disinfection of public water supplies. The goal would be to replace most chlorine compounds with environmentally and economically acceptable substitutes over the next 10–20 years.

The three largest single uses of chlorine are **(1)** plastics (mostly PVC), **(2)** solvents, and **(3)** paper and pulp bleaching. Most of these uses have substitutes and could be phased out, as a small but growing number of companies are doing. Nonchlorine plastics and other materials could replace most uses of PVC, and incineration of PVCs could be prohibited (which would also reduce production of dioxins).

Studies estimate that 60% of the chlorinated organic solvents currently used could be phased out over a decade and replaced with less harmful and affordable substitutes, such as **(1)** soap and water, **(2)** steam cleaning, **(3)** citrus-based solvents (Individuals Matter, p. 281), and **(4)** physical cleaning (including elbow grease and blasting with plastic beads and a pressurized solution of water and baking soda). Such a phaseout has already begun, not just for environmental reasons but also to reduce legal liability for possible effects of such chemicals on workers and consumers.

Using chlorine to bleach wood pulp and paper could be replaced with less harmful processes that rely on oxygen, ozone, chlorine dioxide, and hydrogen peroxide, as several paper companies are doing. By the early 1990s, the German paper industry had achieved totally chlorine-free paper production (which is 30% cheaper than processes using chlorine); the rest of Europe is following Germany's lead.

About 5% of the chlorine produced in the United States is used to purify drinking water (1%) and wastewater from sewage treatment plants (4%). Much of this could be gradually replaced with ozone and other nonchlorine purification processes. Drinking water may still need to be purified by chlorination because it kills harmful microorganisms in water lines.

The remaining 35% of chlorine is used to manufacture a variety of organic and inorganic chemicals, many of them for small and specialized uses. Each of these chemicals could be evaluated to determine which are essential (many pharmaceuticals) and which might be replaced by affordable and environmentally acceptable substitutes.

Producers of chlorine chemicals strongly oppose such threats to their profits. However, if given proper financial incentives and 10–20 years to develop and phase in substitutes, chemical producers could profit from sales of chlorine substitutes and protect themselves from adverse publicity, lawsuits, and cleanup costs.

14-8 HAZARDOUS-WASTE REGULATION IN THE UNITED STATES

What Is the Resource Conservation and Recovery Act? In 1976, the U.S. Congress passed the Resource Conservation and Recovery Act (RCRA, pronounced "RICK-ra"), which was amended in 1984. This law requires the EPA to identify hazardous wastes and set standards for their management, and it provides guidelines and financial aid for states to establish waste management programs. The law also requires all firms that store, treat, or dispose of more than 100 kilograms (220 pounds) of hazardous wastes per month to have a permit stating how such wastes are to be managed.

To reduce illegal dumping, hazardous-waste producers that are granted disposal permits by the EPA must use a cradle-to-grave system to keep track of waste transferred from point of origin to approved off-site disposal facilities. However, the EPA and state regulatory agencies don't have sufficient personnel or money to review the documentation of more than 750,000 hazardous-waste generators and 15,000 haulers each year, let alone detect and prosecute offenders.

What Is the Superfund Act? In 1980, the U.S. Congress passed the Comprehensive Environmental Response, Compensation, and Liability Act, commonly known as the Superfund program. This law (plus later amendments) established a $16.3-billion Superfund financed jointly by federal and state governments and by special taxes on chemical and petrochemical industries (which provide 86% of the funding). The purpose of the Superfund is to identify (Spotlight, p. 378) and clean up abandoned hazardous-waste dump sites such as Love Canal (p. 355) and leaking underground tanks that threaten human health and the environment.

To keep taxpayers from footing most of the bill, cleanups are based on the *polluter-pays principle.* The EPA is charged with **(1)** locating dangerous dump sites, **(2)** finding the potentially liable culprits, **(3)** ordering them to pay for the entire cleanup, and **(4)** suing them if they don't. When the EPA can find no responsible party, it draws money out of the Superfund for cleanup.

To implement the polluter-pays principle, the Superfund legislation considers all polluters of a site to be subject to strict, joint, and several liability. This controversial strategy means that each individual polluter (regardless of its contribution) can be held liable for the entire cost of cleaning up a site if the other parties can't be found or have gone bankrupt.

Currently about 1,375 sites are on a National Priority List for cleanup because they pose a real or potential threat to nearby populations. By June 1998 emergency cleanup had been carried out at virtually all sites and more than 500 sites had been stabilized, with the wastes contained to prevent leakage. Once a site is stabilized, final cleanup (which often involves pumping up and cleaning contaminated groundwater) can take decades. Attempts are being made to find ways to improve the Superfund Act without seriously weakening it (Solutions, p. 379).

The U.S. Office of Technology Assessment and the Waste Management Research Institute estimate that the Superfund list could eventually include at

SPOTLIGHT

Using Honeybees to Detect Toxic Pollutants

Honeybees are being used to detect the presence of toxic chemicals. On their forays from a hive, bees pick up water, nectar, and pollen, particles, and volatile organic compounds that contaminate the air near sites where they visit.

The bees then bring back these materials to the hive. There they fan the air vigorously with their wings to regulate the hive's temperature. This releases and circulates any pollutants they picked up into the air inside the hive.

Scientists have put portable hives, each containing 7,000 to 10,000 bees, near known or suspected hazardous waste sites. A small copper tube attached to the side of each hive pumps air out. The air is analyzed for toxic particulate contaminants such as **(1)** lead and zinc, **(2)** volatile organic contaminants such as perchloroethylene (a toxic degreasing agent and dry-cleaning solvent) and **(3)** pesticide residues.

To find out where the bees have gone to forage for food, a botanist uses a microscope to examine pollen grains to determine what kinds of plants they come from and correlate this with the plants within an area up to 1.6 kilometers (1 mile) of each hive. This approach is much less costly than setting up a number of air pollution monitors around mines, hazardous waste dumps, and other sources of toxic pollutants.

Critical Thinking

Because honeybees can pick up toxic pollutants, should honey be tested for such pollutants before being placed on the market?

least 10,000 priority sites, with cleanup costs of up to $1 trillion, not counting legal fees. Other studies project only 2,000 sites with estimated cleanup costs of $100–165 billion.

Cleaning up toxic U.S. military dumps will cost another $100–200 billion and take at least 30 years. Cleaning up contaminated Department of Energy sites used to make nuclear weapons will cost an additional $200–400 billion and take 30–50 years. In addition, the Department of Interior will need to spend at least $100 billion to clean up 300,000 active and abandoned mine sites, 200,000 sites leased for oil and gas drilling, 3,000 landfills, and 4,200 leaking underground storage tanks on public lands under its jurisdiction. These estimated costs are only for cleanup to prevent future damage; they don't include the health and ecological costs already associated with such wastes.

14-9 SOLUTIONS: ACHIEVING A LOW-WASTE SOCIETY

What Is the Role of Grassroots Action? Bottom-Up Change In the United States, local citizens have worked together to prevent hundreds of incinerators, landfills, or treatment plants for hazardous and radioactive wastes from being built in or near their communities. Opposition has grown as numerous studies have shown that such facilities have traditionally been located in communities populated by African Americans, Asian Americans, Hispanics, and poor whites—cited as an example of *environmental injustice* (Guest Essay, p. 380).

Most members of such groups recognize that health risks from incinerators and landfills, when averaged over the entire country, are quite low. However, they also know that the risks for the people near these facilities are much higher. These people, not the rest of the population, are the ones whose health, lives, and property values are being threatened.

Manufacturers and waste industry officials point out that something must be done with the toxic and hazardous wastes produced to provide people with certain goods and services. If local citizens adopt a "not in my backyard" (NIMBY) approach, the waste still ends up in someone's backyard.

Many citizens don't accept this argument. To them the best way to deal with most toxic or hazardous wastes is to produce much less of them (Guest Essay, p. 362). They agree with many EPA scientists that ultimately there is no such thing as safe disposal of a toxic or hazardous waste. For such materials, their goal is "not in anyone's backyard" (NIABY) or "not on planet Earth" (NOPE). What do you think?

What Can Be Done at the International Level? In 1998 and 1999 delegates from 110 countries met to work on a global treaty to control the world's 12 *persistent organic pollutants* (POPs). These widely used toxic chemicals are insoluble in water and soluble in fatty tissues. This allows them to be concentrated in the fatty tissues of humans and other organisms feeding at high trophic levels in food webs to levels hundreds of thousand times higher than in the general environment (Figures 8-2 and 12-20). They can also be transported long distances by wind and water.

How Can the Superfund Law Be Improved?

SOLUTIONS

Since the Superfund program began, polluters and their insurance companies have been working hard to do away with the polluter-pays principle at the heart of the program and make it mostly a *public-pays* approach.

This strategy, which is working, has three components: **(1)** Deny responsibility (stonewall) in order to tie up the EPA in expensive legal suits for years, **(2)** sue local governments and small businesses to make them responsible for cleanup, both as a delay tactic and to turn local governments and small businesses into opponents of Superfund's strict liability requirements, and **(3)** mount a public relations campaign declaring that toxic dumps pose little threat, that cleanup is too expensive compared with the risks involved, and that Superfund is unfair and is wasteful and ineffective.

The strict joint and several liability of the Superfund law may seem unfair to some, but the authors of the legislation argued that any other liability scheme wouldn't work. If the EPA had to identify and bring an enforcement action against every party liable for a dump site, the agency would be overwhelmed with lawsuits, and the pace of cleanup would be still slower. Administrative costs would also rise sharply, causing the program to become more of a *public-pays program* favored by the polluters.

The EPA also points out that the strict polluter-pays principle in the Superfund Act has been effective in making illegal dump sites virtually a thing of the past—an important form of pollution prevention for the future. It has also forced waste producers who are fearful of future liability claims to reduce their production of such waste and to recycle or reuse much more of what they do generate.

One proposal for reducing lawsuits and speeding cleanup is to set up an $8-billion Environmental Insurance Resolution Fund funded by insurers for a 10-year period. Companies found liable for cleanup of wastes they disposed of before 1986 would be able to use money from this fund rather than going to their individual insurance companies.

Although most experts agree that the Superfund's ultimate goal—permanent cleanup—should be maintained, many argue that sites should be ranked in three general categories: **(1)** sites requiring immediate full cleanup, **(2)** sites considered to pose serious hazard but located sufficiently far from concentrations of people or endangered ecosystems (these sites would receive emergency cleanup and then be isolated by barriers and signs, with more complete cleanup to come later), and **(3)** lower-risk sites requiring only stabilization (capping and containment) and monitoring. Other analysts believe that people and local governments should be involved in cleanup decisions from start to finish.

Other suggestions include **(1)** encouraging nonindustrialized redevelopment of abandoned industrial sites—called *brown fields*—by exempting innocent landowners and clean-up volunteers from superfund liability, and **(2)** exempting innocent individuals, small businesses, and municipalities from superfund liability.

Critical Thinking

Do you support the *polluter-pays principle* used in the original Superfund legislation? Defend your position. Use the library or Internet to find out whether there have been any recent changes in the Superfund law. If so, evaluate these changes and decide which of them you support. Defend your position.

The list of 12 chemicals, called the *dirty dozen,* include DDT and 8 other chlorine-containing persistent pesticides, PCBs, and dioxins. The goals of this treaty are to **(1)** ban or phase out use of these chemicals and **(2)** to detoxify or isolate stockpiles of such chemicals in warehouses and dumps.

How Can We Make the Transition to a Low-Waste Society? According to physicist Albert Einstein, "A clever person solves a problem, a wise person avoids it." To prevent pollution and reduce waste, many environmental scientists urge us to understand and live by four key principles: **(1)** Everything is connected, **(2)** there is no "away" for the wastes we produce, **(3)** dilution is not the solution to most pollution, and **(4)** the best and cheapest way to deal with waste and pollution is to produce less of it and then reuse and recycle most of the materials we use (Figures 14-3 and 14-4). Some actions you can take to reduce your production of solid waste and hazardous waste are listed in Appendix 6.

Unless we learn to depend on fewer virgin resources—by development of secondary materials industries . . . and by redesigning goods, services, and communities—we will continue propelling ourselves toward economic and ecological disaster.

JOHN E. YOUNG AND AARON SACHS

Environmental Justice for All

GUEST ESSAY

Robert D. Bullard

Robert D. Bullard is professor of sociology and director of the Environmental Justice Resource Center at Clark Atlanta University. For more than a decade, he has conducted research in the areas of urban land use, housing, community development, industrial facility siting, and environmental justice. His scholarship and activism have made him one of the leading experts on environmental racism, the systematic selection of communities of color as sites for waste facilities and polluting industries. He is the author of four books and more than three dozen articles, monographs, and scholarly papers that address equity concerns. His book Dumping in Dixie: Race, Class, and Environmental Quality, *2d ed. (Westview Press, 1994), has become a standard text in the field. Other books are* Confronting Environmental Racism *(South End Press, 1993) and* Unequal Protection: Environmental Justice and Communities of Color *(Sierra Club Books, 1994).*

Despite widespread media coverage and volumes written on the U.S. environmental movement, environmentalism and social justice have seldom been linked. Nevertheless, an environmental revolution is now taking shape in the United States that combines the environmental and social justice movements into one framework.

People of color (African Americans, Latinos, Asians, Pacific Islanders, and Native Americans), working-class people, and poor people in the United States suffer disproportionately from industrial toxins, dirty air and drinking water, unsafe work conditions, and the location of noxious facilities such as municipal landfills, incinerators, and toxic-waste dumps. Despite the government's attempts to level the playing field, all communities are not created equal.

The environmental justice movement attempts to dismantle **(1)** exclusionary zoning ordinances, **(2)** discriminatory land-use practices, **(3)** differential enforcement of environmental regulations, **(4)** disparate siting of risky technologies, and **(5)** the dumping of toxic waste on the poor and people of color in the United States and in developing countries.

All communities are not treated as equals when it comes to resolving environmental and public health concerns. Over 300,000 farm workers (over 90% of whom are people of color) and their children are poisoned by pesticides sprayed on crops in the United States. Some 3–4 million children (many of them African Americans or Latinos living in the inner city) are poisoned each year by lead-based paint in old buildings, lead-soldered pipes and water mains, lead-tainted soil contaminated by industry, and air pollutants from smelters.

All communities do not bear the same burden or reap the same benefits from industrial expansion. This is true in the case of the mostly African-American Emelle, Alabama (home of the nation's largest hazardous-waste landfill), Navajo lands in Arizona where uranium is mined, and the 2,000 factories, known as *maquiladores*, located just across the U.S. border in Mexico.

Communities, states, and regions that contain hazardous-waste disposal facilities (importers) receive far fewer economic benefits (jobs) than the geographic locations that generate the wastes (exporters). Nationally,

REVIEW QUESTIONS

1. Define the boldfaced terms used in this chapter.

2. Summarize the Love Canal tragedy.

3. Distinguish between *solid waste* and *municipal solid waste*. What are the major sources of solid waste in the United States? What happens to municipal solid waste in the United States?

4. What is *hazardous waste*? What percentage of the overall hazardous waste produced in the United States is regulated by hazardous-waste laws?

5. Distinguish between the *high-waste* and *low-waste* approaches to management of solid and hazardous waste. According to the U.S. National Academy of Sciences, what should be the five goals of waste management in order of their importance? According to scientists, what percentage of solid and hazardous waste produced could be eliminated through a combination of waste reduction, reuse, and recycling (including composting)?

6. List five reasons that producing less waste and pollution is the best approach. Describe the pollution prevention program implemented by the Minnesota Mining and Manufacturing (3M) company in the United States.

7. List 10 ways to reduce waste and pollution. What two economic changes would probably have to be implemented to encourage a shift to pollution and waste prevention?

8. What is *reuse* and what are three advantages of using this approach for waste management?

9. What is *compost* and how can it be used as a way to deal with solid waste?

10. Distinguish between *primary (closed-loop)* and *secondary (open-loop) recycling*. About what percentage of the municipal solid waste produced in the United States is

60% of African Americans and 50% of Latinos live in communities with at least one uncontrolled toxic-waste site. Three of the five largest hazardous-waste landfills are located in communities that are predominantly African American or Latino.

Environmental justice does not stop at the U.S. border. Environmental injustices exist from the *favelas* of Rio de Janeiro, Brazil, to the shantytowns of Johannesburg, South Africa. Members of the environmental justice movement are also questioning the wasteful and nonsustainable development models being exported to the developing world.

It is no mystery why grassroots environmental justice groups in Louisiana's "Cancer Alley," Chicago's south side, and Los Angeles's East and South Central neighborhoods are attacking the institutions they blame for their underdevelopment, disenfranchisement, and poisoning. Some people see these threats to their communities as a form of genocide.

Grassroots leaders are demanding justice. Residents of communities such as West Dallas and Texarkana (Texas), West Harlem (New York), Rosebud (South Dakota), Kettleman City (California), and Sunrise, Lions, and Wallace (Louisiana) see their struggle for environmental justice as a life-and-death matter. Unfortunately, their stories of environmental racism are not broadcast into the nation's living rooms during the nightly news, nor are they splashed across the front pages of national newspapers and magazines. To a large extent, the communities that are the victims of environmental injustice remain invisible to the larger society.

The environmental justice movement is led, planned, and to a large extent funded by people who are not part of the established environmental community or the "Big 10" environmental organizations. Most environmental justice groups are small and operate with resources generated from the local community.

For too long these groups and their leaders have been invisible and their stories muted. This is changing as these grassroots groups are forcing their issues onto the nation's environmental agenda.

The United States has a long way to go in achieving environmental justice for all its citizens. The membership of decision-making boards and commissions still does not reflect the racial, ethnic, and cultural diversity of the country, and token inclusion of people of color on boards and commissions does not necessarily mean that their voices will be heard or their cultures respected. The ultimate goal of any inclusion strategy should be to democratize the decision-making process and empower disenfranchised people to speak and do for themselves.

Critical Thinking

1. Does your lifestyle help promote or reduce environmental injustice in your community and in society as a whole?

2. Would you work to help prevent polluting factories and hazardous-waste facilities from being located in or near communities made up largely of people of color, working-class people, and poor people?

recycled and composted and what percentage do experts believe could be recycled and composted?

11. What is a *pay-as-you-throw* garbage collection program and how does it encourage recycling?

12. Distinguish between the *centralized recycling* of mixed solid waste and *consumer separation* of solid waste and list the pros and cons of each approach to recycling.

13. List the pros and cons of recycling. According to environmentalists, what are the five major reasons for recycling? What three factors hinder recycling and reuse? List seven ways to encourage more recycling and reuse. Describe Germany's approach to recycling.

14. Summarize the recycling of **(a)** aluminum, **(b)** wastepaper, and **(c)** plastics. List eight advantages of recycling paper. Distinguish between *preconsumer* and *postconsumer paper waste*. Why are so few plastics recycled?

15. Describe Denmark's hazardous-waste detoxification program. What are *bioremediation* and *phytoremediation?*

16. List the pros and cons of dealing with solid and hazardous waste by **(a)** burning it in incinerators, **(b)** burying it in sanitary landfills, **(c)** disposing of liquid hazardous waste in deep underground wells, **(d)** storing hazardous waste in aboveground buildings, and **(e)** exporting solid and hazardous waste to other areas or countries.

17. Describe the hazards of exposure to lead and list eight ways to reduce such exposure.

18. What are *dioxins,* what harm can they cause, and how can we reduce exposure to these hazardous chemicals?

19. What are three major problems with many chlorine-containing compounds? What are the three major uses of chlorine? How can we reduce exposure to harmful chlorine-containing compounds?

20. How is the Resource Conservation and Recovery Act used to deal with the problem of hazardous wastes in the United States?

21. What is the Superfund Act, and what are its strengths and weaknesses?

CRITICAL THINKING

1. Explain why you support or oppose the following:

a. Requiring that all beverage containers be reusable

b. Requiring all households and businesses to sort recyclable materials into separate containers for curbside pickup

c. Requiring consumers to pay for plastic or paper bags at grocery and other stores to encourage the use of reusable shopping bags

2. Would you oppose having a hazardous-waste landfill, waste treatment plant, deep-injection well, or incinerator in your community? Explain. If you oppose these disposal facilities, how do you believe that hazardous waste generated in your community and your state should be managed?

3. Give your reasons for agreeing or disagreeing with each of the following proposals for dealing with hazardous waste:

a. Reducing the production of hazardous waste and encouraging recycling and reuse of hazardous materials by levying on producers a tax or fee for each unit of waste generated

b. Banning all land disposal and incineration of hazardous waste to encourage recycling, reuse, and treatment and to protect air, water, and soil from contamination

c. Providing low-interest loans, tax breaks, and other financial incentives to encourage industries producing hazardous waste to recycle, reuse, treat, destroy, and reduce generation of such waste

d. Banning the shipment of hazardous waste from one country to another country

e. Banning the shipment of hazardous waste from one state to another

***4.** For 1 week, keep a list of the solid waste you throw away. What percentage is materials that could be recycled, reused, or burned for energy? What percentage of the items could you have done without in the first place? Tally and compare the results for your entire class.

***5.** What hazardous and solid wastes do you create? What specific things could you do to reduce, reuse, or recycle as much of these wastes as possible? Which of these things do you actually plan to do? Do you use any of the alternative chemicals shown in Table 14-2?

***6.** Determine whether your school **(a)** has recycling programs, **(b)** sells soft drinks in throwaway cans or bottles, and **(c)** bans release of throwaway helium-filled balloons at sporting events and other activities.

***7.** What happens to solid waste in your community? How much is landfilled? Incinerated? Composted? Recycled? What technology is used in local landfills and incinerators? What leakage and pollution problems have local landfills or incinerators had? Does your community have a recycling program? Is it voluntary or mandatory? Does it have curbside collection? Drop-off centers? Buy-back centers?

***8.** What hazardous wastes are produced at your school? In your community? What happens to these wastes?

***9.** Make a concept map of this chapter's major ideas, using the section heads and subheads and the key terms (in boldface). Look at the inside back cover and in Appendix 5 for information about concept maps.

FURTHER INFORMATION

For more information see Internet Sources and Further Readings at the back of the book. Online resources for this chapter can be found at **www.brookscole.com/biology**

PART IV

SUSTAINING BIODIVERSITY AND ECOLOGICAL INTEGRITY

It is the responsibility of all who are alive today to accept the trusteeship of wildlife and to hand on to posterity, as a source of wonder and interest, knowledge, and enjoyment, the entire wealth of diverse animals and plants. This generation has no right by selfishness, wanton or intentional destruction, or neglect, to rob future generations of this rich heritage. Extermination of other creatures is a disgrace to humankind.

WORLD WILDLIFE CHARTER

15 FOOD RESOURCES

Perennial Crops on the Kansas Prairie

When you think about farms in Kansas, you probably picture seemingly endless fields of wheat or corn plowed up and planted each year. By 2040 the picture might change, thanks to pioneering work at the nonprofit Land Institute near Salina, Kansas (Figure 15-1).

The Institute, headed by plant geneticist Wes Jackson, is experimenting with an ecological approach to agriculture on the Midwestern prairie. The goal is to grow food crops by planting a mix of **(1)** perennial grasses, **(2)** legumes (a source of nitrogen fertilizer), **(3)** sunflowers, **(4)** grain crops, and **(5)** plants that provide natural insecticides in the same field (polyculture).

The Institute's goal is to raise food by mimicking many of the natural conditions of the prairie without losing fertile grassland soil (Figure 13-17). Institute researchers believe that perennial polyculture can be blended with modern agriculture to reduce the growing harmful environmental effects of industrialized agriculture.

Because these plants are perennials, the soil doesn't have to be plowed up and prepared each year to replant them. This requires much less labor than conventional monoculture or diversified organic farms that grow annual crops. It reduces soil erosion because the unplowed soil is not exposed to wind and rain and also reduces pollution caused by chemical fertilizers and pesticides. Because of deep roots such perennials retain more water than annuals and thus reduce the need for irrigation.

If the Institute and similar groups doing such research succeed, within a few decades many people may be eating food made from a mix of perennials such as **(1)** *eastern gamma grass* (a warm-season grass that is a relative of corn with three times as much protein as corn and twice as much as wheat), **(2)** *mammoth wildrye* (a cool-season grass that is distantly related to rye, wheat, and barley), **(3)** *Illinois bundleflower* (a wild nitrogen-producing legume that can enrich the soil and whose seeds may serve as livestock feed), and **(4)** *Maximillian sunflower* (which produces seeds with as much protein as soybeans).

These discoveries will come none too soon. The basic question is, "How can the growing human population produce and distribute enough food to meet everyone's basic nutritional needs without degrading the soil, water, air, and biodiversity that support all food production?"

Figure 15-1 The Land Institute near Salina, Kansas, is a farm, a prairie laboratory, and a school dedicated to changing the way we grow food. It advocates growing a diverse mixture of edible perennial plants to supplement traditional annual monoculture crops. (Terry Evans)

There are two spiritual dangers in not owning a farm. One is the danger of supposing that breakfast comes from the grocery, and the other that heat comes from the furnace.

ALDO LEOPOLD

This chapter addresses the following questions:

- How is the world's food produced?
- What are the world's food problems?
- What can we do to help solve the world's food problems?
- How can we design and shift to more sustainable agricultural systems?

15-1 HOW IS FOOD PRODUCED?

What Three Systems Provide Us with Food? Some Good and Bad News Historically, humans have depended on three systems for their food supply: **(1)** *croplands* (mostly for producing grain), **(2)** *rangelands* (for producing meat from grazing livestock), and **(3)** *oceanic fisheries.*

Some *good news* is that

- Since 1950 there has been a staggering increase in global food production from all three systems.
- This phenomenal growth in food productivity has happened because of technological advances such as **(1)** increased use of tractors and farm machinery and high-tech fishing boats and gear, **(2)** inorganic chemical fertilizers, **(3)** irrigation, **(4)** pesticides, **(5)** high-yield varieties of wheat, rice, and corn, and **(6)** densely populated *feedlots* and enclosed pens for raising cattle, pigs, and chickens and *aquaculture ponds* for raising some types of fish and shellfish.
- Some agricultural scientists believe that such advances can be expanded and improved enough to provide food for at least 8 billion people.

Some *bad but challenging news* is that

- After tripling between 1950 and 1990, meat production from beef and mutton has increased little since 1990, mostly because overgrazing has lowered rangeland productivity in many parts of the world.
- After increasing nearly fivefold between 1950 and 1990, the oceanic fish catch appears to be leveling off, mostly because we have exceeded sustainable yield limits from overfishing.
- Since 1980 there has been little expansion in the area of the world's cropland because **(1)** the best cropland is already being used and **(2)** an increasing amount of existing cropland and rangeland has become less productive, mostly because of environmental degradation from erosion, salinization, waterlogging, and overgrazing.
- Since 1950 most increases in crop productivity have come from greatly increasing grain yields per hectare on existing cropland, but there are signs that we may be approaching the limits of such grain yields.
- There are growing signs that environmental degradation and pollution may limit future food production. Modern industrialized agriculture uses more soil, water, plant, animal, and energy resources and causes more pollution and environmental damage than any other human activity.
- Continuing population growth is putting increasing pressure on agricultural systems and the natural systems and resources (Figure 1-2) that support them.
- To feed the 9 billion people projected by 2054, we must produce and equitably distribute more food than has been produced since agriculture began about 10,000 years ago and do this in an environmentally sustainable manner.

What Plants and Animals Feed the World? The multitude of species of plants and animals (species diversity) and the varieties of plants and animals (genetic diversity) that provide us with food are an important part of the planet's biodiversity. Biologists estimate that even though the earth has perhaps 30,000 plant species with parts that people can eat, only 15 plant and 8 terrestrial animal species supply 90% of our food.

Just three grain crops—wheat, rice, and corn—provide almost half of the calories people consume. These three, and most of our other food crops, are *annuals*, whose seeds must be replanted each year.

Two out of three or the world's people survive primarily on traditional grains (mainly rice, wheat, and corn), mostly because they can't afford meat. As incomes rise, people consume even more grain, but indirectly in the form of meat (mostly beef, pork, and chicken), eggs, milk, cheese, and other products of grain-eating domesticated livestock.

Fish and shellfish are an important source of food for many coastal-dwelling people in developing countries. However, on a global scale fish and shellfish supply less than 1% of the energy and less than 6% of the proteins in the human diet.

What Are the Major Types of Food Production? There are two major types of agricultural systems: industrialized and traditional. **Industrialized agriculture**, or **high-input agriculture**, uses large amounts of fossil fuel energy, water, commercial fertilizers, and pesticides to produce huge quantities of single crops (monocultures) or livestock animals for sale. Practiced on about 25% of all cropland, mostly in developed countries (Figure 15-2), industrialized agriculture has spread since the mid-1960s to some developing countries. **Plantation agriculture**, a form of industrialized agriculture practiced primarily in tropical developing countries, grows cash crops such as bananas, coffee, and cacao, mostly for sale in developed countries.

An increasing amount of livestock production in developed countries is industrialized. This involves bringing large numbers of cattle to densely populated *feedlots*, where they are fattened up for about 4 months before slaughter. Most pigs and chickens in developed countries spend their entire lives in densely populated pens and cages and are fed mostly grain grown on cropland. Livestock production has an enormous environmental impact (Connections, right).

Traditional agriculture consists of two main types, which together are practiced by about 2.7 billion people in developing countries—almost half the people on the earth. **Traditional subsistence agriculture** typically produces only enough crops or livestock for a farm family's survival; in good years there may be

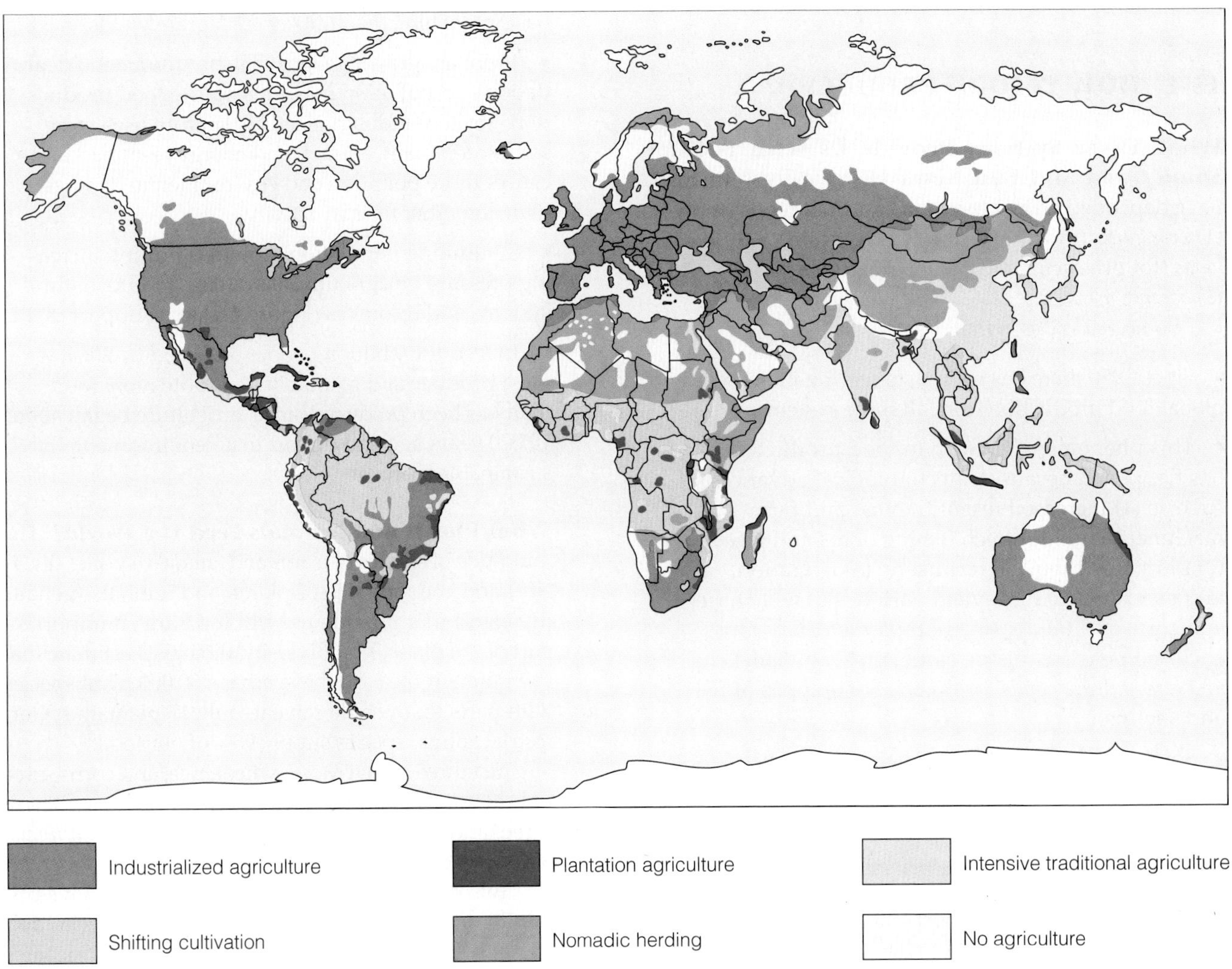

Figure 15-2 Generalized locations of the world's principal types of food production.

Some Environmental Consequences of Meat Production

CONNECTIONS

The meat-based diet of affluent people in developed and developing countries has the following effects on resource use, environmental degradation, and pollution:

- More than half of the world's cropland (19% in the United States) is used to produce livestock feed grain (mostly field corn, sorghum, and soybeans).
- Livestock and fish raised for food consume about 37% of the world's grain production (70% of U.S. prodution).
- Livestock use more than half the water withdrawn each year, mostly to irrigate crops fed to livestock and to wash away manure from livestock pens and feedlots.
- Manure washing off the land or leaking from lagoons used to store animal wastes is a significant source of water pollution that kills fish by depleting dissolved oxygen (Figure 12-18) and contributes to cultural eutrophication (Figure 12-21).
- About 14% of U.S. topsoil loss is directly associated with livestock grazing.
- Globally, overgrazing of sparse vegetation and trampling of the soil by too many livestock is the major cause of desertification in arid and semiarid areas (Figure 13-22).
- Cattle belch out 12–15% of all the methane (a greenhouse gas) released into the atmosphere (Figure 11-2c).
- Some of the nitrogen in commercial inorganic fertilizer used to grow livestock feed is converted to nitrous oxide, another greenhouse gas (Figure 11-2d).
- Nitrogen in manure escapes into the atmosphere as gaseous ammonia (NH_3), a pollutant that contributes to acid deposition (Figure 10-6).
- Livestock in the United States produce 21 times more waste (manure) than is produced by the country's human population. Only about half of this nutrient-rich livestock waste is recycled to the soil.

Some environmentalists have called for reducing livestock production (especially cattle) to reduce its environmental effects and to feed more people. This would decrease the environmental impact of livestock production, but it would not free up much land or grain to feed more of the world's hungry people.

Cattle and sheep that graze on rangeland use a resource (grass) that humans can't eat, and most of this land is not suitable for growing crops. Moreover, because of poverty, insufficient economic aid, and the nature of global economic and food distribution systems, very little if any additional grain grown on land used to raise livestock or livestock feed would reach the world's hungry people.

Critical Thinking

Some environmentalists and health experts call for affluent people to become vegetarians or at least eat less meat as a way to reduce the environmental impact of meat production and to improve human health. Fewer than 500,000 Americans are strict vegetarians (vegans) who eat no meat and meat products. Would you be willing to eat less meat or not eat any meat? Explain.

a surplus to sell or to put aside for hard times. Subsistence farmers use primarily human labor and draft animals. Examples of this type of agriculture include numerous forms of shifting cultivation in tropical forests (Figure 1-18) and nomadic livestock herding. In **traditional intensive agriculture**, farmers increase their inputs of human and draft labor, fertilizer, and water to get a higher yield per area of cultivated land to produce enough food to feed their families and to sell for income. These two forms of small-scale agriculture (carried out mostly by women) produce about 20% of the world's food.

How Have Green Revolutions Increased Food Production? Farmers can produce more food either by **(1)** farming more land or **(2)** getting higher yields per unit of area from existing cropland. Since 1950 most of the increase in global food production has come from increased yields per unit of area of cropland in a process called the **green revolution**.

This process involves three steps: **(1)** developing and planting monocultures (Figure 6-16) of selectively bred or genetically engineered high-yield varieties of key crops such as rice, wheat and corn, **(2)** lavishing fertilizer, pesticides, and water on crops to produce high yields, and **(3)** often increasing the intensity and frequency of cropping. This approach dramatically increased crop yields in most developed countries between 1950 and 1970 in what is considered the *first green revolution* (Figure 15-3).

A *second green revolution* has been taking place since 1967 (Figure 15-3), when fast-growing dwarf varieties of rice and wheat, specially bred for tropical and subtropical climates, were introduced into several developing countries. With sufficient fertile soil and enough fertilizer, water, and pesticides, yields of these new

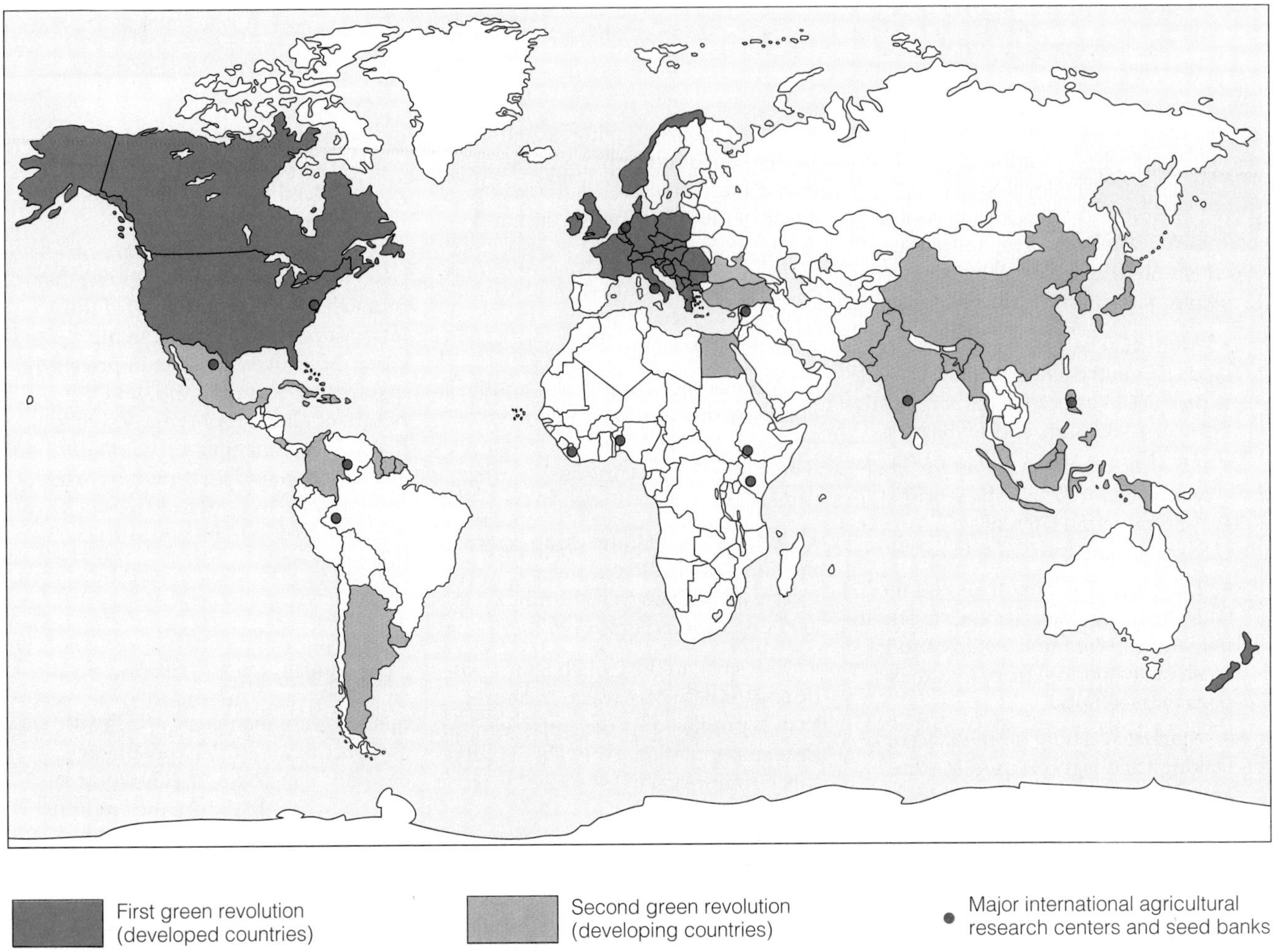

Figure 15-3 Countries whose crop yields per unit of land area increased during the two green revolutions. The first took place in developed countries between 1950 and 1970; the second has occurred since 1967 in developing countries with enough rainfall or irrigation capacity. Several agricultural research centers and gene or seed banks play a key role in developing high-yield crop varieties.

plants (Figure 15-4) can be two to five times those of traditional wheat and rice varieties. The fast growth also allows farmers to grow two or even three crops a year (multiple cropping) on the same land.

Producing more food on less land is also an important way to protect biodiversity by saving large areas of forests, grasslands, wetlands, and mountain terrain from being used to grow food. According to Dennis Avery, without the two major green revolutions the world would have lost wild land equal to the combined land area of the United States, Europe, and Brazil.

These yield increases depend not only on having fertile soil and ample water, but also on extensive use of fossil fuels to run machinery, produce and apply inorganic fertilizers and pesticides, and pump water for irrigation. All told, green-revolution agriculture uses about 8% of the world's oil output.

Figure 15-4 A high-yield, semidwarf variety of rice called IR-8 (left), a part of the second green revolution, was produced by crossbreeding two parent strains of rice: PETA from Indonesia (center) and DGWG from China (right). The shorter and stiffer stalks of the new variety allow the plants to support larger heads of grain without toppling over and increase the benefit of applying more fertilizer. (International Rice Research Institute, Manila)

These high inputs of energy, water, fertilizer, and pesticides on high-yield crop varieties have yielded dramatic results. At some point, however, additional inputs become useless because no more output can be squeezed from the soil and the crop varieties—the *principle of diminishing returns in action.*

In fact, yields may even start dropping because **(1)** the soil erodes, loses fertility, and becomes salty and waterlogged (Figure 13-23), **(2)** underground and surface water supplies become depleted and polluted with pesticides and nitrates from fertilizers, and **(3)** populations of rapidly breeding pests develop genetic immunity to widely used pesticides.

Case Study: Food Production in the United States In 1900 each American farmer produced enough food to feed 7 other people in the United States and abroad. Today, each U.S. farmer feeds 96 other people.

Since 1940 U.S. farmers have more than doubled crop production without cultivating more land. This happened because of industrialized agriculture using green-revolution techniques in a favorable climate on some of the world's most fertile and productive soils. This has also kept large areas of forests, grasslands, wetlands, and easily erodible land from being converted to farmland.

Farming has become *agribusiness* as big companies and larger family-owned farms have taken control of most U.S. food production. Only about 650,000 Americans are full-time farmers. However, about 9% of the population is involved in the U.S. agricultural system, from growing and processing food to distributing it and selling it at the supermarket.

In terms of total annual sales, agriculture is the biggest industry in the United States, bigger than the automotive, steel, and housing industries combined. It generates about 18% of the country's gross national product and 19% of all jobs in the private sector, employing more people than any other industry.

The U.S. agricultural system is highly productive. U.S. farms, with only 0.3% of the world's farm labor force, produce about 25% of the world's food and nearly half of the world's grain exports. In 1998, U.S. agriculture exports reached about $50 billion, more than 7% of all American goods bought abroad.

This industrialization of agriculture was made possible by the availability of cheap energy, most of it from oil. Agriculture consumes about 17% of all commercial energy in the United States each year (Figure 15-5). Most plant crops in the United States provide more food energy than the energy used to grow them. However, if we include livestock as well as crops, the U.S. food production system currently uses about three units of fossil fuel energy to produce one unit of food energy.

Energy efficiency is much lower if we look at the whole U.S. food system. Considering the energy used to grow, store, process, package, transport, refrigerate, and cook all plant and animal food, *about 10 units of nonrenewable fossil fuel energy are needed to put 1 unit of food energy on the table.* By comparison, every unit of energy from human labor in subsistence farming provides at least 1 unit of food energy; with traditional intensive farming, each unit of energy provides up to 10 units of food energy.

What Is Traditional Agriculture? Agrodiversity in Action Traditional farmers in developing countries today grow about 20% of the world's food on about 75% of its cultivated land. Many traditional farmers simultaneously grow several crops on the same plot, a practice known as **interplanting**. Such crop diversity reduces the chance of losing most or all of the year's food supply to pests and other misfortunes.

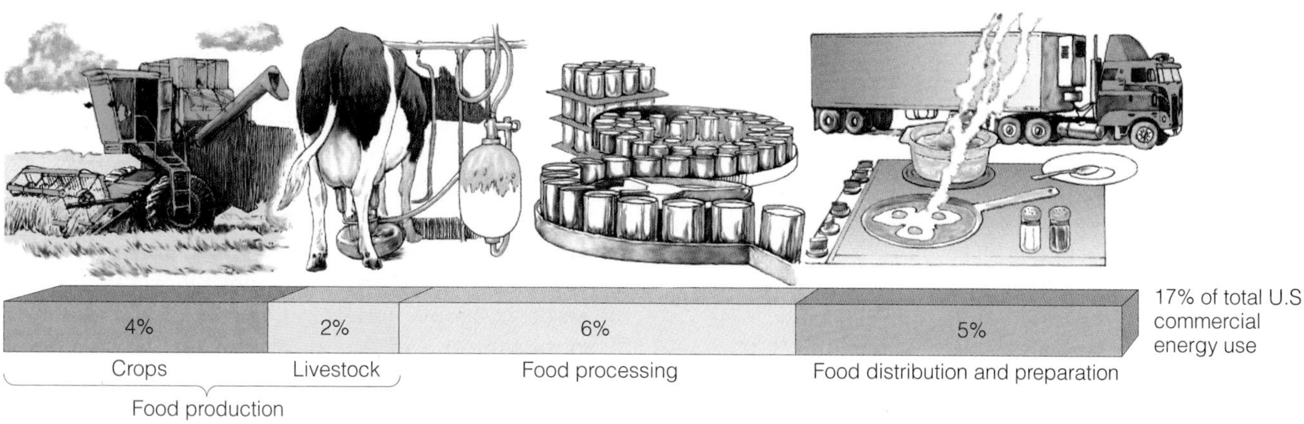

Figure 15-5 In the United States, industrialized agriculture uses about 17% of all commercial energy. On average, a piece of food eaten in the United States has traveled 2,100 kilometers (1,300 miles).

Common interplanting strategies found throughout the world, mostly in developing countries, include the following:

- **Polyvarietal cultivation**, in which a plot is planted with several varieties of the same crop.
- **Intercropping**, in which two or more different crops are grown at the same time on a plot (for example, a carbohydrate-rich grain that uses soil nitrogen alongside a protein-rich legume that puts it back).
- **Agroforestry**, or **alley cropping**, in which crops and trees are planted together (Figure 13-24c). For example, a grain or legume crop can be planted around fruit-bearing orchard trees or in rows between fast-growing trees or shrubs that can be used for fuelwood or for adding nitrogen to the soil.
- **Polyculture**, a more complex form of intercropping in which many different plants maturing at various times are planted together. Root systems at different depths in the soil capture nutrients and moisture efficiently and minimize the need for fertilizer or irrigation water. Year-round plant coverage also protects the soil from wind and water erosion. The presence of various habitats for insects' natural predators means that crops need not be sprayed with insecticides. In addition, weeds have trouble competing with the multitude of crop plants and thus can be removed fairly easily by hand, without herbicides. Crop diversity also provides insurance against bad weather.

Recent ecological research on crop yields of 14 artificial ecosystems found that on average polyculture (with four or five different crop species) produces higher yields per acre than high-input monoculture. This important finding has major implications for development of high-yield sustainable agriculture in developing countries by combining the techniques of traditional high-yield interplanting with modern inputs of fertilizer and irrigation.

15-2 WORLD FOOD PROBLEMS AND CHALLENGES

How Much Has Food Production Increased? Figure 15-6 shows the success story of global agriculture. The *good news* is that between 1950 and 1990, world grain production almost tripled (Figure 15-6, left) and per capita production rose by about 36% (Figure 15-6, right), helping reduce global hunger and malnutrition. During the same period, average food prices adjusted for inflation dropped by 25% and the amount of food traded in the world market quadrupled.

Despite these impressive achievements in food production, here is some *disturbing news*:

- Population growth is outstripping food production and distribution in areas that support about 2 billion people. Since 1978 grain production has lagged behind population growth in 88 developing countries, especially in Africa. Serious food supply problems persist in 34 developing countries, most of them in sub-Saharan Africa.
- Since 1950 global grain production has been rising but its rate of growth has slowed and has leveled off since 1985 (Figure 15-6, left). In addition, per capita grain production declined by 7% between 1985 and 1998 (Figure 15-6, right).
- More than 100 countries regularly import food from the United States, Canada, Australia, Argentina, western Europe, Australia, New Zealand, Thailand, and a few other countries. Any drop in annual grain production by major food-exporting countries (especially the United States) has severe economic and political consequences for food-importing nations. It can also lead to hordes of hungry people and environmental refugees illegally migrating to other countries in a desperate search for food and work.
- To provide the 8 billion people projected by 2028 and the 9 billion projected by 2054 with even a meatless subsistence diet will require that grain production grow by 2% per year, compared with the leveling off of growth per year between 1985 and 1998 (Figure 15-6, left).
- Food production will have to increase even more than 2% a year because projected increasing affluence in many developing countries will raise meat consumption and require higher amounts of grain for animal feed.
- If everyone ate the diet typical of a person in a developed country, with 30–40% of the calories coming from animal products, estimates suggest that the current world agricultural system would support only 2.5 billion people.
- If the 8 billion people projected by 2028 ate the typical American meat-based diet, the annual grain harvest would have to be about 3.2 times the current world's production (Figure 15-6, left).

How Serious Are Undernutrition, Malnutrition, and Overnutrition? We live in a world where one out of six people in developing countries doesn't get enough food and one out of seven in developed countries eats too much. People who are underfed and underweight and those who are overfed and overweight face similar health problems: **(1)** lower life expec-

Brown, L. 1998. "Food Scarcity: An Environmental Wakeup Call." *The Futurist*, vol. 32, no. 1 p. 34(5).

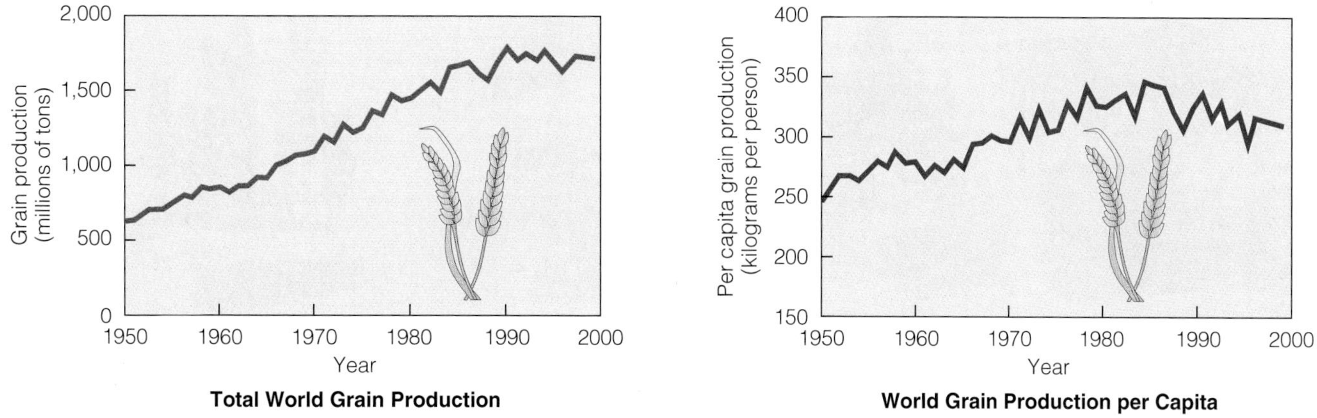

Figure 15-6 Total worldwide grain production of wheat, corn, and rice, and per capita grain production, 1950–99. (Data from U.S. Department of Agriculture, Worldwatch Institute, and UN Food and Agriculture Organization)

tancy, **(2)** increased susceptibility to disease and illness, and **(3)** reduced productivity and life quality.

People who cannot grow or buy enough food to meet their basic energy needs suffer from **undernutrition**. To maintain good health and resist disease, people need not only a certain number of calories, but also the proper amounts of protein (from animal or plant sources), carbohydrates, fats, vitamins, and minerals.

People who are forced to live on a low-protein, high-carbohydrate diet consisting only of grains such as wheat, rice, or corn often suffer from **malnutrition**: deficiencies of protein and other key nutrients. Many of the world's desperately poor people, especially children, suffer from both undernutrition and malnutrition.

The two most common nutritional-deficiency diseases are marasmus and kwashiorkor. *Marasmus* (from the Greek word *marasmos*, "to waste away") occurs when a diet is low in both calories and protein (Figure 1-8). Most victims are either nursing infants of malnourished mothers or children who do not get enough food after being weaned from breast-feeding. If the child is treated in time with a balanced diet, most of these effects can be reversed. In practice, however, relief efforts are often too little or too late to save many children from death.

Kwashiorkor (meaning "displaced child" in a West African dialect) is a severe protein deficiency occurring in infants and children ages 1–3, usually after the arrival of a new baby deprives them of breast milk. The displaced child's diet changes to grain or sweet potatoes, which provide enough calories but not enough protein. If it is caught soon enough, most of the harmful effects can be cured with a balanced diet; otherwise, if the child survives, stunted growth and mental retardation result.

Here's some *good news*. Between 1970 and 1996 the worldwide proportion of people suffering from chronic undernutrition fell from 36% to 14%. Also, despite population growth, the estimated number of chronically malnourished people fell from 940 million in 1970 to 828 million in 1996 (92% of them in Asia and Africa).

Despite this progress, about one of every six people in developing countries (including one of every three children below age 5) is chronically undernourished or malnourished; 87% of them live in Asia (especially India and Bangladesh) and Africa. Such people are disease prone, and adults are too weak to work productively or think clearly. As a result, their children also tend to be underfed and malnourished. If these children survive to adulthood, many are locked in a malnutrition–poverty cycle (Figure 15-7) that can continue for generations.

It is estimated that each year at least 10 million people (some researchers say 18 million), half of them children under age 5, die prematurely from undernutrition, malnutrition, or normally nonfatal diseases such as measles and diarrhea worsened by malnutrition. Because they are spread out and not widely publicized in the media, undernutrition and malnutrition have been described by the United Nations (UN) as a silent and invisible global emergency with a massive impact on children. However, children don't have to die prematurely because of undernutrition and malnutrition (Solutions, p. 393).

Although balanced diets, vitamin-fortified foods, and vitamin supplements have slashed the number of vitamin-deficiency diseases in developed countries, hundreds of millions of cases occur each year in developing countries. For example, an estimated 250,000 million

Hint: Enter the search terms *food supply, environmental* using the Subject Guide.

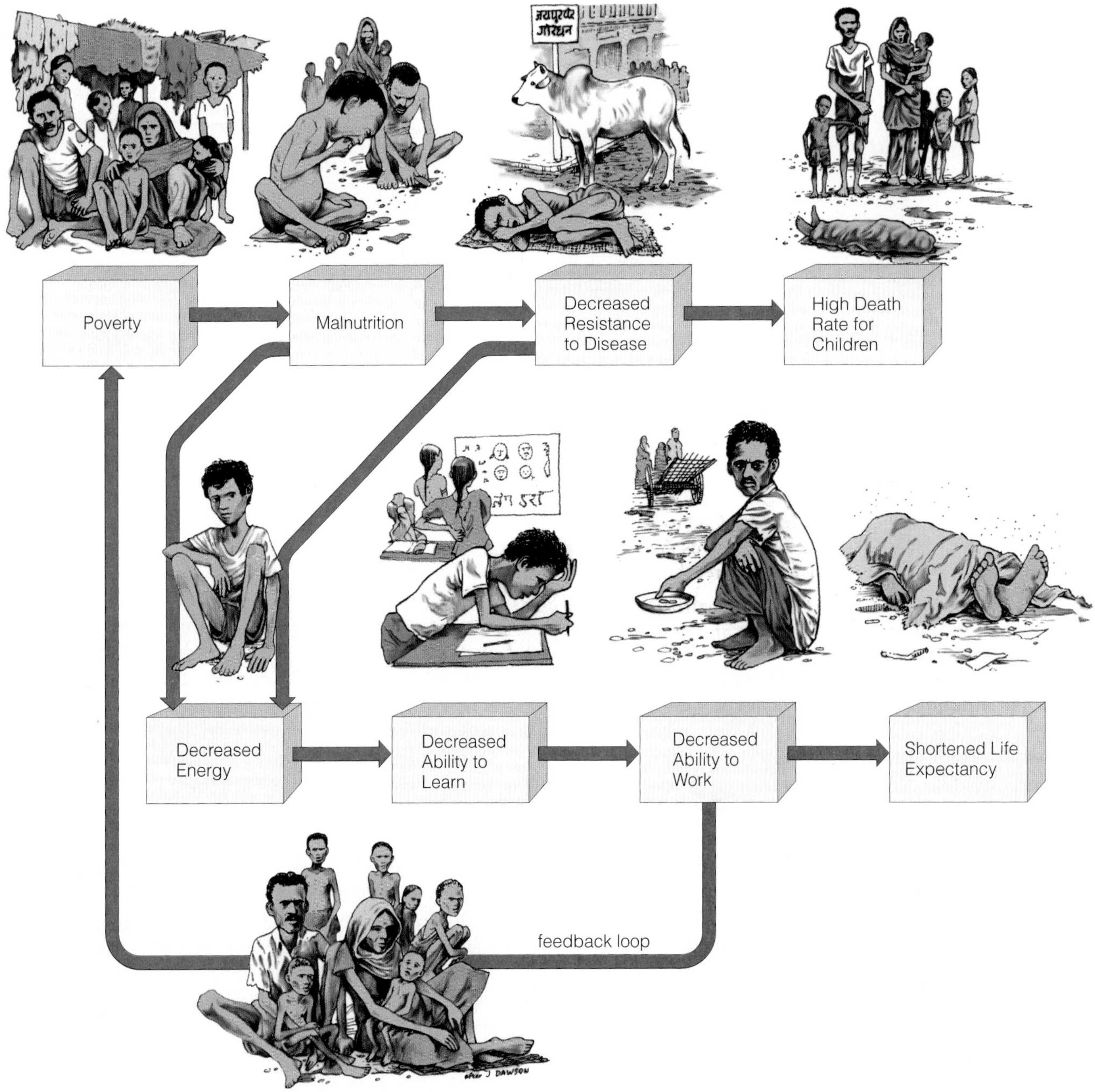

Figure 15-7 Interactions among poverty, malnutrition, and disease form a tragic cycle that tends to perpetuate such conditions in succeeding generations of families.

children in developing countries are deficient in vitamin A. This puts them at risk for blindness and for premature death because even mild vitamin A deficiency impairs the immune system, reducing children's resistance to infectious diseases such as diarrhea and measles.

Other nutritional-deficiency diseases are caused by the lack of certain minerals. For example, too little iron (a component of hemoglobin that transports oxygen in the blood) causes anemia. This mineral deficiency **(1)** causes fatigue, **(2)** makes infection more likely, **(3)** increases a woman's chances of dying in childbirth, **(4)** increases an infant's chances of dying from infection during its first year of life, and **(5)** cripples efforts to improve primary school education because developing brains need adequate iron to learn. According to the World Health Organization (WHO), one out of every three people, mostly women and children in tropical developing countries, are deficient in iron.

According to the WHO, some 1.6 billion people—almost one of every four—get too little iodine in their diet. Lack of iodine can cause goiter. This abnormal enlargement of the thyroid gland in the neck affects 760 million people worldwide and leads to deafness if untreated. Worldwide, an estimated 100 million children born to mothers with severe iodine deficiency suffer from stunted physical growth and severe mental retardation; another 20 million have some degree of brain damage from too little iodine.

In the United States, researchers estimate that 11 million people, including 4 million children, do not get enough to eat. More than 20% of U.S. children live in poverty, more than twice the rate of most other industrialized countries.

Whereas one out of six of the people in developing countries suffers from undernutrition and malnutrition, about one out of seven people in developed countries (one out of three in the United States) suffers from **overnutrition**. It is a condition in which food energy intake exceeds energy use and causes excess body fat (obesity). It can be caused by too much food, too little exercise, or both.

Overnutrition is the second leading cause of preventable deaths after smoking (p. 179), mostly from coronary heart disease, cancer, stroke, and diabetes. A study of thousands of Chinese villagers indicates that the healthiest diet for humans is largely vegetarian, with only 10–15% of calories coming from fat. This is in contrast to the typical meat-based diet, in which 40% of the calories come from fat. The $36 billion Americans spend each year on weight loss programs is almost twice the $19 billion needed to eliminate undernutrition and malnutrition in the world.

Saving Children

SOLUTIONS

According to the United Nations Children's Fund (UNICEF), "Malnourished children who live past childhood face a diminished future. They will become adults with lower physical and intellectual abilities, lower levels of productivity, and higher levels of chronic illness and disability."

UNICEF officials estimate that between one-half and two-thirds of childhood deaths from nutrition-related causes could be prevented at an average annual cost of only $5–10 per child, or only 10–19¢ per week. This life-saving program would involve the following simple measures:

- Immunizing children against childhood diseases such as measles
- Encouraging breast-feeding
- Preventing dehydration from diarrhea by giving infants a mixture of sugar and salt in a glass of water
- Preventing blindness by giving people a vitamin A capsule twice a year at a cost of about 75¢ per person
- Providing family-planning services to help mothers space births at least 2 years apart
- Increasing education for women, with emphasis on nutrition, sterilization of drinking water, and child care

Critical Thinking

How much money (if any) would you be willing to spend each year to help implement such a program for saving children? Why has insufficient money been allocated for such a program?

Do We Produce Enough Food to Feed the World's People? The *good news* is that we produce more than enough food to meet the basic nutritional needs of every person on the earth today. Indeed, if distributed equally, the grain currently produced worldwide is enough to give everyone a meatless subsistence diet.

The *bad news* for those not getting enough food is that food is not distributed equally among the world's people because of differences in soil, climate, political and economic power, and average per capita income throughout the world. Most agricultural experts agree that *the principal cause of hunger is and will continue to be poverty*, which prevents poor people from growing or buying enough food regardless of how much is available.

Moreover, increases in global and per capita food production often hide large differences in food supply and quality among and within countries. For example, despite impressive gains in total and per capita food production since 1970, roughly 40% of India's population suffers from malnutrition because they are too poor to buy or grow enough food to meet their basic needs.

What Are the Environmental Effects of Producing Food? *Agriculture has a greater harmful impact on air, soil, water, and biodiversity resources than any other human activity* (Figure 15-8). Food prices in the United States and many other countries are deceptively low because they do not include the harmful environmental and health costs associated with food production. David Pimentel (Guest Essay, p. 352) has estimated that the harmful costs not included in the prices of food in the United States are $150–200 billion per year. According to a 1990 UN study, degradation of irrigated

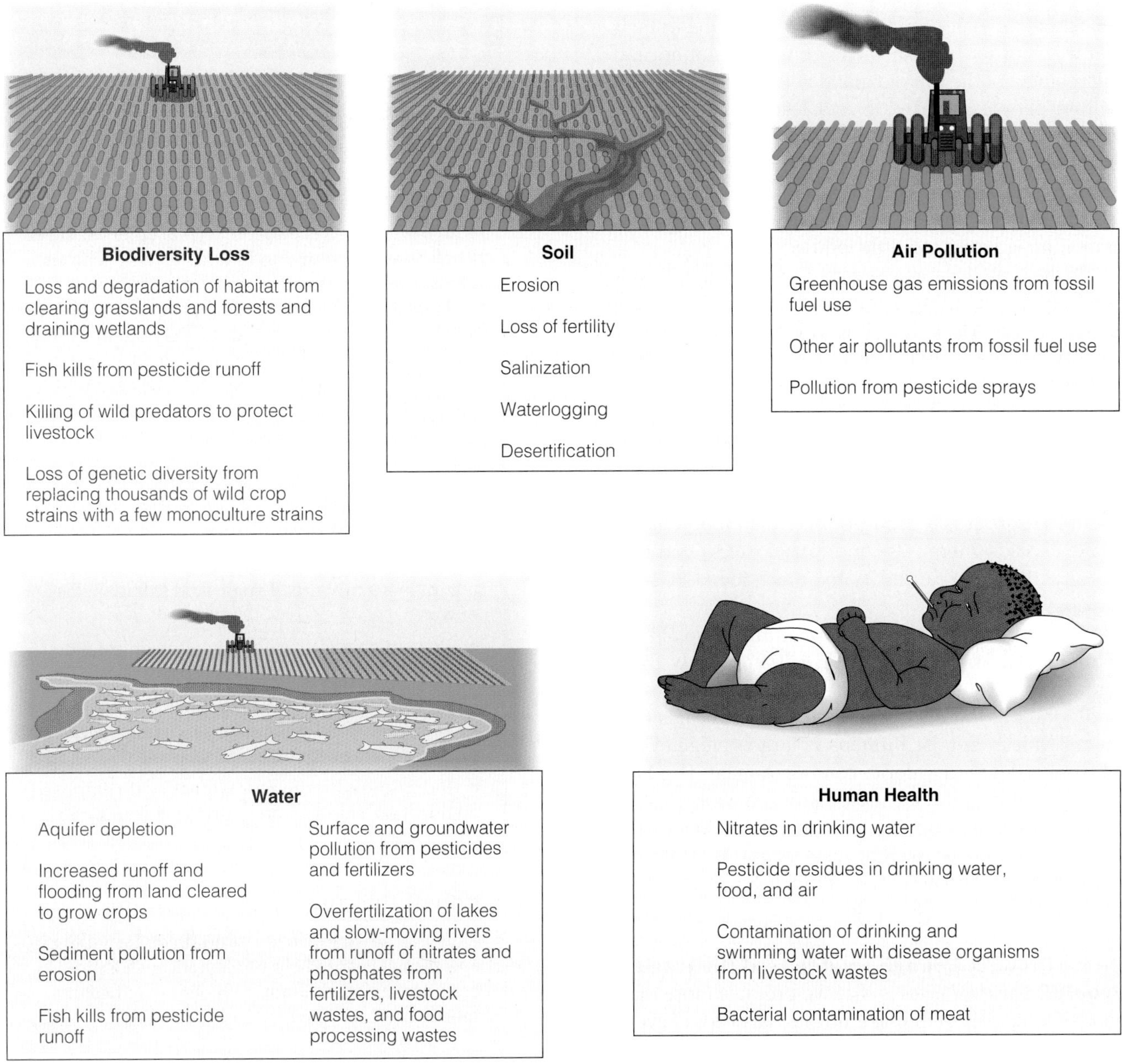

Figure 15-8 Major environmental effects of food production.

cropland, rain-fed cropland, and rangeland now costs the world more than $42 billion a year in lost crop and livestock output; this loss is roughly equal to the annual value of the entire U.S. grain harvest.

Many analysts believe that it is possible to produce enough food to feed the 8 billion people projected by 2025 through new advances in agricultural technology and by spreading the use of existing green-revolution or high-yield techniques. Other analysts disagree. They have serious doubts about the ability of new food production technologies and food distribution systems to keep up with current levels of population growth, mostly because the harmful environmental effects of agriculture will reduce yields (Case Study, right).

According to Worldwatch Institute estimates, between 1945 and 1990, erosion, salinization, waterlogging, desertification, and other forms of environmental degradation eliminated an area of land from global food production equal to the cropland of two Canadas. David Pimentel (Guest Essay, p. 352) estimates that an area of cropland larger than the land area of India is expected to lose most of its agricultural productivity from environmental degradation between 1994 and 2013. Two other environmental constraints that are likely to limit

Can China's Population Be Fed?

CASE STUDY

Since 1970 China has made significant progress in feeding its people and slowing its rate of population growth (Section 9-4). Despite such efforts, between 1999 and 2025 China is projected to add about 300 million people—more than currently live in the United States. With its economy now doubling in size every 8 years and 12 million more people each year, there is growing concern that crop yields may not be able to keep up with demand.

A basic problem is that with 21% of the world's people, China has only 7% of the world's cropland and fresh water, 3% of its forests, and 2% of its oil. China needs to expand its area of cropland. Instead, China is experiencing a drop in available farmland, mostly because of **(1)** lack of suitable cropland, **(2)** massive conversion of limited existing cropland to nonfarm uses as the country industrializes and urbanizes, **(3)** aquifer depletion from irrigation of 60% of its cropland, **(4)** environmental degradation (especially erosion, salinization, and waterlogging of soil), and **(5)** severe air pollution as a result of extensive burning of coal.

China depends on irrigated land to produce 70% of the grain for its population. However, it is diverting increasing amounts of irrigation water to supply the needs of its rapidly growing cities and industries. As aquifers are depleted and some of its rivers run dry, China's rapidly growing demands for water are colliding with its limited supply and its need to increase grain production.

According to Worldwatch Institute projections, China's grain production is likely to fall by at least 20% between 1990 and 2030. Even if China's currently booming economy resulted in no increases in meat consumption, this 20% drop would mean that by 2030 China would need to import more than the world's entire 200 million tons of grain exports in 1993 (roughly half from the United States).

However, if the increased demand for meat led to a rise in per capita grain consumption equal to the current level in Taiwan (one-half the current U.S. level), China would have to import more than the entire current grain output of the United States. In addition to growing more food for export, the United States will need to produce enough food to feed a projected 63 million more Americans by 2025 and 121 million more by 2050.

The Worldwatch Institute warns that if either of these scenarios is correct, no country or combination of countries has the potential to supply even a small fraction of China's potential food supply deficit. This is not even taking into account the huge food deficits that are projected in other parts of the world by 2030, especially in Africa and India.

Other analysts believe that serious and rapidly growing environmental problems may limit China's economic growth and its ability to feed its people. So far China has concentrated on rapid industrialization and devoted little attention to sustainable use of its resources and to reducing pollution and environmental degradation.

However, according to a 1997 study by the International Food Policy Institute, China should be able to feed its population and begin exporting grain again by 2020 if the government invests in expanding irrigation and increasing agricultural research. Recent satellite surveys also show that China has far more potential cropland than previously thought.

If China begins acting now to chart a new course, it has a unique opportunity to show the world how to build a more environmentally sustainable economy and agricultural system over the next few decades.

Critical Thinking

If the scenarios about China's growing dependence on food imports are valid, how might this affect **(a)** world food prices, **(b)** your life, and **(c)** the harmful environmental impacts of food production (Figure 15-8)? What actions, if any, do you suggest for dealing with this potential problem?

food production in the future are **(1)** increased ultraviolet radiation from ozone-layer depletion and **(2)** projected global warming (Chapter 11).

15-3 INCREASING WORLD FOOD PRODUCTION

Is Increasing Crop Yields the Answer? Agricultural experts expect most future increases in food yields per hectare on existing cropland to result from improved strains of plants and from expansion of green-revolution technology to new parts of the world. For example, a new strain of rice developed by the International Rice Research Institute could increase rice yields by as much as 20%. However, the future of such important research may be hampered by drops in funding of the Rice Research Institute from donor nations and foundations.

Scientists are working to create new green revolutions—actually *gene revolutions*—by using genetic engineering and other forms of biotechnology. Over the next 20–40 years they hope to breed high-yield plant strains that **(1)** are more resistant to insects and disease, **(2)** thrive on less fertilizer, **(3)** make their own nitrogen fertilizer, **(4)** do well in slightly salty soils, **(5)** can

Genetically Modified Food

PRO/CON

There is growing controversy over the rapidly increasing use of genetically modified food (GMF). Such food is seen as a potentially sustainable way to solve world food problems by its producers and called potentially dangerous "Frankenfood" by its critics in Europe, where such food has been banned in many countries.

If you live in the United States you probably had some GMF for breakfast, lunch, and dinner. The first genetically modified crop, the Flavr–Savr tomato, went into U.S. supermarkets in 1993. Today more than 60 different U.S. crops, including corn, potatoes, strawberries, and apples, are bioengineered.

By 1998, 25% of corn, 38% of soybeans, and 45% of cotton grown in the United States were genetically modified to **(1)** produce their own pesticides or **(2)** be resistant to herbicides (weedkillers). By 2005, most foods produced in the United States are expected to be genetically modified.

The rapidly growing biotech industry, led by Monsanto, Dow, and DuPont in the United States, sees genetically modified crops (called *transgenics*) as the basis of new biotechnology revolution that will supply more food and reduce farmers' use of pesticides and herbicides. They envision

- Crops genetically engineered to tolerate low-nutrient, dry, or salty soils and expand areas where crops can be grown.
- Crops and trees that grow faster. In 1999, Israeli scientists announced that they had inserted a cellulose-binding gene in plant cells that could accelerate plant and tree growth by up to 50%. This could boost food production and help restore degraded forestland.
- Food genetically modified to provide better nutrition. In 1999, Swiss researchers announced the development of genetically modified rice strains that would add more iron and vitamin A to the human diet.
- Crops that continually release pesticides and herbicide-resistant crops (HRCs). Such crops (already being used) can reduce the need for repeated spraying of toxic pesticides and ideally reduce pesticide costs for farmers (already saving farmers over $1 billion per year). HRCs could also allow increased use of conservation-tillage techniques that reduce or eliminate the need to plow the soil and help reduce soil erosion and water pollution.
- Crops that could reduce water pollution. A new genetically engineered strain of corn is designed to reduce water pollution from the runoff of phosphorus from poultry and hog farms by increasing the amount of phosphorus that such animals can digest.

However, a growing chorus of critics warns of a number of unintended consequences from widespread use of transgenics and GMF. Such potential problems include threats to human health, ecological simplification, and economic problems for the world's poor farmers. Here are some of the concerns:

- Biotech companies talk about feeding the world, but at least two-thirds of all genetically engineered crops now available or in development are designed to increase the sale of pesticides produced by the companies that are selling the genetically engineered seeds.
- Genetic manipulation might enhance the production of natural plant toxins that could contaminate human food. Little research has been done to evaluate this possibility.
- People with allergies could unknowingly be exposed to proteins that they are allergic to. For example, a few years ago researchers found that a Brazil nut gene spliced into soybean genes induced allergies in people sensitive to Brazil nuts.
- The nutritional value of some foods may be lowered by foreign genes. For example, a recent study found that levels of compounds (phytoestrogens) thought to protect against heart disease and possibly some types of cancer were lower in genetically modified soybeans than in traditional strains.
- A safe natural pesticide (*Bacillus thuringensis* or *Bt*) widely used in small doses by organic farmers and in integrated pest management (an important ecologically based form of pest control; Section 16-5) will soon be ineffective. Seed companies have inserted *Bt* genes into corn, potatoes, and cotton, where the pesticide is released continuously by every cell of every plant. This will ensure development of genetic resistance to *Bt* in only a few years and deprive organic farmers of an important tool for making agriculture more sustainable.

withstand drought, and **(6)** use solar energy more efficiently during photosynthesis. However, according to Donald Duvick, former director of research at Pioneer HiBred International (one of the world's largest seed producers), "No breakthroughs are in sight. Biotechnology, while essential to progress, will not produce sharp upward swings in yield potential except for isolated crops in certain situations." There is also considerable controversy over the rapidly growing use of genetically modified crops (Pro/Con, above).

Several factors have limited the success of the green and gene revolutions to date, and may continue to do so.

■ Biologists fear that genes designed to give crops a competitive advantage (such as herbicide tolerance and resistance to insect pests) may be passed on to wild plants, especially weeds. Such hybrid superweeds impervious to herbicides could lower crop productivity. Recent studies indicate that more than 25 of the 30 most agriculturally important plants can hybridize or crossbreed naturally with related and sometimes unrelated wild species. The possibilities for such genetic pollution are especially high in developing countries, which contain a variety of traditional and wild crop strains and weeds. Plans to develop transgenic crops with a variety of toxins in their genes could hasten the diffusion of pesticide-resistant genes to weeds and other wild species.

■ There is concern over harm to nontarget species. For example, in 1999 researchers at Cornell University reported the results of a laboratory experiment in which 44% of the caterpillar (larval) forms of monarch butterflies (Figure 4-3) died or developed abnormally after eating milkweed dusted with pollen from a widely planted bioengineered form of corn containing *Bt* genes. Caterpillars that didn't feed on the pollen did not die. In 1998, a Swiss study found that beneficial insects called lacewings died more quickly if they fed on corn borers reared on *Bt* corn. In Scotland, a toxicologist reported that rats eating potatoes containing insect-resistant genes and proteins suffered damaged immune systems, shrunken brains, and growth problems. Ecologists see such early warnings as red flags pointing to possible unexpected effects of transgenics on various nontarget species and on the natural functioning of ecosystems.

■ A small number of biotech companies may soon have a monopoly on the seed and chemicals needed to raise all of the world's major food crops. For example, in 2000 all of America's soybean farmers depended on a single supplier for their seed and the chemicals needed to allow such seeds to thrive.

■ Farmers increasingly will not be able to reduce their costs and help protect their crops by saving and crossbreeding seeds for replanting. In developing countries, some 1.4 billion farmers still rely almost exclusively on seed saving and crossbreeding as an affordable way to have crop diversity as insurance against changes in growing conditions and pests. Monsanto has taken legal action against 300 farmers in developed countries for replanting the company's patented transgenic seeds. Some biotech seed companies are developing gene protection technology, called "terminator genes," that prevents patented seeds from germinating.

■ It is doubtful that seed companies will develop crop varieties specifically to help the poor grow more food because poor farmers can't afford such seeds and technologies. Thus, hunger among the poor could grow as a result of the transgenics revolution.

■ Increasing dependence on a narrow range of bioengineered crops will reduce genetic diversity even more and hinder the development of more sustainable forms of agriculture based on **(1)** planting a variety of crops (polyculture) and **(2)** using a diverse ecological approach (agroecology) for producing food instead of a narrow chemical approach.

In 1999, researchers at Pennsylvania-based Kimeragem announced a new way to genetically engineer plants in which no foreign gene is inserted. This technique, called *chimeraplasty*, could reduce some of the problems associated with conventional bioengineered crops. It involves inserting a chemical instruction that attaches to a gene and alters the gene in the desired way.

Proponents of GMF argue that its potential benefits far outweigh its potential harmful effects. Critics, mostly ecologists, recognize the potential benefits but say that we know far too little about the potential harm from widespread use of such crops. They call for **(1)** more controlled field experiments, **(2)** more research to better understand the risks, and **(3)** better and more rigorous regulation of this rapidly developing and highly profitable technology.

Critical Thinking

1. Are you for or against the widespread use of transgenic crops? Explain. What are the alternatives?

2. Do you favor much stricter regulation of the use of transgenics and greatly increased research on their possible benefits and harmful effects? If so, how would you accomplish such regulation?

■ Without huge amounts of fertilizer and water, most green-revolution crop varieties produce yields that are no higher (and are sometimes lower) than those from traditional strains; this is why the second green revolution has not spread to many arid and semiarid areas such as much of Africa and Australia (Figure 15-3). Without ample water, good soil, and favorable weather, some new genetically engineered crop strains could fail.

■ Genetically engineered crop strains cost too much for most of the world's subsistence farmers in developing countries. So far, most such crops have been genetically modified as a way to sell more pesticides.

- Continuing to increase inputs of fertilizer, water, and pesticides eventually produces no additional increase in crop yields; the J-shaped curve of crop productivity slows down, reaches its limits, levels off, and becomes an S-shaped curve. At that point, the yield potential for any particular grain in a country depends mostly on that country's **(1)** soil moisture (from rainfall or ability to irrigate), **(2)** day length or latitude, **(3)** temperatures, and **(4)** solar intensity—factors that are difficult to alter.
- Grain yields per hectare are still increasing in many parts of the world, but at a much slower rate. For example, after increasing by about 2.1% a year between 1950 and 1990, the annual increase in grain productivity dropped to about 1% from 1990 to 1998. This has been a key factor in the leveling off of grain production and the decline in per capita grain production since 1985 (Figure 15-6, p. 391).
- It is misleading to use the highest yields achieved for grains such as rice, wheat, and corn in certain countries as the basis for estimating yield increases elsewhere. Lester Brown argues that this is unrealistic because maximum attainable yields vary with variations in temperature, precipitation, day length (based on latitude), solar intensity, and inherent soil fertility.
- Indian economist Vandana Shiva contends that overall gains in crop yields from new green- and gene-revolution varieties may be much lower than claimed. The reason is that the yields are based on comparisons between the output per hectare of old and new *monoculture* varieties, rather than between the even higher yields per hectare for *polyculture* cropping systems and the new monoculture varieties that often replace them.
- In addition to leveling off, yields may even start dropping for a number of reasons: **(1)** the soil erodes, loses fertility, and becomes salty and waterlogged (Figure 13-23), **(2)** underground and surface water supplies become depleted and polluted with pesticides and nitrates from fertilizers, and **(3)** populations of rapidly breeding pests develop genetic immunity to widely used pesticides.

Connections: Will Loss of Genetic Diversity Limit Crop Yields? Some agricultural scientists think that new genetically engineered or crossbred varieties will enable farmers to continue increasing yields of key crops. Other scientists question whether this is possible, primarily because of the environmental impacts of current forms of industrialized agriculture (Figure 15-8) and the accelerating loss of biodiversity, which can limit the genetic raw material needed for future green and gene revolutions.

Scientists can crossbreed varieties of animal and plant life, and genetic engineers can move genes from one organism to another, but they need the genetic materials in the earth's existing species to work with. We are losing much of this genetic diversity as a small number of specially bred monoculture varieties of key crops have replaced thousands of strains of various crops and natural areas have been cleared (Spotlight, p. 400).

Wild varieties of the world's most important plants can be collected and stored in gene or seed banks, agricultural research centers, and botanical gardens. However, space and money severely limit the number of species that can be preserved. Many plants (such as potatoes) cannot be stored successfully as seed in gene banks. Power failures, fires, or unintentional disposal of seeds can also cause irreversible losses.

In addition, because stored seeds don't remain alive indefinitely, periodically they must be planted (germinated) and new seeds collected for storage. Unless this is done, seed banks become seed morgues. Moreover, stored plant species stop evolving; thus, they may be difficult to reintroduce into their native habitats, which may have changed in the meantime.

Because of these limitations, ecologists and plant scientists warn that the only effective way to preserve the genetic diversity of most plant and animal species is to protect representative ecosystems throughout the world from agriculture and other forms of development (Chapter 17).

Will People Try New Foods? Some analysts recommend greatly increased cultivation of less widely known plants to supplement or replace such staples as wheat, rice, and corn. One of many possibilities is the *winged bean,* a protein-rich legume now common only in New Guinea and Southeast Asia (Figure 15-9).

Insects—called *microlivestock*—are also important potential sources of protein, vitamins, and minerals in many parts of the world (Figure 15-10). There are about 450 edible insect species, including black ant larvae (served in tacos in Mexico) and giant waterbugs (crushed into vegetable dip in Thailand). Two basic problems are **(1)** getting farmers to take the financial risk of cultivating new types of food crops and **(2)** convincing consumers to try new foods.

Most crops we depend on are tropical annuals. Each year the land is cleared of all vegetation, dug up, and planted with their seeds. Some plant scientists believe we should rely more on polycultures of perennial crops, which are better adapted to regional soil and climate conditions than most annual crops (p. 384). Using perennials would also **(1)** eliminate the need to till soil and replant seeds each year, **(2)** greatly reduce

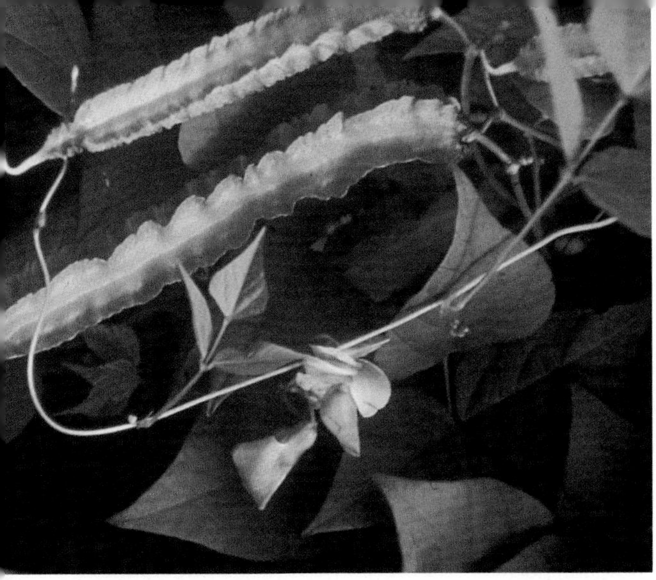

Figure 15-9 The winged bean, a protein-rich annual legume from the Philippines, is one of many currently unfamiliar plants that could become important sources of food and fuel. Its edible winged pods, spinachlike leaves, tendrils, and seeds contain as much protein as soybeans, and its edible roots contain more than four times the protein of potatoes. Its seeds can be ground into flour or used to make a caffeine-free beverage that tastes like coffee. Indeed, this plant produces so many different edible parts that it has been called a supermarket on a stalk. Because of nitrogen-fixing nodules in its roots, this fast-growing plant needs little fertilizer. (Larry Mellichamp/Visuals Unlimited)

Figure 15-10 Insects are important food items in many parts of the world. *Mopani*—emperor moth caterpillars—are among several insects eaten in South Africa. However, this food is so popular that the caterpillars (known as mopane worms) are being overharvested. Kalahari Desert dwellers eat cockroaches, lightly toasted butterflies are a favorite food in Bali, and French-fried ants are sold on the streets of Bogota, Colombia. Most of these insects are 58–78% protein by weight—three to four times as protein-rich as beef, fish, or eggs. (Anthony Bannister/Natural History Photographic Agency)

energy use, **(3)** save water, and **(4)** reduce soil erosion and sediment water pollution. However, widespread use of perennials would reduce the profits of agribusinesses selling annual seeds, fertilizers, and pesticides, which explains why they don't favor this approach.

Is Irrigating More Land the Answer? Between 1950 and 1978 the world's area of irrigated cropland increased almost threefold. Currently, about 40% of the world's food production comes from irrigated land, two-thirds of it in Asia. Irrigated land produces about 70% of the grain harvest in China and 50% in India (compared with 15% in the United States).

According to the UN, about 60% of the extra food needed to sustain a world population of about 8 billion by 2028 must come from irrigated agriculture. According to water expert Sandra Postel, by 2025 the additional water needed for irrigation will be equal to the annual flow of 24 Nile Rivers.

However, since 1978 the world's population has been growing faster than the growth of irrigated agriculture. As a result, the amount of irrigated land per person fell by 7% between 1979 and 1998 and is projected to fall much more between 1998 and 2050 (Figure 15-11). Water available for irrigation is dropping in some crop-growing areas (Chapter 12) and is projected to continue dropping because of

- Chronic water shortages that could affect 2.8 billion people by 2025
- Overpumping of aquifers in the Great Plains of the United States (Case Study, p. 297), most of China (Case Study, p. 395), the Aral Sea area in Central Asia (Case Study, p. 293), and much of India, North Africa, and the Middle East (p. 286)
- Inefficient use of irrigation water
- Increasing urbanization, which puts city dwellers and farmers in competition for limited water supplies

If global warming occurs as projected (Chapter 11), this will disrupt water supplies in many food-growing areas.

The potential for developing new water sources in most parts of the world is so low that further increases in irrigation depend mostly on **(1)** increasing the efficiency of water use for irrigation (Figure 12-13) and in

Shrinking the World's Genetic Plant Library

SPOTLIGHT

According to the UN Food and Agriculture Organization, in 2000 two-thirds of all seeds planted in developing countries were of uniform strains. Such genetic uniformity increases the vulnerability of food crops to pests, diseases, and harsh weather. Many biologists argue that this decreased variability, plus growing extinction of wild plant species, can severely limit the potential of future green and gene revolutions.

In the mid-1970s a valuable wild corn species, the only known perennial strain of corn, was barely saved from extinction. When this strain was discovered in south central Mexico, only a few thousand stalks survived in three tiny patches that were about to be cleared by squatter cultivators and commercial loggers.

Crossbreeding this perennial strain with commercial varieties could reduce the need for yearly plowing and sowing, which would reduce soil erosion and save water and energy. Even more important, this wild corn has a built-in genetic resistance to four of the eight major corn viruses, and it grows better in cooler and damper habitats than established commercial strains. Overall, the economic benefits of cultivating this barely rescued wild plant could total several billion dollars per year.

Critical Thinking

What are the major advantages and disadvantages of relying on a shrinking number of crop varieties? Why do seed companies favor this approach?

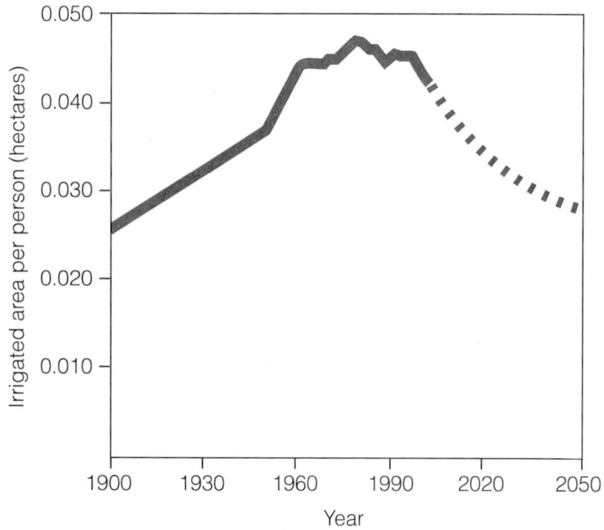

Figure 15-11 World irrigated area of cropland per person, 1900–98, with projections to 2050. (Data from UN Food and Agriculture Organization, U.S. Census Bureau, and the Worldwatch Institute)

industries and cities (Section 12-3), **(2)** switching to crops that use water more efficiently (for example, depending less on rice and more on wheat and sorghum), and **(3)** using water and grain more efficiently in livestock production and encouraging the production of more grain-efficient poultry and aquaculture-raised fish over beef and pork.

Is Cultivating More Land the Answer? Between 1950 and 1998, the area of the world's cropland for growing grain increased by 19%, but global population grew by 132%, seven times as fast. Theoretically, the world's cropland could be more than doubled by clearing tropical forests and irrigating arid land (Figure 15-12). However, many analysts believe that this potential for agricultural expansion is often overestimated because much of the land is marginal land, where cultivation is unlikely to be sustainable.

Clearing rain forests to grow crops and graze livestock can have harmful ecological consequences, as discussed in Section 17-4. Furthermore, most of this cleared land has nutrient-poor soils (Figure 13-17) and cannot support crop growth for more than a couple of years. In addition, potential cropland in savanna and other semiarid land in Africa cannot be used for farming or livestock grazing because of the presence of 22 species of the tsetse fly.

Some researchers hope to develop new methods of intensive cultivation in tropical areas. But other scientists argue that it makes more ecological and economic sense to combine various ancient methods of shifting cultivation (Figure 1-18), followed by fallow periods long enough to restore soil fertility, with various forms of polyculture.

Much of the world's potentially cultivable land lies in dry areas, especially in Australia and Africa. Large-scale irrigation in these areas would require **(1)** large, expensive dam projects with a mixture of beneficial and harmful impacts (Figure 12-6) and **(2)** large inputs of fossil fuel to pump water long distances. Large-scale irrigation could also deplete groundwater supplies by removing water faster than it is replenished. The land would need constant and expensive maintenance to prevent erosion, groundwater contamination, salinization, and waterlogging. Expanding wetland production of rice could also accelerate projected global warming by increasing atmospheric methane emissions (Figure 11-2c).

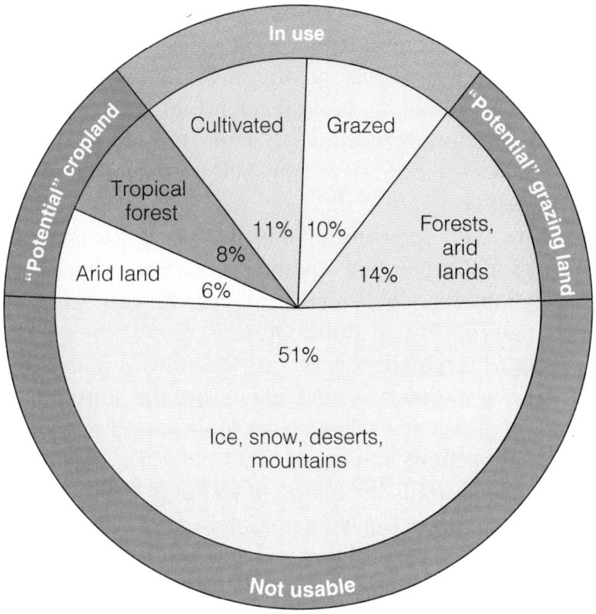

Figure 15-12 Classification of the earth's land. Theoretically, we could double the amount of cropland by clearing tropical forests and irrigating arid lands. However, converting these lands into cropland would **(1)** destroy valuable forest resources, **(2)** reduce the earth's biodiversity, **(3)** affect water quality and quantity, and **(4)** cause other serious environmental problems, usually without being cost-effective.

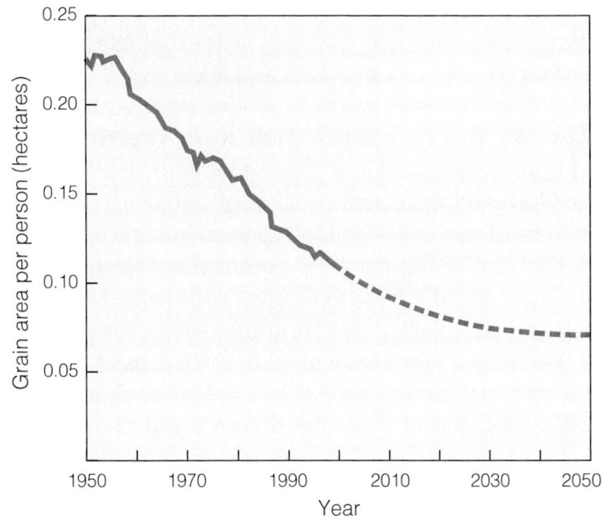

Figure 15-13 Average grain area per person worldwide, 1950–98, with projections to 2050. (Data from U.S. Department of Agriculture and Worldwatch Institute)

Thus, much of the new cropland that could be developed would be on land that is marginal for raising crops, requiring expensive inputs of fertilizer, water, and energy. Furthermore, these potential increases in cropland would not offset the projected loss of almost one-third of today's cultivated cropland due to erosion, overgrazing, waterlogging, salinization, mining, and urbanization.

Even if it is financially feasible, such expansion would reduce wildlife habitats and thus the world's biodiversity and ecological integrity. According to the UN Food and Agriculture Organization (FAO), cultivating all potential cropland in developing countries would reduce forests, woodlands, and permanent pasture by 47%.

In addition to providing wildlife habitats and conserving water and soils, these forests store 20–50 times more carbon as biomass than crops and pasture do. Clearing these forests would release a huge amount of carbon dioxide into the atmosphere and accelerate projected global warming.

For the reasons discussed in this section, *many analysts believe that a major economically profitable and environmentally sustainable expansion of cropland is unlikely over the next few decades.* If this assessment is correct, the world's grainland area per person, which dropped by half between 1950 and 1998 (because population growth grew 7 times faster than grainland expansion), is expected to decline further (Figure 15-13). Then the question is whether crop yields per area of cropland can be increased enough to offset the shrinkage of cropland per person and keep up with projected population growth.

Can We Grow More Food in Urban Areas? Food experts project that people in urban areas could live more sustainably and save money by growing food in empty lots, on rooftops and balconies, and in their own backyards and by raising fish in tanks and sewage lagoons. Currently, about 800 million urban gardens provide about 15% of the world's food, and this amount could be increased. A study by the UN Center for Human Settlements estimated that up to 50% of the total area in many cities in developing countries is vacant public land that could be used to produce food.

According to Worldwatch Institute estimates, at least 200 million people now grow some of their food and provide 800 million people with at least some of their food. Farmers in Accra, Ghana, provide the city with about 90% of its vegetables. In Singapore urban farmers supply 80% of the city's poultry and 25% of its vegetables. Farmers in or near 18 of China's largest cities provide urban dwellers with 85% of their vegetables and more than half of their meat and poultry.

Recycling nutrient-rich animal and human wastes to grow food in urban areas can also greatly reduce water pollution from runoff of plant nutrients (Figure 12-21). In China human waste is treated and sold to farmers as fertilizer. For more than 50 years sewage-fed lagoons (ponds) in Calcutta, India, have provided the city's people with one-tenth of the fish they consume. This wastewater-fed aquaculture system also is a cost-effective way to treat sewage wastes that usually are discharged into nearby rivers and lakes.

Can We Harvest More Fish and Shellfish? In addition to cropland and grazing land, the third major food-producing system consists of **fisheries**: concentrations of particular aquatic species suitable for commercial harvesting in a given ocean area or inland body of water. About 70% of the annual commercial catch of fish and shellfish comes from the ocean. About 99% of this catch is taken from plankton-rich coastal waters, but this vital coastal zone is being disrupted and polluted at an alarming rate (Sections 6-5 and 12-6).

The remainder of the annual catch comes from **(1)** using aquaculture to raise fish in ponds and underwater cages (20%) and **(2)** from inland freshwater fishing from lakes, rivers, reservoirs, and ponds (10%). About one-third of the world fish harvest is not consumed directly by humans and is used primarily as animal feed, fish meal, and oils.

Here is some *good news*. Between 1950 and 1997 the annual commercial fish catch (marine plus freshwater harvest) increased 4.8-fold (Figure 15-14, left). And between 1950 and 1988 the per capita seafood catch more than doubled (Figure 15-14, right).

The *bad news* is that because population grew faster than the annual catch, the per capita catch fell by 6% between 1989 and 1997 (Figure 15-14, right). Because of overfishing, pollution, and population growth, the world's fish catch is not expected to increase significantly and may continue to decline as population growth exceeds the growth of the catch.

Connections: How Are Overfishing and Habitat Degradation Affecting Fish Harvests? Fish are potentially renewable resources as long as the annual harvest leaves enough breeding stock to renew the species for the next year. Ideally, an annual **sustainable yield**—the size of the annual catch that could be harvested indefinitely without a decline in the population of a species—should be established for each species to avoid depleting the stock.

However, determining sustainable yields is difficult. Estimating mobile aquatic populations isn't easy, and sustainable yields shift from year to year because of changes in climate, pollution, and other factors. Furthermore, sustainably harvesting the entire annual surplus of one species may severely reduce the population of other species that rely on it for food—another example of connections in nature.

Overfishing is the taking of so many fish that too little breeding stock is left to maintain numbers; that is, overfishing is a harvest that exceeds the estimated sustainable yield. Prolonged overfishing leads to **commercial extinction**: reduction of a species to the point at which it's no longer profitable to hunt for them. Fishing fleets then move to a new species or a new region, hoping that the overfished species will eventually recover (Connections, right).

According to the UN FAO, *11 of the world's 15 most important oceanic fishing areas have been heavily exploited, overexploited, or depleted.* As a result, about 70% of the world's commercial fish stocks are in decline and in need of urgent management.

According to the U.S. National Fish and Wildlife Foundation, 14 major commercial fish species in U.S. waters (accounting for one-fifth of the world's annual catch and half of all U.S. stocks) are so depleted that even if all fishing stopped immediately it would take up to 20 years for stocks to recover. Ways to manage fisheries to reduce overfishing are discussed in Section 18-5.

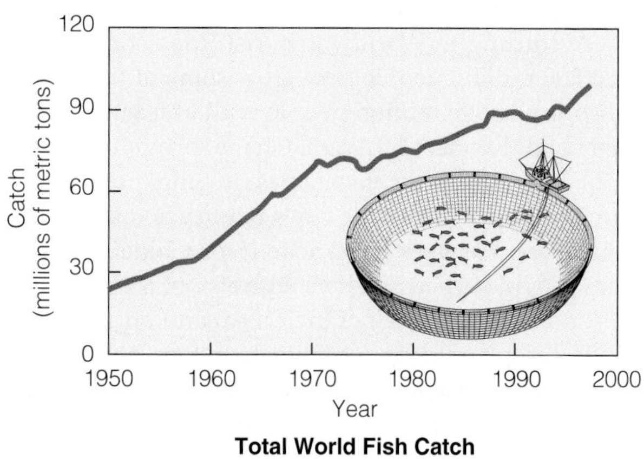

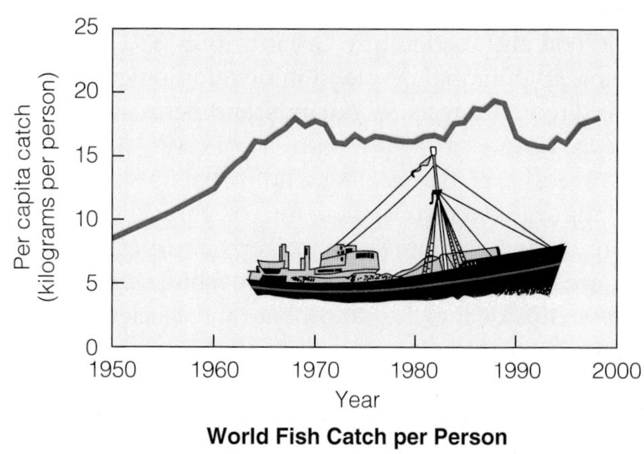

Figure 15-14 World fish catch and world fish catch per person, 1950–97. Worldwide per capita fish catch did not rise much between 1968 and 1989 and has dropped since then. (Data from UN Food and Agriculture Organization and Worldwatch Institute)

CONNECTIONS

Commercial Fishing and the Tragedy of the Commons

Today the commercial fishing industry is dominated by industrial fishing fleets with factory trawlers the size of football fields. They are equipped with satellite positioning equipment, sonar, massive nets, and spotter planes. Most of these industrial fishing fleets can remain at sea for a long time because they have factory ships that can process and freeze their catches.

Between 1975 and 1998, the size of the industrial fishing fleet expanded twice as fast as the rise in catches. As a result, there are now too many boats fishing for a declining number of fish. This leads to overfishing—an example of the tragedy of the commons (p. 14).

A 1998 study by Daniel Pauly warned that the current harvesting of species at increasingly lower trophic levels in ocean food webs decreases chances for recovery of species at the top of ocean food webs. Eventually this can lead to **(1)** abrupt declines in plankton-eating species, **(2)** collapse of marine ecosystems, **(3)** a drop in aquatic biodiversity, and **(4)** a loss of high-quality protein for humans.

Because of the overcapacity of the fishing fleet and overfishing, it costs the global fishing industry about $125 billion a year to catch $70 billion worth of fish. Most of the $54-billion annual deficit of the industry is made up by government subsidies such as fuel tax exemptions, price controls, low-interest loans, and grants for fishing gear.

Critics contend that such subsidies accelerate overfishing. They argue that eliminating these subsidies would reduce the size of the fishing fleet by encouraging free-market competition and would allow some of the economically and biologically depleted stocks to recover.

Critical Thinking

Do you believe that government subsidies for the fishing industry should be eliminated? Explain. How would you feel about eliminating such subsidies if your livelihood depended on fishing?

Degradation and destruction of wetlands, estuaries, coral reefs, salt marshes, and mangroves are also serious and growing threats to populations of fish and shellfish. About 70% of the U.S. fish catch consists of species that depend on increasingly threatened and polluted estuaries for at least part of their life cycle.

In the 1970s and 1980s extensive investment in fishing fleets, aided by government and development agency subsidies, significantly boosted the fish catch (Figure 15-14, left). According to many marine biologists, since 1995 such investments and subsidies have hastened the collapse of ocean fisheries because too many boats are chasing fewer fish. Many analysts believe that global fishing fleets should be shrunk by at least 26% (mostly by eliminating government fishing subsidies) to eliminate overfishing.

An even greater threat than overfishing is the possibility of global climate change over the next 50–100 years. This can warm ocean waters and enhance the harmful effects of habitat degradation, pollution, and ultraviolet radiation (Sections 11-4 and 11-6) on fish catches. Because fish cannot regulate their internal temperature, they are particularly vulnerable to changes in the temperature of their water environments.

Is Aquaculture the Answer? **Aquaculture**, in which fish and shellfish are raised for food, supplies about 20% of the world's commercial fish harvest. Aquaculture production increased 3.3-fold between 1984 and 1997; by 2005 it may account for one-third of the world's fish harvest. China is the world leader in aquaculture (producing almost half of the world's output), followed by India and Japan.

There are two basic types of aquaculture. **Fish farming** involves cultivating fish in a controlled environment, often a pond or tank, and harvesting them when they reach the desired size. **Fish ranching** involves **(1)** holding anadromous species such as salmon (that live part of their lives in fresh water and part in salt water) in captivity for the first few years of their lives (usually in fenced-in areas or floating cages in coastal lagoons and estuaries), **(2)** releasing them, and then **(3)** harvesting the adults when they return to spawn (Figure 15-15).

Species cultivated in developing countries (mostly by inland aquaculture) include carp, tilapia, milkfish, clams, and oysters, all of which feed on phytoplankton and other aquatic plants. In developed countries and some rapidly developing countries in Asia, aquaculture is used mostly to stock lakes and streams with game fish or to raise expensive fish and shellfish such as oysters, catfish, crayfish, rainbow trout, shrimp, and salmon. Aquaculture now produces 90% of all oysters, 40% of all salmon (75% in the United States), and 30% of the shrimp and prawns (50% in the United States) sold in the global marketplace.

Aquaculture has several advantages.

- It is highly efficient and can produce high yields in a small volume of water.

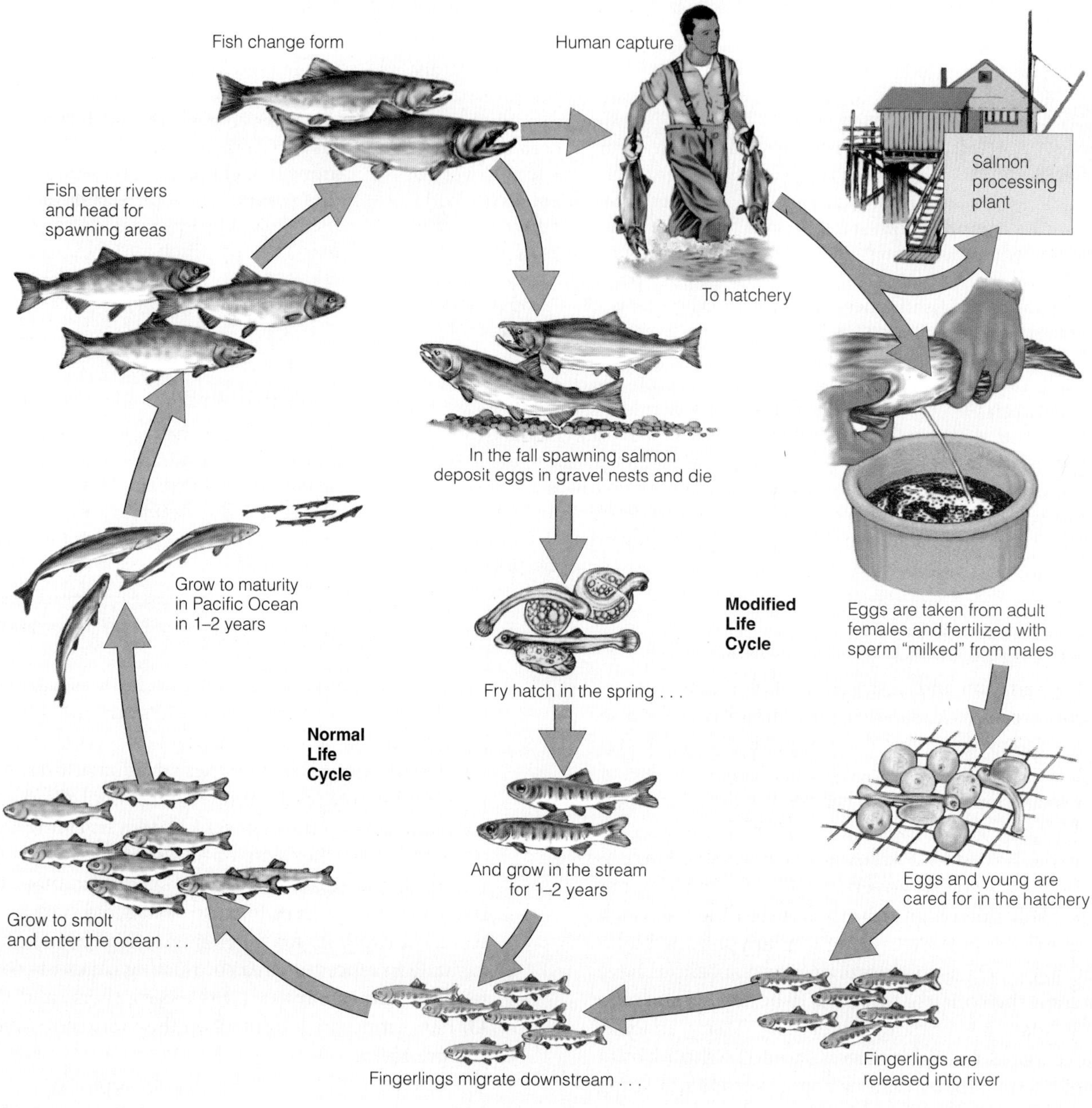

Figure 15-15 Normal life cycle of wild salmon (left) and human-modified life cycle of hatchery-raised salmon (right). Salmon spend part of their lives in fresh water and part in salt water.

- Because little fuel is needed, yields and profits are not closely tied to the price of oil (as they are in commercial marine fishing).

- Crossbreeding and genetic engineering can help increase yields.

- Some analysts project that freshwater and saltwater aquaculture production could double during the next 10 years.

However, the following are problems from aquaculture:

- Fish farms are essentially *aquatic feedlots*, requiring large inputs of land, feed, water, and energy and producing large outputs of wastes, which can contaminate nearby estuaries, surface water, and groundwater.

- Scooping out huge ponds for fish and shrimp farming in some developing countries has destroyed ecologically important mangrove forests (see photo in table of contents, p. XIV).

- Pesticide runoff from nearby croplands can kill fish in aquaculture ponds.
- Dense populations make fish more vulnerable to bacterial and viral infections.
- Escaped farm-raised fish may breed with wild fish and degrade the genetic stock of such species.
- Aquaculture ponds typically last 5 years before they become too contaminated to use. As a result, farmers buy and flood land, extract profits of 40–50% return on their investment, and then move on, leaving barren, salty land that can't be used to raise seafood or crops.

Can We Develop a More Sustainable Approach to Fishery Management? Analysts suggest the following guidelines (many of them politically controversial) for more sustainable management of global fisheries:

- *Shift the burden of proof to the fishing industry by requiring those who profit from harvesting publicly owned resources in territorial waters and the open ocean to show that their harvests are sustainable before allowing them to operate*
- *Set and strictly monitor and enforce quotas for fisheries that are set well below their estimated maximum sustainable yields*
- *Establish and divide up fishing quotas based on* **(1)** *fairness,* **(2)** *local needs and conditions,* **(3)** *the best available scientific data, and* **(4)** *inputs from local communities and fishers*
- *Strengthen commitment to protection of marine biodiversity and integrated coastal management programs that promote both sustainable fishing and the ecological health of marine ecosystems* (Guest Essay, p. 406)
- *Sharply reduce or eliminate fishing subsidies to* **(1)** *shrink the size of global fishing fleets,* **(2)** *encourage free-market competition,* **(3)** *sustain small-scale local fishers,* **(4)** *reduce overfishing, and* **(5)** *allow economically and biologically depleted stocks to recover*
- *Restrict location of fish farms to ensure no net loss of mangrove forests and other threatened coastal environments*
- *Locate fish farms and ranches to prevent displacement of local fishing and spawning grounds*
- *Raise types of aquaculture fish that require little or no fish meal in their diets to reduce the need to divert grain grown on cropland to raise fish*

Ways to manage fisheries to reduce overfishing are discussed in more detail in Section 18-5.

If fish can be harvested more sustainably, this will help feed some people, especially those in coastal areas of developing countries. However, even under the most optimistic projections, harvesting more fish and shellfish will not increase world food supplies significantly because they supply only about 1% of the energy and 6% of the protein in the human diet.

How Do Government Agricultural Policies Affect Food Production? Agriculture is a financially risky business. Whether farmers have a good year or a bad year is determined by factors over which they have little control: weather, crop prices, crop pests and diseases, interest rates, and the global market. Because of the need for reliable food supplies despite fluctuations in these factors, most governments provide various forms of assistance to farmers and consumers.

One approach is to *keep food prices artificially low*. This makes consumers happy but means that farmers may not be able to make a living. Many governments in developing countries keep food prices in cities lower than in the countryside to prevent political unrest. With food prices lower in the cities, more rural people migrate to urban areas, aggravating urban problems and unemployment and increasing the chances of political unrest.

A second approach is to *give farmers subsidies to keep them in business and to encourage them to increase food production*. In developed countries government price supports and other subsidies for agriculture total more than $300 billion per year (including $100 billion per year in the United States). If government subsidies are too generous and the weather is good, farmers may produce more food than can be sold. Then food prices and profits drop because of the surplus and large amounts of food become available for export or food aid to developing countries. This depresses world food prices, which reduces the financial incentive for farmers in developing countries to increase domestic food production. Moreover, the taxes citizens in developed countries pay to provide agricultural subsidies more than offset the lower food prices they enjoy.

A third policy is to *eliminate most or all price controls and subsidies*. This allows market competition to be the primary factor determining food prices and thus the amount of food produced. Some analysts call for phasing out all government price controls and subsidies over, say, 5–10 years and letting farmers respond to market demand. However, these analysts urge that any phaseout of farm subsidies should be coupled with increased aid for the poor and the lower middle class, who would suffer the most from any increase in food prices.

Many environmentalists believe that instead of eliminating all subsidies, we should use them to reward farmers and ranchers who protect the soil, conserve

GUEST ESSAY

Mazunte: A Farming and Fishing Ecological Reserve in Mexico

Alberto Ruz Buenfil

Alberto Ruz Buenfil, an international environmental activist, writer, and performer, is the founder of Huehuecoytl, a land-based ecovillage in the mountains of Mexico. His articles on ecology and alternative living have appeared in publications in the United States, Canada, Mexico, Japan, and Europe. His book, Rainbow Nation Without Borders: Toward an Ecotopian Millennium *(New Mexico: Bear, 1991), has been published in English, Italian, and Spanish.*

The world's eight known species of sea turtles are all officially listed as endangered or threatened. Seven of these species nest on Mexico's Pacific and Atlantic coastlines, making Mexico the world's most important turtle nesting country. The Pacific coasts of southern Mexico, especially the shores of the state of Oaxaca, are the main sites for turtle nesting, reproduction, and conservation; they also contain some of Mexico's last wetland reserves.

Only recently have we begun recognizing the ecological values of wetlands, deltas, and coastal ecosystems [Section 6-5]. They provide habitats for a rich diversity of wildlife, maintain water supplies, protect shorelines from erosion, and play a role in regulating global climate.

Swamps, marshes, and bogs were once considered wastelands to be drained, filled, and turned into "productive land," especially for developing and constructing urban and tourist areas. During the past few decades, the coasts of Oaxaca have not escaped such exploitation, especially after business interests discovered the beaches of Mazunte, Zipolite, San Agustinilo, Puerto Angel, and Puerto Escondido, and the magnificent bays of Hautulco.

The villages of Mazunte and San Agustinilo were founded in the late 1960s, basically to bring in cheap labor from neighboring indigenous villages to provide workers for a slaughterhouse making products from various species of turtles nesting in the area. Members of nearly 200 indigenous farming families became fishermen and employees of the new turtle meat factory, which was fully operational in the 1970s and 1980s. According to some of these workers, nearly 2,000 turtles were killed and quartered every day during those years. At night, dozens of poachers came to collect turtle eggs from their nests.

In the 1980s, this situation came to the attention of two of the first environmental organizations to speak up in defense of species, forests, and natural resources in Mexico. They began denouncing the massacre of the area's turtles, and they pointed out the danger that some of the species might be exterminated. With support from other international organizations, they campaigned for almost 10 years, until a 1990 decree by the president of Mexico made it illegal to exploit the turtles and led to the closing of San Agustinilo's turtle slaughterhouse.

The Mexican government provided some funds, boats, and freezers to compensate for the loss of jobs. However, only about 5% of the indigenous population benefited from this compensation. Since then, most of the people have been living on the verge of starvation, and some have illegally killed protected turtles to survive. What had seemed to be an important environmental victory turned into a nightmare for a large population of indigenous people. Understandably, these people had no use for ecologists or environmentalists.

In 1990, a group called ECOSOLAR A.C. began efforts to change this situation by implementing a plan for sustainable development of the coast of Oaxaca. They were successful in obtaining funding for this project from different national and international institutions. By 1992, the members of this small but effective group had succeeded in

- Making a detailed study of the bioregion, which, with the participation of the local people, is being used to define the possible uses of different areas

water, reforest degraded land, protect and restore wetlands, and conserve wildlife. What do you think?

Another way that governments and private organizations deal with a lack of food production and hunger is through food aid (Pro/Con p. 408).

15-4 SOLUTIONS: SUSTAINABLE AGRICULTURE

What Is Sustainable Agriculture? Evidence suggests that two of the world's major food production systems, rangelands and oceanic fisheries (Figure 15-14), have reached or are near their limits to produce more food. This means that most future growth in food production must come from croplands. However, the total area of cropland is unlikely to expand because of a lack of economically affordable and environmentally sustainable land. Thus, as population continues to grow, the amount of cropland per person for raising grain is expected to decline (Figure 15-13).

Increasing the yields per area of existing cropland may also be limited because of **(1)** a lack of water for irrigation (Figure 15-11), **(2)** reduced genetic diversity, **(3)** a leveling off of yields per hectare, and **(4)** the environmental effects of food production, which degrade existing cropland (Figure 15-8). These *limiting factors* may limit population growth by increasing death rates from lack of food and vulnerability to disease.

Bourne, J. 1999. "The Organic Revolution." Audubon, vol. 101, no. 2, p. 64(1).

- Creating a system of credits to help native inhabitants build better houses, establish small family-run restaurants, and manufacture hammocks for rent or sale to visitors
- Beginning the construction of systems for drainage, water collection, and latrines using low-impact technology and local materials and workers, as well as nurseries for local seeds and facilities for reforestation and wildlife preservation projects
- Working with the community to promote Mazunte as a center for ecotourism where visitors can experience unique ecosystems containing alligators, turtles, and hundreds of bird and fish species

In only 2 years, the native inhabitants of Mazunte and other neighboring communities completely changed their opinion about and perspective on ecology and environmentalists. In May 1992, Mazunte hosted the second annual gathering of Earth Keepers, involving nearly 150 representatives of 35 organizations from 20 different countries.

For 1 week, these specialists shared their practical knowledge with the local people. They used their skills to help the community **(1)** set up an alternative clinic for healing arts, **(2)** establish permaculture (a form of sustainable agriculture) and organic agriculture to improve local hatcheries and home-based plant nurseries, and **(3)** build both biodigestors for producing natural fertilizers from biomass and a village recycling center located at the school. In addition, artistic and cultural activities took place every night in the Center for Biological Investigations, which the people of Mazunte want to turn into Mexico's first Marine Turtles Museum. Run by local people, it will attract and educate visitors from around the world.

However, despite the legal protection of sea turtles in Mexico, illegal killing of turtles and removal of their eggs continues. A few days after the event concluded, the village of Mazunte called a general meeting attended by 150 heads of families to discuss ways to get community members to protect turtle nesting areas instead of illegally killing the turtles for food. Out of that meeting came a "Declaration of Mazunte" requesting that competent higher authorities and the president of Mexico put an immediate end to such destruction, which violates the earlier presidential decree forbidding the annihilation of turtles in Mexico.

The community went on to declare their village and neighboring environments to be Mexico's first *Farming and Fishing Reserve*. Its goals would be to protect the area's forests, water sources, wetlands, wildlife, shores, beaches, and scenic places and to "establish new forms of relationship between humans and nature, for the well-being of today's and tomorrow's generations." This declaration has been presented to the government of Mexico and to many national and international organizations.

Mazunte is taking the lead in demonstrating that cooperation between local people and environmental experts can lead to ecologically sustainable communities that benefit local people and wildlife alike. This model can show farmers and indigenous communities everywhere how to live more sustainably on the earth and turn things around in a short time. It is a message of hope and empowerment for people seeking a better world for themselves and others.

Critical Thinking

1. What lessons that could be applied to your own life have you learned from this essay?

2. Could the rapid change toward sustainability brought about by environmentalists and local people in Mazunte be accomplished in your own community? If so, how? If not, why not?

If these projections are correct, then the two main tools in trying to reduce hunger, malnutrition, poverty, and the harmful environmental effects of agriculture will be **(1)** slowing population growth (Section 9-3) and **(2)** developing and phasing in systems of **sustainable agriculture** (or **low-input agriculture**) over the next three decades.

Most proponents of more sustainable agriculture are not opposed to high-yield agriculture. Instead, they see it as vital for protecting the earth's biodiversity and ecological integrity by reducing the need to cultivate new and often marginal land. They also believe that current agricultural research and economic incentives should be redirected to encourage increases in yield per hectare without depleting or degrading soil, water, and biodiversity.

General guidelines for sustainable agricultural systems suggested by various analysts include the following:

- *Combine traditional high-yield polyculture and modern monoculture methods for growing crops.*
- *Measure the success of food policies not in terms of food production but in terms of human nutrition and health.*
- *Grow more perennial crops* (p. 384).
- *Minimize soil erosion, salinization, and waterlogging.*

Hint: Enter the search term organic farming using the Subject Guide.

How Useful Is International Food Aid?

PRO/CON

Most people view international food aid as a humanitarian effort to prevent people from dying prematurely. However, some analysts contend that giving food to starving people in countries with high population growth rates does more harm than good in the long run. By not helping people grow their own food, they argue, food relief can condemn even greater numbers to premature death from starvation and disease in the future—another example of harmful positive feedback.

Biologist Garrett Hardin (Guest Essay, p. 214) suggests that we use the concept of *lifeboat ethics* to decide which countries get food relief. His basic premise is that there are already too many people in the lifeboat we call earth. Thus, if food relief is given to countries that are not reducing their populations, the effect is to add more people to an already overcrowded lifeboat. According to Hardin, sooner or later the overloaded boat will sink and most of the passengers will drown.

Large amounts of food relief can also depress local food prices, decrease food production, and stimulate mass migration from farms to already overburdened cities. In addition, food relief discourages local and national governments from investing in the rural agricultural development needed to enable farmers to grow enough food for the population.

Another problem is that much food relief does not reach hunger victims. Transportation networks and storage facilities often are inadequate, so some of the food rots or is devoured by pests before it can reach the hungry. Moreover, some officials steal some of the food and sell it for personal profit. In addition, sometimes a portion of the food must be given to officials as bribes for approving the unloading and transporting of the remaining food to the hungry.

Providing food and other forms of aid during famines can also be risky and costly. For example, supplying starving people in Somalia in 1992 required a UN Peacekeeping force that probably cost at least 10 times more than the food that was distributed. This illustrates the strong linkages between environmental security and economic and military security (Guest Essay, p. 282).

Most critics are not against providing aid. However, they believe that such aid should help countries control population growth and grow enough food to feed their populations by using sustainable agricultural methods (Section 15-4). Temporary food relief, they believe, should be given only when there is a complete breakdown of an area's food supply because of natural disaster.

Critical Thinking

Is sending food to famine victims helpful or harmful in the short run? In the long run? Explain. Are there any conditions you would attach to providing such aid? Explain.

- *Reduce destruction of forests, grasslands, and wetlands for producing food by emphasizing increased yields per area of cropland using sustainable methods.*
- *Stabilize aquifers by reducing the rate of water removal to the rate of recharge.*
- *Reduce water waste in irrigation* (Section 12-3).
- *Reduce overfishing by implementing the principles listed on p. 405.*
- *Reduce use and waste of fossil fuels* (Section 20-2).
- *Increase use of organic fertilizers, and of solar, wind, and biomass energy to grow and process crops.*
- *Emphasize biological pest control and integrated pest management* (Section 16-5).
- *Protect existing prime cropland from environmental degradation and conversion to urban or industrial uses* (Section 9-7).
- *Subsidize sustainable farming and phase out government subsidies for unsustainable farming.*
- *Shift to full-cost pricing* (Section 2-3) *that includes the harmful environmental effects of agriculture* (Figure 15-8) *in food prices.*
- *Educate the public about the hidden environmental and health costs they are paying for food and the need to gradually incorporate these costs into market prices.*
- *Reduce food waste.* According to a 1997 U.S. Department of Agriculture study, an estimated 27% of the food produced in the United States (not including crop losses) is wasted by grocery stores, restaurants, and consumers.
- *Greatly increase research on sustainable agriculture and improving human nutrition.*
- *Set up demonstration projects throughout each country so that farmers can see how sustainable agricultural systems work.*
- *Establish training programs in sustainable agriculture for farmers and government agricultural officials and*

encourage the creation of college curricula in sustainable agriculture and human nutrition.

- *Reduce poverty (Section 2-4) so that the poor can have enough money or land to supply their basic food needs.*
- *Provide poor women with access to credit, markets, and technical food-growing advice, as well as education and health care.*
- *Slow population growth to help all of the world's countries reach the more sustainable postindustrial stage of the demographic transition* (Figure 9-19).
- *Integrate agriculture, population, urban and rural, energy, health, climate, water resource, soil resource, land use, pollution, and biodiversity protection policies.*

Can We Make the Transition to Sustainable Agriculture? A growing number agricultural analysts believe that over the next 30 years we must make a transition from unsustainable and environmentally harmful agriculture (Figure 15-8) to more sustainable forms of agriculture.

In developed countries, including the United States, even a partial shift to more environmentally sustainable food production will not be easy. It will be opposed by **(1)** agribusiness, **(2)** successful farmers with large investments in unsustainable forms of industrialized agriculture, and **(3)** specialized farmers unwilling to learn the more demanding art of farming sustainably. It might also be resisted by many consumers unwilling or unable to pay higher prices for food, because full-cost accounting would include agriculture's harmful environmental and health costs in the market prices of food.

Despite such difficulties, many environmentalists believe that a new *eco-agricultural revolution* could take place throughout most of the world over the next 30 years. Whether it does occur is primarily a political and ethical issue. Some actions you can take to help promote sustainable agriculture are listed in Appendix 6.

The need to bring birth rates well below death rates, increase food production while protecting the environment, and distribute food to all who need it is the greatest challenge our species has ever faced.

PAUL AND ANNE EHRLICH

REVIEW QUESTIONS

1. Define the boldfaced terms used in this chapter.

2. What are *perennial crops*? What advantages do they have over conventional annual crops?

3. What are the three main systems that supply our food? List six factors that have led to a huge increase in food production since 1950. List six pieces of bad news about current and future food production.

4. What three crops provide most of the world's food? Distinguish between *industrialized agriculture, plantation agriculture, traditional subsistence agriculture,* and *traditional intensive agriculture.*

5. What is a *green revolution,* and what three steps does it involve? Explain how producing more food on less land can help protect biodiversity. List three factors that could cause yields from green revolution crops to level off or even drop.

6. Describe the nature and importance of the agricultural industry in the United States. How energy-efficient is industrialized agriculture in the United States?

7. What are some of the harmful environmental consequences of meat production? What are some of the advantages of a meat-based diet?

8. Distinguish between *interplanting, polyvarietal cultivation, intercropping, agroforestry,* and *polyculture.* What are some advantages and disadvantages of polyculture?

9. What is the good news about world food production since 1950? What is the bad news about such production?

10. Distinguish between *undernutrition, malnutrition,* and *overnutrition.* What are the two most common nutritional deficiency diseases and how do they differ? What are the effects of nutritional deficiencies of **(a)** vitamin A, **(b)** iron, and **(c)** iodine? About how many chronically malnourished people are there in the world? About how many people die each year from undernutrition, malnutrition, or diseases worsened by malnutrition? List five major ways to reduce sickness and premature death of children from malnutrition. What is the primary cause of hunger in the world?

11. List the major harmful environmental effects of producing food.

12. List the pros and cons of growing more food by **(a)** increasing crop yields through green revolutions, **(b)** genetically engineering crops (transgenics), **(c)** getting people to switch to new types of foods, **(d)** irrigating more cropland, **(e)** cultivating more land, and **(f)** harvesting more fish and shellfish. Why should we be concerned about loss of genetic diversity in planted crops?

13. Summarize the problems with providing China with enough food. How could lack of food in China affect the rest of the world?

14. With respect to obtaining food from fish and shellfish, distinguish between *sustained yield, overfishing,* and *commercial extinction* of a species. About what percentage of the world's fish stocks are in decline because of overfishing and pollution? Distinguish between *fish farming* and *fish ranching.* What are the pros and cons of aquaculture?

15. Explain how commercial fishing can be considered an example of the tragedy of the commons. List eight ways to improve the sustainability of fisheries.

16. List three types of policies governments use to affect food production. List the pros and cons of each approach.

17. List the pros and cons of food aid.

18. What is *sustainable agriculture*? What are some general guidelines for achieving more sustainable agricultural systems?

19. Why is there opposition to switching to more sustainable agriculture? How can such opposition be overcome?

CRITICAL THINKING

1. Summarize the advantages and limitations of each of the following proposals for increasing world food supplies and reducing hunger over the next 30 years: **(a)** cultivating more land by clearing tropical forests and irrigating arid lands, **(b)** catching more fish in the open sea, **(c)** producing more fish and shellfish with aquaculture, and **(d)** increasing the yield per area of cropland.

2. What are the three most important things that you believe should be done to reduce hunger in the country where you live? In the world?

3. Some people argue that starving people could get enough food by eating nonconventional plants and insects; others point out that most starving people don't know what plants and insects are safe to eat and can't take a chance on experimenting when even the slightest illness could kill them. If you had no money to grow or buy food, would you collect and eat protein-rich grasshoppers, moths, beetles, or other insects?

4. The United States is by far the world's largest food exporter. What harmful economic and environmental effects might the United States experience if its ability to supply food for export is exceeded by the demand?

5. Does the United States, with one of the world's highest levels of meat consumption, have a moral responsibility to cut its grain consumption to make more grain available for export to countries whose people get most of their food by direct consumption of grain? Explain.

6. Should all price supports and other government subsidies paid to farmers be eliminated? Explain. Try to consult one or more farmers in answering this question.

7. What might be some drawbacks for relying more on perennial food crops (p. 384)? Why will conventional food-growing companies be opposed to relying more on perennial crop varieties?

8. Should governments phase in agricultural tax breaks and subsidies to encourage farmers to switch to more sustainable farming? Explain.

***9.** If possible, visit both a conventional industrialized farm and an organic or low-input farm. Compare **(a)** soil erosion and other forms of land degradation, **(b)** use and costs of energy, **(c)** use and costs of pesticides and inorganic fertilizer, **(d)** use and costs of natural pest control and organic fertilizer, **(e)** yields per hectare for the same crops, and **(f)** overall profit per hectare for the same crops.

***10.** Use health and other local government records to estimate how many people in your community suffer from undernutrition or malnutrition. Has this problem increased or decreased since 1980? What are the basic causes of this hunger problem, and what is being done to alleviate it? Share the results of your study with local officials and then present your own plan for improving efforts to reduce hunger in your community.

***11.** Make a concept map of this chapter's major ideas, using the section heads and subheads and the key terms (in boldface type). Look at the inside back cover and in Appendix 5 for information about concept maps.

FURTHER INFORMATION

For more information see Internet Sources and Further Readings at the back of the book. Online resources for this chapter can be found at **www.brookscole.com/biology**

16 PROTECTING FOOD RESOURCES: PESTICIDES AND PEST CONTROL

Along Came a Spider

The longest war in human history is our ongoing war against insect pests. This war was declared about 10,000 years ago, when humans first got serious about agriculture. Today we are not much closer to winning this war than we were then.

Some Chinese farmers recently decided to change strategies. Instead of spraying their rice and cotton fields with poisons, they build little straw huts around the fields in the fall.

If this sounds crazy, it's crazy like a fox. These farmers are giving aid and comfort to insects' worst enemy, one that has hunted them for millions of years: spiders (Figure 16-1). The little huts are for hibernating spiders. Protected from the worst of the cold by the huts, far more of the hibernating spiders become active in the spring. Ravenous after their winter fast, they scuttle off into the fields to stalk their insect prey.

Even without human help, the world's 30,000 known species of spiders kill far more insects every year than insecticides do. A typical acre of meadow or woods contains an estimated 50,000 to 2 million spiders, each devouring hundreds of insects per year.

Entomologist Willard H. Whitcomb found that leaving strips of weeds around cotton and soybean fields provides the kind of undergrowth favored by insect-eating wolf spiders (Figure 16-1, right). He also sings the praises of a type of banana spider, which lives in warm climates and can keep a house clear of cockroaches.

The idea of encouraging populations of spiders in fields, forests, and even houses scares some people because spiders have bad reputations. Perhaps because of their eight legs and numerous eyes, spiders look like dangerous space aliens instead of the helpful and mostly harmless creatures they are.

A few species of spiders, such as the black widow, the brown recluse, and eastern Australia's Sydney funnel web, are dangerous to people. However, the vast majority of spider species, including the ferocious-looking wolf spider, don't harm humans. Even the giant tarantula rarely bites people, and its venom is too weak to harm us or other large mammals.

As biologist Thomas Eisner puts it, "Bugs are not going to inherit the earth. They own it now. So we might as well make peace with the landlord." As we seek new ways to coexist with the insect rulers of the planet, we would do well to be sure that spiders are in our corner.

Figure 16-1 Spiders are insects' worst enemies. Most spiders, like the crab spider (left) and the wolf spider (right), found in many parts of the world, are harmless to humans. (Left, James C. Cokendolpher; right, Dan Kline/Visuals Unlimited)

A weed is a plant whose virtues have not yet been discovered.
RALPH WALDO EMERSON

This chapter addresses the following questions:

- What types of pesticides are used?
- What are the pros and cons of using chemicals to kill insects and weeds?
- How well is pesticide use regulated in the United States?
- What alternatives are there to using pesticides?

16-1 PESTICIDES: TYPES AND USES

How Does Nature Keep Pest Populations Under Control? A **pest** is any species that **(1)** competes with us for food, **(2)** invades lawns and gardens, **(3)** destroys wood in houses, **(4)** spreads disease, or **(5)** is simply a nuisance. In natural ecosystems and many polyculture agroecosystems, natural enemies (predators, parasites, and disease organisms) control the populations of 50–90% of pest species (Figure 1-2).

When we simplify natural ecosystems, we upset these natural checks and balances, which keep any one species from taking over for very long. Then we must devise ways to protect our monoculture crops, tree farms, and lawns from insects and other pests that nature once controlled at no charge.

We have done this primarily by developing a variety of **pesticides** (or *biocides*): chemicals to kill organisms we consider undesirable. Common types of pesticides include **(1) insecticides** (insect-killers), **(2)** *herbicides* (weed-killers), **(3)** *fungicides* (fungus-killers), **(4)** *nematocides* (roundworm-killers), and **(5)** *rodenticides* (rat- and mouse-killers).

Humans didn't invent the use of chemicals to repel or kill other species; plants have been producing chemicals to ward off or poison herbivores that feed on them for about 225 million years. This is a neverending, ever-changing process: Herbivores overcome various plant defenses through natural selection; then the plants use natural selection to develop new defenses. The result of these dynamic interactions between predator and prey species is what biologists call *coevolution* (Section 5-2).

What Was the First Generation of Pesticides and Repellents? As the human population grew and agriculture spread, people began looking for ways to protect their crops, mostly by using chemicals to kill or repel insect pests. Sulfur was used as an insecticide well before 500 B.C.; by the 1400s, toxic compounds of arsenic, lead, and mercury were being applied to crops as insecticides. Farmers abandoned this approach in the late 1920s, when the increasing number of human poisonings and fatalities prompted a search for less toxic substitutes. However, traces of these nondegradable toxic metal compounds are still being taken up by tobacco, vegetables, and other crops grown on soil dosed with them long ago.

In the 1600s nicotine sulfate, extracted from tobacco leaves, came into use as an insecticide. In the mid-1800s two more natural pesticides were introduced: **(1)** *pyrethrum*, obtained from the heads of chrysanthemum flowers, and **(2)** *rotenone*, from the root of the derris plant and other tropical forest legumes. These *first-generation pesticides* were mainly natural substances, weapons borrowed from plants that had been at war with insects for eons.

In addition to protecting crops, people have used chemicals (produced by plants) to repel or kill insects in their households, yards, and gardens; they also save money and reduce health hazards associated with using commercial insecticides.

If *ants* come indoors, you can usually persuade them to leave within about 4 days by **(1)** sprinkling repellents such as red or cayenne pepper, crushed mint leaves, or boric acid (with an anticaking agent) along their trails inside a house and **(2)** by wiping off countertops with vinegar. (However, boric acid is poisonous and should not be placed in areas accessible to small children and pets.)

Repel *mosquitoes* by **(1)** planting basil outside windows and doors and **(2)** rubbing a bit of vinegar, basil oil, lime juice, or mugwort oil on exposed skin. You can also reduce mosquito attacks by not using scented soaps or wearing perfumes, colognes, and other scented products outdoors during mosquito season. Researchers have found that the $30 million U.S. consumers spend each year on electric bug zappers to kill mosquitoes is mostly wasted. Only about 3% of the insects they kill on an average night are female mosquitoes—the kind that bite.

You can kill *cockroaches* (Spotlight, p. 121) by **(1)** sprinkling boric acid under sinks and ranges, behind refrigerators, and in cabinets, closets, and other dark, warm places (but not in areas accessible to children and pets) or **(2)** establishing populations of banana spiders. You can also trap roaches by greasing the inner neck of a bottle or large jar with petroleum jelly, filling much of it with raw potato, stale beer, banana skins, or other food scraps (especially fruits), and placing a small ramp leading to it. Placing fresh or dried bay leaves in and around cupboards repels cockroaches.

Repel *flies* by **(1)** planting sweet basil and tansy (a common herb) near doorways and patios and **(2)** hanging a series of polyethylene strips in front of entry doors (like the ones on some grocery-store coolers). You can

Rachel Carson

INDIVIDUALS MATTER

Rachel Carson (Figure 16-2) began her professional career as a biologist for the Bureau of U.S. Fisheries (later to become the U.S. Fish and Wildlife Service). In that capacity, she carried out research on oceanography and marine biology, wrote articles about the oceans and topics related to the environment, and became editor-in-chief of the bureau's publications in 1949.

In 1951, she wrote *The Sea Around Us*, which described in easily understandable terms the natural history of oceans and the harm that humans were doing them. The book was on the best-seller list for 86 weeks, sold more than 2 million copies, was translated into 32 languages, and won a National Book Award.

During the late 1940s and throughout the 1950s, the use of DDT and related compounds—to kill insects that ate food crops, attacked trees, bothered people, and transmitted diseases such as malaria (Figure 8-9)—expanded rapidly.

In 1958, DDT was sprayed to control mosquitoes near the home and private bird sanctuary of Olga Huckins, a good friend of Carson. After the spraying, Huckins witnessed the agonizing deaths of several of her birds, and in distress she asked Carson whether she could find someone to investigate the effects of pesticides on birds and other wildlife.

Carson decided to look into the issue herself and quickly found that almost no independent critical research on the environmental effects of pesticides existed. As a well-trained scientist, Carson surveyed the scientific literature and methodically built her case against the widespread use of pesticides.

In 1962, she published her findings in popular form in *Silent Spring*, an allusion to the silencing of "robins, catbirds, doves, jays, wrens, and scores of other bird voices" because of their exposure to pesticides. She pointed out that "for the first time in the history of the world, every human being is now subjected to dangerous chemicals, from the moment of conception until death."

Carson's book was read by many scientists, politicians, and policy makers and was embraced by the public. However, the chemical industry viewed the book as a serious threat to booming pesticide sales and mounted a $250,000 campaign to discredit Carson. A parade of critical reviewers and industry scientists claimed that her book was full of inaccuracies, made selective use of research findings, and failed to give a balanced account of the benefits of pesticides.

Some critics even claimed that, as a woman, she was incapable of understanding the highly scientific and technical subject of pesticides. Others charged that she was a hysterical woman and a radical nature lover trying to scare the American public in order to sell books.

During this period of intense controversy Carson was suffering from terminal cancer, but she was able to defend her research and strongly counter her critics. She died in 1964—18 months after the publication of *Silent Spring*—without knowing that her efforts were a driving force in the birth of what is now known as the environmental movement in the United States.

Figure 16-2 Biologist Rachel Carson (1907–64) was a pioneer in increasing public awareness of the importance of nature and the threat of pollution. She died without knowing that her efforts were a key in starting today's environmental movement. (©1962 Eric Hartmann/ Magnum Photos)

To environmentalists and most citizens, Rachel Carson is an outstanding example of a dedicated scientist and effective communicator of complex scientific information.

make nontoxic flypaper by applying honey to strips of yellow paper (their favorite color) and hanging it from the ceiling in the center of rooms. You can also hang clusters of cloves.

Keep *fleas* off pets by **(1)** using green dye or flea-repellent soaps, **(2)** feeding them brewer's yeast or vitamin B, **(3)** using flea powders made from eucalyptus, sage, tobacco, wormwood, or vetiver, or **(4)** dipping or shampooing pets in a mixture of water and essential oils such as citronella, cedarwood, eucalyptus, fleabane, sassafras, geranium, clove, or mint. Researchers recently invented a trap that uses green-yellow light to attract fleas to an adhesive-coated surface. Another solution is to put a light over a shallow pan of water before going to bed at night and turn out all other lights, empty the water every morning, and continue for a month. (Fleas are attracted to heat and light but they can't swim.) Desiccant powders, such as Dri-Die, Perma-Guard, or SG-67, can also rid a house of flea infestations by drying up the insects. Diatomaceous earth (or diatom powder) can also be sprinkled on carpets and pets to kill fleas.* This powder consists of the skeletal remains of microscopic algae and kills fleas and many other harmful insects by cutting their outer shells so they lose their body fluids.

Control *lawn weeds* by raising the cutting level of your lawn mower so the grass can grow 8–10 centimeters (3–4 inches) high. This gives it a strong root system that can hinder weed growth; the higher grass also provides habitats for spiders and other insects that eat insect pests. Pull weeds and douse the hole with soap solution or human urine (which is high enough in nitrogen to burn the weed).

What Was the Second Generation of Pesticides? A major pest control revolution began in 1939, when entomologist Paul Mueller discovered that DDT (dichlorodiphenyltrichloroethane), a chemical known since 1874, was a potent insecticide. DDT, the first of the so-called *second-generation pesticides*, soon became the world's most-used pesticide and Mueller received the Nobel prize in 1948. Since 1945 chemists have developed hundreds of synthetic organic chemicals for use as pesticides.

Worldwide, about 2.3 million metric tons (2.5 million tons) of such pesticides are used yearly. About 75% of these chemicals are used in developed countries, but use in developing countries is soaring.

In the United States, about 630 different biologically active (pest-killing) ingredients and about 1,820 inert (inactive) ingredients are mixed to make some 25,000 different pesticide products. About 25% of pesticide use in the United States is for ridding houses, gardens, lawns, parks, playing fields, swimming pools, and golf courses of unwanted pests. According to the U.S. Environmental Protection Agency (EPA), the average lawn in the United States is doused with more than 10 times more synthetic pesticides per hectare than U.S. cropland. Each year, more than 250,000 U.S. residents become ill because of household use of pesticides, and such pesticides are a major source of accidental poisonings and deaths for children under age 5. An EPA study found that at least 47% of U.S. households with children under age 5 were found to have at least one pesticide within reach of small children.

*Diatom powder can be purchased in bulk at stores that sell it for use in swimming pool filters. It is also found in some gardening stores under the name of Permatex. Because it contains fine particles of silicate, you should wear a dust mask when applying this powder to avoid inhaling tiny particles of silicate.

Some pesticides, called *broad-spectrum agents*, are toxic to many species; others, called *selective* or *narrow-spectrum agents*, are effective against a narrowly defined group of organisms. Pesticides vary in their *persistence*, the length of time they remain deadly in the environment (Table 16-1). Most organophosphates (except malathion) are highly toxic to humans and other animals, and they account for most human pesticide poisonings and deaths. In 1962 biologist Rachel Carson warned against relying on synthetic chemicals to kill insects and other species we deem pests (Individuals Matter, p. 414).

16-2 THE CASE FOR PESTICIDES

Proponents of pesticides contend that their benefits outweigh their harmful effects.

Pesticides save human lives. Since 1945 DDT and other chlorinated hydrocarbon and organophosphate insecticides have probably prevented the premature deaths of at least 7 million people from insect-transmitted diseases such as malaria (carried by the *Anopheles* mosquito; Figure 8-9), bubonic plague (rat fleas), typhus (body lice and fleas), and sleeping sickness (tsetse fly).

Pesticides increase food supplies and lower food costs. About 55% of the world's potential human food supply is lost to pests before (35%) or after (20%) harvest. An estimated 37% of the potential U.S. food supply is destroyed by pests before and after harvest; insects cause 13% of these losses, plant pathogens 12%, and weeds 12%. Without pesticides, these losses would be worse, and food prices would rise (by 30–50% in the United States, according to pesticide company officials).

Pesticides increase profits for farmers. Pesticide companies estimate that every $1 spent on pesticides leads to an increase in U.S. crop yields worth approximately $4 (but studies have shown that this benefit drops to about $2 if the harmful effects of pesticides are included).

Pesticides work faster and better than alternatives. Pesticides **(1)** can control most pests quickly and at a reasonable cost, **(2)** have a long shelf life, **(3)** are easily shipped and applied, and **(4)** are safe when handled properly. When genetic resistance occurs, farmers can use stronger doses or switch to other pesticides.

The health risks of pesticides are insignificant compared with their benefits. According to Elizabeth Whelan, director of the American Council on Science and Health (ACSH), which presents the position of the pesticide industry, "The reality is that pesticides, when used in the approved regulatory manner, pose no risk to either farm workers or consumers." Pesticide proponents consider media reports describing pesticide health scares to be distorted science and irresponsible reporting. They also point out that about 99.99% of the pesti-

Blindauer, K. M., et al. 1999. "Environmental Pesticide Illness and Injury." *Journal of Environmental Health,* v61 i10 p9(6).

Table 16-1 Major Types of Pesticides

Type	Examples	Persistence	Biologically Magnified?
Insecticides			
Chlorinated hydrocarbons	DDT, aldrin, dieldrin, toxaphene, lindane, chlordane, methoxychlor, mirex	High (2–15 years)	Yes (Figure 8-2)
Organophosphates	Malathion, parathion, diazinon, TEPP, DDVP, mevingphos	Low to moderate (1–2 weeks), but some can last several years	No
Carbamates	Aldicarb, carbaryl (Sevin), propoxur, maneb, zineb	Low (days to weeks)	No
Botanicals	Rotenone, pyrethrum, and camphor extracted from plants, synthetic pyrethroids (variations of pyrethrum) and rotenoids (variations of rotenone)	Low (days to weeks)	No
Microbotanicals	Various bacteria, fungi, protozoa	Low (days to weeks)	No
Herbicides			
Contact chemicals	Atrazine, simazine, paraquat	Low (days to weeks)	No
Systemic chemicals	2,4-D, 2,4,5-T, Silvex, diruon, daminozide (Alar), alachlor (Lasso), glyphosate (Roundup)	Mostly low (days to weeks)	No
Soil sterilants	Tribualin, diphenamid, dalapon, butylate	Low (days)	No
Fungicides			
Various chemicals	Captan, pentachlorphenol, zeneb, methyl bromide, carbon bisulfide	Most low (days)	No
Fumigants			
Various chemicals	Carbon tetrachloride, ethylene dibromide, methyl bromide	Mostly high	Yes (for most)

cides we consume in our food are natural chemicals produced by plants.

Proponents point out some beneficial changes in pesticides and their use. For example, *safer and more effective pesticides are being developed.* Industry scientists are developing pesticides, such as botanicals and microbotanicals (Table 16-1), that are safer to users and less damaging to the environment. Genetic engineering also holds promise in developing pest-resistant crop strains and genetically altered crops that produce pesticides (Pro/Con, p. 396). However, total research and development and government approval costs for a new pesticide have risen from \$6 million in 1976 to \$80–120 million today. As a result, new chemicals are being developed only for crops such as wheat, corn, and soybeans with large markets.

Many new pesticides are used at very low rates per unit area compared with older products. For example, application amounts per hectare for many new herbicides are 1/100 the rate for older ones.

Scientists continue to search for the ideal pest-killing chemical, which would

- *Kill only the target pest*
- *Harm no other species*
- *Disappear or break down into something harmless after doing its job*
- *Not cause genetic resistance in target organisms*
- *Be more cost-effective than doing nothing*

Unfortunately, no known pesticide meets all these criteria, and most don't even come close.

16-3 THE CASE AGAINST PESTICIDES

How Do Pesticides Cause Genetic Resistance and Kill Natural Pest Enemies? Opponents of widespread use of pesticides believe that their harmful effects outweigh their benefits. The biggest problem is the development of *genetic resistance* to pesticides by pest organisms. Insects breed rapidly (Figure 16-3), and within 5–10 years (much sooner in tropical areas) they

Hint: Enter the search terms *pesticide, surveillance* using the Subject Guide.

Figure 16-3 A boll weevil, just one example of an insect capable of rapid breeding. In the cotton fields of the southern United States, these insects lay thousands of eggs, producing a new generation every 21 days and as many as six generations in a single growing season. Attempts to control the cotton boll weevil account for at least 25% of insecticide use in the United States. However, farmers are now increasing their use of natural predators and other biological methods to control this major pest. (U.S. Department of Agriculture)

TOLES©1992 The Buffalo News. Reprinted with permission of UNIVERSAL PRESS SYNDICATE. All rights reserved.

can develop immunity to pesticides through natural selection (Section 5-2) and come back stronger than before (see cartoon, above). Weeds and plant-disease organisms also become resistant, but more slowly.

Since 1950 at least 520 insect and mite species, 273 weed species, 150 plant diseases, and 10 rodent species (mostly rats) have developed genetic resistance to one or more pesticides. At least 17 insect pest species are resistant to all major classes of insecticides, and several fungal plant diseases are now immune to most widely used fungicides. Because of genetic resistance, most widely used insecticides no longer protect people from insect-transmitted diseases in many parts of the world, leading to resurgences of diseases such as malaria (Figure 8-8).

Another problem is that *broad-spectrum insecticides kill natural predators and parasites that may have been maintaining the population of a pest species at a reasonable level.* With wolf spiders (Figure 16-1, right), wasps, predatory beetles, and other natural enemies out of the way, a rapidly reproducing insect pest species can make a strong comeback only days or weeks after initially being controlled. Although natural predators can also develop genetic resistance to pesticides, most predators can't reproduce as quickly as their insect prey.

Wiping out natural predators can also unleash new pests whose populations the predators had previously held in check, causing other unexpected effects (Connections, p. 175). Currently 100 of the 300 most destructive insect pests in the United States were secondary pests that became major pests through the effects of insecticides.

What Is the Pesticide Treadmill? When genetic resistance develops, pesticide sales representatives usually recommend **(1)** more frequent applications, **(2)** larger doses, or **(3)** a switch to new (usually more expensive) chemicals to keep the resistant species under control. This can put farmers on a **pesticide treadmill**, whereby they pay more and more for a pest control program that often becomes less and less effective. Some biologists also warn that increasing dependence on genetically altered crops (transgenics) can put farmers on a similar *genetic treadmill* (Pro/Con, p. 396).

In 1989 David Pimentel (Guest Essay, p. 352), an expert in insect ecology, evaluated data from more than 300 agricultural scientists and economists and came to the following conclusions:

- Although the use of synthetic pesticides has increased 33-fold since 1942, it is estimated that more of the U.S. food supply is lost to pests today (37%) than in the 1940s (31%). Losses attributed to insects almost doubled (from 7% to 13%) despite a 10-fold increase in the use of synthetic insecticides.
- The estimated environmental, health, and social costs of pesticide use in the United States range from $4 billion to $10 billion per year. The International Food Policy Research Institute puts the estimate much higher, at $100–200 billion per year, or $5–10 in damages for every dollar spent on pesticides.
- Alternative pest control practices (Section 16-5) could halve the use of chemical pesticides on 40 major U.S. crops without reducing crop yields.

- A 50% cut in U.S. pesticide use would cause retail food prices to rise by only about 0.2% but would raise average income for farmers about 9%.

Although pesticides can't be eliminated completely, numerous studies and experience show that their use can be reduced sharply without reducing yields, and in some cases yields would increase. Over the past few years, Sweden has cut pesticide use in half with virtually no decrease in the harvest. Campbell Soup uses no pesticides on tomatoes it grows in Mexico, and yields have not dropped. After a 65% cut in pesticide use on rice in Indonesia, yields increased by 15%.

Where Do Pesticides Go, and How Can They Harm Wildlife? *Pesticides don't stay put.* According to the U.S. Department of Agriculture (USDA), no more than 2% (and often less than 0.1%) of the insecticides applied to crops by aerial spraying (Figure 16-4) or by ground spraying actually reaches the target pests; less than 5% of herbicides applied to crops reaches the target weeds.

Pesticides that miss their target pests end up in the air, surface water, groundwater, bottom sediments, food, and nontarget organisms, including humans and wildlife. Pesticide waste and mobility can be reduced by **(1)** using recirculating sprayers, **(2)** covering spray booms to reduce drift, and **(3)** using rope-wick applicators (which deliver herbicides directly to weeds and reduce herbicide use by 90%). Crops that have been genetically altered to release small amounts of pesticides can help overcome this problem but can also increase the rate of genetic resistance to such pesticides (Pro/Con, p. 396).

Some pesticides can harm wildlife. During the 1950s and 1960s, populations of fish-eating birds such as the osprey, cormorant, brown pelican, and bald eagle plummeted. Research indicated that a chemical derived from DDT, when biologically magnified in food webs (Figure 8-2), made the birds' eggshells so fragile that they could not successfully reproduce. Also hard-hit were such predatory birds as the prairie falcon, sparrow hawk, and peregrine falcon (Figure 16-5), which help control rabbits, ground squirrels, and other crop-eaters.

Since the U.S. ban on DDT in 1972, most of these species have made a comeback. In 1980, however, DDT levels were again rising in the peregrine falcon and the osprey. Scientists believe that the birds may be picking up DDT and other banned pesticides in Latin America, where they winter and where use of such chemicals is still legal. Illegal use of banned pesticides in the United States, as well as airborne drift of long-lived pesticides from developing countries onto U.S. land and surface water, may also play a role. Recently, the EPA found DDT in 99% of the freshwater fish it tested.

Figure 16-4 A crop duster spraying an insecticide on grapevines south of Fresno, California. Aircraft are used to apply about 25% of the pesticides used on U.S. cropland, but only 0.1–2% of these insecticides actually reaches the target pests. To compensate for the drift of pesticides from target to nontarget areas, aircraft apply up to 30% more pesticide than ground-based application does. (National Archives/EPA Documerica)

Each year 20% of U.S. honeybee colonies are wiped out by pesticides, and another 15% are damaged, costing farmers at least $200 million per year from reduced pollination of vital crops. Pesticide runoff from cropland, a leading cause of fish kills worldwide, kills 6–14 million fish each year in the United States. Each year pesticides applied legally to U.S. farmland kills more than 67 million birds. According to the U.S. Fish and Wildlife Service, pesticides menace about 20% of the endangered and threatened species in the United States.

How Can Pesticides Threaten Human Health? According to the World Health Organization and the UN Environment Programme, an estimated 25 million agricultural workers in developing countries are seriously poisoned by pesticides each year, resulting in an estimated 220,000 deaths. Health officials believe that the actual number of pesticide-related illnesses and deaths among the world's farm workers is probably

Figure 16-5 Until 1994 the peregrine falcon was listed as an endangered species in the United States, mostly because DDT caused their young to die before hatching because the eggshells were too thin to protect them. Only about 120 peregrine falcons were left in the lower 48 states in 1975; today there are about 1,990, most of them bred in captivity and then released into the wild in a $2.7-million-per-year recovery program. Because of the success of this program, the peregrine falcon in the United States has been removed from the list of endangered and threatened species. (Hans Reinhard/Bruce Coleman Ltd.)

greatly underestimated because of **(1)** poor records, **(2)** lack of doctors and disease reporting in rural areas, and **(3)** faulty diagnoses.

Farm workers in developing countries are especially vulnerable to pesticide poisoning because three-quarters of all pesticides are applied by hand. In addition, **(1)** educational levels are low, **(2)** warnings are few, **(3)** pesticide regulations are lax or nonexistent, and **(4)** use of protective equipment is rare (especially in the hot and humid tropics). Workers' clothing also spreads pesticide contamination to family members, and it is common in developing countries for children to work or play in fields treated with pesticides. To make matters worse, some families reuse pesticide containers to store food and drinking water.

Some of these same conditions exist for farm workers in parts of the United States and in other developed countries. In the United States, for example, it is estimated that at least 300,000 farm workers suffer from pesticide-related illnesses each year. According to health officials, at least 30,000 of these pesticide poisonings are acute, and typically 25 U.S. farm workers die every year from such poisonings. Every year about 110,000 Americans, mostly children, get sick from home misuse or unsafe storage of pesticides and about 20 die.

DDT and related persistent chlorinated hydrocarbon pesticides have been banned in the United States and most developed countries. However, they were largely replaced by organophosphate pesticides (Table 16-1) that, although considerably less persistent, are much more toxic to humans and have led to increased poisonings and deaths among farm workers.

In 1995 the Environmental Working Group estimated that about 13% of vegetables and fruits consumed in the United States may contain illegal pesticides and levels of approved pesticides above their legally allowed limits. These estimates may be too low because **(1)** U.S. Food and Drug Administration (FDA) inspectors check less than 1% (about 12,000 samples) of domestic and imported food for pesticide contamination, and **(2)** measuring techniques by FDA laboratories can detect only about half of the legal and illegal pesticides in use. Moreover, the FDA's turnaround time for food analysis is so long that the food often has been sold and eaten before contamination is detected. Even when contaminated food is found, the growers and importers are rarely penalized.

In 1987 the EPA ranked pesticide residues in foods as the third most serious environmental health threat in the United States in terms of cancer risk. According to the EPA, approximately 165 of the active ingredients approved for use in U.S. pesticide products are known or suspected human carcinogens. By June 1999 use of only 41 of these chemicals had been banned by the EPA or discontinued voluntarily by manufacturers. A 1993 study of Missouri children revealed a statistically significant correlation between childhood brain cancer and use of various pesticides in the home, including **(1)** flea and tick collars, **(2)** no-pest strips, and **(3)** chemicals used to control pests such as roaches, ants, spiders, mosquitoes, and termites.

A 1995 study also found that children whose yards were treated with pesticides (mostly herbicides) were four times more likely to suffer from cancers of muscle and connective tissues than children whose yards were not treated. The study also found that children whose homes contained pest strips (with dichlorovos) faced 2.5–3 times the risk of leukemia as children whose homes didn't contain such strips.

Some scientists are becoming increasingly concerned about possible **(1)** genetic mutations, **(2)** birth defects, **(3)** nervous system disorders (especially behavioral disorders), and **(4)** effects on the immune and endocrine systems from long-term exposure to low levels of various pesticides (Section 8-2 and Connections, p. 185). Very little research has been conducted on these potentially serious threats to human health.

Accidents and unsafe practices in pesticide manufacturing plants can expose workers, their families, and sometimes the general public to harmful levels of pesticides or toxic chemicals used in their manufacture (Case Study, right).

A Black Day in Bhopal

CASE STUDY

December 2, 1984, will long be a black day for India because on that date the world's worst industrial accident occurred at a Union Carbide pesticide plant in Bhopal, India.

Some 36 metric tons (40 tons) of highly toxic methyl isocyanate (MIC) gas, used in the manufacture of carbamate pesticides, leaked from an underground storage tank. When water entered the tank through leaky valves and corroded pipes, its cooling system failed. This caused the overheated reaction mixture to explode. Once in the atmosphere, some of the toxic MIC was converted to even more deadly hydrogen cyanide gas.

The toxic cloud of gas settled over about 78 square kilometers (30 square miles), exposing up to 600,000 people, many of them illegal squatters living near the plant because they had no other place to go.

According to Indian officials, at least 5,100 people (some say 7,000–15,000, based on the sales of shrouds and cremation wood) were killed; according to a 1996 report by the International Medical Commission on Bhopal, another 50,000 to 60,000 sustained permanent injuries such as blindness or lung damage.

The Indian Supreme Court ordered Union Carbide to pay a $470-million settlement. However, the Indian government challenged the ruling, arguing that the settlement was inadequate. In 1991, the court upheld the settlement amount. In December 1999, survivors and next-of-kin of people killed by the accident filed a lawsuit in a New York U.S. district court, charging the Union Carbide Corporation and its former CEO with violating international law and the fundamental human rights of the victims and survivors of the Bhopal plant accident.

Union Carbide probably could have prevented this tragedy, which cost an estimated $4.1 billion (not including the tragic loss of life and serious health effects) by spending no more than $1 million to improve plant safety.

Critical Thinking

Did Union Carbide behave responsibly or irresponsibly in this matter with respect to **(a)** those killed and injured and **(b)** its stockholders, who saw their stock value rise because of the relatively low settlement the company had to pay? Explain. How could future disasters like this be prevented?

16-4 PESTICIDE REGULATION IN THE UNITED STATES

Is the Public Adequately Protected? The Federal Insecticide, Fungicide, and Rodenticide Act (FIFRA) requires that all commercial pesticides be approved by the EPA for general or restricted use. Pesticide companies first evaluate the biologically active ingredients in their products. EPA officials then review these data. In the late 1970s, the EPA discovered that data used to support registrations for more than 200 pesticide active ingredients had been falsified by a now-defunct laboratory.

When a pesticide is legally approved for use on fruits or vegetables, the EPA sets a *tolerance level* specifying the amount of toxic pesticide residue that can legally remain on the crop when the consumer eats it. In 1993 the National Academy of Sciences recommended that the legally allowed tolerance levels for all active ingredients in pesticides be reduced to one-tenth of their previous levels to help protect infants and children, who are more vulnerable to such chemicals than adults. By early 2000 this had not been done.

Between 1972 and 1999 the EPA banned or severely restricted the use of 57 active pesticide ingredients. The banned chemicals include **(1)** most chlorinated hydrocarbon insecticides, **(2)** several carbamates and organophosphates, and **(3)** the systemic herbicides 2,4,5-T and Silvex (Table 16-1). However, banned or unregistered pesticides may be manufactured in the United States and exported to other countries (Connections, p. 420).

FIFRA required the EPA to reevaluate the more than 600 active ingredients approved for use in pre-1972 pesticide products to determine whether any of them caused cancer, birth defects, or other health risks. However, by 1999—27 years after Congress ordered EPA to evaluate these chemicals—less than 10% of these active ingredients had been evaluated fully. The EPA contends that it hasn't been able to complete this evaluation because of the difficulty and expense of evaluating the health effects of chemicals (Section 8-2).

Environmentalists accuse the EPA of not moving fast enough to protect public safety. However, farm groups and pesticide companies say the EPA is moving too fast. They support legislation requiring the EPA to develop more laboratory data before banning more pesticides.

According to a National Academy of Sciences study, federal laws regulating the use of pesticides in the United States are inadequate and poorly enforced by the EPA, FDA, and USDA. The study also concluded that up to 98% of the potential risk of developing cancer from pesticide residues on food grown in the United States would be eliminated if EPA standards were as strict for pre-1972 pesticides as they are for later ones.

CONNECTIONS

The Circle of Poison

U.S. pesticide companies can make and export to other countries pesticides that have been banned or severely restricted—or never even approved—in the United States. Between 1992 and 1997, U.S. exports of such pesticides (most to developing countries) averaged over 450 metric tons (500 tons) per day. But what goes around can come around.

In what environmentalists call a *circle of poison,* residues of some of these banned or unapproved chemicals can return to the United States in or on imported items such as coffee, cocoa, pineapples, and out-of-season melons, tomatoes, and grapes. More than one-fourth of the produce (fruits and vegetables) consumed in the United States is grown overseas. Persistent pesticides such as DDT can also be carried by winds from other countries to the United States.

Environmentalists have urged Congress—without success—to break this deadly circle. Supporters of pesticide exports argue that such sales increase economic growth and provide jobs. They also contend that if the United States didn't export pesticides, other countries would.

In 1998, more than 100 countries began talks to finalize an international treaty curbing trade in potentially lethal chemicals such as mercury and its compounds and pesticides such as DDT and organophosphates. A related problem is what to do with the more than 91,000 metric tons (100,000 tons) of obsolete pesticide stockpiles in countries around the world.

Critical Thinking

Should U.S. companies be allowed to export pesticides that have been banned, severely restricted, or not registered for use in the United States? Explain.

According to a 1987 study by the National Academy of Sciences, exposure to pesticide residues in food causes 4,000–20,000 cases of cancer per year in the United States. Because roughly 50% of those getting cancer die prematurely, this amounts to about 2,000–10,000 premature deaths per year in the United States from exposure to legally allowed pesticide residues in foods.

Representatives from pesticide companies dispute these findings. Indeed, the food industry denies that anyone in the United States has ever been harmed by eating food that has been grown using pesticides for the past 50 years.

Since 1987 the EPA has been evaluating the potential health effects of the 1,820 so-called *inert* ingredients (such as chlorinated hydrocarbon solvents) that make up 80–99% by weight of pesticide products. It has labeled 100 of them "of known or potential toxicological concern" but has not yet banned their use.

FIFRA allows the EPA to leave inadequately tested pesticides on the market and to license new chemicals without full health and safety data. It also gives the EPA unlimited time to remove a chemical, even when its health and environmental risks are shown to outweigh its economic benefits. The built-in appeals and other procedures often keep a dangerous chemical on the market for up to 10 years. FIFRA is the only major environmental statute that does not allow citizens to sue the EPA for not enforcing the law.

A 1993 study of pesticide safety by the U.S. National Academy of Sciences urged the government to do the following things:

- Make human health the primary consideration for setting limits on pesticide levels allowed in food.
- Collect more and better data on pesticide exposure for different groups, including farm workers, adults, and children.
- Develop new and better test procedures for evaluating the toxicity of pesticides, especially for children.
- Consider cumulative exposures of all pesticides in food and water, especially for children, instead of basing regulations on exposure to a single pesticide.

Progress was made with the passage of the 1996 Food Quality Protection Act:

- It requires new standards for pesticide tolerance levels in foods, based on a reasonable certainty of no harm to human health (defined for cancer as producing no more than one additional cancer per million people exposed to a certain pesticide over a lifetime).
- It requires manufacturers to demonstrate that the active ingredients in their pesticide products are safe for infants and children.
- It allows the EPA to apply an additional 10-fold safety factor to pesticide tolerance levels to protect infants and children.
- It requires the EPA to consider exposure to more than one pesticide when setting pesticide tolerance levels.
- It requires the EPA to develop rules for a program to screen all active and inactive ingredients for their estrogenic and endocrine effects by 1999 (which has not been achieved).

Pesticide control laws in the United States can be improved, but most other countries (especially developing countries) have little protection against the potentially harmful effects of pesticides.

16-5 SOLUTIONS: OTHER WAYS TO CONTROL PESTS

How Can Cultivation Practices Control Pests? The following cultivation practices can be used to help reduce pest damage:

- Changing the type of crop planted in a field each year (crop rotation)
- Planting rows of hedges or trees around fields to hinder insect invasions and provide habitats for their natural enemies (with the added benefit of reduced soil erosion, Figure 13-24d)
- Adjusting planting times so that major insect pests either starve or get eaten by their natural predators
- Growing crops in areas where their major pests do not exist
- Planting trap crops to lure pests away from the main crop
- Switching from vulnerable monocultures to intercropping, agroforestry, and polyculture, which use plant diversity to reduce losses to pests (p. 390)
- Removing diseased or infected plants and stalks and other crop residues that harbor pests
- Using plastic that degrades slowly in sunlight to keep weeds from sprouting between crop rows
- Using vacuum machines to gently remove harmful bugs from plants

With the rise of industrial agriculture and the greatly increased use of synthetic pesticides over the last 40 years, many farmers in developed countries no longer use many of these traditional cultivation methods for controlling pest populations. However, as the problems and expense of using pesticides have risen since 1980, more farmers are returning to these methods.

How Can Genetically Resistant Plants and Crops Help Lower Pest Losses? Plants and animals that are genetically resistant to certain pest insects, fungi, and diseases can be developed. However, resistant varieties usually take a long time (10–20 years) and lots of money to develop by conventional crossbreeding methods. Moreover, insects and plant diseases can develop new strains that attack the once-resistant varieties, forcing scientists to continually develop new resistant strains. Genetic engineering is now helping to speed up this process (Figure 16-6 and Pro/Con, p. 396).

There has been much talk about using genetic engineering to build pest resistance into crops and thus reduce the need for pesticides. However, most biotech companies have been focusing their genetic engineering research on developing crop strains that are resistant to herbicides, a strategy that increases the use and sale of herbicides (Pro/Con, p. 396).

How Can Natural Enemies Help Control Pests? Biological control using predators (Figure 16-1 and 16-7), parasites, and pathogens (disease-causing bacteria and viruses) can be encouraged or imported to regulate pest populations. More than 300 **biological pest control** projects worldwide have been successful, especially in China (p. 411). In Nigeria, crop-duster planes release parasitic wasps instead of pesticides to fight the cassava mealybug; farmers get a $178 return for every $1 they spend on the wasps. In the United States, natural enemies have been used to control more than 70 insect pests and have saved farmers an average of $25 for every $1 invested.

Biological control has several advantages. It focuses on selected target species and is nontoxic to other species, including people. Once a population of natural predators or parasites is established, biological pest control often can be self-perpetuating. Development of genetic resistance is minimized because pest and predator species interact and change together (coevolution).

However, years of research may be needed to understand how a particular pest interacts with its various enemies and to choose the best biological control

Figure 16-6 The results of one example of genetic engineering against pest damage. Both tomato plants were exposed to destructive caterpillars. The normal plant's leaves are almost gone (left), whereas the genetically altered plant (right) shows little damage. (Monsanto)

Figure 16-7 Biological control of pests: An adult convergent ladybug (right) is consuming an aphid (left). (Peter J. Bryant/Biological Photo Service)

agent. Biological agents can't always be mass-produced, and farmers find them slower acting and more difficult to apply than pesticides. Biological agents must also be protected from pesticides sprayed in nearby fields.

Once released into the environment, biological control agents can multiply, cause unpredictable harmful ecological effects, and be impossible to recapture. Some may even become pests themselves; others (such as praying mantises) devour beneficial and pest insects alike.

How Can Biopesticides Help Control Pests? Botanicals such as synthetic pyrethroids (Table 16-1) are an increasingly popular method of pest control. Microbes are also being drafted for insect wars, especially by organic farmers. For example, *Bacillus thuringensis (Bt)* toxin is a registered pesticide sold commercially as a dry powder. Each of the thousands of strains of this common soil bacterium kills a specific pest. Various strains of *Bt* are used by almost all organic farmers as a nonchemical pesticide. The bad news is that genetic resistance is developing to some *Bt* toxins, especially because *Bt* genes have been transferred to widely planted genetically altered crops such as corn, potatoes, and soybeans (Pro/Con, p. 396).

How Can Insect Birth Control, Sex Attractants, and Hormones Help Control Pests? Males of some insect pest species can be **(1)** raised in the laboratory, **(2)** sterilized by radiation or chemicals, and **(3)** released into an infested area to mate unsuccessfully with fertile wild females. Problems with this approach include **(1)** high costs, **(2)** difficulties in knowing the mating times and behaviors of each target insect, **(3)** the large number of sterile males needed, and **(4)** the few species for which this strategy works.

The USDA used the sterile-male approach to essentially eliminate the screwworm fly, a major livestock pest (Figure 16-8), from the southeastern states between 1962 and 1971. To prevent resurgence of this pest, new strains of sterile male flies must be released every few years.

In many insect species, a female that is ready to mate releases a minute amount (typically about one-millionth of a gram) of a chemical sex attractant called a *pheromone*. Whether extracted from insects or synthesized in the laboratory, pheromones can lure pests into traps or attract their natural predators into cropfields (usually the more effective approach). More than 50 companies worldwide sell about 250 pheromones to control pests (Figure 16-9). These chemicals **(1)** attract only one species, **(2)** work in trace amounts, **(3)** have little chance of causing genetic resistance, and **(4)** are not harmful to nontarget species. However, it is costly and time-consuming to identify, isolate, and produce the specific sex attractant for each pest or predator.

Each step in the insect life cycle is regulated by the timely natural release of juvenile hormones (JH) and molting hormones (MH) (Figure 16-10). These chemicals, which can either be extracted from insects or synthesized in the laboratory, can disrupt an insect's normal life cycle, causing the insect to fail to reach maturity and reproduce (Figure 16-11). Insect hormones have the same advantages as sex attractants, but they **(1)** take weeks to kill an insect, **(2)** often are ineffective with large infestations of insects, **(3)** sometimes break down before they can act, and **(4)** must be applied at exactly the right time in the target insect's life cycle. Moreover, they sometimes affect the target's predators and other nonpest species, and they can kill crustaceans if they get into aquatic ecosystems. Finally, like sex attractants, they are difficult and costly to produce.

How Can Hot Water Be Used to Zap Pests? Recently some farmers have begun *zapping pests with hot water* by using a machine that sprays boiling water on crops to kill weeds and insects. So far, the system has worked well on cotton, alfalfa, and potato fields and in citrus groves in Florida, where the machine was invented. The cost is roughly equal to that of using chemical pesticides.

How Can Zapping Foods with Radiation Help Control Pests? Some pests can also be killed by *zapping foods after harvest with gamma radiation*. Such food irradiation extends food shelf life and kills **(1)** insects, **(2)** parasitic worms (such as trichinae in pork), and **(3)** bacteria (such as salmonella, which infects at least 51,000 Americans and kills 2,000 each year, and *E. coli*, which infects more than 20,000 Americans and kills about 250 each year). A food does not become radioactive when it is irradiated, just as exposure to X rays does not make the body radioactive.

Hargrove, T. R. 1999. "Wrangling Over Refuge." *American Scientist*, vol. 86, no. 1, p24(2).

Figure 16-8 Infestation of a steer by screwworm fly larvae in Texas. An adult steer can be killed in 10 days by thousands of maggots feeding on a single wound. (U.S. Department of Agriculture)

Figure 16-9 Pheromones can help control populations of pests, such as the red scale mites that have infested this lemon grown in Florida. (Agricultural Research Service/USDA)

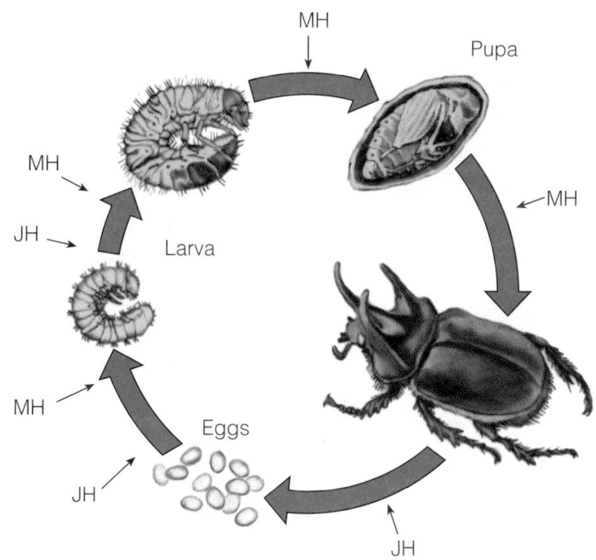

Figure 16-10 For normal insect growth, development, and reproduction to occur, certain juvenile hormones (JH) and molting hormones (MH) must be present at genetically determined stages in the insect's life cycle. If applied at the proper time, synthetic hormones disrupt the life cycles of insect pests and help control their populations.

According to the U.S. FDA and the World Health Organization, more than 2,000 studies show that foods exposed to low doses of ionizing radiation are safe for human consumption. Currently, 37 countries (8 of them in western Europe) allow irradiation of one or more food items. Proponents of this technology contend that irradiation of food is likely to lower health hazards to people and reduce pesticide use and that its potential benefits greatly exceed the risks.

Critics contend that irradiating food destroys some of its vitamins and other nutrients and that the irradiation process may form trace amounts of toxic, possibly carcinogenic compounds in the food. However, proponents of food irradiation point out that fresh foods lose similar amounts of vitamins when they are cooked, canned, or frozen. Critics also contend that more studies are needed to ensure the safety and wholesomeness of irradiated food and that it is too soon to see possible long-term effects, which might not show up for 30–40 years. They also argue that Americans want fresh, wholesome food, not old, possibly less nutritious food made to appear fresh and healthy by irradiation.

According to critics, irradiating food will endanger the health and safety of food irradiation workers and could also increase deaths from botulism poisoning. Current levels of irradiation do not destroy botulinum spores, but they do destroy the bacteria that give off the rotten odor warning us of their presence. Furthermore, some microorganisms could mutate when exposed to radiation, possibly creating new and more dangerous forms of botulism.

The FDA has approved the use of irradiation on poultry, pork, beef, spices, and all domestic fruits and vegetables to delay sprouting, kill insect pests, and slow ripening. However, New York, New Jersey, and Maine have prohibited the sale and distribution of irradiated food, as have Germany, Austria, Denmark, Sweden, Switzerland, Sudan, Singapore, Australia, and New Zealand.

Of course, the best food-processing system in the world can't protect us from ourselves. More than 75% of reported food-poisoning cases in the United States can be traced to improper food handling by consumers or food preparers in restaurants.

How Can Integrated Pest Management Help Control Pests? An increasing number of pest control experts and farmers believe that the best way to control crop pests is a carefully designed **integrated pest management (IPM)** program. In this approach, each crop and

Hint: Enter the search term *biopesticides* using the Subject Guide.

Figure 16-11 A use of hormones to prevent insects from maturing completely, making it impossible for them to reproduce. The stunted tobacco hornworm (left) was fed a compound that prevents production of molting hormones; a normal hornworm is shown on the right. (Agricultural Research Service/USDA)

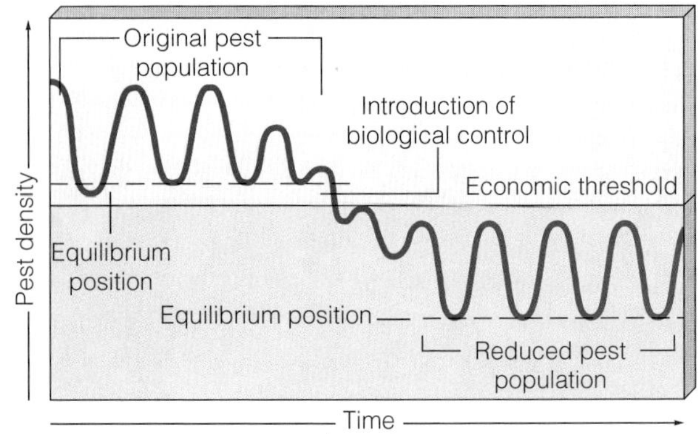

Figure 16-12 The goal of integrated pest management is to keep each pest population just below the size at which it causes economic loss.

its pests are evaluated as parts of an ecological system. Then a control program is developed that includes a mix of cultivation and biological and chemical methods applied in proper sequence and with the proper timing.

The overall aim of IPM is not eradication of pest populations, but reduction of crop damage to an economically tolerable level (Figure 16-12). Fields are carefully monitored; when a damaging level of pests is reached, farmers first use biological methods (natural predators, parasites, and disease organisms) and cultivation controls, including vacuuming up harmful bugs. Small amounts of insecticides (mostly botanicals or microbotanicals) are applied only as a last resort, and different chemicals are used to slow development of genetic resistance and to avoid killing predators of pest species.

In 1986 the Indonesian government **(1)** banned the use of 57 of the 66 pesticides used on rice, **(2)** phased out pesticide subsidies over a 2-year period, and **(3)** used some of the money to help launch a nationwide program to switch to IPM, including a major farmer education program. The results were dramatic: Between 1987 and 1992, **(1)** pesticide use dropped by 65%, **(2)** rice production rose by 15%, and **(3)** more than 250,000 farmers were trained in IPM techniques. By 1993 the program had saved the Indonesian government over $1.2 billion (most from elimination of $120-million-per-year pesticide subsidies)—more than enough to fund its IPM program.

The experiences of countries such as China, Brazil, Indonesia, Australia, and the United States show that a well-designed IPM program can **(1)** reduce pesticide use and pest control costs by 50–90%, **(2)** reduce preharvest pest-induced crop losses by 50%, **(3)** improve crop yields, **(4)** reduce inputs of fertilizer and irrigation water, and **(5)** slow the development of genetic resistance because pests are assaulted less often and with lower doses of pesticides. Thus, IPM is an important form of pollution prevention that reduces risks to wildlife and human health. Consumers Union estimates that if all U.S. farmers practiced IPM by 2020, public health risks from pesticides would drop by 75%.

However, **(1)** IPM requires expert knowledge about each pest situation, **(2)** it is slower acting than conventional pesticides, and **(3)** methods developed for a crop in one area might not apply to areas with even slightly different growing conditions. Although long-term costs typically are lower than the costs of using conventional pesticides, initial costs may be higher.

Despite its promise and growth, widespread use of IPM is hindered by government subsidies of conventional chemical pesticides and by opposition from agricultural chemical companies, whose sales would drop sharply. In addition, farmers get most of their information about pest control from pesticide salespeople (and in the United States from USDA county farm agents), few of whom have adequate training in IPM.

In 1996 a study by the National Academy of Sciences recommended that the United States shift from chemically based approaches to ecologically based approaches to pest management. A growing number of scientists urge the USDA to promote IPM in the United States by **(1)** adding a 2% sales tax on pesticides and using the revenue to fund IPM research and education, **(2)** setting up a federally supported IPM demonstration project on at least one farm in every U.S. county, **(3)** training USDA field personnel and county farm agents in IPM so that they can help farmers use this alternative, **(4)** providing federal and state subsidies, and perhaps government-backed crop insurance, to farmers who use IPM or other approved alternatives to pesticides, and **(5)** gradually phasing out subsidies to farmers who depend almost entirely on pesticides as effective IPM methods are developed for major pest species. Some actions you can take to reduce your use of and exposure to pesticides are listed in Appendix 6.

We need to recognize that pest control is basically an ecological, not a chemical problem.

ROBERT L. RUDD

REVIEW QUESTIONS

1. Define the boldfaced terms used in this chapter.

2. How can spiders be useful in controlling insect pests?

3. What is a *pest*? How are species we consider pests controlled in nature? What happens to this natural pest control service when we simplify ecosystems? How have plants evolved to protect themselves from herbivores?

4. Distinguish between *insecticides, herbicides, fungicides, nematocides,* and *rodenticides.* Distinguish between *first-generation pesticides* and *second-generation pesticides* used by people to combat pests, and give two examples of each type. Distinguish between *broad-spectrum* and *narrow-spectrum pesticides.*

5. Summarize the role of Rachel Carson in alerting people to the potential dangers from widespread use of pesticides.

6. List six benefits of the use of pesticides. List five traits of the ideal pest-killing chemical.

7. Summarize the harmful effects of pesticides in terms of **(a)** development of genetic resistance, **(b)** killing natural predators and parasites that help control pest species, **(c)** creation of new pest species, **(d)** migration of pesticides into the environment, **(e)** effects on wildlife, and **(f)** effects on the human health of pesticide and farm workers and on the general public.

8. What are the *pesticide treadmill* and the *genetic treadmill,* and what causes these effects? What are the alternatives?

9. How are pesticides regulated in the United States? List major weaknesses in these laws. According to the U.S. National Academy of Sciences, what three things should be done to minimize the harmful effects of pesticides?

10. Describe the events that took place at the Dow pesticide plant in Bhopal, India.

11. With respect to pesticides, what is the *circle of poison*? How might this affect you?

12. List nine cultivation practices that can be used to help control pests. Why aren't these methods widely used today?

13. List the pros and cons of the following methods of controlling pests: **(a)** development of plants and crops that are genetically resistant to pests, **(b)** use of crops genetically modified to tolerate higher doses of herbicides or to release pesticides (transgenics), **(c)** biological control, **(d)** biopesticides, **(e)** insect birth control through sterilization, **(f)** insect sex attractants, **(g)** insect hormones (pheromones), **(h)** spraying with hot water, **(i)** irradiation of foods, and **(j)** integrated pest management (IPM).

14. Why do most biologists believe that the primary methods for controlling pest populations should be ecological instead of chemical?

CRITICAL THINKING

1. Should DDT and other pesticides be banned from use in malaria control efforts throughout the world? Explain. What are the alternatives?

2. Do you believe that because essentially all pesticides eventually fail, their use should be phased out or sharply reduced and that farmers should be given economic incentives for switching to IPM? Explain your position.

3. Do you agree or disagree that because DDT and the other banned chlorinated hydrocarbon pesticides pose no demonstrable threat to human health and have saved millions of lives, they should again be approved for use in the United States? Explain.

4. Should certain types of foods used in the United States be irradiated? Explain. If so, should such foods be required to carry a clear label stating that they have been irradiated? Explain.

5. What changes, if any, do you believe should be made in the Federal Insecticide, Fungicide, and Rodenticide Act regulating pesticide use in the United States?

***6.** How are bugs and weeds controlled in your yard and garden? On the grounds of your school? On public school grounds, parks, and playgrounds near your home? Consider organizing an effort to have IPM and organic fertilizers used on school and public grounds. Do the same thing for your yard and garden.

***7.** Make a concept map of this chapter's major ideas, using the section heads and subheads and the key terms (in boldface). Look at the inside back cover and in Appendix 5 for information about concept maps.

FURTHER INFORMATION

For more information see Internet Sources and Further Readings at the back of the book. Online resources for this chapter can be found at **www.brookscole.com/biology**

17 SUSTAINING TERRESTRIAL ECOSYSTEMS: FORESTS, RANGELANDS, PARKS, AND WILDERNESS

How Farmers and Loud Monkeys Saved a Forest

It's early morning in a tropical forest in the Central American country Belize (Figure 9-21). Suddenly, loud roars that trail off into wheezing moans—territorial calls of black howler monkeys (Figure 17-1)—wake up everyone in or near the wildlife sanctuary by the Belize River.

This species is the centerpiece of an experiment recruiting peasant farmers to preserve tropical forests and wildlife. The project is the brainchild of an American biologist, Robert Horwich, who met with villagers and suggested that they establish a sanctuary that would benefit the local black howlers and themselves. He proposed that the farmers leave thin strips of forest along the edges of their fields to provide the howlers with food and a travel route through the sanctuary's patchwork of active garden plots and young and mature forest.

To date, more than 100 farmers have participated, and the 47-square-kilometer (18-square-mile) sanctuary is home for an estimated 1,100 black howlers. The idea has spread to several other villages.

As many as 6,000 ecotourists visit the sanctuary each year to see its loud monkeys and other wildlife. Villagers serve as tour guides, cook meals for the visitors, and lodge tourists overnight in their spare rooms.

As long as excessive ecotourism doesn't disturb the sanctuary ecosystem, this experiment can demonstrate the benefits of integrating ecology and economics, allowing villagers to make money by helping sustain the forest and its wildlife.

Forests, rangelands, parks, wilderness, wetlands (Sections 6–5 and 6–6), and other storehouses of biodiversity are crucial natural resources that are coming under increasing pressure from population growth and economic development. Conservation biologists urge us to use these potentially renewable forms of natural capital more sustainably, as discussed in this chapter.

Figure 17-1 A black howler monkey, shown in a tropical forest in Costa Rica. The populations of six howler monkey species have declined as tropical forests have been cleared. The black howler, like more than two-thirds of the world's 150 known species of primates, is threatened with extinction. (John Lythgoe/ Planet Earth Pictures)

Forests precede civilizations, deserts follow them.
FRANCOIS-AUGUSTE-RENÉ DE CHATEAUBRIAND

This chapter addresses the following questions:

- What are the types of public lands in the United States, and how are they used?
- How should we manage and conserve forest resources?
- Are tropical deforestation and fuelwood shortages serious problems? If so, what can we do about them?
- Why are rangelands important, and how should we manage them?
- What problems do parks face, and how should we manage them?
- Why is wilderness important, and how much wilderness and other biodiversity sanctuaries should we preserve?

17-1 PUBLIC LANDS IN THE UNITED STATES

What Are the Major Types of U.S. Public Lands? No nation has set aside as much of its land—about 42%—for public use, enjoyment, and wildlife as has the United States. Almost one-third of the country's land belongs to every American and is managed for them by the federal government. About 73% of this public land is in Alaska and another 22% is in the Western states (where 60% of all land is public land). These public lands are classified as multiple-use lands, moderately restricted-use lands, and restricted-use lands.

Multiple-Use Lands

- The 156 forests (Figure 17-2) and 20 grasslands of the *National Forest System* are managed by the U.S. Forest Service. These forests are used for logging, mining, livestock grazing, farming, oil and gas extraction, recreation, sport hunting, sport and commercial fishing, and conservation of watershed, soil, and wildlife resources. Off-road vehicles usually are restricted to designated routes.
- *National Resource Lands* in the Western states and Alaska are managed by the Bureau of Land Management (BLM). Emphasis is on providing a secure domestic supply of energy and strategic minerals and on preserving rangelands for livestock grazing under a permit system.

Moderately Restricted-Use Lands

- The 508 *National Wildlife Refuges* (Figure 17-2) are managed by the U.S. Fish and Wildlife Service. Most refuges protect habitats and breeding areas for waterfowl and big game to provide a harvestable supply for hunters; a few protect endangered species from extinction. Sport hunting, trapping, sport and commercial fishing, oil and gas development, mining, logging, grazing, some military activities, and farming are permitted as long as the Department of the Interior finds such uses compatible with the purposes of each unit.

Restricted-Use Lands

- The 378 units of the *National Park System* include 54 major parks (mostly in the West, Figure 17-2) and 324 national recreation areas, monuments, memorials, battlefields, historic sites, parkways, trails, rivers, seashores, and lakeshores managed by the National Park Service. National parks may be used only for camping, hiking, sport fishing, and boating. Motor vehicles are permitted only on roads. In national recreation areas, these same activities, plus sport hunting, mining, and oil and gas drilling, are allowed.
- The 630 roadless areas of the *National Wilderness Preservation System*, which lie within the national parks, national wildlife refuges, and national forests, are managed by the National Park Service (42%), Forest Service (33%), Fish and Wildlife Service (20%), and BLM (5%). These areas are open only for recreational activities such as hiking, sport fishing, camping, nonmotorized boating, and, in some areas, sport hunting and horseback riding. Roads, logging, livestock grazing, mining, commercial activities, and buildings are banned, except when they predate the wilderness designation.

How Should We Manage Public Lands? Because of the resources they contain, there has been intense controversy over how public lands should be used and managed.

Most conservation biologists and many free-market economists believe that the following four principles should govern use of public land:

- Protection of biodiversity and the ecological integrity of vital public lands should be the primary goal.
- No one should be given subsidies or tax breaks for using or extracting resources on public lands. In recent years, the total cost of government subsidies (in public funds spent and taxes and user fees not collected) given to mining (p. 326), logging, and grazing interests using U.S. public lands has exceeded $1 billion per year.
- The American people deserve fair compensation for the use of their property.
- All users or extractors of resources on public lands should be fully responsible for any environmental damage they cause.

Figure 17-2 National forests, national parks, and wildlife refuges managed by the U.S. federal government. (Data from U.S. Geological Survey)

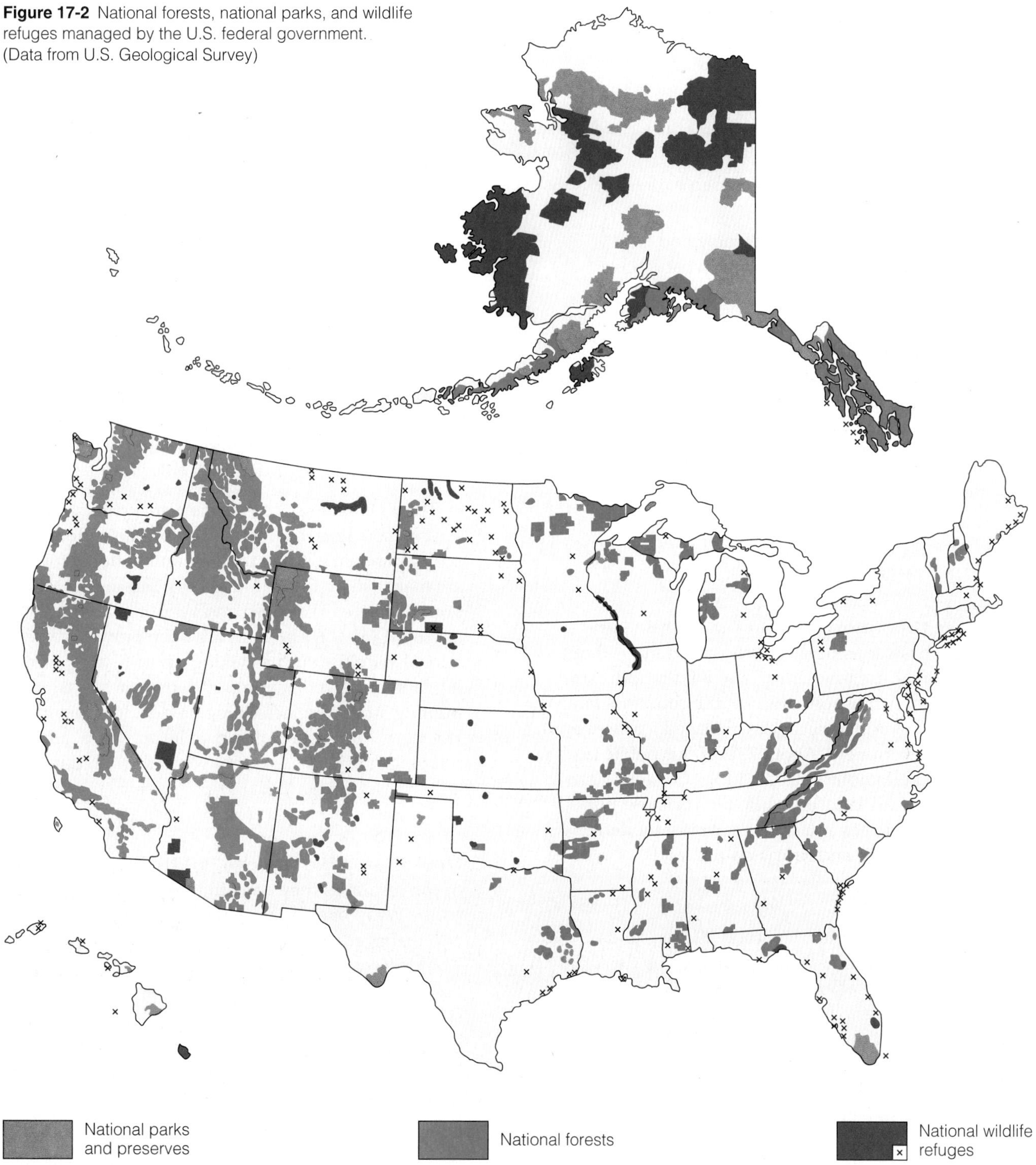

Do you agree or disagree with these principles? Why?

There is strong opposition to these ideas. Economists, developers, and resource extractors tend to view areas of the earth's surface in terms of their **(1)** usefulness in providing mineral and other resources, **(2)** potential for urban development, and **(3)** ability to increase short-term economic growth. Many people with these views have lobbied members of the U.S. Congress to pass laws that would:

- Sell public lands or their resources to corporations or individuals, usually at less than market value
- Give public lands to the states (where these interests often have more influence)

Smith, M. B. 1998. "The Value of a Tree: Public Debates of John Muir and Gifford Pinchot." *The Historian,* v60 n4 p757(22).

■ Slash federal funding for administration of public lands

■ Weaken, waive, or eliminate federal laws that regulate use of public lands

■ Require the government to reimburse private landowners for regulations that decrease the economic value of their land

■ Allow highways to be built through national parks and wilderness areas

■ Redefine protected wetlands so that about half of them would no longer be protected

■ Repeal or seriously weaken the Endangered Species Act.

Do you agree or disagree with such proposals? Explain.

17-2 MANAGING AND SUSTAINING FORESTS

What Are the Major Types of Forests? There are three general types of forests, based primarily on climate: *tropical*, *temperate*, and *polar (boreal)* (Figure 6-10), described in detail in Section 6-4. Since agriculture began about 10,000 years ago, human activities have reduced the earth's forest cover by about one-quarter—from about 34% to 26% of the world's land area.

Forests are potentially renewable resources as long as the rate of cutting and degradation does not exceed the rate of regrowth and biodiversity protection is emphasized. However, forests are being cut or fragmented and degraded faster than they can regenerate in many parts the world, especially in the tropics (Figure 17-3). Deforestation, like agriculture, is ultimately an example of the increasing appropriation of the net primary productivity of natural ecosystems (Figure 4-21) for human needs and wants.

Old-growth forests are uncut forests or regenerated forests that have not been seriously disturbed by human activities or natural disasters for several hundred years. About 70% of the world's remaining old-growth forests are found as boreal forests (Figures 6-9 and 6-20) in Russia and Canada and as tropical forests (Figures 6-9 and 6-17). Two-thirds of Europe's old-growth forests have been cut and in the United States 95–98% have been cut, mostly in the late 1800s and early 1900s.

Old-growth forests provide ecological niches for a multitude of wildlife species (Figure 6-18). These forests also have large numbers of standing dead trees (snags) and fallen logs, which provide habitats for a variety of species. Decay of this dead vegetation returns plant nutrients to the soil (Figure 4-12).

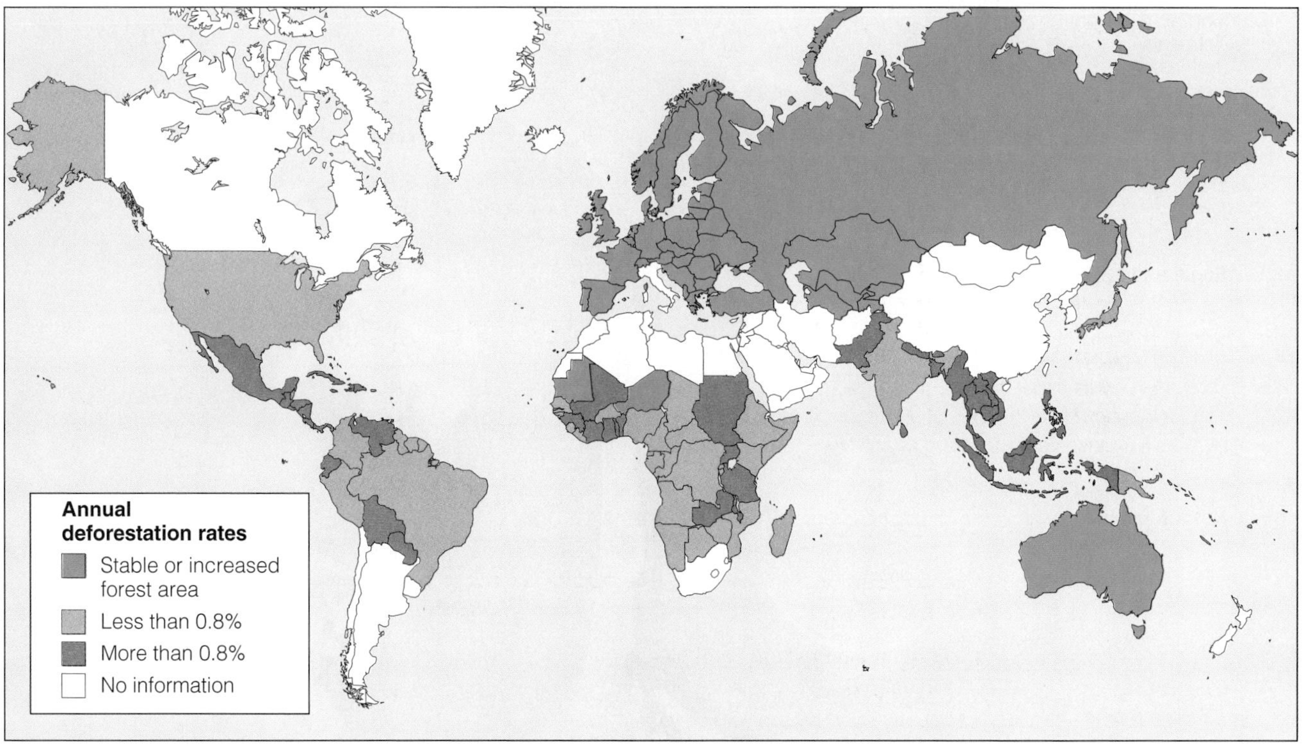

Figure 17-3 Estimated annual rates of deforestation, 1981–90. During this period the area of the world's tropical forest decreased by 8% (an average of 0.8% per year). This rate of deforestation has continued during the 1990s. Currently, global forests shrink at a net rate of at least 113,000 square kilometers (44,000 square miles) per year and an equal amount is degraded each year. (UN Food and Agriculture Organization)

Second-growth forests are stands of trees resulting from secondary ecological succession after the trees in an area have been removed by **(1)** human activities (clear-cutting for timber or conversion to cropland or urban land) or **(2)** natural forces such as fire, hurricanes, or volcanic eruptions (Figure 5-15). Most forests in the United States and other temperate areas are second-growth forests that grew back after virgin forests were cleared for timber or to create farms and ranches that were later abandoned. About 40% of tropical forests are second-growth forests. Some second-growth stands have remained undisturbed long enough to become old-growth forests and others have been cleared and converted to **tree farms** or **plantations** (see photo in table of contents, p. xiii). Such farms are managed tracts with uniformly aged trees of one species that are harvested by clear-cutting as soon as they become commercially valuable. Then they are replanted and clear-cut again on regular cycles.

Tree farms are attempts to produce timber and pulpwood more efficiently by setting back secondary ecological succession (Figure 5-15) to create a greatly simplified ecosystem. This is the same approach used to produce more food per hectare by planting monoculture crops. Although tree farms reduce biodiversity where they are planted, they can help preserve biodiversity elsewhere by reducing pressures to cut down old-growth and second-growth forests.

What Is the Economic and Ecological Importance of Forests? Forests provide many economic benefits. They provide **(1)** lumber for housing, **(2)** biomass for fuelwood, **(3)** pulp for paper, **(4)** medicines, and many other products (Figure 17-4) worth more than $300 billion a year. Many forestlands are also used for mining, grazing livestock, and recreation.

Worldwide, about 55% of the timber cut each year is used as fuelwood or converted to charcoal fuel for heating and cooking, especially in developing countries. About 30% of the world's annual harvest is sawlogs that are converted into building materials such as lumber, plywood, particleboard, and chipboard. Some 16% of the world's annual wood harvest is converted into pulp used in a variety of paper products.

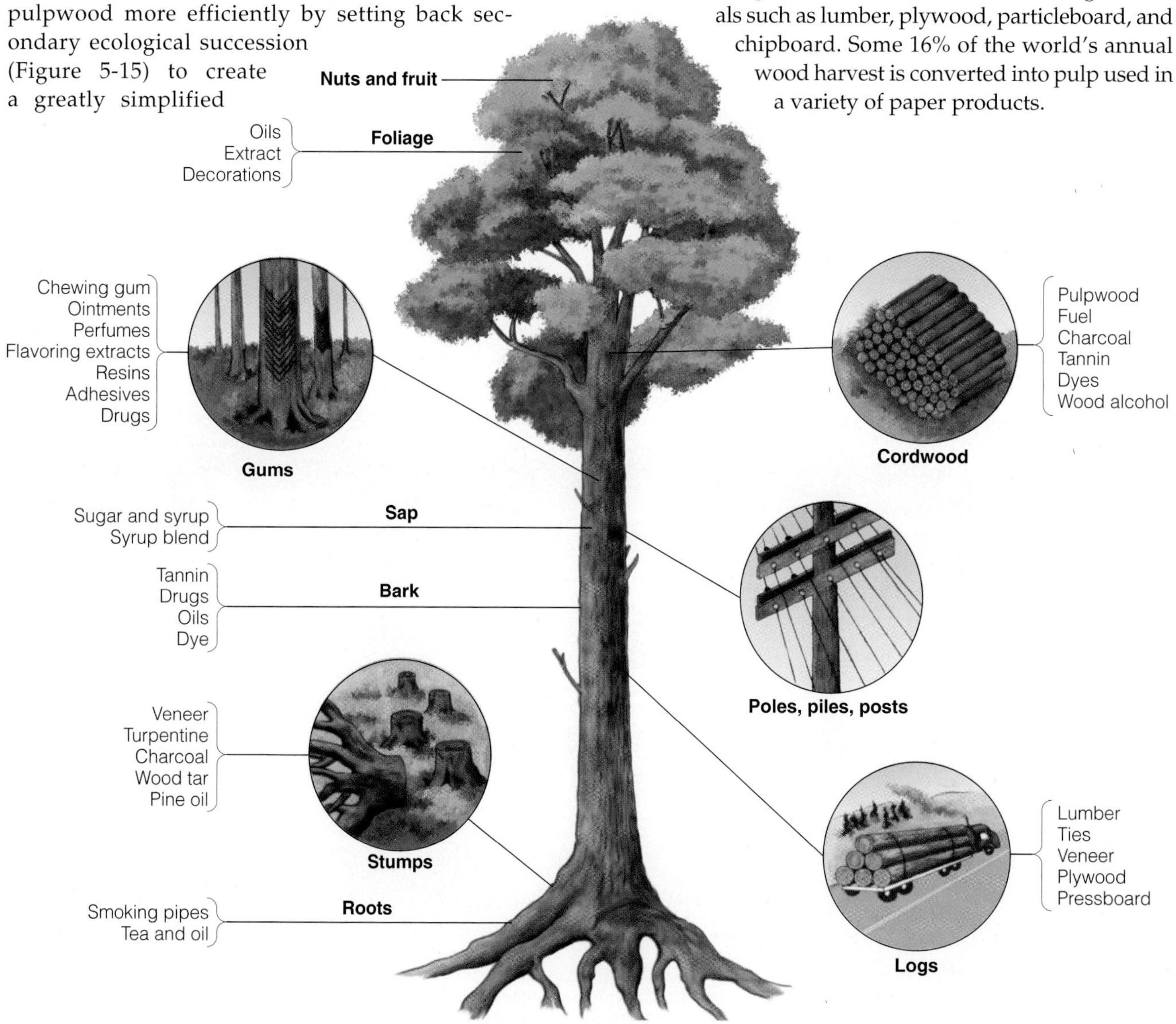

Figure 17-4 Some of the many useful products obtained from trees.

Forests also provide numerous ecological services by

- Supporting food webs and energy flow (Section 4-4) and cycling nutrients (Section 4-5) in forest ecosystems.
- Acting as giant sponges that **(1)** slow runoff, **(2)** hold water that recharges springs, streams, and groundwater, **(3)** regulate the flow of water from mountain highlands to croplands and urban areas, and **(4)** reduce the amount of sediment washing into streams, lakes, and reservoirs by reducing soil erosion.
- Influencing local and regional climate. For example, 50–80% of the moisture in the air above tropical forests comes from trees via transpiration and evaporation. If large areas of these forests are cleared, average annual precipitation drops and the region's climate gets hotter and drier. This process can eventually convert a diverse tropical forest into a sparse grassland or even a desert.
- Providing numerous habitats and niches for wildlife species.
- Buffering us against noise.
- Absorbing air pollutants.

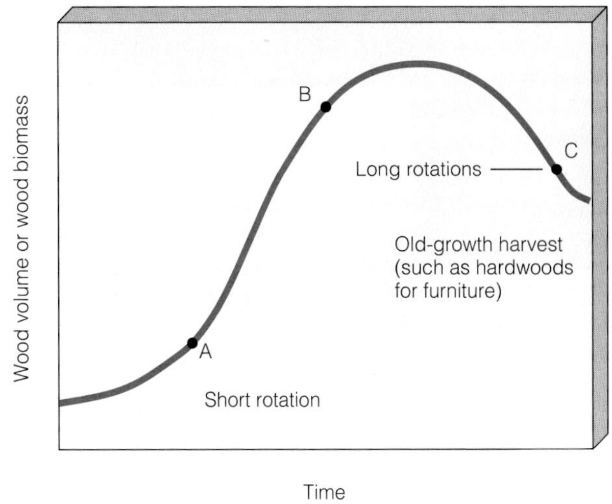

Figure 17-5 Changes in wood volume over various growth–harvest cycles in forest management. Forest management occurs over a cycle of decisions and events called a *rotation*. The most important steps in a rotation include taking an inventory of the site, developing a forest management plan, building roads into the site, preparing the site for harvest, harvesting timber, and regenerating and managing the site until the next harvest.

Planting forests sometimes is suggested as a way to remove carbon dioxide from the atmosphere to help slow projected global warming, but this effect is temporary. Generally, young growing trees remove more carbon dioxide from the atmosphere through photosynthesis than they add through aerobic respiration. However, as they age and their growth slows down, trees are not effective ways to absorb CO_2. Also, when trees die and decay or burn in a fire, the carbon they stored is released back into the atmosphere as CO_2. If CO_2 removal is the goal, then planting and repeatedly harvesting tree plantations would be the more effective approach.

Recently, ecologists and economists have attempted to place a monetary value on the ecosystem services provided by forests and other ecosystems. By estimating the immense economic value of nature's ecological services, they hope to slow down the unsustainable use and degradation of many of the world's forests and other ecosystems (Spotlight, p. 432).

What Are the Major Types of Forest Management? The total volume of wood produced by a particular stand of forest varies as it goes through different stages of growth and ecological succession (Figure 7-15). If the goal is to produce fuelwood or fiber for paper production in the shortest time, the forest usually is harvested on a short rotation cycle, well before the volume of wood produced peaks (point A in Figure 17-5). Harvesting at point B in Figure 17-5 gives the maximum yield of wood per unit of time. If the goal is high-quality wood for fine furniture or veneer, managers use longer rotations to develop larger, older-growth trees (point C in Figure 17-5), whose rate of growth has leveled off and is much lower than that of young trees.

There are two basic forest management systems: even-aged and uneven-aged. With **even-aged management**, trees in a given stand are maintained at about the same age and size. In this approach, sometimes called *industrial forestry*, a biologically diverse natural forest is replaced with a simplified *tree farm* of one or two fast-growing and economically desirable species that can be harvested every 10–100 years, depending on the species. Crossbreeding and genetic engineering can improve the quality and the quantity of tree-farm wood.

Decades of experience with intensive even-aged management of German forests has shown that in a forest in which virtually all of the trees are repeatedly cut and removed, the soil is depleted of nutrients. Trees then have little defense against stresses such as drought, diseases, and pests (Figure 10-9).

In **uneven-aged management**, a variety of tree species in a given stand are maintained at many ages and sizes to foster natural regeneration. Here the goals are **(1)** biological diversity, **(2)** long-term sustainable production of high-quality timber, **(3)** a reasonable economic return, and **(4)** multiple use of the forest for timber, wildlife, watershed protection, and recreation. Mature trees are selectively cut, and the removal of all trees is used only on small patches of species that benefit from such a practice.

How Much Is a Forest Worth?

SPOTLIGHT

In 1997 a group of ecologists, economists, and geographers attempted to estimate how much nature's ecological services (Figure 1-2) are worth. According to this crude appraisal led by ecological economist Robert Costanza of the University of Maryland, the economic value of Mother Nature's bounty ranges from $16 trillion to $54 trillion per year. In other words, each year nature provides us with free goods and services worth somewhere between $2,700 and $9,000 per person.

To make these estimates the researchers divided the earth's surface into 16 biomes (Figure 6-9) and aquatic life zones (they omitted deserts and tundra because of a lack of data). Then they agreed on a list of 17 goods and services provided by nature (Figure 1-2) and sifted through more than 100 studies that attempted to put a dollar value on such services in the 16 different types of ecosystems.

Some analysts believe that such estimates are misleading and dangerous because they put a dollar value on ecosystem services with an infinite value because they are irreplaceable. In the 1970s economist E. F. Schumacher warned that "to undertake to measure the immeasurable is absurd" and is a "pretense that everything has a price." Biologist and conservationist David Ehrenfield's response to this evaluation was, "I am afraid that I don't see much hope for a civilization so stupid that it demands a quantitative estimate of the value of its own umbilical cord."

The researchers admit that their estimates rely on many assumptions and omissions and could easily be too low by a factor of 10 to 1 million or more. For example, their calculations include only estimates of the ecosystem services themselves, not the natural capital that generates them, and omitted the value of nonrenewable minerals and fuels. They also recognize that as the supply of ecosystem services declines their value will rise sharply and that such services can be viewed as having an infinite value.

However, they contend that their estimates are much more accurate than the *zero* value the market usually assigns to these ecosystem services. They hope such estimates will call people's attention to the fact that the earth's ecosystem services (Figure 1-2) are absolutely essential for all humans and their economies (Figure 2-3) and that their economic value is huge. They believe that failure to estimate economic values for ecosystem services helps ensure that such vital life-support services will continue to be degraded for short-term economic gain.

Critical Thinking

Should we put a price on nature's services? Explain. What are the alternatives?

How Are Trees Harvested? Before large amounts of timber can be harvested, roads must be built for access and removal of timber. Even with careful design, logging roads have a number of harmful impacts (Figure 17-6). They include **(1)** increased erosion and sediment runoff into waterways, **(2)** habitat fragmentation and loss of biodiversity, **(3)** exposure of forests to invasion by exotic pests, diseases, and introduced wildlife species, and **(4)** opening of once-inaccessible forests to farmers, miners, ranchers, hunters, and off-road vehicle users. In addition, logging roads on public lands in the United States disqualify the land for protection as wilderness.

Once loggers can reach a forest, they use various methods for harvesting the trees (Figure 17-7). With **selective cutting**, intermediate-aged or mature trees in an uneven-aged forest are cut singly or in small groups (group selection), creating gaps no larger than the height of the standing trees (Figure 17-7a). Selective cutting **(1)** reduces crowding, **(2)** encourages the growth of younger trees, **(3)** maintains an uneven-aged stand of trees of different species, **(4)** allows natural regeneration from the surrounding trees, **(5)** can help protect the site from soil erosion and wind damage, and **(6)** allows a forest to be used for multiple purposes.

Some tree species grow best in full or moderate sunlight in moderate to large clearings. Such sun-loving species usually are harvested by shelterwood cutting, seed-tree cutting, or clear-cutting (in which all trees on a site are removed in a single cut). **Shelterwood cutting** removes all mature trees in two or typically three cuttings over a period of about 10 years (Figure 17-7b). The first cut removes most mature canopy trees, unwanted tree species, and diseased, defective, and dying trees. This opens the forest floor to light but leaves enough mature trees to cast seed and to shelter growing seedlings. Some years later, after enough seedlings take hold, a second cut removes more canopy trees, although some of the best mature trees are left to shelter the young trees. After these are well established, a third cut removes the remaining mature trees, and the even-aged stand of young trees then grows to maturity.

This method **(1)** allows natural seeding from the best seed trees, **(2)** keeps seedlings from being crowded out, **(3)** leaves a fairly natural-looking forest that can serve a variety of purposes until the final cut is made,

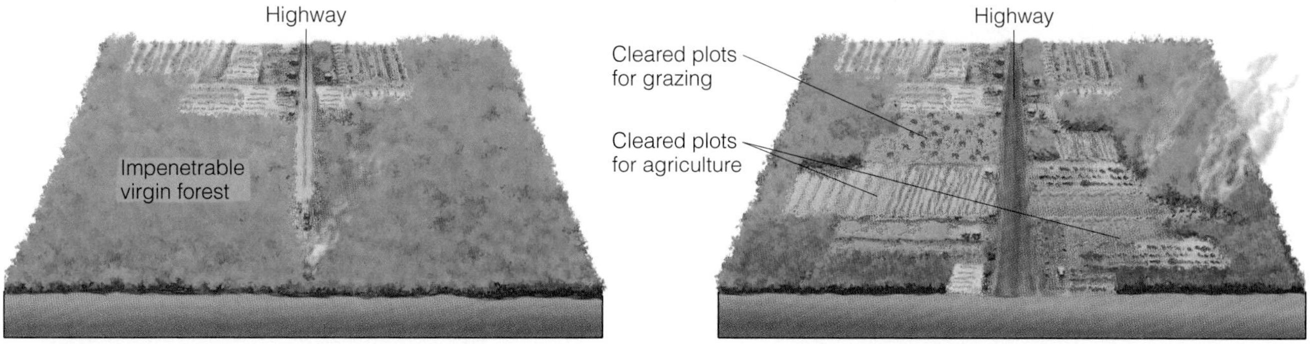

Figure 17-6 Building roads into previously inaccessible tropical (and other) forests paves the way to fragmentation, destruction, and degradation.

(4) helps reduce soil erosion, and **(5)** provides a good habitat for some types of wildlife. Two problems are that **(1)** loggers sometimes take too many trees in the first cut and **(2)** eventually a diverse forest is converted to a mostly monoculture plantation that is clear-cut.

Seed-tree cutting harvests nearly all of a stand's trees in one cutting, leaving a few uniformly distributed seed-producing trees to regenerate the stand (Figure 17-7c). After the new trees become established, the seed trees may be harvested. By allowing several species to grow at once, seed-tree cutting leaves an aesthetically pleasing forest that is useful for recreation, deer hunting, erosion control, and wildlife conservation. Leaving the best trees for seed can also lead to genetic improvement in the new stand.

Clear-cutting is the removal of all trees from an area in a single cutting. The clear-cut area may be a whole stand (Figure 17-7d), a strip, or a series of patches. After all trees are cut, the site usually is reforested **(1)** naturally by seed released by the harvest (or from nearby uncut trees if the clear-cut area is small) or **(2)** artificially as foresters broadcast seed over the site or plant seedlings raised in a nursery.

On the positive side, clear-cutting **(1)** increases timber yield per hectare, **(2)** permits reforesting with genetically improved stocks of fast-growing trees, **(3)** shortens the time needed to establish a new stand of trees, **(4)** requires much less skill and planning than other harvesting methods, and **(5)** usually provides timber companies the maximum economic return in the shortest time. If done carefully and responsibly, clear-cutting often is the best way to harvest tree farms and stands of some tree species that require full or moderate sunlight for growth.

On the negative side, clear-cutting **(1)** leaves forest openings, **(2)** eliminates most recreational value for several decades, **(3)** reduces biodiversity and disrupts the processes of energy flow and chemical cycling that sustain ecosystems by destroying and fragmenting some wildlife habitats, **(4)** makes nearby trees more vulnerable to being blown down by windstorms, and **(5)** leads to severe soil erosion, sediment water pollution, and flooding when done on steep slopes (Figure 12-16).

A variation of clear-cutting that can allow a sustainable timber yield without widespread destruction is **strip cutting**. A strip of trees is clear-cut along the contour of the land, with the corridor narrow enough to allow natural regeneration within a few years (Figure 17-7e). After regeneration, another strip is cut above the first, and so on. This allows a forest to be clear-cut in narrow strips over several decades with minimal damage.

Instead of conventional industrial forestry, conservation biologists and a growing number of foresters call for more sustainable forest management (Solutions, p. 435).

How Can Forests Be Protected from Pathogens and Insects? Insects and tree diseases can damage and even kill trees in forests. A biologically simplified tree farm is particularly vulnerable to attack by pathogens. Examples include **(1)** parasitic fungi such as chestnut blight, Dutch elm disease, dogwood blight, and pine pitch canker disease and **(2)** insects such as bark beetles, Chinese beetles, gypsy moths, spruce budworms, and hemlock wooly adelgids.

To conservation biologists, preserving biodiversity is the best and cheapest defense against tree diseases and insects. Other methods include **(1)** banning imported timber that might introduce harmful new pathogens, **(2)** removing infected trees or clear-cutting infected areas and burning all debris, **(3)** treating diseased trees with antibiotics, **(4)** developing disease-resistant tree species, **(5)** applying pesticides (Section 16-1), and **(6)** using integrated pest management (Section 16-5).

How Do Fires Affect Forest Ecosystems? Intermittent natural fires set by lightning are an important part of the ecological cycle of some types of forests.

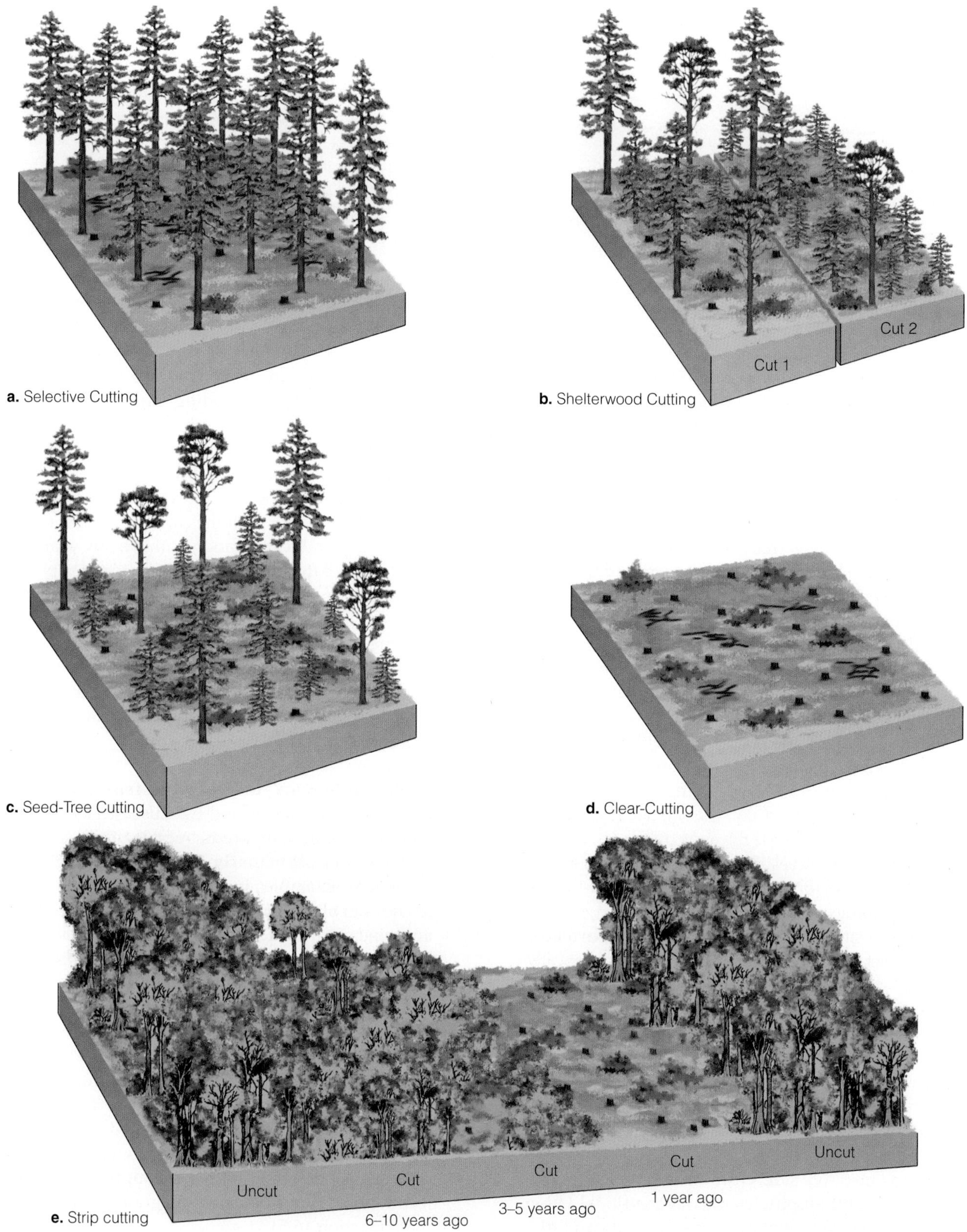

Figure 17-7 Tree harvesting methods.

What Is Sustainable Forestry?

SOLUTIONS

Many conservation biologists and a number of foresters call for sustainable management of forests that emphasizes the following:

- *Recycling more paper to reduce the harvest of pulpwood trees* (Section 14-4)
- *Using fibers from fast-growing plants such as kenaf to make tree-free paper and reducing wood waste* (p. 439)
- *Growing more timber on long rotations,* generally about 100–200 years (point C, Figure 17-5), depending on the species and soil quality
- *Emphasizing selective cutting of individual trees or small groups of most tree species* (Figure 17-7a)
- *Minimizing fragmentation of remaining larger blocks of forest*
- *Using road-building and logging methods that minimize soil erosion and compaction*
- *Practicing strip cutting* (Figure 17-7e), *instead of conventional clear-cutting, and banning clear-cutting (including seed-tree and shelterwood cutting) on land that slopes more than 15°*
- *Leaving most standing dead trees (snags) and fallen timber to maintain diverse wildlife habitats and to be recycled as nutrients* (Figure 4-12)
- *Reducing consumption of wood and paper products*
- *Switching to alternative building materials*
- *Including the estimated ecological and recreational services provided by trees and a forest in estimates of their economic value*

Some foresters have suggested that the United States meet most of its demand for wood and wood products by growing genetically improved trees on carefully managed tree plantations. They contend that this practice would be sustainable if it is done with minimum use of pesticides and commercial fertilizer and carried out mostly on already cleared marginal farmland and pastureland that already has access roads. This would **(1)** reduce the need for the United States to import wood that causes environmental damage in other countries, **(2)** reduce pressures to cut timber in old-growth forests, and **(3)** help protect wildlife habitats.

In 1903 New Zealand adopted such an approach. Today, the country meets all of its domestic wood needs from plantations and has protected about 30% of its original native forest. What do you think about this approach?

Critical Thinking

Do you disagree with these proposals? Explain. If you agree with these proposals, how would you see that they are adopted and enforced?

Some species actually need occasional fires. The seeds of the giant sequoia, the lodgepole pine, and the jack pine, for instance, are released from cones and germinate only after being exposed to intense heat.

Forest ecosystems can be affected by different types of fires. Some, called *surface fires,* usually burn only undergrowth and leaf litter on the forest floor. These fires kill seedlings and small trees but spare most mature trees and allow most wild animals to escape.

Occasional surface fires **(1)** burn away flammable ground material and help prevent more destructive fires, **(2)** release valuable mineral nutrients tied up in slowly decomposing litter and undergrowth, **(3)** increase the activity of underground nitrogen-fixing bacteria, **(4)** stimulate the germination of certain tree seeds, and **(5)** help control pathogens and insects. In addition, some wildlife species such as deer, moose, elk, muskrat, woodcock, and quail depend on occasional surface fires to maintain their habitats and to provide food in the form of vegetation that sprouts after fires.

Some extremely hot fires, called *crown fires,* may start on the ground but eventually burn whole trees and leap from treetop to treetop. They usually occur in forests in which no surface fires have occurred for several decades, allowing dead wood, leaves, and other flammable ground litter to build up. These rapidly burning fires can destroy most vegetation, kill wildlife, and lead to accelerated soil erosion.

Sometimes surface fires go underground and burn partially decayed leaves or peat. Such *ground fires* are most common in northern peat bogs. They may smolder for days or weeks before being detected and are difficult to extinguish.

Protecting forest resources from fire can involve four approaches: **(1)** *prevention,* **(2)** *prescribed burning* (setting controlled ground fires to prevent buildup of flammable material), **(3)** *presuppression* (early detection and control of fires), and **(4)** *suppression* (fighting fires once they have started). Means of forest-fire prevention include **(1)** requiring burning permits, **(2)** closing all or parts of a forest to travel and camping during periods of drought and high fire danger, and **(3)** educating the public. The Smokey Bear educational campaign of the Forest Service and the National Advertising Council, for example, has prevented countless forest fires in the United States, saving many lives and preventing billions of dollars in losses.

However, ecologists contend that prevention can increase the likelihood of highly destructive crown fires because it allows litter to accumulate in some forests. Since 1972 U.S. Park Service policy has been to allow most lightning-caused fires to burn themselves out, as long as they don't threaten human lives, park facilities, private property, or endangered wildlife.

After the fires in parts of Yellowstone National Park and many other fires that broke out in a number of national forests in the West during the summers of 1988 and 1994, some people called for a reversal of the federal let-it-burn policy. However, biologists contend that these fires caused more damage than they should have because the previous policy of fighting all fires in the park had allowed the buildup of flammable ground litter and small plants. They believe that many fires in national parks, national forests, and wilderness areas should be allowed to burn as part of the natural ecological cycle of succession and regeneration (Figure 5-15). What do you think?

How Do Air Pollution and Climate Change Threaten Forests? Forests at high elevations and those downwind from urban and industrial centers are exposed to a variety of air pollutants that can harm trees, especially conifers. Besides doing direct harm, prolonged exposure to multiple air pollutants makes trees much more vulnerable to drought, diseases, and insects (Figure 10-9). The solution is to reduce emissions of the offending pollutants from coal-burning power plants, industrial plants, and motor vehicles (Section 10-6).

In coming decades, an even greater threat to forests (especially temperate and boreal forests) may come from regional climate changes brought about by projected global warming. Ways to deal with projected global warming are discussed in Section 11-5.

17-3 FOREST RESOURCES AND MANAGEMENT IN THE UNITED STATES AND CANADA

What Is the Status of Forests in the United States? Forests cover about one-third of the lower 48 United States and provide habitats for more than 80% of the country's wildlife species. The *good news* is that today forests in the United States are generally bigger and often healthier than they were in 1900, when the country's population was below 100 million.

Most of the early old-growth forests in the lower 48 states were cut between 1800 and 1960 (Figure 17-8). However, many of the old-growth forests that were cleared or partially cleared between 1800 and 1960 have grown back naturally through secondary ecological succession as fairly diverse second-growth (and in some cases third-growth) forests.

Since the mid-1960s a increasing area of the nation's remaining old-growth and fairly diverse second-growth forests has been clear-cut and replaced with tree plantations. Although this doesn't reduce the nation's overall tree cover, conservation biologists argue that it does reduce overall forest biodiversity and ecosystem processes such as energy flow and chemical cycling (ecological integrity).

Others argue that tree farms help preserve overall forest diversity by reducing the pressure to clear-cut more diverse old-growth and second-growth forests. The basic question is, "What percentage of the nation's forests should be converted to tree farms to meet the wood and pulp needs of consumers and still preserve the overall biodiversity and ecological integrity of the nation's forests?" What do you think?

Case Study: How Are U.S. National Forests Managed? About 22% of the commercial forest acreage in the United States is located within the 156 national forests (Figure 17-2) managed by the U.S. Forest Service and occupying about 10% of the U.S. land area. These forests **(1)** serve as grazing lands for more than 3 million cattle and sheep each year, **(2)** support multibillion-dollar mining operations (p. 326), **(3)** supply about 3% of the nation's softwood timber (down from 15% in the 1980s), **(4)** contain a network of roads equal in area to the entire U.S. interstate highway system, and **(5)** receive more recreational visits than any other federal public lands.

The Forest Service is required by law to manage national forests according to the principles of **(1) sustained yield** (which states that potentially renewable tree resources should not be harvested or used faster than they are replenished) and **(2) multiple use** (which says that each of these forests should be managed simultaneously for a variety of uses such as sustainable timber harvesting, recreation, and wildlife conservation).

Environmentalists charge that especially during the 1980s and early 1990s the Forest Service allowed timber harvesting to become the dominant use in most national forests. A major reason for this emphasis on timber cutting is that Congress allows the agency to keep most of the money it makes on timber sales (and thus increase its budget), but any losses are passed on to taxpayers. Local county officials also lobby Congress and encourage Forest Service officials to keep timber harvests high because counties get 25% of the gross receipts from national forests within their boundaries.

A drive through the 122 national forests open to logging, or a boat ride along streams in such forests, does not reveal any emphasis on timber cutting because the Forest Service leaves thin buffers of uncut trees along roads and streams. However, a flight over some of these forests reveals that many of the trees have been clear-cut and many of the forests have been highly fragmented.

Leaving buffer strips along streams is an important way to help reduce runoff of soil into such streams from harvested areas. However, critics of excessive cutting in national forests contend that leaving large areas of forest intact is the best way to prevent stream pollution and loss of biodiversity.

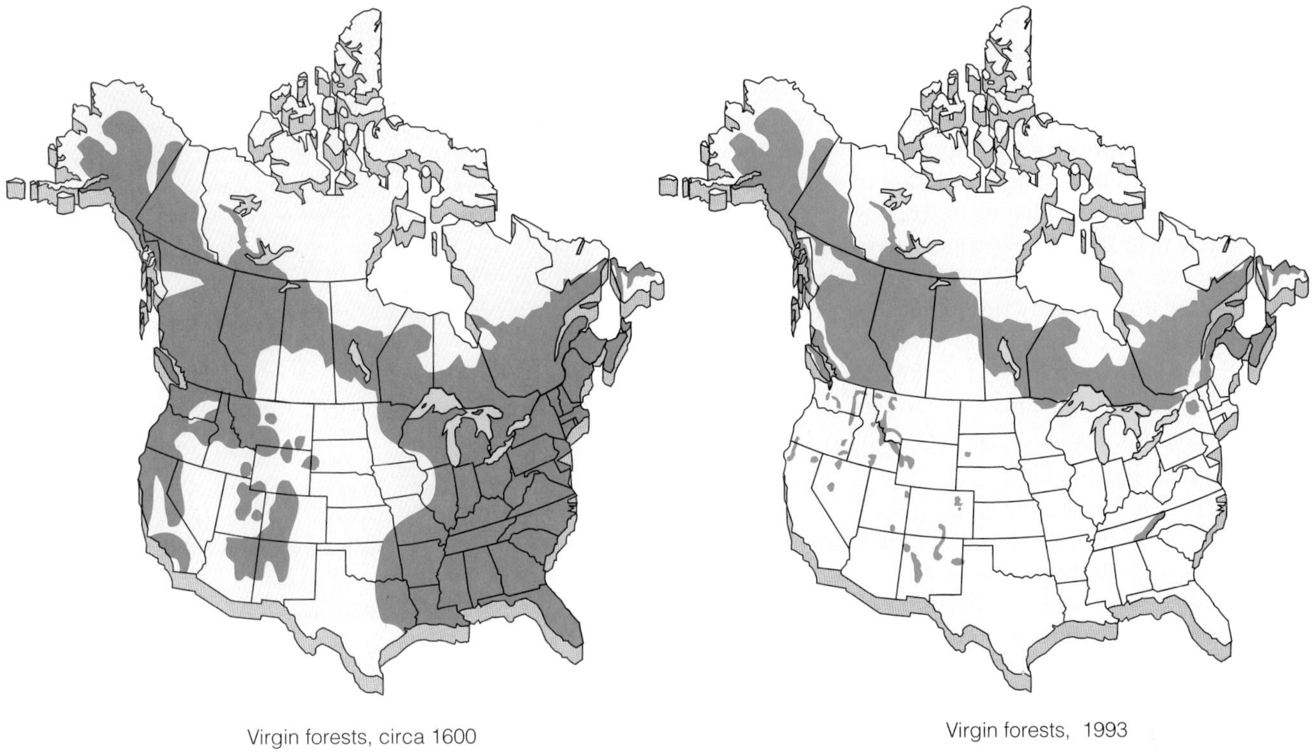

Figure 17-8 Vanishing old-growth forests in the United States and Canada between about 1600 and 1993. (Data from the Wilderness Society, the U.S. Forest Service, and *Atlas Historique du Canada*, vol. 1)

By law, the Forest Service must sell timber for no less than the cost of reforesting the land from which it was harvested. However, the cost of access roads is not included in this price but is provided as a subsidy to logging companies. Usually the companies also get the timber itself for less than they would pay a private landowner for an equivalent amount of timber. According to the Wilderness Society, a lumber company harvesting trees in a national forest pays as little as $2 for 20-meter (65-foot), 100-year-old pine tree that will produce timber products worth thousands of dollars.

Because of such subsidies, timber sales from U.S. federal lands have turned a profit for taxpayers in only 3 of the last 100 years. Studies by the U.S. General Accounting Office have shown that between 1978 and 1998, national forests lost at least $7 billion from below-cost timber sales. According to a 1999 study by the Earth Island Institute, the federal logging program cost taxpayers $1.7 billion in 1997—an estimate confirmed by the U.S. Congressional Research Service.

In 1998 Forest Service chief Mike Dombeck announced a policy designed to shift the primary emphasis of the agency from maximizing timber harvests to putting much greater emphasis on the recreational use of national forests. Between 1988 and 1998, the annual amount of timber cut from national forests dropped by almost 75%. In 1995 recreational use of national forests generated 2.6 million jobs and added $97.8 billion to the national economy. By contrast, logging in these forests added only 76,000 jobs and $3.5 billion. In 1997, the national forests had about 800 million recreation visits, three times the number of such visits to the national parks. (Up from 250 million such visits in 1985.) However, Congress, under pressure from timber companies, may ask legislation to overturn this new policy.

Timber company officials argue that by taking timber from federal lands fairly cheaply, they benefit taxpayers by keeping lumber and paper prices down. Environmentalists respond that below-cost timber prices discourage investments to produce additional timber on private lands and put more pressure for timber cutting on national and state forests.

Timber company officials point out that logging in national forests provides jobs and stimulates economic growth in nearby communities. However, studies show that logging in national forests rarely provides economic stability for local industries and communities. Instead, communities that rely heavily on proceeds from national forest timber sales experience severe economic slumps as the timber is depleted and timber companies move to other areas.

Analysts have proposed a number of reforms in Forest Service policy that would lead to more sustainable use and management of the national forests (Solutions, p. 439).

Case Study: Ecology of Old-Growth Forests in the Northwest In the Pacific Northwest, it typically takes 350 years for an old-growth forest to reach its prime in terms of size and diversity. These forests have a number of important ecological functions. They have the world's largest accumulation of dead standing trees (snags) and fallen dead trees, which decompose slowly over 200–400 years and recycle nutrients in the forest ecosystem (Figure 4-12). For their latitude, these forests are unusually rich in wildlife species, and they act as giant sponges that hold and slowly release enough moisture to help protect against fires and floods and recharge nearby streams and aquifers.

Fragmenting these forests into small patches, or clearing and replacing them with tree plantations and second-growth forests, can disrupt the vital network of relationships among species that help sustain these complex forest communities.

The survival of old-growth forests in the Pacific Northwest is also connected to the survival of the region's salmon stocks (Figure 15-15). Each year the $1.25-billion sport and commercial salmon fishing industry in this region provides jobs for 62,000 people. Several factors—including damming of rivers, pollution, and overfishing—are involved in the decline of salmon stocks. However, a key factor has been excessive logging on public lands throughout the region. Sediment deposited in streams from the logging of old-growth forests and associated road building smothers spawning beds and disrupts the feeding patterns of young salmon.

Pro/Con: Should Remaining Old-Growth Forests on U.S. Public Lands Be Cut or Preserved? To timber company officials, the giant living trees and rotting dead trees in the Pacific Northwest's old-growth forests are valuable resources that could provide jobs and should be harvested for profit, not left alone to please environmentalists. Timber officials claim that protecting large areas of remaining old-growth forests on public lands will cost as many as 53,000 jobs and hurt the economies of logging and milling towns in such areas.

To conservation biologists and environmentalists, however, the remaining old-growth forests on the nation's public lands—owned by all U.S. citizens—are a treasure whose ecological, scientific, aesthetic, and recreational values far exceed the economic value of cutting them down for short-term profit.

Environmentalists have used the threatened northern spotted owl (Figure 17-9) as a way to slow or halt timber harvesting in many remaining old-growth forests in public lands in Oregon, Washington, and northern California. They claim that this species lives almost exclusively in 200-year-old Douglas fir forests found in these areas. However, logging company officials contend that the owls do not require old-growth forest and can instead adapt to younger second-growth forests.

Figure 17-9 The threatened northern spotted owl, which lives almost exclusively in old-growth forests of the Pacific Northwest. Environmentalists have used the owl's threatened status both to save it and to halt or decrease logging in the national forests in which it lives, thereby advancing their wider goal of preserving biodiversity. (Pat & Tom Leeson/ Photo Researchers, Inc.)

The owl and other threatened species in these forests are symbols of the broader clash between **(1)** timber companies who want to clear-cut most remaining old-growth stands in the national forests and **(2)** conservation biologists who want to protect the biodiversity and ecological integrity of these forests, or at least allow only sustainable harvesting in some areas. What do you think?

Case Study: Destruction of Old-Growth Forests in Western Canada With nearly 50% of its land forested (much of it in second-growth forest), Canada has 10% of the world's forests. One of every 10 jobs in Canada is directly related to forestry. Provincial governments own and manage 94% of Canada's forests. Ninety percent of logging in Canada involves clear-cutting.

Critics charge that the Canadian government gives large subsidies to logging companies without providing many jobs for Canadians. Environmentalists and many Canadian citizens were outraged when British Columbia's provincial government awarded MacMillan Bloedel a contract to clear-cut the old-growth forest of Clayoquot (pronounced "clack-what") Sound on the west coast of Vancouver Island. Environmental groups **(1)** exerted intense pressure on government officials, **(2)** staged pub-

How Can U.S. Federal Forest Management Be Improved?

SOLUTIONS

Foresters and conservation biologists have proposed a number of management reforms that would allow national forests in the United States to be used more sustainably without seriously impairing their important ecological functions. These suggestions include the following:

- *Make sustaining biodiversity and ecological integrity the first priority of national forest management policy.*
- *Use full-cost accounting to include the ecological services provided by old-growth forests in the market price of timber harvested from national forests.*
- *Prohibit timber harvesting from all or most of the remaining old-growth forests on public lands.*
- *Make recreation and protection of watershed and wildlife habitats the primary uses of national forests.* The Forest Service estimates that recreational user fees (as low as $3 per day) would generate three times what it earns from timber sales.
- *Sharply reduce all new road building in national forests.*
- *Allow individuals or groups to buy conservation easements that prevent timber harvesting on designated areas of public old-growth forests.* In such *conservation-for-tax-relief swaps*, purchasers would be allowed tax breaks for the funds they put up.
- *Require that timber from national forests be sold at a price that includes the costs of road building, site preparation, and site regeneration, and that all timber sales in national forests yield a profit for taxpayers.*
- *Do not use money from timber sales in national forests to supplement the Forest Service budget* because it encourages overexploitation of timber resources.
- *Eliminate the provision that returns 25% of gross receipts from national forests to counties containing the forests, or base such returns on recreational user-fee receipts only.*
- *Strictly enforce and close loopholes in the current federal ban on exporting raw logs harvested from public lands.*
- *Tax exports of raw logs but not finished wood products* (which provide and create domestic milling and other jobs).
- *Establish federally funded reforestation and ecological restoration projects on degraded public lands.* In addition to the ecological benefits, this will provide jobs for displaced forest harvesting and wood products workers.
- *Provide increased aid and job retraining for displaced workers in the forest harvesting and products industry.*
- *Require use of sustainable forestry methods* (p. 435).

Timber company officials, many county officials, and some Forest Service officials oppose such policies.

Critical Thinking

Explain why you agree or disagree with each of these proposals for reforming federal forest management in the United States.

licized protests, **(3)** subjected the Canadian government to the glare of unfavorable international publicity, and **(4)** called for businesses and governments throughout the world to boycott Canadian timber. In 1995 British Columbia's government announced that it would end clear-cutting in Clayoquot Sound, and in 1999 officials of MacMillan Bloedel signed an agreement not to log in any parcels in the area's unlogged valleys that are larger than 1,000 hectares (2,470 acres).

Environmentalists see this as an important victory that will protect a small area of Canada's forest biodiversity. However, they point out that unsustainable timber harvesting is still occurring in much of the country's forests and call for more of them to be protected.

Solutions: How Can We Reduce Wood Waste and Use Other Fibers to Make Paper? According to the Worldwatch Institute and forestry analysts, up to 60% of the wood consumed in the United States is unnecessarily wasted through construction waste, excess packaging, junk mail, inadequate recycling of paper, and failure to reuse wooden shipping containers. Because only 4% of the total U.S. timber production comes from the national forests, reducing the waste of wood and paper products by only 4% could eliminate the need to remove any timber from the national forests. Then these public lands could be used primarily for recreation and protection of biodiversity and ecological integrity.

Ways to reduce such waste include **(1)** reducing construction waste, **(2)** using laminated boards instead of solid wood, **(3)** using less packaging, **(4)** reducing junk mail, **(5)** recycling more paper to make products with the highest possible content of postconsumer waste, and **(6)** using paper made from fibers obtained from rapidly growing plants other than trees (Solutions, p. 440).

Some have suggested that we can save wood by switching to nonwood home building and remodeling materials such as **(1)** steel studs and beams, **(2)** aluminum and vinyl siding, and **(3)** concrete slabs instead of wooden floor joists. However, this involves shifting from a potentially renewable resource (wood) to nonrenewable resources such as limestone, iron and other mineral products, and crude oil used make these products and plastics.

Analyses show that switching to these nonrenewable resources usually requires more energy for transportation and manufacture and produces more pollution than use of wood and wood products. For example, **(1)** steel framing requires 13 times more

Making Paper from Rice Straw and Kenaf Fibers

SOLUTIONS

Because of deforestation and the rising costs of wood pulp, an increasing number of countries use other sources of fiber for making paper. In China, for example, 60% of the paper is made with nonwood pulp such as rice straw and other agricultural wastes left after harvest.

North America produces enough rice and wheat straw and other agricultural wastes to more than meet all of its paper needs without having to cut down another tree. In 1999, some manufacturers in California announced that they would make paper from rice straw left over from rice production. However, some environmentalists believe that instead of being removed, crop residues such as rice straw should be left to decompose and help fertilize soil and reduce inputs of commercial inorganic fertilizers.

Currently, most of the small amount of treeless paper produced in the United States is made from the fibers of a woody annual plant called *kenaf* (pronounced "kuh–NAHF"; Figure 17-10). The tall, bamboolike stalks of this annual woody plant grow to maturity in only 4–5 months, compared with 7–60 years for pine trees.

Kenaf's other advantages over pulpwood include **(1)** little need to use herbicides because it grows faster than most weeds, **(2)** little use of insecticides because its outer fibrous covering is nearly insect-proof, **(3)** no depletion of soil nitrogen because the plant is a nitrogen fixer, **(4)** use of fewer chemicals and less energy to break kenaf down into fibers used for making paper, and **(5)** less toxic wastewater and lower water treatment costs. The U.S. Department of Agriculture considers kenaf the best renewable fiber to replace trees in paper making and predicts that the market for kenaf will expand rapidly in coming years.

However, kenaf paper currently costs three to five times more than virgin or recycled paper stock, which is why this book is not printed on tree-free kenaf paper. But the price of kenaf should come down as demand increases and producers cut production costs.

Biologists urge caution against the widespread planting of kenaf. They fear that this rapidly growing plant could take over ecosystems, crowd out other plant species, and disrupt food webs. Also, planting large plantations of kenaf may compete with increasingly scarce land need to provide food for the world's growing population.

Most paper will continue to be made from pulpwood. However, use of alternative sources of pulp could be increased greatly by shifting some of the government subsidies given to pulpwood paper producers to these new paper-producing industries.

Critical Thinking

What might be some economic and ecological disadvantages of shifting most paper production to kenaf and other nonpulpwood sources? Do you believe that the ecological and economic benefits of such a shift outweigh the disadvantages? Explain.

Figure 17-10 Kenaf growing in a field in south Texas. This fast-growing plant can be harvested annually and its fibers used to make paper. Its widespread cultivation would decrease the need to grow and harvest trees to make paper. (U.S. Department of Agriculture)

energy than wood framing, **(2)** aluminum framing for walls takes 20 times more energy than wood framing, **(3)** steel floor joists require 50 times more energy than those made of wood, and **(4)** wall-to-wall carpeting is four times more energy intensive than wood floors.

17-4 TROPICAL DEFORESTATION AND THE FUELWOOD CRISIS

How Fast Are Tropical Forests Being Cleared and Degraded? Tropical forests cover about 6% of the earth's land area (roughly the area of the lower 48 states of the United States) and grow in equatorial Latin America, Africa, and Asia (Figure 6-9). Tropical forests come in many varieties (Figure 6-10), including **(1)** rain forests (which receive rainfall almost daily), **(2)** tropical deciduous forests (with one or two dry seasons each year), **(3)** dry and very dry deciduous forests, and **(4)** forests on hills and mountains.

Climatic and biological data suggest that mature tropical forests once covered at least twice as much area as they do today, with most of the destruction occurring since 1950. Satellite scans and ground-level surveys used to estimate forest destruction indicate that large areas of tropical forests are being cut (Figure 17-3). Haiti has lost 99% of its original forest cover, the Philippines 97%, and Madagascar 84%.

There is debate over the current rates of tropical deforestation and degradation because of **(1)** difficulties in interpreting satellite images, **(2)** different ways of defining deforestation and forest degradation, and **(3)** political and economic factors that cause countries to hide or exaggerate deforestation. However, using the best available data, it is estimated that between 1972 and 1999, at least 574,000 square kilometers (222,000 square miles) or 14% of the huge Amazon area was deforested.

In 1998, a team of scientists using satellite images, airplane photography, and ground-level interviews estimated that 44,000 square kilometers (17,000 square miles) of the Amazon were deforested in that year alone. This was three times the official Brazilian estimate. It's estimated that an equivalent area of these forests is seriously degraded and fragmented without being destroyed outright each year. Most attention has focused on the Amazon Basin, but 92% of the once-vast rain forests on Brazil's Atlantic coast and 98% of the coniferous Araucaria forests of the south have been devastated by logging and urban expansion.

Why Should We Care About Tropical Forests? Biologists consider the plight of tropical forests to be a serious global environmental problem, primarily because these forests are home to 50–90% of the earth's terrestrial species (Figure 1-10), most of them still unknown and unnamed. Biologist Edward O. Wilson estimates that by 2022 at least 20% of tropical forest species could be gone, and as many as 50% by 2042, if current rates of tropical deforestation and degradation continue.

Tropical forests have great economic and ecological importance. They supply **(1)** half of the world's annual harvest of hardwood, **(2)** hundreds of food products (including coffee, tea, cocoa, spices, nuts, chocolate, and tropical fruits), and **(3)** materials such as natural latex rubber (Figure 17-11), resins, dyes, and essential oils that can be harvested sustainably. A 1988 study showed that sustainable harvesting of nonwood products such as nuts, fruits, herbs, spices, oils, medicines, and latex rubber in Amazon rain forests over 50 years would generate **(1)** twice as much revenue per hectare as timber production and **(2)** three times as much as cutting them and converting them to grassland for cattle ranching.

The active ingredients for 25% of the world's prescription drugs (worth $100 billion per year worldwide) are derived from plants, most of which grow in tropical rain forests. About 70% of the 3,000 plants identified by the National Cancer Institute as sources of cancer-fighting chemicals come from tropical forests (Figure 17-12). Some tropical tree species can be used for many purposes (Solutions, p. 442).

Most of the original strains of rice, wheat, and corn that supply more than half of the world's food were developed from wild tropical plants. Botanists believe that tens of thousands of plant strains with potential food value await discovery in tropical forests. Despite their economic and health potential, fewer than 1% of the estimated 125,000 flowering plant species in tropical forests (and less than 3% of all the world's 240,000 such species) have been examined closely for possible use as human resources.

Figure 17-11 Rubber tapping is a potentially sustainable use of tropical forests. This 1987 photo shows Brazilian activist Chico Mendez (murdered in 1988 for his efforts to protect these forests) making a cut in a rubber tree. The cut allows milky liquid latex to trickle into a collecting cup. The latex is processed to make rubber, and the scars heal without killing the tree. During the seasons when latex is not flowing, rubber tappers make a living by gathering and selling nuts. (Randall Hyman)

What Causes Tropical Deforestation? Tropical deforestation results from a number of interconnected causes, all of which are related to increasing appropriation of the net primary productivity of the earth's forest for human use as a result of population growth, poverty, and certain government policies that encourage deforestation (Figure 17-13). Population growth and poverty combine to drive subsistence farmers and the landless poor to tropical forests, where they try to grow enough food to survive.

Government subsidies can accelerate deforestation by **(1)** making timber or other resources cheap relative to their full ecological value and **(2)** encouraging the

Figure 17-12 Rosy periwinkle in the threatened tropical forests of Madagascar. Two compounds extracted from this plant have been used to cure most victims of two deadly cancers: lymphocytic leukemia (which was once almost always fatal for children) and Hodgkin's disease (mostly affecting young adults). Worldwide sales of these two drugs are about $180 million per year, none of which is returned to Madagascar. Only a tiny fraction of the world's tropical plant species has been studied for such potential uses, and many will become extinct before we can study them. (Heather Angel/BioFotos)

poor to colonize tropical forests by giving them title to land they clear (as is done in Brazil, Indonesia, and Mexico). International lending agencies encourage developing countries to borrow huge sums of money from developed countries to finance projects such as roads, mines, logging operations, oil drilling, and dams in tropical forests.

The process of degrading a tropical or other type of forest often begins with a road (Figure 17-6), usually cut by logging companies. Once the forest becomes accessible, it can be cleared and increasingly degraded by a number of factors, including

- *Unsustainable forms of small-scale farming.* Instead of practicing various methods of traditional and potentially sustainable shifting cultivation, many poor people migrating to tropical forests practice unsustainable farming that depletes soils and destroys large tracts of forests.
- *Cattle ranching.* Cattle ranches, sometimes supported by government subsidies, often are established on cropland that has been exhausted by small-scale farmers and then abandoned or sold to ranchers. When torrential rains and overgrazing turn the usually thin and nutrient-poor tropical forest soils (Figure 13-17) into eroded wastelands, ranchers move to another area and repeat the destructive process known as *shifting ranching*. In Brazil's Amazon Basin, ranchers (supported by government subsidies averaging $5.6 million per ranch) have claimed large tracts of land and the mineral rights below them.
- *Clearing large areas of tropical forest for raising cash crops.* Large plantations grow crops such as sugarcane, bananas, pineapples, peppers, strawberries, cotton, tea, and coffee, mostly for export to developed countries.

SOLUTIONS

The Incredible Neem Tree

Wouldn't it be nice if there were a single plant that could **(1)** quickly reforest bare land, **(2)** provide fuelwood and lumber in dry areas, **(3)** provide alternatives to toxic pesticides, **(4)** be used to treat numerous diseases, and **(5)** help control population growth? There is: the *neem tree*, a broadleaf evergreen member of the mahogany family.

This remarkable tropical species, native to India and Burma, is ideal for reforestation because it can grow to maturity in only 5–7 years! It grows well in poor soil in semiarid lands such as those in Africa, providing an abundance of fuelwood, lumber, and lamp oil in the process.

It also contains various natural pesticides. Chemicals from its leaves and seeds can repel or kill over 200 insect species, including termites, gypsy moths, locusts, boll weevils, and cockroaches.

Extracts from neem seeds and leaves can be used to fight bacterial, viral, and fungal infections. Indeed, the tree's chemicals have allegedly relieved so many different afflictions that the tree has been called a "village pharmacy." Its twigs are used as an antiseptic toothbrush, and oil from its seeds is used to make toothpaste and soap.

But that's not all. Neem-seed oil evidently acts as a strong spermicide that might be used in producing a much-needed male birth control pill. According to a study by the U.S. National Academy of Sciences, the neem tree "may eventually benefit every person on the planet."

Despite its numerous advantages, ecologists caution against widespread planting of neem trees outside of native range. If introduced as an alien species to other ecosystems, it could take over and displace other species because of its rapid growth and resistance to pests.

Critical Thinking

Assume that you are an elected county official. Someone proposes widespread planting of neem trees on abandoned cropland in the area where you live. Would you favor this proposal? Explain. What tests might you want to run before approving this project?

- *Mining (*Case Study, p. 339*) and oil drilling.*
- *Building dams on rivers that flood large areas of tropical forests.*
- *Commercial logging.* Since 1950 the consumption of tropical timber has risen 15-fold. Japan alone accounts for 53% of the world's tropical timber imports, followed by Europe (32%) and the United

Figure 17-13 Major interconnected causes of the destruction and degradation of tropical forests. All of these causes ultimately are related to the underlying causes of **(1)** population growth, **(2)** poverty, and **(3)** government policies that encourage deforestation.

States (15%). After depleting tropical timber in much of Asia, cutting is now shifting to Latin America and Africa. Timber exports to developed countries contribute significantly to tropical forest depletion and degradation, but over 80% of the trees cut in developing countries are used at home.

- *Increasing forest fires*. Between 1997 and 1999, huge areas of forests were burned in Indonesia, Brazil, Guatemala, Nicaragua, and Mexico, mostly by farmers using fire to prepare fields for planting or cattle-grazing. The resulting highly polluted air **(1)** sickened tens of millions of people, **(2)** killed hundreds, **(3)** caused billions of dollars in damage, and **(4)** released large amounts of carbon dioxide into the atmosphere. In coming years, more fires are expected in Latin America, Africa, and Asia.

Solutions: How Can We Reduce Tropical Deforestation and Degradation? A number of analysts have suggested the following ways to protect tropical forests and use them more sustainably:

- Identify and move rapidly to protect areas of tropical forests that are rich in unique animal and plant species and in imminent danger, called *hot spots* or *critical ecosystems.* So far, Conservation International has identified 25 hot spots where preservation efforts could have maximum benefits, such as the island of Madagascar and the Atlantic forest region of eastern Brazil.

- Mount global efforts to sharply reduce the poverty (Section 2-4) that leads the poor to use forests (and other resources) unsustainably for short-term survival.

- Reduce the flow of the landless poor to tropical forests by slowing population growth and discouraging the poor from migrating to undisturbed tropical forests.

- Establish programs to help new settlers in tropical forests learn how to practice small-scale sustainable agriculture and forestry. (Solutions, right).

- Phase in *full-cost pricing* of forests that includes the ecological services they provide over the next two decades (Section 2-3).

- Phase out government subsidies that encourage unsustainable deforestation and phase in tariffs, use taxes, user fees, and subsidies that favor sustainable forestry and protection of biodiversity.

- Encourage governments to protect large areas of tropical forests. Currently less than 5% of the world's tropical forests are part of parks and preserves, and many of these are protected in name only. In 1998 tiny Suriname (a Dutch colony on South America's northern Caribbean coast with a population 400,000) announced that it was setting aside 12% of its land as a nature reserve instead of turning it over to Asian loggers.

- Use **debt-for-nature swaps** and conservation easements to encourage countries to protect tropical forests. In a *debt-for-nature swap*, participating countries act as custodians for protected forest reserves in return for foreign aid or debt relief (Figure 17-14); in *conservation easements*, a private organization, country, or group of countries compensates other countries for protecting selected forest areas.

- Establish an international system for evaluating and labeling timber produced by sustainable methods.

- Enact and enforce laws to reduce illegal timber cutting. In Cambodia, for example, the estimated amount of money lost to the national treasury because of illegal logging is equal to the entire national budget.

Sustainable Agriculture and Forestry in Tropical Forests

SOLUTIONS

A combination of the knowledge of indigenous peoples, ecological research, and modern technology can show people how to grow crops and harvest timber in tropical forests more sustainably.

One approach would be to show people migrating to forests how to grow crops using forms of more sustainable agroforestry developed by various indigenous peoples throughout the world. The Lacandon Maya Indians of Chiapas, Mexico, for example, have developed a multilayered system of agroforestry that allows them to cultivate up to 75 crop species on 1-hectare (2.5-acre) plots for up to 7 consecutive years. After that another plot must be planted to allow regeneration of the soil in the original plot.

Another approach being used by Yánesha Indians in the lush rain forests of Peru's Palcazú Valley is strip-shelterbelt harvesting (Figure 17-7e) of tropical trees for lumber. They use chain saws to cut widely spaced thin strips—barely visible from an airplane—through the forest. Oxen haul the cut logs to a processing plant, where other tribe members convert them to charcoal fuel, boards, and fence posts. Tribe members market their ecologically harvested wood products in the global marketplace and act as consultants to help other forest dwellers set up similar systems.

Critical Thinking

What applications (if any) might these practices have for growing food and harvesting timber in the United States and other developed counties? How would you encourage use of such approaches for more sustainable use of forests?

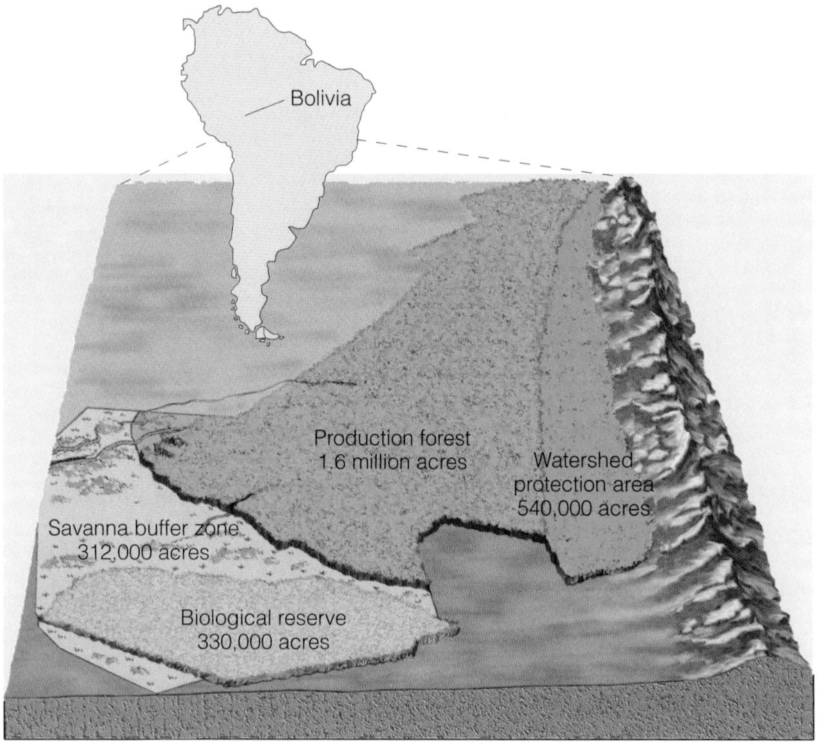

Figure 17-14 Blending economic development and conservation in a 1.5-million-hectare (3.7-million-acre) tract in Bolivia. A U.S. conservation organization arranged a debt-for-nature swap to help protect this land from destructive development.

In Brazil, the government acknowledges that nearly 80% of the timber extracted from the Amazon rain forest is taken illegally, but in 1999 the government rescinded a much-heralded new law to crack down on illegal logging. The government also eliminated most of the $70-million budget for conservation and environmental programs in the Amazon.

- Use gentler methods for harvesting trees. For example, cutting canopy vines (lianas) before felling a tree can reduce damage to neighboring trees by 20–40%, and using the least-obstructed paths to remove the logs can halve the damage to other trees.
- Concentrate peasant farming, tree and crop plantations, and ranching activities on already-cleared tropical forest areas that are in various stages of secondary ecological succession instead of on old-growth forests.
- Mount global efforts to reforest and rehabilitate degraded tropical forests and watersheds (Individuals Matter, p. 446).
- Encourage citizens to work together to protect tropical forests. In Brazil, 500 conservation organizations have formed a coalition to preserve the country's remaining tropical forests. In 1991 Find Fensen, a first-year student at Princeton University, decided to buy and protect some rain forest. By 1993 his efforts to preserve tropical forests had **(1)** spread to 45 campuses, **(2)** attracted contributions from individuals and corporations, and **(3)** resulted in the purchase and protection of hundreds of hectares of rain forest in Belize.
- Reduce the waste and overconsumption of industrial timber, paper, and other resources by consumers, especially in developed countries (Spotlight, p. 447).

How Serious Is the Fuelwood Crisis In Developing Countries? According to the UN Food and Agriculture Organization, in 2000 about 2.7 billion people in 77 developing countries either didn't get enough fuelwood to meet their basic needs or were forced to meet their needs by using wood faster than it is replenished (Figure 17-16).

Besides local deforestation and accelerated soil erosion, fuelwood scarcity has other harmful effects. It places an additional burden on the rural poor, especially women and children, who often must walk long distances searching for firewood. Buying fuelwood or charcoal can take 40% of a poor family's meager income.

An estimated 800 million poor people who can't get enough fuelwood burn dried animal dung and crop residues for cooking and heating. These natural fertilizers thus never reach the soil, cropland productivity is reduced, the land is degraded still further, and hunger and malnutrition increase.

City dwellers in many developing countries burn charcoal because it is much lighter than fuelwood and is thus much cheaper to transport. But burning wood in traditional earthen pits to produce charcoal consumes more than half of the wood's energy. Thus, each city dweller burning charcoal uses twice as much wood for a given amount of energy as a rural dweller who burns firewood. This helps explain the expanding rings of deforested land that surround many cities in developing countries that use charcoal as the major source of fuel.

Solutions: What Can We Do About the Fuelwood Crisis? Developing countries can reduce the severity of the fuelwood crisis by **(1)** planting more fast-growing fuelwood trees or shrubs, **(2)** burning wood more efficiently, and **(3)** switching to other fuels. However, fast-growing tree and shrub species used to establish fuelwood plantations must be selected carefully to prevent harm to local ecosystems.

Kenya's Green Belt Movement

INDIVIDUALS MATTER

In Kenya, Wangari Maathai (Figure 17-15) started the Green Belt Movement. The goals of this highly regarded women's self-help community are to establish tree nurseries, raise seedlings, and plant a tree for each of Kenya's 29 million people. By 1999 the 100,000 members of this grassroots group had planted and protected more than 20 million trees.

The success of this project has sparked the creation of similar programs in more than 30 other African countries. This inspiring leader has said,

> *I don't really know why I care so much. I just have something inside me that tells me that there is a problem and I have got to do something about it. And I'm sure it's the same voice that is speaking to everyone on this planet, at least everybody who seems to be concerned about the fate of the world, the fate of this planet.*

Figure 17-15 Wangari Maathai, the first Kenyan woman to earn a Ph.D. (in anatomy) and to head an academic department (veterinary medicine) at the University of Nairobi, organized the internationally acclaimed Green Belt Movement in 1977. (William Campbell/©TIME magazine)

Eucalyptus trees, for example, are being used to reforest areas threatened by desertification and to establish fuelwood plantations in some parts of the world. Because these species grow fast even on poor soils, this might seem like a good idea, but environmentalists and local villagers see it as an ecological disaster. In their native Australia these trees thrive in areas with good rainfall; when planted in arid areas, however, the trees suck up so much of the scarce soil water that most other plants can't grow. Then farmers don't have fodder to feed their livestock, and groundwater is not replenished.

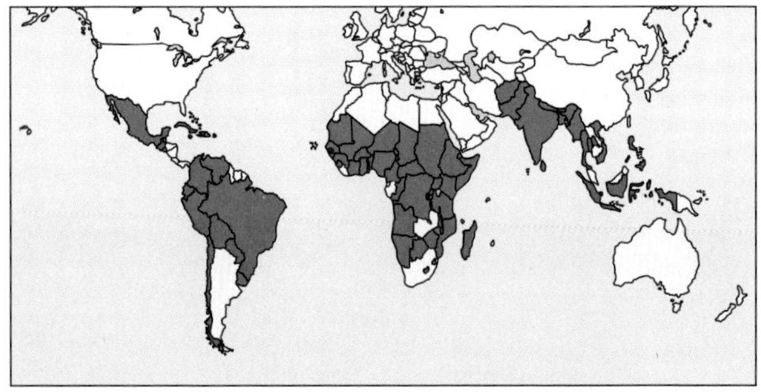

Figure 17-16 Scarcity of fuelwood, 2000. (Data from UN Food and Agriculture Organization)

Eucalyptus trees also deplete the soil of nutrients and produce toxic compounds that accumulate in the soil (because of low rainfall) and inhibit the uptake of nitrogen. In Karnata, India, villagers became so enraged over a government-sponsored project to plant these trees that they uprooted the saplings.

Experience shows that planting projects are most successful when local people participate in their planning and implementation. Programs work best that give village farmers incentives, such as ownership of the land or ownership of any trees they grow on village land. Emphasis should be placed on community woodlots, which are easy to tend and harvest, rather than on creating large fuelwood plantations located far from where the wood is needed.

Another promising method is to encourage villagers to use the sun-dried roots of various gourds and squashes as cooking fuel. These rootfuel plants, which regenerate themselves each year, produce large quantities of burnable biomass per unit area on dry deforested lands. They also help reduce soil erosion and produce an edible seed with a high protein content.

The traditional three-stone fire is a very inefficient way to burn wood, typically wasting about 94% of the wood's energy content. New, cheap, more efficient, less polluting stoves can provide both heat and light while reducing indoor air pollution, a major health threat. The stoves must be easy and cheap to build, and the materials used to make them must be easily accessible locally (Individuals Matter, right).

Cheap and easily made solar ovens (Figure 20-21d, p. 535) that capture sunlight can also be used to reduce wood use and air pollution in sunny and warm areas. However, they often are not accepted by cultures that also use fires for light and heat at night and as centers for social interaction.

Despite encouraging success in some countries (such as China, Nepal, Senegal, and South Korea), most developing countries suffering from fuelwood shortages have inadequate forestry policies and budgets and lack trained foresters. Such countries are cutting trees

Our Ecological Footprints

SPOTLIGHT

In 1996 William Rees and Mathis Wackernagel at the University of British Columbia in Canada introduced the concept of the **ecological footprint**.* It is a measure of the ecological impact of our **(1)** consumption of food, wood products, and other resources, **(2)** use of buildings, roads, garbage dumps, and other things that consume land space, and **(3)** the forests needed to absorb the excess of CO_2 produced by our burning of fossil fuels.

This analysis shows that 5.1 hectares (13 acres) of land are needed to provide the flows of materials and energy that sustain the average U.S. citizen (most people in western Europe require nearly as much). However, if you take all of the available ecologically productive land and divide it by the number of people living today, there are only 1.5 hectares (3.7 acres) for each person.

If this analysis is correct, it tells us that the whole world can never achieve the high-consumption lifestyle of most Americans because there isn't enough ecologically productive land. In fact, supporting today's entire human population in the lifestyle of most Americans (or most western Europeans) would require roughly three additional earths.

Critical Thinking

List three important things we could do to reduce the ecological footprint of the average person in developed countries? What three things would you be willing to do to reduce your own ecological footprint?

*See Mathis Wackernagel and William Rees, *Our Ecological Footprint: Reducing Human Impact on the Earth* (Philadelphia: New Society Publishers, 1996).

New Stoves Help Save India's Forests and Improve Women's Health

INDIVIDUALS MATTER

Since 1984 Lalita Balakrishnan—supported by the All-India Women's Conference (formed in 1926 with its early mission to educate Indian women)—has put together an army of 300,000 women dedicated to spreading efficient and smokeless stoves throughout much of India. These women have helped design, build, and install more than 300,000 smokeless wood stoves, called *chulhas*, across the country.

These stoves cost about $5 and are made of cow-dung, mud, and hay slapped together by hand, with a metal pipe leading to the roof. The Indian government pays one-third of the stoves' cost, and all materials (except the pipe) are provided by the stove buyer. Local women are hired to sell, install, and maintain the stoves.

During the past decade these cheap, fuel-efficient stoves have helped reduce deforestation and saved India more than 182,000 metric tons (200,000 tons) of fuelwood. Replacement of smoky indoor fires with such stoves has also led to improved health for an estimated 1 million Indian women.

In recognition of her efforts, the UN Environment Programme added the name of Lalita Balakrishnan to its 1996 Global 500 Roll of Honor.

for fuelwood and forest products 10–20 times faster than new trees are being planted.

17-5 MANAGING AND SUSTAINING RANGELANDS

What Is Rangeland, and Why Is It Important? Almost half of the earth's ice-free land is **rangeland**: land that supplies forage or vegetation for grazing (grass-eating) and browsing (shrub-eating) animals and is not intensively managed. Most rangelands are grasslands in arid and semiarid areas too dry for nonirrigated crops.

About 42% of the world's rangeland is used for grazing livestock. Much of the rest is too dry, cold, or remote from population centers to support large numbers of livestock. About 34% of the total U.S. land area is rangeland, most of it short-grass prairie in the arid and semiarid western half of the country (Figure 6-9). About 2% of the cattle and 10% of the sheep in the United States graze on public rangelands.

Most rangeland grasses have deep and complex root systems (Figure 13-17) that help anchor the plants. Blades of rangeland grass grow from the base, not the tip. Thus, as long as only its upper half is eaten and its lower half remains, rangeland grass is a renewable resource that can be grazed again and again.

Rangeland has a number of important ecological functions. It provides forage for large numbers of wild herbivores and essential habitats for a variety of wild plant and animal species. Rangelands also act as crucial watersheds that help replenish surface and groundwater resources by absorbing and slowly releasing rainfall.

Figure 17-17 Rangeland: overgrazed (left) and lightly grazed (right). (USDA Soil Conservation Service)

Rangeland is also a valuable resource for recreation. Since 1975 use of U.S. rangeland for activities such as hiking, camping, and hunting has risen dramatically.

What Is Overgrazing? **Overgrazing** occurs when too many animals graze for too long and exceed the carrying capacity of a grassland area. It lowers the productivity of vegetation and changes the number and types of plants in an area. Large populations of wild ruminants can overgraze rangeland during prolonged dry periods. However, most overgrazing is caused by excessive numbers of domestic livestock feeding for too long in a particular area.

Heavy overgrazing compacts the soil, which diminishes its capacity to hold water and to regenerate itself. Overgrazing also converts continuous grass cover into patches of grass and thus exposes the soil to erosion, especially by wind; then woody shrubs such as mesquite and prickly pear cactus take over. Overgrazing is the major cause of desertification in arid and semiarid lands (Figure 13-22). Figure 17-17 shows lightly grazed and overgrazed grassland side by side.

What Is the Condition of the World's Rangelands? *Range condition* usually is classified as *excellent* (containing more than 75% of its potential forage production), *good* (51–75%), *fair* (26–50%), and *poor* (0–25%). Limited data from surveys in various countries indicate that most of the world's rangelands have been degraded to some degree, mostly by desertification (Figure 13-22).

In 1990, 68% of nonarctic U.S. public rangeland was rated by the Bureau of Land Management and the General Accounting Office as being in unsatisfactory (fair or poor) condition, compared with 84% in 1936. Since 1936 U.S. public range area in excellent or good condition has doubled while the area rated as poor has shrunk by half. This is a considerable improvement, but there's still a long way to go. Conservation biologists and some range experts also point out that surveys of U.S. rangeland neglect the damage to vital riparian zones by livestock (Spotlight, p. 450).

How Can Rangelands Be Managed? The primary goal of rangeland management is to maximize livestock productivity without overgrazing rangeland vegetation. The most widely used method to prevent overgrazing is controlling the *stocking rate*—the number of each kind of animal placed in a given area—so that it doesn't exceed an area's carrying capacity. But determining the carrying capacity of a range site is difficult and costly, and carrying capacity varies with factors such as climatic conditions (especially drought), past grazing use, soil type, invasions by new species, kinds of grazing animals, and intensity of grazing.

To prevent overgrazing, both the numbers and the distribution of livestock on a rangeland must be controlled. Ranchers can control livestock distribution by **(1)** fencing off damaged rangeland and riparian zones, **(2)** rotating livestock from one grazing area to another, **(3)** providing supplemental feed at selected sites, and **(4)** situating water holes and salt blocks in strategic places.

A more expensive and less widely used method of rangeland management involves suppressing the growth of unwanted plants (mostly invaders) by herbicide spraying, mechanical removal, or controlled burning. A cheaper and less harmful way to discourage unwanted vegetation is controlled, short-term trampling by large numbers of livestock.

Seeding and applying fertilizer can increase growth of desirable vegetation. Even though this method usually costs too much, reseeding is an excellent way to restore severely degraded rangeland.

Case Study: Livestock and Coyotes Another aspect of rangeland management for livestock is *predator control*. For decades, hundreds of thousands of predators have been shot, trapped, and poisoned by U.S. ranchers, farmers, hunters, and federal officials in a little-known agency called Animal Damage Control (ADC).

Federally subsidized predator control programs have so reduced gray wolf (Case Study, p. 451) and grizzly bear populations that they are now endangered species. Today the U.S. Fish and Wildlife Service spends additional funds to protect and help revive populations of such species.

Coyotes, the main target of livestock predator control programs, are smaller relatives of wolves. The effect of coyotes on livestock, especially sheep, is a controversial issue in the Western United States. Sheep ranchers claim that coyotes kill large numbers of sheep on the open range and should be exterminated. However, many wildlife experts disagree. They agree that coyotes kill some sheep, but contend that their net effect is to increase rangeland vegetation for livestock and wild herbivores by reducing grazing pressure from rodents.

Because coyotes are so prolific, adaptable, and efficient, environmentalists contend that any attempt to control them is doomed to failure. They also argue that it would cost taxpayers much less to pay ranchers for each head of livestock killed by a coyote than to spend millions of dollars each year for predator control.

Environmentalists suggest using a combination of fences, repellents, and trained guard dogs to keep predators away. Such methods have given sheep producers in Kansas one of the country's lowest rates of livestock losses to coyotes.

In 1986 USDA researchers reported that predation can be sharply reduced by penning young lambs and cattle together for 30 days and then allowing them to graze together on the same range. When predators attack, cattle butt and kick them, protecting both themselves and the sheep. Llamas and donkeys are also tough fighters against predators and act in the same way to protect sheep. Livestock raisers have little incentive to use such methods, however, when a local ADC or state predator control official is only a phone call away and takes care of any predator problems at little or no charge.

Case Study: Livestock and U.S. Public Rangeland About 23,000 U.S. ranchers hold what are essentially lifetime permits to graze about 4 million livestock (3 million of them cattle) on BLM and Forest Service rangelands in 16 western states.

Permit holders pay the federal government a grazing fee for this privilege. Since 1981 grazing fees on public rangeland have been set by Congress at only one-fourth to one-eighth the going rate for comparable private land. This means that taxpayers give the roughly 1% of U.S. ranchers with federal grazing permits subsidies amounting to $20–150 million a year, depending on how the costs of federal rangeland management are calculated.

The economic value of a permit is a part of the overall worth of a ranch. It can serve as collateral for a loan and usually is renewed automatically every 10 years, allowing permit holders to treat federal land as part of their ranches. It's not surprising that politically influential permit holders have fought hard to block any change in this system.

The public subsidy does not end with low grazing fees, however. When other costs such as water pipelines, stock ponds, weed control, livestock predator control, clearing of undesirable vegetation, planting of grass for livestock, erosion, and loss of biodiversity are added, it's estimated that taxpayers are providing 23,000 ranchers with an annual subsidy of about $2 billion (an average of $69,000 per rancher) to produce 2% of the country's beef. In 1991 one rancher alone received subsidies worth almost $950,000, which explains why many critics charge that the public lands grazing programs are little more than rancher welfare.

However, ranchers with permits say that grazing fees on public lands should be low because most private rangeland is more productive than public rangeland. Environmentalists contend that this varies with the land involved and that even when this is true, current grazing fees on public rangelands are much too low. What do you think?

Solutions: How Can U.S. Public Rangeland Be Managed More Sustainably? Some environmentalists believe that all commercial grazing of livestock on Western public lands should be phased out over the next 10–15 years. They contend that the water-poor Western rangeland (Figure 12-5) is not a good place to raise cattle and sheep, which require a lot of water.

However, some ranchers have demonstrated that Western rangeland can be grazed sustainably; they make the case that such practices should not be prohibited on public rangeland (Individuals Matter, p. 452). Some environmentalists agree that with proper management, ranching on Western rangeland is a potentially sustainable operation and that encouraging sustainable ranching practices keeps the land from being developed and destroyed. They believe that the following measures can promote more sustainable use of public rangeland:

- *Allow no or only limited grazing on riparian areas.*
- *Ban grazing on rangeland in poor condition.*
- *Use competitive bidding for all grazing permits (similar to the system used to sell timber from natural forests).*
- *Allow individuals and environmental groups to purchase grazing permits and not use the land for grazing.*
- *Until a competitive bidding system is implemented, raise grazing fees to fair market value.*

The problem is that despite their small numbers, Western ranchers with grazing permits wield enough political power to see that measures such as those on the previous page are not enacted by elected officials. There are also efforts in the U.S. Congress to turn management of public rangelands over almost completely to ranchers and exempt them from most environmental laws and any legal challenges. What do you think?

17-6 MANAGING AND SUSTAINING NATIONAL PARKS

How Popular Is the Idea of National Parks? Today, over 1,100 national parks larger than 1,000 hectares (2,500 acres) each are located in more than 120 countries. These parks cover a total area equal to that of Alaska, Texas, and California combined. This important achievement in global conservation was spurred by the creation of the national park system in the United States in 1912.

The U.S. National Park System is dominated by 54 national parks, most of them in the West (Figure 17-2), sometimes called *America's crown jewels.*

These national parks are supplemented by state, county, and city parks. Most state parks are located near urban areas and thus are more heavily used than national parks.

How Are Parks Being Threatened? Parks everywhere are under pressure. In developing countries parks are often invaded by local people who desperately need wood, cropland, and other resources. Poachers kill animals to obtain and sell rhino horns, elephant tusks, and furs. Park services in developing countries typically have too little money and too few personnel to fight these invasions, either by force or by education. In addition, most of the world's national parks are too small to sustain many of the larger animal species.

Popularity is one of the biggest problems of national and state parks in the United States (and other developed countries). Because of increased numbers of roads, cars, and affluent people, annual recreational visits to major national parks increased fourfold (and visits to state parks sevenfold) between 1950 and 1997; visits are projected to increase from 271 million in 1997 to 500 million by 2010.

During the summer, the most popular U.S. national and state parks are often choked with cars and trailers and hour-long entrance backups. They are also plagued by noise, traffic jams, litter, vandalism, poaching, deteriorating trails, polluted water, rundown visitor centers, garbage piles, and crime. Many visitors to heavily used parks leave the city to commune with nature, only to find the parks as noisy, congested, and stressful as where they came from.

U.S. Park Service rangers now spend an increasing amount of their time on law enforcement instead of conservation, management, and education. Since 1976 the number of federal park rangers has not changed while the number of visitors to park units has risen by 85 million; the number of visitors is expected to rise by another 162 million in the next 20 years. Currently, there is one ranger for every 84,200 visitors to the major national parks. Because overworked rangers earn an

SPOTLIGHT

Endangered Riparian Zones

According to some conservation biologists and rangeland experts, overall estimates of rangeland condition obscure severe damage to certain heavily grazed areas, especially vital **riparian zones**: thin strips of lush vegetation along streams.

These zones help prevent floods (and help keep streams from drying out during droughts) by storing and releasing water slowly from spring runoff and summer storms. These centers of biodiversity also provide habitats, food, water, and shade for wildlife in the arid and semiarid Western lands. Studies indicate that 65–75% of the wildlife in the Western United States is totally dependent on riparian habitats.

Because cattle need lots of water, they congregate near riparian zones and feed there until the grass and shrubs are gone. As a result, riparian vegetation is destroyed by trampling and overgrazing. It takes only a few head of cattle to degrade a riparian zone. A 1988 U.S. General Accounting Office report concluded that "poorly managed livestock grazing is the major cause of degraded riparian habitat on federal rangelands."

Riparian areas can be restored by using fencing to restrict access to degraded areas and by developing off-stream watering sites for livestock. Sometimes protected areas can recover in a few years. Despite the threat to soil and wildlife, little has been done to protect or repair riparian zones on U.S. public land, mostly because of political opposition by ranchers and lack of awareness of the problem by the general public.

Critical Thinking

Do you believe that riparian zones on public rangelands in the United States should receive stronger protection? Explain. If so, how would you see that such protection is provided?

CASE STUDY

Who's Afraid of the Big Gray Wolf?

Figure 17-18 The gray wolf is an endangered species in the lower 48 states (except Minnesota, where it is listed as threatened). Efforts to return this species to its former habitat in the Yellowstone National Park area have been vigorously opposed by ranchers, hunters, miners, and loggers, but wolves were reintroduced beginning in 1995. A court case may require that all of these wolves be removed. (Tom J. Ulrich/Visuals Unlimited)

At one time, the gray wolf (Figure 17-18) ranged over most of North America, but between 1850 and 1900, 2 million wolves were systematically shot, trapped, and poisoned by ranchers, hunters, and government employees. The idea was to make the West and the Great Plains safe for livestock and for big-game animals prized by hunters.

This strategy worked. The species is now listed as endangered in all 48 lower states except Minnesota (which has an estimated 2,500 wolves).

According to Yellowstone biologist Douglas Smith, "The gray wolf is a keystone species. You remove it and the effects cascade down to the grasses." Ecologists now recognize the important role these predators once played in parts of the West and the Great Plains by culling herds of bison, elk, caribou, and mule deer of weaker animals, thereby strengthening the genetic pool of the survivors. In recent years, these herds have proliferated, devastating some of the area's vegetation, increasing erosion, and threatening the niches of other wildlife species (Case Study, p. 453).

In 1987, the U.S. Fish and Wildlife Service proposed that gray wolves be reintroduced into the Yellowstone ecosystem, an idea that brought outraged protests. Some objections came from ranchers who feared the wolves would attack their cattle and sheep; one enraged rancher said that the idea was "like reintroducing smallpox." Other protests came from **(1)** hunters who feared that the wolves would kill too many big game animals and **(2)** miners and loggers who worried that the government would force them to cease operations on wolf-populated federal lands.

National Park Service officials promised to trap or shoot any wolves that killed livestock outside designated areas. Ranchers would also be reimbursed from a private fund established by Defenders of Wildlife for any livestock verified to have been killed by wolves. However, these offers fell on deaf ears, and many ranchers and hunters worked hard to unsuccessfully delay or defeat the plan in Congress.

Since 1995 federal wildlife officials have caught gray wolves in Canada and relocated them in Yellowstone National Park and northern Idaho. By 1999 the population of 66 gray wolves released in the Yellowstone system had grown by natural reproduction to nearly 240.

Since 1995 gray wolves that have wandered out of the Yellowstone ecosystem have killed 7 head of cattle and 84 sheep. The Defenders of Wildlife has paid out more than $21,000 to compensate owners for their losses.

Some ranchers and hunters say that they'll take care of the wolves quietly—what they call the "shoot, shovel, and shut up" solution. Meanwhile, there are continuing efforts in Congress to eliminate the program and its funding.

In 1997 a Wyoming federal appeals court judge found the wolf recovery program illegal. He said that the U.S. Fish and Wildlife Service had violated the law by declaring the wolves an "experimental population." That designation denies them the protection of the Endangered Species Act. The judge ordered that the Canadian wolves reintroduced into Yellowstone National Park and central Idaho be removed. However the case was successfully appealed by several wildlife conservation organizations.

Critical Thinking

Do you agree or disagree with the program to reintroduce populations of the gray wolf in the Yellowstone ecosystem? Explain. What are the major drawbacks of this program? Could the money be better spent on other wildlife programs? If so, what programs would you suggest?

The Eco-Rancher

INDIVIDUALS MATTER

Wyoming rancher Jack Turnell is one of a new breed of cowpuncher who gets along with environmentalists. He talks about riparian ecology and biodiversity as fluently as he talks about cattle: "I guess I have learned how to bridge the gap between the environmentalists, the bureaucracies, and the ranching industry."

Turnell grazes cattle on his 32,000-hectare (80,000-acre) ranch south of Cody, Wyoming, and on 16,000 hectares (40,000 acres) of Forest Service land on which he has grazing rights. For the first decade after he took over the ranch, he raised cows the conventional way. Since then, he's made some changes.

Turnell disagrees with the proposals by environmentalists to raise grazing fees and remove sheep and cattle from public rangeland. He believes that if ranchers are kicked off the public range, ranches like his will be sold to developers and chopped up into vacation sites and homes ("ranchettes"), irreversibly destroying the range for wildlife and livestock alike.

At the same time, he believes that ranches can be operated in more ecologically sustainable ways. To demonstrate this, Turnell **(1)** began rotating his cows away from the riparian areas, **(2)** gave up most uses of fertilizers and pesticides, and **(3)** crossed his Hereford and Angus cows with a French breed that tends to congregate less around water. Most of his ranching decisions are made in consultation with range and wildlife scientists, and changes in range condition are carefully monitored with photographs.

The results have been impressive. Riparian areas on the ranch and Forest Service land are lined with willows and other plant life, providing lush habitat for an expanding population of wildlife, including pronghorn antelope, deer, moose, elk, bear, and mountain lions. And this eco-rancher now makes more money because the higher-quality grass puts more meat on his cattle.

average salary of less than $28,000, many leave for better-paying jobs.

Wolves (Case Study, p. 451), bears, and other large predators in and near various parks have all but vanished because of excessive hunting, poisoning by ranchers and federal officials, and the limited size of most parks. As a result, populations of species these predators once helped control have exploded, destroying vegetation and crowding out other species.

Nonnative species have moved into or been introduced into many parks. Wild boars (imported to North Carolina in 1912 for hunting) are threatening vegetation in part of the Great Smoky Mountains National Park. The Brazilian pepper tree has invaded Florida's Everglades National Park. Mountain goats in Washington's Olympic National Park trample native vegetation and accelerate soil erosion. At the same time that some species have moved into parks, other native species of animals and plants (including many threatened or endangered species) are being killed or removed illegally in almost half of U.S. national parks.

Nearby human activities outside many national parks also threaten the ecological health of such parks. Wildlife and recreational values are threatened by mining, logging, livestock grazing, coal-burning power plants, water diversion, and urban development. Polluted air drifts hundreds of kilometers, killing ancient trees in California's Sequoia National Park and blurring the awesome vistas at Arizona's Grand Canyon. According to the National Park Service, air pollution affects scenic views in national parks more than 90% of the time.

That's not all. Mountains of trash wash ashore daily at Padre Island National Seashore in Texas. Water use in Las Vegas threatens to shut down geysers in the Death Valley National Monument. Visitors to Sequoia National Park complain of raw sewage flowing through a parking lot; similar contamination has occurred for years at Kentucky's Mammoth Cave.

Unless a massive ecological restoration project is successful, Florida's Everglades National Park may dry up, making it the country's most endangered national park. Much of the water needed to sustain the park's wildlife has been diverted for crops and cities by a network of canals, levees, spillways, and pumping stations to the north. Nesting birds in the park declined by 90%, from about 90,000 in 1931 to fewer than 9,000 in 1994. All other vertebrates, from deer to turtles, are down 75–95%.

By the 1970s state and federal officials recognized that this massive plumbing project had been a serious ecological blunder. After over 20 years of political haggling, in 1990 state and federal governments agreed on a massive restoration project to undo some of the damage. Whether this 20-year project will work and be funded adequately remains to be seen. However, the high cost of this project is another example of a fundamental lesson from nature: Prevention is the cheapest and best way to go.

Solutions: How Can Park Management in Developing Countries Be Improved? Some park managers, especially in developing countries, are developing *integrated management plans* that combine conservation practices with sustainable development of resources in the park and surrounding areas. In such plans, the inner core and vulnerable areas of each park are protected from development and treated as wilderness. Restricted numbers of people are allowed to use these areas for hiking, nature study, ecological research, and other nondestructive recreational and educational activities.

In buffer areas surrounding the core, controlled commercial logging, sustainable grazing by livestock, and sustainable hunting and fishing by local people are allowed. Money spent by park visitors adds to local income. By involving nearby residents in developing park management and restoration plans, managers and conservation biologists try to help them see the park as a vital resource to protect and sustain rather than ruin. Costa Rica has led the way in protecting some of its remaining parklands and other biodiversity sanctuaries (Solutions, p. 455).

Integrated park management plans look good on paper, but they cannot be carried out without adequate funding and the support of nearby landowners and users. Moreover, in some cases the protected inner core may be too small to sustain the park's larger animal species.

Solutions: How Can Management of U.S. National Parks Be Improved? The 54 major U.S. national parks are managed under the principle of *natural regulation*, as if they were wilderness ecosystems that would sustain themselves if left alone. Many ecologists consider this a misguided policy. Most parks are far too small to even come close to sustaining themselves. Even the biggest ones, such as Yellowstone, cannot be isolated from the harmful effects caused by activities in nearby areas and from destruction from within by exploding populations of some plant-eating species such as elk (Case Study, right).

The U.S. National Park Service has two goals that increasingly conflict: **(1)** to preserve nature in parks and **(2)** to make nature more available to the public. The Park Service tries to accomplish these goals with a $1.5-billion annual budget and a $3.5-billion backlog of maintenance, repairs, and high-priority construction projects at a time when park usage and external threats to the parks are increasing. Private developers continually pressure Congress and the Park Service to turn national parks into high-tech, heavily developed theme or amusement parks instead of the repositories of natural ecosystems and centers for learning about nature they were intended to be.

In 1988 the Wilderness Society and the National Parks and Conservation Association made a number of suggestions for sustaining and expanding the national park system, including the following:

- *Have all user and entrance fees for national park units be used for the management, upkeep, and repair of the national parks.* Currently, such fees go into the general treasury.
- *Require integrated management plans for parks and other nearby federal lands.*
- *Increase the budget for adding new parkland near the most threatened parks.*
- *Increase the budget for buying private lands inside parks.*
- *Locate all new and some existing commercial facilities and visitor parking areas outside parks.*
- *Require private concessionaires who provide campgrounds, restaurants, hotels, and other services for park visitors to* **(1)** *compete for contracts,* **(2)** *pay franchise fees equal to 22% of their gross (not net) receipts, and* **(3)** *pour some of these receipts back into parks.* In 1996, private concessionaires in national parks paid only about 6.7% of their gross receipts in franchise fees to the government. Many large concessionaires have long-term contracts by which they pay the government as little as 0.75% of their gross receipts.
- *Allow concessionaires to lease but not own facilities inside parks.*
- *Provide more funds for park system maintenance and repairs.*
- *Raise fees for park visitors and pour receipts back into parks.*
- *Restrict the number of visitors to crowded park areas.*
- *Increase the number and pay of park rangers.*
- *Survey the condition and types of wildlife species in parks.*
- *Encourage volunteers to give tours and lectures to visitors.*
- *Encourage individuals and corporations to donate money for park maintenance and repair.*

Case Study: What Changes Are Occurring in Yellowstone's Ecosystem? Some ecologists and river experts point to Yellowstone National Park as an ecosystem that is under stress as **(1)** stands of aspens and willows die, **(2)** woody shrubs disappear, **(3)** populations of once abundant animal species dwindle or disappear, and **(4)** stream banks erode.

These scientists point to the explosion of the elk population as the major culprit. They have collected data indicating that between 1968 and 1996, the National Park Service's hands-off management policy of *natural regulation* has been the key factor in allowing the park's elk population to increase from 3,100 to about 20,000.

According to Charles Kay, a wildlife biologist at Utah State University, the rising elk population is consuming tree shoots and other forms of vegetation and disrupting a series of important ecological connections. For example, centuries old stands of aspen, which reproduce by sending up shoots or suckers, have died out as their shoots have been eaten by elk.

Loss of these and other tree species has eliminated food and dam-building materials for beavers, whose numbers have plummeted in the last 60 years. As the beaver populations have dropped, so have their dams and ponds, which trapped eroded silt and slowly built up the park's streambeds. With a stronger flow of water and without aspens, cottonwoods, and willows to hold streambed soil in place, some stream banks have eroded and collapsed.

Since adopting their policy of natural regulation in the early 1970s, park service officials have claimed that the large increase in elk represents the park ecosystem returning to its natural equilibrium state. Many ecologists say this is an outdated idea because recent ecological research has revealed that ecosystems are in a constant state of change and rarely attain an equilibrium point. Park officials also attribute the decline in woody streamside vegetation to climate change or the long-standing policy of suppressing fires within the park. What, if anything, do you think should be done to help improve the ecological health of the Yellowstone ecosystem?

17-7 PROTECTING AND MANAGING WILDERNESS AND OTHER BIODIVERSITY SANCTUARIES

How Much of the Biosphere Should We Protect from Exploitation? Most wildlife and conservation biologists believe that the best way to preserve biodiversity and ecological integrity is through a worldwide network of reserves, parks, wildlife sanctuaries, wilderness, and other protected areas. Currently, about 6% of the world's land area is either strictly or partially protected in more than 20,000 nature reserves, parks, wildlife refuges, and other areas around the globe.

This is an important beginning, but conservation biologists say that to keep biodiversity and ecological integrity from being depleted, a minimum of 10% of the globe's land area should be protected. Moreover, many existing reserves are too small to provide any real protection for the wild species that live on them and many existing preserves receive little protection. Currently, less than 2% of the world's protected areas are at least 10,000 square kilometers (3,900 square miles) in size, a conservative estimate of the minimum habitat needed to maintain viable populations of the largest mammals.

In 1981 the UN Educational, Scientific, and Cultural Organization (UNESCO) proposed that at least one (and ideally five or more) *biosphere reserves* be set up in each of the earth's 193 biogeographical zones. Each reserve should be large enough to prevent gradual species loss and should combine both conservation and sustainable use of natural resources. To date, more than 300 biosphere reserves have been established in 76 countries.

A well-designed biosphere reserve has three zones (Figure 17-19): **(1)** a *core area* containing an important ecosystem that has had little or no disturbance from human activities, **(2)** a *buffer zone* where activities and uses are managed in ways that help protect the core, and **(3)** a second buffer or *transition zone*, which combines conservation and sustainable forestry, grazing, agriculture, and recreation. Buffer zones can also be used for education and research.

Proponents of large reserves say they are the only way to maintain viable populations of large, wide-ranging species (such as panthers, elephants, and grizzly bears). Also, large reserves can **(1)** sustain more species, **(2)** minimize edge effects, and **(3)** provide greater habitat diversity than small preserves. These advantages of large parks or reserves are based on the theory of island biogeography developed by Robert MacArthur and Edward O. Wilson (Connections, p. 133 and Figure 5-17), in which the reserves are assumed to be habitat islands surrounded by expanses of often inhospitable development or other terrain.

In reality, few countries are physically or politically able to set aside and protect large biosphere reserves or other protected areas. Moreover, some research indicates that in some locales several well-placed, medium-sized, isolated reserves may better protect a greater variety of habitat types and more populations of rare species than a single large one. If the population of a species is wiped out in one reserve because of fire, epidemic, or some other disaster, the species might survive in the other reserves.

Conservation biologists also suggest establishing protected habitat corridors between reserves to **(1)** help support more species, **(2)** allow migrations when environmental conditions in a reserve deteriorate, **(3)** reduce loss of genetic diversity from inbreeding, and **(4)** help preserve animals that must make seasonal migrations to obtain food.

However, there are some drawbacks to corridors. They can allow the movement of pest species, disease, and exotic species between reserves. Also, animals mov-

Parks and Ecological Restoration in Costa Rica

SOLUTIONS

Costa Rica (Figure 9-21), smaller in area than West Virginia, was once almost completely covered with tropical forests. Between 1963 and 1983, politically powerful ranching families cleared much of the country's forests to graze cattle, and they exported most of the beef to the United States and western Europe. By 1983, only 17% of the country's original tropical forest remained, and soil erosion was rampant.

Despite such widespread forest loss (which continues today), tiny Costa Rica is a superpower of biodiversity, with an estimated 500,000 species of plants and animals. A single park in Costa Rica is home for more bird species than all of North America.

In the mid-1970s, Costa Rica established a system of national parks and conservation reserves that by 1998 included 12% of its land (6% of it in reserves for indigenous peoples). As a result, Costa Rica now has larger proportion of land devoted to biodiversity conservation than any other country.

One reason for this accomplishment in biodiversity protection was the establishment in 1963 of the Organization of Tropical Studies, a consortium of more than 50 U.S. and Costa Rican universities with the goal of promoting research and education in tropical ecology. The resulting infusion of several thousand scientists has helped Costa Ricans appreciate their country's great biodiversity. It also led to the establishment in 1989 of the National Biodiversity Institute (INBio), a private nonprofit organization set up to survey and catalog the country's biodiversity. This biodiversity conservation strategy has paid off. Today, revenue from tourism (almost two-thirds of it from ecotourists) is the country's largest source of outside income.

However, legal and illegal deforestation threatens this plan because of the country's population growth and poverty which affects 10% of its people. Currently, this small country still loses roughly 400 square kilometers (150 square miles) of primary forest per year—four times the rate of loss in Brazil.

Costa Rica is also the site of one of the world's largest *ecological restoration* projects. In the lowlands of the country's Guanacaste National Park, a small tropical seasonal forest is being restored and relinked to the rain forest on adjacent mountain slopes.

Daniel Janzen, professor of biology at the University of Pennsylvania and a leader in the growing field of rehabilitation and restoration of degraded ecosystems, has helped galvanize international support and has raised more than $10 million for this restoration project.

Janzen's vision is to make the nearly 40,000 people who live near the park an essential part of the restoration of the degraded forest, a concept he calls *biocultural restoration*. By actively participating in the project, local residents reap enormous educational, economic, and environmental benefits. Local farmers have been hired to sow large areas with tree seeds and to plant seedlings started in Janzen's lab.

Students in grade schools, high schools, and universities study the ecology of the park in the classroom and then go on field trips to the park itself. Educational programs for civic groups and tourists from Costa Rica and elsewhere stimulate the local economy.

The project also serves as a training ground in tropical forest restoration for scientists from all over the world. Research scientists working on the project give guest classroom lectures and lead some of the field trips.

Janzen recognizes that in 20–40 years today's children will be running the park and the local political system. If they understand the importance of their environment, they are more likely to protect and sustain its biological resources. He believes that education, awareness, and involvement—not guards and fences—are the best ways to protect ecosystems from unsustainable use.

Critical Thinking

Why do you think Costa Rica has been able to set aside a much larger percentage of its land for national parks and conservation reserves than any other country? Do you agree or disagree that such a plan should be implemented in the country where you live? If so, how would you go about doing this?

ing along corridors may be exposed to greater risks of predation by both natural predators and human hunters. Although most conservation biologists support the development of wildlife corridors between reserves, they recognize the need to evaluate them on a park-by-park basis.

So far, most biosphere reserves fall short of the ideal (Figure 17-19), and too little funding has been provided for their protection and management. An international fund to help developing countries protect and manage bioregions and biosphere reserves would cost about $100 million per year—about what the world's nations spend on arms every 90 minutes.

Because there won't be enough money to protect most of the world's biodiversity, environmentalists believe that efforts should be focused on the most biodiverse countries (Figure 17-20) and threatened species-rich areas within such countries. In the United States, conservation biologists have urged Congress to pass an Endangered Ecosystems or Biodiversity Protection Act as an important step toward surveying and preserving

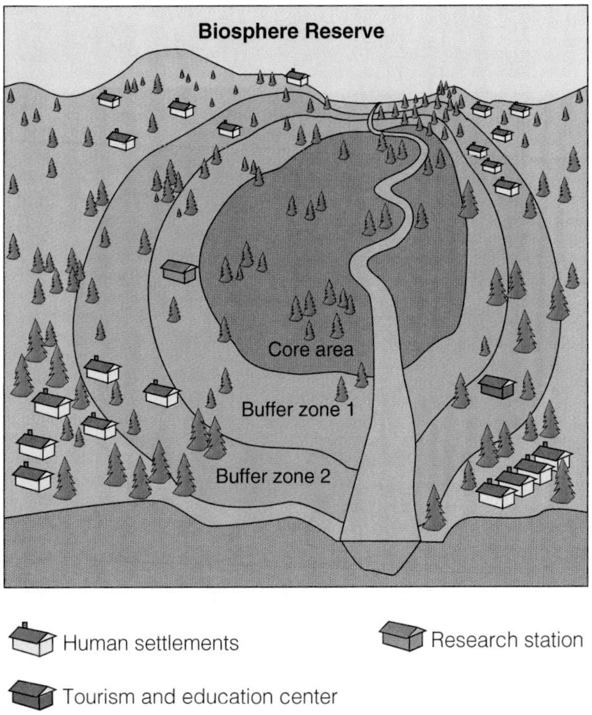

Figure 17-19 Design of a model biosphere reserve. In traditional parks and wildlife reserves, well-defined boundaries keep people out and wildlife in. By contrast, biosphere reserves recognize people's needs for access to sustainable use of various resources in parts of the reserve.

the country's biodiversity. In addition to government action, private groups have helped establish biodiversity and wildlife sanctuaries (Solutions, p. 458).

Most developers and resource extractors and many economists disagree with protecting even the current 6% of the earth's remaining undisturbed ecosystems. They contend that such areas contain valuable resources that would add to economic growth. They view protected areas as "useless" and "empty" because they are not being used to provide mineral and other resources and land for development for humans.

Ecologists and conservation biologists disagree and view protected areas as islands of biodiversity and ecological integrity that are vital parts of the earth's natural capital that sustains all life and economies (Figure 1-2) and are centers of evolution. Instead of being empty or nearly empty of life, these oases of biodiversity are teeming with life. What do you think?

What Is Wilderness and How Much Is Left? Protecting biodiversity and ecological integrity means protecting *wildness*. One way to do this is to protect undeveloped lands from exploitation by setting them aside as wilderness. According to the U.S. Wilderness Act of 1964, **wilderness** consists of areas "where the earth and its community of life are untrammeled by man, where man himself is a visitor who does not remain." U.S. President Theodore Roosevelt summarized what we should do with wilderness: "Leave it as it is. You cannot improve it."

The U.S. Wilderness Society estimates that a wilderness area should contain at least 4,000 square kilometers (1,500 square miles); otherwise it can be affected by air, water, and noise pollution from nearby human activities. Conservation biologists urge that remaining wilderness be protected by law everywhere, focusing first on the most endangered spots in wilderness- and species-rich countries (Figure 17-20). Conservation biologists urge that more wilderness areas be protected in the United States (Case Study, p. 459).

Why Preserve Wilderness? According to wilderness supporters, we need wild places where people can experience the beauty of nature and observe natural biological diversity and where they can enhance their mental and physical health by getting away from noise, stress, and large numbers of people. Wilderness preservationist John Muir advised us,

> *Climb the mountains and get their good tidings. Nature's peace will flow into you as the sunshine into the trees. The winds will blow their freshness into you, and the storms their energy, while cares will drop off like autumn leaves.*

Even those who never use the wilderness may want to know it is there, a feeling expressed by novelist Wallace Stegner:

> *Save a piece of country . . . and it does not matter in the slightest that only a few people every year will go into it. This is precisely its value. . . . We simply need that wild country available to us, even if we never do more than drive to its edge and look in. For it can be a means of reassuring ourselves of our sanity as creatures, a part of the geography of hope.*

Recently some critics have argued that protecting wilderness for its scenic and recreational value for a small number of people is an outmoded concept that keeps some areas of the planet from being economically useful to humans. They also contend that wilderness protection diverts money, energy, and talent from learning how to better manage urban and other landscapes disturbed by human activities.

Scientists argue that this idea is outmoded because it fails to take into account the research that has revealed

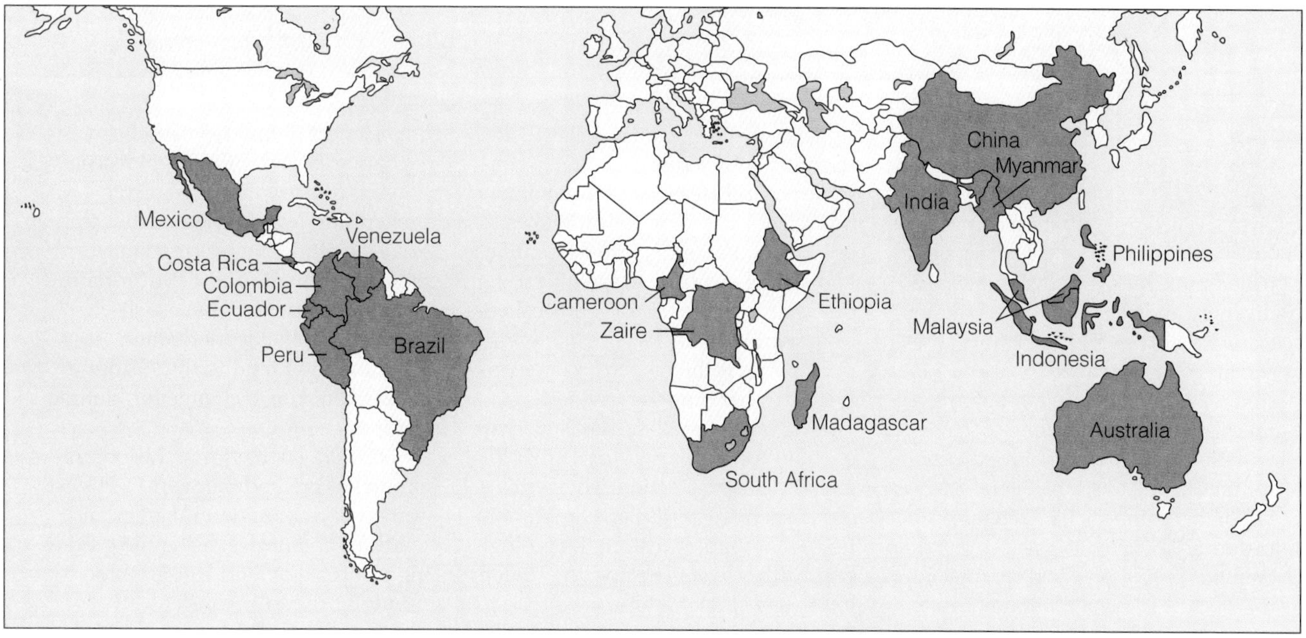

Figure 17-20 The earth's most biodiverse countries. Conservation biologists believe that efforts to preserve repositories of biodiversity as protected wilderness areas should be concentrated in these 17 species-rich countries. The 80% of the world's people who live in developing countries have 80% of the world's biodiversity but only 15% of the world's income and financial wealth. Protecting the world's biodiversity will require considerable financial and scientific help from developed countries. (Data from Conservation International and World Wildlife Fund)

the ecological importance of wilderness areas for all people and all life. To most biologists, *the most important reason for protecting wilderness and other areas from exploitation and degradation is to preserve the biodiversity and ecological integrity they contribute as a vital part of the earth's natural capital* (Figure 1-2). What do you think about this controversy?

Wilderness areas **(1)** provide mostly undisturbed habitats for wild plants and animals, **(2)** protect diverse biomes from damage, and **(3)** provide a laboratory in which we can discover more about how nature works. In other words, wilderness is a biodiversity and wildness bank and an ecoinsurance policy.

Most ecologists and conservation biologists believe there is an urgent need to **(1)** protect at least 10–12% of the earth's surface as wilderness and **(2)** allow some degraded areas to return to a more wild state. They agree that we need to learn much more about how to manage and restore land disturbed by human activities. However, they don't believe that this should be done by decreasing or unraveling wilderness protection, which will lead to even more areas disturbed by human activities. What do you think?

To protect more of the earth's biodiversity and ecological integrity, conservation biologists and other environmental scientists believe that the designation of future wilderness and other protected areas should focus on prairies, lowland forests, and wetlands that are largely absent from the current system. This will bring wilderness advocates into increasing conflict with developers, farmers, ranchers, and resource extractors who view remaining lands in these categories as prime sites for human use.

Some analysts believe that wilderness should be preserved because the wild species it contains have a right to **(1)** exist (or struggle to exist) and **(2)** play their roles in the earth's ongoing saga of biological evolution and ecological processes, without human interference. What do you think about this idea?

How Can We Protect Wilderness? To protect the most popular areas from damage, wilderness managers increasingly designate sites where camping is allowed and limit the number of people using these sites at any one time. Managers have also increased the number of wilderness rangers to patrol vulnerable areas, and they have enlisted volunteers to pick up trash discarded by thoughtless users.

Environmental historian and wilderness expert Roderick Nash suggests that wilderness areas be divided

SOLUTIONS

The Nature Conservancy

Private groups play an important role in establishing wildlife refuges and other protected areas as a major strategy to protect biological diversity and ecological integrity. For example, since 1951 the Nature Conservancy has preserved over 38,000 square kilometers (14,700 square miles) of areas in the United States with unique ecological or aesthetic significance. Outside the United States, it has helped protect 80,000 square kilometers (31,000 square miles) of wildlife habitat. The Nature Conservancy now has over 800,000 members and programs in all 50 states, Latin America, and the Caribbean.

For several years, the Conservancy has been rated among the top conservation organizations. It has one of the lowest overhead rates of any nonprofit organization, and 85% of all contributions go directly to its conservation programs.

The Nature Conservancy began in 1951 when an association of professional ecologists wanted to use their scientific knowledge of nature to conserve natural areas. Since its beginning this science-based organization has used the most sophisticated scientific knowledge available to identify and rank sites that are unique and ecologically significant and whose biodiversity or existence is threatened by development or other human activities.

Once sites are identified, a variety of techniques are used to see that they receive legal protection. Then science-based management plans are drawn up to maintain or restore the ecological health of each site and to provide long-term stewardship.

This conservation organization uses private and corporate donations to maintain a fund of over $150 million, which can be used to buy ecologically important pieces of land or wetlands threatened by development when no other option is available. Sometimes this land is then sold or donated to federal or state agencies.

If it cannot buy land for the protection of habitat, the Conservancy helps landowners obtain tax benefits in exchange for accepting legal restrictions or conservation easements preventing development. Other techniques include long-term management agreements and debt-for-nature swaps. Landowners have also received sizable tax deductions by donating their land to the Nature Conservancy in exchange for lifetime occupancy rights.

Through such efforts, this organization has created the world's largest system of private natural areas and wildlife sanctuaries, using the guiding principle of land conservation through private action.

Critical Thinking

Do you favor this private-group approach to protecting biodiversity over the government approach to protecting public lands and endangered species? Explain. Would you be in favor of selling or giving some public lands to private groups such as the Nature Conservancy? Explain.

into three categories. The easily accessible, popular areas would be intensively managed and have trails, bridges, hiker's huts, outhouses, assigned campsites, and extensive ranger patrols. Large, remote wilderness areas would be used only by people who get a permit by demonstrating their wilderness skills. The third category, biologically unique areas, would be left undisturbed as gene pools of plant and animal species, with no human entry allowed. What do you think about this idea?

What Is the U.S. National Wild and Scenic Rivers System? In 1968 Congress passed the National Wild and Scenic Rivers Act, which allows rivers and river segments with outstanding scenic, recreational, geological, wildlife, historical, or cultural values to be protected in the National Wild and Scenic Rivers System. These waterways are to be kept free of development; they may not be widened, straightened, dredged, filled, or dammed along the designated lengths. The only activities allowed are camping, swimming, nonmotorized boating, sport hunting, and sport and commercial fishing. New mining claims are permitted in some areas, however.

Currently, only 0.2% of the country's 6 million kilometers (3.5 million miles) of waterways along some 150 stretches are protected by the Wild and Scenic River System. In contrast, 17% of the country's wild river length has been tamed and land flooded by dams and reservoirs.

Environmentalists have urged Congress to add 1,500 additional river segments to the system, a goal that is vigorously opposed by some local communities and anti-environmental groups. If that goal is achieved, about 2% of the country's river systems would be protected. Environmentalists also urge that a permanent federal administrative body be established to manage the Wild and Scenic Rivers System and that states develop their own wild and scenic river programs. What do you think?

Some actions you can take to help sustain the earth's biodiversity and ecological integrity are listed in Appendix 6.

The U.S. National Wilderness Preservation System

CASE STUDY

In the United States, preservationists have been trying to save wild areas from development since 1900. On the whole, they have fought a losing battle. Not until 1964 did Congress pass the Wilderness Act, which allowed the government to protect undeveloped tracts of public land from development as part of the National Wilderness Preservation System.

Only about 4.6% of U.S. land area is protected as wilderness, with almost three-fourths of it in Alaska. Only 1.8% of the land area of the lower 48 states is protected, most of it in the West.

Of the 413 wilderness areas in the lower 48 states, only four are larger than 4,000 square kilometers. Furthermore, the present wilderness preservation system includes only 81 of the country's 233 distinct ecosystems. Like the national parks, most wilderness areas in the lower 48 states are habitat islands in a sea of development.

There remain almost 400,000 square kilometers (150,000 square miles) of scattered blocks of public lands that could qualify for designation as wilderness. Conservation biologists would like to see all this land protected as wilderness.

In 1999, U.S. national forests contained more than 60% of the wilderness in the contiguous 48 states. In 1999 President Bill Clinton acted to add 22 million hectares (65 million acres) of roadless but unprotected areas in the national forests to wilderness protection. Such efforts are strongly opposed by timber, mining, ranching, energy, and other interests who want either **(1)** to extract resources from these and most other public lands or **(2)** convert them to private ownership.

Some conservation biologists also urge that *wilderness recovery areas* be created by **(1)** closing and obliterating nonessential roads in large areas of public lands, **(2)** restoring wildlife habitats, **(3)** allowing natural fires to burn, and **(4)** reintroducing species that have been driven from such areas. However, resource developers lobby elected officials and government agencies to build roads into national forests (and other areas being evaluated for inclusion in the wilderness system) so that they can't be designated as wilderness. They also strongly oppose the idea of wilderness recovery areas.

Critical Thinking

1. Do you believe that all or most of the remaining potential wilderness area in the United States (or in the country where you live) should be protected? Explain.

2. Should a large portion of U.S. Forest Service and Bureau of Land Management land be reclassified as wilderness recovery areas, closed off for many uses, and allowed to undergo natural restoration as wildlife habitat? Explain. If so, how would you go about doing this?

We abuse land because we regard it as a commodity belonging to us. When we see land as a community to which we belong, we may begin to use it with love and respect.

ALDO LEOPOLD

REVIEW QUESTIONS

1. Define the boldfaced terms used in this chapter.

2. What are the five major types of public lands in the United States? What are the major uses allowed on each type? Summarize the ways in which developers, resource extractors, and environmentalists disagree about how these public lands should be used and managed.

3. Under what conditions are forests renewable resources? Distinguish between *old-growth forests, second growth forests,* and *tree farms.* Compare tree farms with monoculture food crops. What are the advantages and disadvantages of tree farms?

4. Summarize **(a)** the economic importance of forests and **(b)** the ecological importance of forests. What are the three major economic uses of forests? How useful are forests in helping slow projected global warming by removing excess carbon dioxide from the atmosphere? What are the advantages and disadvantages of putting an estimated price on the value of the ecosystem services provided by forests?

5. Describe the two major types of forest management and list the advantages and disadvantages of each type. Describe five major ways trees are harvested and list the advantages and disadvantages of each type.

6. List major ways suggested by many foresters and conservation biologists to use forests more sustainably.

7. List ways to help protect forests from **(a)** diseases and insects and **(b)** fire. Describe the three major types of forest fires. What are the benefits of fire for some plant and animal species? List the pros and cons of allowing fires in national parks to burn themselves out instead of fighting them.

8. Describe the general extent and condition of forests in the United States today. How are national forests in the United States used? What are the principles of *sustained yield* and *multiple use* that are supposed to guide the use of national forests in the United States? List the pros and cons of making the primary use of national forests **(a)** timber cutting and **(b)** recreation. What are the advantages and disadvantages of giving government subsidies to companies cutting timber in national forests? Summarize reforms in U.S. Forest Service policy that environmentalists and some foresters believe will lead to more sustainable use of these forests. Do you agree or disagree with these proposals? Explain.

9. Summarize some of the ecological services provided by old-growth forests in the Pacific Northwest in the United States. Summarize the controversy over continued cutting of remaining old-growth forests on public lands in the Pacific Northwest. What is your position on this issue? Explain. Summarize the controversy over the cutting of old-growth forests in Canada.

10. List ways to reduce the unnecessary waste of wood. List the pros and cons of making more paper from sources such as rice straw and kenaf fibers. Describe the *ecological footprint concept* and explain its usefulness.

11. How serious is the problem of tropical deforestation? Give economic and ecological reasons for protecting more of the world's **(a)** tropical forests, **(b)** temperate forests, and **(c)** boreal forests. What are the major causes of tropical deforestation and degradation? List ways to reduce tropical deforestation and degradation.

12. Describe efforts in Costa Rica **(a)** to protect some of its forests and **(b)** to restore some of its degraded forests. Describe Kenya's Green Belt movement.

13. What is the *fuelwood crisis* and how serious is this problem? List ways for dealing with the fuelwood crisis.

14. Discuss the pros and cons of widespread planting of neem trees.

15. What is *rangeland,* and what are its major economic and ecological functions? Explain how rangeland grass can be a renewable resource for livestock and wild herbivores. What is *overgrazing,* and what are its major causes? List major ways to manage rangelands to prevent overgrazing. What are *riparian areas,* and why are they important? How can riparian areas be protected? List the pros and cons of wild game ranching as a substitute for cattle ranching.

16. List the pros and cons of widespread killing of coyotes and other predators than can kill sheep. What are some alternatives to this approach?

17. List the pros and cons of providing government subsidies to ranchers holding permits to graze on public lands in the United States. Summarize and evaluate alternatives suggested by some ranchers and conservation biologists for more sustainable management of public rangeland.

18. What are the major threats to national parks in the United States and in other countries? What are the pros and cons of reintroducing species such as wolves into the Yellowstone National Park system in the United States? What is your position on this issue? Explain. Explain how the Yellowstone ecosystem is unraveling.

19. List ways to improve management of national parks in **(a)** developing countries and **(b)** the United States.

20. What is a *biosphere reserve*? What are the pros and cons of **(a)** setting aside large wildlife reserves and **(b)** using corridors to connect wildlife reserves?

21. What is *wilderness,* and why is it important? List the pros and cons of protecting more wilderness. Summarize the role of the Nature Conservancy in protecting ecosystems from development and degradation. What are major ways to manage wilderness areas? What is the U.S. National Wild and Scenic Rivers System?

CRITICAL THINKING

1. Should owners of private property be able to do anything they want with their land? Explain. If not, what specific restrictions on use should be imposed?

2. Should individuals and corporations be compensated financially if they are prevented from using land they own or from extracting resources from land they own because the land is classified by state or federal government as protected wetlands or habitats for endangered or threatened wildlife species? Explain.

3. Should private companies that harvest timber from U.S. national forests continue to be subsidized by federal payments for reforestation and for building and maintaining access roads? Explain.

4. Explain why you agree or disagree with the principles of sustainable forestry listed on p. 435.

5. Explain why you agree or disagree with each of the proposals given in the Spotlight on p. 439 for reforming federal forest management in the United States.

6. Explain why you agree or disagree with each of the proposals given on pp. 444–445 concerning protection of the world's tropical forests.

7. Many Brazilians believe that they have a right to clear their tropical forests for their economic benefit, as industrialized countries have been doing for hundreds of years. They also point out that the industrialized countries are by far the greatest producers of carbon dioxide from burning fossil fuels. Should industrialized countries have the right to tell Brazil to sharply reduce its deforestation to preserve biodiversity and cut CO_2 emissions when such use provides food and jobs for the Brazilian people? Why or why not?

8. Explain why you agree or disagree with each of the proposals given on p. 449 for providing more sustainable use of public rangeland in the United States.

9. Some have suggested that some rangelands could be used to raise wild grazing animals for meat instead of conventional livestock. Others consider it unethical to raise and kill wild herbivores. What do you think? Explain.

10. Explain why you agree or disagree with each of the proposals given on p. 453 concerning the U.S. national park system.

11. Is a protected forest or other area "useless" because it does not add significantly to economic growth? Explain. Describe the usefulness of a forest area in ecological terms.

***12.** Evaluate timber harvesting on private and public lands in your local area. What are the most widely used harvesting methods? Try to document any harmful environmental impacts. Have the economic benefits to the community outweighed any harmful environmental effects?

***13.** Evaluate cattle grazing on private and public rangeland and pastures in your local area. Try to document any harmful environmental impacts. Have the economic benefits to the community outweighed any harmful environmental effects?

***14.** Make a survey of the national, state, and local parks within a 97-kilometer (60-mile) radius of your local community. How widely are they used by local residents and by outside visitors? What is their condition? Develop a plan for their improved management.

***15.** Make a concept map of this chapter's major ideas, using the section heads and subheads and the key terms (in boldface). Look at the inside back cover and in Appendix 5 for information about concept maps.

FURTHER INFORMATION

For more information see Internet Sources and Further Readings at the back of the book. Online resources for this chapter can be found at **www.brookscole.com/biology**

18 SUSTAINING WILD SPECIES

The Passenger Pigeon: Gone Forever

In the early 1800s, bird expert Alexander Wilson watched a single migrating flock of an estimated 2 million passenger pigeons darken the sky for over 4 hours. By 1914 the passenger pigeon (Figure 18-1) had disappeared forever.

How could a species that was once the most common bird in North America become extinct in only a few decades? The answer is humans. The main reasons for the extinction of this species were **(1)** uncontrolled commercial hunting and **(2)** loss of the bird's habitat and food supply as forests were cleared to make room for farms and cities.

Passenger pigeons were good to eat, their feathers made good pillows, and their bones were widely used for fertilizer. They were easy to kill because they flew in gigantic flocks and nested in long, narrow colonies.

Commercial hunters would capture one pigeon alive, sew its eyes shut, and tie it to a perch called a stool. Soon a curious flock would land beside this "stool pigeon," and the birds would then be shot or ensnared by nets that might trap more than 1,000 birds at once.

Beginning in 1858, passenger pigeon hunting became a big business. Shotguns, traps, artillery, and even dynamite were used. Birds were sometimes suffocated by burning grass or sulfur below their roosts. Live birds were even used as targets in shooting galleries. In 1878 one professional pigeon trapper made $60,000 by killing 3 million birds at their nesting grounds near Petoskey, Michigan.

By the early 1880s only a few thousand birds were left. At that point, recovery of the species was doomed because the females laid only one egg per nest. On March 24, 1900, a young boy in Ohio shot the last known wild passenger pigeon. The last passenger pigeon on earth, a hen named Martha after Martha Washington, died in the Cincinnati Zoo in 1914. Her stuffed body is now on view at the National Museum of Natural History in Washington, D.C.

Eventually, all species become extinct or evolve into new species. However, conservation biologists estimate that every day 50–200 species become *prematurely extinct* primarily because of human activities. Studies indicate that this rate of loss of biodiversity is increasing as the human population grows (Figure 1-1), consumes more resources, disturbs more of the earth's land (Figure 1-5), and uses more of the earth's net plant productivity (Figure 4-21) that supports all species.

Figure 18-1 Passenger pigeons, extinct in the wild since 1900. The last known passenger pigeon died in Ohio's Cincinnati Zoo in 1914. (John James Audubon/The New York Historical Society)

Many biologists consider this epidemic of extinction an even more serious problem than depletion of stratospheric ozone and global warming (Chapter 11) because it is happening more quickly and is irreversible. Ways to protect species from premature extinction are the subject of this chapter.

The last word in ignorance is the person who says of an animal or plant: "What good is it?" . . . If the land mechanism as a whole is good, then every part of it is good, whether we understand it or not. . . . Harmony with land is like harmony with a friend; you cannot cherish his right hand and chop off his left.

ALDO LEOPOLD

This chapter addresses the following questions:

- Why should we care about wildlife?
- Are human activities causing a new mass extinction?
- What human activities endanger wildlife?
- How can we prevent premature extinction of species?
- Can we manage game animals more sustainably?
- Can we manage freshwater and marine fish more sustainably?

18-1 THE RISE AND FALL OF SPECIES

What Are Endangered and Threatened Species? Biologists distinguish among three levels of extinction: **(1)** *Local extinction* occurs when a species is no longer found in an area it once inhabited but is still found elsewhere in the world, **(2)** *ecological extinction* occurs when there are so few members of a species left that it can no longer play its ecological roles in the biological communities where it is found, and **(3)** *biological extinction* occurs when a species is no longer found anywhere on the earth. Biological extinction is forever (Figure 18-2).

Species heading toward biological extinction are classified as either *endangered* or *threatened* (Figure 18-3). An **endangered species** has so few individual survivors that the species could soon become extinct over all or most of its natural range. A **threatened species** is still abundant in its natural range but is declining in numbers and is likely to become endangered. Endangered and threatened species are ecological smoke alarms.

Some species have characteristics that make them more vulnerable than others to premature biological extinction (Table 18-1). In general, species more vulnerable to extinction are found in a limited area and have a small population size, a low population density, large body size, a specialized niche (Section 5-4), and a low reproductive rate (K-strategists; Figure 7-8). Such species also tend to **(1)** undergo fairly fixed migrations, **(2)** feed atop long food chains or webs, and **(3)** have high economic values to people.

According to a 1996 joint study by the International Union for the Conservation and Conservation International and a 1999 study by the World Wildlife Fund, **(1)** 34% of the world's fish species (51% of fresh water species), **(2)** 25% of amphibians (Connections, p. 122), **(3)** 25% of mammals (Case Study, p. 469), **(4)** 20% of reptiles, **(5)** 13% of plants (and nearly one of three in the United States), and **(6)** 11% of bird species is under threat of extinction. According to biologists, such estimates probably are low because of lack of data from many areas, including many species-rich tropical countries (Figure 17-20) where areas are being cleared rapidly (Figure 17-3). According to a 1995 study by the Nature Conservancy, about 32% of the 20,500 U.S. animal and plant species studied by scientists are vulnerable to premature extinction, mostly because of human activities.

The populations of many wild species that are not yet in danger of biological extinction have diminished locally or regionally to the point where they are locally or ecologically extinct. Such species may be a better indicator of the condition of entire ecosystems than endangered and threatened species. These *indicator species* can serve as early warnings so that we can prevent species extinction rather than respond to emergencies, often with little chance of success.

Why Preserve Wild Species? If all species eventually become extinct, why should we worry about losing a few more because of our activities? Does it matter that

Passenger pigeon — Great auk — Dodo — Bushy seaside sparrow — *Aepyornis* (Madagascar)

Figure 18-2 Some species that have become extinct largely because of human activities, mostly habitat destruction and overhunting.

Figure 18-3 Species that are endangered or threatened largely because of human activities. Populations of some of the species are recovering. As a result, some endangered species may be relisted as threatened and some threatened species (such as the bald eagle and peregrine falcon) were delisted in 1999.

(1) the passenger pigeon (Figure 18-1), **(2)** the great auk (Figure 18-2), **(3)** the green sea turtle (p. 61), **(4)** the 30–50 remaining Florida panthers (p. 383), or **(5)** some unknown plant or insect in a tropical forest becomes prematurely extinct because of human activities?

Biologists contend that the answer is *yes* because wild species have important scientific, ecological, economic, medical, aesthetic, and recreational values, including the following:

- Studying species can help scientists understand how life has evolved and functions and how it may continue to evolve on this planet.
- Wild species provide many of the ecological services that make up the earth's natural capital (Figure 1-2) and thus are key factors in sustaining the earth's biodiversity and ecological integrity.
- About 90% of today's food crops were domesticated from wild tropical plants (Figure 15-4), and existing wild plant species are needed to develop future crop strains.
- Pollination by birds and insects is essential to many food crops.
- Wild plants and plants domesticated from wild species supply rubber, oils, dyes, fiber, paper, lumber, and many other useful products (Figure 17-4).

- About 3.5 million people in developing countries rely on plant-based medicines for their primary health care.
- At least 40% of all medicines (worth $100 billion per year) owe their existence to the genetic resources of wild plants, mostly from tropical developing countries. This includes 80% of the top 150 prescription drugs used in the United States.
- Plant-derived anticancer drugs save an estimated 30,000 lives per year in the United States (Figure 17-12).
- Wild plants and animals are a source of beauty, wonder, joy, and recreational pleasure for many people. Each year Americans spend over three times more to watch wildlife than they do to watch movies or professional sporting events.
- Wildlife tourism, sometimes called *ecotourism*, is the fastest-growing segment of the global travel industry. Conservation biologist Michael Soulé estimates that one male lion living to age 7 generates $515,000 in tourist dollars in Kenya but only $1,000 if killed for its skin. Similarly, over a lifetime of 60 years, a Kenyan elephant is worth about $1 million in ecotourist revenue. However, there is evidence that excessive and unregulated ecotourism can destroy or degrade fragile areas and promote premature species extinction.

Biophilia

CONNECTIONS

Biologist Edward O. Wilson contends that because of the billions of years of biological connections leading to the evolution of the human species (Figure 5-3), we have an inherent affinity for the natural world, a phenomenon he calls *biophilia* (love of life). He points out that we cannot erase this evolutionary imprint in our genes through a few generations of urban living.

Evidence of this natural affinity for life is seen in the preference most people have for almost any natural scene over one from an urban environment. Given a choice, most people prefer to live in an area where they can see water, grassland, or a forest. More people visit zoos and aquariums than attend all professional sporting events combined.

In the 1970s I was touring the space center at Cape Canaveral in Florida. During our bus ride the tour guide pointed out each of the abandoned multimillion-dollar launch sites and gave a brief history of each launch. Most of us were utterly bored. Suddenly people started rushing to the front of the bus and staring out the window with great excitement. What they were looking at was a baby alligator—a dramatic example of how *biophilia* can triumph over *technophilia*.

Critical Thinking

1. Do you have an affinity for wildlife and wild ecosystems (biophilia)? If so, how do you display this love of wildlife in your daily actions? What patterns of your consumption help destroy and degrade wildlife?

2. Some critics contend that there is no verifiable genetic evidence for Wilson's concept of biophilia and view this idea as opinion, not science. If you have an affinity for wildlife do you believe that it is inherent in your genes or that it is culturally conditioned? Does it matter? Explain.

Is It Ethically Important to Preserve Wild Species? Some people believe that each wild species has an inherent right to exist, or to struggle to exist, which is unrelated to its usefulness to humans.

According to this view, we have an ethical responsibility to protect species from becoming prematurely extinct as a result of human activities. Biologist Edward O. Wilson believes that most people feel obligated to protect other species and the earth's biodiversity because most humans seems to have a natural affinity for nature (Connections, above).

Some people distinguish between the survival rights of plants and those of animals, mostly for what they consider practical reasons. Poet Alan Watts once said that he was a vegetarian "because cows scream louder than carrots."

Other people distinguish among various types of species. They might think little about killing a mosquito, cockroach (Spotlight, p. 121), or rat or ridding the world of disease-causing bacteria. Unless they are strict vegetarians, they might also see no harm in having others kill domesticated animals in slaughterhouses to provide them with meat, leather, and other products. However, these same people might deplore the killing of wild animals such as deer, squirrels, or rabbits.

Some proponents go further and assert that each individual organism, not just each species, has a right to survive without human interference. Others emphasize the importance of preserving the whole spectrum of biodiversity by protecting entire ecosystems rather than individual species or organisms, as discussed in Chapter 17. What is your view?

How Does Background Extinction Differ from Mass Extinction? Evolutionary biologists estimate that more than 99.9% of all the species that have ever existed are now extinct because of a combination of background and mass extinctions.

Each year, a small number of species become extinct naturally at a low rate, a phenomenon called the **background** or **natural rate of extinction**. In contrast, **mass extinction** is a rise in extinction rates above the background level. It is a catastrophic, widespread (often global) event in which large groups of existing species (perhaps 25–70%) are wiped out. Most mass extinctions are believed to result from one or a combination of global climate changes that kill many species and leave behind those able to adapt to the new conditions.

Fossils and geological evidence indicate that the earth's species have experienced five great mass extinctions (20–60 million years apart) during the past 500 million years in which large numbers of species became extinct each year for tens of thousands to millions of years (Figure 5-7). Evidence also shows that these mass extinctions were followed by other periods, called *adaptive radiations*, when the diversity of life increased and spread for 10 million years or more (Figures 5-7 and 5-8). The last mass extinction took place about 65 million years ago, when the dinosaurs became extinct after thriving for 140 million years.

Why Do Most Biologists Believe There Is a New Mass Extinction Crisis? According to a 1998 survey, 70% of the biologists polled believe that we are in the midst of a mass extinction and that this loss of species will pose a major threat to human existence in the 21st

Is There Really an Extinction Crisis?

SPOTLIGHT

Some social scientists and a few biologists question the existence of a human-caused extinction crisis. They cite two problems in estimating species loss: **(1)** We don't know how many species there are (estimates range between 5 million and 100 million, with a best guess of 10-14 million), and **(2)** it is difficult to observe species extinction, especially for species we know little or nothing about.

These critics also point out that biologists contending that we are in the midst of a human-caused extinction crisis are making two questionable assumptions:

- Any loss of habitat also means a net loss in species, usually in some proportion to the amount of habitat lost. For example, when an old-growth forest is cleared, a number of species (many of them with specialized niches) are lost. However, a cleared area that returns by secondary ecological succession to a second-growth forest can still support some of the original species. There may be a net loss of species, but not as high as projected by some estimates.
- The number of species an area holds increases with its geographic size. However, in some types of ecosystems the number of species levels off as size increases. In such cases, a loss of some habitat may not necessarily lead to a net loss of species.

Biologists use models and field data to make estimates of current and future extinction rates. Using such approaches, they estimate the annual loss of tropical forest habitat at about 1.8% per year. Edward O. Wilson and several tropical biologists who have counted species in patches of tropical forest before and after destruction or degradation estimate that this 1.8% loss in habitat results in roughly a 0.5% loss of species.

Do such estimates add up to an extinction crisis? Let us assume, as Wilson and many other biologists do, that a loss of 1 million species over several decades represents an extinction crisis. If we assume the global decline in species to be 0.5% per year, then we will lose 25,000 species per year if there are 5 million species and 500,000 per year if there are 100 million species. If these assumptions are correct, we will lose 1 million species in 40 years if there are 5 million species and in only 2 years with 100 million species.

Let's assume, however, that the estimate of 0.5% species loss per year is too high for the earth as a whole. If it is 0.25% per year, then we will lose 1 million species in 80 years with 5 million species and in 4 years with 100 million species. Even if we halve the estimated species loss again, to 0.125% per year, we can still lose 1 million species within 8 to 160 years, enough to easily qualify the situation as an extinction crisis. According to biodiversity expert Edward O. Wilson, "Clearly, we are in the midst of one of the great extinction spasms of geological history."

Mathematical models recently developed by ecologists indicate a time lag of several generations between habitat loss and extinction, primarily because habitat loss also removes potential colonization sites for the emergence of new species. If these models are correct, biologists may be greatly underestimating the magnitude of the current biotic impoverishment.

Biologists don't contend that their estimates of extinction rates are precise enough to make a firm prediction. Instead, they argue that there is ample evidence that we are destroying and degrading wildlife habitats at an increasing rate and that our actions certainly lead to a significant loss of species, even though the number and rate vary in different parts of the world.

Critical Thinking

Do you believe that we are in the midst of a sixth mass extinction caused mostly by human activities? Explain. If so, list five ways in which you contribute to this loss of biodiversity.

century. In 1999, Peter Raven, president of the International Botanical Congress, reported that, "We are predicting the extinction of about two-thirds of all bird, mammal, butterfly, and plant species by the end of the 21st century."

To these biologists we are not heeding Aldo Leopold's warning about preserving species as we tinker with the earth: "To keep every cog and wheel is the first precaution of intelligent tinkering."

It's difficult to document extinctions, for most go unrecorded. However, fossil and other evidence indicate that since agriculture began about 10,000 years ago, human activities have caused a growing number of species to become extinct (Figure 18-2).

Some conservation biologists estimate that currently 18,000–73,000 species become extinct each year (on average, 50–200 species per day), thousands of times the estimated natural background extinction rate of 3–30 species per year. Even if this estimate of the erosion of biodiversity is 100 times too high, the current extinction rate is many times the estimated rate of background extinction. Scientists expect this rate of extinction to accelerate as the human population grows and takes over more of the planet's wildlife habitats and net primary productivity.

Mass extinctions occurred long before humans evolved (Figure 5-7). However, biologists point out two important differences between the current mass

Table 18-1 Characteristics of Extinction-Prone Species

Characteristic	Examples
Low reproductive rate	Blue whale, polar bear, California condor, Andean condor, passenger pigeon, giant panda, whooping crane
Specialized feeding habits	Everglades kite (eats apple snail of southern Florida), blue whale (krill in polar upwellings), black-footed ferret (prairie dogs and pocket gophers), giant panda (bamboo), koala (certain eucalyptus leaves)
Feed at high trophic levels	Bengal tiger, bald eagle, Andean condor, timber wolf
Large size	Bengal tiger, lion, elephant, Javan rhinoceros, American bison, giant panda, grizzly bear
Limited or specialized nesting or breeding areas	Kirtland's warbler (6- to 15-year-old jack pine trees), whooping crane (marshes), orangutan (only on Sumatra and Borneo), green sea turtle (lays eggs on only a few beaches), bald eagle (prefers forested shorelines), nightingale wren (only on Barro Colorado Island, Panama)
Found in only one place or region	Woodland caribou, elephant seal, Cooke's kokio, many unique island species
Fixed migratory patterns	Blue whale, Kitland's warbler, Bachman's warbler, whooping crane
Preys on livestock or people	Timber wolf, some crocodiles
Behavioral patterns	Passenger pigeon and white-crowned pigeon (nest in large colonies), redheaded woodpecker (flies in the front of cars), Carolina parakeet (when one bird is shot, rest of flock hovers over body), Key deer (forages for cigarette butts along highways—it's a "nicotine addict")

extinction most biologists believe we are bringing about and those of the past:

- It is taking place in only a few decades, rather than over thousands to millions of years (Figure 5-7).
- We are eliminating or degrading many biologically diverse environments (such as tropical forests, tropical coral reefs, wetlands, and estuaries) that in the past have served as evolutionary centers for the 5- to 10-million-year recovery of biodiversity after a mass extinction (Figure 5-7).

In the words of Norman Myers (Guest Essay, p. 282), "Within just a few human generations, we shall—in the absence of greatly expanded conservation efforts—impoverish the biosphere to an extent that will persist for at least 200,000 human generations or twenty times longer than the period since humans emerged as a species." However, a few analysts question whether there is an extinction crisis (Spotlight, p. 467).

Even without extinctions, some ecosystems may lose their ability to support many forms of life because of the disappearance of *local populations* of key organisms. According to a 1997 study by biologist Jennifer Hughes and her colleagues at Stanford University, every year an estimated 16 million local plant and animal populations—an average of 1,800 populations every hour—disappear from the world's tropical forests.

18-2 CAUSES OF DEPLETION AND PREMATURE EXTINCTION OF WILD SPECIES

What Are the Causes of Wildlife Depletion and Premature Extinction? The basic underlying causes of population reduction and extinction of wildlife are

- Human population growth (Figures 1-1 and 1-7)
- Economic systems and policies that fail to value the environment and its ecological services (Figure 1-2), thereby promoting unsustainable exploitation

Ewel, J. et al. 1999. "Deliberate Introductions of Species: Research Needs." *BioScience*, vol. 49 no. 8 p. 619(1).

Bats Are Getting a Bad Rap

CASE STUDY

Despite their variety (950 known species) and worldwide distribution, bats—the only mammals that can fly—have certain traits that make them vulnerable to extinction. Bats reproduce slowly, and many bat species that live in huge colonies in caves and abandoned mines become vulnerable to destruction when people block the passageways or disturb their hibernation.

Bats play important ecological roles. About 70% of all bat species feed on crop-damaging nocturnal insects and other pest species such as mosquitoes, making them the primary control agents for such insects.

In some tropical forests and on many tropical islands, pollen-eating bats pollinate flowers and fruit-eating bats distribute plants throughout tropical forests by excreting undigested seeds. As keystone species, they are vital for maintaining plant biodiversity and for regenerating large areas of tropical forest cleared by human activities. If you enjoy bananas, cashews, dates, figs, avocados, or mangos, you can thank bats.

Many people mistakenly view bats as fearsome, filthy, aggressive, rabies-carrying bloodsuckers. But most bat species are harmless to people, livestock, and crops. In the United States, only 10 people have died of bat-transmitted disease in four decades of recordkeeping; more Americans die each year from falling coconuts.

Because of unwarranted fears of bats and lack of knowledge about their vital ecological roles, several bat species have been driven to extinction. Currently, 26% of the world's bat species, including the ghost bat (Figure 18-3), are listed as endangered or threatened. Conservation biologists urge us to view bats as valuable allies, not as enemies.

Critical Thinking

How do you feel about bats? What would you do if you found bats flying around your yard at night?

- Greater per capita resource use as a result of increasing affluence and economic growth (Figures 1-16 and 1-17)
- Increasing appropriation of the earth's net primary productivity (Figure 4-21) that supports all life for human use.
- Poverty, which pushes many of the poor in developing countries to **(1)** cut forests, **(2)** grow crops on marginal land, **(3)** overgraze grasslands, **(4)** deplete fish species, and **(5)** kill endangered animals for their valuable furs, tusks, or other parts

These underlying causes lead to other more direct causes of the endangerment and premature extinction of wild species (Figure 18-4).

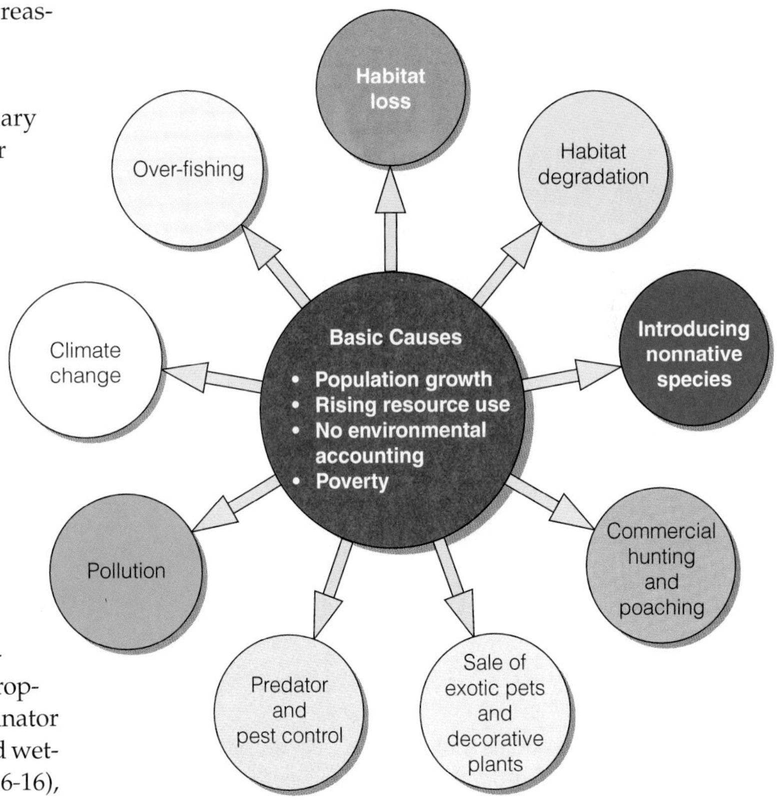

Figure 18-4 Underlying and direct causes of depletion and premature extinction of wild species. The two biggest direct causes of wildlife depletion and premature extinction are **(1)** habitat loss and degradation and **(2)** deliberate or accidental introduction of nonnative species into ecosystems.

What Is the Role of Habitat Loss, Degradation, and Fragmentation? Biologists agree that the greatest threat to wild species is habitat loss (Figure 18-5), degradation, and fragmentation. According to conservation biologists, tropical deforestation (Figure 17-3) is the greatest eliminator of species, followed by destruction of coral reefs and wetlands (Section 6-5), plowing of grasslands (Figure 6-16), and pollution of marine and freshwater habitats (Sections 12-5 and 12-6).

Island species, many of them *endemic species* found nowhere else on earth, are especially vulnerable to extinction. The theory of island biogeography (Figure 5-17) has been used to predict the number

Hint: Enter the search term *native species* using the Subject Guide.

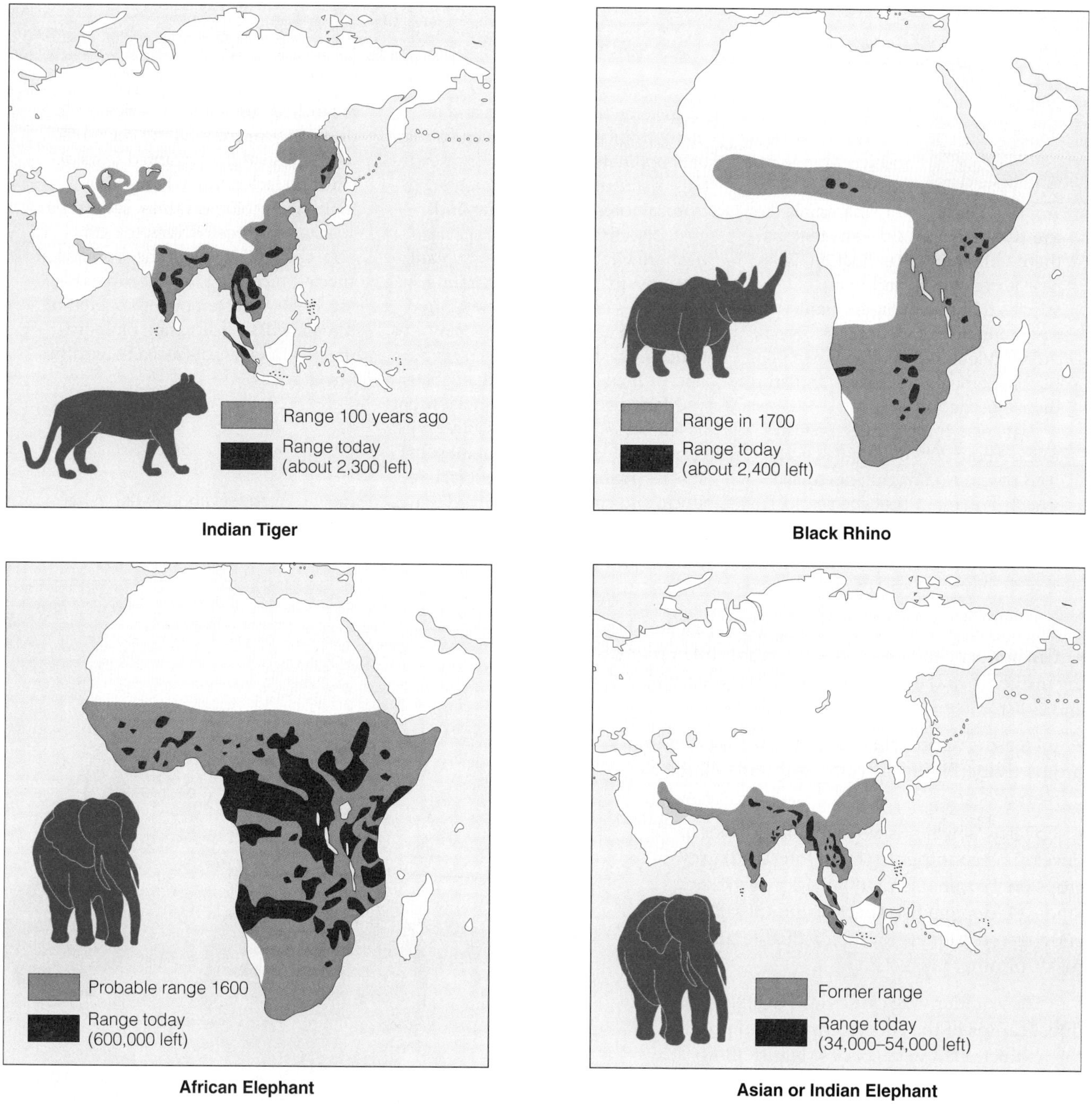

Figure 18-5 Reductions in the ranges of four wildlife species, mostly the result of habitat loss and hunting. What will happen to these and millions of other species when the global human population doubles in the next few decades? (Data from International Union for the Conservation of Nature and World Wildlife Fund)

and percentage of species that would become extinct when habitats on islands are destroyed, degraded, or fragmented. For example, many continuously forested landscapes have been converted into a quilt of small patches of trees criss-crossed with roads, subdivisions, cropfields, and other forms of human land use. This model has been extended from islands to national parks, tropical rain forests, lakes, and nature reserves (Connections, p. 133), which can be viewed as *habitat islands* in an inhospitable sea of unsuitable habitat.

The three main problems caused by habitat fragmentation are **(1)** a decrease in the sustainable population size for many wild species, **(2)** increased surface area (edge), which makes some species more vulnerable to predators, and **(3)** creation of barriers that limit the ability of some species to disperse and colonize new areas, find enough to eat, and find mates. Some species

The Plight of Migrating Birds

CASE STUDY

Migrating bird species face a double habitat problem. Nearly half of the 700 U.S. bird species spend two-thirds of the year in the tropical forests of Central or South America or on Caribbean islands. During the summer they return to North America to breed.

A U.S. Fish and Wildlife study showed that between 1978 and 1987, populations of 44 of the 62 surveyed species of insect-eating, migratory songbirds in North America declined; 20 species experienced drops of 25–45%.

The two main culprits are **(1)** logging of tropical forests in the birds' winter habitats and **(2)** fragmentation of their summer forest habitats in North America. The intrusion of farms, freeways, and suburbs breaks forests into patches. This makes it easier for opossums, skunks, squirrels, raccoons, and blue jays to feast on the eggs and the young of migrant songbirds.

Another serious threat to songbirds from such habitat fragmentation comes from parasitic cowbirds that lay their eggs in the nests of various songbird species and have those birds raise their young for them.

Approximately 68% of the world's 9,600 known bird species are declining in numbers (58%) or are threatened with extinction (11%), mostly because of habitat loss and fragmentation. Conservation biologists view this loss and decline of bird species as an early warning of the greater loss of biodiversity and ecological integrity to come. Birds are excellent environmental indicators because they **(1)** live in every climate and biome, **(2)** respond quickly to environmental changes in their habitats, and **(3)** are easy to track and count.

In addition to serving as indicator species, birds play important ecological roles such as **(1)** helping control populations of insects (including the spruce budworm, gypsy moth, and tent caterpillar, which decimate many tree species) and rodents, **(2)** pollinating a wide variety of flowering plants, and **(3)** spreading plants throughout their habitats by consuming plant seeds and excreting them in their droppings.

Critical Thinking

What three things would you do to help prevent the decline in populations of migrating birds?

of migrating birds face loss, degradation, or fragmentation of their seasonal habitats (Case Study, above).

What Is the Role of Deliberately Introduced Species? After habitat loss and degradation, deliberate or accidental introduction of nonnative or alien species into ecosystems is the biggest cause of animal and plant extinctions.

Deliberate introduction of nonnative species can be a boon or bane to society. We depend heavily on nonnative organisms for ecosystem services, food, shelter, medicine, and aesthetic enjoyment. Over 70% of the world's food comes from nine crops, all of which are cultivated far beyond their natural ranges. According to a 1999 study by David Pimentel (Guest Essay, p. 352), introduced species such as corn, wheat, rice, and other food crops, and cattle, poultry, and other livestock, supply more than 98% of the U.S. food supply at a value of approximately $800 billion per year. Similarly, about 85% of industrial forestry tree plantations use mostly species nonnative to the areas where they are grown. Some deliberately introduced species have helped control pests.

However, some introduced species have no natural predators, competitors, parasites, or pathogens to control their numbers in their new habitats. This can allow them to reduce or wipe out the populations of many native species, making them a form of *biological pollution* (Table 18-2). The kudzu vine is an example of the effects of a deliberately introduced species (Connections, p. 474).

An estimated 1 million *wild hogs* are roaming the state of Florida. They arrived with Spanish conquistadors more than 450 years ago. The state's panthers (see photo, p. 383), bears, and coyotes used to help control the populations of these hogs, but the hogs now outbreed them. They are hogging food from endangered animals, rooting up farm fields, and causing traffic accidents. Game and wildlife officials have had little success in controlling their numbers with hunting and trapping and say there is no way of stopping them.

The United States is home for about 50,000 alien species and the number is increasing. According to a 1999 study by David Pimentel, damages and pest control costs for unwanted species amount to an estimated $138 billion per year—an average of $16 million per hour—and nearly twice the annual budget of New York State.

These are conservative estimates because of **(1)** lack of adequate economic data on many alien species and **(2)** inability to assign monetary values to losses in biodiversity, ecosystem services, and aesthetics. If such costs were included, the overall annual damage from alien species would be several times higher.

About 42% of the species in the United States on the list of endangered and threatened species are there in part because of population declines caused by nonnative species that either compete with or prey on them. In Hawaii, more than 95% of the 282 imperiled plant and bird species are threatened by alien species.

Table 18-2 Damage Caused by Species Imported into the United States			
Name	**Origin**	**Mode of Transport**	**Type of Damage**
Mammals			
European wild boar	Russia	Intentionally imported (1912), escaped captivity	Destroys habitat by rooting; damages crops
Nutria (cat-sized rodent)	Argentina	Intentionally imported, escaped captivity (1940)	Alters marsh ecology; damages levees and earth dams; destroys crops
Mongoose	Africa	Introduced into Puerto Rico and Hawaii in the late 1800s to kill rats in sugarcane plantations	Preys on native ground-nesting birds and on amphibians and reptiles that can be beneficial for pest control; carries pathogenic organisms for rabies and leptospirosis
Birds			
European starling	Europe	Intentionally released (1890)	Competes with native songbirds; damages crops; transmits swine diseases; causes airport nuisance
House sparrow	England	Intentionally released by Brooklyn Institute (1853)	Damages crops; displaces native songbirds; carries 29 diseases that affect people and domestic animals
Aquatic Animals			
Carp	Germany	Intentionally released (1877)	Displaces native fish; uproots water plants; lowers waterfowl population
Sea lamprey	North Atlantic Ocean	Entered Great Lakes via Welland Canal (1829)	Wiped out lake trout, lake whitefish, and sturgeon in Great Lakes
Asian swamp eel	Asia	Recently escaped from home aquariums or fish farms into Florida's Everglades	Displaces native fish; very prolific and hard to capture or kill; expected to spread up the Atlantic Coast as far north as Chesapeake Bay
Walking catfish	Thailand	Imported into Florida	Destroys bass, bluegill, and other fish

Once an introduced species has become established in an ecosystem its wholesale removal is almost impossible—somewhat like trying to get smoke back into a chimney or trying to unscramble an egg. Thus, the best way to limit the harmful impacts of nonnative species is to prevent them from being introduced and becoming established.

What Is the Role of Accidentally Introduced Species? One example of the effects of the accidental introduction of a nonnative species was the invasion of the Great Lakes by zebra and quagga mussels (Case Study, p. 312).

In the late 1930s, extremely aggressive red fire ants were accidentally introduced into the United States in Mobile, Alabama. It is believed that they may have arrived on shiploads of lumber imported from South America or by hitching a ride in the soil-containing ballast of cargo ships.

Without natural predators, these ants have spread rapidly by land and water (they can float) throughout the South from Texas to Florida and as far north as Tennessee and Virginia. So far, the ants' aversion to frost has kept them out of the Midwest and Northwest, but this might not last long because of genetic changes in some of their populations.

Table 18-2 *(continued)*

Name	Origin	Mode of Transport	Type of Damage
Aquatic Animals *(continued)*			
Zebra mussel	Europe	Discharged from ballast of ship near Detroit (1986)	Depletes food supplies of native species; clogs irrigation pipes and water intake systems for water and power plants; fouls beaches, piers, and boat hulls; spreading rapidly
Insects			
Argentine fire ant	Argentina	Probably entered via coffee shipments from Brazil (1918)	Damages crops; destroys native ant species
Camphor scale insect	Japan	Accidentally imported on nursery stock (1920s)	Damaged nearly 200 plant species in Louisiana, Texas, and Alabama
Japanese beetle	Japan	Accidentally imported on irises or azaleas (1911)	Defoliates more than 250 species of trees and other plants, including many of commercial importance
Plants			
Water hyacinth	Central America	Accidentally introduced (1884)	Clogs waterways; shades out other aquatic vegetation
Purple loosestrife	Europe	Imported for ornamental and medicinal purposes	Has overrun large area wetlands in 42 states from Maine to California; eliminated native vegetation; pushed several species of amphibians and butterflies to the brink of extinction
Australian melaleuca tree	Australia	Introduced to south Florida for windbreaks and fence rows (1906)	Has invaded much of the Florida Everglades and displaced many native plant species
Fungus			
Chestnut blight (fungus)	Asia	Accidentally imported on nursery plants (1900)	Killed nearly all eastern U.S. chestnut trees; disturbed forest ecology
Dutch elm disease (fungus)	Europe	Accidentally imported on infected elm timber used for veneers (1930)	Killed millions of elms; disturbed forest ecology

Wherever the fire ant has gone, up to 90% of native ant populations have been sharply reduced or wiped out. Their extremely painful stings have also killed deer fawn, birds, livestock, pets, and at least 80 people allergic to their venom. They have invaded cars and caused accidents by attacking drivers. They have made cropfields unplowable, disrupted phone service and electrical power, and caused some fires by chewing through underground cables. They cost the United States an estimated $10 million per year in damages.

Mass spraying of pesticides in the 1950s and 1960s temporarily reduced populations of fire ants. In the end, however, this chemical warfare hastened the advance of the fire ant by reducing populations of many native ant species. The rapidly multiplying fire ants also developed genetic resistance to heavily used pesticides. A colony cannot be destroyed unless the queen ant is killed.

Researchers at the U.S. Department of Agriculture are evaluating the use of a tiny parasitic fly that lays its

CONNECTIONS

Deliberate Introduction of the Kudzu Vine

In the 1930s, the *kudzu vine* was imported from Japan and planted in the southeastern United States to help control soil erosion. It does control erosion, but it is so prolific and difficult to kill that it engulfs hillsides, trees, abandoned houses and cars, stream banks, patches of forest, and anything else in its path (Figure 18-6). It has spread throughout much of the southern United States and could spread as far north as the Great Lakes by 2040 if projected global warming occurs.

Although kudzu is considered a menace in the United States, Asians use a powdered kudzu starch in beverages, gourmet confections, and herbal remedies for a range of diseases. A Japanese firm has built a large kudzu farm and processing plant in Alabama and ships the extracted starch to Japan.

In an ironic twist, kudzu—which can engulf and kill trees—could eventually help save trees from loggers. Research at Georgia Institute of Technology indicates that kudzu may join kenaf (Figure 17-10) as a source of tree-free paper.

Figure 18-6 Kudzu taking over a field and trees near Lyman, South Carolina. This vine can grow 0.3 meter (1 foot) per day and is now found from East Texas to Florida and as far north as southeastern Pennsylvania and Illinois. Kudzu was deliberately introduced into the United States for erosion control, but can't be stopped by being dug up or burned. Grazing by goats and repeated doses of herbicides can destroy it, but goats and herbicides also destroy other plants, and herbicides can contaminate water supplies. (Angela Lax/Photo Researchers, Inc.)

Critical Thinking

On balance, do you think the potential beneficial uses of the kudzu vine outweigh its harmful ecological effects? Explain.

eggs on the fire ant's body. After the eggs hatch, the larvae eat their way through the ant's head. Before releasing these biological control agents, however, researchers must be sure they will not cause problems for native ant species.

About 50 years ago the *brown tree snake* arrived accidentally on the island of Guam in the cargo of a military plane. Since then this highly aggressive, nocturnal, climbing snake with no natural enemies has eliminated or decimated nine of Guam's native forest bird species. It has also snatched chicken and pets from yards, has attacked babies asleep in cribs, and shorted out power lines.

This light-sensitive snake can select an airplane wheel or outgoing cargo as a hiding place during daylights hours and thus spread to other areas. Since 1981 seven brown tree snakes have been found in Hawaii. Those found alive were captured when discovered. As home for 41% of all endangered birds in the United States, Hawaii has a lot to lose from an invasion by this snake. Another harmful invader is the Formosan termite (Case Study, right).

What Is the Role of Commercial Hunting and Poaching? The international trade in wild plants and animals brings in up to $12 billion a year. Almost half of this trade involves the illegal sale of endangered and threatened species or their parts. Organized crime has moved into wildlife smuggling because of the huge profits involved (second only to drug smuggling). At least two-thirds of all live animals smuggled around the world die in transit.

Worldwide, some 622 species of animals and plants face extinction, mostly because of illegal trade. To poachers, a live mountain gorilla is worth $150,000, a gyrfalcon $120,000, a chimpanzee $50,000, an Imperial Amazon macaw $30,000, and rhinoceros horn (Figure 18-7) as much as $28,600 per kilogram ($13,000 per pound). Excessive hunting brought the American bison to near extinction (Case Study, p. 476).

Despite international protection, the world's total tiger population numbers only about 7,000 (down from about 150,000 in 1950), mostly because of habitat loss

The Termite from Hell

CASE STUDY

Forget killer bees and fire ants. The homeowner's nightmare is the Formosan termite. It is the most voracious, aggressive, and prolific of over 2,000 known termite species. These termites invaded the Hawaiian Islands by 1900 and probably arrived on the mainland of United States during or soon after World War II. They probably arrived in wooden packing materials on military cargo ships that docked in southern ports such as New Orleans, Louisiana, and Houston, Texas.

They can quickly munch through wooden beams and plywood and can even chew through brick. Their huge colonies can contain up to 10 million insects (10 times as many as a nest of native termites).

Over the past decade, the Formosan termite has caused more damage in New Orleans, Louisiana, than hurricanes, floods, and tornadoes combined. In New Orleans virtually every building in every neighborhood has been struck, and the famous French Quarter has one of the most concentrated infestations in the world.

Once confined to Louisiana, they have invaded at least a dozen other states, including Alabama, Florida, Mississippi, North and South Carolina, Texas, and California. They cause at least $1 billion in damage each year.

Most currently used pesticides don't work on these termites. In New Orleans, the U.S. Department of Agriculture is using a variety of techniques all at once in an attempt to control the species in a heavily infested 15-block area of the French Quarter. They hope to develop techniques for battling these invaders elsewhere.

Critical Thinking

What important ecological roles do termites play in nature? If the Formosan termite and other termite species could be eradicated (a highly unlikely possibility), would you favor doing this? Explain.

and poaching for their furs and bones. Bengal tigers are at risk because a tiger fur sells for $100,000 in Tokyo. With the body parts of a single tiger potentially worth as much as $5 million, it is not surprising that illegal hunting has skyrocketed, especially in India. Without emergency action, there may be few or no tigers left in the wild within 20 years.

As more species become endangered, the demand for them on the black market soars, hastening their chances of premature extinction from poaching. Most poachers are not caught, and the money to be made far outweighs the risk of fines and the much smaller risk of imprisonment.

Case Study: How Should We Protect Elephants from Extinction? Habitat loss (Figure 18-5) and legal and illegal trade in elephant ivory (Figure 18-9, p. 477) have reduced wild African elephant numbers from 2.5 million in 1970 to about 500,000 today (plus about 300 in zoos).

Since 1989 this decline has been slowed by an international ban on the sale of ivory from African elephants, but things are not quite that simple. First, the ban on elephant ivory sales has increased the killing of other species with ivory parts such as bull walruses (for their ivory tusks, worth $500–$1500 a pair) and hippos (for their ivory teeth, worth $70 per kilogram). Second, increases in elephant populations in areas where their habitat has shrunk has resulted in widespread destruction of vegetation by these animals. This in turn reduces the niches available for other wild species.

Figure 18-7 Rhinoceros horns are carved into ornate dagger handles that sell for $500–12,000 in Yemen and other parts of the Middle East. In China and other parts of Asia, powdered rhino horn is used for medicinal purposes, particularly as a proven fever reducer and occasionally as an alleged aphrodisiac. All five species of rhinoceros are threatened with extinction because of poachers (who kill them for their horns) and habitat loss. Between 1973 and 1998 the population of African black rhinos (Figure 18-5) dropped from approximately 63,000 to about 2,400. (R. F. Porter/Ardea London)

Near Extinction of the American Bison

CASE STUDY

In 1500, before Europeans settled North America, 30–60 million North American bison grazed the plains, prairies, and woodlands over much of the continent. A single herd on the move might thunder past for hours.

Several Native American tribes depended heavily on bison, and typically they killed only the animals they needed for food, clothing, and shelter. By 1906, however, the once-vast range of the bison had shrunk to a tiny area, and the species had been driven nearly to extinction (Figure 18-8).

How did this happen? First, as settlers moved west after the Civil War, the sustainable balance between Native Americans and bison was upset. Several plains tribes traded bison skins to settlers for steel knives and firearms, so they began killing more bison.

The most relentless slaughter, however, was caused by the new settlers. As railroads spread westward in the late 1860s, railroad companies hired professional bison hunters—including Buffalo Bill Cody—to supply construction crews with meat. Passengers also gunned down bison from train windows for sport, leaving the carcasses to rot.

Commercial hunters shot millions of bison for their hides and tongues (considered a delicacy), leaving most of the meat to rot. "Bone pickers" collected the bleached bones that whitened the prairies and shipped them east to be ground up as fertilizer.

Farmers shot bison because they damaged crops, fences, telegraph poles, and sod houses. Ranchers killed them because they competed with cattle and sheep for grass. The U.S. Army killed at least 12 million bison as part of their campaign to subdue the plains tribes by killing off their primary source of food.

By 1892 only 85 bison were left. They were given refuge in Yellowstone National Park and protected by an 1893 law against the killing of wild animals in national parks.

In 1905, 16 people formed the American Bison Society to protect and rebuild the captive population. Soon thereafter, the federal government established the National Bison Range near Missoula, Montana. Today there are an estimated 200,000 bison, about 97% of them on privately owned ranches.

Research by biologists studying current bison herds indicates that bison were once a keystone species whose grazing activities played an important role in the tallgrass prairie. Some biologists have suggested restoring large herds of bison on public lands in the North American plains as a way of restoring the ecological integrity of remaining tracts of tallgrass prairie. This idea has been strongly opposed by ranchers with permits to graze cattle and sheep on federally managed lands.

Critical Thinking

Explain why you agree or disagree with the proposal to reestablish large numbers of American bison on public lands in the North American plains.

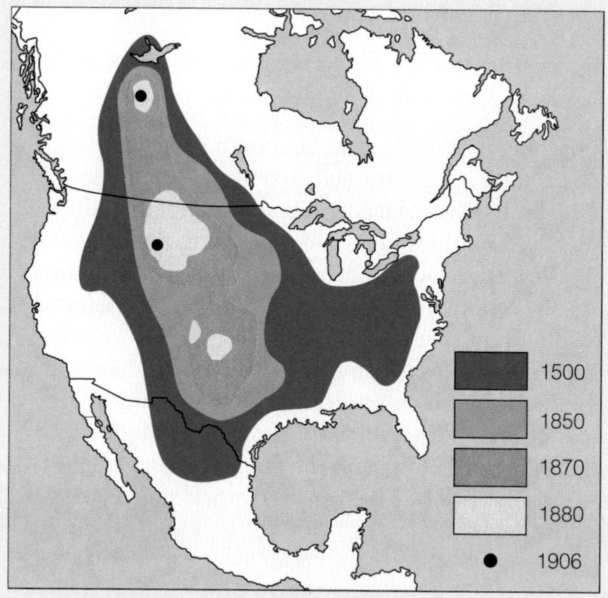

Figure 18-8 The historical range of the bison shrank severely between 1500 and 1906, mostly because of unregulated and deliberate overhunting.

Some wildlife conservationists and leaders of several southern African nations have called for a partial lifting of the elephant ivory ban in areas where elephant populations are not endangered. Ivory from the sustainable culling of elephants in these areas would be marketed in a way that certifies that it was obtained legally. Most of the profits would go to local people.

In 1997 the Convention on International Trade in Endangered Species (CITES) voted to let Zimbabwe, Botswana, and Namibia sell some of their stockpiled ivory (most confiscated from poachers) under strict monitoring. Profits from the sale are used for wildlife conservation and rural development programs.

Some wildlife conservationists oppose such sales. They believe that the best way to protect African elephants from extinction is to continue the ivory ban, which discourages poaching by keeping the price of ivory down. What do you think should be done?

What Is the Role of Predator and Pest Control? People try to exterminate species that compete with them for food and game animals. For example, U.S.

Figure 18-9 Vultures are feeding on this elephant carcass, poached in Tanzania for its ivory tusks, which are used to make jewelry, piano keys, ornamental carvings, and art objects. (E. R. Degginger)

fruit farmers exterminated the Carolina parakeet around 1914 because it fed on fruit crops. The species was easy prey because when one member of a flock was shot, the rest of the birds hovered over its body, making themselves easy targets (Table 18-1).

African farmers kill large numbers of elephants to keep them from trampling and eating food crops. Many ranchers, farmers, and hunters in the United States support the killing of coyotes (p. 448), wolves (Case Study, p. 451), and other species that can prey on livestock and on species prized by game hunters.

Since 1929 U.S. ranchers and government agencies have poisoned 99% of North America's prairie dogs because horses and cattle sometimes step into the burrows and break their legs. This has also nearly wiped out the endangered black-footed ferret (Figure 18-3, about 600 left in the wild), which preyed on the prairie dog.

What Is the Role of the Market for Exotic Pets and Decorative Plants? The global legal and illegal trade in wild species for use as pets is a huge and very profitable business. However, for every live animal captured and sold in the pet market, an estimated 50 other animals are killed.

About 25 million U.S. households have exotic birds as pets, 85% of them imported. Over 60 bird species, mostly parrots, are endangered or threatened because of this wild bird trade. Collectors of exotic birds may pay $10,000 for a threatened hyacinth macaw (Figure 18-3) smuggled out of Brazil; however, during its lifetime, a single macaw left in the wild might yield as much as $165,000 in tourist income. A 1992 study suggested that keeping indoor pet birds for more than 10 years doubles a person's chances of getting lung cancer.

Other wild species whose populations are depleted because of the pet trade include amphibians, reptiles, mammals, and tropical fish (taken mostly from the coral reefs of Indonesia and the Philippines). Divers catch tropical fish by using plastic squeeze bottles of cyanide to stun them. For each fish caught alive, many more die. In addition, the cyanide solution kills the coral animals that create the reef, which is a center for marine biodiversity (Figures 6-28 and 6-29).

Some exotic plants, especially orchids and cacti, are endangered because they are gathered (often illegally) and sold to collectors to decorate houses, offices, and landscapes. A collector may pay $5,000 for a single rare orchid. A single rare mature crested saguaro cactus can earn cactus rustlers as much as $15,000.

What Are the Roles of Climate Change and Pollution? A potential problem for many species is fairly rapid changes in climate over several decades, accelerated by deforestation and emissions of heat-trapping gases into the atmosphere (Sections 11-3 and 11-4). Wildlife in even the best-protected and best-managed terrestrial reserves and ocean sanctuaries could be depleted in a few decades if projected changes in climate take place.

Toxic chemicals, such as pesticides (Figure 8-2), degrade wildlife habitats and kill some terrestrial plants and animals and aquatic species (Sections 12-5, 12-6, and 16-3). According to the U.S. Fish and Wildlife Service, pesticides menace about 20% of the endangered and threatened species in the United States. Some aquatic wild species are also threatened by acid deposition (Section 10-3) and ozone depletion (Section 11-6).

What Is the Role of Overfishing in Reducing Aquatic Biodiversity? According to the UN Food and Agriculture Organization, at least 60% of the world's 200 commercially valuable fish species are either overfished or fished to the limit, mostly because too many fishing boats are pursuing too few fish—another example of the tragedy of the commons (Connections, p. 14).

Consider the plight of the Atlantic bluefin tuna. In 1980 there were about 1 million of these fish. Now there are only about 20,000. Before they disappeared from overfishing, a single large Atlantic bluefin tuna sold in a Tokyo, Japan, fish market for about $83,000.

Various methods for harvesting fish are shown in Figure 18-10. Some fishing boats, called *trawlers*, catch bottom-dwelling fish and shellfish by dragging a funnel-shaped net held open at the neck along the ocean bottom. Newer trawling nets are large enough to swallow 12 jumbo jets in a single gulp, and even larger ones are on the way.

The large mesh of the net allows most small fish to escape but can capture and kill other species, such as seals and endangered and threatened sea turtles. Only the large fish are kept, with most of the fish and

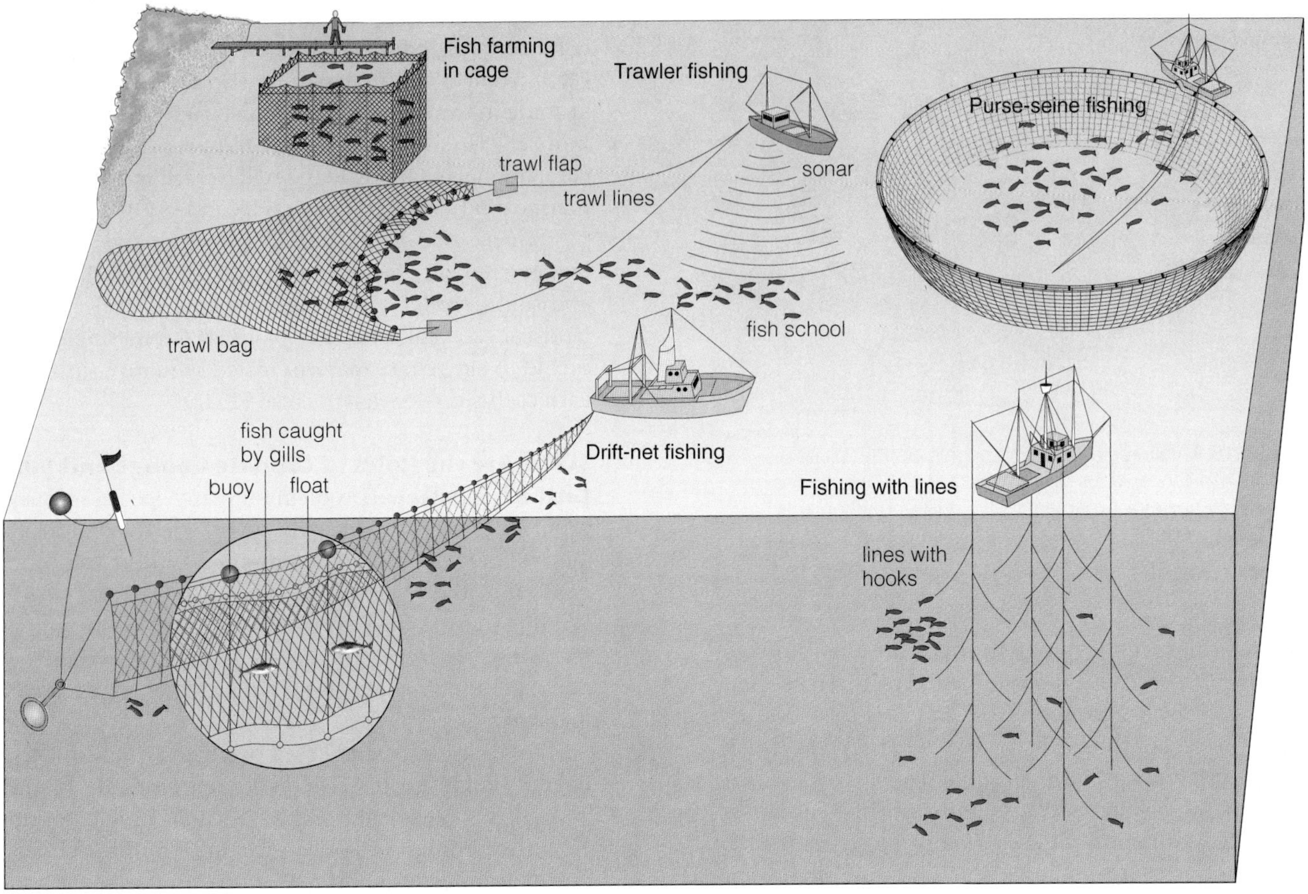

Figure 18-10 Major commercial fishing methods used to harvest various marine species.

other aquatic species thrown back into the ocean either dead or dying.

Thousands of trawlers scrape and disturb an area of ocean bottom that is the size of Brazil and India combined and is about 100 times larger than area of forests clear-cut each year. According to a 1998 study by oceanographers, this scraping is devastating to life on the seafloor, removing organisms that provide shelter for little fish eaten by larger fish that the trawlers are trying to catch. These ocean floor communities can take decades or centuries to recover.

Marine scientist Carl Safina says, "Bottom trawling is akin to harvesting corn with bulldozers." According to marine scientist Elliot Norse, "Bottom trawling is probably the largest human-caused disturbance to the biosphere."

Surface-dwelling species such as tuna, which feed in schools near the surface or in shallow areas, are often caught by *purse-seine fishing* (Figure 18-10). After a school is found, it's surrounded by a large purse-seine net. The net is then closed like a drawstring purse to trap the fish, and the catch is hauled aboard. Nets used to capture yellowfin tuna in the eastern tropical Pacific Ocean have also killed large numbers of dolphins, which swim on the surface above schools of the tuna.

Another method for catching open-ocean fish species such as swordfish, tuna, and sharks (Case Study, p. 123) is *longlining*, in which fishing vessels put out lines up to 130 kilometers (80 miles) long, hung with thousands of baited hooks. This practice has caused such a severe drop in the population of Atlantic swordfish that many restaurant owners and chefs have agreed to stop serving this fish and many consumers are refusing to order it. Long lines also hook pilot whales, dolphins, and endangered sea turtles (Figure 18-11, pp. 480–481).

In addition to trawling, another serious threat to marine biodiversity is *drift-net fishing (*Figure 18-10). Each net descends as much as 15 meters (50 feet) below the surface and is up to 65 kilometers (40 miles) long. Almost anything that comes in contact with these nearly invisible "curtains of death" becomes entangled.

Since 1992, use of drift nets longer than 2.5 kilometers (1.6 miles) has been banned in international waters to help reduce overfishing and reduction of marine biodiversity. However, the practice continues because **(1)** there is no effective way to monitor activi-

Baker, B. 1999. "Spending on the Endangered Species Act: Too Much or Not Enough?" *BioScience*, v49 i4 p279(1).

ties of fishing fleets over vast ocean areas and **(2)** fines for violations are small.

How Serious Are the Threats to the Biodiversity of Freshwater and Marine Systems? According to a number of aquatic scientists, the economic pressures to catch more and more fish pose a serious threat to freshwater and marine diversity. Most people are unaware how vulnerable freshwater environments are to environmental degradation and depletion. Although only 1% of the earth's surface is fresh water, these aquatic systems are home for 12% of all known animal species (including 41% of the known fish species).

According to the UN Food and Agriculture Organization and the World Wildlife Fund, about 34% of all known fish species and 51% of freshwater fish species are at risk of becoming extinct. According to a recent study by the World Resources Institute, coastal developments threaten marine biodiversity along roughly half of the world's coasts. The greatest threats occur in Europe (86% of coastal waters at high or moderate risk) and Asia (69% threatened).

The major causes of declines in fish populations are **(1)** habitat loss and degradation (mostly from pollution), **(2)** invasions by nonnative species, and **(3)** overharvesting. Although loss and degradation of forests get most of the attention, freshwater habitats are even more serioously altered by human activities. More than 40,000 large dams (Figure 12-6) and hundreds of thousands of smaller barriers alter the seasonal flows, temperatures, and sediment loads, and other river properties to which native fish are adapted. For example, in the heavily altered Columbia River basin, 29 of 50 native fish species are either extinct or endangered mostly because of habitat alteration. Flood control levees (Figure 12-14, middle) disconnect rivers from their floodplains and thus eliminate wetlands and backwaters that are important spawning grounds for fish. Pollution from industries, farms, and urban areas further reduces and degrades fish habitat (Chapter 12).

Populations of many marine species are also in decline. The most obvious signs of this degradation of marine life include **(1)** dramatic declines in many of the world's fish stocks, mostly because of overfishing, **(2)** destruction and bleaching of coral reefs (Figure 6-28, top), **(3)** massive die-offs of dolphins and seals, **(4)** loss of mangrove forests, **(5)** increasing incidences of huge blooms of toxic red tides, and **(6)** pollution and sedimentation.

Overfishing leads in most cases to *commercial extinction,* which is usually only a temporary depletion of fish stocks, as long as depleted areas are allowed to recover. However, large-scale fish harvesting can lead to serious depletion and extinction of species such as sea turtles (Figure 18-11) and other marine mammals that are unintentionally caught.

18-3 SOLUTIONS: PROTECTING WILD SPECIES FROM DEPLETION AND EXTINCTION

How Can We Protect Wildlife and Biodiversity? The three basic approaches to protecting wildlife and protecting biodiversity emphasize **(1)** ecosystems, **(2)** species, or **(3)** wildlife management. The *ecosystem approach*, discussed in Chapter 17, aims to **(1)** preserve balanced populations of species in their native habitats, **(2)** establish legally protected wilderness areas and wildlife reserves, and **(3)** eliminate or reduce the populations of nonnative species.

Biologists consider protecting ecosystems to be the best way to preserve biological diversity and natural ecosystem processes such as chemical cycling and energy flow (ecological integrity). Two basic problems are that **(1)** fully or partially protected wildlife sanctuaries make up only about 6% of the world's land area and **(2)** the human population is expected to grow by 3 billion between 1999 and 2054.

The *species approach* is based on protecting endangered species by **(1)** identifying them, **(2)** giving them legal protection, **(3)** preserving and managing their critical habitats, **(4)** propagating them in captivity, and **(5)** reintroducing them into suitable habitats.

The *wildlife management approach* manages game species for sustained yield by **(1)** using laws to regulate hunting, **(2)** establishing harvest quotas, **(3)** developing population management plans, and **(4)** using international treaties to protect migrating game species such as waterfowl. The species and wildlife management approaches are discussed in the remainder of this chapter.

How Can International Treaties Help Protect Endangered Species? Several international treaties and conventions help protect endangered or threatened wild species. One of the most far-reaching is the 1975 Convention on International Trade in Endangered Species (CITES). This treaty, now signed by 145 countries, lists almost 700 species that cannot be commercially traded as live specimens or wildlife products because they are endangered or threatened.

However, enforcement of this treaty is spotty, convicted violators often pay only small fines, and member countries can exempt themselves from protecting any listed species. Furthermore, much of the estimated $6-billion annual illegal trade in wildlife and wildlife products goes on in countries that have not signed the treaty. Centers of illegal animal trade are Singapore, Argentina, Indonesia, Spain, Taiwan, and Thailand.

How Can Laws Help Protect Endangered Species? The United States controls imports and exports of

Hint: Enter the search term *endangered species act* using the Subject Guide.

Figure 18-11 Major species of sea turtles that have roamed the seas for 150 million years, showing their relative adult sizes. Nearly half of the world's sea turtle species are becoming endangered because of **(1)** loss or degradation of beach habitat (where they come ashore to lay their eggs), **(2)** legal and illegal taking of their eggs (Guest Essay, p. 406), **(3)** their increased use as sources of food and medicinal ingredients, and **(4)** unintentional capture by commercial fishing boats.

endangered wildlife and wildlife products through two important laws. The *Lacey Act of 1900* prohibits transporting live or dead wild animals or their parts across state borders without a federal permit. The *Endangered Species Act of 1973* (amended in 1982 and 1988) makes it illegal for Americans to import or trade in any product made from an endangered or threatened species unless it is used for an approved scientific purpose or to enhance the survival of the species.

The Endangered Species Act (ESA) authorizes the National Marine Fisheries Service (NMFS) to identify and list endangered and threatened ocean species; the U.S. Fish and Wildlife Service (USFWS) identifies and lists all other endangered and threatened species. These species cannot be hunted, killed, collected, or injured in the United States.

Any decision by either agency to add or remove a species from the list must be based on biology only, not on economic or political considerations. However, economic factors can be used in deciding whether and how to protect endangered habitat and in developing recovery plans for listed species. The act also forbids federal agencies to carry out, fund, or authorize projects that would either jeopardize an endangered or threatened species or destroy or modify the critical habitat it needs to survive. On private lands, fines and even jail sentences can be imposed to ensure protection of the habitats of endangered species.

Between 1973 and 1999, the number of U.S. species included in the official endangered and threatened list increased from 92 to almost 1,200 species (of which approximately 60% are plants and 40% animals).This is only a fraction of the more than 6,500 U.S. species that the Nature Conservancy estimates are at risk. Each year about 50–85 species are added to the list. The Secretary of the Interior generally is required to designate and protect the *critical habitat* needed for the survival and recovery of each listed species.

Getting listed is only half the battle. Next, the USFWS or the NMFS is supposed to prepare a plan to help the species recover. By 1999 final recovery plans had been developed and approved for about 75% of the endangered or threatened U.S. species, but about half of those plans exist only on paper.

The ESA requires that all commercial shipments of wildlife and wildlife products enter or leave the country through one of nine designated ports. Few illegal shipments are confiscated (Figure 18-12) because the 60 USFWS inspectors can examine only about one-fourth of at least 90,000 shipments that enter and leave the United States each year. Even if caught, many violators are not prosecuted, and convicted violators often pay only a small fine.

In 1982 Congress amended the ESA to allow the Secretary of the Interior to use *habitat conservation plans* (HCPs) to strike a compromise between the interests of private landowners and the interests of endangered and threatened wildlife. With an HCP, landowners, developers, or loggers are allowed to destroy some critical habitat or even kill a certain number of endangered or threatened species on private land in exchange for taking steps to protect that species. Such protective measures might include **(1)** setting aside a part of the species' habitat as a preserve, **(2)** paying to relocate the species to another suitable habitat, or **(3)** paying money to have the government buy suitable habitat elsewhere.

Figure 18-12 Confiscated products made from endangered species. Because of a scarcity of funds and inspectors, probably no more than one-tenth of the illegal wildlife trade in the United States is discovered. The situation is even worse in most other countries. (Steve Hillebrand/ U.S. Fish and Wildlife Service)

Once the plan is approved it cannot be changed even if new data show that the plan is inadequate to protect a species and help it recover.

This approach is supported by some wildlife conservationists. However, there is growing concern that such plans are being developed too quickly without enough scientific evaluation of the effects of such plans on a species' recovery.

In 1999, the U.S. Fish and Wildlife Service approved two new approaches that remove barriers in the ESA that have kept private landowners from protecting threatened or endangered species on their property. One of the new approaches allows *safe harbor agreements* in which landowners voluntarily agree to take specified steps to restore, improve, or maintain habitat for threatened or endangered species located on their land. In return, landowners get **(1)** technical help from local conservation agencies, **(2)** government assurances that the land, water, or other natural resources involved will not face future restrictions once the agreement is over, and **(3)** assurances that after the agreement has expired landowners can return the property to its original condition without penalty.

Under the second approach, landowners enter into voluntary *candidate conservation agreements* with the government. In this prevention approach, landowners agree to take specific steps to help conserve a species whose population is declining but is not yet listed as endangered or threatened. Participating landowners receive technical help and assurances that no additional resource use restrictions will be imposed on the land covered by the agreement if the species is listed as endangered or threatened in the future.

In 1998, Secretary of the Interior Bruce Babbitt announced that the U.S. Fish and Wildlife Service would remove (delist) 17 animals and 12 plants from the endangered and threatened species list over 2 years, subject to final approval by federal biologists. Some would be downgraded from endangered to threatened and others removed from the law's protection. Candidates for this delisting include well-known species such as the bald eagle, peregrine falcon, brown pelican, and gray wolf, as well as lesser-known species such as Pahrump poolfish and the Missouri bladder-pod plant.

Should the Endangered Species Act Be Weakened? Since 1995 there have been efforts to weaken the ESA by **(1)** making protection of endangered species on private land voluntary, **(2)** having the government pay landowners if it forces them to stop using part of their land to protect endangered species, **(3)** making it harder and more expensive to list newly endangered species by requiring government wildlife officials to navigate through a series of hearings and peer-review panels, **(4)** giving the Secretary of the Interior the power to permit a listed species to become extinct without trying to save it, **(5)** allowing the Secretary of the Interior to give any state, county, or landowner permanent exemption from the law, with no requirement for public notification or comment, and **(6)** prohibiting the public from commenting on or bringing lawsuits on any changes in habitat conservation plans for endangered species.

Should the Endangered Species Act Be Strengthened? Most conservation biologists and wildlife scientists contend that the ESA has not been a failure (Spotlight, right). They also refute the charge that the ESA has caused severe economic losses. For example, between 1979 and 1992 only 69 (or only 0.05%) of more than 145,000 projects evaluated by the USFWS were blocked or canceled as a result of the ESA.

Moreover, the act does allow for economic concerns. By law, a decision to list a species must be based solely on science. But once a species is listed, economic considerations can be weighed against species protection in protecting critical habitat and in designing and implementing recovery plans. The act also allows a special Cabinet-level panel, called the "God Squad," to exempt any federal project from having to comply with the act if the economic costs are too high. In addition, the act allows the government to issue permits and exemptions to landowners with listed species living on their property and to use HCPs to bargain with private landowners.

Most biologists and wildlife conservationists believe that we should develop a new system to protect and sustain biological diversity and ecological integrity based on three principles: **(1)** Find out what species and

Accomplishments of the Endangered Species Act

SPOTLIGHT

Critics of the ESA call it an expensive failure because only a few species have been removed from the endangered list.

Most biologists strongly disagree that the act has been a failure, for several reasons. *First*, species are listed only when they are already in serious danger of extinction. This is like setting up a poorly funded hospital emergency room that takes only the most desperate cases, often with little hope for recovery, and then saying it should be shut down because it has not saved enough patients.

Second, it takes decades for most species to become endangered or threatened. Thus, it should not be surprising that it usually takes decades to bring a species in critical condition back to the point where it can be removed from the list. Expecting the ESA (which has been in existence only since 1973) to quickly repair the biological depletion of many decades is unrealistic.

The most important measure of the law's success is that the conditions of almost 40% of the listed species are stable or improving. A hospital emergency room taking only the most desperate cases yet stabilizing or improving the condition of 40% of its patients would be considered an astounding success.

Third, the federal endangered species budget was only $93 million in 1999 (up from $23 million in 1993). This funding is equal to about what it takes to build 3.2 kilometers (1.5 miles) of urban interstate highway, or about 36¢ a year per U.S. citizen. One C-17 transport airplane costs $300 million—more than three times what was spent on protecting endangered species in the United States during 1999.

To most biologists, it's amazing that so much has been accomplished in stabilizing or improving the condition of almost 40% of the listed species on a shoestring budget. What do you think?

Critical Thinking

Explain why you agree or disagree each of the proposals made on p. 482 to **(a)** weaken the ESA and **(b)** strengthen the act.

ecosystems we have, **(2)** locate and protect the most endangered ecosystems and species, and **(3)** give private landowners financial carrots (tax breaks and write-offs), technical help, and assurances of no future requirements (safe harbor and candidate conservation agreements) for helping protect endangered species and ecosystems. Do you believe that the ESA should be weakened or strengthened?

Should We Try to Protect All Endangered and Threatened Species? Because of limited funds, information, and trained personnel, only a few endangered and threatened species can be saved. Many wildlife experts suggest that the limited funds available for preserving threatened and endangered wildlife be concentrated on species that **(1)** have the best chance for survival, **(2)** have the most ecological value to an ecosystem, and **(3)** are potentially useful for agriculture, medicine, or industry.

Others oppose such ideas on ethical grounds or contend that currently we don't have enough biological information to make such evaluations. Proponents argue that **(1)** in effect we are already deciding by default which species to save and that **(2)** despite limited knowledge, the selective approach is more effective and a better use of limited funds than the current one. What do you think?

Sometimes legislation enacted to save some species puts others at risk. For example, in 1973 California legislators attempted to protect the state's remaining mountain lions (also known as cougars or pumas) by outlawing mountain lion hunting for trophies. In 1990, voters passed a referendum banning all killing of these animals.

As a result, the lion's numbers have risen to levels not seen since the 1800s. To find enough food some of the cougars have been killing and eating the Sierra Nevada bighorn sheep, whose number is now about 100. The situation became so bleak that in 1999 the U.S. Fish and Wildlife gave emergency protection to the bighorn sheep. This allows federal officials to kill or remove cougars that might pose a threat to the sheep.

Another case of unexpected consequences involves California sea lions. The Marine Mammal Protection Act of 1972 bans killing of this fish-eating species. Now the growing population of these sea lions is devouring endangered salmon and steelhead trout in Oregon and Washington.

How Can Wildlife Refuges and Other Protected Areas Help Protect Endangered Species? Since 1903, when President Theodore Roosevelt established the first U.S. federal wildlife refuge at Pelican Island, Florida, the National Wildlife Refuge System has grown to 508 refuges (Figure 17-2). About 85% of the area included in these refuges is in Alaska.

Over three-fourths of the refuges are wetlands for protection of migratory waterfowl. About 20% of the

Should Oil and Gas Development Be Allowed in the Arctic National Wildlife Refuge?

PRO/CON

The Arctic National Wildlife Refuge on Alaska's North Slope (Figure 18-13) contains more than one-fifth of all the land in the National Wildlife Refuge System and has been called the crown jewel of the system. The refuge's coastal plain, its most biologically productive part, is the only stretch of Alaska's arctic coastline not open to oil and gas development.

For years, U.S. oil companies have been working to change this because they believe that this coastal area might contain oil and natural gas deposits. They argue that such exploration is needed to increase U.S. oil and natural gas supplies and to reduce dependence on oil imports.

Conservation biologists oppose this proposal and urge Congress to designate the entire coastal plain as wilderness. They cite Department of Interior estimates that there is only a 19% chance of finding as much oil there as the United States consumes every 6 months. Even if the oil does exist, environmentalists do not believe the potential degradation of any portion of this irreplaceable wilderness area would be worth it, especially when improvements in energy efficiency would save far more oil at a much lower cost (Sections 20-1 and 20-2).

Oil company officials claim **(1)** that they have developed Alaska's Prudhoe Bay oil fields without significant harm to wildlife and **(2)** that the area they seek to open to oil and gas development is less than 1.5% of the entire coastal plain region. To environmentalists, however, the huge 1989 oil spill from the tanker *Exxon Valdez* in Alaska's Prince William Sound (Figure 12-25) casts doubt on the first claim. Moreover, a study leaked from the U.S. Fish and Wildlife Service in 1988 revealed that oil drilling at Prudhoe Bay has caused much more air and water pollution than was anticipated before drilling began in 1972.

A 1988 EPA study also found that "violations of state and federal environmental regulations and laws are occurring at an unacceptable rate" in portions of the Prudhoe Bay area in which oil fields and facilities have been developed.

Critical Thinking

Do you believe that exploration and removal of oil from this wildlife refuge should be allowed? Why or why not?

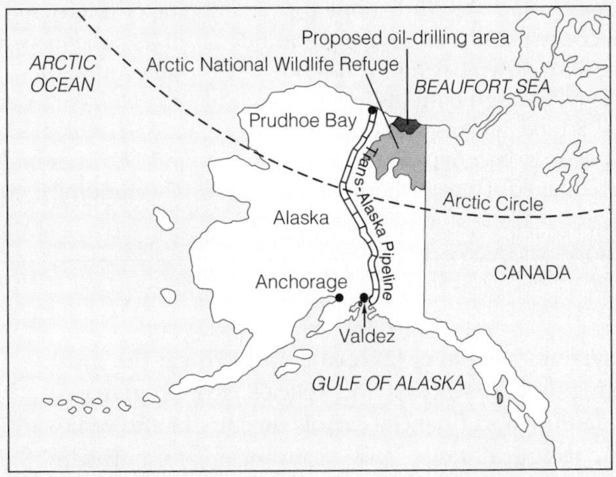

Figure 18-13
Proposed oil-drilling area in Alaska's Arctic National Wildlife Refuge. (Courtesy U.S. Fish and Wildlife Service)

species on the U.S. endangered and threatened list have habitats in the refuge system, and some refuges have been set aside for specific endangered species. These have helped Florida's key deer, the brown pelican, and the trumpeter swan to recover.

Conservation biologists urge the establishment of more refuges for endangered plants. They also urge Congress and state legislatures to allow abandoned military lands that contain significant wildlife habitat to become national or state wildlife refuges. Private groups have also helped establish biodiversity and wildlife sanctuaries (Solutions, p. 458).

So far Congress has not established guidelines for management within the National Wildlife Refuge System, as it has for other public lands. A 1990 report by the General Accounting Office found that activities considered harmful to wildlife occur in nearly 60% of the nation's wildlife refuges. In addition, a 1986 USFWS study estimated that one federal refuge in five is contaminated with chemicals from old toxic-waste dump sites and runoff from nearby agricultural land. There is much controversy over whether to allow oil and gas development in Alaska's Arctic National Wildlife Refuge (Pro/Con, above).

Can Gene Banks and Botanical Gardens Help Save Most Endangered Species? Botanists preserve genetic information and endangered plant species by storing their seeds in gene or seed banks: refrigerated, low-humidity environments (p. 398). Scientists urge that many more such banks be established, especially in developing countries. But some species can't be preserved in gene banks, and maintaining the banks is very expensive.

The world's 1,500 botanical gardens and arboreta maintain at least 40,000 plant species. However, these sanctuaries have too little storage capacity and too little funding to preserve most of the world's rare and threatened plants.

Can Zoos Help Protect Most Endangered Species? Zoos and animal research centers are increasingly being used to preserve some individuals of critically endangered animal species, with the long-term goal of reintroducing the species into protected wild habitats.

Two techniques for preserving such species are egg pulling and captive breeding. *Egg pulling* involves collecting wild eggs laid by critically endangered bird species and then hatching them in zoos or research centers. In *captive breeding*, some or all wild individuals of a critically endangered species are captured for breeding in captivity, with the aim of reintroducing the offspring into the wild. Other techniques for increasing the populations of captive species include **(1)** artificial insemination, **(2)** surgical implantation of eggs of one species into a surrogate mother of another species (embryo transfer), **(3)** use of incubators, and **(4)** cross-fostering, in which the young of a rare species are raised by parents of a similar species.

Captive breeding programs at zoos in Phoenix, Arizona, and San Diego and Los Angeles in California temporarily saved the nearly extinct Arabian oryx (Figure 18-3), a large antelope that once lived throughout the Middle East. By 1960 it had been hunted nearly to extinction in the wild (with only about 48 left) by people riding in Jeeps and helicopters and wielding rifles and machine guns. Between 1980 and 1996, the return of oryx bred in captivity to protected habitats in the Middle East allowed the population to reach about 400. However, since 1996 poachers have been capturing these animals for sale to private zoos. As a result, between 1996 and 1999 the wild population was reduced to about 100 and the species was again facing extinction.

Endangered species now being bred in captivity in the United States and returned to the wild include the **(1)** peregrine falcon (Figure 18-3, removed from the endangered species list in 1999), **(2)** black-footed ferret (Figure 18-3), and **(3)** golden lion tamarin (Figure 18-3). However, most reintroductions fail because of **(1)** lack of suitable habitat, **(2)** inability of individuals bred in captivity to survive in the wild, or **(3)** renewed overhunting or capture of some returned species (such as the Arabian oryx).

Efforts to maintain populations of endangered species in zoos and research centers are limited by lack of space and money. The captive population of each species must number 100 to 500 individuals to avoid extinction through accident, disease, or loss of genetic diversity through inbreeding. Recent genetic research indicates that 10,000 or more individuals are needed for an endangered species to maintain its capacity for biological evolution.

It is estimated that today's zoos and research centers have space to preserve healthy and sustainable populations of only 925 of the 2,000 large vertebrate species that could vanish from the planet. According to one estimate, if all of the space in U.S. zoos were used for captive breeding, only about 100 species of large animals could be sustained on a long-term basis.

Some critics see zoos as prisons for once-wild animals. They also contend that zoos foster the notion that we don't need to preserve large numbers of wild species in their natural habitats.

Whether one agrees or disagrees with this position, conservation biologists point out that zoos and botanical gardens are not a biologically or economically feasible solution for most of the world's current endangered species and the much larger number expected to become endangered over the next few decades.

18-4 WILDLIFE MANAGEMENT

How Can Wildlife Populations Be Managed? **Wildlife management** involves manipulating wildlife populations (especially game species) and their habitats for their welfare and for human benefit.

In the United States, funds for state game management programs come from the sale of hunting and fishing licenses and from federal taxes on hunting and fishing equipment. Two-thirds of the states also have provisions on state income tax returns that allow individuals to contribute money to state wildlife programs.

Only 10% of all government wildlife dollars are spent to study or benefit nongame species, which make up nearly 90% of the country's wildlife species. Since the passage of the Wildlife Restoration Act in 1937 there has been a large increase in the number of game animals (such as white-tailed deer, wild turkeys, Rocky Mountain elk, and Pronghorn antelope) sought by many sport hunters.

The first step in wildlife management is to make an often controversial decision about which species are to be managed in a particular area. Ecologists and conservation biologists emphasize preserving biodiversity and ecological integrity, wildlife conservationists are concerned about endangered species, birdwatchers want the greatest diversity of bird species, and hunters want large populations of game species. In the United States and other developed countries, most wildlife management is devoted to producing surpluses of game animals and birds for hunters.

After goals have been set, the wildlife manager must develop a management plan. Ideally, this is based on the principles of **(1)** ecological succession (Figures 5-14 and 5-15), **(2)** wildlife population dynamics (Figures 7-3, 7-4, 7-5, 7-6, and 7-7), and **(3)** an understanding of the cover, food, water, space, and other habitat requirements of each species to be managed. The manager must also consider the number of potential hunters, their likely success rates, and the regulations for preventing excessive hunting.

This information is difficult, expensive, and time-consuming to obtain. It involves much educated guesswork and trial and error, which is why wildlife management is as much an art as a science. Management plans must also be sensitive to political pressures from conflicting groups and to budget constraints.

How Can Vegetation and Water Supplies Be Manipulated to Manage Wildlife? Wildlife managers can encourage the growth of plant species that are the preferred food and cover for a particular animal species by controlling the ecological succession of the vegetation in an area.

Wildlife species can be classified into four types, according to the stage of ecological succession at which they are most likely to be found: early successional, midsuccessional, late successional, and wilderness (Figure 5-16). *Early successional species* find food and cover in weedy pioneer plants that invade an area that has been disturbed, whether by human activities or natural phenomena (fires, volcanic activity, or glaciation).

Midsuccessional species are found around abandoned croplands and partially open areas created by **(1)** logging of small stands of timber, **(2)** controlled burning, and **(3)** clearing of vegetation for roads, firebreaks, oil and gas pipelines, and electrical transmission lines. Such openings of the forest canopy **(1)** promote the growth of vegetation favored by midsuccessional mammal and bird species and **(2)** increase the amount of edge habitat, where two communities such as a forest and a field come together. This transition zone allows animals such as deer to feed on vegetation in clearings and then quickly escape to cover in the nearby forest.

Late successional species rely on old-growth and mature forests to produce the food and cover on which they depend. These animals require the protection of moderate-sized, old-growth forest refuges.

Wilderness species flourish only in undisturbed, mature vegetation communities, such as large areas of old-growth forests, tundra, grasslands, and deserts. They can survive only in large wilderness areas and wildlife refuges.

Various types of habitat management can be used to attract a desired species and encourage its population growth. Examples are **(1)** planting seeds, **(2)** transplanting certain types of vegetation, **(3)** building artificial nests, and **(4)** deliberately setting controlled, low-level ground fires to help control vegetation. Wildlife managers often create or improve ponds and lakes in wildlife refuges to provide water, food, and habitat for waterfowl and other wild animals.

How Useful Is Sport Hunting in Managing Wildlife Populations? Most developed countries use sport hunting laws to manage populations of game animals. Licensed hunters are allowed to hunt only during certain portions of the year so as to protect animals during their mating season. Limits are set on the size, number, and sex of animals that can be killed, as well as on the number of hunters allowed in a given area.

Close control of sport hunting is difficult. Accurate data on game populations may not exist and may cost too much to get. People in communities near hunting areas, who benefit from money spent by hunters, may seek to have hunting quotas raised.

There is controversy over sport hunting. Proponents argue that without carefully regulated sport hunting, deer and other large game animals exceed the carrying capacity of their habitats and destroy vegetation they and other species need. For example, between 1900 and 1997, the estimated population of white-tailed deer in the United States increased from 500,000 to 30 million. Many of these deer **(1)** invade subdivisions and eat shrubs and home gardens, **(2)** raid farmers' fields and orchards, **(3)** threaten rare plants and animals in some areas, **(4)** spread Lyme disease carried by deer ticks to humans, and **(5)** are involved in more than 500,000 deer-vehicle collisions each year that kill more Americans than any other wild animal, injure thousands, and cost at least $1.1 billion. Wildlife biologist William Porter estimates that without hunting, the U.S. deer population would be five times its current level and cause extensive ecological damage.

According to its proponents sport hunting also **(1)** provides recreational pleasure for millions of people (16 million in the United States), **(2)** stimulates local economies, and **(3)** provides money through sales of hunting licenses and taxes on firearms and ammunition (more than $1.7 billion since 1937) that is used to buy, restore, and maintain wildlife habitats and to support wildlife research in the United States. Environmental groups such as the Sierra Club and Defenders of Wildlife support carefully controlled sport hunting as a way of preserving biological diversity by helping to prevent depletion of other native species of plants and animals.

However, some individuals and groups, including the Humane Society, oppose sport hunting. They argue

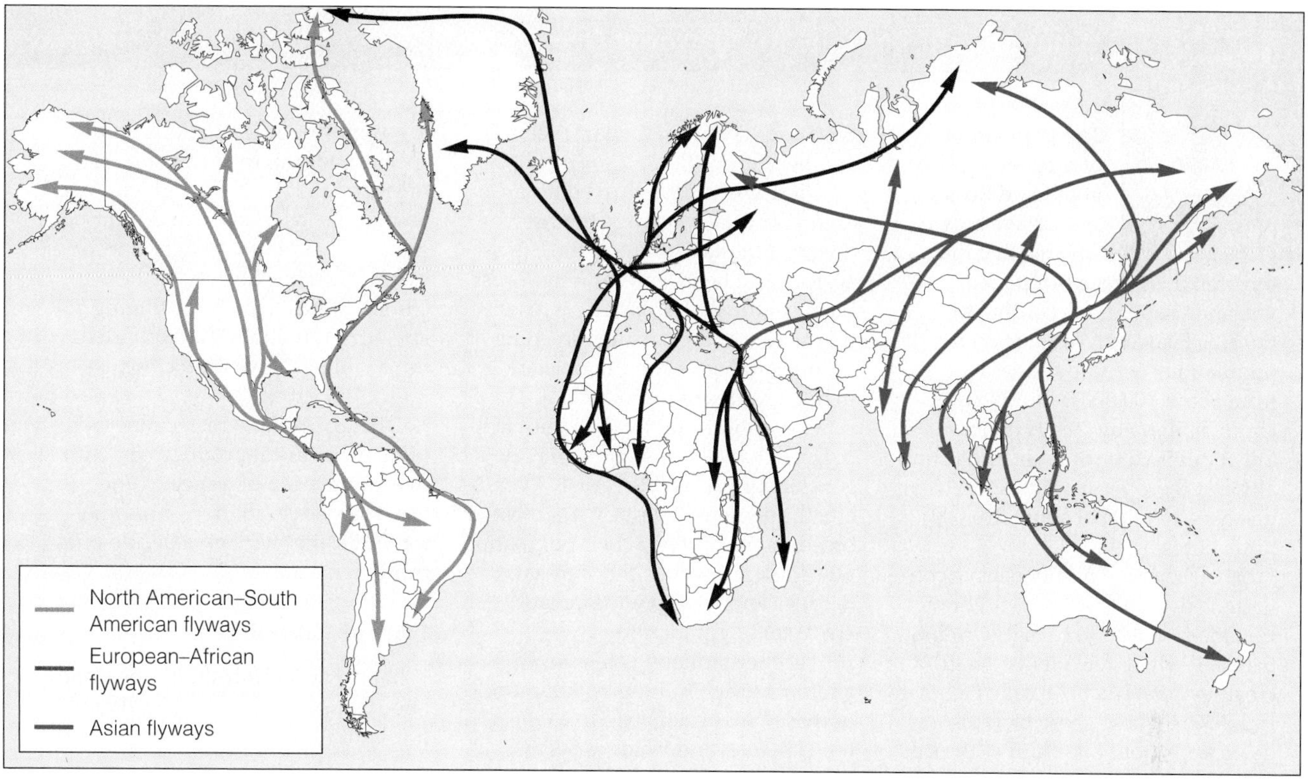

Figure 18-14 Major flyways used by migratory birds, mostly waterfowl. Each route has a number of subroutes.

(1) that it inflicts unnecessary pain and suffering on wild animals (most of which are not killed to supply food humans need for survival) and **(2)** that game managers create a surplus of game animals by deliberately eliminating their natural predators (such as wolves) and then claim that the surplus must be harvested by hunters to prevent habitat degradation or starvation of the game. Instead of eliminating natural predators, say opponents, wildlife managers should reintroduce them to reduce the need for sport hunting (Cast Study, p. 451).

Supporters of hunting point out that **(1)** populations of many game species (for example, deer) are so large that predators such as wolves cannot possibly control them, and **(2)** because most wildlife habitats are fragmented, introduction of predators can lead to the loss of nearby livestock. However, critics of hunting contend that deer (which make up only 2% of the 200 million animals hunters kill in the United States each year) are being used as a smoke screen argument to allow killing of many other game species that don't threaten vegetation. What do you think?

How Can Populations of Migratory Waterfowl Be Managed? Migratory birds—including ducks, geese, swans, and many songbirds (Case Study, p. 471)—make north–south journeys from one habitat to another each year, usually to find food, suitable climate, and other conditions necessary for reproduction. Such bird species use many different north–south routes called **flyways**, but only about 15 are considered major routes (Figure 18-14).

Some countries along such flyways have entered into agreements and treaties to protect crucial habitats needed by such species, both along their migration routes and at each end of their journeys. However, the populations of migrating snow geese have risen to the point where they are causing serious ecological damage (Connections, p. 488).

Wildlife officials manage waterfowl by regulating hunting, protecting existing habitats, and developing new habitats, including artificial nesting sites, ponds, and nesting islands. More than 75% of the federal wildlife refuges in the United States are wetlands used by migratory birds. Local and state agencies and private conservation groups such as Ducks Unlimited, the Audubon Society, and the Nature Conservancy (Solutions, p. 458) have also established waterfowl refuges.

Since 1934 the Migratory Bird Hunting and Conservation Stamp Act has required waterfowl hunters to buy a duck stamp each season they hunt. Revenue from these sales goes into a fund to buy land and easements for the benefit of waterfowl.

What Should Be Done About Snow Geese?

CONNECTIONS

The populations of the lesser snow goose and Ross's geese have grown so large that they threaten their arctic tundra and wetlands habitats. According to the U.S. Fish and Wildlife Service, the populations of these species rose from about 800,000 in the 1960s to about 5 billion by 1999. Hunters kill about 500,000 of these birds annually, but their population is still growing at a rate of about 250,000 birds per year.

Their massive numbers have converted vast tracts of arctic tundra in Canada and Alaska to highly saline, bare soil where few plants can grow, creating what has been called a botanical desert. These geese have also destroyed about one-third of the salt marshes along the Hudson and James Bays and have heavily damaged another third.

These geese have thrived because each year they can gorge themselves on the abundant grains in the U.S. agricultural heartland from the Great Plains to the Gulf Coast. As a result, they arrive at their winter breeding grounds in better condition, leading to higher rates of reproduction.

Two options for dealing with this problem have been suggested. One is to let nature take its course. Eventually the geese will destroy their winter breeding ground and their population will crash. However, this may take decades and lead to widespread ecosystem degradation.

Another strategy is to allow and encourage hunters to kill more of these birds each year. In 1999, the U.S. Fish and Wildlife Service announced plans to allow 24 states to ease hunting restrictions on the species, hoping to at least double the number killed. Though endorsed by many conservation and hunting groups, this plan was denounced by the Humane Society of the United States, which is bringing a lawsuit saying that the expanded hunt violates laws protecting migratory birds.

This problem, which connects high agricultural productivity with overpopulation of a species and ecosystem degradation, is another reminder that we can never do one thing in nature.

Critical Thinking

What do you think should be done about snow goose overpopulation? Explain.

18-5 FISHERY MANAGEMENT AND PROTECTING MARINE BIODIVERSITY

How Can Freshwater Fisheries Be Managed and Sustained? Managing freshwater fish involves encouraging populations of commercial and sport fish species and reducing or eliminating populations of less desirable species. It is usually done by regulating time and length of fishing seasons and the number and size of fish that can be taken.

Other techniques include **(1)** building reservoirs and farm ponds and stocking them with fish, **(2)** fertilizing nutrient-poor lakes and ponds, **(3)** protecting and creating spawning sites, **(4)** protecting habitats from buildup of sediment and other forms of pollution, **(5)** removing debris, **(6)** preventing excessive growth of aquatic plants from cultural eutrophication (excessive inputs of plant nutrients from human activities, Figure 12-21), and **(7)** building small dams to control water flow.

Predators, parasites, and diseases can be controlled by **(1)** improving habitats, **(2)** breeding genetically resistant fish varieties, and **(3)** using antibiotics and disinfectants judiciously. Hatcheries can be used to restock ponds, lakes, and streams with prized species such as trout and salmon, and entire river basins can be managed to protect such valued species as salmon (Figure 15-15).

How Can Marine Fisheries Be Managed? By international law, a country's offshore fishing zone extends to 370 kilometers (200 nautical miles, or 230 statute miles) from its shores. Foreign fishing vessels can take certain quotas of fish within such zones, called *exclusive economic zones*, but only with a government's permission. Ocean areas beyond the legal jurisdiction of any country are known as the *high seas*. Any limits on the use of the living and mineral common-property resources in these areas are set by international maritime law and international treaties.

Managers of marine fisheries use several techniques to help prevent overfishing and commercial extinction and allow depleted stocks to recover. Fishery commissions, councils, and advisory bodies (with representatives from countries or states using a fishery) can **(1)** set annual quotas, **(2)** establish rules for dividing the allowable catch among the participating countries or states, **(3)** limit fishing seasons, and **(4)** regulate the type of fishing gear that can be used to harvest a particular species.

As voluntary associations, however, fishery commissions don't have any legal authority to compel their members to follow their rules. Furthermore, it is very difficult to estimate the sustainable yields of various marine species. Experience has shown that many fishery com-

Privatizing Fishing Rights in Public Waters

SPOTLIGHT

In 1996, the National Marine Fisheries Service (NMFS) initiated a controversial system of individual transferable quotas (ITQs), in which the government gives each fishing vessel owner a specified percentage of the total allowable catch for each fishery in public waters. The largest quotas go to those who have historically taken the most fish, and ITQ owners are permitted to buy, sell, or lease their quotas like private property.

The NMFS argued that because of its limited resources it could not effectively monitor and help manage fish stocks in U.S. public waters. It suggested that privatizing such fish stocks would be a more effective way to prevent overfishing as a result of the tragedy of the commons.

On the surface, this example of free-market environmentalism sounds like a good idea, but most environmentalists generally oppose this approach for several reasons. *First*, it would in effect transfer ownership of publicly owned fisheries to the private commercial fishing industry but still use taxpayer dollars to pay for NMFS oversight of this newly privatized resource.

Second, the scheme would shift fishing quotas from small-scale operators to large-scale operators (whose unsustainable harvesting caused most of the problem in the first place), as has happened with the ITQ system now used in New Zealand. Small fishers would be squeezed out because they don't have the capital to buy the ITQs from others.

Third, the New Zealand experience shows that the ITQ system increases poaching and sales of illegally caught fish on the black market by small-scale fishers who receive no quota or too small a quota to make a living.

Critical Thinking

Do you believe that privatizing fishing in publicly owned waters by using transferable quotas is a good or a bad idea? Explain.

missions, most of them dominated by fishing industry representatives, have not prevented overfishing.

A major problem is that the world's fishing fleets are about 2.5 times larger than needed to ensure a sustainable catch. However, the world's 13 major fishing nations agreed in 1999 to develop plans to reduce their distant-water fishing fleets by 2005. This would be done primarily by reducing fishing subsidies that are one of the major driving forces of fleet overcapacity and overfishing (Connections, p.403).

Suggestions to reduce overfishing in U.S. waters include the following:

- Gradually phase out government subsidies of the fishing industry, such as low-interest loans and direct subsidies for boats and fishing operations.
- Impose fees for harvesting fish and shellfish from publicly owned and managed offshore waters, similar to fees now imposed for grazing and logging on U.S. public lands.
- Give each fishing vessel owner a marketable quota for removing a certain amount of fish (Spotlight, above).
- Establish no-fishing marine areas to allow depleted fish species to recover, such as the 13 marine reserves New Zealand has established over the past 20 years.

Why Is It Difficult to Protect Marine Biodiversity? It is difficult to protect marine biodiversity for several reasons:

- Shore-hugging species are adversely affected by coastal development and the accompanying massive inputs of sediment and other wastes from land (Figure 6-23).
- Much of the damage is not visible to most people.
- The seas are viewed by many people as an inexhaustible resource, capable of absorbing an almost infinite amount of waste and pollution.
- Most of the world's ocean area lies outside the legal jurisdiction of any country and is thus an open-access resource, subject to overexploitation because of the tragedy of the commons.
- There are no effective international agreements to protect biodiversity in the open seas.

18-6 CASE STUDY: THE WHALING INDUSTRY

Why Is Whaling an Example of the Tragedy of the Commons? *Cetaceans* are an order of mostly marine mammals ranging in size from the 0.9-meter (3-foot) porpoise to the giant 15- to 30-meter (50- to 100-foot) blue whale. They are divided into two major groups: toothed whales and baleen whales (Figure 18-15).

Toothed whales, such as the porpoise, sperm whale, and killer whale (orca), bite and chew their food, and feed mostly on squid, octopus, and other marine animals. *Baleen whales*, such as the blue, gray, humpback,

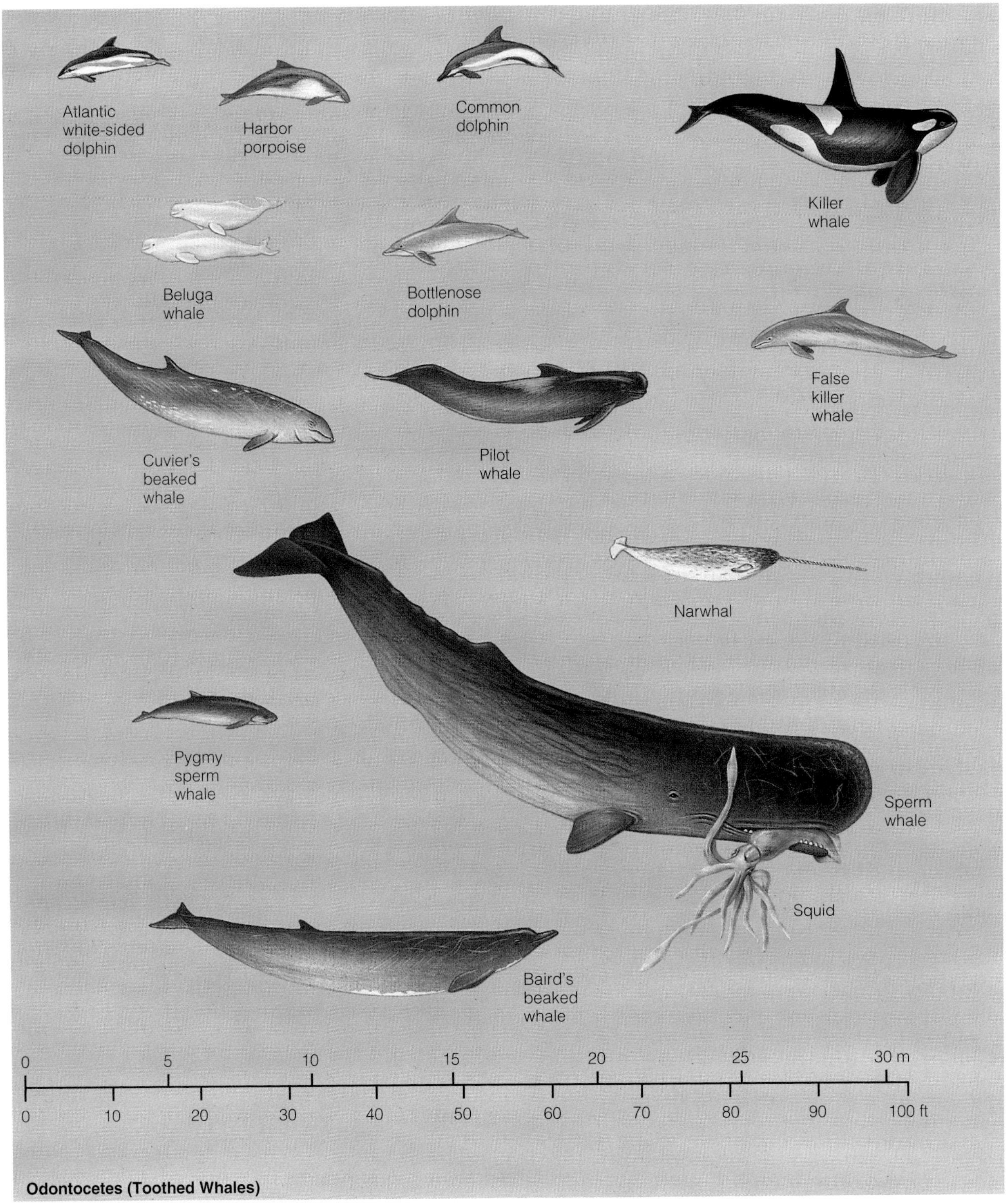

Figure 18-15 Examples of cetaceans, which can be classified as baleen whales and toothed whales.

Humpback whale
Bowhead
whale
Right
whale
Minke
whale
Blue
whale
Feeding
on krill
Fin whale
Sei whale
Gray whale
Mysticetes (Baleen Whales)

and finback, are filter feeders. Instead of teeth, they have several hundred horny plates made of baleen, or whalebone, that hang down from their upper jaw. These plates filter plankton from the seawater, especially tiny shrimplike krill (Figure 4-15). Baleen whales are the most abundant group of cetaceans.

Whales are fairly easy to kill because of their large size and their need to come to the surface to breathe. Mass slaughter has become very efficient with the use of fast ships, harpoon guns, and inflation lances (which pump dead whales full of air and make them float).

Whale harvesting—mostly in international waters, a commons open to all resource extractors—has followed the classic pattern of a tragedy of the commons. Between 1925 and 1975, whalers killed an estimated 1.5 million whales; this overharvesting drove the populations of 8 of the 11 major species to commercial extinction, the point where it no longer paid to hunt and kill them.

Case Study: Near Extinction of the Blue Whale In addition to commercial extinction, the populations of some prized species were reduced to the brink of biological extinction. A prime example is the endangered blue whale (Figure 18-15), the world's largest animal. Fully grown, it's more than 30 meters (100 feet) long—longer than three train boxcars—and weighs more than 25 elephants. The adult has a heart the size of a Volkswagen Beetle car, and some of its arteries are so big that a child could swim through them.

Blue whales spend about 8 months of the year in Antarctic waters. There they find an abundant supply of krill, which they filter daily by the trillions from seawater. During the winter, they migrate to warmer waters, where their young are born.

Before commercial whaling began, an estimated 200,000 blue whales roamed the Antarctic Ocean. Today, the species has been hunted to near biological extinction for its oil, meat, and bone. Its decline was caused by a combination of prolonged overharvesting and certain natural characteristics of blue whales. Their huge size made them easy to spot. They were caught in large numbers because they grouped together in their Antarctic feeding grounds. They also take 25 years to mature sexually and have only one offspring every 2–5 years, a reproductive rate that makes it difficult for the species to recover once its population falls beneath a certain threshold.

Blue whales haven't been hunted commercially since 1964 and have been classified as an endangered species since 1975. Despite this protection, some marine biologists believe that too few blue whales—an estimated 3,000-10,000—remain for the species to recover and avoid extinction.

How Have Whale Populations Been Managed? In 1946, the International Whaling Commission (IWC) was established to regulate the whaling industry by setting annual quotas to prevent overharvesting and commercial extinction. However, these quotas often were based on inadequate data or were ignored by whaling countries. Without any powers of enforcement, the IWC has been unable to stop the decline of most whale species.

In 1970, the United States stopped all commercial whaling and banned all imports of whale products. Under intense pressure from environmentalists and governments of many countries and led by the United States, the IWC has imposed a moratorium on commercial whaling since 1986. As a result, the estimated number of whales killed commercially worldwide dropped from 42,480 in 1970 to about 1,100 in 1999. Some whaling has continued because Japan, Norway, Peru, Iceland, and the former Soviet Union have exempted themselves from the moratorium or have quit the IWC.

Should the International Ban on Whaling Be Lifted? Whaling is a traditional part of the economies and cultures of countries such as Japan, Iceland, and Norway. To many people in these cultures hunting whales is similar to the legalized hunting of deer or elk. These and other whaling nations believe that the international ban on commercial whaling should be lifted, arguing that the moratorium is based on emotion, not science.

Whaling proponents also say that fears of extinction for some species are outdated. According to IWC estimates, the population of minke whales (which Japan hunts) now numbers about 1 million and that of pilot whales about 1.4 million. They see no scientific reason for not resuming controlled hunting of these species.

Most environmentalists disagree. Some argue that whales are peaceful, intelligent, sensitive, and highly social mammals that pose no threat to humans and should not be killed. Others fear that opening the door to any commercial whaling may eventually lead to widespread harvests of most whale species by weakening current international disapproval and sanctions against commercial whaling. Whaling opponents also **(1)** point to the failure of the IWC (with no enforcement powers) to prevent the past commercial extinction of most whale species in the past and **(2)** contend that comparing whale hunting to deer and elk hunting is misleading because populations of deer and elk in the United States are large and in no danger of becoming endangered or extinct. What do you think?

In 1994 the IWC voted 23 to 1 to establish a permanent whale sanctuary in the Antarctic Ocean. However, it is almost impossible to monitor and prevent illegal whaling in this huge sanctuary. Moreover, Japan and Russia filed objections to the IWC decision and Japan announced its intentions to continue commercial whaling in the sanctuary.

Opponents of commercial whaling have urged the U.S. Congress to impose sanctions against countries defying the IWC ban on commercial whaling. They also urge consumers to **(1)** support economic boycotts of all imports from offending nations, **(2)** avoid using air and sea vessels owned by such countries, and **(3)** cease traveling to offending countries. By 1994 a boycott of Norwegian fish products in the United Kingdom and Germany had cost Norway more than $30 million in canceled contracts. What is your view about such boycotts?

Are We a Wise Species? We named ourselves the wise (*sapiens*) species. Most conservation biologists believe that in the next few decades, we will learn whether we are indeed a wise species—whether we have the wisdom to learn from and work with nature to protect ourselves and other species. Some actions you can take to help protect wildlife and preserve biodiversity are listed in Appendix 6.

A greening of the human mind must precede the greening of the earth. A green mind is one that cares, saves, and shares. These are the qualities essential for conserving biological diversity now and forever.

M. S. SWAMINATHAN

REVIEW QUESTIONS

1. Define the boldfaced terms used in this chapter.

2. Describe the factors that led to the extinction of the passenger pigeon in the United States.

3. Distinguish between *local, ecological,* and *biological extinction* of a species. Distinguish between *endangered* and *threatened species.* List characteristics that make species vulnerable to biological extinction.

4. Give scientific, ecological, economic, medical, aesthetic, recreational, and ethical reasons for not allowing wild species to become prematurely extinct because of human activities.

5. What is *biophilia*? Do you agree with this hypothesis? Explain.

6. Distinguish between *background* and *mass extinction.* According to fossil and geological evidence, how many mass extinctions have occurred during the past 500 million years? After each mass extinction, about how long did it take for adaptive radiations to again increase the diversity of life? Give three reasons why most biologists believe that we are in the midst of sixth mass extinction caused by human activities. Do you agree with this hypothesis? Explain.

7. What are five underlying causes of population reduction and extinction of wild species? Describe how each of the following factors contributes to the premature biological extinction of species as a result of human activities and give an example of a species affected by each factor: **(a)** habitat loss and degradation, **(b)** habitat fragmentation, **(c)** deliberately introduced species, **(d)** accidentally introduced species, **(e)** commercial hunting and illegal hunting (poaching), **(f)** predator and pest control, **(g)** the legal and illegal market for exotic pets and decorative plants, **(h)** climate change and pollution, and **(i)** overfishing.

8. Describe the factors leading to the near extinction of the American bison.

9. Why are many island species especially vulnerable to extinction? Why are some migrating birds vulnerable to premature extinction? What are some important ecological roles of bats and why are they vulnerable to extinction?

10. Describe the following methods used for harvesting fish and explain and how each can lead to overfishing and depletion of other species: **(a)** trawling, **(b)** purse-seine fishing, **(c)** longlining, and **(d)** drift-net fishing.

11. What are the three basic approaches to protecting wildlife and biodiversity?

12. List the benefits and limitations of protecting species using **(a)** the Convention on International Trade in Endangered Species (CITES) and **(b)** the Endangered Species Act (ESA) in the United States.

13. List reasons why you believe that the Endangered Species Act has been a failure or a success. List measures that would strengthen and weaken the ESA in the United States. Which do you favor? Why?

14. Distinguish between *habitat conservation plans, safe harbor agreements,* and *candidate conservation plans* used as ways to help implement the Endangered Species Act.

15. Should we try to protect all endangered and threatened species? Explain. If not, what guidelines would you use?

16. Summarize the pros and cons of using the following to help protect endangered species: **(a)** wildlife refuges, **(b)** gene banks and botanical gardens, and **(c)** zoos and animal research centers.

17. What is *wildlife management*, and what are the major steps in managing wildlife? Describe how each of the following methods can be used to help manage wildlife and list the limitations of each approach: **(a)** manipulation of vegetation and water supplies, **(b)** sport hunting, and **(c)** controlling populations of migratory waterfowl. Give the pros and cons of sport hunting.

18. Describe methods for managing **(a)** freshwater fish species and **(b)** marine fish species and list the limitations of the approaches used. List methods for reducing overfishing in U.S. waters.

19. List five reasons why protecting marine biodiversity is so difficult.

20. Distinguish between *toothed whales* and *baleen whales* and give an example of each type. Describe how international whaling is an example of the tragedy of the commons. Describe the factors leading to the near extinction of the blue whale.

21. What is the International Whaling Commission, and why has it not been successful in preventing the commercial extinction of almost all commercially hunted whale species? Should the international ban on whaling be lifted? Explain.

CRITICAL THINKING

1. Discuss your gut-level reaction to the following statement: "Eventually all species become extinct. Thus, it doesn't really matter that the passenger pigeon is extinct and that the blue whale, the whooping crane, the California condor, and the world's remaining species of rhinoceros and tigers are endangered mostly because of human activities." Be honest about your reaction and give arguments for your position.

2. **(a)** Do you accept the ethical position that each *species* has the inherent right to survive without human interference, regardless of whether it serves any useful purpose for humans? Explain. **(b)** Do you believe that each *individual* of an animal species has an inherent right to survive? Explain. Would you extend such rights to individual plants and microorganisms? Explain.

3. Your lawn and house are invaded by fire ants, which can cause painful bites. What would you do?

4. Should whaling nations be allowed by the International Whaling Commission to resume a limited annual harvest of minke whales as long as this species' estimated numbers don't decline significantly? Explain. Would you extend this to other commercially valuable whale species if their populations increase significantly? Explain. How would you monitor and enforce annual commercial whaling quotas?

5. Which of the following statements best describes your feelings toward wildlife: **(a)** As long as it stays in its space, wildlife is OK; **(b)** as long as I don't need its space, wildlife is OK; **(c)** I have the right to use wildlife habitat to meet my own needs; **(d)** when you've seen one redwood tree, fox, elephant, or some other form of wildlife you've seen them all, so lock up a few of each species in a zoo or wildlife park and don't worry about protecting the rest; **(e)** wildlife should be protected.

6. List your three favorite species. Examine why they are your favorites. Are they cute and cuddly-looking, like the giant panda and the koala? Do they have human-like qualities, like apes or penguins that walk upright? Are they large, like elephants or blue whales? Are they beautiful, like tigers and monarch butterflies? Are any of them plants? Are any of them species such as bats, sharks, snakes, or spiders that most people are afraid of? Are any of them microorganisms that help keep you alive? Reflect on what your choice of favorite species tells you about your attitudes toward most wildlife.

7. Environmental groups in a heavily forested state want to restrict logging in some areas to save the habitat of an endangered squirrel. Timber company officials argue that the well-being of one type of squirrel is not as important of the well-being of the many families affected if the restriction causes them to lay off thousands of workers. If you had the power to decide this issue, what would you do and why? Can you come up with a compromise?

8. If you lived in the suburbs and deer invaded your yards and ate the shrubs and vegetables in your organic garden, what would you do?

***9.** Make a log of your own consumption of all products for a single day. Relate your level and types of consumption to the decline of wildlife species and the increased destruction and degradation of wildlife habitats in the United States, in tropical forests, and in aquatic ecosystems.

***10.** Identify examples of habitat destruction or degradation in your community that have had harmful effects on the populations of various wild plant and animal species. Develop a management plan for the rehabilitation of these habitats and wildlife.

***11.** Make a concept map of this chapter's major ideas, using the section heads and subheads and the key terms (in boldface). Look at the inside back cover and in Appendix 5 for information about concept maps.

FURTHER INFORMATION

For more information see Internet Sources and Further Readings at the back of the book. Online resources for this chapter can be found at **www.brookscole.com/biology**

PART V

ENERGY RESOURCES

A country that runs on energy cannot afford to waste it.

BRUCE HANNON

19 NONRENEWABLE ENERGY RESOURCES

Bitter Lessons from Chernobyl

Chernobyl is a chilling word recognized around the globe as the site of a major nuclear disaster (Figure 19-1). On April 26, 1986, a series of explosions in one of the reactors in a nuclear power plant in Ukraine (then part of the Soviet Union) blew the massive roof off the reactor building and flung radioactive debris and dust high into the atmosphere. A huge radioactive cloud spread over much of Belarus, Russia, the Ukraine, and other parts of Europe and eventually encircled the planet. Here are some consequences of this disaster, caused by poor reactor design and human error:

- In 1998 the Ukrainian Health Ministry put the official death toll from the accident at 3,576. However, Greenpeace Ukraine estimates that by 1995 the total death toll from the accident was about 32,000.
- According to the United Nations, almost 400,000 people have been forced to leave their homes, probably never to return. Most were not evacuated until 10 days or more after the accident.
- According to a recent UN report, some 160,000 square kilometers (62,000 square miles) of the former Soviet Union remain contaminated with radioactivity.
- The 9 million people (about one-third of them children) still living in this area (and especially the almost 400,000 evacuated environmental refugees) live under great stress.
- Over half a million people were exposed to dangerous levels of radioactivity. Between 1986 and 1997 the rate of thyroid cancer (mostly among children) in Ukraine and nearby Belarus was more than 10 times the 1986 level and 30 times higher among those evacuated from the Chernobyl area. In Belarus, where 70% of the radiation was deposited, the World Health Organization says that thyroid cancer rates among children are 100 times preaccident levels.
- Government officials say that the total cost of the accident will reach at least $358 billion, many times more than the value of all the nuclear electricity that has ever been generated in the former Soviet Union.

The environmental refugees evacuated from the Chernobyl region had to leave their possessions behind and say good-bye to lush, green wheat fields and blossoming apple trees, to land their families had farmed for generations, to cows and goats that would be shot because the grass they ate was radioactive, and to their radioactivity-poisoned cats and dogs. They will not be able to return. Today the Chernobyl power plant remains one of the most dangerous places on the earth.

Chernobyl taught us that a major nuclear accident anywhere is a nuclear accident everywhere.

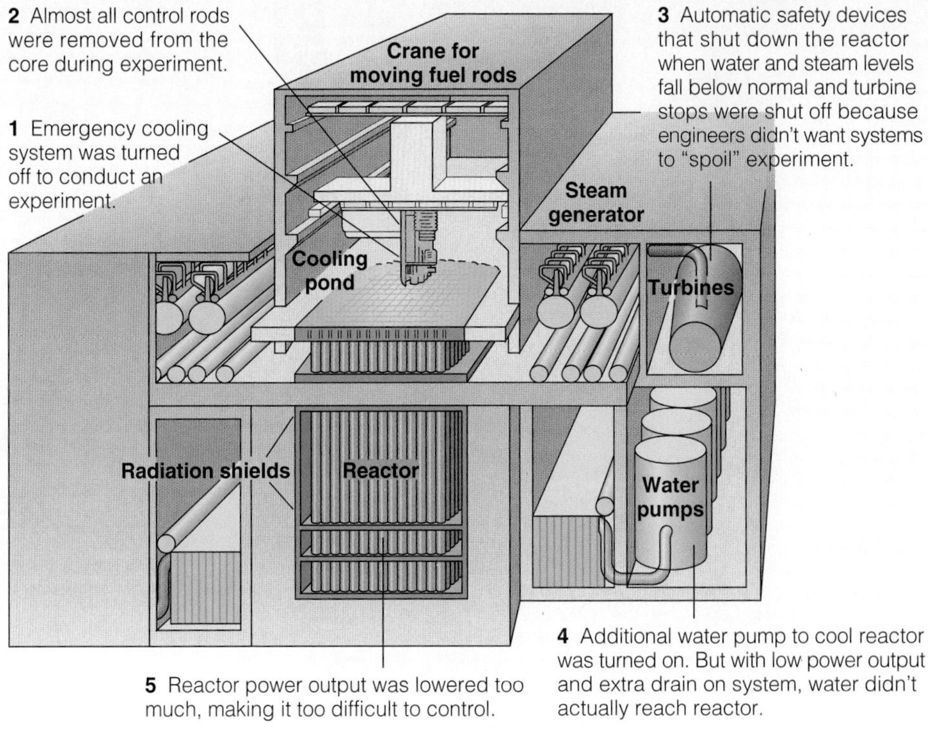

Figure 19-1 Major events leading to the Chernobyl nuclear power-plant accident on April 26, 1986, in the former Soviet Union. The accident happened because **(1)** engineers turned off most of the reactor's automatic safety and warning systems (to keep them from interfering with an unauthorized safety experiment) **(2)** inadequate safety design of the reactor (no secondary containment shell, as in Western-style reactors, no emergency core cooling system to prevent reactor core meltdown, and **(3)** a design flaw that leads to unstable operation at low power). After the reactor exploded, crews exposed themselves to lethal levels of radiation to put out fires and encase the shattered reactor in a hastily constructed concrete tomb. This 20-story concrete tomb is now sagging and full of holes that allow water to seep in and radioactive dust to drift out. Building a new tomb for the reactor will cost at least $1.5 billion—money the Ukrainian government doesn't have.

We are an interdependent world and if we ever needed a lesson in that, we got it in the oil crisis of the 1970s.

ROBERT S. MCNAMARA

This chapter addresses the following questions:

- How should we evaluate energy alternatives?
- What are the advantages and disadvantages of oil?
- What are the advantages and disadvantages of natural gas?
- What are the advantages and disadvantages of coal?
- What are the advantages and disadvantages of conventional nuclear fission, breeder nuclear fission, and nuclear fusion?

19-1 EVALUATING ENERGY RESOURCES

What Types of Energy Do We Use? *Some 99% of the energy used to heat the earth and all of our buildings comes directly from the sun.* Without this input of essentially inexhaustible solar energy, the earth's average temperature would be –240°C (–400°F), and life as we know it would not exist. This direct input of solar energy also produces several other forms of renewable energy: **(1)** wind, **(2)** falling and flowing water (hydropower), and **(3)** biomass (solar energy converted to chemical energy stored in chemical bonds of organic compounds in trees and other plants).

Commercial energy sold in the marketplace makes up the remaining 1% of the energy we use to supplement the solar input. Most commercial energy comes from extracting and burning mineral resources obtained from the earth's crust, primarily nonrenewable, carbon-containing fossil fuels (Figure 19-2).

About 81% of the commercial energy consumed in the world comes from nonrenewable energy resources (75% from fossil fuels and 6% from nuclear power, Figure 19-3, left). In developing countries the most important supplemental source of energy is potentially renewable biomass (mostly fuelwood and charcoal made from fuelwood), the main sources of energy for heating and cooking for roughly half the world's population. Within a few decades one-fourth of the world's population in developed countries may face an oil shortage, but half the world's population in developing countries already faces a fuelwood shortage (Figure 17-16).

The United States is the world's largest user of energy. With only 4.6% of the population, it uses 24% of the world's commercial energy. In contrast, India, with 17% of the world's people, uses only about 3% of the world's commercial energy. About 92% of the commercial energy used in the United States comes from nonrenewable energy resources (85% from fossil fuels and 7% from nuclear power, Figure 19-3, right). Energy use per person in Canada and the United States is about twice as high as in Japan, Germany, France, and the United Kingdom and at least 100 times as high as that in China and India.

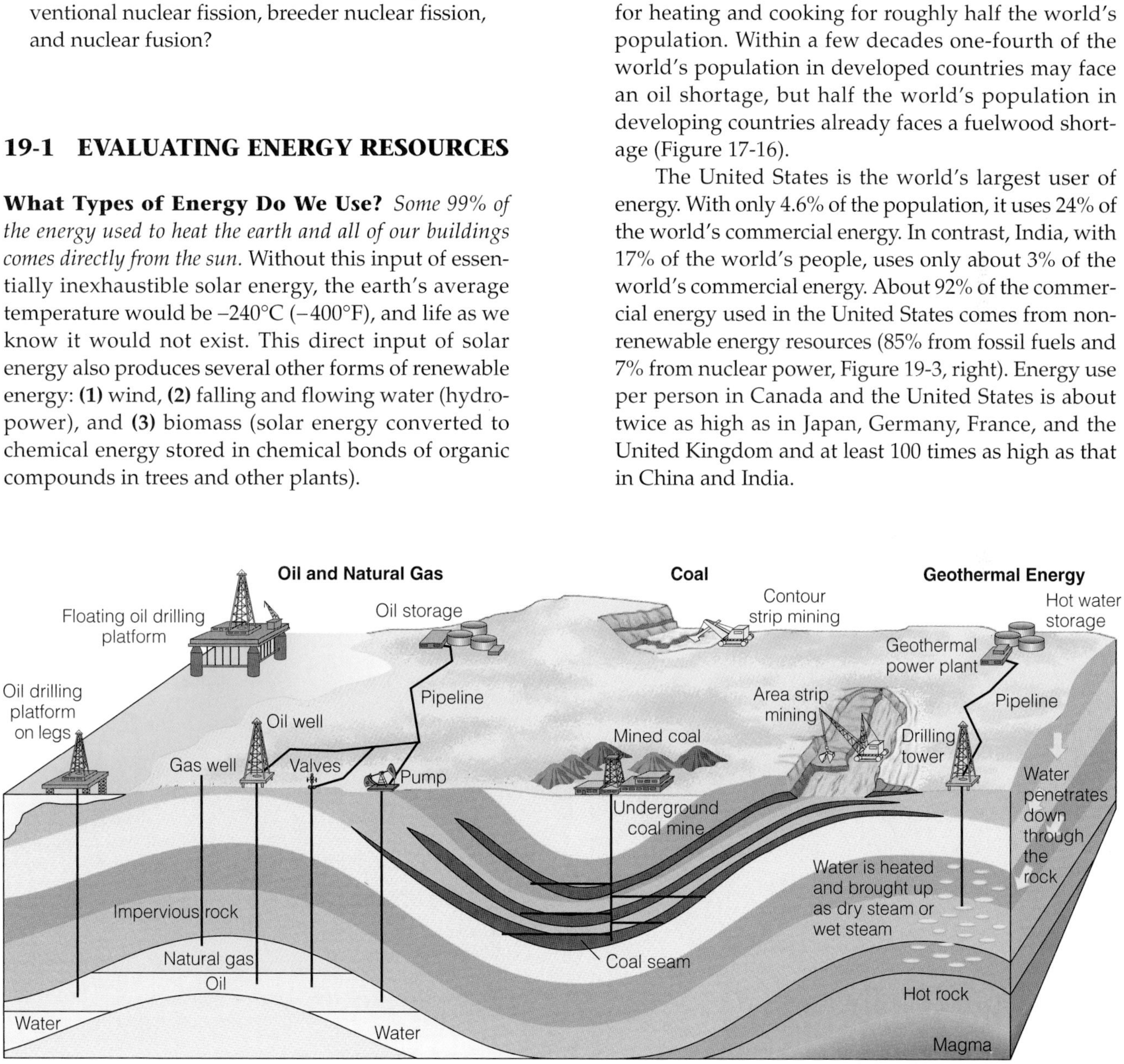

Figure 19-2 Important commercial energy resources from the earth's crust are geothermal energy, coal, oil, and natural gas. Uranium ore is also extracted from the crust and then processed to increase its concentration of uranium-235, which can be used as a fuel in nuclear reactors to produce electricity.

How Should We Evaluate Energy Resources? Our current dependence on nonrenewable fossil fuels is the primary cause of air and water pollution, land disruption, and projected global warming. Moreover, affordable oil, the most widely used energy resource in developed countries, probably will be economically depleted within 40–80 years and must be replaced by other energy resources.

What is our best immediate energy option? The general consensus is to cut out unnecessary energy waste by improving energy efficiency, as discussed in Chapter 20. There is disagreement about our next best energy option. Some say we should get much more of the energy we need from renewable energy sources such as the **(1)** sun, **(2)** wind, **(3)** flowing water, **(4)** biomass, **(5)** heat stored in the earth's interior, and **(6)** hydrogen gas, as discussed in Chapter 20.

Others say we should burn more coal and synthetic liquid and gaseous fuels made from coal because it is our most plentiful energy resource. Many believe that fairly abundant and clean-burning natural gas is the answer, especially as a transition fuel to a new solar age built around improved energy efficiency and renewable energy. Some think nuclear power is the answer. These nonrenewable energy options are evaluated in this chapter.

Experience shows that it usually takes at least 50 years and huge investments to phase in new energy alternatives (Figure 19-4). Thus, we must plan for and begin the shift to a new mix of energy resources now. To do so involves answering the following questions for *each* energy alternative:

- How much of the energy source will be available in the near future (the next 15–25 years) and in the long term (the next 25–50 years)?
- What is this source's net energy yield?
- How much will it cost to develop, phase in, and use this energy resource?
- How will extracting, transporting, and using the energy resource affect the environment? The key question may no longer be when will we run out of carbon-based fossil fuels, but how will burning these fuels affect the earth's climate (Chapter 11) and human health (Section 10-5)?

What Is Net Energy? The Only Energy That Really Counts It takes energy to get energy. For example, oil must be **(1)** found, **(2)** pumped up from beneath the ground, **(3)** transported to a refinery and converted to useful fuels (such as gasoline, diesel fuel, and heating oil), **(4)** transported to users, and **(5)** burned in furnaces and cars before it is useful to us. Each of these steps uses energy, and the second law of energy (Section 3-7) tells us that each time we use energy to perform a task, some of it is always wasted and is degraded to low-quality energy.

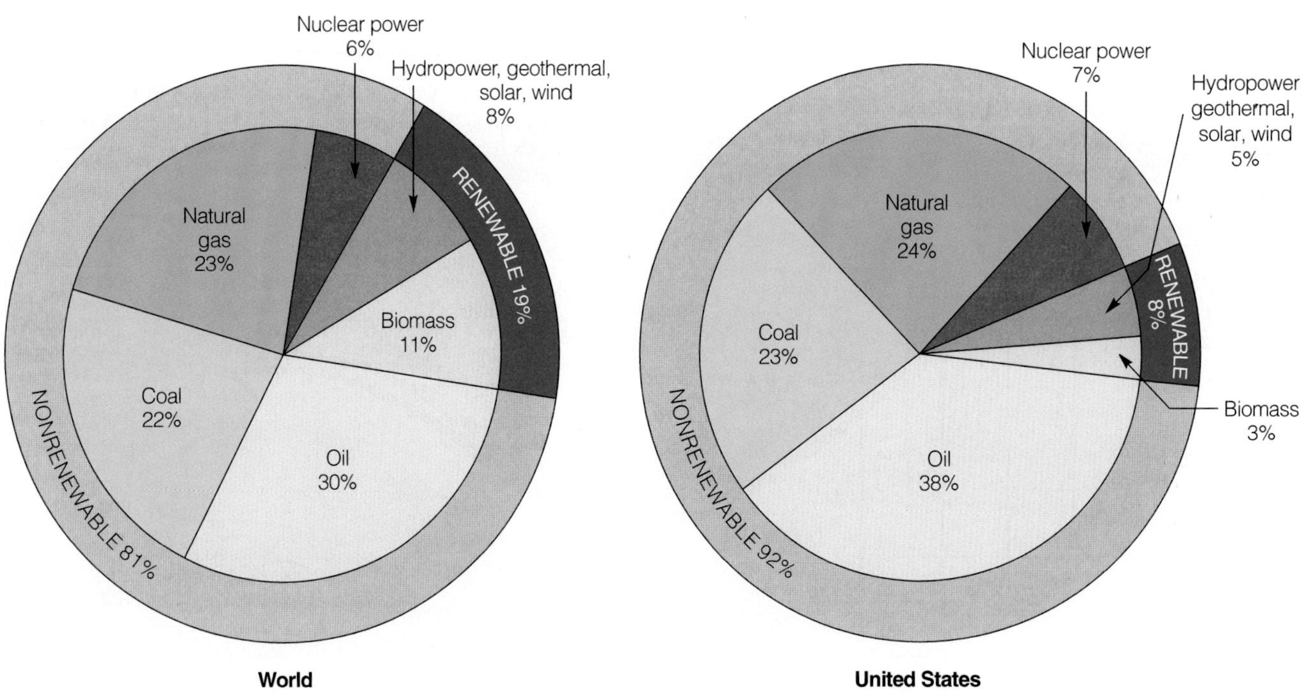

Figure 19-3 Commercial energy use by source for the world and the United States in 1998. Commercial energy amounts to only 1% of the energy used in the world; the other 99% comes from the sun and is not sold in the marketplace. (Data from U.S. Department of Energy, British Petroleum, and Worldwatch Institute)

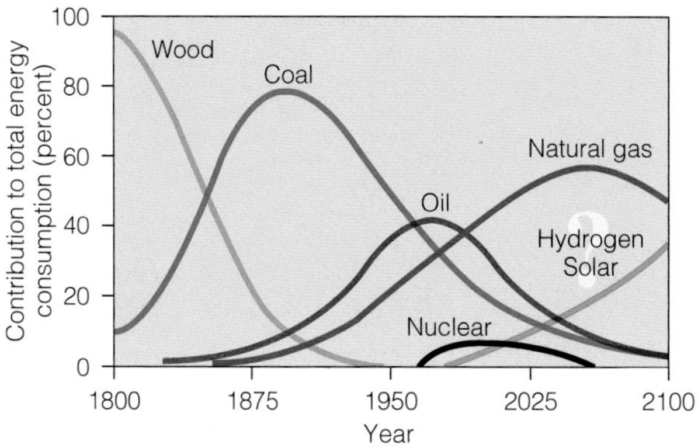

Figure 19-4 Shifts in the use of commercial energy resources in the United States since 1850, with projected changes to 2100. Shifts from wood to coal and then from coal to oil and natural gas have each taken about 50 years. Affordable oil is expected to be running out within 40–80 years, and burning fossil fuels is the primary cause of air pollution and projected warming of the atmosphere. For these reasons, most analysts believe we must make a new shift in energy resources over the next 50 years. (Data from U.S. Department of Energy)

The usable amount of high-quality energy available from a given quantity of an energy resource is its **net energy**: the total useful energy available from the resource over its lifetime minus the amount of energy **(1)** used (the first law of energy), **(2)** automatically wasted (the second law of energy), and **(3)** unnecessarily wasted in finding, processing, concentrating, and transporting it to users.

Net energy is like your net spendable income (your wages minus taxes and other deductions). For example, suppose that for every 10 units of energy in oil in the ground we have to use and waste 8 units of energy to find, extract, process, and transport the oil to users. Then we have only 2 units of *useful energy* available from every 10 units of energy in the oil.

We can express net energy as the ratio of useful energy produced to the useful energy used to produce it. In the example just given, the *net energy ratio* would be 10/8, or approximately 1.25. The higher the ratio, the greater the net energy yield. When the ratio is less than 1, there is a net energy loss.

Figure 19-5 shows estimated net energy ratios for various types of space heating, high-temperature heat for industrial processes, and transportation. Currently, oil has a high net energy ratio because much of it comes from large, accessible deposits such as those in the Middle East. When those sources are depleted, the net energy ratio of oil will decline and prices will rise. Then more money and high-quality fossil-fuel energy will be needed to find, process, and deliver new oil from **(1)** widely dispersed small deposits, **(2)** deposits buried deep in the earth's crust, or **(3)** deposits located in remote areas.

Conventional nuclear energy has a low net energy ratio because large amounts of energy are required to extract and process uranium ore, convert it into a usable nuclear fuel, and build and operate power plants. In addition, more energy is needed to dismantle the plants after their 15–40 years of useful life and to store the resulting highly radioactive wastes for thousands of years.

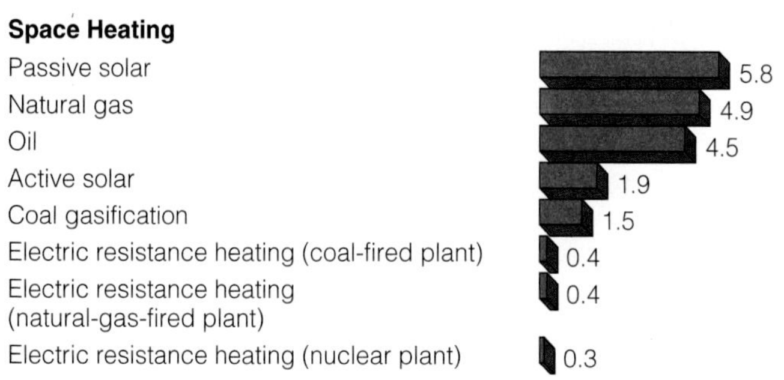

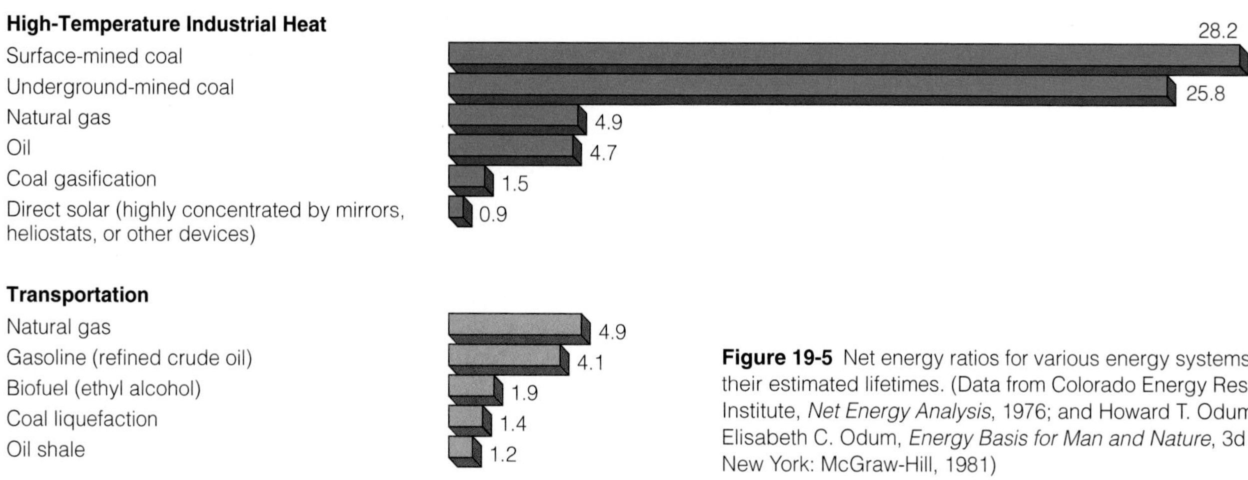

Figure 19-5 Net energy ratios for various energy systems over their estimated lifetimes. (Data from Colorado Energy Research Institute, *Net Energy Analysis*, 1976; and Howard T. Odum and Elisabeth C. Odum, *Energy Basis for Man and Nature*, 3d ed., New York: McGraw-Hill, 1981)

19-2 OIL

What Is Crude Oil, and How Is It Extracted and Processed? **Petroleum**, or **crude oil** (oil as it comes out of the ground) is a fossil fuel produced by the decomposition of dead organic matter from plants and animals that has been buried deeply and subjected to high temperatures and pressures over millions of years. Typically this gooey, smelly liquid consists mostly of hydrocarbons, with small amounts of sulfur, oxygen, and nitrogen impurities.

Crude oil and natural gas are often trapped together under a dome deep within the earth's crust (Figure 19-2). The crude oil is dispersed in pores and cracks in underground rock formations, somewhat like water in a sponge.

Primary oil recovery involves drilling a well and pumping out the oil that flows by gravity into the bottom of the well. After the flowing oil has been removed, water can be injected into nearby wells to force some of the remaining heavy oil to the surface, a process known as **secondary oil recovery**.

On average, producers get only about 35% of the oil out of a reservoir by primary and secondary recovery before they abandon it because the *heavy oil* that remains is too difficult or expensive to recover.

As oil prices rise, it may become economical to remove about 10–25% of this heavy oil by **enhanced**, or **tertiary, oil recovery**. Steam or CO_2 gas can be used to force some of the heavy oil into the well cavity for pumping to the surface. However, enhanced oil recovery is expensive: It takes the energy in one-third of a barrel of oil to retrieve each barrel of heavy oil. Researchers hope to use bacteria to increase tertiary recovery by 10–25% at a lower cost.

Most crude oil travels by pipeline to a *refinery*, where it is heated and distilled in gigantic columns to separate it into components with different boiling points (Figure 19-6). Some of the products of oil distillation, called **petrochemicals**, are used as raw materials in industrial organic chemicals, pesticides, plastics, synthetic fibers, paints, medicines, and many other products.

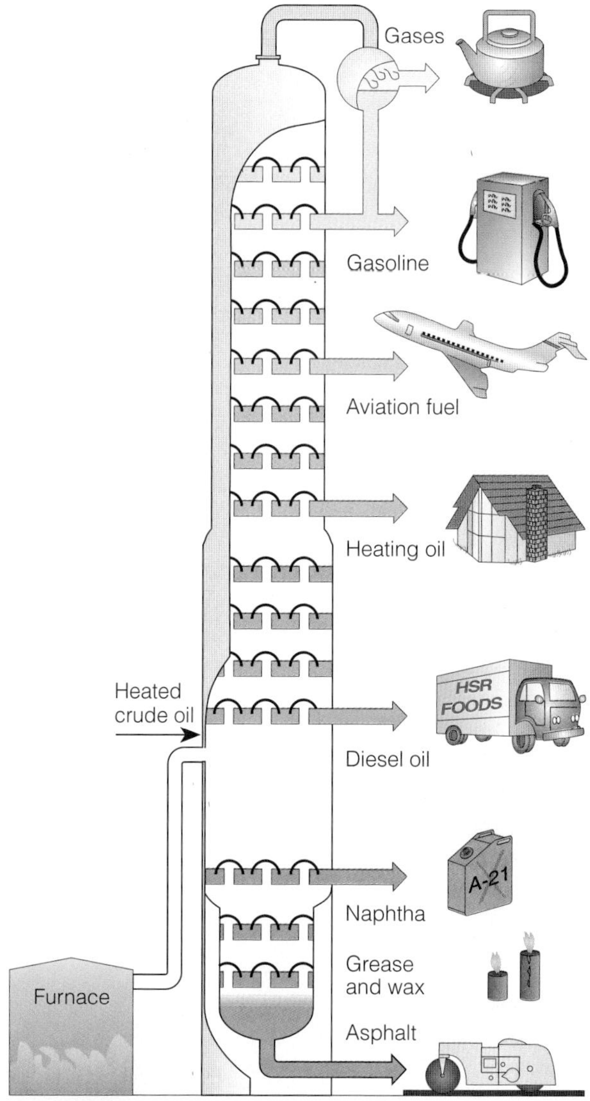

Figure 19-6 Refining crude oil. Components are removed at various levels, depending on their boiling points, in a giant distillation column. The most volatile components with the lowest boiling points are removed at the top of the column.

Who Has the World's Oil Supplies? Oil *reserves* are identified deposits from which oil can be extracted profitably at current prices with current technology. The 13 countries that make up the Organization of Petroleum Exporting Countries (OPEC)* have 67% of the world's reserves, which explains why OPEC is expected to have long-term control over world oil supplies and prices. Saudi Arabia, with 26%, has the world's largest known crude oil reserves, followed by Iraq with 10%.

Geologists believe that the politically volatile Middle East contains the majority of the world's undiscovered oil. However, some additional oil is expected to be found from increased exploration in China, Mexico, and other developing countries.

Currently, the United States has only 2.3% of the world's oil reserves but uses nearly 30% of the oil

*OPEC was formed in 1960 so that developing countries with much of the world's known and projected oil supplies could get a higher price for this resource. Today its members are Algeria, Ecuador, Gabon, Indonesia, Iran, Iraq, Kuwait, Libya, Nigeria, Qatar, Saudi Arabia, United Arab Emirates, and Venezuela.

Campbell, C. J. 1998. "Running Out of Gas: This Time the Wolf Is Coming." *The National Interest*, no. 51 p. 47(9).

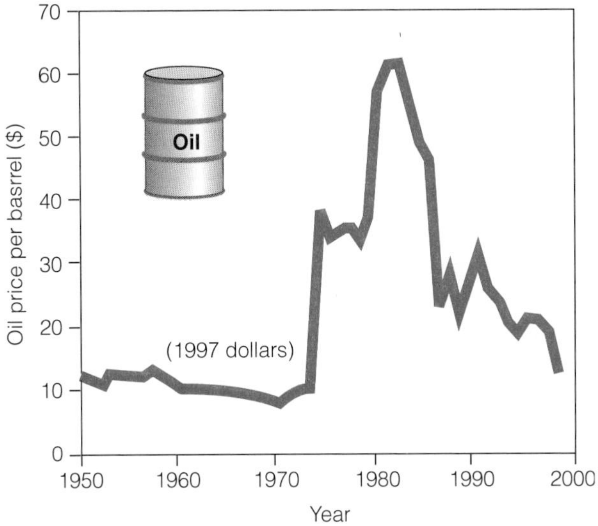

Figure 19-7 Inflation-adjusted price of oil, 1950–98. When adjusted for inflation, oil costs about the same as it did in 1975. Although low oil prices have stimulated economic growth, they have discouraged improvements in energy efficiency and increased use of renewable energy resources. (Data from U.S. Department of Energy and Department of Commerce)

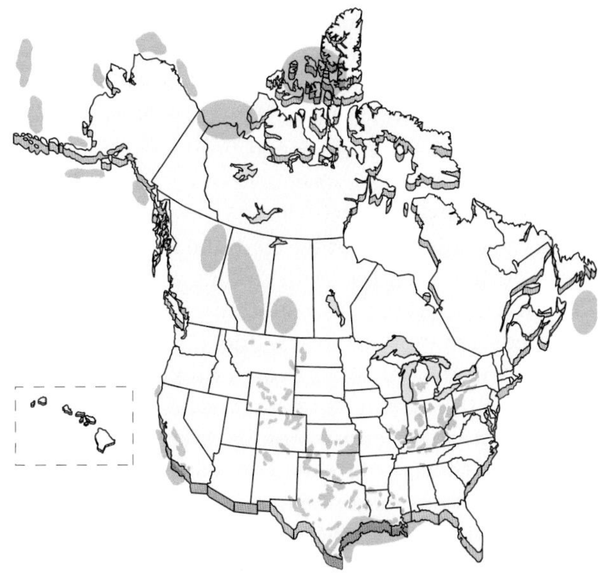

Figure 19-8 Locations of the largest crude oil and natural gas fields in the United States and Canada. Little *new* oil and natural gas is expected to be found in the United States. (Data from Council on Environmental Quality)

extracted worldwide each year (68% of it for transportation) mostly because oil is cheap (Figure 19-7). Figure 19-8 shows the locations of the largest crude oil fields in the United States and Canada. Despite an upsurge in exploration and test drilling, U.S. oil extraction has declined since 1985.

In 1998, the United States imported about 40% of the oil it used (up from 36% in 1973), mostly because of declining oil reserves and increased oil use. By 2010 the United States could be importing 72% or more of the oil it uses. In 1998 oil imports cost the United States about $47 billion—an average of $129 million per day.

This dependence on imported oil (about half of it from OPEC countries) and the likelihood of much higher oil prices within 10–20 years could drain the United States and other major oil-importing nations of vast amounts of money. Economists warn that this could lead to severe inflation and widespread economic recession, perhaps even a major depression.

When Iraq invaded Kuwait in the summer of 1990, the United States and other developed countries went to war, mostly to protect their access to oil in the Middle East, especially from Saudi Arabia. As Worldwatch Institute researchers Christopher Flavin and Nicholas Lenssen put it, "Not only is the world addicted to cheap oil, but the largest liquor store is in a very dangerous neighborhood."

The U.S. military presence in the Middle East costs at least $55 billion per year, which adds about $80 to each barrel of oil the United States imports from this area. When these military costs are added to the roughly $20-per-barrel market price of oil, Americans in 1999 were paying at least $100 per barrel for oil.

How Long Will Oil Supplies Last? The world's oil's reserves (Figure 13-7) may be 80% depleted within 35–84 years, depending on how rapidly it is used. The remaining 20% is considered too expensive to extract (Figure 1-12). At the current rate of consumption, global oil reserves will last at least 44 years. Undiscovered oil that is thought to exist might last another 20–40 years.

Instead of remaining at the current level, however, global oil consumption is projected to increase by about 25% between 1999 and 2010. This will hasten depletion of global oil reserves, with global oil production projected to begin declining sometime between 2010 and 2020. At today's consumption rate, U.S. oil reserves will be depleted in about 24 years (less if consumption increases as projected); potential reserves might yield an additional 24 years of production.

Some analysts argue that rising oil prices (when oil consumption exceeds oil production) will stimulate exploration and lead to enough new reserves to meet future demand through the next century. Other analysts argue that such optimistic projections about future oil supplies ignore the consequences of exponentially increasing consumption of oil.

Even assuming that we continue to use crude oil at the current rate, **(1)** Saudi Arabia, with the largest known crude oil reserves, could supply all the world's oil needs for only 10 years, and **(2)** the estimated reserves under Alaska's North Slope (the largest ever found in North America) would meet current world demand for only 6 months or U.S. demand for 3 years. In short, just to keep on using oil at the *current* rate and not run out, we must discover and add to global oil reserves the equivalent of a new Saudi Arabian supply *every 10 years*.

According to the U.S. Geological Survey, global discovery of large oil fields peaked in 1962 and has been declining since then. In the United States, Alaskan oil production peaked in 1988 and has been declining since then.

Burning any fossil fuel releases carbon dioxide into the atmosphere and can thus enhance projected global warming. Figure 19-9 compares the relative amounts of CO_2 emitted per unit of energy by the major fossil fuels and nuclear power. In the United States, coal-burning electric power plants produce about 35% of the annual CO_2 emissions and gasoline-burning vehicles produce about 31% of such emissions.

Figure 19-10 lists the pros and cons of conventional oil as an energy resource. According to a 1998 study by the International Center for Technology Assessment, **(1)** the environmental, health, and social costs of using oil are estimated to be at least $200 billion per year, and **(2)** federal tax subsidies for the U.S. oil industry are at least $7.6 billion per year, plus $55–96 billion per year to protect U.S. access to global oil reserves (mostly in the Middle East). Suppose these government subsidies were phased out and the harmful environmental effects of oil (Figure 13-12) were included in its market price (full-cost pricing, Section 2-3). Analysts project that oil would become so expensive that it probably would be replaced by improved energy efficiency and by a variety of less harmful and cheaper renewable energy resources (also evaluated using full-cost pricing).

How Useful Are Heavy Oils from Oil Shale and Tar Sands? **Oil shale** is a fine-grained rock (Figure 19-11) that contains a solid, waxy mixture of hydrocarbon compounds called **kerogen**. After being removed by surface or subsurface mining, the shale is crushed and heated above ground in a retort to vaporize the kerogen (Figures 19-11 and 19-12). The kerogen vapor is condensed, forming heavy, slow-flowing, dark-brown **shale oil**. Before shale oil can be sent by pipeline to a refinery, it must be heated to increase its flow rate and processed to remove sulfur, nitrogen, and other impurities.

The shale oil potentially recoverable from U.S. deposits, mostly on federal lands in Colorado, Utah, and Wyoming, could probably meet the country's crude oil demand for 41 years at current use levels. Canada,

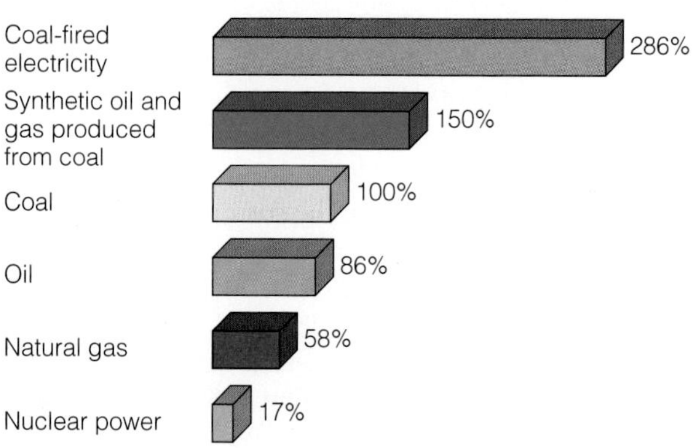

Figure 19-9 Carbon dioxide emissions per unit of energy produced by various fuels, expressed as percentages of emissions produced by coal.

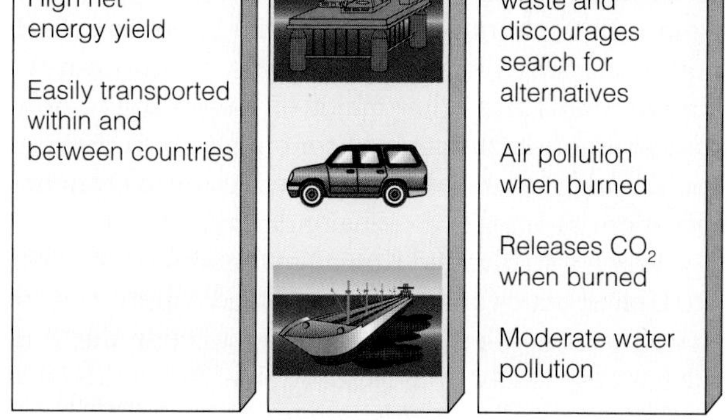

Figure 19-10 Advantages and disadvantages of using conventional oil as an energy resource.

Figure 19-11 Oil shale and the shale oil extracted from it. Big U.S. oil shale projects have been canceled because of excessive cost. (U.S. Department of Energy)

China, and several republics in the former Soviet Union also have large oil shale deposits. Estimated potential global supplies of shale oil are 200 times larger than estimated global supplies of conventional oil.

Tar sand (or oil sand) is a mixture of clay, sand, water, and **bitumen** (a gooey, black, high-sulfur heavy oil). It is usually removed by surface mining. It is then heated with pressurized steam until the bitumen fluid softens and floats to the top. The bitumen is then purified and chemically upgraded into a synthetic crude oil suitable for refining (Figure 19-13).

The world's largest known deposits of tar sands, the Athabasca Tar Sands, lie in northern Alberta, Canada. Currently, these deposits supply about 21% of Canada's oil needs. These deposits could supply all of Canada's projected oil needs for about 33 years at its current consumption rate, but they would last the world only about 2 years. Other large deposits of tar sands are in Venezuela, Colombia, and parts of the former Soviet Union. Figure 19-14 lists the pros and cons of using heavy oil from oil shale and tar sand as energy resources.

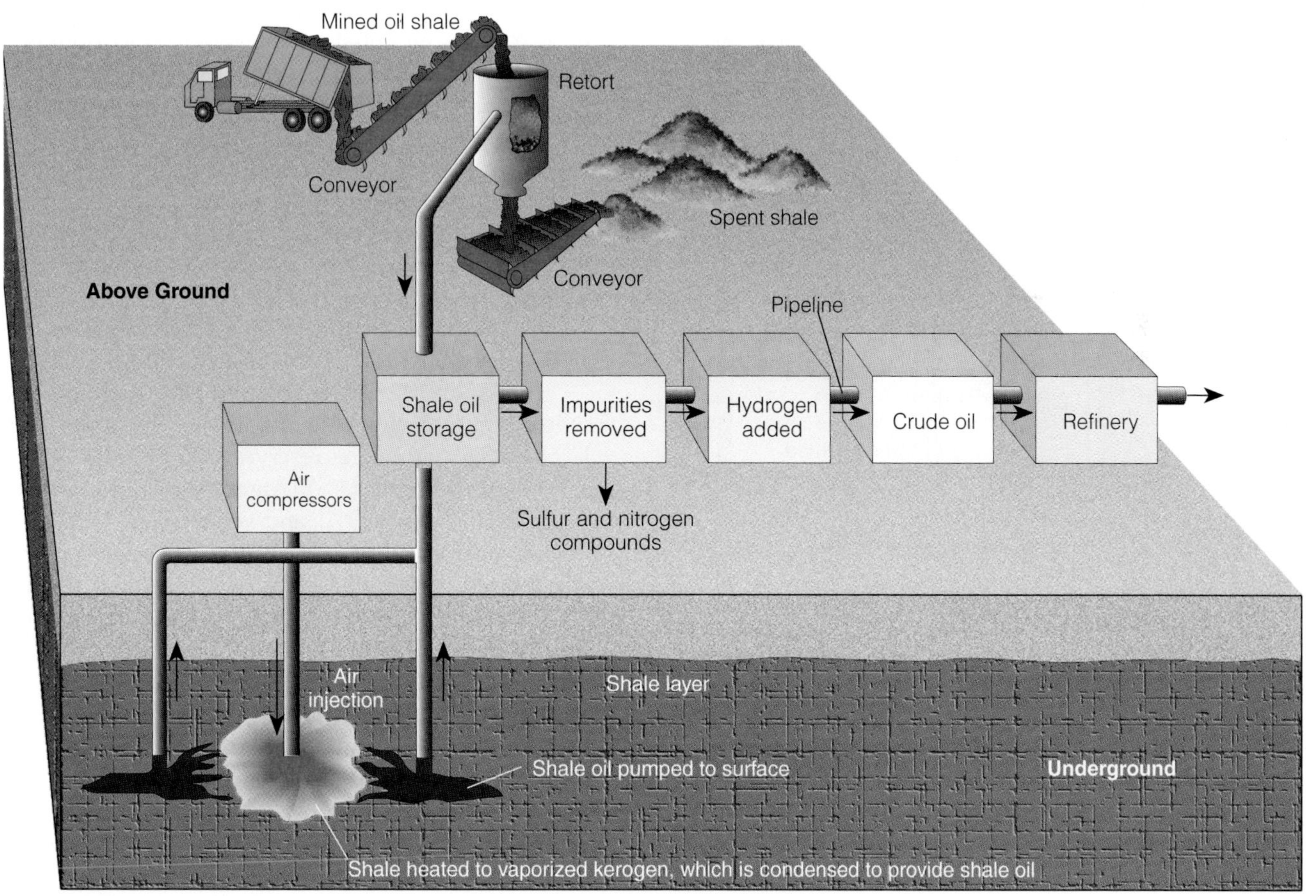

Figure 19-12 Above ground and underground (*in situ*) methods for producing synthetic crude oil from oil shale. The net energy yield is fairly low because it takes the energy from almost half a barrel of conventional oil to extract, process, and upgrade one barrel of shale oil to synthetic crude oil.

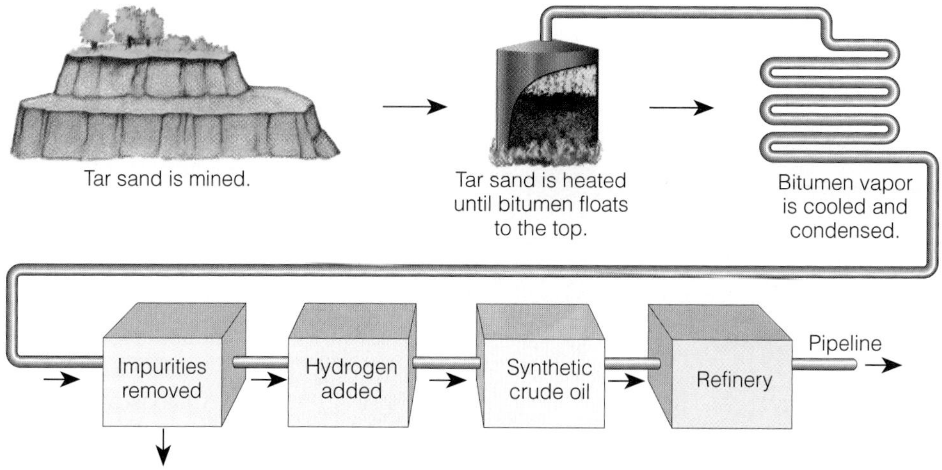

Figure 19-13 Generalized summary of how synthetic crude oil is produced from tar sand. The net energy yield is low because it takes the energy in almost one-half a barrel of conventional oil to extract and process one barrel of bitumen and upgrade it to synthetic crude oil.

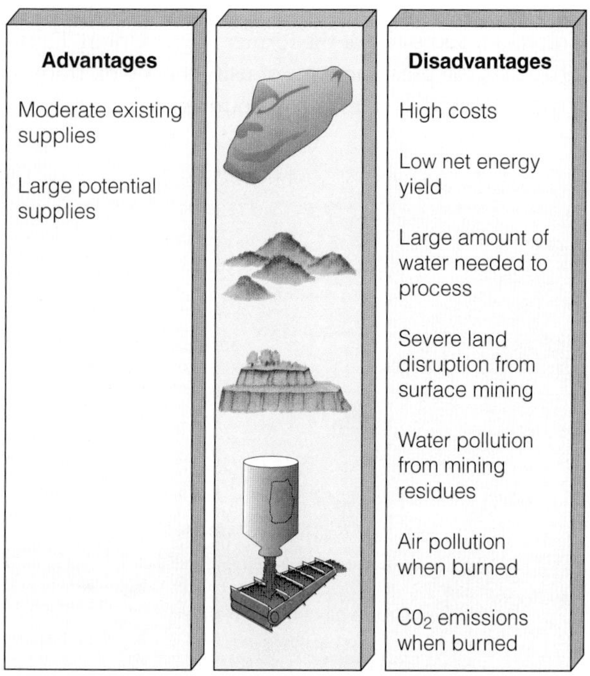

Figure 19-14 Advantages and disadvantages of using heavy oils from oil shale and tar sand as energy resources.

19-3 NATURAL GAS

What Is Natural Gas? In its underground gaseous state, **natural gas** is a mixture of **(1)** 50–90% by volume of methane (CH_4), the simplest hydrocarbon, **(2)** smaller amounts of heavier gaseous hydrocarbons such as ethane (C_2H_6), propane (C_3H_8), and butane (C_4H_{10}), and **(3)** highly toxic hydrogen sulfide (H_2S), a by-product of naturally occurring sulfur in the earth.

Conventional natural gas lies above most reservoirs of crude oil (Figure 19-2). *Unconventional natural gas* is found by itself in other underground sources. It is not yet economically feasible to get natural gas from such unconventional sources, but the extraction technology is being developed rapidly.

When a natural gas field is tapped, propane and butane gases are liquefied and removed as **liquefied petroleum gas (LPG)**. LPG is stored in pressurized tanks for use mostly in rural areas not served by natural gas pipelines. The rest of the gas (mostly methane) is **(1)** dried to remove water vapor, **(2)** cleansed of poisonous hydrogen sulfide and other impurities, and **(3)** pumped into pressurized pipelines for distribution. At a very low temperature of –184°C (–300°F), natural gas can be converted to **liquefied natural gas (LNG)**. This highly flammable liquid can then be shipped to other countries in refrigerated tanker ships.

Russia and Kazakhstan have almost 40% of the world's natural gas reserves. Other countries with large known natural gas reserves are Iran (15%), Qatar (5%), Saudi Arabia (4%), Algeria (4%), the United States (3%), Nigeria (3%), and Venezuela (3%). Geologists expect to find more natural gas, especially in unexplored developing countries. Most U.S. natural gas reserves are located in the same places as crude oil (Figure 19-8).

How Long Will Natural Gas Supplies Last? At the current consumption rate, known reserves and undiscovered, potential reserves of conventional natural gas in the United States are projected to last 65–80 years. World reserves are expected to last at least 125 years at the current consumption rate. It is estimated that conventional supplies of natural gas, plus unconventional supplies available at higher prices, will last at

Loeb, P. 1997. "Shear Madness." *U.S. News & World Report*, vol. 123, no. 6 p.26(9).

Advantages	Disadvantages
Ample supplies (125–200 years)	Releases CO_2 when burned
High net energy yield	Leaks of methane (a greenhouse gas)
Low cost (with huge subsidies)	Shipped across ocean as highly explosive LNG
Less air pollution than other fossil fuels	Sometimes burned off and wasted at wells because of low price
Lower CO_2 emissions than other fossil fuels	
Moderate environmental impact	
Easily transported by pipeline	
Good fuel for fuel cells and gas turbines	

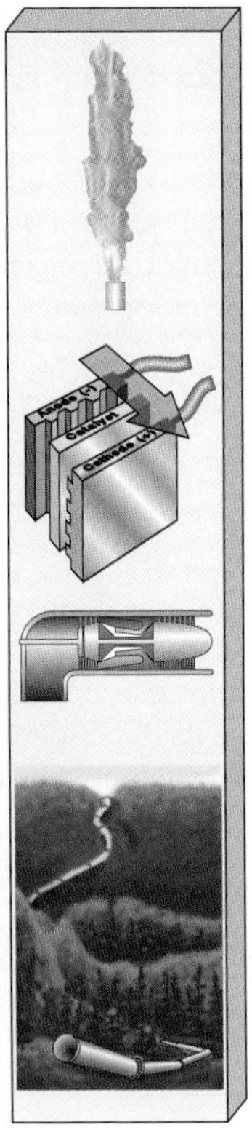

Figure 19-15 Advantages and disadvantages of using natural gas as an energy resource.

least 200 years at the current consumption rate and 80 years if usage rates rise 2% per year.

What Is the Future of Natural Gas? Figure 19-15 lists the pros and cons of using natural gas as an energy resource. In new *combined-cycle natural gas systems,* natural gas burned in gas turbines (that work like jet engines) can produce electricity much more efficiently and cheaply than burning coal or oil or using nuclear power. Some utilities plan to convert their aging coal plants into combined-cycle natural gas plants. These systems can also provide backup power for solar energy and wind power systems. Smaller combined-cycle gas units being developed could supply all of the heat and electricity needs of an apartment building or office building.

Because of its advantages over oil, coal, and nuclear energy, some analysts see natural gas as the best fuel to help make the transition to improved energy efficiency and more renewable energy over the next 50 years.

19-4 COAL

What Is Coal and How Is It Extracted and Processed? **Coal** is a solid fossil fuel formed in several stages as buried plant remains are subjected to intense heat and pressure over many millions of years (Figure 19-16). Coal is mostly carbon (40–98% depending on the type), with much smaller amounts of water (0.2–1.2%), sulfur (0.2–2.5%), and trace amounts of radioactive materials found in the earth. As coal ages its carbon content increases and its water content decreases (Figure 19-16). Anthracite, which is about 98% carbon, is the most desirable type of coal because of its high heat content and low sulfur content. However, because it takes much longer to form it is less common and therefore more expensive.

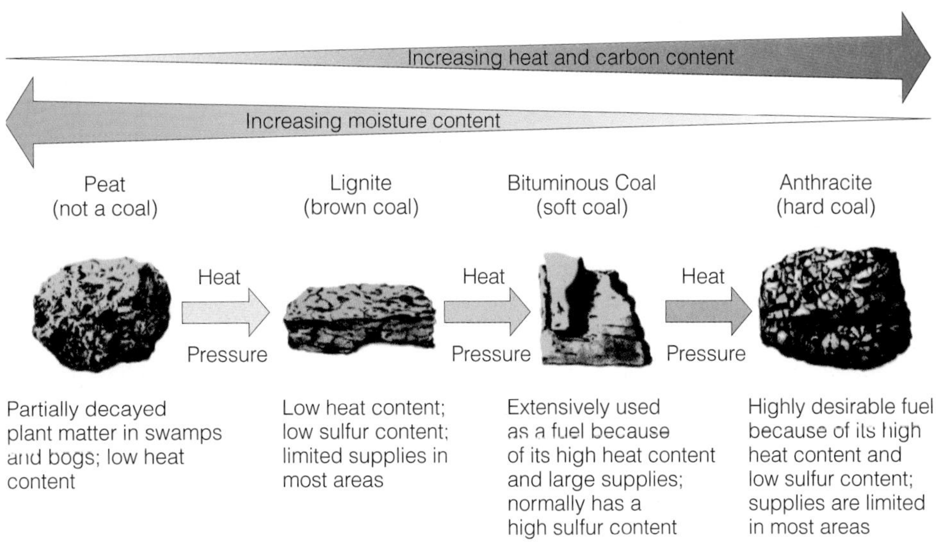

Figure 19-16 Stages in the formation of coal over millions of years. Peat is a soil material made of moist, partially decomposed organic matter. Lignite and bituminous coal are sedimentary rocks, whereas anthracite is a metamorphic rock.

Coal Burning and Destructive Synergies in China

CONNECTIONS

China has huge quantities of coal (11% of the world's reserves) and is the world's largest producer of coal, accounting for 25% of global output. Currently, the country burns coal to provide 72% of its commercial energy and has enough coal to last 300 years at current consumption rates.

Mostly because of coal burning, China now produces 10% of the world's emissions of CO_2 each year. By 2025, China is projected to emit more CO_2 than the current combined total of the United States, Canada, and Japan by relying on coal to provide 68% of its energy and on oil for 25% of its energy.

Several factors interact synergistically to cause China to burn more coal than it needs to. One is inefficient use of energy. About 66% of the energy used in China is unnecessarily wasted, compared to about 43% in the United States (which is in turn only half as energy-efficient as Japan and most western European countries).

Another factor is that in China drinking water usually must be boiled because it is too contaminated to drink. This uses more energy, much of it produced by the inefficient burning of coal in homes.

For China's economic future, and perhaps for the world's climate, many analysts believe that developed countries should help China reduce its dependence on coal and improve public health by **(1)** improving energy efficiency, **(2)** increasing its use of natural gas, nuclear power, and potentially large supplies of renewable energy, and **(3)** implementing a nationwide program to provide safe drinking water.

Critical Thinking

1. How might your life be affected if China continues to burn coal as its primary fuel?

2. What role, if any, should the United States and other developed countries play in helping China reduce its dependence on coal? Should such help be confined to transferring energy efficiency and renewable energy technologies or should it also include aid in building nuclear power plants?

Some coal is extracted underground (Figure 19-2) by miners working in tunnels and shafts. Such *subsurface mining* is very labor intensive and is one of the world's most dangerous occupations because of accidents and black lung disease (caused by prolonged inhalation of coal dust particles).

When coal lies close to the earth's surface it is extracted by *surface mining*. Bulldozers and huge earth-moving machines remove soil and rock, known as *overburden*, to expose underlying coal deposits. *Area strip mining* is used on fairly flat terrain (Figure 13-9), and *contour strip mining* is used on hilly or mountainous terrain (Figure 13-10). Thick beds of coal fairly near the surface are removed by digging a deep pit to remove the coal—a form of surface mining called *open-pit mining*.

After the coal is removed it is transported (usually by train) to a processing plant, where it is broken up, crushed, and then washed to remove impurities. The coal is then dried and shipped (again usually by train) to users, mostly power plants and industrial plants.

How Is Coal Used and Where Are the Largest Supplies? Coal provides about 22% of the world's commercial energy (23% in the United States). It is used to generate 62% of the world's electricity (53% in the United States) and to make 75% of its steel. The remainder of the world's electricity is supplied by hydroelectric dams (20%), nuclear energy (17%), and other alternatives such as wind and solar energy (1%). China, which gets 72% of its energy from coal, is the world's largest user of coal, followed closely by the United States.

About 66% of the world's proven coal reserves and 85% of the estimated undiscovered coal deposits are located in the United States, the former Soviet Union, and China (Connections, above). Figure 19-17 shows the locations of major coal fields in the United States and Canada.

How Long Will Coal Supplies Last? Coal is the world's most abundant fossil fuel. *Identified* world reserves of coal should last at least 225 years at current usage rates, and 65 years if usage rises 2% per year. The world's *unidentified* coal reserves are projected to last about 900 years at the current consumption rate and 149 years if the usage rate increases 2% per year.

Identified U.S. coal reserves should last about 300 years at the current consumption rate; unidentified U.S. coal resources could extend those supplies for perhaps 100 years, at a much higher average cost.

What Is the Future of Coal? Figure 19-18 lists the pros and cons of using coal as an energy resource. Coal is very abundant but it has the highest environmental

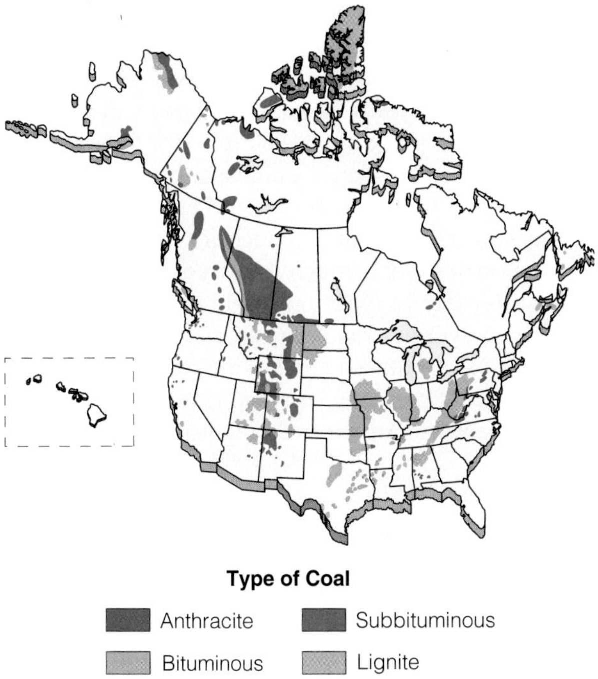

Figure 19-17 Locations of major coal fields in the United States and Canada. Wyoming, West Virginia, and Kentucky are the three largest coal-producing states in the United States. (Data from U.S. Council on Environmental Quality)

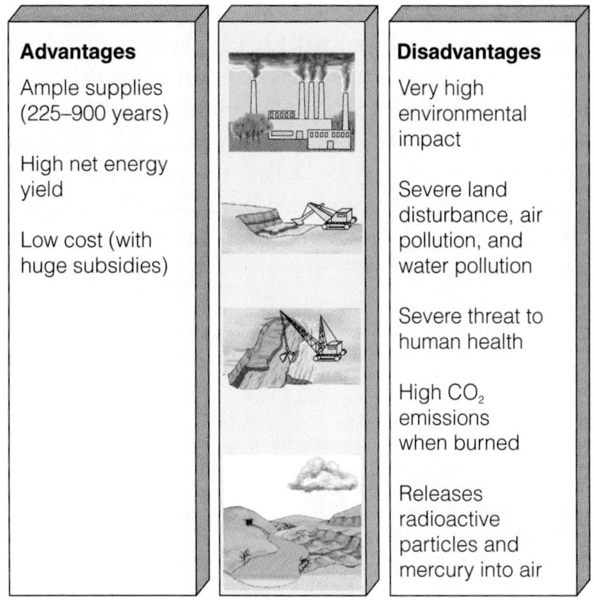

Figure 19-18 Advantages and disadvantages of using coal as an energy resource.

impact from **(1)** land disturbance (Figures 13-9, 13-10, and 19-2), **(2)** soil erosion, **(3)** air pollution, **(4)** water pollution (Figure 13-13), and **(5)** CO_2 emissions (Figure 19-9) of any fossil fuel. Surface-mined land can be restored (except in arid and semiarid areas) but this is expensive and is not done in many countries.

Without expensive air pollution control devices and other strategies (Figures 10-14 and 10-15), burning coal produces more air pollution per unit of energy than any other fossil fuel. Burning coal also releases thousands of times more radioactive particles into the atmosphere per unit of energy produced than does a normally operating nuclear power plant. According to a 1998 Environmental Protection Agency (EPA) study, the most threatening toxic material produced by coal-burning power plants is mercury and currently there is no effective way to cut these emissions.

Burning coal is also one of the greatest threats to human health. Each year in the United States alone, air pollutants from coal burning **(1)** kill thousands of people (with estimates ranging from 65,000 to 200,000), **(2)** cause at least 50,000 cases of respiratory disease, and **(3)** result in several billion dollars of property damage. However, new ways, such as *fluidized-bed combustion* (Figure 19-19), have been developed to burn coal more cleanly and efficiently.

Burning coal releases about 29% more carbon dioxide per unit of energy than oil and 80% more than natural gas. While providing about 22% of the world's energy, burning coal accounts for about 43% of the world's annual emissions of carbon dioxide.

Recent research done for the U.S. Department of Energy concluded that the *environmental costs* of producing electricity were **(1)** 5.7¢ per kilowatt-hour for coal, **(2)** 5.0¢ for nuclear, **(3)** 2.7¢ for oil, **(4)** 1¢ for natural gas, **(5)** under 0.7¢ for biomass, **(6)** less than 0.4¢ for solar cells, and **(7)** under 0.1¢ for wind and geothermal. Suppose that **(1)** taxes (per unit of energy produced) added these coal's environmental costs to its market price and **(2)** government subsidies for coal were phased out. Coal use would decline and become so expensive that it probably would be replaced for most uses by a combination of improved energy efficiency and cheaper and less environmentally harmful renewable energy resources.

Many analysts project a decline in coal use over the next 40–50 years because of **(1)** its high CO_2 emissions (Figure 19-9) and harmful health effects and **(2)** the availability of safer and increasingly cheaper renewable energy resources. By removing government subsidies for coal over the past 15 years, Belgium, France, Spain, the United Kingdom, and Japan have collectively cut coal use in half. However, world coal subsidies still amount to an estimated $63 billion, with about half of these subsides in developed countries.

What Are the Pros and Cons of Converting Solid Coal into Gaseous and Liquid Fuels? Solid coal can be converted into **synthetic natural gas (SNG)** by **coal gasification** (Figure 19-20) or into a liquid fuel such as methanol (Table 10-4) or synthetic gasoline by **coal liquefaction**. Figure 19-21 lists the pros and cons of using these *synfuels* produced from coal. Most analysts expect synfuels to play only a minor role as an energy resource in the next 30–50 years.

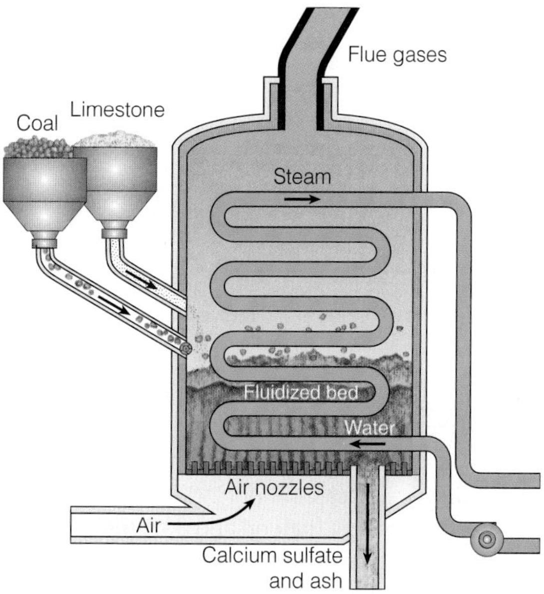

Figure 19-19 Fluidized-bed combustion of coal. A stream of hot air is blown into a boiler to suspend a mixture of powdered coal and crushed limestone. This method removes most of the sulfur dioxide, sharply reduces emissions of nitrogen oxides, and burns the coal more efficiently and cheaply than conventional combustion methods.

19-5 NUCLEAR ENERGY

What Happened to Nuclear Power? Studies indicate that U.S. utility companies began developing nuclear power plants in the late 1950s for three reasons: **(1)** The Atomic Energy Commission (which had the conflicting roles of promoting and regulating nuclear power) promised them that nuclear power would produce electricity at a much lower cost than coal and other alternatives (with one early proponent touting nuclear power as "too cheap to meter"), **(2)** the government paid about a quarter of the cost of building the first group of commercial reactors, and **(3)** Congress passed the Price–Anderson Act, which protects the U.S. nuclear industry and utilities from significant liability to the general public in case of accidents.

In the 1950s researchers predicted that by the year 2000 at least 1,800 nuclear power plants would supply 21% of the world's commercial energy (25% in the United States) and most of the world's electricity. By 1999, after over 40 years of development, enormous government subsidies, and an investment of $2 trillion, 429 commercial nuclear reactors in 32 countries were producing 6% of the world's commercial energy and 16% of its electricity. Little growth in nuclear power is projected, and its capacity is expected to decline between 2003 and 2020 as existing plants wear out and are retired (decommissioned) or become too costly to run.

In western Europe, plans to build more new nuclear power plants have come to a halt. Some countries such as Sweden and Germany are planning to phase out their use of nuclear power.

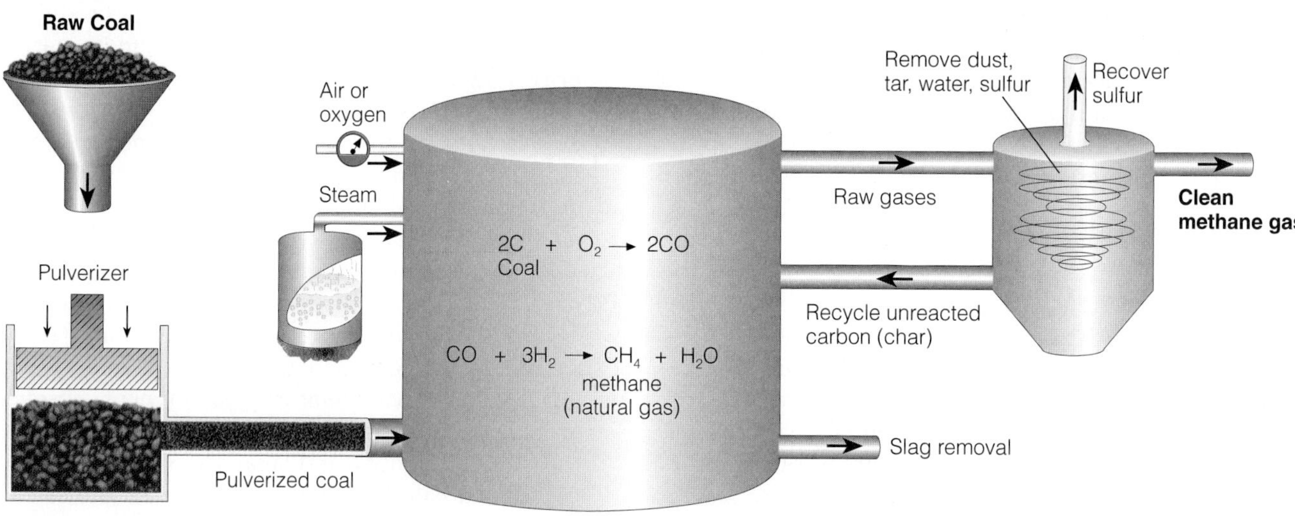

Figure 19-20 Coal gasification. Generalized view of one method for converting solid coal into synthetic natural gas (methane).

An exception is France, which gets about 78% of its electricity from nuclear power. However, France has more electricity than it needs (Guest Essay, p. 528), and the heavily subsidized government agency that builds and operates France's nuclear plants has lost money for 24 years, accumulating a $30 billion debt. In 1999, France adopted a moratorium on further construction of nuclear power plants.

In Asia 26 new nuclear plants are under construction, but a slowdown in such construction is taking place in some countries. In Japan, which gets 35% of its electricity from nuclear power, public opposition has been so intense that only two sites for new nuclear plants have been approved since 1979. However, China, which gets 1% of its electricity from nuclear power, plans to build more than 50 new plants by 2020 as a way to reduce its dependence on highly polluting coal (Connections, p. 506). However, many analysts doubt that China will reach these goals because of the high costs of nuclear power compared to abundant alternatives such as natural gas, hydropower and solar and wind energy.

In the United States, no new nuclear power plants have been ordered since 1978, and all 120 plants ordered since 1973 have been canceled. In 1999, the 104 licensed commercial nuclear power plants in the United States generated about 20% of the country's electricity. This percentage is expected to decline over the next two decades as existing plants wear out and are retired (decommissioned) or become too costly to run.

Between 1996 and 1999, U.S. power companies closed five reactors before their normal lifetimes were up because they were too expensive to operate. Wall Street analysts and research firms expect that as many as one-third of U.S. reactors could be shut down for economic reasons by 2005. Seven of Canada's 21 reactors have been shut down and are unlikely to operate again.

The major reasons for the failure of nuclear power to grow as projected are **(1)** multibillion-dollar construction cost overruns, **(2)** high operating costs, **(3)** more malfunctions than expected, **(4)** poor management, **(5)** public concerns about safety after the Chernobyl (p. 496) and Three Mile Island (Case Study, p. 512) accidents, and **(6)** public concerns about costs and radioactive waste disposal.

Advantages

- Large potential supply
- Vehicle fuel

Disadvantages

- Low to moderate net energy yield
- Higher cost than coal
- High environmental impact
- Increased surface mining of coal
- High water use
- Higher CO_2 emissions than coal

Figure 19-21 Advantages and disadvantages of using synthetic natural gas (SNG) and liquid synfuels produced from coal.

How Does a Nuclear Fission Reactor Work? To evaluate the pros and cons of nuclear power, we first must know how a nuclear power plant and its accompanying nuclear fuel cycle work. When the nuclei of atoms such as uranium-235 and plutonium-239 are split by neutrons in a nuclear fission chain reaction, energy is released and converted mostly into high-temperature heat (Figure 3-11). In the reactor of a nuclear power plant the rate of fission can be controlled, and the heat generated can produce high-pressure steam, which spins turbines that generate electricity.

Light-water reactors (LWRs) like the one diagrammed in Figure 19-22 produce about 85% of the world's nuclear-generated electricity (100% in the United States). An LWR has the following key parts:

- *Core* containing 35,000–40,000 long, thin fuel rods, each packed with fuel pellets (about one-third the size of a cigarette).
- *Uranium oxide fuel* consisting of about 97% nonfissionable uranium-238 and 3% fissionable uranium-235. To create a suitable fuel, the concentration of uranium-235 in the ore is increased (enriched) from 0.7% (its natural concentration in uranium ore) to 3% by removing some of the uranium-238.
- *Control rods*, which are moved in and out of the reactor core to absorb neutrons and thus regulate the rate of fission and amount of power the reactor produces.
- *Moderator*, which slows down the neutrons emitted by the fission process so that the chain reaction can be kept going. This is a material such as **(1)** liquid water (75% of the world's reactors, called *pressurized water reactors*), **(2)** solid graphite (20% of reactors), or **(3)** heavy water (D_2O, 5% of reactors). Graphite-moderated reactors, including the ill-fated one at Chernobyl (Figure 19-1), can also produce fissionable plutonium-239 for nuclear weapons.

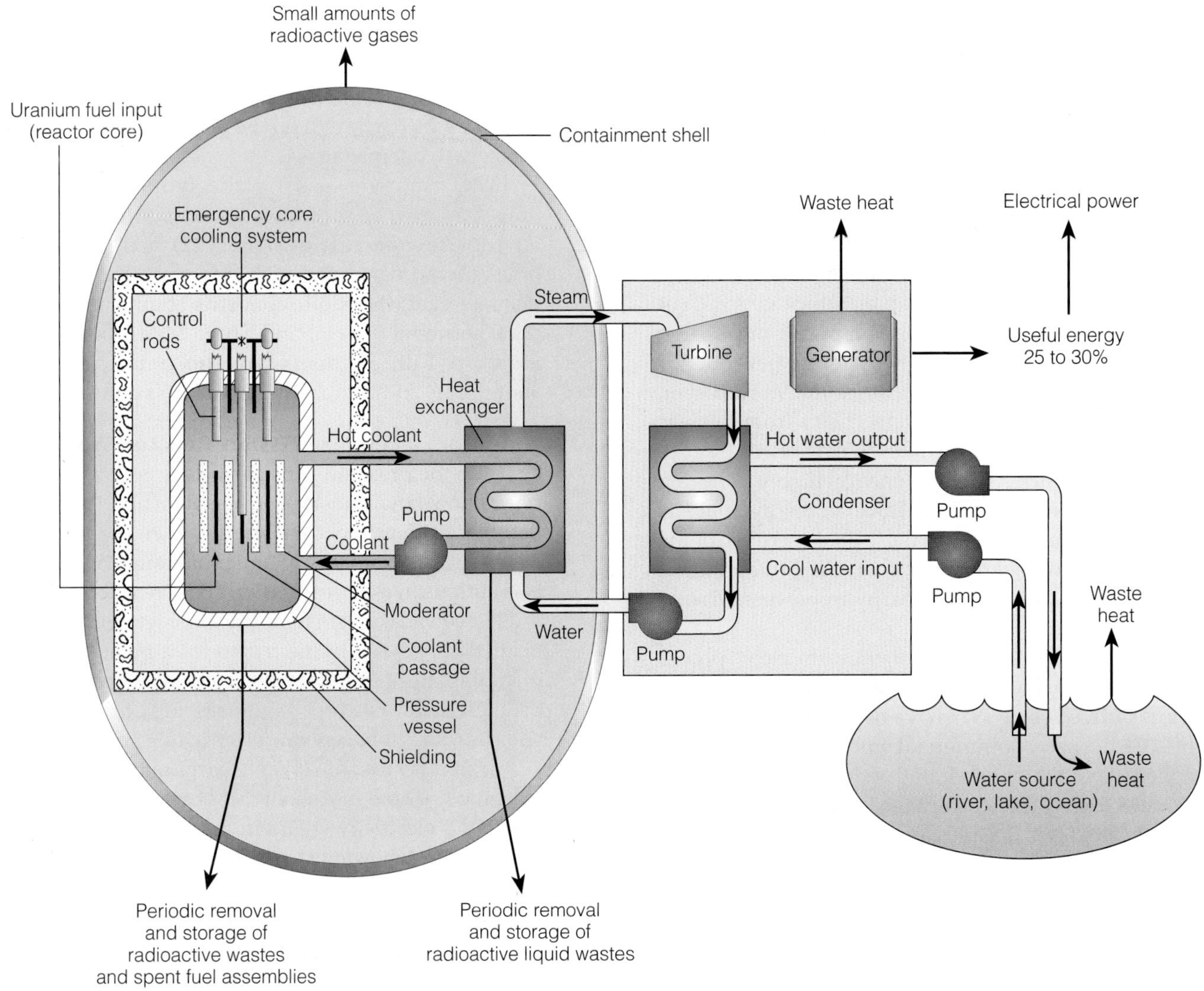

Figure 19-22 Light-water–moderated and –cooled nuclear power plant with a pressurized water reactor. The ill-fated Chernobyl nuclear reactor (Figure 19-1) in Ukraine was a graphite-moderated reactor with less extensive safety features than most reactors in the rest of the world.

- *Coolant,* usually water, which circulates through the reactor's core to remove heat (to keep fuel rods and other materials from melting) and to produce steam for generating electricity.

Nuclear power plants, each with one or more reactors, are only one part of the nuclear fuel cycle (Figure 19-23). In evaluating the safety and economic feasibility of nuclear power, we need to look at this entire cycle, not just the nuclear plant itself.

What Are the Pros and Cons of Nuclear Power? Using nuclear power to produce electricity has some important advantages, especially when compared to coal. However, the disadvantages of nuclear power have essentially halted the growth of this energy resource. These advantages and disadvantages are summarized in Figure 19-24. Several of the major problems with nuclear power are discussed further in the remainder of this chapter.

How Safe Are Nuclear Power Plants and Other Nuclear Facilities? Because of the built-in safety features, the risk of exposure to radioactivity from nuclear power plants in the United States (and presumably in most other developed countries) is *extremely low* compared to other environmental and health risks (Figures 8-10 and 8-11). However, a partial or complete meltdown or explosion is possible, as accidents at the Chernobyl (Figure 19-1) nuclear power plant in Ukraine and the Three Mile Island plant in Pennsylvania (Case Study, p. 512) have taught us.

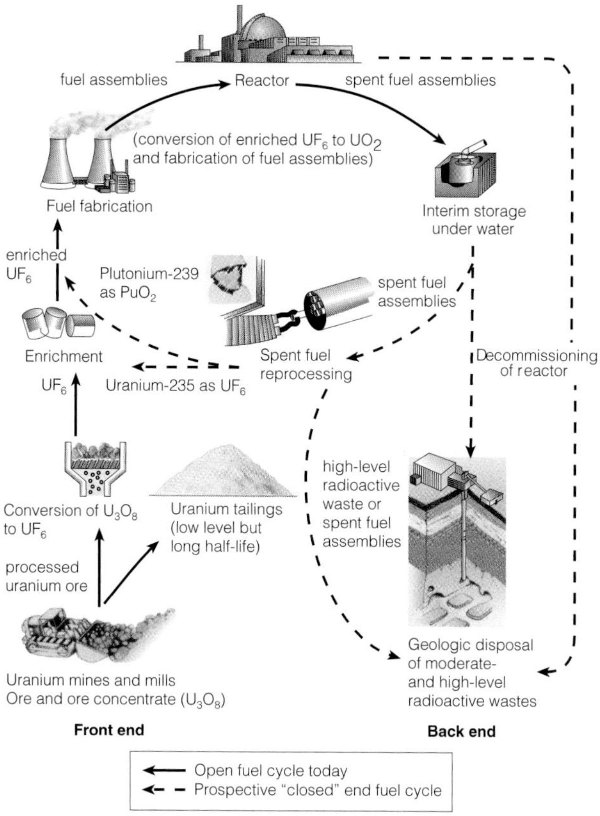

Figure 19-23 The nuclear fuel cycle. As long as it is operating properly, little pollution or radioactivity is released from a nuclear power plant (although the potential health, environmental, and economic consequences from a major accident are significant). Each of the other steps in the nuclear fuel cycle can release significant amounts of toxic chemicals and radioactive materials into the environment.

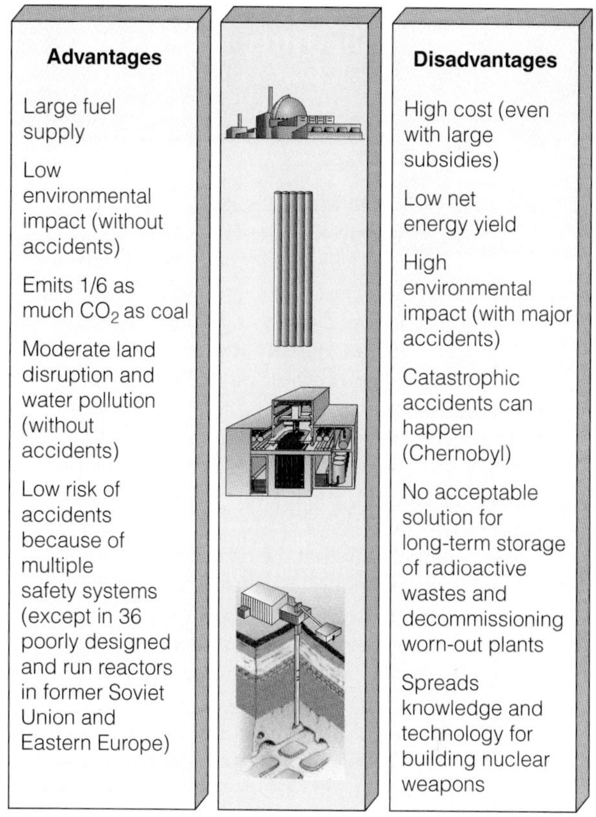

Figure 19-24 Advantages and disadvantages of using nuclear power to produce electricity. This evaluation includes the entire nuclear fuel cycle (Figure 19-23).

The Nuclear Regulatory Commission (NRC) estimates that there is a 15–45% chance of a complete core meltdown at a U.S. reactor during the next 20 years. The NRC also found that 39 U.S. reactors have an 80% chance of containment shell failure from a meltdown or an explosion of gases inside the containment structures.

Nuclear scientists and government officials throughout the world urge the shutdown of 36 poorly designed and poorly operated nuclear reactors in some republics of the former Soviet Union and in eastern Europe. However, without considerable economic aid from the world's developed countries, it is unlikely that these potentially dangerous plants will be closed and replaced with safer nuclear or nonnuclear alternatives.

All French pressurized water nuclear reactors are built using a cost-saving standardized design. However, plants built with this generic design have suffered from a growing number of faulty steam generator problems and cracked reactor vessel heads, which must be replaced at great cost to minimize the chances of a serious accident.

In 1999 an accident caused by human error in a Japanese nuclear fuel processing plant **(1)** exposed 49 workers to high levels of radioactivity as they attempted to stop an out-of-control chain reaction and **(2)** released some radiation into the nearby community. This worst nuclear accident in Japanese history increased public concern about the country's nuclear industry, already plagued by smaller accidents, plant shutdowns, radiation leaks, and cover-up attempts.

In the United States, there is widespread public distrust in the ability of the NRC and the Department of Energy (DOE) to enforce nuclear safety in commercial (NRC) and military (DOE) nuclear facilities. Congressional hearings in 1987 uncovered evidence that high-level NRC staff members **(1)** destroyed documents and obstructed investigations of criminal wrongdoing by utilities, **(2)** suggested ways utilities could evade commission regulations, and **(3)** provided utilities and their contractors with advance notice of "surprise" inspections. In 1996 George Galatis, a respected senior nuclear engineer, said, "I believe in nuclear power but after seeing the NRC in action I'm convinced a serious accident is not just likely, but inevitable. . . . They're asleep at the wheel."

Since 1986 government studies and once-secret documents have revealed that for decades most of the

The Three Mile Island Nuclear Power-Plant Accident

CASE STUDY

On March 29, 1979, the number 2 reactor at the Three Mile Island (TMI) nuclear plant near Harrisburg, Pennsylvania, lost its coolant water because of a series of mechanical failures and human operator errors not anticipated in safety studies.

The reactor's core became partially uncovered and about 50% of it melted and fell to the bottom of the reactor. Unknown amounts of radioactive materials escaped into the atmosphere, 50,000 people were evacuated, and another 50,000 fled the area on their own.

Most analysts say that the estimated radiation released was not enough to cause deaths or cancers. However, there is controversy over how much radiation was released and there has been an increase in certain cancers among people who live near the plant.

A 1991 study concluded that the excess cancers may have been caused by immune system damage from stress related to the accident. However, a 1997 study concluded that the increased cancer rates were caused by radiation released from the plant.

Partial cleanup of the damaged TMI reactor, lawsuits, and payment of damage claims have cost $1.2 billion so far, almost twice the reactor's $700-million construction cost. Banks and other lending institutions have little interest in financing new U.S. power plants because the TMI accident showed that utility companies could lose more than $1 billion in equipment and cleanup costs, even without any established harmful effects on public health.

Critical Thinking

The nuclear power industry claims that the TMI accident showed that its safety systems work because no deaths were known to have been caused by the accident. Other analysts dispute this claim and argue that the accident was a wake-up call about the potential dangers of nuclear power plants. What is your position? Explain.

nuclear weapon production facilities supervised by the DOE have been operated with gross disregard for the safety of their workers and the people nearby. Since 1957 these facilities have **(1)** released huge quantities of radioactive particles into the air and **(2)** dumped tons of radioactive waste and toxic substances into flowing creeks and leaking pits without informing the public. According to former senator John Glenn of Ohio, "We are poisoning our own people in the name of national security."

The nuclear power industry claims that nuclear power plants in the United States have not killed anyone. However, according to U.S. National Academy of Sciences estimates, U.S. nuclear plants cause 6,000 premature deaths and 3,700 serious genetic defects each year. If correct, this annual death toll is much smaller than the 65,000 to 200,000 deaths per year caused by coal-burning plants in the United States. However, critics point out the estimated annual deaths from both of these types of plants are unacceptable, given the availability of much less harmful alternatives.

What Do We Do with Low-Level Radioactive Waste? Each part of the nuclear fuel cycle (Figure 19-23) produces low-level and high-level solid, liquid, and gaseous radioactive wastes with various half-lives (Table 3-1). Wastes classified as *low-level radioactive wastes* give off small amounts of ionizing radiation and must be stored safely for 100–500 years before decaying to safe levels.

From the 1940s to 1970, most low-level radioactive waste produced in the United States (and most other countries) was put into steel drums and dumped into the ocean; the United Kingdom and Pakistan still dispose of their low-level wastes in this way.

Today, low-level waste materials from commercial nuclear power plants, hospitals, universities, industries, and other producers in the United States are put in steel drums and shipped to the two remaining regional landfills run by federal and state governments (both scheduled for closure). Attempts to build new regional dumps for low-level radioactive waste using improved technology (Figure 19-25) have met with fierce public opposition.

Some scientists believe that low-level radioactive waste should be stored in carefully designed aboveground buildings that could capture and contain any leaks. They urge that all such buildings (or low-level waste dumps) be located at nuclear power-plant sites, which produce more than half of these wastes (as much as 80% in some states).

They point out that nuclear power-plant sites **(1)** have been carefully evaluated to meet strict geological standards, **(2)** have the personnel with the expertise and equipment to manage the low-level wastes, and **(3)** are strictly regulated by the federal government. Locating low-level waste at plant sites would also reduce the need to transport most such wastes long distances. Utility companies and the nuclear power industry oppose such proposals because they would then be financially and legally responsible for the wastes they produce. What do you think about this proposal?

What Should We Do with High-Level Radioactive Waste? *High-level radioactive wastes* give off large

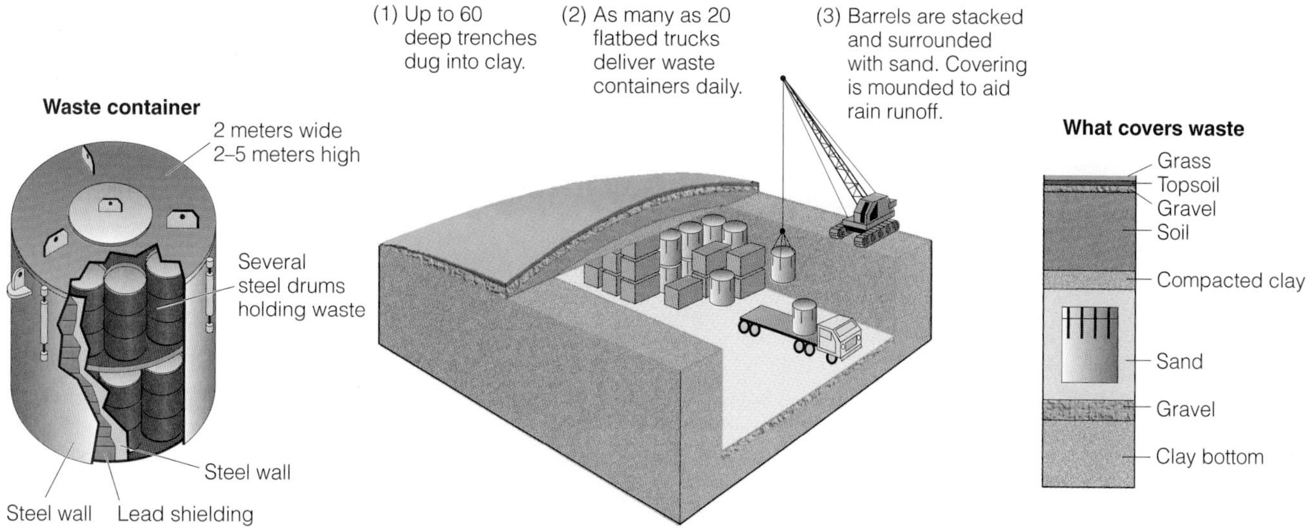

Figure 19-25 Proposed design of a state-of-the-art low-level radioactive waste landfill. According to the EPA, all landfills eventually leak. The best-designed landfills may not leak for several decades, after which the companies running them would no longer be liable for leaks and problems; any problems would be passed on to taxpayers and future generations. (Data from U.S. Atomic Industrial Forum)

amounts of ionizing radiation for a short time and small amounts for a long time. Such wastes must be stored safely for at least 10,000 years and about 240,000 years if plutonium-239 is not removed by reprocessing. Most high-level radioactive wastes are spent fuel rods from commercial nuclear power plants and an assortment of wastes from plants that produce plutonium and tritium for nuclear weapons.

After 50 years of research, scientists still don't agree on whether there is any safe method of storing these wastes. Some scientists believe that the long-term safe storage or disposal of high-level radioactive wastes is technically possible. Others disagree, pointing out that it is impossible to demonstrate that any method will work for the 10,000–240,000 years of fail-safe storage needed for such wastes. Here are some of the proposed methods and their possible drawbacks:

- *Bury it deep underground* (Figure 19-26 and Case Study, p. 515). This favored strategy is under study by all countries producing nuclear waste. However, according to a 1990 report by the U.S. National Academy of Sciences, "Use of geological information—to pretend to be able to make very accurate predictions of long-term site behavior—is scientifically unsound."
- *Shoot it into space or into the sun.* Costs would be very high and a launch accident, such as the explosion of the space shuttle *Challenger*, could disperse high-level radioactive wastes over large areas of the earth's surface. This strategy has been abandoned for now.
- *Bury it under the antarctic ice sheet or the Greenland ice cap.* The long-term stability of the ice sheets is not known. They could be destabilized by heat from the wastes, and retrieval of the wastes would be difficult or impossible if the method failed. This strategy is prohibited by international law and has been abandoned for now.
- *Dump it into descending subduction zones in the deep ocean* (Figure 13-4, middle). Wastes might eventually be spewed out somewhere else by volcanic activity, containers might leak and contaminate the ocean before being carried downward, and retrieval would be impossible if the method did not work. This method is being evaluated by a group of 10 countries, but current international agreements and U.S. laws forbid dumping radioactive waste beneath the seas.
- *Bury it in thick deposits of mud on the deep ocean floor in areas that tests show have been geologically stable for 65 million years.* The waste containers would eventually corrode and release their radioactive contents. This approach is banned under current international agreements and U.S. law.
- *Change it into harmless, or less harmful, isotopes.* Currently there is no way to do this. Even if a method were developed, **(1)** costs probably would be very high and **(2)** the resulting toxic materials and low-level (but very long-lived) radioactive wastes would still need to be disposed of safely.

How Widespread Are Contaminated Radioactive Sites? In 1992 the EPA estimated that as many as 45,000 sites in the United States may be contaminated with radioactive materials, 20,000 of them belonging to the DOE and the Department of Defense. According to the DOE it will cost taxpayers at least $230 billion

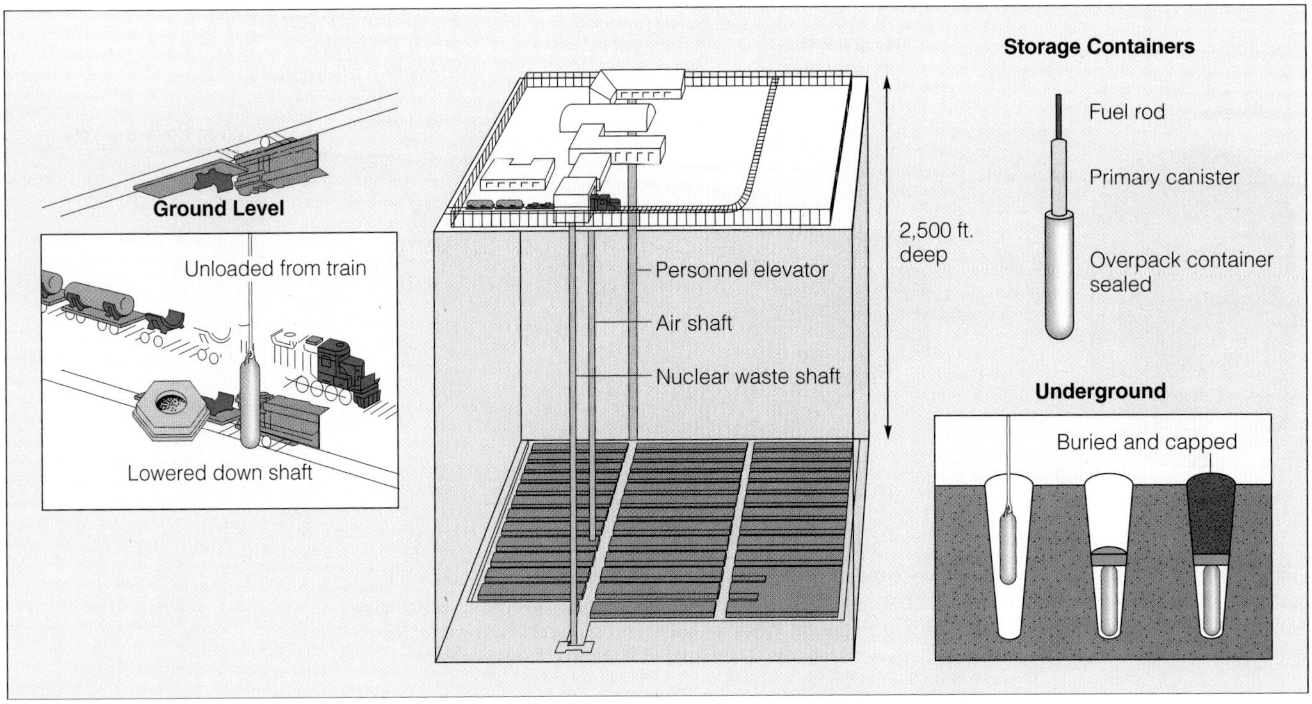

Figure 19-26 Proposed general design for deep-underground permanent storage of high-level radioactive wastes from commercial nuclear power plants in the United States. (Source: U.S. Department of Energy)

over the next 75 years to clean up these facilities (with some estimating that the price will range from $400 to $900 billion).

Some critics question spending so much money on a problem that is ranked by scientific advisers to the EPA as a low-risk ecological problem and not among the top high-risk health problems (Figure 8-10). The radioactive contamination situation in the United States pales in comparison to the post–Cold War legacy of nuclear waste and contamination in the republics of the former Soviet Union (Spotlight, p. 516).

What Can We Do with Worn-Out Nuclear Plants? After approximately 15–40 years of operation, a nuclear reactor becomes dangerously contaminated with radioactive materials, and many of its parts are worn out. Then the plant must be *decommissioned* or retired (the last step in the nuclear fuel cycle, Figure 19-23) by **(1)** dismantling it and storing its large volume of highly radioactive materials in high-level nuclear waste storage facilities (Figure 19-26), **(2)** putting up a physical barrier and setting up full-time security for 30–100 years before the plant is dismantled, or **(3)** enclosing the entire plant in a tomb that must last for several thousand years.

At least 228 large commercial reactors worldwide (20 in the United States) are scheduled for retirement between 2000 and 2012. By 2030 all U.S. reactors will have to be retired, based on the life of their current operating licenses, and many may be retired early for safety or financial reasons.

U.S. utilities have been setting aside funds for decommissioning more than 100 reactors, but to date these funds fall far short of the estimated costs. If sufficient funds are not available for decommissioning, the balance of the costs could be passed along to ratepayers and taxpayers.

What Is the Connection Between Nuclear Reactors and the Spread of Nuclear Weapons? Today the United States and at least 14 other countries sell nuclear power technology in the international marketplace. Information, components, and materials used to build and operate such reactors can be used to produce fissionable isotopes such as uranium-235 and plutonium-239 (the explosive material in nuclear weapons).

We now live in a world with enough nuclear weapons to kill everyone on the earth 40 times over—20 times if current nuclear arms reduction agreements are carried out. Currently 60 countries—1 of every 3 in the world—either have nuclear weapons or the knowledge and ability to build them, with the fuel and knowledge coming mostly from research and commercial nuclear reactors. Dismantling thousands of Russian and American nuclear warheads increases the threat from the resulting huge amounts of bomb-grade plutonium-239 that must be safeguarded.

Underground Storage of High-Level Radioactive Wastes in the United States

CASE STUDY

In 1999, the U.S. Department of Energy's controversial $2-billion Waste Isolation Pilot Plant (WIPP) began storing high-level radioactive wastes from nuclear weapon production facilities. It is built on government-owned land near Carlsbad, New Mexico, and took 19 years to build and put into operation.

In 1985 the DOE announced plans to build a repository for underground storage of high-level radioactive wastes from commercial nuclear reactors (Figure 19-26) on federal land in the Yucca Mountain desert region, 160 kilometers (100 miles) northwest of Las Vegas, Nevada. The facility, which is expected to cost at least $26 billion, is scheduled to open by 2010. It may never open, mostly because of rock faults, a nearby active volcano, and 36 active earthquake faults on the site itself.

Fed up with the delay, some U.S. utility companies are trying to develop a temporary, aboveground, monitored retrievable storage (MRS) facility for commercial high-level radioactive waste on the Mescalero Apache reservation in New Mexico. Because of the sovereignty of Native American nations, it is easier to site a radioactive storage facility on tribal lands than anywhere else in the United States. New Mexico officials have threatened to block the MRS facility, claiming it would be too risky to transport spent fuel across the country to the facility. Even if a permit is granted, the site would not be open until 2003 at the earliest.

Regardless of the storage method, most citizens strongly oppose the location of a low-level or high-level nuclear waste disposal facility anywhere near them. Even if the problem is technically solvable, it may be politically unacceptable.

Critical Thinking

1. What do you believe should be done with high-level radioactive wastes? Explain.

2. Would you favor having nuclear waste transported by truck or train through the area where you live?

Can We Afford Nuclear Power? Experience has shown that nuclear power is an extremely expensive way to boil water to produce electricity, even when huge government subsidies partially shield it from free-market competition with other energy sources.

Costs rose dramatically in the 1970s and 1980s because of unanticipated safety problems and stricter regulations after the Three Mile Island and Chernobyl accidents. For example, in the United States the cost of a new nuclear power plant rose from $200 to $3,000 per kilowatt between 1972 and 1985. During the 1990s the cost of other new and clean electric power technologies such as natural gas turbines fell sharply. Such turbines cost about $700 per kilowatt, compared to $1,000 for coal power plants and $3,000 for nuclear power plants. In 1995 the World Bank said that nuclear power is too costly and risky. In the 1980s Great Britain tried to sell its government-run nuclear power plants to private interests, but no one would buy them or even accept them for free. U.S. utility companies have also tried to sell some of their reactors at bargain-basement prices. By 1999 they had found buyers for only two reactors.

Abandoned reactor projects have cost U.S. utility investors over $100 billion since the mid-1970s, bankrupting U.S. utilities in Washington and New Hampshire. *Forbes* business magazine has called the failure of the U.S. nuclear power program "the largest managerial disaster in U.S. business history."

The U.S. nuclear industry hopes to persuade the federal government and utility companies to build hundreds of smaller second-generation plants using standardized designs, which they claim are safer and can be built more quickly (in 3–5 years). These so-called *advanced light water reactors (ALWRs)* have built-in *passive safety* features that are supposed to make explosions or the release of radioactive emissions virtually impossible. However, according to *Nucleonics Week*, an important nuclear industry publication, "Experts are flatly unconvinced that safety has been achieved—or even substantially increased—by the new designs."

What Should Be the Future of Conventional Nuclear Power? Some analysts call for phasing out all or most government subsidies and tax breaks for nuclear power and using such funds to subsidize the development of technologies for improving energy efficiency, natural gas turbines, and forms of renewable energy (Chapter 20).

Other analysts argue that we should continue government funding of research and development and pilot plant testing of new reactor designs. They argue that we need to keep the nuclear option available for use in the future if improved energy efficiency and renewable energy options fail to **(1)** keep up with electricity demands and **(2)** reduce CO_2 emissions to acceptable levels. What do you think?

The Soviet Nuclear-Waste Legacy

SPOTLIGHT

Much of the land in various parts of the former Soviet Union is dotted with **(1)** areas severely contaminated by nuclear accidents, **(2)** 26 operating nuclear power plants with flawed and unsafe designs, **(3)** nuclear-waste dump sites, **(4)** radioactive waste-processing plants, and **(5)** contaminated nuclear test sites.

Mayak, a plutonium production facility in southern Russia, has spewed into the atmosphere 2.5 times as much radiation as Chernobyl, mostly from the explosion of a nuclear-waste storage tank in 1957. About 2,600 square kilometers (1,000 square miles) of the surrounding area is still contaminated with radioactive wastes so dangerous that no one can live there.

Large quantities of radioactive materials from the Mayak facility were dumped into nearby Lake Karachay between 1949 and 1967. Today the lake is so radioactive that standing on its shores for about an hour would be fatal.

Around the shores of Novaya Zemyla, an island off the coast of Russia, 18 defunct nuclear-powered submarines were dumped (at least three of them loaded with nuclear fuel), along with as many as 17,000 containers of radioactive waste. If released by corrosion, radioactivity from these submarines and containers threatens Russian, Norwegian, Finnish, and perhaps Swedish citizens in the area. In 1993, Greenpeace caught Russians dumping massive amounts of radioactive waste from nuclear submarines into the Sea of Japan.

Critical Thinking

Because of serious economic problems, there is little money for cleanup of radioactive sites in parts of the former Soviet Union. Should the United States and other developed nations provide technical and financial aid to help deal with this problem? Explain.

Is Breeder Nuclear Fission a Feasible Alternative? Some nuclear power proponents urge the development and widespread use of **breeder nuclear fission reactors**, which generate more nuclear fuel than they consume by converting nonfissionable uranium-238 into fissionable plutonium-239. Because breeders would use over 99% of the uranium in ore deposits, the world's known uranium reserves would last at least 1,000 years, and perhaps several thousand years.

However, if the safety system of a breeder reactor fails, the reactor could lose some of its liquid sodium coolant, which ignites when exposed to air and reacts explosively if it comes into contact with water. This could cause a runaway fission chain reaction and perhaps a nuclear explosion powerful enough to blast open the containment building and release a cloud of highly radioactive gases and particulate matter. Leaks of flammable liquid sodium can also cause fires, as has happened with all experimental breeder reactors built so far.

In addition, existing experimental breeder reactors produce plutonium so slowly that it would take 100–200 years for them to produce enough plutonium to fuel a significant number of other breeder reactors. In 1994, the United States ended government-supported research for breeder technology after providing about $9 billion in research and development funding.

In December 1986, France opened a commercial-size breeder reactor. It was so expensive to build and operate that after spending $13 billion the government closed it permanently in 1998. Because of this experience, other countries have abandoned their plans to build full-size commercial breeder reactors.

Is Nuclear Fusion a Feasible Alternative? Scientists hope that someday controlled nuclear fusion will provide an almost limitless source of high-temperature heat and electricity. Research has focused on the D–T nuclear fusion reaction, in which two isotopes of hydrogen—deuterium (D) and tritium (T)—fuse at about 100 million degrees (Figure 3-12).

After 50 years of research and huge expenditures of mostly government funds, controlled nuclear fusion is still in the laboratory stage, with none of the approaches tested so far producing more energy than they use. In 1989 two chemists claimed to have achieved deuterium–deuterium (D–D) nuclear fusion at room temperature using a simple apparatus, but subsequent experiments have not substantiated their claims.

If researchers can eventually get more energy out of nuclear fusion than they put in, the next step would be to build a small fusion reactor and then scale it up to commercial size, an extremely difficult engineering problem. The estimated cost of a commercial fusion reactor is several times that of a comparable conventional fission reactor.

Proponents contend that with greatly increased federal funding, a commercial nuclear fusion power plant might be built by 2030, but many energy experts don't expect nuclear fusion to be a significant energy source until 2100, if then.

We are embarked on the beginning of the last days of the Age of Oil.

MIKE BOWLIN (CEO ARCO OIL)

REVIEW QUESTIONS

1. Define the boldfaced terms used in this chapter.

2. Describe the causes and major effects of the Chernobyl nuclear power plant accident.

3. Where does most of the energy we use come from? What percentages of the energy we use come from nonrenewable and from renewable energy in **(a)** the world and **(b)** the United States?

4. How long does it usually take to phase in a new energy alternative? What four questions should we try to answer about each energy resource?

5. What is *net energy,* and why is it important in evaluating an energy resource?

6. What is *petroleum* or *crude oil*? Distinguish between *primary, secondary,* and *tertiary recovery* of crude oil. What happens to crude oil at a refinery? What are *petrochemicals*? Who has most of the world's oil reserves? What percentage of the world's oil reserves is found in the United States? What percentage of the world's oil does the United States use? What percentage of the oil used in the United States is imported? How long are affordable known and projected supplies of conventional oil expected to last in **(a)** the world and **(b)** the United States?

7. What are the pros and cons of using oil as an energy resource?

8. What are *oil shale* and *kerogen,* and how can these substances be converted to shale oil? What are the pros and cons of using shale oil as an energy resource?

9. What is *tar sand,* and how can it be converted to synthetic crude oil? What are the pros and cons of using oil from tar sand as an energy resource?

10. What is *natural gas*? Who has most of the world's reserves of natural gas? How long are known and projected supplies of natural gas expected to last in **(a)** the world and **(b)** the United States? What is a combined-cycle natural gas turbine system, and why is its use rising rapidly? What are the pros and cons of using natural gas as an energy resource?

11. What is coal? How is it extracted from the earth? How is coal used? What three countries have the largest reserves of coal? How long are known and projected supplies of coal expected to last **(a)** in the world and **(b)** in the United States?

12. How does heavy dependence on coal in China affect **(a)** the people of China and **(b)** the world?

13. What are the pros and cons of using coal as an energy resource?

14. What are the pros and cons of converting solid coal into gaseous and liquid fuels?

15. What five factors contribute to the leveling off of the use of nuclear power plants to produce electricity?

16. Describe how a nuclear fission reactor works. What are the five major components of a light-water nuclear reactor, and what role does each play? What is the nuclear fuel cycle?

17. How safe are nuclear power plants? Describe the Three Mile Island nuclear power plant accident in Pennsylvania.

18. What is being done with low-level radioactive waste produced by the nuclear fuel cycle? What are the options for dealing with high-level radioactive waste produced by the nuclear fuel cycle? Currently, what is being done with such waste in the United States? How widespread are contaminated radioactive waste sites in **(a)** the United States and **(b)** the former Soviet Union?

19. What are the three options for retiring (decommissioning) nuclear power plants? How are nuclear power plants connected with the spread of nuclear weapons throughout much of the world?

20. Why are the costs of nuclear power so high? What is the likely financial future of nuclear power?

21. What are the pros and cons of using conventional nuclear fission as an energy resource?

22. What are the pros and cons of using **(a)** breeder nuclear fission and **(b)** nuclear fusion as an energy resource?

CRITICAL THINKING

1. Just to continue using oil at the current rate (not the projected higher exponential increase in its annual use) and not run out, we must discover and add to global oil reserves the equivalent of a new Saudi Arabia supply (the world's largest) *every 10 years*. Do you believe this is possible? If not, what effects might this have on your life? On a child or grandchild you might have?

2. List five things you can do to help reduce your dependence on oil and resources derived from oil (such as gasoline and most plastics; see Appendix 6). Which of these things do you actually plan to do?

3. The United States now imports more than half of the oil it uses and could be importing 70% of its oil by 2010. What do you believe are the five best ways to reduce this dependence?

4. **(a)** Should air pollution emission standards for *all* new and existing coal-burning plants be tightened significantly? Explain. **(b)** Do you favor an energy strategy based on greatly increased use of coal-burning plants to produce electricity? Explain. What are the alternatives?

5. Explain why you agree or disagree with each of the following proposals made by the U.S. nuclear power industry:

a. A large number of new, better-designed nuclear fission power plants should be built in the United States to reduce dependence on imported oil and slow down projected global warming.

b. Federal subsidies to the commercial nuclear power industry (already totaling $1 trillion) should be continued so that it does not have to compete in the open marketplace with other energy alternatives receiving no, or smaller, federal subsidies.

c. A comprehensive program for developing the breeder nuclear fission reactor should be developed and funded largely by the federal government.

d. Current federal subsidies for developing nuclear fusion should be increased.

6. If you had to choose, would you rather live next door to a coal-burning power plant or a nuclear power plant? Explain.

***7.** How is the electricity in your community produced? How has the cost of that electricity changed since 1970 compared to general inflation? Do your community and your school have a plan for improving energy efficiency? If so, what is this plan, and how much money has it saved during the past 10 years? If there is no plan, develop one and present it to the appropriate officials.

***8.** Write a two-page scenario of what your life might be like without oil. Compare and discuss the scenarios developed by members of your class.

***9.** Make a concept map of this chapter's major ideas, using the section heads and subheads and the key terms (in boldface). Look at the inside back cover and in Appendix 5 for information about concept maps.

FURTHER INFORMATION

For more information see Internet Sources and Further Readings at the back of the book. Online resources for this chapter can be found at **www.brookscole.com/biology**

20 ENERGY EFFICIENCY AND RENEWABLE ENERGY

The Energy-Efficiency and Renewable-Energy Revolution

Energy experts Hunter and Amory Lovins (Guest Essay, p. 528) have built a large, passively heated, superinsulated, partially earth-sheltered home in Snowmass, Colorado (Figure 20-1), where winter temperatures can drop to −40°C (−40°F).

This structure, which also houses the research center for the Rocky Mountain Institute, an office used by 40 people, gets 99% of its space and water heating and 95% of its daytime lighting from the sun. It uses one-tenth the usual amount of electricity for a structure of its size. Total energy savings repaid the cost of its energy-saving features after 10 months and are projected to pay for the entire facility in 40 years.

Superinsulating windows, already here, mean that a house can have large numbers of windows without much heat loss in cold weather or heat gain in hot weather. Thinner insulation material now being developed will allow roofs and walls to be insulated far better than in today's best superinsulated houses.

In energy-efficient houses of the near future, sensors will monitor indoor temperatures, sunlight angles, and the occupants' locations. Then small microprocessors will send heat or cooled air where it is needed and open or close window shutters to let solar energy in or reduce heat loss from windows at night and on cloudy days.

A small but growing number of people in developed and developing countries are getting their electricity from *solar cells* that convert sunlight directly into electricity. They can be attached like shingles to a roof or applied to window glass as a coating. Solar-cell prices are high but are falling rapidly.

As the price of producing electricity from solar cells drops, we could see a *solar–hydrogen revolution*. Electricity produced by large banks of solar cells could be passed through water to make hydrogen gas (H_2), which could be used to fuel vehicles, industries, and buildings. Another solution is to burn hydrogen in energy-efficient *fuel cells* that produce electricity to run cars and heat buildings and water.

These are only a few of the components of the *energy-efficiency and renewable-energy revolution* that many analysts believe will help us make the transition to more sustainable societies over the next 40–50 years.

Figure 20-1 The Rocky Mountain Institute in Colorado. This facility is a home and a center for the study of energy efficiency and sustainable use of energy and other resources. It is also an example of energy-efficient passive solar design. (Robert Millman/Rocky Mountain Institute)

If the United States wants to save a lot of oil and money and increase national security, there are two simple ways to do it: Stop driving Petropigs and stop living in energy sieves.

Amory B. Lovins

This chapter addresses the following questions:

- What are the advantages and disadvantages of **(1)** improving energy efficiency, **(2)** using solar energy to heat buildings and water and to produce electricity, **(3)** using flowing water and solar energy stored as heat in water to produce electricity, **(4)** using wind to produce electricity, **(5)** burning plants and organic waste (biomass) for heating buildings and water, for producing electricity, and for transportation (biofuels), **(6)** producing hydrogen gas and using it to make electricity, heat buildings and water, and propel vehicles, and **(7)** and extracting heat from the earth's interior (geothermal energy)?
- What are the best energy options?

20-1 THE IMPORTANCE OF IMPROVING ENERGY EFFICIENCY

What Is Energy Efficiency? You may be surprised to learn that *84% of all commercial energy used in the United States is wasted* (Figure 20-2). About 41% of this energy is wasted automatically because of the degradation of energy quality imposed by the second law of energy. However, about 43% is wasted unnecessarily, mostly by **(1)** using fuel-wasting motor vehicles, furnaces, and other devices and **(2)** by living and working in leaky, poorly insulated, poorly designed buildings.

People in the United States unnecessarily waste as much energy as two-thirds of the world's population consumes. Improvements in energy efficiency since the oil embargo in 1973 have cut U.S. energy bills by a whopping $275 billion a year. However, according to energy expert Amory Lovins (Guest Essay, p. 528), unnecessary energy waste in the United States still costs about $300 billion per year (an average of $570,000 per minute)—more than the entire $281-billion military budget in 1999.

Japan and most developed countries waste much less energy than the United States, but most developing countries waste a higher percentage of the energy they use than the United States. Reducing such energy waste has a number of advantages (Figure 20-3).

One way to reduce energy waste is to increase the efficiency of the energy conversion devices we use. **Energy efficiency** is the percentage of total energy input that does useful work (is not converted to low-quality, essentially useless heat) in an energy conversion system.

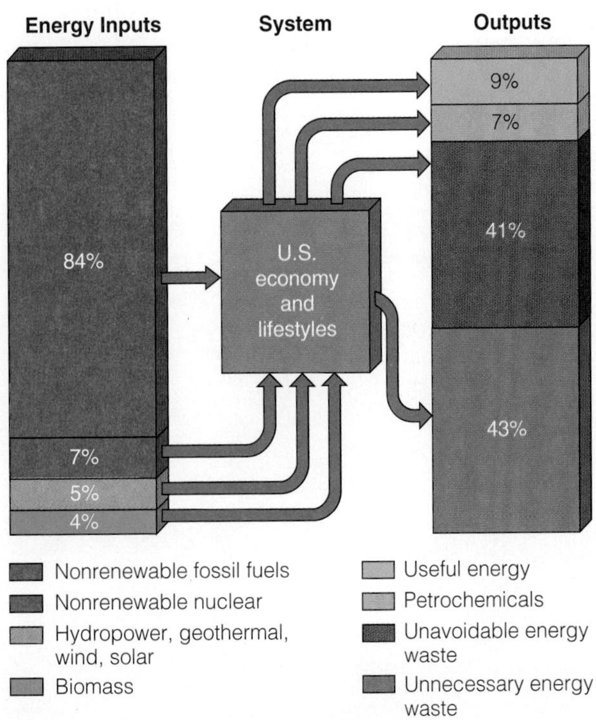

Figure 20-2 Flow of commercial energy through the U.S. economy. Note that only 16% of all commercial energy used in the United States ends up performing useful tasks or being converted to petrochemicals; the rest is either automatically and unavoidably wasted because of the second law of energy (41%) or wasted unnecessarily (43%).

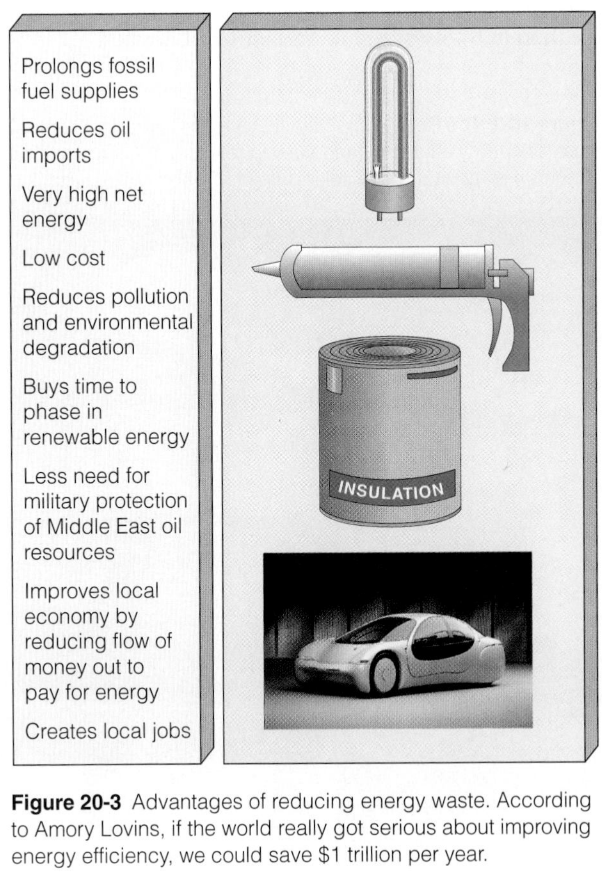

Figure 20-3 Advantages of reducing energy waste. According to Amory Lovins, if the world really got serious about improving energy efficiency, we could save $1 trillion per year.

The energy conversion devices we use vary in their energy efficiencies (Figure 20-4). We can save energy and money by buying the most energy-efficient home heating systems, water heaters, cars, air conditioners, refrigerators, computers, and other appliances. Some energy-efficient models may cost more initially, but in the long run they usually save money by having a lower **life cycle cost**: initial cost plus lifetime operating costs.

The net efficiency of the entire energy delivery process for a space heater, water heater, or car is determined by the efficiency of each step in the energy conversion process. For example, the sequence of energy-using (and energy-wasting) steps involved in using electricity produced from fossil or nuclear fuels is

Extraction → Transportation → Processing → Transportation to power plant → Electric generation → Transmission → End use

Figure 20-5 shows the net energy efficiency for heating two well-insulated homes: one with electricity produced at a nuclear power plant, transported by wire to the home, and converted to heat (electric resistance heating), the other heated passively, with an input of direct solar energy through high-efficiency windows facing the sun, with heat stored in rocks or water for slow release.

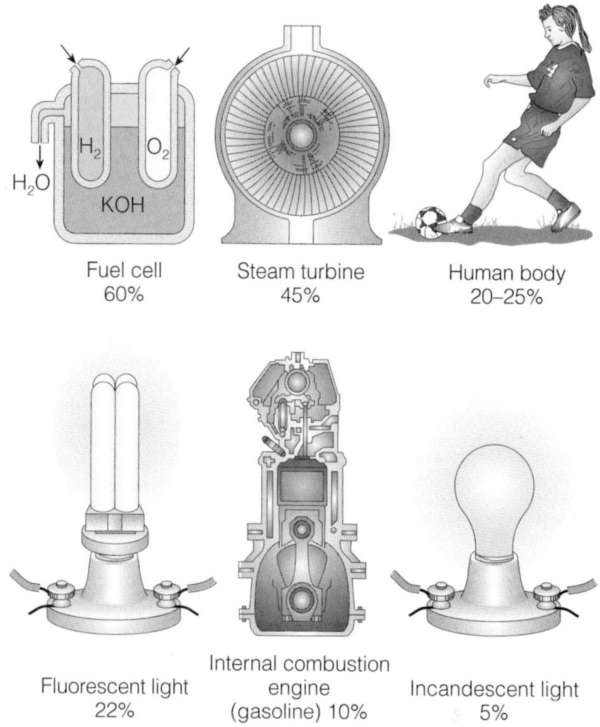

Figure 20-4 Energy efficiency of some common energy conversion devices.

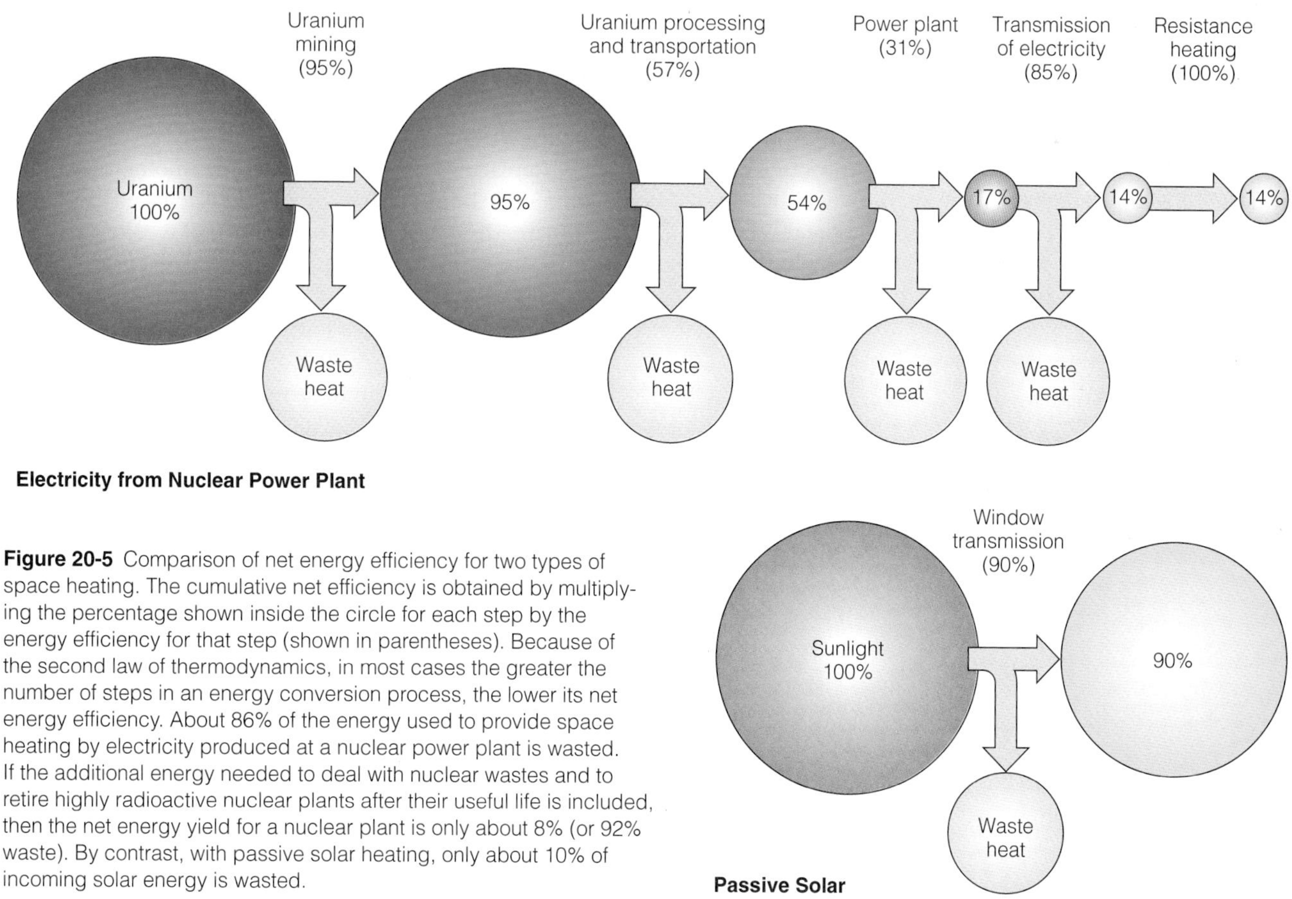

Figure 20-5 Comparison of net energy efficiency for two types of space heating. The cumulative net efficiency is obtained by multiplying the percentage shown inside the circle for each step by the energy efficiency for that step (shown in parentheses). Because of the second law of thermodynamics, in most cases the greater the number of steps in an energy conversion process, the lower its net energy efficiency. About 86% of the energy used to provide space heating by electricity produced at a nuclear power plant is wasted. If the additional energy needed to deal with nuclear wastes and to retire highly radioactive nuclear plants after their useful life is included, then the net energy yield for a nuclear plant is only about 8% (or 92% waste). By contrast, with passive solar heating, only about 10% of incoming solar energy is wasted.

This analysis shows that the process of **(1)** converting the high-quality energy in nuclear fuel to high-quality heat at several thousand degrees in the power plant, **(2)** converting this heat to high-quality electricity, and **(3)** using the electricity to provide low-quality heat for warming a house to only about 20°C (68°F) is very wasteful of high-quality energy. Burning coal or any fossil fuel at a power plant to supply electricity for space heating is also inefficient. It is much less wasteful to **(1)** collect solar energy from the environment, **(2)** store the resulting heat in stone or water, and, **(3)** if necessary, use a small backup heating system to raise its temperature slightly to provide space heating or household hot water.

Physicist and energy expert Amory Lovins (Guest Essay, p. 528) points out that using high-quality electrical energy to provide low-quality heating for living space or household water is like using a chain saw to cut butter or a sledgehammer to kill a fly. As a rule, he suggests that we not use high-quality energy to do a job that can be done with lower-quality energy (Figure 3-9).

The logic of this rule is illustrated by looking at the costs of providing heat using various fuels. According to the Department of Energy, in 1997, the average price of obtaining 250,000 kilocalories (1 million Btu) for heating either space or water in the United States was $5.32 using a heat pump (useful in warm to moderate climate areas only), $6.83 using natural gas, $7.56 using kerosene, $8.64 using oil, $11.82 using propane, and $20.19 using electricity (resistance heating).

Perhaps the three least efficient energy-using devices in widespread use today are **(1)** incandescent light bulbs (which waste 95% of the energy input), **(2)** vehicles with internal combustion engines (which waste 86–90% of the energy in their fuel), and **(3)** nuclear power plants producing electricity for space heating or water heating (which waste 86% of the energy in their nuclear fuel, and probably 92% when the energy needed to deal with radioactive wastes and retired nuclear plants is included). Energy experts call for us to replace these devices or greatly improve their energy efficiency over the next few decades.

20-2 WAYS TO IMPROVE ENERGY EFFICIENCY

How Can We Use Waste Heat? Could we save energy by recycling energy? No. The second law of energy tells us that we cannot recycle energy. However, we can slow the rate at which waste heat flows into the environment when high-quality energy is degraded.

For a house, the best way to do this is to **(1)** insulate it thoroughly, **(2)** eliminate air leaks, and **(3)** equip it with an air-to-air heat exchanger to prevent buildup of indoor air pollutants. Many homes in the United States (and in most other countries) are so full of leaks that their heat loss in cold weather and heat gain in hot weather are equivalent to having a large window-size hole in the wall of the house (Figure 3-1).

In office buildings and stores, waste heat from lights, computers, and other machines can be collected and distributed to reduce heating bills during cold weather. During hot weather, the collected heat can vented outdoors to reduce cooling bills.

How Can We Save Energy in Industry? There are a number of ways to save energy and money in industry. One is **cogeneration**, or *combined heat and power (CHP)* systems, the production of two useful forms of energy (such as steam and electricity) from the same fuel source. Waste heat from coal-fired and other industrial boilers can produce steam that spins turbines and generates electricity at roughly half the cost of buying it from a utility company. Conventional electricity generation by coal-burning or nuclear power plants has an average efficiency of about 32%. But by recovering wasted energy, cogeneration systems have an efficiency of up to 80% and reduce carbon dioxide emissions from coal-burning plants by two-thirds.

Cogeneration has been widely used in western Europe for years, and its use in the United States and China is growing. In Germany, small cogeneration units that run on natural gas or liquefied petroleum gas (LPG) supply restaurants, apartment buildings, and houses with all their energy. In 4–5 years they pay for themselves in saved fuel and electricity.

Replacing energy-wasting electric motors is another strategy. About 60–70% of the electricity used in U.S. industry drives electric motors, most of which are run at full speed with their output throttled to match the task—somewhat like driving with the gas pedal to the floor and the brake engaged. Each year a heavily used electric motor consumes 10 times its purchase cost in electricity—equivalent to using $200,000 worth of gasoline each year to fuel a $20,000 car. According to Amory Lovins, it would be cost-effective to scrap virtually all such motors and replace them with adjustable-speed drives. The costs would be paid back in 1–3 years, depending on how the motors were used.

Energy can also be saved by **(1)** switching to high-efficiency lighting, **(2)** recycling, reusing, and repairing products instead of using energy to make them from virgin resources, and **(3)** changing how electric utility companies are regulated (Solutions, p. 524).

How Can We Save Energy in Transportation? According to most energy analysts, the best way to save

Figure 20-6 The Ultralite prototype car built by General Motors. It is fast and much safer than existing cars. When driven at 81 kph (50 mph), it gets 43 kpl (100 mpg). It has a 220-kph (135-mph) top speed, is equipped with four air bags, and easily carries four adults. Its body is made from lightweight carbon-fiber composite plastics that won't dent, scratch, or corrode. With 200 fewer parts than a conventional engine, it costs about $400 less to make and costs consumers less in maintenance expenses. (General Motors)

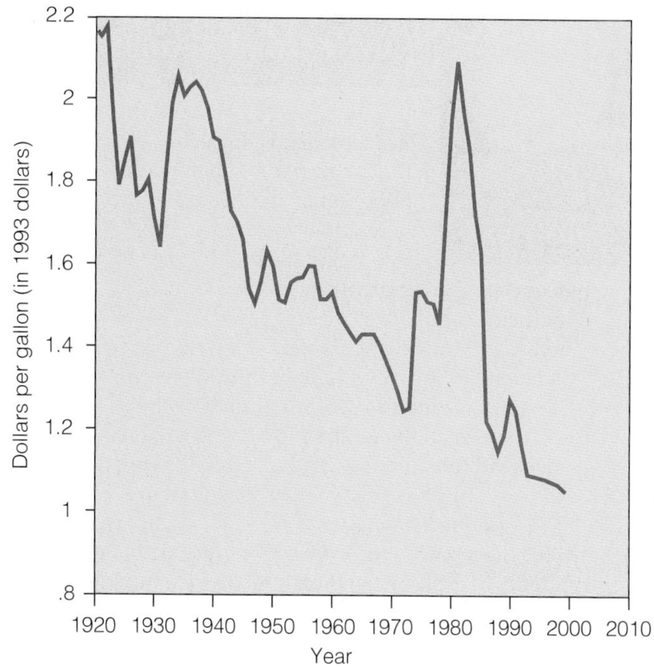

Figure 20-7 Real price of gasoline (in 1993 dollars), 1920–99. When the price is adjusted for inflation, U.S. consumers are paying less for gasoline than at any time since 1920. This underpricing of gasoline encourages energy waste and pollution by failing to include in its market price the many harmful social and environmental costs of gasoline, which consumers ultimately end up paying anyway. (U.S. Department of Energy)

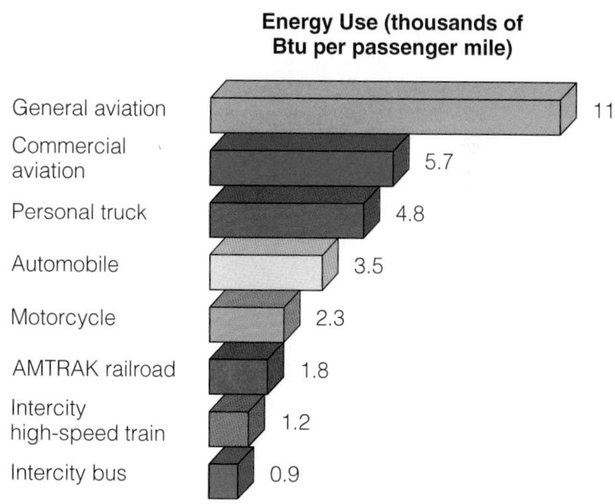

Figure 20-8 Energy use of various types of domestic passenger transportation; the lower the energy use, the greater the efficiency. (U.S. Department of Energy)

energy (especially oil) and money in transportation is to *increase the fuel efficiency of motor vehicles*.

Between 1973 and 1985, the average fuel efficiency doubled for new American cars and rose 37% for all passenger cars on the road because of government-mandated standards. However, since 1985 automakers have successfully opposed any increase in the government standards. Actually, the average fuel efficiency of new vehicles has fallen by about 5% in recent years because of the popularity of **(1)** sport utility vehicles (SUVs), minivans, and light trucks (subject to much lower mileage standards than cars) and **(2)** larger, less efficient autos.

Since 1985 at least 10 automobile companies have made nimble and peppy prototype cars that meet or exceed current safety and pollution standards, with fuel efficiencies of 29–59 kilometers per liter (67–138 miles per gallon) (Figure 20-6). If such cars were mass-produced, their slightly higher costs would be more than offset by their fuel savings. The problem is that there is little consumer interest in fuel-efficient cars mostly because the inflation-adjusted price of gasoline today in the United States is the lowest it has been since 1920 (Figure 20-7).

To encourage consumers to buy energy-efficient vehicles, Amory Lovins has suggested a system of revenue-neutral rebates and "freebates" for motor vehicles. Buyers of a new vehicle would pay a fee or receive a rebate depending on its fuel efficiency. The fees on inefficient vehicles would be used to pay for the rebates on efficient ones. What do you think of this idea?

Another way to save energy is to *shift to more energy-efficient ways to move people* (Figure 20-8) *and freight*. More freight could be shifted from trucks and planes to more energy-efficient trains and ships. With improved aerodynamic design, turbocharged diesel engines, and radial tires, new transport trucks can be 50% more fuel efficient than today's conventional trucks.

The Negawatt Revolution

SOLUTIONS

Traditionally, U.S. electric utility companies have made more money by increasing the demand for electricity. This encourages the building of often unnecessary power plants to send electricity to inefficient appliances, heating and cooling systems, and industrial plants.

A small but growing number of utility companies in the United States and elsewhere are trying to reverse this wasteful process by reducing the demand for electricity and avoiding having to finance and build expensive new power plants.

This approach is known as *demand-side management* or the *negawatt revolution*. To reduce demand, utilities give customers **(1)** cash rebates for buying efficient lights and appliances, **(2)** free home energy audits, **(3)** low-interest loans for home weatherization or industrial retrofits, and **(4)** lower rates to households or industries meeting certain energy-efficiency standards.

To make demand-side management feasible, government regulators must allow utility investors to make a reasonable return on their money, based on the amount of energy the utilities save. Such a policy allows utility companies to profit by shifting their emphasis from megawatts to energy-saving negawatts. Amory Lovins (Guest Essay, p. 528) says we should "think of a 16-watt compact-fluorescent replacement for a 75-watt incandescent bulb, for example, as a 59-negawatt power plant."

However, since 1993 such programs have been cut back sharply because the U.S. electric power industry is being restructured and deregulated to help increase competition.

Critical Thinking

Do you believe that regulators should allow utility companies to make a profit for their investors based on how much electricity they save? Explain.

Are Electric Cars the Answer? Conventional battery-powered *electric cars* might help reduce dependence on oil, especially for urban commuting and short trips. Electric vehicles are extremely quiet, need little maintenance, and can accelerate rapidly.

The cars themselves produce no air pollution, but using coal and nuclear power plants to produce the electricity needed to recharge their batteries produces air pollution and nuclear wastes—something called *elsewhere pollution*. If solar cells or wind turbines could be used for recharging the car's batteries, CO_2 and other air pollution emissions would be virtually eliminated.

On the negative side, today's electric cars **(1)** can travel only 81–161 kilometers (50–100 miles) before needing a 3- to 8-hour recharge (although a new device may reduce recharge time to 10–20 minutes) and **(2)** batteries must be replaced about every 48,000 kilometers (30,000 miles) at a cost of at least $2,000. This plus the costs for daily recharging and a buying a recharger ($700–3,500) means that today's electric cars have twice the operating cost of gasoline-powered cars. Because of high costs and a lack of consumer interest, in 1999 major car companies abandoned their production of electric cars.

Are Hybrid and Fuel Cell Cars the Answer? There is growing interest in developing *ecocars* or *supercars* that could eventually get 64–128 kilometers per liter (150–300 miles per gallon). One type of ecocar uses **(1)** a small hybrid electric–internal combustion engine that runs on gasoline or some other liquid fuel and **(2)** a small battery (recharged by the internal combustion engine) to provide the energy needed for acceleration and hill climbing (Figure 20-9). In 1999 Toyota and Honda began selling the first generation of fuel-efficient hybrid engine cars.

Amory Lovins (Guest Essay, p. 528) argues that future models of ecocars on the drawing boards will sell "not just because they are more than 5 times as efficient and more than 100 times cleaner than current cars, but also because they will be markedly superior in comfort, quietness, performance, durability, beauty, and safety" (Figure 20-6).

Another type of ecocar is an electric vehicle that uses *fuel cells* (Figure 20-10). Fuel cells consist of two electrodes immersed in a solution (electrolyte) that conducts electricity. They produce electricity by combining hydrogen and oxygen ions, typically from hydrogen and oxygen gas supplied as fuel for the cell (Figure 20-10). Such cars are 50–60% efficient (compared to 10–14% efficiency for gasoline-powered vehicles).

How Can We Save Energy in Buildings? The 110-story, twin-towered World Trade Center in New York City is a monument to energy waste: It uses as much electricity as a city of 100,000 people for about 53,000 employees.

In contrast, Atlanta's 13-story Georgia Power Company building uses 60% less energy than conventional office buildings of the same size. The largest surface of the building faces south to capture solar energy. Each floor extends out over the one below it, blocking out the higher summer sun to reduce air conditioning costs

Figure 20-9 General features of a hybrid car powered by a *hybrid gas/electric engine.* The batteries are recharged by a small internal combustion engine, thus reducing the need for heavy banks of batteries and solving the problem of the limited range of conventional electric cars. The bodies of future models of such cars will probably be made of lightweight composite plastics that offer more protection in crashes, don't need to be painted, don't rust, can be recycled, and have fewer parts than conventional cars. (Concept information from DaimlerChrysler, Ford, Honda, and Toyota)

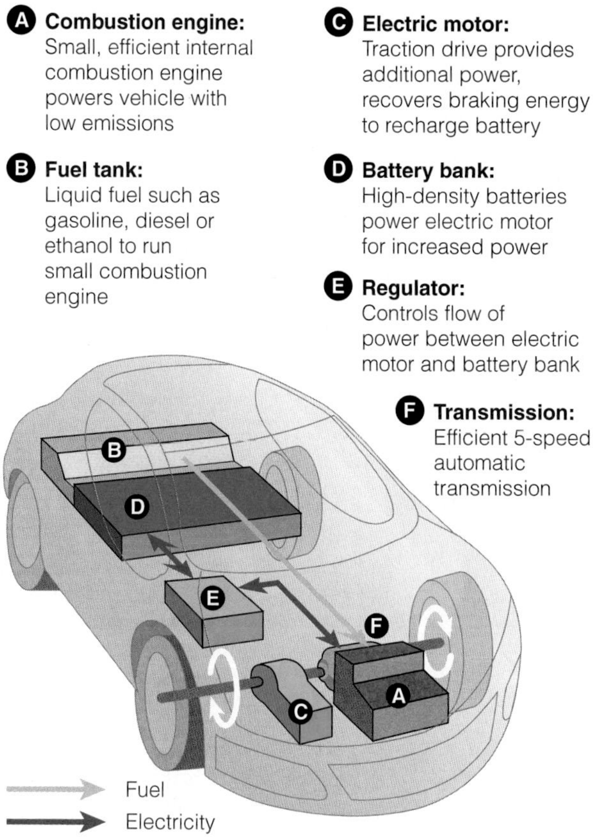

but allowing warming by the lower winter sun. Energy-efficient lights focus on desks rather than illuminating entire rooms. The Georgia Power model and other existing cost-effective commercial building technologies could **(1)** reduce energy use by 75% in buildings, **(2)** cut carbon dioxide emissions in half, and **(3)** in the United States save more than $130 billion per year in energy bills.

There are a number of ways to improve the energy efficiency of buildings, some of them discussed in the opening of this chapter (p. 519). One is to build more *superinsulated houses* (Figure 20-11). Although such houses typically cost 5% more to build than conventional houses of the same size, this extra cost is paid back by energy savings within 5 years and can save a homeowner $50,000–100,000 over a 40-year period.

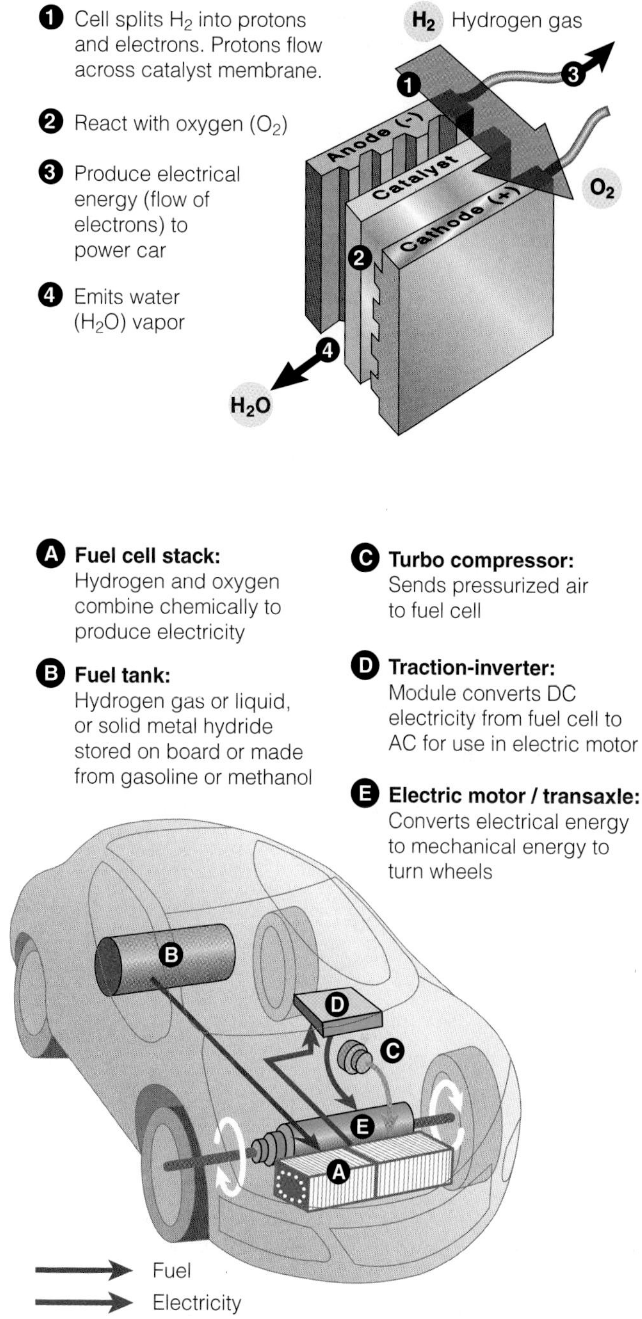

Figure 20-10 General features of an electric car powered by a *fuel cell* running on hydrogen gas. Such cars will **(1)** be almost pollution-free, emitting only water vapor and small amounts of nitrogen oxides and no CO_2, and **(2)** should get at least twice the mileage of comparable gasoline-powered cars. Several automobile companies have developed prototypes and are working to get costs down and improve hydrogen fuel-storage systems. Early models could be on the road by 2010. (Concept information from DaimlerChrysler, Ford, Ballard, Toyota, and Honda)

Since the mid-1980s there has been growing interest in building superinsulated houses called *straw-bale houses* with walls consisting of compacted bales of certain types of straw (available at a low cost almost

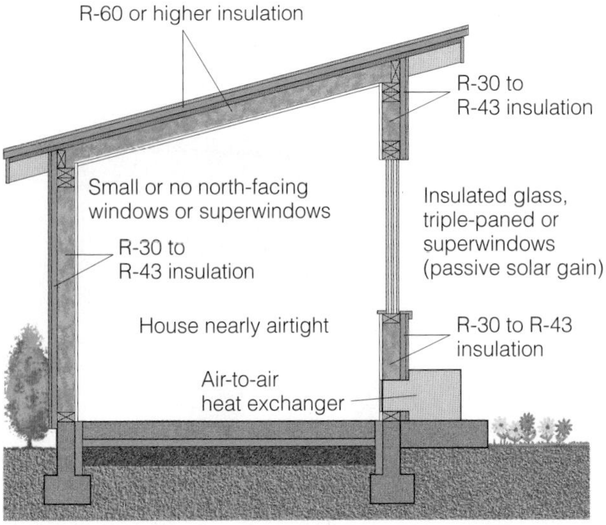

Figure 20-11 Major features of a superinsulated house. Such a house is so heavily insulated and so airtight that heat from direct sunlight, appliances, and human bodies warms it, with little or no need for a backup heating system. An air-to-air heat exchanger prevents buildup of indoor air pollution.

Figure 20-12 An energy-efficient, environmentally healthy, and affordable Victorian style strawbale house designed and built by Alison Gannett in Crested Butte, Colorado. The top photo is during construction and the bottom photo is the completed house. Depending on the thickness of the bales, plastered strawbale walls have an insulating value of R-35 to R-60, compared to R-12 to R-19 in a conventional house. (The R value is a measure of resistance to heat flow.) (Alison Gannett)

everywhere) covered with plaster or adobe (Figure 20-12).* By 1999 there were more than 1,000 such homes built or under construction in the United States. Using straw, an *annually* renewable agricultural residue often burned as a waste product, for the walls reduces the need for wood and thus slows deforestation. The main problem is getting banks and other money lenders to recognize the potential of this type of housing and provide homeowners with construction loans.

Another way to save energy is to *use the most energy-efficient ways to heat houses* (Figure 20-13). The most energy-efficient ways to heat space are **(1)** a superinsulated house, **(2)** passive solar heating, **(3)** a high-efficiency (85–98% efficient) natural gas furnace, and **(4)** heat pumps in warm climates (but not in cold climates because at low temperatures they automatically switch to costly electric resistance heating). The most wasteful and expensive way is to use electric resistance heating with the electricity produced by a coal-fired or nuclear power plant.

The energy efficiency of existing houses and buildings can be improved significantly by adding insulation, plugging leaks, and installing energy-saving windows and lighting. About one-third of heated air in U.S. homes and buildings escapes through closed windows and holes and cracks (Figure 3-1)—equal to the energy in all the oil flowing through the Alaska pipeline every year. During hot weather these windows also let heat in, increasing the use of air conditioning. Replacing all windows in the United States with low-E (low-emissivity) windows would cut these expensive losses by two-thirds and reduce CO_2 emissions.

Simply wrapping a water heater in a $20 insulating jacket can save a homeowner $45 a year and reduce CO_2 emissions. Leaky ducts allow 20–30% of a home's heating and cooling energy to escape, and draw unwanted moisture and heat into the home. Careful sealing can reduce this loss. Some designs for new homes eliminate the need for duct sealing by keeping the ducts inside the home's thermal envelope so that escaping hot or cool air leaks into the living space. In Chicago,

*For information on strawbale houses, see Steen et al., *The Straw Bale House* (White River Junction, Vt.: Chelsea Green, 1994); and GreenFire Institute, 1509 Queen Anne Ave. North, #606, Seattle, WA 98109, 206-284-7470.

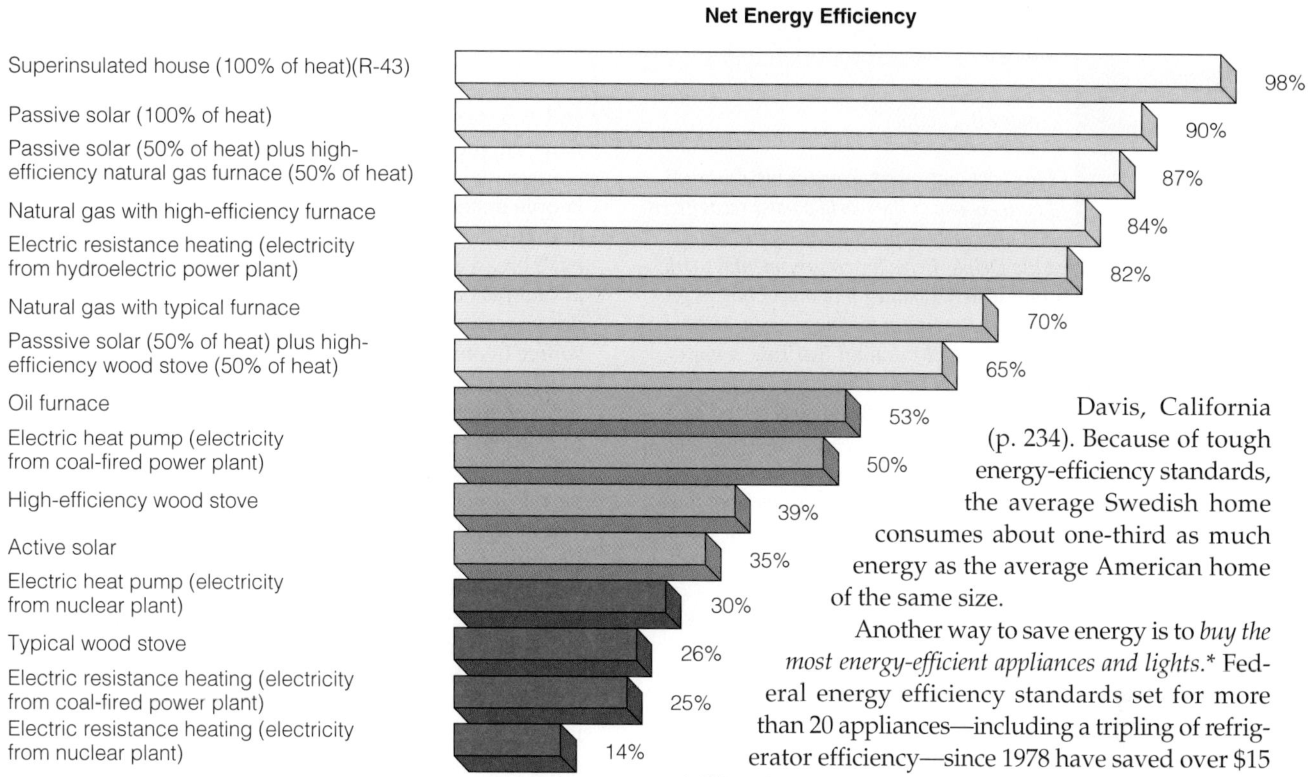

Figure 20-13 Net energy efficiencies for various ways to heat an enclosed space such as a house. (Data from Howard T. Odum)

Illinois, a retrofit of a 20-year-old office tower reduced energy use by 75% while increasing lighting, comfort, and productivity.

An energy-efficient way to heat hot water for washing and bathing is to use tankless instant water heaters (about the size of bookcase loudspeakers) fired by natural gas or LPG. These devices, widely used in many parts of Europe, heat the water instantly as it flows through a small burner chamber and provide hot water only when it is needed. A well-insulated, conventional natural gas or LPG water heater is also fairly efficient (although all conventional natural gas and electric resistance heaters waste energy by keeping a large tank of water hot all day and night and can run out after a long shower or two).

The most inefficient and expensive way to heat water for washing and bathing is to use electricity produced by any type of power plant. A $425 electric water heater can cost $5,900 in energy over its 15-year life, compared to about $2,900 for a comparable natural gas water heater over the same period.

Setting higher energy-efficiency standards for new buildings would also save energy. Building codes could require that all new houses use 60–80% less energy than conventional houses of the same size, as has been done in Davis, California (p. 234). Because of tough energy-efficiency standards, the average Swedish home consumes about one-third as much energy as the average American home of the same size.

Another way to save energy is to *buy the most energy-efficient appliances and lights.** Federal energy efficiency standards set for more than 20 appliances—including a tripling of refrigerator efficiency—since 1978 have saved over $15 billion in energy costs in the United States and are expected to save U.S. consumers some $56 billion between 1990 and 2015. Developing these standards cost only $50 million, yielding a benefit-to-cost ratio for the government's investment of 2,600 to 1.

Energy-efficient lighting could save U.S. businesses $15–20 billion per year in electricity bills. Replacing a standard incandescent bulb with an energy-efficient compact fluorescent bulb (Figure 20-14) saves about $48–70 per bulb over its 10-year life. Thus, replacing 25 incandescent bulbs in a house with energy-efficient fluorescent bulbs saves $1,250–1,750. Students in Brown University's environmental studies program showed that the school could save more than $40,000 per year just by replacing the incandescent light bulbs in exit signs with compact fluorescents. So far less than 10% of U.S. homes use these energy- and money-saving bulbs, which pay for themselves in 2–4 years.

If all U.S. households used the most efficient frost-free refrigerator now available, 18 large (1,000-megawatt) power plants could close. Microwave ovens can cut electricity use for cooking by 25–50% (but not if used for defrosting food). Clothes dryers with moisture

*Each year the American Council for an Energy-Efficient Economy (ACEEE) publishes a list of the most energy-efficient major appliances mass-produced for the U.S. market. A copy can be obtained from the council at 1001 Connecticut Ave. NW, Suite 801, Washington, DC 20036. Each year they also publish *A Consumer Guide to Home Energy Savings*, available in bookstores or from the ACEEE.

GUEST ESSAY

Technology Is the Answer (But What Was the Question?)

Amory B. Lovins

Physicist and energy consultant Amory B. Lovins is one of the world's most recognized and articulate experts on energy strategy. In 1989, he received the Delphi Prize for environmental work; in 1990 the Wall Street Journal *named him one of the 39 people most likely to change the course of business in the 1990s. He is research director at Rocky Mountain Institute, a nonprofit resource policy center that he and his wife, Hunter, founded in Snowmass, Colorado, in 1982. He has served as a consultant to more than 200 utilities, private industries, and international organizations, and to many national, state, and local governments. He is active in energy affairs in more than 35 countries and has published several hundred papers and a dozen books on energy strategies and policies.*

The answers you get depend on the questions you ask. It is fashionable to suppose that we're running out of energy and that the solution is obviously to get lots more of it. But asking how to get more energy begs the question of how much we need and what are the cheapest and least environmentally harmful ways to meet these needs.

How much energy it takes to make steel, run a sewing machine, or keep ourselves comfortable in our houses depends on how cleverly we use energy, and the more it costs, the smarter we seem to get. It is now cheaper, for example, to double the efficiency of most industrial electric motor drive systems than to fuel existing power plants to make electricity. (*Just this one saving can more than replace the entire U.S. nuclear power program.*) We know how to make lights five times as efficient as those currently in use and how to make household appliances that give us the same work as now, but use one-fifth as much energy (saving money in the process).

Ten automakers have made good-sized, peppy, safe prototype cars averaging 29–59 kilometers per liter (67–138 miles per gallon) and within a decade automakers could have cars getting 64–128 kpl (150–300 mpg) on the road if consumers demanded such cars. We know today how to make new buildings (and many old ones) so heat-tight (but still well ventilated) that they need essentially no outside energy to maintain comfort year-round, even in severe climates. In fact, I live and work in one [Figure 20-1].

These energy-saving measures are uniformly cheaper than going out and getting more energy. However, the old view of the energy problem included a worse mistake than forgetting to ask how much energy we needed: It sought more energy, in any form, from any source, at any price—as if all kinds of energy were alike.

Just as there are different kinds of food, so there are many different forms of energy, whose different prices and qualities suit them to different uses [Figure 3-9]. After all, there is no demand for energy as such; nobody wants raw kilowatt-hours or barrels of sticky black goo. People instead want energy services: comfort, light, mobility, hot showers, cold beverages, and the ability to bake bread and make cement. We ought therefore to start at that part of the energy problem and ask, "What tasks do we want energy for, and what amount, type, and source of energy will do each task most cheaply?"

Electricity is a particularly high-quality, expensive form of energy. An average kilowatt-hour delivered in the United States in 1998 was priced at about 6.7¢, equivalent to buying the heat content of oil costing $111 per barrel—over six times oil's average world price in 1998. The average cost of electricity from nuclear plants (including construction, fuel, and operating expenses) that began operation in 1988 was 13.5¢ per kilowatt-hour in 1998, equivalent on a heat basis to buying oil at about $216 per barrel.

Such costly energy might be worthwhile if it were used only for the premium tasks that require it, such as lights, motors, electronics, and smelters. But those special uses—only 8% of all delivered U.S. energy needs—are already met twice over by today's power stations. Two-fifths of our electricity is already spilling over into uneconomic, low-grade uses such as water heating, space heating, and air conditioning. Yet no matter how efficiently we use electricity (even with heat pumps), we can never get our money's worth on these applications.

Thus, *supplying more electricity is irrelevant to the energy problem we have.* Even though electricity accounts for almost all the federal energy research-and-development budget and for at least half the national energy investment, it is the wrong kind of energy to meet our needs economically. Arguing about what kind of new power station to build—coal, nuclear, solar—is like shopping for the best buy in antique Chippendale chairs to burn in your stove, or for brandy to put in your car's gas tank. *It is the wrong approach.*

Indeed, *any kind of new power station is so uneconomical that if you have just built one, you will save the country money by writing it off and never operating it.* Why? Because its additional electricity can be used only for low-temperature heating and cooling (the premium "electricity-specific" uses being already filled up) and is the most expensive way of supplying those services. Saving electricity (creating negawatts) is much cheaper than making it.

The real question is, "What is the cheapest way to do low-temperature heating and cooling?" The answer is weather-stripping, insulation, heat exchangers, greenhouses, superwindows (which have as much insulating value as the outside wall of a typical house), window shades and overhangs, trees, and so on. These measures generally cost about half a penny to 2¢ per kilowatt-hour; the running costs *alone* for a new nuclear plant will be nearly 4¢ per kilowatt-hour, so it's cheaper not to run it. In fact, under the crazy U.S. tax laws, the extra saving from not having to pay the plant's future subsidies is probably so big that by shutting the plant down society can also recover the capital cost of having built it!

If we want more electricity, we should get it from the cheapest sources first. In approximate order of increasing price, these include

- Converting to efficient lighting equipment. This would save the United States electricity equal to the output of 120 large power plants, plus $30 billion a year in fuel and maintenance costs.
- Using more efficient electric motors to save up to half the energy used by motor systems. This would save electricity equal to the output of another 150 large power plants and repay the cost in about a year.
- Eliminating pure waste of electricity, such as lighting empty offices.
- Displacing the electricity now used for water heating and for space heating and cooling with good architecture, weatherization, insulation, and passive and some active solar techniques.
- Making appliances, smelters, and the like cost-effectively efficient.

Just these five measures can quadruple U.S. electrical efficiency, making it possible to run today's economy with no changes in lifestyles and using no power plants, whether old or new or fueled with oil, gas, coal, or uranium. We would need only the present hydroelectric capacity, readily available small-scale hydroelectric projects, and a modest amount of wind power.

If we still wanted more electricity, the next cheapest sources would include industrial cogeneration, combined heat-and-power plants, low-temperature heat engines run by industrial waste heat or by solar ponds, filling empty turbine bays and upgrading equipment in existing big dams, modern wind machines or small-scale hydroelectric turbines in good sites, steam-injected natural-gas turbines, and perhaps recent developments in solar cells with waste heat recovery.

It is only after we had clearly exhausted all these cheaper opportunities that we would even consider building a new central power station of any kind—the slowest and costliest known way to get more electricity (or to save oil).

To emphasize the importance of starting with energy end uses rather than energy sources, consider a sad little story from France. In the mid-1970s, energy conservation planners in the French government found that their biggest need for energy was to heat buildings, and that even with good heat pumps, electricity would be the costliest way to do this. So they had a fight with their nationalized utility; they won, and electric heating was supposed to be discouraged or even phased out because it was so wasteful of money and fuel.

Meanwhile, down the street, the energy supply planners (who were far more numerous and influential in the French government) said, "Look at all that nasty imported oil coming into our country! We must replace that oil. Oil is energy. We need some other source of energy. Voilà! Reactors can give us energy; we'll build nuclear reactors all over the country." But they paid little attention to who would use that extra energy and no attention to relative prices.

Thus these two groups of the French energy establishment went on with their respective solutions to two different, indeed contradictory, French energy problems: *more energy of any kind* versus *the right kind to do each task in the most inexpensive way.* It was only in 1979 that these conflicting perceptions collided. The supply-side planners suddenly realized that the only thing they would be able to *sell* all that nuclear electricity for would be electric heating, which they had just agreed not to do.

Every industrial country is in this embarrassing position (especially if we include as "heating" air conditioning, which just means heating the outdoors instead of the indoors). Asking *what we think the energy problem is* is not an academic abstraction; *it determines what we buy*. It is the most fundamental source of disagreement about energy policy.

Supply-oriented planners think the problem boils down to whether to build coal or nuclear power stations (or both). Energy-use planners realize that *no* kind of new power station can be an economic way to meet the needs for using electricity to provide low- and high-temperature heat and for the vehicular liquid fuels that are 92% of our energy problem.

So if we want to provide energy services at a price we can afford, let's get straight what question our technologies are supposed to answer. This means asking what we really need the energy *for*!

Critical Thinking

1. The author argues that building more nuclear, coal, or other electrical power plants to supply electricity for the United States is unnecessary and wasteful. Summarize the reasons for this conclusion, and give your reasons for agreeing or disagreeing with this viewpoint.

2. Do you agree or disagree that increasing the supply of energy, instead of improving energy efficiency, is the wrong answer to energy problems? Explain.

sensors cut energy use by 15%, and front-loading washers use 50% less energy than top-loading models but cost no more. New microwave clothes dryers use 15% less energy than conventional electric dryers and 28% less energy than gas units while virtually eliminating wrinkles.

Why Aren't We Doing More to Reduce Energy Waste? With such an impressive array of benefits (Figure 20-3), why isn't there more emphasis on improving energy efficiency? Major reasons include the following:

- A glut of low-cost, underpriced fossil fuels. As long as such energy is artificially cheap because its harmful environmental costs are not included in its market prices, people are more likely to waste it and not make investments in improving energy efficiency.
- Huge government subsidies and tax breaks to promote the use of energy-wasting power plants, motor vehicles, and buildings and few economic incentives for consumers and businesses to invest in improving energy efficiency.
- Lack of information about the availability of energy-saving devices and the amount of money such items can save consumers using *life cycle cost* analysis.

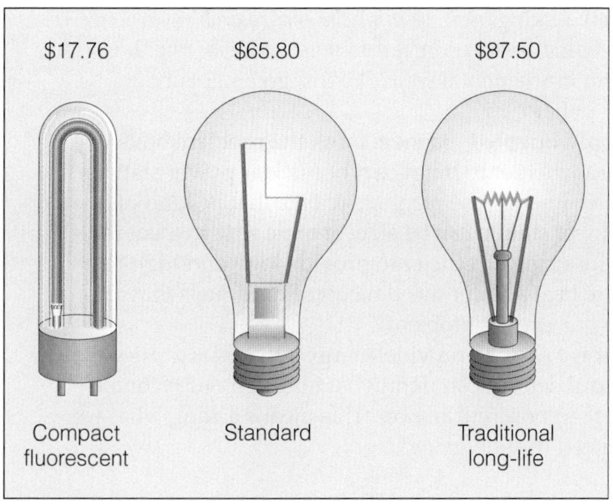

Figure 20-14 Cost of electricity for comparable light bulbs used for 10,000 hours. Because conventional incandescent bulbs are only 5% efficient and last only 750–1,500 hours, they waste enormous amounts of energy and money, and they add to the heat load of houses during hot weather. Socket-type fluorescent light use one-fourth as much electricity as conventional bulbs. Although these bulbs cost about $10–20 per bulb, they last 10–20 times longer than conventional incandescent bulbs used the same number of hours, and save considerable money (compared with incandescent bulbs) over their long life. Recently, smaller compact-fluorescent bulbs (costing about $6) that fit into ordinary light fixtures and ones that can be dimmed have been developed. (Data from Electric Power Research Institute)

20-3 DIRECT USE OF SOLAR ENERGY FOR HEAT AND ELECTRICITY

What Are the Advantages of Solar Energy? The Renewable-Energy Revolution In 1994 Shell International Petroleum in London projected that renewable energy, especially solar, will account for 50% of world energy production by 2050. Analysts project that if current subsidies for nonrenewable fossil fuels and nuclear energy are phased out, a combination of **(1)** improvements in energy efficiency and **(2)** a mix of renewable energy from the sun, wind, flowing water, biomass, and earth's internal heat could meet virtually all of the energy needs of the United States by 2050. In 1999, renewable energy provided only about 8% of the commercial energy used in the United States (Figure 19-3). About 4% of this energy came from hydropower, 3% from biomass and geothermal, and 1% from wind and solar energy.

Developing renewable energy resources would **(1)** save money, **(2)** create two to five times more jobs per unit of electricity produced than coal and nuclear power plants, **(3)** eliminate the need for oil imports, **(4)** cause much less pollution and environmental degradation per unit of energy used, and **(5)** increase military, economic, and environmental security. In the United States, wind farms and power plants using geothermal, wood (biomass), hydropower, and high-temperature solar energy (with natural gas backup) can already produce electricity more cheaply than can new nuclear power plants, and with far fewer federal subsidies (Figure 20-15).

How Can We Use Solar Energy to Heat Houses and Water? Buildings and water can be heated by solar energy using two methods: passive and active (Figure 20-16). A **passive solar heating system** captures sunlight directly within a structure (Figures 20-16 and 20-17) and converts it into low-temperature heat for space heating. Energy-efficient windows, greenhouses, and sunspaces face the sun to collect solar energy by direct gain. Walls and floors of concrete, adobe, brick, stone, salt-treated timber, and water in 55-gallon drums store much of the collected solar energy as heat and release it slowly throughout the day and night. A small backup heating system such as a vented natural gas or propane heater may be used but is not necessary in many climates.

Engineer and builder Michael Sykes has designed a solar envelope house that is heated and cooled passively by solar energy and the slow storage and release of energy by massive timbers and the earth beneath the house (Figure 20-18). The front and back of this house are double walls of heavy timber impregnated with salt to increase the wood's ability to store heat. The space between the two walls, plus the basement, forms a convection loop or envelope around the inner shell of the house. In summer, roof vents release heated air from

Flavin, C. and M. O'Meara. 1998. "Solar Power Markets Boom." *World Watch,* vol. 11, no. 5 p. 23(5).

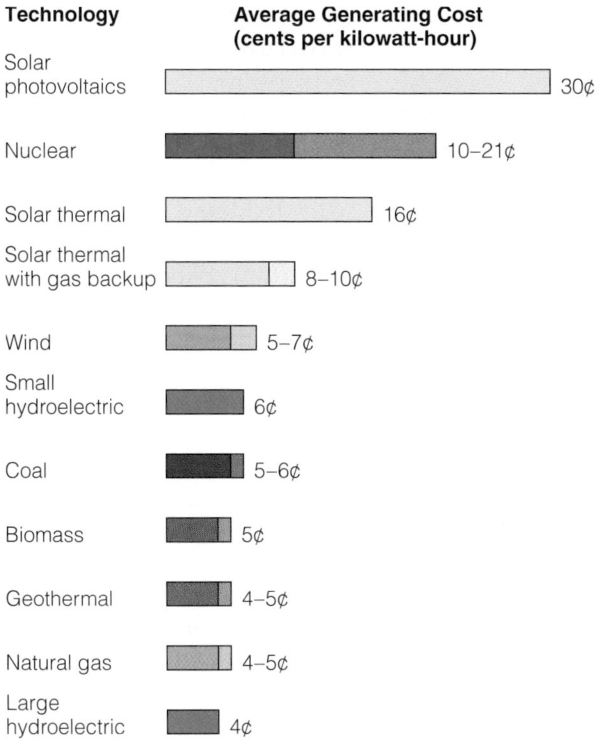

Figure 20-15 Generating costs of electricity per kilowatt-hour by various technologies in 1995. (A kilowatt-hour or kwh can light a 100-watt bulb for about 8 hours or a typical hair dryer for about an hour.) Between 2000 and 2005, costs per kilowatt-hour for wind are expected to decrease to 4–5¢, solar thermal with gas assistance 5–6¢, solar photovoltaic 6–12¢ (compared to about 20¢ per kwh in 1998), natural gas 3–4¢, and coal 4–5¢; nuclear power is expected to remain at 10–21¢. Costs for the other technologies shown in this figure are expected to remain about the same. (Data from U.S. Department of Energy, Council for Renewable Energy Education, Investor Responsibility Research Center, California Energy Commission, and Worldwatch Institute)

the convection loop throughout the day; at night, these roof vents, with the aid of a fan, draw air into the loop, passively cooling the house. The interior temperature of the house typically stays within 2° of 21°C (70°F) year-round without any conventional cooling or heating system. In cold or cloudy climates, a small wood stove or vented natural gas heater in the basement can be used as a backup to heat the air in the convection loop.

On a life cycle cost basis, good passive solar and superinsulated design is the cheapest way to heat a home or small building in regions where ample sunlight is available more than 60% of daylight hours (Figure 20-19). Such a system usually adds 5–10% to the construction cost, but the life cycle cost of operating such a house is 30–40% lower. The typical payback time for passive solar features is 3–7 years. Roof-mounted passive solar water heaters can supply all or most of the hot water for a typical house in suitable areas.

In an **active solar heating system**, collectors absorb solar energy and a fan or a pump supplies part of a building's space-heating or water-heating needs (Figure 20-16, right). Several connected collectors usually are mounted on the roof with an unobstructed exposure to the sun. Some of the heat can be used directly, and the rest can be stored in insulated tanks containing rocks, water, or a heat-absorbing chemical for release as needed. Active solar collectors can also supply hot water. In Colombia, South America, an ecopioneer has spent 30 years building a solar village in an inhospitable wet savanna infested with mosquitoes and with soil so toxic that it could support only tough grass (Individuals Matter, p. 534). Figure 20-20 (p. 534) lists the major advantages and disadvantages of using passive or active solar energy for heating buildings.

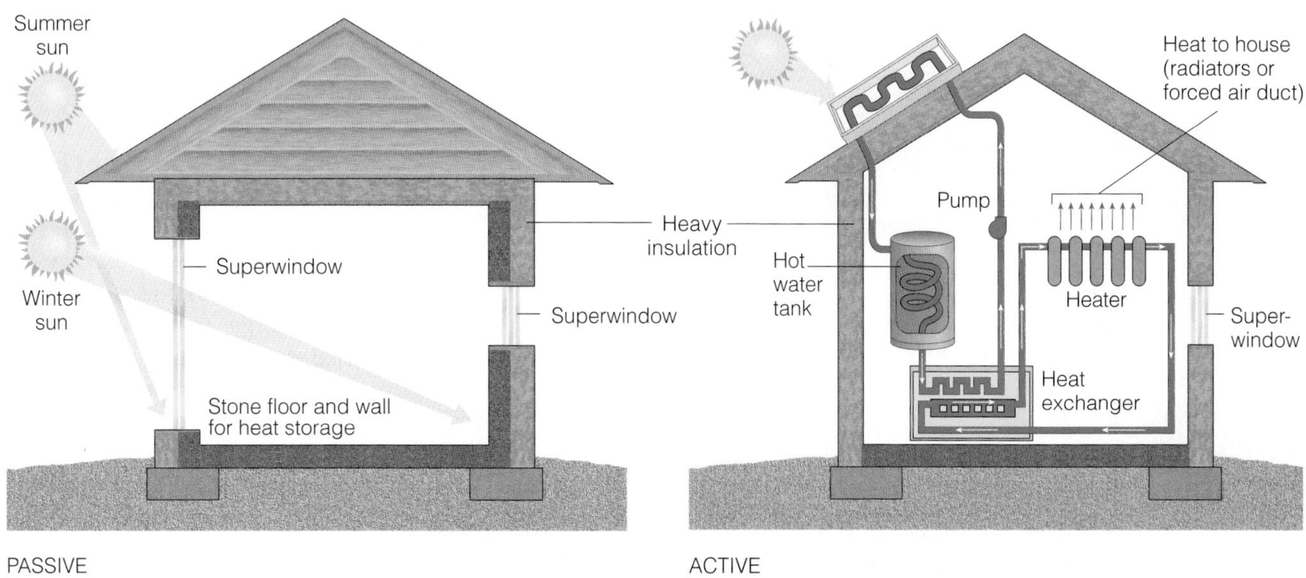

Figure 20-16 Passive and active solar heating for a home.

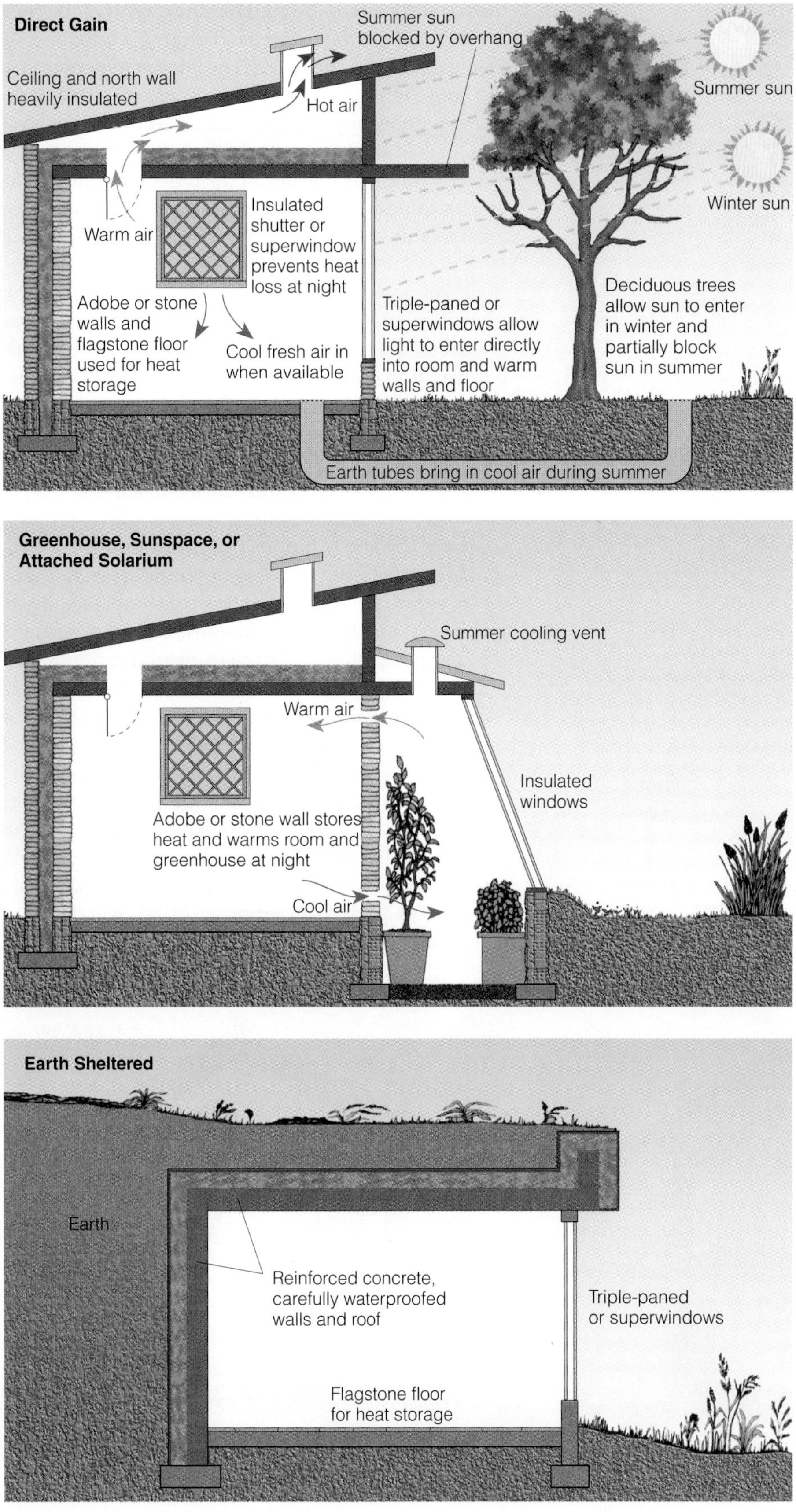

Figure 20-17 Three examples of passive solar design for houses.

How Can We Cool Houses Naturally? Superinsulation and superinsulated windows help make buildings cooler. Passive cooling can also be provided by blocking the high summer sun with deciduous trees, window overhangs, or awnings (Figure 20-17, top). Windows and fans can be used to take advantage of breezes and keep air moving. A reflective insulating foil sheet suspended in the attic can block heat from radiating down into the house. Solar-powered evaporative air conditioners have been developed but so far are too expensive for residential use.

Earth tubes can also be used for indoor cooling (Figure 20-17, top). At a depth of 3–6 meters (10–20 feet), the soil temperature stays at about 5–13°C (41–55°F) all year long in cold northern climates, and about 19°C (67°F) in warm southern climates. Several earth tubes, simple plastic (PVC) plumbing pipes with a diameter of 10–15 centimeters (4–6 inches), buried about 0.6 meter (2 feet) apart and 3–6 meters deep, can pipe cool and partially dehumidified air into an energy-efficient house at a cost of a few dollars per summer.* People allergic to pollen and molds should add an air purification system, but this is also necessary with a conventional cooling system.

How Can We Use Solar Energy to Generate High-Temperature Heat and Electricity? Several so-called *solar thermal* systems collect and transform radiant energy from the sun into high-temperature thermal energy (heat), which can be used directly or converted to electricity (Figure 20-21). In one such *central receiver system*, called a *power tower*, huge

*They work. I used them in a passively heated and cooled office/home for 15 years.

Figure 20-18 Solar envelope house that is heated and cooled passively by solar energy and the earth's thermal energy. This patented Enertia® design developed by Michael Sykes needs no conventional heating or cooling system in most areas. It comes in a precut kit engineered and tailored to the buyer's design goals. Sykes plants 50 trees for each one used in his house kits and his design has received both the Department of Energy's Innovation Award and the North Carolina Governor's Energy Achievement Award. (Enertia® Building Systems, Rt. 1, Box 67, Wake Forest, NC 27587; Web site: http://enertia.com)

arrays of computer-controlled mirrors called *heliostats* track the sun and focus sunlight on a central heat-collection tower (Figure 20-21a).

A government-subsidized power tower system, called Solar Two, began operating in the California desert in 1996. This experimental plant cost about eight times more to build than a coal-fired plant and produces electricity at about twice the cost of a coal-fired plant. Supporters say the construction and operating costs will drop as the technology spreads. Moreover, if the estimated environmental cost of coal-fired electricity is included, the electricity cost from Solar Two is comparable to that from coal.

In a *solar thermal plant* or *distributed receiver system*, sunlight is collected and focused on oil-filled pipes running through the middle of curved solar collectors (Figure 20-21b). This concentrated sunlight can generate temperatures high enough for industrial processes or for producing steam to run turbines and generate electricity. At night or on cloudy days, high-efficiency natural gas turbines supply backup electricity as needed. In California's Mojave Desert, such a solar thermal system with a natural gas turbine backup system has produced power much more cheaply than nuclear power plants (Figure 20-15).

Another type of distributed receiver system uses *parabolic dish collectors* (that look somewhat like TV satellite dishes) instead of parabolic troughs. These collectors can track the sun along two axes and generally are more efficient than troughs. A pilot plant is now being built in northern Australia. The U.S. Department of Energy projects that early in the 21st century parabolic dishes with gas turbine backup should be able to produce electrical power for 6¢ per kilowatt-hour.

Another promising approach for intensifying incoming solar energy about 80,000 times is a *nonimaging optical solar concentrator*. With this technology, the sun's rays are allowed to scramble instead of being focused on a particular point (Figure 20-21c). Because of its high efficiency and ability to generate extremely high temperatures, nonimaging concentrators may make solar energy practical for widespread industrial and commercial use within a decade.

Inexpensive solar cookers can focus and concentrate sunlight and cook food, especially in rural villages in sunny developing countries. They can be made by fitting an insulated box big enough to hold three or four pots, with a transparent, removable top (Figure 20-21d). Solar cookers

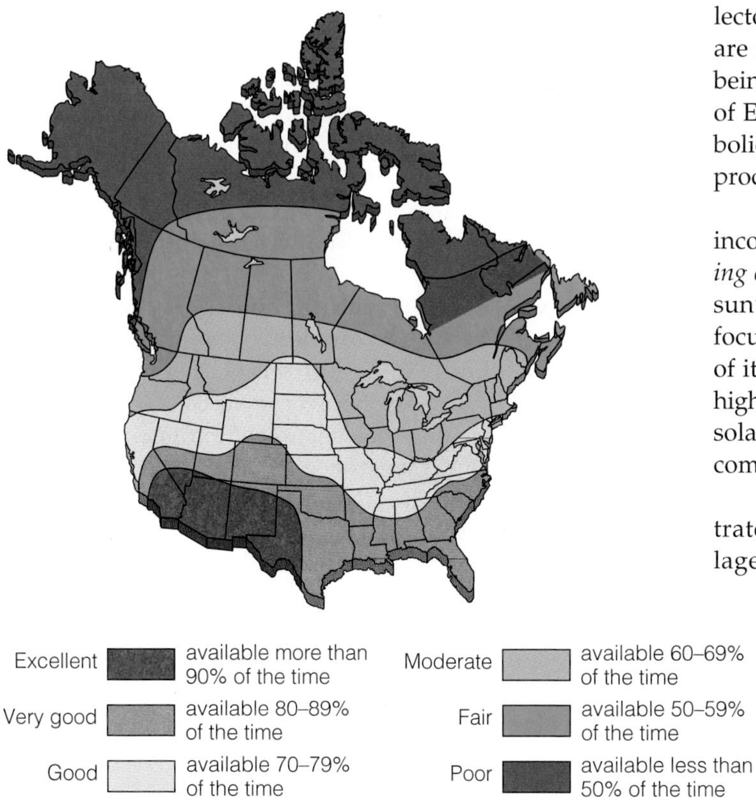

Figure 20-19 Availability of solar energy in the continental United States and Canada. (Data from U.S. Department of Energy and National Wildlife Federation)

Gaviotas: A Solar Village in Colombia*

INDIVIDUALS MATTER

In 1965 ecopioneer and visionary Paolo Lugari began developing an ecologically sustainable solar village called Gaviotas in Columbia east of the Andes.

When he first saw this area he thought that if people can live in this inhospitable area, they can live anywhere. With the help of professors and students from the universities in Bogotá—16 hours away by jeep—he set out to build a sustainable ecovillage.

They learned how to make a soil cement for use in building dams, water pipes for irrigation, and buildings by mixing 14 parts of the area's toxic soil with 1 part cement. They attached water pumps to seesaws powered by kids having a good time.

They invented cheap solar water heaters that could work even on cloudy days and helped finance development of the village by building and selling them to thousands of people in Bogotá. They also produced electricity by designing ultralight windmills to catch the area's mild but steady winds.

They built a village hospital heated by the sun and cooled by the wind for the area's several hundred inhabitants. People in rural areas for hundreds of miles around came there for medical care and sent their children to school there.

The villagers found enough nontoxic soil along the riverbanks to grow mangoes, cassava, and cashews. They got protein from fish in the river and cattle that could eat the area's tough grass.

They planted hundreds of pines, which have grown into forests. The villagers tap gum oozing from the pines and use solar energy to distill it into turpentine and a valuable resin used in paints, cosmetics, and medicines. The huge market for these products provided a new industry for Gaviotas.

Dropping needles from the pines built up the soil, which then spouted with hundreds of new kinds of plants. Now a rain forest that once grew in the area is returning from seeds carried by birds and roots emerging from the river edges.

Members of this ecovillage are willing to teach all comers about what they have learned about living more sustainably by tapping into the earth's sun, water, and wind. Their simple and affordable technologies, purposely unpatented, are spreading throughout many developing countries.

Critical Thinking

Why is it apparently more difficult to develop such sustainable communities in most developed countries? Give three important ways to overcome this problem.

*For more details, see Alan Weisman, *Gaviotas: A Village to Reinvent the World* (White River Junction, Vt.: Chelsea Green Publishing, 1998).

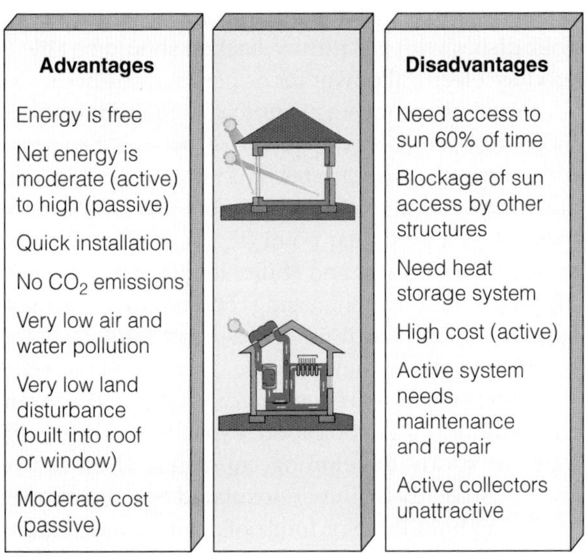

Figure 20-20 Advantages and disadvantages of heating a house with passive or active solar energy.

reduce deforestation for fuelwood, the time and labor needed to collect firewood, and indoor air pollution from smoky fires. Figure 20-22 lists the advantages and disadvantages of concentrating solar energy to produce high-temperature heat or electricity.

How Can We Produce Electricity with Solar Cells? The PV Revolution Solar energy can be converted directly into electrical energy by **photovoltaic (PV) cells**, commonly called **solar cells** (Figure 20-23). Sunlight falling on a solar cell, a transparent silicon wafer thinner than a sheet of paper, releases a flow of electrons when it strikes silicon atoms, creating an electrical current.

Because a single solar cell produces only a tiny amount of electricity, many cells are wired together in a panel. Several panels are wired together to produce the amount of electricity needed. The direct current (DC) electricity produced can be stored in batteries and used directly or converted to conventional alternating current

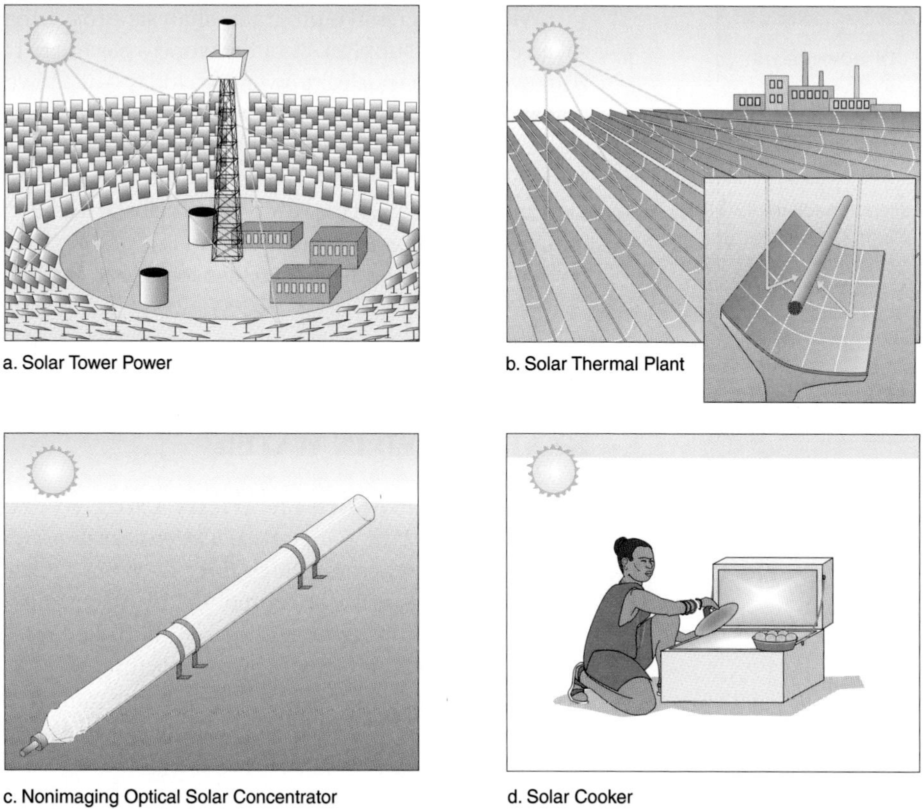

Figure 20-21 Several ways to collect and concentrate solar energy to produce high-temperature heat and electricity are in use. Today, solar plants are used mainly to supply reserve power for daytime peak electricity loads, especially in sunny areas with a high demand for air conditioning.

(AC) electricity. Currently, a separate inverter is used to convert the DC to AC, but new panels being built by Solarex will be equipped with built-in converters.

Traditional-looking solar-cell roof shingles and photovoltaic panels that resemble metal roofs are now available, reducing the cost of solar-cell installations by saving on roof costs. A Japanese company offers a home that comes with a solar-cell roof and can be assembled in 8 hours.

A California company is working with utilities to install prototype solar-cell panels directly over the rubber or plastic roofing membranes of flat-roofed apartment and office buildings. A German company is testing a *solar-electric window* that incorporates solar cells into a semitransparent glazing that simultaneously generates electricity and provides filtered light during daylight hours. Expandable banks of solar cells can also produce electricity at a small power plant, using efficient turbines burning natural gas to provide backup power when the sun isn't shining.

In New York City, the Durst Building at 4 Times Square incorporates green design on a grand scale. This 48-story, $500-million tower uses **(1)** low-E windows, **(2)** energy-efficient lighting, **(3)** PV panels built into the walls on the southern and eastern sides, **(4)** two large fuel cells on the roof burning natural gas to supplement daytime electricity and provide all of the building's electricity at night, **(5)** recycled building materials, and **(6)** separate waste chutes to facilitate recycling by tenants.

Oberlin College in Ohio has recently built a new environmental studies center incorporating major elements of *green design*, including use of passive solar energy and solar cells (Figure 20-24). David Orr (head of Oberlin's Environmental Studies Program, Guest Essay, p. 232) designed the building with the help of more than 250 students, faculty, and townspeople.

In 1999, BP Amoco announced plans to spend $50 million on a pilot plan to reduce CO_2 emissions by using 400 solar panels as a supplemental source of power for 200 of its service stations in 11 countries. This will make BP one of the largest single users of solar power. The company is also installing a solar-cell system at three new office buildings in the United Kingdom and is sponsoring rural solar electric projects in Africa and South America.

By 2005, electricity produced by solar cells could drop to 6–12¢ per kilowatt-hour (compared to about 20¢ per kwh in 1999), making it cost-competitive with most other technologies (Figure 20-15). If new solar-cell technologies can be brought to the marketplace, the production, sale, and installation of photovoltaics (already a $2-billion-per-year business) will be one of the world's largest and fastest-growing businesses. In 1999 researchers announced the development of a new breed of solar cells. These cells are thinner than a human hair and could be created with materials costing only a few pennies. Mass production and technological advances could bring costs down as low as 4¢ per kilowatt-hour by 2020. Figure 20-25 lists the advantages and disadvantages of solar cells.

Solar cells are an ideal technology for providing electricity to 2 billion people in rural areas in most developing countries, which have not committed to expensive large-scale centralized nuclear and fossil fuel power plants and electric-line transmission systems. With financing from the World Bank, India (the world's number-one market for solar cells) is installing solar-cell systems in 38,000 villages, and Zimbabwe is bringing solar electricity to 2,500 villages. Houston-based Enron Corporation announced

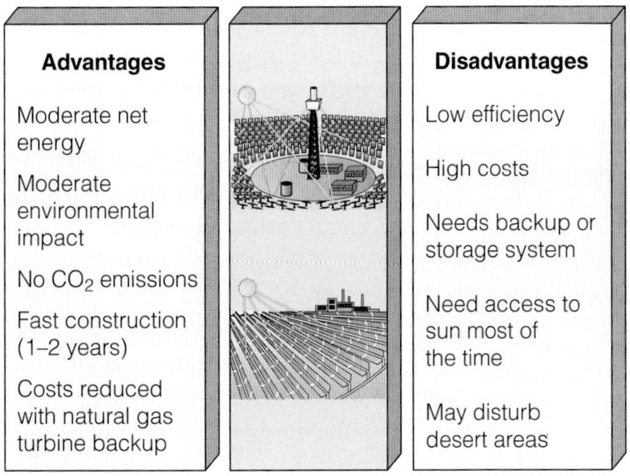

Figure 20-22 Advantages and disadvantages of using solar energy to generate high-temperature heat and electricity. According to the Worldwatch Institute, solar power plants installed on a total land area equal to that of Panama or South Carolina could provide as much electricity as the entire world now uses, but costs may be high.

plans to build large, grid-connected solar photovoltaic power plants in the desert regions of China, India, and the United States. In 1999, Shell Oil (which owns two PV companies) launched a solar electrification project in South Africa that will provide 50,000 homes in impoverished areas with a solar panel and battery.

In 1998, the U.S. Department of Energy launched a program to encourage utilities, the solar industry, and governments to work together to install solar-cell energy systems on 1 million roofs in the United States by 2010. In Japan the government is giving incentives for contractors to build 70,000 residential homes with solar-cell roof tiles.

With an aggressive program, analysts project that solar cells could supply 17% of the world's electricity by 2010—as much as nuclear power does today—at a lower cost and much lower risk. With a strong push from governments and private investors, by 2050 solar cells (which could easily be a $100-billion-a-year business) could provide as much as 25% of the world's electricity (at least 35% in the United States). Currently, Japan is spending three times more on solar-cell research than the United States is.

20-4 PRODUCING ELECTRICITY FROM MOVING WATER AND FROM HEAT STORED IN WATER

How Can We Produce Electricity Using Hydropower Plants? In a *large-scale hydropower project*, a high dam is built across a large river to create a reservoir. Some of the water stored in the reservoir is allowed to flow through huge pipes at controlled rates, spinning turbines and producing electricity.

In a *small-scale hydropower project*, a low dam with no reservoir (or only a small one) is built across a small stream. Output in small systems can vary with seasonal changes in stream flow.

Falling water can also produce electricity in *pumped-storage hydropower systems*. When demand is low (usually at night), pumps using surplus electricity from a conventional power plant pump water from a lake or a reservoir to another reservoir at a higher elevation. When a power company temporarily needs more electricity, water in the upper reservoir is released, flows through turbines, and generates electricity on its return to the lower reservoir.

Hydroelectric power supplies about 20% of the world's electricity (99% in Norway, 75% in New Zealand,

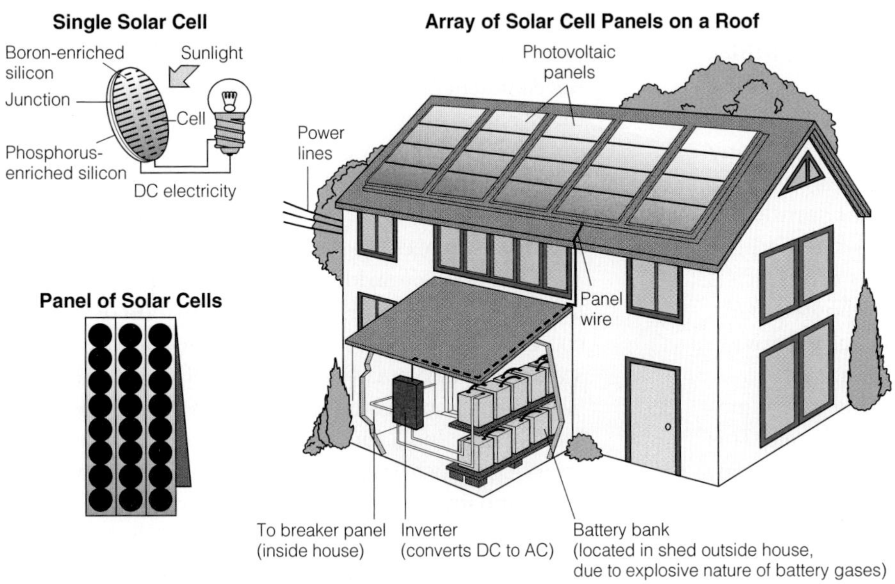

Figure 20-23 Photovoltaic (solar) cells can provide electricity for a house. Small and easily expandable arrays of such cells can provide electricity for urban villages throughout the world without large power plants or power lines. Massive banks of such cells can also produce electricity at a small power plant. As the price of such electricity drops, usage will increase dramatically. The racks of solar-cell panels shown in this diagram can now be incorporated into the roof in the form of solar-cell roof shingles or PV panel roof systems that look like metal roofs.

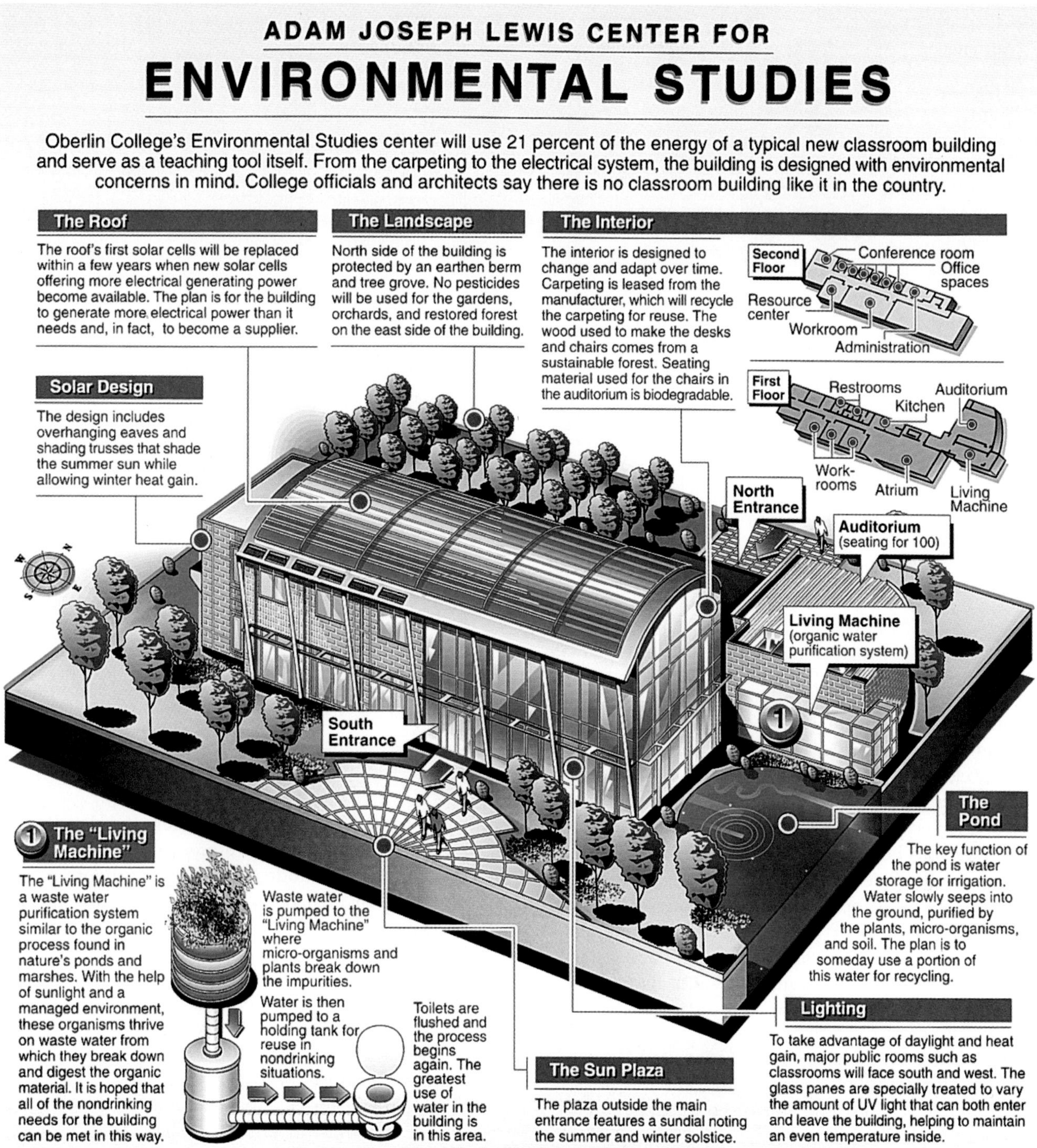

Figure 20-24 Major elements of green design in the Adam Joseph Lewis Center for Environmental Studies at Oberlin College in Ohio. There were no waste products as a result of the construction and no use of toxic materials. The building is the brainchild of David Orr (head of Oberlin's Environmental Studies program) and a number of students, faculty, and townspeople. (Illustration by James Owen. Courtesy of Oberlin College)

Advantages

- Fairly high net energy
- Work on cloudy days
- Quick installation
- Easily expanded or moved
- No CO_2 emissions
- Low environmental impact
- Last 20–30 years
- Low land use (if on roof)

Disadvantages

- Need access to sun
- Sun access can be blocked
- Low efficiency
- Need electricity storage system or backup
- High costs (but should be competitive in 5–15 years)
- DC current must be converted to AC

Figure 20-25 Advantages and disadvantages of using solar cells to produce electricity. Theoretically, all U.S. electricity needs could be met by placing solar cells on only 0.6% of U.S. land (and on only half this area if solar-cell efficiencies increase as projected from about 12% to 21%).

50% in developing countries, 25% in China, and 10% in the United States) and 6% of its total commercial energy (4% in the United States). Figure 20-26 lists the advantages and disadvantages of using hydropower plants to produce electricity.

Because of increasing concern about the environmental and social consequences of large dams (Figure 12-6), there has been growing pressure on the World Bank and other development agencies to stop funding new large-scale hydropower projects. Small-scale projects eliminate most of the harmful environmental effects of large-scale projects, but they can **(1)** threaten recreational activities and aquatic life, **(2)** disrupt the flow of wild and scenic rivers, and **(3)** destroy wetlands.

Is Producing Electricity from Tides and Waves a Useful Option? Twice a day in high and low tides, water that flows into and out of coastal bays and estuaries can spin turbines to produce electricity (Figure 20-27). Two large tidal energy facilities are currently operating, one at La Rance in France and the other in Canada's Bay of Fundy. However, most analysts expect tidal power to make only a tiny contribution to world electricity supplies. There are few suitable sites, and construction costs are high.

The kinetic energy in ocean waves, created primarily by wind, is another potential source of electricity (Figure 20-27). Most analysts expect wave power to make little contribution to world electricity production, except in a few coastal areas with the

Advantages

- Moderate to high net energy
- High efficiency (80%)
- Low cost electricity
- Long life span
- No CO_2 emissions during operation
- May provide flood control below dam
- Provides water for year round irrigation

Disadvantages

- High construction costs
- High environmental impact
- High CO_2 emission from biomass decay in shallow tropical reservoirs
- Floods natural areas
- Converts land habitat to lake habitat
- Danger of collapse
- Uproots people
- Decreases fish harvest below dam
- Decreases flow of natural fertilizer (silt) to land below dam

Figure 20-26 Advantages and disadvantages of using large dams and reservoirs to produce electricity.

right conditions (such as western England). Construction costs are moderate to high and the net energy yield is moderate, but equipment could be damaged or destroyed by saltwater corrosion and severe storms.

How Can We Produce Electricity from Heat Stored in Water? Japan and the United States have been evaluating the use of the large temperature differences (between the cold, deep waters and the sun-warmed surface waters) of tropical oceans for producing electricity. If economically feasible, this would be done in *ocean thermal energy conversion* (OTEC) plants anchored to the bottom of tropical oceans in suitable sites (Figure 20-27). However, most energy analysts believe that the large-scale extraction of energy from ocean thermal gradients may never compete economically with other energy alternatives. Despite 50 years of work, the technology is still in the research-and-development stage.

Saline solar ponds, usually located near inland saline seas or lakes in areas with ample sunlight, can be used to produce electricity (Figure 20-27). Heat accumulated during the day in the denser bottom layer can be used

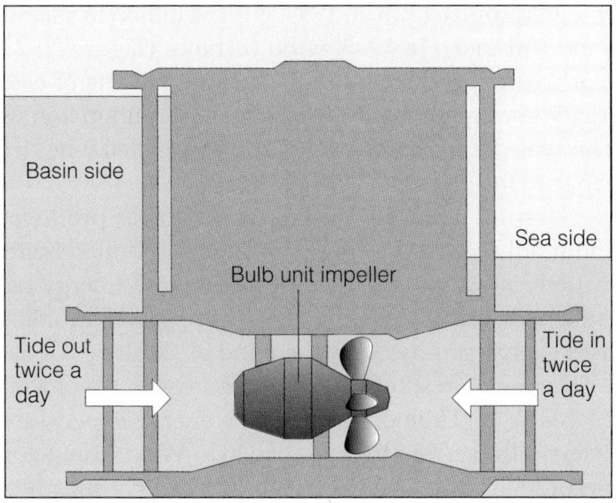

Tidal Power Plant

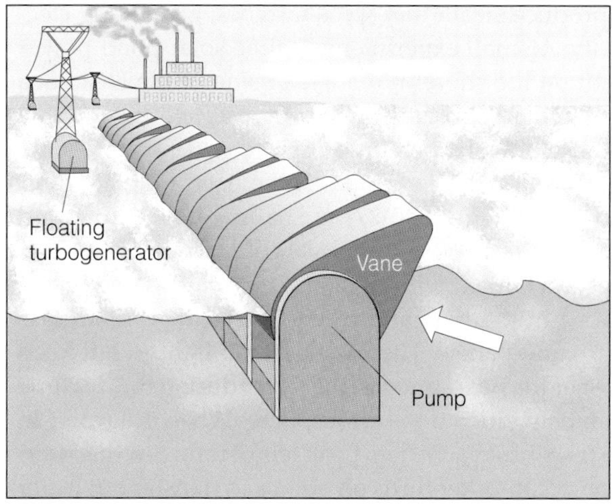

Wave Power Plant

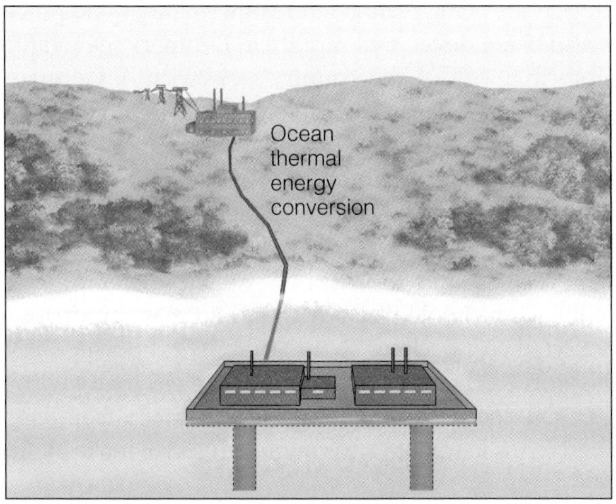

Ocean Thermal Power Plant - Overview

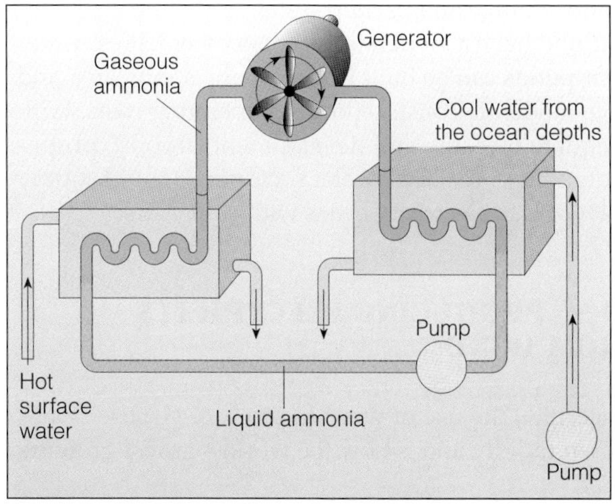

Enlargement of Ocean Thermal Power Plant

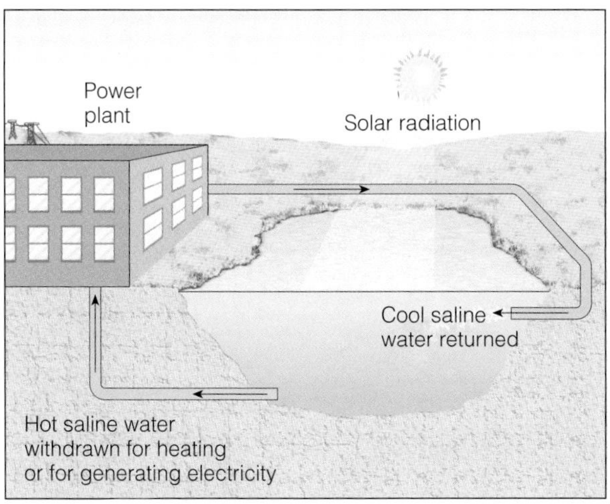

Saline Water Solar Pond

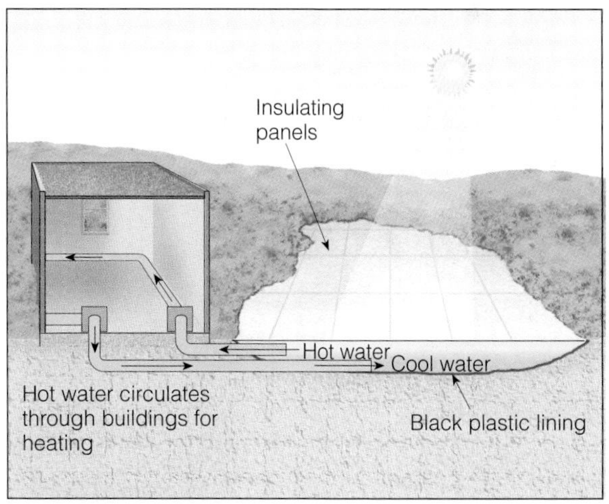

Freshwater Solar Pond

Figure 20-27 Ways to produce electricity from moving water and to tap into solar energy stored in water as heat. None of these systems are expected to be significant sources of energy in the near future.

to produce steam that spins turbines, generating electricity. A small experimental saline solar pond power plant on the shore of the Israeli side of the Dead Sea operated for several years but was closed in 1989 because of high operating costs.

Freshwater solar ponds can be used to heat water and space (Figure 20-27). A shallow hole is dug and lined with concrete. A number of large, black plastic bags, each filled with several centimeters of water, are placed in the hole and then covered with fiberglass insulation panels. The panels let sunlight in but keep most of the heat stored in the water during the daytime from being lost to the atmosphere. When the water in the bags has reached its peak temperature in the afternoon, a computer turns on pumps to transfer hot water from the bags to large, insulated tanks for distribution.

Saline and freshwater solar ponds require no energy storage and backup systems, emit no air pollution, and have a moderate net energy yield. Freshwater solar ponds can be built in almost any sunny area and have moderate construction and operating costs. With adequate research and development support, proponents believe that freshwater solar ponds could supply 3–4% of U.S. electricity needs within 10 years.

20-5 PRODUCING ELECTRICITY FROM WIND

Since 1980 the use of wind to produce electricity has grown rapidly and is now the world's fastest-growing energy resource (35% in 1998, with $2 billion in sales of wind turbines). In 1998, wind turbines (Figure 20-28) worldwide produced about 9,600 megawatts of electricity, enough to meet the needs of 3.4 million homes. This was double the capacity in 1995 and 960 times the capacity in 1980.

Despite its rapid growth, wind power produced about only 1% of the energy used in the United States in 1999. However, the U.S. Department of Energy has launched a program designed to have 5% of the country's energy produced by the wind by 2020.

Figure 20-29 shows the potential areas for use of wind power in the United States, which energy expert John Berger calls the "Saudi Arabia of wind." Wind farms (clusters of 20–100 wind turbines, Figure 20-28, right and photo on p. 495) provide about 1.5% of California's electricity—enough to power about 300,000 homes. Sizable wind-farm projects are being planned in 12 other states. Wind power is also being developed rapidly in Denmark (generating more than 8% of the country's electricity), Germany, Spain, and India (the world's number-two market for wind energy); China is a potential wind superpower.

Wind power (with much lower government subsidies) has a significant cost advantage over nuclear power (Figure 20-15) and at 4–6¢ per kilowatt-hour has become competitive with coal-fired power plants in many places. With new technological advances and mass production, projected cost declines should make wind power one of the world's cheapest ways to produce electricity within the next decade. In the long run, electricity from large wind farms in remote areas might be used to make hydro-

Figure 20-28 Wind turbines can be used to produce electricity individually or in clusters called wind farms. In 1999, wind turbines (many of them in automated wind farms, photo on p. 495) produced about $1.2 billion worth of electricity, enough to meet the needs of 3.4 million homes.

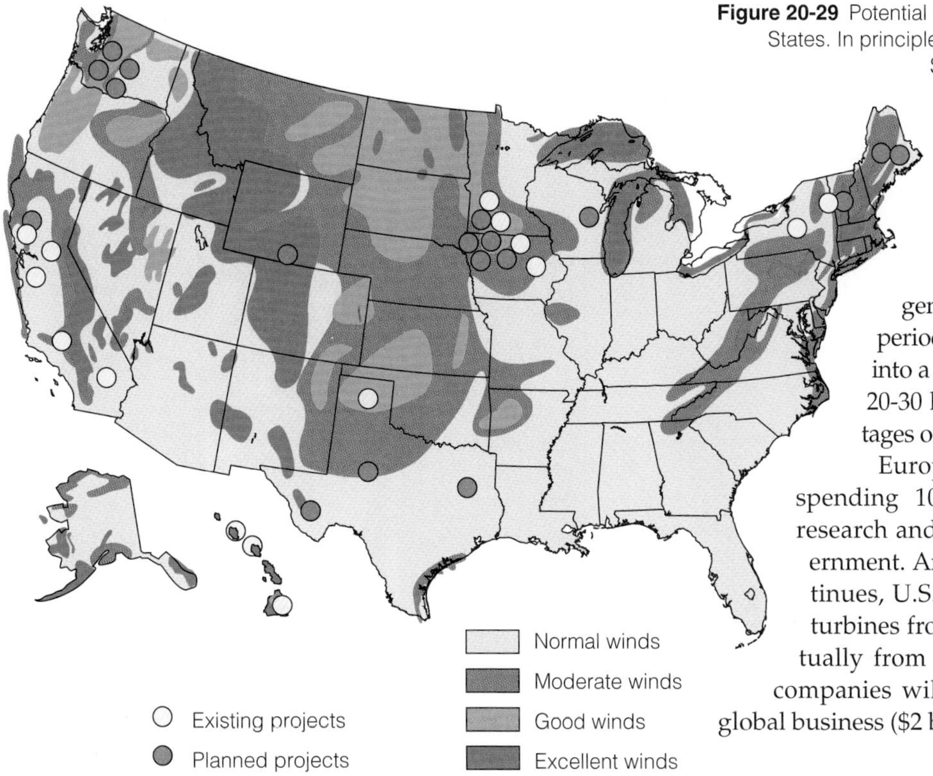

Figure 20-29 Potential for use of wind power in the United States. In principle, all the power needs of the United States could be provided by exploiting the wind potential of just three states: North Dakota, South Dakota, and Texas. (Data from U.S. Department of Energy)

gen gas from water during off-peak periods. The hydrogen could then be fed into a pipeline and storage system. Figure 20-30 lists the advantages and disadvantages of using wind to produce electricity.

European governments are currently spending 10 times more for wind energy research and development than the U.S. government. Analysts warn that if this trend continues, U.S. consumers will be buying wind turbines from European countries (and eventually from China and India) and American companies will lose out on a rapidly growing global business ($2 billion in 1998).

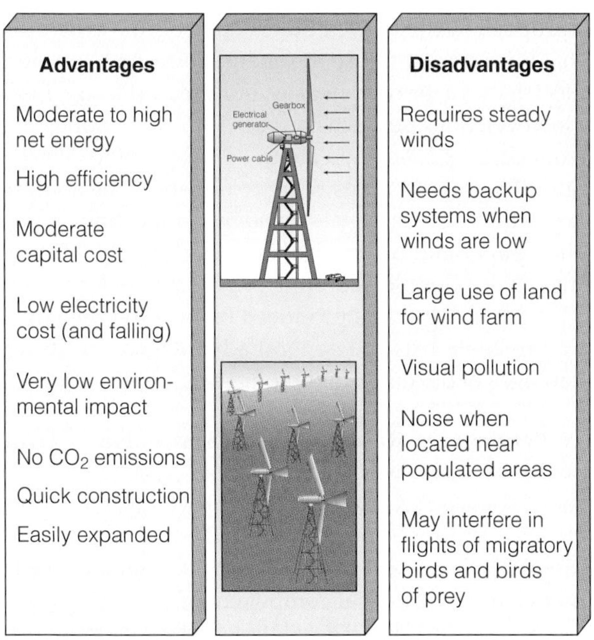

Figure 20-30 Advantages and disadvantages of using wind to produce electricity. Wind power experts project that by 2050 wind power could supply more than 10% of the world's electricity and 10–25% of the electricity used in the United States.

20-6 PRODUCING ENERGY FROM BIOMASS

How Useful Is Burning Solid Biomass? Many forms of *biomass*—organic matter in plants produced through photosynthesis—can be burned directly as a solid fuel or converted into gaseous or liquid **biofuels** (Figure 20-31).

The burning of wood and manure to heat buildings and cook food supplies about 11% of the world's energy (4–5% in Canada and almost 3% the United States) and about 35% of the energy used in developing countries. Almost 70% of the people living in developing countries heat their homes and cook their food by burning wood or charcoal. However, about 2.7 billion people in developing countries cannot find (or are too poor to buy) enough fuelwood to meet their needs (Figure 17-16). Sweden leads the world in burning wood as an energy source, mostly for district heating plants.

Biomass can also be burned to drive turbines and produce electricity. About 2% of the electricity in the United States is produced from biomass—twice as much as from wind power. The U.S. government has a goal of tripling the use of biomass energy from 3% to 9% of the country's energy use between 1999 and 2010.

One way to produce biomass fuel is to plant, harvest, and burn large numbers of fast-growing trees (especially cottonwoods, poplars, sycamores, willows, and leucaenas), shrubs, perennial grasses, and water hyacinths in *biomass*

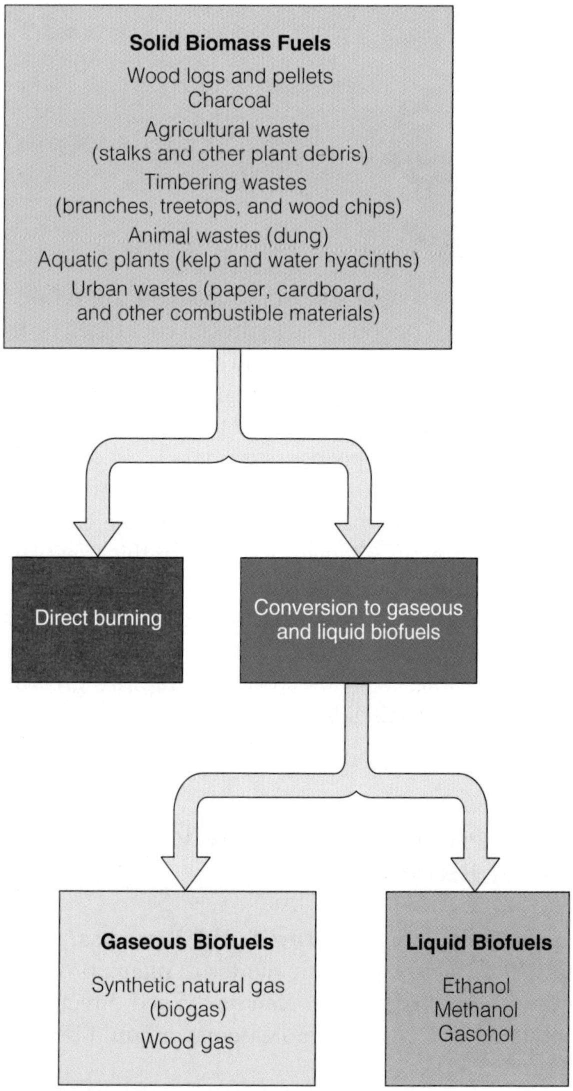

Figure 20-31 Principal types of biomass fuel.

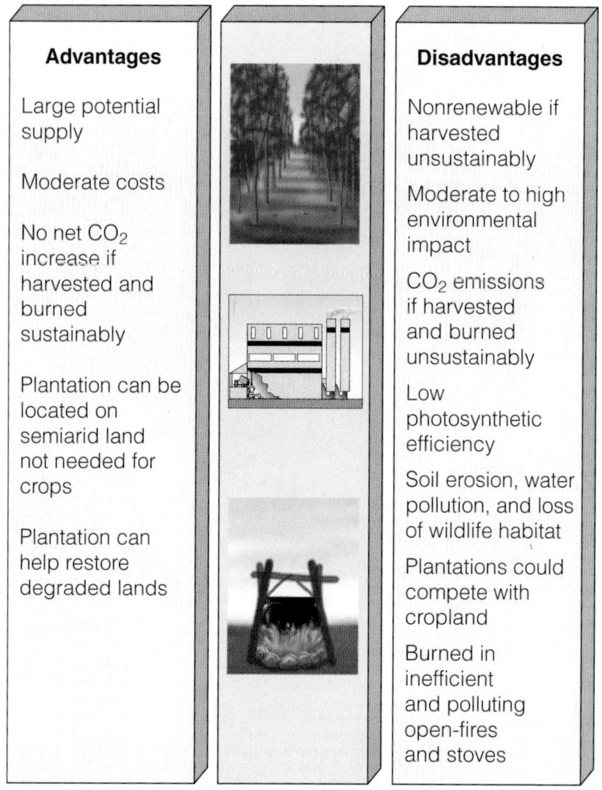

Figure 20-32 General advantages and disadvantages of burning solid biomass as a fuel.

plantations. However, if the plants are harvested faster than they are replenished, this approach can cause soil erosion and water pollution and reduce wildlife habitats.

In agricultural areas, crop residues (such as sugarcane residues, rice husks, cotton stalks, and coconut shells) and animal manure can be collected and burned or converted into biofuels. Burning *bagasse*—the residue left after sugarcane harvesting and processing—supplies about 10% of the electricity in Hawaii and in Brazil.

This approach makes sense when crop residues are burned near areas where the residues are produced. Otherwise, it takes too much energy to collect, dry, and transport the residues to power plants. Some ecologists argue that it makes more sense to use animal manure as a fertilizer and to use crop residues to feed livestock, retard soil erosion, and fertilize the soil.

An increasing number of cities in Japan, western Europe, and the United States have built incinerators that burn trash and use the energy released to produce electricity or to heat nearby buildings (Figure 14-7). However, this approach has been limited by opposition from citizens concerned about emissions of toxic gases and disposal of the resulting toxic ash. Some analysts argue that more energy is saved by composting or recycling paper and other organic wastes than by burning them. For example, recycling paper saves at least twice as much energy as is produced by burning the paper. Figure 20-32 lists the general advantages and disadvantages of burning solid biomass as a fuel.

Is Producing Gaseous and Liquid Fuels from Solid Biomass a Useful Option? Various forms of solid biomass can be converted by bacteria and various chemical processes into gaseous and liquid biofuels (Figure 20-31). Examples include **(1)** *biogas*, a mixture of 60% methane (the principal component of natural gas) and 40% carbon dioxide, **(2)** *liquid ethanol* (ethyl, or grain, alcohol), and **(3)** *liquid methanol* (methyl, or wood alcohol).

In China, anaerobic bacteria in more than 6 million *biogas digesters* convert plant and animal wastes into methane fuel for heating and cooking. These simple devices can be built for about $50 including labor. After

INDIVIDUALS MATTER

A New Car Fuel That Is Mostly Water

It seems too good to be true, but a breakthrough fuel that is more than half water could soon power cars, trucks, buses, farm machinery, train locomotives, generators, boilers in power plants, and aircraft running on gasoline, diesel fuel, or fuel oil.

This new, milky-looking fuel, called A-55, was developed by Reno, Nevada, inventor Rudolf Gunnerman. It consists of 55% water and 45% naphtha (a cheap-to-produce by-product of petroleum distillation, Figure 19-6). It also contains **(1)** lubricants, **(2)** antirust and antifreeze agents, and **(3)** small amounts of a blending agent developed by Gunnerman (which attracts the normally repellent water and oily naphtha to one another).

In 1994 Gunnerman and diesel equipment giant Caterpillar formed Advanced Fuels, a joint venture to test and market the new fuel. So far in every test the new fuel tops conventional gasoline and diesel fuel as a clean, safe, and cheap fuel that can be used in almost any internal combustion engine.

In 1995, Nevada certified A-55 as a clean alternative fuel. A-55 is especially attractive to the trucking industry, facing federal crackdowns on dirty diesel engine emissions.

Since 1996 the fuel has been used successfully in a city bus and a power generator in Reno, in generators at Caterpillar's Illinois plants, and in public and private vehicles in Sacramento, California. In 1997, the federal government approved A-55 for legal purchase in the United States and stated that engine modifications for its use are legal as long as they don't boost air pollution emissions.

Tests so far show that using A-55 cuts emissions of nitrogen oxides (NO_x) by 50% and particulate emissions by up to 50%. A-55 is virtually immune to fire and explosions and can be stored in aboveground tanks without risking explosions. This makes it a much safer fuel for school buses and urban transit buses.

Converting a car to use A-55 is a simple $300 job involving adjusting fuel injectors and timing. Modifying a truck or bus to use the fuel takes 2 hours and costs less than $1,000. Once that's done, the vehicle could use either the water-based fuel or conventional gasoline or diesel fuels.

Car and diesel truck drivers could see their fuel costs cut in half. That's because the refining process for producing naphtha is faster and cheaper than producing gasoline and diesel fuel. Using naphtha instead of gasoline and diesel fuel eliminates up to 90% of the air pollutants produced by oil refineries.

Gunnerman came to the United States from Germany in 1949 at age 21 with $20 in his pocket. After several failed ventures he became a multimillionaire from an environmentally friendly invention that converted wood waste into wood pellets to fuel wood stoves.

Instead of retiring in luxury, Gunnerman (who turned 71 in 1999) has used most of his $25-million fortune to develop a clean, cheap fuel and help leave the planet in better shape for future generations. You can find out the latest news about this fuel by visiting the company's Web site at www.a-55.com/power.html.

the biogas has been separated, the solid residue is used as fertilizer on food crops or, if contaminated, on trees. When they work, biogas digesters are very efficient. However, they are also slow and unpredictable, a problem that could be corrected with development of more reliable models.

According to the U.S. Department of Energy, gasifying biomass and burning it in advanced turbines could produce electricity costing 4.5¢ per kilowatt-hour by 2010. Shell Oil projects that by 2010 biomass could provide 5% of the world's electricity (worth $20 billion).

Some analysts believe that liquid ethanol and methanol produced from biomass could replace gasoline and diesel fuel when oil becomes too scarce and expensive. *Ethanol* can be made from sugar and grain crops (sugarcane, sugar beets, sorghum, and corn) by fermentation and distillation. Gasoline mixed with 10–23% pure ethanol makes *gasohol*, which can be burned in conventional gasoline engines and is sold as super unleaded or ethanol-enriched gasoline.

Another alcohol, *methanol*, is made mostly from natural gas but can also be produced at a higher cost from wood, wood wastes, agricultural wastes (such as corncobs), sewage sludge, garbage, and coal. The advantages and disadvantages of using ethanol, methanol, and several other fuels as alternatives to gasoline are summarized in Table 10-4. Recently a new water-based gasoline fuel has entered the picture and is currently undergoing extensive testing (Individuals Matter, above).

Projections about the future role of biomass energy vary widely, with this resource providing 14–40% of the world's energy by 2050 depending on assumptions about biomass gasification, use in energy-efficient fuel cells, and establishment of biomass plantations on degraded forestland and idle cropland. Whether such a projection becomes reality depends on **(1)** the availability of large areas of productive land and adequate water (resources that may be in short supply in coming decades) and **(2)** the ability to minimize the harmful environmental effects of large-scale production of biomass. Biomass is also limited by low photosynthetic efficiency.

Currently, much of the world's potentially renewable biomass is being exploited in nonrenewable and unsustainable ways. According to a 1997 analysis by David Pimentel and two other researchers, "Large-scale biofuel production is not an alternative to the current use of oil and is not even an advisable option to cover a significant fraction of it."

20-7 THE SOLAR–HYDROGEN REVOLUTION

What Can We Use to Replace Oil? Good-bye Oil and Smog, Hello Hydrogen When oil is gone (or when what's left costs too much to use), what will we use to fuel vehicles, industry, and buildings? Some scientists say the fuel of the future is hydrogen gas (H_2) (Table 10-4).

When hydrogen gas burns in air, it combines with oxygen gas in the air and produces nonpolluting water vapor and some nitrogen oxides. This eliminates most of the air pollution problems we face today and greatly reduces the threats from global warming.

There is very little hydrogen gas around, but we can get it from something we have plenty of: *water.* Water can be split by electricity into gaseous hydrogen and oxygen (Figure 20-33).

What's the Catch? If you think using hydrogen as an energy source sounds too good to be true, you're right. Several problems must be solved to make hydrogen one of our primary energy resources, but scientists are making rapid progress in finding solutions.

One problem is that it takes energy (and thus money) to produce this marvelous fuel. Hydrogen can be produced by reacting natural gas and steam to form carbon dioxide and hydrogen or by gasifying coal, but the cost is about twice as much as producing gasoline with the same fuel value. We could use electricity from coal-burning and nuclear power plants to split water and produce hydrogen, but this subjects us to the harmful environmental effects associated with using these fuels (Chapter 19), and it costs more than the hydrogen fuel is worth.

Most proponents of hydrogen gas believe that if we are to get its very low pollution benefits, the energy to produce the gas from water must come from the sun, in the form of electricity generated by sources such as hydropower, solar thermal and solar-cell power plants, and wind farms. If scientists and engineers can learn how to use sunlight to decompose water cheaply enough, they will set in motion a *solar–hydrogen revolution* over the next 50 years and change the world as much as the agricultural and industrial revolutions did. An important step in this direction occurred in 1998 when researchers at the U.S. National Renewable Energy Laboratory created a combined *photovoltaic–photoelectrochemical cell* that uses sunlight to split water into hydrogen and oxygen at an efficiency of 12.4%, a spectacular advance according to prominent researchers in this field.

Currently, using solar energy to produce hydrogen gas is too costly, but the costs of using solar energy to produce electricity are coming down. Moreover, if the health and environmental costs of using gasoline were

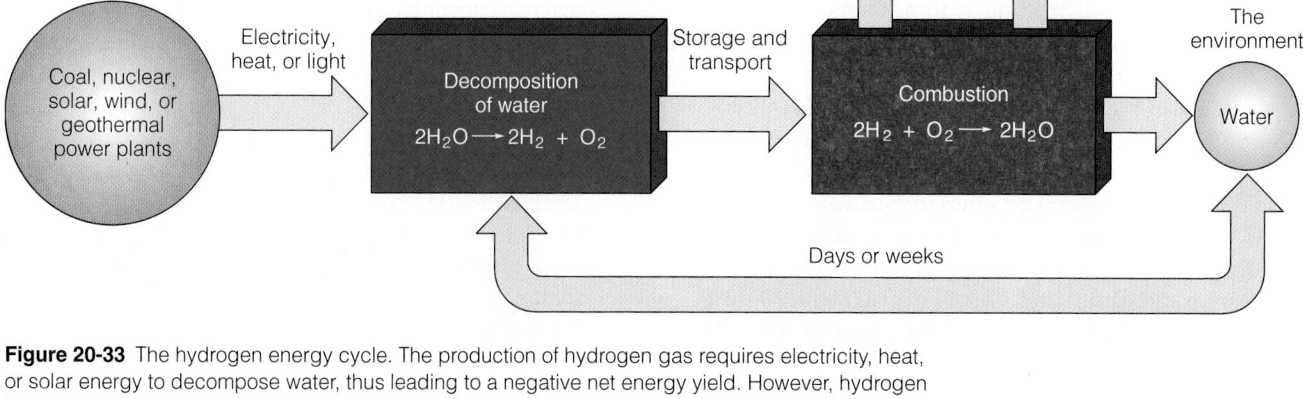

Figure 20-33 The hydrogen energy cycle. The production of hydrogen gas requires electricity, heat, or solar energy to decompose water, thus leading to a negative net energy yield. However, hydrogen is a clean-burning fuel that can replace oil, other fossil fuels, and nuclear energy. Using solar energy to produce hydrogen from water could also eliminate most air pollution and greatly reduce the threat of global warming.

included in its market prices (through gasoline taxes), it would cost at least $1 per liter ($4 per gallon), roughly the price in Japan and in many European countries. At this price, hydrogen produced by any form of solar energy would be competitive.

Hydrogen for vehicles could initially be produced from natural gas. Blends of natural gas and hydrogen produced from solar sources could then be phased in gradually as reserves of natural gas fuel become depleted. Hydrogen produced by solar-cell power plants (in sunny, mostly desert areas) could be carried by pipeline wherever it was needed.

Hydrogen gas can be stored in a pressurized tank, metal powders, and activated carbon that absorbs gaseous hydrogen and releases it when heated for use as a fuel. Unlike gasoline, compounds and powders containing absorbed hydrogen will not explode or burn if a vehicle's tank is ruptured in an accident. However, it's difficult to store enough hydrogen gas in a car as a compressed gas or a metal powder for it to run very far, a problem similar to the one electric cars face today. Scientists and engineers are seeking solutions to this problem.

Another possibility is to power a car with a *fuel cell* (Figure 20-4 and 20-10) in which hydrogen and oxygen gas combine to produce electrical current. Fuel cells produce no air pollution and have high energy efficiencies 65–95%—several times the efficiency of conventional gasoline-powered engines and electric cars.

A number of prototype fuel-cell systems for cars, buses, homes, and buildings are being tested and evaluated. Currently, fuel cells are expensive and heavy, and they produce energy at a cost equivalent to gasoline at $2.50 per gallon, but this could change with more research and mass production. Several prototype buses in Chicago, Illinois, are running on hydrogen.

Two U.S. companies are developing residential fuel cells that use methane to produce hydrogen fuel for the cell. These units are about the size of a dishwasher, should cost less than $4,000, and could be on the market by 2002. A single unit could supply all the heating, cooling, cooking, refrigeration, and electrical needs of a home and provide hydrogen fuel for one or more cars at an affordable price.

In 1997, a panel of technology experts projected that fuel cells could supply as much as 30% of the world's electricity by 2017. In 1999, DaimlerChrysler and Shell announced plans to turn the tiny country of Iceland into the world's first "hydrogen economy," eventually replacing the gasoline and diesel engines on all of its cars, buses, and fishing vessels with hydrogen. Figure 20-34 lists the pros and cons of using hydrogen as an energy resource.

What's Holding Up the Solar–Hydrogen Revolution? Politics and economics, not a lack of promising technology, are the main factors holding up a more rapid transition to a solar–hydrogen age. Phasing out dependence on fossil fuels and phasing in new solar–hydrogen technologies over the next 40–50 years involves convincing investors and energy and automobile companies with strong vested interests in fossil fuels to risk a lot of capital on hydrogen.

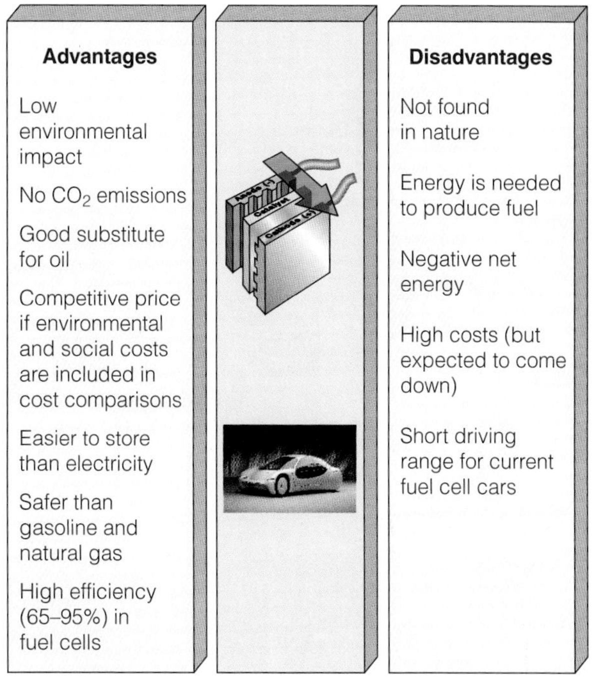

Figure 20-34 Advantages and disadvantages of using hydrogen as a fuel for vehicles and for providing heat and electricity.

It also involves convincing governments to put up some of the money for developing hydrogen energy (as they have done for decades for fossil fuels and nuclear energy). Currently the U.S. government spends about 60 times more on petroleum research and development than on hydrogen research and development.

20-8 GEOTHERMAL ENERGY

How Can We Tap the Earth's Internal Heat? Going Underground Heat contained in underground rocks and fluids is an important source of energy. Over millions of years, this **geothermal energy** from earth's mantle (Figure 13-2) has been transferred to underground concentrations of **(1)** *dry steam* (steam with no water droplets), **(2)** *wet steam* (a mixture of steam and water droplets), and **(3)** *hot water* trapped in fractured or porous rock at various places in earth's crust. All three types are currently being used to heat space and water, and in some cases to produce electricity (Figure 19-2).

If such geothermal sites are close to the surface, wells can be drilled to extract the dry steam, wet steam

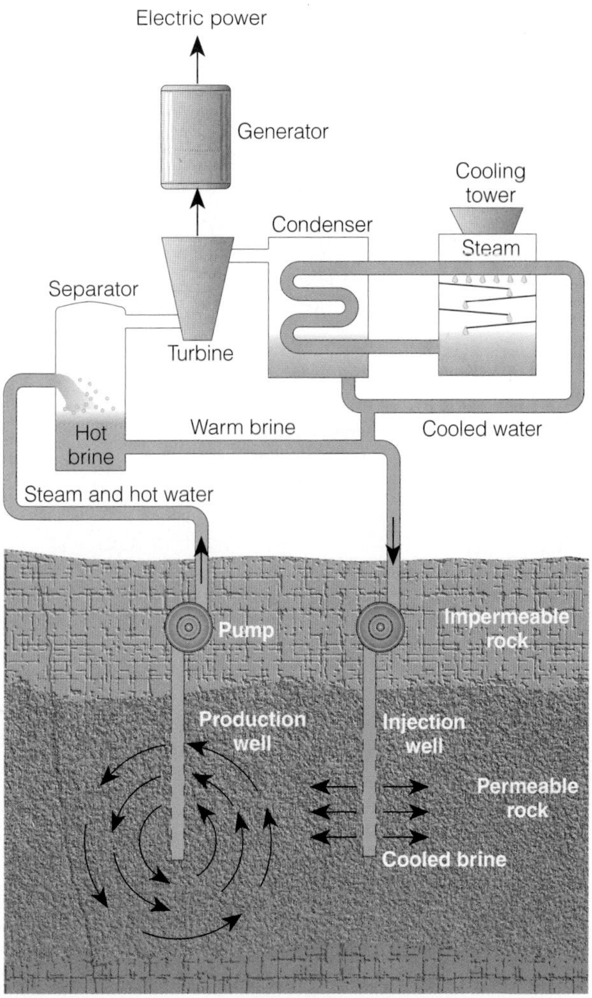

Figure 20-35 Tapping the earth's heat or geothermal energy in the form of wet steam to produce electricity.

(Figure 20-35), or hot water. This thermal energy can be used for space heating and to produce electricity or high-temperature heat for industrial processes.

Geothermal reservoirs containing dry steam, wet steam, or hot water can be depleted if heat is removed faster than it is renewed by natural processes. Thus, geothermal resources can be nonrenewable on a human time scale, but the potential supply is so vast that it is usually classified as a potentially renewable energy resource.

Currently, about 22 countries (most of them in the developing world) are extracting energy from geothermal sites to produce about 1% of the world's electricity. The United States accounts for 44% of the geothermal electricity generated worldwide, with most of the favorable sites in California, Hawaii, Nevada, and Utah. The Philippines is the world's second largest user, followed by Mexico. Reykjavik, Iceland, whose skies were once blackened by chimney smoke from coal burning, now gets most of its heat from geothermal energy and is considered to be one of the world's cleanest cities.

The world's largest operating geothermal system is a steam-driven complex called The Geysers, north of San Francisco, California. Its first geothermal well was drilled in 1924. Now the area contains 26 power plants producing enough power for a city of 1.3 million people. However, heat is being withdrawn from this geothermal site at about 80 times faster than it is being replenished, converting this potentially renewable resource to a nonrenewable source of energy. In 1999 Santa Monica, California, became the first city in the world to get all of its electricity from renewable energy in the form of geothermal energy.

Three other virtually nondepletable or potentially renewable sources of geothermal energy are **(1)** *molten rock* (magma), **(2)** *hot dry-rock zones*, where molten rock that has penetrated the earth's crust heats subsurface rock to high temperatures, and **(3)** low- to moderate-temperature *warm-rock reservoir deposits*, which could be used to preheat water and run heat pumps for space heating and air conditioning. Research is being carried out to see whether hot dry-rock zones, which can be found almost anywhere if one can drill deep enough, can provide affordable geothermal energy.

Figure 20-36 lists the pros and cons of using geothermal energy. Currently, the cost of tapping geothermal energy is too high for all but the most concentrated and accessible sources.

Advantages	Disadvantages
Very high efficiency	Scarcity of suitable sites
Moderate net energy at accessible sites	Depleted if used too rapidly
Lower CO_2 emissions than fossil fuels	CO_2 emissions
Low cost at favorable sites	Moderate to high local air pollution
Moderate environmental impact	Noise and odor (H_2S)

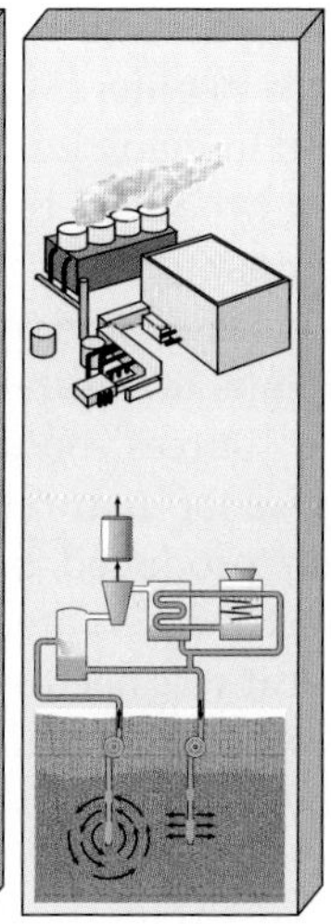

Figure 20-36 Advantages and disadvantages of using geothermal energy for space heating and to produce electricity or high-temperature heat for industrial processes.

Kammen, D. M. 1999. Bringing Power to the People. *Environment,* v41 i5 p10(1).

20-9 SOLUTIONS: A SUSTAINABLE ENERGY STRATEGY

What Are the Best Energy Alternatives? We have a variety of nonrenewable and renewable energy resources, each with certain advantages and disadvantages. Many scientists and energy experts who have evaluated these energy alternatives have come to the following general conclusions:

- *The best alternatives are a combination of improved energy efficiency and greatly increased use of locally available renewable energy resources.*
- *Future energy alternatives probably will have low-to-moderate net energy yields and moderate to high development costs.*
- *Because there is not enough financial capital to develop all energy alternatives, projects must be chosen carefully.*
- *We cannot and should not depend mostly on a single nonrenewable energy resource such as oil, coal, natural gas, or nuclear power.*

What Role Does Economics Play in Energy Resource Use? Governments use three basic economic and political strategies to stimulate or dampen the short-term and long-term use of a particular energy resource.

The first approach is *allowing all energy resources to compete in a free market without any government interference.* This is rarely politically feasible because of well-entrenched government intervention into the marketplace in the form of subsidies, taxes, and regulations. Furthermore, the free-market approach, with its emphasis on short-term profit, inhibits development of new energy resources, which can rarely compete economically in their early stages without government support.

The second approach is *keeping energy prices artificially low to encourage use of selected energy resources*. In the United States (and most other countries), the energy marketplace is distorted by huge government subsidies and tax breaks that make the prices of fossil fuels and nuclear power artificially low and help perpetuate the use of these energy resources, even when better and less costly alternatives are available (Figure 20-15). This creates an uneven economic playing field that **(1)** encourages energy waste and rapid depletion of a nonrenewable energy resource and **(2)** discourages the development of energy alternatives that are not getting at least the same level of subsidies and tax breaks.

For example, the U.S. fossil fuel industry gets almost $20 billion a year in government subsidies, compared to only about $200 million for renewable energy. In developing countries the fossil fuel industry receives subsidies of at least $101 billion per year, compared to extremely small subsidies for renewable energy.

The third approach is *keeping energy prices artificially high to discourage use of a resource.* Governments can raise the price of an energy resource by withdrawing existing tax breaks and other subsidies or by adding taxes on its use. This **(1)** increases government revenues, **(2)** encourages improvements in energy efficiency, **(3)** reduces dependence on imported energy, and **(4)** decreases use of an energy resource that has a limited future supply.

Many economists favor *increasing taxes on fossil fuels* as a way to reduce air and water pollution and slow global warming. The tax revenues would be used to **(1)** improve energy efficiency, **(2)** encourage use of renewable energy resources, and **(3)** provide energy assistance to the poor and lower-middle-class. Some economists believe that the public might accept these higher taxes if income or payroll taxes were lowered as gasoline or other fossil fuel taxes were raised. What do you think?

How Can We Develop a More Sustainable Energy Future? Analysts have suggested a variety of strategies for making the transition to a more sustainable energy future based on improving **(1)** energy efficiency, **(2)** increasing use of renewable energy, and **(3)** decreasing the harmful environmental and health effects of using energy (Figure 20-37).

Energy experts estimate that implementing a mix policies such as those shown in Figure 20-37 over the next two decades could **(1)** save money, **(2)** create a net gain in jobs, **(3)** slow projected global warming, and **(4)** sharply reduce air and water pollution. Some actions you can take toward achieving a more sustainable energy future are listed in Appendix 6.

The chief feature of the *sustainable energy revolution* over the next 40–50 years is likely to be significant *decentralization,* analogous to the computer industry's shift from large, centralized mainframes to increasingly smaller, widely dispersed PCs.

Two-way communication systems using fiber-optic cables and microprocessors will allow utility companies to monitor and control individual fuel cells, solar-cell rooftop systems, wind turbines, cogenerators, air conditioners, and water heaters of their customers. These control systems can **(1)** turn such devices on or off (to save energy), **(2)** withdraw and buy excess power from customers, and **(3)** sell them power as needed.

Energy analysts project that by the year 2050 the decentralized global energy system is likely to be based mostly on improved energy efficiency and much greater use of renewable forms of energy such as wind power, solar cells, fuel cells, cogenerators, and solar-produced hydrogen. Companies and investors trying

Hint: Enter the search term *alternative energy* using the Subject Guide.

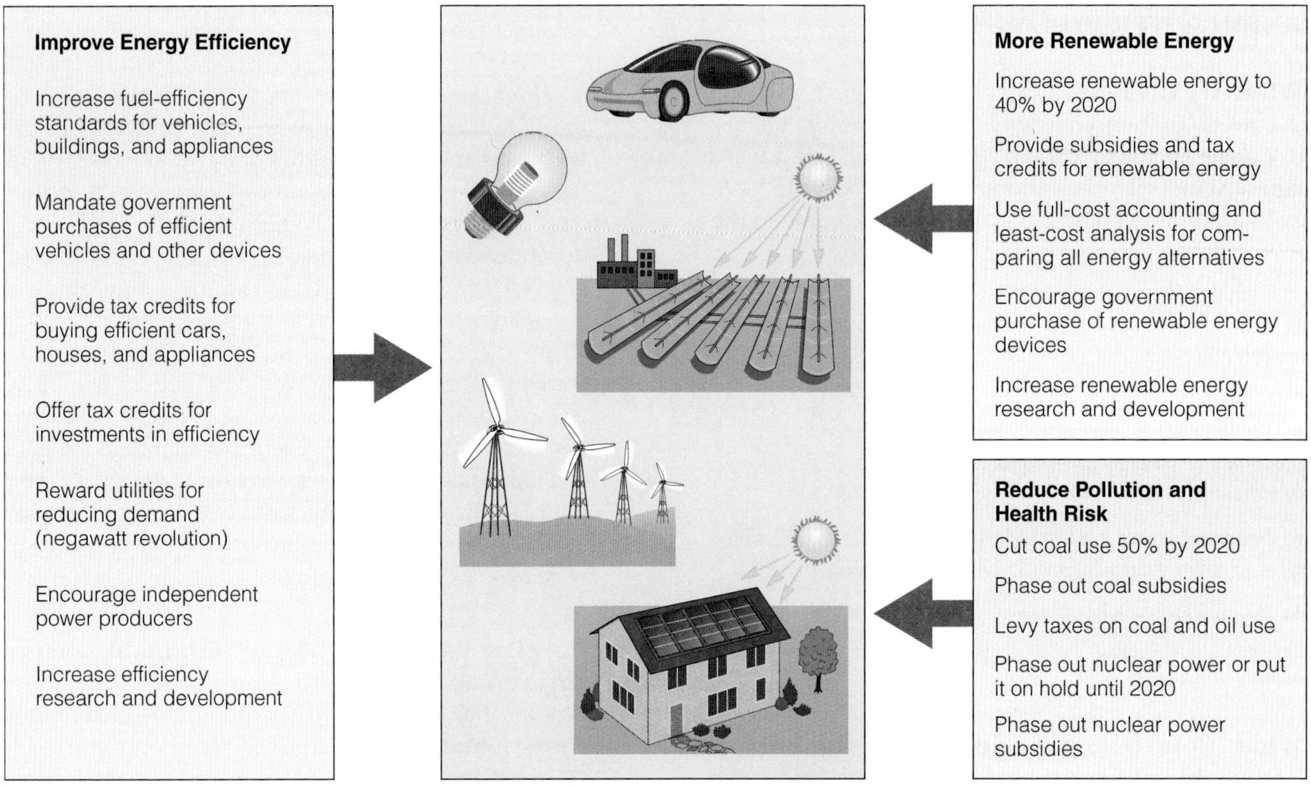

Figure 20-37 Solutions. Some ways various analysts have suggested to attain a more sustainable energy future.

to preserve large, centralized coal-burning and nuclear power plants may find themselves saddled with expensive technological dinosaurs.

In the long run, humanity has no choice but to rely on renewable energy. No matter how abundant they seem today, eventually coal and uranium will run out.

DANIEL DEUDNEY AND CHRISTOPHER FLAVIN

REVIEW QUESTIONS

1. Define the boldfaced terms found in this chapter.

2. How much of the energy used in the United States is wasted? What percentage of this is wasted because of the second law of energy and what percentage is wasted unnecessarily? What is *energy efficiency*? What is *life cycle cost*? What are three of the least efficient energy-using devices?

3. Explain why we can't recycle energy. List three ways to slow down the flow of heat from **(a)** a house and **(b)** an office building.

4. What are the advantages of saving energy?

5. What is *cogeneration*, and how efficient is it compared to producing electricity by a conventional coal-burning or nuclear power plant? List four other ways to save energy in industry.

6. What do most experts believe is the best way to save energy in transportation?

7. List the pros and cons of using **(a)** battery-powered electric cars, **(b)** hybrid cars, and **(c)** fuel-cell cars.

8. Describe how we can save energy in houses by using **(a)** superinsulated houses and **(b)** straw-bale houses. What are the four most efficient ways to heat a house? Describe ways to make an existing house more energy efficient. What are the most efficient and least efficient ways to heat water for washing and bathing?

9. List three reasons why there is little emphasis on saving energy in the United States, despite its important benefits.

10. List five major benefits of relying more on renewable energy.

11. Distinguish between a *passive solar heating system* and an *active solar heating system* and list the pros and cons of each.

12. Describe three ways to cool houses naturally.

13. Distinguish between the following solar systems used to generate high-temperature heat and electricity: **(a)** power tower, **(b)** solar thermal plant, **(c)** parabolic dish collection system, **(d)** nonimaging optical solar concentrator, and **(e)** solar stoves. List the advantages and disadvantages of concentrating solar energy to produce high-temperature heat or electricity.

14. What is a *solar cell*? List the advantages and disadvantages of using solar cells to produce electricity.

15. List the advantages and disadvantages of using hydropower to produce electricity.

16. List the advantages and disadvantages of using the following systems for storing heat in water to produce electricity: **(a)** ocean thermal energy conversion (OTEC), **(b)** saline solar ponds, and **(c)** freshwater solar ponds.

17. List the advantages and disadvantages of using wind to produce electricity.

18. List the advantages and disadvantages of **(a)** burning solid biomass as a source of energy and **(b)** producing gaseous and liquid fuels from solid biomass.

19. What is the *solar–hydrogen revolution*? List the advantages and disadvantages of using hydrogen as a source of energy. What's holding up the solar–hydrogen revolution?

20. List the advantages and disadvantages of using geothermal energy to produce heat and electricity.

21. What are four conclusions energy experts have reached about possible future energy alternatives?

22. Summarize the three different economic approaches that can be used to stimulate or dampen the use of a particular energy resource. Which approach do you favor?

23. What are major ways to help make the transition to a more sustainable energy future?

CRITICAL THINKING

1. A home builder installs electric baseboard heat and claims that "it's the cheapest and cleanest way to go." Apply your understanding of the second law of energy (thermodynamics) to evaluate his claim.

2. Someone tells you that we can save energy by recycling it. How would you respond?

3. What are the five most important things an individual can do to save energy at home and in transportation (see Appendix 6)? Which, if any, of these do you currently do? Which, if any, do you plan to do?

4. Should the United States (or the country where you live) institute a crash program to develop solar photovoltaic cells and solar-produced hydrogen fuel? Explain.

5. Congratulations. You have just won $150,000 to build a house of your choice anywhere in the country. What type of house would you build? Where would you locate it? What types of materials would you use? What types of materials would you be sure *not* to use? How would you heat and cool your house? How would you heat your water? Considering fuel and energy efficiency, what sort of lighting, stove, refrigerator, washer, and dryer would you use? Which of these appliances could you do without? Would you use a dishwasher? A garbage disposal? A trash compactor? Suppose you learn that it will cost $200,000 to build the type of energy-efficient, passive solar house you want. Considering lifetime energy costs, would it make sense for you to borrow $50,000 to complete your house or to save the $50,000 by cutting back on energy-saving and solar design features?

6. Explain why you agree or disagree with the following proposals by various energy analysts:

a. Federal subsidies for all energy alternatives should be eliminated so that all energy choices can compete in a true free-market system.

b. All government tax breaks and other subsidies for conventional fuels (oil, natural gas, coal), synthetic natural gas and oil, and nuclear power (fission and fusion) should be removed and replaced with subsidies and tax breaks for improving energy efficiency and developing solar, wind, geothermal, and biomass energy alternatives.

c. Development of solar and wind energy should be left to private enterprise and receive little or no help from the federal government, but nuclear energy and fossil fuels should continue to receive large federal subsidies.

d. To solve present and future U.S. energy problems, all we need to do is find and develop more domestic supplies of oil, natural gas, and coal, and increase dependence on nuclear power.

e. A heavy federal tax should be placed on gasoline and imported oil used in the United States.

f. By 2020, the United States should phase out all nuclear power plants.

7. Explain why you agree or disagree with the proposals suggested in Figure 20-37 by various analysts as ways to promote a more sustainable energy future.

8. Congratulations. You have just been put in charge of the world. List the five most important features of your energy policy.

***9.** Make an energy use study of your school or dorm and use the findings to develop an energy-efficiency improvement program. Present your plan to school officials.

***10.** Make a concept map of this chapter's major ideas, using the section heads and subheads and the key terms (in boldface). Look at the inside back cover and in Appendix 5 for information about concept maps.

FURTHER INFORMATION

For more information see Internet Sources and Further Readings at the back of the book. Online resources for this chapter can be found at **www.brookscole.com/biology**

APPENDIX 1

PUBLICATIONS, ENVIRONMENTAL ORGANIZATIONS, AND FEDERAL AND INTERNATIONAL AGENCIES

PUBLICATIONS

The following publications can help you keep well informed and up-to-date on environmental and resource problems. Subscription prices, which tend to change, are not given.

American Biology Teacher Journal of the National Association of Biology Teachers, 11250 Roger Bacon Dr., Rm. 319, Reston, VA 22090; Tel: 703-471-1134

American Forests American Forestry Association, P.O. Box 2000, Washington, DC 20013; Tel: 202-955-4500, E-mail: member@amfor.org, Web: www.amfor.org

American Journal of Alternative Agriculture 9200 Edmonston Rd., Suite 117, Greenbelt, MD 20770; Web: www.haw-iaa.org/ajaa.htm

Amicus Journal Natural Resources Defense Council, 40 W. 20th St., New York, NY 10011; Tel: 212-727-2700, Web: www.igc.apc.org/nrdc/nrdc/nrdc/eami-cus/back/maga.html

Annual Review of Energy Department of Energy, Forrestal Building, 1000 Independence Ave. SW, Washington, DC 20585; Web: www.eia.doe.gov/emeu/plugs/plaer96.html

Audubon National Audubon Society, 700 Broadway, New York, NY 10003-9501; Tel: 212-979-3000, Fax: 212-979-3188, Web: www.magazine.audubon.org

BioCycle Journal of Waste Recycling Biocycle, 419 State Ave., Emmaus, PA 19049; Tel: 610-967-4135, E-mail: Biocycle@jgpress.com, Web: www.grn.com/news/home/biocycle/index-t.html

BioScience American Institute of Biological Sciences, Central Station, P.O. Box 27417, Washington, DC 20077-0038; Tel: 202-628-1500 or 800-992-2427, Fax: 202-628-1509, E-mail: bioscienc@aibs.org, Web: www.aibs.org/bio-sciencemagazine/index.html

Conservation Biology Blackwell Scientific Publications, Commerce Place, 350 Maine St., Malden, MA 02148-5018; Tel: 888-661-5800, Fax: 617-388-8255, Web: www.conbio.rice.edu/scb/journal

Conservation Ecology Ecological Society of America, 2010 Massachusetts Ave., Suite 400 NW, Washington, DC 20036; Tel: 202-833-8773, Fax: 202-833-8775, E-mail: ques-tions@consecol.org, Web: www.consecol.org/Journal

Demographic Yearbook Department of International Economic and Social Affairs, Statistical Office, United Nations Publishing Service, United Nations, New York, NY 10017; Tel: 617-253-2889, Fax: 617-258-6779

Discover 114 Fifth Ave., New York, NY 10011-5690; Tel: 800-829-9132, E-mail: let-ters@discover.com, Web: www.discover.com

Earth Island Journal Earth Island Institute, 300 Broadway, Suite 28, San Francisco, CA 94133-3312; Tel: 415-788-2666, Fax: 415-788-7324, E-mail: earthisland@earthisland.org, Web: www.earthisland.org/eijournal/jour-nal.html

The Ecologist MIT Press Journals, 55 Hayward St., Cambridge, MA 02142; Web: www.gn.apc.org/ecologist

Ecology Ecological Society of America, Dr. Duncan T. Patten, Center for Environmental Studies, Arizona State University, Tempe, AZ 85281; Tel: 602-965-3979

EcoNet Institute for Global Communication, 18 De Boom St., San Francisco, CA 94107; Tel: 415-422-0220, Web: www.igc.org/igc/econet

E Magazine 28 Knight St., Norwalk, CT 06881; Tel: 203-854-5559, Fax: 203-866-0602, E-mail: axgm65a@prodigy.com, Web: www.emagazine.com

Endangered Species Update School of Natural Resources, University of Michigan, Ann Arbor, MI 48109-1115; Tel: 313-763-3243, Fax: 313-936-2195, E-mail: jfwatson@umich.edu, Web: www.eelink.umich.edu/ESupdate/ESupda te.about.html

Energy Efficiency and Renewable Energy Clearinghouse P.O. Box 3048, Merrifield, VA 22116; Tel: 800-363-3732, Fax: 1-703-893-0400, E-mail: doe.erec@nciinc.com, Web: www.eren.doe.gov/consumerinfo

Environment Heldref Publications, 1319 Eighteenth St. NW, Washington, DC 20036-1802; Tel: 800-365-9753

Environment Abstracts Congressional Information Services, Inc., 4520 East[[endash]]West Hwy., Bethesda, MD 20814-3389; Tel: 800-638-8380 or 301-654-1550, Fax: 301-654-4033 (in most libraries)

Environmental Abstracts Annual Bowker A & I Publishing, 121 Chanlon Rd., New Providence, NJ 07974

Environmental Communicator North American Association for Environmental Education, Suite 400, 1255 23rd St. NW, Washington, DC 20037; Tel: 202-884-8912, Fax: 202-202-884-8701, E-mail: ashotkin@aed.org, Web: www.eelink.umich.edu/COMMUNICATOR

Environmental Ethics Center for Environmental Philosophy, P.O. Box 310980, University of North Texas, Denton, TX 76203-0980; Tel: 940-565-2727, Fax: 565-4439, E-mail: ee@unt.edu, Web: www.cep.unt.edu/enethics.html

Environmental Health Perspectives: Journal of the National Institute of Environmental Health Sciences Government Printing Office, Washington, DC 20402; Web: www.ehpnet1.niehs.nih.gov/docs/month-ly.html

Environmental News Web Sites **(1)** *CNN-Nature*: www.cnn.com/NATURE/; *Earth Times Daily Edition*: www.earthtimes.org/cur-rentedition.htm; **(2)** *Earthvision News Page*: www.204.255.211.112/ColdFusion/news_top 10.cfm; *EnviroLink News Service*: www.envi-rolink.org/environews; *Enviro-Newsbrief (EPA)*: www.epa.gov/natlibra/hqirc/enb/-current.htm; **(3)** *Environmental News Network*: www.enn.com/features; **(4)** *Lycos Environment News*: www.ens.lycos.com; **(5)** *MSNBC*: www.msnbc.com/news/ENVI-RONMENT_Front.asp; **(6)** *Reuters World Environment News (Planet Ark)*: www.plane-tark.org/dailynewshome.cfm; **(7)** *Tiempo Global Warming Newswatch*: www.cru.uea.ac.uk/tiempo/-newswatch/index.htm

Everyone's Backyard Center for Health, Environment, and Justice, P.O. Box 6806, Falls Church, VA 22216; Tel: 703-237-2249, Fax: 703-237-8389, E-mail: CCHW@essen-tial.org, Web: www.essential.org.cchw

Greenpeace Magazine Greenpeace USA, 1436 U St. NW, Washington, DC 20009

Green Teacher Box 1431, Lewiston, NY 14092; Tel: 416-960-1244, Fax: 416-925-3474, E-mail: greentea@web.net, Web: web.net/~greentea/address.html

Hydrogen and Fuel Cell Letter Grinnell St., P.O. Box 14, Rhinecliff, NY 12574-0014; Tel: 914-876-5988, Fax: 914-876-7599, E-mail: hfcletter@mhv.net, Web: www.ttcorp.com/nha/thl/index.htm

International Environmental Affairs University Press of New England, 171/2 Lebanon St., Hanover, NH 03755

International Rivers Network 1847 Berkeley Way, Berkeley, CA 94703; Tel: 510-848-1155, Fax: 510-848-1008, E-mail: irn@irn.org, Web: www.irn.org

International Wildlife National Wildlife Federation, 8925 Leesburg Pike, Vienna, VA

22184; Tel: 703-790-4000, Web: www.207.87.15.60/nwf/intlwild

Issues in Science and Technology National Academy of Sciences, 2101 Constitution Ave. NW, Washington, DC 20077-5576; Tel: 213-883-6325, Web: www.utdallas.edu/research/issues

Journal of the American Public Health Association Subscriptions Dept. 5037, Washington, DC 20061-5037; Tel: 202-789-5662, Fax: 202-789-5661, Web: www.apha.org/news/publications/journal/subscribejourn.html

Journal of Forestry Society of American Foresters, 5400 Grosvenor Lane, Bethesda, MD 20814; Tel: 301-897-8720, Fax: 301-897-8720, E-mail: safweb@safnet.org, Web: www.safnet.org/market/storeperiodicals.htm

Journal of Pesticide Reform NCAP Office, 1249 Willamette St., Eugene, OR 97401; Tel: 541-344-5044, Fax: 514-344-6923, E-mail: info@pesticide.org, Web: www.efn.org/~ncap

Journal of Soil and Water Conservation Soil and Water Conservation Society, 7515 NE Ankeny Rd., Ankeny, IA 50021; Tel: 515-289-2331 or 800-843-7645, Web: www.swcs.org

Journal of Sustainable Agriculture 10 Alice St., Binghampton, NY 13904-1580; Web: www.bubl.ac.uk/journals/agr/jsusagr/index.html

Journal of Wildlife Management Wildlife Society, 5410 Grosvenor Lane, Bethesda, MD 20814; Tel: 301-897-9770, Web: www.wildlife.org/publications.html

National Geographic National Geographic Society, P.O. Box 2895, Washington, DC 20077-9960; Tel: 202-857-7000, Web: www.nationalgeographic.com/ngm/index.html

National Library for the Environment National Institute for the Environment Online Library. Web: www.cnie.org

National Parks National Parks and Conservation Association, 1776 Massachusetts Ave. NW, Washington, DC 20036; Tel: 800-NAT-PARKS or 202-223-6722, Fax: 202-659-0650, E-mail: npca@npca.org, Web: www.npca.org/magazine

National Wildlife National Wildlife Federation, 1400 16th St. NW, Washington, DC 20036; Tel: 202-790-4000, Web: www.207.87.15.60/nwf/natlwild

Nature Subscriptions Dept., P.O. Box 5055, Brentwood, TN 37024-5055; Tel: 800-524-0384, Fax: 615-377-0525, E-mail: subscriptions@natureny.com, Web: www.nature.com/home.html

Nature Conservancy Nature Conservancy, 1815 N. Lynn St., Arlington, VA 22209; Tel: 703-841-5300, Fax: 703-841-1283, Web: www.tnc.org/magazine/main.html

New Bottom Line: Strategic Perspectives on Business and Environment (online journal) Gil Friend and Associates, 48 Shattuck Square, Suite 103, Berkeley, CA 94704; Tel: 510-548-7904, Fax: 510-849-2341, E-mail: gfriend@eco-ops.com, Web: www.iisd.ca/business/jump.asp?cur-page=naturalconcept.htm&ref=http://www.eco-ops.com/eco-ops/nbl/index.subject.html

The New Scientist 1151 Wardour St., London, W1V 4Bn, UK; Tel: +44(0)171 331 2701, Web: www.newscientist.com/home.html

Nucleus Union of Concerned Scientists, Two Brattle Square, Cambridge, MA 02238-9105; Tel: 617-547-5552, Fax: 617-864-9405, E-mail: ucs@ucsusa.org, Web: www.ucsusa.org/publications/index.html

Organic Gardening Rodale Press, Inc., 33 E. Minor St., Emmaus, PA 18049; Web: www.organicgardening.com

Our Planet United Nations Environment Programme (U.S.), 2 United Nations Plaza, Room 803, New York, NY 10017

Pesticides and You National Coalition Against the Misuse of Pesticides, 701 E St. SE, Suite 200, Washington, DC 20003; Tel: 202-543-5450, Fax: 202-543-4791, E-mail: ncamp@ncamp.org, Web: www.ncamp.org/newsletter.html

Pollution Abstracts Cambridge Scientific Abstracts, 7200 Wisconsin Ave., Bethesda, MD 20814 (in many libraries)

Population Bulletin Population Reference Bureau, 1875 Connecticut Ave. NW, Suite 520, Washington, DC 20009; Tel: 202-483-1100, Web: www.prb.org

Rachel's Environment and Health Weekly Environmental Research Foundation, P.O. Box 5036, Annapolis, MD 21403-7036; Tel: 410-263-1584, Fax: 410-293-8944, E-mail: erf@rachel.org

Rocky Mountain Institute Newsletter 1739 Snowmass Creek Rd., Snowmass, CO 81654-9199; Tel: 970-927-3851, Fax: 970-927-3420, E-mail: emalia@rmi.org, Web: www.rmi.org/newsletters/currentnewsletter/index.html

Science American Association for the Advancement of Science, 1200 New York Ave. NW, Washington, DC 20005; Tel: 202-326-6400, Web: www.aaas.org/#anchor442146

Science News Science Service, Inc., 1719 N St. NW, Washington, DC 20036; Tel: 202-785-2255, Web: www.sciencenews.org

Scientific American 415 Madison Ave., New York, NY 10017-1111; Tel: 800-333-1199, Fax: 712-755-7118, Web: www.ciam.com/marketplace/mrktsubscription.html#link1

SEJournal Society of Environmental Journalists, 7904 Germantown Ave., Philadelphia, PA 19118; Tel: 215-247-9710, Web: www.sej.org/sejournal/index.html

Sierra 85 Second St., San Francisco, CA 94105; Tel: 415-977-5500, Fax: 415-977-5799, E-mail: information@sierraclub.org, Web: www.sierraclub.org

Solar Today 2400 Central Ave., G-1, Boulder, CO 80301; Tel: 303-443-3130, Fax: 303-443-3212, E-mail: ases@ases.org, Web: www.ases.org/solartoday/index.html

State of the Environment OECD Publications and Information Center, 1750 Pennsylvania Ave., Suite 1207, Washington, DC 20006 (published annually); Web: www.electrade.gfi.fr/cgi-bin/OECDBookShop.storefront

State of the World Worldwatch Institute, 1776 Massachusetts Ave. NW, Washington, DC 20036-1904; Tel: 202-452-1999, Fax: 202-296-7365 (published annually)

Statistical Yearbook Department of International Economic and Social Affairs, Statistical Office, United Nations Publishing Service, United Nations, New York, NY 10017

Technology Review P.O. Box 489, Mount Morris, IL 61504; Tel: 800-877-5320, Fax: 815-734-1127, E-mail: trsubscriprions@mit.edu, Web: www.techreview.com

Vital Signs: The Trends That Are Shaping Our Future Worldwatch Institute, 1776 Massachusetts Ave. NW, Washington, DC 20036-1904; Tel: 202-452-1999, Fax: 202-296-7365, Web: www.worldwatch.org (published annually)

Waste Not Work on Waste USA, 82 Judson, Canton, NY 13617

Wild Earth Cenozoic Society, Inc., P.O. Box 455, Richmond, VT 05477; Tel: 802-434-4077

Wilderness The Wilderness Society, 900 17th St. NW, Washington, DC 20006-2596; Tel: 202-833-2300, Web: www.wilderness.org

Wildlife Conservation New York Zoological Society, 185th St. & Southern Blvd., Bronx, NY 10460; Tel: 718-220-5100, Fax: 718-220-7114

World Rainforest Report Rainforest Action Network, 3221 Pine St., Suite 500, San Francisco, CA 94104; Tel: 415-398-4404, Fax: 415-389-2732, Web: www.ran.org

World Resources World Resources Institute, 10 G St. NE, Suite 800, Washington, DC 20002; Tel: 202-729-7600, Fax: 202-729-7610, Web: www.wri.org (published every 2 years)

World Rivers Review International Rivers Network, 1847 Berkeley Way, Berkeley, CA 94703; Tel: 510-848-1155, Fax: 510-848-1008, E-mail: irn@igc.apc.org

World Watch Worldwatch Institute, 1776 Massachusetts Ave. NW, Washington, DC 20036-1904; Tel: 202-452-1999, Fax: 202-296-7365, E-mail: wwpub@worldwatch.org, Web: www.worldwatch.org/mag/index.html

Worldwatch Papers Worldwatch Institute, 1776 Massachusetts Ave. NW, Washington, DC 20036-1904; Tel: 202-452-1999, Fax: 202-296-7365, Web: www.worldwatch.org

YES: A Journal of Positive Futures P.O. Box 10818, Bainbridge Island, WA 98110-0818; Tel: 800-937-4451, Fax: 206-842-5208, Web: www.futurenet.org/new.html

ENVIRONMENTAL AND RESOURCE ORGANIZATIONS

For a more detailed list of national, state, and local organizations, see **(1)** *Conservation Directory* (published annually by the National Wildlife Federation, 1400 16th St. NW, Washington, DC 20036), **(2)** *Your Resource Guide to Environmental Organizations* (Irvine, CA: Smiling Dolphin Press,

1991), **(3)** *National Environmental Organizations* (published annually by U.S. Environmental Directories, Inc., P.O. Box 65156, St. Paul, MN 55165), and **(4)** *World Directory of Environmental Organizations* (published by the California Institute of Public Affairs, P.O. Box 10, Claremont, CA 91711). Also see Environmental Organizations Web site (www.econet.apc.org/econet/en.orgs.html).

Adopt-a-Stream Foundation Northwest Stream Center, 600 12th St. SE, Everett, WA 98208-6353; Tel: 424-316-8592, Fax: 424-338-1423, E-mail: aasf@streamkeeper.org, Web: www.streamkeeper.org

African Wildlife Foundation 1400 Sixteenth St. NW, Suite 120, Washington, DC 20036; Tel: 202-939-3333, Fax: 939-3332, E-mail: awfwash@igc.apc.org, Web: www.awf.org/home.html

Alliance for Chesapeake Bay 6600 York Rd., Baltimore, MD 21212; Tel: 410-377-6270, Fax: 410-377-7144, E-mail: mail@acb-online.org, Web: www.acb-online.org/index.htm

Alliance to Save Energy 1200 18th St. NW, Suite 900, Washington, DC 20036; Tel: 202-857-0666, E-mail: info@ase.org, Web: www.ase.org

American Association for the Advancement of Science 1200 New York Ave. NW, Washington, DC 20005; Tel: 202-326-6400, Web: www.aaas.org

American Cetacean Society P.O. Box 1391, San Pedro, CA 90733-1391; Tel: 310-548-6279, Fax: 310-548-6950, E-mail: acs@pobox.com, Web: www.acsonline.org

American Council for an Energy-Efficient Economy 1001 Connecticut Ave. NW #801, Washington, DC 20036; Tel: 202-429-8873, E-mail: ACEEE@ix.netcom.com, Web: www.aceee.org

American Forests (formerly American Forestry Association) P.O. Box 2000, Washington, DC 20013; Tel: 202-955-4500, Web: www.amfor.org

American Hydrogen Association P.O. Box 15075, Phoenix, AZ 85060; Tel: 480-827-7915, Fax: 602-967-6601, Web: www.clean-air.org/index.html

American Institute of Biological Sciences, Inc. 71444 I St. NW, Suite 200, Washington, DC 20005; Tel: 202-628-1500, Fax: 202-628-1509, E-mail: jkolber@aibs.org, Web: www.aibs.org/core/index.html

American Public Health Association 801 I St. NW, Washington, DC 20001-3710; Tel: 202-777-2742, E-mail: comments@apha.org, Web: www.apha.org

American Rivers 1025 Vermont Ave. NW, Suite 720, Washington, DC 20005; Tel: 202-347-7550, Fax: 202-347-9240, Web: www.amrivers.org

American Solar Energy Society 2400 Central Ave, G-1, Boulder, CO 80301; Tel: 303-443-3130, Fax: 303-443-3212, E-mail: ases@ases.org, Web: ases.org/index.html

American Water Resources Association 950 Herndon Parkway, Suite 300, Herndon, VA 20170-5531; Tel: 703-904-1255, Fax: 703-904-1228, E-mail: awrahq@aol.com, Web: www.awra.org

American Wind Energy Association 122 C St. NW, Fourth Floor, Washington, DC 20001; Tel: 202-383-2500, Fax: 202-383-2505, E-mail: windmail@awea.org, Web: www.awea.org

Bat Conservation International P.O. Box 162603, Austin, TX 78716; Tel: 800-538-BATS, Web: www.batcon.org

Bio-Integral Resource Center P.O. Box 7414, Berkeley, CA 94707; Tel: 510-524-2567, Fax: 524-2567, Web: www.igc.apc.org/birc

Campus Ecology National Wildlife Federation, 8925 Leesburg Pike, Vienna, VA 22184; Tel: 800-588-1650, Web: www.nwf.org/nwf/campus

Center for Health, Environment, and Justice P.O. Box 6806, Falls Church, VA 22040; Tel: 703-237-2249, Fax: 703-237-8389, Web: www.essential.org/cchw

Center for Science in the Public Interest 1875 Connecticut Ave. NW, Suite 300, Washington, DC 20009; Tel: 202-332-9110, Fax: 202-265-4594, Web: www.cspinet.org

Chesapeake Bay Foundation 162 Prince George St., Annapolis, MD 21401; Tel: 410-268-8816, Web: www.baylink.org/bay-watch.html

Clean Water Action Project 4455 Connecticut Ave. NW, Suite A300, Washington, DC 20008-2238; Tel: 202-895-0420, Fax: 202-895-0438, E-mail: CWA@Clean Water Action.org, Web: www.cleanwateraction.org

Coalition for Environmentally Responsible Economies (CERES) 11 Arlington St., 6th Floor, Boston, MA 02116-3411; Tel: 617-247-0700, Fax: 617-267-5400, E-mail: schiller@ceres.org, Web: www.ceres.org

Common Cause 1250 Connecticut Ave. NW, Washington, DC 20036; Tel: 202-833-1200, E-mail: 75300.3120@compuserve.com, Web: www.commoncause.org

Conservation International 2501 M St. NW, Suite 200, Washington, DC 20037; Tel: 800-429-5660, Fax: 202-887-0193, Web: www.conservation.org

Critical Mass Energy Project Public Citizen 1600 20th St. NW, Washington, DC 20009; Tel: 202-546-4996, E-mail: cmep@citizen.org, Web: www.citizen.org/CMEP

Cultural Survival 96 Mount Auburn St., Cambridge, MA 02138; Tel: 617-441-5400, Fax: 617-441-5417, E-mail: csinc@cs.org, Web: www.cs.org/main.html

Defenders of Wildlife 1101 14th St. NW, Suite 1400, Washington, DC 20005; Tel: 202-682-9400, Fax: 202-682-1331, E-mail: information@defenders.org, Web: www.defenders.org

Ducks Unlimited One Waterfowl Way, Memphis, TN 38120; Tel: 901-758-3825, Fax: 901-758-3850, Web: www.ducks.org

Earth Island Institute 300 Broadway, Suite 28, San Francisco, CA 94133; Tel: 415-788-3666, Fax: 415-788-7324, Web: www.earthisland.org

Environmental Defense Fund, Inc. 257 Park Ave. South, New York, NY 10010; Tel: 212-505-2100, Web: www.edf.org

Environmental Law Institute 1616 P St. NW, #200, Washington, DC 20036; Tel: 202-939-3248, Fax: 202-939-3868, E-mail: eckl@eli.org, Web: www.eli.org

Florida Solar Energy Center 1679 Clearlake Rd., Cocoa, FL 32922; Tel: 407-638-1000, Fax: 407-638-1010, Web: www.fsec.ucf.edu

Forest Service Employees for Environmental Ethics P.O. Box 11615, Eugene, OR 97403; Tel: 541-484-2692, Fax: 541-484-3004, E-mail: afseee@afseee.org, Web: www.afseee.org

Friends of the Earth 1025 Vermont Ave. NW, 3rd Floor, Washington, DC 20005; Tel: 202-783-7400, Fax: 202-783-0444, E-mail: foe@foe.org, Web: www.foe.org

Greenpeace, Canada 250 Dundas St. W, Suite 605, Toronto, Ontario M5T 2Z5; Tel: 416-597-8408, Fax: 416-597-8422, Web: www.greenpeacecanada.org

Greenpeace, USA 1436 U St. NW, Washington, DC 20009; Tel: 800-326-0959 or 202-462-1177, Fax: 202-462-4507, Web: www.greenpeaceusa.org

Habitat for Humanity 121 Habitat St., Americus, GA 31709-3498; Tel: 912-924-6935, E-mail: public_info@habitat.org, Web: www.habitat.org

Humane Society of the United States 2100 L St. NW, Washington, DC 20037; Tel: 202-452-1100, Fax: 301-258-3077, Web: www.hsus.org

Institute for Alternative Agriculture 9200 Edmonston Rd., Suite 117, Greenbelt, MD 20770-1551; Tel: 301-441-8777, Fax: 301-220-0164, E-mail: hawiaa@access.digex.net, Web: www.hawiaa.org

Institute for Earth Education Cedar Cove, Box 115, Greenville, WV 24945; Tel: 304-832-6404, Fax: 304-832-6077, E-mail: iee1@aol.com, Web: www.eartheducation.org

Institute for Local Self-Reliance 2425 18th St. NW, Washington, DC 20009-2096; Tel: 202-232-4108, Fax: 202-332-0463, E-mail: ilsr@itp.apc.org, Web: www.ilsr.org

International Institute for Energy Conservation 750 First St. NE, #940, Washington, DC 20002; Tel: 202-842-3388, Fax: 202-842-1565, E-mail: iiec@igc.apc.org, Web: www.iiec.org

International Planned Parenthood Federation 120 Wall St., 9th Floor, New York, NY 10005; Tel: 212-248-6400, Fax: 212-248-4221, E-mail: info@ippfwhr.org, Web: www.ippfwhr.org

International Rivers Network 1847 Berkeley Way, Berkeley, CA 94703; Tel: 510-848-1155, Fax: 510-848-1008, E mail: irn@igc.apc.org, Web: www.irn.org

International Society for Ecological Economics P.O. Box 1589, Solomons, MD 20688; Tel: 410-326-0794, Fax: 410-326-7354, E-mail: beckman@cbl.umces.edu, Web: isee.aibs.org

IUCN: The World Conservation Union 128 Rue Mauverney CH-1196 Gland, Switzerland (World Headquarters) or 1630 Connecticut Ave NW, 3rd Floor,

Washington, DC 20009; Tel: 202-387-IUCN (4826), Fax: 202-387-4823, E-mail: postmaster@iucnus.org, Web: www.iucn.org and www.iucnus.org

Izaak Walton League of America 707 Conservation Lane, Gaithersburg, MD 20878; Tel: 301-548-0150 or 800-IKE-LINE (453-5463), Fax: 301-548-0146, E-mail: general@iwla.org, Web: www.iwla.org

Land Institute 2440 E. Well Water Rd., Salina, KS 67401; Tel: 785-823-5376, Fax: 785-823-8728, E-mail: the land@midkan.com, Web: www.midkan.com/theland

Land Trust Alliance 1319 F St. NW, Suite 501, Washington, DC 20004; Tel: 202-638-4725, Fax: 202-638-4730, Web: www.lta.org

League of Conservation Voters 1707 L St. NW, Suite 750, Washington, DC 20036; Tel: 202-785-8683, Fax: 202-835-0491, E-mail: lcv@lcv.org, Web: www.lcv.org

League of Women Voters of the U.S. 1730 M St. NW, Washington, DC 20036; Tel: 800-249-VOTE or 202-429-1965, Fax: 202-429-0854, Web: www.lwv.org

Mineral Policy Center 1612 K St. NW, Suite 808, Washington, DC 20006; Tel: 202-887-1872, Fax: 202-887-1875, E-mail: mpc@mineralpolicy.org, Web: www.mineralpolicy.org

National Audubon Society 700 Broadway, New York, NY 10003-9501; Tel: 212-979-3000, Fax: 212-979-3188, E-mail: webmaster@list.audubon.org, Web: www.audubon.org/nas

National Coalition Against the Misuse of Pesticides 701 E St. SE, Suite 200, Washington, DC 20003; Tel: 202-543-5450, Fax: 202-543-4791, E-mail: ncamp@ncamp.org, Web: www.igc.org/ncamp

National Environmental Health Association 720 S. Colorado Blvd., South Tower 970, Denver, CO 80246-1925; Tel: 303-756-9090, Fax: 303-691-9490, E-mail: staff@neha.org, Web: www.neha.org

National Environmental Trust Tel: 202-887-8800, E-mail: netinfo@envirotrust.com, Web: www.envirotrust.com

National Geographic Society 1145 17th St. NW, Washington, DC 20036-4688; Tel: 800-647-5463 or 202-857-7000, Web: www.nationalgeographic.com

National Hydrogen Association 1800 M St. NW, Suite 300, Washington, DC 20036-5802; Tel: 202-223-5547, Fax: 202-223-5537, Web: www.ttcorp.com/nha

National Park Foundation 1101 17th St. NW, Suite 1102, Washington, DC 20036; Tel: 202-785-4500, Web: www.nationalparks.org

National Parks and Conservation Association 1776 Massachusetts Ave. NW, Suite 200, Washington, DC 20036; Tel: 800-NAT-PARKS or 202-223-6722, Fax: 202-659-0650, E-mail: natparks@aol.com, Web: www.npca.org

National Recycling Coalition 1727 King St., Suite 105, Alexandria, VA 22314-2720; Tel: 703-683-9025, Fax: 703-683-9026, Web: www.earthsystems.org

National Science Teachers Association 1840 Wilson Blvd., Arlington, VA 22201-3000; Tel: 703-7100, Fax: 703-243-7177, E-mail: maiser@nsta.org, Web: www.nsta.org

National Wildlife Federation 8925 Leesburg Pike, Vienna, VA 22184; Tel: 703-790-4000, E-mail: feedback@nws.org, Web: www.nwf.org

Natural Resources Defense Council 40 W. 20th St., New York, NY 10011; Tel: 212-727-2700; and 1350 New York Ave. NW, Suite 300, Washington, DC 20005; Tel: 202-783-7800, E-mail: nrdcinfo@igc.apc.org, Web: www.nrdc.org

Nature Conservancy 4245 North Fairfax Drive, Suite 100, Arlington, VA 22203-1606; Tel: 703-841-5300, Fax: 703-841-1283, Web: www.tnc.org

North American Association for Environmental Education 1825 Connecticut Ave. NW, Suite 800, Washington, DC 20009-5708; Tel: 202-884-8912, Fax: 202-884-8455, Web: www.naaee.org

Permaculture Institute of North America 4649 Sunnyside Ave. N, Seattle, WA 98103

Pesticide Action Network 965 Mission St. #514, San Francisco, CA 94103; Web: www.igc.org/panna

Physicians for Social Responsibility 1101 14th St. NW, Suite 700, Washington, DC 20005; Tel: 202-898-0150, Fax: 202-898-0172, E-mail: psrnatl@psr.org, Web: www.psr.org

Planet/Drum Foundation P.O. Box 31251, San Francisco, CA 94131; Tel: 415-285-6556, Fax: 415-285-6563, E-mail: planetdrum@igc.apc.org, Web: www.planetdrum.org

Planned Parenthood Federation of America 810 Seventh Ave., New York, NY 10019; Tel: 212-541-7800, Fax: 212-261-4560, Web: www.plannedparenthood.org

Population Action International (formerly Population Crisis Committee) 1120 19th St. NW, Suite 550, Washington, DC 20036-3605; Tel: 202-659-1833, Fax: 202-293-1795, E-mail: pai@popact.org, Web: www.populationaction.org

Population Council One Dag Hammarskjold Plaza, New York, NY 10017; Tel: 212-339-0500, Fax: 212-755-6052, E-mail: pubinfo@popcouncil.org, Web: www.popcouncil.org

Population–Environment Balance 2000 P St. NW, Suite 210, Washington, DC 20036-5915; Tel: 202-955-5700, Fax: 202-955-6161, E-mail: uspop@balance.org, Web: www.balance.org

Population Institute 107 2nd St. NE, Washington, DC 20002; Tel: 800-787-0038 or 202-544-3300, Fax: 544-0068, E-mail: web@populationinstitute.org, Web: www.populationinstitute.org

Population Reference Bureau 1875 Connecticut Ave. NW, Suite 520, Washington, DC 20009-5728; Tel: 202-483-1100, Fax: 202-328-3937, E-mail: popref@prb.org, Web: www.igc.apc.org

Public Citizen 1600 20th St. NW, Washington, DC 20009; Tel: 202-588-1000, Web: www.citizen.org

Rainforest Action Network 3221 Pine St., Suite 500, San Francisco, CA 94104; Tel: 415-398-4404, Fax: 415-389-2732, Web: www.ran.org

Rainforest Alliance 65 Bleecker St., New York, NY 10012; Tel: 212-677-1900, Fax: 212-677-2187, E-mail: canopy@cdp.apc.org, Web: www.rainforest-alliance.org

Resources for the Future 1616 P St. NW, Washington, DC 20036; Tel: 202-328-5000, Fax: 202-939-3460, E-mail: info@rrf.org, Web: www.rff.org

Rocky Mountain Institute 1739 Snowmass Creek Rd., Snowmass, CO 81654-9199; Tel: 970-927-3851, Fax: 970-927-4178, E-mail: emalia@rmi.org, Web: www.rmi.org

Scientists' Institute for Public Information 355 Lexington Ave., New York, NY 10017; Tel: 212-661-9110

Sea Shepherd Conservation Society P.O. Box 628, Venice, CA 90294; Tel: 310-301-7325, Fax: 310-574-3161, E-mail: seashepherd@seashepherd.org, Web: www.seashepherd.org

Sierra Club 85 Second St., San Francisco, CA 94105; Tel: 415-977-5500, Fax: 415-977-5799, E-mail: information@sierraclub.org, Web: sierraclub.org

Smithsonian Institution 1000 Jefferson Dr. SW, Washington, DC 20560; Tel: 202-357-2700, Web: www.si.edu

Society for Conservation Biology Alice Blandin, *Conservation Biology*, University of Washington, Box 351800, Seattle, WA 98195-1800; Tel: 206-616-4054, Fax: 206-543-3041, E-mail: conbio@u.washington.edu, Web: www.conbio.rice.edu

Society for Ecological Restoration 1207 Seminole Hwy., Madison, WI 53711; Tel: 608-262-9547, Fax: 608-262-8557, E-mail: ser@vms2.macc.wisc.edu, Web: www.ser.org

Soil and Water Conservation Society 7515 NE Ankeny Rd., Ankeny, IA 50021-9764; Tel: 515-289-2331 or 800-863-7645, Fax: 515-289-1227, E-mail: swcs@swcs.org, Web: www.swcs.org

Solid Waste Association of North America P.O. Box 7219, Silver Spring, MD 20907-7219; Tel: 301-585-2898, Fax: 301-589-7068, E-mail: info@swana.org, Web: www.swana.org

Student Conservation Association Inc. P.O. Box 550, Charlestown, NH 03603-0550; Tel: 603-543-1700, Fax: 603-543-1828, E-mail: debw@sca-inc.org, Web: www.sca-inc.org

Student Environmental Action Coalition (SEAC) P.O. Box 31909, Philadelphia, PA 19104-0609; Tel: 215-222-4711, Fax: 215-222-2896, E-mail: seac@seac.org, Web: www.seac.org

Survival International 11-15 Emerald St., London WC1N 3QL, England; Tel: 0171-242-1441, Fax: 0171-242-1771, E-mail: survival@gn.apc.org, Web: www.survival.org.uk

Union of Concerned Scientists Two Brattle Square, Cambridge, MA 02238-9105;

Tel: 617-547-5552, Fax: 617-864-9405, E-mail: ucs@ucsusa.org, Web: www.ucsusa.org

United Nations Population Fund 220 East 42nd St., New York, NY 10017; Tel: 212 297-5020, Fax: 212-557-6416, Web: www.unfpa.org

U.S. Public Interest Research Group 215 Pennsylvania Ave. SE, Washington, DC 20003; Tel: 202-546-9707, Web: www.pirg.org

The Wilderness Society 900 17th St. NW, Washington, DC 20006-2596; Tel: 800-THE-WILD, Web: www.wilderness.org

Wildlife Society 5410 Grosvenor Lane, Suite 200, Bethesda, MD 20814-2197; Tel: 301-897-9770, Fax: 301-530-2471, E-mail: tws@wildlife.org, Web: www.wildlife.org

Work on Waste 82 Judson St., Canton, NY 13617; Tel: 315-375-9200, Fax: 315-379-0448, E-mail: wastenot@northnet.org

World Rainforest Movement International Coordination, Ricardo Carrere, Instituto del Tercer Mundo, Jackson 1136, Montevideo 11200, Uruguay; Tel: +598 2 4096192, Fax: +598 2 4019222, E-mail: wrm@chasque.apc.org, Web: www.wrm.org.uy

World Resources Institute 10 G St. NE, Suite 800, Washington, DC 20002; Tel: 202-729-7600, Fax: 202-729-7610, Fax: 202-729-7610, E-mail: lauralee@wri.org, Web: www.wri.org/wri

Worldwatch Institute 1776 Massachusetts Ave. NW, Washington, DC 20036-1904; Tel: 202-452-1999, Fax: 202-296-7365, E-mail: worldwatch@worldwatch.org, Web: www.worldwatch.org

World Wildlife Fund 1250 24th St. NW, Washington, DC 20037; Tel: 800-225-5993 or 202-293-4800, Web: www.worldwildlife.org

Zero Population Growth 1400 16th St. NW, Suite 320, Washington, DC 20036; Tel: 800-POP-1956 or 202-332-2200, Fax: 202-332-2302, E-mail: info@zpg.org, Web: www.zpg.org

FEDERAL, SCIENTIFIC, AND INTERNATIONAL AGENCIES

Agency for International Development (USAID) State Building, 320 21st St. NW, Washington, DC 20523-0016; Tel: 202-647-1850, Fax: 202-647-8321, Web: www.info.usaid.gov

Bureau of Land Management U.S. Department of Interior, 1620 L St. NW, Rm. 5600, Washington, DC 20240; Tel: 202-208-3801, Web: www.blm.gov

Bureau of Reclamation Washington, DC 20240; Web: www.usbr.gov/main

Congressional Research Service 101 Independence Ave. SW, Washington, DC 20540-7000; Web: www.lcweb.loc.gov/crsinfo

Conservation and Renewable Energy Inquiry and Referral Service P.O. Box 8900, Silver Spring, MD 20907; Tel: 800-523-2929

Department of Agriculture 14th St. and Independence Ave. SW, Washington, DC 20250; Tel: 202-720-8732, Web: www.usda.gov

Department of Commerce Herbert C. Hoover Bldg., Rm. 5610, 15th & Constitution Ave. NW, Washington, DC 20230; Tel: 202-219-3605, Web: www.doc.gov

Department of Energy Forrestal Building, 1000 Independence Ave. SW, Washington, DC 20585; Web: www.doe.gov and www.eren.doe.gov

Department of Health and Human Services 200 Independence Ave. SW, Washington, DC 20585; Web: www.os.dhhs.gov

Department of Housing and Urban Development 451 7th St. SW, Washington, DC 20410; Tel: 202-755-5111, Web: www.hud.gov

Department of the Interior 1849 C St. NW, Washington, DC 20240; Tel: 202-208-3100, Web: www.usga.gov.doi

Department of Transportation 400 7th St. SW, Washington, DC 20590; Tel: 202-366-4000, Web: www.dot.gov

Energy Information Administration Dept. of Energy, National Energy Information Center, Forrestal Bldg., Washington, DC 20585; Tel: 202-586-8800, Web: www.eia.doe.gov

Environmental Protection Agency 401 M St. SW, Washington, DC 20460; Tel: 202-260-2090, Web: www.epa.gov

Federal Energy Regulatory Commission 888 First St. NE, Washington, DC 20426; Web: www.ferc.fed.us

Food and Agriculture Organization (FAO) of the United Nations 101 22nd St. NW, Suite 300, Washington, DC 20437; Web: www.fao.org

Food and Drug Administration Department of Health and Human Services, 5600 Fishers Lane, Rockville, MD 20857; Tel: 410-433-1544, Web: www.fda.gov

Forest Service P.O. Box 96090, Washington, DC 20090-6090; Tel: 202-205-0957, Web: www.fs.fed.us

Great Lakes Commission The Argus II Bldg., 400 Fourth St., Ann Arbor, MI 48103-4816; Tel: 313-665-9135, Fax: 313-665-4370, Web: www.glc.org

International Whaling Commission The Red House, 135 Station Rd., Histon, Cambridge CB4 4NP, England (0223 233971); Web: www.marine.gov.uk/iwc.htm

Mine Safety and Health Administration Ballston Tower 3, 4015 Wilson Blvd., Arlington, VA 22203; Tel: 703-235-1452, Web: www.msha.gov

National Academy of Sciences 2101 Constitution Ave. NW, Washington, DC 20418; Web: www.nas.edu

National Aeronautics and Space Administration 400 Maryland Ave. SW, Washington, DC 20546; Web: www.nasa.gov

National Cancer Institute 9000 Rockville Pike, Bethesda, MD 20892; Web: www.nci.nih.gov

National Center for Atmospheric Research P.O. Box 3000, Boulder, CO 80307; Web: www.ucar.edu

National Lead Information Center Tel: 1-800-LEAD-FYI, Web: www.nsc.org/ehc/lead.htm

National Marine Fisheries Service U.S. Dept. of Commerce, Silver Spring Metro Center 1, 1335 East[[endash]]West Hwy., Silver Spring, MD 20910; Tel: 301-713-2239, Web: www.nmfs.gov

National Oceanic and Atmospheric Administration Herbert C. Hoover Bldg., Rm. 5128, 14th & Constitution Ave. NW, Washington, DC 20230; Tel: 202-482-3384, Web: www.noaa.gov

National Park Service Interior Bldg., P.O. Box 37127, Washington, DC 20013-7127; Tel: 202-208-4747, Web: www.nps.gov

National Renewable Energy Laboratory 1617 Cole Blvd., Golden, CO 80401; Web: www.nrel.gov

National Science Foundation 4201 Wilson Blvd., Arlington, VA 22230; Tel: 703-306-1234, Web: www.nsf.gov

National Technical Information Service U.S. Department of Commerce, 5285 Port Royal Rd., Springfield, VA 22161; Web: www.ntis.gov

Natural Resource Conservation Service USDA, 14th and Independence Ave. SW, P.O. Box 2890, Washington, DC 20013; Tel: 202-720-3210, Web: www.nrcs.usda.gov

Nuclear Regulatory Commission Washington, DC 20555; Tel: 301-415-7000, Web: www.nrc.gov

Occupational Safety and Health Administration Department of Labor, 200 Constitution Ave. NW, Washington, DC 20210; Web: www.osha.gov

Office of Ocean and Coastal Resource Management 1825 Connecticut Ave. NW, Suite 700, Washington, DC 20235; Tel: 202-208-2553, Web: www.wave.nos.noaa.gov/ocrm

Office of Surface Mining Reclamation and Enforcement Interior South Bldg., 1951 Constitution Ave. NW, Washington, DC 20240; Web: www.eia.doe.gov

Organization for Economic Cooperation and Development (U.S. Office) 2001 L St. NW, Suite 700, Washington, DC 20036; Web: www.oecd.org

Soil Conservation Service See *Natural Resource Conservation Service*

United Nations 1 United Nations Plaza, New York, NY 10017; Web: www.unol.org

United Nations Environment Programme Regional North American Office, United Nations Rm. DC 20803, New York, NY 10017; and 1889 F St. NW, Washington, DC 20006; Web: www.unep.org

U.S. Fish and Wildlife Service Department of the Interior, Washington, DC 20240; Web: www.fws.gov

U.S. Geological Survey National Center, Reston, VA 22092; Tel: 703-648-4000, Web: www.usgs.gov

World Bank 1818 H St. NW, Washington, DC 20433; Web: www.worldbank.org

APPENDIX 2

UNITS OF MEASUREMENT

LENGTH

Metric

1 kilometer (km) = 1,000 meters (m)
1 meter (m) = 100 centimeters (cm)
1 meter (m) = 1,000 millimeters (mm)
1 centimeter (cm) = 0.01 meter (m)
1 millimeter (mm) = 0.001 meter (m)

English

1 foot (ft) = 12 inches (in)
1 yard (yd) = 3 feet (ft)
1 mile (mi) = 5,280 feet (ft)
1 nautical mile = 1.15 miles

Metric-English

1 kilometer (km) = 0.621 mile (mi)
1 meter (m) = 39.4 inches (in)
1 inch (in) = 2.54 centimeters (cm)
1 foot (ft) = 0.305 meter (m)
1 yard (yd) = 0.914 meter (m)
1 nautical mile = 1.85 kilometers (km)

AREA

Metric

1 square kilometer (km^2) = 1,000,000 square meters (m^2)
1 square meter (m^2) = 1,000,000 square millimeters (mm^2)
1 hectare (ha) = 10,000 square meters (m^2)
1 hectare (ha) = 0.01 square kilometer (km^2)

English

1 square foot (ft^2) = 144 square inches (in^2)
1 square yard (yd^2) = 9 square feet (ft^2)
1 square mile (mi^2) = 27,880,000 square feet (ft^2)
1 acre (ac) = 43,560 square feet (ft^2)

Metric-English

1 hectare (ha) = 2.471 acres (ac)
1 square kilometer (km^2) = 0.386 square mile (mi^2)
1 square meter (m^2) = 1.196 square yards (yd^2)
1 square meter (m^2) = 10.76 square feet (ft^2)
1 square centimeter (cm^2) = 0.155 square inch (in^2)

VOLUME

Metric

1 cubic kilometer (km^3) = 1,000,000,000 cubic meters (m^3)
1 cubic meter (m^3) = 1,000,000 cubic centimeters (cm^3)
1 liter (L) = 1,000 milliliters (mL) = 1,000 cubic centimeters (cm^3)
1 milliliter (mL) = 0.001 liter (L)
1 milliliter (mL) = 1 cubic centimeter (cm^3)

English

1 gallon (gal) = 4 quarts (qt)
1 quart (qt) = 2 pints (pt)

Metric-English

1 liter (L) = 0.265 gallon (gal)
1 liter (L) = 1.06 quarts (qt)
1 liter (L) = 0.0353 cubic foot (ft^3)
1 cubic meter (m^3) = 35.3 cubic feet (ft^3)
1 cubic meter (m^3) = 1.30 cubic yards (yd^3)
1 cubic kilometer (km^3) = 0.24 cubic mile (mi^3)
1 barrel (bbl) = 159 liters (L)
1 barrel (bbl) = 42 U.S. gallons (gal)

MASS

Metric

1 kilogram (kg) = 1,000 grams (g)
1 gram (g) = 1,000 milligrams (mg)
1 gram (g) = 1,000,000 micrograms (μg)
1 milligram (mg) = 0.001 gram (g)
1 microgram (μg) = 0.000001 gram (g)
1 metric ton (mt) = 1,000 kilograms (kg)

English

1 ton (t) = 2,000 pounds (lb)
1 pound (lb) = 16 ounces (oz)

Metric-English

1 metric ton (mt) = 2,200 pounds (lb) = 1.1 tons (t)
1 kilogram (kg) = 2.20 pounds (lb)
1 pound (lb) = 454 grams (g)
1 gram (g) = 0.035 ounce (oz)

ENERGY AND POWER

Metric

1 kilojoule (kJ) = 1,000 joules (J)
1 kilocalorie (kcal) = 1,000 calories (cal)
1 calorie (cal) = 4,184 joules (J)

Metric-English

1 kilojoule (kJ) = 0.949 British thermal unit (Btu)
1 kilojoule (kJ) = 0.000278 kilowatt-hour (kW-h)
1 kilocalorie (kcal) = 3.97 British thermal units (Btu)
1 kilocalorie (kcal) = 0.00116 kilowatt-hour (kW-h)
1 kilowatt-hour (kW-h) = 860 kilocalories (kcal)
1 kilowatt-hour (kW-h) = 3,400 British thermal units (Btu)
1 quad (Q) = 1,050,000,000,000,000 kilojoules (kJ)
1 quad (Q) = 2,930,000,000,000 kilowatt-hours (kW-h)

TEMPERATURE CONVERSIONS

Fahrenheit (°F) to Celsius (°C): °C = ((°F − 32.0) ÷ 1.80)
Celsius (°C) to Fahrenheit (°F): °F = ((°C × 1.80) + 32.0)

APPENDIX 3

MAJOR U.S. RESOURCE CONSERVATION AND ENVIRONMENTAL LEGISLATION

GENERAL

National Environmental Policy Act of 1969 (NEPA)
International Environmental Protection Act of 1983
Environmental Education Act of 1990

ENERGY

Energy Policy and Conservation Act of 1975
National Energy Act of 1978, 1980
National Appliance Energy Conservation Act of 1987
Energy Policy Act of 1992

WATER QUALITY

Water Quality Act of 1965
Water Resources Planning Act of 1965
Federal Water Pollution Control Acts of 1965, 1972
Ocean Dumping Act of 1972
Safe Drinking Water Act of 1974, 1984, 1996
Water Resources Development Act of 1986
Clean Water Act of 1977, 1981, 1987
Ocean Dumping Ban Act of 1988
Great Lakes Critical Programs Act of 1990
Oil Spill Prevention and Liability Act of 1990

AIR QUALITY

Clean Air Act of 1963, 1965, 1970, 1977, 1990
Asbestos Hazard and Emergency Response Act of 1986
Pollution Prevention Act of 1990

NOISE CONTROL

Noise Control Act of 1965
Quiet Communities Act of 1978

RESOURCES AND SOLID WASTE MANAGEMENT

Mining Act of 1872
Solid Waste Disposal Act of 1965
Resource Recovery Act of 1970
Resource Conservation and Recovery Acts (RCRA) of 1976, 1989
Marine Plastic Pollution Research and Control Act of 1987
Waste Reduction Act of 1990

TOXIC SUBSTANCES

Hazardous Materials Transportation Act of 1975
Toxic Substances Control Act (TOSCA) of 1976
Resource Conservation and Recovery Acts (RCRA) of 1976, 1984
Comprehensive Environmental Response, Compensation, and Liability (Superfund) Act of 1980, 1986, 1990
Low-Level Radioactive Policy Act of 1980
Nuclear Waste Policy Act of 1982

PESTICIDES

Food, Drug, and Cosmetics Acts of 1938, 1954, 1958
Federal Insecticide, Fungicide, and Rodenticide Control Act of 1972, 1988
Food Quality Protection Act of 1996

WILDLIFE CONSERVATION

Lacey Act of 1900
Migratory Bird Treaty Act of 1918
Migratory Bird Conservation Act of 1929
Migratory Bird Hunting Stamp Act of 1934
Pittman-Robertson Act of 1937
Anadromous Fish Conservation Act of 1965
Fur Seal Act of 1966
Species Conservation Act of 1966, 1969
National Wildlife Refuge System Act of 1966, 1976, 1978
Marine Mammal Protection Act of 1972
Marine Protection, Research, and Sanctuaries Act of 1972
Endangered Species Act of 1973, 1982, 1985, 1988
Whale Conservation and Protection Study Act of 1976
Fishery Conservation and Management Act of 1976, 1978, 1982
Fish and Wildlife Improvement Act of 1978
Fish and Wildlife Conservation Act of 1980 (Nongame Act)

LAND USE AND CONSERVATION

Homestead Act of 1862
Taylor Grazing Act of 1934
Soil Conservation Act of 1935
Wilderness Act of 1964
Land and Water Conservation Act of 1965
Multiple Use Sustained Yield Act of 1968
Wild and Scenic Rivers Act of 1968
National Trails System Act of 1968
National Coastal Zone Management Act of 1972, 1980
Forest Reserves Management Act of 1974, 1976
Forest and Rangeland Renewable Resources Act of 1974, 1978
Federal Land Policy and Management Act of 1976
National Forest Management Act of 1976
Soil and Water Conservation Act of 1977
Surface Mining Control and Reclamation Act of 1977
Antarctic Conservation Act of 1978
Public Rangelands Improvement Act of 1978
Endangered American Wilderness Act of 1978
Alaskan National Interests Lands Conservation Act of 1980
Coastal Barrier Resources Act of 1982
Food Security Act of 1985
Emergency Wetlands Resources Act of 1986
Coastal Development Act of 1990
California Desert Protection Act of 1994

APPENDIX 4

TACTICS OF THE ANTI-ENVIRONMENTAL MOVEMENT

Whether you are for or against the environmental movement, it is useful to understand the following legal and political tactics being used against this movement.* Start by *establishing an enemy to create fear and divert people's attention away from the real issues.* Label environmentalists as antibusiness, antipeople, antireligious, scaremongering radical extremists who are **(1)** crippling the economy, **(2)** hurting small business owners, **(3)** costing taxpayers billions of dollars in unnecessary regulations, **(4)** robbing small landowners of the right to do what they want with their land, **(5)** trying to lock up the natural resources on public lands, and **(6)** threatening jobs, national security, and traditional values.

When accused of using wild exaggeration and smear tactics against environmentalists to raise money and spread distrust, hate, and fear, Ron Arnold, one of the leaders of the anti-environmental Wise-Use movement (p. 47-8), said, "Facts don't really matter. In politics, perception is reality."

Here are some other tactics used against the environmental movement.

Weaken and intimidate. Infiltrate and spy on environmental groups to learn their plans and to create dissent and mistrust. Slap individual activists and small grassroots groups with nuisance lawsuits. Intimidate and threaten individual scientists, scientific writers, and mainstream scientific professional groups that stand in the way of anti-environmental goals by lawsuits and threats of lawsuits.

Threaten or use violence. If environmental activists become too effective, harass them with phone calls threatening their lives or the lives of their family members. Try to have them fired, kill their pets, trash their homes and offices, cut phone lines, slash tires, sabotage their cars, plant drugs in their cars and notify the police, burn their houses or businesses, assault them, fire warning shots into their houses or cars—all things that have happened to environmentalists in the United States in recent years.** Former U.S. Interior Secretary James Watt publicly declared in 1990, "If the troubles from the environmentalists cannot be solved in the jury box or the ballot box, perhaps the cartridge box should be used."

Influence public opinion. Commission books and articles, establish magazines, set up think tanks, and hire PR firms to deny that environmental problems exist or are serious. Use fax and letter-writing campaigns against corporations that provide financial support for TV or radio programs critical of environmentally harmful business practices or anti-environmental concerns. Flood schools with free pamphlets, videos, and other teaching materials that promote your position.

Exploit the built-in limitations of science and public ignorance about the nature of science to sow seeds of doubt. Mislead the public by saying that claims against your position have not been absolutely scientifically proven. If that doesn't work, imply that scientific findings that are contrary to your position should not be taken seriously because science can only put forth theories (Section 3-1). These tactics should drive scientists and environmentalists crazy and undermine public support of environmental causes.

Urge the president and Congress to require that all government environmental decisions and regulations be evaluated primarily by cost–benefit and risk–benefit analysis. Then try to load the evaluation panels with experts who support your position. If this doesn't work, challenge the analyses in courts. The built-in limitations (Spotlight, p. 37), information gaps, and vulnerable assumptions of many such analyses could take years to unravel. All of this would be achieved without trying to repeal existing environmental laws (a politically unpopular cause).

Get Congress to require rigorous scientific proof for all environmental decision-making and at the same time cut funding for such scientific research.

Support unenforceable legislation and regulations. This undermines confidence in government, splits up opposition groups, and helps convince the public that real change is impossible.

When unfavorable environmental laws are passed, urge legislators not to fund the laws. Passing the cost of federally unfunded environmental mandates on to financially strapped states and cities will generate intense political pressure against such laws by citizens, mayors, and governors. Because Congress is unlikely to come up with funding for these laws, they probably will be weakened, unenforced, or repealed.

Require federal and state governments to compensate property owners whenever environmental, health, safety, or zoning laws limit how the owners can use their property or decrease its financial value. This will make such laws and regulations too expensive to enforce and will keep new ones from being enacted.

Attach anti-environmental laws as riders to key pieces of legislation (such as annual budget funding) that a president will have a difficult time vetoing.

Divide and conquer. Keep people and interest groups fighting with one another so they can't get together on vital issues that threaten the status quo (such as election campaign financing; Solutions, p. 48).

CRITICAL THINKING

What do you think about the use of such tactics? Proponents of such tactics contend that they learned many of them from environmental and civil rights groups. They argue that they are merely exercising their rights under the U.S. Constitution. What do you think?

---

*For more details see the following references found in the Further Readings for Chapter 2 on pp. A18-20: **(1)** Arnold and Gottlieb, 1993; **(2)** Beder, 1997; **(3)** Ehrlich and Ehrlich, 1996; **(4)** Fagin and Lavelle, 1996; **(5)** Flatteau, 1998; **(6)** Gelbspan, 1999; **(7)** Greenpeace, 1993; **(8)** Helvarg, 1994; **(9)** Korten, 1995; **(10)** Libby, 1999; **(11)** Rowell, 1997; **(12)** Selcraig, 1998; **(13)** Stauber and Rampton, 1995; **(14)** Switzer, 1997; and **(15)** Wilkinson, 1998. See also the following web sites: **(1)** Anti-environmental myths at http://www.members.aol.com/jimn469897/myths.htm; **(2)** The Way Things Really Are: Debunking Rush Limbaugh on the Environment at http://www.maui.net/~jstarh/rush/html; **(3)** The Way Things Aren't at http://www.fairorg/press-releases/limbaugh-debates-reality.html; **(4)** Pesticide Industry Propaganda at http://www.ewg.org/pub/home/reports/myths/industry_myths.html; and **(5)** Junk science at http://www.junkscience.com/news/gelbspan.html

**For details see Helvarg, 1994.

APPENDIX 5

MAPPING CONCEPTS AND CONNECTIONS*

This textbook emphasizes the connections between environmental principles, problems, and solutions. One way to organize the material in various chapters and to understand such connections is to map the concepts in a particular chapter or portion of a chapter. Inside the back cover you will find a general concept map showing how the parts of this entire textbook are related. At the end of each chapter you will find a question asking you to make a chapter concept map. This appendix provides you with information about how to create such maps. Here are the basic steps in making a concept map:

1. List the key general and specific concepts by indenting specific concepts under more general concepts. Basically, make an outline of the material using section heads, subsection heads, and key terms.
2. Arrange the clusters of concepts with the more general concepts at the top of a page and the more specific concepts at the bottom of the page.
3. Circle key concepts, draw lines connecting them, and write labels on the lines that describe linkages between the concepts.
4. Draw in crosslinking connections (lines that relate various parts of the map) and write labels on the line, describing these connections.
5. Study the map to see how you might improve it. You might add clustering concepts and connections of your own. You might eliminate detail to make the big picture stand out. You might add more details to clarify certain concepts. You might add or alter broad crosslinking connections on the map as you think and learn more about the material.

Let's go through these steps using the concept of living systems discussed in Chapter 6 as an example.

1. List the appropriate general and specific concepts. Note that Section 7-6 is about land systems (called biomes) and Sections 6–5 and 6–6 are about aquatic systems. Use the list of section titles, subtitles, words in bold and italics, key drawings, and other terms that you think are particularly important. An outline of these two sections in which specific concepts are indented under more general ones would look something like Figure 1.
2. Arrange clusters of the concepts in Figure 1 with the more general ones at the top and more specific ones at the bottom. Concepts of the same general level should be at about the same height on the map. Figure 2 shows what this might look like. You might be able to do this on a single sheet of paper turned sideways, or you might find it easier to use large sheets of paper. Use a pencil to draw the map so that you can make changes. Note that the diagram has been simplified by eliminating some of the concepts in Figure 1 such as latitude and altitude, various types of temperate grasslands, and tropical scrub forests.
3. Circle key concepts, draw lines connecting them, and write labels on the lines that describe linkages between the concepts. You might decide to add or delete some of the concepts to make the map more logical, clear, and helpful to you. Figure 3 shows one way to do this. Note that in this figure the subcategories of deserts, grasslands, and forests are put together to improve clarity.
4. Draw in crosslinks that relate various parts of the map. Figure 4 illustrates one possibility. At this point you are trying to understand how the various concepts are connected to and interact with one another. Note that many of the human systems concepts interact positively or negatively with ecosystems and natural resource concepts (in terms of human value judgment). Note that Chapter 5 focuses on natural systems. As you learn more about how human systems interact with natural systems in this chapter and in other chapters, you should be able to add additional crosslinkages.
5. Study your map and try to improve it. You might choose to eliminate excessive detail or add additional concepts and linkages to improve clarity. You might also find it useful to rearrange the position of concepts vertically or horizontally to eliminate crowding or to reduce confusion over too many connecting links. In Figure 5 note that the general concepts of *fresh water* and *salt water* have been added, and aquatic systems has been divided into two subcategories based on salinity (dissolved salt content) instead of four. To improve clarity, the concepts of *moving* and *still* under *fresh water* have also been added because these characteristics greatly affect the types of organisms living in such systems. Finally, concepts have been connected with some of those in Chapter 4 by including details about photosynthetic and chemosynthetic productivity because these activities form the basis of the food web relationships in terrestrial and aquatic systems.

Remember the following things about concept maps:

- They are particularly useful for showing the big picture and for clarifying particularly difficult concepts. They can also be used in studying for tests to help you sort out what is important and how concepts are related.
- They take time to make. Sometimes they are easy and seem almost to write themselves. Other times, they can be difficult and frustrating to draw.
- There is no single "best" map. Rarely do concept maps developed by different individuals look alike.
- They are works in progress that should change and improve over time as you get better at making them and develop a deeper understanding of how things are connected.
- Comparing your maps with those developed by others is a good way to learn how to improve your mapmaking skills and to come up with more connections.
- Making them is challenging and fun.

*This appendix was developed by **Jane Heinze-Fry** with assistance from G. Tyler Miller, Jr.

LAND SYSTEMS (Biomes)
- climate
  - precipitation
  - temperature
  - latitude
  - altitude
- deserts
  - tropical
  - temperate
  - cold
  - semidesert
- grasslands
  - tropical
    - savanna
  - temperate
    - tall-grass
    - short-grass
    - pampas
    - veld
    - steppes
  - polar (arctic tundra)
    - permafrost
- forests
  - tropical
    - rain
    - deciduous
    - scrub
  - temperate deciduous
  - evergreen coniferous

WATER SYSTEMS
- ocean
  - coastal
    - coral reefs
    - estuaries
    - wetlands
    - barrier islands
    - beaches
      - rocky shore
      - barrier beach
- open sea
  - euphotic zone
  - bathyal zone
  - abyssal zone
- freshwater lakes
  - eutrophic
  - oligotrophic
  - mesotrophic
  - littoral zone
  - limnetic zone
  - profundal zone
  - benthic zone
  - thermal stratification
    - thermoclines
    - turnover
- freshwater streams
  - surface water
  - runoff
  - watershed
- inland wetlands
  - year-round
  - seasonal

Figure 1 List of concepts for Living Systems map.

LIVING SYSTEMS

Land Systems (Biomes)
- climate
  - precipitation
  - temperature
  - ~~latitude~~
  - ~~altitude~~

Water Systems

ocean

- deserts
  - tropical
  - temperate
  - cold
  - ~~semidesert~~
- grasslands
  - tropical
    - ~~savanna~~
  - temperate
    - ~~tall grass~~
    - ~~short grass~~
    - ~~pampas~~
    - ~~veld~~
    - ~~steppes~~
  - polar (arctic tundra)
    - ~~permafrost~~
- forests
  - tropical
    - rain
    - deciduous
    - ~~scrub~~
  - temperate deciduous
  - evergreen coniferous
- coastal
  - coral reefs
  - estuaries
  - wetlands
  - barrier islands
  - beaches
    - rocky shore
    - barrier beach
- open sea
  - euphotic zone
  - bathyal zone
  - abyssal zone
- freshwater lakes
  - eutrophic
  - oligotrophic
  - mesotrophic
  - littoral zone
  - limnetic zone
  - profundal zone
  - benthic zone
  - thermal stratification
    - thermoclines
    - turnover
- inland wetlands
  - year-round
  - seasonal
- freshwater streams
  - ~~surface water~~
  - ~~runoff~~
  - ~~watershed~~

Figure 2 Clusters of concepts are arranged in a hierarchy from more general at the top to more specific at the bottom.

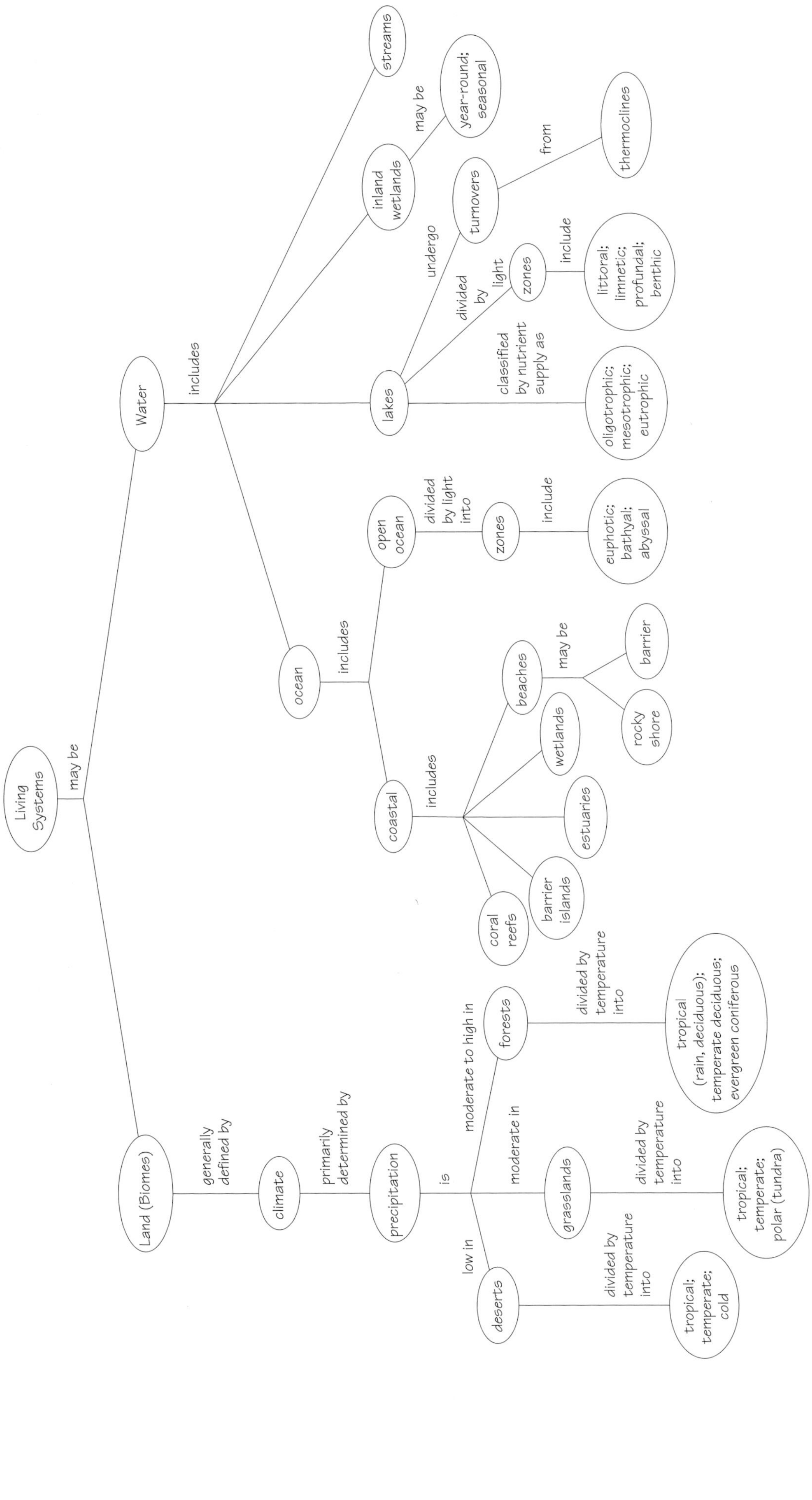

Figure 3 Concepts are circled and linkages are drawn.

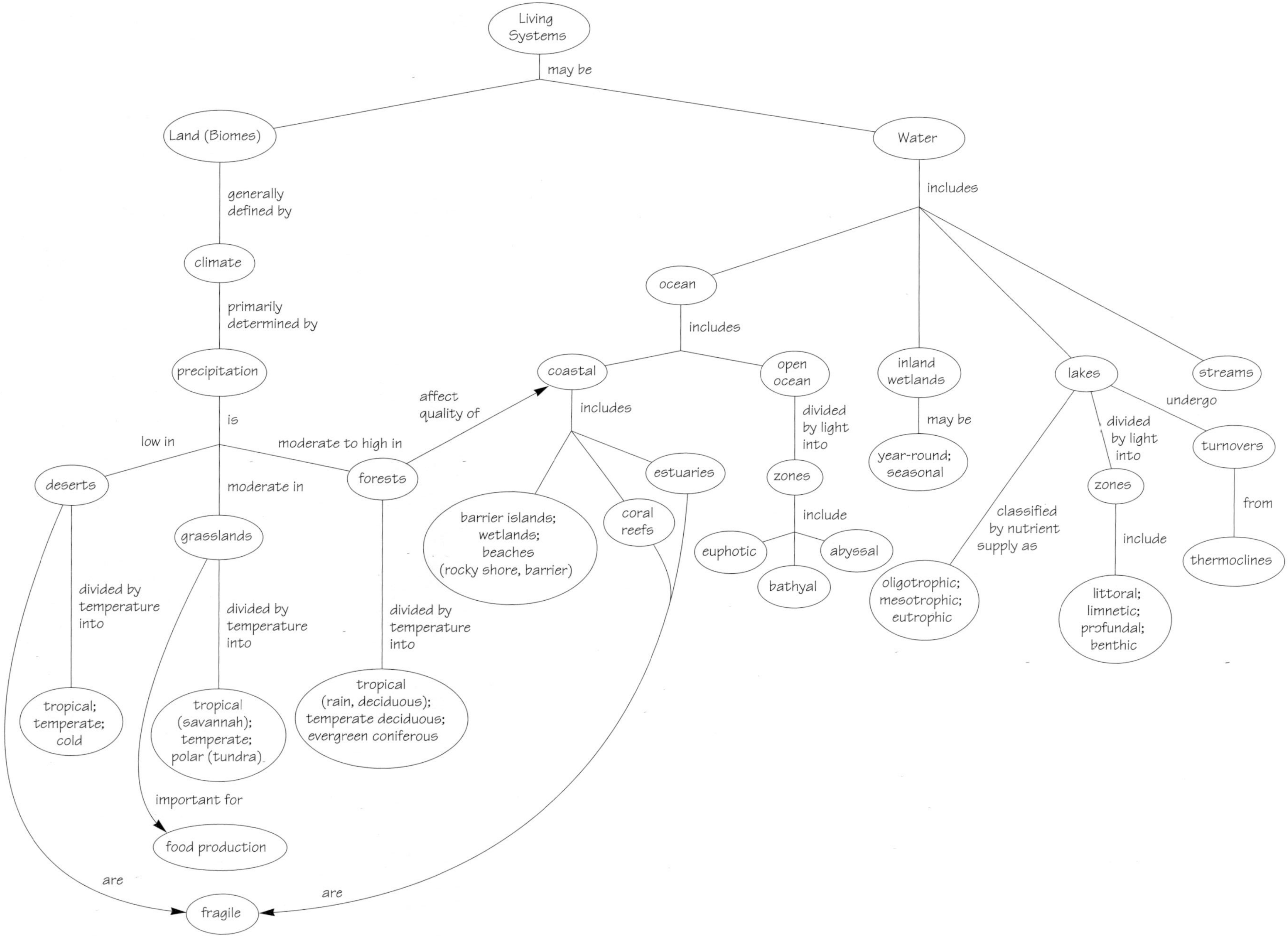

Figure 4 Crosslinkages are drawn and map is improved.

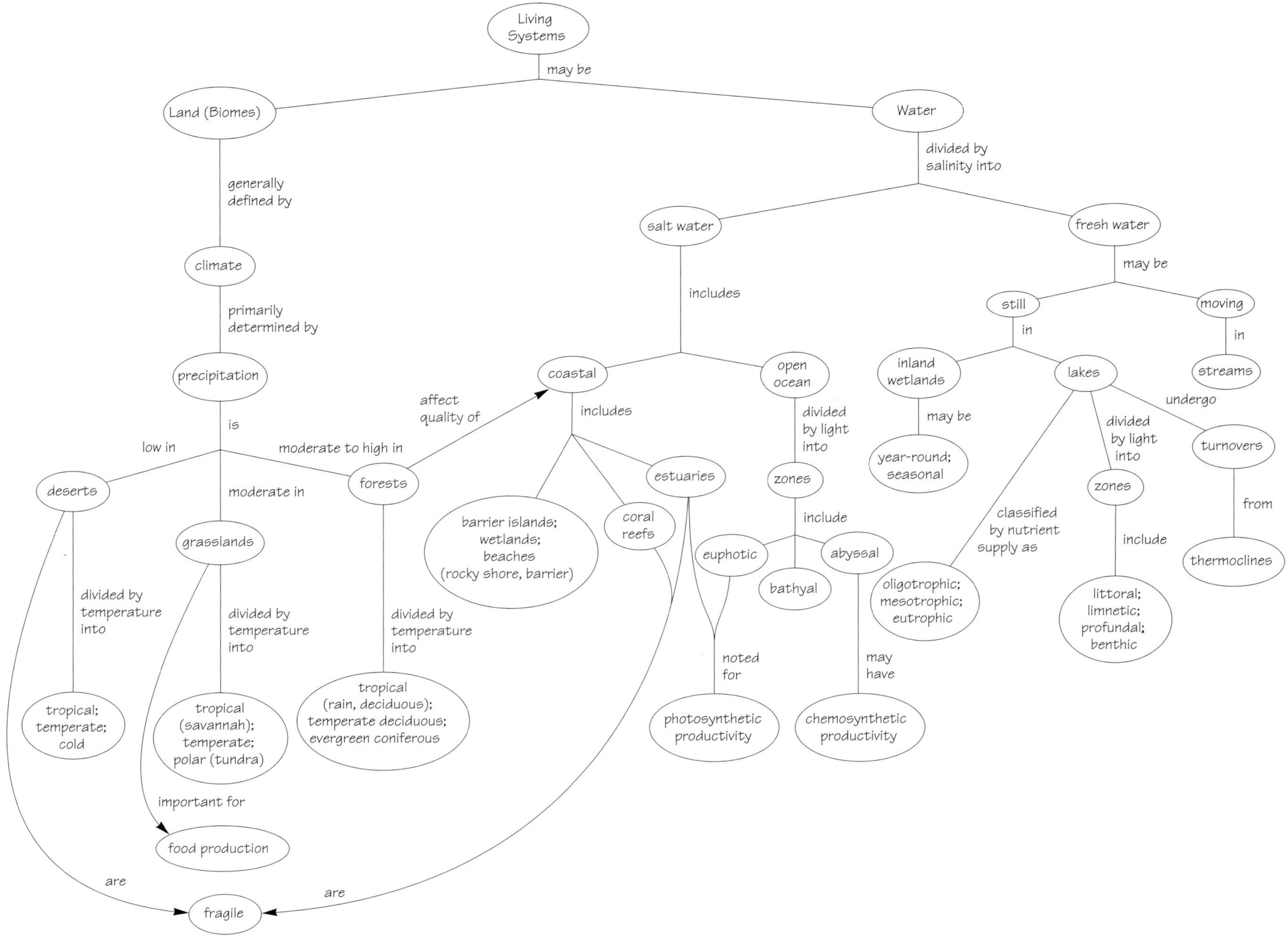

Figure 5 Map is further refined. New hierarchical clusters may emerge.

APPENDIX 6

INDIVIDUALS MATTER: WORKING WITH THE EARTH

This is a list of things individuals can do based on suggestions from a wide variety of environmentalists. It is not meant to be a list of things you must do but a list of actions you might consider. Don't feel guilty about the things you are not doing. Start by picking the ones that you are willing to do and that you feel will have the most impact. Many of these suggestions are controversial. Carefully evaluate each action to see whether it fits in with your beliefs. Based on practicality and my own beliefs, I do some of these things. Each year I look over this list and try to add several new items.

PRESERVING BIODIVERSITY AND PROTECTING THE SOIL

- *Develop a plan for the sustainable use of any forested area you own.*
- *Plant trees on a regular basis and take care of them.*
- *Reduce use of wood and paper products, recycle paper products, and buy recycled paper products.*
- *Don't buy furniture, doors, flooring, window frames, paneling, or other products made from tropical hardwoods such as teak or mahogany.* Look for the Good-Wood seal given by Friends of the Earth, or consult the *Wood User's Guide* by Pamela Wellner and Eugene Dickey (San Francisco: Rainforest Action Network, 1991).
- *Don't wood and paper products produced by cutting remaining old-growth forests in the tropics and elsewhere.* Information on such products can be obtained from the Rainforest Action Network, Rainforest Alliance, and Friends of the Earth (Appendix 1).
- *Help rehabilitate or restore a degraded area of forest near your home.*
- *Don't buy furs, ivory products, items made of reptile skin, tortoiseshell jewelry, and materials endangered or threatened animal species.*
- *When building a home, save all the trees possible.* Require that the contractor **(1)** disturb as little soil as possible, **(2)** set up barriers that catch any soil eroded during construction, and **(3)** save and replace any topsoil removed instead of hauling it off and selling it.
- *Landscape areas not used for gardening with a mix of wildflowers, herbs (for cooking and for repelling insects), low-growing ground cover, small bushes, and other forms of vegetation natural to the area.*
- *Set up a compost bin and use it to produce soil conditioner for yard and garden plants.*
- *Before you embark on a planned ecotour find out in writing what precautions are being taken to reduce the tour's impact on local ecosystems, customs, and cultures and what percentage of the gross profits go to local people in the form of jobs and donations.*

PROMOTING SUSTAINABLE AGRICULTURE AND REDUCING PESTICIDE USE

- *Waste less food.* An estimated 27% of all food produced in the United States is wasted.
- *Eat lower on the food chain* by reducing or eliminating meat consumption to reduce its environmental impact.
- *If you have a dog or a cat, don't feed it canned meat products.* Balanced-grain pet foods are available.
- *Reduce the use of pesticides on agricultural products by asking grocery stores to stock fresh produce and meat produced by organic methods.*
- *Grow some of your own food using organic farming techniques and drip irrigation to water your crops.*
- *Compost your food wastes.*
- *Think globally, eat locally.* Whenever possible, eat food that is locally grown and in season. This **(1)** supports your local economy, **(2)** gives you more influence over how the food is grown (by either organic or conventional methods), **(3)** saves energy required to transport food over long distances, and **(4)** reduces the fossilfuel use and pollution. If you deal directly with local farmers, you can also save money.
- *Give up the idea that the only good bug is a dead bug.* Recognize that insect species keep most of the populations of pest insects in check and that full-scale chemical warfare on insect pests wipes out many beneficial insects.
- *Don't insist on perfect-looking fruits and vegetables.* These are more likely to contain high levels of pesticide residues.
- *Use pesticides in your home only when absolutely necessary, and use them in the smallest amount possible.*
- *Don't become obsessed with having the perfect lawn.* About 40% of U.S. lawns are treated with pesticides. These chemicals can cause **(1)** headaches, **(2)** dizziness, **(3)** nausea, **(4)** eye trouble, and **(5)** even more acute effects in sensitive people (including children who play on treated lawns and in parks).
- *If you hire a lawn care company, use one that relies only on organic methods, and get its claims in writing.*

SAVING ENERGY AND REDUCING OUTDOOR AIR POLLUTION

- *Reduce use of fossil fuels.* Drive a car that gets at least 15 kilometers per liter (35 miles per gallon), join a carpool, and use mass transit, walking, and bicycling as much as possible. This reduces emissions of CO_2 and other air pollutants, saves energy and money, and can improve your health.
- *Plant and care for trees to help absorb* CO_2. During its lifetime, the average tree absorbs enough CO_2 to offset the amount produced by driving a car 42,000 kilometers (26,000 miles).
- *Insulate new or existing houses heavily, caulk and weatherstrip to reduce air infiltration and heat loss, and use energy-efficient windows.* Add an air-to-air heat exchanger to minimize indoor air pollution.
- *Obtain as much heat and cooling as possible from natural sources*, especially sun, wind, geothermal energy, and trees.
- *Buy the most energy-efficient homes, lights, cars, and appliances available. Evaluate them only in terms of lifetime cost.*
- *Turn down the thermostat on water heaters*

to 43-49C (110–120F), and insulate hot water pipes.

- *Lower the cooling load on an air conditioner by increasing the thermostat setting, installing energy-efficient lighting, using floor and ceiling fans, and using whole-house window or attic fans to bring in outside air (especially at night, when temperatures are cooler).*

REDUCING EXPOSURE TO INDOOR AIR POLLUTANTS

- *Test for radon and take corrective measures as needed.*
- *Install air-to-air heat exchangers or regularly ventilate your house by opening windows.*
- *At the beginning of the winter heating season, test your indoor air for formaldehyde when the house is closed up.* To locate a testing laboratory in your area, write to Consumer Product Safety Commission, Washington, DC 20207, or call 301-492-6800.
- *Don't buy furniture and other products containing formaldehyde. Use low-emitting formaldehyde or nonformaldehyde building materials.*
- *Reduce indoor levels of formaldehyde and other toxic gases by growing certain house plants.* Examples are the spider or airplane plant (removes 96% of carbon monoxide), aloe vera (90% of formaldehyde), banana (89% of formaldehyde), elephant ear philodendron (86% of formaldehyde), ficus (weeping fig, 47% of formaldehyde), golden porthos (67% of formaldehyde and benzene and 75% of carbon monoxide), Chinese evergreen (92% of toluene and 81% of benzene), English ivy (90% of benzene), peace lily (80% of benzene and 50% of trichloroethylene), and Janet Craig (corn plant, 79% of benzene). (Toxic removal figures indicate percentage of toxin removed by one plant in a 24-hour period in a 3.4-cubic-meter 12-cubic-foot space). Plants should be potted with a mixture of soil and granular charcoal (which absorbs organic air pollutants).
- *Consider not using carpeting and using wood or linoleum floors instead.* New synthetic carpeting releases vapors from more than 100 volatile organic compounds. New and old carpeting is a haven for microbes (many of them highly allergenic), dust, and traces of lead and pesticides brought in by shoes.
- *Remove your shoes before entering your house.* This reduces inputs of dust, lead, and pesticides.
- *Test your house or workplace for asbestos fiber levels and for any crumbling asbestos materials if it was built before 1980.* Don't buy a pre-1980 house without having its indoor air tested for asbestos and lead. To get a free list of certified asbestos laboratories that charge $25–50 to test a sample call the EPA's Toxic Substances Control Hotline at 202-554-1404.
- *Don't store gasoline, solvents, or other volatile hazardous chemicals inside a home or attached garage.*
- *Don't use aerosol spray products and commercial room deodorizers or air fresheners.*
- *If you smoke, do it outside or in a closed room vented to the outside.*
- *Make sure that wood-burning stoves, fireplaces, and kerosene- and gas-burning heaters are properly installed, vented, and maintained.* Install carbon monoxide detectors in all sleeping areas.

SAVING WATER

- *For existing toilets, reduce the amount of water used per flush* by putting into each tank a tall plastic container weighted with a few stones or buying and installing a toilet dam.
- *Install water-saving toilets that use no more than 6 liters (1.6 gallons) per flush.*
- *Flush toilets only when necessary.* Consider using the advice found on a bathroom wall in a drought-stricken area: "If it's yellow, let it mellow; if it's brown, flush it down."
- *Install water-saving showerheads and flow restrictors on all faucets.* If a 3.8-liter (1-gallon) jug can be filled by your showerhead in less than 15 seconds, you need a more efficient fixture.
- *Check frequently for water leaks in toilets and pipes, and repair them promptly.* A toilet must be leaking more than 940 liters (250 gallons) *per day* before you can hear the leak. To test for toilet leaks, add a water-soluble vegetable dye to the water in the tank, but don't flush. If you have a leak, some color will show up in the bowl's water within a few minutes.
- *Turn off sink faucets while brushing teeth, shaving, or washing.*
- *Wash only full loads of clothes;* if smaller loads must be used, use the lowest possible water-level setting.
- *When buying a new washer, choose one that uses the least amount of water and that fills up to different levels for loads of different sizes.* Front-loading clothes models use less water and energy than comparable top-loading models.
- *Use automatic dishwashers for full loads only.* Also, use the short cycle and let dishes air dry to save energy and money.
- *When washing many dishes by hand, don't let the faucet run.* Instead, use one filled dishpan or sink for washing and another for rinsing.
- *Keep one or more large bottles of water in the refrigerator rather than running water from the tap until it gets cold enough for drinking.*
- *Don't use a garbage disposal system—a large user of water.* Instead, consider composting your food wastes.
- *Wash a car from a bucket of soapy water, and use the hose for rinsing only. use a commercial car wash that recycles its water.*
- *Sweep walks and driveways instead of hosing them off.*
- *Reduce evaporation losses by watering lawns and gardens in the early morning or evening, rather than in the heat of midday or when it's windy.*
- *Use drip irrigation and mulch for gardens and flower beds.* Better yet, landscape with native plants adapted to local average annual precipitation so that watering is unnecessary.

REDUCING WATER POLLUTION

- *Use manure or compost instead of commercial inorganic fertilizers to fertilize garden and yard plants.*
- *Use biological methods or integrated pest management to control garden, yard, and household pests.*
- *Use low-phosphate, phosphate-free, or biodegradable dishwashing liquid, laundry detergent, and shampoo.*
- *Don't use water fresheners in toilets.*
- *Use less harmful substances instead of commercial chemicals for most household cleaners* (Table 14-2, p. 361).
- *Don't pour pesticides, paints, solvents, oil, antifreeze, or other products containing harmful chemicals down the drain or onto the ground.* Contact your local health department about disposal.
- *If you get water from a private well or suspect that municipal water is contaminated, have it tested by an EPA-certified laboratory for lead, nitrates, trihalomethanes, radon, volatile organic compounds, and pesticides.*
- *If you have a septic tank, have it cleaned out every 3–5 years by a reputable contractor so that it won't contribute to groundwater pollution.*

REDUCING SOLID WASTE AND HAZARDOUS WASTE

- *Buy less by asking yourself whether you really need a particular item.*
- *Buy things that are reusable, recyclable, or compostable, and be sure to reuse, recycle,*

and compost them.

- *Buy beverages in refillable glass containers instead of cans or throwaway bottles. Urge companies and legislators to make refillable plastic (PET) bottles available in the United States.*
- *Use reusable plastic or metal lunch boxes and metal or plastic garbage containers without throwaway plastic liners* (unless such liners are required for garbage collection).
- *Carry sandwiches and store food in the refrigerator in reusable containers instead of wrapping them in aluminum foil or plastic wrap.*
- *Use rechargeable batteries and recycle them when their useful life is over.* In 1993 Rayovac began selling mercury-free, rechargeable alkaline batteries (Renewal batteries) that outperform conventional nickel–cadmium rechargeable batteries.
- *Carry groceries and other items in a reusable basket, a canvas or string bag, or a small cart.*
- *Use sponges and washable cloth napkins, dish towels, and handkerchiefs instead of paper ones.*
- *Stop using throwaway paper and plastic plates,cups, and eating utensils, and other disposable items when reusable or refillable versions are available.*
- *Buy recycled goods, especially those made by primary recycling, and then make an effort to recycle them.* If you're not buying recycled materials, you're not recycling.
- *Reduce the amount of junk mail you get.* Do this (as several million Americans have done) at no charge by contacting the Mail Preference Service, Direct Marketing Association, 11 West 42nd St., P.O. Box 3681, New York, NY 10163-3861 (212-768-7277) and asking that your name not be sold to large mailing-list companies. Of the junk mail you do receive, recycle as much of the paper as possible.
- *Buy products in concentrated form whenever possible.*
- *Choose items that have the least packagingor better yet, no packaging* ("nude products").
- *Don't buy helium-filled balloons that end up as litter. Urge elected officials and school administrators to ban balloon releases except for atmospheric research and monitoring.*
- *Lobby local officials to set up a community composting program.*
- *Use pesticides and other hazardous chemicals (Table -1) only when absolutely necessary and in the smallest amount possible.*
- *Use less hazardous (and usually cheaper) cleaning products* (Table 14-2, p. 361).
- *Don't dispose of hazardous chemicals by flushing them down the toilet, pouring them down the drain, burying them, throwing them into the garbage, or dumping them down storm drains.* Consult your local health department or environmental agency for safe disposal methods.

COMMUNICATING WITH ELECTED OFFICIALS

- Find out their names and addresses. Then write, call, fax, or e-mail them. Contact a senator by writing: The Honorable ___, U.S. Senate, Washington, DC 20510; Tel: 202-224-3121, Web site: http://thomas.loc.gov. Contact a representative by writing: The Honorable ___, U.S. House of Representatives, Washington, DC 20510; Tel: 202-225-3121 Web site: http://thomas.loc.gov; Contact the president by writing: President ___, The White House, 1600 Pennsylvania Ave. NW, Washington, DC 20500; Tel 202-456-1414 Comment line: 202-456-1111 Fax: 202-456-2461 e-mail: president www.whitehouse.gov.
- When you write a letter or e-mail, **(1)** use your own words, **(2)** be brief and courteous, address only one issue, and **(3)** ask the elected official to do something specific (such as cosponsoring, supporting, or opposing certain bills). Give reasons for your position, explain its effects on you and your district, try to offer alternatives, share any expert knowledge you have, and ask for a response. Be sure to include your name and return address.
- After your representatives have cast votes* supporting your position, send them a short note of thanks.
- Call and ask to speak to a staff member who works on the issue you are concerned about: for the White House, 202-456-1414 (Web site: www.whitehouse.gov); for the U.S. Senate, 202-224-3121; for the House of Representatives, 202-225-3121. The Web site address for the U.S. Congress is http://thomas.loc.gov.
- Once a desirable bill is passed, call or write to urge the president not to veto it. Urge the members of the appropriations committee to appropriate enough money to implement the law—a crucial decision.
- Monitor and influence action at the state and local levels, where all federal and state laws are either ignored or enforced. As Thomas Jefferson said, "The execution of laws is more important than the making of them."
- Get others who agree with your position to contact their elected officials.

---

*Each year the League of Conservation Voters (1707 L St. NW, Suite 750, Washington, DC 20036; Tel: 202-785-8683, Web: www.lcv.org) publishes an *Environmental Scorecard* that rates all members of Congress on how they voted on environmental issues.

INTERNET SOURCES AND FURTHER READINGS

INTERNET

For a list of useful Internet sites, hyperlinks to material in this book, and other helpful and interesting information, please visit our Web site at **www.brookscole.com/biology**

GENERAL SOURCES OF ENVIRONMENTAL INFORMATION

Readings

Ashworth, William. 1991. *The Encyclopedia of Environmental Studies.* New York: Facts on File.

Beacham, W., ed. 1993. *Beacham's Guide to Environmental Issues and Sources.* 5 vols. Washington, D.C.: Beacham.

British Petroleum. Annual. *BP Statistical Review of World Energy.* New York: BP America.

Brown, Lester R., et al. Annual. *State of the World.* New York: Norton.

Brown, Lester R., et al. Annual. *Vital Signs.* New York: Norton.

Katz, Michael, and Dorothy Thornton. 1997. *Environmental Management Tools on the Internet: Assessing the World of Environmental Information.* Delray Beach, Calif.: St. Lucie.

Kurland, Daniel J., and Jane Heinze-Fry. 1996. *Introduction to the Internet.* Belmont, Calif.: Wadsworth. Available as a supplement for use with this book.

Lean, Geoffrey, and Don Hinrichsen, eds. 1994. *Atlas of the Environment.* 2d ed. New York: HarperCollins.

Marien, Michael, ed. 1996. *Environmental Issues and Sustainable Futures: A Critical Guide to Recent Books, Reports, and Periodicals.* Bethesda, Md.: World Future Society.

Paehilke, Robert, ed. 1995. *Conservation and Environmentalism: An Encyclopedia.* New York: Garland.

Population Reference Bureau. Annual. *World Population Data Sheet.* Washington, D.C.: Population Reference Bureau.

Schupp, Jonathan F. 1995. *Environmental Guide to the Internet.* Rockville, Md.: Government Institutes, Inc.

Seager, Joni, et al. 1995. *The New State-of-the-Earth Atlas.* New York: Simon & Schuster.

United Nations. Annual. *Demographic Yearbook.* New York: United Nations.

United Nations Children's Fund (UNICEF). Annual. *The State of the World's Children.* New York: UNICEF.

United Nations Environment Programme (UNEP). Annual. *State of the Environment.* New York: UNEP.

United Nations Population Fund. Annual. *The State of World Population.* New York: United Nations Population Fund.

U.S. Bureau of the Census. Annual. *Statistical Abstract of the United States.* Washington, D.C.: U.S. Bureau of the Census.

World Health Organization (WHO). Annual. *World Health Statistics.* Geneva, Switzerland: WHO.

World Resources Institute and International Institute for Environment and Development. Biannual. *World Resources.* New York: Basic Books.

1 / ENVIRONMENTAL PROBLEMS, THEIR CAUSES, AND SUSTAINABILITY

Internet

For a list of useful Internet sites, hyperlinks to material in this book, and other helpful and interesting information, please visit our Web site at **www.brookscole.com/biology**

Readings

Atkisson, Alan. 1999. *Believing Cassandra.* White River Junction, Vt.: Chelsea Green.

Ayres, Ed. 1999. *God's Last Offer: Negotiating for a Sustainable Future.* New York: Four Walls Eight Windows.

Bailey, Ronald, ed. 1995. *The True State of the Planet.* New York: Free Press.

Barney, Gerald G., et al. 1993. *Global 2000 Revisited: What Shall We Do?* Arlington, Va.: Millennium Institute.

Bast, Joseph L., et al. 1994. *Eco Sanity: A Common-Sense Guide to Environmentalism.* Lantham, Md.: Madison Books.

Bell, Simon, and Stephen Morse, 1998. *Sustainability Indicators.* East Haven, Conn.: Earthscan.

Brown, Lester R., and Hal Kane. 1994. *Full House: Reassessing the Earth's Population Carrying Capacity.* New York: Norton.

Burger, Joanna, and Michael Gochfeld. 1998. "The Tragedy of the Commons: 30 Years Later." *Environment,* vol. 40, no. 10, 4–13, 26–27.

Chivian, Eric, et al. 1993. *Critical Condition: Human Health and the Environment.* Cambridge, Mass.: MIT Press.

Commoner, Barry. 1994. *Making Peace with the Planet.* Rev. ed. New York: New Press.

Daily, Gretchen, ed. 1997. *Nature's Services: Society's Dependence on Natural Ecosystems.* Washington, D.C.: Island Press.

Dunn, Terry L. 1997. *Guide to Global Environmental Issues.* Golden, Colo.: Fulcrum.

Easterbrook, Gregg. 1995. *A Moment on the Earth: The Coming Age of Environmental Optimism.* New York: Viking.

Ehrlich, Paul R., and Anne H. Ehrlich. 1990. *The Population Explosion.* New York: Doubleday.

Ehrlich, Paul R., and Anne H. Ehrlich. 1991. *Healing the Planet.* Reading, Mass.: Addison-Wesley.

Ehrlich, Paul R., and Anne H. Ehrlich. 1996. *Betrayal of Science and Reason: How Anti-Environmental Rhetoric Threatens Our Future.* Washington, D.C.: Island Press.

Flattau, Edward. 1998. *Tracking the Charlatans: Countering the Eco-Bashers.* New York: Global Horizons.

Gore, Al. 1992. *Earth in the Balance: Ecology and the Human Spirit.* Boston: Houghton Mifflin.

Hammond, Allen. 1998. *Which World? Scenarios for the 21st Century.* Washington, D.C.: World Resources Institute.

Hardin, Garrett. 1968. "The Tragedy of the Commons." *Science,* vol. 162, 1243–48.

Hardin, Garrett. 1993. *Living Within Limits: Ecology, Economics, and Population Taboos.* New York: Oxford University Press.

Hardin, Garrett. 1998. "Extensions of the Tragedy of the Commons." *Science,* vol. 279, 682–83.

Hawken, Paul, Amory Lovins, and Hunter Lovins. 1999. *Natural Capitalism.* Boston: Little Brown.

Hinrichsen, Don. 1999. "6,000,000,000 Consumption Machines." *International Wildlife,* September/October, 22–29.

IUCN. 1997. *Conservation and the Future: Trends and Options toward the Year 2025.* Washington, D.C.: Island Press (IUCN).

Keating, Michael. 1997. *Canada and the State of the Planet: The Social, Economic, and Environmental Trends That Are Shaping Our Lives.* New York: Oxford University Press.

Lehr, Jay, ed. 1994. *Rational Readings on Environmental Concerns.* New York: Van Nostrand Reinhold.

Lubchenco, Jane. 1998. "Entering the Century of the Environment: A New Social Contract for Science." *Science,* vol. 279, 491-97.

Markley, Oliver W., and Walter R. McCuan, eds. 1996. *21st Century Earth:*

Opposing Viewpoints. Boston: Greenhaven.

Mazur, Laurie Ann, ed. 1994. *Beyond the Numbers: A Reader on Population, Consumption, and the Environment.* Washington, D.C.: Island.

Meadows, Donella. 1991. *The Global Citizen.* Washington, D.C.: Island Press.

Meadows, Donella H., et al. 1992. *Beyond the Limits: Confronting Global Collapse, Envisioning a Sustainable Future.* White River Junction, Vt.: Chelsea Green.

Mitchell, Bruce, ed. 1995. *Resource and Environmental Management in Canada: Addressing Conflict and Uncertainty.* New York: Oxford University Press.

Myers, Norman, ed. 1993. *Gaia: An Atlas of Planet Management.* Garden City, N.Y.: Anchor/Doubleday.

Myers, Norman. 1996. *Ultimate Security: The Environmental Basis of Political Stability.* New York: Norton.

Myers, Norman, and Julian Simon. 1994. *Scarcity or Abundance? A Debate on the Environment.* New York: Norton.

Quinn, Daniel. 1992. *Ishmael.* New York: Bantam/Turner.

Rees, William, and Mathis Wackernagel. 1995. *Our Ecological Footprint.* Philadelphia: New Society.

Sadler, A. E. 1996. *The Environment: Opposing Viewpoints.* San Diego, Calif.: Greenhaven.

Simon, Julian L., ed. 1995. *The True State of the Planet.* Cambridge, Mass.: Basil Blackwell.

Simon, Julian L. 1996. *The Ultimate Resource 2.* Princeton, N.J.: Princeton University Press.

Suzuki, David. 1994. *Time to Change.* Toronto: Stoddard.

United Nations. 1997. *Global Environmental Outlook.* New York: Oxford University Press.

Wagner, Travis. 1993. *In Our Backyard: A Guide to Understanding Pollution and Its Effects.* New York: Van Nostrand Reinhold.

West, Bernadette. 1995. *The Reporter's Environmental Handbook.* New Brunswick, N.J.: Rutgers University Press.

2 / ECONOMICS, POLITICS, ETHICS, AND SUSTAINABILITY

Internet

For a list of useful Internet sites, hyperlinks to material in this book, and other helpful and interesting information, please visit our Web site at **www.brookscole.com/biology**

Readings

Aaseng, Nathan. 1994. *Jobs vs. the Environment: Can We Save Both?* Hillside, N.J.: Enslow.

Adler, Jonathan. 1996. *Environmentalism at the Crossroads: Green Activism in America.* Washington, D.C.: Capital Research Center.

Andersen, Terry, and Donald R. Leal. 1991. *Free-Market Environmentalism.* Boulder, Colo.: Westview.

Anderson, Alison. 1997. *Media, Culture, and the Environment.* New Brunswick, N.J.: Rutgers University Press.

Anderson, Ray. 1999. *Mid-Course Correction: Toward a Sustainable Enterprise.* White River Junction, Vt.: Chelsea Green.

Arnold, Ron, and Alan Gottlieb. 1993. *Trashing the Economy: How Runaway Environmentalism Is Wrecking America.* Bellevue, Wash.: Free Enterprise Press.

Ashworth, William. 1995. *The Economy of Nature: Rethinking the Connections Between Ecology and Economics.* Boston: Houghton Mifflin.

Athansiou, Tony. 1996. *Divided Planet: The Ecology of Rich and Poor.* Boston: Little, Brown.

Ausbel, Kenny. 1997. *Restoring the Earth: Visionary Solutions from the Bioneers.* Tiburon, Calif.: H.J. Kramer.

Ayers, Robert U. 1998. *Turning Point: The End of the Growth Paradigm.* New York: St. Martin's.

Barnet, Richard J., and John Cavanagh. 1994. *Global Dreams: Imperial Corporations and the New World Order.* New York: Simon & Schuster.

Beder, Sharon. 1997. *Global Spin: The Corporate Assault on Environmentalism.* White River Junction, Vt.: Chelsea Green.

Bennett, Steven J., et al. 1994. *Corporate Realities and Environmental Truths.* New York: Wiley.

Berry, Thomas. 1988. *The Dream of the Earth.* San Francisco: Sierra Club Books.

Berry, Wendell. 1990. *What Are People For?* Berkeley, Calif.: North Point Press.

Bhat, V. N. 1996. *The Green Corporation: The Next Competitive Advantage.* Westport, Conn.: Greenwood.

Bossel, Hartmut. 1998. *Paths to a Sustainable Future.* New York: Cambridge University Press.

Bowers, C. A. 1997. *The Culture of Denial: Why the Environmental Movement Needs a Strategy for Reforming Universities and Schools.* Albany: State University of New York.

Bramwell, Anna. 1994. *The Fading of the Greens: The Decline of Environmental Politics in the West.* New Haven, Conn.: Yale University Press.

Brockmeier, Jons, et al. 1998. *Greenspeak: A Study of Environmental Discourse.* Thousand Oaks, Calif.: Sage.

Brower, Michael, and Warren Leon. 1999. *The Consumer's Guide to Environmental Choices.* New York: Three Rivers.

Brown, Lester R. 1999. "Threshold: Early Signs of an Environmental Awakening." *World Watch,* Mar./Apr., 13–22.

Brown, Lester R., and Christopher Flavin. 1999. "A New Economy for a New Century." In Lester R. Brown et al., *State of the World 1999.* Washington, D.C.: Worldwatch Institute, pp. 3–21.

Bryant, Bunyan, ed. 1995. *Environmental Justice: Issues, Policies, and Solutions.* Washington, D.C.: Island Press.

Bullard, Robert D., ed. 1994. *Unequal Protection: Environmental Justice and Communities of Color.* San Francisco: Sierra Club Books.

Cairncross, Frances. 1995. *Green, Inc.: A Guide to Business and the Environment.* Washington, D.C.: Island Press.

Caldwell, Lynton K. 1997. *International Environmental Policy: From the Twentieth to the Twenty-First Century.* Durham, N.C.: Duke University Press.

Callicott, J. Baird. 1994. *Earth's Insights (A Survey of Ecological Ethics from the Mediterranean Basin to the Australian Outback).* Berkeley: University of California Press.

Carley, Michael, and Philippe Sparpens. 1998. *Sharing the Wealth: Sustainable Living and Global Equity in the 21st Century.* East Haven, Conn.: Earthscan.

Carson, Rachel. 1962. *Silent Spring.* Boston: Houghton Mifflin.

Chetlow, Marian R., and Daniel C. Esty. 1997. *Thinking Ecologically: The Next Generation of Environmental Policy.* New Haven, Conn.: Yale University Press.

Cobb, Clifford, et al. 1995. *The Genuine Progress Indicator.* San Francisco: Redefining Progress.

Cobb, John B., Jr. 1994. *Is It Too Late? A Theology of Ecology.* Rev. ed. Denton, Tex.: Environmental Ethics Books.

Cohen, Joel E., and David Tilghman. 1996. "Biosphere 2 and Biodiversity: The Lessons So Far." *Science,* vol. 274, 1150–51.

Cohen, Michael J. 1989. *Connecting with Nature: Creating Moments That Let Earth Teach.* Eugene, Ore.: World Peace University Press.

Cohn, Susan. 1995. *Green at Work: Finding a Business Career That Works for the Environment.* Rev. ed. Washington, D.C.: Island Press.

Conca, Ken, and Geoffrey D. Dabelko, eds. 1998. *Green Planet Blues: Environmental Politics from Stockholm to Tokyo.* Boulder, Colo.: Westview.

Costanza, Robert, et al. 1997. *An Introduction to Ecological Economics: The Science and Management of Sustainability.* Boca Raton, Fla.: Lewis.

Creighton, Sara H. 1998. *Greening the Ivory Tower.* Cambridge, Mass.: MIT Press.

Crocker, David A., and Toby Linden, eds. 1997. *The Ethics of Consumption.* Savage, Md.: Rowman & Littlefield.

Daily, Gretchen C., ed. 1997. *Nature's Services: Societal Dependence on Natural Ecosystems.* Washington, D.C.: Island Press.

Daly, Herman E. 1996. *Beyond Growth: The Economics of Sustainable Development.* Boston: Beacon.

Daly, Herman E., and John B. Cobb, Jr. 1994. *For the Common Good: Redirecting the Economy Toward Community, the Environment, and a Sustainable Future.* 2d ed. Boston: Beacon.

Davies, J. Clarence, and Jan Mazurek. 1997. *Regulating Pollution: Does the U.S. System Work?* Washington, D.C.: Resources for the Future.

Day, Martyn. 1997. *Environmental Action: A Citizen's Guide.* Chicago: Pluto.

Devall, Bill, and George Sessions. 1985. *Deep Ecology: Living as if Nature Mattered.* Salt Lake City, Utah: Gibbs H. Smith.

Dowie, Mark. 1995. *Losing Ground: American Environmentalism at the Close of the Twentieth Century.* Cambridge, Mass.: MIT Press.

Doyle, Kevin. 1999. *The Complete Guide to Environmental Careers in the 21st Century.* New York: Environmental Careers Organization.

Durning, Alan T. 1992. *How Much Is Enough? The Consumer Society and the Earth.* New York: Norton.

Durning, Alan T. 1999. *Green-Collar Jobs: Working in the New Northwest.* Seattle, Wash. Northwest Environment Watch.

Durning, Alan Their, and Yoram Bauman. 1998. *Tax-Shift: How to Help the Economy, Improve the Environment, and Get the Tax Man Off Our Backs.* Seattle, Wash. Northwest Environment Watch.

Eagen, David J., and David W. Orr. 1994. *The Campus and Environmental Responsibility.* San Francisco: Jossey-Bass.

Earth Island Books. 1993. *The Case Against Free Trade: GATT, NAFTA, and the Globalization of Corporate Power.* Berkeley, Calif.: North Atlantic Books.

Eden, Sally. 1996. *Environmental Issues and Business: Implications of a Changing Agenda.* New York: Wiley.

Ehrenfeld, David. 1993. *Beginning Again: People and Nature in the New Millennium.* New York: Oxford University Press.

Ehrlich, Paul R., and Anne H. Ehrlich. 1996. *Betrayal of Science and Reason: How Anti-Environmental Rhetoric Threatens Our Future.* Washington, D.C.: Island Press.

Eisenberg, Evan. 1998. *The Ecology of Eden.* New York: Knopf.

Elderige, Niles. 1995. *Dominion.* New York: Henry Holt.

Elgin, Duane. 1993. *Voluntary Simplicity: Toward a Way of Life That Is Outwardly Simple, Inwardly Rich.* Rev. ed. New York: William Morrow.

Fagin, Dan, and Marianne Lavelle. 1996. *Toxic Deception.* Secaucus, N.J.: Carol Publishing.

Fanning, Odum. 1995. *Opportunities for Environmental Careers.* Lincolnwood, Ill.: VCM Career Horizons.

Farber, Daniel, ed. 1998. *The Struggle for Ecological Democracy: Environmental Justice Movements in the United States.* New York: Guilford.

Ferkiss, Victor. 1993. *Nature, Technology, and Society: Cultural Roots of the Current Environmental Crisis.* New York: New York University Press.

Fiorino, Daniel J. 1995. *Making Environmental Policy.* Berkeley: University of California Press.

Flattau, Edward. 1998. *Tracking the Charlatans: Countering the Eco-Bashers.* New York: Global Horizons.

Foreman, Christopher H., Jr. 1998. *The Promise and Peril of Environmental Justice.* Washington, D.C.: Brookings Institution.

Foreman, Dave. 1991. *Confessions of an Eco-Warrior.* New York: Harmony Books.

Foster, John Bellamy. 1995. *The Vulnerable Planet: A Short Economic History of the Planet.* New York: Monthly Review Press.

Frank, Robert H. 1999. *Luxury Fever: Why Money Fails to Satisfy in an Era of Excess.* New York: Free Press.

French, Hilary E. 1993. *Costly Trade-Offs: Reconciling Trade and the Environment.* Washington, D.C.: Worldwatch Institute.

French, Hilary F. 1998. *Investing in the Future: Harnessing Private Capital Flows for Environmentally Sustainable Development.* Washington, D.C.: Worldwatch Institute.

Gay, Kathlyn. 1994. *Pollution and the Powerless: The Environmental Justice Movement.* New York: F. Watts.

Gelbspan, Ross. 1998. *The Heat is On: The Climate Crisis, The Cover-up, and the Prescription.* Reading, Mass.: Perseus.

Gottlieb, Robert. 1993. *Forcing the Spring: The Transformation of the Environmental Movement.* Washington, D.C.: Island Press.

Goudy, John, and Sabine O'Hara. 1995. *Economic Theory for Environmentalists.* New York: St. Lucie.

Graham, Mary. 1999. *The Morning After Earth Day: Practical Environmental Politics.* Washington, D.C.: Brookings Institution.

Greenpeace. 1993. *The Greenpeace Guide to Anti-Environmental Organizations.* Berkeley, Calif.: Odonian.

Greider, William. 1992. *Who Will Tell the People? The Betrayal of American Democracy.* New York: Simon & Schuster.

Hallman, David G. 1994. *Ecotheology: Voices from South and North.* New York: Orbis.

Hammond, Allen. 1998. *Which World? Scenarios for the 21st Century.* Washington, D.C.: World Resources Institute.

Hardin, Garrett. 1978. *Exploring New Ethics for Survival.* 2d ed. New York: Viking.

Hardin, Garrett. 1996. *Stalking the Wild Taboo.* Petrosky, Mich.: Social Contract Press.

Hargrove, Eugene C. 1989. *Foundations of Environmental Ethics.* Englewood Cliffs, N.J.: Prentice Hall.

Hawken, Paul. 1994. *The Ecology of Commerce.* New York: HarperCollins.

Hawken, Paul, Amory Lovins, and Hunter Lovins. 1999. *Natural Capitalism.* Boston: Little Brown.

Hayden, Tom. 1996. *The Lost Gospel of the Earth: A Call for Renewing Nature, Spirit, and Politics.* San Francisco: Sierra Club Books.

Helvarg, David. 1994. *The War Against the Greens.* San Francisco: Sierra Club Books.

Henderson, Hazel. 1996. *Building a Win–Win World: Life Beyond Global Economic Warfare.* San Francisco: Berrett-Koehler.

Hoffman, Andrew J. 1997. *From Heresy to Dogma: An Institutional History of Corporate Environmentalism.* San Francisco: New Lexington.

Hunter, J. Robert. 1997. *Simple Things Won't Save the Earth.* Austin: University of Texas Press.

Irvine, Sandy. 1989. *Beyond Green Consumerism.* London: Friends of the Earth.

Jacob, Jeffrey. 1997. *New Pioneers: The Back-to-the-Land Movement and the Search for Sustainability.* University Park: Pennsylvania State University Press.

Jacobs, Michael. 1991. *The Green Economy: Environment, Sustainable Development, and Politics.* New York: Pluto.

Johnson, Huey D. 1995. *Green Plans: Greenprint for Sustainability.* Lincoln: University of Nebraska Press.

Jordan, Carl F. 1998. *Working with Nature: Resource Management for Sustainability.* Williston, Vt.: Harwood.

Kamieniecki, Sheldon, et al. 1997. *Flashpoints in Environmental Policymaking: Controversies in Achieving Sustainability.* Albany: State University of New York Press.

Kellert, Stephen R. 1996. *The Value of Life: Biological Diversity and Human Society.* Washington, D.C.: Island Press.

Keniry, Julian. 1994. *Ecodemia: Profiles of Environmentally Smart Green Campuses.* Carbondale: Southern Illinois University Press.

Klenig, John. 1991. *Valuing Life.* Princeton, N.J.: Princeton University Press.

Kline, Benjamin. 1997. *First Along the River: A Brief History of the U.S. Environmental Movement.* San Francisco: Acada.

Korten, David. 1995. *When Corporations Rule the World.* West Hartford, Conn.: Kumarian.

Korten, David C. 1999. *The Post Corporate World: Life After Capitalism.* West Hartford, Conn.: Kumarian.

Krishnan, Rajaram, et al., eds. 1995. *A Survey of Ecological Economics.* Washington, D.C.: Island Press.

Landy, Marc K., et al. 1994. *The Environmental Protection Agency: Asking the Wrong Questions, from Nixon to Clinton.* New York: Oxford University Press.

Learner, Steve. 1997. *Eco-Pioneers: Practical Visionaries Solving Environmental Problems.* Cambridge, Mass.: MIT Press.

Lemons, John, et al., eds. 1998. *Ecological Sustainability and Integrity: Concepts and Approaches.* Boston: Kluwer.

Leopold, Aldo. 1949. *A Sand County Almanac.* New York: Oxford University Press.

Leslie, John. 1996. *The End of the World: The Science and Ethics of Human Extinction.* New York: Routledge.

Lewis, Martin W. 1992. *Green Delusions: An Environmentalist Critique of Radical Environmentalism.* Durham, N.C.: Duke University Press.

Libby, Ronald T. 1999. *Eco-Wars: Political Campaigns and Social Movements.* New York: Columbia University Press.

List, Peter C. 1993. *Radical Environmentalism: Philosophy and Tactics.* Belmont, Calif.: Wadsworth.

MacDonald, Mary. 1998. *Agendas for Sustainability: Environment and Development into the 21st Century.* New York: Routledge.

Maguire, Daniel C., et al. 1998. *New Ethics for a Small Planet.* Albany: State University of New York Press.

Makower, Joel. 1993. *The E-Factor: The Bottom-Line Approach to Environmentally Responsible Business.* New York: Times Books.

Mander, Jerry, and Edward Goldsmith, eds. 1996. *The Case Against the Global Economy.* San Francisco: Sierra Club Books.

McKibben, Bill. 1997. *Hope, Human, and Wild: Treading Lightly on the Earth.* St. Paul, Minn.: Hungry Mind.

Mehta, Shekhar, et al. 1997. *Controlling Pollution: Incentives and Regulation.* Thousand Oaks, Calif.: Sage.

Milbrath, Lester W. 1989. *Envisioning a Sustainable Society.* Albany: State University of New York Press.

Milbrath, Lester W. 1995. *Learning to Think Environmentally While There Is Still Time.* Albany: State University of New York Press.

Myers, Norman. 1996. *Ultimate Security: The Environmental Basis of Political Stability.* New York: Norton.

Myers, Norman, and Jennifer V. Kent. 1998. *Perverse Subsidies: Tax $s Undercutting Our Economies and Environments Alike.* Winnipeg, Canada: International Institute for Sustainable Development.

Nabhan, Gary Paul, and Stephen Trimble. 1994. *The Geography of Childhood: Why Children Need Wild Places.* Boston: Beacon.

Naess, Arne. 1989. *Ecology, Community, and Lifestyle.* New York: Cambridge University Press.

Nagpal, Tanvi, and Camilla Foltz, eds. 1995. *A World of Difference: Giving Voices to Visions of Sustainability.* Washington, D.C.: World Resources Institute.

Nash, Roderick. 1988. *The Rights of Nature: A History of Environmental Ethics.* Madison: University of Wisconsin Press.

National Academy of Sciences. 1993. *Greening Industrial Ecosystems.* Washington, D.C.: National Academy Press.

National Academy of Sciences. 1995. *Assigning Economic Value to Natural Resources.* Washington, D.C.: National Academy Press.

National Academy of Sciences. 1997. *The Industrial Green Game: Implications for Environmental Design and Management.* Washington, D.C.: National Academy Press.

National Wildlife Federation. 1995. *Ecodemia: Campus Environmental Stewardship at the Turn of the 21st Century.* Washington, D.C.: National Wildlife Federation.

Ophuls, William, and A. Stephen Boyan, Jr. 1992. *Ecology and the Politics of Scarcity Revisited: The Unraveling of the American Dream.* San Francisco: W.H. Freeman.

O'Riordan, ed. 1997. *Ecotaxation.* New York: St. Martin's.

O'Riordan, Timothy, and James Cameron, eds. 1994. *Interpreting the Precautionary Principle.* East Haven, Conn.: Earthscan.

Orr, David. 1992. *Ecological Literacy.* Ithaca: State University of New York Press.

Orr, David. 1994. *Earth in Mind: On Education, Environment, and the Human Prospect.* Washington, D.C.: Island Press.

Paepke, C. Owen. 1993. *The Evolution of Progress: The End of Economic Growth and the Beginning of the Human Transformation.* New York: Random House.

Palmer, Joy. 1997. *Environmental Education in the 21st Century: Theory, Practice, Progress, and Promise.* New York: Routledge.

Pearce, David. 1995. *Blueprint 4: Sustaining the Earth.* East Haven, Conn.: Earthscan.

Peterson, Tarla Rai. 1997. *Sharing the Earth: The Rhetoric of Sustainable Development.* Columbia: University of South Carolina Press.

Piasecki, Bruce W. 1995. *Corporate Environmental Strategy: The Avalanche of Change Since Bhopal.* New York: Wiley.

Ponting, Clive. 1992. *A Green History of the World. The Environment and the Collapse of Great Civilizations.* New York: St. Martin's.

Porter, Gareth, and Janet Welsh Brown. 1994. *Global Environmental Politics.* 2d ed. Boulder, Colo.: Westview.

President's Council on Sustainable Development. 1996. *Sustainable America: A New Consensus for the Future.* Washington, D.C.: U.S. Government Printing Office.

Prugh, Thomas, Robert Costanza, and Herman Daly. 1999. *The Local Politics of Global Sustainability.* Washington, D.C.: Island Press.

Prugh, Thomas, et al., eds. 1999. *National, Capital and Economic Survival.* 2d ed.

Quinn, Daniel. 1992. *Ishmael.* New York: Bantam/Turner.

Raskin, Jamin B., and John Bonifaz. 1994. *The Wealth Primary.* Washington, D.C.: Center for Responsive Politics.

Rasmussen, Larry. 1998. *Earth Community, Earth Ethics.* Maryknoll, N.Y.: Orbis.

Rees, William, and Mathis Wackernagel. 1995. *Our Ecological Footprint.* Philadelphia: New Society.

Repetto, Robert, et al. 1992. *Green Fees: How a Tax Shift Can Work for the Environment and the Economy.* Washington, D.C.: World Resources Institute.

Rich, Bruce. 1994. *Mortgaging the Earth.* Boston: Beacon.

Rifkin, Jeremy. 1991. *Biosphere Politics: A New Consciousness for a New Century.* New York: Crown.

Rocky Mountain Institute. 1996. *Economic Renewal Guide.* Snowmass, Colo.: Rocky Mountain Institute.

Rolston, Holmes, III. 1994. *Conserving Natural Values.* New York: Columbia University Press.

Romm, Joseph J. 1993. *Defining National Security: The Nonmilitary Aspects.* New York: Council on Foreign Relations Press.

Roodman, Alan M. 1998. *The Natural Wealth of Nations: Harnessing the Market for the Environment.* New York: Norton.

Roodman, David M. 1996. *Paying the Piper: Subsidies, Politics, and the Environment.* Washington, D.C.: Worldwatch Institute.

Roodman, David M. 1997. *Getting the Signals Right: Tax Reform to Protect the Environment and the Economy.* Washington, D.C.: Worldwatch Institute.

Rosenbaum, Walter A. 1998. *Environmental Politics and Solutions.* Washington, D.C.: Congressional Quarterly Press.

Rosenblatt, Roger. 1999. *Consuming Desires: Consumption, Culture, and the Pursuit of Happiness.* Washington, D.C.: Island Press (Shearwater).

Rowell, Andrew. 1997. *Green Backlash: Subversion of the Environmental Movement.* New York: Routledge.

Rubin, Charles T. 1994. *The Green Crusade: Rethinking the Roots of Environmentalism.* New York: Free Press.

Ruckelshaus, William D., and Karl Hausker, eds. 1998. *The Environmental Protection System in Transition: Toward a More Desirable Future.* Washington, D.C.: Center for Strategic and International Studies.

Ryan, John C., and Alan Thein. 1997. *Stuff: The Secret Lives of Everyday Things.* Seattle, Wash.: Northwest Environmental Press.

Sachs, Aaron. 1995. *Eco-Justice: Linking Human Rights and the Environment.* Washington, D.C.: Worldwatch Institute.

Sanjor, William. 1992. *Why the EPA Is Like It Is and What Can Be Done About It.* Washington, D.C.: Environmental Research Foundation.

Saunders, Tedd, and Loretta McGovern. 1994. *The Bottom Line of Green Is Black.* New York: HarperCollins.

Schmidheiny, Stephan. 1992. *Changing Course: A Global Business Perspective on Development and the Environment.* Cambridge, Mass.: MIT Press.

Schor, Juliet B. 1999. *The Overspent American: Why We Want What We Don't Need.* New York: HarperPerennial.

Schumacher, E. F. 1973. *Small Is Beautiful: Economics as if People Mattered.* New York: Harper & Row.

Seidel, Peter. 1998. *Invisible Walls: Why We Ignore the Damage We Inflict on the Planet and Ourselves.* New York: Prometheus.

Selcraig, Bruce. 1998. "Reading, 'Riting, and Ravaging: The Three R's Brought to You by Corporate America and the Far Right." *Sierra,* May/June, 60–65, 86–88.

Sessions, George, ed. 1994. *Deep Ecology for the Twenty-First Century.* Boston: Shambhala.

Shabecoff, Philip. 1993. *A Fierce Green Fire: The American Environmental Movement.* New York: Hill & Wang.

Shabecoff, Philip. 1996. *A New Name for Peace: International Environmentalism, Sustainable Development, and Democracy.* Hanover, N.H.: University of New England Press.

Smith, April, and the Student Environmental Action Coalition. 1993. *Campus Ecology: A Guide to Assessing Environmental Quality and Creating Strategies for Change.* Los Angeles, Calif.: Living Planet.

Smith, Fraser, ed. 1997. *Environmental Sustainability: Practical Global Implications.* Boca Raton, Fla.: St. Lucie.

Smith, Joseph, et al. 1998. *The Bankruptcy of Economics: Ecology, Economics, and the Sustainability of the Earth.* New York: St. Martin's.

Stauber, John C., and Sheldon Rampton. 1995. *Toxic Sludge Is Good for You: Lies, Damn Lies, and the Public Relations Industry.* Madison, Wis.: PR Watch.

Student Conservation Association. 1997. *The Guide to Graduate Environmental Programs.* Washington, D.C.: Island Press.

Suzuki, David. 1998. *The Sacred Balance: Rediscovering Our Place in Nature.* New York: Prometheus.

Swimme, Brian, and Thomas Berry. 1992. *The Universe Story: From the Primordial Flaring Forth to the Ecozoic Era.* San Francisco: HarperCollins.

Switzer, Jacqueline. 1997. *Green Backlash: The History and Politics of Environmental Opposition in the United States.* Boulder, Colo.: Lynne Rienner.

Switzer, Jacqueline, and Gary Byner. 1998. *Environmental Politics: Domestic and Global Dimensions.* New York: St. Martin's.

Thomashow, Mitchell. 1995. *Ecological Identity: Becoming a Reflective Environmentalist.* Cambridge, Mass.: MIT Press.

Tietenberg, Tom. 1992. *Environmental and Resource Economics.* 3d ed. Glenview, Ill.: Scott, Foresman.

Tokar, Brian. 1997. *Earth for Sale: Reclaiming Ecology in the Age of Corporate Greenwash.* Boston: South End Press.

Tucker, Mary Evelyn, and John A. Grim, eds. 1994. *Worldviews and Ecology: Religion, Philosophy, and the Environment.* Maryknoll, N.Y.: Orbis.

Turner, R. Kerry, et al. 1994. *Environmental Economics: An Elementary Introduction.* Baltimore: Johns Hopkins University Press.

Van DeVeer, Donald, and Christine Pierce. 1994. *The Environmental Ethics and Policy Book: Philosophy, Ecology, Economics.* Belmont, Calif.: Wadsworth.

Watson, Paul. 1994. *Ocean Warrior.* Marina del Ray, Calif.: Key Porter Books.

Wege, Peter M. 1999. *Economicology: The Eleventh Commandment.* Grand Rapids, Mich.: Center for Environmental Study, Grand Rapids Community College.

Werbach, Adam. 1997. *Act Now, Apologize Later.* New York: Cliff Street Books.

Westra, Laura. 1997. *Living in Integrity: A Global Effort to Restore a Fragmented Earth.* Savage, Md.: Rowman & Littlefield.

Wilkinson, Todd. 1998. *Science Under Siege: The Politicians' War on Nature and Truth.* New York: Johnson.

World Resources Institute. 1993. *A New Generation of Environmental Leadership: Action for the Environment and the Economy.* Washington, D.C.: World Resources Institute.

Worster, Donald. 1993. *The Wealth of Nature: Environmental History and the Ecological Imagination.* New York: Oxford University Press.

Zimmerman, Michael. 1995. *Science, Nonscience, and Nonsense: Approaching Environmental Literacy.* Baltimore, Md.: Johns Hopkins University Press.

3 / SCIENCE, SYSTEMS, MATTER, AND ENERGY

Internet

For a list of useful Internet sites, hyperlinks to material in this book, and other helpful and interesting information, please visit our Web site at **www.brookscole.com/biology**

Readings

Bauer, Henry H. 1992. *Scientific Literacy and the Myth of the Scientific Method.* Urbana: University of Illinois Press.

Carey, Stephen S. 1998. *A Beginner's Guide to Scientific Method.* 2d ed. Belmont, Calif.: Wadsworth. (Available as a supplement for use with this book.)

Heinze-Fry, Jane. 1996. *An Introduction to Critical Thinking.* Belmont, Calif.: Wadsworth. Available as a supplement for use with this book.

Kuhn, Thomas S. 1970. *The Structure of Scientific Revolutions.* 2d ed. Chicago: University of Chicago Press.

Lubchenco, Jane. 1998. "Entering the Century of the Environment: A New Social Contract for Science." *Science*, vol. 279, 491–97.

McCain, Garvin, and Erwin M. Segal. 1998. *The Game of Science.* 5th ed. Pacific Grove, Calif.: Brooks/Cole. Available as a supplement for use with this book.

McComas, William. 1996. "Ten Myths of Science: Reexamining What We Think We Know." *School Science and Mathematics*, vol. 86, 10–16.

Miller, G. Tyler, Jr., and David G. Lygre. 1991. *Chemistry: A Contemporary Approach.* 3d. ed. Belmont, Calif.: Wadsworth.

Rifkin, Jeremy. 1989. *Entropy: Into the Greenhouse World: A New World View.* New York: Bantam.

Zimmerman, Michael. 1995. *Science, Nonscience, and Nonsense: Approaching Environmental Literacy.* Baltimore, Md.: Johns Hopkins University Press.

4 / ECOSYSTEMS AND HOW THEY WORK: CONNECTIONS IN NATURE

Internet

For a list of useful Internet sites, hyperlinks to material in this book, and other helpful and interesting information, please visit our Web site at **www.brookscole.com/biology**

Readings

Begon, Michael, John L. Harper, and Colin R. Townsend. 1996. *Ecology: Individuals, Populations, and Communities.* 3d ed. Boston: Blackwell Scientific Publications.

Berenbaum, May R. 1995. *Bugs in the System: Insects and Their Impact on Human Affairs.* Reading, Mass.: Addison-Wesley.

Berner, Elizabeth K., and Robert A. Berner. 1996. *Global Environment: Water, Air, and Geochemical Cycles.* Upper Saddle River, N.J.: Prentice Hall.

Bradbury, Ian K. 1991. *The Biosphere.* New York: Belhaven.

Chapin, F. S. Terry, and Pamela A. Matson. 1998. *Principles of Ecosystem Ecology.* New York: Springer-Verlag.

Colinvaux, Paul A. 1992. *Ecology.* 2d ed. New York: Wiley.

Daily, Gretchen, ed. 1997. *Nature's Services: Society's Dependence on Natural Ecosystems.* Washington, D.C.: Island Press.

Dodson, Stanley I., et al. 1998. *Ecology.* New York: Oxford University Press.

Ehrlich, Anne H., and Paul R. Ehrlich. 1987. *Earth.* New York: Franklin Watts.

Elsom, Derek. 1992. *Earth: The Making, Shaping, and Working of a Planet.* New York: Macmillan.

Krebs, Charles J. 1994. *Ecology: : The Experimental Analysis of Distribution & Abundance.* 4th ed. New York: HarperCollins.

Lovelock, James. 1988. *The Ages of Gaia.* New York: Norton.

Lubchenco, Jane. 1998. "Entering the Century of the Environment: A New Social Contract for Science." *Science*, vol. 279, 491–97.

Mackenzie, Fred T. 1998. *Our Changing Planet.* 2d ed. Englewood Cliffs, N.J.: Prentice Hall.

Macmillan Publishing. 1992. *The Way Nature Works.* New York: Macmillan.

Odum, Eugene P. 1997. *Ecology: A Bridge Between Science and Society.* Sunderland, Mass.: Sinauer.

Odum, Eugene P. 1998. *Ecological Vignettes: Ecological Approaches to Dealing with Human Predicaments.* Williston, Vt.: Harwood.

Rickleffs, Robert E. 1997. *The Economy of Nature: A Textbook in Basic Ecology.* 4th ed. New York: W.H. Freeman.

Smil, Vaclav. 1997. *Cycles of Life: Civilization and the Biosphere.* New York: Scientific American Library.

Smil, Vaclav. 1997. "Global Population and the Nitrogen Cycle." *Scientific American*, July, 76–81.

Smith, Robert L. 1998. *Elements of Ecology.* 4th ed. New York: Harper & Row.

Tudge, Colin. 1991. *Global Ecology.* New York: Oxford University Press.

Vitousek, Peter M., Paul R. Ehrlich, Anne H. Ehrlich, and Pamela Matson. 1986. "Human Appropriation of the Products of Photosynthesis." *BioScience*, June, 368–73.

5 / EVOLUTION, BIODIVERSITY, AND COMMUNITY PROCESSES

Internet

For a list of useful Internet sites, hyperlinks to material in this book, and other helpful and interesting information, please visit our Web site at **www.brookscole.com/biology**

Readings

Allègre, Claude J., and Stephen H. Schneider. 1994. "The Evolution of the Earth." *Scientific American,* October, 66–75.

Baskin, Yvonne. 1997. *The Work of Nature: How the Diversity of Life Sustains Us.* Washington, D.C.: Island Press.

Blaustein, Andrew R., and David W. Wake. 1995. "The Puzzle of Declining Amphibians." *Scientific American,* April, 52–57.

Brown, Bruce, and Lane Morgan. 1990. *The Miracle Planet.* Edison, N.J.: W.H. Smith.

Elderidge, Niles. 1991. *The Miner's Canary: Unraveling the Mysteries of Extinction.* Englewood Cliffs, N.J.: Prentice Hall.

Elderidge, Niles. 1995. *Dominion.* New York: Henry Holt.

Leakey, Richard, and Roger Lewin. 1995. *The Sixth Extinction: Patterns of Life and the Future of Humankind.* New York: Doubleday.

Levin, Simon. 1999. *Fragile Dominion: Complexity and the Commons.* New York: Perseus.

Lovelock, James E. 1988. *The Ages of Gaia: A Biography of Our Living Earth.* New York: Norton.

Lunine, Jonathan I. 1998. *Earth: Evolution of a Natural World.* New York: Columbia University Press.

Margulis, Lynn, and Lorraine Olendzenski. 1992. *Environmental Evolution: Effects of the Origin and Evolution of Life on Planet Earth.* Cambridge, Mass.: MIT Press.

Myers, Norman. 1997. "Mass Extinction and Evolution." *Science,* vol. 278, 597–98.

Pimentel, David, et al. 1997. "Economic and Environmental Benefits of Biodiversity." *BioScience,* vol. 47, no. 11, 747–57.

Potts, Rick. 1997. *Humanity's Descent: The Consequences of Ecological Instability.* New York: Avon.

Reaka-Kudla, Marjorie I., Don E. Wilson, and Edward O. Wilson, eds. 1997. *Biodiversity II: Understanding and Protecting Our Biological Resources.* Washington, D.C.: Joseph Henry.

Ridley, Mark. 1993. *Evolution.* Boston: Blackwell Scientific.

Wilson, Edward O. 1992. *The Diversity of Life.* Cambridge, MA: Belknap Press of Harvard University.

6 / CLIMATE, WEATHER, AND BIODIVERSITY

Internet

For a list of useful Internet sites, hyperlinks to material in this book, and other helpful and interesting information, please visit our Web site at **www.brookscole.com/biology**

Readings

Aber, John, and Jerry Melito. 1991. *Terrestrial Ecosystems.* Philadelphia: Saunders.

Ahrens, C. Donald. 1994. *Meteorology Today: An Introduction to Weather, Climate, and the Environment.* 5th ed. Belmont, Calif.: Wadsworth.

Akin, Wallace E. 1991. *Global Patterns: Climate, Vegetation, and Soils.* Norman: University of Oklahoma Press.

Allan, J. David. 1995. *Stream Ecology: Structure and Function of Running Waters.* New York: Chapman & Hall.

Attenborough, David, et al. 1989. *The Atlas of the Living World.* Boston: Houghton Mifflin.

Bailey, Robert G. 1998. *The Ecosystem Geography of the Oceans and Continents.* New York: Springer-Verlag.

Cappuccino, Naomi, and Peter W. Price, eds. 1995. *Population Dynamics: New Approaches and Syntheses.* New York: Academic Press.

Cook, L. M. 1991. *Genetic and Ecological Diversity.* New York: Chapman & Hall.

Dean, Cornelia. 1999. *Against the Tide: The Battle for America's Beaches.* New York: Columbia University Press.

Denniston, Derek. 1995. *High Priorities: Conserving Mountain Ecosystems and Cultures.* Washington, D.C.: Worldwatch Institute.

Dugan, Patrick, ed. 1993. *Wetlands in Danger: A World Conservation Atlas.* New York: Oxford University Press.

Elder, Danny, and John Pernetta, eds. 1991. *Oceans.* London: Michael Beazley.

Eldridge, Niles. 1991. *The Miner's Canary: Unraveling the Mysteries of Extinction.* New York: Prentice Hall.

Garrison, Tom. 1998. *Essentials of Oceanography.* 2d ed. Belmont, Calif.: Wadsworth.

Goldman, C., and A. Horne. 1994. *Limnology.* New York: McGraw-Hill.

Goldsmith, Edward, et al. 1990. *Imperiled Planet: Restoring Our Endangered Ecosystems.* Cambridge, Mass.: MIT Press.

Golley, Frank B., ed. 1983. *Tropical Rain Forest Ecosystems: Structure and Function.* New York: Elsevier.

Hare, Tony, ed. 1994. *Habitats.* New York: Macmillan.

Hinrichsen, Don. 1997. *The World's Coastal Seas: Trends, Threats, and Strategies.* Washington, D.C.: Island Press.

Hutchinson, G. Evelyn. 1993. *A Treatise on Limnology.* New York: Wiley.

Lincare, Edward, and Bart Geerts. 1997. *Climate and Weather Explained.* New York: Routledge.

Lovelock, James E. 1991. *Healing Gaia: Practical Medicine for the Planet.* New York: Random House.

MacArthur, Robert H. 1972. *Geographical Ecology.* San Francisco: Harper & Row.

National Academy of Sciences. 1995. *Wetlands: Characteristics and Boundaries.* Washington, D.C.: National Academy Press.

Niering, William A. 1991. *Wetlands of North America.* Charlottesville, Va.: Thomasson-Grant.

Nisbet, E. G. 1991. *Living Earth.* San Francisco: HarperCollins.

Pilkey, Orin H., Jr., and William J. Neal, eds. 1987. *Living with the Shore.* Durham, N.C.: Duke University Press.

Reaka-Kudla, Marjorie, Don E. Wilson, and Edward O. Wilson, eds. 1996. *Biodiversity II: Understanding and Protecting Our Biological Resources.* New York: John Henry.

Sapp, Jan. 1999. *What Is Natural? Coral Reef Crisis.* New York: Oxford University Press.

Simon, Noel. 1995. *Nature and Danger: Threatened Habitats and Species.* New York: Oxford University Press.

Stevenson, Robert F., and Frank H. Talbot. 1993. *Oceans.* Emmaus, Pa.: Rodale.

Thorne-Miller, Boyce. 1999. *The Living Ocean: Understanding and Protecting Marine Biodiversity. 2d ed.* Washington, D.C.: Island Press.

Tudge, Colin. 1991. *The Environment of Life.* New York: Oxford University Press.

Vileisis, Ann. 1997. *Discovering the Unknown Landscape: A History of America's Wetlands.* Washington, D.C.: Island Press.

Weber, Michael L., and Judith A. Gradwohl. 1995. *The Wealth of Oceans: Environment and Development on Our Ocean Planet.* New York: Norton.

Weber, Peter. 1993. *Abandoned Seas: Reversing the Decline of the Oceans.* Washington, D.C.: Worldwatch Institute.

Whitmore, T. C. 1990. *An Introduction to Tropical Rainforests.* New York: Oxford University Press.

Wilson, E. O., ed. 1988. *Biodiversity.* Washington, D.C.: National Academy Press.

Wilson, E. O. 1992. *The Diversity of Life.* Cambridge, Mass.: Harvard University Press.

7 / POPULATION DYNAMICS, CARRYING CAPACITY, AND CONSERVATION BIOLOGY

Internet

For a list of useful Internet sites, hyperlinks to material in this book, and other helpful and interesting information, please visit our Web site at **www.brookscole.com/biology**

Readings

Andrewartha, H. G., and L. C. Birch. 1986. *The Ecological Web: More on the Distribution and Abundance of Animals.* Chicago: University of Chicago Press.

Botkin, Daniel. 1990. *Discordant Harmonies: A New Ecology for the Twenty-First Century.* New York: Oxford University Press.

Cappuccino, Naomi, and Peter W. Price, eds. 1995. *Population Dynamics: New Approaches and Syntheses.* New York: Academic Press.

Gordon, David G. 1996. *The Compleat Cockroach: A Comprehensive Guide to the Most Despised (and Least Understood)*

Creature on Earth. Berkeley, Calif.: Ten Speed.

Holt, Robert D., George R. Robinson, and Michael S. Gaines. 1995. "Vegetation Dynamics in an Experimentally Fragmented Landscape." *Ecology,* vol. 76, 1610–24.

Hunter, Malcolm L. Jr., 1996. *Fundamentals of Conservation Biology.* Cambridge, Mass.: Blackwell Scientific.

MacArthur, Robert H., and E. O. Wilson. 1967. *The Theory of Island Biogeography.* Princeton, N.J.: Princeton University Press.

Meffe, Gary K., and C. Ronald Carroll. 1994. *Principles of Conservation Biology.* Sunderland, Mass.: Sinauer.

Orians, Gordon H. 1990. "Ecological Concepts of Sustainability." *Environment,* November, 10–39.

Pimm, Stuart L. 1992. *The Balance of Nature?* Chicago: University of Chicago Press.

Primack, Richard B. 1995. *A Primer of Conservation Biology.* Sunderland, Mass.: Sinauer.

Rapport, David J., and Walter G. Whitford. 1999. "How Ecosystems Respond to Stress." *BioScience,* March, 193–203.

Rickleffs, Robert E. 1996. *The Economy of Nature: A Textbook of Ecology.* New York: W.H. Freeman.

Vitousek, Peter M., et al. 1997. "Human Domination of Earth's Ecosystems." *Science,* vol. 277, 494–99.

8 / RISK, TOXICOLOGY, AND HUMAN HEALTH

Internet

For a list of useful Internet sites, hyperlinks to material in this book, and other helpful and interesting information, please visit our Web site at **www.brookscole.com/biology**

Readings

American Council on Science and Health. 1997. *Cigarettes: What the Warning Label Doesn't Tell You.* Amherst, N.Y.: Prometheus.

Baggs, Sydney, and Joan C. Baggs. 1997. *The Healthy House: Creating a Safe, Healthy, and Environmentally Friendly Home.* New York: Harper Collins.

Bartecchi, Carl E., et al. 1995. "The Global Tobacco Epidemic." *Scientific American,* May, 44–47.

Chivian, Eric, et al. 1993. *Critical Condition: Human Health and the Environment.* Cambridge, Mass.: MIT Press.

Cohen, Mark N. 1989. *Health and the Rise of Civilization.* New Haven, Conn.: Yale University Press.

Colborn, Theo, et al. 1996. *Our Stolen Future.* New York: Dutton.

Dadd-Redalla, Debra. 1994. *Sustaining the Earth: Choosing Consumer Products That Are Environmentally Safe for You.* New York: Hearst Books.

Davies, Clarence, ed. 1996. *Comparing Environmental Risks: Tools for Setting Government Guidelines.* Washington, D.C.: Resources for the Future.

Dixon, Bernard. 1994. *Power Unseen: How Microbes Rule the World.* San Francisco: W.H. Freeman.

Drlica, Karl L. 1996. *Double-Edged Sword: The Promises and Risks of the Genetic Revolution.* New York: Addison-Wesley.

Emsley, John. 1994. *The Consumers' Good Chemical Guide: A Jargon-Free Guide to the Chemicals of Everyday Life.* New York: W.H. Freeman.

Environmental Protection Agency. 1987. *Unfinished Business: A Comparative Assessment of Environmental Problems.* Washington, D.C.: Environmental Protection Agency.

Environmental Protection Agency. 1990. *Reducing Risk: Setting Priorities and Strategies for Environmental Protection.* Washington, D.C.: Environmental Protection Agency.

Fisher, Jeffrey A. 1994. *The Plague Makers: How We Are Creating Catastrophic New Epidemics and What We Must Do to Avert Them.* New York: Simon & Schuster.

Fox, Michael W. 1992. *Superpigs and Wondercorn: The Brave New World of Biotechnology and Where It All May Lead.* New York: Lyons & Burford.

Freudenburg, William R. 1988. "Perceived Risk, Real Risk: Social Science and the Art of Probabilistic Risk Assessment." *Science,* vol. 242, 44–49.

Freudenthal, Ralph I., and Susan L. Freudenthal. 1989. *What You Need to Know to Live with Chemicals.* Greens Farms, Conn.: Hill & Garnett.

Garrett, Laurie. 1994. *The Coming Plague: Newly Emerging Diseases in a World Out of Balance.* New York: Farrar, Straus & Giroux.

Graham, John D., and Jonathan Baert Weiner, eds. 1995. *Risk vs. Risk: Tradeoffs in Protecting Health and the Environment.* Cambridge, Mass.: Harvard University Press.

Harris, John. 1992. *Wonderwoman and Superman: The Ethics of Human Biotechnology.* New York: Oxford University Press.

Harris, Stephen L. 1990. *Agents of Chaos: Earthquakes, Volcanoes, and Other Natural Disasters.* Missoula, Mont.: Mountain Press.

Harte, John, et al. 1992. *Toxics A to Z: A Guide to Everyday Pollution Hazards.* Berkeley: University of California Press.

Kamarin, M. A. 1988. *Toxicology: A Primer on Toxicology Principles and Applications.* Boca Raton, Fla.: Lewis.

Kitcher, Philip. 1996. *The Lives to Come: The Genetic Revolution and Human Possibilities.* New York: Simon & Schuster.

Klaassen, Curtis D., et al. 1999. *Cassarett and Doull's Toxicology: The Basic Science of Poisons.* New York: Macmillan.

Kluger, Richard. 1997. *Ashes to Ashes: America's Hundred-Year Cigarette War, the Public Health, and the Unabashed Triumph of Philip Morris.* New York: Vintage.

Kovach, Robert L. 1995. *Earth's Fury: An Introduction to Natural Hazards and Disasters.* Englewood Cliffs, N.J.: Prentice Hall.

Knudson, Mary. 1998. "The Hunt Is On for New Ways to Overcome Bacterial Resistance." *Technology Review,* January/February, 22–29.

Krimsky, Sheldon. 1999. *Hormonal Chaos: The Scientific and Social Origins of the Environmental Endocrine Hypotheses.* Baltimore, MD.: Johns Hopkins University Press.

Lappé, Marc. 1995. *The Evolving Threat of Drug-Resistant Disease.* San Francisco: Sierra Club Books.

Laudan, Larry. 1997. *Danger Ahead: The Risks You Really Face on Life's Highway.* New York: Wiley.

Levy, Stuart B. 1998. "The Challenge of Antibiotic Resistance." *Scientific American,* March, 46–53.

Merrell, Paul, and Carol Van Strum. 1990. "Negligible Risk or Premeditated Murder?" *Journal of Pesticide Reform,* vol. 10, Spring, 20–22.

Moeller, Dade W. 1997. *Environmental Health.* Rev. ed. Cambridge, Mass.: Harvard University Press.

Montague, Peter, ed. *Rachel's Environment and Health Weekly.* (P.O. Box 5036, Annapolis, MD 21403-7036). (A two-page weekly newsletter explaining health, risk, and environmental issues in an easily understandable manner.)

Moore, Gary S. 1999. *Living with the Earth: Concepts in Environmental Health Science.* Boca Raton, Fla.: Lewis.

Myers, Norman. 1997. "Development, Environment, and Health." *Environment and Development Economics,* vol. 2, no. 1, 367–70.

Nadakavukaren, Anne. 1995. *Our Global Environment: A Health Perspective.* 4th ed. New York: Waveland.

National Academy of Sciences. 1992. *Eat for Life: The Food and Nutrition Board's Guide to Reducing Your Risk of Chronic Disease.* Washington, D.C.: National Academy Press.

National Academy of Sciences. 1996. *Understanding Risk: Informing Decisions in a Democratic Society.* Washington, D.C.: National Academy Press.

Ottoboni, M. Alice. 1991. *The Dose Makes the Poison: A Plain-Language Guide to Toxicology.* 2d ed. New York: Van Nostrand Reinhold.

Piller, Charles. 1991. *The Fail-Safe Society.* New York: Basic Books.

Platt, Anne. 1996. *Infecting Ourselves: How Environmental and Social Disruptions Trigger Disease.* Washington, D.C.: Worldwatch Institute.

Platt, Rutherford, et al. 1999. *Disasters and Democracy: The Politics of Extreme Natural Events.* Washington, D.C.: Island Press.

Preston, Richard. 1994. *The Hot Zone.* New York: Random House.

Regenstein, Lewis G. 1993. *Cleaning Up America the Poisoned.* Washington, D.C.: Acropolis Books.

Rifkin, Jeremy. 1998. *The Biotech Century: Harnessing the Gene and Remaking the World.* New York: Tarcher/Plenum.

Rodricks, Joseph V. 1992. *Calculated Risks: Understanding the Toxicity and Human Health Risks of Chemicals in the Environment.* New

York: Cambridge University Press.

Steingraber, Sandra. 1997. *Living Downstream: An Ecologist Looks at Cancer and the Environment.* Reading, Mass.: Addison-Wesley.

Suzuki, David, and Peter Knudtson. 1989. *Genethics: The Clash Between the New Genetics and Human Values.* Cambridge, Mass.: Harvard University Press.

Tenner, Ward. 1996. *Why Things Bite Back: Technology and the Revenge of Unintended Consequences.* New York: Knopf.

VanDoren, Peter. 1999. *Chemicals, Cancer, and Choices.* Washington, D.C.: Cato Institute.

Vitousek, Peter M., et al. 1997. "Human Domination of Earth's Ecosystems." *Science*, vol. 277, 494–99.

Whelan, Elizabeth M. 1993. *Toxic Terror: The Truth Behind the Cancer Scares.* 2d ed. Buffalo, N.Y.: Prometheus.

Wildavsky, Aaron. 1995. *But Is It True? A Citizen's Guide to Environmental Health and Safety Issues.* Cambridge, Mass.: Harvard University Press.

9 / THE HUMAN POPULATION: GROWTH AND DISTRIBUTION

Internet

For a list of useful Internet sites, hyperlinks to material in this book, and other helpful and interesting information, please visit our Web site at **www.brookscole.com/biology**

Readings

Aaberley, Doug, ed. 1994. *Futures by Design: The Practice of Ecological Planning.* Philadelphia: New Society.

Abernathy, Virginia D. 1993. *Population Politics: The Choices That Shape Our Future.* New York: Plenum.

Arizpe, Lourdes, et al., eds. 1994. *Population and Environment: Rethinking the Debate.* Boulder, Colo.: Westview.

Ashford, Lori S. 1995. "New Perspectives on Population: Lessons from Cairo." *Population Bulletin*, vol. 50, no. 1, 1–44.

Badshah, Akhtar A. 1997. *Our Urban Future: Paradigms for Equity and Sustainability.* London: Zed.

Barna, George. 1992. *The Invisible Generation: Baby Busters.* New York: Barna Research Group.

Beatley, Timothy. 1999. *Green Urbanism: Learning from European Cities.* Washington, D.C.: Island Press.

Berg, Peter, et al. 1989. *A Green City Program for San Francisco Bay Area Cities and Towns.* San Francisco: Planet/Drum Foundation.

Bouvier, Leon F., and Lindsey Grant. 1994. *How Many Americans? Population, Immigration, and the Environment.* San Francisco: Sierra Club Books.

Brower, Michael. 1994. *Population Complications: Understanding the Population Debate.* Cambridge, Mass.: Union of Concerned Scientists.

Brown, Lester R. 1994. *Who Will Feed China?* New York: Norton.

Brown, Lester R., Gary Gardner, and Brian Halweil. 1998. *Beyond Malthus: Sixteen Dimensions of the Population Problem.* Washington, D.C.: Worldwatch Institute.

Brown, Lester R., and Hal Kane. 1994. *Full House: Reassessing the Earth's Population Carrying Capacity.* New York: Norton.

Cadman, D., and G. Payne, eds. 1990. *The Living City: Towards a Sustainable Future.* New York: Routledge.

Calthorpe, Peter. 1993. *The Next American Metropolis: Ecology, Community, and the American Dream.* Princeton, N.J.: Princeton Architectural Press.

Carlson, Daniel, et al. 1995. *At Road's End: Transportation and Land Use Choices for Communities.* Washington, D.C.: Island Press.

Corbett, Michael, and Judy Corbett. 1999. *Learning from Village Homes in Davis, California.* Washington, D.C.: Island Press.

Day, Lincoln H. 1992. *The Future of Low-Birthrate Populations.* New York: Routledge.

DeVita, Carol J. 1996. "The United States at Mid-Decade." *Population Bulletin*, vol. 50, no. 4, 1–48.

Donaldson, Peter J., and Amy Ong Tsui. 1990. "The International Family Planning Movement." *Population Bulletin*, vol. 43, no. 3, 1–42.

During, Alan Thein, and Christopher D. Crowther. 1997. *Misplaced Blame, The Real Roots of Population Growth.* Seattle: Northwest Environment Watch.

Ehrlich, Paul R., Anne H. Ehrlich, and Gretchen C. Daily. 1995. *The Stork and the Plow: The Equity Answer to the Human Dilemma.* New York: Putnam.

Ezcurra, Exequiel, and Marisa Mazari-Hriart. 1996. "Are Megacities Viable? A Cautionary Tale from Mexico City." *Environment*, vol. 38, no. 1, 6–15, 25–34.

Fodor, Eben. 1998. *Better Not Bigger: How to Take Care of Urban Growth and Improve Your Community.* Philadelphia: New Society.

Gordon, Deborah. 1991. *Steering a New Course: Transportation, Energy, and the Environment.* Washington, D.C.: Island Press.

Grant, Lindsey. 1996. *Juggernaut: Growth on a Finite Planet.* Santa Ana, Calif.: Seven Locks.

Hall, Charles A. H., et al. 1995. "The Environmental Consequences of Having a Baby in the United States." *Wild Earth*, Summer, 78–87.

Hardin, Garrett. 1993. *Living Within Limits: Ecology, Economics, and Population Taboos.* New York: Oxford University Press.

Hardin, Garrett. 1995. *Immigration Reform: Avoiding the Tragedy of the Commons.* Washington, D.C.: Federation for American Immigration Reform.

Hardin, Garrett. 1999. *The Ostrich Factor: Our Population Myopia.* New York : Oxford University Press.

Harrison, Paul. 1992. *The Third Revolution: Environment, Population, and a Sustainable World.* New York: I. B. Tauris.

Hart, John. 1992. *Saving Cities, Saving Money: Environmental Strategies That Work.* Washington, D.C.: Resource Renewal Institute.

Hartmann, Betsy. 1995. *Reproductive Rights and Wrongs: The Global Politics of Population Control and Contraceptive Choice.* Boston: South End Press.

Haupt, Arthur, and Thomas T. Kane. 1997. *The Population Handbook.* 4th ed. Washington, D.C.: Population Reference Bureau.

Hohm, Charles F., ed. 1995. *Population: Opposing Viewpoints.* San Diego, Calif.: Greenhaven.

Hollingsworth, William J. 1996. *Ending the Explosion: Population Policies and Ethics.* Santa Ana, Calif.: Seven Locks.

Kay, Jane Holtz. 1997. *Asphalt Nation: How the Automobile Took Over America and How We Can Take It Back.* New York: Crown.

Leisinger, Klaus M., and Karin Schmitt. 1994. *All Our People: Population Policy with a Human Face.* Washington, D.C.: Island Press.

Lowe, Marcia D. 1991. *Shaping Cities: The Environmental and Human Dimensions.* Washington, D.C.: Worldwatch Institute.

Lowe, Marcia D. 1994. *Back on Track: The Global Rail Revival.* Washington, D.C.: Worldwatch Institute.

Lutz, Wolfgang. 1994. "The Future of World Population." *Population Bulletin*, vol. 46, no. 2, 1–43.

Lyle, John T. 1993. *Regenerative Design for Sustainable Development.* New York: Wiley.

Lyman, Francesca. 1997. "Twelve Gates to the City." *Sierra*, May/June, 29–35.

MacKenzie, James J., et al. 1992. *The Going Rate: What It Really Costs to Drive.* Washington, D.C.: World Resources Institute.

Makower, Joel. 1992. *The Green Commuter.* Washington, D.C.: National Press.

Martin, Philip, and Elizabeth Midgley. 1994. "Immigration to the United States: Journey to an Uncertain Destination." *Population Bulletin*, vol. 49, no. 2, 1–47.

Mazur, Laurie Ann, ed. 1994. *Beyond the Numbers: A Reader on Population Consumption, and the Environment.* Washington, D.C.: Island Press.

McFalls, Joseph A., Jr. 1998. "Population: A Lively Introduction." *Population Bulletin*, vol. 53, no. 3, 1–48.

McHarg, Ian L. 1992. *Designing with Nature.* Reprinted ed. New York: Wiley.

McHarg, Ian L., and Frederick R. Steiner, eds. 1998. *To Heal the Earth: Selected Writings of Ian L. McHarg.* Washington, D.C.: Island Press.

McKibben, Bill. 1995. *Hope, Human and Wild: True Stories of Living Lightly on the Earth.* Boston: Little, Brown.

McKibben, Bill. 1998. *Maybe One: A Personal and Environmental Argument for Single-Child Families.* New York: Simon & Schuster.

Moffett, George D. 1994. *Critical Masses: The Global Population Challenge.* New York: Viking.

Nadis, Steve, and James J. MacKenzie. 1993. *Car Trouble.* Boston: Beacon.

National Academy of Sciences. 1997. *The New Americans: Economic, Demographic, and*

Fiscal Effects of Immigration. Washington, D.C.: National Academy Press.

Peterson, Peter G. 1999. *How the Coming Age Wave Will Transform America and the World.* New York: Times Books.

Population Reference Bureau. 1997. *Population Handbook.* 4th ed. Washington, D.C.: Population Reference Bureau.

Population Reference Bureau. Annual. *World Population Data Sheet.* Washington, D.C.: Population Reference Bureau.

Rabinovitch, Jonas, and Josef Leitman. 1996. "Urban Planning in Curitiba." *Scientific American,* March, 46–53.

Register, Richard. 1992. *Ecocities.* Berkeley, Calif.: North Atlantic Books.

Rocky Mountain Institute. 1997. *Green Development: Integrating Ecology and Real Estate.* Snowmass, Colo.: Rocky Mountain Institute.

Ryn, Sin van der, and Stuart Cowan. 1995. *Ecological Design.* Washington, D.C.: Island Press.

Simon, Julian L. 1996. *The Ultimate Resource 2.* Princeton, N.J.: Princeton University Press.

Steele, James. 1997. *Sustainable Architecture: Principles, Paradigms, and Case Studies.* New York: McGraw-Hill.

Tien, H. Yuan, et al. 1992. "China's Demographic Dilemmas." *Population Bulletin,* vol. 47, no. 1, 1–44.

Todd, John, and Nancy Jack Todd. 1993. *From Ecocities to Living Machines: Precepts for Sustainable Technologies.* Berkeley, Calif.: North Atlantic Books.

Todd, John, and George Tukel. 1990. *Reinhabiting Cities and Towns: Designing for Sustainability.* San Francisco: Planet/Drum Foundation.

Tsui, E. 1999. *Evolutionary Architecture: Nature as a Basis for Design.* New York: Wiley.

United Nations. 1991. *Consequences of Rapid Population Growth in Developing Countries.* New York: United Nations.

United Nations. 1992. *Long-Range World Population Projections: Two Centuries of Population Growth, 1950–2150.* New York: United Nations.

United Nations Population Division. 1991. *World Urbanization Prospects.* New York: United Nations.

Visaria, Leela, and Pravin Visaria. 1995. "India's Population in Transition." *Population Bulletin,* vol. 50, no. 3, 1–51.

Walter, Bob, et al., eds. 1992. *Sustainable Cities: Concepts and Strategies for Eco-City Development.* Los Angeles: Eco-Home Media.

Wann, David. 1994. *Biologic: Environmental Protection by Design.* 2d ed. Boulder, Colo.: Johnson Books.

Wann, David. 1995. *Deep Design: Pathways to a Livable Future.* Washington, D.C.: Island Press.

White, Rodney R. 1994. *Urban Environmental Management. Environmental Change and Urban Design.* New York: Wiley.

Zero Population Growth. 1990. *Planning the Ideal Family: The Small Family Option.* Washington, D.C.: Zero Population Growth.

Zuckerman, Wolfgang. 1991. *End of the Road: The World Car Crisis and How We Can Solve It.* White River Junction, Vt.: Chelsea Green.

10 / AIR AND AIR POLLUTION

Internet

For a list of useful Internet sites, hyperlinks to material in this book, and other helpful and interesting information, please visit our Web site at **www.brookscole.com/biology**

Readings

Alleman, James E., and Brooke T. Mossman. 1997. "Asbestos Revisited." *Scientific American,* July, 70–75.

Ayres, Harvard, et al., eds. 1998. *An Appalachian Tragedy: Air Pollution and Tree Death in the Eastern Forests of North America.* San Francisco: Sierra Club Books.

Bridgman, Howard. 1991. *Global Air Pollution: Problems for the 1990s.* New York: Belhaven.

Bryner, Gary. 1992. *Blue Skies, Green Politics: The Clean Air Act of 1990.* Washington, D.C.: Congressional Quarterly Press.

Cohen, Bernie. 1988. *Radon: A Homeowner's Guide to Detection and Control.* Mt. Vernon, N.Y.: Consumer Report Books.

Cole, Leonard D. 1994. *Element of Risk: The Politics of Radon.* New York: Oxford University Press.

Elsom, Derek. 1987. *Atmospheric Pollution: Causes, Effects, and Control Policies.* Cambridge, Mass.: Basil Blackwell.

Goklany, Indur. 1999. *Clearing the Air: The Real Story of the War on Air Pollution.* Washington, D.C.: Cato Institute.

Hedin, Lars O., and Gene E. Likens. 1996. "Atmospheric Dust and Acid Rain." *Scientific American,* December, 88–92.

Leslie, G. B., and F. W. Linau. 1994. *Indoor Air Pollution.* New York: Cambridge University Press.

Little, Charles E. 1997. *The Dying of the Trees: The Pandemic in America's Forests.* New York: Penguin.

MacKenzie, James J., and Mohamed T. El-Ashry. 1990. *Air Pollution's Toll on Forests and Crops.* Washington, D.C.: World Resources Institute.

McCormick, John. 1997. *Acid Earth: The Politics of Acid Pollution.* 3d ed. East Haven, Conn.: Earthscan.

Munton, Don. 1998. "Dispelling the Myths of the Acid Rain Story." *Environment,* vol. 40, no. 6, 4–7, 29–34.

National Academy of Sciences. 1995. *Health Effects of Exposure to Radon: Time for Reassessment?* Washington, D.C.: National Academy Press.

Soroos, Marvin S. 1997. *The Endangered Atmosphere: Preserving a Global Commons.* Columbia: University of South Carolina Press.

Steidlmeier, Paul. 1993. "The Morality of Pollution Permits." *Environmental Ethics,* vol. 15, 133–50.

Wagner, Travis. 1993. *In Our Backyard: A Guide to Understanding Pollution and Its Effects.* New York: Van Nostrand Reinhold.

Warde, John. 1997. *The Healthy Home Handbook: All You Need to Know to Rid Your Home of Health and Safety Hazards.* New York: Times Books.

11 / GLOBAL WARMING AND OZONE LOSS

Internet

For a list of useful Internet sites, hyperlinks to material in this book, and other helpful and interesting information, please visit our Web site at **www.brookscole.com/biology**

Readings

American Forestry Association. 1993. *Forests and Global Warming.* Vol. 1. Washington, D.C.: American Forestry Association.

Balling, Robert C., Jr. 1992. *The Heated Debate: Greenhouse Predictions Versus Climate Reality.* San Francisco: Pacific Research Institute for Public Policy.

Bates, Albert K. 1990. *Climate in Crisis: The Greenhouse Effect and What We Can Do.* Summertown, Tenn.: The Book Publishing Company.

Benedick, Richard Eliot. 1991. *Ozone Diplomacy: New Directions in Safeguarding the Planet.* Cambridge, Mass.: Harvard University Press.

Bernard, Harold W., Jr. 1993. *Global Warming Unchecked: Signs to Watch For.* Bloomington: University of Indiana Press.

Brown, Paul. 1997. *Global Warming: Can Civilization Survive?* London: Blandford.

Charlson, Robert J., and Tom M. L. Wigley. 1994. "Sulfate Aerosol and Climate Change." *Scientific American,* February, 48–57.

Christianson, Gale E. 1998. *Greenhouse: The 200-Year Story of Global Warming.* New York: Walker.

Council for Agricultural Science and Technology. 1992. *Preparing U.S. Agriculture for Global Climate Change.* Ames, Iowa: Council for Agricultural Science and Technology.

Fermann, Gunnar, ed. 1997. *International Politics of Climate Change: Key Issues and Critical Actors.* Oslo, Norway: Scandinavian University Press.

Flavin, Christopher, and Odil Tunali. 1996. *Climate of Hope: New Strategies for Stabilizing the World's Atmosphere.* Washington, D.C.: Worldwatch Institute.

Flavin, Christopher, and Seth Dunn. 1997. *Rising Sun, Gathering Winds: Policies to Stabilize Climate and Strengthen Economies.* Washington, D.C.: Worldwatch Institute.

Gates, David M. 1993. *Climate Change and Its Biological Consequences.* Sunderland, Mass.: Sinauer.

Gelbspan, Ross. 1998. *The Heat is On: The Climate Crisis, The Cover-up and the Prescription.* Reading, Mass.: Perseus.

Graedel, T. C., and Paul J. Cutzen. 1995. *Atmosphere. Climate, and Change.* New York: W.H. Freeman.

Hileman, Bette. 1999. "Case Grows for Climate Change." *Chemistry & Engineering News,* August 9, 16–23.

Houghton, J. T., 1997. *Global Warming: The*

Complete Briefing. New York: Cambridge University Press.

Intergovernmental Panel on Climate Change (IPCC). 1990. *Climate Change: The IPCC Assessment*. New York: Cambridge University Press.

Intergovernmental Panel on Climate Change (IPCC). 1995. *The Supplementary Report to the IPCC Scientific Assessment*. New York: Cambridge University Press.

IUCN Global Change Programme. 1993. *Impact of Climate Change on Ecosystems and Species*. Washington, D.C.: Island Press.

Karplus, Walter J. 1992. *The Heavens Are Falling: The Scientific Prediction of Catastrophes in Our Time*. New York: Plenum.

Kunzig, Robert, and Carl Zimmer. 1998. "Carbon Cuts and Techno-Fixes: 10 Things to Do About the Greenhouse Effect." *Technology Review*, June, 61–71.

Lee, Henry, ed. 1995. *Shaping Responses to Climate Change*. Washington, D.C.: Island Press.

Leffell, David J., and Douglas E. Brash. 1996. "Sunlight and Skin Cancer." *Scientific American*, July, 52–59.

Leggett, Jeremy, ed. 1994. *The Climate Time Bomb: Signs of Climate Change from the Greenpeace Database*. Amsterdam, Netherlands: Stichting Greenpeace Council.

Lovins, Amory B., et al. 1989. *Least-Cost Energy: Solving the CO_2 Problem*. 2d ed. Snowmass, Colo.: Rocky Mountain Institute.

Lyman, Francesca, et al. 1990. *The Greenhouse Trap: What We're Doing to the Atmosphere and How We Can Slow Global Warming*. Washington, D.C.: World Resources Institute.

Makhijani, Arjun, and Kevin Gurney. 1995. *Mending the Ozone Hole: Science, Technology, and Policy*. Cambridge, Mass.: MIT Press.

McKibben, Bill. 1989. *The End of Nature*. New York: Random House.

Mintzer, Irving, and William R. Moomaw. 1991. *Escaping the Heat Trap: Probing the Prospects for a Stable Environment*. Washington, D.C.: World Resources Institute.

Mintzer, Irving, et al. 1990. *Protecting the Ozone Shield: Strategies for Phasing Out CFCs During the 1990s*. Washington, D.C.: World Resources Institute.

Moore, Thomas G. 1998. *Climate of Fear: Why We Shouldn't Worry About Global Warming*. Washington, D.C.: Cato Institute.

National Academy of Sciences. 1990. *Confronting Climate Change*. Washington, D.C.: National Academy Press.

National Academy of Sciences. 1990. *Sea Level Change*. Washington, D.C.: National Academy Press.

National Academy of Sciences. 1992. *Global Environmental Change*. Washington, D.C.: National Academy Press.

National Audubon Society. 1990. *CO_2 Diet for a Greenhouse Planet: A Citizen's Guide to Slowing Global Warming*. New York: National Audubon Society.

Natural Resources Defense Council et al. 1999. *Kingpins of Carbon: How Fossil Fuel Producers Contribute to Global Warming*. New York: Natural Resources Defense Council.

Office of Technology Assessment. 1993. *Preparing for an Uncertain Climate*. Washington, D.C.: U.S. Government Printing Office.

Oppenheimer, Michael, and Robert H. Boyle. 1990. *Dead Heat: The Race Against the Greenhouse Effect*. New York: Basic Books.

Peters, Robert L., and Thomas E. Lovejoy, eds. 1992. *Global Warming and Biological Diversity*. New Haven, Conn.: Yale University Press.

Philander, S. George. 1998. *Is the Temperature Rising?: The Uncertain Science of Global Warming*. Princeton, N.J.: Princeton University Press.

Ray, Dixie Lee, and Lou Guzzo. 1993. *Environmental Overkill: Whatever Happened to Common Sense?* Washington, D.C.: Regnery Gateway.

Revkin, Andrew. 1992. *Global Warming: Understanding the Forecast*. New York: Abbeville.

Roan, Sharon L. 1989. *Ozone Crisis:. The 15-Year Evolution of a Sudden Global Emergency*. New York: Wiley.

Romm, Joseph J. 1999. *Cool Companies: How the Best Businesses Boost Profits and Productivity by Cutting Greenhouse Gas Emissions*. Washington, D.C.: Island Press.

Rowland, F. Sherwood, and Mario J. Molina. 1994. "Ozone Depletion: 20 Years After the Alarm." *Chemistry & Engineering News*, August 15, 8–13.

Ryan, John. 1999. *Over our heads: A Look at Global Climate*. Seattle, Wash.: Northwest Environmental Watch.

Schneider, David. 1997. "Rising Seas.".*Scientific American*, March, 112-17.

Singer, S. Fred and Frederich Seitz. 1999. *Hot Talk Cold Science: Global Warming's Unfinished Debate*. Cincinatti, Ohio: Seven Hills.

Turekian, Karl K. 1996. *Global Environmental Change: Past, Present, and Future*. Upper Saddle River, N.J.: Prentice Hall.

UN Environment Programme, World Meteorological Organization, U.S. National Aeronautics and Space Administration, and National Oceanic and Atmospheric Administration. 1994. *International Assessment of Ozone Depletion*. Washington, D.C.: UN Environment Programme.

Woodwell, George M., and Fred T. Mackenzie, eds. 1995. *Biotic Feedbacks in the Global Climatic System. Will the Warming Feed the Warming?* New York: Oxford University Press.

12 / WATER RESOURCES AND WATER POLLUTION

Internet

For a list of useful Internet sites, hyperlinks to material in this book, and other helpful and interesting information, please visit our Web site at **www.brookscole.com/biology**

Readings

Abel, P. D. 1997. *Water Pollution Biology*. 2d ed. London: Taylor & Francis.

Abramovitz, Janet N. 1996. "Sustaining Freshwater Ecosystems." In Lester R. Brown et al., *State of the World 1996*. New York: Norton, 60–77.

Adler, Robert W., et al. 1993. *The Clean Water Act Twenty Years Later*. Washington, D.C.: Island Press.

Allaby, Michael. 1992. *Water: Its Global Nature*. New York: Facts on File.

Ashworth, William. 1986. *The Late, Great Lakes: An Environmental History*. New York: Knopf.

Biswas, Asit K., et al. 1997. *Core and Periphery: A Comprehensive Approach to Middle Eastern Water*. New York: Oxford University Press.

Bolling, David M. 1994. *How to Save a River: A Handbook for Citizen Action*. Washington, D.C.: Island Press.

Burger, Joanna. 1997. *Oil Spills*. Rutgers, N.J.: Rutgers University Press.

Christopher, Thomas. 1994. *Water-Wise Gardening: America's Backyard Revolution*. New York: Simon & Schuster.

Clarke, Robin. 1993. *Water: The International Crisis*. Cambridge, Mass.: MIT Press.

Colborn, Theodora E., et al. 1989. *Great Lakes, Great Legacy?* Washington, D.C.: Conservation Foundation.

Costner, Pat, and Glenna Booth. 1986. *We All Live Downstream: A Guide to Waste Treatment That Stops Water Pollution*. Berkeley, Calif.: Bookpeople.

Doppelt, Bob, et al. 1993. *Entering the Watershed: A New Approach to Save America's River Ecosystems*. Washington, D.C.: Island Press.

Edmonson, W. T. 1991. *The Uses of Ecology: Lake Washington and Beyond*. Seattle: University of Washington Press.

El-Ashry, Mohamed, and Diana C. Gibbons, eds. 1988. *Water and Arid Lands of the Western United States*. Washington, D.C.: World Resources Institute.

Environmental Protection Agency. 1990. *Citizen's Guide to Ground-Water Protection*. Washington, D.C.: Environmental Protection Agency.

Falkenmark, Malin, and Carl Widstand. 1992. "Population and Water Resources: A Delicate Balance." *Population Bulletin*, vol. 47, no. 3, 1–36.

Gardner, Gary. 1995. "From Oasis to Mirage: The Aquifers That Won't Replenish." *World Watch*, May/June, 30–36, 40–41.

Gardner, Gary. 1998. "Recycling Human Waste: Fertile Ground or Toxic Legacy?" *World Watch*, January/February, 28–34.

Gleick, Peter H. 1993. *Water in Crisis: A Guide to the World's Fresh Water Resources*. Berkeley, Calif.: Pacific Institute for Studies in Development, Environment, and Security.

Gleick, Peter H. 1994. "Water, War and Peace in the Middle East." *Environment*, April, 6–41.

Gleick, Peter H. 1999. *The World's Water*

1998–1999: The Biennial Report on Freshwater Resources. Washington, D.C.: Island Press.

Hansen, Nancy R., et al. 1988. *Controlling Nonpoint-Source Water Pollution.* New York: National Audubon Society and the Conservation Society.

Hileman, Bette. 1997. "*Pfiesteria* Health Concerns Realized." *Chemical & Engineering News*, October 13, 14–15.

Hillel, Daniel. 1995. *Rivers of Eden: The Struggle for Water and the Quest for Peace in the Middle East.* New York: Oxford University Press.

Hinrichsen, Don. 1998. *Coastal Waters of the World: Trends, Threats, and Strategies.* Washington, D.C.: Island Press.

Horton, Tom, and William Eichbaum. 1991. *Turning the Tide: Saving the Chesapeake Bay.* Washington, D.C.: Island Press.

Hundley, Norris, Jr. 1992. *The Great Thirst: Californians and Water, 1770s–1990s.* Berkeley: University of California Press.

Ives, J. D., and B. Messeric. 1989. *The Himalayan Dilemma: Reconciling Development and Conservation.* New York: Routledge.

Jewell, William J. 1994. "Resource-Recovery Wastewater Treatment." *American Scientist*, vol. 82, 366–75.

Keeble, John. 1991. *Out of the Channel: The Exxon Valdez Oil Spill in Prince William Sound.* New York: HarperCollins.

Knopman, Debra S., and Richard A. Smith. 1993. "20 Years of the Clean Water Act." *Environment*, vol. 35, no. 1, 17–20, 34–41.

Lewis, Scott A. 1995. *Guide to Safe Drinking Water.* San Francisco: Sierra Club Books.

McCully, Patrick. 1996. *Silenced Rivers: The Ecology and Politics of Large Dams.* London: Zed.

McCutcheon, Sean. 1992. *Electric Rivers: The Story of the James River Project.* New York: Paul & Co.

Micklin, Philip P., and William D. Williams. 1996. *The Aral Sea Basin.* New York: Springer-Verlag.

Mitchell, Bruce, ed. 1990. *Integrated Water Management.* New York: Belhaven.

Munasinghe, Mohan. 1992. *Water Supply and Environmental Management.* Boulder, Colo.: Westview.

National Academy of Sciences. 1992. *Water Transfers in the West: Efficiency, Equity, and the Environment.* Washington, D.C.: National Academy Press.

National Academy of Sciences. 1997. *Valuing Ground Water.* Washington, D.C.: National Academy.

Natural Resources Defense Council. 1989. *Ebb Tide for Pollution: Actions for Cleaning Up Coastal Waters.* Washington, D.C.: Natural Resources Defense Council.

Opie, John. 1993. *Ogallala: Water for a Dry Land.* Lincoln: University of Nebraska Press.

Outwater, Alice. 1996. *Water: A Natural History.* New York: Basic Books.

Palmer, Tim. 1994. *Lifelines: The Case for River Conservation.* Washington, D.C.: Island Press.

Patrick, Ruth, et al. 1992. *Surface Water Quality: Have the Laws Been Successful?* Princeton, N.J.: Princeton University Press.

Pearce, Fred. 1992. *The Dammed: Rivers, Dams, and the Coming World Water Crisis.* London: Bodley Head.

Pimentel, David, et al. 1997. "Water Resources: Agriculture, the Environment, and Society." *BioScience*, vol. 47, no. 2, 97–106.

Platt, Anne. 1995. "Dying Seas." *World Watch*, January/February, 10–19.

Platt, Anne. 1996. "Water-Borne Killers." *World Watch*, March/April, 28–35.

Postel, Sandra, et al. 1996. "Human Appropriation of Renewable Fresh Water." *Science*, vol. 271, 785–87.

Postel, Sandra. 1997. *Last Oasis: Facing Water Scarcity.* 2d ed. New York: Norton.

Postel, Sandra. 1999. *Pillar of Sand: Can the Irrigation Miracle Last?* New York: Norton.

Qing, Dai. 1994. *Yangtse! Yangtze! Debate Over the Three Gorges Project.* East Haven, Conn.: Earthscan.

Reisner, Marc, and Sara Bates. 1990. *Overtapped Oasis: Reform or Revolution for Western Water.* Washington, D.C.: Island Press.

Rocky Mountain Institute. 1990. *Catalog of Water-Efficient Technologies for the Urban/Residential Sector.* Snowmass, Colo.: Rocky Mountain Institute.

Sierra Club Defense Fund. 1989. *The Poisoned Well: New Strategies for Groundwater Protection.* Washington, D.C.: Island Press.

Starr, Joyce R. 1991. "Water Wars." *Foreign Policy*, Spring, 12–45.

Stauffer, Julie. 1998. *The Water Crisis: Constructing Solutions to Freshwater Pollution.* East Haven, Conn.: Earthscan.

Steingraber, Sandra. 1997. *Living Downstream: An Ecologist Looks at Cancer and the Environment.* Reading, Mass.: Addison-Wesley.

Wagner, Travis. 1993. *In Our Backyard: A Guide to Understanding Pollution and Its Effects.* New York: Van Nostrand Reinhold.

Weber, Michael L., and Judith A. Gradwohl. 1995. *The Wealth of Oceans: Environment and Development on Our Ocean Planet.* New York: Norton.

Weber, Peter. 1993. *Abandoned Seas: Reversing the Decline of the Oceans.* Washington, D.C.: Worldwatch Institute.

Wilkinson, Charles F. 1992. *Crossing the Next Meridian: Land, Water, and the Future of the West.* Washington, D.C.: Island.

Worster, Donald. 1985. *Rivers of Empire: Water, Aridity, and the Growth of the American West.* New York: Pantheon.

13 / MINERALS AND SOIL

Internet

For a list of useful Internet sites, hyperlinks to material in this book, and other helpful and interesting information, please visit our Web site at **www.brookscole.com/biology**

Readings

Brady, Nyle C. 1989. *The Nature and Properties of Soils.* 10th ed. New York: Macmillan.

Brown, Bruce, and Lane Morgan. 1990. *The Miracle Planet.* Edison, N.J.: W.H. Smith.

Campbell, Stu. 1990. *Let It Rot: The Gardener's Guide to Composting.* Pownal, Vt.: Storey Communications.

Crosson, Pierre. 1997. "Will Erosion Threaten Agricultural Productivity?" *Environment*, vol. 39, no. 8, 4–9, 29–31.

Donahue, Roy, et al. 1990. *Soils and Their Management.* 5th ed. Petaluma, Calif.: Inter Print.

Eggert, R. G., ed. 1994. *Mining and the Environment.* Washington, D.C.: Resources for the Future.

Ernst, W. G. 1990. *The Dynamic Planet.* New York: Columbia University Press.

Gardner, Gary, and Payal Sampat. 1999. "Forging a Sustainable Materials Economy." In Lester R. Brown et al., *State of the World 1999.* New York: Norton, pp. 41–59.

Grainger, Alan. 1990. *The Threatening Desert: Controlling Desertification.* East Haven, Conn.: Earthscan.

Hillel, Daniel. 1992. *Out of the Earth: Civilization and the Life of the Soil.* New York: Free Press.

Hodges, Carroll A. 1995. "Mineral Resources, Environmental Issues, and Land Use." *Science*, vol. 268, 1305–11.

Little, Charles E. 1987. *Green Fields Forever: The Conservation Tillage Revolution in America.* Washington, D.C.: Island Press.

Mainguet, Monique. 1991. *Desertification: Natural Background and Human Mismanagement.* New York: Springer-Verlag.

Mollison, Bill. 1990. *Permaculture.* Washington, D.C.: Island Press.

National Academy of Sciences. 1993. *Soil and Water Quality: An Agenda for Agriculture.* Washington, D.C.: National Academy Press.

Paddock, Joe, et al. 1987. *Soil and Survival: Land Stewardship and the Future of American Agriculture.* San Francisco: Sierra Club Books.

Pimentel, David, ed. 1993. *World Soil Erosion and Conservation.* New York: Cambridge University Press.

Pimentel, David, et al. 1995. "Environmental and Economic Costs of Soil Erosion and Conservation Benefits." *Science*, vol. 267, 1117–23.

Ripley, Earle A., et al. 1996. *Environmental Effects of Mining.* Delray Beach, Fla.: St. Lucie.

Wild, Alan. 1993. *Soils and the Environment: An Introduction.* New York: Cambridge University Press.

Young, John E. 1992. *Mining the Earth.* Washington, D.C.: Worldwatch Institute.

Young, John E., and Aaron Sachs. 1994. *The Next Efficiency Revolution: Creating a Sustainable Materials Revolution.* Washington, D.C.: Worldwatch Institute.

Youngquist, Walter. 1990. *Mineral Resources and the Destinies of Nations.* Portland, Ore.: National Book Press.

Youngquist, Walter. 1997. *GeoDestinies: The Inevitable Control of Earth Resources Over Nations and Individuals.* Portland, Oreg.: National Book Press.

14 / SOLID AND HAZARDOUS WASTE

Internet

For a list of useful Internet sites, hyperlinks to material in this book, and other helpful and interesting information, please visit our Web site at **www.brookscole.com/biology**

Readings

Ackerman, Frank. 1997. *Why Do We Recycle? Markets, Values, and Public Policy.* Washington, D.C.: Island Press.

Alexander, Judd H. 1993. *In Defense of Garbage.* Westport, Conn.: Praeger.

Allen, Robert. 1992. *Waste Not, Want Not.* East Haven, Conn.: Earthscan.

Barnett, Harold C. 1994. *Toxic Debts and the Superfund Dilemma.* Chapel Hill: University of North Carolina Press.

Bullard, Robert D., ed. 1993. *Confronting Environmental Racism: Voices from the Grassroots.* Boston: South End Press.

Bullard, Robert D. 1994. *Dumping in Dixie: Race, Class, and Environmental Quality.* 2d ed. Boulder, Colo.: Westview.

Bullard, Robert D., ed. 1994. *Unequal Protection: Environmental Justice and Communities of Color.* San Francisco: Sierra Club Books.

Calow, Peter, 1997. *Controlling Environmental Risks from Chemicals: Principles and Practice.* New York: Wiley.

Carless, Jennifer. 1992. *Taking Out the Trash: A No-Nonsense Guide to Recycling.* Washington, D.C.: Island Press.

Centers for Disease Control and Prevention. 1991. *Preventing Lead Poisoning in Young Children.* Atlanta, Ga.: Centers for Disease Control and Prevention.

Clayton, Tony, et al. 1998. *Policies for Cleaner Technology.* East Haven, Conn.: Earthscan.

Cohen, Gary, and John O'Connor. 1990. *Fighting Toxics: A Manual for Protecting Family, Community, and Workplace.* Washington, D.C.: Island Press.

Commoner, Barry. 1994. *Making Peace with the Planet.* Rev. ed. New York: New Press.

Connett, Paul H. 1992. "The Disposable Society." In F. H. Bormann and Stephen R. Kellert, eds. *Ecology, Economics, Ethics.* New Haven, Conn.: Yale University Press, 99–122.

Connett, Paul, and Ellen Connett. 1994. "Municipal Waste Incineration: Wrong Question, Wrong Answer." *The Ecologist,* January/February, vol. 24, 14–20.

Denison, Richard A., and John Ruston. 1997. "Recycling Is Not Garbage." *Technology Review,* October, 55–60.

Devito, Stephen C., and Roger L. Garrett, eds. 1996. *Designing Safe Chemicals: Green Chemistry for Pollution Prevention.* Washington, D.C.: American Chemical Society.

Durning, Alan Thein. 1992. *How Much Is Enough? The Consumer Society and the Future of the Earth.* New York: Norton.

Environmental Protection Agency. 1992. *The Consumer's Handbook for Reducing Solid Waste* Washington, D.C.: Environmental Protection Agency.

Frosch, Robert A. 1995. "Industrial Ecology." *Environment,* December, 16–37.

Gardner, Gary. 1997. *Recycling Organic Waste: From Urban Pollutant to Farm Resource.* Washington, D.C.: Worldwatch Institute.

Gertler, Nicholas, and John R. Ehrenfeld. 1996. "A Down-to-Earth Approach to Clean Production." *Technology Review,* February/March, 48–54.

Gibbs, Lois. 1995. *Dying from Dioxin.* Boston: South End Press.

Gibbs, Lois. 1997. *Love Canal: The Story Continues.* Philadelphia, Penn.: New Society.

Gordon, Ben, and Peter Montague. 1989. *Zero Discharge: A Citizen's Toxic Waste Manual.* Washington, D.C.: Greenpeace.

Gottlieb, Robert, ed. 1995. *Reducing Toxics: A New Approach to Policy and Industrial Decisionmaking.* Washington, D.C.: Island Press.

Graedel, T. E., and B. R. Allenby. 1995. *Industrial Ecology.* Englewood Cliffs, N.J.: Prentice Hall.

Harte, John, et al. 1991. *Toxics A to Z: A Guide to Everyday Pollution Hazards.* Berkeley: University of California Press.

Hilz, Christoph. 1993. *The International Toxic Waste Trade.* New York: Van Nostrand Reinhold.

Hird, John A. 1994. *Superfund: The Political Economy of Environmental Risk.* Baltimore: Johns Hopkins University Press.

Hofrichter, Richard, ed. 1993. *Toxic Struggles: The Theory and Practice of Environmental Justice.* Philadelphia: New Society.

Kane, Hal. 1996. "Shifting to Sustainable Industries." In Lester R. Brown et al., *State of the World 1996.* Washington, D.C.: Worldwatch Institute, 152–67.

Kessel, Irene, and John T. O'Connor. 1997. *Getting the Lead Out: The Complete Resource on How to Prevent and Cope with Lead Poisoning.* New York: Plenum.

Lappé, Marc. 1991. *Chemical Deception: The Toxic Threat to Health and Environment.* San Francisco: Sierra Club.

League of Women Voters. 1993. *The Garbage Primer: A Handbook for Citizens.* New York: Lyons & Burford.

League of Women Voters. 1993. *A Plastic Waste Primer.* New York: Lyons & Burford.

Mazmanian, Daniel, and David Morrell. 1992. *Beyond Superfailure: America's Toxics Policy for the 1990s.* Boulder, Colo.: Westview.

McLenighan, Valjean. 1991. *Sustainable Manufacturing: Saving Jobs, Saving the Environment.* Chicago: Center for Neighborhood Technology.

Minnesota Mining and Manufacturing. 1988. *Low- or Non-Pollution Technology Through Pollution Prevention.* St. Paul, Minn.: 3M Company.

Mowrey, Marc, and Tim Redmond. 1993. *Not in Our Backyard.* New York: William Morrow.

Moyers, Bill. 1990. *Global Dumping Ground: The International Traffic in Hazardous Waste.* Cabin John, Md.: Seven Locks.

National Academy of Sciences. 1994. *The Greening of Industrial Ecosystems.* Washington, D.C.: National Academy of Engineering.

Needleman, Herbert L., and Philip J. Landrigan. 1995. *Raising Children Toxic Free.* New York: Avon Books.

Nemerow, Nelson L. 1995. *Zero Pollution for Industry: Waste Minimization Through Industrial Complexes.* New York: Wiley.

Office of Technology Assessment. 1992. *Green Products by Design.* Washington, D.C.: Government Printing Office.

Pellerano, Maria B. 1995. *How to Research Chemicals: A Resource Guide.* Annapolis, Md.: Environmental Research Foundation.

Platt, Brenda, et al. 1991. *Beyond 40 Percent: Record-Setting Recycling and Composting Programs.* Washington, D.C.: Island Press.

Puckett, Jim. 1994. "Disposing of the Waste Trade: Closing the Recycling Loophole." *The Ecologist,* vol. 24, no. 2, 53–58.

Rathje, William, and Cullen Murphy. 1992. *Rubbish! The Archaeology of Garbage: What Our Garbage Tells Us About Ourselves.* San Francisco: HarperCollins.

Regenstein, Lewis G. 1993. *Cleaning Up America the Poisoned.* Washington, D.C.: Acropolis Books.

Sachs, Aaron. 1995. *Eco-Justice: Linking Human Rights and the Environment.* Washington, D.C.: Worldwatch Institute.

Setterberg, Fred, and Lonny Shavelson. 1993. *Toxic Nation: The Fight to Save Our Communities from Chemical Contamination.* New York: Wiley.

Stapleton, Richard M. 1994. *Lead Is a Silent Hazard.* New York: Walker.

Stein, Kathy. 1998. *Beyond Recycling: A Reuser's Guide.* Santa Fe, N.M.: Clear Light Publishers.

Theodore, Louis, and Young C. McGuinn. 1992. *Pollution Prevention.* New York: Van Nostrand Reinhold.

Thomas, William. 1995. *Scorched Earth: The Military's Assault on the Environment.* Philadelphia: New Society.

Thornton, Joe. 1997. *The PVC Lifecycle: Dioxin from Cradle to Grave.* Washington, D.C.: Greenpeace USA.

Wagner, Travis. 1993. *In Our Backyard: A Guide to Understanding Pollution and Its Effects.* New York: Van Nostrand Reinhold.

Walsh, Edward J., et al. 1997. *Don't Burn It Here: Grassroots Challenges to Trash Incineration.* University Park: Pennsylvania State University Press.

Whelan, E. M. 1993. *Toxic Terror: The Truth Behind the Cancer Scares.* Buffalo, N.Y.: Prometheus.

Whitaker, Jennifer Seymour. 1994. *Salvaging the Land of Plenty: Garbage and the American Dream.* New York: William Morrow.

Young, John E. 1991. *Discarding the Throwaway Society.* Washington, D.C.: Worldwatch Institute.

Young, John E. 1995. "The Sudden New Strength of Recycling." *World Watch,* July/August, 20–25.

15 / FOOD RESOURCES

Internet

For a list of useful Internet sites, hyperlinks to material in this book, and other helpful and interesting information, please visit our Web site at **www.brookscole.com/biology**

Readings

Aberley, Doug. 1993. *Greening the Garden: A Guide to Sustainable Growing.* Philadelphia: New Society.

Aliteri, Miguel A. 1995. *Agroecology: The Scientific Basis of Alternative Agriculture.* 2d ed. Boulder, Colo.: Westview.

Ausubel, Kenny. 1994. *Seeds of Change.* New York: HarperCollins.

Avery, Dennis T. 1997. "Saving Nature's Legacy Through Better Farming." *Issues in Science and Technology*, Fall, 59–64.

Bardach, John E., ed. 1997. *Sustainable Aquaculture.* New York: Wiley.

Berrill, Michael. 1997. *The Plundered Seas: Can the World's Fish Be Saved?* San Francisco: Sierra Club Books.

Berry, Wendell. 1990. *Nature as Measure.* Berkeley, Calif.: North Point.

Berstein, Henry, et al., eds. 1990. *The Food Question: Profits Versus People.* East Haven, Conn.: Earthscan.

Bongaarts, John. 1994. "Can the Growing Human Population Feed Itself?" *Scientific American*, March, 36–42.

Boyd, Claude E., and Jason W. Clay. 1998. "Shrimp Aquaculture and the Environment." *Scientific American*, June, 58–65.

Brookfield, Harold, and Christine Padoch. 1994. "Appreciating Biodiversity: A Look at the Dynamism and Diversity of Indigenous Farming Practices." *Environment*, June, 6–42.

Brown, Lester R. 1994. *Who Will Feed China?* New York: Norton.

Brown, Lester R. 1996. *Tough Choices: Facing the Challenge of Food Scarcity.* New York: Norton.

Brown, Lester R., and Brian Halwell. 1998. "China's Water Shortage Could Shake World Food Security." *WorldWatch*, July/August, 10–21.

Brown, Lester R., and Hal Kane. 1994. *Full House: Reassessing the Earth's Population Carrying Capacity.* New York: Norton.

Carroll, C. Ronald, et al. 1990. *Agroecology.* New York: McGraw-Hill.

Coleman, Elliot. 1992. *The New Organic Grower's Four-Season Harvest.* White River Junction, Vt.: Chelsea Green.

Dunning, Alan B., and Holly W. Brough. 1991. *Taking Stock: Animal Farming and the Environment.* Washington, D.C.: Worldwatch Institute.

Edwards, Clive A., et al., eds. 1990. *Sustainable Agricultural Systems.* Ankeny, Iowa: Soil and Water Conservation Society.

Evans, L. T. 1998. *Feeding the Ten Billion: Plants and Population Growth.* New York: Cambridge University Press.

Faeth, Paul. 1995. *Growing Green: Enhancing the Economic and Environmental Performance of U.S. Agriculture.* Washington, D.C.: World Resources Institute.

Fowler, Cary, and Pat Mooney. 1990. *Shattering: Food, Politics, and the Loss of Genetic Diversity.* Tucson: University of Arizona Press.

Fukuoka, Masanobu. 1985. *The Natural Way of Farming: The Theory and Practice of Green Philosophy.* New York: Japan Publications.

Gardner, Gary. 1996. *Shrinking Fields: Cropland Loss in a World of Eight Billion.* Washington, D.C.: Worldwatch Institute.

Goering, Peter, et al. 1993. *From the Ground Up: Rethinking Industrial Agriculture.* Atlantic Highlands, N.J.: Zed.

Gordon, R. Conway, and Edward R. Barbier. 1990. *After the Green Revolution: Sustainable Agriculture for Development.* East Haven, Conn.: Earthscan.

Halweil, Brian. 1999. "The Emperor's New Crops." *WorldWatch*, July/August, 21–29.

Jacobson, Michael, et al. 1991. *Safe Food: Eating Wisely in a Risky World.* Washington, D.C.: Planet Earth.

Jeavons, John, and Carol Cox. 1993. *Lazy-Bed Gardening: The Quick and Dirty Guide.* Willits, Calif.: Ecology Action.

Kane, Hal. 1993. "Growing Fish in Fields." *World Watch*, September/October, 20–27.

Klopenburg, Jack, Jr., and Beth Burrows. 1996. "Biotechnology to the Rescue: Twelve Reasons Why Biotechnology Is Incompatible with Sustainable Agriculture." *The Ecologist*, vol. 26, no. 2, 61–67.

Kneen, Brewester. 1999. *Farmaggedon: Food and the Culture of Biotechnology.* Philadelphia: New Society.

Lappe, Francis M. 1999. *World Hunger: Twelve Myths.* New York: Grove.

League of Women Voters. 1991. *U.S. Farm Policy: Who Benefits? Who Pays? Who Decides?* Washington, D.C.: League of Women Voters.

Logsdon, Gene. 1994. *At Nature's Pace: Farming and the American Dream.* New York: Pantheon.

Mann, Charles. 1997. "Reseeding the Green Revolution." *Science*, vol. 277, 1038–43.

McGinn, Anne. 1998. *Rocking the Boat: Conserving Fisheries and Protecting Jobs.* Washington, D.C.: Worldwatch Institute.

Mollison, Bill. 1990. *Permaculture: A Practical Guide for a Sustainable Future.* Washington, D.C.: Island Press.

Montague, Peter. 1999. "Against the Grain: Genetically Modified Foods." *Yes! A Journal of Positive Futures*, Summer, 46–48.

Motavalli, Jim. 1998. "The Trouble with Meat: Dangerous Diseases Emerge in a Planet Eating High on the Food Chain." *E Magazine*, May/June, 29–35.

National Academy of Sciences. 1992. *Marine Aquaculture: Opportunities for Growth.* Washington, D.C.: National Academy Press.

National Academy of Sciences. 1993. *Soil and Water Quality: An Agenda for Agriculture.* Washington, D.C.: National Academy Press.

National Academy of Sciences. 1993. *Sustainable Agriculture and the Environment in the Humid Tropics.* Washington, D.C.: National Academy Press.

Pauly, Daniel et al. 1998. "Fishing Down Marine Food Webs." *Science*, vol. 279, 860–63.

Pimentel, David, and Marcia Pimentel, eds. 1996. *Food, Energy, and Society.* Niwot, Colo: University Press of Colorado.

Postel, Sandra L. 1998. "Water for Food Production: Will There Be Enough in 2025?" *BioScience*, vol. 48, no. 8, 629–637.

Postel, Sandra. 1999. *Pillar of Sand: Can the Irrigation Miracle Last?* New York: Norton.

Pretty, Jules N. 1995. *Regenerating Agriculture: Policies and Prospects for Sustainability and Self-Reliance.* East Haven, Conn.: Earthscan.

Prosterman, Roy L., et al. 1996. "Can China Feed Itself?" *Scientific American*, November, 90–96.

Raeburn, Paul. 1995. *The Last Harvest: The Genetic Gamble That Threatens to Destroy American Agriculture.* New York: Simon & Schuster.

Rifkin, Jeremy. 1992. *Beyond Beef: The Rise and Fall of the Cattle Culture.* New York: Dutton.

Rissler, Jane, and Margaret Mellon. 1996. *The Ecological Risks of Engineered Crops.* Cambridge, Mass.: MIT Press.

Robbins, John. 1992. *May All Be Fed: Diet for a New World.* New York: William Morrow.

Rosegrant, Mark W., and Robert Livernash. 1996. "Growing More Food, Doing Less Damage." *Environment*, vol. 38, no. 7, 6–11, 28–30.

Serageldin, Ismail. 1999. "Biotechnology and Food Security in the 21st Century." *Science*, vol. 285, 387–89.

Shiva, Vandana. 1991. *The Violence of the Green Revolution.* Atlantic Highlands, N.J.: Zed.

Smith, Katherine R. 1995. "Time to 'Green' U.S. Farm Policy." *Issues in Science and Technology*, Spring, 71–78.

Soil and Water Conservation Society. 1990. *Sustainable Agricultural Systems.* Ankeny, Iowa: Soil and Water Conservation Society.

Solbrig, Otto T., and Dorothy Solbrig. 1994. *So Shall You Reap: Farming and Crops in Human Affairs.* Washington, D.C.: Island Press.

Soule, Judith D., and Jon Piper. 1992. *Farming in Nature's Image: An Ecological Approach to Agriculture.* Washington, D.C.: Island Press.

Swanson, Louis E., and Frank B. Clearfield, eds. 1994. *Agricultural Policy and the Environment: Iron Fist or Open Hand?* Ankeny, Iowa: Soil and Water Conservation Society.

Tamsey, Geoff, and Tony Worsley. 1995. *The Food System.* East Haven, Conn.: Earthscan.

United Nations Development Fund. 1996. *Urban Agriculture: Food, Jobs, and Sustainability.* New York: United Nations.

Weber, Peter. 1994. *Net Loss: Fish, Jobs, and the Marine Environment.* Washington, D.C.: Worldwatch Institute.

Welch, Ross M. 1997. "Toward a 'Greener' Revolution." *Issues in Science and Technology*, Fall, 50–58.

Yang, Linda. 1990. *The City Gardener's Handbook: From Balcony to Backyard*. New York: Random House.

16 / PROTECTING FOOD RESOURCES: PESTICIDES AND PEST CONTROL

Internet

For a list of useful Internet sites, hyperlinks to material in this book, and other helpful and interesting information, please visit our Web site at **www.brookscole.com/biology**

Readings

Benbrook, Charles M., et al. 1996. *Pest Management at the Crossroads*. Yonkers, N.Y.: Consumer's Union.

Bormann, F. Herbert, et al. 1993. *Redesigning the American Lawn*. New Haven, Conn.: Yale University Press.

Briggs, Shirley, and the Rachel Carson Council. 1992. *Basic Guide to Pesticides: Their Characteristics and Hazards*. Washington, D.C.: Taylor & Francis.

Care, James R., and Maureen K. Hinkle. 1994. *Integrated Pest Management: The Path of a Paradigm*. Washington, D.C.: National Audubon Society.

Carson, Rachel. 1962. *Silent Spring*. Boston: Houghton Mifflin.

Cassels, Jamie. 1993. *The Uncertain Promise of Law: Lessons from Bhopal*. Toronto: University of Toronto Press.

Dinham, Barbara. 1993. *The Pesticide Hazard: A Global Health and Environmental Audit*. New York: Zed.

Flint, Mary L., and Steve H. Dreistadt. 1998. *Natural Enemies Handbook: The Illustrated Guide to Biological Pest Control*. Los Angeles: University of California Integrated Pest Management Program.

Friends of the Earth. 1990. *How to Get Your Lawn and Garden Off Drugs*. Ottawa, Ontario: Friends of the Earth.

Gustafson, David. 1993. *Pesticides in Drinking Water*. New York: Van Nostrand Reinhold.

Heylin, Michael, ed. 1991. "Pesticides: Costs Versus Benefits." *Chemical & Engineering News*, January 7, 27–56.

Horn, D. J. 1988. *Ecological Approach to Pest Management*. New York: Guilford.

Jenkins, Virginia Scott. 1994. *The Lawn: A History of an American Obsession*. Washington, D.C.: Smithsonian Institute Press.

Marquardt, Sandra. 1989. *Exporting Banned Pesticides: Fueling the Circle of Poison*. Washington, D.C.: Greenpeace.

National Academy of Sciences. 1993. *Pesticides in the Diet of Infants and Children*. Washington, D.C.: National Academy Press.

National Academy of Sciences. 1996. *Ecologically Based Pest Management: New Solutions for a New Century*. Washington, D.C.: National Academy Press.

Natural Resources Defense Council. 1993. *After Silent Spring: The Unsolved Problems of Pesticide Use in the United States*. New York: Natural Resources Defense Council.

Pimentel, David, ed. 1997. *Techniques for Reducing Pesticide Use: Economic and Environmental Benefits*. New York: Wiley.

Pimentel, David, et al. 1992. "Environmental and Economic Cost of Pesticide Use." *BioScience*, vol. 42, no. 10, 750–60.

Platt, Anne. 1996. "IPM and the War on Pests." *World Watch*, March/April, 21–27.

Wargo, J. 1996. *Our Toxic Legacy: How Science and Law Fail to Protect Us From Pesticides*. New Haven Conn.: Yale University Press.

Wiles, Richard, and Christopher Campbell. 1993. *Pesticides in Children's Food*. Washington, D.C.: Environmental Working Group.

Winston, Mark L. 1997. *Nature Wars: People vs. Pests*. Cambridge, Mass.: Harvard University Press.

Yepsen, Roger B., Jr. 1987. *The Encyclopedia of Natural Insect and Pest Control*. Emmaus, Pa.: Rodale.

17 / SUSTAINING TERRESTRIAL ECOSYSTEMS: FORESTS, RANGELANDS, PARKS, AND WILDERNESS

Internet

For a list of useful Internet sites, hyperlinks to material in this book, and other helpful and interesting information, please visit our Web site at **www.brookscole.com/biology**

Readings

Abramovitz, Janet N. 1998. *Taking a Stand: Cultivating a New Relationship with the World's Forests*. Washington, D.C.: Worldwatch Institute.

Abramovitz, Janet N., and Ashley T. Mattoon. 1999. "Reorienting the Forest Products Economy." In Lester R. Brown, et al., *State of the World 1999*. New York: Norton, pp. 60–77.

Anderson, Patrick. 1989. "The Myth of Sustainable Logging: The Case for a Ban on Tropical Timber Imports." *The Ecologist*, vol. 19, no. 5, 166–68.

Aplet, Greg, et al., eds. 1993. *Defining Sustainable Forestry*. Washington, D.C.: Island Press.

Balch, Si, et al. 1998. *The Land Ethic: Meeting the Needs for the Land and Its Resources*. Bethesda, Md.: Society of American Foresters.

Baskin, Yvonne. 1997. *The Work of Nature: How the Diversity of Life Sustains Us*. Washington, D.C.: Island Press.

Beatley, Timothy. 1994. *Ethical Land Use: Principles of Policy and Planning*. Baltimore: Johns Hopkins University Press.

Booth, Douglas E. 1993. *Valuing Nature: The Decline and Preservation of Old Growth Forests*. Lantham, Md.: University Press of America.

Boucher, Norman. 1995. "Back to the Everglades." *Technology Review*, August/September, 25–35.

Brush, Stephen B., and Doreen Stabinsky. 1995. *Valuing Local Knowledge: Indigenous People and Intellectual Property Rights*. Washington, D.C.: Island Press.

Callenbach, Ernest. 1995. *Bring Back the Buffalo: A Sustainable Future for America's Great Plains*. Washington, D.C.: Island Press.

Callicott, J. Baird, ed. 1999. *Beyond the Land Ethic*. Albany: State University of New York Press.

Callicott, J. Baird, 1999. *In Defense of the Land Ethic*. Albany: State University of New York Press.

Callicott, J. Baird, and Eric T. Fregfoyle, eds. 1999. *For the Health of the Land: Unpublished Essays and Other Writings by Aldo Leopold*. Washington D.C.: Island Press (Shearwater).

Carroll, Mathew S. 1995. *Community and the Northwestern Logger*. Boulder, Colo.: Westview.

Chagnon, Napoleon A. 1992. *Yanomamo: The Last Days of Eden*. New York: Harcourt Brace Jovanovich.

Chapin, F. Stuart, et al. 1998. "Ecosystem Consequences of Changing Biodiversity." *BioScience*, vol. 48, no. 1, 45–51.

Daily, Gretchen, ed. 1997. *Nature's Services: Society's Dependence on Natural Ecosystems*. Washington, D.C.: Island Press.

Defenders of Wildlife. 1996. *A Status Report on America's Vanishing Habitat and Wildlife*. Washington, D.C.: Defenders of Wildlife.

Dekker-Robinson, Donna L., and William J. Libby. 1998. "American Forest Policy: Global Ethical Tradeoffs." *BioScience*, vol. 48, no. 6, 471–77.

Denniston, Derek. 1995. *High Priorities: Conserving Mountain. Ecosystems and Cultures*. Washington, D.C.: Worldwatch Institute.

Devall, Bill, ed. 1993. *Clearcut: The Tragedy of Industrial Forestry*. San Francisco: Sierra Club Books/Earth Island.

Dietrich, William. 1993. *The Final Forest: The Last Great Trees of the Pacific Northwest*. New York: Penguin.

DiSilvestro, Roger L. 1993. *Reclaiming the Last Wild Places: A New Agenda for Biodiversity*. New York: Wiley.

Dobson, Andrew. 1998. *Conservation and Biodiversity*. New York: Scientific American Library.

Drengson, Alan, and Duncan Taylor, eds. 1997. *Ecoforestry: The Art & Science of Sustainable Forest Use*. Gabriola Island, Canada: New Society.

Durning, Alan Thein. 1992. *Guardians of the Land: Indigenous Peoples and the Health of the Earth*. Washington, D.C.: Worldwatch Institute.

Durning, Alan Thein. 1992. *Saving the*

Forests: What Will It Take? Washington, D.C.: Worldwatch Institute.

Foreman, Dave, and Howie Wolke. 1989. *The Big Outside.* Tucson, Ariz.: Nedd Ludd Books.

Friends of the Earth. 1992. *The Rainforest Harvest: Sustainable Strategies for Saving Tropical Forests.* Washington, D.C.: Friends of the Earth.

Frome, Michael. 1992. *Regreening the National Parks.* Tucson: University of Arizona Press.

Goldsmith, Edward et al. 1990. *Imperiled Planet: Restoring our Endangered Ecosystems.* Cambridge, Mass.: MIT Press.

Halvorson, William L., and Gary E. Davis, eds. 1996. *Science and Ecosystem Management in the National Parks.* Tucson, Ariz.: University of Arizona Press.

Heady, Harold F., and R. Dennis Child. 1994. *Rangeland Ecology and Management.* Boulder, Colo.: Westview.

Hendee, John, et al. 1991. *Principles of Wilderness Management.* Golden, Colo.: Fulcrum.

Hess, Karl, Jr. 1992. *Visions Upon the Land: Man and Nature on the Western Range.* Washington, D.C.: Island Press.

Hirt, Paul W. 1994. *A Conspiracy of Optimism: Management of the National Forests Since World War Two.* Lincoln: University of Nebraska Press.

Holloway, Marguerite. 1993. "Sustaining the Amazon." *Scientific American,* July, 90–99.

Hunter, Malcolm L., Jr. 1990. *Wildlife, Forests, and Forestry: Principles of Managing Forests for Biodiversity.* Englewood Cliffs, N.J.: Prentice Hall.

Jacobs, Lynn. 1992. *Waste of the West: Public Lands Ranching.* Tucson, Ariz.: Lynn Jacobs.

Johnson, Nels, and Brice Cabarle. 1993. *Looking Ahead: Sustainable Natural Forest Management in the Humid Tropics.* Washington, D.C.: World Resources Institute.

Kellert, Stephen R. 1996. *The Value of Life: Biological Diversity and Human Society.* Washington, D.C.: Island Press.

Kelly, David, and Gary Braasch. 1988. *Secrets of the Old Growth Forest.* Salt Lake City, Utah: Peregrine Smith.

Knight, Richard L., and Sarah F. Bates, eds. 1994. *A New Century for Natural Resources Management.* Washington, D.C.: Island Press.

Kimmins, James P. 1997. *Forest Ecology: A Foundation for Sustainable Management.* 2d ed. Upper Saddle River, N.J.: Prentice Hall.

Kohn, Kathryn, and Jerry M. Franklin. 1997. *Creating a Forestry for the 21st Century.* Washington, D.C.: Island Press.

Kuletz, Valerie L. 1998. *The Painted Desert: Environmental Ruin in the American West.* New York: Routledge.

Lansky, Mitch. 1992. *Beyond the Beauty Strip: Saving What's Left of Our Forests.* Gardiner, Maine: Tilbury House.

Leopold, Aldo. 1949. *A Sand County Almanac.* New York: Oxford University Press.

Little, Charles E. 1995. *The Dying of the Trees: The Pandemic in America's Forests.* New York: Viking.

Loomis, John D. 1993. *Integrated Public Lands Management: Principles and Applications to National Forest, Parks, Wildlife Refuges, and BLM Lands.* New York: Columbia University Press.

Lowry, William R. 1994. *The Capacity for Wonder: Preserving National Parks.* Washington, D.C.: The Brookings Institution.

Mahony, Rhona. 1992. "Debt-for-Nature Swaps: Who Really Benefits?" *The Ecologist,* vol. 22, no. 3, 97–103.

Manning, Richard. 1993. *Last Stand.* New York: Penguin.

Maser, Chris, et al. 1994. *Sustainable Forestry: Philosophy, Science and Economics.* San Francisco: Rainforest Action Network.

Maybury-Lewis, David. 1992. *Millennium: Tribal Wisdom and the Modern World.* New York: Viking.

Meffe, Gary K., and C. Ronald Carroll. 1997. *Principles of Conservation Biology* 2d ed. Sunderland, Mass.: Sinauer.

Miller, Kenton, and Laura Tangley. 1991. *Trees of Life: Saving Tropical Forests and Their Biological Wealth.* Boston: Beacon.

Myers, Norman. 1993. *The Primary Source: Tropical Forests and Our Future.* New York: Norton.

Naar, Jon, and Alex J. Naar. 1993. *This Land Is Your Land: A Guide to North America's Endangered Ecosystems.* New York: Harper.

Nash, Roderick. 1982. *Wilderness and the American Mind.* 3d ed. New Haven, Conn.: Yale University Press.

National Academy of Sciences. 1996. *Biodiversity II. Understanding and Protecting Our Biological Resources.* Washington, D.C.: National Academy Press.

National Parks and Conservation Association. 1994. *Our Endangered Parks: What You Can Do to Protect Our National Heritage.* San Francisco: Foghorn.

National Park Service. 1992. *National Parks for the 21st Century.* Washington, D.C.: National Park Service.

Nester, William R. 1997. *The War for America's Resources.* New York: St. Martin's.

Newman, Arnold. 1990. *The Tropical Rainforest: A World Survey of Our Most Valuable Endangered Habitats.* New York: Facts on File.

Norse, Elliot A. 1990. *Ancient Forests of the Pacific Northwest.* Washington, D.C.: Island Press.

Noss, Reed F. 1999. *A Citizen's Guide to Ecosystem Management.* Richmond, Vt.: Wild Earth.

Noss, Reed F., and Allen Y. Cooperrider. 1994. *Saving Nature's Legacy: Protecting and Restoring Biodiversity.* Washington, D.C.: Island Press.

Oelschlager, Max. 1991. *The Idea of Wilderness from Prehistory to the Age of Ecology.* New Haven, Conn.: Yale University Press.

Office of Technology Assessment. 1992. *Combined Summaries: Technologies to Sustain Tropical Forest Resources and Biological Diversity.* Washington, D.C.: Office of Technology Assessment.

Panayotou, Theodore, and Peter S. Ashton. 1992. *Not by Timber Alone: Economics and Ecology for Sustaining Tropical Forests.* Washington, D.C.: Island Press.

Patterson, Alan. 1990. "Debt for Nature Swaps and the Need for Alternatives." *Environment,* vol. 32, no. 10, 5–13, 31–32.

Perlin, John. 1989. *A Forest Journey: The Role of Wood in the Development of Civilization.* New York: Norton.

Perlman, Dan L., and Glenn Adelson. 1997. *Biodiversity: Exploring Values and Priorities in Conservation.* Malden, Mass.: Blackwell Science.

Pimentel, David, et al. 1992. "Conserving Biological Diversity in Agricultural/Forestry Systems." *BioScience,* vol. 42, no. 5, 354–62.

Pimentel, David, et al. 1997. "Economic and Environmental Benefits of Biodiversity." *BioScience,* December, 747–57.

Primack, Richard B. 1998. *Essentials of Conservation Biology* 2d. ed. Sunderland, Mass.: Sinauer.

Revkin, Andrew. 1990. *The Burning Season: The Murder of Chico Mendes and the Fight for the Amazon.* Boston: Houghton Mifflin.

Riebsame, William E. 1996. "Ending the Range Wars?" *Environment,* vol. 38, no. 4, 4–29.

Rietbergen, Simon, ed. 1994. *The Earthscan Reader in Tropical Forestry.* San Francisco: Rainforest Action Network.

Robbins, Jim. 1994. *Last Refuge: The Environmental Showdown in the American West.* San Francisco: HarperCollins.`

Robinson, Gordon. 1987. *The Forest and the Trees: A Guide to Excellent Forestry.* Washington, D.C.: Island Press.

Romme, William H., and Don G. Despain. 1989. "The Yellowstone Fires." *Scientific American,* vol. 261, no. 5, 37–46.

Runte, Alfred. 1990. *Yosemite: The Embattled Wilderness.* Lincoln: University of Nebraska Press.

Runte, Alfred. 1991. *Public Lands, Public Heritage: The National Forest Idea.* Niwot, Colo.: Roberts Rinehart.

Rush, James. 1991. *The Last Tree: Reclaiming the Environment in Tropical Asia.* Boulder, Colo.: Westview.

Ryan, John C. 1992. *Life Support: Conserving Biological Diversity.* Washington, D.C.: Worldwatch Institute.

Sauer, Leslie J., 1998. *The Once and Future Forest: A Guide to Forest Restoration Strategies.* Washington, D.C.: Island Press.

Savory, Allan, and Jody Butterfield. 1999. *Holistic Management: A New Framework for Decision Making.* Washington, D.C.: Island Press.

Sedjo, Roger A., and Daniel Botkin. 1997. "Using Forest Plantations to Spare National

Forests." *Environment*, vol. 39, no. 10, 12–20, 29–30.

Shiva, Vandana. 1993. *Monocultures of the Mind: Perspectives on Biodiversity and Biotechnology.* Atlantic Highlands, N.J.: Zed.

Shiva, Vandana. 1997. *Biopiracy: The Plunder of Nature and Knowledge.* Boston: South End Press.

Simon, Noel. 1995. *Nature in Danger: Threatened Habitats and Species.* New York: Oxford University Press.

Snow, Donald, et al., eds. 1997. *The Next West: Public Lands, Community, and Economy in the American West.* Washington, D.C.: Island Press.

Teitel, Martin. 1992. *Rain Forest in Your Kitchen: The Hidden Connection Between Extinction and Your Supermarket.* Washington, D.C.: Island Press.

Tennebaum, David. 1995. "The Greening of Costa Rica." *Technology Review*, October, 42–52.

Thompson, Claudia. 1992. *Recycled Papers: The Essential Guide.* Cambridge, Mass.: MIT Press.

Tobin, Richard J. 1990. *The Expendable Future: U.S. Politics and the Protection of Biodiversity.* Durham, N.C.: Duke University Press.

Vandermeer, John, and Ivette Perfecto. 1995. *Breakfast of Biodiversity: The Truth About Rain Forest Destruction.* Oakland, Calif.: Food First.

Wellner, Pamela, and Eugene Dickey. 1991. *The Wood User's Guide.* San Francisco: Rainforest Action Network.

Whitmore, T.C. 1990. *An Introduction to Tropical Rainforests.* New York: Oxford University Press.

Wilderness Society. 1991. *Keeping It Wild: A Citizen Guide to Wilderness Management.* Washington, D.C.: Wilderness Society.

Wilson, E. O., ed. 1988. *Biodiversity.* Cambridge, Mass.: Harvard University Press.

Wilson, E. O. 1992. *The Diversity of Life.* Cambridge, Mass.: Harvard University Press.

World Commission on Forests and Sustainable Development. 1999. *Our Forests . . . Our Future.* New York: Cambridge University Press.

World Wildlife Fund. 1999. *Living Planet: Preserving Edens of the Earth.* New York: Crown.

Worster, Donald. 1994. *An Unsettled Country: Changing Landscapes of the American West.* Albuquerque: University of New Mexico Press.

Yaffee, Steven Lewis. 1994. *The Wisdom of the Spotted Owl: Policy Lessons for a New Century.* Washington, D.C.: Island Press.

Zaslowsky, Dyan, and T. H. Watkins. 1994. *These American Lands: Parks, Wilderness, and the Public Lands.* Rev. ed. Washington, D.C.: Island Press.

Zuckerman, Seth. 1991. *Saving Our Ancient Forests.* Federalsburg, Md.: Living Planet.

18 / SUSTAINING WILD SPECIES

Internet

For a list of useful Internet sites, hyperlinks to material in this book, and other helpful and interesting information, please visit our Web site at **www.brookscole.com/biology**

Readings

Ackerman, Diane. 1991. *The Moon by Whalelight.* New York: Random House.

Barker, Rocky. 1993. *Saving All the Parts: Reconciling Economics and the Endangered Species Act.* Washington, D.C.: Island Press.

Berrill, Michael. 1997. *The Plundered Seas: Can the World's Fish Be Saved?* San Francisco: Sierra Club Books.

Bolen, Eric G., and William J. Robinson. 1998. *Wildlife Ecology and Management.* Englewood Cliffs, N.J.: Prentice Hall.

Boo, Elizabeth. 1990. *Ecotourism: The Potential and the Pitfalls.* Vols. 1, 2. Washington, D.C.: World Wildlife Fund.

Botsford, Louis W., et al. 1997. "The Management of Fisheries and Marine Ecosystems." *Science*, vol. 277, 509–15.

Bright, Chris. 1995. "Understanding the Threat of Bioinvasions." In Lester R. Brown et al., *State of the World 1996.* Washington, D.C.: Worldwatch Institute, 95–113.

Caughley, Graeme, and A. R. E. Sinclair. 1994. *Wildlife Ecology and Management.* Cambridge, Mass.: Basil Blackwell.

Clark, Tim W., et al., eds. 1994. *Endangered Species Recovery: Finding the Lessons, Improving the Process.* Washington, D.C.: Island.

Cox, George W. 1993. *Conservation Ecology.* Dubuque, Iowa: Wm. C. Brown.

Credlund, Arthur G. 1983. *Whales and Whaling.* New York: Seven Hills Books.

Croke, Vicki. 1997. *The Modern Ark: The Story of Zoos: Past, Present, Future.* New York: Scribner.

DeBlieu, Jan. 1991. *Meant to Be Wild: The Struggle to Save Endangered Species Through Captive Breeding.* New York: Fulcrum.

DiSilvestro, Roger L. 1992. *Rebirth of Nature.* New York: Wiley.

Eldredge, Niles. 1991. *The Miner's Canary.* New York: Prentice Hall.

Ewel, John J., et al. 1999. "Deliberate Introductions of Species: Research Needs." *BioScience*, vol. 49, no. 8, 619–30.

Galan, Mark. 1997. *There's Still Time: The Success of the Endangered Species Act.* Washington, D.C.: National Geographic Society.

Gilbert, Frederick F., and Donald G. Dodds. 1992. *The Philosophy and Practice of Wildlife Management.* 2d ed. Malabar, Fla.: Robert E. Krieger.

Gray, Gary G. 1993. *Wildlife and People: The Human Dimensions of Wildlife Ecology.* Champaign: University of Illinois Press.

Grumbine, R. Edward. 1992. *Ghost Bears: Exploring the Biodiversity Crisis.* Washington, D.C.: Island.

Hargrove, Eugene C. 1992. *The Animal Rights/Environmental Ethics Debate.* Ithaca: State University of New York.

Howes, Chris. 1997. *The Spice of Life: Biodiversity and the Extinction Crisis.* London: Blandford.

Kaufman, Les, and Kenneth Mallay. 1993. *The Last Extinction.* 2d ed. Cambridge, Mass.: MIT Press.

Kellert, Stephen R. 1997. *Kinship to Mastery: Biophilia in Human Evolution and Development.* Washington, D.C.: Island.

Kellert, Stephen R., and E. O. Wilson, eds. 1995. *The Biophilia Hypothesis.* Washington, D.C.: Island.

Klenig, John. 1992. *Valuing Life.* Princeton, N.J.: Princeton University Press.

Kohm, Kathryn A., ed. 1990. *Balancing on the Brink of Extinction: The Endangered Species Act and Lessons for the Future.* Washington, D.C.: Island.

Leakey, Richard, and Roger Lewin. 1995. *The Sixth Extinction: Patterns of Life and the Future of Humankind.* New York: Doubleday.

Livingston, John A. 1981. *The Fallacy of Wildlife Conservation.* Toronto: McClelland & Stewart.

Luoma, Jon. 1987. *A Crowded Ark: The Role of Zoos in Wildlife Conservation.* Boston: Houghton Mifflin.

Maehr, David S. 1997. *The Florida Panther: Life and Death of a Vanishing Carnivore.* Washington, D.C.: Island.

Mann, Charles C., and Mark L. Plummer. 1995. *Noah's Choice: The Future of Endangered Species.* New York: Knopf.

Myers, Norman. 1994. *A Wealth of Wild Species: Storehouse for Human Welfare.* Boulder, Colo.: Westview.

Myers, Norman. 1997. "Mass Extinction and Evolution." *Science*, vol. 278, 597–98.

Nash, Roderick F. 1988. *The Rights of Nature: A History of Environmental Ethics.* Madison: University of Wisconsin Press.

National Academy of Sciences. 1995. *Science and the Endangered Species Act.* Washington, D.C.: National Academy of Sciences.

National Academy of Sciences. 1999. *Sustaining Marine Fisheries.* Washington, D.C.: National Academy of Sciences.

Office of Technology Assessment. 1989. *Oil Production in the Arctic National Wildlife Refuge.* Washington, D.C.: Government Printing Office.

Primack, Richard B. 1993. *Essentials of Conservation Biology.* Sunderland, Mass.: Sinauer.

Reisner, Marc. 1991. *Game Wars: The Undercover Pursuit of Wildlife Poachers.* New York: Viking.

Roe, Frank G. 1970. *The North American Buffalo.* Toronto: University of Toronto Press.

Schaller, George. 1993. *The Last Panda.* Chicago: University of Chicago Press.

Soulé, Michael E. 1991. "Conservation: Tactics for a Constant Crisis." *Science*, vol. 253, 744–50.

Stearns, Beverly P., and Stephen C. Stearns. 1999. *Watching from the Edge of Extinction*. New Haven, Conn.: Yale University Press.

Terborgh, John. 1992. "Why American Songbirds Are Vanishing." *Scientific American*, May, 98–104.

Terborgh, John. 1999. *Requiem for Nature*. Washington, D.C.: Island Press.

Thorne-Miller, Boyce. 1999. *The Living Ocean: Understanding and Protecting Marine Biodiversity*. Washington, D.C.: Island Press.

Tudge, Colin. 1992. *Last Animals at the Zoo: How Mass Extinction Can Be Stopped*. Washington, D.C.: Island Press.

Tuttle, Merlin D. 1988. *America's Neighborhood Bats: Understanding and Learning to Live in Harmony with Them*. Austin: University of Texas Press.

Tuxill, John. 1998. *Losing Strands in the Web of Life: Vertebrate Declines and the Conservation of Biological Diversity*. Washington, D.C.: Worldwatch Institute.

Ward, Peter, 1995. *The End of Evolution: Mass Extinctions and the Preservation of Biodiversity*. New York: Bantam.

Wilder, Robert J., et al. 1999. "Saving Marine Biodiversity." *Issues in Science and Technology*, Spring, 57–64.

World Resources Institute et al. 1992. *Global Biodiversity Strategy: Guidelines for Action to Save, Study, and Use Earth's Biotic Wealth Sustainably and Equitably*. Washington, D.C.: World Resources Institute.

19 / NONRENEWABLE ENERGY RESOURCES

Internet

For a list of useful Internet sites, hyperlinks to material in this book, and other helpful and interesting information, please visit our Web site at **www.brookscole.com/biology**

Readings

Ahearne, John F. 1993. "The Future of Nuclear Power." *American Scientist*, vol. 81, 24–35.

Ballonoff, Paul. 1997. *Energy: The Never-Ending Crisis*. Washington, D. C.: Cato Institute.

Beck, Peter. 1994. *Prospects and Strategies for Nuclear Power: Global Boon or Dangerous Diversion?* East Haven, Conn.: Earthscan.

Campbell, Colin J. 1999. *The Coming Oil Crisis*. Santa Cruz, Calif.: EcoSystems.

Campbell, Colin J., and Jean H. Laherrère. 1998. "The End of Cheap Oil." *Scientific American*, March, 78–83.

Chernousenko, Vladimir M. 1991. *Chernobyl: Insight from the Inside*. New York: Springer-Verlag.

Cohen, Bernard L. 1990. *The Nuclear Energy Option: An Alternative for the '90s*. New York: Plenum.

Department of Energy Annual. *Annual Energy Outlook*. Washington, D.C.: U.S. Energy Information Administration, Department of Energy.

Dunn, Seth. 1999. "King Coal's Weakening Grip on Power." *WorldWatch*, September/October, 10–19.

Flavin, Christopher, and Nicholas Lenssen. 1994. "Reshaping the Power Industry." In Lester Brown et al., *State of the World 1994*. New York: Norton, 61–80.

Ford, Daniel F. 1986. *Meltdown*. New York: Simon & Schuster.

George, Richard L. 1998. "Mining for Oil." *Scientific American*, March, 84–88.

Gershey, Edward L., et al. 1993. *Low-Level Radioactive Waste: From Cradle to Grave*. New York: Van Nostrand Reinhold.

Golay, Michael W., and Neil E. Todreas. 1990. "Advanced Light-Water Reactors." *Scientific American*, April, 82–89.

Gould, Jay M., and Benjamin A. Goldman. 1991. *Deadly Deceit: Low-Level Radiation, High-Level Cover-Up*. New York: Four Walls Eight Windows.

Hollister, Charles D., and Steven Nadis. 1998. "Burial of Radioactive Waste Under the Seabed." *Scientific American*, January, 60–65.

Huizenga, John R. 1992. *Cold Fusion: The Scientific Fiasco of the Century*. Rochester, N.Y.: University of Rochester Press.

Lenssen, Nicholas. 1991. *Nuclear Waste: The Problem That Won't Go Away*. Washington, D.C.: Worldwatch Institute.

Lenssen, Nicholas. 1993. "All the Coal in China." *Worldwatch*, March/April, 22–28.

Lenssen, Nicholas, and Christopher Flavin. 1996. "Meltdown." *World Watch*, May/June, 23–31.

MacKenzie, James J. 1996. "Heading Off the Permanent Oil Crisis." *Issues in Science and Technology*, Summer, 48–54.

May, John. 1990. *The Greenpeace Book of the Nuclear Age*. New York: Pantheon.

Murray, Raymond L., and Judith Powell, eds. 1994. *Understanding Radioactive Waste*. 4th ed. Columbus, Ohio: Battelle.

National Academy of Sciences. 1991. *Nuclear Power: Technical and Institutional Options for the Future*. Washington, D.C.: National Academy Press.

National Academy of Sciences. 1991. *Undiscovered Oil and Gas Resources*. Washington, D.C.: National Academy Press.

National Academy of Sciences. 1995. *Coal: Energy for the Future*. Washington, D.C.: National Academy Press.

Oppenheimer, Ernest J. 1990. *Natural Gas, the Best Energy Choice*. New York: Pen & Podium.

President's Commission on the Accident at Three Mile Island. 1979. *Report of the President's Commission on the Accident at Three Mile Island*. Washington, D.C.: Government Printing Office.

Read, Piers Paul. 1993. *Ablaze: The Story of Chernobyl*. New York: Random House.

Rhoades, Richard. 1993. *Nuclear Renewal: Common Sense About Energy*. New York: Viking.

Savchenko, V. K. 1995. *The Ecology of the Chernobyl Catastrophe*. Pearl River, N.Y.: Parthenon.

Sierra Club. 1991. *Kick the Oil Habit: Choosing a Safe Energy Future for America*. San Francisco: Sierra Club Books.

Squillace, Mark. 1990. *Strip Mining Handbook*. Washington, D.C.: Friends of the Earth.

Sthcherbak, Yuri M. 1996. "Ten Years of the Chernobyl Era." *Scientific American*, April, 44–49.

Toke, David. 1995. *Energy and Environment: The Political and Economic Debate*. Boulder, Colo.: Westview.

Union of Concerned Scientists. 1990. *Safety Second: The NRC and America's Nuclear Power Plants*. Bloomington: Indiana University Press.

Whipple, Chris G. 1996. "Can Nuclear Waste Be Stored Safely at Yucca Mountain?" *Scientific American*, June, 72–79.

Yeargin, Daniel. 1990. *The Prize: The Epic Quest for Oil, Money, and Power*. New York: Simon & Schuster.

Youngquist, Walter L. 1998. *Spending Our Genetic Inheritance—Then What?* Portland, Oreg.: Halcyon House.

Zorpette, Glenn. 1996. "Hanford's Nuclear Wasteland." *Scientific American*, May, 88–97.

20 / ENERGY EFFICIENCY AND RENEWABLE ENERGY

Internet

For a list of useful Internet sites, hyperlinks to material in this book, and other helpful and interesting information, please visit our Web site at **www.brookscole.com/biology**

Readings

American Council for an Energy Efficient Economy. Annual. *The Most Energy-Efficient Appliances*. Washington, D.C.: American Council for an Energy Efficient Economy.

American Council for an Energy Efficient Economy. 1991. *Energy Efficiency and Environment: Forging the Link*. Washington, D.C.: American Council for an Energy Efficient Economy.

American Council for an Energy Efficient Economy. 1999. *Combined Heat and Power: Capturing Wasted Energy*. Washington, D.C.: American Council for an Energy Efficient Economy.

Anderson, Victor. 1993. *Energy Efficiency Policies*. New York: Routledge.

Berger, John. 1997. *Charging Ahead: The Business of Renewable Energy and What It Means to America*. New York: Henry Holt.

Berman, Daniel M., and John T. O'Connor. 1996. *Who Owns the Sun? People, Politics, and the Struggle for a Solar Economy*. River Junction, Vt.: Chelsea Green.

Blackburn, John O. 1987. *The Renewable Energy Alternative: How the United States and the World Can Prosper Without Nuclear*

Energy or Coal. Durham, N.C.: Duke University Press.

Brower, Michael. 1992. *Cool Energy: The Renewable Solution to Global Warming.* Cambridge, Mass.: Union of Concerned Scientists.

Carless, Jennifer. 1993. *Renewable Energy: A Concise Guide to Green Alternatives.* New York: Walker.

Cole, Nancy, and P. S. Skerrett. 1995. *Renewables Are Ready: People Creating Energy Solutions.* White River Junction, Vt.: Chelsea Green.

Field, Frank R. III, and Joel P. Clark. 1997. "A Practical Road to Lightweight Cars." *Technology Review*, January, 28–36.

Flavin, Christopher. 1996. "Power Shock: The Next Energy Revolution." *World Watch*, January/February, 10–19.

Flavin, Christopher, and Seth Dunn. 1997. *Rising Sun, Gathering Winds: Policies to Stabilize the Climate and Strengthen Economies.* Washington, D.C.: Worldwatch Institute.

Flavin, Christopher, and Nicholas Lenssen. 1994. *Power Surge: Guide to the Coming Energy Revolution.* New York: Norton.

Gever, John, et al. 1991. *Beyond Oil: The Threat to Food and Fuel in Coming Decades.* Boulder: University of Colorado Press.

Giampietro, Mario et al. 1998. "Feasibility of Large-Scale Biofuel Production." *BioScience*, vol. 47, no. 9, 587–600.

Gipe, Paul. 1995. *Wind Energy Comes of Age.* New York: Wiley.

Gipe, Paul. 1999. *Wind Energy Basics: A Guide to Small and Micro Wind Systems.* White River Junction, Vt.: Chelsea Green.

Harland, Edward. 1994. *Eco-Renovation: The Ecological Home Improvement Guide.* White River Junction, Vt.: Chelsea Green.

Hart, David. 1997. *Hydrogen Power.* London: FT Energy Publishing.

Kozloff, Keith, and Roger C. Dower. 1993. *A New Power Base: Renewable Energy Policies for the Nineties and Beyond.* Washington, D.C.: World Resources Institute.

Lovins, Amory B. 1990. *The Negawatt Revolution.* Snowmass, Colo.: Rocky Mountain Institute.

Lovins, Amory B., and L. Hunter Lovins. 1995. "Reinventing the Wheels." *The Atlantic Monthly*, January, 75–94.

MacKenzie, James J. 1994. *The Keys to the Car: Electric and Hydrogen Vehicles for the 21st Century.* Washington, D.C.: World Resources Institute.

McKeown, Walter. 1991. *Death of the Oil Age and the Birth of Hydrogen America.* San Francisco: Wild Bamboo Press.

Nansen, Ralph. 1995. *Sun Power: The Global Solution for the Coming Energy Crisis.* Ocean Shores, Wash.: Ocean Press.

National Academy of Sciences. 1992. *Automotive Fuel Efficiency: How Far Can We Go?* Washington, D.C.: National Academy Press.

National Renewable Energy Laboratory. 1999. *Choices for a Brighter Future: Perspectives on Renewble Energy.* Springfield, VA.: National Technical Information Service.

Office of Technology Assessment. 1990. *Replacing Gasoline: Alternative Fuels for Light-Duty Vehicles.* Washington, D.C.: Government Printing Office.

Office of Technology Assessment. 1991. *Improving Automobile Fuel Economy: New Standards, New Approaches.* Washington, D.C.: Government Printing Office.

Office of Technology Assessment. 1992. *Building Energy Efficiency.* Washington, D.C.: Government Printing Office.

Ogden, Joan M., and Robert H. Williams. 1989. *Solar Hydrogen: Moving Beyond Fossil Fuels.* Washington, D.C.: World Resources Institute.

Pimentel, David, et al. 1994. "Environmental and Social Costs of Biomass Energy." *BioScience*, February, 89–93.

Potts, Michael. 1993. *The Independent Home: Living Well with Power from the Sun, Wind, and Water.* White River Junction, Vt.: Chelsea Green.

Rocky Mountain Institute. 1995. *Community Energy Workbook.* Snowmass, Colo.: Rocky Mountain Institute.

Rocky Mountain Institute. 1995. *Homemade Energy: How to Save Energy and Dollars in Your Home.* Snowmass, Colo.: Rocky Mountain Institute.

Roodman, David M., and Nicholas Lenssen. 1995. *A Building Revolution: How Ecology and Health Concerns Are Transforming Construction.* Washington, D.C.: Worldwatch Institute.

Sperling, Daniel. 1996. "The Case for Electric Vehicles." *Scientific American*, November, 54–59.

Steen, Athena Swentzell, et al. 1994. *The Straw Bale House.* White River Junction, Vt.: Chelsea Green.

Tenenbaum, David. 1995. "Tapping the Fire Down Below." *Technology Review*, January, 38–47.

Vale, Brenda, and Robert Vale. 1991. *Green Architecture.* Boston: Little, Brown.

Weisman, Alan. 1998. *Gaviotas: A Village to Reinvent the World.* White River Junction, Vt.: Chelsea Green.

Weizacker, Ernst von, and Amory and Hunter Lovins. 1997. *Factor Four: Doubling Wealth, Halving Resource Use.* Snowmass, Colo.: Rocky Mountain Institute.

Wells, Malcolm. 1991. *How to Build an Underground House.* Brewster, Mass.: Malcolm Wells.

Wouk, Victor. 1997. "Hybrid Electric Vehicles." *Scientific American*, October, 70–74.

Yeang, Ken. 1995. *Designing with Nature: The Ecological Basis for Architectural Design.* New York: McGraw-Hill.

Zweibel, Ken. 1990. *Harnessing Solar Power: The Challenge of Photovoltaics.* New York: Plenum.

GLOSSARY

abiotic Nonliving. Compare *biotic*.

absolute humidity Amount of water vapor found in a certain mass of air (usually expressed as grams of water per kilogram of air). Compare *relative humidity*.

acclimation Adjustment to slowly changing new conditions. Compare *threshold effect*.

acid deposition The falling of acids and acid-forming compounds from the atmosphere to the earth's surface. Acid deposition is commonly known as *acid rain*, a term that refers only to wet deposition of droplets of acids and acid-forming compounds.

acid rain See *acid deposition*.

acid solution Any water solution that has more hydrogen ions (H^+) than hydroxide ions (OH^-); any water solution with a pH less than 7. Compare *basic solution*, *neutral solution*.

active solar heating system System that uses solar collectors to capture energy from the sun and store it as heat for space heating and heating water. Liquid or air pumped through the collectors transfers the captured heat to a storage system such as an insulated water tank or rock bed. Pumps or fans then distribute the stored heat or hot water throughout a dwelling as needed. Compare *passive solar heating system*.

adaptation Any genetically controlled structural, physiological, or behavioral characteristic that helps an organism survive and reproduce under a given set of environmental conditions. It usually results from a beneficial mutation. See *biological evolution*, *differential reproduction*, *mutation*, *natural selection*.

adaptive radiation Period of time (usually millions of years) during which numerous new species evolve to fill vacant and new ecological niches in changed environments, usually after a mass extinction.

adaptive trait See *adaptation*.

advanced sewage treatment Specialized chemical and physical processes that reduce the amount of specific pollutants left in wastewater after primary and secondary sewage treatment. This type of treatment usually is expensive. See also *primary sewage treatment*, *secondary sewage treatment*.

aerobic respiration Complex process that occurs in the cells of most living organisms, in which nutrient organic molecules such as glucose ($C_6H_{12}O_6$) combine with oxygen (O_2) and produce carbon dioxide (CO_2), water (H_2O), and energy. Compare *photosynthesis*.

age structure Percentage of the population (or number of people of each sex) at each age level in a population.

agricultural revolution Gradual shift from small, mobile hunting and gathering bands to settled agricultural communities in which people survived by learning how to breed and raise wild animals and to cultivate wild plants near where they lived. It began 10,000–12,000 years ago. Compare *environmental revolution*, *hunter-gatherers*, *industrial revolution*, *information revolution*.

agroforestry Planting trees and crops together.

air pollution One or more chemicals in high enough concentrations in the air to harm humans, other animals, vegetation, or materials. Excess heat and noise can also be considered forms of air pollution. Such chemicals or physical conditions are called air pollutants. See *primary pollutant*, *secondary pollutant*.

alien species See *nonnative species*.

allele Slightly different molecular form found in a particular gene.

alley cropping Planting of crops in strips with rows of trees or shrubs on each side.

alpha particle Positively charged matter, consisting of two neutrons and two protons, that is emitted as a form of radioactivity from the nuclei of some radioisotopes. See also *beta particle*, *gamma rays*.

altitude Height above sea level. Compare *latitude*.

anaerobic respiration Form of cellular respiration in which some decomposers get the energy they need through the breakdown of glucose (or other nutrients) in the absence of oxygen. Compare *aerobic respiration*.

ancient forest See *old-growth forest*.

animal manure Dung and urine of animals that can be used as a form of organic fertilizer. Compare *green manure*.

animals Eukaryotic, multicelled organisms such as sponges, jellyfishes, arthropods (insects, shrimp, lobsters), mollusks (snails, clams, oysters, octopuses), fish, amphibians (frogs, toads, salamanders), reptiles (turtles, lizards, alligators, crocodiles, snakes), birds, and mammals (kangaroos, bats, cats, rabbits, elephants, whales, porpoises, monkeys, apes, humans). See *carnivores*, *herbivores*, *omnivores*.

annual Plant that grows, sets seed, and dies in one growing season. Compare *perennial*.

aquaculture Growing and harvesting of fish and shellfish for human use in freshwater ponds, irrigation ditches, and lakes, or in cages or fenced-in areas of coastal lagoons and estuaries. See *fish farming*, *fish ranching*.

aquatic Pertaining to water. Compare *terrestrial*.

aquatic life zone Marine and freshwater portions of the ecosphere. Examples include freshwater life zones (such as lakes and streams) and ocean or marine life zones (such as estuaries, coastlines, coral reefs, and the deep ocean).

aquifer Porous, water-saturated layers of sand, gravel, or bedrock that can yield an economically significant amount of water.

arable land Land that can be cultivated to grow crops.

arid Dry. A desert or other area with an arid climate has little precipitation.

asexual reproduction Reproduction in which a mother cell divides to produce two identical daughter cells that are clones of the mother cell. This type of reproduction is common in single-celled organisms. Compare *sexual reproduction*.

atmosphere The whole mass of air surrounding the earth. See *stratosphere*, *troposphere*.

atom Minute unit made of subatomic particles that is the basic building block of all chemical elements and thus all matter; the smallest unit of an element that can exist and still have the unique characteristics of that element. Compare *ion*, *molecule*.

atomic number Number of protons in the nucleus of an atom. Compare *mass number*.

autotroph See *producer*.

background extinction Normal extinction of various species as a result of changes in local environmental conditions. Compare *mass extinction*.

bacteria Prokaryotic, one-celled organisms. Some transmit diseases. Most act as decomposers and get the nutrients they need by breaking down complex organic compounds in the tissues of living or dead organisms into simpler inorganic nutrient compounds.

barrier islands Long, thin, low offshore islands of sediment that generally run parallel to the shore along some coasts.

basic solution Water solution with more hydroxide ions (OH^-) than hydrogen ions (H^+); water solution with a pH greater than 7. Compare *acid solution, neutral solution.*

beneficiation Separation of an ore mineral from the waste mineral material (gangue). See *tailings.*

beta particle Swiftly moving electron emitted by the nucleus of a radioactive isotope. See also *alpha particle, gamma rays.*

bioaccumulation An increase in the concentration of a chemical in specific organs or tissues at a level higher than would normally be expected. Compare *biomagnification.*

biodegradable Capable of being broken down by decomposers.

biodegradable pollutant Material that can be broken down into simpler substances (elements and compounds) by bacteria or other decomposers. Paper and most organic wastes such as animal manure are biodegradable but can take decades to biodegrade in modern landfills. Compare *degradable pollutant, nondegradable pollutant, slowly degradable pollutant.*

biodiversity Variety of different species (*species diversity*), genetic variability among individuals within each species (*genetic diversity*), and variety of ecosystems (*ecological diversity*).

biofuel Gas or liquid fuel (such as ethyl alcohol) made from plant material (biomass).

biogeochemical cycle Natural processes that recycle nutrients in various chemical forms from the nonliving environment to living organisms, and then back to the nonliving environment. Examples are the carbon, oxygen, nitrogen, phosphorus, sulfur, and hydrologic cycles.

biological amplification See *biomagnification.*

biological community See *community.*

biological diversity See *biodiversity.*

biological evolution Change in the genetic makeup of a population of a species in successive generations. If continued long enough, it can lead to the formation of a new species. Note that populations—not individuals—evolve. See also *adaptation, differential reproduction, natural selection, theory of evolution.*

biological oxygen demand (BOD) Amount of dissolved oxygen needed by aerobic decomposers to break down the organic materials in a given volume of water at a certain temperature over a specified time period.

biological pest control Control of pest populations by natural predators, parasites, or disease-causing bacteria and viruses (pathogens).

biomagnification Increase in concentration of DDT, PCBs, and other slowly degradable, fat-soluble chemicals in organisms at successively higher trophic levels of a food chain or web. Compare *bioaccumulation.*

biomass Organic matter produced by plants and other photosynthetic producers; total dry weight of all living organisms that can be supported at each trophic level in a food chain or web; dry weight of all organic matter in plants and animals in an ecosystem; plant materials and animal wastes used as fuel.

biome Terrestrial regions inhabited by certain types of life, especially vegetation. Examples are various types of deserts, grasslands, and forests.

biosphere Zone of earth where life is found. It consists of parts of the atmosphere (the troposphere), hydrosphere (mostly surface water and groundwater), and lithosphere (mostly soil and surface rocks and sediments on the bottoms of oceans and other bodies of water) where life is found. Also called the *ecosphere.*

biotic Living organisms make up the biotic parts of ecosystems. Compare *abiotic.*

biotic potential Maximum rate at which the population of a given species can increase when there are no limits on its rate of growth. See *environmental resistance.*

birth rate See *crude birth rate.*

bitumen Gooey, black, high-sulfur, heavy oil extracted from tar sand and then upgraded to synthetic fuel oil. See *tar sand.*

breeder nuclear fission reactor Nuclear fission reactor that produces more nuclear fuel than it consumes by converting nonfissionable uranium-238 into fissionable plutonium-239.

broadleaf deciduous plants Plants such as oak and maple trees that survive drought and cold by shedding their leaves and becoming dormant. Compare *broadleaf evergreen plants, coniferous evergreen plants.*

broadleaf evergreen plants Plants that keep most of their broad leaves year-round. Examples are the trees found in the canopies of tropical rain forests. Compare *broadleaf deciduous plants, coniferous evergreen plants.*

calorie Unit of energy; amount of energy needed to raise the temperature of 1 gram of water 1°C (unit on Celsius temperature scale). See also *kilocalorie.*

cancer Group of more than 120 different diseases, one for each type of cell in the human body. Each type of cancer produces a tumor in which cells multiply uncontrollably and invade surrounding tissue.

capitalism See *pure market economic system.*

carbon cycle Cyclic movement of carbon in different chemical forms from the environment to organisms and then back to the environment.

carcinogen Chemicals, ionizing radiation, and viruses that cause or promote the development of cancer. See *cancer, mutagen, teratogen.*

carnivore Animal that feeds on other animals. Compare *herbivore, omnivore.*

carrying capacity (K) Maximum population of a particular species that a given habitat can support over a given period of time.

cell Smallest living unit of an organism. Each cell is encased in an outer membrane or wall and contains genetic material (DNA) and other parts to perform its life function. Organisms such as bacteria consist of only one cell, but most of the organisms we are familiar with contain many cells. See *eukaryotic cell, prokaryotic cell.*

centrally planned economy See *pure command economic system.*

CFCs See *chlorofluorocarbons.*

chain reaction Multiple nuclear fissions, taking place within a certain mass of a fissionable isotope, that release an enormous amount of energy in a short time.

chemical One of the millions of different elements and compounds found naturally and synthesized by humans. See *compound, element.*

chemical change Interaction between chemicals in which there is a change in the chemical composition of the elements or compounds involved. Compare *nuclear change, physical change.*

chemical evolution Formation of the earth and its early crust and atmosphere, evolution of the biological molecules necessary for life, and evolution of systems of chemical reactions needed to produce the first living cells. These processes are believed to have occurred about 1 billion years before biological evolution. Compare *biological evolution.*

chemical formula Shorthand way to show the number of atoms (or ions) in the basic structural unit of a compound. Examples are H_2O, NaCl, and $C_6H_{12}O_6$.

chemical reaction See *chemical change.*

chemosynthesis Process in which certain organisms (mostly specialized bacteria) extract inorganic compounds from their environment and convert them into organic nutrient compounds without the presence of sunlight. Compare *photosynthesis.*

chlorinated hydrocarbon Organic compound made up of atoms of carbon, hydrogen, and chlorine. Examples are DDT and PCBs.

chlorofluorocarbons (CFCs) Organic compounds made up of atoms of carbon, chlorine, and fluorine. An example is Freon-12 (CCl_2F_2), used as a refrigerant in refrigerators and air conditioners and in making plastics such as Styrofoam. Gaseous CFCs can deplete the ozone layer when they slowly rise into the stratosphere and their chlorine atoms react with ozone molecules. Use of these molecules is being phased out.

chromosome A grouping of various genes and associated proteins in plant and animal cells that carry certain types of genetic information. See *genes.*

clear-cutting Method of timber harvesting in which all trees in a forested area are removed in a single cutting. Compare *selective cutting, seed-tree cutting, shelterwood cutting, strip cutting.*

climate Physical properties of the troposphere of an area based on analysis of its weather records over a long period (at least 30 years). The two main factors determining an area's climate are *temperature*, with its seasonal variations, and the amount and distribution of *precipitation.* Compare *weather.*

climax community See *mature community.*

coal Solid, combustible mixture of organic compounds with 30–98% carbon by weight, mixed with various amounts of water and small amounts of sulfur and nitrogen compounds. It is formed in several stages as the remains of plants are subjected to heat and pressure over millions of years.

coal gasification Conversion of solid coal to synthetic natural gas (SNG).

coal liquefaction Conversion of solid coal to a liquid hydrocarbon fuel such as synthetic gasoline or methanol.

coastal wetland Land along a coastline, extending inland from an estuary that is covered with salt water all or part of the year. Examples are marshes, bays, lagoons, tidal flats, and mangrove swamps. Compare *inland wetland.*

coastal zone Warm, nutrient-rich, shallow part of the ocean that extends from the high-tide mark on land to the edge of a shelflike extension of continental land masses known as the continental shelf. Compare *open sea.*

coevolution Evolution in which two or more species interact and exert selective pressures on each other that can lead each species to undergo various adaptations. See *evolution, natural selection.*

cogeneration Production of two useful forms of energy, such as high-temperature heat or steam and electricity, from the same fuel source.

commensalism An interaction between organisms of different species in which one type of organism benefits and the other type is neither helped nor harmed to any great degree. Compare *mutualism.*

commercial extinction Depletion of the population of a wild species used as a resource to a level at which it is no longer profitable to harvest the species.

commercial inorganic fertilizer Commercially prepared mixtures of plant nutrients such as nitrates, phosphates, and potassium applied to the soil to restore fertility and increase crop yields. Compare *organic fertilizer.*

common-property resource Resource that people are normally free to use; each user can deplete or degrade the available supply. Most are potentially renewable and are owned by no one. Examples are clean air, fish in parts of the ocean not under the control of a coastal country, migratory birds, gases of the lower atmosphere, and the ozone content of the upper atmosphere (stratosphere). See *tragedy of the commons.*

community Populations of all species living and interacting in an area at a particular time.

community development See *ecological succession.*

competition Two or more individual organisms of a single species (*intraspecific competition*) or two or more individuals of different species (*interspecific competition*) attempting to use the same scarce resources in the same ecosystem.

compost Partially decomposed organic plant and animal matter that can be used as a soil conditioner or fertilizer.

compound Combination of atoms, or oppositely charged ions, of two or more different elements held together by attractive forces called chemical bonds. Compare *element.*

concentration Amount of a chemical in a particular volume or weight of air, water, soil, or other medium.

condensation nuclei Tiny particles on which droplets of water vapor can collect.

coniferous evergreen plants Cone-bearing plants (such as spruces, pines, and firs) that keep some of their narrow, pointed leaves (needles) all year. Compare *broadleaf deciduous plants, broadleaf evergreen plants.*

coniferous trees Cone-bearing trees, mostly evergreens, that have needle-shaped or scalelike leaves. They produce wood known commercially as softwood. Compare *deciduous plants.*

consensus science Scientific data, models, theories, and laws that are widely accepted by scientists considered experts in the area of study. These results of science are very reliable. Compare *frontier science.*

conservation Sensible and careful use of natural resources by humans. People with this view are called conservationists.

conservationists See *conservation.*

conservation biology Multidisciplinary science created to deal with the crisis of maintaining the genes, species, communities, and ecosystems that make up earth's biological diversity. Its goals are to investigate human impacts on biodiversity and to develop practical approaches to preserving biodiversity and ecological integrity.

conservation-tillage farming Crop cultivation in which the soil is disturbed little (minimum-tillage farming) or not at all (no-till farming) to reduce soil erosion, lower labor costs, and save energy. Compare *conventional-tillage farming.*

constancy Ability of a living system, such as a population, to maintain a certain size. Compare *inertia, resilience.* See *homeostasis.*

consumer Organism that cannot synthesize the organic nutrients it needs and gets its organic nutrients by feeding on the tissues of producers or of other consumers; generally divided into *primary consumers* (herbivores), *secondary consumers* (carnivores), *tertiary (higher-level) consumers, omnivores,* and *detritivores* (decomposers and detritus feeders). In economics, one who uses economic goods.

contour farming Plowing and planting across the changing slope of land, rather than in straight lines, to help retain water and reduce soil erosion.

controlled burning Deliberately set, carefully controlled surface fires to reduce flammable litter and decrease the chances of damaging crown fires.

conventional-tillage farming Crop cultivation method in which a planting surface is made by plowing land, breaking up the exposed soil, and then smoothing the surface. Compare *conservation-tillage farming.*

convergent plate boundary Area where earth's lithospheric plates are pushed together. See *subduction zone.* Compare *divergent plate boundary, transform fault.*

coral reef Formation produced by massive colonies containing billions of tiny coral animals, called polyps, that secrete a stony substance (calcium carbonate) around themselves for protection. When the corals die, their empty outer skeletons form layers that cause the reef to grow. They are found in the coastal zones of warm tropical and subtropical oceans.

core Inner zone of the earth. It consists of a solid inner core and a liquid outer core. Compare *crust, mantle.*

cost–benefit analysis Estimates and comparison of short-term and long-term costs (losses) and benefits (gains) from an economic decision. If the estimated benefits exceed the estimated costs, the decision to buy an economic good or provide a public good is considered worthwhile.

critical mass Amount of fissionable nuclei needed to sustain a nuclear fission chain reaction.

crop rotation Planting a field, or an area of a field, with different crops from year to year to reduce depletion of soil nutrients. A plant such as corn, tobacco, or cotton, which removes large amounts of nitrogen from the soil, is planted one year. The next year a legume such as soybeans, which adds nitrogen to the soil, is planted.

crown fire Extremely hot forest fire that burns ground vegetation and tree tops. Compare *controlled burning, ground fire, surface fire.*

crude birth rate Annual number of live births per 1,000 people in the population of a geographic area at the midpoint of a given year. Compare *crude death rate.*

crude death rate Annual number of deaths per 1,000 people in the population of a geographic area at the midpoint of a given year. Compare *crude birth rate.*

crude oil Gooey liquid consisting mostly of hydrocarbon compounds and small amounts of compounds containing oxygen, sulfur, and nitrogen. Extracted from underground accumulations, it is sent to oil refineries, where it is converted to heating oil, diesel fuel, gasoline, tar, and other materials.

crust Solid outer zone of the earth. It consists of oceanic crust and continental crust. Compare *core, mantle.*

cultural eutrophication Overnourishment of aquatic ecosystems with plant nutrients (mostly nitrates and phosphates) because of human activities such as agriculture, urbanization, and discharges from industrial plants and sewage treatment plants. See *eutrophication.*

cyanobacteria Single-celled, prokaryotic, microscopic organisms. Before being reclassified as monera, they were called blue-green algae.

DDT Dichlorodiphenyltrichloroethane, a chlorinated hydrocarbon that has been widely used as a pesticide but is now banned in some countries.

death rate See *crude death rate.*

debt-for-nature swap Agreement in which a certain amount of foreign debt is canceled in exchange for local currency investments that will improve natural resource management or protect certain areas in the debtor country from harmful development.

deciduous plants Trees, such as oaks and maples, and other plants that survive during dry seasons or cold seasons by shedding their leaves. Compare *coniferous trees, succulent plants.*

decomposer Organism that digests parts of dead organisms and cast-off fragments and wastes of living organisms by breaking down the complex organic molecules in those materials into simpler inorganic compounds and then absorbing the soluble nutrients. Most of these chemicals are returned to the soil and water for reuse by producers. Decomposers consist of various bacteria and fungi. Compare *consumer, detritivore, producer.*

deforestation Removal of trees from a forested area without adequate replanting.

degradable pollutant Potentially polluting chemical that is broken down completely or reduced to acceptable levels by natural physical, chemical, and biological processes. Compare *biodegradable pollutant, nondegradable pollutant, slowly degradable pollutant.*

degree of urbanization Percentage of the population in the world, or a country, living in areas with a population of more than 2,500 people (higher in some countries). Compare *urban growth.*

democracy Government by the people through their elected officials and appointed representatives. In a *constitutional democracy*, a constitution provides the basis of government authority and puts restraints on government power through free elections and freely expressed public opinion.

demographic transition Hypothesis that countries, as they become industrialized, have declines in death rates followed by declines in birth rates.

depletion time How long it takes to use a certain fraction—usually 80%—of the known or estimated supply of a nonrenewable resource at an assumed rate of use. Finding and extracting the remaining 20% usually costs more than it is worth.

desalination Purification of salt water or brackish (slightly salty) water by removing dissolved salts.

desert Biome in which evaporation exceeds precipitation and the average amount of precipitation is less than 25 centimeters (10 inches) a year. Such areas have little vegetation or have widely spaced, mostly low vegetation. Compare *forest, grassland.*

desertification Conversion of rangeland, rain-fed cropland, or irrigated cropland to desertlike land, with a drop in agricultural productivity of 10% or more. It usually is caused by a combination of overgrazing, soil erosion, prolonged drought, and climate change.

detritivore Consumer organism that feeds on detritus, parts of dead organisms, and cast-off fragments and wastes of living organisms. The two principal types are *detritus feeders* and *decomposers.*

detritus Parts of dead organisms and cast-off fragments and wastes of living organisms.

detritus feeder Organism that extracts nutrients from fragments of dead organisms and their cast-off parts and organic wastes. Examples are earthworms, termites, and crabs. Compare *decomposer.*

deuterium (D; hydrogen-2) Isotope of the element hydrogen, with a nucleus containing one proton and one neutron, and a mass number of 2.

developed country Country that is highly industrialized and has a high per capita GNP. Compare *developing country.*

developing country Country that has low to moderate industrialization and low to moderate per capita GNP. Most are located in Africa, Asia, and Latin America. Compare *developed country.*

dew point Temperature at which condensation occurs for a given amount of water vapor.

dieback Sharp reduction in the population of a species when its numbers exceed the carrying capacity of its habitat. See *carrying capacity.*

differential reproduction Phenomenon in which individuals with adaptive genetic traits produce more living offspring than do individuals without such traits. See *natural selection.*

dioxins Family of 75 different chlorinated hydrocarbon compounds formed as unwanted by-products in chemical reactions involving chlorine and hydrocarbons, usually at high temperatures.

discount rate The economic value a resource will have in the future compared with its present value.

dissolved oxygen (DO) content Amount of oxygen gas (O_2) dissolved in a given volume of water at a particular temperature and pressure, often expressed as a concentration in parts of oxygen per million parts of water.

divergent plate boundary Area where earth's lithospheric plates move apart in opposite directions. Compare *convergent plate boundary, transform fault.*

DNA (deoxyribonucleic acid) Large molecules in the cells of organisms that carry genetic information in living organisms.

dose The amount of a potentially harmful substance an individual ingests, inhales, or absorbs through the skin. Compare *response.* See *dose–response curve, median lethal dose.*

dose–response curve Plot of data showing effects of various doses of a toxic agent on a group of test organisms. See *dose, median lethal dose, response.*

doubling time The time it takes (usually in years) for the quantity of something growing exponentially to double. It can be calculated by dividing the annual percentage growth rate into 70.

drainage basin See *watershed.*

dredge spoils Materials scraped from the bottoms of harbors and streams to maintain shipping channels. They are often contaminated with high levels of toxic substances that have settled out of the water. See *dredging.*

dredging Type of surface mining in which chain buckets and draglines scrape up sand, gravel, and other surface deposits covered with water. It is also used to remove sediment from streams and harbors to maintain shipping channels. See *dredge spoils.*

drift-net fishing Catching fish in huge nets that drift in the water.

drought Condition in which an area does not get enough water because of lower-than-normal precipitation or higher-than-normal temperatures that increase evaporation.

dust dome Dome of heated air that surrounds an urban area and traps pollutants, especially suspended particulate matter. See also *urban heat island.*

dust plume Elongation of a dust dome by winds that can spread a city's pollutants hundreds of kilometers downwind.

earth capital See *natural capital.*

earthquake Shaking of the ground resulting from the fracturing and displacement of rock, which produces a fault, or from subsequent movement along the fault.

earth resources See *natural capital.*

earth-sustaining economy Economic system in which the number of people and the quantity of goods are maintained at a constant level. This level is ecologically sustainable over time and meets at least the basic needs of all members of the population.

earth-wisdom revolution See *environmental revolution.*

earth-wisdom society See *low-throughput society.*

earth-wisdom worldview Beliefs that **(1)** nature exists for all of the earth's species, not just for us; and we are not in charge of the rest of nature; **(2)** there is not always more, and it's not all for us; and **(3)** some forms of economic growth are beneficial and some are harmful and our goals should be to design economic and political systems that encourage earth-sustaining forms of growth and discourage or prohibit earth-degrading forms; and **(4)** our success depends on learning to cooperate with one another and with the rest of nature instead of trying to dominate and manage earth's life-support systems primarily for our own use. Compare *planetary management worldview.*

ecological diversity The variety of forests, deserts, grasslands, oceans, streams, lakes, and other biological communities interacting with one another and with their nonliving environment. See *biological diversity.* Compare *genetic diversity, species diversity.*

ecological efficiency Percentage of energy transferred from one trophic level to another in a food chain or web.

ecological footprint A measure of the ecological impact of the human **(1)** consumption of food, wood products, and other resources, **(2)** use of buildings, roads, garbage dumps, and other things that consume land space, and **(3)** the forests needed to absorb the CO_2 produced by burning fossil fuels.

ecological health The degree to which an area's biodiversity and ecological integrity remain intact. See *biological diversity, ecological integrity.*

ecological integrity The conditions and natural processes (such as the cycling of matter and flow of energy through ecosystems) that **(1)** generate and maintain biodiversity and **(2)** allow evolutionary change as a key mechanism for adapting to changes in environmental conditions.

ecological land-use planning Method for deciding how land should be used; development of an integrated model that considers geological, ecological, health, and social variables.

ecologically sustainable development Development in which the total human population size and resource use in the world (or in a region) are limited to a level that does not exceed the carrying capacity of the existing natural capital and is therefore sustainable. Compare *economic development, economic growth, sustainable development.*

ecological niche Total way of life or role of a species in an ecosystem. It includes all physical, chemical, and biological conditions a species needs to live and reproduce in an ecosystem. See *fundamental niche, realized niche.*

ecological population density Number of individuals of a population per unit area of habitat. Compare *population density.*

ecological resource Anything required by an organism for normal maintenance, growth, and reproduction. Examples include habitat, food, water, and shelter. Compare *economic resource.*

ecological succession Process in which communities of plant and animal species in a particular area are replaced over time by a series of different and often more complex communities. See *primary succession, secondary succession.*

ecology Study of the interactions of living organisms with one another and with their nonliving environment of matter and energy; study of the structure and functions of nature.

economic decision Deciding **(1)** what goods and services to produce, **(2)** how to produce them, how much to produce, and **(3)** how to distribute them to people.

economic depletion Exhaustion of 80% of the estimated supply of a nonrenewable resource. Finding, extracting, and processing the remaining 20% usually costs more than it is worth; may also apply to the depletion of a potentially renewable resource, such as a species of fish or tree.

economic development Improvements in the efficiency of resource use so that the same or greater output of goods and services is produced with smaller throughputs of natural, manufactured, and human capital. Compare *ecologically sustainable development, economic growth, sustainable development.*

economic growth Increase in the real value of all final goods and services produced by an economy; an increase in real GNP. Compare *ecologically sustainable development, economic development, sustainable development.*

economic resource Anything obtained from the environment (the earth's life-support systems) to meet human needs and wants. Examples include food, water, shelter, manufactured goods, transportation, communication, and recreation. Compare *ecological resource.*

economic resources Natural resources, capital goods, and labor used in an economy to produce material goods and services. See *natural capital, human capital, manufactured capital.*

economic system Method that a group of people uses to **(1)** choose what goods and services to produce, **(2)** how to produce them, **(3)** how much to produce, and **(4)** how to distribute them to people. See *mixed economic system, pure command economic system, pure market economic system.*

economy System of production, distribution, and consumption of economic goods.

ecosphere Earth's collection of living organisms interacting with one another and their nonliving environment (energy and matter) throughout the world; all earth's ecosystems. Also called the *biosphere.*

ecosystem Community of different species interacting with one another and with the chemical and physical factors making up its nonliving environment.

ecosystem services Natural services or natural capital that support life on the earth and are essential to the quality of human life and to the functioning of the world's economies. See *natural capital.*

electromagnetic radiation Forms of kinetic energy traveling as electromagnetic waves. Examples are radio waves, TV waves, microwaves, infrared radiation, visible light, ultraviolet radiation, X rays, and gamma rays. Compare *ionizing radiation, nonionizing radiation.*

electron (e) Tiny particle moving around outside the nucleus of an atom. Each electron has one unit of negative charge (–) and almost no mass. Compare *neutron, proton.*

element Chemical, such as hydrogen (H), iron (Fe), sodium (Na), carbon (C), nitrogen (N), or oxygen (O), whose distinctly different atoms serve as the basic building blocks of all matter. There are 92 naturally occurring elements. Another 26 have been made in laboratories. Two or more elements combine to form compounds that make up most of the world's matter. Compare *compound.*

endangered species Wild species with so few individual survivors that the species could soon become extinct in all or most of its natural range. Compare *threatened species.*

energy Capacity to do work by performing mechanical, physical, chemical, or electrical tasks or to cause a heat transfer between two objects at different temperatures.

energy efficiency Percentage of the total energy input that does useful work and is not converted into low-quality, usually useless heat in an energy conversion system or process. See *energy quality, net energy.*

energy quality Ability of a form of energy to do useful work. High-temperature heat and the chemical energy in fossil fuels and nuclear fuels are concentrated high-quality energy. Low-quality energy such as low-temperature heat is dispersed or diluted and cannot do much useful work. See *high-quality energy, low-quality energy.*

enhanced oil recovery Removal of some of the heavy oil left in an oil well after primary and secondary recovery. Compare *primary oil recovery, secondary oil recovery.*

environment All external conditions and factors, living and nonliving (chemicals and energy), that affect an organism or other specified system during its lifetime.

environmental degradation Depletion or destruction of a potentially renewable resource such as soil, grassland, forest, or wildlife by using it faster than it is naturally replenished. If such use continues, the resource can become nonrenewable (on a human time scale) or nonexistent (extinct). See also *sustainable yield.*

environmental ethics Our beliefs about what is right or wrong environmental behavior.

environmental resistance All the limiting factors jointly acting to limit the growth of a population. See *biotic potential, limiting factor.*

environmental revolution Cultural change involving halting population growth and altering lifestyles, political and economic systems, and the way we treat the environment so that we can help sustain the earth for ourselves and other species. This involves working with the rest of nature by learning more about how nature sustains itself. See *earth-wisdom worldview*. Compare *agricultural revolution, hunter-gatherers, industrial revolution, information revolution*.

environmental science Study of how we and other species interact with one another and with the nonliving environment (matter and energy). It is a physical and social science that integrates knowledge from a wide range of disciplines, including physics, chemistry, biology (especially ecology), geology, geography, resource technology and engineering, resource conservation and management, demography (the study of population dynamics), economics, politics, sociology, psychology, and ethics. In other words, it is a study of how the parts of nature and human societies operate and interact—a study of connections and interactions in nature.

environmental worldview How individuals think the world works, what they think their role in the world should be, and what they believe is right and wrong environmental behavior (environmental ethics). See *earth-wisdom worldview, planetary management worldview, spaceship-earth worldview*.

epidemiology Study of the patterns of disease or other harmful effects from toxic exposure within defined groups of people to find out why some people get sick and some do not.

epiphytes Plants that use their roots to attach themselves to branches high in trees, especially in tropical forests.

erosion Process or group of processes by which loose or consolidated earth materials are dissolved, loosened, or worn away and removed from one place and deposited in another. See *weathering*.

estuary Partially enclosed coastal area at the mouth of a river where its fresh water, carrying fertile silt and runoff from the land, mixes with salty seawater.

eukaryotic cell Cell containing a *nucleus*, a region of genetic material surrounded by a membrane. Membranes also enclose several of the other internal parts found in a eukaryotic cell. Compare *prokaryotic cell*.

eutrophication Physical, chemical, and biological changes that take place after a lake, estuary, or slow-flowing stream receives inputs of plant nutrients—mostly nitrates and phosphates—from natural erosion and runoff from the surrounding land basin. See *cultural eutrophication*.

eutrophic lake Lake with a large or excessive supply of plant nutrients, mostly nitrates and phosphates. Compare *mesotrophic lake, oligotrophic lake*.

evaporation Conversion of a liquid into a gas.

even-aged management Method of forest management in which trees, sometimes of a single species in a given stand, are maintained at about the same age and size and are harvested all at once. Compare *uneven-aged management*.

evergreen plants Plants that keep some of their leaves or needles throughout the year. Examples are ferns and cone-bearing trees (conifers) such as firs, spruces, pines, redwoods, and sequoias. Compare *deciduous plants, succulent plants*.

evolution See *biological evolution*.

exhaustible resource See *nonrenewable resource*.

exotic species See *nonnative species*.

exploitation competition Situation in which two competing species have equal access to a specific resource but differ in how quickly or efficiently they exploit it. See *interference competition, interspecific competition*.

exponential growth Growth in which some quantity, such as population size or economic output, increases by a fixed percentage of the whole in a given time period; when the increase in quantity over time is plotted, this type of growth yields a curve shaped like the letter *J*. Compare *linear growth*.

external benefit Beneficial social effect of producing and using an economic good that is not included in the market price of the good. Compare *external cost, full cost*.

external cost Harmful social effect of producing and using an economic good that is not included in the market price of the good. Compare *external benefit, full cost*.

externalities Social benefits ("goods") and social costs ("bads") not included in the market price of an economic good. See *external benefit, external cost*. Compare *full cost, internal cost*.

extinction Complete disappearance of a species from the earth. This happens when a species cannot adapt and successfully reproduce under new environmental conditions or when it evolves into one or more new species. Compare *speciation*. See also *endangered species, threatened species*.

family planning Providing information, clinical services, and contraceptives to help people choose the number and spacing of children they want to have.

famine Widespread malnutrition and starvation in a particular area because of a shortage of food, usually caused by drought, war, flood, earthquake, or other catastrophic events that disrupt food production and distribution.

feedback loop Circuit of sensing, evaluating, and reacting to changes in environmental conditions as a result of information fed back into a system; it occurs when one change leads to some other change, which eventually reinforces or slows the original change. See *negative feedback loop, positive feedback loop*.

feedlot Confined outdoor or indoor space used to raise hundreds to thousands of domesticated livestock. Compare *rangeland*.

fermentation See *anaerobic respiration*.

fertilizer Substance that adds inorganic or organic plant nutrients to soil and improves its ability to grow crops, trees, or other vegetation. See *commercial inorganic fertilizer, organic fertilizer*.

first law of energy See *first law of thermodynamics*.

first law of human ecology We can never do merely one thing. Any intrusion into nature has numerous effects, many of which are unpredictable.

first law of thermodynamics In any physical or chemical change, no detectable amount of energy is created or destroyed, but in these processes energy can be changed from one form to another; you can't get more energy out of something than you put in; in terms of energy quantity, you can't get something for nothing (there is no free lunch). This law does not apply to nuclear changes, in which energy can be produced from small amounts of matter. See also *second law of thermodynamics*.

fishery Concentrations of particular aquatic species suitable for commercial harvesting in a given ocean area or inland body of water.

fish farming Form of aquaculture in which fish are cultivated in a controlled pond or other environment and harvested when they reach the desired size. See also *fish ranching*.

fish ranching Form of aquaculture in which members of a fish species such as salmon are held in captivity for the first few years of their lives, released, and then harvested as adults when they return from the ocean to their freshwater birthplace to spawn. See also *fish farming*.

fissionable isotope Isotope that can split apart when hit by a neutron at the right speed and thus undergo nuclear fission. Examples are uranium-235 and plutonium-239.

floodplain Flat valley floor next to a stream channel. For legal purposes the term is often applied to any low area that has the potential for flooding, including certain coastal areas.

flows See *throughputs*.

flyway Generally fixed route along which waterfowl migrate from one area to another at certain seasons of the year.

food chain Series of organisms in which each eats or decomposes the preceding one. Compare *food web*.

food web Complex network of many interconnected food chains and feeding relationships. Compare *food chain*.

forest Biome with enough average annual precipitation (at least 76 centimeters, or 30 inches) to support growth of various species of trees and smaller forms of vegetation. Compare *desert, grassland*.

fossil fuel Products of partial or complete decomposition of plants and animals that occur as crude oil, coal, natural gas, or heavy oils as a result of exposure to heat and pressure in earth's crust over millions of years.

See *coal, crude oil, natural gas.*

fossils Skeletons, bones, shells, body parts, leaves, seeds, or impressions of such items that provide recognizable evidence of organisms that lived long ago.

Freons See *chlorofluorocarbons.*

freshwater life zones Aquatic systems where water with a dissolved salt concentration of less than 1% by volume accumulates on or flows through the surfaces of terrestrial biomes. Examples are **(1)** *standing* (lentic) bodies of fresh water such as lakes, ponds, and inland wetlands and **(2)** *flowing* (lotic) systems such as streams and rivers. Compare *biomes.*

frontier science Preliminary scientific data, hypotheses, and models that have not been widely tested and accepted. Compare *consensus science.*

full cost Cost of a good when its internal costs and its estimated short- and long-term external costs are included in its market price. Compare *external cost, internal cost.*

fundamental niche The full potential range of the physical, chemical, and biological factors a species can use if there is no competition from other species. See *ecological niche.* Compare *realized niche.*

fungi Eukaryotic, mostly multicelled organisms such as mushrooms, molds, and yeasts. As decomposers, they get the nutrients they need by secreting enzymes that speed up the breakdown of organic matter in the tissue of other living or dead organisms. Then they absorb the resulting nutrients.

fungicide Chemical that kills fungi.

Gaia hypothesis Hypothesis that earth is alive and can be considered a system that operates and changes by feedback of information between its living and nonliving components.

game species Type of wild animal that people hunt or fish for, for sport and recreation and sometimes for food.

gamma rays A form of ionizing electromagnetic radiation with a high energy content emitted by some radioisotopes. They readily penetrate body tissues. See also *alpha particles, beta particles.*

gangue Waste or undesired material in an ore. See *ore.*

GDP See *gross domestic product.*

gene mutation See *mutation.*

gene pool The sum total of all genes found in the individuals of the population of a particular species.

generalist species Species with a broad ecological niche. They can live in many different places, eat a variety of foods, and tolerate a wide range of environmental conditions. Examples are flies, cockroaches, mice, rats, and human beings. Compare *specialist species.*

genes Coded units of information about specific traits that are passed on from parents to offspring during reproduction. They consist of segments of DNA molecules found in chromosomes.

genetic adaptation Changes in the genetic makeup of organisms of a species that allow the species to reproduce and gain a competitive advantage under changed environmental conditions. See *differential reproduction, evolution, mutation, natural selection.*

genetic diversity Variability in the genetic makeup among individuals within a single species. See *biodiversity.* Compare *ecological diversity, species diversity.*

geographic isolation Separation of populations of a species for fairly long times into areas with different environmental conditions.

geothermal energy Heat transferred from the earth's underground concentrations of **(1)** dry steam (steam with no water droplets), **(2)** wet steam (a mixture of steam and water droplets), or **(3)** hot water trapped in fractured or porous rock.

global warming Warming of the earth's atmosphere as a result of increases in the concentrations of one or more greenhouse gases. See *greenhouse effect, greenhouse gases.*

GNP See *gross national product.*

grassland Biome found in regions where moderate annual average precipitation (25 to 76 centimeters, or 10 to 30 inches) is enough to support the growth of grass and small plants, but not enough to support large stands of trees. Compare *desert, forest.*

greenhouse effect A natural effect that traps heat in the atmosphere (troposphere) near the earth's surface. Some of the heat flowing back toward space from the earth's surface is absorbed by water vapor, carbon dioxide, ozone, and several other gases in the lower atmosphere (troposphere) and then radiated back toward the earth's surface. If the atmospheric concentrations of these greenhouse gases rise and are not removed by other natural processes, the average temperature of the lower atmosphere will increase gradually.

greenhouse gases Gases in the earth's lower atmosphere (troposphere) that cause the greenhouse effect. Examples are carbon dioxide, chlorofluorocarbons, ozone, methane, water vapor, and nitrous oxide.

green manure Freshly cut or still-growing green vegetation that is plowed into the soil to increase the organic matter and humus available to support crop growth. Compare *animal manure.*

green revolution Popular term for introduction of scientifically bred or selected varieties of grain (rice, wheat, maize) that, with high enough inputs of fertilizer and water, can greatly increase crop yields.

gross domestic product (GDP) Total market value in current dollars of all goods and services produced *within a country* for final use, usually during a year. Compare *gross national product.*

gross national product (GNP) Total market value in current dollars of all goods and services produced by an economy for final use usually during a year. Compare *gross domestic product.*

gross primary productivity The rate at which an ecosystem's producers capture and store a given amount of chemical energy as biomass in a given length of time. Compare *net primary productivity.*

ground fire Fire that burns decayed leaves or peat deep below the ground surface. Compare *crown fire, surface fire.*

groundwater Water that sinks into the soil and is stored in slowly flowing and slowly renewed underground reservoirs called aquifers; underground water in the zone of saturation, below the water table. Compare *runoff, surface water.*

gully reclamation Restoring land suffering from gully erosion by seeding gullies with quick-growing plants, building small dams to collect silt and gradually fill in the channels, and building channels to divert water away from the gully.

habitat Place or type of place where an organism or population of organisms lives. Compare *ecological niche.*

habitat fragmentation Breakup of a habitat into smaller pieces, usually as a result of human activities.

half-life Time needed for one-half of the nuclei in a radioisotope to emit its radiation. Each radioisotope has a characteristic half-life, which may range from a few millionths of a second to several billion years.

hazard Something that can cause injury, disease, economic loss, or environmental damage. See also *risk.*

hazardous chemical Chemical that can cause harm **(1)** because it is flammable or explosive, **(2)** can irritate or damage the skin or lungs (such as strong acidic or alkaline substances) or **(3)** cause allergic reactions of the immune system (allergens). See *toxic chemical.*

hazardous waste Any solid, liquid, or containerized gas that **(1)** can catch fire easily, **(2)** is corrosive to skin tissue or metals, **(3)** is unstable and can explode or release toxic fumes, or **(4)** has harmful concentrations of one or more toxic materials that can leach out. See also *toxic waste.*

heat Total kinetic energy of all the randomly moving atoms, ions, or molecules within a given substance, excluding the overall motion of the whole object. This form of kinetic energy flows from one body to another when there is a temperature difference between the two bodies. Heat always flows spontaneously from a hot sample of matter to a colder sample of matter. This is one way to state the second law of thermodynamics. Compare *temperature.*

herbicide Chemical that kills a plant or inhibits its growth.

herbivore Plant-eating organism. Examples are deer, sheep, grasshoppers, and zooplankton. Compare *carnivore, omnivore.*

heterotroph See *consumer.*

high-input agriculture See *industrialized agriculture.*

high-quality energy Energy that is organized or concentrated and has great ability to perform useful work. Examples are high-temperature heat and the energy in electricity, coal, oil, gasoline, sunlight, and nuclei of uranium-235. Compare *low-quality energy.*

high-quality matter Matter that is organized and concentrated and contains a high concentration of a useful resource. Compare *low-quality matter.*

high-throughput society The situation in most advanced industrialized countries, in which ever-increasing economic growth is sustained by maximizing the rate at which matter and energy resources are used, with little emphasis on pollution prevention, recycling, reuse, reduction or unnecessary waste, and other forms of resource conservation. Compare *low-throughput society, matter-recycling society.*

high-waste society See *high-throughput society.*

homeostasis Maintenance of favorable internal conditions in a system despite fluctuations in external conditions. See *constancy, inertia, resilience.*

host Plant or animal on which a parasite feeds.

human capital Physical and mental talents of people used to produce, distribute, and sell an economic good. Compare *manufactured capital, natural capital, solar capital.*

humus Slightly soluble residue of undigested or partially decomposed organic material in topsoil. This material helps retain water and water-soluble nutrients, which can be taken up by plant roots.

hunter–gatherers People who get their food by gathering edible wild plants and other materials and by hunting wild animals and fish. Compare *agricultural revolution, environmental revolution, industrial revolution, information revolution.*

hydrocarbon Organic compound of hydrogen and carbon atoms. The simplest hydrocarbon is methane (CH_4), the major component of natural gas.

hydroelectric power plant Structure in which the energy of falling or flowing water spins a turbine generator to produce electricity.

hydrologic cycle Biogeochemical cycle that collects, purifies, and distributes the earth's fixed supply of water from the environment to living organisms, and then back to the environment.

hydropower Electrical energy produced by falling or flowing water. See *hydroelectric power plant.*

hydrosphere The earth's **(1)** liquid water (oceans, lakes and other bodies of surface water, and underground water), **(2)** frozen water (polar ice caps, floating ice caps, and ice in soil known as permafrost), and **(3)** small amounts of water vapor in the atmosphere. See also *hydrologic cycle.*

identified resources Deposits of a particular mineral-bearing material of which the location, quantity, and quality are known or have been estimated from direct geological evidence and measurements. Compare *undiscovered resources.*

igneous rock Rock formed when molten rock material (magma) wells up from the earth's interior, cools, and solidifies into rock masses. Compare *metamorphic rock, sedimentary rock.* See *rock cycle.*

immature community Community at an early stage of ecological succession. It usually has a low number of species and ecological niches and cannot capture and use energy and cycle critical nutrients as efficiently as more complex, mature ecosystems. Compare *mature community.*

immigrant species Species that migrate into an ecosystem or are deliberately or accidentally introduced into an ecosystem by humans. Some of these species are beneficial, whereas others can take over and eliminate many native species. Compare *indicator species, keystone species, native species.*

immigration Migration of people into a country or area to take up permanent residence.

indicator species Species that serve as early warnings that a community or ecosystem is being degraded. Compare *immigrant species, keystone species, native species.*

industrialized agriculture Using large inputs of energy from fossil fuels (especially oil and natural gas), water, fertilizer, and pesticides to produce large quantities of crops and livestock for domestic and foreign sale. Compare *subsistence farming.*

industrial revolution Use of new sources of energy from fossil fuels and later from nuclear fuels, and use of new technologies, to grow food and manufacture products. Compare *agricultural revolution, environmental revolution, hunter-gatherers, information revolution.*

industrial smog Type of air pollution consisting mostly of a mixture of sulfur dioxide, suspended droplets of sulfuric acid formed from some of the sulfur dioxide, and a variety of suspended solid particles. Compare *photochemical smog.*

inertia Ability of a living system to resist being disturbed or altered. Compare *constancy, resilience.*

infant mortality rate Number of babies out of every 1,000 born each year that die before their first birthday.

infiltration Downward movement of water through soil.

information revolution Use of new technologies such as the telephone, radio, television, and computers to enable people to deal with more and more information more rapidly. Compare *agricultural revolution, environmental revolution, hunter-gatherers, industrial revolution.*

inland wetland Land away from the coast, such as a swamp, marsh, or bog, that is covered all or part of the time with fresh water. Compare *coastal wetland.*

inorganic fertilizer See *commercial inorganic fertilizer.*

input Matter, energy, or information entering a system. Compare *output, throughput.*

input pollution control See *pollution prevention.*

insecticide Chemical that kills insects.

integrated pest management (IPM) Combined use of biological, chemical, and cultivation methods in proper sequence and timing to keep the size of a pest population below the size that causes economically unacceptable loss of a crop or livestock animal.

intercropping Growing two or more different crops at the same time on a plot. For example, a carbohydrate-rich grain that depletes soil nitrogen and a protein-rich legume that adds nitrogen to the soil may be intercropped. Compare *monoculture, polyculture, polyvarietal cultivation.*

interference competition Situation in which one species limits access of another species to a resource, regardless of whether the resource is abundant or scarce. See *exploitation competition, interspecific competition.*

intermediate goods See *manufactured capital.*

internal cost Direct cost paid by the producer and the buyer of an economic good. Compare *external cost.*

interplanting Simultaneously growing a variety of crops on the same plot. See *agroforestry, intercropping, polyculture, polyvarietal cultivation.*

interspecific competition Members of two or more species trying to use the same limited resources in an ecosystem. See *competition, competitive exclusion, exploitation competition, interference competition.*

intertidal zone The area of shoreline between low and high tides.

intrinsic rate of increase (r) Rate at which a population could grow if it had unlimited resources. Compare *environmental resistance.*

inversion See *thermal inversion.*

invertebrates Animals that have no backbones. Compare *vertebrates.*

ion Atom or group of atoms with one or more positive (+) or negative (–) electrical charges. Compare *atom, molecule.*

ionizing radiation Fast-moving alpha or beta particles or high-energy radiation (gamma rays) emitted by radioisotopes. They have enough energy to dislodge one or more electrons from atoms they hit, forming charged ions in tissue that can react with and damage living tissue. Compare *nonionizing radiation.*

isotopes Two or more forms of a chemical element that have the same number of protons but different mass numbers due to different numbers of neutrons in their nuclei.

***J*-shaped curve** Curve with a shape similar to that of the letter *J*; can represent prolonged exponential growth.

kerogen Solid, waxy mixture of hydrocarbons found in oil shale rock. When the rock is heated to high temperatures, the kerogen is vaporized. The vapor is condensed, purified, and then sent to a refinery to produce gasoline, heating oil, and other products. See also *oil shale, shale oil.*

keystone species Species that play roles affecting many other organisms in an ecosystem. Compare *immigrant species, indicator species, native species.*

kilocalorie (kcal) Unit of energy equal to 1,000 calories. See *calorie.*

kilowatt (kw) Unit of electrical power equal to 1,000 watts. See *watt.*

kinetic energy Energy that matter has because of its mass and speed or velocity. Compare *potential energy.*

K-strategists Species that produce a few, often fairly large offspring but invest a great deal of time and energy to ensure that most of those offspring reach reproductive age. Compare *r-strategists.*

kwashiorkor Type of malnutrition that occurs in infants and very young children when they are weaned from mother's milk to a starchy diet low in protein. See *marasmus.*

labor See *human capital.*

lake Large natural body of standing fresh water formed when water from precipitation, land runoff, or groundwater flow fills a depression in the earth created by **(1)** glaciation, **(2)** earth movement, **(3)** volcanic activity, or **(4)** a giant meteorite. See *eutrophic lake, mesotrophic lake, oligotrophic lake.*

landfill See *sanitary landfill.*

land-use planning Process for deciding the best present and future use of each parcel of land in an area. See *ecological land-use planning.*

late successional plant species Mostly trees that can tolerate shade and form a fairly stable complex forest community. Compare *midsuccessional plant species.*

latitude Distance from the equator. Compare *altitude.*

law of conservation of energy See *first law of thermodynamics.*

law of conservation of matter In any physical or chemical change, matter is neither created nor destroyed, but merely changed from one form to another; in physical and chemical changes, existing atoms are rearranged into different spatial patterns (physical changes) or different combinations (chemical changes).

law of conservation of matter and energy In any nuclear change, the total amount of matter and energy involved remains the same.

law of tolerance The existence, abundance, and distribution of a species in an ecosystem are determined by whether the levels of one or more physical or chemical factors fall within the range tolerated by the species. See *threshold effect.*

LD_{50} See *median lethal dose.*

LDC See *developing country.*

leaching Process in which various chemicals in upper layers of soil are dissolved and carried to lower layers and, in some cases, to groundwater.

less developed country (LDC) See *developing country.*

life cycle cost Initial cost plus lifetime operating costs of an economic good. Compare *full cost.*

life expectancy Average number of years a newborn infant can be expected to live.

limiting factor Single factor that limits the growth, abundance, or distribution of the population of a species in an ecosystem. See *limiting factor principle.*

limiting factor principle Too much or too little of any abiotic factor can limit or prevent growth of a population of a species in an ecosystem, even if all other factors are at or near the optimum range of tolerance for the species.

linear growth Growth in which a quantity increases by some fixed amount during each unit of time. Compare *exponential growth.*

liquefied natural gas (LNG) Natural gas converted to liquid form by cooling to a very low temperature.

liquefied petroleum gas (LPG) Mixture of liquefied propane (C_3H_8) and butane (C_4H_{10}) gas removed from natural gas and used as a fuel.

lithosphere Outer shell of the earth, composed of the crust and the rigid, outermost part of the mantle outside of the asthenosphere; material found in earth's plates. See *crust, mantle.*

loams Soils containing a mixture of clay, sand, silt, and humus. Good for growing most crops.

logistic growth Exponential population growth when the population is small and results in a steady decrease in population growth with time as the population approaches the carrying capacity.

low-input agriculture See *sustainable agriculture.*

low-quality energy Energy that is disorganized or dispersed and has little ability to do useful work. An example is low-temperature heat. Compare *high-quality energy.*

low-quality matter Matter that is disorganized, dilute, or dispersed, or contains a low concentration of a useful resource. Compare *high-quality matter.*

low-throughput society Society based on working with nature by **(1)** recycling and reusing discarded matter; **(2)** preventing pollution; **(3)** conserving matter and energy resources through reducing unnecessary waste and use; **(4)** not degrading renewable resources; **(5)** making things that are easy to recycle, reuse and repair; **(6)** not allowing population size to exceed the carrying capacity of the environment; and **(7)** preserving biodiversity and ecological integrity. Compare *high-throughput society, matter-recycling society.*

low-waste society See *low-throughput society.*

LPG See *liquefied petroleum gas.*

macroevolution Long-term, large-scale evolutionary changes among groups of species. Compare *microevolution.*

magma Molten rock below the earth's surface.

malnutrition Faulty nutrition, caused by a diet that does not supply an individual with enough protein, essential fats, vitamins, minerals, and other nutrients needed for good health. Compare *overnutrition, undernutrition.*

mangrove swamps Swamps found on the coastlines in warm tropical climates. They are dominated by mangrove trees, any of about 55 species of trees and shrubs that can live partly submerged in the salty environment of coastal swamps.

mantle Zone of the earth's interior between its core and its crust. Compare *core, crust.* See *lithosphere.*

manufactured capital Manufactured items made from natural capital and used to produce and distribute economic goods and services bought by consumers. These include tools, machinery, equipment, factory buildings, and transportation and distribution facilities. Compare *human capital, natural capital, solar capital.*

manure See *animal manure, green manure.*

marasmus Nutritional-deficiency disease caused by a diet that does not have enough calories and protein to maintain good health. See *kwashiorkor, malnutrition.*

market equilibrium State in which sellers and buyers of an economic good agree on the quantity to be produced and the price to be paid.

mass The amount of material in an object.

mass extinction A catastrophic, widespread, often global event in which major groups of species are wiped out over a short time compared to normal (background) extinctions. Compare *background extinction.*

mass number Sum of the number of neutrons (n) and the number of protons (p) in the nucleus of an atom. It gives the approximate mass of that atom. Compare *atomic number.*

mass transit Buses, trains, trolleys, and other forms of transportation that carry large numbers of people.

matter Anything that has mass (the amount of material in an object) and takes up space. On the earth, where gravity is present, we weigh an object to determine its mass.

matter quality Measure of how useful a matter resource is, based on its availability and concentration. See *high-quality matter, low-quality matter.*

matter-recycling society Society that emphasizes recycling the maximum amount of all resources that can be recycled. The goal is to allow economic growth to continue without depleting matter resources and without producing excessive pollution and environmental degradation. Compare *high-throughput society, low-throughput society.*

mature community Fairly stable, self-sustaining community in an advanced stage of ecological succession; usually has a diverse array of species and ecological niches; captures and uses energy and cycles critical chemicals more efficiently than simpler, immature communities. Compare *immature community.*

maximum sustainable yield See *sustainable yield.*

MDC See *developed country.*

median lethal dose (LD_{50}) Amount of a toxic material per unit of body weight of test animals that kills half the test population in a certain time.

megacities Cities with 10 million or more people.

meltdown The melting of the core of a nuclear reactor.

mesosphere Third layer of the atmosphere; found above the stratosphere. Compare *stratosphere, troposphere.*

mesotrophic lake Lake with a moderate supply of plant nutrients. Compare *eutrophic lake, oligotrophic lake.*

metabolism Ability of a living cell or organism to capture and transform matter and energy from its environment to supply its needs for survival, growth, and reproduction.

metamorphic rock Rock produced when a preexisting rock is subjected to high temperatures (which may cause it to melt partially), high pressures, chemically active fluids, or a combination of these agents. Compare *igneous rock, sedimentary rock.* See *rock cycle.*

metastasis Spread of malignant (cancerous) cells from a cancer to other parts of the body.

microevolution The small genetic changes a population undergoes. Compare *macroevolution.*

microorganisms Organisms such as bacteria that are so small that they can be seen only by using a microscope.

midsuccessional plant species Grasses and low shrubs that are less hardy than early successional plant species. Compare *late successional plant species.*

mineral Any naturally occurring inorganic substance found in the earth's crust as a crystalline solid. See *mineral resource.*

mineral resource Concentration of naturally occurring solid, liquid, or gaseous material in or on the earth's crust, in a form and amount such that extracting and converting it into useful materials or items is currently or potentially profitable. Mineral resources are classified as *metallic* (such as iron and tin ores) or *nonmetallic* (such as fossil fuels, sand, and salt).

minimum-tillage farming See *conservation-tillage farming.*

mixed economic system Economic system that falls somewhere between pure market and pure command economic systems. Virtually all of the world's economic systems fall into this category, with some closer to a pure market system and some closer to a pure command system. Compare *pure command economic system, pure market economic system.*

mixture Combination of one or more elements and compounds.

model An approximate representation or simulation of a system being studied.

molecule Combination of two or more atoms of the same chemical element (such as O_2) or different chemical elements (such as H_2O) held together by chemical bonds. Compare *atom, ion.*

monera See *bacteria, cyanobacteria.*

monoculture Cultivation of a single crop, usually on a large area of land. Compare *polyculture, polyvarietal cultivation.*

more developed country (MDC) See *developed country.*

multiple use Use of an ecosystem such as a forest for a variety of purposes such as timber, wildlife, watershed protection, and recreation.

municipal solid waste Solid materials discarded by homes and businesses in or near urban areas. See *solid waste.*

mutagen Chemical or form of ionizing radiation that causes inheritable changes (mutations) in the DNA molecules in the genes found in chromosomes. See *carcinogen, mutation, teratogen.*

mutation A random change in DNA molecules making up genes that can yield changes in anatomy, physiology, or behavior in offspring. See *mutagen.*

mutualism Type of species interaction in which both participating species generally benefit. Compare *commensalism.*

native species Species that normally live and thrive in a particular ecosystem. Compare *immigrant species, indicator species, keystone species.*

natural capital The earth's natural resources and processes that sustain us and other species. Compare *human capital, manufactured capital, solar capital.*

natural gas Underground deposits of gases consisting of 50–90% by weight methane gas (CH_4) and small amounts of heavier gaseous hydrocarbon compounds such as propane (C_3H_8) and butane (C_4H_{10}).

natural greenhouse effect Heat buildup in the troposphere because of the presence of certain gases, called greenhouse gases. Without this effect, the earth would be nearly as cold as Mars and life as we know it could not exist. There is considerable evidence that we are enhancing this natural effect by excess additions of greenhouse gases from human activities.

natural ionizing radiation Ionizing radiation in the environment from natural sources.

natural radioactive decay Nuclear change in which unstable nuclei of atoms spontaneously shoot out particles (usually alpha or beta particles) or energy (gamma rays) at a fixed rate.

natural rate of extinction See *background extinction.*

natural recharge Natural replenishment of an aquifer by precipitation, which percolates downward through soil and rock. See *recharge area.*

natural resource capital See *natural capital.*

natural resources Nutrients and minerals in the soil and deeper layers of the earth's crust, water, wild and domesticated plants and animals, air, and other resources produced by the earth's natural processes. Compare *human capital, manufactured capital, solar capital.* See *natural capital.*

natural selection Process by which a particular beneficial gene (or set of genes) is reproduced in succeeding generations more than other genes. The result of natural selection is a population that contains a greater proportion of organisms better adapted to certain environmental conditions. See *adaptation, biological evolution, differential reproduction, mutation.*

negative feedback loop Situation in which a change in a certain direction provides information that causes a system to change less in that direction. Compare *positive feedback loop.*

negawatt A watt of electrical power saved by improving energy efficiency. See *watt.*

nematocide Chemical that kills nematodes (roundworms).

net economic welfare (NEW) Measure of annual change in quality of life in a country. It is obtained by subtracting the value of all final products and services that decrease the quality of life from a country's GNP.

net energy Total amount of useful energy available from an energy resource or energy system over its lifetime, minus the **(1)** amount of energy used (the first energy law), **(2)** automatically wasted (the second energy law), and **(3)** unnecessarily wasted in finding, processing, concentrating, and transporting it to users.

net primary productivity Rate at which all the plants in an ecosystem produce net useful chemical energy; equal to the difference between the rate at which the plants in an ecosystem produce useful chemical energy (primary productivity) and the rate at which they use some of that energy through cellular respiration. Compare *primary productivity.*

neutral solution Water solution containing an equal number of hydrogen ions (H^+) and hydroxide ions (OH^-); water solution with a pH of 7. Compare *acid solution, basic solution.*

neutron (n) Elementary particle in the nuclei of all atoms (except hydrogen-1). It has a relative mass of 1 and no electric charge. Compare *electron, proton.*

niche See *ecological niche.*

nitrogen cycle Cyclic movement of nitrogen in different chemical forms from the environment to organisms and then back to the environment.

nitrogen fixation Conversion of atmospheric nitrogen gas into forms useful to plants by lightning, bacteria, and cyanobacteria; it is part of the nitrogen cycle.

noise pollution Any unwanted, disturbing, or harmful sound that **(1)** impairs or interferes with hearing, **(2)** causes stress, **(3)** hampers concentration and work efficiency, or **(4)** causes accidents.

nondegradable pollutant Material that is not broken down by natural processes. Examples are the toxic elements lead and mercury. Compare *biodegradable pollutant, degradable pollutant, slowly degradable pollutant.*

nonionizing radiation Forms of radiant energy such as radio waves, microwaves, infrared light, and ordinary light that do not have enough energy to cause ionization of atoms in living tissue. Compare *ionizing radiation.*

nonnative species Species that migrate into an ecosystem or are deliberately or accidentally introduced into an ecosystem by humans. Compare *native species.*

nonpersistent pollutant See *degradable pollutant.*

nonpoint source Large or dispersed land areas such as crop fields, streets, and lawns that discharge pollutants into the environment over a large area. Compare *point source.*

nonrenewable resource Resource that exists in a fixed amount (stock) in various places in the earth's crust and has the potential for renewal only by geological, physical, and chemical processes taking place over hundreds of millions to billions of years. Examples are copper, aluminum, coal, and oil. We classify these resources as exhaustible because we are extracting and using them at a much faster rate than they were formed. Compare *potentially renewable resource.*

nontransmissible disease A disease that is not caused by living organisms and does not spread from one person to another. Examples are most cancer, diabetes, cardiovascular disease, and malnutrition. Compare *transmissible disease.*

no-till farming See *conservation-tillage farming.*

nuclear change Process in which nuclei of certain isotopes spontaneously change, or are forced to change, into one or more different isotopes. The three principal types of nuclear change are natural radioactivity, nuclear fission, and nuclear fusion. Compare *chemical change, physical change.*

nuclear energy Energy released when atomic nuclei undergo a nuclear reaction such as the spontaneous emission of radioactivity, nuclear fission, or nuclear fusion.

nuclear fission Nuclear change in which the nuclei of certain isotopes with large mass numbers (such as uranium-235 and plutonium-239) are split apart into lighter nuclei when struck by a neutron. This process releases more neutrons and a large amount of energy. Compare *nuclear fusion.*

nuclear fusion Nuclear change in which two nuclei of isotopes of elements with a low mass number (such as hydrogen-2 and hydrogen-3) are forced together at extremely high temperatures until they fuse to form a heavier nucleus (such as helium-4). This process releases a large amount of energy. Compare *nuclear fission.*

nucleus Extremely tiny center of an atom, making up most of the atom's mass. It contains one or more positively charged protons and one or more neutrons with no electrical charge (except for a hydrogen-1 atom, which has one proton and no neutrons in its nucleus).

nutrient Any food or element an organism must take in to live, grow, or reproduce.

nutrient cycle See *biogeochemical cycle.*

oil See *crude oil.*

oil shale Fine-grained rock containing various amounts of kerogen, a solid, waxy mixture of hydrocarbon compounds. Heating the rock to high temperatures converts the kerogen into a vapor that can be condensed to form a slow-flowing heavy oil called shale oil. See *kerogen, shale oil.*

old-growth forest Virgin and old, second-growth forests containing trees that are often hundreds, sometimes thousands of years old. Examples include forests of Douglas fir, western hemlock, giant sequoia, and coastal redwoods in the western United States. Compare *second-growth forest, tree farm.*

oligotrophic lake Lake with a low supply of plant nutrients. Compare *eutrophic lake, mesotrophic lake.*

omnivore Animal that can use both plants and other animals as food sources. Examples are pigs, rats, cockroaches, and people. Compare *carnivore, herbivore.*

open-pit mining Removal of minerals such as gravel, sand, and metal ores by digging them out of the earth's surface and leaving an open pit.

open sea The part of an ocean that is beyond the continental shelf. Compare *coastal zone.*

ore Part of a metal-yielding material that can be economically and legally extracted at a given time. An ore typically contains two parts: the ore mineral, which contains the desired metal, and waste mineral material (gangue).

organic farming Producing crops and livestock naturally by using organic fertilizer (manure, legumes, compost) and natural pest control (bugs that eat harmful bugs, plants that repel bugs, and environmental controls such as crop rotation) instead of using commercial inorganic fertilizers and synthetic pesticides and herbicides.

organic fertilizer Organic material such as animal manure, green manure, and compost, applied to cropland as a source of plant nutrients. Compare *commercial inorganic fertilizer.*

organism Any form of life.

other resources Identified and undiscovered resources not classified as reserves.

output Matter, energy, or information leaving a system. Compare *input, throughput.*

output pollution control See *pollution cleanup.*

overburden Layer of soil and rock overlying a mineral deposit; removed during surface mining.

overconsumption Situation in which some people consume much more than they need at the expense of those who cannot meet their basic needs and at the expense of the earth's present and future life-support systems.

overfishing Harvesting so many fish of a species, especially immature fish, that there is not enough breeding stock left to replenish the species, such that it is not profitable to harvest them.

overgrazing Destruction of vegetation when too many grazing animals feed too long and exceed the carrying capacity of a rangeland area.

overnutrition Diet so high in calories, saturated (animal) fats, salt, sugar, and processed foods, and so low in vegetables and fruits that the consumer runs high risks of diabetes, hypertension, heart disease, and other health hazards. Compare *malnutrition, undernutrition.*

oxygen-demanding wastes Organic materials that are usually biodegraded by aerobic (oxygen-consuming) bacteria, if there is enough dissolved oxygen in the water. See also *biological oxygen demand.*

ozone depletion Decrease in concentration of ozone (O_3) in the stratosphere. See *ozone layer.*

ozone layer Layer of gaseous ozone (O_3) in the stratosphere that protects life on earth by filtering out harmful ultraviolet radiation from the sun.

PANs Peroxyacyl nitrates. Group of chemicals found in photochemical smog.

parasite Consumer organism that lives on or in and feeds on a living plant or animal, known as the host, over an extended period of time. The parasite draws nourishment from and gradually weakens its host; it may or may not kill the host. See *parasitism.*

parasitism Interaction between species in which one organism, called the parasite, preys on another organism, called the host, by living on or in the host. See *host, parasite.*

parts per billion (ppb) Number of parts of a chemical found in one billion parts of a particular gas, liquid, or solid.

parts per million (ppm) Number of parts of a chemical found in one million parts of a particular gas, liquid, or solid.

passive solar heating system System that captures sunlight directly within a structure and converts it into low-temperature heat for space heating or for heating water for domestic use without the use of mechanical devices. Compare *active solar heating system.*

pathogen Organism that produces disease.

PCBs See *polychlorinated biphenyls.*

per capita GNP Annual gross national product (GNP) of a country divided by its total population. See *gross national product, real per capita GNP.*

perennial Plant that can live for more than 2 years. Compare *annual.*

permafrost Perennially frozen layer of the soil that forms when the water there freezes. It is found in arctic tundra.

permeability The degree to which underground rock and soil pores are interconnected with each other, and thus a measure of the degree to which water can flow freely from one pore to another. Compare *porosity.*

perpetual resource See *renewable resource.*

persistence How long a pollutant stays in the air, water, soil, or body. See also *inertia.*

persistent pollutant See *slowly degradable pollutant.*

pest Unwanted organism that directly or indirectly interferes with human activities.

pesticide Any chemical designed to kill or inhibit the growth of an organism that people consider to be undesirable. See *fungicide, herbicide, insecticide.*

pesticide treadmill Situation in which the cost of using pesticides increases while their effectiveness decreases, mostly because the pest species develop genetic resistance to the pesticides.

petrochemicals Chemicals obtained by refining (distilling) crude oil. They are used as raw materials in the manufacture of most industrial chemicals, fertilizers, pesticides, plastics, synthetic fibers, paints, medicines, and many other products.

petroleum See *crude oil.*

pH Numeric value that indicates the relative acidity or alkalinity of a substance on a scale of 0 to 14, with the neutral point at 7. Acid solutions have pH values lower than 7, and basic or alkaline solutions have pH values greater than 7.

phosphorus cycle Cyclic movement of phosphorus in different chemical forms from the environment to organisms and then back to the environment.

photochemical smog Complex mixture of air pollutants produced in the lower atmosphere by the reaction of hydrocarbons and nitrogen oxides under the influence of sunlight. Especially harmful components include ozone, peroxyacyl nitrates (PANs), and various aldehydes. Compare *industrial smog.*

photosynthesis Complex process that takes place in cells of green plants. Radiant energy from the sun is used to combine carbon dioxide (CO_2) and water (H_2O) to produce oxygen (O_2) and carbohydrates (such as glucose, $C_6H_{12}O_6$) and other nutrient molecules. Compare *aerobic respiration, chemosynthesis.*

photovoltaic cell (solar cell) Device in which radiant (solar) energy is converted directly into electrical energy.

physical change Process that alters one or more physical properties of an element or a compound without altering its chemical composition. Examples are changing the size and shape of a sample of matter (crushing ice and cutting aluminum foil) and changing a sample of matter from one physical state to another (boiling and freezing water). Compare *chemical change, nuclear change.*

phytoplankton Small, drifting plants, mostly algae and bacteria, found in aquatic ecosystems. Compare *plankton, zooplankton.*

pioneer community First integrated set of plants, animals, and decomposers found in an area undergoing primary ecological succession. See *immature community, mature community.*

pioneer species First hardy species, often microbes, mosses, and lichens, that begin colonizing a site as the first stage of ecological succession. See *ecological succession, pioneer community.*

planetary management worldview Beliefs that **(1)** we are the planet's most important species; **(2)** there is always more, and it's all for us; **(3)** all economic growth is good, more economic growth is better, and the potential for economic growth is limitless; and **(4)** our success depends on how well we can understand, control, and manage earth's life-support systems for our own benefit. See *spaceship-earth worldview.* Compare *earth-wisdom worldview.*

plankton Small plant organisms (phytoplankton) and animal organisms (zooplankton) that float in aquatic ecosystems.

plantation agriculture Growing specialized crops such as bananas, coffee, and cacao in tropical developing countries, primarily for sale to developed countries.

plants (plantae) Eukaryotic, mostly multicelled organisms such as algae (red, blue, and green), mosses, ferns, flowers, cacti, grasses, beans, wheat, rice, and trees. These organisms use photosynthesis to produce organic nutrients for themselves and for other organisms feeding on them. Water and other inorganic nutrients are obtained from the soil for terrestrial plants and from the water for aquatic plants.

plates Various-sized areas of the earth's lithosphere that move slowly around with the mantle's flowing asthenosphere. Most earthquakes and volcanoes occur around the boundaries of these plates. See *lithosphere, plate tectonics.*

plate tectonics Theory of geophysical processes that explains the movements of lithospheric plates and the processes that occur at their boundaries. See *lithosphere, plates.*

point source A single identifiable source that discharges pollutants into the environment. Examples are the **(1)** smokestack of a power plant or an industrial plant, **(2)** drainpipe of a meat-packing plant, **(3)** chimney of a house, or **(4)** exhaust pipe of an automobile. Compare *nonpoint source.*

poison A chemical that in one dose kills exactly 50% of the animals (usually rats and mice) in a test population (usually 60 to 200 animals) within a 14-day period. See *median lethal dose.*

politics Process through which individuals and groups try to influence or control government policies and actions that affect the local, state, national, and international communities.

pollutant A particular chemical or form of energy that can adversely affect the health, survival, or activities of humans or other living organisms. See *pollution.*

pollution An undesirable change in the physical, chemical, or biological characteristics of air, water, soil, or food that can adversely affect the health, survival, or activities of humans or other living organisms.

pollution cleanup Device or process that removes or reduces the level of a pollutant after it has been produced or has entered the environment. Examples are automobile emission control devices and sewage treatment plants. Compare *pollution prevention.*

pollution prevention Device or process that prevents a potential pollutant from forming or entering the environment, or that sharply reduces the amounts entering the environment. Compare *pollution cleanup.*

polychlorinated biphenyls (PCBs) Group of 209 different toxic, oily, synthetic chlorinated hydrocarbon compounds that can be biologically amplified in food chains and webs.

polyculture Complex form of intercropping in which a large number of different plants maturing at different times are planted together. See also *intercropping.* Compare *monoculture, polyvarietal cultivation.*

polyvarietal cultivation Planting a plot of land with several varieties of the same crop. Compare *intercropping, monoculture, polyculture.*

population Group of individual organisms of the same species living within a particular area.

population change An increase or decrease in the size of a population. It is equal to (Births + Immigration) – (Deaths + Emigration).

population density Number of organisms in a particular population found in a specified area.

population dispersion General pattern in which the members of a population are arranged throughout its habitat.

population distribution Variation of population density over a particular geographic area. For example, a country has a high population density in its urban areas and a much lower population density in rural areas.

population dynamics Major abiotic and biotic factors that tend to increase or decrease the population size and age and sex composition of a species.

population size Number of individuals making up a population's gene pool.

porosity Percentage of space in rock or soil occupied by voids, whether the voids are isolated or connected. Compare *permeability.*

positive feedback loop Situation in which a change in a certain direction provides information that causes a system to change further in the same direction. This can lead to a runaway or vicious cycle. Compare *negative feedback loop.*

potential energy Energy stored in an object because of its position or the position of its parts. Compare *kinetic energy.*

potentially renewable resource Resource that can be replenished fairly rapidly (hours to several decades) through natural processes. Examples are trees in forests, grasses in grasslands, wild animals, fresh surface water in lakes and streams, most groundwater, fresh air, and fertile soil. If such a resource is used faster than it is replenished, it can be depleted and converted into a nonrenewable resource. Compare *nonrenewable resource* and *renewable resource.* See also *environmental degradation.*

poverty Inability to meet basic needs for food, clothing, and shelter.

ppb See *parts per billion.*

ppm See *parts per million.*

precipitation Water in the form of rain, sleet, hail, and snow that falls from the atmosphere onto the land and bodies of water.

predation Situation in which an organism of one species (the predator) captures and feeds on parts or all of an organism of another species (the prey).

predator Organism that captures and feeds on parts or all of an organism of another species (the prey).

predator–prey relationship Interaction between two organisms of different species in which one organism, called the *predator,* captures and feeds on parts or all of another organism, called the *prey.*

prey Organism that is captured and serves as a source of food for an organism of another species (the predator).

primary consumer Organism that feeds on all or part of plants (herbivore) or on other producers. Compare *detritivore, omnivore, secondary consumer.*

primary oil recovery Pumping out the crude oil that flows by gravity or under gas pressure into the bottom of an oil well. Compare *enhanced oil recovery, secondary oil recovery.*

primary pollutant Chemical that has been added directly to the air by natural events or human activities and occurs in a harmful concentration. Compare *secondary pollutant.*

primary sewage treatment Mechanical treatment of sewage in which large solids are filtered out by screens and suspended solids settle out as sludge in a sedimentation tank. Compare *advanced sewage treatment, secondary sewage treatment.*

primary succession Sequential development of communities in a bare area that has never been occupied by a community of organisms. Compare *secondary succession.*

probability A mathematical statement about how likely it is that something will happen.

producer Organism that uses solar energy (green plant) or chemical energy (some bacteria) to manufacture the organic compounds it needs as nutrients from simple inorganic compounds obtained from its environment. Compare *consumer, decomposer.*

prokaryotic cell Cell that doesn't have a distinct nucleus. Other internal parts are also not enclosed by membranes. Compare *eukaryotic cell.*

protists (protista) Eukaryotic, mostly single-celled organisms such as diatoms, amoebas, some algae (golden brown and yellow-green), protozoans, and slime molds. Some protists produce their own organic nutrients through photosynthesis. Others are decomposers and some feed on bacteria, other protists, or cells of multicellular organisms.

proton (p) Positively charged particle in the nuclei of all atoms. Each proton has a relative mass of 1 and a single positive charge. Compare *electron, neutron.*

pure capitalism See *pure market economic system.*

pure command economic system System in which all economic decisions are made by the government or some other central authority. Compare *mixed economic system, pure market economic system.*

pure market economic system System in which all economic decisions are made in the market, where buyers and sellers of economic goods interact freely, with no government or other interference. Compare *mixed economic system, pure command economic system.*

pyramid of biomass Diagram representing the biomass, or total dry weight of all living organisms, that can be supported at each trophic level in a food chain or food web. See *pyramid of energy flow, pyramid of numbers.*

pyramid of energy flow Diagram representing the flow of energy through each trophic level in a food chain or food web. With each energy transfer, only a small part (typically 10%) of the usable energy entering one trophic level is transferred to the organisms at the next trophic level. Compare *pyramid of biomass, pyramid of numbers.*

pyramid of numbers Diagram representing the number of organisms of a particular type that can be supported at each trophic level from a given input of solar energy at the producer trophic level in a food chain or food web. Compare *pyramid of biomass, pyramid of energy flow.*

radiation Fast-moving particles (particulate radiation) or waves of energy (electromagnetic radiation). See *alpha particles, beta particles, gamma radiation.*

radioactive decay Change of a radioisotope to a different isotope by the emission of radioactivity.

radioactive isotope See *radioisotope.*

radioactive waste Radioactive waste products of nuclear power plants, research, medicine, weapon production, or other processes involving nuclear reactions. See *radioactivity.*

radioactivity Nuclear change in which unstable nuclei of atoms spontaneously shoot out "chunks" of mass, energy, or both, at a fixed rate. The three principal types of radioactivity are gamma rays and fast-moving alpha particles and beta particles.

radioisotope Isotope of an atom that spontaneously emits one or more types of radioactivity (alpha particles, beta particles, gamma rays).

rain shadow effect Low precipitation on the far side (leeward side) of a mountain when prevailing winds flow up and over a high mountain or range of high mountains. This creates semiarid and arid conditions on the leeward side of a high mountain range.

rangeland Land that supplies forage or vegetation (grasses, grasslike plants, and shrubs) for grazing and browsing animals and is not intensively managed. Compare *feedlot.*

range of tolerance Range of chemical and physical conditions that must be maintained for populations of a particular species to stay alive and grow, develop, and function normally. See *law of tolerance.*

real GNP Gross national product adjusted for inflation. Compare *per capita GNP, gross national product, real per capita GNP.*

realized niche Parts of the fundamental niche of a species that are actually used by that species. See *ecological niche, fundamental niche.*

real per capita GNP Per capita GNP adjusted for inflation.

recharge area Any area of land allowing water to pass through it and into an aquifer. See *aquifer, natural recharge.*

recycling Collecting and reprocessing a resource so that it can be made into new products. An example is collecting aluminum cans, melting them down, and using the aluminum to make new cans or other aluminum products. Compare *reuse.*

reforestation Renewal of trees and other types of vegetation on land where trees have been removed; can be done naturally by seeds from nearby trees or artificially by planting seeds or seedlings.

relative humidity The amount of water vapor in a certain mass of air, expressed as a percentage of the maximum amount it could hold at that temperature. Compare *absolute humidity.*

renewable resource An essentially inexhaustible resource on a human time scale. Solar energy is an example. See *potentially* *renewable resource.* Compare *nonrenewable resources.*

replacement-level fertility Number of children a couple must have to replace themselves. The average for a country or the world is usually slightly higher than 2 children per couple (2.1 in the United States and 2.5 in some developing countries) because some children die before reaching their reproductive years. See also *total fertility rate.*

reproduction Production of offspring by one or more parents.

reproductive isolation Long-term geographic separation of members of a particular sexually reproducing species.

reproductive potential See *biotic potential.*

reserves Resources that have been identified and from which a usable mineral can be extracted profitably at present prices with current mining technology. See *identified resources, undiscovered resources.*

reserve-to-production ratio Number of years reserves of a particular nonrenewable mineral will last at current annual production rates. See *reserves.*

resilience Ability of a living system to restore itself to original condition after being exposed to an outside disturbance that is not too drastic. See *constancy, inertia.*

resource Anything obtained from the living and nonliving environment to meet human needs and wants. It can also be applied to other species. Compare *ecological resource, economic resource.*

resource partitioning Process of dividing up resources in an ecosystem so that species with similar requirements (overlapping ecological niches) use the same scarce resources at different times, in different ways, or in different places. See *ecological niche, fundamental niche, realized niche.*

respiration See *aerobic respiration.*

response The amount of health damage caused by exposure to a certain dose of a harmful substance or form of radiation. See *dose, dose–response curve, median lethal dose.*

reuse To use a product over and over again in the same form. An example is collecting, washing, and refilling glass beverage bottles. Compare *recycling.*

riparian zones Thin strips and patches of vegetation that surround streams. They are very important habitats and resources for wildlife.

risk The probability that something undesirable will happen from deliberate or accidental exposure to a hazard. See *risk analysis, risk assessment, risk–benefit analysis, risk management.*

risk analysis Identifying hazards, evaluating the nature and severity of risks (*risk assessment*), using this and other information to determine options and make decisions about reducing or eliminating risks (*risk management*), and communicating information about risks to decision makers and the public (*risk communication*).

risk assessment Process of gathering data and making assumptions to estimate short- and long-term harmful effects on human health or the environment from exposure to hazards associated with the use of a particular product or technology. See *risk, risk–benefit analysis.*

risk–benefit analysis Estimate of the short- and long-term risks and benefits of using a particular product or technology. See *risk.*

risk communication Communicating information about risks to decision makers and the public. See *risk, risk analysis, risk–benefit analysis.*

risk management Using risk assessment and other information to determine options and make decisions about reducing or eliminating risks. See *risk, risk analysis, risk–benefit analysis, risk communication.*

rock Any material that makes up a large, natural, continuous part of the earth's crust. See *mineral.*

rock cycle Largest and slowest of the earth's cycles, consisting of geologic, physical, and chemical processes that form and modify rocks and soil in the earth's crust over millions of years.

rodenticide Chemical that kills rodents.

r-strategists Species that reproduce early in their life span and produce large numbers of usually small and short-lived offspring in a short period of time. Compare *K-strategists.*

rule of 70 Doubling time (in years) = 70/percentage growth rate. See *doubling time, exponential growth.*

runoff Fresh water from precipitation and melting ice that flows on the earth's surface into nearby streams, lakes, wetlands, and reservoirs. See *surface runoff, surface water.* Compare *groundwater.*

rural area Geographic area in the United States with a population of less than 2,500. The number of people used in this definition may vary in different countries. Compare *urban area.*

salinity Amount of various salts dissolved in a given volume of water.

salinization Accumulation of salts in soil that can eventually make the soil unable to support plant growth.

saltwater intrusion Movement of salt water into freshwater aquifers in coastal and inland areas as groundwater is withdrawn faster than it is recharged by precipitation.

sanitary landfill Waste disposal site on land in which waste is spread in thin layers, compacted, and covered with a fresh layer of clay or plastic foam each day.

scavenger Organism that feeds on dead organisms that either were killed by other organisms or died naturally. Examples are vultures, flies, and crows. Compare *detritivore.*

science Attempts to discover order in nature and use that knowledge to make predictions about what should happen in nature. See *consensus science, frontier science, scientific data, scientific hypothesis, scientific law, scientific methods, scientific model, scientific theory.*

scientific data Facts obtained by making observations and measurements. Compare *model, scientific hypothesis, scientific law, scientific methods, scientific theory.*

scientific hypothesis An educated guess that attempts to explain a scientific law or certain scientific observations. Compare *model, scientific data, scientific law, scientific methods, scientific theory.*

scientific law Description of what scientists find happening in nature over and over in the same way, without known exception. See *first law of thermodynamics, second law of thermodynamics, law of conservation of matter.* Compare *scientific data, scientific hypothesis, scientific methods, scientific model, scientific theory.*

scientific methods The ways scientists gather data and formulate and test scientific hypotheses, models, theories, and laws. See *model, scientific data, scientific hypothesis, scientific law, scientific theory.*

scientific model A simulation of complex processes and systems. Many are mathematical models that are run and tested using computers.

scientific theory A well-tested and widely accepted scientific hypothesis. Compare *scientific data, scientific hypothesis, scientific law, scientific methods, scientific model.*

secondary consumer Organism that feeds only on primary consumers. Most secondary consumers are animals, but some are plants. Compare *detritivore, omnivore, primary consumer.*

secondary oil recovery Injection of water into an oil well after primary oil recovery to force out some of the remaining, usually thicker, crude oil. Compare *enhanced oil recovery, primary oil recovery.*

secondary pollutant Harmful chemical formed in the atmosphere when a primary air pollutant reacts with normal air components or other air pollutants. Compare *primary pollutant.*

secondary sewage treatment Second step in most waste treatment systems, in which aerobic bacteria break down up to 90% of degradable, oxygen-demanding organic wastes in wastewater. This is usually done by bringing sewage and bacteria together in trickling filters or in the activated sludge process. Compare *advanced sewage treatment, primary sewage treatment.*

secondary succession Sequential development of communities in an area in which natural vegetation has been removed or destroyed but the soil is not destroyed. Compare *primary succession.*

second-growth forest Stands of trees resulting from secondary ecological succession. Compare *old-growth forest, tree farm.*

second law of energy See *second law of thermodynamics.*

second law of thermodynamics In any conversion of heat energy to useful work, some of the initial energy input is always degraded to a lower-quality, more dispersed, less useful energy, usually low-temperature heat that flows into the environment; you can't break even in terms of energy quality. See *first law of thermodynamics*.

sedimentary rock Rock that forms from the accumulated products of erosion and in some cases from the compacted shells, skeletons, and other remains of dead organisms. Compare *igneous rock, metamorphic rock*. See *rock cycle*.

seed-tree cutting Removal of nearly all trees on a site in one cutting, with a few seed-producing trees left uniformly distributed to regenerate the forest. Compare *clear-cutting, selective cutting, shelterwood cutting, strip cutting*.

selective cutting Cutting of intermediate-aged, mature, or diseased trees in an uneven-aged forest stand, either singly or in small groups. This encourages the growth of younger trees and maintains an uneven-aged stand. Compare *clear-cutting, seed-tree cutting, shelterwood cutting, strip cutting*.

septic tank Underground tank for treatment of wastewater from a home in rural and suburban areas. Bacteria in the tank decompose organic wastes and the sludge settles to the bottom of the tank. The effluent flows out of the tank into the ground through a field of drain pipes.

sexual reproduction Reproduction in organisms that produce offspring by combining sex cells or *gametes* (such as ovum and sperm) from both parents. This produces offspring that have combinations of traits from their parents. Compare *asexual reproduction*.

shale oil Slow-flowing, dark brown, heavy oil obtained when kerogen in oil shale is vaporized at high temperatures and then condensed. Shale oil can be refined to yield gasoline, heating oil, and other petroleum products. See *kerogen, oil shale*.

shelterbelt See *windbreak*.

shelterwood cutting Removal of mature, marketable trees in an area in a series of partial cuttings to allow regeneration of a new stand under the partial shade of older trees, which are later removed. Typically, this is done by making two or three cuts over a decade. Compare *clear-cutting, seed-tree cutting, selective cutting, strip cutting*.

shifting cultivation Clearing a plot of ground in a forest, especially in tropical areas, and planting crops on it for a few years (typically 2–5 years) until the soil is depleted of nutrients or the plot has been invaded by a dense growth of vegetation from the surrounding forest. Then a new plot is cleared and the process is repeated. The abandoned plot cannot successfully grow crops for 10–30 years. See also *slash-and-burn cultivation*.

slash-and-burn cultivation Cutting down trees and other vegetation in a patch of forest, leaving the cut vegetation on the ground to dry, and then burning it. The ashes that are left add nutrients to the nutrient-poor soils found in most tropical forest areas. Crops are planted between tree stumps. Plots must be abandoned after a few years (typically 2–5 years) because of loss of soil fertility or invasion of vegetation from the surrounding forest. See also *shifting cultivation*.

slowly degradable pollutant Material that is slowly broken down into simpler chemicals or reduced to acceptable levels by natural physical, chemical, and biological processes. Compare *biodegradable pollutant, degradable pollutant, nondegradable pollutant*.

sludge Gooey mixture of toxic chemicals, infectious agents, and settled solids, removed from wastewater at a sewage treatment plant.

smelting Process in which a desired metal is separated from the other elements in an ore mineral.

smog Originally a combination of smoke and fog but now used to describe other mixtures of pollutants in the atmosphere. See *industrial smog, photochemical smog*.

soil Complex mixture of inorganic minerals (clay, silt, pebbles, and sand), decaying organic matter, water, air, and living organisms.

soil conservation Methods used to reduce soil erosion, prevent depletion of soil nutrients, and restore nutrients already lost by erosion, leaching, and excessive crop harvesting.

soil erosion Movement of soil components, especially topsoil, from one place to another, usually by wind, flowing water, or both. This natural process can be greatly accelerated by human activities that remove vegetation from soil.

soil horizons Horizontal zones that make up a particular mature soil. Each horizon has a distinct texture and composition that varies with different types of soils.

soil permeability Rate at which water and air move from upper to lower soil layers. Compare *porosity*.

soil porosity See *porosity*.

soil profile Cross-sectional view of the horizons in a soil.

soil structure How the particles that make up a soil are organized and clumped together. See also *soil permeability, soil texture*.

soil texture Relative amounts of the different types and sizes of mineral particles in a sample of soil.

solar capital Solar energy from the sun reaching earth. Compare *natural capital*.

solar cell See *photovoltaic cell*.

solar collector Device for collecting radiant energy from the sun and converting it into heat. See *active solar heating system, passive solar heating system*.

solar energy Direct radiant energy from the sun and a number of *indirect* forms of energy produced by the direct input. Principal indirect forms of solar energy include wind, falling and flowing water (hydropower), and biomass (solar energy converted into chemical energy stored in the chemical bonds of organic compounds in trees and other plants).

solid waste Any unwanted or discarded material that is not a liquid or a gas. See *municipal solid waste*.

spaceship-earth worldview View of the earth as a spaceship: a machine that we can understand, control, and change at will by using advanced technology. See *planetary management worldview*. Compare *earth-wisdom worldview*.

specialist species Species with a narrow ecological niche. They may be able **(1)** to live in only one type of habitat, **(2)** tolerate only a narrow range of climatic and other environmental conditions, or **(3)** use only one or a few types of food. Compare *generalist species*.

speciation Formation of two species from one species as a result of divergent natural selection in response to changes in environmental conditions; usually takes thousands of years. Compare *extinction*.

species Group of organisms that resemble one another in appearance, behavior, chemical makeup and processes, and genetic structure. Organisms that reproduce sexually are classified as members of the same species only if they can actually or potentially interbreed with one another and produce fertile offspring.

species diversity Number of different species and their relative abundances in a given area. See *biodiversity*. Compare *ecological diversity, genetic diversity*.

species equilibrium model See *theory of island biogeography*.

spoils Unwanted rock and other waste materials produced when a material is removed from the earth's surface or subsurface by mining, dredging, quarrying, and excavation.

S-shaped curve Leveling off of an exponential, *J*-shaped curve when a rapidly growing population exceeds the carrying capacity of its environment and ceases to grow.

stability Ability of a living system to withstand or recover from externally imposed changes or stresses. See *constancy, inertia, resilience*.

stewardship View that because of our superior intellect and power or because of our religious beliefs, we have an ethical responsibility to manage and care for domesticated plants and animals and the rest of nature. Compare *earth-wisdom worldview, planetary management worldview*.

stratosphere Second layer of the atmosphere, extending about 17–48 kilometers (11–30 miles) above the earth's surface. It contains small amounts of gaseous ozone (O_3), which filters out about 99% of the incoming harmful ultraviolet (UV) radiation emitted by the sun. Compare *troposphere*.

stream Flowing body of surface water. Examples are creeks and rivers.

strip cropping Planting regular crops and close-growing plants, such as hay or nitrogen-fixing legumes, in alternating rows or bands to help reduce depletion of soil nutrients.

strip cutting A variation of clear-cutting in which a strip of trees is clear-cut along the contour of the land, with the corridor narrow enough to allow natural regeneration within a few years. After regeneration, another strip is cut above the first, and so on. Compare *clear-cutting, seed-tree cutting, selective cutting, shelterwood cutting.*

strip mining Form of surface mining in which bulldozers, power shovels, or stripping wheels remove large chunks of the earth's surface in strips. See *surface mining.* Compare *subsurface mining.*

subatomic particles Extremely small particles—electrons, protons, and neutrons—that make up the internal structure of atoms.

subduction zone Area in which oceanic lithosphere is carried downward (subducted) under the island arc or continent at a convergent plate boundary. A trench ordinarily forms at the boundary between the two converging plates. See *convergent plate boundary.*

subsidence Slow or rapid sinking down of part of the earth's crust that is not slope-related.

subsistence farming Supplementing solar energy with energy from human labor and draft animals to produce enough food to feed oneself and family members; in good years there may be enough food left over to sell or put aside for hard times. Compare *industrialized agriculture.*

subsurface mining Extraction of a metal ore or fuel resource such as coal from a deep underground deposit. Compare *surface mining.*

succession See *ecological succession.*

succulent plants Plants, such as desert cacti, that survive in dry climates by having no leaves, thus reducing the loss of scarce water. They store water and use sunlight to produce the food they need in the thick fleshy tissue of their green stems and branches. Compare *deciduous plants, evergreen plants.*

sulfur cycle Cyclic movement of sulfur in different chemical forms from the environment to organisms and then back to the environment.

superinsulated house House that is heavily insulated and extremely airtight. Typically, active or passive solar collectors are used to heat water, and an air-to-air heat exchanger is used to prevent buildup of excessive moisture and indoor air pollutants.

surface fire Forest fire that burns only undergrowth and leaf litter on the forest floor. Compare *crown fire, ground fire.*

surface mining Removing soil, subsoil, and other strata, and then extracting a mineral deposit found fairly close to the earth's surface. Compare *subsurface mining.*

surface runoff Water flowing off the land into bodies of surface water.

surface water Precipitation that does not infiltrate the ground or return to the atmosphere by evaporation or transpiration. See *runoff.* Compare *groundwater.*

survivorship curve Graph showing the number of survivors in different age groups for a particular species.

sustainability Ability of a system to survive for some specified (finite) time. See *sustainable system.*

sustainable agriculture Method of growing crops and raising livestock based on organic fertilizers, soil conservation, water conservation, biological control of pests, and minimal use of nonrenewable fossil-fuel energy.

sustainable development Forms of economic growth and activities that do not deplete or degrade natural resources on which present and future economic growth and life depend. Compare *economic growth, economic development, ecologically sustainable development.*

sustainable earth Situation in which the earth's supplies of resources and the processes that make up natural capital are used and maintained over a specified period in ways that don't deplete or degrade these resources and processes.

sustainable living Taking no more potentially renewable resources from the natural world than can be replenished naturally and not overloading the capacity of the environment to cleanse and renew itself by natural processes.

sustainable society A society that manages its economy and population size without doing irreparable environmental harm by overloading the planet's ability to absorb environmental insults, replenish its resources, and sustain human and other forms of life over a specified period, usually hundreds to thousands of years. During this period, it satisfies the needs of its people without depleting natural capital and thereby jeopardizing the prospects of current and future generations of humans and other species.

sustainable system A system that survives and functions over some specified (finite) time; a system that attains its full expected lifetime.

sustainable yield (sustained yield) Highest rate at which a potentially renewable resource can be used without reducing its available supply throughout the world or in a particular area. See also *environmental degradation.*

symbiosis Any intimate relationship or association between members of two or more species. See *symbiotic relationship.*

symbiotic relationship Species interaction in which two kinds of organisms live together in an intimate association. Members of the participating species may be harmed by, benefit from, or be unaffected by the interaction. See *commensalism, interspecific competition, mutualism, parasitism, predation.*

synergistic interaction Interaction of two or more factors or processes so that the combined effect is greater than the sum of their separate effects.

synergy See *synergistic interaction.*

synfuels Synthetic gaseous and liquid fuels produced from solid coal or sources other than natural gas or crude oil.

synthetic natural gas (SNG) Gaseous fuel containing mostly methane produced from solid coal.

system A set of components that function and interact in some regular and theoretically predictable manner.

tailings Rock and other waste materials removed as impurities when waste mineral material is separated from the metal in an ore.

tar sand Deposit of a mixture of clay, sand, water, and varying amounts of a tar-like heavy oil known as bitumen. Bitumen can be extracted from tar sand by heating. It is then purified and upgraded to synthetic crude oil. See *bitumen.*

technology Creation of new products and processes intended to improve our efficiency, chances for survival, comfort level, and quality of life. Compare *science.*

temperature Measure of the average speed of motion of the atoms, ions, or molecules in a substance or combination of substances at a given moment. Compare *heat.*

temperature inversion See *thermal inversion.*

teratogen Chemical, ionizing agent, or virus that causes birth defects. See *carcinogen, mutagen.*

terracing Planting crops on a long, steep slope that has been converted into a series of broad, nearly level terraces with short vertical drops from one to another that run along the contour of the land to retain water and reduce soil erosion.

terrestrial Pertaining to land. Compare *aquatic.*

tertiary (higher-level) consumers Animals that feed on animal-eating animals. They feed at high trophic levels in food chains and webs. Examples are hawks, lions, bass, and sharks. Compare *detritivore, primary consumer, secondary consumer.*

tertiary oil recovery See *enhanced oil recovery.*

tertiary sewage treatment See *advanced sewage treatment.*

theory of evolution Widely accepted scientific idea that all life-forms developed from earlier life-forms. Although this theory conflicts with the creation stories of many religions, it is the way biologists explain how life has changed over the past 3.6–3.8 billion years and why it is so diverse today.

theory of island biogeography The number of species found on an island is determined by a balance between two factors: the **(1)** *immigration rate* (of species new to the island) from other inhabited areas and **(2)** *extinction rate* (of species established on the island). The model predicts that at some point the rates of immigration and extinction will reach an equilibrium point that determines the island's average number of different species (species diversity).

thermal inversion Layer of dense, cool air trapped under a layer of less dense, warm air. This prevents upward-flowing air currents from developing. In a prolonged inversion, air pollution in the trapped layer may build up to harmful levels.

thermocline Zone of gradual temperature decrease between warm surface water and colder deep water in a lake, reservoir, or ocean.

threatened species Wild species that is still

abundant in its natural range but is likely to become endangered because of a decline in numbers. Compare *endangered species.*

threshold effect The harmful or fatal effect of a small change in environmental conditions that exceeds the limit of tolerance of an organism or population of a species. See *law of tolerance.*

throughput Rate of flow of matter, energy, or information through a system. Compare *input, output.*

throwaway society See high-*throughput society.*

time delay Time lag between the input of a stimulus into a system and the response to the stimulus.

tolerance limits Minimum and maximum limits for physical conditions (such as temperature) and concentrations of chemical substances beyond which no members of a particular species can survive. See *law of tolerance.*

total fertility rate (TFR) Estimate of the average number of children that will be born alive to a woman during her lifetime if she passes through all her childbearing years (ages 15–44) conforming to age-specific fertility rates of a given year. In simpler terms, it is an estimate of the average number of children a woman will have during her childbearing years.

totally planned economy See *pure command economic system.*

toxic chemical Chemical that is fatal to humans in low doses, or fatal to over 50% of test animals at stated concentrations. Most are neurotoxins, which attack nerve cells. See *carcinogen, hazardous chemical, mutagen, teratogen.*

toxicity Measure of how harmful a substance is.

toxicology Study of the adverse effects of chemicals on health.

toxic waste Form of hazardous waste that causes death or serious injury (such as burns, respiratory diseases, cancers, or genetic mutations). See *hazardous waste.*

traditional intensive agriculture Producing enough food for a farm family's survival and perhaps a surplus that can be sold. This type of agriculture requires higher inputs of labor, fertilizer, and water than traditional subsistence agriculture. See *traditional subsistence agriculture.*

traditional subsistence agriculture Production of enough crops or livestock for a farm family's survival and, in good years, a surplus to sell or put aside for hard times. Compare *traditional intensive agriculture.*

tragedy of the commons Depletion or degradation of a potentially renewable resource to which people have free and unmanaged access. An example is the depletion of commercially desirable species of fish in the open ocean beyond areas controlled by coastal countries. See *common-property resource.*

transform fault Area where the earth's lithospheric plates move in opposite but parallel directions along a fracture (fault) in the lithosphere. Compare *convergent plate boundary, divergent plate boundary.*

transmissible disease A disease that is caused by living organisms (such as bacteria, viruses, and parasitic worms) and can spread from one person to another by air, water, food, or body fluids (or in some cases by insects or other organisms). Compare *nontransmissible disease.*

transpiration Process in which water is **(1)** absorbed by the root systems of plants, **(2)** moves up through the plants, **(3)** passes through pores (stomata) in their leaves or other parts, and **(4)** then evaporates into the atmosphere as water vapor.

tree farm Site planted with one or only a few tree species in an even-aged stand. When the stand matures it is usually harvested by clear-cutting and then replanted. These farms normally are used to grow rapidly growing tree species for fuelwood, timber, or pulpwood. See *even-aged management.* Compare *old-growth forest, second-growth forest, uneven-aged management.*

tree plantation See *tree farm.*

trophic level All organisms that are the same number of energy transfers away from the original source of energy (for example, sunlight) that enters an ecosystem. For example, all producers belong to the first trophic level, and all herbivores belong to the second trophic level in a food chain or a food web.

troposphere Innermost layer of the atmosphere. It contains about 75% of the mass of earth's air and extends about 17 kilometers (11 miles) above sea level. Compare *stratosphere.*

true cost See *full cost.*

undernutrition Consuming insufficient food to meet one's minimum daily energy requirement for a long enough time to cause harmful effects. Compare *malnutrition, overnutrition.*

undiscovered resources Potential supplies of a particular mineral resource, believed to exist because of geologic knowledge and theory, although specific locations, quality, and amounts are unknown. Compare *identified resources, reserves.*

uneven-aged management Method of forest management in which trees of different species in a given stand are maintained at many ages and sizes to permit continuous natural regeneration. Compare *even-aged management.*

upwelling Movement of nutrient-rich bottom water to the ocean's surface. This can occur far from shore but usually occurs along certain steep coastal areas where the surface layer of ocean water is pushed away from shore and replaced by cold, nutrient-rich bottom water.

urban area Geographic area with a population of 2,500 or more. The number of people used in this definition may vary, with some countries setting the minimum number of people at 10,000–50,000.

urban growth Rate of growth of an urban population. Compare *degree of urbanization.*

urban heat island Buildup of heat in the atmosphere above an urban area. This heat is produced by the large concentration of cars, buildings, factories, and other heat-producing activities. See also *dust dome.*

urbanization See *degree of urbanization.*

vertebrates Animals with backbones. Compare *invertebrates.*

volcano Vent or fissure in the earth's surface through which magma, liquid lava, and gases are released into the environment.

water cycle See *hydrologic cycle.*

waterlogging Saturation of soil with irrigation water or excessive precipitation so that the water table rises close to the surface.

water pollution Any physical or chemical change in surface water or groundwater that can harm living organisms or make water unfit for certain uses.

watershed Land area that delivers the water, sediment, and dissolved substances via small streams to a major stream (river).

water table Upper surface of the zone of saturation, in which all available pores in the soil and rock in the earth's crust are filled with water.

watt Unit of power, or rate at which electrical work is done. See *kilowatt.*

weather Short-term changes in the temperature, barometric pressure, humidity, precipitation, sunshine, cloud cover, wind direction and speed, and other conditions in the troposphere at a given place and time. Compare *climate.*

weathering Physical and chemical processes in which solid rock exposed at the earth's surface is changed to separate solid particles and dissolved material, which can then be moved to another place as sediment. See *erosion.*

wetland Land that is covered all or part of the time with salt water or fresh water, excluding streams, lakes, and the open ocean. See *coastal wetland, inland wetland.*

wilderness Area where the earth and its community of life have not been seriously disturbed by humans and where humans are only temporary visitors.

wildlife All free, undomesticated species. Sometimes the term is used to describe only free, undomesticated species of *animals.*

wildlife management Manipulation of populations of wild species (especially game species) and their habitats for **(1)** human benefit, **(2)** the welfare of other species, and the preservation of threatened and **(3)** endangered wildlife species.

wildlife resources Species of wildlife that have actual or potential economic value to people.

wild species Species found in the natural environment.

windbreak Row of trees or hedges planted to partially block wind flow and reduce soil erosion on cultivated land.

wind farm Cluster of small to medium-sized wind turbines in a windy area to capture wind energy and convert it into electrical energy.

worldview How individuals think the world works and what they think their role in the world should be. See *earth-wisdom worldview, planetary management worldview, spaceship-earth worldview*.

zero population growth (ZPG) State in which the birth rate (plus immigration) equals the death rate (plus emigration) so that the population of a geographic area is no longer increasing.

zone of saturation Area where all available pores in soil and rock in the earth's crust are filled by water. See *water table*.

zoning Regulating how various parcels of land can be used.

zooplankton Animal plankton. Small floating herbivores that feed on plant plankton (phytoplankton). Compare *phytoplankton*.

INDEX

Note: Page numbers in **boldface type** indicate major discussions or definitions of key terms. Page numbers followed by italicized letters *f* and *t* indicate figures and tables, respectively.